W9-CWT-648

# Biophysical Science —A Study Program

# Biophysical Science —A Study Program

*Planned and Edited by*

**J. L. Oncley**
*Editor-in-Chief*

F. O. Schmitt
R. C. Williams
M. D. Rosenberg
R. H. Bolt

for the
Biophysics and Biophysical Chemistry Study Section
of the National Institutes of Health
Public Health Service
United States Department of Health, Education, and Welfare

Published in the

**Reviews of Modern Physics, January and April, 1959**

and by

**John Wiley & Sons, Inc., New York**

SECOND PRINTING, AUGUST, 1960

Library of Congress Catalog Card Number: 59–10786

PRINTED IN THE UNITED STATES OF AMERICA

# Acknowledgments

The Study Program in Biophysical Science, Boulder, Colorado, 20 July–16 August 1958, was organized and conducted by the Biophysics and Biophysical Chemistry Study Section of the National Institutes of Health, Public Health Service, United States Department of Health, Education, and Welfare, under research grant RG-5048 from that agency. About 120 senior research contributors and selected younger scientists took part in this program, which was held in an environment conducive to informal interchange of ideas.

The general objective of the Study Program was to aid and encourage the further blending of concepts and methods of physical science with those of life science in the investigation of biological problems. The Study Section conceived this activity as an experiment in the exchange of information and ideas among representatives of different branches of science. The core of the Study Program was a carefully integrated series of about 60 lectures, constituting compact summaries of certain key problems and critical evaluations of recent advances. The lecture material provided a framework for other activities such as planned workshop sessions, spontaneous discussion groups, and library study.

The results of the experiment reinforced the view of the Study Section that the same lecture material, appropriately edited and published, would also provide a timely research guide and a useful base on which to build new courses and seminars in biophysical science. Thus, the values of this effort would be made available to a much broader audience than could be accommodated in the informal atmosphere desired for the Study Program.

The method of publication reflects the objectives of the Study Program. Presentation in the *Reviews of Modern Physics* brings the material before the community of physicists, who may find interest and stimulation here. In this respect, the publication is a sequel to "Borderland Problems in Biology and Physics" by John R. Loofbourow, which appeared in the *Reviews of Modern Physics* during 1940. Simultaneous publication in book form by John Wiley and Sons, Inc., who have prepared the illustration drawings, brings the material to the attention of scientists in all fields and countries. Through the combined publication procedure, the results of the Study Program are being disseminated promptly, to a wide audience, at the lowest practicable cost.

The entire Study Section devoted long hours and much thought to the formulation of the Study Program. Encouragement and help were given by former members T. Hill and G. B. B. M. Sutherland. An invaluable aid in formulating the program was the experience gained from the "Conference on Certain Fundamental Aspects of Biophysical Science," conducted by the Study Section in January 1958 at Bethesda, Maryland, and planned by a committee: P. H. Abelson, R. C. Williams, and H. Neurath, chairman.

The detailed planning and final decisions regarding subject matter of the summer Study Program were mainly the responsibility of a committee: R. H. Bolt, J. L. Oncley, M. D. Rosenberg, R. C. Williams, and F. O. Schmitt, chairman. Valuable help was provided also by H. Fernández-Morán, A. Rich, C. Levinthal, and D. F. Waugh.

Able technical editorial assistance was given by a considerable number of the Study Program participants, especially R. L. Baldwin, M. S. Blois, Jr., A. S. Brill, L. S. Frishkopf, M. J. Glimcher, M. H. Goldstein, Jr., P. R. Gross, R. G. Hart, R. I. Henkin, M. Kasha, G. Kegeles, M. S. Meselson, H. H. Pattee, Jr., F. W. Putnam, S. Shulman, I. W. Sizer, and I. Tincoco, Jr. In the revision of manuscripts, special assistance was given by M. V. Edds, Jr., and W. Bloom.

This publication would not have been possible without the devoted and expert assistance of Jean Doty. She has carried much of the burden of editing, proofreading, and coordination of manuscript preparation. We are grateful to G. M. Kolodny and W. Scheider for collecting bibliographic material and checking references; to G. F. Gove for undertaking the difficult job of indexing; to M. E. Platin for extensive secretarial and administrative assistance.

The combined publication was made possible by the active interest and cooperation of the editor of the *Reviews of Modern Physics*, E. U. Condon, who also was a participant in the Study Program. We are indebted to Ruth Bryans, Publication Manager of the American Institute of Physics, and staff; and to many other friends in scientific institutions and publishing houses who have advised and encouraged us in this undertaking. We are indebted also to the many authors, societies, and publishers for the use of figures, tables, and other illustrative material; specific acknowledgments are given where they appear in the papers.

The Study Program participants, in addition to those named elsewhere in these acknowledgments or listed as authors of papers, were: B. Altshuler, W. T. Astbury, L. G. Augenstine, J. J. Baruch, E. D. Becker, S. A. Bernhard, H. B. Callen, F. D. Carlson, W. R. Carroll, N. R. Davidson, D. R. Davies, E. A. Edelsack, M. Eden, A. V. Engström, D. J. Fluke, F. L. Friedman, H. E. von Gierke, J. Z. Hearon, D. F. Hornig, S. H. Hutner, H. S. Kaplan, N. O. Kaplan, J. Kraut, U. Liddel, L. Lorand, E. F. MacNichol, Jr., P. Morrison, L. J. Mullins, J. R. Platt, R. Podolsky, E. Ribi, S. J. Singer, R. L. Sinsheimer, H. W. Smith, M. R. Stetten, L. Szilard, E. W. Taylor, T. E. Thompson, C. H. Townes, J. Towsend, B. L. Vallee, A. C. Young.

To all participants, we express our gratitude for critical discussions of the scientific material and cogent suggestions regarding its presentation. It is they as a group who have vitalized this account of horizons in biophysical science.

## CONTENTS

### BIOPHYSICAL SCIENCE

### A STUDY PROGRAM

## CONTENTS

# Biophysical Science—A Study Program

Physical and chemical approaches to problems in biology have become increasingly productive in recent years. Major advances in the understanding of life processes have been made through research in such specialties as biophysical chemistry, molecular biology, biophysics, and electrophysiology. Continuing progress will require an ever more perceptive study of the interactions of matter, energy, and information in biological systems.

This publication grew out of a special activity, designed to aid and stimulate the further blending of the concepts and methods of physics and chemistry with those of the life sciences in the study of biological problems. The papers in this volume were presented in the Study Program in Biophysical Science, held in Boulder, Colorado, during the summer of 1958.

A living system is a self-perpetuating combination of atoms, organized in a highly specific manner and interacting with its environment in such a way as to metabolize, reproduce, grow, and adapt. The Study Program was designed to present biological problems as "viewed through physical spectacles and investigated by physical ideas and methods."* Biological processes can be examined, for example, in terms of atomic structures, energy levels, and binding forces. The engineer might prefer to specify the design and operation in terms of components and "black boxes." The chemist discusses molecular configurations and kinetic and thermodynamic details of biochemical reactions. These and other modes for describing biological processes are found in this volume.

The Study Program included a number of "case studies" in which topics were chosen to point up the need and value of physical approaches at all levels of biological organization. Particular emphasis centered upon muscle and nerve, where considerable information is available at many of these levels. The series of papers related to nerve, for example, encompasses the molecular organization of the nerve fiber, the nature of the nerve impulse, nerve metabolism, sensory performance of organisms, receptor mechanisms, and integrated responses of higher neural centers. Another series deals with genetics, replication, and synthesis of nucleic-acid and protein macromolecules. Here again, the topics range from descriptions of the biological processes of replication, through physical and biochemical details of relevant structures, to mathematical considerations in the coding of information.

The subjects were selected with emphasis on fundamental concepts, to broaden the base of common language and understanding among a heterogeneous group of biologists, physicists, and chemists. This emphasis restricted the range of topics that could be presented within the limited duration of the Study Program. Also, the interests of those who organized the program are unavoidably reflected in its contents.

This particular choice of topics is not intended to define or delimit the field of biophysics, but, rather, to illustrate the power and success of the integrated physical-biological approach. Future conferences and publications of this kind will undoubtedly emphasize alternative subjects such as, for example, those at higher levels of biological complexity.

The arrangement of the material presented in the Study Program was chosen to provide a logical unfolding of the selected subjects, along with a continuous injection of background material. The initial chapters give many of the fundamental concepts, whereas the later chapters make use of developments introduced in the intervening chapters. Since references may be needed to supplement the necessarily abridged discussions in the papers, a Selected General Bibliography is given on pages 2 to 4. The entries are divided into 10 subject categories, together with a designation of those papers that are most closely related to each category.

Biophysics and Biophysical Chemistry Study Section

P. H. Abelson
J. W. Beams
R. H. Bolt
J. D. Ferry
I. Fuhr
D. E. Goldman
I. Gray
J. D. Hardy
H. K. Hartline
J. G. Kirkwood
H. Neurath
J. L. Oncley
M. D. Rosenberg
H. A. Sober
R. C. Williams
F. O. Schmitt,
*Chairman*

October, 1958
Cambridge, Massachusetts

* A. V. Hill, "Why Biophysics?" *Lectures on the Scientific Basis of Medicine* (Athlone Press, London, 1956), Vol. IV, p. 1. Reprinted in Science **124**, 1233 (1956).

# Selected General Bibliography

**Cellular Biology** (related papers: 1–3, 47, 48)

1. Frey-Wyssling, A., *Submicroscopic Morphology of Protoplasm* (Elsevier Publishing Company, New York, 1953).
2. Giese, A. C., *Cell Physiology* (Saunders Company, Philadelphia, 1957).
3. Conference on Tissue Fine Structure, Biophys. Biochem. Cytol. **2**, Suppl., 1–448 (1956).
4. McElroy, W. D., and Glass, B., editors, *The Chemical Basis of Heredity* (The Johns Hopkins Press, Baltimore, Maryland, 1957).
5. Mellors, R. C., *Analytical Cytology* (McGraw-Hill Book Company, Inc., New York, 1955).
6. Schmidt, W. J., *Die Doppelbrechung von Karyoplasma, Zytoplasma, und Metaplasma* (Verlage Borntraeger, Berlin, 1937).
7. Schmitt, F. O., Doty, P., Hall, C. E., Williams, R. C., and Weiss, P. A., Symposium on Biomolecular Organization and Life-Processes, Proc. Natl. Acad. Sci. U. S. **42**, 789–830 (1956).
8. Willier, B. H., Weiss, P. A., and Hamburger, V., editors, *Analysis of Development* (Saunders Company, Philadelphia, 1955).
9. Wilson, E. B., *The Cell in Development and Heredity* (The Macmillan Company, New York, 1925).

**Physical and Chemical Characteristics of Macromolecules** (related papers: 4–15, 23).

1. Chargaff, E., and Davidson, J. N., editors, *The Nucleic Acids* (Academic Press, Inc., New York, 1955).
2. Davidson, J. N., *The Biochemistry of the Nucleic Acids* (Methuen Company, London; and John Wiley and Sons, Inc., New York, 1957), third edition.
3. Edsall, J. T., and Wyman, J., *Biophysical Chemistry* (Academic Press, Inc., New York, 1958), Vol. I.
4. Eirich, F. R., editor, *Rheology: Theory and Applications* (Academic Press, Inc., New York, 1956–1958).
5. Flory, P. J., *Principles of Polymer Chemistry* (Cornell University Press, Ithaca, New York, 1953).
6. Fruton, J. S., and Simmonds, S., *General Biochemistry* (John Wiley and Sons, Inc., New York, 1958), second edition.
7. Katchalsky, A., "Problems in the physical chemistry of polyelectrolytes," J. Polymer Sci. **12**, 159–184 (1954).
8. Mark, H., and Tobolsky, A. V., *Physical Chemistry of High Polymeric Systems* (Interscience Publishers, Inc., New York, 1950), second edition.
9. Neuberger, A., editor, *Symposium on Protein Structure* (John Wiley and Sons, Inc., New York; and Methuen Company, London, 1958).
10. Neurath, H., and Bailey, K., editors, *The Proteins: Chemistry, Biological Activity and Methods* (Academic Press, Inc., New York, 1953).
11. Springall, H. D., *The Structural Chemistry of Proteins* (Butterworths Scientific Publications, London, 1954).

**Energy Transfer and Biochemical Synthesis** (related papers: 16–22, 24, 25)

1. Arnold, W., and Sherwood, H. K., "Are chloroplasts semiconductors?" Proc. Natl. Acad. Sci. U. S. **43**, 105–114 (1957).
2. Bassham, J. A., and Calvin, M. C., *The Path of Carbon in Photosynthesis* (Prentice-Hall, Inc., Englewood Cliffs, New Jersey, 1957).
3. Chance, B., and Williams, G. R., "The respiratory chain and oxidative phosphorylation," Advances in Enzymol. **17**, 65–134 (1956).
4. Gutfreund, H., "The nature of entropy and its role in biochemical processes," Advances in Enzymol. **11**, 1–33 (1951).
5. Karreman, G., and Steele, R. H., "On the possibility of long distance energy transfer by resonance in biology," Biochim. et Biophys. Acta **25**, 280–291 (1957).
6. Krebs, H. A., and Kornberg, A. L., *Energy Transformations in Living Matter* (Springer-Verlag, Berlin, 1957).
7. Lehninger, A. L., "Oxidative phosphorylation" in *The Harvey Lectures 1953–1954*, The Harvey Society of New York, Series XLIX (Academic Press, Inc., New York, 1955), pp. 176–215.
8. Lehninger, A. L., Wadkins, C. L., Cooper, C., Devlin, T. M., and Gamble, J. L., Jr., "Oxidative phosphorylation," Science **128**, 450–456 (1958).
9. Reid, C., *Excited States in Chemistry and Biology* (Academic Press, Inc., New York, 1957).
10. Sogo, P. B., and Tolbert, B. M., "Nuclear and electron paramagnetic resonance and its application to biology," in *Physics*, J. H. Lawrence and C. A. Tobias, editors (Academic Press, Inc., New York, 1957), Vol. V, pp. 1–35.

**Genetics and Replication of Proteins and Nucleic Acids** (related papers: 26–31)

1. Cherry, C., editor, *Information Theory* (Butterworths Scientific Publications, London, 1956).
2. Cohn, M., and Monod, J., "Specific inhibition and induction of enzyme biosynthesis," in *Adaptation in Microorganisms*, R. Davies and E. F. Gale, editors (Cambridge University Press, Cambridge, England, 1953), pp. 132–149.

3. "Viruses," Cold Spring Harbor Symposia Quant. Biol. **18**, (1953).
4. Dulbecco, R., "Interaction of viruses and animal cells—a study of facts and interpretations," Physiol. Rev. **35**, 301–335 (1955).
5. Feinstein, A., *Foundations of Information Theory* (McGraw-Hill Book Company, Inc., New York, 1958).
6. Hershey, A. D., "Bacteriophages as genetic and biochemical systems," Advances in Virus Research **4**, 25–62 (1957).
7. Jacob, F., *Les Bactéries Lysogènes et la Notion de Provirus* (Masson et Cie., Paris, 1954).
8. McElroy, W. D., and Glass, B., editors, *A Symposium on the Chemical Basis of Heredity* (The Johns Hopkins Press, Baltimore, Maryland, 1957).
9. Pontecorvo, G., "Genetic formulation of gene structure and gene action," Advances in Enzymol. **13**, 121–149 (1952).
10. Shannon, C. E., and Weaver, W., *The Mathematical Theory of Communication* (University of Illinois Press, Urbana, Illinois, 1949).
11. Spiegelman, S., and Campbell, A. M., "The Significance of induced enzyme formation," in *Currents in Biochemical Research*, D. E. Green, editor (Interscience Publishers, Inc., New York, 1956), pp. 115–161.
12. Stanier, R. Y., "Enzymatic adaptation in bacteria," Ann. Rev. Microbiol. **5**, 35–56 (1951).

**Biological Effects of Radiation** (related papers: 32–35)

1. Bacq, Z. M., and Alexander, P., *Fundamentals of Radiobiology* (Academic Press, Inc., New York, 1955).
2. Hollaender, A., editor, *Radiation Biology*. Vol. I: High Energy Radiation (1954); Vol. II: Ultraviolet and Related Radiations (1955); Vol. III: Visible and Near-visible Light (1956) (McGraw-Hill Book Company, Inc., New York, 1954–1956).
3. Lea, D. E., *Actions of Radiations on Living Cells* (Cambridge University Press, New York, 1955), second edition.
4. Nickson, J. J., editor, *Symposium on Radiobiology* (John Wiley and Sons, Inc., New York, 1952).

**Molecular Organization and Function** (related papers: 36–40)

1. "Active transport and secretion," Symp. Soc. Exptl. Biol. **8**, (1954).
2. Brachet, J., *Biochemical Cytology* (Academic Press, Inc., New York, 1957).
3. Clarke, H. T., editor, *Ion Transport Across Membranes* (Academic Press, Inc., New York, 1954).
4. Conference on Tissue Fine Structure, J. Biophys. Biochem. Cytol. **2**, Suppl. (1956).
5. DeRobertis, E. D. P., Nowinski, W. W., and Saez, F. A., *General Cytology* (Saunders Company, Philadelphia, 1954).
6. Engström, A., and Finean, J. B., *Biological Ultrastructure* (Academic Press, Inc., New York, 1958).
7. Frey-Wyssling, A., *Macromolecules in Cell Structure* (Harvard University Press, Cambridge, 1957).
8. Mellors, R. C., editor, *Analytical Cytology* (McGraw-Hill Book Company, Inc., New York, 1955).
9. "Mitochondria and other cytoplasmic inclusions," Symp. Soc. Exptl. Biol. **10**, (1957).
10. Oster, G., and Pollister, A. W., editors, *Physical Techniques in Biological Research*. Vol. III: Cells and Tissues (Academic Press, Inc., New York, 1956).
11. Schmitt, F. O., "Tissue ultrastructure analysis: polarized light method," in *Medical Physics*, O. Glasser, editor (Yearbook Publishers, Chicago, 1944), pp. 1586–1591.

**Connective Tissue and Muscle** (related papers: 41–46)

1. "Fibrous proteins and their biological significance," Symp. Soc. Exptl. Biol. **9**, (1955).
2. Gustavson, K. H., *The Chemistry and Reactivity of Collagen* (Academic Press, Inc., New York, 1956).
3. O'Flaherty, F., editor, *The Chemistry and Technology of Leather* (Rheinhold Publishing Corporation, New York, 1956).
4. Peters, R., Hill, A. V., Huxley, A. F., Huxley, H. E., Wilkie, D. R., Bailey, K., Perry, S. V., Needham, D. M., Pantin, C. F. A., Gray, J., Katz, B., *et al.*, "Physiology of voluntary muscle," Brit. Med. Bull. **12**, 161–235 (1956).
5. Randall, J. T., "Observations on the collagen system," J. Intern. Soc. Leather Trades' Chemists **38**, 362–387 (1954).
6. Schmitt, F. O., "Macromolecular interaction patterns in biological systems," Proc. Am. Phil. Soc. **100**, 476–486 (1956).
7. Schmitt, F. O., "Giant molecules in cells and tissues," Scientific American **197**, 204–216 (1957).
8. Stainsby, G., editor, *Recent Advances in Gelatin and Glue Research*, (Pergamon Press, Inc., New York, 1958).
9. Szent-Györgyi, A., *Chemistry of Muscular Contraction* (Academic Press, Inc., New York, 1951), second edition.
10. Tunbridge, R. E., editor, *Connective Tissue: A C.I.O.M.S. Symposium* (Blackwell Scientific Publications, Oxford, 1957).

**Nerve** (related papers: 49–51)

1. Fernández-Morán, H., "The submicroscopic structure of nerve fibers," Progr. in Biophys. and Biophys. Chem. **4**, 112–147 (1954).
2. Geren, B. B., "Structural studies of the formation of the myelin sheath in peripheral nerve fibers," in *The Fourteenth Symposium of the Society for the Study of Growth and Development*, D. Rudnick, editor (Princeton University Press, Princeton, New Jersey, 1956), pp. 213–219.
3. Hodgkin, A. L., "Ionic movements and electrical activity in giant nerve fibres," Proc. Roy. Soc. (London) **148B**, 1–37 (1958).
4. Katz, B., "Microphysiology of the neuro-muscular junction. A physiological 'quantum of action' at the myoneural junction," Bull. Johns Hopkins Hosp. **102**, 275–296 (1958).
5. Palay, S. L., "Structure and function in the neuron," in *Progress in Neurobiology and Neurochemistry*, S. R. Korey and J. I. Nurnberger, editors (Paul B. Hoeber, New York, 1956), Vol. I, pp. 64–82.
6. Richter, D., editor, *Metabolism of the Nervous System* (Pergamon Press, New York, 1957).
7. Schmitt, F. O., and Geschwind, N., "The axon surface," Progr. in Biophys. and Biophys. Chem. **8**, 165–215 (1957).
8. Shedlovsky, T., editor, *Electrochemistry in Biology and Medicine* (John Wiley and Sons, Inc., New York, 1955).
9. Waelsch, H., *Ultrastructure and Cellular Chemistry of Neural Tissue* (Paul B. Hoeber, New York, 1957).
10. Weiss, P., *Genetic Neurology* (The University of Chicago Press, Chicago, 1950).

**Sensory Reception and Signal Processing** (related papers: 52–58)

1. Adrian, E. D., Bremer, F., and Jasper, H. H., editors, *Brain Mechanisms and Consciousness* (Blackwell Scientific Publications, Oxford, 1954).
2. Broadbent, D. E., *Perception and Communication* (Pergamon Press, New York, 1958).
3. Bruner, J. S., "Neural mechanisms in perception," Psychol. Rev. **64**, 340–358 (1957).
4. Cherry, C., editor, *On Human Communication* (John Wiley and Sons, Inc., New York, 1957).
5. Eccles, J. C., *The Physiology of Nerve Cells* (The Johns Hopkins Press, Baltimore, Maryland, 1957).
6. Geldard, F., *The Human Senses* (John Wiley and Sons, Inc., New York, 1953).
7. Granit, R., *Receptors and Sensory Perception* (Yale University Press, New Haven, 1955).
8. Magoun, H. W., *The Waking Brain* (Charles C. Thomas, Springfield, Illinois, 1958).
9. Miller, G. A., "Human memory and the storage of information," Inst. Radio. Engrs. Trans. PGIT **2**, 129–137 (1956).
10. von Neumann, J., *The Computer and the Brain* (Yale University Press, New Haven, 1958).
11. de Vries, Hl., "Physical aspects of the sense organs," Progr. in Biophys. and Biophys. Chem. **6**, 207–264 (1956).
12. Wiener, N., *Nonlinear Problems in Random Theory* (Technology Press, Cambridge; and John Wiley and Sons, Inc., New York, 1958).

**Specificity in the Chemical Control of Biological Systems** (related papers: 59–61)

1. Blum, H. F., *Time's Arrow and Evolution* (Princeton University Press, Princeton, New Jersey, 1951).
2. Ferguson, J. H., "General properties of blood: the formed elements," pp. 503–520; "Coagulation of blood," pp. 540–556; in *Textbook of Physiology*, J. F. Fulton, editor (Saunders Company, Philadelphia, 1949), sixteenth edition.
3. Gurd, F. R. N., "The specificity of metal-protein interactions," in *Ion Transport Across Membranes*, H. T. Clarke, editor (Academic Press, Inc., New York, 1954), pp. 246–272.
4. Henderson, L. J., *Fitness of the Environment* (The Macmillan Company, New York, 1927).
5. Klotz, I. M., "Protein interactions," in *The Proteins: Chemistry, Biological Activity and Methods*, H. Neurath and K. Bailey, editors (Academic Press, Inc., New York, 1953), Vol. I, part B, pp. 727–806.
6. Landsteiner, K., *The Specificity of Serological Reactions* (Harvard University Press, Cambridge, 1945), revised edition.
7. Pauling, L., and Itano, H., editors, *Molecular Structure and Biological Specificity* (American Institute of Biological Sciences, Washington, D. C., Publication No. 2, 1957).
8. Waugh, D. F., "Protein-protein interactions," Advances in Protein Chem. **9**, 326–437 (1954).
9. White, A., Handler, P., Smith, E. L., and Stetten, DeW., Jr., "Biochemistry of the endocrine glands," in *Principles of Biochemistry*, Part 6 (McGraw-Hill Book Company, Inc., New York, 1954), pp. 867–997.
10. Wintrobe, M. M., *Clinical Hematology* (Lea and Febiger, Philadelphia, 1956), fourth edition.

# 1
# Molecular Biology and the Physical Basis of Life Processes

Francis O. Schmitt

*Department of Biology, Massachusetts Institute of Technology, Cambridge 39, Massachusetts*

PRESENT-DAY research in the life sciences is characterized by unprecedented advances in the biochemical and biophysical analysis of the mechanisms underlying life processes. Biochemistry has passed from the early phase, in which the determination of the chemical composition of complex biomolecules was its major task, through the period of investigation of the detailed processes of intermediary metabolism, whereby "energy-rich" compounds become available to provide the energy for the building, maintenance, and repair of the organismic machinery. Biochemistry is turning its attention more and more to a study of the ways in which this available energy is coupled with physiological processes and particularly with biosynthetic reactions, as in the synthesis of the more critical biomolecules such as the proteins and nucleic acids and their derivatives. In these studies, biochemists require, in addition to the molecules that react in a given process, also an enzyme or enzymes whose role it is to catalyze the reaction in a very specific manner. Biochemistry thus has become to some extent the organic chemistry of enzyme-catalyzed reactions. But it has become obvious also that, although each energy-yielding or energy-requiring reaction in the cell is catalyzed by a specific enzyme, reactions in the cell involve enzymes organized in specific *patterns* or structural arrays. This organization or ordering of enzymes permits cycles of reactions to occur that would be highly improbable or impossible if the individual steps in the reaction were catalyzed by individual enzymes randomly distributed in the cell. The biochemist is now striving to isolate not only individual enzymes required for particular reactions but also specifically organized groups or "assemblies" of enzymes. These are found readily-available in partial systems of cell particulates such as the mitochondria and microsomes that can be isolated from fragmented cells and purified *in vitro* by differential centrifugation. With such subcellular, organized *systems* of enzymes it has become possible to achieve spectacular syntheses in the test tube of proteins, nucleic acids, and other critical biomolecules. With the help of isotopic tracers, it is possible also to determine the ways in which individual groups are shuttled about in these biosyntheses.

Biophysics, which is rapidly developing to the status of a major branch of the life sciences, is following the pattern set by biochemistry, and biophysicists are learning to operate in close harmony with biochemists. Just as biochemists had to determine first the composition and structural chemistry of the complex biomolecules, so in the early development of the field, biophysicists have to determine first, with the aid of crystallographic and physicochemical methods, the detailed configuration of the molecular chains of which the complex macromolecules are constructed. It will then be necessary to investigate the forces between the macromolecules and the smaller molecules in the protoplasmic environment and between various types of macromolecules. By now, it is fully apparent that the interaction properties of the macromolecules crucial for life processes depend upon the highly specific manner in which the covalent chains are structured and integrated with each other by secondary bonds within individual macromolecules. It is equally apparent that these interaction properties also spontaneously tend to form supermacromolecular aggregates whose stability and interaction properties depend sensitively upon their chemical environment of small molecules and ions. These supermacromolecular aggregates are, in fact, not only the active machinery by which muscles contract, glands secrete, and energy is mobilized by spatially organized enzymes in mitochondria, but also the means by which the genetic material is segregated and recombined with mathematical precision and by which the genetic coding is preserved and transmitted to posterity. Such specificity of interaction properties underlies the process by which the coded biochemical and biophysical instructions are passed from the germ cells to the various tissue cells, causing them to develop into the mature organism and to maintain and repair the organism until processes of aging and degenerative disease disturb the nice balance of interactions sufficiently to cause death.

Thus biophysics, like biochemistry, has to reckon with hierarchies of organization and with the properties that are characteristic of systems no less complex than those provided by living organisms at each particular level of organizational complexity, *viz.*, molecular, macromolecular, subcellular, cellular, supercellular, organismic, and superorganismic.

It may be apparent that some of the foregoing is based on prognostication of advances presently "in the works" or still to come, as well as being a statement likely to be made by the historian of mid-twentieth century life science. However, the characterization makes it clear that there are two rather different ways of investigating life processes, both of them highly important. One way is primarily analytical. It is illustrated by the biochemist's search for pure enzymes which specifically catalyze individual reactions. This leads not only to an understanding of the individual reactions, but also eventually to the discovery that these reactions constitute complex cycles of metabolic or biosynthetic reactions. The analytical method is characterized also by the biophysicist's striving to isolate fibrous macromolecules in monodisperse condition and to determine the physicochemical properties and intramolecular-chain configuration of these macromolecules. It is inevitable that such investigations must precede successful attempts to determine how the macromolecules interact to cause biological function. The analytical method has made possible most of the striking advances in the biophysical and biochemical sciences which, together with the parent science of general physiology, have come to be known as "molecular biology."

The hallmark of the alternative approach, which may be called the "organismic" or "systems" approach, is to allow the cell, tissue, organism, or groups of organisms to remain intact in their normal environment and to observe the interaction properties of these intact entities, their adaptation to their environment and their tendency to form still more highly organized entities or systems with new properties, many of which are seemingly unpredictable from a knowledge of less complex systems.

Most of the current advances in biophysical science, and certainly the major portion of the present Study Program in Biophysical Science, involve primarily the analytical approach of molecular biology. This may consist in the isolation and examination of particular cellular constituents at the molecular level; in the demonstration of the role of free radicals or "excitons" in particular cell processes; or in the application of polyelectrolyte theory in an understanding of muscle contraction, to cite a few examples. Actually, the tendency of physicists who have become biologists or biophysicists has been to search for simplified models of complex biological processes and, by a sophisticated study of such models, to discover fundamental new principles in biology. In this approach, the numerous complexities and vast areas of ignorance that lie between the simplified model system and the end biological process under study are characteristically and purposely avoided or neglected. It is recognized that there is a formidable "black box" between the molecular effectors, as studied in the model system, and the final behavior of the cells or organism under study. It is quite possible that this approach may lead to important breakthroughs in biological theory. One of the best illustrations of its effectiveness is the phage-microorganism studies and the modern biophysical approach to genetics generally. This subject forms an important part of this Study Program. These advances have not been limited to genetics, narrowly conceived as concerned alone with mechanisms of inheritance, but also have illuminated some of the basic biosynthetic processes by which the genetic determiners guide cellular differentiation and regulation. Thus, the contents of the "black box" are also being scrutinized wherever possible.

Actually, much of the substance of present-day analytical biology would, in these terms, be considered "black box" science—i.e., concerned with the intermediary detailed mechanisms as well as with eventual "basic" mechanisms. However, particularly in the framework of a Study Program such as this, it is important that the dynamic, adaptive, regulatory, homeostatic properties of organisms and cells be kept clearly in mind, as well as the more readily studied properties of simpler partial systems. For such purposes, textbook descriptions of the "typical" cell are of little help. Rather, the living cells themselves must be studied, and it is the purpose of the first papers in the Study Program to demonstrate some of these properties.

In carrying out my own assignment of discussing the role of cell constituents in life processes, I should like to sketch very briefly the present situation in cellular biology in the light of some of the great accomplishments and controversies of the past. It may be hoped that this not only will provide much needed perspective concerning the "systems" or organismic properties of living organisms but also may stimulate some to consider the vast opportunities in theoretical biology, a field that is almost nonexistent at the present time, at least if judged by the standards of theoretical physics. Of necessity, such a theoretical biology must deal not only with the properties of cellular constituents but also with the properties of the organism as a whole. It must seek to identify and characterize those properties of living systems which, like the quantum of action, are ingrained in nature itself and must be assumed as a principle of operation of the living system.

## HISTORICAL PERSPECTIVES

In characterizing the search for the physicochemical basis of life processes, it is pertinent to mention a few concepts and controversies as well as several of the major theories and principles. In these one may trace, from the very beginnings of Greek thought, the thread of the conflict between the atomistic or analytical and the holistic or organismic (molecular biology contrasted with organismic or evolutionary biology).

### Epigenesis *vs* Preformation

The principle of epigenesis—namely, that the *potentialities* rather than the miniature structures of the adult organism themselves reside in the egg or sperm and that, in development, the tissues and adult structures arise by an orderly series of structural and chemical changes—was firmly grasped by Aristotle and, in the seventeenth century, by William Harvey. However, in the seventeenth and eighteenth centuries, there arose speculative controversies about the possibility of "preformation"—i.e., the possibility that the egg or sperm contains within it an embryo fully formed in miniature, that in development the miniature organs unfold into the mature form. Many of the great scientists of the day adhered to this patently impossible concept, differing among themselves chiefly as to whether the sperm (Leeuwenhoek, Boerhaave, Leibnitz) or the egg (Swammerdam, Malpighi, Haller, Bonnet) contained the miniature structures. This kind of speculation was terminated by the demonstration by the early embryologists of the progressive manner in which new structures are formed from previously existing simpler systems. The investigations of the continuity of genetic determiners, as well as the pluripotential properties of the cells of the early embryo, were enriched greatly by the ingenious experimental work in developmental biology in the last few decades.

Obviously, a certain amount of basic information must be carried in the egg and sperm, and this is currently regarded as being the "coding" or instructions of how to construct the structures rather than being microeditions or templates of the structures themselves. It must be remembered that, in addition to the genetic coding which must be preserved intact in the germ cells for transmission to the progeny, there is also the coding that goes to every cell of the body and that directs the chemical and structural differentiation of the various tissues and organs during development. For example, certain cells of the developing embryo must be instructed to become differentiated from their fellows as limb buds and these in turn must cause the orderly development not merely of the skeleton and muscles of the limb but also the appropriate innervation from the developing nervous system, as well as the various secondary patterns of organization of the limb such as the color and patterns of feathers, hair, and so on.

Great emphasis in recent years has been placed on the mechanism by which deoxyribonucleic-acid (DNA) macromolecules uncoil, replicate, and recombine, in explanation of the nature and action of the genetic determiners, and on the manner in which ribonucleic acid (RNA) is coded by the DNA and is, in turn, made the code for the biosynthesis of proteins and other complex biomolecules in the cytoplasm. There is, however, little definitive evidence presently available concerning the manner in which this coding is regulated, modulated, activated, and inhibited so as to make possible the adaptive, pattern-directed reactions that characterize the normal processes of development, maintenance, and repair. There is a tendency among biophysicists to take all of this more or less for granted as being primarily a problem of complexity—properties of the black box. But this may well turn out to be as much a part of the essential problem of life as the definitive coding and biosynthetic processes as such. We may, in fact, be dealing here with a bit of the essence of the biological problem generally—i.e., the difficulty in arranging not only to provide primary, secondary, and derivative codes for particular reactions, but also to see to it that the proper substrate and enzyme molecules are ready at hand at the right places and all at the right moment so as to make possible orderly development, regulation, maintenance, and repair. As Delbrück* has indicated, there is still a great gap between the DNA map and the genetic map; probably also between many other complex biological processes and the macromolecular coding that apparently directs the process. The nature of this regulatory, adaptive, feedback mechanism also is discussed in this Study Program, though regrettably not nearly so thoroughly as the importance of the subject demands.

## Cell Theory

Formulation of the cell theory in 1838–1840 by Schleiden and Schwann proved to be one of the greatest generalizations in biology, ranking in importance with the theory of evolution, because it made possible the explanation of biological process in terms of the cells of which the organism is composed. The rapid rise of physiology and of pathology as well as an understanding of the special significance of the germ cells followed as soon as the concept of the cell as the unit of life was fully apprehended.

In the century that followed, it became obvious that the idea of the cell as the unit of life must be modified. Any particular cell is what it is by virtue of its relationship to all the other cells in the body. The organism is an ideal federation of cells, each with a high degree of specialization, division of labor, and a corresponding mutual interdependence. It is the organism that is alive in the strict sense of the word, though individual constituent cells may readily be removed and cultivated *in vitro* under appropriate conditions.

## Theory of Organic Evolution

Mention is made of this most important hypothesis primarily to emphasize the fact that it was the result of naturalistic observations on *whole* organisms in *whole* communities. It was developed quite independent of the then current tendency to focus attention on the cells and their parts. Darwin was, in fact, singularly misleading and ineffective as an analytical "molecular biologist" (witness his theory of pangenesis and his hypothetical "gemmules"). However, the evolutionary theory strongly reinforced the need to explain the mechanism of inheritance and provided a challenge to molecular biologists for more than a century to come.

Mendel's work provided a beautifully clear set of basic relationships. But, unfortunately, they were soon

* M. Delbrück, "Atomic physics in 1910 and molecular biology in 1957," a lecture delivered at the Massachusetts Institute of Technology in connection with the Karl Taylor Compton Lectures by Professor Niels Bohr.

forgotten and had to be reintroduced at the turn of the century.

### Theories of the Fundamental Structure of Protoplasm as the Physical Basis of Life

Soon after the formulation of the cell theory, attention was focused upon the protoplasm (so called by Purkinje in 1840) as the physical basis of life.† The search was on to discover what is actually alive in the cell. With the rapid development of optical instrumentation, methods of "preserving" (fixing), sectioning, and staining tissues, there was ushered in the golden age of descriptive morphology. Investigators were in hot pursuit of the "living subcellular particles," the ultimate living units.

To discoveries made during this very active period one owes the development of the knowledge of some of the most fundamental cell phenomena. In this active expansion of descriptive cytology from 1870 to 1890, three general theories were proposed for the fundamental structure of protoplasm as the physical basis of life, as follows:

#### *1. The Fibrillar Theory ("One-Dimensional" Arrays)*

According to this view, proposed most succinctly by Flemming (1882), "The essential energies on which life depends have their seat in fibrillae." Heidenhain's book *Plasma und Zelle* abounds in examples of fibrous structures in cells (muscle, mitotic apparatus, cilia, flagella, sperm tails). Two fibrous types were distinguished: the unbranched (filar) and the branched (reticular). The fibrils are immersed in a clear "hyaloplasm" (Leydig).

#### *2. Alveolar or Foam Theory (Two-Dimensional, Membrane-Limited Structures)*

The chief proponent of this theory, Bütschli (1878–1892), believed that protoplasm is a diphasic system containing a continuous clear, viscous fluid (hyaloplasm) and a discontinuous phase of microsomes (*sic*) which are probably lamellar in life. In this theory, fibers were ignored. However, since fibers undoubtedly do exist, Strasburger (1892) proposed a compromise. Protoplasm contains two plasms: a "trophoplasm" which is membranous or alveolar and whose function is metabolic and nutritive, and a "kinoplasm" which is fibrillar and which mediates movement or contraction of any kind.

#### *3. Granule Theory*

Between 1880 and 1890, Altmann exhaustively described certain granules ("plastidules") in protoplasm which he regarded as the organic units of the cell, as the cells are the organic units of the organism. He considered the granules to be "living elementary organisms (bioplasts)." In 1890, he coined the phrase "*omne granulum e granulo*," paraphrasing Virchow's earlier (1855) dictum "*omnis cellula e cellula.*"

However, as is now known, the cytologists were being led astray by their insistence that the structures they observed in fixed tissues were present *as such* in the living cells. A halt was called, particularly by W. B. Hardy, at the turn of the century, to the speculative race by descriptive morphologists to discover the physical basis of life, through observation of structures in fixed and stained tissues. He showed that many of the structures claimed by cytologists to be the physical basis of life are, in fact, coagulation artifacts. He was able to imitate many of them in artificial systems subjected to precipitation under controlled conditions.

J. Loeb's trail-blazing work on the physicochemical and colligative properties of proteins helped remove the needless complexity with which the early colloid enthusiasts had cloaked their studies of protoplasm.

The startling work of Langmuir, Harkins, Rideal, Adam, and Gorter on molecular orientation in mono- and multi-molecular films laid a basis for an understanding of the properties of interfacial films and factors stabilizing polyphasic systems. Studies of the paracrystalline, mesomorphic state, beginning with the work of Lehmann on fluid crystals, led to an understanding of molecular ordering in anisotropic systems.

The return to the study of *living* cells by means of micromanipulation (Peterfi, Chambers) and by tissue culture (Harrison, Levi, Weiss) helped offset the eclipse into which descriptive morphology had been sent by the skeptical, physicochemically minded physiologists, and revealed new and important properties of cells and tissues.

With the introduction of electron microscopy, and particularly since the development of techniques for examining tissues in ultrathin (100 to 500 A) sections by high-resolution (*ca* 10 A) electron microscopy, we seem to have entered upon a new golden age of *ultracytology*. From a study of tissues fixed by only a few types of fixatives (because of limitations imposed by plastic embedding and other preparative techniques as well as because of the intrinsic qualities of preservation of the cells), descriptive papers in ever-increasing number (and length) are being produced as electron-optical equipment is made easy for the nonphysicist to operate and is being installed in more and more laboratories the world over.

As in the analogous period of development of light microscopy in the last century, the new flood of electron-optical observations has brought with it important new discoveries concerning the structure of protoplasm and its subcellular, "submicroscopic" particulates or organelles. There is also danger that one may again be misled in matters of interpretation because of the fixation

† For an excellent review of these early studies, the reader is referred to the book of E. B. Wilson, *The Cell in Development and Heredity* (The Macmillan Company, New York, 1925).

artifacts that are now being observed at a hundred times the magnification available to the morphologists of the last century. However, the danger of this is greatly reduced because other biophysical and biochemical advances make it possible to obtain independent evidence concerning the molecular organization of protoplasm.

### CURRENT CONCEPTS OF CELLULAR ULTRASTRUCTURE IN RELATION TO FUNCTION

Electron microscopy has revealed highly significant new facts about each of the three types of protoplasmic structures that figured prominently in early theories of histology: the granular, the membranous, and the fibrous structures. For present purposes, it suffices to indicate briefly the upshot of the more significant of these discoveries.

By appropriate fragmentation or maceration of cells, followed by differential centrifugation and purification, it is possible to prepare, in separate fractions, certain of the cellular organelles believed vital by the early cytologists. Their biochemical properties and ultrastructure also may be studied in one and the same preparation.

In this way, it was shown that mitochondria are the "power plants" or energy source of the cell by which, through oxidative phosphorylation, phosphate-bond energy is made available through the splitting of adenosine triphosphate (ATP), the metabolic fuel of the cell.

The most characteristic aspect of mitochondrial organization is the internal layered structure. Biochemical studies show that both the outer enclosing membrane and the internal lamellae contain the enzymes required for the citric-acid cycle. From the work of Lehninger and others, it seems possible that the various individual enzymes involved in such biochemical cycles occur as clusters or "assemblies" anchored in or upon the membranous structures. The direct demonstration of such assemblies and of the molecular organization of the layers now becomes a challenging problem in electron-optical technique. Thus, although mitochondria play a vital metabolic role in protoplasmic function, they are far from being the elementary units of life that Altmann pictured them to be.

In the presence of appropriate substrate molecules, cofactors, and activators, fragments of macerated cells have proved capable of rather spectacular biosynthesis of peptides, steroids, and other complex biomolecules. Apparently, sufficient amounts of enzymatically active cell surface are retained in such preparations to produce biosynthesis. This has proved to be a very valuable biochemical tool particularly when used in conjunction with isotopic tracers.

Similar results were obtained with the so-called "microsome" preparations from fragmented cells subjected to differential centrifugation. The microsomes occur in the fraction lighter than that containing the mitochondria. Electron-microscopic examination shows these microsomes to be fragments of the membrane-limited structures in the cell ("endoplasmic reticulum," "cytomembranes," "ergastoplasm") which, when associated with RNA-rich particles ("ribosomes"), are thought to constitute the biosynthetic center of the cell. Electron-microscopic examination of very thin sections of actively synthesizing cells suggests that the cytoplasm may be divided into two phases by such membrane-limited structures. One phase is continuous with the space between the nuclear membranes and perhaps also with the extracellular fluid by means of micropores in the limiting cell membrane. This phase contains a relatively homogeneous fluid material which doubtless contains biosynthetized materials. The second phase contains the mitochondria and the RNA-rich granules (ribosomes), which are the biochemical and structural apparatus of biosynthesis. Thus, there may actually be two plasms in cytoplasm (one of which might perhaps be called trophoplasm) as adherents of the alveolar theory supposed. However, the membranes that separate the two are themselves metabolically very active ones which, by partitioning off regions of the cytoplasm, facilitate the biosynthetic process.

Other membrane-limited structures are also of fundamental significance. Of these, perhaps only two need be mentioned: the Golgi system and the limiting envelope or plasma membrane of the cell.

The Golgi "apparatus," which was considered by cytologists to play an essential role in secretion processes, came under severe criticism as a possible fixation artifact. However, electron-microscopic observations indicate that it is a multilayered structure frequently seen in the secretogenous zone of the cell. It is supposed by some to be the "wrapping and packaging" department of the cell in which secretion granules and other particulates may be enclosed in a membranous wrapping which, by fusion with the cell membrane, allows the granules to traverse the plasma membrane.

The plasma membrane serves not only to enclose the cell and to direct the molecular traffic into and out of the cell but also, presumably, to mount the biochemical mechanism by which solutes, such as sodium ions, may be transported or "pumped" against an activity gradient. Electron-microscopic observations indicate that such membranes have a structure similar to that of the typical unit membrane in cytoplasm generally. This consists of two dense lines separated by a less dense area. The total thickness is 60 to 70 A and may consist of a bimolecular layer of mixed lipids with a monolayer of protein or other material on either interface. As in the case of mitochondrial membranes, it remains to be seen whether or not improved techniques can reveal the presence of specialized, possibly enzymatic, structure having a role in transport or other biochemical processes mediated at the cell surface.

One comes finally to the fibrous types of arrays which are ubiquitous in protoplasm, being particularly conspicuous in the connective tissue, in the mitotic apparatus of dividing cells, and in tissues such as muscle where rapid and reversible interaction of the macromolecular lattice causes contraction or produces tension.

Since the early pioneering work in the twenties and thirties, particularly that of Astbury, many fibrous proteins have been isolated and investigated with respect to their structural and physicochemical properties. It has been found that these elongate macromolecules manifest a high degree of specificity in their ability to aggregate in ordered arrays, to change the pattern of interaction with change in the chemical environment, and possibly to change the intramolecular configuration (helical, supercoiled, and uncoiled, random chains). This matter is dealt with in detail in this Study Program in the case of certain proteins such as collagen and certain muscle proteins which manifest a specific band structure in the electron microscope and by means of which their aggregation states can be determined.

The production of fibrous lattices in cells and tissues apparently depends upon such builtin properties of the macromolecules. There are no microworkmen to go about in the cells joining up the proper molecular beams of the scaffolding according to some predetermined blueprint! Rather, after the elongate macromolecules have been synthesized, and possibly activated by preliminary enzyme action, they automatically polymerize and aggregate with each other according to their builtin specificity, to form the structure or microlattice appropriate in each protoplasmic situation. Thus, the genetic code may be thought of as arranging for the biosynthesis of the complex macromolecules of many kinds, but interaction of the macromolecules under specified conditions occurs spontaneously. In many instances, for example, characteristic fibrogenesis can occur *in vitro*.

Thus, fibrous arrays are essential, not only to provide structural and contractile elements but also, as in the case of the nucleic acids, to provide a means of arranging specific linear sequences of determiners corresponding to particular chemical codes. However, even the most enthusiastic exponent of DNA and viruses as linear codes, would hardly ascribe to them the properties of life such as was suggested by Flemming and the adherents of the fiber school.

Thus, as the specialized biophysical and biochemical functions of morphologically identifiable components, such as the mitochondria, ribosomes, Golgi system, and other membrane-limited structures, come to be better understood (chiefly by a study of partial systems *in vitro*), the importance of the organization of the cell as a whole becomes more and more evident. It is this highly coordinated temporal and spatial ordering that provides the dynamic, regulatory, homeostatic, and adaptive properties of the cell which are the essence of life. Somehow, the biosynthetic capabilities of the cell are regulated and geared into the energy-yielding reactions of metabolism so as to produce the substances that may be required not just locally within the synthesizing cell but by cells at remote loci in the organism. One has only to think of the complex ordering that controls the processes of growth and development in which the individual cells seem to be governed by master plans or codes (morphogenetic "fields" of general or local specificity) but in which the individual cells obviously must be constantly producing substances which make it possible for them to react meaningfully to the general and local fields. Indeed, the very existence of the field depends upon such processes in the individual cells. Yet the master plan (genetic coding) is characteristic of the entire organism as a whole, not only during development and growth but in all of the processes of aging.

One sees illustrated here in especially impressive fashion a general property that seems to characterize living systems at all levels of complexity and that has in it an indeterminacy which is of fundamental significance. Somehow, the appropriate reactions occur at the right places and at the right time, in individual cells and in the organism as a whole, to produce the eventual result preordained in the coding of the DNA transmitted in the germ cells.

It is easy to see, therefore, why it is difficult at the present stage of scientific development to deal with problems of this nature at the molecular level with precise methods of physics and chemistry, and at the same time to characterize with equal precision the ordering and the reactions that govern the over-all "directive" behavior of the organism as a whole.

To a certain degree, the difference in mode of attack is revealed in the various papers in this Study Program and is reflected also by the professional orientation of those whose efforts are primarily in the area of "molecular biology" as compared with those dealing with the larger, systems aspects of cell, tissue, and organismic function.

# 2
# Cellular Dynamics

PAUL WEISS
*The Rockefeller Institute, New York 21, New York*

BIOLOGY is now in an exciting phase marked by the confluence of different disciplines of research in the attack on focal problems. But the enthusiasm raised by the spectacular results of combined physical and chemical approaches to biology sometimes has outraced people's ability to keep pace conceptually with the technical developments. As a result, we frequently try to fit our questions to the very limited answers which our fragmentary knowledge has been able to provide, instead of boldly facing the much broader questions posed by living systems and phrasing them in such a way that still more penetrating answers may be obtained in the future. In order to do this, one needs to focus on the real living objects, rather than on the somewhat fictitious and oversimplified models that one is prone to formulate, as intended targets for physical and chemical attack. Models are necessary, but they must bear more than a coincidental resemblance to the real object if they are to serve as meaningful aids to analysis.

As the result of this extensive use of overly simple models, notions about the cell have become at times slightly vague and unrealistic. It would be presumptuous in one single chapter to try to do more than to just give a few illustrative examples of what the real cell is like. The best that can be hoped is to show the change that has occurred in our thinking about the cell from the static to the dynamic—that is, from static organization to organized behavior. Much of the knowledge of what the cell is has come from ruling out erroneous conceptions of what it is not. Progress has come from narrowing the margin of error. By being exposed to a few examples of the living cell in action, the reader can judge for himself whether or not his mental picture of the cell corresponds to the real thing.

The first example deals with one of the most prominent characteristics of the cell—its shape. Figure 1(a) shows a textbook picture of a particular cell found in the cerebellum. One sees that the cell body has elaborate ramifications. This is the way one usually learns about a cell—through pictures in a book; and since the picture looks the same in all of the thousands of copies of a textbook of microscopic anatomy, one forms the notion that all such cells are like tin soldiers stamped out according to a standard pattern. Thus, the mental habit of cell form as something static and rigid becomes engrained. The truth is, however, that no two cells are ever strictly alike, nor is any one cell quite the same at different times of its life history. It is this history which a static textbook picture fails to reveal. To stress this fact, Fig. 1(b) shows by comparison a Chinese brush drawing of some shrub. Here common experience tells us that the bush has not been stamped out in the shape in which one finds it. It has grown into that shape from seed. So, what one sees as pattern in the shrub is merely the residual record of prior activities of that particular protoplasmic system. In other words, shape is simply an index of antecedent processes by which that shape has come about.

Something else is lacking, however, in both of the pictures besides the account of prior events. The objects are portrayed against a blank background as if they were in a vacuum. Again, in the case of the plant, one knows from daily life how vital the invisible air is for its existence as a provider of chemical necessities.

Now, in the case of the cell, the medium is involved even more intricately: it provides not only chemical components for nutriment, but also a physical framework that integrates the separate cells into a structural continuum. The existence of this continuum usually remains unrecognized because staining techniques deliberately leave the substratum out of sight. Yet, to the classical morphologist only seeing was believing and, what is worse, not seeing amounted to not believing. This attitude is undergoing radical change.

Processes as such are not visible. What is visible is a constellation of elements at different stages of the process. Visible form, a pattern at any one stage, must be viewed as the product of antecedent formative processes. The cell thus appears as a system of highly complex, but ordered, molecular populations grouped in a hierarchy of supramolecular complexes, in constant interactions among themselves and with their environment (that is, the space beyond the cell border), leading to features, some permanent, others transitory, the visible expression of which is recorded as shape. If the environmental conditions are reasonably constant for a group of cells of the same type, the behavioral history of the latter will be reasonably similar so as to end up

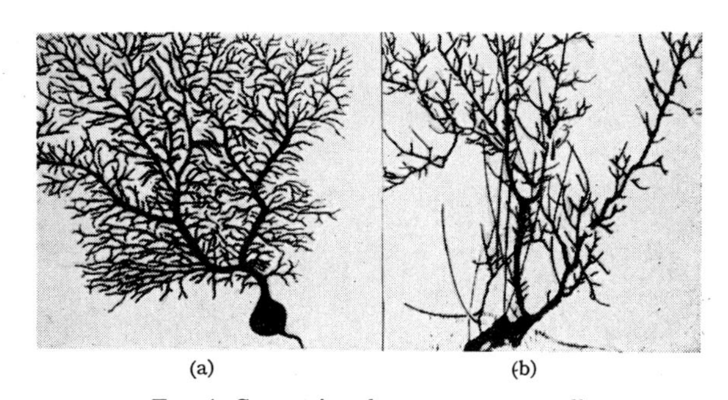


FIG. 1. Comparison between a nerve cell (a) and a plant (b).

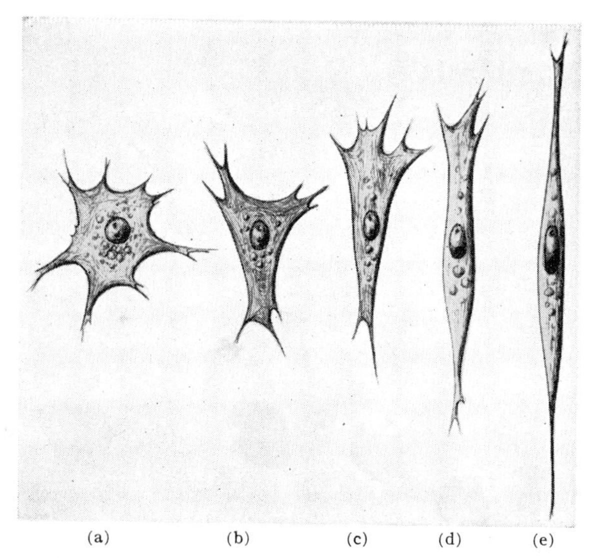


FIG. 2. Samples of shapes assumed by cells of the same connective-tissue cell strain in tissue culture.

with reasonably similar and classifiable shapes of the sort that have made it possible for sciences of microscopic anatomy and microscopic pathology to develop. As soon as there is a change in the conditions, the behavioral response of the cell likewise changes and the familiar shape derived from normal standard conditions ceases to be a diagnostic sign.

A most dramatic illustration of this situation is seen when cells are taken out of their normal site in an organism and transferred into an extraneous medium in tissue culture. As an example, a time-lapse phase-contrast microcinematograph* of human-liver cells spread on glass in horse serum (film made in my laboratory by A. Cecil Taylor and Albert Bock) shows up impressively the lack of fixity and the incessant reshuffling of cell content and contour. No static description can do justice to this vivid record of ever-changing activity. These liver cells in culture look quite different from those one would be used to seeing in stained sections through an intact liver. Except for shape, however, they still possess most of the essential properties of liver cells. In a third type of setting, for instance, in suspension, they would assume still other shapes.

Thus, one realizes that there is no way of getting a fully valid description of a cell except by studying its behavior under as wide a spectrum of conditions as is feasible. Cells of different kinds behave differently. While the transfer to tissue culture alters their morphological expressions markedly, they do retain their constitutional distinctions of behavior.

In conclusion, one is led to the thesis that cell shape is the result of a distinctive behavioral reaction of a living cell to its environment.

To make this concrete, consider a specific example. A colony of fibroblast cells cultured in dilute blood plasma yields a wide spectrum of shapes. The same cell can appear in any of the series of forms pictured in Fig. 2, ranging from the bipolar spindle at one extreme [Fig. 2(e)] to the multipolar star at the other [Fig. 2(a)]. The shape thus depends upon the number of directions in which the cell border shows radial extensions. The tips of these extensions are the active mobile organs of the cell. They push outward and thus distort the originally rounded surface of the cell. Evidently, if there are only two processes tugging in opposite directions, the cell body is drawn out between them into the shape of a spindle [Fig. 2(e)]. If there are three major protrusions, the cell assumes a tricornered shape [Fig. 2(b)], and with even more processes along its circumference it approaches more and more a star shape [Fig. 2(a)]

Must one accept this spectrum merely as a given descriptive fact, or can it be explained causally in the way physical systems are treated? The answer is that, to a certain extent, the whole series can be expressed in terms of a single function derived from a study of cellular behavior. Cells do not live in a structural vacuum as is the illusion created by standard histological preparations, which, by stressing only those features which happen to be stained, obliterate the structural continuum within which the cells reside. In tissue culture in a blood-plasma clot, for instance, this continuum is provided by a network of fibrin fibers—aggregates of molecular chains of varying diameter from submicroscopic to microscopic dimensions, the meshes of the network being filled with serum (Fig. 3). It is in this fibrous jungle that these cells live and move, applying themselves to the interfaces between the fibers and the liquid medium as to a trellis. As was mentioned before, the shape of these cells is determined by the number of protrusions from their surface. One can go one step further and prove that the number of such processes, in turn, is a function of the fibrous constitution of the medium.

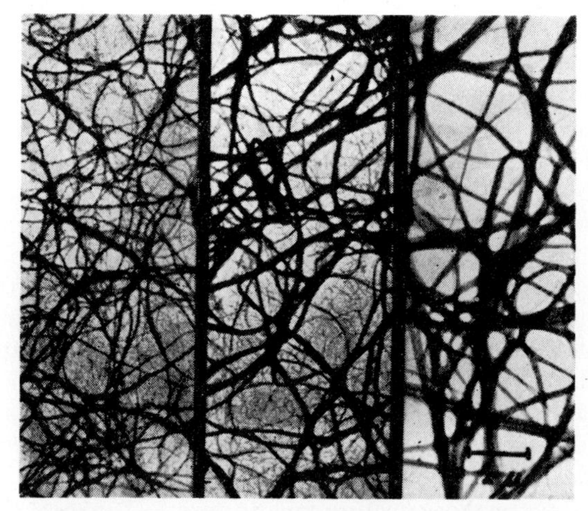

FIG. 3. Electron micrograms of plasma clots coagulated at different $p$H values (from left to right: alkaline, neutral acid).

* The motion pictures referred to in the text were shown at the Study Program in Biophysical Science in conjunction with the lecture on which this article is based.

The relevant interaction is between the cell surface and the fibers in its microenvironment. To understand such surface reactions, one must give up in the first place the outdated notion that the cell surface is a sort of static cellophane-like bag. This may be true of some specialized cell types, for instance, the cellulose membrane of a plant cell, the capsule of a bacterial cell, or the envelope of a red blood corpuscle. But in most types of cells, the surface is far from stable and is by no means of identical composition and state all over the cell. In tissue culture, this state of disequilibrium manifests itself in the continual thrusting forth and withdrawal of surface processes at the expense of cellular energy, showing great variations of the contractile force along the surface. Temporarily weaker points along the surface thus become outlets for thrusts. Such microleaks or "herniations" may occur at random or they may be determined systematically by outside factors, of which one of the most important is the encounter of a fibrous-liquid interface with the cell surface. The fiber contact, in a sense, pricks the cell surface locally. The strength of the resulting strain can be shown to vary with the size of the fibers. Hydrodynamic, viscous, and elastic competition for outflow favors fibers which have a larger diameter (Fig. 4). Consequently, the prevalence of a few major protrusions over minor ones may be expected to be the greater, the larger the average fiber size is in the medium.

These predictions have been tested (jointly with B. Garber) by culturing cells in plasma clots containing fibrin fibers of different average dimensions. The average diameter of such fibers is a function both of *p*H (Fig. 3) and of plasma concentration, larger fibers being formed at either lower *p*H or higher plasma concentrations. It actually was found, in line with expectations, that the ratio of bipolar cells (few processes) over multipolar stellate forms (many processes) increased as a steady linear function as the plasma concentration was raised

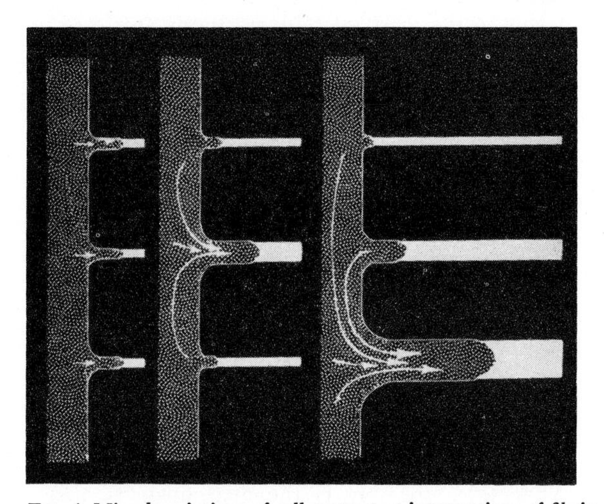

FIG. 4. Microherniations of cell content at intersections of fibrin fibers of various sizes with cell surface. Arrows indicate protoplasmic outflow.

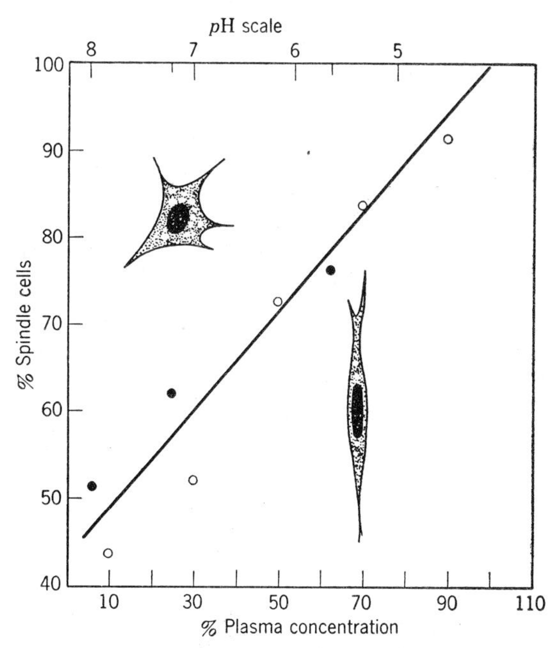


FIG. 5. Distribution of cell shapes in populations cultured in plasma clots of different average fiber sizes.

or as the *p*H during clotting was lowered (Fig. 5). Other criteria of cell shape, such as the ratios of length over width of the cell bodies and of the cell nuclei, showed correspondingly systematic changes. In other words, the whole gamut of shapes displayed by this particular cell strain could be written in a single formula derived from insight into the mechanisms by which deformations into one or another shape come about. The point to stress is that one gets further by studying realistically the formative process rather than by dwelling upon pictorial samples of forms already achieved.

At the same time, it must be stressed that the formula is a probabilistic one, for to predict just what any individual cell will look like is impossible because of the accidental nature of the details of its surroundings. The microclimate and the microenvironment of the individual cell are unique, unknown in each particular instance, and this establishes a certain degree of variance for the actual expressions within each cellular system which is built into its nature.

In a different medium, the same cell strain would give different responses. For instance, these cells, when suspended in a liquid medium without interlaced fibers, would manifest the inequalities along their surfaces by blunt herniations rather than by the pointed protrusions noted along filaments. As a result, such cells appear to be blistering and boiling along their surfaces as can be seen in cinematographs of unattached single cells, and is well known from the loose cells in the late stages of cell division. Conversely, cells of different tissue types would show morphological responses different from the strain just exemplified. The more a cell is given to producing internal structures that serve as a cytoskele-

FIG. 6. Effect of regionally varying degrees of tensions on the organization of a fibrin network and, through it, on the morphology and orientation of enclosed cells.

ton—such as in sperm cells, higher forms of protozoans, or muscle fibers—the less will its shape be codetermined by physical constellations in its environment, although even in the most extreme case the polarization of the axiated system of the cell is presumably still a response to external gradients or to other inhomogeneities of the environment. There is an enormous task laid out here for future detailed investigations on the physical factors involved in cytogenesis and morphogenesis.

As a further example of the complexities involved, one can again cite experiments with connective-tissue cells in tissue culture. A long time ago, I became preoccupied with the role of external factors in guiding the oriented movements of cells. In 1927, I succeeded in orienting cells in tissue culture by applying stretch to the blood-plasma clot; the cells assumed a common orientation along the lines of tension. The analysis of this phenomenon over the years has led to the following summary conclusion: The primary effect is on the orientation of molecular chains which become aligned in the direction of the stretch. When cells are contained in such a medium, their shape—that is, behavioral deviation from a sphere—simply reflects the degree of structural organization of the underlying submicroscopic fibrous network. Where all fibers run parallel and, therefore, where one direction only is open to the cells, they naturally become bipolar, all pointing in the same direction (Fig. 6). With decreasing amounts of stretch and correspondingly less rigorous orientation of the fibrin, cell forms grade over from the strictly determined spindles at one extreme to the only probabilistically describable cells of the multipolar sort mentioned in the earlier example. In the present case, however, the average shape does not vary from clot to clot in accordance with the average constitution of the clot, but varies locally within the same clot in accordance with a systematic variation of an extrinsic factor, namely, stretch.

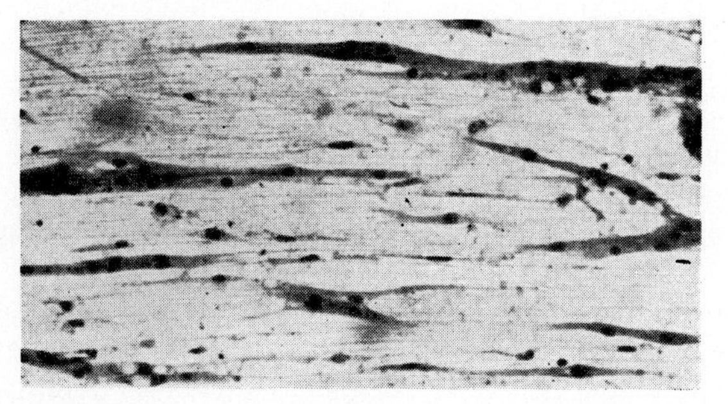

FIG. 7. Elongation of erstwhile-round connective-tissue cells placed on a sheet of parallel collagen fibers from the interior of a fish scale.

However, the immediately relevant thing for the orientation of the cells is the orientation of the fibrous pathways which they are bound to follow, and it is immaterial that, in those earlier experiments, the agent for producing such fibrous orientation was stretch. Even though the normal organism frequently uses stretch in that capacity, other forces also are at work to produce oriented fiber patterns, which likewise act as guides for cells. The inside of a fish scale, for instance, contains layers of collagen fibers beautifully arrayed in parallel lines, and when loose spherical cells are deposited on such a sheet they immediately assume spindle shapes, with the axes strictly aligned along the fibrous substratum (Fig. 7). Therefore, to carry the analysis further, one must study why a round cell on a linear track becomes correspondingly deformed.

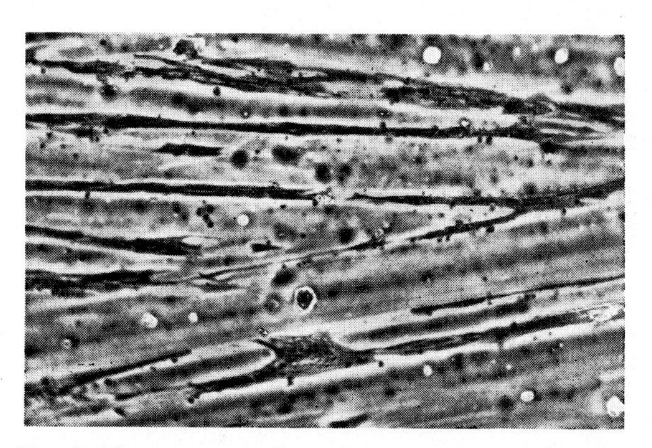

FIG. 8. Phase-contrast photomicrogram of cells which have become elongated along streaks of silicone paste on glass (round white and dark splotches are clusters of cells over nonadhesive silicone where they have been unable to attach themselves).

A clue to the mechanism operating in this case may be obtained from the following experiment. A glass slide first is streaked with silicone paste or with cholesterol. A loose cell suspension then is placed on top in an appropriate liquid medium. As the cell surface does not adhere to cholesterol but does adhere to glass, the cell becomes drawn out along the striplets of bare glass; hence, it is oriented parallel to the streaks (Fig. 8). Shape is determined here by the differential adhesion of different parts of the cell surface to the substratum. An even more impressive example of this kind is provided by cells which have been seeded out on glass scored with a microlathe (experiments performed with A. Cecil Taylor). These cells become deformed from a spherical to an elongate shape in the direction of the microgrooves (Fig. 9).

What then precisely is the mechanism of this response? To understand it, consider an older experiment in which spindle-shaped cells were made to spread out flat when they were forced to splash against a smooth surface from above. Such a surface offers the cell innumerable directions in which to radiate at the same time. Thus, the periphery of the cell actually expands concentrically, flattening the cell into a disk (Fig. 10). This transformation is accompanied by profound changes in function—the cell becomes phagocytic—and in the distribution of intracellular materials, further emphasizing the intimate dependence of chemical activity on the physical constellation in a cell. Similarly, a freshly isolated cell set down on scored glass spreads out at first in all directions, but only those sectors of the cell border which lie in a linear direction of the substratum retain a foothold and continue to advance. The other sectors are retracted and become consolidated as cell flanks.

Each cell surface thus acquires radically different properties at the ends and along the flank. The two "engines," one at either end, remain active and by extending in opposite directions simply draw the cell out into an elongate form. Such elongation again has further consequences for the cell. For instance, the mitotic spindle preparatory to cell division mostly will become aligned with the long axis of the cell so as to make elongation a major factor in the orientation of cell growth. But because of the opposing tugs, to which this elongate cell is subjected from its two active ends, there is no net forward movement. Such a cell simply shuttles back and forth about a stationary position, comparable to Brownian motion, depending upon which one of the two ends happens to have the upper hand at the moment.

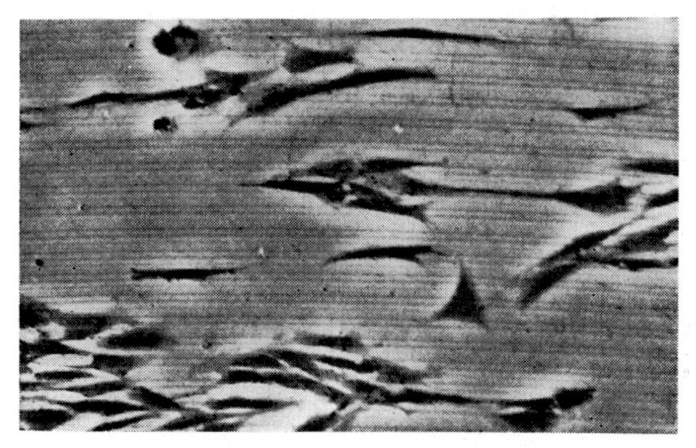

FIG. 9. Cells which have become elongated along the microgrooves of scored glass.

Thus, although the foregoing has brought some deeper understanding of cellular orientation, it tells nothing about the mechanism of cellular locomotion, which remains one of the basic unsolved biophysical problems. Something is known about those cases in which cells have special locomotor organs, such as cilia or flagella, but when it comes to cells moving with their free, unstable surfaces devoid of structural specializations that could be related to motility, ignorance is profound. Cyclosis in plant cells, the gliding of slime molds, the shift of a sheet of skin to cover a raw wound, the invasion of tissues by metastasizing cancer cells, or the penetration of leucocytes through capillary walls to converge on a focus of infection—in none of these cases is it known just how the cell achieves these movements, except that it is suspected that gelation-solvation cycles or contraction-relaxation alternations may somehow be involved. This is one of the most neglected areas of physical approaches to biology. Not only is the mode of locomotion shrouded in ignorance, but also there is equal uncertainty about the reasons why a free cell, which can extend in many directions, often advances steadily in one direction to the exclusion of others. As was just said, a cell left to its own devices in an isotropic environment strays at random with no net dislocation.

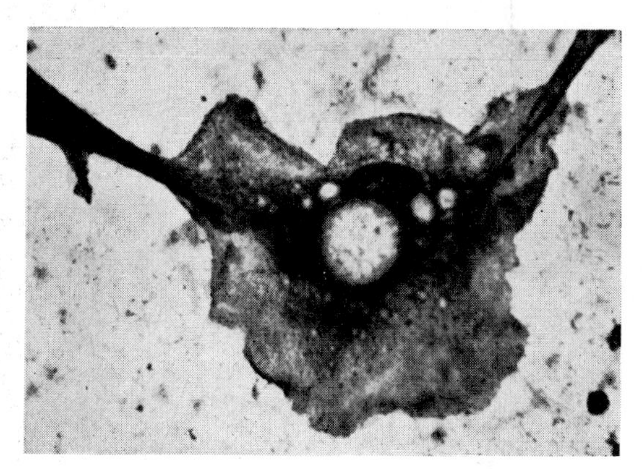

FIG. 10. Spindle cell expanding into a flat disk on impact with glass surface.

There have been many theories and speculations about the directive movement of free cells. Once again, one can illustrate how progress has come from eliminating among such competing concepts the ones ruled out by factual analysis. Turning again to the sample object of cells in tissue culture, contrary to the scattered population of stationary isolated cells discussed before, the cells of a solid fragment of tissue explanted into culture behave quite differently. They move in droves from the explanted piece into the empty medium, giving rise to the well-known phenomenon of "peripheral outgrowth." Why do they move centrifugally? For a long time, I had considered this question synonymous with that of cell orientation and had invoked the fibrous guide rails as explanations of "oriented movement." But, as was just explained, orientation and displacement are two different things, and for the displacement there has been no crucial explanation. "Tropisms" and gradients of various kinds have been proposed to explain the phenomenon. It has been assumed that cells respond either positively or negatively to differential concentrations of hypothetical "attractive" or "repelling" substances emanating from point sources, even though it has never been possible to demonstrate just how a cell could translate such directional cues into actual convection towards or away from the source. Recent observations on our tissue-culture strain have, however, turned up a wholly different story, and it is this.

The only way to get locomotion of a cell with two motor engines at opposite ends is to stop one of the engines, at least for a while. Then the remaining engine can tow the cell away without opposition. This is precisely what happens in tissue cultures whenever free tips of two cells make accidental contact with each other: the colliding ends become temporarily paralyzed. A wave of retraction runs over the affected processes, which become partly detached from the substratum, and the two cells thus come under the exclusive pull from their remaining motile ends facing away from each other. Thus, they move apart in opposite directions. After some time, the paralyzed ends gradually recover their motility and again take hold on the ground, so that the cells are stalled once more, but at a greater distance from each other.

Extrapolating this process to a cell population with a gradient of densities, such as a tissue explant, it is evident that, of a pair of cells moving apart, the one shifting peripherally has a lower probability of encountering another cell than has the one which moves toward the explant. Statistically, this leads to a prevalence of outward migration even though cells are actually free to move in any direction. Eventually, a situation obtains in which the cells have become so widely spaced that random collisions are no longer likely; at this point further migration ceases. Thus, the only gradient which plays any role is a gradient of population density. The phenomenon is formally comparable to the diffusion of molecules from regions of higher to lower concentration, except that one deals with the random collisions not of molecules but of complex entities which may be treated as units for the purposes of description. Again, a close observation of the behavior of the real living object has brought answers far more concrete than what could be anticipated from such generalities as "attractive" or "repulsive" forces between cells. Parenthetically, it should be stressed that the type of contact separation between cells mentioned above is characteristic of the species of connective-tissue cells here described, as well as of several similar strains, but it does not apply to other cell strains, for instance of the epithelial variety, where reaction of two cells on contact can be just the opposite—namely, their drawing much closer together, provided they are both of the same kind. Further details on this behavior are given by Weiss (p. 449).

As still another instance of the dangers inherent in dealing with the living cell in terms of broad generalities, one may once more cite the behavior of cells of diverse shapes observed in plasma clots of different *p*H or concentrations. As was reported before, such cells have a variable number of processes, each of which now may be thought of as a train behind an "engine." In such cultures, it is possible to calculate the average rates of advance of the various cell types in a given direction by dividing the total distance spanned in a given period by the time elapsed. Since such cells move neither steadily nor in a straight course, however, any such average value for "rate of advance" is wholly unrepresentative of the true velocity of cell motility. As plasma concentration is increased, the cells tend to become bipolar. This means, inevitably, that the average rate of locomotion increases simply because the course is less tortuous and the cell is stalled less frequently by simultaneous divergent pulls from multiple processes to which multipolar cells are subjected. Consequently, in the lower concentration range, where most cells are multipolar, there is a progressive increase of the average rate as the number of processes declines toward the liminal value of two (Fig. 11). Once the great majority of cells have attained bipolarity, the average rate of advance remains constant, expressing more nearly the true velocity. By comparing Fig. 11 with Fig. 5, it can be seen that, at a plasma concentration above 50%, where the "rate" curve levels off, more than three-fourths of the cell population are actually bipolar. Just as one cannot tell the true speed of a railroad train if the times of departure and arrival at the terminal are the only data available and the frequency and duration of station stops on the way are not taken into account, so the mere establishment of the average rate of locomotion of a cell under various conditions has little practical meaning. Yet, the literature is full of examples in which such average rates have been used to assess the effects of a variety of agents or drugs on cells, without due attention to the effects of these agents on the medium, which may alter the whole setting in which

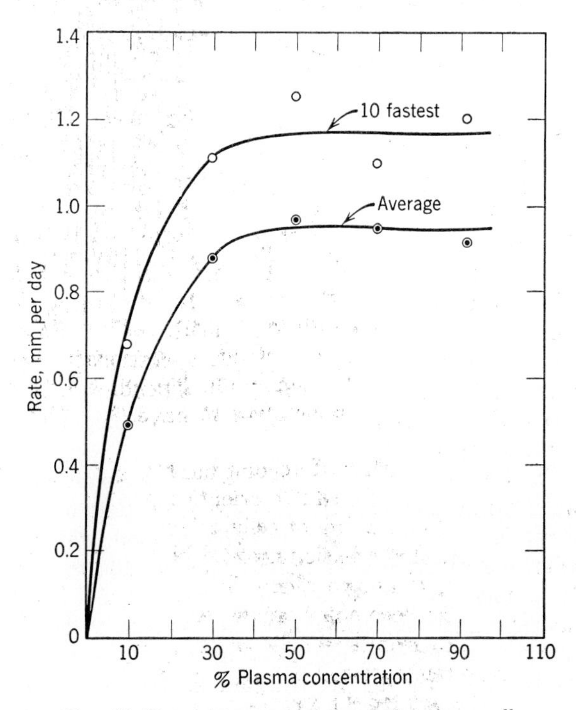


FIG. 11. Rate of progress of connective-tissue cells in plasma clots of different constitution.

the cells move. Without knowledge of these effects, no meaningful comparisons can be made. This illustrates some of the hazards of operating with *average* quantities when one deals with systems such as cells, the composition and behavior of which are inhomogeneous in space and time.

The foregoing discussion has illustrated how a controlled modification of the medium can indirectly modify cell morphology and behavior, including locomotion. It has explained the orienting effect of tensions on the fiber systems of the medium; how such structures evoke conforming organization in the cell population residing in them; and how cell-to-cell interaction and, in the last analysis, population dynamics govern cell locomotion. In this chain of events, an outside experimenter applying tensions to a culture medium appeared as the primary agent. This, of course, immediately raises the question as to what factors serve this organizing function within the living body. The sole agency of the body is its own cells and it was comforting, therefore, to find that cells, by their own activity, can create the type of orderly structural patterns in the intercellular spaces which had been imitated crudely by extraneous tensions. The cells engender in their own environment physical conditions and orderly restraints which then in turn play back on them as guides and regulators of their own behavior. Thus, a further step of complexity is added to the picture. There are innumerable examples, mostly poorly understood, all showing how, through an enormous variety of mechanisms, the same basic principle is served; which is that the cell population, through its products and interactions, sets up conditions modifying the behavior of the enclosed cells, and thus often leading to new settings and interactions which may cause further alterations of the cells, and so forth, in sequences of interactions of ever increasing complexity.

To illustrate this, consider a piece of tissue embedded in a fibrous network and let the boundary of the tissue

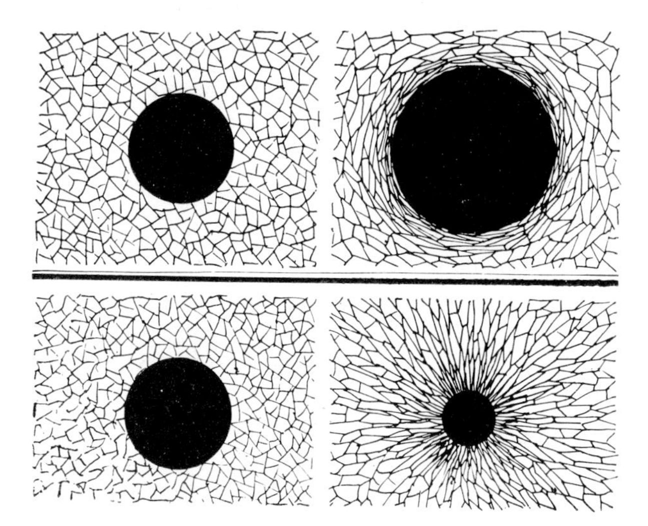

FIG. 12. Effect of an expanding (top) or contracting (bottom) center on the architecture of the surrounding meshwork.

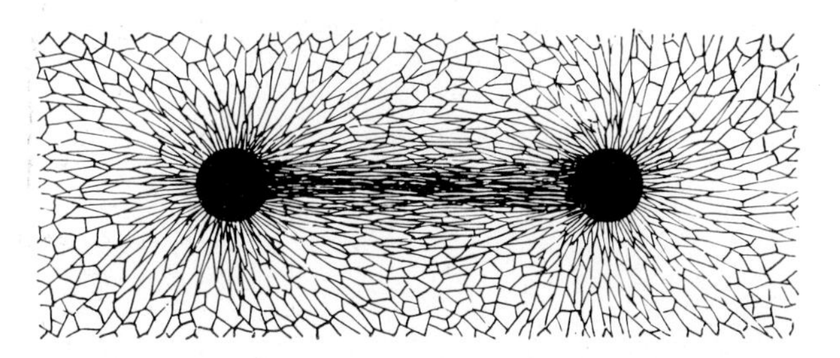

FIG. 13. Effect of two simultaneously contracting centers on a common meshwork.

expand, as happens, for instance, in a vesicle or tube or cyst swelling by the secretion of liquid into its closed lumen. Evidently, as shown in Fig. 12, top, the meshes of the fibrous medium become circumferentially compressed so as to assume a predominantly tangential orientation around the fragment. Cells happening upon such a territory would obviously be forced to circle round and round producing an envelope or tunic. Many of the connective-tissue sheaths and capsules in the meshwork of the body owe their origin to this mechanism. By contrast, if somewhere in the meshwork there arises an area which contracts, the fibrous components will be gathered toward that center in a predominantly radial orientation (Fig. 12, bottom). Cells in such zones then become likewise disposed radially. This type of effect obtains frequently in the vicinity of rapidly proliferating cell groups. It seems that chemical agents, as yet undefined, are being discharged by such cells which cause intensive synaeresis of the surrounding colloids, which means condensation of the fibrous components with loss of bound water, the resulting local shrinkage being much greater than the gain of mass by cell growth. In other words, a purely *scalar* change in a piece of tissue—increase or decrease of volume—can, through the intermediary of a fibrous continuum, translate itself into marked *vectorial* effects, establishing well-defined geometrical and structural patterns. A further degree of ordered complexity is introduced if two or more cell masses are explanted together in a common clot (Fig. 13). Through their joint constriction effects, they deflect the fibrous meshes of the intervening medium into a line connecting the active centers, and this, of course, constitutes a path for direct cell traffic between them. The cells growing out from the two centers simply follow the submicroscopic bridge which has been laid down for them automatically by the tension-engendering chemical activity of their sources (Fig. 14). Here is one primitive example of how chemical action can translate itself into physical organization.

Extrapolating briefly from these model systems of living cells, one may assume that the same sort of intimate interdependence shown here for the microscopic dimension repeats itself both in submicroscopic and in higher supracellular dimensions. One is led to the conclusion that there is a tie between physical structure and chemical activity which is indissoluble and which

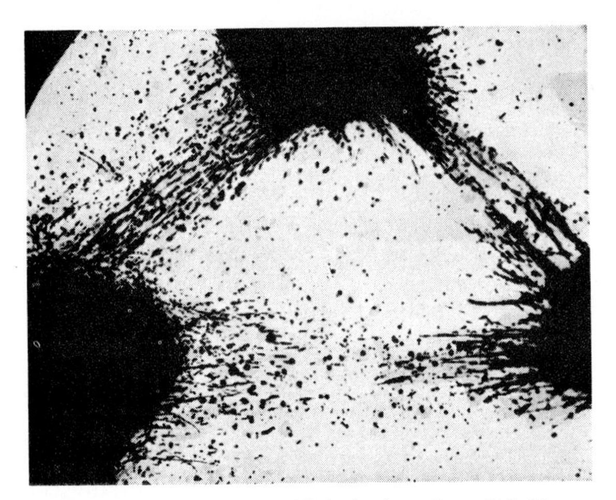

Fig. 14. Automatically established triangular cell-bridge connections between three separate tissue cultures (dark masses) in a common blood-plasma membrane.

it is not enough to assert, but which has to be explored systematically. Where, in order to study a particular metabolic reaction in isolation, the biochemist provides optimum conditions for that process, the cell produces the unique prerequisites for that same reaction only in certain strictly confined localities. It does it through the metabolic products of other chemical reactions going on in another equally confined sample of cell space. Thus, one system feeds another and, in turn, is fed by a third, in a vast system of mutually interdependent, symbiotic, and harmonized partial reactions subsidiary one to another. This interdependence, characteristic of the living state, is what I once termed "molecular ecology." As a field of investigation, it barely has reached infancy. Yet its significance is pointed up by the fact that, in the living cell, the various biochemical partial systems coexist in a common space, without preformed rigid partitions, but rather compartmentalize themselves by structural effects of their own activities of the sort exemplified in crude and elementary fashion by the case just cited. Physical structure and physicochemical conditions limit the types of enzymatic and other chemical reactions that can go on in a given spot —for instance, along interfaces of fibrillar, lamellar, or corpuscular systems—while the resulting reactions, in turn, modify the physical substratum; and by continual interplay of this sort between physical structure and chemical action, the cell system passes first through its progressive developmental transformations and then is stabilized in the steady state of maturity. To some extent, physical structure then is frozen into static arrangements of cytoskeletons, but even then physical structures still are regenerated continuously by cellular activity, leaving at least part of the cellular system in a state of incessant development and self-renewal.

A realistic concept of cell behavior also must take into account the limits set to interactions between distant parts by the formation of compartments within compartments. A diagram of the organism (Fig. 15) would represent it as a system of concentric shells, with the gene in the center enclosed in the chromosome, which is enclosed in the nucleus, which is enclosed in the cytoplasm, which forms part of a tissue, which forms part of the organism, which is surrounded by the external environment; for simplicity, cytoplasmic particles are omitted. No outer agent can influence any of the inner shells except through the mediation of the shells in between, which may or may not modify that factor during its inward passage. Conversely, products of inner systems may not reach outer shells as such, but may be significantly screened and altered in transit. The arrows in the diagram indicate the complex network of relations that one must bear in mind. Oversimplified mental pictures form when, for instance, the simple statement is heard that a gene "controls" a particular feature of the organism, without allowing that it can do so only through interactions with the outer shells, which in themselves have become progressively modified in their long developmental history by countless chains of interactions with other shells, including, of course, the innermost, the gene. Since it has been my assignment to sketch the living cell as it truly is in all its highly ordered complexity, I feel compelled to caution against the illusion that a simple statement, such as, "a gene controls a character" reflects any similar degree of simplicity in the phenomenon covered by the statement.

A final example projects the principle of interaction between physical structure and chemical activity upward into the realm of supracellular order, a field so baffling in its problems that many an investigator prefers to look the other way when he encounters them. The example is chosen from an almost diagrammatically simple object which, because of this, holds at least some promise of more penetrating analysis by the combined physical and chemical tools now at one's disposal. It refers to the origin of the internal architecture of cartilage. Cartilage is formed by groups of cells producing

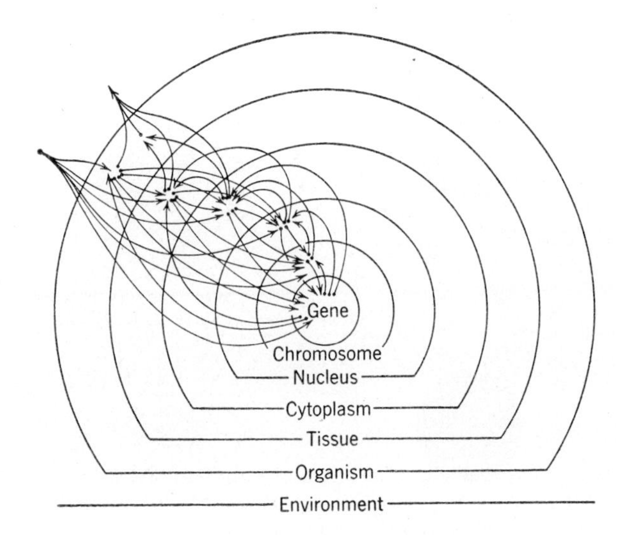


Fig. 15. Diagram of network of possible interactions in an organism.

ground substance between them. That ground substance has been identified partly as a mucopolysaccharide—chondroitin sulphuric acid—and collagen. But cartilage is not just chemical substances; it assumes characteristic shapes and configurations, depending upon its sites, owing to developmental processes under genic control. The nature of the problem becomes clear when one contemplates that the peculiar pattern of convolutions of one's earlobes, which are but covered cartilage, is an individual inherited characteristic. How do such patterns form? Cartilage of the limb is different from cartilage around the eye; the former grows in compact, whorl-shaped masses, the latter develops as a flat plate. It has long been known that, if the cell group that is destined to form limb cartilage is reared in tissue culture, it will grow according to the standard limb pattern, giving rise to recognizable skeletal elements. Similarly, the author observed years ago (jointly with R. Amprino) that the precursor cells of the scleral cartilage around the eye, after explantation as an intact group, go on to form cartilage in the shape of a plate. Evidently, in either case, the explanted tissue complex contained some physical properties that guided the cells contained in it into their respective typical arrangements and growth patterns. The crucial property differentiating between limb and eye cartilage had to be thought of as inherent in the block of tissue as a whole, the further development *in vitro* merely amplifying some distinctive architectural pattern already present in the tissue fragments at the time of their isolation from the embryo.

To test this conjecture, the author (with A. Moscona) recently resorted to a technique that permits one to disintegrate a tissue into its constituent cells and thereby to destroy any intercellular structures and supracellular arrangements that could have acted as guides for subsequent development. Cells can be separated from one another and from their matrix by trypsin, washed in a loose suspension, and then seeded out in a tissue culture where they can aggregate into random clusters. More about the manner of aggregation is reported in another article (Weiss, p. 449). When a piece of prospective limb cartilage and a piece of prospective eye cartilage were dissociated in this manner into their component cells and the cells of each type were permitted then to aggregate in tissue culture and to continue with their actual cartilage-forming activities, it turned out that the cells from the limb site produced cartilage of the typical whorl-shaped massive limb pattern (Fig. 16, top), while the cells that had originated at the eye site produced the plate-shaped laminated structure to which they would have given rise prior to their disaggregation (Fig. 16, bottom). In other words, the "blueprint" of the architecture of the group performance is ingrained in each individual cell and not just carried over into the culture by the block of tissue as a whole. Now, how to convert this figure of speech into concrete terms is problematical. But one may speculate that the architecture lies in the specific

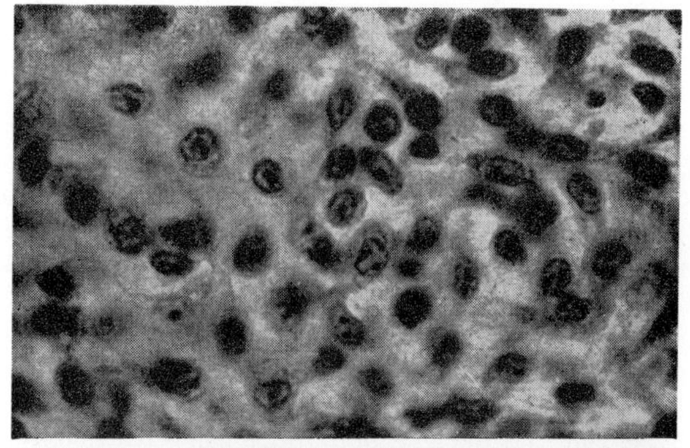

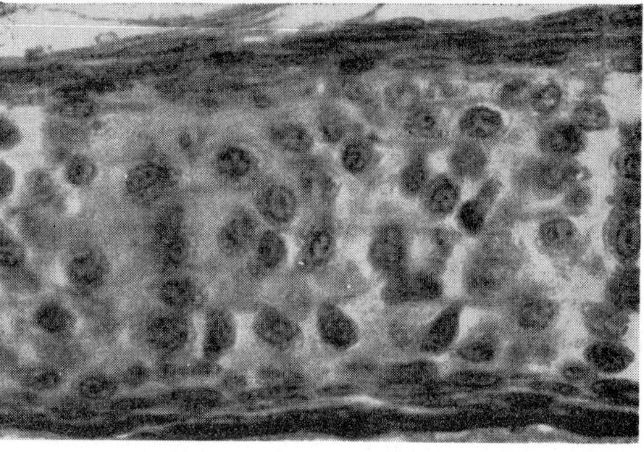

FIG. 16. Cross sections through cartilages developed in tissue culture from dispersed and reaggregated precartilage cells from a prospective limb (top) and from a prospective eye coat (bottom) of the chick embryo, forming whorl-shaped and lamellar structures, respectively.

ground substances secreted by the cells. One would have to suppose that each cell type secretes a complex ground substance of a distinctive pattern of macromolecular stacking or crystallinity of such a kind that it would predispose a planar array of the mass in the form of layers in one case, and a more massive isotropic arrangement in the form of three-dimensional whorls in the other. The cells then would dispose themselves in conformity with these structural patterns of ground substance for which they had furnished the elements in the first place.

This is sheer conjecture. No facts are known that would either support or contradict it. Yet this very uncertainty helps to point up the immensity of ignorance in matters of supracellular organization. If one is convinced that such higher-order organization is to be explained solely in terms of cellular dynamics, then one must raise one's sights to the domain where those phenomena occur, which is no longer the intracellular microcosm.

The study of tissue architecture, used here as an example, may be one of the easier inroads into the maze of perplexing problems presented by the orderliness of the complex living organism, as against the relative disorder and simplicity of its shattered fragments, on which one preferably concentrates, mostly after homogenizing them either physically or conceptually.

The lessons of this story repeat themselves as the viewpoint is shifted from the organization of the body in its structural aspects to the supracellular coordination and ordering systems of its functional activity, whether this be the homeostatic maintenance of blood composition, the integrative action of the nervous system in behavior, or the mobilization of defense and repair mechanisms in response to pathological disturbances. Of course, the operative tools in all of these performances are the individual component cells, which in turn operate through physically ordered subsystems in incessant chemical interaction. But unless one remains cognizant of the fact that the level of the organism is reached from the level of molecular biology not in one single jump over the conceptual gap customarily bridged by the word "organization," and unless one learns to think in terms of a hierarchy of ordered systems, each one with some degree of identity and stability even on supramolecular levels, one obscures rather than enucleates the problems to which thinking and research must be directed. The most impressive feature of a cell is not the constant flux, reshuffling, and variability of its population of molecules and particles (of which our cinematographs presented at the Study Program have given a clear expression), but the fact that, in spite of this ever-present change, each cell remains so remarkably invariant in its total behavior; that indeed, as an entity each behaves so much like millions of other entities equally variable in inner detail, that one comes to recognize them as essentially alike. Such relative invariance of the whole presupposes the harmonious subordination of the behavior of the parts to the conditions of the collective group. It presupposes that the free interactions among the subunits are subject to restraints, the nature and direction of which vary adaptively with the state of the system as a whole. I can think of no more propitious introduction to a biophysics program than by re-emphasizing that the restraints in question are not the sole province of chemistry, especially the stoichiometric branch, but that they include crucial restraints of *physical* nature, with nonrandom molecular arrays a most prominent and analytically most promising feature.

Awareness of the existence of these problems in no way detracts from the pragmatic value of analyzing each elementary component process in its own right and of using whatever models, however simplified, may be found to be constructive as aids to understanding. But it is precisely the phenomenal advances made in the study of the partial and isolated system, in contrast to the dearth of information on the *organized interactions of such systems in the living cell and organism*, that lead me to conclude with a plea for a more balanced effort and more extensive occupation with the latter problems than there are at present. Cellular and tissue dynamics are population dynamics of species of molecules, of cellular subunits, and of cells. Such group dynamics cannot be derived solely from looking at the members of the population in isolation, but only from a study of the ecology and technology of their behavior in the group. It is hoped that this modest effort at presenting the case of the living cell not only has given a fair introduction to the problem but also has shown that practical techniques for its eventual solution are at hand, with physical and chemical approaches indissolubly interwoven.

## BIBLIOGRAPHY

P. Weiss in *Chemistry and Physiology of Growth*, A. Parpart, editor (Princeton University Press, Princeton, New Jersey, 1949), p. 135.

P. Weiss and B. Garber, Proc. Natl. Acad Sci. U. S. **38**, 264 (1952).

P. Weiss, J. Embryol. Exptl. Morphol. **1**, 181 (1953).

P. Weiss in *Analysis of Development*, B. H. Willier, P. Weiss, and V. Hamburger, editors (Saunders Company, Philadelphia, 1955), p. 346.

P. Weiss, J. Cellular Comp. Physiol. **49**, Suppl. 1, 105 (1957).

P. Weiss, Intern. Rev. Cytol. **7**, 391 (1958).

# 3
# Cellular Responses

William Bloom
*Committee on Biophysics, The University of Chicago, Chicago 37, Illinois*

THIS paper is concerned with some aspects of the organization of cells, particularly as seen with the light microscope, and the results of some experiments on dividing cells in tissue culture as shown by time-lapse motion pictures. The paper is illustrated by portions of a few key frames from the motion pictures.

There have been several important periods in the study of cellular organization. One of them was the period of descriptive histology, embryology, and cellular pathology of the nineteenth century. In it a variety of important processes in the development and function of cells were portrayed largely, but not exclusively, on the basis of dead material. This was followed by a period of extreme skepticism in which a number of investigators claimed that nearly everything seen after fixation and staining was artifact; they believed that only what could be seen in the living cell existed in it. There were even claims that nuclei could not be seen in living cells and hence were artifact. Actually, nuclei are easily demonstrable during life. This period of skepticism was helpful in some respects but it also did great harm because some of its claims were too destructively critical. One of the rewards of living in the present period is to see that many of the so-called artifacts of the previous century have been shown to exist in living cells by our more modern methods of investigation. I remember distinctly the first time I saw a living cell with the newly developed phase-contrast microscope. What I saw was like an iron-hematoxylin stained cell moving around under the microscope and offering complete verification, at the light-microscope level, of the existence of certain structures previously known mainly in preparations of dead cells.

As our methods of morphologic, chemical, and physical analysis become more powerful and as the study of the minute structure of living systems reaches ever closer to the molecular level, it becomes more and more obvious that the old problem of what is artifact in our observational material is still with us.

I shall show a series of newt cells cultured in diluted chicken plasma. In the time-lapse movies, constituents of a typical cell are shown as it goes through the process of cell division (mitosis). In the film, the reality becomes apparent of some structures that to some of you may have been only words. Three cells show the effects of either ionizing or ultraviolet, total cell irradiation. The succeeding five dividing cells demonstrate some of the effects of localized irradiation of small parts of them. In the cells which were irradiated throughout their extent, several abnormalities appear, while in those irradiated locally, the number of abnormalities is much smaller. The comparison of local *versus* total cell irradiation gives a little insight into some aspects of the machinery of cell division, a problem which Raymond Zirkle and the author began to work on some eight or nine years ago with the aid of our proton microbeam.[1] A few years later, Robert Uretz began to work with us in our study of mitosis; the simple ultraviolet microbeam, which he devised, has been of great help.[2,3] With it, one can aim directly at a structure and watch it as it is being irradiated.[4,5] Others in our laboratory, especially E. W. Taylor and R. P. Perry, have also contributed to our knowledge of what happens in dividing cells. Zirkle[6] has recently reviewed the methods and results of microbeam irradiation which started with the work of Tschachotin with ultraviolet light in 1912.[7]

In our proton microbeam, the protons start with an energy of 2.0 Mev but lose perhaps a quarter of this as they pass through two layers of mica, each 5 $\mu$ thick. One of these seals the delivery opening of the Van de Graaff generator; the other is the cover slip for the culture. Although there is a small percentage of scatter, if the culture is essentially in contact with the microaperture, 85 to 90% of the protons bombard an area about 2.5 $\mu$ in diameter, and practically all of the protons pass within a circle of 8 to 10 $\mu$ diameter at the target. (The indicated variations reflect slight differences in approximation of the mica cover slip to the microaperture.) The scatter is not important with small numbers, but obviously becomes serious with hundreds of thousands. With a more powerful generator, it would have been possible to deliver a single proton into a target of 1 $\mu^2$, but the smallest number we have been able to deliver with our generator is a burst of about 20.

The Uretz microbeam makes use of a reflecting objective which brings all wavelengths of visible and ultraviolet light into approximately the same plane. This device has an important advantage over the proton microbeam in that the size and shape of the area to be irradiated can be varied easily from circles 2 $\mu$ in diameter to much larger ones, or to slits, or to irregular shapes of various dimensions. We can also compare the effects of microbeams of monochromatic and heterochromatic ultraviolet light with those of the proton microbeam.

The machinery by which a plant or animal cell becomes two cells is probably much the same in all, although much more is known about the process in nucleated cells than in bacteria. Much of what is shown here has been known in the descriptive sense since about 1880 in the work of Strassburger and especially of Flemming, among others. But we know very little more today than these early workers knew about the causal

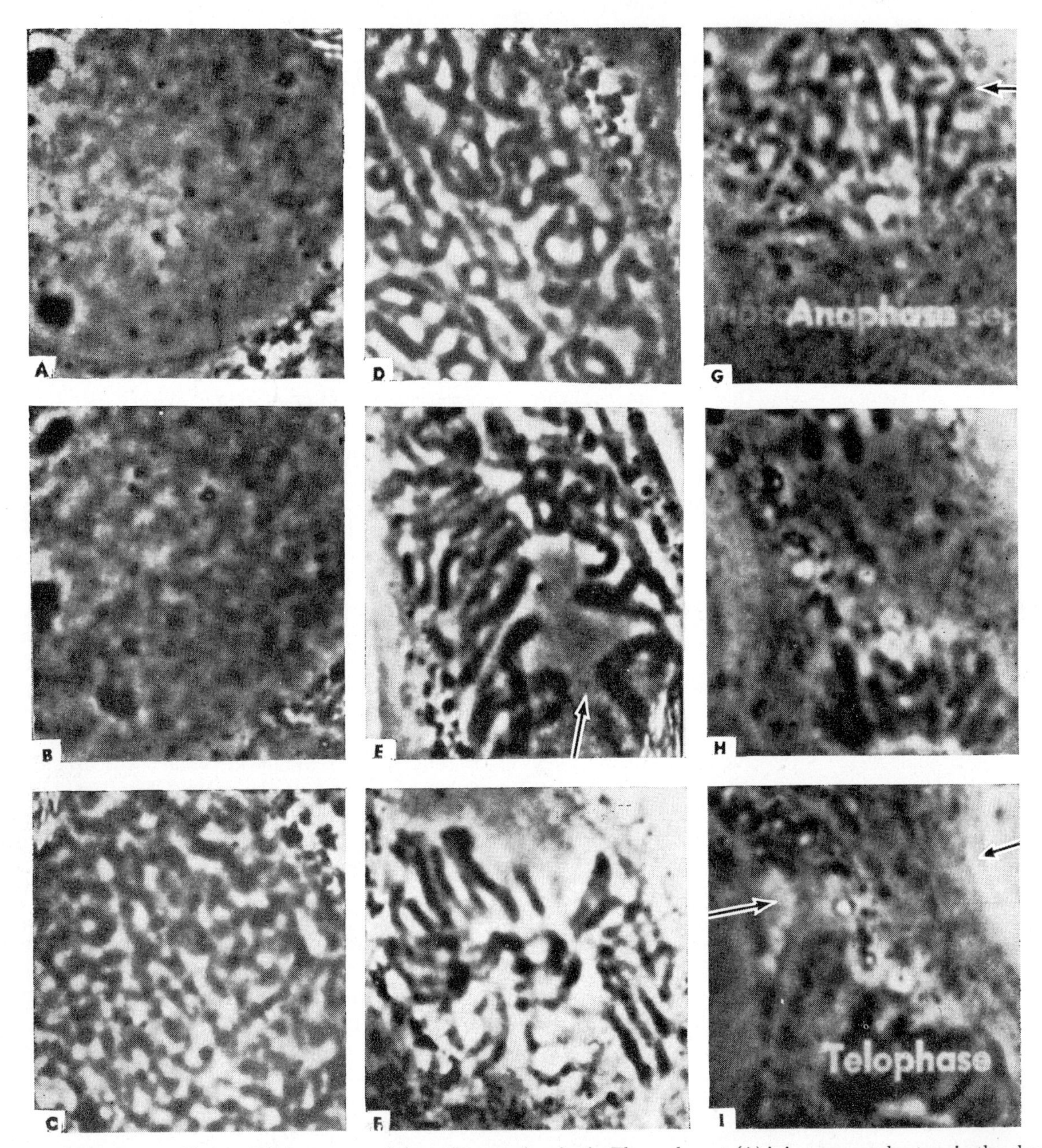


FIG. 1. This sequence [(A) to (I)] shows some of the main stages in mitosis. The nucleus at (A) is in a very early stage in the change of a relatively nongranular nucleus into chromosomes. The two large bodies are nucleoli. At (B), the cell has moved slightly and three nucleoli are visible. The nuclear granules begin to form vague cords which are much more definite at (C) and especially so at (D) where they have become typical chromosomes. (E) shows the change in arrangement of chromosomes from prophase into metaphase at (F). In (G), each chromosome is dividing into two daughter chromosomes (as at arrow) which are beginning to move to opposite poles of the cell (anaphase). At (H), the chromosomes have completely separated and in (I) reconstruction of the daughter nucleoli is beginning and the cell body is constricting (arrows). The arrow in (E) points to a centriole.

mechanisms involved in this complicated, meticulous process by which the two daughter cells inherit equal amounts of genetic material.

In the films, the area of the cells shown is approximately 40 by 60 $\mu$, and the motion pictures are speeded up about 100 times. When present, the clock indicates the lapse of time. Figures 1 to 9 are from the time-lapse motion picture shown at the presentation of the paper, photographed with medium dark, phase-contrast microscopy; 1400×. All cells are from newt heart explanted in a mixture of one part chicken plasma and nine parts amphibian Tyrode solution. The division process in these cells at 70°F may take as long as 6 or 7 hr or as little as 4 hr. These differences, at the same temperature, are unexplained.

## UNTREATED CELL IN MITOSIS

In cell 1, a typical mitosis, one can see tiny fat droplets surrounding the nuclear membrane, the filamentous and granular mitochondria, the nucleus with large nucleoli, and numerous, irregular small granules [Fig. 1(A)]. The author used to question whether the movement of granules in the cytoplasm was Brownian motion, since one sees it only in the speeded-up time-lapse movies, and not by direct observation of the living cells. But Taylor studied this phenomenon and convinced him that it was Brownian motion.

As the film continues, the small granules in the nucleus become more numerous and darker and gradually form long snake-like bodies—the chromosomes [Figs. 1(A)–1(D)]. They are about 2 $\mu$ thick and some 30 $\mu$ or more long. The nucleoli disappear suddenly and we do not see what happened to them [between Figs. 1(B) and 1(D)]. The nuclear membrane also disappears and about the time that it does the chromosomes begin a continuous, irregular movement. Warren Lewis, who has spent many years studying living cells, called this the "Dance of the Chromosomes."

The cell contains two other important small structures called centrioles, one above and another below the group of chromosomes. The poles of the spindle are connected with the centrioles. The area with practically no granules is part of the developing spindle [Fig. 1(E)]. After some time, a pale stripe appears within each chromosome and extends through its length. Suddenly, the chromosomes divide along this stripe and the daughter chromosomes (or chromatids), each a half of an original chromosome, move apart [Fig. 1(G)] and form the two separate daughter nuclei [Figs. 1(H) and 1(I)]. Then the cell body constricts, the whole mass separating into two new cells, each with a nucleus. The temporary excrescences on the surfaces of the cell are normal when they are not more extensive than in this cell. In the irradiated cells, which are seen next, they may be much larger, more numerous, more violent in movement, and have longer persistence.

The process of cell division goes on in our bodies hundreds of millions of times a day in order to replace those cells which are normally lost for various reasons. In the vast majority of cases, these mitoses are executed as meticulously as in this cell.

## RESPONSES TO TOTAL CELL IRRADIATION

The next three cells show a number of different phenomena as a result of total cell irradiation. Cell 2 received 200 r of 200-kv x-rays. The film starts with the cell in an early stage of mitosis. As a consequence of the irradiation, an abnormality develops instead of the orderly cell division which occurred in the untreated cell. Chromosomes start to separate but the separation is not complete [Fig. 2(B)]. Then cell constriction occurs about the mass of chromosomes [Fig. 2(C)] and forms a dumbbell-shaped nucleus connected by a long bridge of chromosomal material extending between the two daughter cells [Fig. 2(D)].

Cell 3 received 350 r of 200-kv x-rays. Instead of producing a relatively simple malformation, as produced by 200 r in the previous cell, several abnormalities occur. The chromosomes make an abortive attempt to separate into their daughter chromosomes [Fig. 3(A)], violent "bubbling" appears on the cell surface, and two constrictions of cytoplasm occur [Fig. 3(B)]. We have seen bubbling go on for days in cytoplasm pinched off as in this case. Nuclear material now separates into two irregularly sized groups [Fig. 3(C)], but at one point there is a tiny connection between them, and the smaller group of chromosomes spurts over into the larger one, making a bi-lobed nucleus.

Cell 4 received ultraviolet light over its whole body. Despite the marked abnormalities which result, the cell will eventually give rise to daughter cells which are able to exist for a time at least in tissue culture. Immediately after being irradiated, some of the motion of structures within the cell ceases, but some of it gradually returns. For a short period, there is an irregular arrangement of chromosomes and then terrific bubbling movements begin to appear [Fig. 4(B)]. The chromosomes and cytoplasmic granules move in and out of the bubbles. We call this "type 4 bubbling," the most extreme degree. The term "bubbles" is a misnomer as they are protuberances of cytoplasm and not bubbles; their nature is completely unknown. Through it all, the chromosomes stick together, more or less. The bubbling motion gradually begins to subside. Then the chromosomes separate into one large and one small group which move apart and a constriction of the cytoplasm develops [Fig. 4(C)]. As nuclear reconstruction begins, it becomes apparent that the two nuclear masses are connected by a thin chromosomal bridge.

## RESPONSES TO PARTIAL CELL IRRADIATION

In the next 5 cells, very small areas of nucleus or cytoplasm were irradiated by protons or ultraviolet light.

There is a widely held view in the radiobiological literature that chromosomes are damaged secondarily as a result of irradiation of the cytoplasm. The experiments on the next two cells clearly refute this claim for the newt cells in our culture. In cell 5, the area indicated by cross-hairs [Fig. 5(A)], about 14 $\mu$ from the nearest chromosomes (arrows), was irradiated with 28 300 protons. The intensity of irradiation of this bit of the cell is enormous when one realizes that each proton equals 600 rep. Nevertheless, the resulting mitosis is perfectly normal [Figs. 5(B) and 5(C)].

Cell 6 received 60 protons localized over a small area of two chromosomes [see cross-hairs and arrowhead in Fig. 6(A)]. The daughter chromosomes on the nonirradiated side separate cleanly, while those on the irradiated side are stuck together. The daughter chromo-

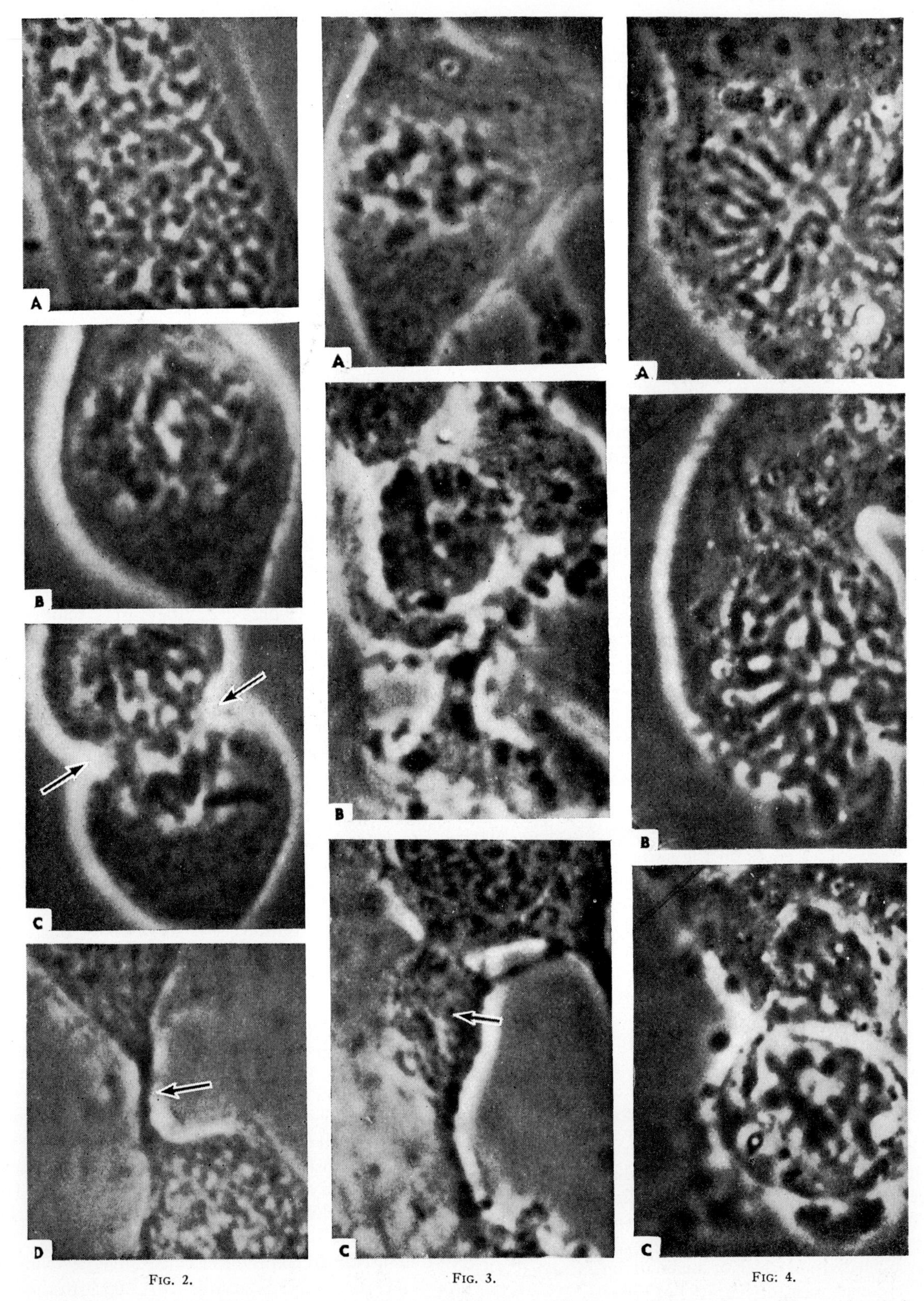


FIG. 2. FIG. 3. FIG. 4.

FIG. 2. This cell received 200 r of 200-kv x-rays. (A) is immediately after irradiation and shows no change. (B) was taken during the abortive anaphase. This is the maximum that the daughter chromosomes have separated. At (C), the constriction of the cell (arrows) includes the chromosomal mass and produces a dumbbell-shaped nucleus (D) with a thin bridge (arrow) connecting two main portions of the reconstructing nuclear material. (*Note.*—Captions for Figs. 3 and 4 are on page 25.)

somes move apart as a group, much as in an opening hinge pinioned on the irradiated parts of the chromosomes [Fig. 5(B)]. This produces a temporary chromosomal bridge [Fig. 6(C)]. The difference between the failure of the 28 300 protons delivered to cytoplasm to produce any visible change in the chromosomes of the previous cell is in sharp contrast to the effect of 60 protons delivered locally to chromosomes in this cell.

Areas 8 $\mu$ in diameter of the next three cells were irradiated with ultraviolet light. The first of these cells (number 7) is in early mitosis (prophase). The spindle has not yet formed and, as a result of cytoplasmic irradiation, will not form. There is an irregular arrangement of chromosomes characteristic of this stage. After the localized, cytoplasmic, ultraviolet irradiation, the irregular configuration persists [Fig. 7(B)]. The chromosomes shift around from time to time and there are pulsations and temporary extrusions of the cell surface. After several hours, a regrouping of the chromosomes occurs suddenly. They do not give rise by longitudinal splitting to a double number of daughter chromosomes, as in normal mitosis. Instead, haphazard numbers of *whole* chromosomes aggregate into two groups which separate one from another [Fig. 7(C), arrows]. Then the cytoplasm constricts and two daughter cells are formed. Such cells obviously differ completely in their genetic complements from each other and from those cells arising from a normal mitosis.

In the next cell (number 8), it is possible to follow the aggregation of the two groups of chromosomes more easily than in the previous one wherein cytoplasmic irradiation had prevented the appearance of the spindle. In this cell, which is in the mid-phase of mitosis, the spindle is fully developed. The area to be irradiated is shown at the cross-hairs [Fig. 8(A)]. After irradiation of a small part of the cytoplasm (8 $\mu$) with ultraviolet light, the spindle breaks down and the chromosomes lose their characteristic, butterfly-like arrangement, becoming irregularly disposed in the center of the cell [Figs. 8(B) and 10]. A pale longitudinal line becomes visible in each chromosome and marks the usual line of cleavage of each chromosome into its two daughter chromosomes (chromatids). The pale line disappears, however, before the whole chromosomes aggregate into two irregular groups [Fig. 8(C), arrows]. In cells so treated, the number of chromosomes in each new nucleus may vary from 9 to 15 or even more, instead of 22 which is characteristic of this species. Constriction occurs and the daughter cells move apart.

Since, in the experiments on cells 7 and 8 (and nearly 100 others similarly treated), irregularly sized groups of whole chromosomes move apart in the absence of the spindle, it would seem that the spindle serves primarily as a mechanism by which *daughter chromosomes* are guided as they are moved apart by unknown forces.

From the days of Flemming on, there have been many theories about the machinery of mitosis. Some of the models, such as those based on rubber bands, have some resemblance to what is seen in some stages of mitosis, but they contribute nothing to an understanding of the forces determining the processes in mitosis. It has been claimed that the daughter chromosomes become separated by electrostatic forces or by a localized imbibition of water, or that the spindle is made of a contractile material which pulls them apart. Although cognizant of many claims to the contrary, it is probably fair to say that there has been no delineation at all of the nature or amount of the forces involved in mitosis.

Nor do we have any idea of the nature of the force which moves the irregular groups of whole chromosomes apart in the absence of the spindle. It is known that, in the early stages of mitosis, the two centrioles separate and take position at the apices of the spindle, and that there is a temporary attraction between the centrioles and the kinetochores of some of the chromosomes. Then, this attraction ceases and all the chromosomes move into a position midway between the two centrioles (metaphase). In the next stage of division (anaphase), the daughter chromosomes separate from their mates and move with their kinetochores foremost toward the centrioles. In those cells in which the spindle is abolished, we assume that, during the separation of the groups of whole chromosomes, those chromosomes close to each of the centrioles go with it to a daughter cell.

The last cell (number 9) shows a different phenomenon after localized irradiation (8 $\mu$ in diameter) of chromosomes with ultraviolet light [Fig. 9(A)]. A dramatic change in the refractive index of the chromosomes appears in the irradiated area and produces what we call a "pale" spot [Fig. 9(B), arrows]. This change does not appear if much more than half of the chromosome mass is irradiated. However, irradiation of as little as 2 $\mu$ of a chromosome will produce it. The mass of chromosomes immediately around the pale area becomes sticky and cannot separate into chromatids, while those some distance away can. From this there results an irregular, lobated single nucleus with malformed chromosomes and a pale spot in the center [Fig. 9(C)].

In fixed preparations, the pale area also appears pale with most of the usual stains. A similar appearance

---

FIG. 3. This cell was irradiated with 350 r of 200-kv x-rays. In it the anaphase movement did not progress as much as in the previous one [compare with Fig. 2(B)]. Shortly after (A), the cell begins to show violent "bubbling" and multiple constrictions. As a result, portions of cytoplasm with or without chromosomal material become separated. At (C), the uppermost part of the picture shows a newly formed daughter nucleus and attached below it is an irregular chromosomal mass (arrow).

FIG. 4. This cell received a 2.5-min exposure of a germicidal ultraviolet lamp 2.5 cm from the culture. At the time of irradiation, the cell showed typical metaphase configuration. At (A), the chromosomes show an irregular rosette appearance. A few minutes later, the cell begins to undergo violent bubbling and at (B) the chromosomal mass is being extruded into one of the bubbles. Later, there are multiple constrictions (C).

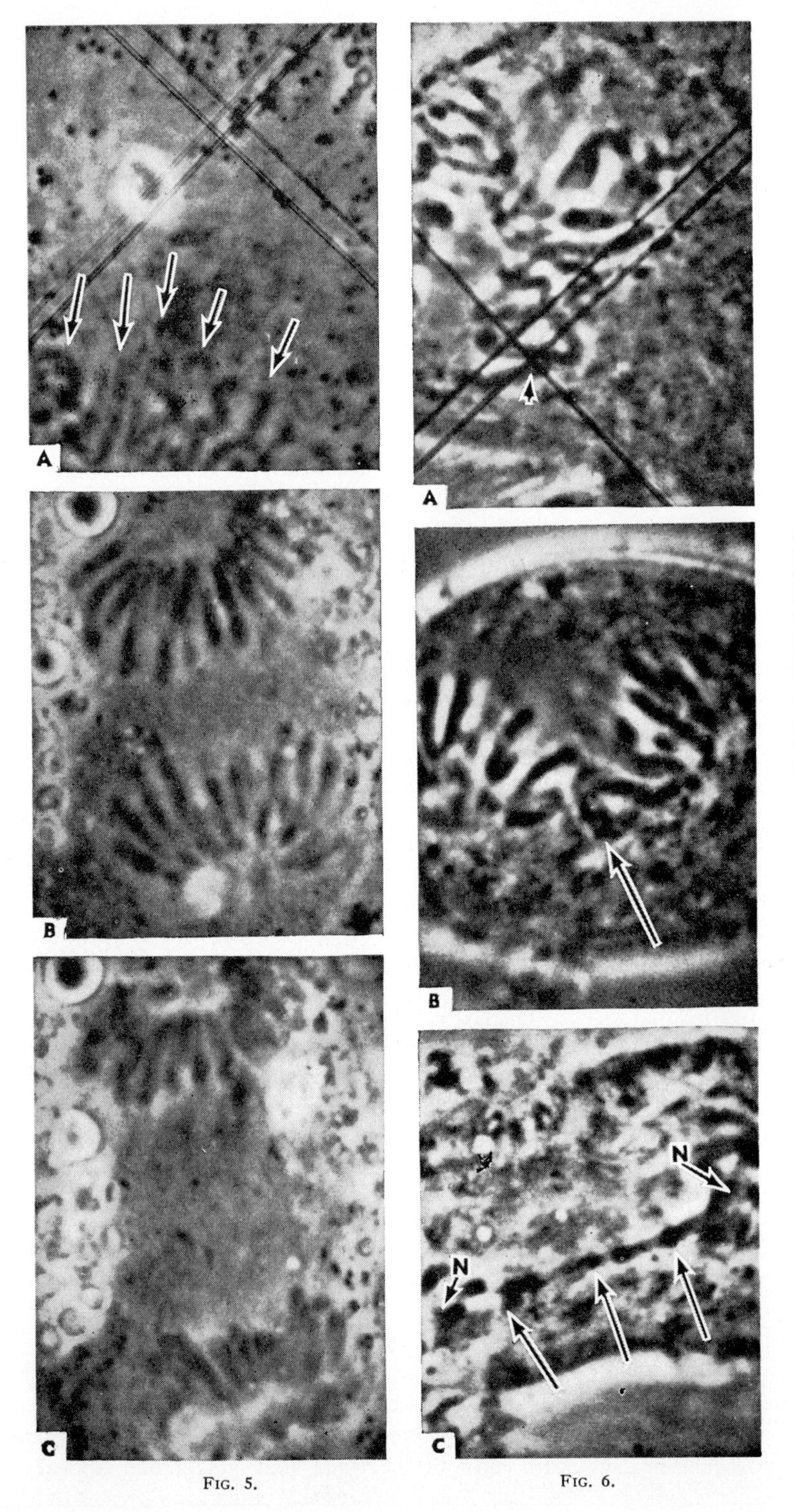


Fig. 5. Fig. 6.

Figs. 5 and 6 show the difference in effect of localized irradiation of cytoplasm *versus* localized irradiation of chromosomes in dividing cells. The cell shown in Fig. 5 received 28 300 protons at the area indicated by the double cross-hairs. The nearest chromosomes are indicated by the arrows. This cell divided into two by a perfectly normal mitosis as shown in (B) and (C). Figure 6 shows a totally different picture as a result of irradiating small parts of two chromosomes (indicated by cross-hairs and arrow) with 60 protons during metaphase. As shown in Fig. 6(B), the chromosomes separated normally into daughter chromatids except those at the point of irradiation where, as indicated by the arrow, the chromosomes are stuck together. As a result, the two daughter nuclei (N,N) are joined by a chromosomal bridge indicated by the three arrows.

results from the application of the Feulgen stain for deoxyribonucleic acid.[8] This suggests that, in the development of the pale spot, the nucleic-acid moiety of the irradiated part of the chromosomes may have been so altered that it does not respond to the stain or has actually left the area. Ultraviolet-absorption studies of the pale spot by Perry show a diminished absorption at 260 m$\mu$ (Fig. 11).[8]

Zirkle has spent much time in the past few years determining the relative effectiveness of the following wavelengths (225, 240, 250, 260, 270, 280, and 300 m$\mu$) on paling. He found the action spectrum to have the

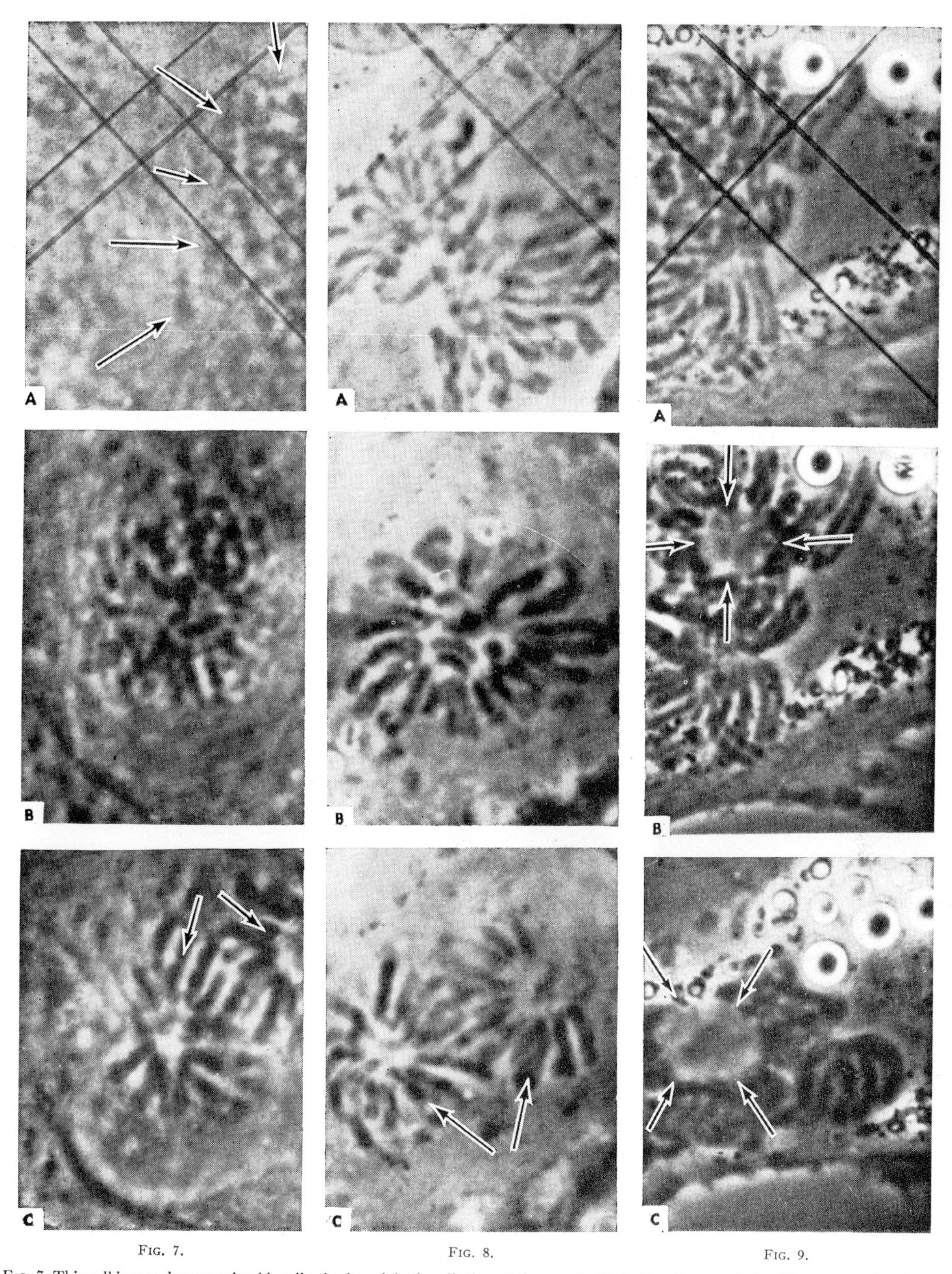


Fig. 7. Fig. 8. Fig. 9.

Fig. 7. This cell in prophase received localized, ultraviolet irradiation at the area indicated by the cross-hairs. The nearest chromosomes are shown by the arrows. The spindle did not form and the chromosomes occupied an irregular clumped area (B) and later separated as whole chromosomes into two unequal groups, indicated by the arrows in (C).

Fig. 8. The cell, in metaphase with a clearly marked spindle, received localized, ultraviolet irradiation of the area indicated by the cross-hairs. The spindle disappeared promptly and the chromosomes show a haphazard clumping (B) much like that in 7(B). In cell 8, as in the previous one, whole chromosomes separate into two daughter clumps, shown by the arrows in (C).

Fig. 9. This cell, in metaphase, received ultraviolet, localized irradiation of an area of chromosomes 8 $\mu$ in diameter, indicated by the cross-hairs. Within a minute, the chromosomes showed a typical "paling" reaction (B). The cell attempted to divide but was unable to execute a complete anaphase as the chromosomes immediately around the pale area were very sticky. There resulted a single irregularly lobed nucleus with a paled spot (arrows) in (C).

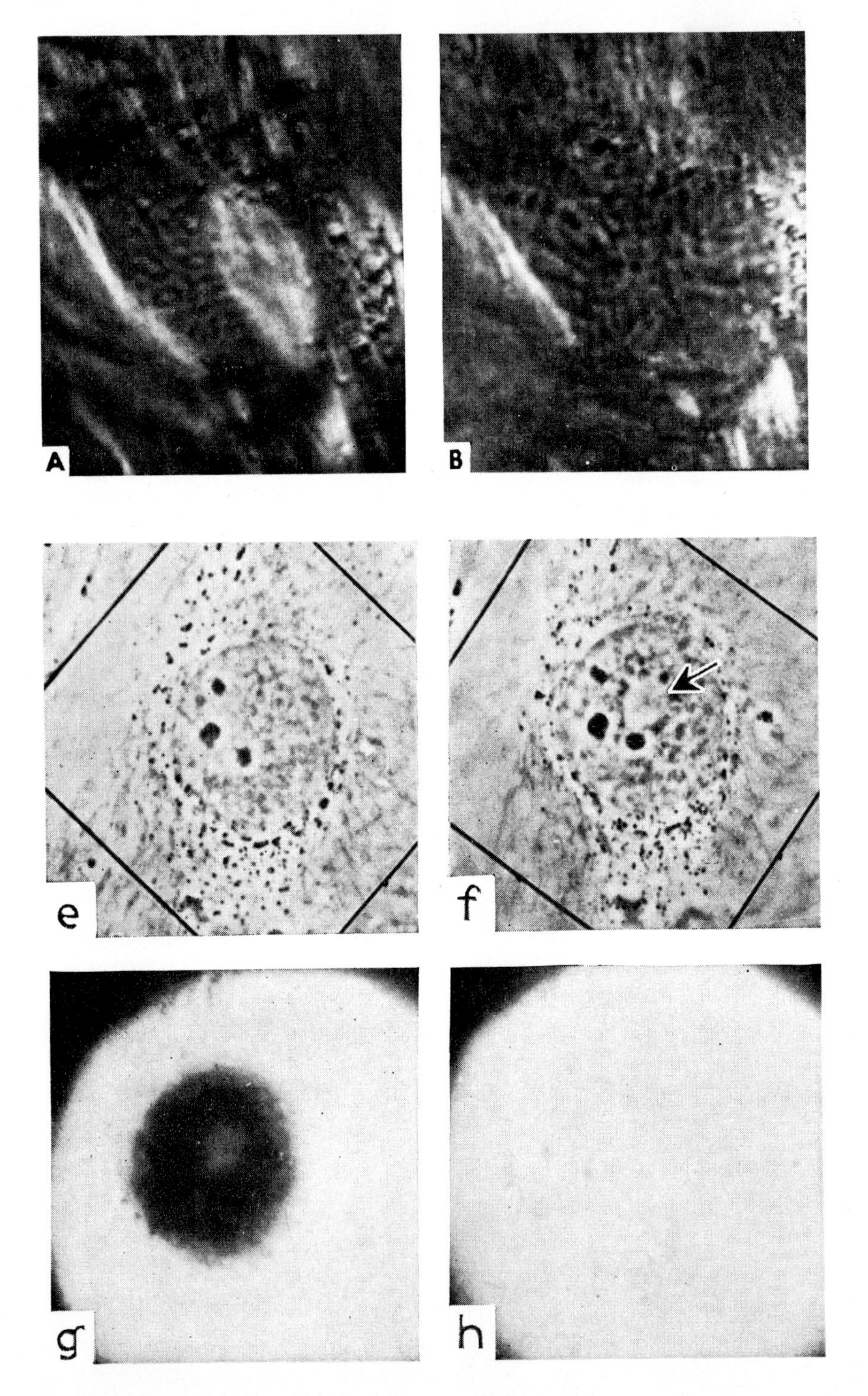


FIG. 10. Newt cell in metaphase as seen with polarized light. (A) The spindle is prominent before irradiation. (B) The spindle has disappeared shortly after localized, ultraviolet irradiation of cytoplasm. Courtesy R. B. Uretz.

FIG. 11. Newt cell in prophase. (e) Before irradiation; (f) shortly after ultraviolet irradiation of 4-$\mu$ diam circle of nucleus. The paled area at the place of irradiation is shown by the arrow. (g) After irradiation, photographed by 260 m$\mu$ light, and (h) photographed by 310 m$\mu$ light. The diminished absorption of the paled area at 260 m$\mu$ is obvious (after R. P. Perry)[8].

general absorption spectrum typically shown by proteins containing aromatic amino acids. There is a minimum at 250 m$\mu$ and a broad maximum between 260 and 280 m$\mu$.

The relative effectiveness of the two types of irradiation is roughly as follows: Irradiation with ultraviolet light of an 8-$\mu$ spot of achromosome will require 5 sec to make it sticky and about 10 sec to make it pale, while spindle destruction will need about 150 sec of irradiation of an 8-$\mu$ spot of cytoplasm. Chromosomes irradiated locally with 2 Mev protons become sticky with 20 protons and pale with tens of thousands. It seems to require a million or more protons delivered to a small area of cytoplasm to destroy the spindle; we have not done many such experiments and this is a very approximate figure.

## SUMMARY

(1) Irradiation of dividing newt cells in culture with 200-kv x-rays or with ultraviolet light produces few or many abnormalities depending on the amount of radiation applied.

(2) Localized irradiation of chromosomes with a few tens of protons makes them sticky. But no abnormalities in the mitotic process result when a small area of cytoplasm is irradiated with 28 300 protons.

(3) Localized irradiation of cytoplasm with ultraviolet light prevents formation of the spindle or destroys it if it has already formed. In such cells, groups of whole chromosomes (not chromatids) move apart, usually in inequal numbers, to form the daughter nuclei of the dividing cell.

(4) The spindle thus seems to provide a mechanism for guiding the chromatids in anaphase, since, in the absence of a spindle, groups of whole chromosomes can move apart and form the nuclei for the daughter cells.

(5) Localized ultraviolet irradiation of chromosomes changes the index of refraction at the affected area. Such pale spots stain very faintly for DNA with the Feulgen method and absorb very little at 260 mμ.

## BIBLIOGRAPHY

[1] R. E. Zirkle and W. Bloom, Science, **117**, 487 (1953).

[2] R. B. Uretz, W. Bloom, and R. E. Zirkle, Science, **120**, 197 (1954).

[3] R. B. Uretz and R. P. Perry, Rev. Sci. Instr. **28**, 861 (1957).

[4] W. Bloom, R. E. Zirkle, and R. B. Uretz, Ann. N. Y. Acad. Sci. **59**, 503 (1955).

[5] R. B. Uretz and R. E. Zirkle, Biol. Bull. **109**, 370 (1955).

[6] R. E. Zirkle, Advances in Biol. and Med. Phys. **5**, 103 (1957).

[7] S. Tschachotin, *Handbuch der biol. Arbeitsmethoden* (Urban and Schwarzenberg, Berlin, 1938), Division 5, Part 10, p. 877.

[8] R. P. Perry, Exptl. Cell. Research **12**, 546 (1957).

[9] W. Bloom and R. J. Leider, Radiation Research **3**, 214 (1955).

# 4

# Chemical Characterization of Proteins, Carbohydrates, and Lipids

J. L. Oncley

*Department of Biological Chemistry, Harvard University Medical School, Boston 15, Massachusetts*

## PROTEINS

POLYMER molecules of high molecular weight but with relatively simple repeating units of one or two kinds are well known today. Proteins are naturally occurring polymers in which amino-acid residues are joined through peptide bonds and certain types of cross-linking bonds. They vary widely in physical and biological properties—from the insoluble keratin of hair to small soluble hormones like insulin—and are essential components of all living matter.

It has been found that only certain amino-acid residues—those from the "natural amino acids"—are present in proteins. These amino-acid residues are shown in Table I, and can be considered in three categories. The first category represents the neutral amino-acid residues—residues which do not have acidic or basic properties over the normal *p*H range. In the second category are the acidic and the basic amino-acid residues, listed in order of decreasing acidity or increasing basicity (as indicated by the $pK_a$ values). A third category represents amino-acid residues found only in certain proteins. Neglecting this last category of rarely found residues, 21 amino-acid residues are listed. The number of fundamental residues can be reduced to 18 if one considers the cystinyl residue as the oxidation product of two cysteinyl residues, and asparaginyl and glutaminyl residues as simple amide derivatives of aspartyl and glutamyl residues.

All of the residues except prolyl are derived from α-amino acids. The prolyl residue is derived from an α-imino acid (i.e., an N-substituted α-amino acid). An important feature of these 18 or 21 natural amino-acid residues is that in proteins they all seem to occur only in the L-configuration. The naturally occurring (protein) forms of all of the amino acids have been related to L-glyceraldehyde, and recent studies by Bijvoet *et al.*[1] have established the correctness of the Fischer convention which assumed the absolute configuration illustrated in Fig. 1. To date, no significant concentrations of D-amino-acid residues have been found in any of the proteins that have been studied.*

The peptide bond, proposed independently by Fischer and Hofmeister in 1902, is the principal linkage of the various amino-acid residues making up protein structures. It is formed by the splitting-out of water between the α-amino group of one amino acid and the carboxyl group of the next, with the formation of an amide linkage between the two residues:

$$\underset{R_1}{NH_2—CH—COOH} + \underset{R_2}{NH_2—CH—COOH}$$

$$= \underset{R_1}{NH_2—CH}—CO—\underset{R_2}{NH—CH}—COOH + H_2O.$$

Peptide formulas are customarily written with the free α-amino group at the left, and are named as substitution products of the amino acid furnishing the free α-carboxyl group.

The peptide bond is fairly stable, but its hydrolysis is catalyzed by acid, base, or various proteolytic enzymes. The peptide bond of the various amino-acid residues differ in stability during acid or base catalyzed hydrolysis, and there is a high degree of residue specificity in the enzymatic hydrolysis (see Neurath, p. 185). It is possible to form amide linkages involving the ε-amino group of lysine or the ω-carboxyl group of aspartic or glutamic acid. These other amide bonds, with properties very much like the peptide bond, have not been observed to form an important part of any protein structure. On the other hand, considerable amounts of glutathione, a dipeptide amide with the structure γ-glutamyl-cysteinyl-glycine, are found in many organisms, and poly-glutamic acid of certain bacteria contains the γ-glutamyl amide bond. Figure 2, from a more comprehensive discussion of protein structure by Low,[2] shows the detailed configuration of a fully extended polypeptide chain. The amide group, —CO—NH—, is planar, and the bond distances observed in various peptides ($\alpha C—C' = 1.53$ A, $C'—O = 1.24$ A, $C'—N = 1.32$ A, $N—\alpha C = 1.47$ A, with angles $N—\alpha C—C' = 110°$, $\alpha C—C'—N = 114°$, $O—C'—N = 125°$, and $C'—N—\alpha C = 123°$) indicate that the αC—N bond has about 40% double-bond character.[3]

A number of other bonds in addition to those making up the peptide linkage are found in most proteins. The disulfide linkage, where two cysteinyl residues are

* However, it might be argued that this statement needs a little more proof. Too often one assumes without proof that the amino acids are all in the L-configuration, particularly if one is working with small amounts of the amino acids. Since many peptide antibiotics have been found to contain the D-forms of these natural amino acids (as well as other amino-acid residues), the complete absence of the D-configuration in all proteins might be questioned.

TABLE I. Natural amino acids.

| Residue symbol | (a) Residue name<br>(b) Amino-acid name (trivial)<br>(c) Amino-acid name (systematic) | Residue formula |
|---|---|---|
| | | Amino-acid residues usually found in proteins |
| | | Neutral |
| -gly- | (a) glycyl<br>(b) glycine<br>(c) aminoacetic acid | $-CO-CH_2$<br>\|<br>NH<br>\| |
| -ala- | (a) alanyl<br>(b) alanine<br>(c) α-aminopropionic acid | $-CO-CH-CH_3$<br>\|<br>NH<br>\| |
| -val- | (a) valyl<br>(b) valine<br>(c) α-aminoisovaleric acid | $CH_3$<br>/<br>$-CO-CH-CH$<br>\| \\<br>NH $CH_3$<br>\| |
| -leu- | (a) leucyl<br>(b) leucine<br>(c) α-aminoisocaproic acid | $CH_3$<br>/<br>$-CO-CH-CH_2-CH$<br>\| \\<br>NH $CH_3$<br>\| |
| -ileu- | (a) isoleucyl<br>(b) isoleucine<br>(c) α-amino-β-methylvaleric acid | $CH_2-CH_3$<br>/<br>$-CO-CH-CH$<br>\| \\<br>NH $CH_3$<br>\| |
| -phe- | (a) phenylalanyl<br>(b) phenylalanine<br>(c) α-amino-β-phenylpropionic acid | $CH-CH$<br>// \\\\<br>$-CO-CH-CH_2-C$ $CH$<br>\| \\ /<br>NH $CH=CH$<br>\| |
| -pro- | (a) prolyl<br>(b) proline<br>(c) pyrrolidine-2-carboxylic acid | $-CO-CH-CH_2$<br>\| \\<br>\| $CH_2$<br>\| /<br>$N-CH_2$<br>\| |
| -try- | (a) tryptophanyl<br>(b) tryptophan<br>(c) α-amino-β-indolylpropionic acid | H<br>C<br>/ \\\\<br>$-CO-CH-CH_2-C$ — — $C$ $CH$<br>\| ‖ ‖ \|<br>NH $HC$ $C$ $CH$<br>\\ / \\ //<br>N C<br>H H |
| -ser- | (a) seryl<br>(b) serine<br>(c) α-amino-β-hydroxypropionic acid | $-CO-CH-CH_2OH$<br>\|<br>NH<br>\| |
| -thr- | (a) threonyl<br>(b) threonine<br>(c) α-amino-β-hydroxybutyric acid | OH<br>/<br>$-CO-CH-CH$<br>\| \\<br>NH $CH_3$<br>\| |
| -met- | (a) methionyl<br>(b) methionine<br>(c) α-amino-γ-methylthiobutyric acid | $-CO-CH-CH_2-CH_2-S-CH_3$<br>\|<br>NH<br>\| |

TABLE I.—*Continued.*

| Residue symbol | (a) Residue name (b) Amino-acid name (trivial) (c) Amino-acid name (systematic) | Residue formula | |
|---|---|---|---|
| -cyS-* \| -cyS- | (a) cystinyl (b) cystine (c) β-β′-dithiobis (α-aminopropionic acid) | —CO—CH—$CH_2$—S—S—$CH_2$—CH—NH— (first CH bonded below to NH—; second CH bonded below to CO—) | |
| $NH_2$ \| -asp- | (a) asparaginyl (b) asparagine (c) α-aminosuccinamic acid | —CO—CH—$CH_2$—CO—$NH_2$ (CH bonded below to NH—) | |
| $NH_2$ \| -glu- | (a) glutaminyl (b) glutamine (c) α-aminoglutaramic acid | —CO—CH—$CH_2$—$CH_2$—CO—$NH_2$ (CH bonded below to NH—) | |
| | **Amino-acid residues usually found in proteins** | | |
| | **Acidic (ionized form)** | | |
| -OH-* | terminal carboxyl | —$O^-$ (—$CO_2^-$) | $pK_a$=3.6–4.0 |
| -asp- | (a) aspartyl (b) aspartic acid (c) aminosuccinic acid | —CO—CH—$CH_2$—$CO_2^-$ (CH bonded below to NH—) | $pK_a$=3.9–4.7 |
| -glu- | (a) glutamyl (b) glutamic acid (c) α-aminoglutaric acid | —CO—CH—$CH_2$—$CH_2$—$CO_2^-$ (CH bonded below to NH—) | $pK_a$=3.9–4.7 |
| -tyr- | (a) tyrosyl (b) tyrosine (c) α-amino-β-(*p*-hydroxyphenyl) propionic acid | —CO—CH—$CH_2$—$C_6H_4$—$O^-$ (benzene ring: C, HC=CH, C—O⁻, HC=CH; CH bonded below to NH—) | $pK_a$=8.5–10.9 |
| -cySH-* | (a) cysteinyl (b) cysteine (c) α-amino-β-mercaptopropionic acid | —CO—CH—$CH_2$—$S^-$ (CH bonded below to NH—) | $pK_a$=*ca* 10 |
| | **Basic (ionized form)** | | |
| -his- | (a) histidyl (b) histidine (c) α-amino-β-imidazolyl-propionic acid | —CO—CH—$CH_2$—C==CH (imidazole ring: C==CH, $^+$N, N, C—H; CH bonded below to NH—) | $pK_a$=6.4–7.0 |
| -H-* | terminal amino | —$H_2^+$ (—$NH_3^+$) | $pK_a$=7.4–8.5 |
| -lys- | (a) lysyl (b) lysine (c) α-ε-diaminocaproic acid | —CO—CH—$CH_2$—$CH_2$—$CH_2$—$CH_2$—$NH_3^+$ (CH bonded below to NH—) | $pK_a$=8.5–10.9 |
| -arg- | (a) arginyl (b) arginine (c) α-amino-γ-guanidino-valeric acid | —CO—CH—$CH_2$—$CH_2$—NH—CH=$NH_2^+$ (first CH bonded below to NH—; second CH bonded below to $NH_2$) | $pK_a$=11.9–13.3 |

TABLE I.—*Continued.*

| Residue symbol | (a) Residue name<br>(b) Amino-acid name (trivial)<br>(c) Amino-acid name (systematic) | Residue formula | |
|---|---|---|---|
| | **Amino-acid residues occasionally found in proteins** | | |
| | **Neutral** | | |
| -hypro- | (a) hydroxyprolyl<br>(b) hydroxyproline<br>(c) 4-hydroxypyrroli-dine-2-carboxylic acid<br>(Found only in collagen) | 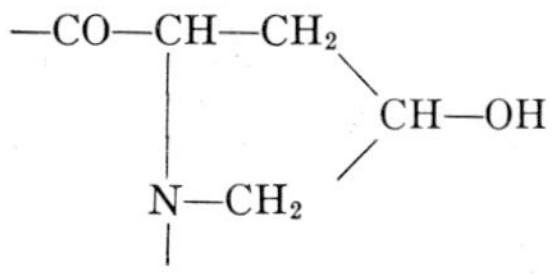 —CO—CH—$CH_2$ (ring: CH—$CH_2$—CH—OH—$CH_2$—N)<br>N—$CH_2$<br>\| | |
| | **Acidic (ionized form)** | | |
| | (a) diiodotyrosyl<br>(b) diiodotyrosine<br>(c) 3,5-diiodotyrosine<br>(Found only in marine organisms and thyroglobulin) | H I<br>C—C<br>—CO—CH—$CH_2$—C C—$O^-$<br>NH C=C<br>\| H I | $pK_a = 6.5$ |
| | (a) dibromotyrosyl<br>(b) dibromotyrosine<br>(c) 3,5-dibromotyrosine<br>(Found only in marine organisms) | H Br<br>C—C<br>—CO—CH—$CH_2$—C C—$O^-$<br>NH C=C<br>\| H Br | $pK_a = ca\ 7$ |
| | (a) thyroxyl<br>(b) thyroxine<br>(c) 3,3′,5,5′-tetra-iodothyronine<br>(Found only in thyroglobulin) | 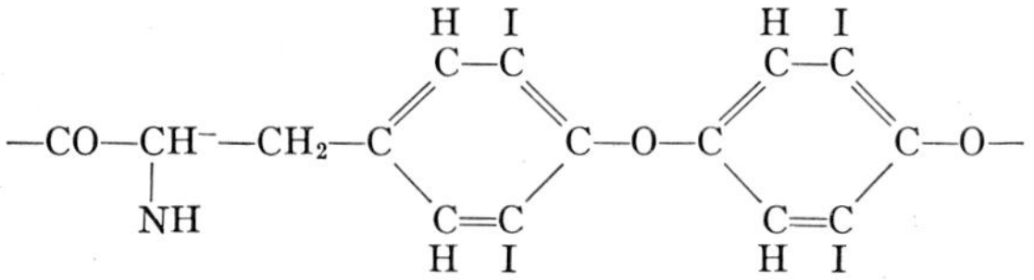 H I H I<br>C—C C—C<br>—CO—CH⁻—$CH_2$—C C—O—C C—O—<br>NH C=C C=C<br>H I H I | $pK_a = ca\ 6.5$ |
| | **Basic (ionized form)** | | |
| -hylys- | (a) hydroxylysyl<br>(b) hydroxylysine<br>(c) α, -diamino-hydroxy-caproic-acid<br>(Found only in collagen) | OH<br>\|<br>—CO—CH—$CH_2$—$CH_2$—CH—$CH_2$—$NH_3^+$<br>NH<br>\| | $pK_a = ca\ 11$ |

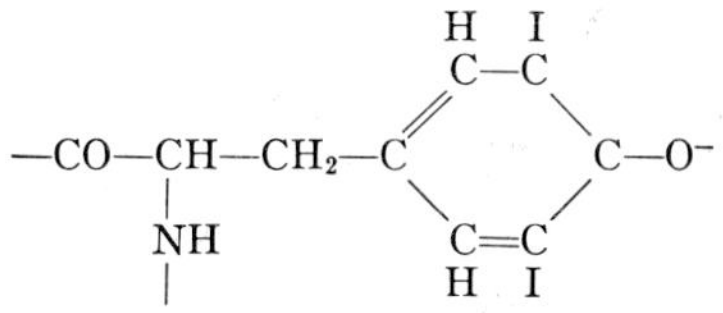

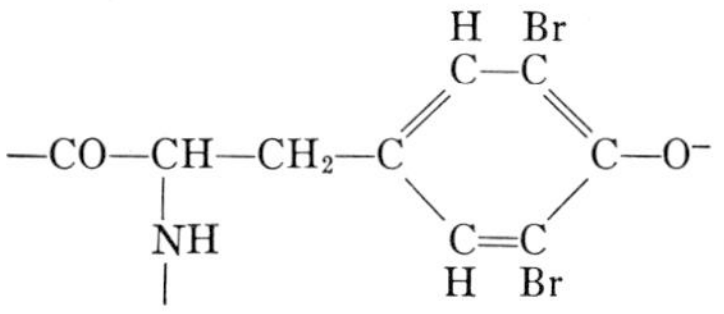

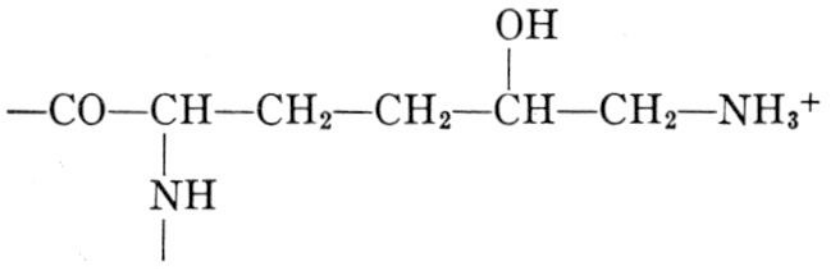

oxidized to form the disulfide linkage,

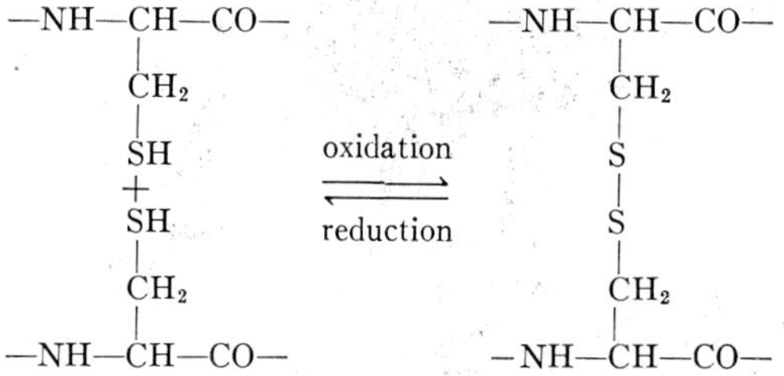

```
—NH—CH—CO—                    —NH—CH—CO—
     |                              |
     CH2                            CH2
     |                              |
     SH         oxidation           S
     +         ⇌                    |
     SH         reduction           S
     |                              |
     CH2                            CH2
     |                              |
—NH—CH—CO—                    —NH—CH—CO—
```

or

```
—cySH—     oxidation    —cyS—
   +          ⇌            |
—cySH—     reduction    —cyS— ,
```

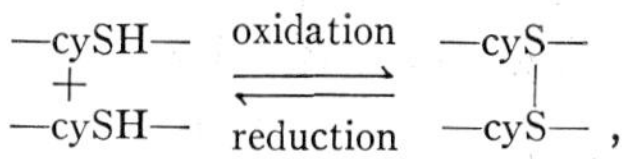

is found in all but a few proteins. This linkage may be between two otherwise separate peptide chains, or it may be an additional intrachain link. Both of these types of disulfide bonds are found in the insulin molecule, shown diagrammatically in Fig. 3. Calvin[4] has recently discussed the geometry of such C—S—S—C linkages. The actual spatial arrangement of a disulfide bond is shown in Fig. 4. The intrachain disulfide of insulin leads to a "link" in the polypeptide chain. This link contains six amino-acid residues (20 atoms), and it is of great interest to find that links of exactly the same number of residues are found in the peptide hormones oxytocin, arginine vasopressin, and lysine vasopressin (see Stetten, p. 563).

Recent studies by Perlmann[5] have done much to elucidate the various types of crosslinkages which involve phosphoric-acid residues. Orthophosphate —O—$PO_2^-$—O— and pyrophosphate —O—$PO_2^-$—

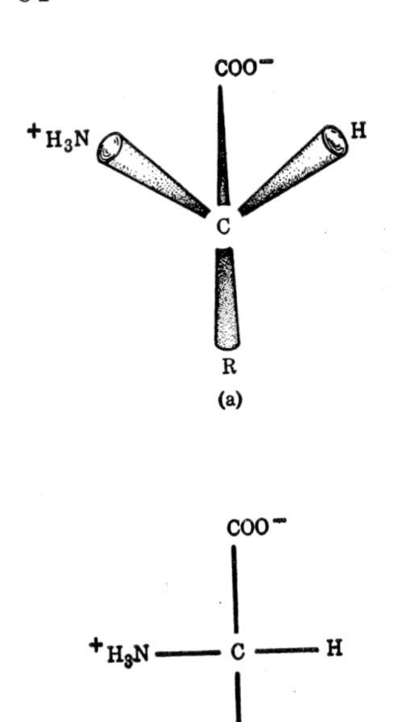

FIG. 1. Absolute configuration of the L-amino acids. (a) Geometrical representation. The $COO^-$, $NH_3^+$, and H groups are represented on a face plain of a tetrahedral carbon atom, with the R-group at the opposite apex. (b) Conventional chemical representation. (c) Photograph of a three-dimensional model of L-tryptophan.

$O—PO_2^-—O—$ crosslinkages have been demonstrated to occur in pepsin and in α-casein, respectively. The phospho-amide crosslinkage $—O—PO_2^-—NH—$ has also been found in α-casein, and the $—O—PO_3^=$ terminal residue is present in α-casein and in pepsin. The $—NH—PO_3^=$ terminal residue may also occur in phosphoproteins, but has not so far been demonstrated. These phosphoric-acid linkages would seem to involve seryl and threonyl residues in the case of linkage to oxygen, and possibly arginyl and lysyl residues in the case of linkage to nitrogen.

The hydrogen bond also seems to be an element in intra- and inter-peptide chain crosslinkage. Intrachain hydrogen bonding of the type —C—O—H—N— has been proposed by a number of workers, and makes up the important element of the Pauling α-helix, discussed in a later paper (see Rich, p. 50). Hydrogen bonding of the same type, but between different peptide chains, is the important crosslinking element in the pleated-sheet structures of Pauling and others, and in various double- and triple-stranded structures [for example,

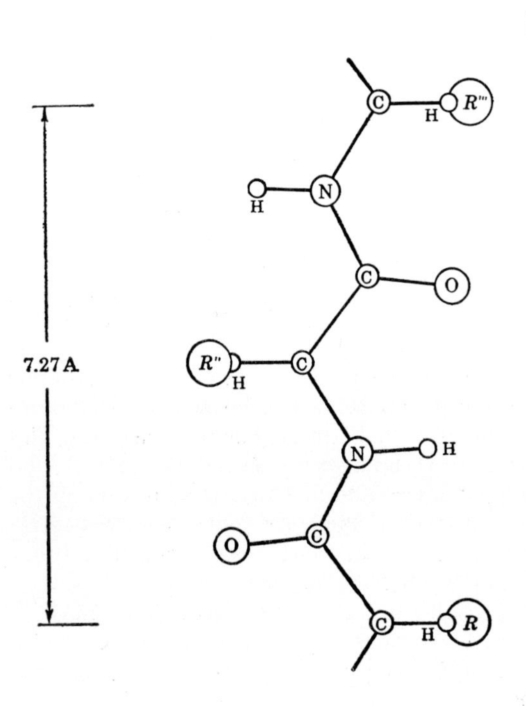


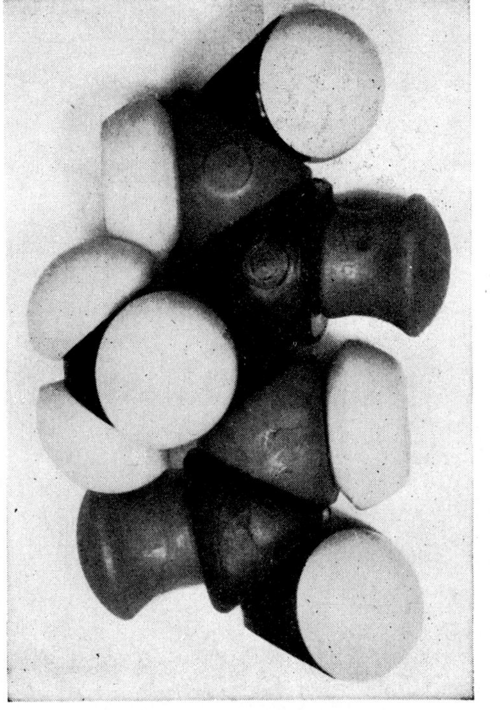

FIG. 2. Photograph of scale model of extended polypeptide chain. The R- and R'''-groups lie at the rear of this model [from B. W. Low in *The Proteins*, H. Neurath and K. Bailey, editors (Academic Press, Inc., New York, 1953), Vol. I, p. 259].

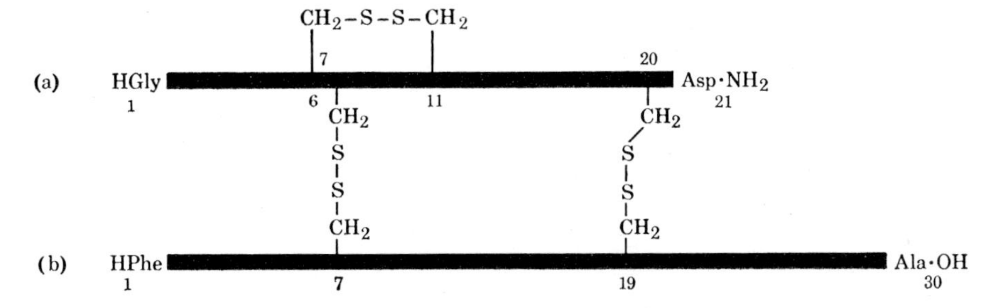


FIG. 3. Outline of the structure of insulin, showing the intrapeptide- and interpeptide-chain disulfide linkages. The numbers refer to the sequence of residues, as shown in Fig. 6 [from B. W. Low and J. T. Edsall in *Currents in Biochemical Research,* 1956, D. E. Green, editor (Interscience Publishers, Inc., New York, 1956), p. 379].

collagen (Rich, p. 58)]. Hydrogen bonds may also exist between carboxyl, amino, and tyrosyl residues. Although the energies involved in any one hydrogen bond are low (2000 to 6000 cal/mole, and even much lower if one considers the differences in energy between peptide-peptide hydrogen bonds and the same groups hydrogen bonded to water[6]), they are of great importance in structures where large numbers of hydrogen bonds can operate cooperatively. Linderstrøm-Lang[7] has reviewed an important method for the estimation of the extent of hydrogen bonding in a protein structure. (A further discussion of the hydrogen bond is found in the article by Orgel, p. 100.)

Another type of secondary bonding, leading both to intra- and inter-peptide chain crosslinkages, results from the action of van der Waals forces between the hydrocarbon-like residues. Waugh[6] has recently called attention to the large "nonpolar side-chain volume" contributed by such residues, and has stressed the point that allowance must be made for the fact that water must first be removed from contact with such groups before they can interact, and that this effect makes the interaction energies considerably larger than would be calculated solely on the basis of van der Waals attractive energies. It seems likely that the magnitude of such attractive forces within the protein molecule is quite comparable to those contributed by hydrogen bonding, and certain positions of large nonpolar residues on the peptide chains might lead to the stability of structures not compatible with the α-helix configuration. These considerations are treated in the later contribution by Waugh (p. 84).

Electrostatic forces between positively and negatively charged groups of the protein molecule can give rise both to attractive and to repulsive forces. The magnitude of such electrostatic forces depends upon the ionic strength of the solution. Most of the physicochemical studies of solutions of globular proteins (see Doty, p. 61) do not show a great dependence upon the ionic strength if the net charge of the protein is not too great, and can probably be interpreted as indicating that electrostatic forces are usually not of great importance in determining the conformation of the molecule. In the case of certain elongated protein molecules, the electrostatic forces may be of more importance in accounting for the molecular conformation, and it must be remembered that only electrostatic forces are likely to account for interactions acting over moderately large distances. Recent studies by Tanford have helped in the consideration of these electrostatic forces.[8]

Another type of intrachain linkage that may exist in proteins is the thiazoline ring, found in the peptide antibiotic bacitracin A. This linkage can be formed by a rearrangement of a peptide bond adjacent to a cysteinyl residue:

```
—NH—CH(R)—CO—NH—CH—CO--
                 |
                 CH2
                 |
                 SH

—NH—CH(R)—C=N—C—CO—
           |    |
           S————CH2
```

The same type of rearrangement could occur with a peptide bond adjacent to a seryl residue, to form an oxazoline ring where the sulfur atom is replaced by oxygen. Neither of these ring structures has been shown to exist in proteins.

The behavior of amino acids, peptides, and proteins is strongly dependent on the ionic or dipolar-ionic con-

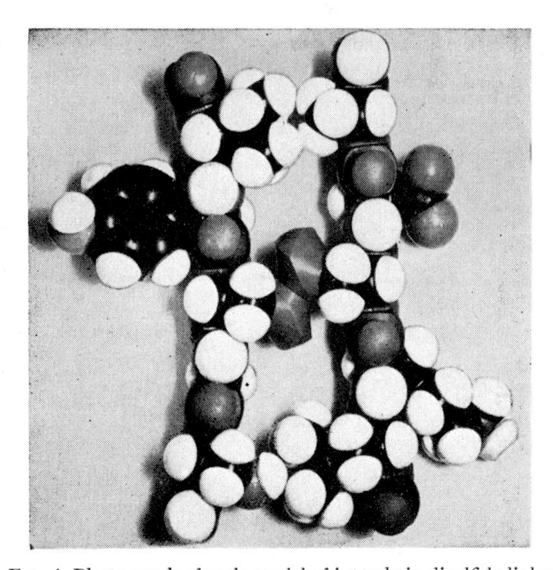

FIG. 4. Photograph of scale model of interchain disulfide linkage with antiparallel chain directions. The peptide chain on the left has the sequence -CO-leu-lys-cyS-asp-ala-NH- (bottom to top), and the chain on the right has the sequence -CO-val-tyr-cyS-ala-ser-NH- (top to bottom). The two peptide chains are about 6.3 A apart at the disulfide bond [from B. W. Low in *The Proteins,* H. Neurath and K. Bailey, editors (Academic Press, Inc., New York, 1953), Vol. I, p. 258].

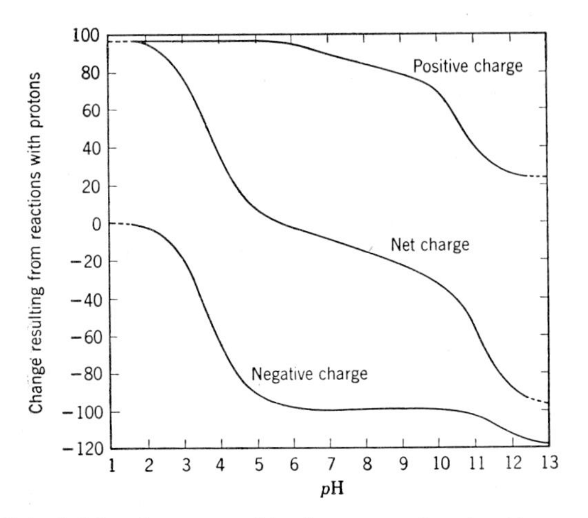


FIG. 5. Titration curve of bovine serum albumin. Also represented is the total positive charge and the total negative charge. These values are computed for 1 C-terminal carboxyl residue, 99 aspartyl and glutamyl residues, 19 tyrosyl residues, 1 cysteinyl residue, 16 histidyl residues, 1 N-terminal amino residue, 57 lysyl residues, and 22 arginyl residues.

figuration of these molecules. Each amino acid contains $\alpha$-$NH_3^+$ and $\alpha$-$COO^-$ groups, and the acidic or basic amino acids also contain an additional charged group. When amino acids are combined in peptides or proteins, the $\alpha$-$NH_3^+$ and $\alpha$-$COO^-$ groups are lost through the formation of the peptide bond, and only the terminal $NH_3^+$ and -$COO^-$ residues can become ionized. But the acidic and basic groups of the amino-acid residues of category two (Table I) remain capable of being ionized, and it is primarily these residues which give the dipolar and ionic properties to the protein molecule. Titration of a protein solution with strong acid or base allows a calculation of the number of available acidic and basic groups and an estimation of their *p*K's. Such a titration curve for bovine serum albumin calculated from the measurements of Tanford[9] is shown in Fig. 5. It can be seen that, when the net charge of the protein is zero, there are approximately 94 positively charged groups (basic groups of histidyl, terminal amino, lysyl, and arginyl residues) and an equal number of negatively charged groups (carboxyl groups of terminal carboxyl, aspartyl, and glutamyl residues). The properties of a protein are greatly influenced by the presence of these many charged groups, and the dipole moment of the uncharged protein is a measure of the symmetry of the distribution of these charges. Studies of the titration curves of many proteins have indicated that most of the acidic and basic amino-acid residues are available for reaction. [Rice (p. 69) discusses some of the difficulties in the analysis of titration curves of proteins and other polyelectrolytes.] In certain instances, however, some of these residues appear to be unavailable for reaction unless the protein is taken to extremes of *p*H.[8] Two well-established examples of this behavior are seen in hemoglobin and in ribonuclease. In hemoglobin, about 36 groups (probably 18 carboxyl groups of aspartyl and glutamyl residues, and 18 $\epsilon$-amino groups of lysyl residues) are found to be unreactive at neutral *p*H values,[10] whereas in ribonuclease, 3 of the 6 tyrosyl residues are found to be unreactive at their normal *p*K.[11]

One of the most exciting developments concerning the structure of proteins has been the evolution of methods which have led to the identification of the sequence of amino-acid residues in the peptide chains of a number of proteins. These developments have been made possibly largely through the work of Sanger and his colleagues who blazed the way by determining the sequence in insulin, a protein of about 6000 molecular weight.[12] Sanger's approach was to develop a method for the identification and estimation of the N-terminal residues of proteins and peptides. The reaction of the N-terminal residue with 1.2,4-fluorodinitrobenzene (FDNB) leads to the yellow dinitrophenyl (DNP) compound,

F
$O_2N$⟨benzene ring⟩ $NO_2$ $+ H_2N—CH—CO—prot. \rightarrow$ (R on CH)

NH—CH—CO—prot. (R on CH) on $O_2N$⟨benzene ring⟩ $NO_2$ $\xrightarrow{HCl}$ NH—CH—COOH on $O_2N$⟨benzene ring⟩ $NO_2$

This reaction is carried out under mild (slightly alkaline) conditions where the peptide bonds are quite stable. Acid hydrolysis of the resulting DNP compound leads to a mixture of the various amino acids present in the protein and the DNP-amino acid from the N-terminal residue. The DNP-amino acids are reasonably stable under the conditions of acid hydrolysis, and can usually be obtained in good yield. There are certain DNP-amino acids, for example DNP-proline and DNP-cysteine, which are considerably less stable than others, however, and substantial corrections must be made for the destruction of such DNP-amino acids during hydrolysis.

If the DNP-protein is only partially hydrolyzed, DNP-peptides can be isolated, and subsequent complete hydrolysis of these purified DNP-peptides reveals the nature of the amino-acid residues and the N-terminal residue; whereas a partial hydrolysis of the purified DNP-peptide can lead to the arrangement of the residues in the N-terminal peptide. By this method, the four or five residues adjoining the N-terminal residue can be arranged in sequence. In the studies on insulin, the N-terminal sequences DNP-phe-val-asp-glu- and DNP-gly-ileu-val-glu-glu- were identified in this way. Also, this same method of attack has been used to determine the residue sequence in various purified peptides obtained from either acid or enzymatic hydrolysis of the protein. After a large number of such peptides were completely identified (about 65 in the case of the *B*-chain of insulin), it was found that there

TABLE II. Peptides identified in hydrolyzates of fraction B of oxidized insulin.[a]

| |
|---|
| *Dipeptides from acid and alkaline hydrolyzates* |
| H-phe-val-OH H-his-leu-OH H-his-leu-OH H-ala-leu-OH H-gly-glu-OH H-thr-pro-OH |
| H-val-asp-OH H-leu-cy$SO_3H$-OH H-leu-val-OH H-leu-val-OH H-glu-arg-OH H-lys-ala-OH |
| H-asp-glu-OH H-cy$SO_3H$-gly-OH H-val-glu-OH H-val-cy$SO_3H$-OH H-arg-gly-OH |
| H-glu-his-OH H-ser-his-OH H-glu-ala-OH H-cy$SO_3H$-gly-OH H-gly-phe-OH |
| *Tripeptides from acid and alkaline hydrolyzates* |
| H-phe-val-asp-OH H-leu-cy$SO_3H$-gly-OH H-ala-leu-tyr-OH H-gly-glu-arg-OH H-pro-lys-ala-OH |
| H-val-asp-glu-OH H-ser-his-leu-OH H-tyr-leu-val-OH |
| H-glu-his-leu-OH H-leu-val-glu-OH H-leu-val-cy$SO_3H$-OH |
| H-his-leu-cy$SO_3H$-OH H-val-glu-ala-OH H-val-cy$SO_3H$-gly-OH |
| *Higher peptides from acid and alkaline hydrolyzates* |
| H-phe-val-asp-glu-OH H-ser-his-leu-val-glu-OH H-tyr-leu-val-cy$SO_3H$-OH H-thr-pro-lys-ala-OH |
| H-phe-val-asp-glu-his-OH H-ser-his-leu-val-glu-ala-OH H-leu-val-cy$SO_3H$-gly-OH |
| H-glu-his-leu-cy$SO_3H$-OH H-his-leu-val-glu-OH |
| H-his-leu-cy$SO_3H$-gly-OH H-leu-val-glu-ala-OH |
| H-ser-his-leu-val-OH |
| *Sequences deduced from above peptides* |
| H-phe-val-asp-glu-his-leu-cy$SO_3H$-gly- -tyr-leu-val-cy$SO_3H$-gly- -thr-pro-lys-ala |
| -ser-his-leu-val-glu-ala- -gly-glu-arg-gly- |
| *Peptides identified in peptic hydrolyzate* |
| H-phe-val-asp-glu-his-leu-cy$SO_3H$-gly-ser-his-leu-OH H-leu-val-cy$SO_3H$-gly-glu-arg-gly-phe-OH |
| H-val-glu-ala-leu-OH H-tyr-thr-pro-lys-ala-OH |
| H-his-leu-cy$SO_3H$-gly-ser-his-leu-OH |
| *Peptides identified in chymotryptic hydrolyzate* |
| H-phe-val-asp-glu-his-leu-cy$SO_3H$-gly-ser-his-leu-val-glu-ala-leu-tyr-OH H-tyr-thr-pro-lys-ala-OH |
| H-leu-val-cy$SO_3H$-gly-glu-arg-gly-phe-phe-OH |
| *Peptides identified in tryptic hydrolyzate* |
| H-gly-phe-phe-tyr-thr-pro-lys-ala-OH |
| *Structure of the B-(phenylalanyl terminal) chain of insulin* |
| H-phe-val-asp-glu-his-leu- (cyS-) -gly-ser-his-leu-val-glu-ala-leu-tyr-leu-val- (cyS-) -gly-glu-arg-gly-phe-phe-tyr-thr-pro-lys-ala-OH |

[a] Taken from F. Sanger, Advances in Protein Chem. **7**, 56 (1952).

was only one sequence which would fit all of the experimental results, assuming that there was a single polypeptide chain of about 30 residues. It was further assumed that no rearrangements of the residues had occurred during the hydrolyses.

The attack outlined in the foregoing is applicable only to single peptide chains devoid of interchain linkages, such as the disulfide link described earlier. Since insulin and most other proteins contain interchain and/or intrachain disulfide linkages, these must be broken before the sequence studies can be undertaken by the foregoing methods. In the case of insulin, oxidation of all of the disulfide linkages to sulfonic-acid groups was carried out with performic acid. This reagent thus converts cystine to cysteic acid (H—cy$SO_3H$—OH), a strong acid. It also converts methionine to the corresponding sulfone, and tryptophan to unidentified products. Since insulin contains no methionine or tryptophan, performate oxidation was capable of converting the insulin to two chains, called the *A*-(glycine terminal) chain and the *B*-(phenylalanine terminal) chain, with each half-cystine residue (-cyS-) converted to a cysteic-acid residue (-cy$SO_3H$-). The *A*- and *B*-chains were then purified, and the sequences of amino-acid residues in each of the two peptide chains were determined by the method outlined in the following. Table II records a number of the peptides obtained from such a study of the *B*-chain, and Fig. 6 shows the entire sequence for the insulin molecule. The acid hydrolysis used in the determination of the amino-acid sequence in insulin caused the liberation of six moles of ammonia, originally present as the amide groups of the

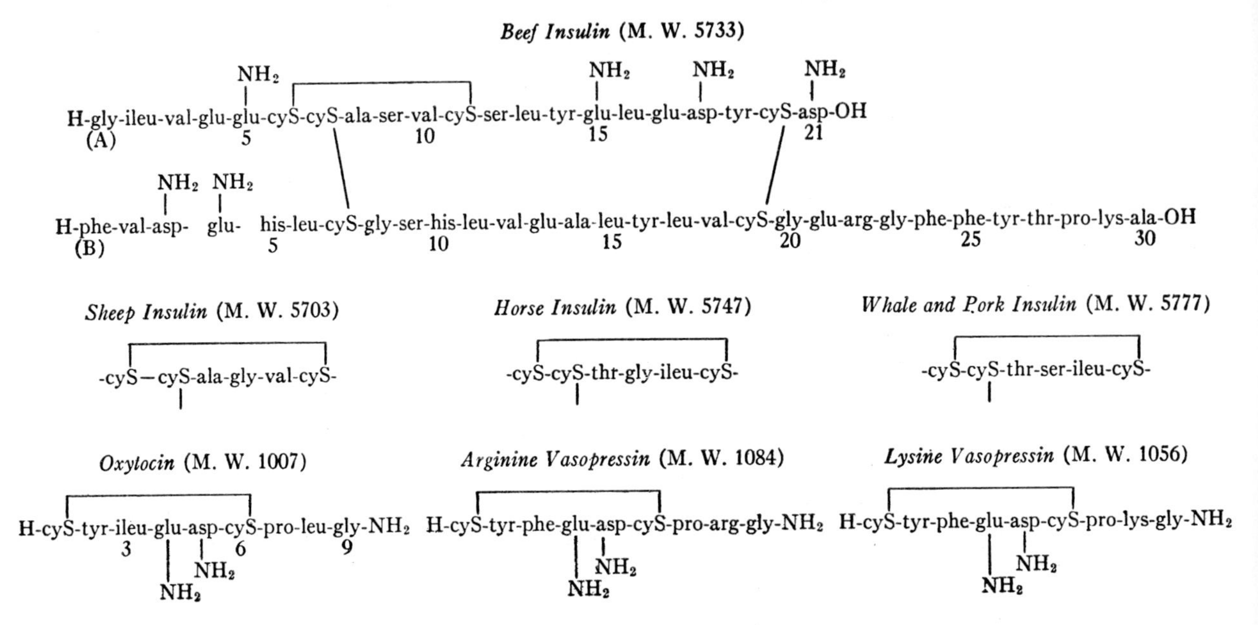


FIG. 6. Amino-acid sequences in insulin and some hormone peptides.

asparaginyl and glutaminyl residues. Reduction of the insulin by lithium borohydride ($LiBH_4$) in tetrahydrofuran converted the $\omega$-carboxyl groups to primary alcohol groups, and acid hydrolysis of the reduced insulin then showed that three asparaginyl residues, three glutaminyl residues, and four glutamyl residues were present in insulin. In order to locate these amide groups on the individual glutamic-acid residues, two methods were used which depend on the fact that enzymatic hydrolysis does not break the $\omega$-amide linkage. The purified peptides obtained after enzymatic hydrolysis could then be tested for amide content, either by studies of their electrophoretic behavior (since the amide derivatives had one negative charge less than the corresponding $\omega$-acid derivatives), or by studies of the extent of ammonia liberation during their acid hydrolysis. In this way, it was found that residues 5, 15, 18, and 21 of the *A*-(glycyl terminal) chain and residues 3 and 4 of the *B*-chain were present in the amide form.

In order to find the distribution of the disulfide bridges, Sanger submitted intact insulin to partial enzymatic hydrolysis. The resulting cystine-containing peptides were then purified and sequences of these peptides were determined. This gave Sanger the final data required to completely identify the sequences and cross-linkages in beef insulin,[13] and resulted in the formula given in Fig. 6. The intrachain disulfide linkage found in the *A*-(glycyl terminal) chain is of special interest. The half-cystinyl residue at position 6 is combined with the position 11 half-cystinyl residue through a disulfide linkage. Studies on horse, sheep, whale, and pork insulin indicate sequences similar to those found in beef insulin, except for the amino acids in positions 8, 9, and 10 of the *A*-(glycyl terminal) chain. These residue differences are all within the hexapeptide disulfide link. Figure 6 also shows the sequences of the insulins of the other species studied. Whale and pork insulin are seen to be identical.

As mentioned in the discussion of the intrachain disulfide linkage, three peptide hormones isolated from the anterior pituitary and synthesized by du Vigneaud —oxytocin, arginine vasopressin, and lysine vasopressin —all contain a similar hexapeptide disulfide link. The formulas of these peptides are shown in Fig. 6 also, and the possible significance of this structure in terms of hormone structure is discussed by Stetten (p. 563).

The methods developed by Sanger and his colleagues for the study of the amino-acid sequence in insulin have subsequently been applied by a number of laboratories to the elucidation of the structure of several other proteins. Other methods of attack involving replacement of the DNP compounds by other types of derivatives have been developed, and the stepwise freeing of N-terminal and C-terminal residues by enzymatic hydrolysis catalyzed by aminopeptidases and carboxypeptidase has proved to be fruitful. Methods have also been developed to replace the performate oxidation procedure for the elimination of disulfide bonds, and these can be applied in the study of proteins containing tryptophan and other easily oxidized amino acids. A short summary of these methods for amino-acid sequence determinations can be found in a recent review by Anfinsen and Redfield.[14]

Figures 7–9 summarize some of the more complete amino-acid sequence studies on other proteins and large peptides. The most complete amino-acid sequence study of a series of related hormones is the work on the melanocyte-stimulating hormones (MSH) and adrenocorticotropic hormones (ACTH). Figure 7 records the

*Beef Melanotropin (Seryl-β-MSH)* (M.W. 2134)

H-asp-ser-gly-pro-*tyr*-lys-*met-glu-his-phe-arg-try-gly*-ser-*pro*-pro-lys-asp-OH
5 10 15 18

*Pork β-Melanocyte-Stimulating Hormone (Glutamyl-β-MSH)* (M.W. 2176)

H-asp-glu-gly-pro-*tyr*-lys-*met-glu-his-phe-arg-try-gly*-ser-*pro*-pro-lys-asp-OH
5 10 15 18

*Pork α-Melanocyte-Stimulating Hormone (α-MSH)* (M.W. 1665)

$CH_3$-CO-ser-*tyr*-ser-*met-glu-his-phe-arg-try-gly*-lys-*pro*-val-$NH_2$
5 10 13

*Sheep α-Corticotropin (α-ACTH)* (M.W. 4540)*

H-ser-*tyr*-ser-*met-glu-his-phe-arg-try-gly*-lys-*pro*-val-gly-lys-lys-arg-arg-pro-val-
5 10 15 20

$NH_2$ (on glu-33)

-lys-val-tyr-pro-ala-gly-glu-asp-asp-glu-ala-ser-glu-ala-phe-pro-leu-glu-phe-OH
21 25 30 35 39

| ACTH Preparation | Residue No. 25 26 27 28 29 30 31 32 33 |
|---|---|
| Sheep α-corticotropin | -ala-gly-glu-asp-asp-glu-ala-ser-glu- ($NH_2$ on 33) |
| Pork corticotropin-A (Cyanamid) | -asp-gly-ala-glu-asp-glu-leu-ala-glu- ($NH_2$ on 30) |
| Pork corticotropin-A (Armour) | -gly-ala-glu-asp-asp-glu-leu-ala-glu- |

FIG. 7. Amino-acid sequences in melanocyte-stimulating and adrenocorticotropic hormones. Sheep α-corticotropin contains one more amide group, probably in position 27, 28, or 29. Beef α-corticotropin appears from preliminary structural investigations to be identical with sheep α-ACTH. All of these ACTH preparations seem to have no loss of biological potency when residues 29 to 39 (11 residues) are removed by limited enzyme hydrolysis.

results of extensive studies on these hormones in some five or six laboratories. Recent reviews of this work have been presented by Li,[15–17] Harris,[18] and by Anfinsen and Redfield.[14] The amino-acid sequences of all of the MSH and ACTH materials studied to date have shown almost identical sequences . . . *-tyr-*. X . *-met-glu-his-phe-arg-tyr-gly-*. Y . *-pro-*. . . . This sequence has been shown in italics in Fig. 7, and is seen to occur at different distances from the N-terminal residue in the β-MSH preparations (residues 5–15) and in α-MSH and the various ACTH preparations (residues 2–12). The residues X and Y in this unique sequence are seen to be lysyl or seryl, and seryl or lysyl, respectively (see Stetten, p. 563). The two pork corticotropins-A, as prepared in the Cyanamid and in the Armour Laboratories, are known to differ with respect to the total content of amide groups (none for the Armour product, and one for the Cyanamid product). The sequence differences in residues 25–28 of the two corticotropin-A preparations have not been definitely shown to indicate different sequences in the two preparations, since technical difficulties may have occurred in the sequence determination. The sheep α-corticotropin preparation of Li differs from the corticotropin-A, in that Li's preparation contains two amide groups (one is glutaminyl residue 33, and one not yet located, but probably in position 27, 28, or 29), and also one more seryl and one less leucyl residue. It may be noted that the component amino acids in positions 25–28 are the same in all three ACTH preparations, but differ in order. The different location of the amide residue (position 33 in sheep α-corticotropin and 30 in pork corticotropin-A) appears to represent a real difference in sequence. Preliminary structural studies by Li have indicated that the amino-acid sequence for beef α-corticotropin is identical with that for sheep α-corticotropin.[17]

The two β-MSH preparations are seen to differ only in the second residue, seryl for the beef hormone and glutamyl for the pork hormone. These materials are, therefore, often referred to as seryl-β-MSH and glutamyl-β-MSH. The acetylated amino group of the N-terminal seryl residue in α-MSH has only recently been identified,[18] and its presence is rather unusual. The presence of this group, as well as of the amide group of the C-terminal valyl residue in α-MSH, causes a drastic change in the hormonal activity of this compound as compared with that of the longer α-ACTH peptide.[18] This effect, as well as other aspects of the relation between the structure of these hormones, and their physiological activities, is discussed by Stetten (p. 563).

The amino-acid sequence in glucagon, a small protein involved in glucose metabolism, has recently been determined and is shown in Fig. 8. This protein, like the ACTH and MSH preparations, contains no cystinyl residues. Unlike these hormones, however, the sulfur-containing methionyl residue is present in glucagon. The relationship between glucagon and insulin in glucose metabolism is discussed by Stetten (p. 563). The

H-his-ser-glu($NH_2$)-gly-thr-phe-thr-ser-asp-tyr-ser-lys-tyr-leu-asp- (5, 10, 15)

-ser-arg-arg-ala-glu($NH_2$)-asp-phe-val-glu($NH_2$)-try-leu-met-asp($NH_2$)-thr-OH (20, 25, 29)

FIG. 8. Amino-acid sequence in glucagon (M. W. 3647).

sequence arrangements in all of the unrelated proteins so far evaluated show no more similarity than could be expected from simple probability considerations, and none of them indicates that repeating sequences are important elements in the structure of these small globular proteins. Whether or not repeating sequences occur in the larger globular proteins is as yet unknown. Some strong evidence of repeating elements has, however, been found in certain fibrous-protein structures, and in histone and protamine, proteins found associated with nucleic acid in the nucleoproteins.[14]

Ribonuclease is the most complicated protein in which an almost complete amino-acid sequence has been determined. This enzyme, containing 124 amino-acid residues and four disulfide bonds (molecular weight 12 000), contains more than twice the number of residues found in insulin. The presently known sequence as shown in Fig. 9 has recently been discussed by Hirs *et al.*[19] and by Anfinsen.[14,20] About 23 residues remain to be located definitely in the sequence. No unusual partial sequences seem to occur, with the possible exception of a number of repeats (-ala-ala-ala- in positions 4–7, and -met-met- in positions 29–30). These partial sequences can be compared with the unusual sequence -phe-phe-tyr- in the *B*-chain of insulin (positions 24–26).

Short, partial-residue sequences have been determined for a large number of proteins. A few of the more interesting partial sequences are reported in Figs. 10 and 11. The partial sequence shown for beef cytochrome-*c* (Fig. 10) is interesting in that this sequence contains the residues to which the porphyrin-*c* prosthetic group is attached. The exact attachment of the porphyrin-*c* residue may be either that shown in Fig. 10, or that of an otherwise identical structure where the porphyrin-*c* residue is rotated about the plane of the page by 180°. The corresponding partial sequences in a series of cytochrome-*c* preparations from sources other than beef are also shown in Fig. 10. Here, alanyl residues are sometimes replaced by seryl or glutamyl residues, lysyl residues by arginyl residues, valyl residues by lysyl residues, and glutaminyl residues by threonyl residues. The italicized sequence *-cyS-*.A.-.B.*-cyS-his-thr-val-glu-* has been found to occur in this order in each of the six cytochrome-*c* molecules. This study, involving work by Theorell, Tuppy, and others, has been reviewed by Tuppy[21] and by Anfinsen and Redfield.[14] The partial sequences shown for lysozyme and for human serum albumin (Fig. 11) are very incomplete. Many disulfide bonds are known to be present in these molecules, and serum albumin is somewhat unusual in that it contains a single cysteinyl residue. Many other partial sequences have been found for egg-white lysozyme,[14,22] but a study of these sequences shows that certain of them must be spurious, and it has not been possible to locate the various partial sequences in relation one to another. Anfinsen and Redfield[14] suggest that the study of lysozyme illustrates the limit of structural information that can be obtained by acid hydrolysis alone, and that more specific degradative methods must be applied before the complete amino-acid sequence can be obtained.

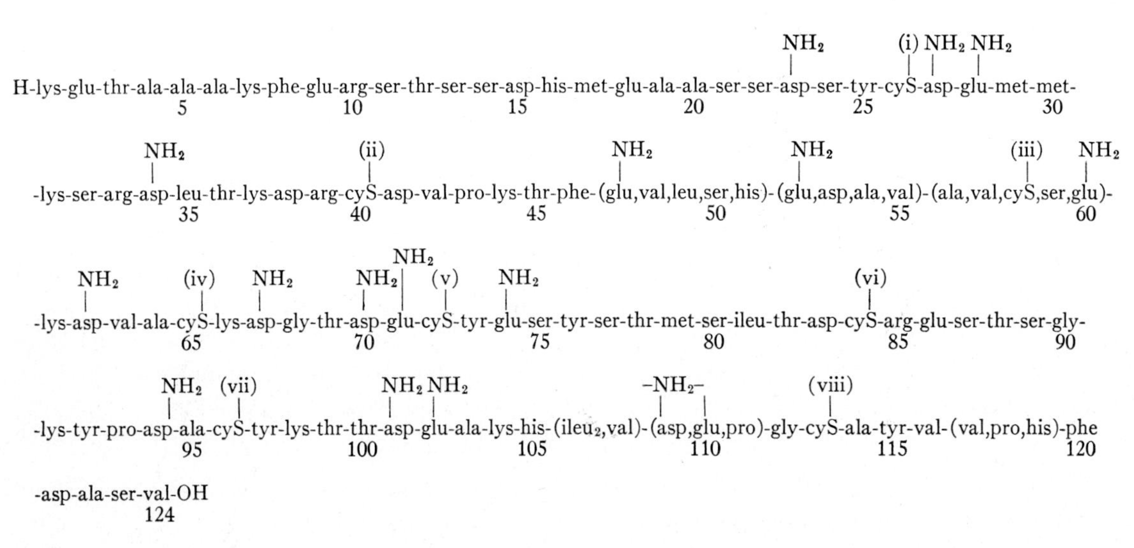


FIG. 9. Amino-acid sequence in ribonuclease (M. W. 12 000). Where the composition and position of a group of residues is known, but not the relative positions of the individual residues within the group, the listing of the group is enclosed in parentheses. The half-cystinyl residues are identified by roman numerals. Work of Anfinsen[20] has shown that the four disulfide bridges in ribonuclease are paired as follows: i–vi, iii–vii, ii–viii and iv–v (after studies by C. H. W. Hirs *et al.*[19]).

Other species of serum albumin have been studied, and it has been shown that bovine serum albumin, while containing the same N-terminal residue (H-asp-), has threonyl as the second residue and ends with a different C-terminal sequence, possibly -(ala, leu, thr, val, ser)-ala-OH.[14] Porter[23] has obtained a fragment of the bovine serum-albumin molecule after mild chymotryptic hydrolysis which seems to represent about one-fifth of the entire molecule (molecular weight about 12 000).

*Beef Cytochrome-c* (M.W. 13 100)

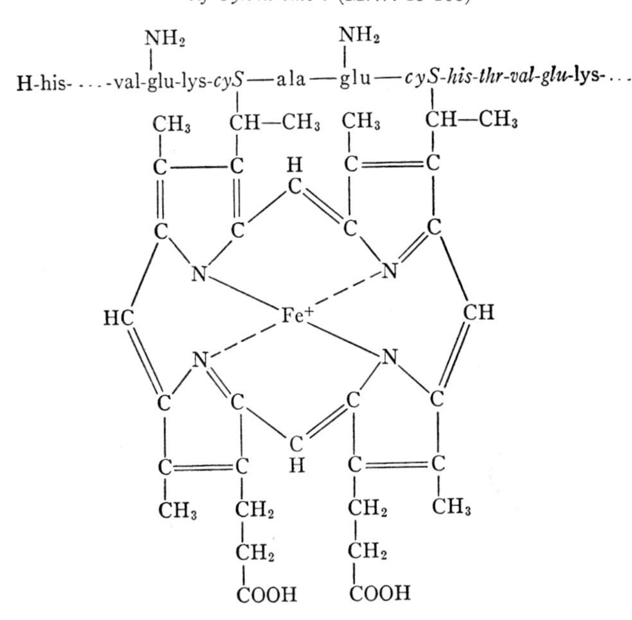


| *Preparation* | *Observed residue sequence* |
|---|---|
| Beef | -val-glu-lys-*cyS*-ala-glu-*cyS-his-thr-val-glu*-lys- |
| Horse and pork | -lys-*cyS*-ala-glu-*cyS-his-thr-val-glu*-lys- |
| Salmon | -val-glu-lys-*cyS*-ala-glu-*cyS-his-thr-val-glu*- |
| Chicken | -val-glu-lys-*cyS*-ser-glu-*cyS-his-thr-val-glu*- |
| Silkworm | -val-glu-arg-*cyS*-ala-glu-*cyS-his-thr-val-glu*- |
| Yeast | -phe-lys-thr-arg-*cyS*-glu-leu-*cyS-his-thr-val-glu*- |

FIG. 10. Partial amino-acid sequence in cytochrome-*c*. The exact attachment of the porphyrin-*c* residue may be either the residue shown or the structure where the porphyrin-*c* residue is rotated by 180° [from H. Tuppy in *Symposium on Protein Structure*, A. Neuberger, editor (Methuen and Company, Ltd., London; John Wiley and Sons, Inc., New York, 1958), p. 71].

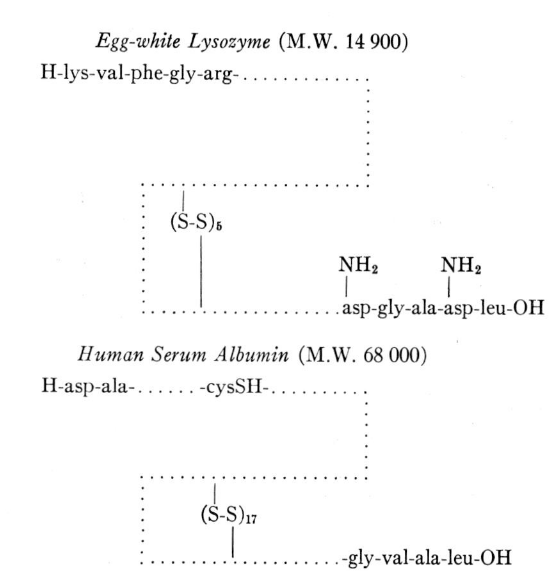


FIG. 11. Partial amino-acid sequences in lysozyme and serum albumin [from C. B. Anfinsen and R. R. Redfield, Advances in Protein Chem. **11**, 1 (1956)].

This fragment appeared to have very nearly the same configuration as part of the original albumin molecule, since it effectively inhibited the reaction of a specific antiserum to bovine serum albumin with its antigen (see Kauzmann, p. 549). This fragment contained a single N-terminal amino acid (H-phe-), the unique cysteinyl residue, and a single disulfide bond.

Further structural studies of amino-acid sequences are reported in other contributions to the Paris *Symposium on Protein Structure*.[24] Considerable sequence data exist for many other proteins, notably papain, growth hormone, trypsin, chymotrypsin, pepsin, hemoglobin, and tobacco mosaic virus.

It is necessary to always bear in mind the difficulties in the isolation of purified and homogeneous components in their native state. Proteins occur in nature as complex mixtures, and are often found in the presence of highly charged polysaccharides and/or other macromolecules. These naturally occurring colloidal mixtures are often enclosed by membranes of varying stability. The extraction of a particular protein from such a mixture is always difficult, and is often impossible without the use of procedures which may cause permanent changes in the configuration or even in the covalent linkages of the resulting protein preparation. A number of useful methods for the isolation and purification of such protein systems are currently available for the separation of these components. These methods include precipitation and differential extraction by high concentrations of salts or organic solvents, adsorption and partition between immiscible solvents, as well as such physical methods as electrophoresis and ultracentrifugation. A discussion of these methods cannot be undertaken here, but it is important to remember that the purified components isolated by these methods may

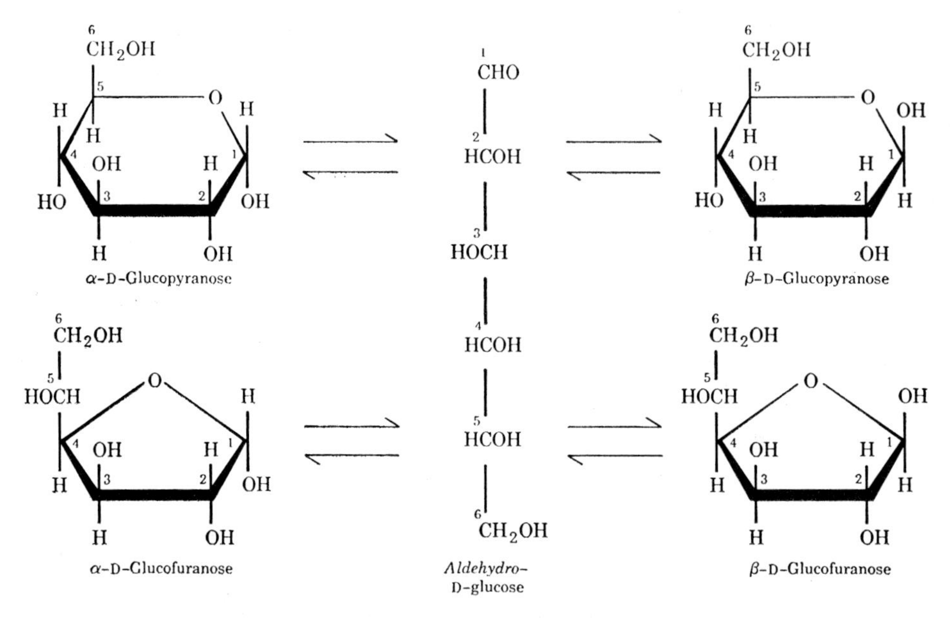


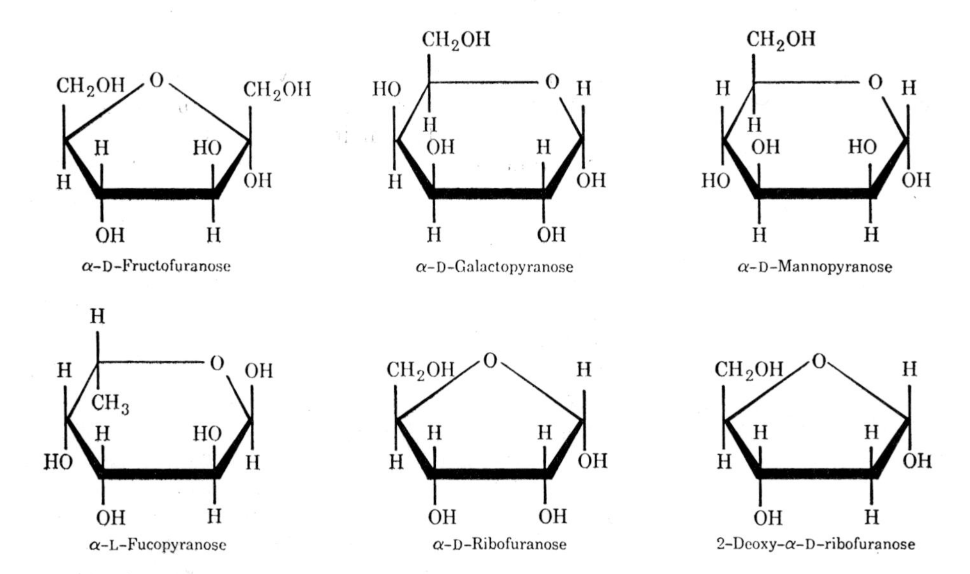


FIG. 12. Configuration of some monosaccharides.

not be truly homogeneous, and artifacts may be introduced by the isolation procedures.

Tests for the homogeneity of the purified components must be evaluated critically. Physical methods are most useful; they include studies of behavior in the ultracentrifuge and in electrophoresis, measurements of solubility (akin to the phase-rule methods of constant melting or boiling points as applied in the case of smaller molecules), and determination of the distribution coefficient between various solvents. Chemical methods showing the constancy of composition (in regard to the individual amino-acid residues, for example) are also of much value. Biological assay methods measuring activities ascribed to various components and quantitatively evaluating the activities of the purified components in terms of the activity of the native tissues are of special importance.

## CARBOHYDRATES

Carbohydrates comprise another of the major groups of naturally occurring organic materials. They are often found in combination with proteins as glycoproteins and mucoproteins, especially in the higher animals. They also occur in combination with lipids to give cerebrosides, and in combination with purines and phosphoric acid to give nucleotides and nucleic acids. They form the structural elements of plants, bacteria, and certain animals in the forms of cellulose, chitin, and other high molecular-weight polysaccharides. They provide a means for the storage of chemical energy in the forms of amylose and amylopectin in plant starches, and glycogen in animals.

Although most monosaccharides are fairly stable when in the crystalline condition, they undergo many

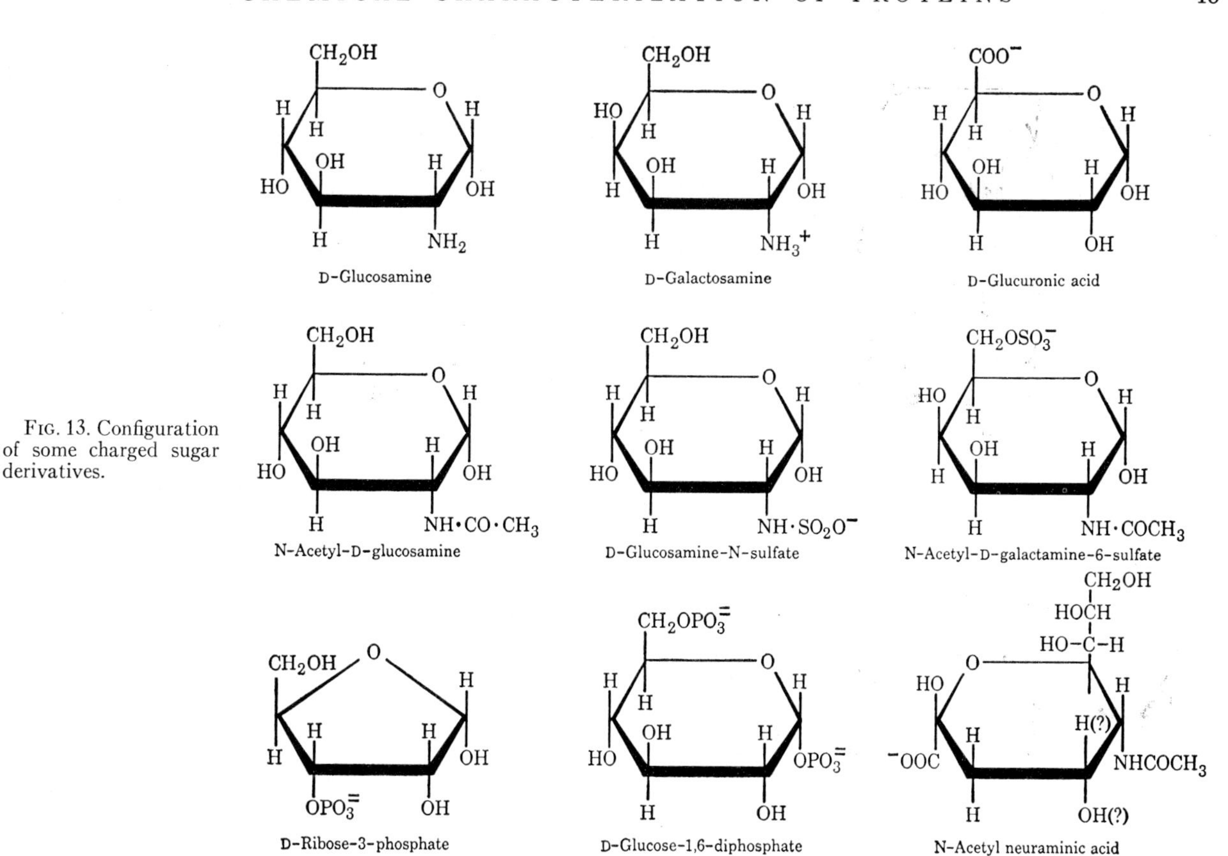


FIG. 13. Configuration of some charged sugar derivatives.

transformations when dissolved in water, particularly in the presence of acids or bases. Although the chemical formulas of the monosaccharides are often shown in a linear form, much evidence suggests that, in the main, they exist in the form of five- and six-membered rings called the furanose and pyranose forms. Each of these forms exists as $\alpha$- and $\beta$-isomers, termed anomers, and interconversions between anomers and between ring isomers occur even under the mildest possible conditions of acidity and temperature. One of the simplest methods for demonstrating and studying the equilibria between the various forms is by the measurement of changes of optical rotation with time ("mutarotation"), which can easily be observed with freshly prepared sugar solutions. These various equilibria (for D-glucose) are illustrated in Fig. 12. The linear forms are shown in the Fischer formulation where the carbon atoms must be thought of as tetrahedrons, bonded together by angles projecting into the page and with the H— and —OH groups projecting out from the page. The ring forms shown in the Haworth formulation are to be viewed as a perspective representation with the heavy bonds extending out from the page. Side chains in the Haworth formulas are written according to the Fischer convention. The rings, shown here as planar, are oversimplified, since the valence angles in a single coplanar ring would be appreciably greater than those in a "strainless" structure having valence angles of 109°. The usual conformation of the pyranose ring is probably a chain form, but the free sugars in solution probably can occur to some extent in any of the possible conformations, since there is a spontaneous equilibrium with the aldehyde form. Derivatives in which the ring is fixed (because of substitution on C-1) probably are stabilized in the chair form.[25]

The aldehydo-hexoses such as D-glucose and D-galactose contain four asymmetric carbon atoms, while the keto-hexoses such as D-fructose and aldehydo-pentoses such as D-ribose contain three. Thus, many optical isomers exist (16 for the aldehydo-hexoses and 8 for the keto-hexoses and aldehydo-pentoses), and a number of these are found in various natural products. The ring structures contain an additional center of asymmetry at carbon atom 1, and the anomers of each ring structure are usually designated $\alpha$- when the hydroxyl group of carbon atom 1 is *cis*- to the hydroxyl group at carbon atom 2, and $\beta$- when these two hydroxyl groups are *trans*-. In the case of 2-deoxy-D-ribose, the anomeric forms are designated like the D-ribose forms. $\alpha$-D-mannose does not follow the (*cis*-) (*trans*-) rule given in the foregoing, and the commonly used nomenclature of Hudson is to be found in Pigman's review.[25] The designation L- indicates that the asymmetric carbon atom most remote from the reference group (e.g., aldehyde, keto, carboxyl, etc.) has the same configuration as L-glyceraldehyde (see Fig. 1). The L- and D-forms differ in the configuration of all asymmetric carbon atoms. In addition to the three monosaccharides

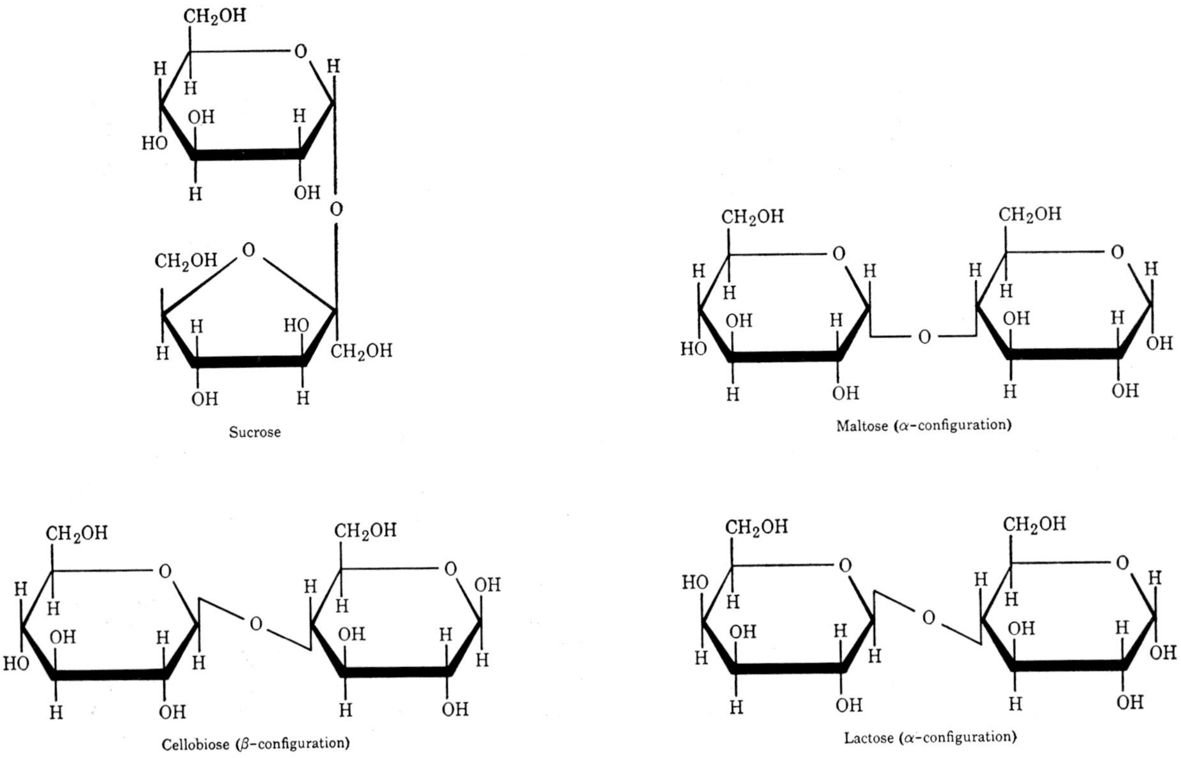

Sucrose

Maltose ($\alpha$-configuration)

Cellobiose ($\beta$-configuration)

Lactose ($\alpha$-configuration)

FIG. 14. Configuration of some disaccharides.

mentioned above, those most commonly found in bacterial and animal sources include D-galactose (like D-glucose with the configuration of C-4 reversed) and D-mannose (like D-glucose with configuration of C-2 reversed). Three other monosaccharides of major interest include L-fucose (6-deoxy-L-galactose), L-rhamnose (6-deoxy-L-mannose), and 2-deoxy-D-ribose, where the designation -deoxy- indicates that the appropriate —OH is replaced by —H.

Haworth formulas for most of the aforementioned monosaccharides are shown in Fig. 12. Each of these sugars would exist in the same sorts of conformations as are shown for D-glucose, and the ring structures commonly found in oligo- and poly-saccharides are portrayed in the diagrams. Only the $\alpha$-anomer is depicted. The relative amounts of the four-ring conformations which will exist at equilibrium in aqueous solutions of the various monosaccharides vary considerably. In the case of neutral solutions of D-glucose, nearly two-thirds of the sugar seem to exist as $\beta$-glucopyranose, and just over one-third as $\alpha$-glucopyranose. The two glucofuranose forms and the *aldehydo*-glucose form are thought to exist in minor amounts. In the case of D-mannose, over two-thirds exist as $\alpha$-mannopyranose and just under one-third as $\beta$-mannopyranose. Solutions of D-ribose probably contain considerable quantities of the furanose and *aldehydo-* forms, and the mutarotation is complex, and exhibits a minimum.

There is also a number of biologically important derivatives of the monosaccharides, a few of which are listed in Fig. 13. Certain of these derivatives contain ionizable groups. Thus, glucosamine and galactosamine provide primary amine groups with a $pK_a$ near 7.8. Glucuronic acid and galacturonic acid provide carboxyl groups with $pK_a$ about 3.3; N-acetyl-neuraminic acid (sialic acid) has a stronger carboxyl group with $pK_a$ 2.7; and the phosphate esters have two potential hydrogen ions, with $pH_a$ values from 0.9 to 1.5 and from 5.9 to 6.3. The sulfate group of the amino-N-sulfate and the sulfate esters have a very strongly acidic hydrogen with $pK_a < 1$. The phosphate esters of the monosaccharides are products of intermediary metabolism, discussed in a number of other papers (see Lehninger, p. 136; Calvin, p. 147; Roberts, p. 170; Meister, p. 210).

The monosaccharide units of disaccharides may be alike as in maltose and cellobiose, or different as in sucrose and lactose. The hydrolysis of maltose and of cellobiose yields two molecules of D-glucose, whereas sucrose yields D-glucose and D-fructose, and lactose yields D-glucose and D-galactose. The monosaccharide residues are combined through an oxygen bridge of the hemiacetal hydroxyl to a second hydroxyl from another residue. The possible number of combinations of two monosaccharide units is large, since there are usually four or five free hydroxyl groups in the monosaccharide, and the oxygen bridge can come from either the $\alpha$- or $\beta$-anomer of the other sugar unit. Figure 14 shows the

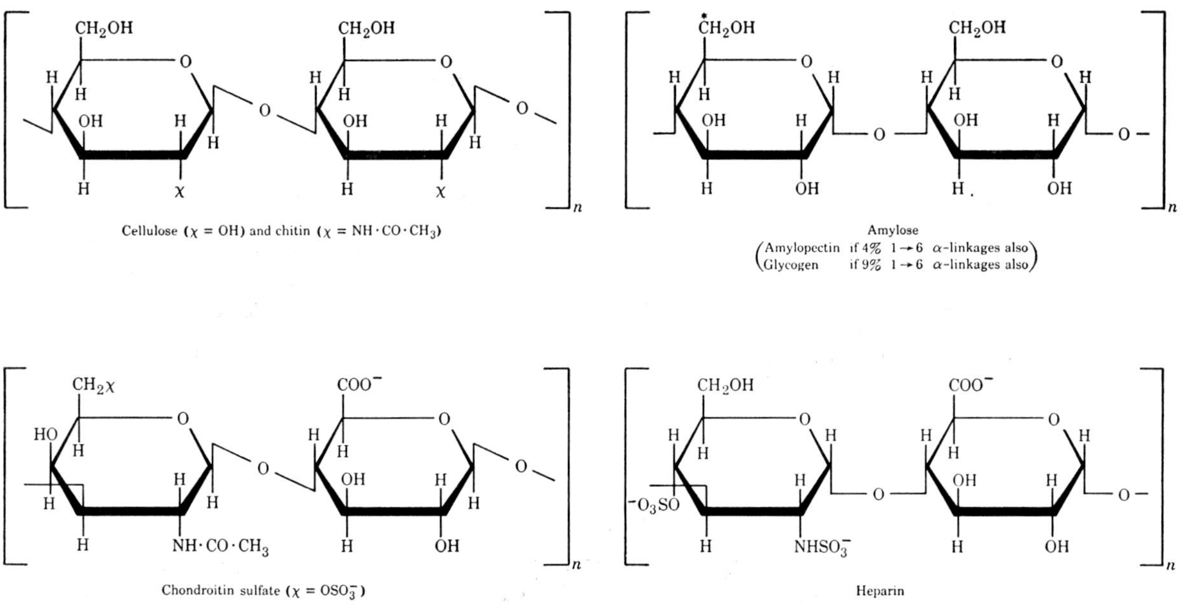

Cellulose ($\chi$ = OH) and chitin ($\chi$ = NH·CO·$CH_3$)

Amylose
(Amylopectin if 4% 1→6 α-linkages also)
(Glycogen if 9% 1→6 α-linkages also)

Chondroitin sulfate ($\chi$ = $OSO_3^-$)

Heparin

FIG. 15. Structural elements in selected polysaccharides.

structure of four common disaccharides, and illustrates some of the linkages possible. In sucrose, the two hemiacetyl hydroxyls are comblned through an oxygen bridge between the two anomeric carbons, with the D-glucose being in the α-pyranose configuration, and the D-fructose in the β-furanose form. In lactose, the hemiacetyl hydroxyl (C-1) of β-D-galactopyranose bridges to the C-4 hydroxyl of D-glucopyranose. Both maltose and cellobiose contain 1 → 4 oxygen bridges between D-glucopyranose units, but in maltose there is an α-linkage while in cellobiose there is a β-linkage.

There are numerous polysaccharides of biological importance, with monosaccharide units or their derivatives as the polymer unit, often repeating either a single unit, ...A-A-A-A-..., or alternating two units, ...A-B-A-B-A-B-.... The type of glycosidic linkage between these units has the same diversity as was found in the disaccharides. There are certain of the polysaccharides that show a more complex structure owing to branching of the polymer chain by means of linkages through a third hydroxyl group of the polymer unit. Usually, the naturally occurring polysaccharides contain many hundreds of residues. The lability of the linkages makes the isolation and purification of undegraded materials difficult, and many polysaccharide preparations thus show molecular weights and polydispersity not characteristic of the native product.

Cellulose and amylose provide two of the simplest polysaccharide structures (Fig. 15). Both of these polymers are made up of D-glucose residues in the pyranose conformation, with 1 → 4 linkages between the units. In cellulose, the pyranose ring has the β-configuration and each linkage is like that in the disaccharide cellobiose; in amylose, the rings are in the α-configuration and each glycosidic linkage is like that in maltose. Chitin resembles cellulose, and is a polymer of N-acetyl-D-glucosamine units linked by a 1 → 4 β-glycosidic bond. This 1 → 4 β-glycosidic linkage leads to polymers of low solubility, whereas the 1 → 4 α-glycosidic linkage gives polymers of much higher solubility, owing to a spiraling of the macromolecule in a helix-like fashion induced by this type of bond.

Amylopectin and glycogen have a basic structure like amylose, but about five percent of the residues branch by means of 1 → 6 α-glycosidic bonds in the amylopectins, and about nine percent in the glycogens. The branched structure leads to solutions of much lower viscosity than would occur with amylose macromolecules of the same molecular weight. Detailed structures of glycogen and amylopectin have been established by the use of specific enzymes which differentiate between the 1 → 4 and 1 → 6 linkages, and Fig. 16 shows a representation of a segment of a glycogen molecule as indicated by such studies.

The aforementioned polysaccharides have been made up of uncharged polymer units. A more reactive class of polymers is made up from sugar derivatives. Less is known of their detailed structure, because the high charge density in the polymer makes the application of physicochemical methods much more difficult, and because the modifying groups are split from the polymerizing unit by many of the same reagents that are used to degrade the polysaccharide. Among the more important materials of this type, one finds hyaluronic acid, heparin, and chondroitin sulfate. These highly reactive polysaccharides are found in various animal tissues, and the structures now thought most likely are recorded in Fig. 15. None of these structures has been

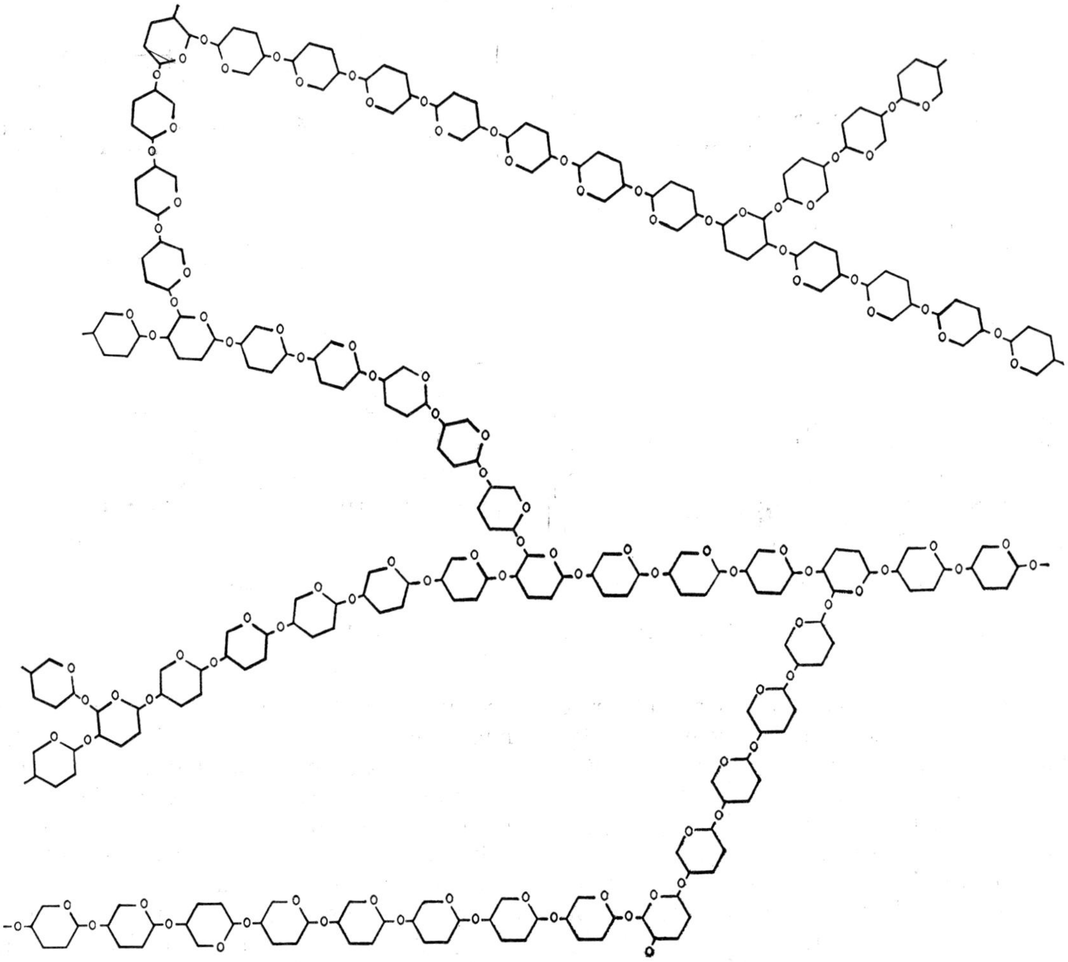

FIG. 16. Representation of a segment of a glycogen or amylopectin macromolecule.

definitely established at the present time. Other polysaccharides of this general type are to be found in bacterial systems, but none of these has structures that have been completely elucidated.

In the glycoproteins and mucoproteins, one finds large amounts of carbohydrate combined with protein.[26,27] Thus, the $\alpha_1$-acid glycoprotein, a homogeneous crystallized material contains about 42% of carbohydrate with the following composition (expressed as moles per mole of glycoprotein, molecular weight 45 000): N-acetyl-D-glucosamine, 32; N-acetyl-nuraminic acid, 16; D-galactose, 18; D-mannose, 18; L-fucose, 2. Recent studies by Eylar indicate that most of this carbohydrate moiety can be removed in association with a small peptide fragment containing glutamyl, seryl, threonyl, and isoleucyl residues. Earlier studies by E. Smith showed that the smaller carbohydrate moiety of serum $\gamma$-globulin could be removed in association with a peptide fragment containing glutamyl, aspartyl, and seryl residues. These interesting results indicate that covalent linkages exist between the carbohydrate residues and certain of the aforementioned amino-acid residues.

## LIPIDS

The classification "lipids" is taken to embrace the "fatty acids," all actual or potential esters of fatty acids, and often includes other materials soluble in "fat solvents," such as triterpenes, carotenoids, and fat-soluble vitamins. This large field is only touched upon here, since a recent short review by Lovern,[28] as well as the comprehensive reference book by Deuel[29] provide adequate background material. The simple fundamental lipid structures are often found to occur in complex structures of intermediate molecular weight, but to date no really high molecular-weight lipids have been discovered. On the other hand, these complex lipid molecules are often found in association with polysaccharides and/or proteins to form high molecular-weight materials.[30]

Although complex lipids often contain no charged groups (e.g., triglycerides, cholesterol, cholesterol

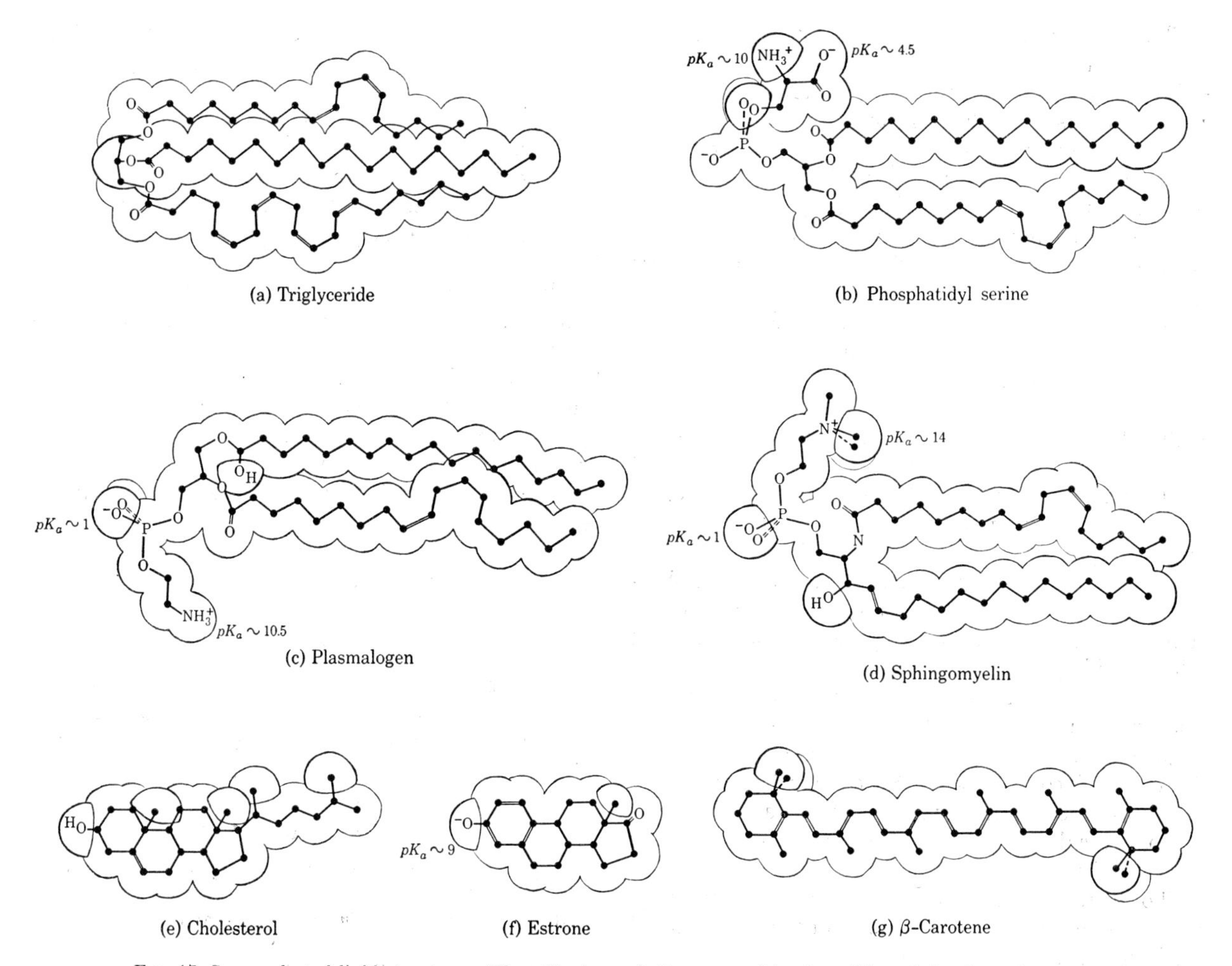


FIG. 17. Some selected lipid structures. The $pK_a$ shown indicates roughly the acidity of the charged groups.

esters, etc.), there are lipid structures containing primary amine, tertiary amines, phosphoric-acid groups, phenolic hydroxyl groups, etc. It can easily be imagined that the lipids owe much of their importance in biological systems to the fact that they represent classes of materials which can bridge the gap from water soluble to water insoluble phases without necessitating a sharp discontinuity. They are often thought to be concentrated in biological membranes and interfaces. Lipids also provide the most compact form of storage of chemical energy.

The term "fatty acid" is not easy to define accurately, but is usually taken to include all of the straight-chain members of the acetic-acid series of carboxylic acids, and a number of naturally occurring unsaturated straight-chain acids (e.g., oleic, linoleic, arachidonic acids, etc.). The esters of these fatty acids include compounds with glycerol to form triglycerides, with α-glycerylphosphoric acid to form phosphatidic acids, with α-glyceryl phosphoryl esters to form lecithins, phosphatidyl ethanolamines, and phosphatidyl serines; with sphingosine to form sphingomyelin and cerebrosides; and with inositol N-phosphate to form phosphoinositides; etc. A few typical lipid structures involving certain of the lipid residues mentioned above are shown in Fig. 17. In (a) a typical triglyceride is shown, with arachidonic acid (4 double bonds), stearic acid (saturated), and linolenic acid (2 double bonds) joined to glycerol. The three fatty-acid residues are shown in their extended conformation, with the stearic-acid residue above the other two. The double bonds in the unsaturated fatty-acid residues are usually found in the *cis*-configuration. Phosphatidyl serine (b) represents a typical phospholipid, and is shown with one residue of linolenic acid and one of palmitic acid (saturated), joined through a glycerol residue to phosphoserine. Three ionizable groups are seen here—the strongly acidic phosphoric-acid residue ($pK_a$ about 1), the weakly acidic carboxyl of serine ($pK_a$ about 4.5), and the weakly basic amino residue of serine ($pK_a$ about 10). In other phospholipids the serine residue may be replaced by ethanolamine ($HO-CH_2-CH_2-NH_3^+$) or by choline [$HO-CH_2-CH_2-N^+(CH_3)_3$]. In such phospholipids, no $pK_a$ near 4.5 would be expected, and the basic group would be found to have a $pK$ value of about 10 in the case of ethanolamine and above 14 with choline.

The plasmalogens [Fig. 17(c)] are glycerophosphatides containing an active group (shown here as the $-OH$ on carbon-1 of the stearic aldehyde residue) which reacts as an aldehyde group. The structure shown is one of several which have been proposed, and illustrates stearic-aldehyde and oleic-acid (1 double bond) residues joined to glycerol, which in turn is linked to phosphoethanolamine. This structure contains two ionizable residues—the phosphoric-acid group ($pK_a$ about 1) and the amine group ($pK_a$ about 10.5). Palmitic aldehyde sometimes replaces stearic aldehyde, and choline often replaces ethanolamine in typical plasmalogens. Other common types of lipids contain the lipid base sphingosine, $CH_3-(CH_2)_{12}-CH=CH-CH(OH)-CH(NH_2)-CH_2OH$. Illustrated here is a typical sphingomyelin [Fig. 17(d)] with a linolenic-acid residue joined through an amide linkage to the $-NH-$ of sphingosine, and phosphocholine joined through an ester linkage to the terminal hydroxyl of sphingosine. This molecule would have ionizable groups in the phosphoric-acid and the choline residues ($pK_a$ about 1 and above 14). The cerebrosides represent another typical sphingolipid type (not illustrated). Their structure is similar to sphingomyelin, but in place of phosphocholine a sugar residue, often D-galactose (as a pyranose ring), is joined to the terminal hydroxyl of sphingosine through a glycoside linkage. Such a cerebroside would not contain ionizable groups, but the many hydroxyl groups of galactose would serve as active sites for reaction with other molecules.

Cholesterol [Fig. 17(e)] is frequently found in animal tissues, either as the free alcohol (as illustrated here) or as the ester, in which the cholesterol hydroxyl is linked to a fatty acid. Neither cholesterol nor cholesterol esters contain ionizable groups, and the solubility properties of these molecules resemble those of the triglycerides. Estrone [Fig. 17(f)], a typical female sex hormone, has a ring structure much like cholesterol except that the first (*A*) ring contains three double bonds, the second ring is saturated, and the C-18 methyl is lacking. Estrone contains a ketonic oxygen at C-17, and a phenolic hydroxide residue ($pK_a$ about 9) at C-3.

$\beta$-Carotene [Fig. 17(g)] is a highly unsaturated hydrocarbon, occurring in many plants and green leaves, and also in animal systems. It contains a highly conjugated system of eleven double bonds, responsible for the intense color and the reactivity of the hydrocarbon. An alcohol (vitamin A) and an aldehyde (vitamin A aldehyde) are of great importance in visual pigments and contain a hydrocarbon chain similar to one-half $\beta$-carotene. Two closely related hydrocarbons, $\alpha$- and $\gamma$-carotene usually occur with $\beta$-carotene. The double bonds in the carotenoids usually are found to be in the *trans*-configuration (in contrast to the situation in the fatty-acid residues). The carotenoid hydrocarbons, along with other terpene-type compounds, appear to be built up from repeating isoprene units $[CH_2=C(CH_3)-CH=CH_2]$. Another hydrocarbon, squalene, $[CH_3-C(CH_3)=CH-CH_2-CH_2-C(CH_3)=CH-CH_2-CH_2-C(CH_3)=CH-CH_2-]_2$, is found in large amounts in shark-liver oil, and also is an intermediate for the synthesis of cholesterol in mammalian systems.

The mode of attachment of these lipid structures to the peptide moiety of a lipoprotein is poorly understood.[30] Denaturation of the protein and extraction with lipid solvents is usually sufficient to completely break the lipoprotein complex, so that it would seem that, if any covalent linkages are present, they must be very labile. A number of lipo-polysaccharides are known, and lipids are found to be firmly linked with some of the glycoproteins. It may be that the carbohydrate moiety of such complex macromolecules provides a covalent attachment for the lipid molecules. Much of the lipid material now thought to occur in cells as "free lipid" will, in the near future, be found to have intimate molecular binding to protein, glycoprotein, or carbohydrate cellular components.

**ACKNOWLEDGMENTS**

The author is greatly indebted to Dr. Margaret J. Hunter, Dr. Martha L. Ludwig, Dr. Donald F. H. Wallach, and Dr. Colin Green for their constructive criticism regarding this manuscript.

**BIBLIOGRAPHY**

[1] J. M. Bijvoet, A. F. Peerdeman, and A. J. van Bommel, Nature **168**, 271 (1950).

[2] B. W. Low in *The Proteins*, H. Neurath and K. Bailey, editors (Academic Press, Inc., New York, 1953), Vol. I., p. 235.

[3] L. Pauling in *Symposium on Protein Structure*, A. Neuberger, editor (Methuen and Company, Ltd., London; John Wiley and Sons, Inc., New York, 1958), p. 17.

[4] M. Calvin, Federation Proc. **13**, 697 (1954).

[5] G. E. Perlmann, Advances in Protein Chem. **10**, 1 (1955).

[6] D. F. Waugh, Advances in Protein Chem. **9**, 325 (1954).

[7] K. Linderstrøm-Lang in *Symposium on Protein Structure*, A. Neuberger, editor (Methuen and Company, Ltd., London; John Wiley and Sons, Inc., New York, 1958), p. 23.

[8] C. Tanford in *Symposium on Protein Structure*, A. Neuberger, editors (Methuen and Company, Ltd., London; John Wiley and Sons, Inc., New York, 1958), p. 35.

[9] C. Tanford, S. A. Swanson, and W. S. Shore, J. Am. Chem. Soc. **77**, 6414 (1955).

[10] J. Steinhardt and E. M. Zaiser, Advances in Protein Chem. **10**, 152 (1955).

[11] C. Tanford, J. D. Hauerstein, and D. G. Rands, J. Am. Chem. Soc. **77**, 6409 (1955).

[12] F. Sanger, Advances in Protein Chem. **7**, 1 (1952).

[13] F. Sanger, *Currents in Biochemical Research*, 1956, D. E. Green, editor (Interscience Publishers, Inc., New York, 1956).

[14] C. B. Anfinsen and R. R. Redfield, Advances in Protein Chem. **11**, 1 (1956).

[15] C. H. Li, Advances in Protein Chem. **11**, 101 (1956).

[16] *Ibid.*, **12**, 270 (1957).

[17] C. H. Li in *Symposium on Protein Structure*, A. Neuberger, editor (Methuen and Company, Ltd., London; John Wiley and Sons, Inc., New York, 1958), p. 302.

[18] J. I. Harris in *Symposium on Protein Structure*, A. Neuberger, editor (Methuen and Company, Ltd., London; John Wiley and Sons, Inc., New York, 1958), p. 333.

[19] C. H. W. Hirs, W. H. Stein, and S. Moore in *Symposium on Protein Structure*, A. Neuberger, editor (Methuen and Company, Ltd., London; John Wiley and Sons, Inc., New York, 1958), p. 211.

[20] C. B. Anfinsen in *Symposium on Protein Structure*, A Neuberger, editor (Methuen and Company, Ltd., London; John Wiley and Sons, Inc., New York, 1958), p. 223.

[21] H. Tuppy in *Symposium on Protein Structure*, A. Neuberger, editor (Methuen and Company, Ltd., London; John Wiley and Sons, Inc., New York, 1958), p. 66.

[22] P. Jollés, J. Jollés-Thaureaux, and C. Fromageot in *Symposium on Protein Structure*, A. Neuberger, editor (Methuen and Company, Ltd., London; John Wiley and Sons, Inc., New York, 1958), p. 277.

[23] R. R. Porter in *Symposium on Protein Structure*, A. Neuberger, editor (Methuen and Company, Ltd., London; John Wiley and Sons, Inc., New York, 1958), p. 290.

[24] A. Neuberger, editor, *Symposium on Protein Structure* (Methuen and Company, Ltd., London; John Wiley and Sons, Inc., New York, 1958).

[25] W. Pigman, *The Carbohydrates* (Academic Press, Inc., New York, 1957).

[26] K. Meyer, Advances in Protein Chem. **2**, 249 (1945).

[27] G. E. W. Wolstenholme and M. O'Connor, editors, *Chemistry and Biology of Mucopolysaccharides, Ciba Symposium* (J. and A. Churchill, London, 1958).

[28] J. A. Lovern, *The Chemistry of Lipids of Biochemical Significance* (Methuen and Company, Ltd., London; John Wiley and Sons, Inc., New York, 1955).

[29] H. J. Deuel, Jr., *The Lipids* (Interscience Publishers, Inc., New York, 1951), Vols. I–III.

[30] E. Chargaff, Advances in Protein Chem. **1**, 1 (1944).

# 5
# Molecular Configuration of Synthetic and Biological Polymers

ALEXANDER RICH

*Department of Biology, Massachusetts Institute of Technology, Cambridge 39, Massachusetts*

IT is possible to make a rough division in biological systems between large molecules and small molecules. They may be regarded as having quite different roles. Small molecules are actively involved in metabolic cycles, and they provide most of the energy needed for the activity of a biological system. The large molecules often have quite a different role in that their configuration—the three-dimensional organization of the molecule—has a particularly important part in determining the function of the molecule. Indeed, on the molecular level, structure and function are closely related.

Almost all large molecules that are found in biological systems are polymers; that is, they are formed by the linear association of certain fundamental repeating units. This review discusses some of the general structural features of these polymeric molecules, starting with some synthetic polymers which illustrate certain features of structural organization, and then continuing on to a group of more biologically significant synthetic and naturally occurring polymers.

In recent years, there has been an increasing awareness of the widespread occurrence of helical polymer molecules, both from natural and synthetic sources. This is not surprising, since the helix is the most general configuration for a regular polymer. The position of adjacent residues along a helix can be specified by a translation distance and an angular rotation. The most familiar example of a helix is the twofold screw rotation in which adjoining residues are related by an angular rotation of 180°; however, this is a specialized configuration and is discussed in the following.

Most molecular structure determinations are the results of x-ray diffraction studies. For diffraction work on polymers, a very useful expression has been obtained by Cochran *et al.*,[1] who derived the Fourier transform for helical molecules. Their analysis utilizes the helical symmetry in the molecule and expresses the Fourier transform in terms of Bessel-function contributions to various layer lines. In this way, the position and intensity of the x-ray scattering by helical molecules can be computed relatively directly. Their analysis has been programed into computing machines in order to facilitate the testing of a variety of proposed helical molecular models.[2]

## SYNTHETIC POLYMERS

Some understanding of the factors which determine the configuration of polymers can be gained from studies of the simpler synthetic polymers. It is illuminating to compare the configuration of polyethylene $-(CH_2)_n-$ with that of polytetrafluoroethylene $-(CF_2)_n-$ (known commercially as Teflon). The polyethylene molecule [Fig. 1(a)] consists of a series of $CH_2$ groups organized in a zigzag chain in which all of the carbon atoms lie in one plane. The distance between alternate $CH_2$ groups is 2.54 A. If, however, the hydrogen atoms are replaced by fluorine atoms, there is a change in configuration as is shown in Fig. 1(b).[3] Instead of lying in a planar zigzag, the carbon atoms in the backbone are now

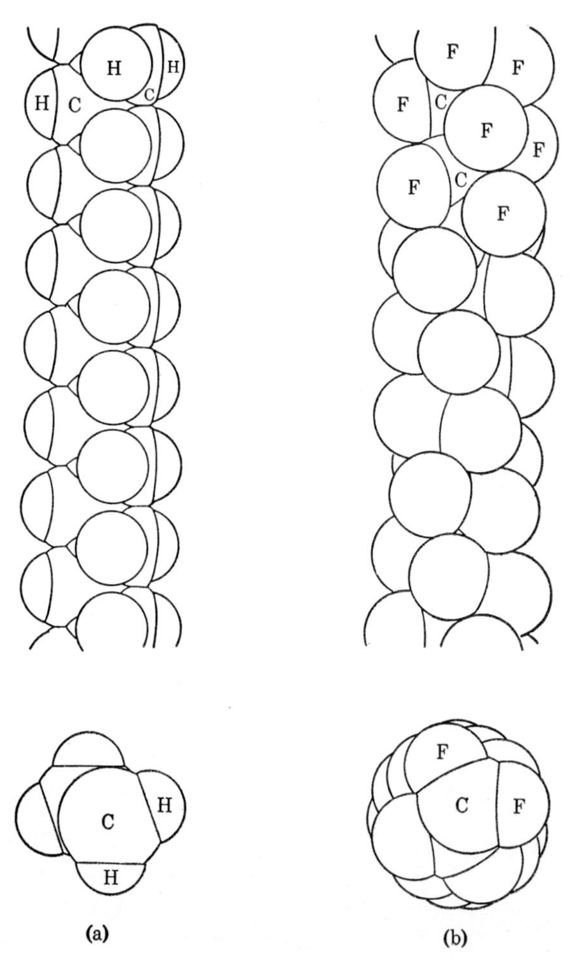


FIG. 1. (a) Planar zigzag chain formed by polyethylene (side and end views). The hydrogen atoms are not in van der Waals contact so that the carbon backbone remains fully extended. (b) Helical-chain configuration formed by the fluorocarbon molecule (side and end views). The bulkier fluorine atoms are in van der Waals contact and force the backbone chain of carbon atoms into a helical configuration [from C. L. Bunn and E. R. Howells, Nature **174**, 549 (1954)].

twisted. In the hydrocarbon chain, adjacent residues are related by a rotation of 180°; in the fluorocarbon chain, this angle has changed to 166°, and the molecule has a helical appearance. The fluorocarbon chain assumes this form as a result of the steric restraints imposed on it by the bulkier fluorine atoms. The hydrogen atoms with a van der Waals radius of 1.1 A are so small that they are not in contact in polyethylene. However, the fluorine atoms with a van der Waals radius of 1.35 A are too bulky to lie in a planar zigzag, and a helical twist is introduced into the carbon backbone in order to fit each fluorine atom into the groove between the two fluorine atoms on adjacent residues. This fairly small structural change brings about a considerable alteration in the properties of the two materials. Looking down the axis of the fluorocarbon molecule [Fig. 1(b)], one can see that the polymer is rounded in cross section. This rounded contour is responsible for the first-order phase transition which occurs near 20°C, and is undoubtedly owing to the rotation of these molecules in the crystalline state about their molecular axes. The hydrocarbon molecule, with its flattened cross section, does not have a phase change of this sort. This is an example of the effect of atomic size on molecular configuration and chemical properties.

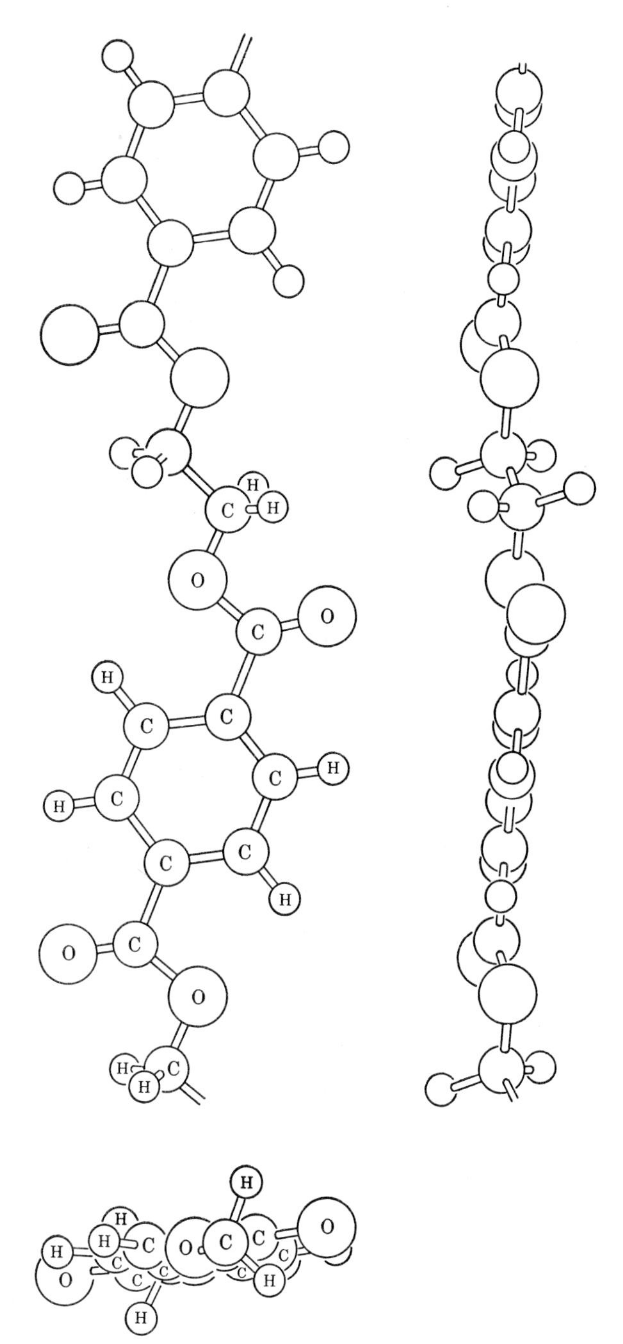


FIG. 2. Configuration of the polyethylene-terephthalate molecule. The unsaturated benzene groups lie nearly in the same plane so that the molecule has a ribbon-like configuration [from R. P. Daubeny, C. W. Bunn, and L. Brown, Proc. Roy. Soc. (London) **A226**, 531 (1954).

Another feature of interest is illustrated by the polymer molecule polyethylene terephthalate (known commercially as Dacron or Terylene). The repeating molecular unit contains a benzene ring as shown in Fig. 2. Succeeding residues lie almost in the same plane, hence the molecule has a ribbon-like configuration. Figure 3 illustrates how these molecules are packed in the crystal.[4] The flat, unsaturated rings lie largely on top of each other. In this orientation, they are stabilized considerably by the van der Waals interaction which is large for the highly polarizable $\pi$-electron system of the ring. This stabilization undoubtedly contributes to the high melting point (266°C) observed in the fiber. This type of interaction occurs in biological polymers and is most frequent in the nucleic acids where the large unsaturated ring systems usually pile on top of each other. These are discussed more thoroughly in the chapter devoted to the nucleic acids (p. 191).

When polymers contain electronegative atoms such as nitrogen and oxygen, there is often an opportunity for hydrogen bonds to be important in determining the configuration of the molecule. An example of this type of polymer is seen in polyhexamethylene adipamide, known as 66 nylon.[5] The structure of the $\alpha$-crystal illustrated in Fig. 4 is most clearly understood by noting that the molecules form a sheet structure in which adjacent units on the "$a$" face of the unit cell are held together by the horizontal hydrogen bonds which are shown as dotted lines. This polymer is of particular interest, of course, because its residues are joined by peptide bonds (−NH−CO−) as are the amino-acid residues in a polypeptide or in a protein.

Hydrogen bonds represent fairly weak or secondary forces between atoms. Nonetheless, they are often of considerable importance in determining the configuration of polymeric systems, because the total number of bonds involved is of the same order of magnitude as the number of monomer units in the polymer molecule. This is usually a large number so that the total stabiliz-

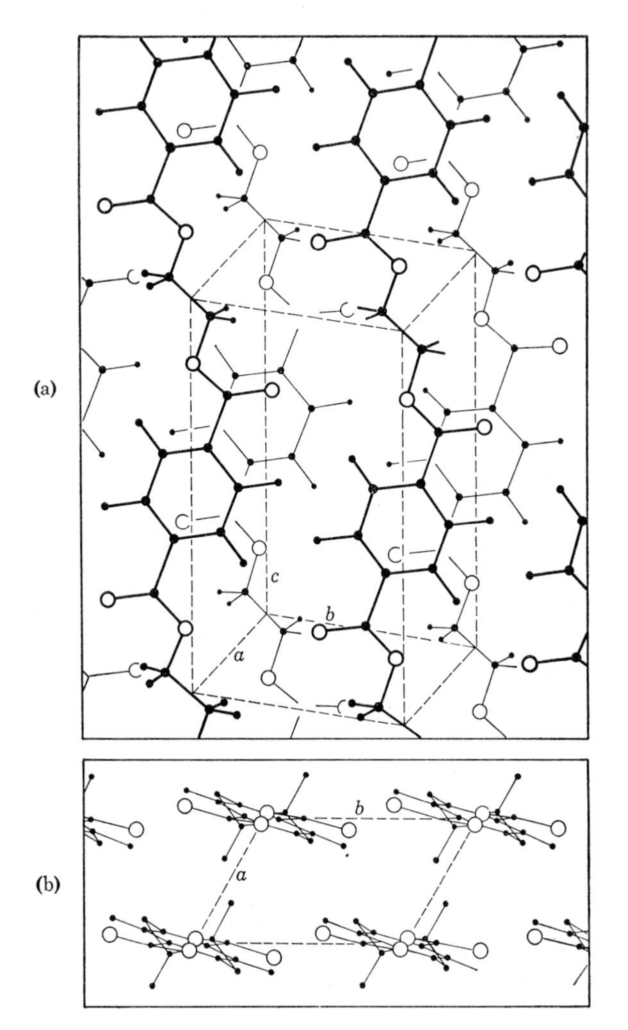


FIG. 3. The arrangement of polyethylene-terephthalate molecules in the crystal. (a) View perpendicular to the fiber axis. (b) View down the fiber axis. The larger dots represent carbon; smaller dots, hydrogen; open circles, oxygen [from R. P. Daubeny, C. W. Bunn, and L. Brown, Proc. Roy. Soc. (London) **A226**, 531 (1954)].

ing energy per covalently-linked molecule may become considerable.

## BIOLOGICAL POLYMERS

In all of the polymer structures described in the foregoing, there is one regular repeating unit and the configuration of the molecule is the resultant of systematic and repeating steric effects, van der Waals stabilization, electrostatic forces, and hydrogen bonding. The biological polymers, however, have an additional element of subtlety. In the proteins, the peptide backbone is the repeating unit, while the sugar phosphate groups form the backbone in the nucleic acids. In addition to these repeating units, however, there is also a variety of side groups attached periodically in some definite sequence to the polymer backbone. Hence, in addition to the systematic, repeating interactions described in the foregoing, the configuration of the molecule must be dependent also upon the nature of these side groups and the sequence in which they occur.

In discussing the molecular configuration of biological polymers, one usually describes the results of structural investigations carried out in the solid state. This description is useful, since the solid-state structure often is retained in solution or in the living cell.

The biological polymers may be listed as follows:

(a) The nucleic-acid polymers. These are discussed at length in another chapter (Rich, p. 191). However, it should be mentioned that, in this polymer, each repeating residue has a negative charge located on the phosphate group. Thus, the configuration of a nucleic-acid molecule represents a compromise in which the internally repelling electrostatic forces are balanced by the cohesive forces in the molecule.

(b) The carbohydrate polymers. Only a few such as cellulose and chitin have a regular configuration with the polymer chains organized systematically. The ab-

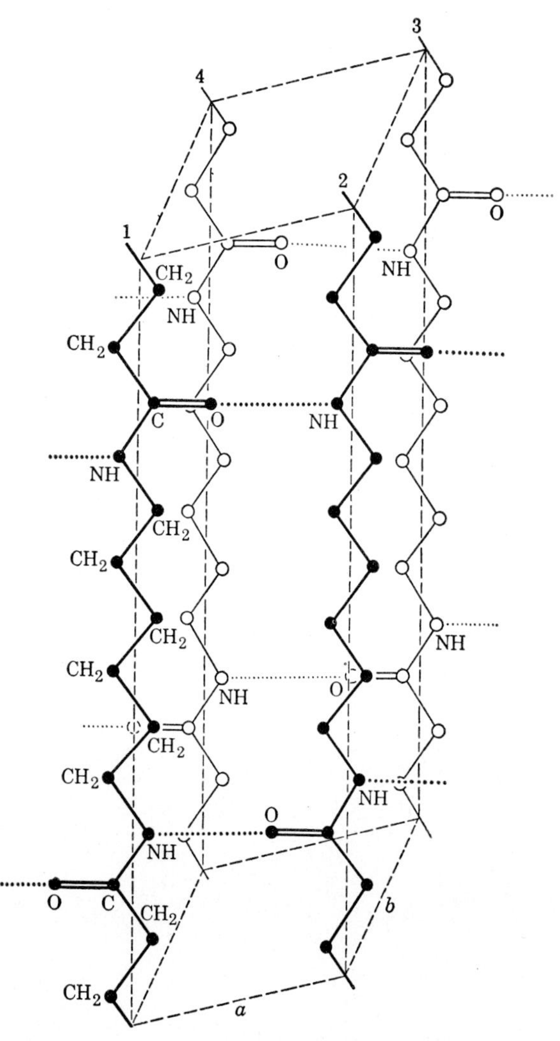


FIG. 4. The molecular structure of the $\alpha$-crystal of 66 nylon. The molecules are hydrogen-bonded into sheets in the plane of the page [from C. W. Bunn and E. V. Garner, Proc. Roy. Soc. (London) **A189**, 39 (1947)].

sence of a highly ordered configuration in most carbohydrate polymers is probably an indication that their biological function does not require any elaborate three-dimensional organization. Instead, it may depend on the stereochemistry of the sugar-to-sugar linkage. For example, the amorphous particles of glycogen stored in the cell are apparently just a convenient depot from which glucose residues can be removed by enzymatic action when they are needed. These structures are not discussed here.

(c) Proteins. Perhaps the most varied biological polymer is the polyamino-acid chain from which proteins are built. The repeating chemical unit is a single peptide bond with one α-carbon atom between each bond in contrast to the six carbon atoms between peptide bonds in 66 nylon. In the proteins, however, about twenty different types of side chains can be attached to the single α-carbon atom. The nature of these side chains has a profound effect in determining molecular configuration (see following).

(d) Lipids. There are no lipid polymers, even though there are several lipid molecules with extremely long side chains which are almost polymeric in nature.

It is interesting to note that there are apparently no mixed polymers in biological systems. Thus, one never finds a polymer chain involving both nucleotides and amino acids or sugar residues. If such mixed polymers existed, they might be unable to interact systematically through their backbones, and hence would lack an important element which gives rise to configurational stability.

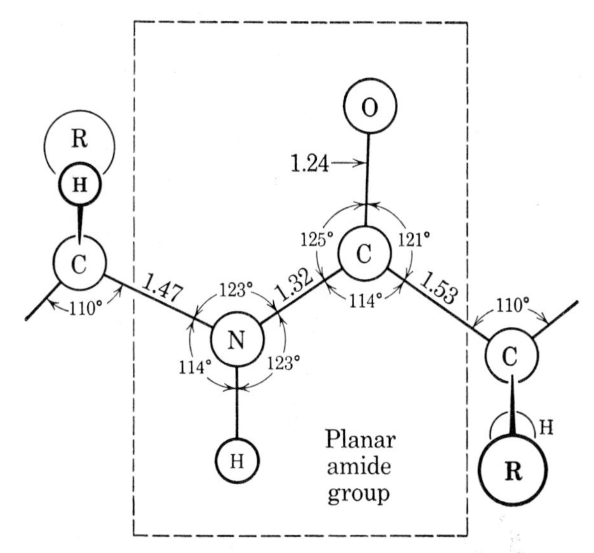


FIG. 5. Fundamental dimensions of polypeptide chains as derived from x-ray analyses of amino acids and simple peptides [from R. B. Corey and L. Pauling, Rend. ist. lombardo sci. **89**, 10 (1955)].

## POLYAMINO ACIDS

Present knowledge of the structure of the simpler polypeptides and proteins has its foundation in the extensive work of Pauling and Corey and their collaborators who worked out the crystal structures of a number of amino acids and small polypeptides. From their work, several important conclusions can be drawn.

(1) The dimensions of bond lengths and bond angles of the peptide group were obtained to an accuracy of ±0.02 A and ±2°.[6–8] They found a similarity in the dimensions of the peptide group regardless of the particular amino acids involved in the linkage.[9] These are shown in Fig. 5. The carbon-oxygen distance in the carbonyl group was found to be 1.24 A, even though the sum of the double-bond covalent radii is 1.21 A. In a similar way, the carbon-nitrogen bond distance in the amide group is 1.32 A, which is less than the sum of the single-bond radii (1.47 A). This points to approximately 40% double-bond character in the C—N bond, and 60% double-bond character in the carboxyl group owing to resonance between the two.

(2) As a direct consequence of the double-bond character in the C—N bond, the amide group is expected to have a planar configuration. Experimental results amply confirm this expectation. It has been estimated that the distortion energy is about 1 kcal/mole for a deviation of 10° from planarity and about 3.5 kcal/mole for a 20° distortion.[6,7] In all of the peptide structures determined, the peptide group is found planar to within a few degrees.

(3) All of the small peptides now examined show a *trans*-configuration for the amide group. This is shown in Fig. 5 where the α-carbon atoms are attached to the planar amide group on opposite sides of the C—N bond. This conclusion from crystal-structure work is also supported by dipole moment studies.[10] Using infrared absorption, Badger and Rubaclava[11] have shown that the *trans*-configuration is more stable than the *cis*- by more than 2 kcal/mole.

(4) Another feature of the peptide group is the fact that it has a hydrogen atom attached to the nitrogen and is, therefore, capable of forming a hydrogen bond. All of the crystal structures analyzed were characterized by forming a maximum number of N—H···O hydrogen bonds. Most of the hydrogen bonds are linear to within a few degrees and have a length of 2.79±0.12 A.[6,7]

Thus, the polypeptide backbone consists of a series of fairly rigid planar groups which may or may not be able to rotate relative to each other, depending upon the nature of the side groups attached to the α-carbon atom. This rotational barrier which may be as large as 3 kcal/mole is an additional important factor in determining the ultimate configuration of the molecules. Pauling and Corey first pointed out that a relatively small number of configurations will meet all of the four criteria described above. Thus, one can expect a fairly small number of stable polypeptide configurations. The following paragraphs review several of the more important ones.

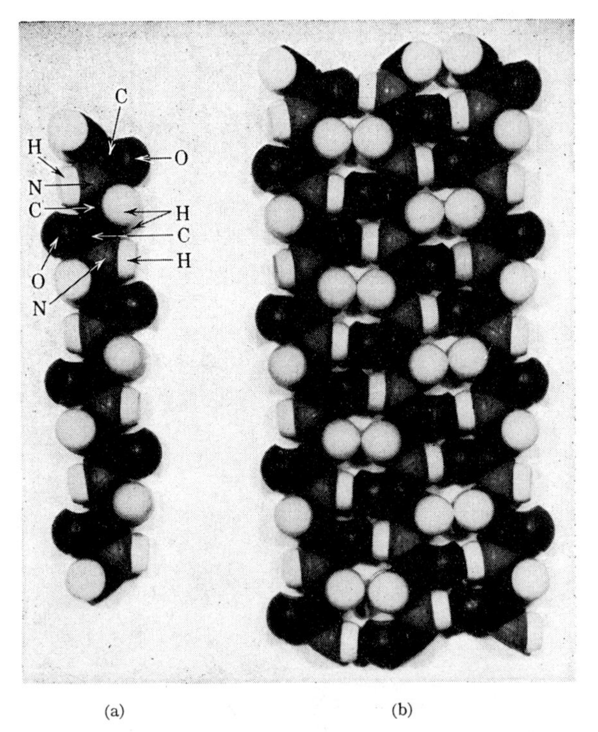


FIG. 6. A planar sheet composed of fully-extended polyglycine chains. The atoms are represented as solid out to their van der Waals radii. The hydrogen atoms attached to nitrogen have a depression in them for use in illustrating hydrogen bonding with carbonyl groups (>C=O). (a) Single extended polypeptide chain. (b) Antiparallel arrangement of polyglycine chains to form a sheet. Note that the hydrogen atoms in adjoining chains are in van der Waals contact. This would prevent the formation of this sheet from polypeptides other than polyglycine.

## Extended Polypeptide Configurations

The simplest configuration of a polypeptide chain is one in which the polypeptide chain is extended so that all of the planar peptide units tend to lie in a common plane. In this extended form, alternate amino-acid residues have carboxyl groups pointing in the same direction, and adjacent residues are related by a translation of 3.6 A and a rotation of 180°. The usual extended form of polyglycine has a configuration which is close to that described in the foregoing. When fully extended, the polyglycine chain should have a repeat distance of 7.2 A. However, the fiber-axis repeat distance, obtained experimentally, is somewhat less[12]; hence, the chains are almost but not fully extended. An extended configuration is shown in Fig. 6, a photograph of atom models that are designed to show all of the van der Waals radii for the various atoms.[6] In Fig. 6, a single extended chain of polyglycine is shown also. This structure is helical in the same sense that polyethylene [Fig. 1(a)] is helical, with a rotation angle of 180°. Chains of polyglycine can be hydrogen bonded to make a sheet structure, as shown in Fig. 6(b), where the chains alternate in direction. Such a configuration in which the successive planar peptide residues are almost in the same plane could exist in polyglycine only and not in any other polypeptide, because steric interference would result in the presence of a bulkier side chain. This can be seen most clearly by noting that the hydrogen atoms in the sheet structure are in contact. Thus, as illustrated in Fig. 6, it would be impossible to have a larger β-carbon atom attached in the position of the hydrogen atoms.

It is possible, however, to make a sheet structure utilizing hydrogen bonds in the plane of the sheet by rotating the peptide units until the β-carbon atoms are no longer in contact as they are in the polyglycine configuration of Fig. 6. This configuration is important in the structure of silk fibroin.

## Structure of Silk Fibroin

There is a rotational barrier between adjacent planar peptide units in the polypeptide chain, and, as a consequence, only a limited number of configurations are stable. One of the earliest to be investigated critically by Pauling and Corey was a semi-extended form in which adjacent planar peptide groups were related to each other by a translation of approximately 3.5 A and a rotation of 180°. A drawing of the configuration, looking down the plane of the peptide units, is illustrated in Fig. 7(a). In this configuration, the β-carbon atoms emerge from alternate sides of the polypeptide backbone, and the backbone itself has a zigzag rather than a fully extended form. The most significant feature of this configuration is the orientation of the C=O and N—H groups which line in one plane and are perpendicular to Fig. 7(a). These groups are oriented so that all hydrogen bonds can be formed by aligning a series of these polypeptide chains side by side. This has been done in Fig. 7(b) to form what is called the "antiparallel chain pleated sheet" structure. Alternate chains run in opposite directions, and in this way the successive amino and carboxyl groups are hydrogen-bonded to chains on either side of the original chain. By working carefully with molecular models, it is possible to show that the identity period along the fiber axis is 7.00 A, whereas the lateral distance between equivalent chains (alternate chains) is 9.50 A.

There are several interesting features in the antiparallel chain pleated sheet structure. The β-carbon atoms protrude at right angles from the plane of the sheet on alternate sides for successive residues along one chain. Furthermore, the β-carbon atoms are in phase on either side of the sheet; that is, they all appear at the same points along the fiber axis. Thus, there are large grooves between adjacent rows of β-carbon atoms.

The common form of silk fibroin is obtained from the silkworm *Bombyx mori*, and it has been the subject of x-ray diffraction investigations for many years. Most recently, Marsh *et al.*[13] have carried out an extensive and careful investigation and have arrived at a structure for this form of silk fibroin based on the antiparallel chain pleated sheet structure. Although the diffraction pattern is not simple, most of the reflections can be

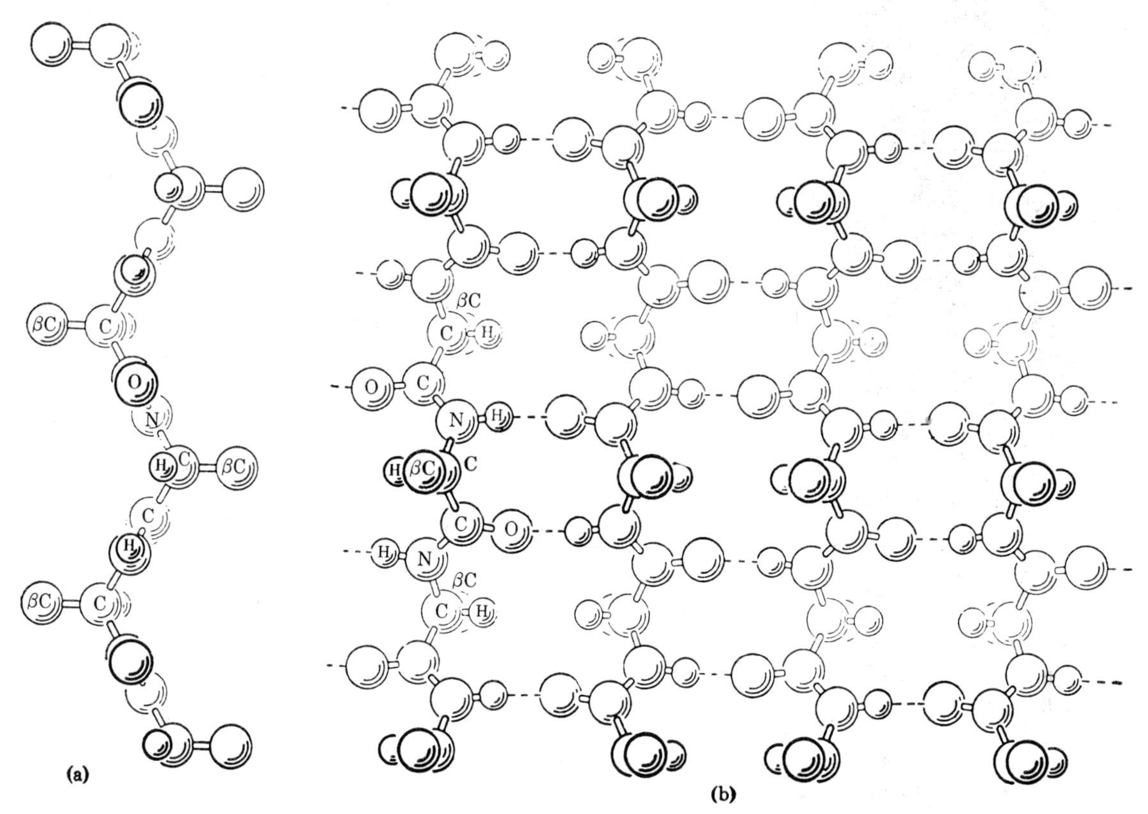


FIG. 7. (a) A drawing of a single polypeptide chain in the configuration found in the antiparallel chain pleated sheet. The chain is viewed in the direction of the hydrogen bonds and, therefore, along the plane of the sheet. Note the zigzag arrangement produced by alternate planar peptide residues. (b) A drawing of the antiparallel chain pleated sheet. In a given polypeptide chain, alternate β-carbon atoms are found on the same side of the pleated sheet [from R. E. Marsh, R. B. Corey, and L. Pauling, Biochim. et Biophys. Acta **16**, 1 (1955)].

accounted for by a simple unit cell which is orthogonal with dimensions 9.20 and 9.40 A at right angles to the fiber and a fiber-axis repeat of 6.97 A. The similarity between two of these dimensions observed experimentally and the pleated-sheet dimensions (7.00 A along the polypeptide chain and 9.50 A between equivalent alternate chains) suggested to these investigators that the antiparallel chain pleated sheet structure might serve as the structural basis of silk fibroin. Experiments carried out with rolled and flattened samples of silk fibroin produce the double orientation usually associated with layered structures. In this way, they were able to show that the 9.20-A reflection represented the repeat distance between molecular layers in silk fibroin. To deduce the structure, they had to arrange the layers of the sheet structure such that an identity repeat of 9.2 A was obtained.

One of the most remarkable features of silk fibroin is its unusual chemical composition. The simplest of the amino acids, glycine, constitutes almost one-half (44%) of the total number of amino acids present, while the next simplest amino acids (alanine and serine) comprise almost 40% of the residues. This suggested to the investigators that a simplified model of the structure might contain glycine residues with the alanine and serine residues in between at alternate sites along the polypeptide chain. In this way, all of the side groups protruding from one side of the sheet are from glycine residues and, therefore, consist simply of hydrogen atoms. The other alternate sites are occupied by other side chains—mainly alanine which has one methyl group or serine which has $CH_2OH$ as a side chain. When the

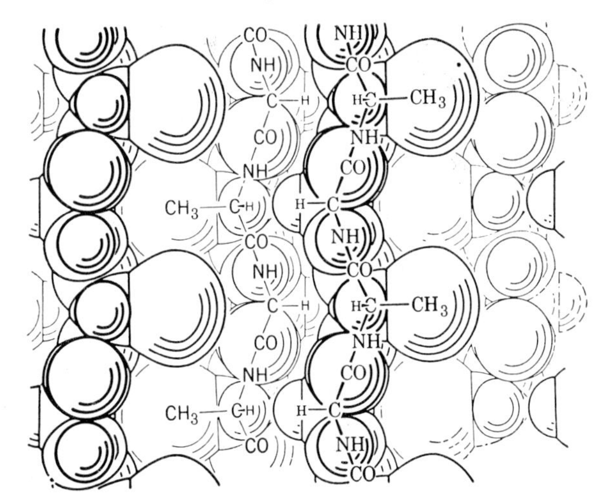


FIG. 8. A packing drawing of the structure of silk fibroin (*Bombyx mori*) viewed parallel to the pleated sheet. The fiber axis is vertical in this view [from R. E. Marsh, R. B. Corey, and L. Pauling, Biochim. et Biophys. Acta **16**, 1 (1955)].

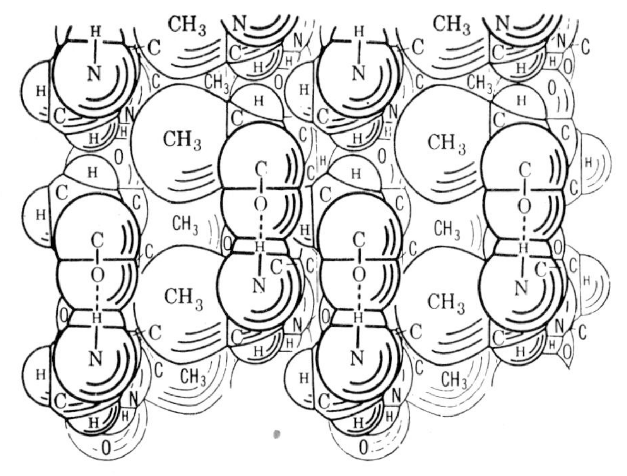

FIG. 9. A packing drawing of the structure of silk fibroin (*Bombyx mori*) viewed along the fiber axis. The hydrogen-bonding sheets are vertical in this view [from R. E. Marsh, R. B. Corey, and L. Pauling, Biochim. et Biophys. Acta **16**, 1 (1955)].

side chains are distributed in this manner, it is possible to pack pairs of sheets together so that the bulky side chains are located between the two sheets. In this way, the $\beta$-carbon atoms of one sheet protrude into the space between the $\beta$-carbon atoms from the other sheet. The distance from one sheet to the next is 5.7 A when they are packed in this way. However, when these double-sheet structures are stacked together, the glycine side chains are in contact and the distance from sheet to sheet over the glycine contacts is 3.5 A. Hence, the identity repeat for such a head-to-head, tail-to-tail stacking of sheets is $5.7+3.5=9.2$ A. This is, of course, exactly what is observed in the simple unit cell derived from the x-ray diffraction study of silk fibroin.

A packing drawing of the structure of silk fibroin is shown in Figs. 8 and 9. In Fig. 8, the view is perpendicular to the fiber axis and parallel to the pleated sheets. This view shows how the bulky methyl side chains are packed together with adjacent sheets filling alternate positions along the fiber axis, and how the small hydrogen side chains of the glycine residues fit together. Figure 9 is a view down the fiber axis. In this view, the hydrogen-bonding between the chains can be seen in the vertical direction, and the alternate packing arrangement of side chains is also clearly shown.

The reflections calculated from a simplified structural model of this type agree very well with the intensities obtained from the x-ray diffraction study of silk fibroin. This model does not account for all of the amino acids found in silk, including many with bulky side chains. They act to perturb this simplified model and produce some of the minor complications present in the diffraction pattern. Most of the properties of silk, however, can be understood from the simplified unit of structure.

## Coiled Helical Configurations

The extended form of polyglycine shown in Fig. 6 is called polyglycine I and was the first form studied crystallographically by x-ray methods. In 1955, Bamford and his colleagues[14] described a second form of polyglycine which they called polyglycine II. The struc-

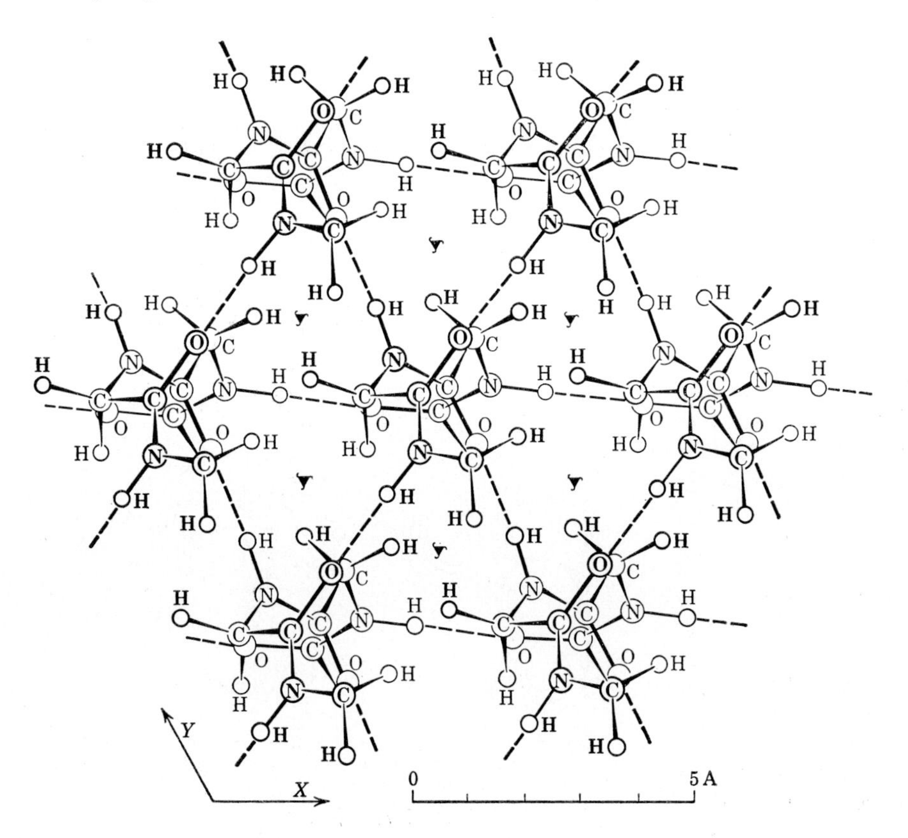


FIG. 10. The crystal structure of polyglycine II as viewed along the axis of the helical chains. The dashed lines represent hydrogen bonds [from F. H. C. Crick and A. Rich, Nature **176**, 780 (1955)].

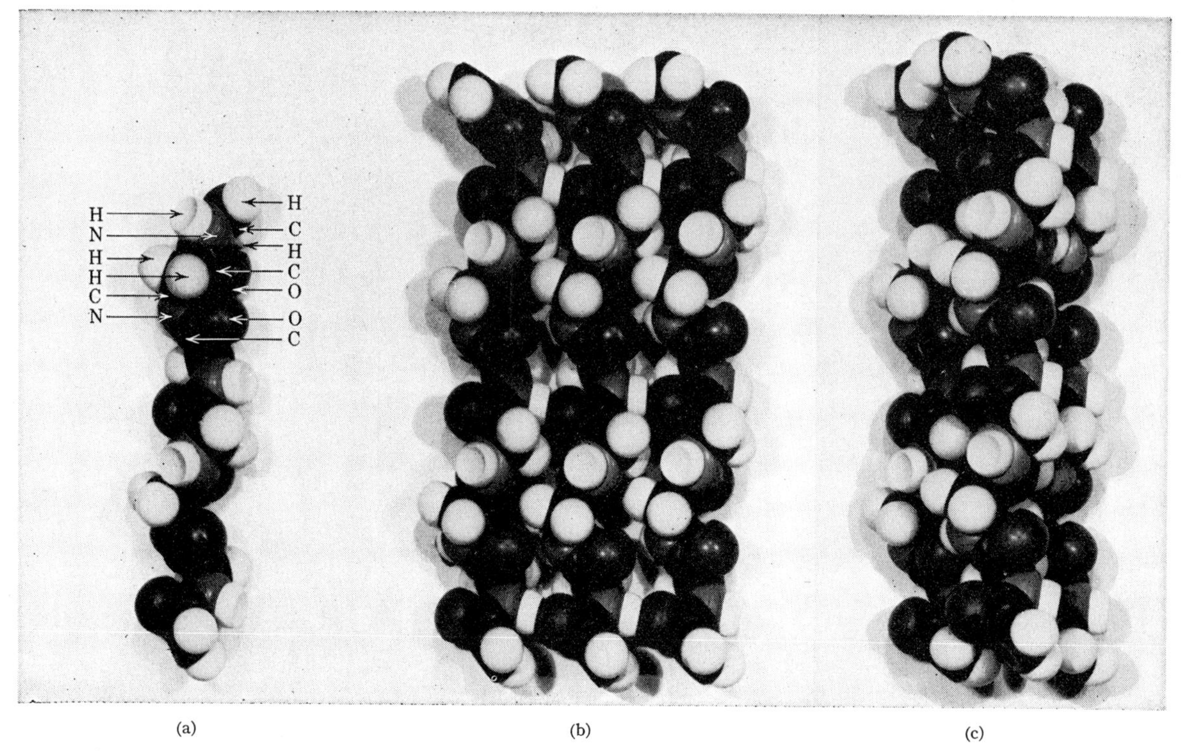


FIG. 11. (a) Molecular model of a single polyglycine chain arranged in the helical form found in polyglycine II. There are three peptide residues per turn in this form. (b) Three chains hydrogen bonded together to make a planar part of the polyglycine-II lattice. (c) Three chains packed in triangular form as found in the polyglycine-II lattice. The hydrogen bonds connecting these three chains are partly obscured.

ture of polyglycine II has been worked out by Crick and Rich,[15] and it is a particularly simple one. In it, all of the polypeptide chains are parallel, and each one is organized into a threefold screw axis so that adjacent residues are related by a translation of 3.1 A and rotation of 120°. The chains are packed in an hexagonal array, each chain being hydrogen bonded to six neighbors. The hydrogen bonds lie roughly perpendicular to the screw axis, and run in several directions and not solely in one plane as in the polyglycine sheet structure described in the foregoing. Figure 10 is a view of this structure down the fiber axis. In it, seven polypeptide chains are hexagonally packed with dashed lines indicating the linear hydrogen bonds which hold the adjoining chains together. The hydrogen-bond distance is 2.76 A, and the spacing between the polypeptide chains is 4.8 A. Since the translation along the axis is 3.1 A per residue, the axial repeat is 9.3 A because of the threefold screw symmetry. In this configuration, the polypeptide chains can fulfill all of the criteria developed by Pauling for stable configurations of polypeptide chains. It should be noted, however, that the polyglycine-II lattice could be formed only by a polymer of glycine. If bulkier side chains were present, the molecules would not be able to hydrogen bond, since the chains are already densely packed with only hydrogen atoms attached to the $\alpha$-carbon atoms. A side view of the polyglycine-II chain structure is seen in Fig. 11, using molecular models which show the atomic van der Waals distances. On the left [Fig. 11(a)] is a single polyglycine chain arranged in a threefold screw axis. In the center [Fig. 11(b)], three of these chains are hydrogen-bonded to make one plane through the polyglycine-II lattice. It can be seen that only one-third of the hydrogen bonds is utilized in forming a sheet. The remaining two-thirds of the hydrogen bonds are used in holding the sheets together. Thus, in order to form all of the hydrogen bonds, a three-dimensional network must be built up in polyglycine II in contrast to the two-dimensional sheet structure seen in polyglycine I. In Fig. 11(c), the same three polyglycine chains are present as are in Fig. 11(b), but they are arranged so that they are no longer planar. One polyglycine chain is lying on top of the other two so that the three are hydrogen bonded. In this form, they represent a group of three polyglycine chains obtained by selecting a triangle of three close-packed chains (see Fig. 10). It should be pointed out that there are two ways of selecting a group of these three polyglycine chains. This can be seen most readily by observing the directions of the hydrogen bonds (N—H···O) which hold together the group of three. In one group, they are clockwise and, in the other, they are counterclockwise. This difference in the two groups is a consequence of the space-group symmetry, since the chains are arranged about a threefold screw axis and not about a sixfold screw axis.

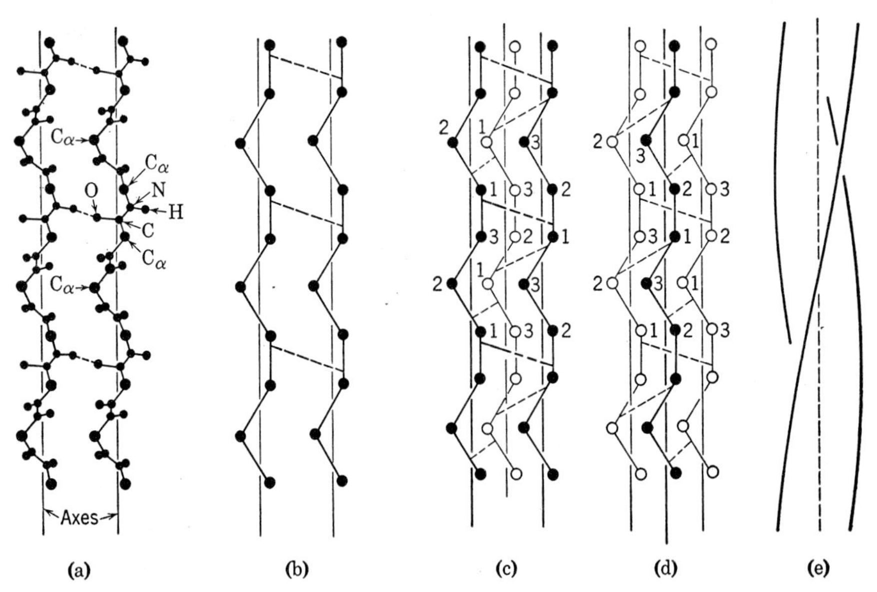


FIG. 12. A schematic drawing which illustrates how the collagen molecule may be generated from the polyglycine-II lattice. (a) Two molecular chains from polyglycine II. (b) The same chains showing only the α-carbon atoms (solid circles) and hydrogen bonds (dashed lines). (c) A third chain is added behind the two chains of (b). This generates collagen I. (d) The third chain is added in front of the two chains found in (b). This generates collagen II. (e) In the collagen molecule itself, the three molecular axes are coiled as shown in the diagram. The numbers on (c) and (d) correspond to the three kinds of sites found on the collagen molecule. Site number 1 is usually occupied by glycine [from A. Rich and F. H. C. Crick in *Recent Advances in Gelatine and Glue Research* (Pergamon Press, London, 1957), p. 1].

## Structure of Collagen

Just as the structure of the antiparallel pleated sheet serves as a model for deriving the structure of silk fibroin, the polyglycine-II lattice can be used to derive the structure of collagen. Collagen, found in the form of elongate fibrils, is the major structural element of skin and connective tissue in the animal kingdom. One of its most unusual features is the chemical composition. One-third of its residues is glycine, and it has, in addition, a large number of pyrrolidine rings. Proline plus hydroxyproline make up about 24% of the residues. Because of the high glycine content, it is perhaps not surprising that there is a relationship between the polyglycine-II lattice and the collagen molecule. This relationship can be seen most clearly in Fig. 12, which shows a schematic development starting from the polyglycine-II lattice and ending in the collagen molecule.[16] In Fig. 12(a), two parallel polyglycine chains are sketched with the hydrogen bonds joining them (dotted lines). In Fig. 12(b), only the α-carbon atoms of the glycine residues are shown, with short solid lines schematically indicating the peptide units connecting them. It is seen that these polypeptide chains coil around the two axes and are held together by the hydrogen bonds (dashed lines). The collagen molecule may be derived from three polyglycine chains such as is shown in Figs. 12(c) and 12(d). In Fig. 12(c), a third polypeptide chain is added behind the two shown in Fig. 12(b), whereas in Fig. 12(d), the third chain is added in front of the original two. In this way, two groups of three polypeptide chains are created which are prototypes of two models, collagen I and collagen II. These correspond to the two different ways of collecting groups of three molecules as shown in the end view of the polyglycine-II lattice (Fig. 10). In the collagen molecule, however, the axes around which the polypeptide chains coil are no longer straight as in Figs. 10(a)–10(d), but are coiled slightly as shown in Fig. 12(e). Thus, the collagen molecule is regarded as a coiled-coil structure.

In silk fibroin, the spacings between the sheets in the molecule are uneven; the closer spacing (3.5 A) arises because there are only the side chains of glycine between the sheets. Similarly, in the collagen molecule, the three polypeptide chains can come close together, because the glycine residues are located near the center of the group of three chains, whereas the other two-thirds of the residues are projecting outward from the center of the molecule. Thus, two-thirds of the residues can have any side chains on them. And, what is more significant, they can accommodate pyrrolidine rings without producing any steric difficulties.

It is generally agreed that the collagen-II model is the most likely form for the collagen molecule,[17–19] and in it pyrrolidine rings can be found on two of the externally situated sites, while the third internal site must have a glycine residue. This explains why all collagens are found with 33% glycine in their composition. Figure 13 shows a form of the collagen-II molecule where two-thirds of the sites are filled with pyrrolidine rings such that a repeating sequence is used in each chain. In it, the sequence glycine-proline-hydroxyproline is used, since this is a common constituent of collagen digests. This model of collagen was made using the polypeptide chains of Fig. 11(c), except that the axes of the polyglycine chains are now coiled rather than straight. As seen in Fig. 13, the collagen-molecule model is elongated, and has a distinct helical groove running down it which arises from the absence of a bulky side chain on the glycine site. The helical ridge on the surface of the molecule arises because the bulky side chains, in this case pyrrolidine rings, all lie closely packed. In the collagen molecule, however, there is a variety of amino acids found in that ridge, including many which have polar side chains. These are undoubtedly important in stabilizing collagen by hydrogen bonding, as well as in forming electrostatic bonds from one molecule to its nearest neighbor.

In the polyglycine-II lattice, the fundamental screw operation used in generating the lattice is a translation of 3.1 A and a rotation of 120°. Because collagen is a coiled-coil molecule formed from this lattice, the fundamental screw operation in generating the collagen helix is a translation of 2.86 A and a rotation of 108°. Of course, the asymmetric unit in the collagen molecule now consists of a group of three amino acids in contrast to the single amino acid in the polyglycine chain. In collagen, the coiling of the three polypeptide chains about one another is produced by the steric interaction of the $\beta$-carbon atoms which are present in the collagen molecule but which are absent in the polyglycine-II lattice.

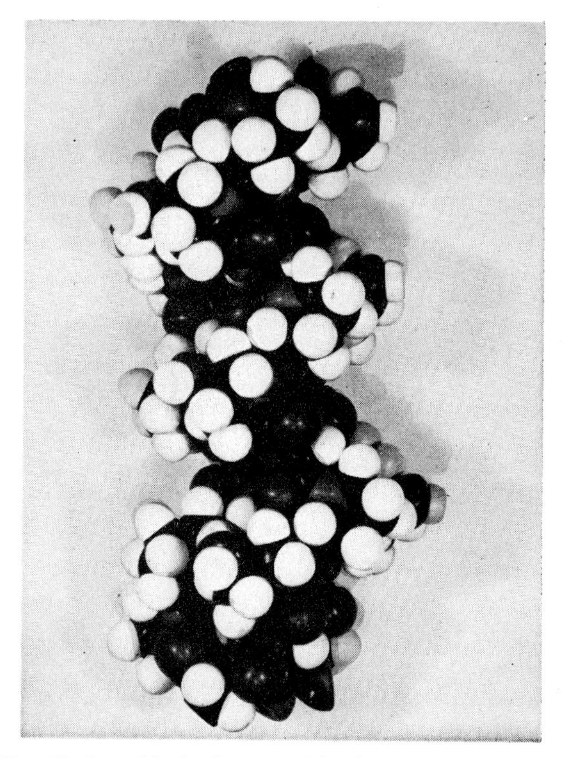

FIG. 13. A model of collagen II. The three polypeptide chains of Fig. 11(c) have been modified by the addition of proline and hydroxyproline residues to sites 2 and 3. This causes the three chains to coil about as shown in Fig. 12(e).

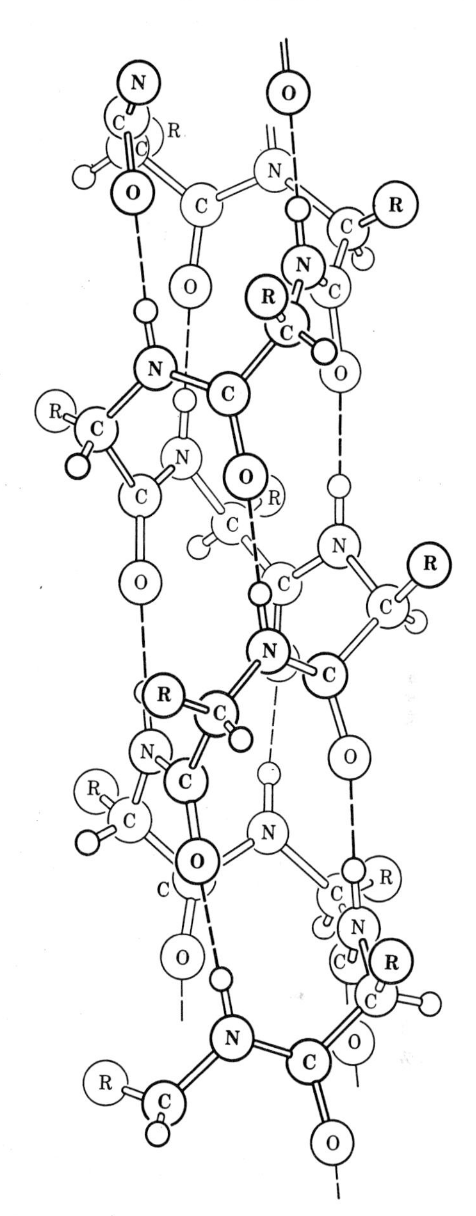

FIG. 14. A drawing of the $\alpha$-helix [from R. B. Corey and L. Pauling, Rend. ist. lombardo sci. **89**, 10 (1955).

## The $\alpha$-Helix

An important feature of the structure of silk fibroin is the fact that alternate residues are glycine units along the polypeptide chain. This permits the sheets to pack very closely. Similarly, in the collagen molecule, an important feature is the fact that every third residue is glycine, thereby allowing the three chains to come close together so that they can hydrogen bond. In a sense, both of these structures are very specialized and are a consequence of a restricted amino-acid composition. The best example of a polypeptide configuration which can be formed with minimal or no compositional restrictions is the $\alpha$-helix.[20] In the $\alpha$-helix (Fig. 14), the peptide units are oriented so that their planes are tangent to a cylinder around the axis of the helix. In this way, hydrogen bonds can be formed which are parallel to the axis, and thus serve to hold together the various turns of the helix. The fundamental operations for generating the α-helix are a rotation of 100° around the axis and a translation of 1.5 A along the axis. Each amide group is connected by a hydrogen bond to the third amide group from it along the peptide chain. There are 3.6 residues per turn of the helix, and the total rise of the helix per turn is 5.4 A. The helix formed in this way is packed very firmly, and there is no open space in the center.

Of all of the configurations discussed hereto, the α-helix is by far the most important. It is a completely general structure in that it will accommodate any amino-acid side chain. When proline residues are incorporated into the structure, the helix axis changes direction, but this incorporation probably does not destroy the organization of the helix. This structure has been found in a variety of synthetic polypeptides,[21,22] and there is evidence for its existence in many globular proteins as well as hair, muscle, wool, and myosin.[23,24] The α-helix should produce a reflection at a spacing of 1.5 A, which is the translation distance along the fiber axis for each amino-acid residue. Perutz[23] has found this reflection in a variety of proteins, and this has provided strong inferential evidence for the α-helix.

It has been suggested that the α-keratin proteins such as hair contain helices twisted around one another to form coiled-coil structures.[25,26] Thus, these molecules may form by hexagonal packing into cable-like structures involving several strands which coil around each other. Structures of this sort give reasonable agreement with the observed x-ray diffraction photograph obtained from hair.

The importance of the α-helix lies in its extreme generality. Knowledge of the detailed configuration of globular proteins, as discussed by Kendrew (p. 94), is just beginning to be advanced. It seems quite likely that the α-helix will form an important feature of many of these molecules. Significantly, the α-helix buries the systematically repeating part of the polypeptide chain, using it to form a supporting fabric of great stability. In so doing, it exposes to the external environment all of the side chains with their great diversity. Thus, the α-helix enables a protein structure to maximize its variability and, thereby, produce the considerable chemical versatility of the proteins.

## SUMMARY

On considering the molecular structure of biological polymers, the proteins and the nucleic acids, one can see that they are more complex than are synthetic polymers. Nonetheless, there are many common structural features in all of these materials. The helix is seen as the most general form for organizing successive residues along a polymer chain. The helix is found in many forms, ranging from fully extended to tightly coiled molecules. In all of these structures, space is filled efficiently, thereby maximizing the amount of van der Waals stabilization. Most synthetic polymers achieve a limited degree of structural complexity, as for example, coiling into a helical form. Materials of biological origin, by contrast, often go beyond this in complexity. They may form coiled-coils, as in collagen or in the α-keratin structure. On even a higher level, these polymers may fold in a very complicated fashion to produce globular molecules in which the interactions of the different side chains play a critical part in maintaining configurational stability.

## BIBLIOGRAPHY

[1] W. Cochran, F. H. C. Crick, and V. Vand, Acta Cryst. **5**, 581 (1952).
[2] D. R. Davies and A. Rich, Acta Cryst. (to be published).
[3] C. W. Bunn and E. R. Howells, Nature **174**, 549 (1954).
[4] R. P. Daubeny, C. W. Bunn, and C. J. Brown, Proc. Roy. Soc. (London) **A226**, 531 (1954).
[5] C. W. Bunn and E. V. Garner, Proc. Roy. Soc. (London) **A189**, 39 (1947).
[6] R. B. Corey and L. Pauling, Proc. Roy. Soc. (London) **B141**, 10 (1953).
[7] R. B. Corey and L. Pauling, Rev. Sci. Instr. **24**, 621 (1953).
[8] L. Pauling and R. B. Corey, Fortschr. Chem. org. Naturstoffe **8**, 310 (1954).
[9] R. B. Corey and L. Pauling, Rend. ist. lombardo sci. **89**, 10 (1955).
[10] S. Mizushima, T. Simanouti, S. Nagakura, K. Kuratani, M. Tsuboi, H. Baba, and O. Fujioka, J. Am. Chem. Soc. **72**, 3490 (1950).
[11] R. M. Badger and H. Rubalcava, Proc. Natl. Acad. Sci. U. S. **40**, 12 (1954).
[12] W. T. Astbury, Nature **163**, 722 (1949).
[13] R. E. Marsh, R. B. Corey, and L. Pauling, Biochim. et Biophys. Acta **16**, 1 (1955).
[14] C. H. Bamford, L. Brown, E. M. Cant, A. Elliott, W. E. Hanby, and B. R. Malcolm, Nature **176**, 396 (1955).
[15] F. H. C. Crick and A. Rich, Nature **176**, 780 (1955).
[16] F. H. C. Crick and A. Rich in *Recent Advances in Gelatine and Glue Research* (Pergamon Press, London, 1957), p. 20.
[17] A. Rich and F. H. C. Crick, Nature **176**, 915 (1955).
[18] P. M. Cowan, S. McGavin, and A. C. T. North, Nature **176**, 1062 (1955).
[19] G. N. Ramachandran, Nature **177**, 710 (1956).
[20] L. Pauling, R. B. Corey, and H. R. Branson, Proc. Natl. Acad. Sci. U. S. **37**, 205 (1951).
[21] L. Pauling and R. B. Corey, Proc. Natl. Acad. Sci. U. S. **37**, 241 (1951).
[22] C. H. Bamford, L. Brown, A. Elliott, W. E. Hanby, and I. F. Trotter, Proc. Roy. Soc. (London) **B141**, 49 (1953).
[23] M. F. Perutz, Nature **167**, 1053 (1951).
[24] W. T. Astbury, Proc. Roy. Soc. (London) **B141**, 1 (1953).
[25] F. H. C. Crick, Nature **170**, 882 (1952).
[26] L. Pauling and R. B. Corey, Nature **171**, 59 (1953).

# 6
# Physicochemical Characterization of Macromolecules

Paul Doty

*Department of Chemistry, Harvard University, Cambridge 38, Massachusetts*

ALTHOUGH the x-ray diffraction method appears to be the only means capable of yielding the complete three-dimensional structure of large macromolecules (p. 94), an investigation of macromolecular structures in solution is important for two reasons. In a number of cases, the macromolecule is either not crystallizable or insufficiently crystallizable; consequently, x-ray diffraction methods cannot be applied. In the other cases, where x-ray structure determination is possible, physicochemical examination is necessary to show whether or not the configuration characteristic of the solid state remains intact in solution.

It is essential, therefore, that one find out from the physical methods as much as possible about these molecules in solution. These methods are complementary to the x-ray investigation which goes on in the solid state. In addition, they furnish unique information such as molecular weight and molecular-weight distribution. In general, the more that one can do to bridge these two approaches, the better.

Nature seems to deal primarily with polymeric structures in the production of macromolecules. Fortunately, there is not the chaos that might be present should all possible covalent bonds be joined together to make an infinite variety of large macromolecules. There has been, in fact, a great sorting out so that only three basic polymeric chains have been selected for wide use in living systems. Evolved over many millions of years are the polysaccharides, the polypeptide, and the nucleic-acid chains. These sometimes are tied together by cross-links, but this generally does not mar the basic simplicity of the long-chain structures which may, or may not, be folded in a particular way. It is these polymer chains, subject to many specific perturbations, that we wish to examine in the greatest detail that is practical.

## SPECIAL PROBLEMS IN CHARACTERIZING MACROMOLECULES

One must consider at the outset three special and constantly recurring problems which arise in the characterization of macromolecules and have no simple counterparts in the field of small molecules. First to be mentioned is the fact that there may be a distribution of molecular weight. This does not usually occur in globular proteins, but it does in most other biological macromolecules. Consequently, the molecular weight is no longer a unique and simple number. One must deal instead with chains of varying lengths, but which are otherwise homogeneous. Each physical method that can be used for molecular-weight determination reflects a particular average of the molecular-weight distribution and as a result this average must be specified.

There are two widely used averages. The first of these is the number-average molecular weight—a very democratic form of averaging wherein each molecule is counted with a weight of one regardless of the actual weight. The second, the weight-average, corresponds not so much to democracy as to a more primitive arrangement wherein the importance of the molecule influences its count in accordance with its actual physical weight. The ratio of one of these to the other is a rough measure of the breadth of the molecular-weight distribution. This quality of having a distribution of molecular weight is sometimes referred to as "polydispersity."

Also to be considered is the problem of the specification of molecular size and shape. There are a variety of situations which arise from the basic polymeric chains. At one extreme, there can be found no periodic internal structure to the macromolecule whatsoever. It will become a randomly coiled molecule which is a rather common but degenerate form, since it does not have any specific configuration that is generally thought to be required for the functioning of many biological macromolecules. At the opposite extreme, many species of the three types of biological macromolecules can form helices (see Rich, p. 50), most commonly made of one, two, or three molecular strands, in which atoms are grouped in a periodic, inflexible array. Now, it is clear that the specification of size and shape must be treated differently in these two cases. For the case of the random coil, one will wish to measure the volume of space occupied by the molecule or some effective radius; in the case of the compact helical structure, one obviously will measure the length of the molecule and its mass, and this ratio of length to mass will characterize the particular helix. Therefore, the kind of parameter used to characterize the size of a macromolecule depends very much upon the form the molecule presents.

The final feature, which arises here and which has no counterpart with small molecules, is a space-filling concept in solution. If one had to deal only with dilute solutions of small compact spheres, then these particles could be considered as spaced at random with no overlapping and only rarely making contact. However, if these same molecules were not compact spheres, but random coils of the same mass, they would be expanded to such an extent that they would be continuously overlapping. This feature is maximal in the case of the random-coil form, but it is significant for rigid, asymmetric macromolecules as well. The consequence of this is that, if physical measurements are to be used to characterize the individual properties of these molecules, the system must be diluted down to the point where molecules are clearly separated one from another. This requires a very high dilution which, in turn, puts a very severe requirement upon the sensitivity of the physical method to be employed. A number of physical methods which otherwise might be very attractive fail on this account. They are not sufficiently sensitive to respond to the very high dilutions necessary to allow the individual expression of the intrinsic properties of the single macromolecule.

## THERMODYNAMIC METHODS

Keeping in mind these three special problems, a survey is undertaken now of what appear to be the most practical and general methods for obtaining quantitative information about the anatomical character of biological macromolecules. It is of interest to note that all of these methods have their origins in the work of eminent physicists at the turn of the century. The contributions of Einstein and Boltzmann appear in a number of places and our indebtedness to them is very great.

### Osmotic Pressure

The thermodynamic methods, three in number, are considered first. Perhaps the widest known is that of osmotic pressures. It is one property of solution which, as it happens, gives a macroscopic response to the dilute macromolecular solutions that is sufficiently large to be practical. The osmotic pressure is a measure of the number of molecules of solute per unit volume in solution. Therefore, if the weight of the material per unit volume in solution is known, and if one measures the number by means of osmotic pressure, a number-average molecular weight can be obtained.

This was perceived in the latter part of the last century, but only recently has the exact and rigorous relation of the concentration-dependent behavior of osmotic pressure to that of gas pressure been demonstrated. This is important, because, in each one of these methods, particular attention must be paid to the way in which the physical property depends upon concentration, and so, in each case, measurements must be carried out on a series of dilute solutions and extrapolated to an infinitely dilute solution in order to isolate the property of the individual molecule.

In all of these thermodynamic methods, the concentration dependence bears a close analogy to that of the pressure in a gas. Thus, the pressure-volume relation of an imperfect gas has the following form

$$PV/nRT = 1 + B(n/V) + C(n/V)^2 + \cdots, \tag{1}$$

where $B$ and $C$ are the second and third virial coefficients. The analogy with the concentration dependence of osmotic pressure is exact. Osmotic pressure, $\pi$, replaces gas pressure, $P$, and concentration (mass per unit volume) divided by molecular weight, $c/M$, replaces number of moles per unit volume, $n/V$. In this way,[1] one obtains the relation

$$\frac{\pi}{cRT} = \frac{1}{M} + \frac{B}{M^2}c + \frac{C}{M^3}c^2 + \cdots. \tag{2}$$

At sufficient dilution, the terms in $c^2$ and higher will be negligible. Under these conditions, a plot of $\pi/cRT$ against $c$ yields a straight line whose intercept is the reciprocal of the number of average molecular weight and whose slope is equal to $B/M^2$.

In small molecules of gases, the virial coefficient often is thought of in terms of excluded volume—that is, the amount of space not available for the center of one molecule because of the volume of another molecule. In the domain of macromolecular solutions, this effect can become very much larger because of the extra-large volume-filling capacity of some forms of macromolecules. However, if the molecules have an attraction for each other, this over-all effect becomes smaller and even can become negative immediately before phase separation. On the other hand, if there is a large electric repulsion, the over-all effect can become very, very large making it impossible to establish the curve with precision.

The range in which the osmotic-pressure method can be applied is from molecular weights of about 10 000 upward to a value which depends upon the precision desired—usually, a few hundred thousand. The lower limit is set by membrane permeability.

As an illustration, some measurements on collagen solutions[2] are shown in Fig. 1. The molecular weight obtained from the reciprocal of the intercept at the concentration axis is 310 000, and the scatter shows the uncertainty is of the order of 15%. This is about the upper limit for practical use of this method in aqueous solutions. The second virial coefficient can be evaluated from the slope, and, in this case, is found to arise almost entirely from the excluded volume effects.

### Light Scattering

Light scattering, the second thermodynamic method, is considered next. Concentrating only upon its thermo-

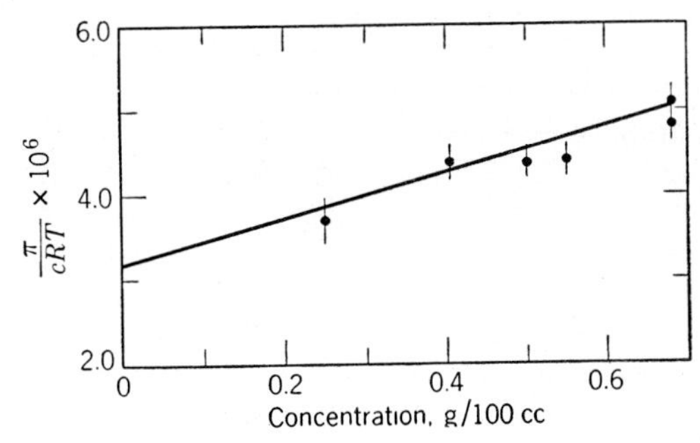


FIG. 1. Osmotic pressure of collagen solutions at 2°C.

dynamic aspect, compounded, of course, by the central electrostatic features which go along with dipole scattering, one finds again a close relation to the solution of the corresponding problem in gases.

Rayleigh's analysis of the problem of light scattering from dilute gases, worked out nearly a century ago, took the following form. The removal of light from an incident beam by scattering may be expressed quantitatively by means of the turbidity, $\tau$, defined as

$$I = I_0^{e^{-\tau l}}, \tag{3}$$

where $I_0$ and $I$ are the intensities of a parallel beam of light before and after passing through a length, $l$, of the gas. Alternatively, the scattering can be characterized by measuring its intensity at any angle $\theta$, $i_\theta$, and a distance, $r$. When $\theta$ is 90°, $\tau$ and $i_{90}$ are related in the following manner:

$$\tau = \frac{16\pi}{3} \frac{i_{90} r^2}{I_0} = \frac{16\pi}{3} R_{90}, \tag{4}$$

where $R_{90}$ represents the reduced form of the scattering intensity and is known as the Rayleigh ratio. This ratio was shown to be for any angle

$$R_\theta = \frac{2\pi^2 (n-1)^2}{\lambda^4} \frac{1}{\nu} (1+\cos^2\theta), \tag{5}$$

where $n$ is the refractive index of the gas, $\lambda_0$ the wavelength of light in vacuum and $\nu$ the number of molecules per cc.

The relevance of this to scattering from a solution of macromolecules is due to Einstein[3] and Debye.[4] At very great dilution, the same results hold except that the refractive index of the solvent, $n_0$, replaces unity in Eq. (5). This then can be rearranged to give:

$$R_{90} = \frac{2\pi^2 n_0^2 [(n-n_0)/c]^2}{N\lambda_0^4} cM = KcM. \tag{6}$$

From this it can be seen that the scattering as measured by $R_{90}$ is proportional to the molecular weight (the weight average) and the proportionality factor is experimentally determinable.

This result holds for a collection of independent scattering centers. In real macromolecular solutions, the space-filling property, previously referred to, produces a concentration-dependent correlation between scattering centers. Einstein showed this could be taken into account in a manner that now can be expressed as follows:

$$K\frac{c}{R_{90}} = \frac{1}{M} + 2\frac{B}{M^2} c. \tag{7}$$

One sees here a complete analogy with osmotic pressure except for a factor of 2 and the type of molecular-weight average that is obtained. Again, scattering measurements must be made at a number of concentrations and upon extrapolation yield the desired intercept.

In contrast to osmotic pressure, the sensitivity of this method increases with molecular weight. However, when the molecular size becomes significant with respect to the wavelength of the light, the angular dependence of the scattered light is affected in a manner discussed in the next section, and this must be taken into account. Finally, it should be noted that, since osmotic pressure and light scattering provide different molecular-weight averages, their combined results provide a measure of the molecular-weight distribution.

### Sedimentation Equilibrium

The third and final thermodynamic method is the various exploitations of sedimentation equilibrium that are found to be useful in the study of macromolecules. It is very likely that this will turn out to be the most versatile of all the tools which one can have for the physical investigation of macromolecules. The key to these methods is found in the high speed of the ultracentrifuge, which allows one to observe optically the distribution of macromolecules in a cell in which the gravitational field is increased to the order of 100 000 times that of gravity. The versatility of the modern ultracentrifuge allows one to carry out a number of quite different kinds of measurements. In one case, one may allow the field to be exerted on the molecules in the solution until they have redistributed themselves at equilibrium.[5] This would then correspond with the distribution of gas molecules in the earth's atmosphere, and the analysis of the situation can be readily adapted to yield an average molecular weight of the macromolecules distributed in the much greater field provided by the ultracentrifuge.

The basis of determining the molecular weight in this way can be seen if one recalls the relation between the pressure $p$ and the height $h$ in the earth's gravitational field $g$. That is,

$$\frac{dp}{p} = -\frac{Mg}{RT} dh. \tag{8}$$

In the ultracentrifuge cell, the pressure is replaced by the concentration, the molecular mass $M$ is corrected for the buoyancy owing to the solvent by multiplying by $(1-\bar{v}p)$, where $\bar{v}p$ is the ratio of solvent to solute density, $h$ is replaced by $x$, the distance from the center of rotation, and $g$ is replaced by the centrifugal force $w^2x$. These exchanges lead to

$$\frac{dc}{c} = \frac{M(1-\bar{v}p)w^2 x dx}{RT}. \tag{9}$$

If there is only one macromolecular species present, its molecular weight can be determined if the concentration is known at only two places in the cell. When a distribution of molecular weights exists, its weight-average molecular weight can be obtained most readily.[6] Again, the observed molecular weight is found to be dependent upon the concentration in a manner governed by the second virial coefficient.[6] Consequently, the reciprocal of the apparent value must be plotted against concentration and the true value of $M_w$ determined from the reciprocal of the intercept at zero concentration.

Now, in actual practice this method is not widely used because a number of days are required to attain the equilibrium distribution of the macromolecules in the solution in most cases. However, the smaller the molecular weight of the macromolecule, the sooner such distribution is attained; consequently, it is practical for molecules whose molecular weights are only a few thousand. Waugh[7] has pioneered in devising a particular kind of ultracentrifuge cell containing a retracting partition, which greatly facilitates this kind of application. By determining the concentration of the two parts, above and below the partition, one can obtain the molecular weight.

Within the last three years, an alternative version of the sedimentation-equilibrium method has come into use that promises to have wide utility. It is the Archibald approach to equilibrium method.[8] Kegeles[9,10] has been chiefly responsible for recognizing and developing its potential. Its basis is both novel and simple. The condition on which the general method of sedimentation equilibrium is based is that at equilibrium the net flux of solute species across any plane within the solution and perpendicular to the radius is zero. Archibald pointed out about 10 years ago that this condition of no net transport through a boundary perpendicular to the direction of sedimentation was always valid at the bottom and top of the solution, that is, at the meniscus and the cell bottom. Consequently, very soon after speed is attained, one can determine the initial redistribution of solute near the meniscus, or near the bottom of the cell; and from this one can derive the weight-average molecular weight by use of Eq. (9), together with appropriate extrapolation. Operationally, one now has precisely the opposite situation, a very short run instead of a very long run.

This method appears to be applicable over a wide range of molecular weight, and fortunately it does well in the region of 1000 to 10 000 where most other methods fail. As one illustration of this, I have chosen some work that we have done on determining the molecular weight of polypeptides and correlating this with intrinsic viscosity.

Before the Archibald method was available, it had been possible to show that the weight-average molecular weights of poly-γ-benzyl-L-glutamate were related to the intrinsic viscosity in the expected manner. That is, on a double logarithmic plot, a linear relation was found as indicated by the open circles in Fig. 2. We were unable to carry measurements below 20 000, however, and extrapolation was precarious. When the Archibald results were obtained (filled circles), the proper extension down to molecular weights of 1000 could be made as shown.[11]

## INTERFERENCE OF SCATTERED LIGHT AND MOLECULAR SIZE

If the macromolecules are quite small as compared to the wavelength of light, and if the incident light is vertically polarized, the scattering will be the same at all angles; that is, $R_\theta = R_0$. However, for larger molecules with dimensions exceeding 200 or 300 A, interference arises from light scattered from different parts of the same particle. As a consequence, the scattering is diminished, the effect increasing with the scattering angle $\theta$. The effect vanishes at zero angle and, as a consequence, measurements over the accessible angular range (usually 30 to 135°C) can be extrapolated to give $R_0$ which is required for molecular-weight determination. The character of the angular dependence, however, is of particular interest because it reflects the distribution of matter within the scattering article. The angular dependence can be separated from the concentration dependence because the following relation is generally valid

$$\frac{Kc}{R_\theta} = \frac{1}{MP(\theta)} + 2Bc. \tag{10}$$

That is, the angular dependence can be shown to enter only as a function, $P(\theta)$, known as the particle-scattering factor equivalent to the square of the structure factor in x-ray diffraction. The experimental determination of $P(\theta)$ is used to obtain information of dimensions and shape through the relation

$$P(\theta) - 1 - \frac{\rho^2}{3}\left[\frac{4\pi \sin(\theta/2)}{\lambda'}\right]^2 + \cdots, \tag{11}$$

where $\lambda'$ is the wavelength of light in solution and $\rho$ is

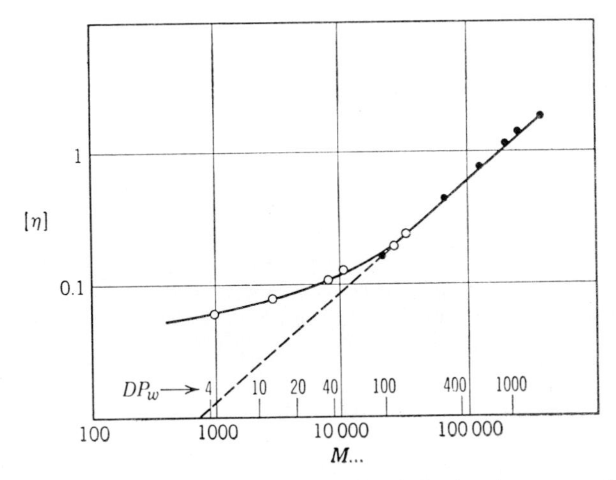


FIG. 2. A double logarthmic plot of intrinsic viscosity measured in dichloroacetic acid against weight-average molecular weight of poly-γ-benzyl-L-glutamate.

radius of gyration of the macromolecule. This radius of gyration is related to the dimensions of simply shaped particles. For example, the length of a rod-like molecule is $(12)^{\frac{1}{2}}\rho$. The root-mean-square end-to-end distance $(\langle r^2 \rangle)^{\frac{1}{2}}$ of a randomly coiled polymer is given by $(6)^{\frac{1}{2}}\rho$. For quite large macromolecules, higher terms in Eq. (11) become significant in such a way that shape, as well as the size, can be independently determined or at least estimated.

As an illustration of these methods, one can consider a particular fraction of cellulose nitrate.[12] Osmotic-pressure measurements plotted according to Eq. (1) yielded a value of 234 000 for $M_n$. Light-scattering measurements showed substantial angular dependence. A double extrapolation against concentration and $\sin^2(\theta/2)$ in accordance with Eqs. (10) and (11), known as a Zimm plot, showed that $M_w = 400\,000$ and $(\langle r^2 \rangle)^{\frac{1}{2}} - 1500$ A (Fig. 3). This indicates a rather broad molecular-weight distribution for a sample that has already been fractionated and the size is indicative of a quite extended random coil. This is evident when the following is considered. Each monomer unit has a molecular weight of 294 and a length of 5.15 A. Hence, the number of units (degree of polymerization) making up a chain of 400 000 molecular weight is 1350. Completely extended, this would be 6950 A in length. By comparing this with $(\langle r^2 \rangle)^{\frac{1}{2}} = 1500$ A, the extent of coiling is readily visualized. If there were no hindrances to rotation at the glycosidic linkages, the value of $(\langle r^2 \rangle)^{\frac{1}{2}}$ would be several times smaller. The extent to which this molecule is extended because of steric and potential hindrances to rotation is the highest yet found for single chains. One can appreciate that this "stiffness" is put to good use in the biological role of this material since this will not only give rigidity to the crystallites of cellulose, but will make firm the amorphous regions in between the crystallites. Moreover, this natural tendency toward rod-like behavior in localized regions of the chain is effective in lowering the entropy of melting, and thereby contributes to the high melting point and hence the negligible solubility of cellulose.

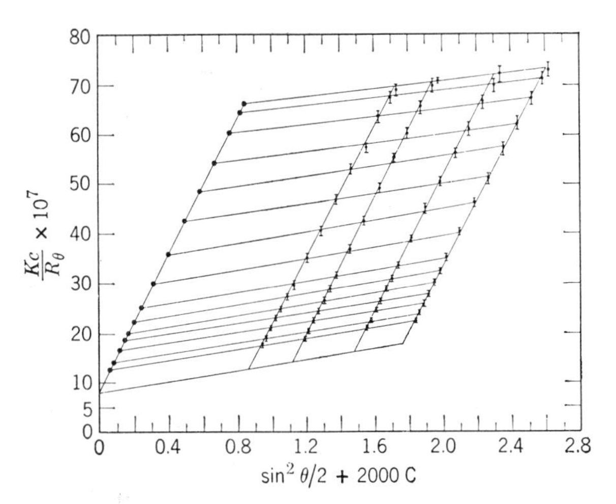


FIG. 3. Cellulose-nitrate fraction Ab in acetone at 25°C.

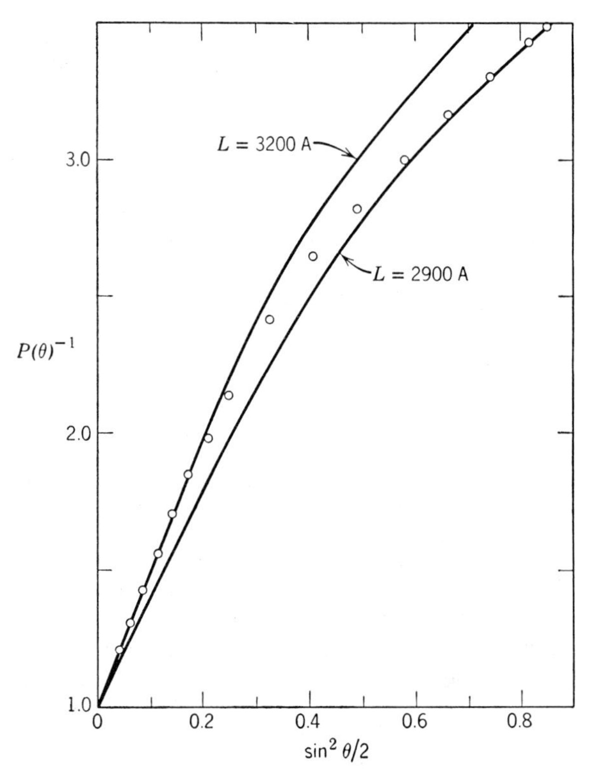


FIG. 4. Reciprocal particle-scattering factor of tobacco mosaic virus solution: sample *A*-4, experimental points,—theoretical scattering curves.

The Zimm plot for tobacco mosaic virus is shown in Fig. 4. It can be seen that the experimental points are bounded by the theoretical scattering curves for rods (having negligible diameter) of 2900 and 3200 A. Boedtker and Simmons[13] were able to conclude from these data that the length was 3000±50 A. In contrast to the complete linearity of the plot for the randomly coiled cellulose derivative in the previous figure, one finds here a pronounced downward curvature characteristic of rod-shaped scattering elements. From the molecular weight of 39.5 million obtained, from the intercept and the length and density, an effective diameter of 150 A can be computed. Thus, the shape and dimensions of this virus are completely determined from light-scattering measurements.

An application of light-scattering studies to collagen[2] is shown in Figs. 5 and 6. In the former, the Zimm plot for the native-collagen molecule in solution is shown. The downward curvature indicates a rod-like shape and the quantitative interpretation of the results show that the molecule is 3000 A long and has a molecular weight of 360 000. The diameter in this case is only 13.5 A and, as described by Rich (p. 50), consists of only three polypeptide chains in a helical arrangement. As in other cases which are dealt with in the paper by Rich, this macromolecule is essentially a one-dimensional crystallite. Hence, it should melt on raising the temperature.

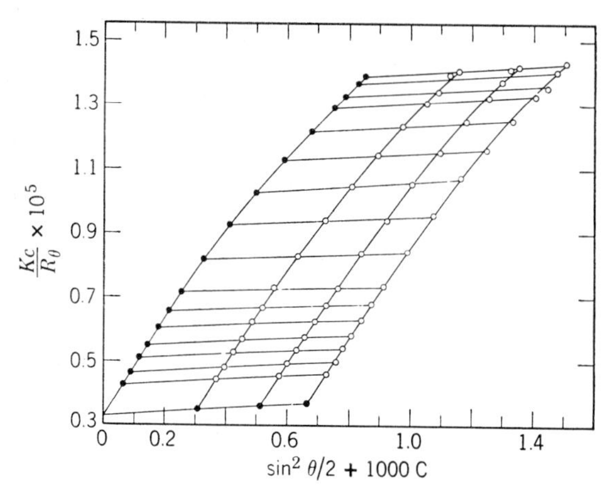


FIG. 5. Collagen (ichthyocol) in citrate buffer (*p*H 37) at 15°C. $M_w$=310 000; $P$=955 A; $L$=3300 A; $M/L$=94.

Upon melting, the secondary bonding holding the three chains together disappears and the "melted" or denatured state should consist of the three polypeptide chains in randomly coiled configurations. This configurational transition can be observed directly in light scattering. In Fig. 6, the reciprocal envelopes are plotted before and after heating. The intercept is seen to be nearly three times higher indicating a correspondingly lower molecular weight, and the slope is likewise greatly diminished indicating a much smaller spatial extent. The curvature is also gone, consistent with the assumption of a randomly coiled configuration.

Light-scattering studies of deoxyribonucleic acid ((DNA) have been particularly useful in showing the nature of these molecules in solution. These macromolecules are so highly extended in space that very great dilution is required in their physical characterization, and as a consequence few methods could be properly applied. A typical light-scattering result is shown in Fig. 7. The slope relative to the intercept is seen to be much greater than those encountered before. For this case, its interpretation as a coiled molecule leads to an estimate of 7000 A for the average end-to-end length and a molecular weight of 8 000 000. The contour length of this molecule is about 40 000 A on the basis of the Watson-Crick model. Thus, it is only very modestly curved and the mild downward curvature reflects its intermediate status between a coiled and rod-like shape. Actually, the size of this molecule is greater than present light-scattering techniques can handle because the extrapolation should be based on measurements down to 5°C in order to eliminate the effects of polydispersity on the final answer. Thus far, however, the errors have not been serious because the polydispersity has been such as to justify a linear extrapolation from the higher angles, but this good fortune cannot be expected to hold in general.

The use of the angular dependence of scattered light in this manner requires only the ratio of particle size to wavelength to be such that the interference is first order. This situation holds in the region of x-rays as well, and as a consequence the scattering of long x-rays can lead to the assignment of dimensions of the order of hundreds of A, just as light scattering deals with dimensions of a few thousands A. Time does not permit an examination of this, but reference should be made to the work of Beeman[14] and Luzatti[15] in this connection.

## HYDRODYNAMIC METHODS

The motions of a macromolecule in solution can be resolved into those of translation and rotation. If the motion is simply that of diffusion, one terms it translational or rotatory diffusion and characterizes it by a diffusion constant $D$ for translation and $\Theta$ for rotation. If the motion arises from an imposed gradient, two other situations occur. If the imposed force is that of a gravitational gradient as in the ultracentrifuge, sedimentation is observed and one characterizes it by a sedimentation constant $s$. If the imposed force arises from a hydrodynamic gradient, the molecule is caused to rotate with a definite bias instead of at random. The dissipation of energy that this produces gives rise to an increase in viscosity over that of the solvent. The fractional increase is known as the specific viscosity, and, when divided by concentration and extrapolated to zero concentration, it becomes known as the intrinsic viscosity $[\eta]$.

These four types of motion are all the result of a certain amount of resistance to an applied force (or couple). This resistance is hydrodynamic in nature and arises from the size and shape of the molecule. On very general grounds, it can be argued that for a given molecule the resistance can be characterized by a frictional factor $f$, which is the same in all four cases. Consequently, two developments are possible. Theoretical

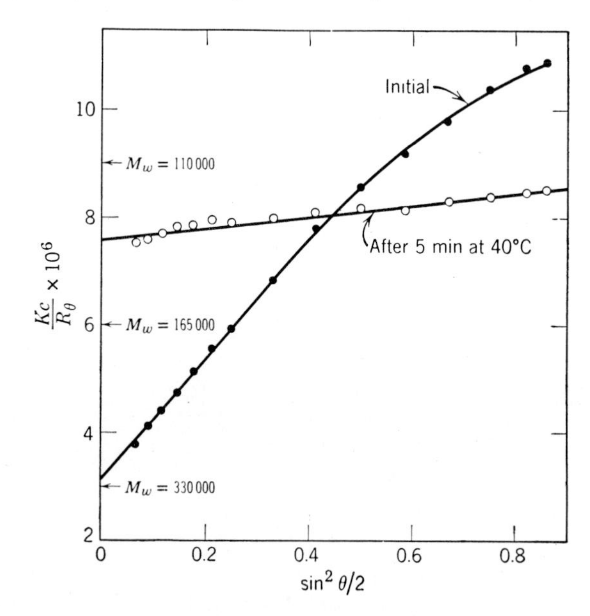


FIG. 6. Denaturation of collagen in solutions. ● = before denaturation; ○ = after denaturation.

investigations can aim at calculating the frictional factor for various macromolecular models as a function of dimensions and provide, in this way, a basis of obtaining dimensions from measuring one or more of the foregoing quantities. On the other hand, the measurement of any two quantities provides at least the possibility of eliminating $f$ and obtaining the molecular weight. This is illustrated in the well-known Svedberg equation. Here, the expressions for $D$ and $s$ are combined to eliminate $f$ as shown.

$$D = kT/f, \qquad s = M(1-\bar{v}\rho)Nf, \tag{12}$$

$$M = \frac{sRT}{D(1-\bar{v}\rho)}. \tag{13}$$

Thus, if the partial specific volume $\bar{v}$ and the solvent density $\rho$ are determined, the molecular weight can be obtained from measurements of $s$ and $D$.

The difficult and time-consuming nature of the measurement of $D$, particularly for chain-like and rod-like macromolecules, has led to a search for other means of achieving the same result by using the much more easily measured quantity, the intrinsic viscosity $[\eta]$, in its place. This can be done rigorously for ellipsoids, making use of the work of Perrin and Simha. The result is

$$M = \left[\frac{s[\eta]^{\frac{1}{3}}\eta_0 N}{\beta(1-\bar{v}\rho)}\right]^{\frac{3}{2}}, \tag{14}$$

where the constant $\beta$ is a slowly varying quantity dependent only upon the axial ratio. Its values range from 2.12 for spheres to 3.50 for infinite axial ratio. The axial ratio can always be determined from $[\eta]$ with enough precision to permit the correct choice of $\beta$.

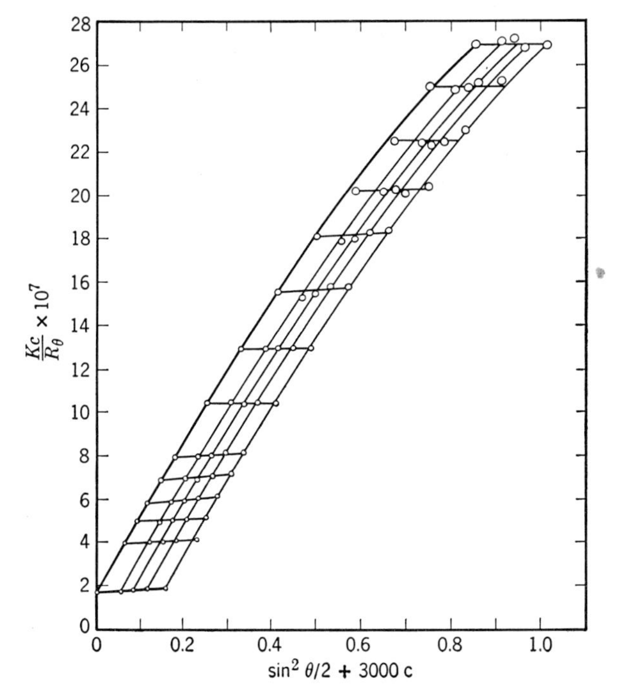


FIG. 7. Light scattering of DNA.

The usefulness of intrinsic-viscosity determinations is by no means limited to Eq. (14). Quite the contrary. Its application stems from a series of impressive theoretical derivations that have been made relating it to size and shape. The first of these was owing to Einstein who showed that the intrinsic viscosity of compact (unpenetrated by solvent) spheres in solution was 2.5 times their specific volume in units of cc/g. This value represents a lower limit: almost all values observed for macromolecules are much larger. For example, 4 to 10 for a number of globular proteins, 10 to 100 for a number of proteins known to be asymmetric, and extremely high values in the case of collagen (1200) and deoxyribonucleic acid (7000). Later theories showed that these higher values could all be accounted for in terms of two effects: asymmetry of the macromolecule, or solvent immobilization. The latter effect could arise from hydration owing to the binding of water (or other solvent) at specific sites on the macromolecule or by loosely incorporating it with the domain of the macromolecule through swelling. Thus, a typical randomly coiled polymer pervades a domain that is perhaps 100 times or more greater than it would occupy in a compact form; yet, the solvent in this domain moves with the polymer and is, therefore, hydrodynamically immobilized. As a consequence, the intrinsic viscosity measures this volume of immobilized material, and, in a rough way, its numerical value expresses in units of cc/g the ratio of the volume of the domain to the mass of the polymer molecule. Extremely useful developments in the last few years have led to the recognization of the relation between intrinsic viscosity (and sedimentation constant as well) and the mean end-to-end length, $(\langle r^2\rangle)^{\frac{1}{2}}$, of polymer chains.[16] As a consequence, the size of such polymer molecules can be determined from the intrinsic viscosity, if their molecular weight is known. Moreover, the large variations in intrinsic viscosity observed when the solvent or temperature is changed are seen to be the result of swelling and shrinking of the polymer coils. Finally, the previously observed empirical relation between intrinsic viscosity and molecular weight for a series of homologous polymers,

$$[\eta] = KM^a, \tag{15}$$

becomes understandable from this point of view. The exponent in this relation is a constant with a value between 0.5 and 1.0 for any given polymer-solvent system. These limiting values correspond to tightly coiled molecules with many intramolecular contacts (0.5) and to very highly swollen molecules (1.0), with most actual cases distributed in between.

In the case of rigid particles, the exaltation of the intrinsic viscosity arises from the increased energy dissipation resulting from the end-over-end rotations in solution. The analysis of this problem, particularly by Simha,[17] showed that the intrinsic viscosity was a func-

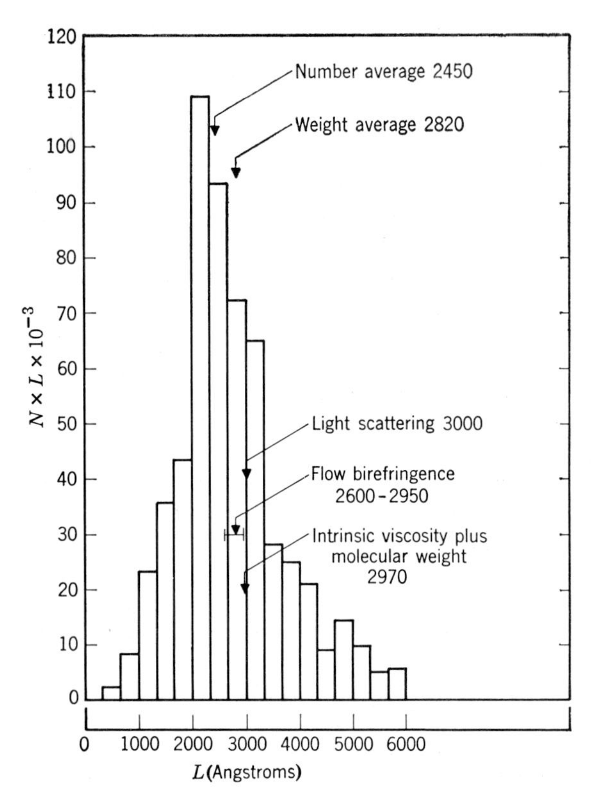


FIG. 8. Weight distribution of lengths of ichthyocol macromolecules as measured from electron micrographs compared with results from other methods. Total number represented is 238 [from C. E. Hall and P. Doty, J. Am. Chem. Soc. **80**, 1269 (1958)].

tion only of the axial ratio of the particle and that this dependence rapidly approached that of the square of the axial ratio. In many typical proteins, one is faced with the possibility that the intrinsic viscosity reflects both molecular asymmetry and moderate amounts of hydration. The resolution of this problem has been worked out by Oncley.[18]

The final point to make on intrinsic viscosity is about its ease of measurement relative to the other three hydrodynamic properties.

The determination of the rotatory-diffusion constant from streaming birefringence of flow is a rather specialized technique and is limited to rigid particles. In these cases, provided the length of the molecules exceeds a few hundred angstroms, the length and often the length distribution can be measured with considerable accuracy.

As a single illustration of the methods discussed in this section, as well as a demonstration of their self-consistency, the results on soluble collagen are presented in Table I.[2] The intrinsic viscosity was 1150 cc/g, the sedimentation constant was 2.96 Svedbergs and the distribution of rotatory-diffusion constants corresponded to a range in length from 2500 to 2950 A.

This brief discussion of the physical determination of the characteristics of macromolecules would not be complete without the mention of a new technique that has, within the last two years, become practical for asymmetric, rigid macromolecules. This is the perfection of the direct viewing of such macromolecules in the electron microscope by Hall.[18] Thus, nucleic acids, some proteins, and even simple polypeptides in the α-helical configuration have been measured and found to be in agreement with the results of other physical methods applied in solution.[19] As one illustration of this, shown in Fig. 8 is the weight distribution of the lengths of collagen molecules compared with the average values shown in Table I. The agreement is seen to be quite good. It is possible that the distribution was somewhat broadened as a result of some damage in spraying and shadowing the electron-microscope preparation, but the effect has been modest. F. O. Schmitt deals with the way in which these macromolecules are united in collagen fibrils and the extent to which the dimensions found here are compatible with his studies (p. 349).

TABLE I.

| Method | Mol. Wt. | Length A | Diameter A |
|---|---|---|---|
| Osmotic pressure ($M_n$) | 310 000 | ... | ... |
| Light scattering ($M_w$) | 345 000 | 3100 | 13.0 |
| Intrinsic viscosity and $M_w$ | ... | 2970 | 13.6 |
| Sedimentation and viscosity | 300 000 | ... | 12.8 |
| Flow birefringence and viscosity | 350 000 | 2900 | 13.5 |

## BIBLIOGRAPHY

[1] The rigorous derivation of Eq. (2) may be found in the work of W. G. MacMillan, Jr., and J. E. Mayer, J. Chem. Phys. **13**, 276 (1945).
[2] H. Boedtker and P. Doty, J. Am. Chem. Soc. **78**, 4267 (1956).
[3] A. Einstein, Ann. Physik **33**, 1275 (1910).
[4] P. Debye, J. Phys. & Colloid Chem. **51**, 18 (1947).
[5] T. Svedberg and K. Pedersen, *The Ultracentrifuge* (The Clarendon Press, Oxford, 1940).
[6] M. Wales, J. W. Williams, J. O. Thompson, and R. H. Ewart, J. Phys. & Colloid Chem. **52**, 983 (1948).
[7] D. A. Yphantis and D. F. Waugh, J. Phys. Chem. **60**, 630 (1956).
[8] W. J. Archibald, J. Phys. Chem. **51**, 1204 (1947).
[9] S. M. Klainer and G. Kegeles, J. Phys. Chem. **59**, 952 (1955).
[10] S. M. Klainer and G. Kegeles, Arch. Biochem. Biophys. **63**, 247 (1956).
[11] J. C. Mitchell, A. E. Woodward, and P. Doty, J. Am. Chem. Soc. **79**, 3955 (1957).
[12] A. M. Holtzer, H. Benoit, and P. Doty, J. Phys. Chem. **58**, 624 (1954).
[13] H. Boedtker and N. S. Simmons, J. Am. Chem. Soc. **80**, 2550 (1958).
[14] H. N. Ritland, P. Kaesberg, and W. W. Beeman, J. Chem. Phys. **18**, 1237 (1950).
[15] M. Champagne, V. Luzzati, and A. Nicolaieff, J. Am. Chem. Soc. **80**, 1002 (1958).
[16] P. J. Flory, *Principles of Polymer Chemistry* (Cornell University Press, Ithaca, New York, 1953), p. 595.
[17] R. Simha, J. Phys. Chem. **44**, 25 (1940).
[18] C. Hall, J. Biophys. Biochem. Cytol. **2**, 625 (1956).
[19] C. Hall and P. Doty, J. Am. Chem. Soc. **80**, 1269 (1958).

# 7
# Polyelectrolytes

STUART A. RICE

*Department of Chemistry and Institute for the Study of Metals, The University of Chicago, Chicago 37, Illinois*

## I. INTRODUCTION

THE past decade has witnessed an extensive development in two directions of the theory of electrolyte solutions. On the one hand, the investigations of Mayer[1] and of Kirkwood and Poirier[2] finally have established from first principles the validity of the Debye-Huckel limiting law. These studies provide a formal framework for extending the theory of electrolytes to concentrations higher than the traditional "slightly polluted water," and they provide insight into the nature of the approximations which make valid the use of the Poisson-Boltzmann equation in dilute solution. The recent works of Onsager and Kim,[3] of Fuoss and Onsager,[4] of Falkenhagen and Kelbg,[5] and of Pitts[6] have extended, within the framework of the Poisson-Boltzmann equation, the theory of transport processes to include the effects of finite ion size. While it is doubtful if the validity of the Poisson-Boltzmann equation permits further development, the inclusion of finite ion size does permit the description of dissipative processes, in terms of an ion-size parameter, to be extended to much higher concentrations than heretofore, and it suggests a more satisfactory interpretation of the varying behavior of different strong electrolytes. Further extension of the theory of dissipative processes probably requires the more powerful tools of the Mayer theory combined with the general statistical theory of transport.[7]

On the other hand, there has been an almost explosive development of the theory of very highly charged macromolecular ions. Whereas, prior to 1940, there existed almost no quantitative theory of colloid solutions, and whereas the measurements of Kern represented the only investigations of flexible polyelectrolytes,[8] there exists now a number of theories for the equilibrium properties of these substances, and there have been several attempts at the theory of transport. In all instances, the theory of polyelectrolytes has been developed within the framework of the Poisson-Boltzmann equation; in some cases, the analysis uses a linearized form, while in others the analysis adopts different approximations equivalent to the assumption that the domain occupied by the polyion has zero net charge. Although there are many difficulties associated with the current theory, and, though qualitative or semiquantitative agreement with experiment is all that can be attained at present, it may be asserted safely that the broad features of the basic physical processes responsible for the observed phenomena have been delineated and that the major problems are quantitative rather than qualitative. In this article, the theory of polyelectrolyte solutions is reviewed briefly, with emphasis on the basic physical processes. The available experimental data suitable for comparison with the theory, and the directions which future developments are likely to take are mentioned only cursorily, owing to limitations of space.

## II. GENERAL SURVEY AND QUALITATIVE PICTURE

The fundamental theory of rigid, impenetrable macroions first was published in 1948 by Verwey and Overbeek[9] in their well-known monograph; it has been refined continuously by Levine,[10–12] by Booth,[13] and by Kirkwood and co-workers.[14,15] Briefly, the model adopted for the interaction between particles represents the macroion as a rigid sphere with a specified surface-charge density. The high surface charge tends to force the counterions to cluster close to the surface of the macroion, creating a thin double layer which shields the surface charges from similar charges on other colloidal particles. The net repulsive force between two macroions usually is obtained from a suitable solution of the Poisson-Boltzmann equation, although more sophisticated methods also have been used.[14–16] The fact that the macroion is rigid and impenetrable simplifies the problem to the extent that almost no features are intrinsically different from the low molecular weight, strong electrolyte case. There are, however, qualitative differences in the relative importance of the various contributions.

For colloidal ions with one type of charge, one may imagine the free energy of building the double layer to consist of four contributions. First, if the charge is due to adsorption or desorption of ions from the solution, there is an intrinsic chemical potential change due to the dissociation reaction. To this must be added the sum of the chemical potentials of all of the species in solution, the difference in self-energy (interaction of central ion and atmosphere) of an adsorbed ion from a free ion, and the electrical free energy of charging up and assembling the double layer. When considering the interaction of two macroions, it is important to note that the overlap of double layers causes a redistribution of small ions in the solution with the net result that the mutual interactions of the small ions are changed. Thus, in addition to the expected screened Coulomb force between macroions, there is a force arising from the change in interaction energy of the small ions as they redistribute under the influence of the overlapping double layers.

When consideration is given to the possibility that charges of both signs may exist on the surface of the molecule, a new mechanism of interaction between macroions becomes apparent.[16] If there is a net difference in the numbers of positive and negative charges, at the isoelectric point there will be a number of uncharged groups otherwise identical in chemical character with some of the charged groups. For one of the types of charge, there exists under these conditions a large number of different possible arrangements of the charges on the surface of the molecule. A number of these arrangements is of approximately equal electrostatic energy and one may expect the charges to fluctuate in occupation of first one and then another set of sites. If two such molecules are brought close to each other, the charges on one ion tend to polarize the charges on the other in such a manner as to separate like charges. This is possible if there exist surface-charge distributions of roughly equal energy and the resultant decreased repulsion is, of course, equivalent to a net attraction, since the net charge of each particle is zero. It has been postulated that this fluctuation force is operative in enzyme reactions.[17]

Another fundamental difference between macroions and small ions is that, in the former case, consideration must be given to the mutual interactions of charges on the surface of the ion. The magnitude of these interactions depends upon the geometric arrangement of the charges, upon the dielectric constant, and upon the salt concentration, and readily may be ascertained experimentally from the titration curve. Early attempts to characterize this electrostatic energy made the crude approximation of smearing out the discrete-charge structure and applying the Debye-Huckel theory in substantially its original form. Refinements, such as allowing the ion to be partially penetrable by solvent, alter the model only slightly.[18] On the other hand, direct calculation[14,15] of the electrostatic energy for discrete-charge distributions reveals a marked dependence of the titration curve upon the details of the distribution of charges, as was to be expected. Physically, this corresponds to the fact that neighboring acid and base groups tend to become zwitterions, whereas neighboring groups of the same kind resist ionization because of the large repulsive interactions created by such ionizations. Consequently, the amount of work required to ionize a group depends markedly upon the group environment, and a smeared-out charge distribution gives an erroneous, even if fortuitously numerically accurate, picture of the molecule.

In the case of flexible polyelectrolytes, attention has been focussed almost entirely on the interactions between the charges of the polyion and on the relationship between configurational and thermodynamic properties and ionic strength, degree of neutralization, and so forth. Interactions between polyions only now are being investigated in detail.

A basic step in the study of linear polyelectrolytes was taken in 1948 by Kuhn, Künzle, and Katchalsky,[19] who described the configuration of a polyion as that of a chain which was randomly coiled with the restriction that its mean end-to-end distance be such as to minimize the sum of the configurational and electrostatic free energies. This simple model was able to predict qualitatively the size changes accompanying changes in charge on a polymer. This treatment was approximate, however, in that it assumed the charges on the polymer to have an interaction energy determined only by the end-to-end distance, rather than by all of the distances between charges on the chain. Moreover, it was not shown that a screened Coulomb potential was the proper expression to use in deriving the interaction energy between a pair of fixed charge in a solution containing mobile ions.

In the same year, Hermans and Overbeek[20] proposed a model in which the polymer was regarded as a porous spherical-charge distribution, whose radius was to be varied to minimize the total free energy. By using a linearized form of the Poisson-Boltzmann equation, they computed the electrostatic free energy without using the screened Coulomb potential. Although the Hermans-Overbeek model avoids explicit consideration of the chain configuration, it predicts the size changes of polyions moderately well. But neither of the models discussed thus far was conspicuously successful in interpreting the titration curves of polyelectrolytes.

Two years later, Huizenga, Grieger, and Wall[21] reported experiments which, for the first time, fully showed the importance of ion binding in polyelectrolyte solutions. They found that up to 60 percent of the sodium ions counter to a carboxylate polymer moved with it in electrophoresis. Since these polymers are thought to be approximately free draining under the conditions of this experiment, one concludes that the counterions are not merely within the volume of the polymer, but are relatively tightly bound to it.

The foregoing work, and contributions by Flory,[22] by Kimball, Cutler, and Samelson,[23] by Osawa, Imai, and Kagawa,[24] and by others, led to the proposal of a model[25] which was new in that (a) the interactions between charges of the polymer influenced its local configurational properties as well as its end-to-end distance, (b) these interactions affected the tendency toward ionization at each ionizable group, making the various ionizable groups of a polymer interdependent, rather than independent, as heretofore postulated, and (c) the interactions were large enough to cause important amounts of binding of counterions, even though negligible amounts of binding would occur if the ionizable groups did not interact. In implementing this model, it is necessary to allow for the interrelations among its several features. In particular, to determine the mean size, degree of ionization, and degree of binding of a polyion in a solution of given *p*H, counterion concentrations, and ionic strength, one would determine first the free energy as a function of all six of these variables, and then minimize this with respect to the first three, which are not controllable externally. The expression for the free energy must reflect the fact that not all chain configurations of the same end-to-end distance will have the same electrostatic interaction, nor will all arrangements of the same total number of charges be equienergetic for a given chain configuration. A further discussion of these and related points is presented after a more detailed definition of the molecular model.

To specify the model more completely, consider a weak polyacid, each functional group of which can be un-ionized (zero charge), ionized ($-1$ charge), or ion paired (zero charge). It is assumed that an intrinsic "chemical" free energy change can be assigned to the conversion of an ionizable group, from one to another of these charge states, so that, ignoring interactions among the charges, the ionization and binding would be describable by equilibrium constants. The binding is thus treated phenomenologically. The ion pairs introduced here may be described as "site bound," so as to distinguish them from ions merely required to be near the polymer for the maintenance of electrical neutrality.

The chain configurations available to the polymer are assumed to be just those accessible to an otherwise similar uncharged polymer, so that the effect of the electrostatic interactions is to alter the distribution among these configurations. The chain configurations are described in the manner first proposed by Kuhn,[26] by regarding the polymer as a series of rigid links connected by universal joints. The lengths of the links are chosen so as to allow, as well as possible, for all charge-independent forces restricting the short-range flexibility of the chain. The Kuhn model makes all of the accessible configurations of equal nonelectrostatic energy, so that the charge interactions completely determine the manner in which various configurations are weighted. Thus, the charge interactions have the two following effects: they influence the equilibria among un-ionized, ionized, and bound functional groups; they affect the configurational distribution. These interactions, of course, are to be calculated, keeping in mind that the space between the charges is occupied by an ionic solution, and hence the calculation depends upon the ionic strength and upon other properties of the solution.

In accordance with the over-all approach sketched in the third preceding paragraph, one turns first to the formulation of the free energy when the mean size and degrees of ionization and binding are specified. It is convenient to begin by assigning to the polymer a specific chain configuration, and by specifying which particular functional groups are to be un-ionized, ionized, and ion paired. One may consider then the free energy of this visualizable, but unattainable, "state" of the polymer. On a completely microscopic scale, this state is really a large number of states, because the distributions of the small mobile ions of the system, the energy levels of the various species, etc., are not specified. Thus, it is relevant to consider the *free energy* of this state. Each change in the configuration or in the status of the function groups results in a new state, with its associated free energy. These states, together with their free energies, may be taken as the starting point for a statistical mechanical calculation of the over-all free energy when only the *numbers* of ionized, un-ionized, and bound groups, and the *average* configuration are specified.

The free energy of a state of the type outlined above consists only of the intrinsic free energies of the functional groups plus the electrostatic interaction free energy. The configuration of the chain does not contribute by virtue of the Kuhn chain model. The only problem is in the calculation of the interaction free energy for the charge distribution of each state. It has been shown[27] that, in spite of the presence of fixed charges, the Debye-Huckel potential can be used for this purpose. The demonstration involves approximations not much more restrictive than those of the ordinary Debye-Huckel theory.

The methods outlined above provide, at least in principle, a way to calculate the free energy of a state of specified configuration and charge distribution. At this point, mathematical complexity forces an approximation. We choose to calculate exactly for each individual state only the electrostatic interaction between nearest neighboring functional groups along the chain, and to approximate the remainder of the interaction under the assumption that the charge is distributed evenly along the chain. This approximation has the effect of preserving the strong influence that a charged group has upon the ionization and upon the binding at nearest neighboring groups, while making tractable the calculation of the less sensitive dependence on the interactions which promote configurational expansion. To be more explicit, it has just been assumed that, if the net charge is specified, the distribution of that charge along the chain is independent of the chain configuration, and that the free-energy difference among configurations is to be calculated assuming the charge to be uniformly distributed along the chain.

The remainder of the over-all free energy now separates into two parts: (a) the configurational free energy of a uniformly charged chain, and (b) the distributional free energy of a chain of charged and uncharged sites with near-neighbor interaction. The computation of (a) is essentially a biased random-walk problem, while (b) is formally equivalent to the calculation of the spin distribution in a one-dimensional Ising lattice. The details of both of these computations are described elsewhere.[25] Qualitatively, it is apparent that increasing the charge on a chain tends to restrict its configurational freedom, or to increase its configurational free energy. One would expect likewise that computation (b) might result in weighting heavily only those charge distributions in which not very many more charges than necessary are nearest neighbors.

When the free energy, as computed by the methods outlined above, is minimized with respect to the number of ionized and ion-paired groups, it is found that the electrostatic interactions cause the net charge of the polymer to be much less than if no such interactions existed. This means that the free-energy increase associated with forming un-ionized groups or ion pairs is more than compensated for by the reduction in electrostatic free energy, until many more than the "normal" number of ion pairs and un-ionized groups have been created. The effect is much larger than one might suppose at first, inasmuch as the charges are frequently close enough together such that the effective dielectric constant of the region between them is far lower than that of normal water. Moreover, as the number of charged groups increases, the vast majority of the possible distributions of the charges along the chain becomes extremely unfavorable energetically, so that the free energy contains a large negative entropy term. The interaction free energy thus makes it progressively more difficult to remove successive protons from a weak polyacid, spreading the titration curve in the manner experimentally observed. When the protons finally are removed, many of them are replaced by counterions having little intrinsic affinity for ion pairing.

The configurational expansion accompanying neutralization of a polyacid is calculable by minimizing the free energy with respect to its average end-to-end distance. It is found that the binding and the reduced ionization cause the expansion as predicted for the polymer to be in much better agreement with experiment than if the binding or the reduced ionization were ignored. In addition, the amount of binding predicted to satisfy the titration and configuration properties is consistent with the amount of site binding observed in electrophoresis.[21]

The model as set forth in the preceding section may be applied to polyelectrolyte gels[28] which may be analogues of some of the functions of cell membranes. The basic concepts are the same as those already described, but the details are significantly affected by the crosslinking of the gels. The specific systems for which the theory was developed are strong electrolyte ion exchangers in equilibrium with various kinds of monatomic counter ions. Examples of such systems are the sulfonated crosslinked polystyrene exchangers in equilibrium with mixtures of alkali metal ions. There is nothing, however, in the theory restricting the analysis to this case, and the general concepts are applicable to any charged gels or membranes.

Consider, therefore, a crosslinked gel containing a number of functional groups which can be ionized ($-1$ charge) or which can be paired with either one of two types of counter ion to give an electrically neutral ion pair. As before, it is assumed that the intrinsic free energy associated with the formation of each type of ion pair has a value dependent upon the kind of counter ion and upon its concentration in the external solution. The ion-exchange selectivity is introduced by assuming different intrinsic free energies for the formation of different kinds of ion pairs. Exchange selectivity results only, however, when the conditions are such that significant numbers of ion pairs can be formed.

Just as for linear polyelectrolytes, it is desired to compute the free energy for a given configuration and charge distribution of the gel. As pointed out, the Debye-Hückel screened Coulomb potential may be used to calculate the electrostatic interactions. The crosslinked systems, however, present problems not encountered when dealing with linear polyelectrolytes. Many exchange resins are crosslinked sufficiently such that there are reasonably well-defined regions inside and outside the exchanger, respectively. The interior region is characterized by different ion concentrations and by the presence of enough organic matter to cause the solution to be quite different from that outside. The analysis confirming the use of the screened Coulomb potential[27] shows that the screening constant to be used must involve the proper dielectric constant for the interior region, and must recognize that the space occupied by organic matter is not accessible to the mobile ions. The bulk concentrations should be used, as the different ion concentrations within the resin are allowed for in the derivation of the screened Coulomb potential.

The large concentration differences between regions internal and external to an ion-exchange resin have led many investigators to include an "osmotic" contribution to the free energy when describing an ion-exchange resin on a phenomenological basis. The present model, however, is entirely molecular in character, and it would be incorrect to include such a contribution in addition to the direct calculation of the detailed interactions. The concentration differences are merely a consequence of the forces already considered. The point of view presented here removes a conceptual difficulty in treating resins of low crosslinking. For, if an osmotic description is used in preference to a detailed model, ambiguity arises when the degree of crosslinking is diminished to the point where the regions inside and outside the resin cease to be well defined.

Although there is in principle a means of determining the free energy at every specified configuration and charge distribution, to simplify the calculation approximations again must be made. The difference in configurational properties of linear systems and gels suggests that the approximations used here be somewhat different from those of the preceding section. In a gel, many of the spatially near-neighboring functional groups are not near neighbors along the polymer chains, so that it is definitely necessary to consider interactions between groups which are not closely connected. In fairly concentrated gels, near neighbors to a given group in general are distributed so that most motions at constant gel volume do not affect the electrostatic energy seriously, with some pairs of charges approaching more closely as others are carried further apart. This is a radically different state of affairs from that of a linear polymer, where at a constant end-to-end distance the electrostatic energy depends critically upon the local coiling of the chain. It is assumed, therefore, without gross inconsistency, that all gel configurations of the same volume are electrostatically equivalent, and that the interaction between the charge state of the gel and its configuration depends only upon the gel volume. For the purpose of calculating the electrostatic interactions, the functional groups of the gel are assumed to be on a lattice whose scale is a function of the gel volume. The distributional free energy of the gel then can be calculated subject to the same sort of approximations as enters into the simple cell theories of liquids.

The calculation of the gel free energy is complete when the configurational free energy and the electrostatic free energy are each characterized as a function of gel volume. As pointed out above, the electrostatic interactions may be handled by the methods of the cell liquid theory. The analogy is exact if ion-paired groups are identified as the holes of the liquid theory, and the charges as occupied cells. The details have been given elsewhere.[28] The configurational free energy is simply that of an uncharged gel, so that the theory of rubber elasticity may be applied. We have found it convenient to use the theory in the form given by Flory.[26]

From the free-energy function determined in the manner just described, one can relate the external solution composition to the volume and to the ion pairing in a resin. The model differs from the traditional approaches in that the only source of selectivity enters through the intrinsic ion-pairing constants. It has been assumed that the size of the ions enters in no other way. Another difference is the assumption that there is an equilibrium between paired and free charged groups, with only the paired groups contributing to a selective effect. This concept leads to the conclusion that exchange selectivity is increased when conditions are such that ion pairing increases. Since the ion pairing occurs only when necessary to reduce strong electrostatic fields, one may understand why resins of high exchange capacity should be more selective than resins of lower capacity. Likewise, resins unable to reduce their electrostatic interactions through expansion because of high crosslinking are more selective than loosely crosslinked resins which easily can swell.

Flexible polyelectrolytes have been analyzed here in terms of ion-pair formation. There is some controversy on this point though the author believes the available evidence supports the concept of ion pairs. Further discussion of this topic may be found elsewhere.[29]

The preceding description of flexible polyelectrolytes may be extended easily to polyampholytes.[30] No new considerations are necessary, excepting those already discussed for rigid polyampholytes—i.e., fluctuation attraction, local charge structure and its effect on titration behavior, and possible ion binding. However, for flexible polymers, it must be remembered that the internal molecular configurations are subject to the forces arising from these phenomena, a complication absent in the case of rigid ions. A quantitative formulation of the models as discussed in this section is considered next.

## III. SUMMARY OF THE MATHEMATICAL THEORY OF RIGID POLYIONS

Consider as the first and simplest case a rigid polyion of arbitrary shape in a volume $v$ of electrolyte.[9–12] For simplicity, attention is restricted to the case where the charge on the polyion is due to the adsorption or dissociation of groups of one species of charge. The existence of adsorbed ions also implies the existence of a dissociation equilibrium which may be characterized, of course, by the change in chemical potential per ion adsorbed. If all of the ions are discharged, the Helmholtz free energy may be written

$$A_0(n)=\sum_{i=0}^{r} N_i\mu_i(0)+a_0(n)$$
$$a_0(n)=\int_s a_0^*(\nu)dS, \qquad (1)$$

where $a_0(n)$ is the change in free energy when $n$ discharged ions are adsorbed on the polyion, $\mu_i(0)$ is the chemical potential of an ion of species $i$ when there are no adsorbed ions, $\nu$ is the surface density of adsorbed ions, and the integration is over the surface of a macromolecule. Each element of volume-containing ions interacts not only with the charges on the polyion surface, but also with other parts of the medium external to the macromolecule. If $\tau_i$ is the chemical potential of ionic species $i$,

$$\tau_i=\mu_i(0)+\chi_i(1)+q_i\int_0^1 \phi_i(\lambda,n)d\lambda, \qquad (2)$$

where $\lambda$ is the fractional charge on an ion (a hypothetical charging process is used to calculate the change in free energy due to electrostatic forces), $\chi_i(1)$ is the free energy arising from the interaction of the fully charged ($\lambda=1$) central ion and its ionic atmosphere, and $\phi_i(n)$ is the average electrostatic potential in the bulk of the solution at very large distances from the polyion. If the average electrostatic potential at the point $r$ is $\Psi_i(r)$, one has

$$\Psi_i(r)=\psi_i(r)+\phi_i(n), \qquad (3)$$

with $\psi_i(r)$ the change in average potential experienced by ion $i$ in being brought from the bulk of the medium to the point $r$. The total change in Helmholtz free energy due to the excess charge and hence excess electrostatic interaction of the double layer over that in the uniform bulk medium easily is seen to be

$$A_e=\int_0^1 d\lambda\left[\sum_0^r q_i\int_v n_i\psi_i dv+q_i\int_s \nu(s)\psi_1 dS\right], \quad (4)$$

where it is assumed that the surface charge is due to species 1. Note that Eq. (4) is of the simple form, charge multiplied by potential, with $n_i(r)$ the volume density of ions of species $i$ at $r$ and $\nu(s)$ the corresponding surface-charge density. The total free energy is then (solvent is represented as species zero)

$$A(n)=N_0\mu_0(0)+\sum_1^r N_i\tau_i +a_0(n)-n\chi_1^{\text{bulk}}(1)+n\chi_1^{\text{surface}}(1)+A_e(n). \quad (5)$$

To determine the equilibrium number of adsorbed ions, the free energy represented in Eq. (5) must be minimized with respect to $n$, giving the result

$$\chi_1^{\text{bulk}}(1)-\chi_1^{\text{surface}}(1)=\frac{\partial a_0}{\partial n}+\frac{\partial A_e}{\partial n} +\frac{\partial}{\partial n}\sum_1^r q_i\int_0^1[\phi(\lambda,n)-\phi(\lambda,0)]d\lambda, \quad (6)$$

with the last term accounting for the change in composition and hence mean potential of the bulk medium when $n$ ions are adsorbed.

Consider now the interaction of two polyions, fixed in position a distance $R$ apart. The free energy of this system differs from the preceding only in that the relevant mean potentials must be obtained as a function of $R$. One may write

$$A(n,R)=N_0\mu_0(0)+\sum_1^r N_i\tau_i +2a_0(n)-2n(\chi_1^{\text{bulk}}(1)-\chi_1^{\text{surface}}(1))+A_e(n,R), \quad (7)$$

where $A_e(n,R)$ is to be computed now for two double layers, and the force between the macroions is

$$\left(\frac{\partial A(n,R)}{\partial R}\right)_{n=n_{\text{eq}}}=\frac{\partial}{\partial R}\left[A_e(n,R) +\sum_1^r q_i\int_0^1[\phi(\lambda,n)-\phi(\lambda,0)]d\lambda\right]_{n=n_{\text{eq}}}. \quad (8)$$

The derivatives are to be evaluated along the equilibrium-adsorption isotherm which may be a function of the interparticle separation. As before, the second term on the right-hand side of Eq. (8) is the correction arising from the change in mutual energy of the small ions of the solution. To implement this general result, the right-hand side of Eq. (8) must be evaluated. Solution of the linearized Poisson-Boltzmann equation gives the following complicated results[10–12]: The average potential of each particle may be shown to be

$$\langle\Psi_1\rangle=\frac{nq_1}{Da(1+\kappa a)}\left\{1+\frac{\exp\left[-\kappa a\left(\frac{R}{a}-2\right)\right]}{(R/a)(1+\kappa a)}\right\}, \quad (9)$$

where $a$ is the radius of a spherical macroion and $\kappa$ the usual screening parameter. The interaction energy has two terms,

$$\delta W_1(\lambda,R)=\frac{\lambda q_1}{Da(1+\lambda\kappa a)^2}\frac{\exp\left[-\lambda\kappa a\left(\frac{R}{a}-2\right)\right]}{(R/a)} \quad (10)$$

and

$$\delta W_2(R)=-\frac{5\kappa_0 q_1^2 q_2}{4D^2akT}\cdot\frac{a}{R(\kappa a)^5}\left[\frac{1}{\left(\frac{R}{a}-2\right)}\left\{\frac{R}{a}+1-\frac{2}{\left(\frac{R}{a}-2\right)}+\frac{2}{\left(\frac{R}{a}-2\right)^2}\right\}\right.$$
$$-\left(\frac{R}{a}+2\right)\exp\left(\frac{R}{a}-2\right)\left\{Ei\left(\frac{R}{a}-2\right)-Ei\left[\left(\frac{R}{a}-2\right)(1+\kappa a)\right]\right\}$$
$$\left.-\frac{\exp\left[-\kappa a\left(\frac{R}{a}-2\right)\right]}{(R/a)-2}\left\{\frac{2}{[(R/a)-2]^2}+\frac{2(\kappa a+1)}{[(R/a)-2]}+\frac{[(R/a)-2)]}{1+\kappa a}+(\kappa a-1)^2+2+\frac{(\kappa a)^5\left(\frac{R}{a}-2\right)}{5(1+\kappa a)^5}\right\}\right], \quad (11)$$

where

$$Ei(x)=\int_x^\infty\frac{e^{-u}}{u}du. \quad (12)$$

For very large $R$,

$$\delta W_2(R) \sim -\frac{30\kappa_0 q_1^2 q_2}{D^2 akT(\kappa a)^5}\left(\frac{a}{R}\right)^6 \qquad (13)$$

and both forces are repulsive.

Direct numerical integration of the complete Poisson-Boltzmann equation leads to potential distributions differing quantitatively but not qualitatively from that in Eqs. (10) and (11). Graphical and numerical results can be found in the paper by Hoskin and Levine.[10–12]

At most distances, $\delta W_1 \gg \delta W_2$, so that the dominant repulsive potential is given by Eq. (10), but at very large distances the term of Eq. (12) is dominant. The potential, Eq. (10), was considered by Verwey and Overbeek[9] and shown to lead to quantitative description of the stability of lyophobic colloids with respect to changes in ionic strength, coagulation, etc. For further details, the reader is referred to their book.

A rather different study of the nature of the repulsive interactions was undertaken by Steiner.[31] Because of the strength of the repulsive forces, there is considerable ordering of the polyions in solution. Consequently, by inversion of the angular distribution of scattered light from suitably chosen colloidal solutions (i.e., those that exhibit diffraction due to interparticle interference), the radial-distribution function may be determined experimentally. Steiner has carried out such a program for AgI sols in which the charge is due to the adsorption of $I^-$. Figure 1 shows the radial-distribution function deduced by Steiner (reproduced from his Ph.D. dissertation). The structure of the solution alters with dilution and ionic strength in the expected manner. It is found that the strong repulsive forces prevent the particles from approaching closer than 1500 A at low ionic strengths. The maximum in the distribution function occurs at about 3000 A, so the long range of the repulsive force is not unreasonable. Further, the shift in position of the maximum in the distribution function is approximately of the magnitude expected from the concentration change, and the addition of a quantity of bivalent cation produces an effect roughly equivalent to that of one hundred times its concentration of monovalent cation, in accordance with the Schulze-Hardy rule.

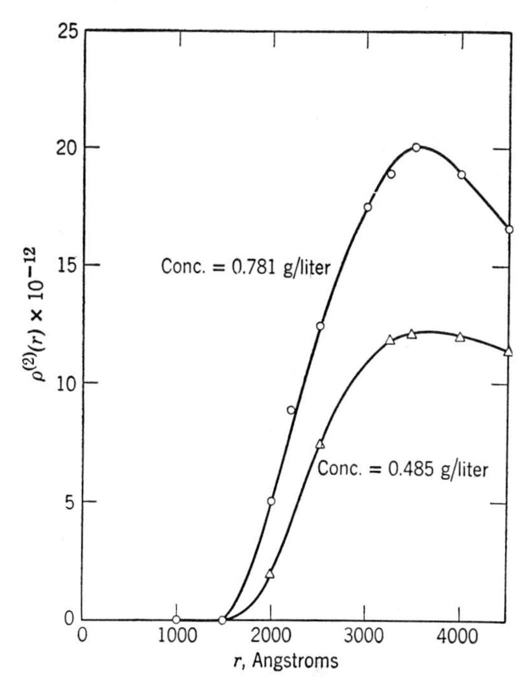


FIG. 1. A typical radial-distribution function for AgI sol.

Quantitative examination of the data reveals, however, deviations from the theoretical predictions. From the shape of the radial-distribution function, it is known that the forces are repulsive. To determine the nature of these forces, one may plot $\log(V/kT)$ against $\log R$ (appropriate to a repulsion proportional to $R^{-n}$) and $\log(RV)$ against $R$ (appropriate to a repulsion of the form $e^{-kr}/R$). Neither one of these plots is at all linear, nor does either one appear to be rectifiable by the addition or subtraction of linear parts. It is possible, even probable, that this deviation is due to the known polydispersity of size among the colloidal particles. Certainly, though the results are fragmentary, the repulsive forces calculated by Levine and by Verwey and Overbeek are not of the form found in experiments of Steiner. A discrepancy other than that due to polydispersity would be expected to show up in the radial-distribution function owing in part to the linearization of the Poisson-Boltzmann equation. Still another possible source of error is the neglect of third-body polyions in the calculation of the potential of mean force, since, in the solutions investigated, the range of the force is great enough to involve simultaneous interaction of more than two particles. Despite the apparent sophistication of the mathematical treatment, it is evident that much work remains to be done before quantitative agreement with experiment is attained.

### IV. RIGID POLYAMPHOLYTES[14–16]

Polyampholytes contain both positive and negative charges on the surface of the macromolecule. It is fairly obvious that, far from the isoelectric point when charge of only one sign is present, the considerations presented in the previous section suffice to describe the interactions. At the isoelectric point where the average net charge is zero, conditions are rather different.

Consider two polyampholytes with their centers of mass separated by the distance $R$. The interaction between the two molecules may be written

$$V = \sum_{i=1}^{Z_1}\sum_{j=1}^{Z_2} \frac{\eta_i^{(1)}\eta_j^{(2)}q^2}{DR_{ij}^{(12)}} \frac{\exp[-\kappa(R_{ij}^{(12)} - a_{12})]}{1+\kappa a_{12}}, \qquad (14)$$

where there are $Z_1$ and $Z_2$ sites on polyions one and two, respectively, $q$ is the magnitude of the charge, $R_{ij}^{(12)}$ is the separation of site $i$ on molecule one from site $j$ on molecule 2, and $\eta_i^{(1)}$ and $\eta_j^{(2)}$ are occupation variables which can have the values zero for an uncharged site and plus or minus unity for positive and negative charges on the specified site. Now the potential of mean force is defined by

$$\exp(-W/kT)=\langle\exp(-V/kT)\rangle, \tag{15}$$

where the average is taken as unweighted and with molecules one and two fixed. It may be shown readily that the potential of mean force may be expressed in terms of unweighted averages of the potential energy $V$.[32] We use the truncated form,

$$W(R)=\langle V\rangle-\frac{1}{2kT}(\langle V^2\rangle-\langle V\rangle^2), \tag{16}$$

correct to terms of order $(kT)^{-1}$. At the isoelectric point, the average net charge is zero and the first term vanishes, as do all powers of $\langle V\rangle$. Now define

$$\begin{aligned} q_i^{(1)}&=q\eta_i^{(1)}=\langle q_i^{(1)}\rangle+\Delta q_i^{(1)} \\ \Delta q_i^{(1)}&=q[\eta_i^{(1)}-\langle\eta_i^{(1)}\rangle]=q\Delta\eta_i^{(1)}. \end{aligned} \tag{17}$$

The use of Eq. (14) in Eq. (16) gives

$$W(R)=-\frac{q^4}{2kTD^2R^2}\sum_{i=1}^{z_1}\sum_{l=1}^{z_1}\sum_{k=1}^{z_2}\sum_{s=1}^{z_2}\langle\eta_i^{(1)}\eta_l^{(1)}\rangle\langle\eta_k^{(2)}\eta_s^{(2)}\rangle \times\frac{e^{-2\kappa(R-a)}}{(1+\kappa a)^2}\left\langle\frac{R^2}{R_{ik}^{(12)}R_{ls}^{(12)}}\right\rangle, \tag{18}$$

where the last factor accounts for differences in charge-charge separation due to distribution of charges on the surface of the molecule. With the substitution of Eq. (17) into Eq. (18), the potential of mean force becomes (setting $\langle R^2/R_{ik}^{(12)}R_{ls}^{(12)}\rangle=1$, valid at large separations)

$$W(R)=-\frac{q^4}{2kTD^2R^2}\sum_{i,l=1}^{z_1}\sum_{k,s=1}^{z_2}[\langle\eta_i^{(1)}\eta_l^{(1)}\rangle-\langle\eta_i^{(1)}\rangle\langle\eta_l^{(1)}\rangle] \times[\langle\eta_k^{(2)}\eta_s^{(2)}\rangle-\langle\eta_k^{(2)}\rangle\langle\eta_s^{(2)}\rangle]\frac{e^{-2\kappa(R-a_{12})}}{(1+\kappa a_{12})^2}. \tag{19}$$

This term represents the interactions between the fluctuating charges and fluctuating multipoles of both molecules. The charge fluctuations may be obtained easily by differentiation of the Grand Partition Function, as is seen later. To calculate the excess chemical potential, one uses the relation

$$\frac{100M_1}{NkT}\left(\frac{\partial\mu_1^e}{\partial c_1}\right)_{T,p,\kappa_0}=4\pi\left[\int_0^\infty R^2[g_{10}(R)-1]dR-\int^\infty R^2[g_{11}(R)-1]dR\right], \tag{20}$$

where, in the usual notation,

$$\begin{aligned} \mu_1&=\mu_1^0(T,p)+kT\ln c_1+\mu_1^e \\ \mu_1^0(T,p)&=\lim_{c\to0}[\mu_1-kT\ln c_1] \\ g_{10}(R)&=\exp\left[-\frac{W^{(10)}}{kT}\right] \\ g_{11}(R)&=\exp\left[-\frac{W^{(11)}}{kT}\right] \end{aligned} \tag{21}$$

with $W^{(10)}$ and $W^{(11)}$ the potentials of mean force between a macromolecule and a solvent molecule and between two macromolecules, respectively. The dimensions of a solvent molecule are small relative to those of the macromolecule in almost all cases, and the solvent plays the role of a dielectric continuum. One finds, therefore, after substitution and integration, that

$$\frac{\mu_1^e}{kT}=\frac{c_1}{100M_1}\left[\frac{7\pi Na_{12}^3}{6}-\frac{\pi Nq^4\langle\Delta\eta_i^{(1)2}\rangle}{(DkT)^2\kappa_0(1+\kappa_0a_{12})^2}\right], \tag{22}$$

where the macroion contribution to the ionic strength is neglected.

At high salt concentrations, $\kappa_0$ is large and the second term is negligible relative to the first term. Under these conditions, the excess chemical potential is positive, representing the net repulsive forces which result in an excluded volume. When the salt concentration is very low, the second term dominates and the excess chemical potential is negative. Careful light-scattering experiments have verified Eq. (22) quantitatively.[33] The magnitude of the fluctuating charge may be obtained independently from the titration curve of the polyampholyte and is in complete agreement with the values determined from the independent light-scattering experiments.[34]

A very interesting application of this theory has been presented by Kirkwood[17] in a discussion of enzyme kinetics. Suppose the enzymatic reaction can be described by the Michaelis-Menten relation

$$-\frac{d[S]}{dt}=\frac{k_3[E][S]}{K_m+[S]} \tag{23}$$

in which $[S]$ and $[E]$ are the substrate and enzyme concentrations, $K_m$ the Michaelis-Menten constant, and $k_3$ the intrinsic rate constant for the enzymatic reaction. Now the potential of mean force is the work required to bring a pair of molecules from infinite separation to the distance $R$. If the standard free energy of formation of a complex is $\Delta G^0$, then this is equal to the potential of mean force at contact; i.e.,

$$\Delta G^0=W(a), \tag{24}$$

and, therefore,

$$kT\ln K=-W(a), \tag{25}$$

where $K$ is the equilibrium constant for the formation of the complex. Let the complex now undergo reaction through the intermediary of a transition state with free energy of activation $\Delta G^{\ddagger}$. Thus,

$$\Delta G^{\ddagger} = W^{\ddagger}(a^{\ddagger}) - W(a). \quad (26)$$

Suppose the substrate molecule is specifically bound to the active site of the enzyme by local forces with potential $W_0$, in which the protein does *not* participate. In the activated state, the potential is denoted by $W_0^{\ddagger}$, the subscript zero again indicating that protein forces are omitted. Under the influence of the protein, the local forces are different and the corresponding potentials are denoted $W'$ and $W^{\ddagger}$. If $K_m$ and $K_m^0$ and $k_3$ and $k_3^0$ are the Michaelis-Menten constants and intrinsic rates of reaction with and without protein participation, then

$$kT \ln \frac{K_m}{K_m^0} = W' - W_0,$$

$$kT \ln \frac{k_3}{k_3^0} = \Delta W_0^{\ddagger} - \Delta W^{\ddagger}, \quad (27)$$

where the intrinsic rate of decomposition of $ES$ into reactants is assumed to be small relative to $k_3$.

If the protein contains $Z$ groups with acid dissociation constants $K_i$, and if the substrate molecule has dipole moments $\mu'$ and $\mu^{\ddagger}$ in the normal and activated states, then, neglecting all forces except those arising from charge fluctuations leads to

$$W - W_0 = -\frac{1}{2kT} \sum_1^Z \frac{K_i[H^+]}{([H^+]+K_i)^2} \frac{q^2\mu^2 \cos^2\gamma_i}{D_e^2 R_i^4}, \quad (28)$$

where $\gamma_i$ is the angle between the dipole moment of the substrate molecule and the radius vector $\mathbf{R}_i$ from site $i$, $D_e$ is the effective dielectric constant, and the first factor arises from the evaluation of $\langle \eta_i^2 \rangle$ as is seen in a subsequent section.[30] Note that electrostatic interactions between charges on the protein again are neglected.

Since the fluctuation potentials fall off as $R^{-4}$, one may simplify Eq. (28) by assuming that only nearest neighboring groups are effective. If there are $Z_\alpha$ nearest neighboring groups with dissociation constants $K_\alpha$,

$$W - W_0 = -\frac{Z_\alpha q^2 \mu^2}{4D_e^2 r_\alpha^4 kT} \frac{K_\alpha[H^+]}{([H^+]+K_\alpha)^2}$$

$$r_\alpha^{-4} = \frac{2}{\nu_\alpha} \sum \cos^2\gamma_i / R_i^4. \quad (29)$$

From Eqs. (29) and (27),

$$\log \frac{k_3}{k_3^0} = \frac{Z_\alpha q^2(\mu^{\ddagger 2} - \mu'^2)}{4D_e^2 k^2 T^2 r_\alpha^4} \frac{K_\alpha[H^+]}{([H^+]+K_\alpha)^2}. \quad (30)$$

Note that, for this simplified model, $k_3$ has a maximum when the $p$H is equal to $pK_\alpha$. Further, the $p$H dependence predicted by the model is a symmetric bell-shaped curve. If the electrostatic interactions between charges on the surface of the protein are accounted for, then the curve may be skewed since the proton population on the molecular surface, and hence the free energy of activation, are not symmetric functions of the $p$H. Kirkwood has made a crude comparison of the theory with some experimental data of Bergmann and Fruton on the effect of $p$H on the pepsin hydrolysis of carbobenzoxy-1-glutamyl-1-tyrosine. The maximum rate occurs at $p$H 4, thereby identifying the participating groups as carboxylates. With an estimated dielectric constant and an estimated dipole moment difference, the fall in rate to one-half of its value in an interval of one *pH* unit is accounted for if there are about ten carboxyl groups at an average distance of 5 A from the adsorption site. This picture is qualitatively reasonable. It should be stressed that we have considered only one of many possible contributory forces operative in enzyme reactions and it is not to be implied that the charge-fluctuation mechanism is the only or even the most important of the possibilities. The charge-fluctuation mechanism, however, does provide a physical picture in qualitative accord with experiment.

It should be apparent by this time that the calculation of the properties of an isolated polyion eventually requires the computation of the mutual electrostatic energy of the charges on the molecule. Until very recently, it has been customary to calculate this electrostatic free energy for spherical macroions by assuming the charge to be uniformly distributed over the surface, and the macromolecule to be impenetrable by the small ions of the bulk medium. The result obtained is,

$$A_{\text{unif}}^{\text{sphere}} = \frac{q^2 Z^2}{2Da}\left[1 - \frac{\kappa a}{1+\kappa a}\right], \quad (31)$$

where $Zq$ is the net charge on the ion. Aside from the questions of penetrability which introduce no basically new features,[18] the assumption of a uniform charge distribution is grossly inaccurate. The calculation of the electrostatic interaction energy for an arbitrary distribution of positive and negative charges is extremely difficult. Basically, the method consists in the direct counting of all configurations with given total charge and weighting each configuration with the appropriate Boltzmann factor. Fortunately, for small numbers of charges, there are many configurations of equal energy, and this degeneracy simplifies the calculation. With the energy zero taken as the completely discharged protein, the work required to charge a polyampholyte sphere may be shown to be[14,15]

$$A_{\text{discrete}}^{\text{sphere}}=\frac{q^2}{2a}\sum_{k=1}^{Z}\sum_{l=1}^{Z}\eta_k\eta_l(\mathfrak{A}_{kl}-B_{kl})-\frac{q^2}{2a}\sum_{k=1}^{Z}\sum_{l=1}^{Z}\eta_k\eta_l C_{kl}, \quad (32)$$

where

$$\mathfrak{A}_{kl}=\frac{a}{D_i r_{kl}},$$

$$B_{kl}=\frac{1}{D_i}\sum_{n=0}^{\infty}\frac{(n+1)(D-D_i)}{(n+1)D+nD_i}\rho_{kl}^n P_n(\cos\vartheta_{kl}),$$

$$C_{kl}=\frac{1}{D}\left[\frac{\kappa a}{1+\kappa a}+\sum_{n=1}^{\infty}\frac{2n+1}{2n-1}\left\{\frac{D}{(n+1)D+nD_i}\right\}^2 \times\frac{(\kappa a)^2\rho_{kl}^n P_n(\cos\vartheta_{kl})}{\dfrac{K_{n+1}}{K_{n-1}}+\dfrac{n(D-D_i)}{(n+1)D+nD_i}\cdot\dfrac{(\kappa a)^2}{4n^2-1}}\right], \quad (33)$$

$$\rho_{kl}=r_k r_l/a^2,$$

$$K_n(\kappa a)=\sum_{s=0}^{n}\frac{2^s n!(2n-s)!}{s!(2n)!(n-s)!}(\kappa a)^s,$$

with $P_n(\cos\vartheta_{kl})$ as the Legendre polynomial of order $n$, $r_k$, $r_l$, $r_{kl}$, and $\vartheta_{kl}$ are, respectively, the distances of charges $k$ and $l$ from the center of the polyion, the distance between charges $k$ and $l$, and the angle between the vectors $\mathbf{r}_k$ and $\mathbf{r}_l$. To this electrostatic energy must be added the chemical-potential change due to the intrinsic change in free energy on ionization. This total free energy then represents a protein with specified charge distribution, and to obtain the relevant macroscopic free energy, an average over the charge distribution must be performed. The theory has been applied to models of proteins in which the charges were variously located at the vertices of a cube and dodecahedron.[14,15] The results of these laborious calculations may be summarized as follows. The titration curve depends markedly upon the depth of the charges below the surface and upon the distribution of charges. As expected, the interaction energy is smallest for a regular, uniform distribution, larger for a random distribution, and very large if charges are crowded at one end of the molecule and the charge distribution made nonuniform. Arguments can be offered which suggest that the charge groups in a protein are uniformly about one angstrom below the surface.[14,15] All deviations from the electrostatic energy given in Eq. (31) can be accounted for in terms of the charge distribution on the molecule. Great care must be used in interpreting variations in titration curve with configurational changes in the macromolecule.

## V. FLEXIBLE POLYELECTROLYTES

In turning to a consideration of the properties of flexible polyelectrolytes, a complicating feature appears. This is the change in mean dimensions of the molecule when the ionic strength or charge density is altered. As stated earlier, the change in configuration is caused by a readjustment between the contractile forces due to the Brownian motion of the stretched chain, a force of entropic origin, and the repulsive forces due to the electric charges. Using the procedures discussed in the preceding sections, it may be shown that the Helmholtz free energy of an independent polyion immersed in a medium of small ions is of the following form[25]:

$$A=A_1+A_2+A_3+A_4 \quad (34)$$

with

$$A_1=Z\alpha\mu_{H^+}+Z\alpha(1-f)\mu_{C^+}$$

$$A_2=-ZkT\ln\tfrac{1}{2}\left(\lambda e^{-\chi/kT}+1+\left[(\lambda e^{-\chi/kT}+1)^2-4\lambda(e^{-\chi/kT}-1)\right]^{\frac{1}{2}}\right)+Z\alpha' kT\ln\lambda-\frac{q^2\kappa Z\alpha'}{2D} \quad (35)$$

$$A_3=-Z\alpha(kT\ln K_a+\mu_{H^+}{}^0)+Z\alpha f(kT\ln K_s{}^0+\mu_{C^+}{}^0)-TS_m$$

$$A_4=-kT\ln Q^0-Z_k kT\ln\frac{\int e^{-u(\gamma)/kT}d\Omega}{4\pi}.$$

In Eqs. (35), $\alpha'$ is the degree of ionization of the macromolecule and $f$ the fraction of dissociated sites occupied by bound-ion pairs. The degree of neutralization then is related to the degree of ionization by the conservation condition $\alpha'=\alpha(1-f)$, and the fraction of bound sites referred to the total number of sites of the polymer will be $\alpha f$. $A_1$ describes the free energy of the free hydrogen ions and counterions eligible to take part in binding and neutralization phenomena. $A_2$ represents the electrical free energy of the net charge of the polyion, regarding the ion pairs as uncharged sites and thus includes the interaction energy between portions of the net charge of the polymer and the entropy associated with the mixture of charged and uncharged sites along the polymer chain. $A_3$ includes the chemical free energy associated with the state of ionization and binding of the polymer, computed from a reference state in which the polyion is in its completely undissociated form. Also included in $A_3$ is the free energy of mixing of the bound ion pairs and un-ionized groups among the uncharged sites of the chain. Finally, the last contribution, $A_4$, is the intrinsic free energy of the polymeric skeleton, a function of its configuration. Of the remaining undefined symbols, $S_m$ is the entropy of mixing uncharged ion-paired and un-ionized groups, $K_a$ and $K_s^0$ are the intrinsic dissociation constants of the acid- and ion-paired groups, $Z_k$ is the number of Kuhn statistical elements in the randomly coiled chain, and $u(\gamma)$ is the interaction potential between neighboring statistical elements making an angle $\gamma$ at their junction.

From Eqs. (34) and (35), the equilibrium root-mean-square end-to-end separation of the chain and the extent of counterion binding may be evaluated by straightforward minimization. The agreement between experiment and theory is semiquantitative, as can be seen in Figs. 2 and 3 which show the amount of binding and the expansion of sodium polyacrylate and sodium carboxymethyl cellulose, respectively. Lengthy discussion of the suitability of the ion-pairing concept can be found elsewhere, and space limitations prevent any amplification of these arguments. Suffice it to say that this author believes the concept to be useful and at least partially supported by experiment, whereas other explanations (such as ion trapping in the region where $e\psi/kT>1$) of the intimate association of counterions, with the polyion cannot explain all the relevant data.[29]

It is apparent that the permeability of the open network representing the expanded polymer coil permits the penetration of electrolyte and the consequent reduction of the electrostatic interactions between polyions. Several approaches[27,35–37] to the solution of the Poisson-Boltzmann equation have been published. Despite widely differing approximations, the conclusion of all investigators is that the domain occupied by the polyion is essentially neutral and polyion-polyion interactions, therefore, are dominated by the excluded volume-type interactions characteristic of neutral polymers. This is in complete agreement with the data of Schneider and Doty.[38]

From the microscopic point of view, there is a very important difference between solutions of polyions and solutions of ordinary electrolytes. In an ordinary electrolyte solution, the medium is essentially homogeneous, and the work required to bring an ion from infinity to **r** does not depend upon **r** (always neglecting boundary effects). In a solution of macroions, the internal region of any polyion is, roughly speaking, a solution of much higher ionic strength than the external medium. Thus, the work required to bring a charge to the point **r** depends upon whether **r** is within or without the volume occupied by the macroion. The extra work required in this case corresponds exactly to the difference in ion-atmosphere interaction corresponding to the two different concentrations of electrolyte.[27] Because of counterion binding, the potential inside a polyion is actually much smaller than would appear to be the case. In fact, marked deviations from a linearized Poisson-Boltzmann equation are expected only in the vicinity of the unit charges of the polymer, and these deviations are no more serious than those in solutions of small electrolytes. Solution of the linearized Poisson-Boltzmann equation (with ion pairing) shows that the internal and external interactions are differently screened, corresponding to the crude picture of a drop of concentrated electrolyte suspended in dilute electrolyte. Of course, when the charge density is small, or the polymer highly extended, the different screening lengths approach equality. Nevertheless, there are systems, such as polyelectrolyte gels, for which the difference is of great importance. The conclusions sketched above also may be arrived at from the diametrically opposed assumption that, to a first approximation, the polymeric volume is electrically neutral, and from calculating the small potential difference between the interior and the bulk medium from a Donnan equilibrium condition. That the same conclusions are reached when two such widely differing approximations are used lends support to the conclusions as general results.

While the forces which are operative in polyampho-

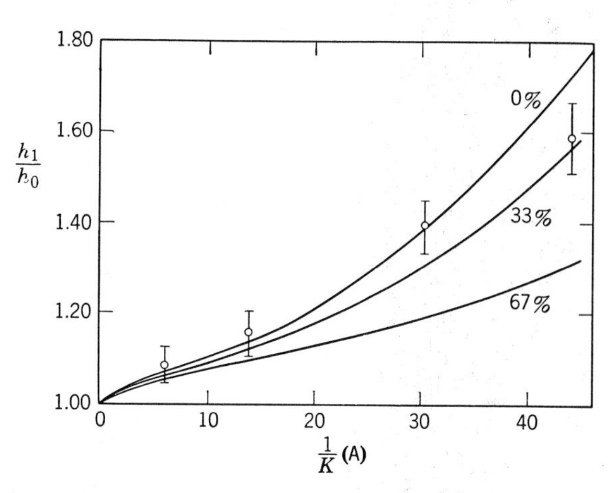


FIG. 3. Expansion of sodium carboxymethyl cellulose as a function of ionic strength. The curves are calculated assuming the amounts of counterion binding indicated. The circles are the experimental points of Schneider and Doty.[38]

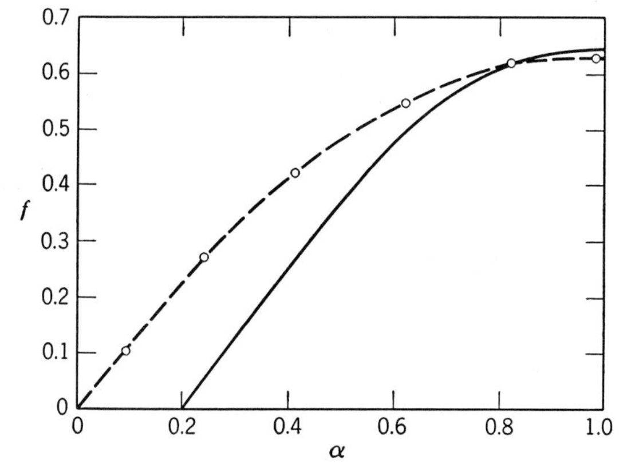


FIG. 2. Fraction of counterions bound to sodium polyacrylate as a function of the degree of neutralization. Solid line is theoretical curve. The dashed curve is from the experimental data of Huizenga, Grieger, and Wall.[21]

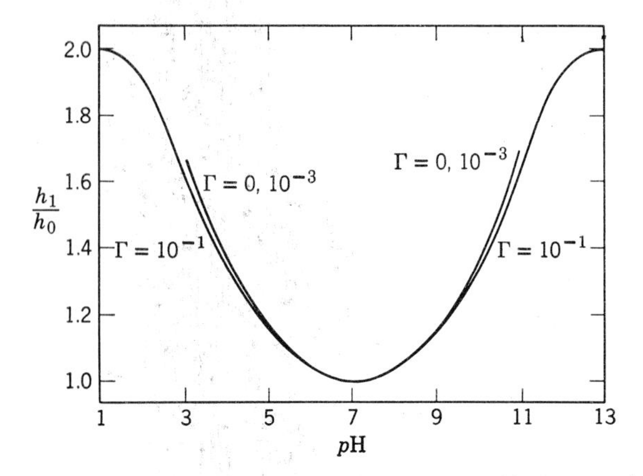


FIG. 4. Expansion as a function of $p$H for several ionic strengths for a regularly alternating copolymer; $K_1^0=10^{-5}$, $K_2^0=10^{-9}$, $K_3^0=K_4^0=10$, charge spacings 6 and 8 A for nearest and next nearest neighboring groups. For further details, see reference 20.

lyte systems are qualitatively no different from those in purely acidic or basic polyelectrolytes, significant new problems arise because of the inhomogeneity of the polymer skeleton. Obviously, far from the isoelectric point, the polyampholyte behaves as an ordinary polyelectrolyte. On the other hand, in the nearly isoelectric region where appreciable numbers of positive and negative charges can coexist, these complications dominate the situation.

One may proceed[30] to calculate the titration curve and other thermodynamic properties by exactly the same technique as used before, taking cognizance of both the existence of two types of charge and the skeletal distribution. At the isoelectric point, it is no longer true that all sites capable of bearing a charge are the same. Even though there is no net charge on the polymer, forces acting between the elements may arise from the two following distinct sources: (a) fluctuations of the charge of each statistical element about its mean value; (b) correlations in charge distribution because of the polymerization statistics.

From the properties of the Grand Partition Function, it may be readily shown that

$$\langle(q-\bar{q})^2\rangle = -\frac{1}{2.303Z_\kappa}\frac{1-\alpha_1'}{1-\alpha_1}\left(\frac{\partial \bar{Q}}{\partial pH}\right)_{a_{x^+},a_{y^-}} \tag{36}$$

for the mean-square charge on a statistical element due to fluctuations and for the hydrogen-ion activity at the isoelectric point,

$$a_{H^+} = \left[\frac{K_1^0K_2^0}{1+\dfrac{a_{x^+}}{a_{H^+}}\dfrac{K_1^0}{K_3^0}-\dfrac{a_{y^-}}{a_{H^+}}\dfrac{K_1^0}{K_4^0}}\right]^{\frac{1}{2}}. \tag{37}$$

The evaluation of the degree of dissociation of acid groups $\alpha_1$, of binding to acid groups, $\alpha_1'=\alpha_1(1-f)$, and the total charge $\bar{Q}$, requires specific consideration of skeletal structure. The dissociation constants $K_1^0$, $K_2^0$, $K_3^0$, $K_4^0$ are for the acid, base, acid-counterion, and base-counterion pairs.

Consider the simplest possible case, when the acid and base groups alternate regularly along the polymer skeleton. It is found that:

(a) The counterion binding does not cause a large change in the titration curve near the isoelectric point. This is due to the stabilizing influence of adjacent opposite charges. When the charge is predominantly of one sign, the polyampholyte behaves essentially like a one-component polyion.

(b) The expansion of the polyion is not strongly dependent upon the ionic strength at a given $p$H. This is due to the fact that as the ionic strength increases, tending to weaken the force expanding the polymer, the charge which can be supported on it also increases, since the group interactions are decreased. This latter tends to counterbalance the first effect.

(c) Since the binding of counterions is small, the effective net charge is essentially identical with the number of molecules of acid or base added to the solution. The effects of added salts are to shift the $p$H at any given degree of neutralization towards the isoelectric point, creating a curve characteristic of weaker acidic and basic groups. The primary source of this shift in the titration curve is due to electrostatic interaction among segments of the polymer and not to binding. This effect is, of course, similar to the observed effects in proteins.

(d) Effects of correlations due to polymerization statistics vanish identically, and those due to charge fluctuations are very small. These results are depicted in Figs. 4–6.

As a more general case, consider an equimolar polyampholyte, but with statistically distributed monomeric groups. This is a much more difficult problem. To

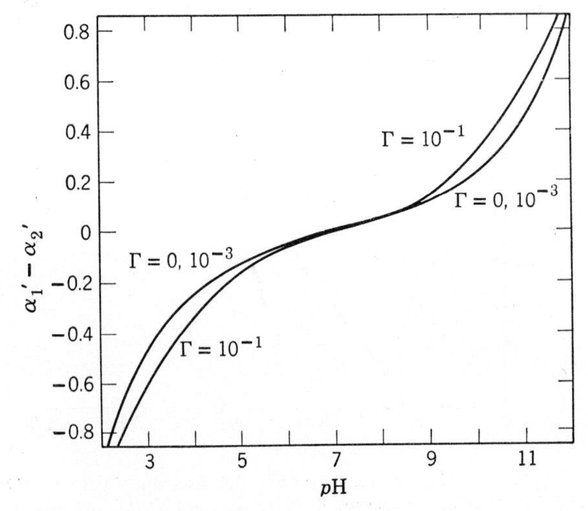


FIG. 5. Titration curve for the same regularly alternating polyampholyte as in Fig. 4.

proceed, the polymer is characterized by the number of groups of a given type which occur in a row. It must be recognized that those charge-bearing groups adjacent to groups of the same type have a tendency to remain uncharged as compared to groups surrounded by oppositely charged sites. If the polymer is regarded as a succession of acidic and basic groups, it is clear that, near the isoelectric point, almost all of the end groups are charged. Thus, the interior groups of the sequences of monomers may be regarded as independent to a good approximation. The problem now has been reduced to a combinatorial question of how many ways one can distribute a certain number of charges over the polymer.

The results of this computation are as follows:

(a) The titration curve has a smaller variation near the isoelectric point in *p*H with changes in the net charge of the polymer than has the corresponding regularly alternating polyampholyte. This means that the slope of the titration curve at the isoelectric point is steeper, and that the effect is due to the considerable number of charges situated in the interior of sequences and adjacent to charges of the same sign. These are, therefore, relatively loosely bound and a much smaller change in the chemical potential of the hydrogen ion results from their removal than from the removal of an equal number of the charges of a regularly alternating polyampholyte. The inverse statement is perhaps clearer; i.e., it takes a smaller change in *p*H to effect a change in charge for the randomly distributed monomers.

(b) A wide range of behavior is possible due to variations in structure. One obvious extreme is the block

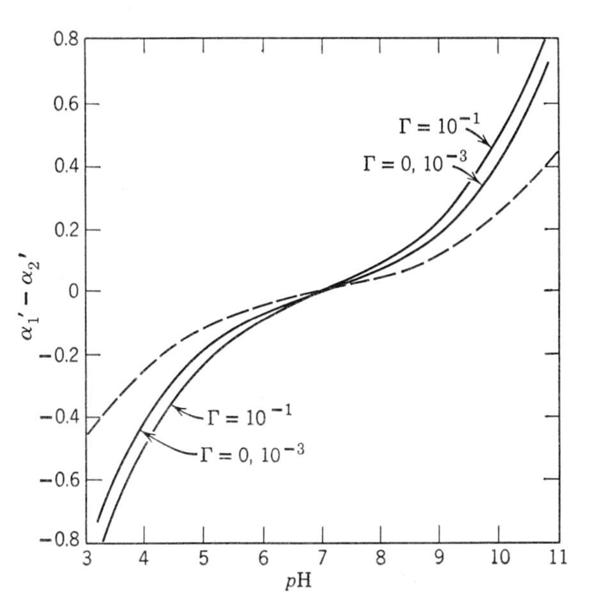


Fig. 6. Comparison of titration curves for a regularly alternating (broken line) and a randomly distributed (solid line) copolymer. The constants are the same as in Fig. 4.

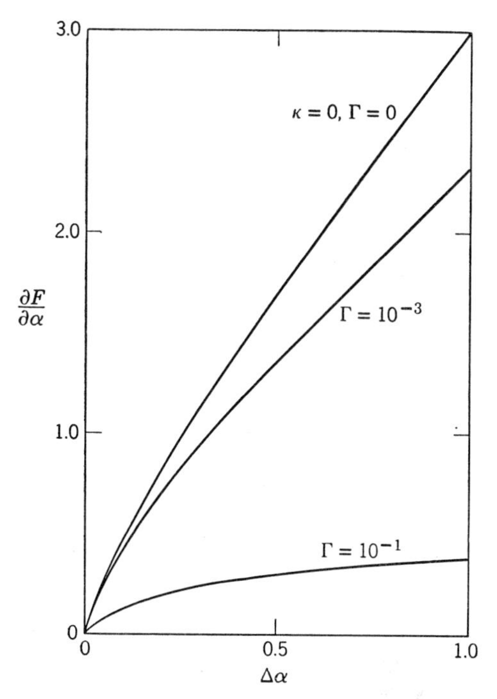


Fig. 7. Rate of change of expansion free energy with degree of neutralization as a function of net charge. The copolymer is the same as in Fig. 4.

copolymer with all acid groups at one end and base groups at the other end of the molecule.

(c) An increase in ionic strength, resulting in a decrease in interaction between charge groups, causes a shift of the titration curve in the direction of the behavior expected of independent acidic and basic groups.

(d) The configurational properties are almost identical with the corresponding regular copolymer of equivalent net segment charge.

(e) Charge fluctuation interactions are small due to the equimolarity of acidic and basic groups. This effect is large only when the numbers of one type of group exceed the other.

The properties of the regular and statistical polyampholytes are compared in Figs. 6 and 7.

## VI. HELIX-COIL TRANSITIONS AS INFLUENCED BY ELECTROSTATIC FORCES[39]

The fact that the electrostatic energy of a molecule depends upon the molecular configuration suggests that electrostatic forces can be the motive source for molecular transitions. Consider a simple model wherein several isomeric states of a molecule may exist, characterized by different electrostatic energies. For simplicity, assume that the distribution of charges and uncharged sites is random in all isomeric states and that ion binding effects may be neglected. Further, for simplicity, assume the reaction to proceed in an all-or-none fashion. If the electrostatic energy is computed

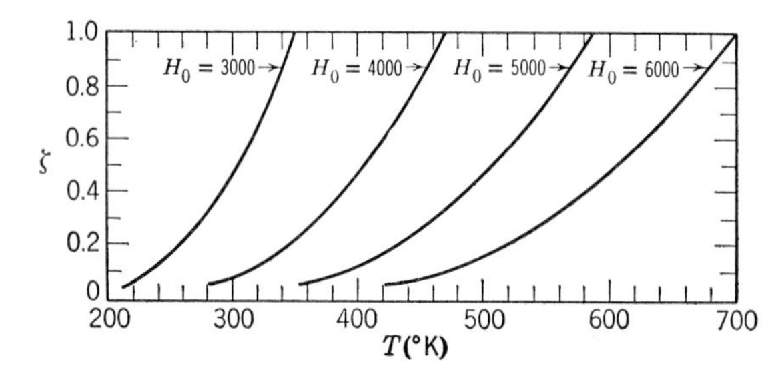


FIG. 8. Equilibrium fraction of bonds broken in DNA as a function of temperature. The broadening can be appreciably reduced by different choices of the parameters. For further details, see references 39 and 44–46.

for smeared-out charges, it is readily shown that

$$pH - \log\frac{\alpha'}{1-\alpha'} = 0.868 \frac{\alpha'}{Z} \frac{W_h^0 + W_c^0 \Theta \exp[-(\alpha')^2 \Delta W^0]}{1+\exp[-(\alpha')^2 \Delta W^0]} \quad (38)$$

with

$$\Delta W^0 = W_c^0 - W_h^0$$
$$\Theta = Q_c^0/Q_h^0$$
$$\frac{U_h^{\text{elec}}}{ZkT} = (\alpha')^2 W_h^0 \quad (39)$$
$$\frac{U_c^{\text{elec}}}{ZkT} = (\alpha')^2 W_c^0,$$

which defines the pair-interaction parameters $W_h^0$ and $W_c^0$, and $Q_c^0$ and $Q_h^0$ are the partition functions for the coil and helix without electrostatic interactions. The pair-interaction parameters are[40,41]

$$W_h^0 = \frac{Z\rho}{kT} \frac{4\pi^2 b^2 l_0}{D} \frac{K_0(\kappa b)}{\kappa b K_1(\kappa b)}$$
$$W_c^0 = \frac{Z^2 q^2}{DkT}\left[\frac{6}{\kappa h_0^2} - \left(\frac{8}{3\pi}\right)^{\frac{1}{2}} \frac{h_1}{2}\left(\frac{6}{\kappa h_0^2}\right)^2\right] \quad (40)$$

with $l_0$ the length of a segment, $b$ the radius of the helix, $K_0$ and $K_1$ are Bessel functions of the second kind and $\rho$ is the charge density on the helix surface. Correspondingly, $h_0$ and $h_1$ are the root-mean-square end-to-end separations of the coil in the absence and in the presence of electrostatic interactions, respectively. Equation (38) does not apply to any real helix-coil transition because of the assumption that the reaction does not involve intermediate states, an approximation known to be poor.[42–45] For the more general case, detailed calculations for the helix-coil transition in the thermal denaturation of DNA predict equilibrium bond breakage as a function of temperature as indicated in Fig. 8.[39] These calculations are for the case when electrostatic interactions are omitted. The effect of charge-charge interaction is to lower the enthalpy required to break bonds, $H_0$. In Fig. 8, this corresponds to moving from right to left across the lines of constant $H_0$. In Fig. 9 is a titration curve for DNA at two temperatures showing that the transition occurs titrimetrically at the same place as it is observed to occur viscometrically, calorimetrically, and by several other techniques.[46] The treatment can be made much more sophisticated by the introduction of discrete charges, ion binding, differences in charge between isomeric states, and other effects previously discussed, but the qualitative results are the same. A qualitative change does occur when intermediate states between pure helix and pure random coil are considered. The reader is referred elsewhere for details.[39,44]

## VII. SOME BRIEF REMARKS

Nowhere in this review are transport properties considered, a subject of great intrinsic interest. The treatment of, say, the conductance is even more complicated than of some equilibrium property due to the loss of elements of symmetry. Some progress has been made by Booth,[13] by Hermans,[47] and by Overbeek,[48] and the reader is referred to their papers.

In closing, the author wishes to emphasize that electrostatic forces only have been considered. In any

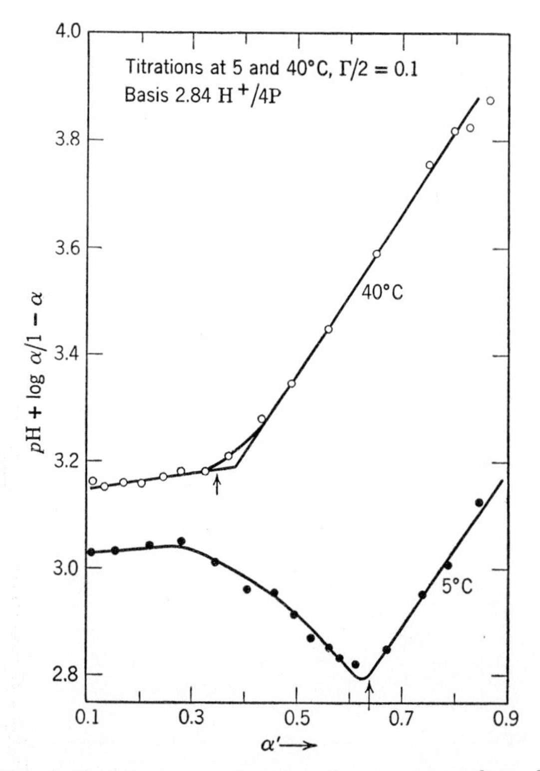


FIG. 9. Titration curve of sodium deoxypentosenucleate. The arrows indicate the points at which the helix-coil transition occurs calorimetrically and viscometrically. Note that we have plotted $pH+\log\alpha/(1-\alpha)$ instead of $pH-\log\alpha/(1-\alpha)$. This serves to sharpen the appearance of the transition.

real system, many other forces are operative. To name only two, consideration must be given to hydrogen bonds and to steric hindrances. Thus, in any real system, one must account *simultaneously* for the electrostatic effects discussed and for all other relevant phenomena. Separation of these from the electrostatic part is arbitrary and not always justified. It should, therefore, be borne in mind that the applicability of the theory discussed herein to any real system must be made with forethought and with circumspection.

## BIBLIOGRAPHY


[1] J. E. Mayer, J. Chem. Phys. **18**, 1426 (1950).
[2] J. G. Kirkwood and J. Poirier, J. Phys. Chem. **58**, 591 (1954).
[3] L. Onsager and S. Kim, J. Phys. Chem. **61**, 198 (1957).
[4] R. M. Fuoss and L. Onsager, J. Phys. Chem. **61**, 668 (1957).
[5] H. Falkenhagen, M. Leist, and G. Kelbg, Ann. Physik **11**, 51 (1952).
[6] E. Pitts, Proc. Roy. Soc. (London) **A217**, 43 (1953).
[7] J. G. Kirkwood, J. Chem. Phys. **14**, 51 (1946).
[8] See for example, P. Doty and G. Ehrlich, Ann. Rev. Phys. Chem. **3**, 81 (1952).
[9] E. J. Verwey and J. Th. G. Overbeek, *Theory of the Stability of Lyophobic Colloids* (Elsevier Publishing Company, New York, 1948).
[10] S. Levine, Proc. Roy. Soc. (London) **A170**, 165 (1939).
[11] S. Levine, Phil. Mag. **41**, 53 (1950).
[12] N. E. Hoskin and S. Levine, Phil. Trans. Roy. Soc. (London) **A248**, 449 (1956).
[13] F. Booth, Proc. Roy. Soc. (London) **A203**, 514, 533 (1950).
[14] J. G. Kirkwood and J. Mazur, J. Polymer Sci. **9**, 519 (1952).
[15] C. Tanford and J. G. Kirkwood, J. Am. Chem. Soc. **79**, 5333 (1957).
[16] J. G. Kirkwood and J. B. Shumaker, Proc. Nat. Acad. Sci. U. S. **38**, 863 (1952).
[17] J. G. Kirkwood, Discussions Faraday Soc. **20**, 78 (1955).
[18] C. Tanford, J. Phys. Chem. **59**, 788 (1955).
[19] W. Kuhn, O. Künzle, and A. Katchalsky, Helv. Chim. Acta **31**, 1994 (1948).
[20] J. Hermans and J. Th. G. Overbeek, Rec. trav. chim. **67**, 761 (1948).
[21] J. Huizenga, P. F. Grieger, and F. T. Wall, J. Am. Chem. Soc. **72**, 2636 (1950).
[22] P. J. Flory, J. Chem. Phys. **21**, 162 (1953).
[23] G. Kimball, M. Cutler, and H. Samelson, J. Phys. Chem. **56**, 57 (1952).
[24] F. Osawa, N. Imai, and I. Kagawa, J. Polymer Sci. **13**, 93 (1954).
[25] F. E. Harris and S. A. Rice, J. Phys. Chem. **58**, 725, 733 (1954).
[26] See, for example, P. J. Flory, *Principles of Polymer Chemistry* (Cornell University Press, Ithaca, 1953).
[27] F. E. Harris and S. A. Rice, J. Chem. Phys. **25**, 955 (1956).
[28] S. A. Rice and F. E. Harris, Z. physik. Chem. (Frankfort) **8**, 207 (1956).
[29] S. A. Rice, J. Am. Chem. Soc. **78**, 5247 (1956).
[30] S. A. Rice and F. E. Harris, J. Chem. Phys. **24**, 326, 336 (1956).
[31] R. Steiner, "Light scattering studies," Ph.D. dissertation, Department of Chemistry, Harvard University (1950).
[32] See for example, T. L. Hill, *Statistical Mechanics* (McGraw-Hill Book Company, Inc., New York, 1956).
[33] S. N. Timasheff, H. M. Dintzis, J. G. Kirkwood, and B. Coleman, J. Am. Chem. Soc. **79**, 782 (1957).
[34] S. Lowey, Ph.D. dissertation, Department of Chemistry, Yale University (1957).
[35] S. Lifson, J. Chem. Phys. **26**, 700, 727, 1356 (1957).
[36] F. T. Wall and J. Berkowitz, J. Chem. Phys. **26**, 114 (1957).
[37] M. Nagasawa and I. Kagawa, Bull. Chem. Soc. Japan. **30**, 961 (1957).
[38] N. S. Schneider and P. Doty, J. Phys. Chem. **58**, 762 (1954).
[39] S. A. Rice and A. Wada, J. Chem. Phys. **29**, 233 (1958).
[40] A. Katchalsky and S. Lifson, J. Polymer Sci. **11**, 409 (1953).
[41] T. L. Hill, Arch. Biochem. Biophys. **57**, 229 (1955).
[42] B. Zimm and J. Bragg, J. Chem. Phys. **28**, 1246 (1958).
[43] J. Gibbs and E. DiMarzio, J. Chem. Phys. **28**, 1247 (1958).
[44] B. H. Zimm, S. A. Rice, and A. Wada (to be published).
[45] S. A. Rice, A. Wada, and E. P. Geiduschek, Discussions Faraday Soc. **25**, 130 (1958).
[46] J. Sturtevant, S. A. Rice, and E. P. Geiduschek, Discussions Faraday Soc. **25**, 138 (1958).
[47] J. Hermans, J. Polymer Sci. **18**, 527 (1955).
[48] J. Th. G. Overbeek and D. Stigter, Rec. trav. chim. **75**, 543 (1956).

# 8
# Proteins and Their Interactions

David F. Waugh

*Department of Biology, Massachusetts Institute of Technology, Cambridge 39, Massachusetts*

THIS article, early in the series devoted to the soluble (globular) proteins, discusses briefly some of the properties of the native proteins, the main characteristics of protein denaturation, and a few examples illustrating specificity and complexity in the interactions between proteins. A model is first presented as a means for describing and correlating, in a general way, structure and properties.

At this point, one accepts at once the facts that globular (soluble) proteins are condensed structures, that they have many of the characteristics of molecules and, from their diffraction patterns, that portions of the main or backbone chain of atoms are arranged with considerable regularity. Of the possible arrangements for the latter, that of a helix has been most attractive. For some time, the $\alpha$-helix of Pauling and Corey[1] has provided a singular foundation for discussion. Several elegant helices were discovered by Pauling and Corey by assuming (1) that the planarity of the CONH grouping around the peptide bond should be preserved, and by assuming (2) the formation of a maximal number of hydrogen bonds between CO and NH groups separated by intervening peptide groups along the main chain. In the $\alpha$-helix, a compact coil of sufficiently high density results. The direction in which one finds the first carbon atom of the side chains for amino acids, other than proline or hydroxy proline, is indicated on the plan of Fig. 1 by the symbol R. A single turn of the $\alpha$-helix, in which van der Waals atomic radii are included to describe the outer limit of the plan, is shown in Fig. 2. For purposes of discussion, this plan is distorted into a circle of radius equal to 3.4 A. To this circle, one must attach the side chains. An analysis of a variety of proteins, undertaken a few years ago,[2] showed that the average side-chain extension was 5.1 A. Thus, the distance from the helix center to the end of the average extended side chain would be about 8.5 A.

At this stage, certain properties of the side chains should be examined since both internal structure and protein interactions are strongly dependent on these properties. Figure 3, taken from a review by Low,[3] shows the variations in size and configuration of the amino acids themselves. The side-chain configurations may be visualized by subtracting from each approximately the equivalent of glycine. The nonpolar amino acids have been grouped in the upper left, the upper center are those containing hydroxyl groups, while acidic and basic side-chain groups are in the upper and the lower right, respectively. Table I gives lengths and volumes for twenty common amino-acid side chains. An examination of the second column, which gives extended length, reveals that side chains vary from 1.5 A for glycine to 8.8 A for arginine. The volumes vary from 5.1 $A^3$ for glycine to 175.5 $A^3$ for tryptophan. Figure 4 shows that, in general, there is a closely linear relationship between side-chain length and volume except for the larger nonpolar side chains of valine, leucine, isoleucine, phenylalanine, tyrosine, and tryptophan. All of these are significantly *shorter* for the volumes occupied. I feel that the reason for this situation is to prevent, under ordinary circumstances, interactions of the larger nonpolar side chains. Thus, an alteration in structure generally is necessary to allow these side chains to come into contact.

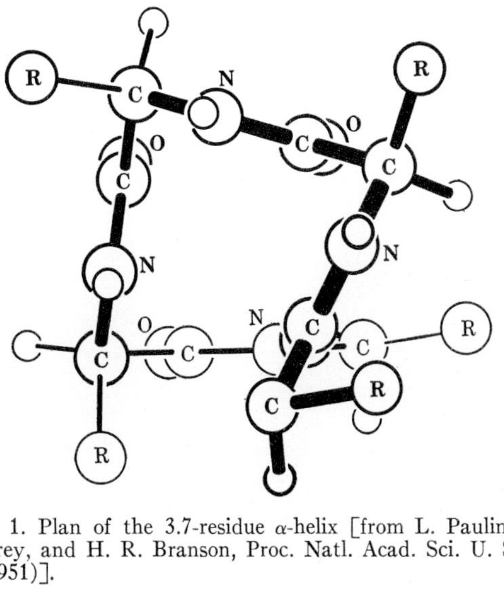

Fig. 1. Plan of the 3.7-residue $\alpha$-helix [from L. Pauling, R. B. Corey, and H. R. Branson, Proc. Natl. Acad. Sci. U. S. **37**, 205 (1951)].

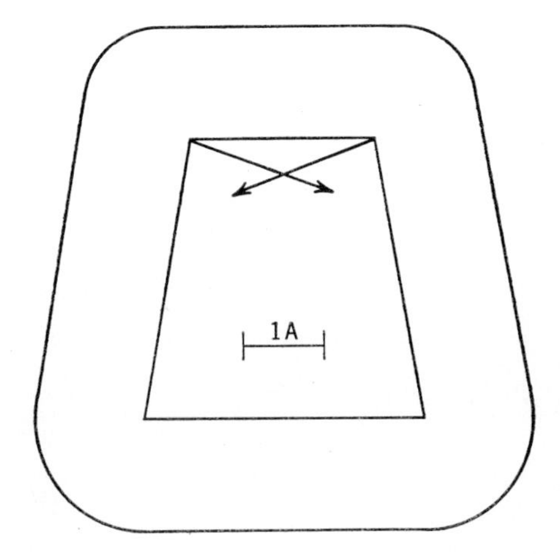


Fig. 2. Plan view of one turn of the $\alpha$-helix including van der Waals radii [from D. F. Waugh, Advances in Protein Chem. **9**, 326 (1954)].

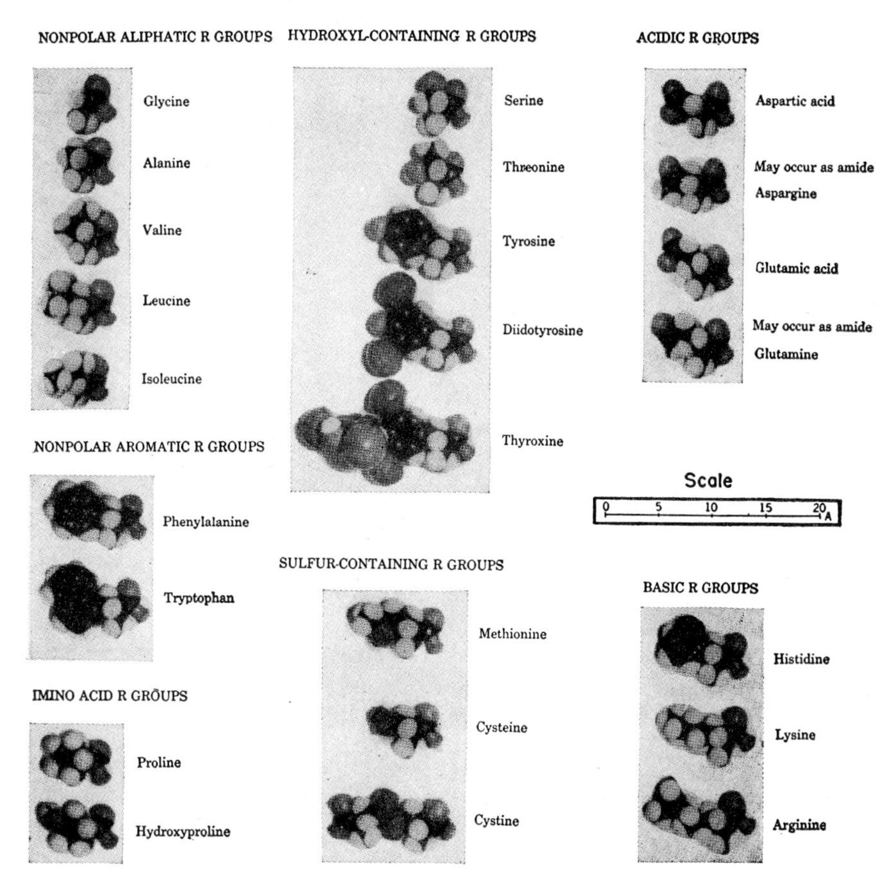


FIG. 3. Packing models of amino acids as dipolar ions [from B. W. Low in *The Proteins*, H. Neurath and K. Bailey, editors (Academic Press, Inc., New York, 1953), Vol. I, p. 235].

The coiled main chain is viewed as a tube, flat-sided in the case of the $\alpha$-helix, having protuberances of varying shape and volume. The shapes and spatial positions of the ring-containing side chains are fixed except for the freedom permitted by rotation around the bonds of the $CH_2$ group which, except for proline and hydroxy proline, joins the ring structure to the main chain. The positions of the protuberances will depend on the amino-acid sequence and the precise nature of the main-chain coil through that position.

In many, if not most proteins, the polypeptide chain occurs in segments or folds, and in the native molecule these segments or folds are approximated and bonded laterally. Certain of the lateral bonds may be covalent bonds, the primary focus of attention in this respect being the disulfide bonds of cystine. Others will be ionic interactions and secondary valence interactions such as hydrogen bonds and short-range van der Waals forc s. Comments about the interactions of chain segments in producing the protein molecule are, I believe, applicable with some modification to the interactions between protein molecules.

Figure 5 gives a diagrammatic representation of a cross section through a four-chain molecule. This molecule is constructed on the assumption that, at least for short distances, the chains are in the form of straight columns and that the center-to-center distance between chains is about 10 A, a value heretofore generally ac-

TABLE I. Characteristics of amino-acid side chains.

| Amino acid | Length max, A | Volume, A³ | Amino acid | Length max, A | Volume, A³ |
|---|---|---|---|---|---|
| Aspartic acid | 5.0 | 58.4 | Glycine | 1.5 | 5.1 |
| Asp. (amide) | 5.1 | 65.4 | Alanine | 2.8 | 32.2 |
| Glutamic acid | 6.3 | 85.5 | Serine | 3.8 | 36.0 |
| Glu. (amide) | 6.4 | 92.5 | Threonine | 4.0 | 63.1 |
| Arginine | 8.8 | 125.7 | Methionine | 6.9 | 112.1 |
| Histidine | 6.5 | 89.0 | Valine | 4.0 | 86.3 |
| Lysine | 7.7 | 121.0 | Leucine | 5.3 | 113.4 |
| Cystine ½ | 2.9 | 52.8 | Isoleucine | 5.3 | 113.4 |
| Cysteine | 4.3 | 57.9 | Phenylalanine | 6.9 | 136.6 |
| Tyrosine | 7.7 | 138.8 | Tryptophan | 8.1 | 175.5 |

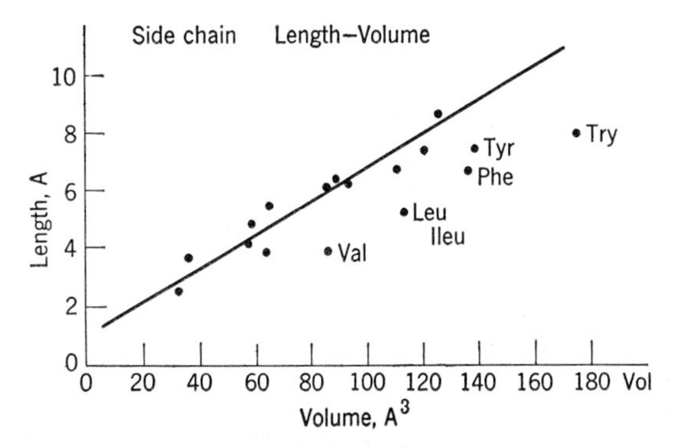


FIG. 4. Side-chain length *vs* volume.

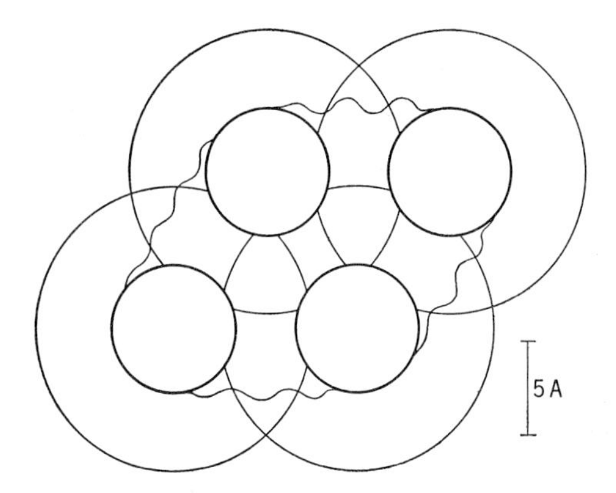


FIG. 5. Schematic representation of the interaction of four helices to produce an internal volume (enclosed by wavy lines) in which side chains are closely packed. Heavy circles are main-chain domains and light circles give maximum extension of average side chain [from D. F. Waugh, Advances in Protein Chem. 9, 326 (1954)].

cepted from crystallographic data. It is probable that the axes of the helices are not linear and that the center-to-center distance is not always 10 A, but the diagrammatic representation is examined for what it suggests as to the way in which the side chains are packed in the protein molecule.

Figure 5 shows heavy circles which are the main-chain domain, and light circles which give the average maximum side-chain extension calculated from the amino-acid composition of insulin. The wavy lines demarcate an internal volume, where side chains are closely packed, from surface regions where side chains are brought into contact with other side chains usually through interactions with other molecules. The side-chain volume enclosed within the wavy lines is expected to be crowded, and thus any regularity in the positioning of the main chain would necessitate a careful selection of the size, shape, and interaction properties of the side chains present on each of the four contributing helices. The precision of this selection is appreciated when an examination is made of the amount of unoccupied space in the protein molecule. By unoccupied space is meant a space which is too small to accept a water molecule or a larger space which is not accessible to the solvent. The maximum unoccupied space may be assessed as follows.

First, from a comparison of the measured specific volume of a protein with that calculated by Traubes' rules from the amino-acid composition, McMeekin and Marshall[4] find that the two values are in excellent agreement for a variety of proteins, as shown in Table II. As Edsall[5] has pointed out, however, the measured value should be about 3.5% lower than the calculated value because of the electrostriction of water around charged groups. The difference one may attribute to unoccupied space.

Second, Linderstrøm-Lang[6,7] has shown for several proteins that enzymatic cleavage of the first few peptide bonds leads to a volume decrement in excess of that predicted from electrostriction of water around the new charged groups produced. The excess volume decrement is about 3% of the molar volume.

The internal volume of the four-chain molecule shown in Fig. 5 is about one-third of the total volume; thus, if all of the unoccupied space occurs in the internal volume, about 10% of the latter will be unoccupied.

A rough calculation, based on the average side-chain volume and the frequency with which side chains are expected to occur in the internal volume, also suggests that roughly 10% of the internal volume will be unoccupied.[2]

The quantity of unoccupied space calculated in the foregoing is expected to be a maximum value. For example, disruption of the protein structure may alter the structure of the surrounding solvent so that a decrease in volume occurs. If such is the case, the unoccupied space in the internal volume would be less than the 10% already suggested, and as a result the side chains in the internal volume would necessarily be more closely packed. Linderstrøm-Lang[6] has pointed out that the protein molecule has some forced structure which makes it occupy more space than the unfolded elements. This forced structure is, I feel, the result of a compromise between uniformity in main-chain configuration and perfection in side-chain packing.

Here are a few additional remarks about the internal volume. In native proteins, it is the rule—rather than the exception—to find that a portion of groups such as sulfhydryl, disulfide, phenol, imidazol, etc., appear to be altered strikingly either in their physical or in their chemical properties. These have been referred to as hidden groups.

Interest in hidden groups has generally been associated with interest in the more general problems of protein denaturation (see the following), for denaturation, reversible or otherwise, gives at least a portion of the hidden groups a freedom to interact in more customary fashion.[8–10] Thus, such groups in native proteins may not react chemically as would be expected from

TABLE II. Specific volumes of proteins [from T. L. McMeekin and K. Marshall, Science 116, 142 (1952)].

| Protein | Observed volume cc/g | Calculated volume cc/g |
|---|---|---|
| Ribonuclease | 0.709 | 0.703 |
| Lysozyme | 0.722 | 0.717 |
| Fibrinogen (human) | 0.725 | 0.723 |
| α-casein | 0.728 | 0.725 |
| Chymotrypsinogen | 0.73 | 0.734 |
| Serum albumin (bovine) | 0.734 | 0.734 |
| Insulin | 0.735 | 0.724 |
| β-casein | 0.741 | 0.743 |
| Ovalbumin | 0.745 | 0.738 |
| Hemoglobin (horse) | 0.749 | 0.741 |
| β-lactoglobulin | 0.751 | 0.746 |
| Edestin | 0.744 | 0.719 |
| Botulinus toxin | 0.75 | 0.736 |
| Gelatin | 0.682 | 0.707 |

TABLE III. Average side-chain nonpolarity and nonpolar side-chain frequencies.

| Protein | Av side chain Equiv. $CH_2$ | Nonpolar class Equiv. $CH_2$ | Nonpolar class Occurrence frequency |
|---|---|---|---|
| Collagen | 1.33 | 4.06 | 0.21 |
| Rabbit tropomyosin | 2.15 | 4.32 | 0.25 |
| Ribonuclease | 2.16 | 4.33 | 0.26 |
| Rabbit myosin | 2.28 | 4.41 | 0.28 |
| Calf-liver histone | 2.29 | 4.13 | 0.31 |
| Chymotrypsinogen | 2.29 | 4.32 | 0.33 |
| Human fibrinogen | 2.31 | 4.53 | 0.32 |
| Ovalbumin | 2.33 | 4.32 | 0.36 |
| Bovine serum albumin | 2.37 | 4.74 | 0.32 |
| Edestin | 2.37 | 4.31 | 0.33 |
| Horse myoglobin | 2.39 | 4.32 | 0.36 |
| β-lactoglobin | 2.41 | 4.24 | 0.37 |
| FSH | 2.44 | 4.27 | 0.39 |
| Horse hemoglobin | 2.46 | 4.19 | 0.43 |
| Rat-sarcoma histone | 2.47 | 4.08 | 0.39 |
| Ox insulin | 2.52 | 4.27 | 0.45 |
| Human γ-globulin | 2.54 | 4.32 | 0.40 |
| Ox-growth hormone | 2.57 | 4.48 | 0.39 |
| Calf-thymus histone | 2.59 | 4.21 | 0.40 |
| α-casein | 2.59 | 4.38 | 0.40 |
| Human serum albumin | 2.61 | 4.33 | 0.40 |
| β-casein | 2.77 | 4.08 | 0.49 |
| Average | 2.36 | 4.30 | 0.35 |

the behavior of the same groups in smaller molecules, the ultraviolet absorption spectra may differ from those determined with less complicated molecules, and the acceptance of protons by basic groups may be retarded. Most of these effects must be associated with that portion of the molecular structure capable of bringing side chains into juxtaposition and of shielding either through secondary valence interactions or through steric hindrance. In a four-chain molecule, approximately half of the side chains will be involved in interactions within the internal volume, the other half will, to a large extent, define the configuration and surface properties of the molecule.

Examination of a series of proteins reveals that the amino acids are distributed as follows: 0.2 to 0.5 can carry charges (are, therefore, in the groups arginine, histidine, and lysine which may carry positive charge; aspartic acid, glutamic acid, cysteine, and tyrosine which may carry negative charge). The smaller amino acids such as glycine, alanine, threonine, and methionine make up 0.13 to 0.5 of the side chains, and the larger nonpolar amino acids of valine, leucine, isoleucine, proline, phenylalanine, tryptyophane, and tyrosine contribute 0.21 to 0.47 of the side chains. There appear to be no striking correlations between the average occurrences of various types of side chains, other than that as the frequency of one type increases, the frequencies of the other types decrease. However, the average side-chain volume may be kept relatively constant. This is suggested by the fact that the average side-chain volumes for a group of five proteins varied over a range of only 6% (Table V in reference 2).

The interactions of proteins with each other must, to a considerable extent, depend upon the characteristics of side chains although segments or regions of exposed main-chain groups might also be involved. A brief summary of the interaction characteristics of charged groups, hydrogen bonds, and van der Waals forces has been given.[2] Of particular interest are the following special possibilities. Kirkwood and Shumaker[11] show that the mobile protons on a group of particles, at *p*H values close to the isoelectric point and at low ionic strength, will effect charge patterns which will produce a net attractive force. At ionic strengths in the physiological range (~0.15) and at *p*H values removed by more than a *p*H unit from the isoelectric point, interactions of charges are expected to give rise to repulsion and the hydration of charged groups to a barrier which will modify the interactions of those groups dependent upon close approach, such as hydrogen-bond-forming groups and nonpolar groups.

The formation of an interprotein or intraprotein hydrogen bond is usually an exhange reaction involving the interacting groups and water. As shown by Pauling and Pressman[12] and by Schellman,[13] the resulting interaction energy is about 1 kcal/mole; thus, several such bonds, acting alone, would be required to give a stable protein-protein interaction.

The interactions of a few of the larger nonpolar groups with the concomitant formation of new hydrogen bonds in the released water, would suffice to form a stable intermolecular linkage. This conclusion stems from the fact that the association of hydrocarbon chains in an aqueous environment liberates about the same energy as the condensation of the hydrocarbon from the vapor state—namely, about 1.2 kcal/mole of $CH_2$ groups. Thus, the interaction, under the proper conditions, of a single $CH_2$ group may be equivalent to the formation of an interprotein hydrogen bond.[2] Important also, from the standpoint of internal structure and stability as well as of interaction, is the insensitivity of the nonpolar interaction itself to the *p*H or ionic strength of an aqueous environment.

The average, large, nonpolar side chain is the equivalent of about 4.2 $CH_2$ groups. Table III, which lists the frequency of occurrence of large nonpolar side chains for a group of proteins, shows an average value of 0.35, the minimum for corpuscular proteins being 0.25 for highly soluble tropomyosin and 0.49 for β-casein. It is clear that the interaction of a small fraction of the total number of the larger nonpolar side chains (e.g., 5 to 10 side chains) would be sufficient to produce an interaction product which would be stable over a wide range in *p*H near the isoelectric point. Some proteins, like serum albumin, are soluble at their isoelectric points and most are soluble to an appreciable extent at *p*H values one or two units away from the isoelectric point. It is thus apparent that the native-protein structure (or family of structures) positions most of the groups capable of attraction so that they are shielded; in other words, so that the interactions of attractive groups are opposed by energy-requiring processes such as (a) the approximation of groups carrying like charges, (b) the removal of water of hydration from dipolar or charged groups, (c) the breaking of hydrogen bonds between water and protein groups, and (d) a distortion of the helix (secondary) structure or side-chain interactions (tertiary structure[7]), etc. In this connection, it is significant that the nonpolar side chains are shorter per unit volume than other side chains and that the charged side chains, particularly those carrying positive charges, are somewhat longer. The addition of water of hydration to charged side chains would cause all of these to project beyond the values given in Table I.

The native protein is extracted by using as mild as possible a set of physical and chemical conditions, experience being required to define the sets of conditions which can be considered mild for a given protein. The protein is characterized, so far as is possible, also using mild conditions, the end result being a description of the protein in terms of its physical, chemical, and biological properties. If severe conditions of treatment are chosen (such as *p*H, temperature, the presence of urea, guanidine, salicylate, etc.), the properties of the native protein change. One of the most striking alterations is in solubility, and initially the term denaturation was based on the fact that almost all proteins become insoluble after a variety of severe treatments. In general, denaturation should not involve hydrolysis of protein covalent bonds. The importance of and interest in denaturation are indicated by the sequence of reviews which have treated this subject.[8–10,14,15]

It is instructive to examine a typical experiment, involving denaturation by alkali, in which case experience dictates an appropriate choice of *p*H, ionic strength, and temperature.

Figure 6, in the left-hand column, indicates that the native protein under ordinary conditions has a low net charge and that establishing denaturing conditions first produces a native protein of high net charge (upper left). With time, the properties of the protein population alter according to first-order kinetics to give reversibly denatured molecules also of high net charge. Under the conditions of denaturation, the protein usually remains in solution, aggregation and precipitation being prevented by the electrostatic energy barrier to close approach. That alterations in structure have occurred is demonstrated by taking an aliquot of the protein solution under denaturing conditions and, by diluting with an appropriate buffer, rapidly returning to a *p*H near the isoelectric point where the native protein is soluble. The fraction of the protein which has been denatured will precipitate. Once a precipitate is allowed to form, the precipitate is insoluble over a broad *p*H range including alkaline *p*H values where the denatured protein remains in solution on careful downward adjustment of the denaturing *p*H. Not only is this general, but if the precipitate is tested with respect to time, the limits of *p*H over which insolubility is observed are broadened, either indicating an increasing number of participating bonding groups or indicating that slow rearrangements take place which make existing group participations more effective. The insolubility of denatured proteins is attributed by many primarily to nonpolar interactions, for, since the driving energy is actually the formation of new water-hydrogen bonds, these are the group interactions which are particularly insensitive to changes in *p*H and ionic strength. Clearly, the conformations of the denatured molecules do not preserve the balance between attractive group interactions and energy-requiring processes discussed in the foregoing in connection with the native protein, alterations in the structures of the latter generally producing local patches of attractive groups which are no longer shielded.

That the denatured protein has a structure different from that of the native protein may also be shown under the denaturing conditions—for example, by changes in ultraviolet absorption, optical rotation, chemical reactivity of particular groups, and by applying techniques designed to examine size, shape, and hydration. Occasionally, determinations of biological activity may also be made under the conditions of denaturation.

If the protein is denatured as described—i.e., when the molecules are prevented from interacting by hydration and charge repulsion—a slow or stepwise return to conditions where the protein is native may lead to a reversal of denaturation, that is, to a return of solubility, biological activity, normal optical activity, etc.

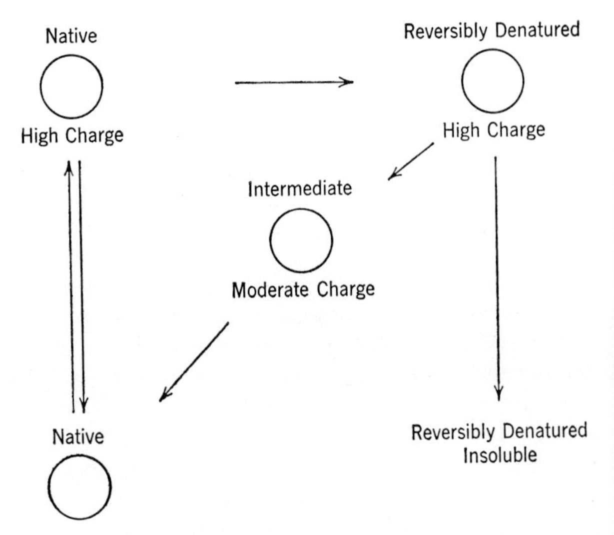


FIG. 6. Scheme showing the manipulations involved in demonstrating reversible denaturation by alkali.

A situation frequently encountered is illustrated by the central diagram of Fig. 6. An aliquot of the solution under denaturing conditions (reversibly denatured, high charge) is brought to a critical *p*H intermediate between the denaturing *p*H and a lower *p*H where the denatured protein will precipitate. At this intermediate *p*H, the molecules are somewhat expanded, the behavior of the system suggesting that a compromise has been reached: progressively below this narrowly defined *p*H, activation energy barriers to recovery increase (and thus effectively freeze the conformations of the denatured states), while above this *p*H, conformation changes corresponding to denaturation are increasingly favored. Clearly, the set of conformations which are established with time at the intermediate *p*H will, on reducing the *p*H further, tend to produce conformations corresponding to the native state. For the protein fibrinogen, denaturation is carried out at *p*H 12.4, ionic strength 0.15, and 0°C. Denaturation by a slow reaction is largely complete in 60 min. The critical intermediate *p*H is 10.8 and a treatment time of 16 hr at 0°C is necessary to effect a recovery of native properties. As is so often the case, however, the recovered protein resembles native protein in many gross aspects—for example, clottability after treatment with thrombin—but differs in detail—for example, in optical activity or solubility at *p*H values close to the isoelectric point. After reversal of denaturation, few proteins have been shown to be identical in all respects with the original native protein.

For many proteins, conditions may be chosen so that an equilibrium exists between the native and denatured states. For example, the enzyme trypsin[14,16,17] at a *p*H near *p*H 11 and 50°C shows a distribution between native and denatured states, for, on adding salt and rapidly altering the *p*H to *p*H 7, a fraction of the protein precipitates while a fraction remains soluble and exhibits enzymatic activity. The distribution is sensitive to temperature, the properties of the system suggesting a true equilibrium, a fact which allows a determination of the free energy, enthalpy, and entropy of the reaction. The data[18] for soybean trypsin inhibitor are shown in Table IV (see also reference 14). It is quite clear that a relatively small change in free energy is the result of a balance between large positive changes in enthalpy and entropy. Essentially the same result is obtained when other proteins showing temperature-sensitive denaturation equilibria are examined (trypsin, chymotrypsinogen, pepsin, luciferase; see reference 14). The

TABLE IV. Soybean trypsin inhibitor $pH=3$ $T=40°C$ [from M. Kunitz, J. Gen. Phys. **32**, 241 (1948)].

| | Thermodynamic | Rate parameters: Forward reaction | Rate parameters: Back reaction |
|---|---|---|---|
| $\Delta F$, kcal/mole | 1 | 25.4 | 24.4 |
| $\Delta H$, kcal/mole | 57.3 | 55.3 | − 1.9 |
| $\Delta S$, cal/mole/deg | 180.0 | 99.5 | −85 |

balance between large entropy and enthalpy changes (of either sign) to yield small free-energy changes is quite apparent also in the rate-defining parameters of the forward- and backward-activation reactions. So far, denaturation reactions have been too complicated to be analyzed in precise structure terms. The present concepts involve mechanisms in which there are:

(1) Alterations in proton association and attending water of hydration.

(2) Expansion (contraction) effects with concomitant (a) increasing (decreasing) entropies, and (b) increasing (decreasing) heats.

(3) Alterations in tertiary (side-chain) and secondary (main-chain coil) structure leading to an altered availability of surface side chains.

In the foregoing, reference has been made to the fact that interaction generally stabilizes the denatured state, the precipitate of denatured protein becoming more insoluble with time. Precipitation, however, does not necessarily lead to irreversible changes. R. H. Hartley and I have denatured fibrinogen by alkali and precipitated the denatured protein by rapid alteration in *p*H. The precipitate was recovered by centrifugation and redissolved in alkali at *p*H 12.4. It was then brought to *p*H 10.8 and allowed to stand for several hours at 0°C after which the *p*H was returned to *p*H 7. The majority of the protein was then found to be soluble and clottable with thrombin, although it differed somewhat from native fibrinogen in its precipitation at the isoelectric point.

So far, those interactions have been considered which, I feel, are generally accepted as involving, relatively, a disorganization of structure and thus the distribution of side chains. A well-organized site of interaction is clearly indicated in interactions between antibody molecules and hapten groups as antigens, a subject which has been extensively studied by Pauling, Campbell, Pressman, and their associates. For example, a protein such as serum albumin is coupled with the diazonium salt of a molecule to give substitutions such as azobenzenearsonates, (hydroxyphenyl azo) benzoates, azobenzoates, etc. Antibodies, found in the $\gamma$-globulin fraction of the plasma proteins, appear when these modified proteins are injected into animals. The antigens are found to combine relatively specifically with the coupled azo group.[19]

A variety of studies has been made of the combination of hapten with antibody and of the way in which various types of molecules inhibit the formation of an antigen-antibody precipitate.[20,21] The inhibiting action is most effective when the structure of the hapten is duplicated and becomes less effective as the inhibiting molecule differs in the positions of the groups attached to the benzene ring (the substituent groups), in the sizes of these groups (assuming similar chemical affinity), and in chemical affinity. The conclusion has been reached that the main-chain coils of the antibody molecule are arranged so that they enclose a cavity or slot which accommodates the hapten group and follows the contours of the latter within a few tenths of an Ångström unit. Interaction has been shown to involve whatever forces are permissible from the properties of the hapten: nonpolar forces if the hapten has no polar or dipolar groups, hydrogen bonds where donor (acceptor groups) are present on the hapten, and charge interactions when such can be made.

What data are available suggest that the free-energy change accompanying a reaction such as

$$\mathrm{Ag+AgAb \leftrightarrow Ag_2Ab}$$
$$\mathrm{Ab\ (ppt) \leftrightarrow Ab\ (sol.)}$$

is small and about 5 to 10 kcal/mole. There are uncertainties concerning changes in entropy and heat. Further considerations are given to antigen-antibody interactions in a later article by Kauzmann (p. 549).

The combination of a hapten-antigen and an antibody can be considered a "point" combination, unless the antibody is directed against the protein surface surrounding the hapten group. In other words, the interacting molecules may have some freedom of rotation around a single axis going through the locus of combination.

Interactions having a specificity which leads to the formation of complexes whose molecular subunits are held in particular positions and orientations with respect to each other are known to occur and some to have particular biological significance. Among these one would place at once the association reactions of insulin[22–24] and α-chymotrypsin,[25] the remarkable series of interactions leading to the high molecular-weight hemocyanins,[26,27] and the interactions of the caseins which lead to the spontaneous formation of stable colloidal micelles.[28,29] The interactions just mentioned lead to symmetrical complexes of limited size, the last two presumably so that they may contribute a relatively small increment in viscosity to the fluid media in which they are transported.

Other interactions lead essentially to extended structures of unlimited size, thus to fibrils. They are commonly referred to as globule-fibril (GF) transformations. Examples of proteins in this group are insulin,[30] tropomyosin,[31] actin,[32] fibrin,[33] and collagen, the latter being treated extensively by F. O. Schmitt (p. 349). These materials are arranged roughly in an order which reflects the increasing difficulty experienced in obtaining the globular (soluble) form of the protein. In addition, tissues such as collagen may contain nonprotein molecules which aid, somewhere in the hierarchy of aggregation, in producing the tissue structure: the tandon or fascia, etc.

The interactions of insulin serve as a specific example which illustrate a more general experience. As described by Oncley (p. 30), the insulin monomer is a unit of $M=5733$ crosslinked covalently by two disulfide linkages. The maximum, net positive charge at low *p*H values is 4 units/monomer, a value which should be obtained at *p*H$\sim$1, and can be obtained since insulin is stable in acid solution. The maximum, net negative charge would be 8 charges/monomer, although the instability of insulin in alkali and the high *p*H at which the two arginine side chains lose their protons would reduce this value to 14 for most experimental conditions. The association reactions of insulin have been carried out at *p*H 2 where each monomer carries an average, net positive charge of 3.5 units. The monomers of $M=5733$ form dimers of $M=11\,466$ so readily that the 5733 monomer is observed only in dilute solutions, after chemical modification, or in solvents such as acetic acid and pyridine (see Yphantis and Waugh[34] for references). Thus, the dimer of $M\sim12\,000$ is usually considered the interacting unit. These will associate readily in pairs or in groups of three.[22–24] Summarizing statements concerning reversible association would be (1) the electrostatic work involved in bringing molecules together is balanced against short-range attractive forces (hydrogen bonds, nonpolar interactions); (2) at *p*H=2, for pairing, $\Delta F_2=-4.93$ kcal/mole, $\Delta H_2=-8.1$ kcal/mole, and $\Delta S_2=-12.1$ cal/mole/deg; and (3) the entropy decrease of $-12$ eu is far less than the 122 eu which would be expected. Doty and Myers[23] point out that the freeing of water molecules from an association with polar groups would provide the requisite positive entropy change and suggest that 24 water molecules would be sufficient.

Insulin is also one of the smaller proteins which exhibits the GF transformation. It is chosen for discussion here since its kinetic and mechanistic complications may well foreshadow similar complications in other systems, just as the GF transformation with insulin was one of the first to be described. A recent summary has been given.[30]

The insulin fibril forms under conditions where the dimer is the prevalent form, i.e., at *p*H 2. Heating at 80 to 100°C causes a spontaneous transformation into a population of fibrils, the most numerous, and largest, of which are about 200 A in diameter and many thousands of Ångströms long. The reaction goes essentially to completion but is reversible in the sense that, under alkaline conditions, the association product, the fibril, disaggregates to yield insulin. Recovery of insulin after disaggregation is evidence that the insulin molecule does not undergo extensive unfolding in the process of forming fibrils. Stronger evidence, and evidence which gives a clue to the mechanism of fibril formation, comes from seeding experiments. In these, preformed fibrils or fibril segments are seeded into insulin solutions at *p*H 2.0. While such solutions alone are stable for long periods of time at temperatures of 20°C or below, the seeded fibrils or segments recruit insulin from solution and grow in the process. Structurally, the new portions of the fibrils obtained after fibril growth at lower temperatures appear to be identical with those which are formed at 80 to 100°C.

When a solution of insulin is heated and the transformation of insulin into fibrils is plotted as a first-order reaction (Fig. 7), the resulting curves typically have a lag period which is followed by an essentially linear rise. The extent of the lag period is determined markedly by the initial insulin concentration, as is the slope of the near-linear portion of the curve.

A comparison of the reaction kinetics observed (a) with the growth of seeded fibrils and (b) in the absence of seeding suggests that the fibril is first initiated by a nucleation reaction which involves the cooperative effects of $p$-interacting units (dimers) according to Eq. (1), where the differential is the rate of nucleation, $k_1$ is a rate constant, and $C$ is the insulin

$$dn/dt = k_1 C^p \tag{1}$$

concentration. The value of $p$ is near 3. After initiation, when the fibril has achieved a reasonable size, the fibril grows as a function of its surface area and the free

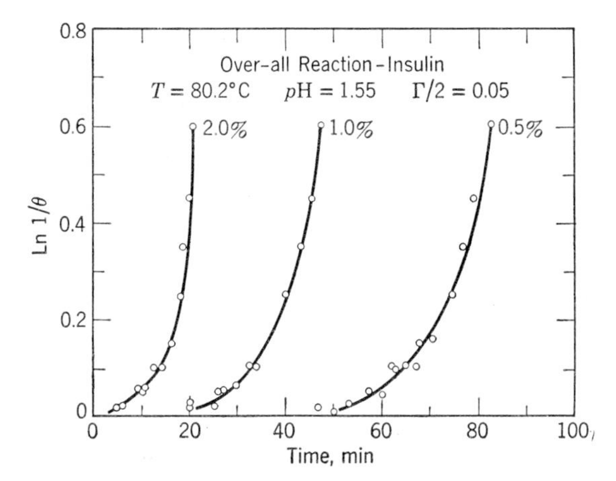


FIG. 7. Kinetics of the transformation of insulin into insulin fibrils. The percent protein has been indicated for each curve. $\theta$ represents the fraction of insulin remaining at time $t$ [from D. F. Waugh, J. Cellular Comp. Physiol. 49, Suppl. 1, 145 (1957)].

insulin concentration [Eq. (2)]. The cooperative effect established during nucleation is perpetuated, for the surfaces (particularly the ends) of the fibril present to the entering interaction unit the correct cooperative configuration necessary for bonding.

$$-(dc/dt) = k_2 \text{ area } C. \tag{2}$$

One particularly interesting consequence of this type of mechanism is that the majority of the fibrils are initiated during the lag period and those which are initiated during the first few minutes of the lag period dominate the reaction in the sense that they are responsible for removing the majority of the insulin. Thus, as shown in Fig. 8 which shows the numbers of fibrils in successive groups plotted against the radius of each group, the fibril population at the end of the reaction appears to be relatively homogeneous.

It is clear that, at least over a portion of its development, the fibril axial ratio increases with fibril mass.

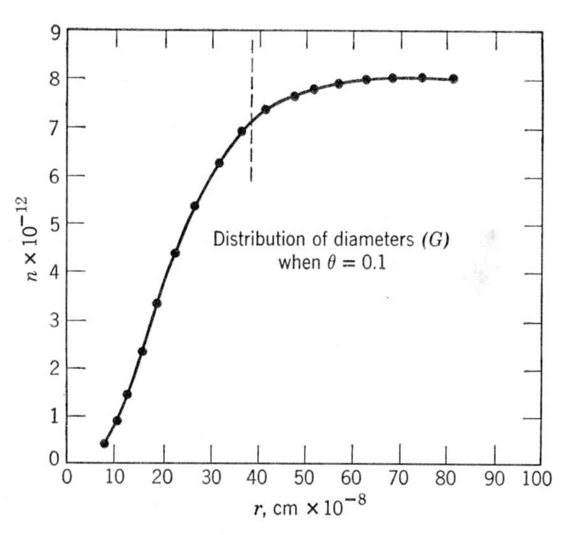


FIG. 8. Distribution of insulin-fibril diameters ($G$) near the end of a reaction involving 2% insulin when $\theta = 0.1$ [from D. F. Waugh, J. Cellular Comp. Physiol. 49, Suppl. 1, 145 (1957)].

That an asymmetric aggregate should form when charged molecules link, without specification of surface structure, has been pointed out by Rees.[35] This is owing to the fact that the electrostatic-potential barrier is lowest at the ends of the dimer or other asymmetric unit. However, the cooperative effect itself will exert a strong directional influence. This is illustrated in Fig. 9 which is a possible arrangement for the insulin-fibril nucleus. The stable structure is formed when any fourth soluble unit is added to the correct previous group of three. Thereafter, units are expected to add most frequently in a manner which perpetuates this stable

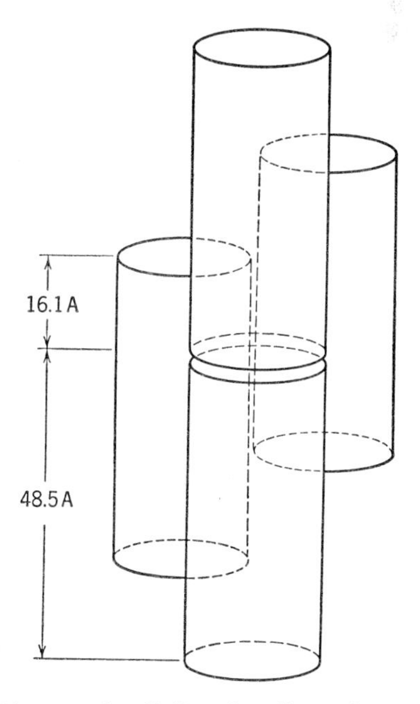


FIG. 9. Diagram of a fibril nucleus illustrating a cooperative interaction [from D. F. Waugh, J. Cellular Comp. Physiol. 49, Suppl. 1, 145 (1957)].

structure—consequently, the aggregate elongates in the direction of the cylinder axis of Fig. 9.

A variety of chemically modified insulins will also form insulin fibrils[36]: esterified and acetylated insulins, and those having nonpolar or polar groups added through coupling of insulin with substituted benzene-diazonium chlorides. Several different solvents also were used including concentrated urea, organic acids, and ethanol. The results of such studies led to the conclusion that the associations of nonpolar groups were primarily responsible for linking molecules. During the course of these investigations, it was observed also that the introduction of essentially nonpolar groups, for example by reacting benzene- or tolyl-diazonium chloride with the protein, led to insolubility. As the number of groups is increased, the progression is towards larger and more-stable polymers until the insulin itself becomes insoluble. A series of fibrils with increasing stability to disaggregation by alkali is formed also from these insulins.

We have been studying another system in which there occur both specificity of interaction and cooperative effects, the end result being a complex in which secondary valence and charge interactions arrive at a remarkable mutual satisfaction. Reference is made to the interactions of $\alpha_s$- and $\kappa$-caseins.[28,29] Here, $\alpha_s$-casein is a phosphoprotein of $M = 23\,000$ and has a phosphorus content of 1%. Physically, this protein appears to be a single coil 210 A long and 16 A in diameter. In the absence of calcium, the protein is a highly soluble polymer. The polymer becomes quite insoluble in the presence of calcium. $\kappa$-casein is also apparently a single coil about 150 A long, which contains sulfur but little phosphorus. It forms soluble condition-insensitive polymers of 13.5*S* in the presence or absence of calcium. When mixtures are made containing molecular ratios of $3\alpha_s$ to $1\kappa$, the original polymers disappear and a stoichiometric $\alpha_s$-$\kappa$-casein complex forms. This complex can form in the absence of calcium and under conditions where both of the monomers carry a high, net negative charge. One must assume that here, as in other cases, secondary valence forces are responsible for the association. In this connection, an examination of the amino-acid compositions of all of the caseins reveals a high-occurence frequency for the larger nonpolar amino acids mentioned at the start.

The $\alpha_s$-$\kappa$-complex, although it exists at all temperatures between 0 and 37°C, is unstable at the lower temperatures. This is apparent from the results obtained on adding calcium (to give 0.05 M). At the lower temperature, a Ca-$\alpha_s$-caseinate precipitate forms immediately, leaving $\kappa$-casein in solution. At the higher temperature, the complex is stabilized and will now engage in further aggregation to form stable colloidal particles, namely, micelles typical of milk. Once formed, the Ca-$\alpha_s$-$\kappa$-casein complex is unusually stable to heat, changes in *p*H, etc. We propose for it a structure in which the $\alpha_s$-casein molecules are oriented axially around a $\kappa$-casein molecule. At higher temperatures, secondary valence interactions hold the $\alpha_s$-casein molecules in positions such that pairs of phosphorus groups attached to adjacent $\alpha_s$-casein monomers are in juxtaposition and can be linked through the introduction of a calcium ion. The precision with which the phosphate groups are positioned is probably dependent upon the conformations of the interactants, for, at lower temperatures, the instability of the complex suggests that the phosphorus groups cannot be crosslinked by calcium. If each phosphorus is able to accept a calcium ion, stabilization through Ca-phosphate crosslinking would be absent, and, since Ca-$\alpha_s$-caseinate is relatively insoluble at low temperatures, it would precipitate and thus lead to a dissociation of complexes.

Another association which apparently does not involve covalent crosslinks, but does involve specificity of interaction and units of a higher asymmetry than any mentioned so far, is the collagen fibril. According to Boedtker and Doty,[37] the molecule is about 3000 A long and 13.6 A in diameter (hydrated), and is unusual in that it is made up of three polypeptide chains intercoiled to produce a coiled coil. The molecule contains unusually high percentages of hydroxyproline and glycine, and the occurrence frequency for larger nonpolar amino acids, including proline and hydroxyproline, is low (0.31). The collagen molecule denatures in the unusually low temperature range of 25 to 35°C during which process the molecule falls apart irreversibly into its constituent chains, which then fold into more symmetrical structures. A cooperative effect occurs when collagen molecules are linked together to form protofibrils, filaments, etc. (F. O. Schmitt, p. 349). That the cooperative effect exists is shown by the striking difference between the temperature of molecular denaturation and the temperature of thermal shrinkage, the uniform cross-striations, and the diffraction patterns of collagen, etc. Another result of the cooperative effect in collagen is the remarkably high tensile strength of wet collagen fibers. The latter is about 60 kg mm$^2$ for intact collagen and may be about 80% of this value for fibers reconstituted from collagen gels. These high tensile strengths, in the absence of intermolecular covalent bonds, must be the result of molecular overlapping which allows a maximum of interaction in the formation of secondary valence attractions.

## BIBLIOGRAPHY


[1] L. Pauling, R. B. Corey, and H. R. Branson, Proc. Natl. Acad. Sci. U. S. **37**, 205 (1951).

[2] D. F. Waugh, Advances in Protein Chem. **9**, 326 (1954).

[3] B. W. Low in *The Proteins*, H. Neurath and K. Bailey, editors (Academic Press, Inc., New York, 1953), Vol. I, p. 235.

[4] T. L. McMeekin and K. Marshall, Science **116**, 142 (1952).

[5] J. T. Edsall in *The Proteins*, H. Neurath and K. Bailey, editors (Academic Press, Inc., New York, 1953), Vol. I, p. 549.

[6] K. Linderstrøm-Lang, Cold Spring Harbor Symposia Quant. Biol. **14**, 117 (1950).

[7] K. Linderstrøm-Lang in *Lane Medical Lectures, 1951* (Stanford University Press, Stanford, California, 1952).

[8] H. Neurath, J. P. Greenstein, F. W. Putnam, and J. O. Erickson, Chem. Revs. **34**, 157 (1944).
[9] M. L. Anson, Advances in Protein Chem. **2**, 361 (1945).
[10] J. Steinhardt and E. M. Zaiser, Advances in Protein Chem. **10**, 151 (1955).
[11] J. G. Kirkwood and J. B. Shumaker, Proc. Natl. Acad. Sci. U. S. **38**, 863 (1952).
[12] L. Pauling and D. Pressman, J. Am. Chem. Soc. **67**, 1003 (1945).
[13] J. A. Schellman, Compt. rend. trav. Lab. Carlsberg. Sér. chim. **29**, 223 (1955). See also, W. F. Harrington and J. A. Schellman, *ibid.* **30**, 21 (1956).
[14] R. Lumry and H. Eyring, J. Phys. Chem. **58**, 110 (1954).
[15] F. W. Putman in *The Proteins*, H. Neurath and K. Bailey, editors (Academic Press, Inc., New York, 1953), Vol. I, p. 893.
[16] A. S. Stern, Ergeb. Enzymforsch. **7**, 1 (1938).
[17] M. L. Anson and A. E. Mirsky, J. Gen. Physiol. **17**, 393 (1934).
[18] M. Kunitz, J. Gen. Physiol. **31**, 241 (1948).
[19] K. Landsteiner, *Specificity of Serological Reactions* (Harvard University Press, Cambridge, 1945), revised edition.
[20] D. H. Campbell and N. Bulman in *Fortschritte der Chemie organischer Naturstoffe,* L. Zechmeister, editor (Springer-Verlag, Wien, 1952), Vol. IX, p. 443.
[21] D. Pressman in *Molecular Structure and Biological Specificity,* L. Pauling and H. Itano, editors (American Institute of Biological Sciences, Publication No. 2, Baltimore, Maryland, 1957), p. 1.
[22] J. L. Oncley, E. Ellenbogen, D. Giflin, and F. R. N. Gurd, J. Phys. Chem. **56**, 85 (1952).
[23] P. Doty and G. E. Myers, Discussions Faraday Soc. **13**, 51 (1953).
[24] R. F. Steiner, Arch. Biochem. Biophys. **44**, 120 (1953).
[25] G. W. Schwert and S. Kaufman, J. Biol. Chem. **190**, 807 (1952).
[26] I. Ericsson-Quensel and T. Svedberg, Biol. Bull. **71**, 498 (1936).
[27] A. Tiselius and F. L. Horsfall, Jr., J. Exptl. Med. **69**, 83 (1939).
[28] D. F. Waugh and P. H. Von Hippel, J. Am. Chem. Soc. **78**, 4576 (1956).
[29] D. F. Waugh, Discussions Faraday Soc. **25**, 186 (1958).
[30] D. F. Waugh, J. Cellular Comp. Physiol. **49**, Suppl. 1, 145 (1957).
[31] T. C. Tsao and K. Bailey, Discussions Faraday Soc. **13**, 145 (1953).
[32] S. V. Perry, Physiol. Rev. **36**, 1 (1956).
[33] H. Scheraga and M. Laskowski, Jr., Advances in Protein Chem. **12**, 1 (1957).
[34] D. A. Yphantis and D. F. Waugh, Biochim. et Biophys. Acta **26**, 218 (1957).
[35] A. L. G. Rees, J. Phys. & Colloid Chem. **55**, 1340 (1951).
[36] D. F. Waugh, D. F. Wilhelmson, S. L. Commerford, and M. L. Sackler, J. Am. Chem. Soc. **75**, 2592 (1953).
[37] H. Boedtker and P. Doty, J. Am. Chem. Soc. **78**, 4267 (1956).

9

# Three-Dimensional Structure of Globular Proteins

JOHN C. KENDREW

*Cavendish Laboratory, The University of Cambridge, Cambridge, England*

PRECEDING articles (Oncley, p. 30; Rich, p. 50; Waugh, p. 84) have presented protein molecules in various ways. With an eye to the properties they wished to describe, the authors have used conceptual schemes ranging from the straight polypeptide chain to the helical configurations, and finally to the unextended, bundled-up shape of the globular proteins. These different configurational aspects must be studied in different ways. Chemical methods are used to study the sequence of amino-acid residues in the polypeptide chain. Indications of the helical nature of the chain can be obtained, for example, by measuring optical rotations. Finally, the three-dimensional configuration of a globular protein, the detailed twisting and folding of a polypeptide chain to form a bundle of characteristic shape, can be determined only by x-ray diffraction.

Proteins are, of course, much more complicated than the molecules usually studied by the x-ray method. Use of this method began in 1913 with Bragg's determination of the structure of sodium chloride, which has just two atoms in its molecule. The most complicated complete structure so far determined—vitamin $B_{12}$, studied by D. Hodgkin—has 90 some atoms in its molecule (exclusive of hydrogens). Even the smallest proteins are much larger than vitamin $B_{12}$. Their great size and complexity would, in fact, soon discourage the crystallographer, if the importance of the problem were not so great; for it is not difficult to see that in such highly complicated molecules a knowledge of the three-dimensional configuration is vitally important for an understanding of physicochemical properties or biological function.

X-rays are scattered by electrons, and the amount of scattering of x-rays at any point in a crystal is proportional to the electron density at that point. A crystal is essentially a three-dimensional periodically repeating structure. The electron density plotted as a function of distance along a crystal axis is, therefore, periodic and can be analyzed into a series of harmonics, just as a musical note can be analyzed into its fundamental and overtone components. To carry the analogy further, a complicated sound can be reconstituted by a so-called Fourier synthesis or putting together of its component waves (Fourier components). In order to synthesize a musical note, one needs certain items of information about each of the harmonics: their frequencies, their amplitudes, and also their phases, that is, how much they are out of step with one another.

In x-ray diffraction, one must generalize this sort of argument into three dimensions. The complete diffraction pattern of a crystalline protein consists of a regular three-dimensional array of spots, any plane of which can be sampled by means of an x-ray photograph (Fig. 1). The great usefulness of the diffraction pattern lies in the fact that, for any crystal, each of the spots corresponds to a separate Fourier component of the periodically repeating electron-density distribution of that crystal. The complete distribution is the sum of all such waves, each having a characteristic amplitude, wavelength and direction relative to the crystal axes, and phase. The length and direction of the wave can be determined directly from the position of the spot on the photograph and the geometry of the x-ray camera, and the amplitude can easily be inferred from the blackness of the spot on the photograph. It is the determination of the remaining parameter, the relative phase of the spot, which presents the biggest difficulties in x-ray crystallography, for no direct physical method of phase determination is available.

In the study of simpler structures, phases are usually found by trial-and-error methods. The crystallographer guesses the atomic positions before he begins the analysis. If his guess is good enough, he can predict the phase relationships and then begin a mathematical refinement process in which he gradually distorts the proposed model until the diffraction pattern predicted from it matches that obtained experimentally. He does this by minimizing certain error functions which express the discrepancies between the observed and predicted pat-

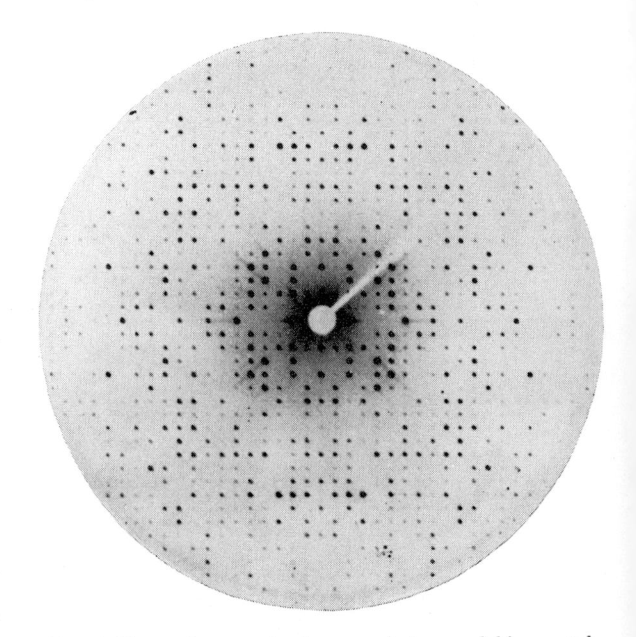

FIG. 1. X-ray photograph of sperm-whale myoglobin crystal.

terns. In general, the method can yield any one of a number of solutions, and the correct one will be found only if the initial guess was a fairly good one. For complex molecules, such as proteins, the method has been found to be useless, for the number of possible solutions in such cases is so great that there is negligible probability that any initial guess as to the positions of all of the atoms is sufficiently near the truth to be refined into the correct solution.

Fortunately, there are a few tricks that the crystallographer can use for determining phases in more complicated structures. One such trick, the method of isomorphous replacement, is to replace an atom or small group in the molecule by a very heavy atom or group. If the replacement causes an appreciable change in the diffraction pattern; if the positions of the heavy groups in the unit cell are known; and if the replacement causes no distortion of the original crystal structure, it is then possible to determine phases by comparing diffraction patterns before and after replacement.

The isomorphous-replacement method was first used successfully in protein crystallography in 1953, when M. F. Perutz applied it to horse hemoglobin. This protein molecule contains sulfhydryl groups which can react specifically with reagents such as *p*-chloromercuribenzoic acid (PCMB), and Perutz found a definite change in the diffraction pattern when the reaction took place. For the method to work, it is essential that only a small number of heavy groups are attached to each molecule and that they attach at specific sites. When these conditions are met, one can establish the positions of the heavy atoms in the unit cell by straightforward methods. From the weights of the heavy groups and from their positions, one can calculate their vectorial contribution (i.e., both phase and amplitude) to the harmonic wave represented by each of the spots in the diffraction pattern. Corresponding waves for the protein alone, and for protein plus heavy atom, are represented by vectors of known (experimentally determined) amplitudes, but unknown phase angles; however, since the vectorial difference between them (i.e., the contribution of the heavy groups) is completely determined, one can solve for their phase angles geometrically by finding an arrangement in which the two vectors of specified length form a triangle with the difference vector. None of these operations requires any guesswork.

Perutz determined some phase angles for hemoglobin in this way and proceeded to calculate a two-dimensional Fourier synthesis. He confined his attention to a small class of reflections present in the diffraction pattern of hemoglobin (and most other crystals) where the symmetry of the crystal restricts the phase angles to values of either 0 or $\pi$ radians. When the Fourier synthesis is carried out with just these reflections, a plane projection of the three-dimensional electron-density distribution is obtained. By working in two

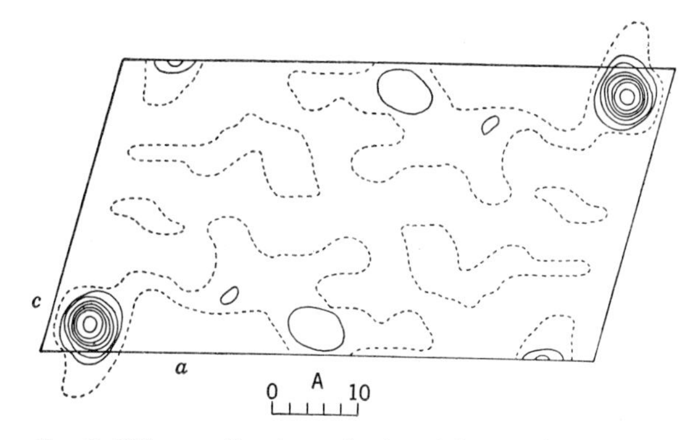


FIG. 2. Difference Fourier projection of the complex of mercury diammine with myoglobin. The unit cell contains two protein molecules; the two peaks indicate that one mercury atom is attached to each molecule.

dimensions instead of three, he achieved a twofold simplification: only a few of the reflections had to be considered, and the phases of these were relatively easy to determine. It is obviously easier to determine a variable with just two possible values than it is to determine one which may have any value.

By putting other heavy atoms in different positions in the molecule, Perutz was able to redetermine the same phases by an independent set of calculations. The results were the same, showing that his projection of hemoglobin was quite certainly correct; unfortunately, however, it proved impossible to interpret in terms of chemical structure. The unit cell is about 40 atoms thick, and in the projection all of these atoms came on top of one another, so that the features of the molecule obscured one another in a hopeless confusion. Obviously, the only way of getting useful information was to extend this analysis into three dimensions.

I now turn to our own studies of myoglobin, which like those of hemoglobin began with a two-dimensional analysis. We chose myoglobin for our study because it is a small protein, having a molecular weight of only 18 000. The molecule consists of 153 amino-acid residues, all arranged, so far as we know, in a single polypeptide chain to which is attached one heme group. This group is common to hemoglobin and myoglobin, and has the property of reversible combination with oxygen. While hemoglobin is used for transporting oxygen in the blood stream, myoglobin is intracellular and its function is to act as a temporary store of oxygen. Though myoglobin is present in all mammals, it is especially abundant in animals such as whales which spend a great amount of time under water and must store oxygen for long periods.

There are several heavy groups which form good isomorphous-replacement compounds with myoglobin. In order to find out whether a heavy group has attached to specific sites in the myoglobin crystal, one takes an x-ray diffraction picture of the protein with heavy group added and another of the protein alone. One then, as it were, subtracts one diffraction pattern from the other

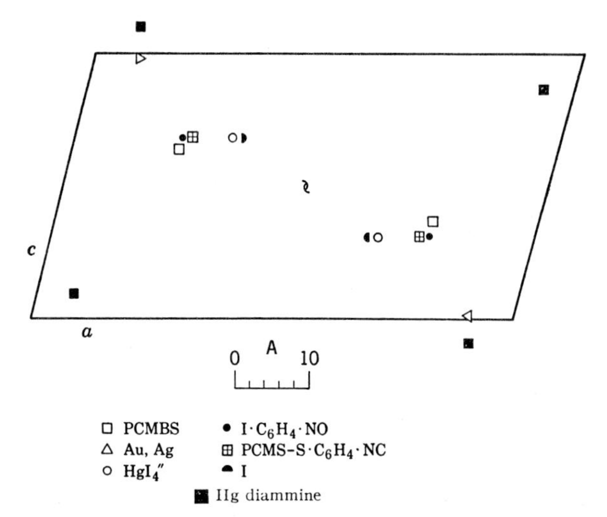


FIG. 3. Sites at which heavy atoms have been attached to the myoglobin molecule. In each case, these are two sites per unit cell, corresponding to one site per molecule.

and carries out a Fourier synthesis in such a way that only the contribution of the heavy group appears in the projected electron-density contour map. If a good isomorphous replacement has been achieved, this two-dimensional difference Fourier synthesis of the unit cell will show a few high peaks and very little else (Fig. 2). A nonspecific absorption of the heavy group or a deformation of the protein would cause the appearance of significant "background" features in the map.

We find, by the foregoing criterion, that myoglobin forms good isomorphous-replacement compounds with mercuri-iodide ion, gold-tetrachloride ion, silver ion, *p*-chloromercuribenzene sulfonate, and mercury diammine. The gold-tetrachloride and silver ions appear to attach at the same site in the molecule. The other groups all go to different sites (Fig. 3). In most of these replacements, we have no idea as to the chemical nature of the attachment, and indeed there are several unexpected features of our results. Thus *p*-chloromercuribenzene sulfonate is known as a reagent for sulfhydryl groups, yet it combines quite specifically to myoglobin which has no free sulfhydryl groups. We do not understand this, nor do we understand why the positively charged silver ion and the negatively charged gold-tetrachloride ion go to the same site in the molecule. There is a great need for some good chemical work on the rationale of combining heavy atoms with proteins. Our approach, however, has simply been to try various reagents and judge our success from the x-ray pictures.

In two-dimensional work, only one replacement compound is really needed, but it is useful to have others because a double check is then possible. For our two-dimensional Fourier projection, we have checked the phases of all of the reflections out to a resolution of about 2 A. We get the same answer with each of four isomorphous compounds, except for a few reflections where the heavy atom contribution is so small that we cannot be sure of its direction. Our two-dimensional Fourier projection, cross-checked four times, was therefore as reliable as Perutz's hemoglobin projection but just as unintelligible (Fig. 4). We were unable to deduce anything about the structure of the myoglobin molecule by looking at it, although now that we have a three-dimensional synthesis we are able to understand in retrospect all of the features of the projection.

We have already noted some of the factors that make it more difficult to work in three dimensions than in two. Besides these difficulties, it turns out that in three dimensions the phases cannot be determined unambiguously if there is only one replacement compound available. On attempting to draw the vector triangle, one finds that there are two possible phase angles consistent with the data for each reflection. However, if the calculations can be done for each of two different replacement compounds, the method yields two pairs of answers of which one from each pair should agree, and these give the correct value of the phase angle. In practice, because of experimental errors and other difficulties, two replacement compounds are hardly sufficient for determining all of the phases. Three compounds or even more are highly desirable.

Before embarking on our three-dimensional synthesis, we had some difficult decisions to make. The amount of work required for such a project goes up very steeply with the degree of resolution sought in the final Fourier synthesis. For myoglobin, we estimated that in order to get a resolution of 6 A one would have to solve the phases of about 400 reflections, while for a resolution of 2 A one would have to treat some 10 000 reflections; to discern the individual atoms in myoglobin, one would need a resolution of at least 1.5 A, and that corresponds to about 20 000 reflections. At this point, the diffraction pattern of a myoglobin crystal begins to fade away altogether. In an ordinary crystal, the diffraction pattern goes on further than this, but a protein crystal is more disordered, and it will probably be touch-and-go as to whether or not we can resolve directly the individual atoms in protein crystals by any method. At the

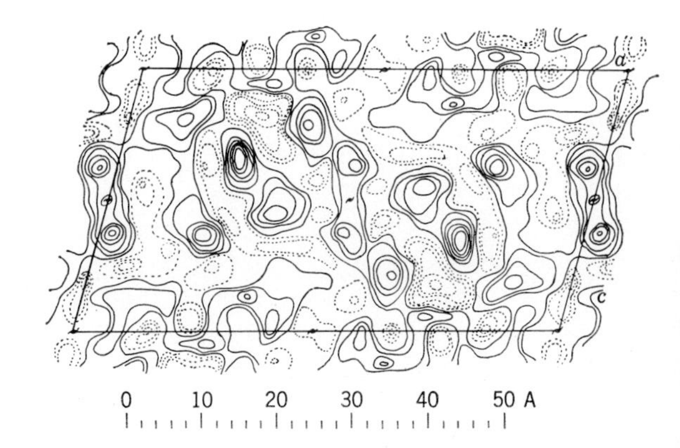


FIG. 4. Two-dimensional Fourier projection of the myoglobin unit cell. The highest peaks in the projection are owing to the iron atoms of the heme group.

FIG. 5. Three-dimensional Fourier synthesis of the myoglobin unit cell. Some of the rod-like polypeptide chains can be seen.

present time, however, these theoretical limits of resolution are of academic interest only, and in fact, because the amount of labor goes up so rapidly with increasing resolution, we decided in the first instance to try out the isomorphous-replacement method at the lowest resolution that was likely to give us useful and interesting structural information.

How were we to decide what resolution would be useful? Here we had to bring in preconceived notions about protein structure, and we adopted a working hypothesis which would nowadays be accepted by most workers as being at least plausible—that $\alpha$-helices are somehow involved in the protein molecule. There is some experimental evidence that this is the case, though it is not of an absolutely conclusive kind. An $\alpha$-helix is a very dense object, and at low resolution it should appear as a rod of electron density about 1 electron/$A^3$. The side chains sticking out from the helix are more open structures, and they should appear as regions of density about 0.29 electron/$A^3$, surrounding the central rod. $\alpha$-helices would be expected to pack together so that each is about 9 to 10 A from its neighbors. It was decided, therefore, to calculate the first Fourier synthesis at 6 A resolution, when the $\alpha$-helices, if present, should be clearly distinguishable.

The process by which we finally arrived at the Fourier synthesis is not detailed here. Much of the calculation was done with a high-speed computer. The method is analogous to that used in two dimensions; it requires no guesswork, and the phases of all reflections can be cross-checked. We are, therefore, confident that our three-dimensional electron-density distribution is essentially correct, although it is somewhat blurred by its low resolution, and must contain a small background because of experimental error.

When the Fourier synthesis was finished, we had the problem of making a suitable geometrical representation of this three-dimensional object. We first plotted a series of electron-density contour maps for 16 parallel planes at different levels in the unit cell, and then traced the contours of these onto transparent sheets of Lucite, so that by stacking two or more of them together one could see how the contour surfaces were arranged in three dimensions (Fig. 5). At this point, we could see the two differently oriented myoglobin molecules which comprise the unit cell of the crystal, their positions being related to one another by the symmetry of the crystal (a screw dyad axis). We could also see rod-like features of high density within the molecules, and indeed except for a few regions of uncertainty we could follow a continuous path of high electron density throughout the entire molecule. Thus, we were able to make a model

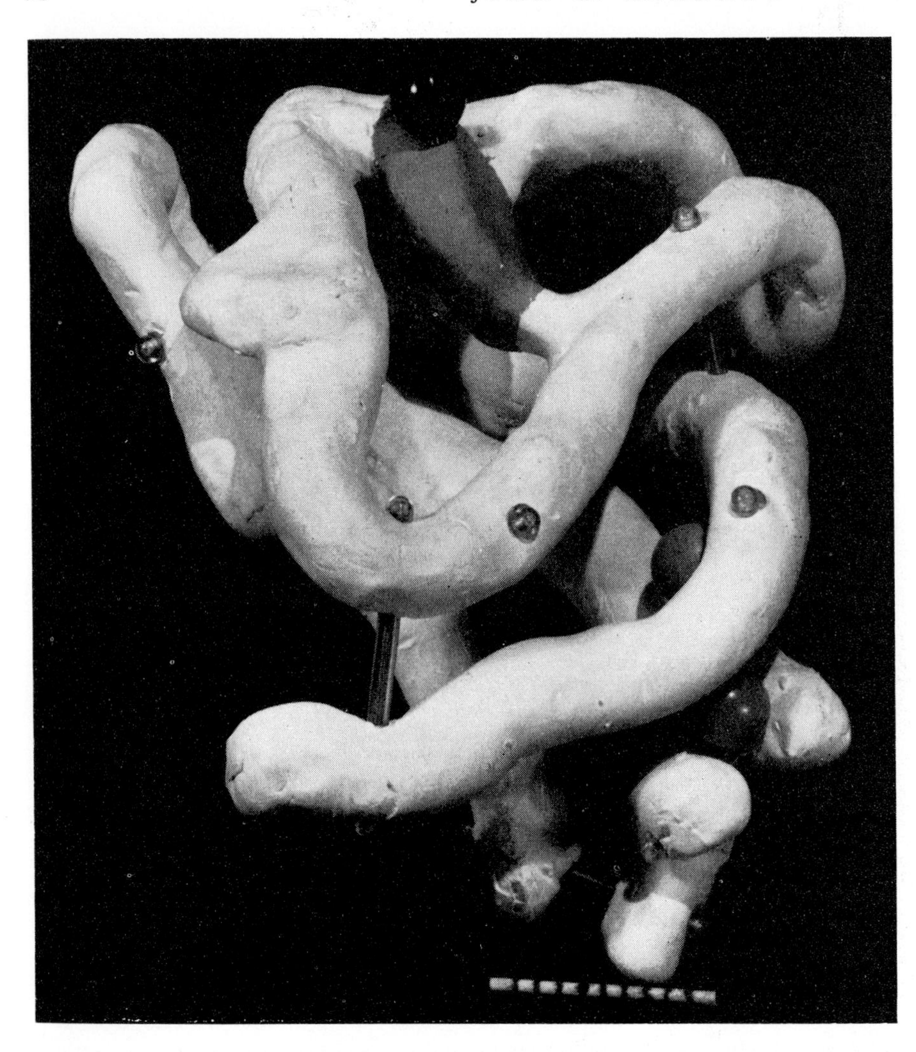

FIG. 6. Model of the myoglobin molecule. The grey disk seen nearly edge-on is the heme group; the small spheres are heavy atoms attached to the molecule. The marks on the scale are 1 A apart.

to represent the protein molecule consisting of one long rod of modeling clay or thermo-setting plastic, molded to the shape of the rod of high electron density. The rod is irregularly coiled to form a bundle measuring about 45 A by 35 A by 25 A. On looking at the model, one is impressed by its total lack of symmetry, and by the absence of parallel lengths of rod (Fig. 6).

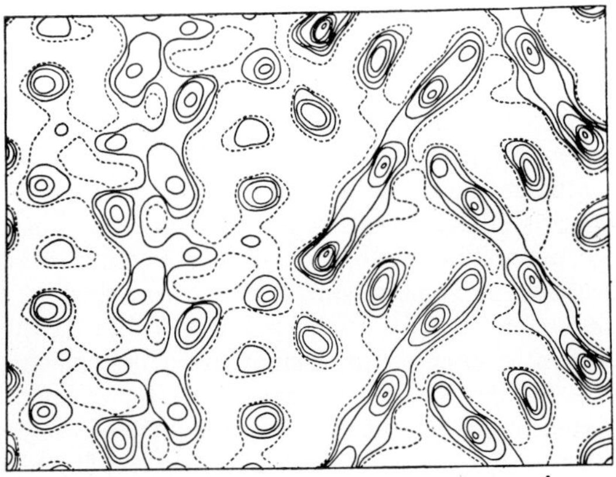

2b

0 A 10

Section parallel to [20$\bar{1}$] at $x = 0$

FIG. 7. Section through the three-dimensional Fourier synthesis showing, on the right, straight rods 40 A long.

The rod has segments over which it is nearly straight for distances of 20, 30, or even 40 A (Fig. 7). Adjacent segments of rods are always at least 9 to 10 A apart. The electron density at the rod axis is about 1 electron/$A^3$. Thus, in all respects, the rod seems to fit the specifications of an $\alpha$-helix, and we have little doubt that it is in fact the polypeptide chain. We cannot yet be sure, however, whether the configuration of the chain is that of an $\alpha$-helix, though it must be something fairly closely resembling this in gross dimensions.

From chemical evidence, the molecule is thought to consist of just one chain, and that is the way we represent it in our solid models. But, in following the high-density streak through the molecule, one finds regions where its path cannot be traced unambiguously. These are the places where we think the rod is sharply bent. The presence of a bend would imply an interruption in the very dense helical structure, and would explain our difficulty in tracing it through such a region. A simple calculation shows that, in any case, the entire chain cannot be coiled as tightly into an $\alpha$-helix, for the total length of rod appearing in our model is 300 A, whereas if the molecule consisted entirely of $\alpha$-helix, the rod would be about 230 A long. If we make the obviously oversimplified assumption that part of the molecule has the $\alpha$-helix form, and that the rest of it is fully extended, then the measured length of the rod requires that about 70% of it be in the helical form. This figure agrees with other estimates made by the optical-rotation method and by the deuterium-exchange method.

The heme group of myoglobin can also be seen in our Fourier synthesis. This group was easy to identify because of its iron atom, by far the heaviest atom in the protein molecule. It produced a peak of electron density 50% higher than any other feature in the molecule. Though we cannot resolve the porphyrin-ring system, the heme group appears flattened in one direction, and this direction corresponds very well with what we already knew of the orientation of the heme group from optical and magnetic studies of myoglobin crystals. In order to make our identification of this group more certain, we treated the protein with a reagent, *p*-iodophenyl hydroxylamine, which is known to react with the myoglobin heme group, and it turned out that the electron density was raised at the appropriate place in the molecule (Fig. 8).

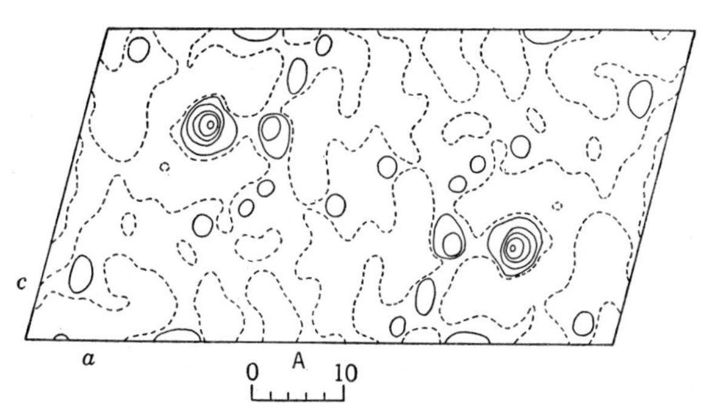


FIG. 8. Difference Fourier projection of the *p*-iodophenyl hydroxylamine derivative of myoglobin. The peaks are caused by the iodine atoms, and they lie near the heme groups in projection.

We are now collecting further data in order to extend the resolution of our model to 2 A. At this resolution, our rod, if it is truly a helix, should show up as a hollow tube with spiral walls like a spring. We might even be able to recognize some of the larger side chains, such as those containing benzene rings. The resolution of individual atoms is still a long way off, if indeed it is attainable.

When considering the future of protein crystallography, one must remember that a full three-dimensional analysis of a protein crystal is a very long and tedious process. In that respect, it is like the determination of the amino-acid sequence of a protein. Neither is likely to be embarked on lightly, yet it is only by means of these now difficult and tedious methods that we will ever be able to put into concrete form the vague concepts concerning configurational specificity in proteins. Edmundson at the Rockefeller Institute has recently begun an amino-acid sequence analysis of myoglobin, and his work taken together with the x-ray results at higher resolution may eventually lead to a detailed picture of the configuration of this one protein. There are, however, very many other interesting proteins for which the same sort of dual approach would be highly warranted, and it is very important to try to develop short-cut methods, but as yet nobody has much idea what form they will take.

The author desires to express his appreciation of the work done by Dr. J. Kraut and Dr. R. G. Hart in preparing this manuscript for publication.

**BIBLIOGRAPHY**

F. H. C. Crick and J. C. Kendrew, Advances in Protein Chem. **12**, 133 (1957).

M. M. Bluhm, G. Bodo, H. M. Dintzis, and J. C. Kendrew, Proc. Roy. Soc. (London) **A246**, 369 (1958).

J. C. Kendrew, G. Bodo, H. M. Dintzis, R. G. Parrish, H. Wyckoff, and D. C. Phillips, Nature **181**, 662 (1958).

# 10
# The Hydrogen Bond

LESLIE E. ORGEL
*Department of Theoretical Chemistry, University Chemical Laboratory,*
*The University of Cambridge, Cambridge, England*

IT is an experimental fact that a hydrogen atom attached to an electronegative atom in one molecule can cause an interaction with another molecule which also contains an electronegative group, or with a different electronegative group of the same molecule. This interaction is referred to as hydrogen bonding.[1] The order of hydrogen-bond strength is F>O>N≫Cl,C. There is no clear-cut lower limit to the strength of hydrogen bonds, those to chlorine and carbon atoms in particular giving rise to very little energy of interaction.

## THERMOCHEMISTRY

The heat of formation of the strongest hydrogen bonds is about 10 kcal, but this value is achieved only in polymers of HF. The strongest hydrogen bonds between groups of biological interest are between oxygen atoms, and the heat of formation of these is rarely greatly in excess of 5 kcal. Hydrogen bonds between nitrogen and oxygen are usually somewhat weaker and bonds between pairs of nitrogen atoms somewhat weaker again. These relations apply only to general trends; it is not suggested that every O—O hydrogen bond is stronger than every N—O or N—N bond.

## STEREOCHEMISTRY[2]

The lengths of hydrogen bonds vary considerably, but there is a general rule that the stronger the bond, the shorter its length. Oxygen bonds may have a length as short as 2.5 A, although those occurring in biochemical systems are usually somewhat longer than this, perhaps 2.7 A. There is no sharp upper limit to the length of a hydrogen bond, but if the O—O distance is much greater than 3.0 A, there is very little interaction. Similar considerations apply to N—O and N—N hydrogen bonds.

In almost all hydrogen-bonded systems, the proton does not lie symmetrically between the atoms bonded. The exceptions to this rule are the $[HF_2]^-$ ion and the hydrogen-maleate ion, but in systems of biochemical interest the hydrogen will always be associated definitely with one or the other of the atoms bonded.

The position of the hydrogen atom relative to the atom to which it is most strongly bound is always close to that which it would occupy if no hydrogen bond were formed. On the other hand, its position relative to the second atom involved in the bond is much more variable. Thus, both of the arrangements shown in Fig. 1 are encountered in certain proteins and in peptides. It is interesting that x-ray studies show that, when the general stereochemistry of a molecule makes it impossible for the hydrogen atom of a hydrogen bond both to lie on the line joining the atoms bonded and to occupy its normal position relative to its nearest neighbor, it is the second requirement which is satisfied. Thus, in salicylic acid the hydrogen atom lies well off the line joining the two oxygen atoms.[3]

## SPECTROSCOPIC PROPERTIES

Hydrogen bonding has a profound influence, both on the infrared and the nuclear-resonance spectrum of a hydrogen-bonded system. Here it is mentioned only that these techniques are of great importance in studies of hydrogen bonding.

## EFFECT OF HYDROGEN BONDING ON CHEMICAL PROPERTIES

In general, the effect of hydrogen bonding on the donor group is in the same direction as the effect of ionization, but less extreme. In a similar way, the acceptor group behaves in the same manner as it does when a proton is added, but again the effect is less marked. An example of this is salicylic acid which is a much stronger acid than one would have anticipated if no hydrogen bond were formed. The effect of the hydroxyl proton on the carboxylic-acid group is like that which would be produced by the addition of an extra proton and, therefore, leads to an easy loss of the carboxylic-acid proton.

## THEORIES ABOUT THE HYDROGEN BOND

The earliest theories of hydrogen bonding postulated either an electrostatic or a covalent interaction, usually the former.[4] In the electrostatic theories, it is argued that, since the fluorine oxygen and nitrogen atoms are very electronegative, a proton attached to them must carry a positive charge. Thus, if a second molecule containing an electronegative and consequently negatively charged atom is available, the hydrogen atom will be attracted to it as shown below.

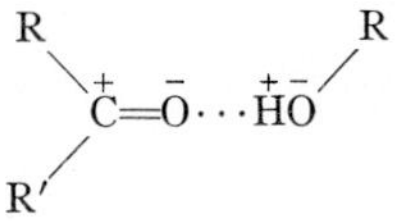

$$(R)(R')\overset{+}{C}{=}\overset{-}{O}\cdots\overset{+}{H}\overset{-}{O}{-}R$$

Naturally, the strength of the bond will be greater if the positive charge on the hydrogen atom is large and if the negative charge on the acceptor is also large. This explains why the strongest bonds are formed by fluorine, the most electronegative element, and the next strongest by oxygen, etc. Rough calculations show that the energies of formation of hydrogen bonds are not incompatible with a simple electrostatic theory, while the covalent theory could not be tested quantitatively.

More-recent detailed calculations[4] show that the situation is intermediate between those suggested by the electrostatic and covalent theories. It seems that weak hydrogen bonds are always entirely electrostatic in origin, but that, as the hydrogen bond becomes stronger and shorter, the covalent contribution to bonding increases. In typical hydrogen bonds occurring in biological systems, it seems unlikely that the covalent character of the bonds will ever exceed about 10%.

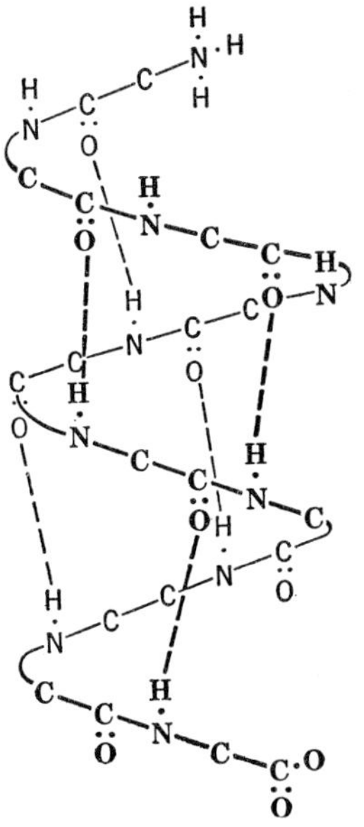

(a)

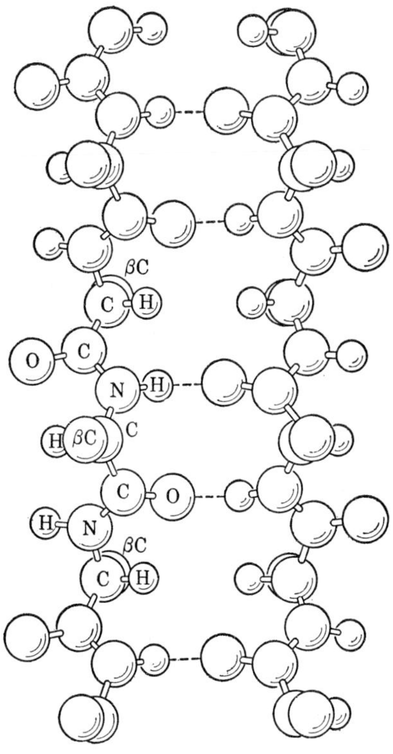

(b)

Fig. 1. Hydrogen-bond arrangement in (a) α-helix and (b) sheet structure.

### PARTICULAR HYDROGEN-BONDING CONFIGURATIONS OF BIOLOGICAL INTEREST

There is one arrangement of atoms which leads to particular stable associations between pairs of molecules. It is the one present in carboxylic acids, amides, guanidine, etc.

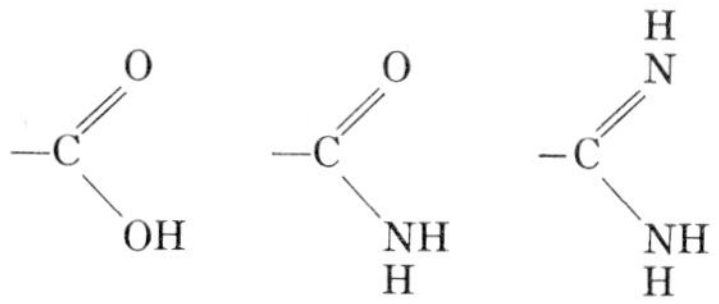

This arrangement is particularly advantageous since it allows for the formation of two bonds rather than one between molecules, e.g.,

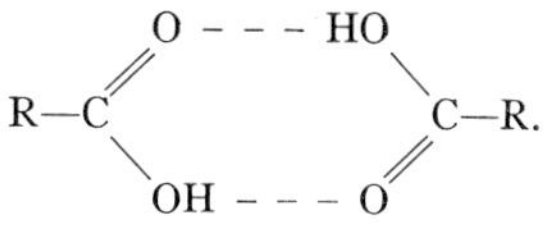

We believe that it must be one of the most important contributors to the hydrogen-bond stabilization by side chains in proteins, etc., and it is also the basis of the Watson-Crick[5] base-pairing scheme, e.g., in the base pair adenine-thymine.

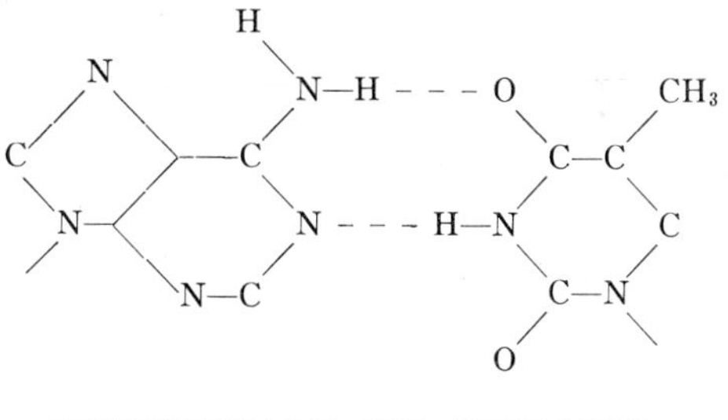

### ELECTROSTATIC AND HYDROGEN-BOND INTERACTIONS

In proteins, one important interaction is that between charged amino groups and charged carboxylate ions. If the two ions are sufficiently close to be in contact, it is not possible to distinguish between electrostatic and hydrogen-bond contributions to the energy of interaction. Thus, it is only in the case of interaction between widely separated charged groups that a clear case of electrostatic (without hydrogen bond) interaction can be recognized.

### HYDROGEN BONDING IN AQUEOUS SOLUTION

The heats of formation of hydrogen bonds in the gas phase give one little information about their stability in aqueous solution, because in this case one is concerned with the difference in the hydrogen-bonding energy of the solute and solvent, firstly when the solute molecules are bound together and secondly when they are separated. Thus, the heat obtained on forming hydrogen bonds in solution is usually less than that obtained in the gas phase. Most simple single hydrogen bonds are split almost completely in water except when the solute is present at very high concentration. The amide group, however, because of its great hydrogen-bonding power, can form moderately stable dimers in aqueous solution.

Schellman[6] has made a detailed study of the dimerization of urea in aqueous solution. Although his assumptions are open to a good deal of doubt, the final value for $\Delta H$, the heat of formation of an O−N hydrogen bond in aqueous solution of −1500 cal is probably roughly right. In the author's view, this figure is probably a little too high. Schellman then went on to calculate the degree of stability of the $\alpha$-helix in aqueous solution. He was obliged to make a number of questionable assumptions and so, instead of obtaining a firm value for the free energy of formation, he could give only a rather wide range of possible values. His work is of great interest because it shows that in aqueous solution the $\alpha$-helix must be on the border of stability so that side-chain interactions may be critical in determining whether a particular protein exists in the $\alpha$-helix form or not. Even though the numerical values which he obtains are probably not very reliable, this conclusion—which was reached before much of the evidence which is described by Doty (p. 61) was obtained—is of very great importance.

Schellman discussed end effects in some detail and showed that the $\alpha$-helix should be stable only provided the chain size exceeds a certain lower limit. Using reasonable values for the parameters involved, the critical length came out to between 8 and 15 units. Experimentally, there does seem to be some evidence for this conclusion.

## PROTEINS IN AQUEOUS SOLUTION

The available experimental evidence discussed in detail in a later paper [Doty (p. 107)] shows that most proteins are only partially present in the $\alpha$-helix configurations, the rest presumably being in some less regular structure. This raises a point of great importance to all discussions of the physical properties of proteins in aqueous solution, namely, the question of whether or not proteins are present in equilibrium configurations. It seems at least possible that, in fact, proteins are present in frozen configurations formed during the peeling-off of the molecule from its template. If so, many of the arguments from the structure of simple synthetic polypeptides may not apply directly to proteins. If this is true, any structural information about proteins also should give useful information about their mode of formation. In particular, it may be that certain sections of proteins which can be removed without effecting the enzymatic activity are present only in order to allow the protein to fold up in the right way. Of course, there are also many other possible explanations of the same effects. Experiments with synthetic copolymers containing, for example, appreciable quantities of glutamine or asparagine might help to solve this problem.

## BIBLIOGRAPHY

[1] See L. C. Pauling, *The Nature of the Chemical Bond* (Cornell University Press, Ithaca, New York, 1940), for much introductory material.

[2] J. Donohue, J. Phys. Chem. **56**, 502 (1952).

[3] W. Cochran, Acta Cryst. **6**, 260 (1953).

[4] C. A. Coulson and U. Danielsson, Arkiv Fysik **8**, 245 (1954).

[5] F. H. C. Crick and J. D. Watson, Proc. Roy. Soc. (London), **A223**, 80 (1954).

[6] J. A. Schellman, Compt. rend trav. lab. Carlsberg. Sér. chim. **29**, 223, 230 (1955).

# 11
# Forces between Macromolecules

Walter H. Stockmayer

*Department of Chemistry, Massachusetts Institute of Technology, Cambridge 39, Massachusetts*

In earlier articles, special attention has been paid to electrostatic forces (see Rice, p. 69), steric forces (see Waugh, p. 84), and hydrogen bonds (see Orgel, p. 100) in biological systems. Here, the attempt is made to round out the discussion of various types of intermolecular action[1] and to take up several special applications.

One recalls first the nature of the mutual potential energy of two simple and chemically inert molecules, such as two argon atoms. Instantaneously, this energy is the sum of the Coulombic interactions involving all of the electrons and nuclei. However, one cannot follow the electronic motion in detail; all that is required is simply the average force between the nuclei, which move much more slowly. In fact, little error is made[2] by assuming the nuclei to be stationary while quantum-mechanical averages over the electronic motions are computed. The results of such theoretical calculations, reinforced by experimental evidence from properties of gases and molecular crystals, lead to a potential energy given approximately by relations such as:

$$\phi(r)=Ar^{-n}\quad -Cr^{-6}, \tag{1a}$$

or

$$\phi(r)=B\exp(-\alpha r)\quad -Cr^{-6}, \tag{1b}$$

where $r$ is the internuclear distance and the other symbols are constants.* The positive (repulsive) term in these expressions is the steric energy resulting from direct overlap of the two atomic electron clouds, and the negative term is the dipole-dipole dispersion or London energy. The familiar curve for the function $\phi(r)$ is shown in Fig. 1(a).

Two further points concerning these intermolecular potential energies should be mentioned.

(a) In a system of many molecules, the total potential energy is to a good approximation the sum of those for all possible pairs; thus, useful predictions of the behavior of bulk matter can be made in principle with the aid of appropriate statistical-mechanical and kinetic theories, from a knowledge of the potential energy between just two isolated molecules. Without this fortunate result, progress would be difficult indeed.

(b) The most important features of the $\phi(r)$ curve—namely, the equilibrium distance and the potential energy corresponding thereto (the minimum in $\phi$)—are not given either by steric or by London forces alone, but result from a *balance* between them. This truism warns that, in general, one should not hope to single out one unique type of interaction (and the classification into "types" is really somewhat arbitrary) as solely responsible for a particular effect.†

One of the simplest applications of a knowledge of $\phi(r)$ is found in the theory of slightly imperfect gases[3] for which the equation of state can be written

$$PV/RT=1+B(T)/V+\cdots, \tag{2}$$

where $P$ is the pressure and $V$ the molar volume. The so-called *second virial coefficient*, $B(T)$, is given for monatomic gases (except for the lightest molecules at very low temperatures) by

$$B(T)=2\pi N_0\int_0^\infty [1-\exp(-\phi(r)/kT)]r^2dr, \tag{3}$$

where $N_0$ is Avogadro's number and $k$ Boltzmann's constant. At temperatures low enough so that the depth of the potential well is at least comparable to $kT$, the quantity $B(T)$ can be thought of as the negative of an equilibrium constant describing binary clustering of the molecules.

On turning to the interactions of two molecules *in solution*, as one inevitably must in most biological considerations, it is extremely important to recognize that the space intervening between the molecules is no longer empty, but is filled with solvent and perhaps other solutes. Detailed treatments of the interactions are now much more difficult, but Eqs. (2) and (3) may be retained if $P$ now is defined to be the *osmotic pressure*, $V$ the volume of solution per mole of *solute*, and if $\phi(r)$ in Eq. (3) is replaced by a quantity $W(r,P,T)$, called the *potential of mean force*.[4] The physical meaning of the latter quantity is easily stated: It is the work that would be required to pull apart the two solute molecules from a separation $r$ to an infinite separation when these are immersed not in a vacuum (as for the potential energy $\phi$), but in a very large quantity of the solvent, the molecules of which are permitted to perform all motions consistent with equilibrium at the stated temperature and pressure. Thus, $W$ depends (through the properties of the solvent) upon pressure and temperature, and its dependence upon the separation $r$ reflects the structure of the solvent.

The simplest and most familiar example of such a potential of mean force is the Coulombic energy

$$W(r,P,T)=e_ie_j/rD(P,T) \tag{4}$$

between two point charges, $e_i$, $e_j$, separated by a distance $r$ in a medium of dielectric constant $D$.

* For pairs of simple molecules, $n$ is about 12, or $\alpha$ about $3\times10^8$ cm$^{-1}$.

† An apparent exception is given by the Coulombic interactions in very dilute ionic solutions, but even here a short-range steric energy is essential to insure stability.

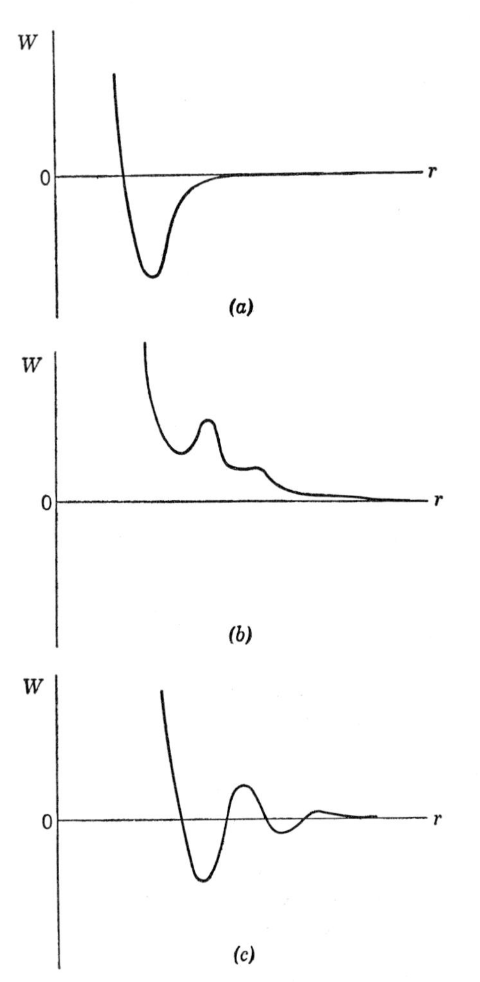


FIG. 1. Potentials of mean force $W(r)$ between two nonpolar spherical molecules as a function of their separation $r$: (a) *in vacuo;* (b) in good solvent; (c) in poor solvent.

Outside the domain of very dilute electrolyte solutions, such simple and general expressions for $W$ as Eq. (4) are no longer found. Possible curves of $W(r)$ for two uncharged nonpolar solute molecules in a good and in a poor solvent, respectively, are drawn in Figs. 1 (b) and 1 (c). The oscillatory nature of these curves is imposed by the structure of the solvent, but it is seen that the damping is quite large. Thus, many approximate theories of liquids and solids can be constructed in which molecules are considered to interact only if they are nearest neighbors.[4]

The important role of the solvent has already been emphasized in connection with hydrogen bonding (see Orgel, p. 100). Perhaps unfortunately for students of both inorganic and biophysical chemistry, the structure of liquid water is probably uniquely complicated. As discussed in detail by Bernal and Fowler,[5] a large residuum of the characteristic, tetrahedrally coordinated structure of ice may be considered to persist in liquid water at ordinary temperatures. The orientations of the water molecules are far from random, and they oscillate torsionally with rather small amplitudes, instead of rotating freely. The introduction of solute molecules or ions into water produces large perturbations in the structure surrounding each solute particle, roughly out to about the third shell of neighbors. Despite considerable effort,[6,7] a complete quantitative theory of these effects still is lacking. However, the student of macromolecular behavior rarely needs an absolute theory, since his task is to infer the behavior of large molecules, given all necessary information about small ones.

Steric effects between macromolecules, in general, may be considered in terms of the well-known covalent and van der Waals radii of the constituent atoms.[8] Naive expectations may sometimes be misleading, as illustrated by the measurements by Chinai and coworkers[9] of the random-coil dimensions in dilute solution of a series of synthetic polymethacrylates

$$(-CH_2-\underset{\substack{|\\ COOR}}{\overset{\substack{CH_3\\ |}}{C}}-)_n$$

in suitably ideal ("theta") solvents.[10] The effective bond length, instead of increasing monotonically as R is varied from methyl to *n*-octyl, passes through a minimum at the *n*-butyl ester. Clearly, a detailed examination of the structure of the polymer-chain skeleton and its surrounding substituent groups and solvent molecules is necessary to the understanding of such behavior.

Turning now to a more detailed consideration of the London dispersion forces,[1,11] one first recalls that they may be considered as arising out of the interaction between instantaneously unsymmetrical electrical-charge distributions in the molecules concerned. Although a noble-gas atom such as argon has, on the average, a spherically symmetric charge cloud, an instantaneous snapshot of the atom, if such were possible, would reveal almost always an unsymmetrical array, with a dipole moment, and also, in general, higher moments of the charge distribution. If this unsymmetrically distributed set of charges is brought close to another atom, its field polarizes the latter into an unsymmetrical condition. The result of this mutual perturbation of the charge distribution is an attractive potential energy whose principal part comes from the interaction of the instantaneous dipoles. London's general formula obtained by second-order perturbation theory, is (for the case of two identical molecules)

$$\phi_{\text{dis}}(r) = -\frac{3e^4\hbar^4}{4m^2r^6}\sum_k \frac{f_{0k}^2}{(E_k - E_0)^3}, \tag{5}$$

where $e$ and $m$ are electronic charge and mass, $\hbar$ is the Planck constant divided by $2\pi$, $r$ is the internuclear distance, $E_0$ and $E_k$ are the electronic energies of the isolated molecule in the general state and the $k$th excited state, respectively, and $f_{0k}$ is the so-called *oscillator strength* of the $0-k$ transition, closely related to the intensity of absorption or emission of the corresponding radiation.

It is seen that the dispersion energy is always attractive, and that it varies inversely as the sixth power of the intermolecular separation, in contrast to a dependence upon $r^{-3}$ for two dipoles in fixed specified orientations. This indicates the second-order nature of the effect, as does the presence in Eq. (5) of properties related to the excited states of the molecules.

The quantities required for directly computing the sum in Eq. (5) are rarely available. However, useful estimates of dispersion forces can be made by appealing to a similar expression for the (low-frequency) molecular polarizability which contains a similar sum. If it is

$$\alpha=\frac{e^2\hbar^2}{m}\sum_k\frac{f_{0k}}{(E_k-E_0)^2}, \qquad (6)$$

assumed (sometimes apparently without too great error) that the largest part of each sum is contributed by a single excited state, elimination between Eqs. (5) and (6) leads to the much simpler result

$$\phi_{\text{dis}}(r)=-\frac{3}{4}\frac{\alpha^2}{r^6}(E_k-E_0), \qquad (7)$$

in which the energy difference $(E_k-E_0)$ is of the order of magnitude of an ionization potential (say, 10 to 20 v).

The corresponding expression for the London dipole energy between two *unlike* molecules[1] shows that no great specificity is to be sought in these interactions. A rough rule, often embodied in approximate theories of solutions, is that the force between a pair of unlike molecules is the geometric mean of those between the related pairs of similar molecules, at the same value of $r$.

Equation (5) suggests that highly colored molecules, which have at least one large oscillator strength at a rather small excitation energy, should exert especially strong dispersion forces. This expectation seems to be supported[12] by the fact that many dyestuffs are extensively dimerized in solution at concentrations as low as $10^{-4}$ molar, often in spite of Coulombic repulsion between ionic charges (cf. Fig. 2). The interaction between lyophobic colloid particles (see Rice, p. 69, and Verwey and Overbeek[13]) is qualitatively similar, Coulombic repulsion at large separation giving way to London attraction at smaller separations until overlap repulsion produces the stable minimum.

Unless extensive arrays of conjugated double bonds or condensed aromatic rings are found in large molecules, their mutual London interactions can be regarded as the sum of contributions from all possible pairs of localized centers, which may be taken as the individual atoms or the bonds. This additivity is well substantiated in the properties of the normal paraffins, for example, and [recall Eq. (6)] also is found in refractivity. When nonlocalized electron orbitals are permitted by the molecular structure, the simple London formula, Eq. (5), is inadequate and unusual orientational dependence of the forces is found,[14,15] but the magnitudes are not unusually great (e.g., compare benzene with cyclohexane, or hexatriene with $n$-hexane).

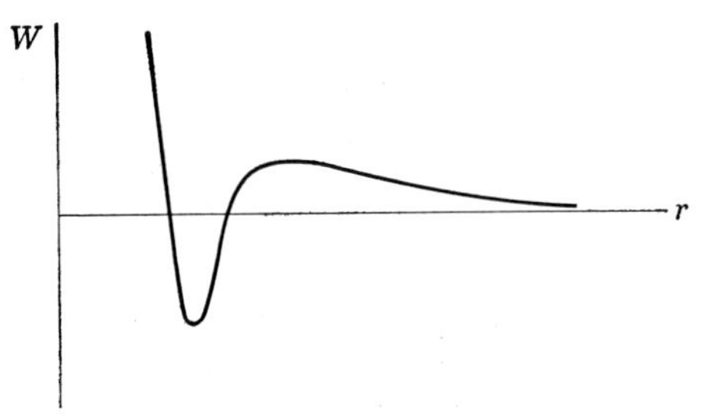


FIG. 2. Potential of mean force between two similarly charged molecules or lyophobic colloid particles. At large separations, Coulombic repulsion is dominant. As $r$ decreases, this gives way in turn to London attraction and steric repulsion.

It has been suggested[16] that very strong and highly specific attractive forces can operate between identical macromolecules. Consider two identical and widely separated molecules, one in its ground state and one in some excited state. If these approach each other, the fundamental indistinguishability of the molecules prevents the identification of one of the two as the excited one. In familiar language, there is resonance[8] between two identical structures, with resultant (first-order) perturbation of the energy levels. The resonance energy (dipole term) varies as $r^{-3}$ and can be either attractive or repulsive[1]; but, statistically, the former is favored at any finite temperature.

Resonance forces are highly specific indeed, for, if the excitation energies of the two molecules do not match rather closely, the resonance energy is negligible; but clearly, they cannot contribute much to the stability of a system unless appreciable numbers of the molecules are in excited states to begin with. This, in turn, requires low values of the excitation energies. The excited *electronic* states of all known macromolecules do not satisfy this requirement. Excited *vibrational* states of heavy molecules, on the other hand, are more generously populated but they do not make very large contributions to the forces (compare the magnitudes of the so-called "atomic" and "electronic" polarizabilities of molecules[17]), because only small changes of charge distribution attend the vibrational motions. It is concluded, therefore, that, although specific dipole resonance forces between identical molecules are certainly capable of existence, their magnitudes must be negligible in all practical cases. However, their role in energy-transfer processes involving excited states is very important.[18]

Another class of attractive force operates between electron acceptors and donors (Lewis acids and bases). These so-called "charge transfer" forces[19] contribute to the stability of many colored complexes (e.g., benzene-iodine) and, thus, have interesting consequences for spectroscopy, but they are probably too rare to be accorded major attention in biology.

These views regarding the forces between biological macromolecules are summarized by stating that, in general, they do not differ qualitatively from those between simple molecules, and that the most important specific interactions are steric and Coulombic.

Turning now to applications, consider first the stability of micelles formed by aggregation of soap ions, which have long hydrocarbon tails and ionic heads. Whether the micelles are spherical [compare casein (see Waugh, p. 84)] or lamellar (as lipid membranes), the ions arrange themselves so that their charges remain at the micelle surface, where counterions from the surrounding medium congregate preferentially to complete an electrical double layer. The potential of mean force between unchanged hydrocarbon molecules in water is strongly attractive, so that the micelles would grow indefinitely large (as do droplets of uncharged hydrocarbon in water), except for the counteracting mutual repulsion between the ionic heads of the ions. More specifically, the standard free energy of formation $\Delta F^\circ$ of a micelle containing $n$ ions is the sum of a negative (favorable) bulk term directly proportional to $n$ (arising from the net attraction between the hydrocarbon tails), and of a positive (unfavorable) surface term varying more rapidly‡ than $n$ (arising from the electrostatic repulsion between the ionic heads). The resultant curve of $\Delta F^\circ$ against $n$ displays a minimum defining the most stable micelle size. Quantitative development of these ideas[20,21] accounts satisfactorily for the sizes and size distributions of soap micelles.

In systems of more than two components, electrostatic repulsion is not the only possible size-regulating mechanism. For example, the size of stable oil-in-water emulsion droplets is controlled by the available quantity of a third component, the surface-active agent, which need not be an electrolyte. Casein micelles (Waugh, p. 84) are characterized by large size and unusual stability toward added electrolytes, and may well involve the simultaneous action of several size-regulating effects.

Overbeek[22] suggests that lipid molecules naturally form flat membranes rather than spherical micelles because of the bulkiness of their hydrophobic parts (two hydrocarbon chains per glyceride unit), and that there is no need, therefore, to postulate a more specific biological mechanism of membrane synthesis. He also accounts for the order of magnitude of the observed spacing (50 to 100A) between multiple layers of such membranes (as in a myelin sheath) as a consequence of balance between London attraction and electrostatic repulsion.

The formation of gels or gelatinous precipitates involves a somewhat different balance of forces. Here, the interactions between the major portions of the primary particles, frequently chain- or rod-like macromolecules, are insufficient to cause aggregation and even may be weakly repulsive, but they are overshadowed by strong attractions between specific local sites. The resulting aggregate has a network structure and usually is extensively swollen by the solvent medium.

‡ The precise dependence of the electrostatic term upon $n$ is complicated, and involves the charge and concentration of the counterions forming the double layer. If the counterion concentration is *very* low, this term varies as $n^{5/3}$ for spherical micelles or as $n^{3/2}$ for lamellar micelles.

The mathematical theory of gelation[23] is similar to that of explosions and of certain classical gambling problems. It may seem obvious perhaps that networks of unlimited size can form from primary molecules only if each carries an average of more than two reactive sites. The nature of the attraction between these sites need not be of any special type. Thus, in synthetic thermosetting resins,[23] the bonds are covalent and essentially permanent, while in gelatin[24] they are presumed to be the hydrogen bonds which stabilize the triple-stranded helical collagen structure. In antigen-antibody reactions,[25] a variety of possibilities exists.

If each of the primary molecules has a large number of reactive sites, only a very small attraction suffices to produce giant network structures. For example, if each primary molecule had 1000 reactive groups, a solution of about one percent concentration would gel even if the association equilibrium constant between two groups were only about $10^{-2}$ liter/mole. One can see that this mechanism affords the possibility of producing large effects (e.g., gelation greatly reduces the mobility of large particles) at small cost of free energy.

## BIBLIOGRAPHY


1 J. O. Hirschfelder, C. F. Curtiss, and R. B. Bird, *Molecular Theory of Gases and Liquids* (John Wiley and Sons, Inc., New York, 1954), particularly Chaps. 1, 12, 13, and 14.

2 M. Born and J. R. Oppenheimer, Ann. Physik **84**, 457 (1927).

3 See reference 1, Chap. 3.

4 T. L. Hill, *Statistical Mechanics* (McGraw-Hill Book Company, Inc., New York, 1956).

5 J. D. Bernal and R. H. Fowler, J. Chem. Phys. **1**, 515 (1933).

6 R. W. Gurney, *Ionic Processes in Solution* (McGraw-Hill Book Company, Inc., New York, 1953).

7 H. S. Frank and W.-Y. Wen, Discussions Faraday Soc. **24**, 133 (1958).

8 L. Pauling, *Nature of the Chemical Bond* (Cornell University Press, Ithaca, New York, 1948), second edition.

9 S. N. Chinai, J. Polymer Sci. **25**, 413 (1957).

10 P. J. Flory, *Principles of Polymer Chemistry* (Cornell University Press, Ithaca, New York, 1953), Chaps. 12–14.

11 H. Margenau, Revs. Modern Phys. **11**, 1 (1939).

12 E. Rabinowitch and L. F. Epstein, J. Am. Chem. Soc. **63**, 69 (1941).

13 E. J. W. Verwey and J. Th. G. Overbeek, *Theory of the Stability of Lyophobic Colloids* (Elsevier Publishing Company, Amsterdam, 1948).

14 F. London, J. Phys. Chem. **46**, 305 (1942).

15 C. A. Coulson and P. L. Davies, Trans. Faraday Soc. **48**, 777 (1952).

16 H. Jehle, J. Chem. Phys. **18**, 1150 (1950).

17 J. H. Van Vleck, *Electric and Magnetic Susceptibilities* (Oxford University Press, London, 1932).

18 C. Reid, *Excited States in Chemistry and Biology* (Academic Press, Inc., New York, 1957).

19 R. S. Mulliken, J. Am. Chem. Soc. **74**, 811 (1952).

20 P. Debye, J. Phys. & Colloid Chem. **53**, 1 (1949).

21 D. Stigter and J. Th. G. Overbeek, *Gas/Liquid and Liquid/Liquid Interface,* Proceedings of the Second International Congress of Surface Activity (Butterworths Scientific Publications, London, 1957), p. 311.

22 J. Th. G. Overbeek (private communication, 1958).

23 See reference 10, Chap. 9.

24 P. Doty, Proc. Natl. Acad. Sci. U. S. **42**, 791 (1956).

25 S. J. Singer, Proc. Natl. Acad. Sci. U. S. **41**, 1041 (1955).

# 12
# Configurations of Biologically Important Macromolecules in Solution

Paul Doty

*Department of Chemistry, Harvard University, Cambridge 38, Massachusetts*

ALMOST all biologically important macromolecules are particular chemical and structural modifications of three basic, polymeric-chain structures. These polymeric chains are the polysaccharides, the polypeptides, and the polynucleotides. They are better known in their more particular forms; for example, cellulose and amylose in the first instance, fibrous and globular proteins in the second, and deoxyribonucleic and ribonucleic acid in the third.

A polysaccharide chain such as cellulose exists as a completely extended chain in the crystalline, fibrous form and as such has a unique configuration. In solution, however, the limited rotational freedom permitted at the juncture of each repeating unit results in the chain being flexible and, because of internal Brownian motion, it undergoes continuous, worm-like changes in configuration. This state is referred to as a randomly coiled configuration. Such a configuration is devoid of any fixed relationship between pairs of residues, and hence is said to have no secondary structure. Since the fixed configuration in the crystalline form and the randomly coiled configuration in solution are typical of the situation found in most synthetic polymers as well, these have been widely studied and are well understood. Consequently, they are not further dealt with here.

The polypeptides and polynucleotides offer a sharp contrast in configurational properties to the polysaccharides and all other known polymeric chains. This uniqueness lies in the ability of these two types of chains to form hydrogen-bonded helical configurations consisting on one, two, or three chains which are stable in aqueous solution and hence in the cellular environment. Each of these helical configurations is unique and is equivalent to one-dimensional crystallites in that they consist of a periodic arrangement of the repeating-chain units along the helical axis. In this way unique, stereospecific relations in the individual macromolecule can be maintained while the macromolecule itself moves about in solution. The randomly coiled configuration exhibited in solution by all other macromolecules does not have this property. Consequently, it is evident that the preservation of unique configurations by polypeptides and polynucleotides in solution offers a basis of biological specificity.

These unique configurations of individual macromolecules are similar to the crystalline state in several respects, in addition to their having a one-dimensional periodic order. Most important is the implication that they will have a melting point; that is, a temperature will exist at which the supporting hydrogen-bonded structure will undergo a transition to the equivalent of the liquid state. For the macromolecule, the liquid state is simply the randomly coiled configuration already described. Thus, it is not surprising that transitions are found in these macromolecules that can have a sharpness approaching a phase transition. These are generally known as helix-coil transitions, and their study in the present case is of interest because their location reveals the relative stability of the unique configurations upon which one's attention becomes focused.

In the last few years, the study of these helical structures has greatly benefited from the possibility of making pure polypeptides and pure polynucleotides—pure in the sense that the repeating units are identical rather than differentiated as they are in the naturally occurring counterparts, proteins and nucleic acids. As a consequence, the properties of the purely helical forms could be carefully studied and the means of detecting such forms in the naturally occurring materials were thereby greatly sharpened.

It is against this background[1] that some studies of the configurations of polypeptides and their relation to protein structure are examined briefly, and following that a similar look is taken at the corresponding situation in polynucleotides and nucleic acids. The configurations have been established in solution by the use of the methods outlined earlier (Doty, p. 61). In addition, optical-rotatory dispersion and ultraviolet spectroscopy have been found widely useful in detecting the configuration established by the physical methods.

## POLYPEPTIDES AND PROTEINS

In 1951, shortly after the proposal of the α-helix by Pauling and Corey,[2,3] Perutz[4] showed by means of x-ray diffraction that a few fibrous proteins in crystalline form did contain this configuration in unspecified amounts. However, it was not possible to extend this method to globular proteins.

In 1953, E. R. Blout and the author initiated a program of synthesis and characterization of synthetic polypeptides: this had as one of its aims the direct testing of whether or not polypeptides could take up unique configurations such as the α-helix in solution. In 1954, we found that poly-γ-benzyl-L-glutamate could exist in two configurations, the α-helix and the solvated, randomly coiled chain, depending on the solvent. Moreover, we found that the two forms showed a substantial difference in specific rotation similar in sign and magnitude

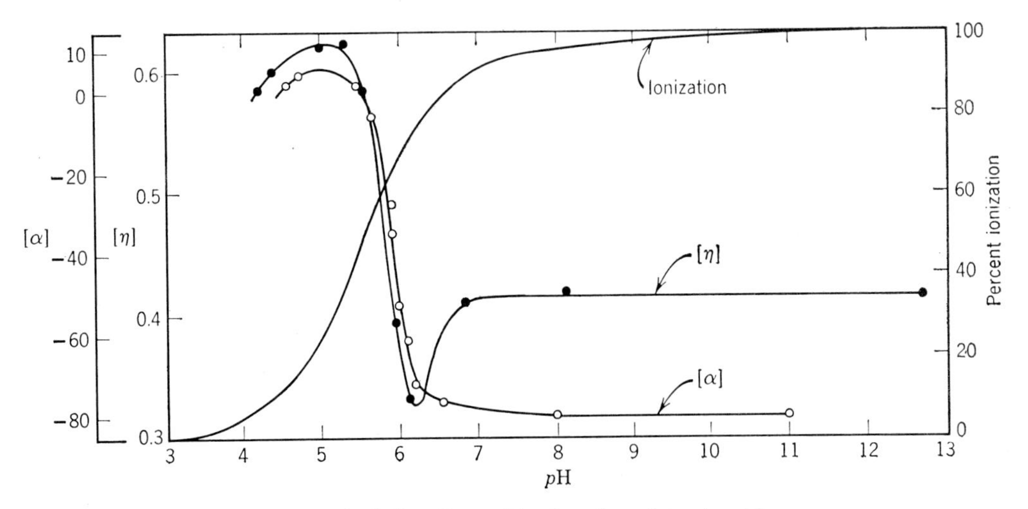


FIG. 1. The helix-coil transition in poly-L-glutamic acid.

to the difference between native and denatured proteins.[5] This fitted nicely with the suggestion made in 1955 by Cohen[6] that a specific main-chain configuration may be the cause of the difference in optical rotation between native and denatured states of proteins.

In 1956, more-detailed evidence was presented in support of the assignment of configuration in solution,[7] and it was shown that a sharp transition could be observed between the two configurations.[8] More to the point, however, was the investigation of the rotatory dispersion (i.e., the wavelength dependence of specific rotation), because this showed that, whereas the dispersion for the randomly coiled form was of the simple (or Drude) type found in most organic compounds containing asymmetric carbon, the dispersion of the $\alpha$-helical form was anomalous.[8] At the same time, poly-L-glutamic acid was investigated in aqueous solution, and found to be helical in acid solution and randomly coiled in neutral and alkaline solutions.[9,10] This provided additional data on a system that more closely resembled proteins in aqueous solution.

It is perhaps useful to examine the case of poly-L-glutamic acid in some detail. Measurement of its intrinsic viscosity, sedimentation constant, and molecular weight[10] at *p*H values below *p*H 5 (where the carboxyl group was largely non-ionized) showed that it had the rod-like form and mass-to-length ratio of the $\alpha$-helix. As the *p*H was raised, however, substantial changes in all of the physical properties of this polypeptide were observed[9] (Fig. 1), and the interpretation of these changes indicated that the polypeptide has gone over to a random-coil form. This can be considered as a melting process brought about by the electrostatic repulsion. That is, owing to the ionization of the carboxyl groups resulting in turn from the increase in *p*H, there is imposed upon the original helical structure an outer helix of negative charges; the repulsion between these charges is more than the hydrogen-bonding frame can withstand; and the $\alpha$-helix structure breaks apart. One can observe this transition by measuring changes in the viscosity, for example, corresponding to the change in the shape of the molecule. However, the most generally useful of all of the changes accompanying this transition is the change in optical rotation. In this case, one observes a change of optical rotation from +4 to −80°C accompanying the helix-coil transition. With this nearly instantaneous means of detecting the configuration, it is easy to show that this is a reversible transition.

It is an old observation that the specific rotation of proteins falls* upon denaturation.[11] During the last decade, much specific support has been given to this generalization by quantitative studies, particularly by Kauzmann.[12] These studies show that the specific rotation, $[\alpha]_D$, for the majority of proteins lies between −30 and −60°C and that these values fall to approximately −100°C upon denaturation. Now, if the values of $[\alpha]_D$ observed for the helix and coil forms of poly-L-glutamic acid are adjusted to the mean residue weight of proteins (151 for sodium glutamate to 115 for proteins), the result is that a value of about +5°C would be expected for a protein in the completely helical form and about −105°C for a protein in the completely random-coil form. Thus, the observed specific rotation of proteins is compatible with 40 to 70% of their residues being in the $\alpha$-helical configuration, and the specific rotation of denatured proteins corresponds to that of the randomly coiled configuration as previously surmised.[13] Naturally, this proposal requires extensive testing, and much of this is under way or has been done.

Since it is well known that proteins denature upon heating, one would expect the helical form to break up as the temperature is raised. The data in Fig. 2 show that this does occur if one uses the optical-rotation

* In this article, changes in specific rotation are always described with reference to the absolute value, not the magnitude as is customary. This is necessary since we now have both positive and negative values of specific rotation to discuss in contrast to earlier times when the only measured values of proteins were negative.

method to gain a measure of helix content.[9] At *p*H 4.1 and temperatures below 40°C, the optical rotation is nearly constant. Above 40°C, the optical rotation begins to decrease, showing the start of the helix-coil transition. It melts out nearly all of the way at *p*H 4.6. One can pick up the rest of the transition by going to *p*H 5.0 where it melts out completely.

It is possible that the stability of the α-helix in poly-L-glutamic acid resides not so much in the hydrogen-bonding framework of the peptide units as in the pairwise interaction of carboxyl groups. This is the kind of hydrogen bonding that is always pointed out as being quite strong, and, if one makes models of the α-helix configuration of polyglutamic acid, one sees that uncharged glutamic residues can pair very nicely on the surface of the helix. Therefore, it is of great interest to see if other polypeptides that do not have this possibility also show the same phenomena.

Poly-L-lysine has an amino group on the side chain. This amino group is in the charged form ($-NH_3^+$) at *p*H values below 9.5, and becomes uncharged ($-NH_2$) above *p*H 10.5. Thus, one would expect that, if the helix is only marginally stabile, it would exist only at high *p*H values and would melt out as one lowered the *p*H. Current experiments show just this.[14] Again, there is about a 90°C change in optical rotation upon passing through the helix-coil transition. Physicochemical investigations were used to show that it exists as a pure α-helix at high *p*H.

If one examines the state of charge of the poly-L-lysine molecule when the helix begins to melt out in a quantitative fashion, one sees that this helix is considerably weaker than the poly-L-glutamic-acid helix. However, the difference is not enough to prevent the poly-L-lysine helix from being stable in aqueous solution. From this and a few other examples, it can be concluded that the α-helical configuration is stable in aqueous solution in the absence of pronounced electrostatic repulsions or other disrupting influences.

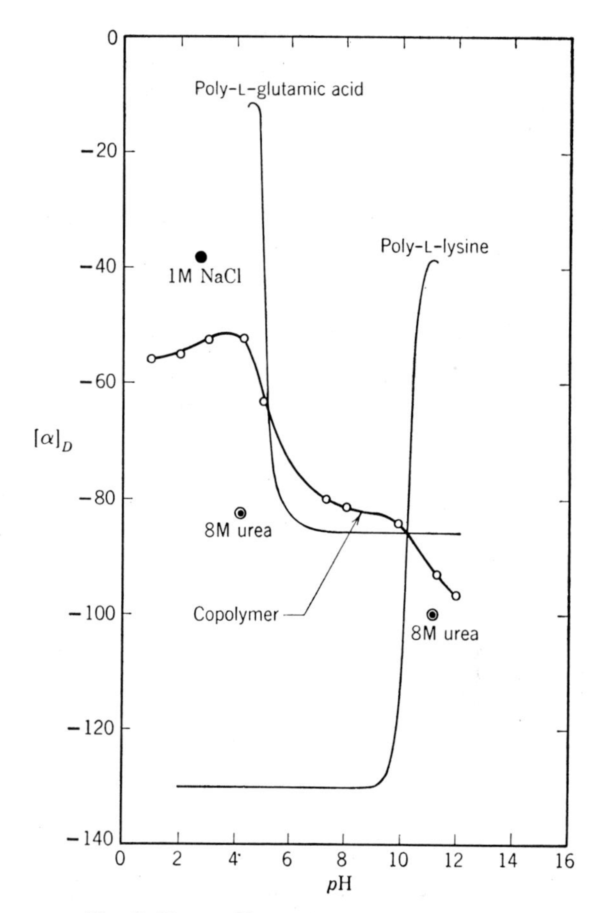


FIG. 3. The specific rotation as a function of *p*H for copoly-L-lysine-L-glutamic acid.

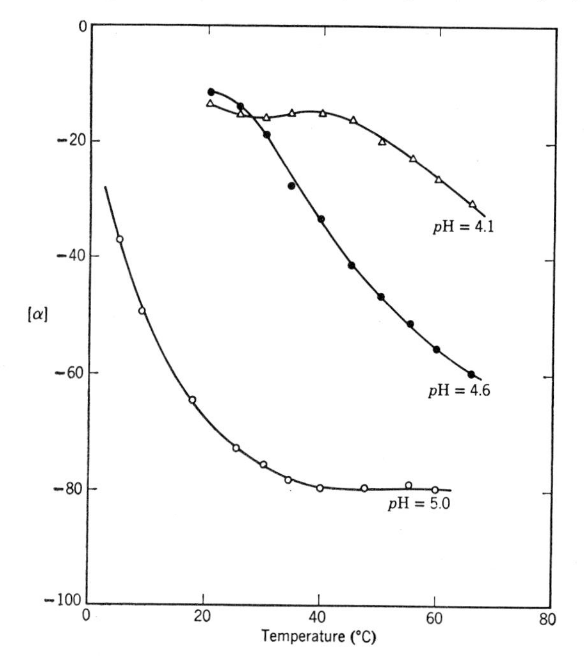


FIG. 2. Optical rotation of poly-L-glutamic acid as a function of temperature as several values of *p*H.

An interesting point associated with the nature of these helix-coil transitions is the question of whether or not they have an "all-or-none" character. That is, at the midpoint of the transition are half of the molecules in each configuration or are the residues in each molecule partitioned between these two forms? One of several answers indicating that the latter was the case was obtained in the following way. The intrinsic viscosity increases much more strongly with molecular weight in the helical form than in the random-coiled form. Thus, there is one molecular weight for which the viscosity is the same in both configurations. By choosing a sample of this molecular weight and carrying it through the transition, the intrinsic viscosity would be expected to remain unchanged if the transition were of the all-or-none type. This was not observed in several cases.[10,14,15] Consequently, the transition must be viewed as one that proceeds via intermediate states in which individual molecules have interspersed helical and nonhelical regions.

It is of some interest to consider the behavior of a copolymer composed of equal amounts of L-glutamic acid and L-lysine. The optical rotation, which again is taken as a measure of helix constant for this copolymer, is shown in Fig. 3.[16] These results can be compared with those previously reported for poly-L-glutamic acid and poly-L-lysine which are also shown. The copolymer exhibits an intermediate behavior here; that is, it is about 50% helical at acid *p*H, and not at all helical at alkaline *p*H. This is interpreted as resulting from the difference in intrinsic stability of the glutamic-acid and lysine residues in supporting the helical structure. When the lysine residues are uncharged, there appears to be no helix, because the lysine residues cannot compensate for the electrical repulsion arising from the 50% negative charges that exist here owing to the ionized glutamic-acid residues. At neutral *p*H, one finds about 15 or 20% helix.

## ROTATORY DISPERSION AND HELICAL CONTENT OF POLYPEPTIDES AND PROTEINS

Although one could proceed to estimate the fraction of residues (helical content) in the helical configuration from the specific rotation of proteins in the manner indicated in the foregoing, one would be failing to take advantage of another aspect of the situation that, upon proper analysis, offers an independent means of estimating the helical content and in addition offers a much more perceptive view of the problem. Soon after the first observations of the helix-coil transition in polypeptides, the rotatory dispersion (wavelength dependence of the specific rotation) of the two forms was measured[8] and found to be simple Drude dispersion for the random-coil form and a complex dispersion for the helical form. The complex dispersion could be fitted in a manner expected for coupled oscillators and originally suggested by Moffitt.[17] This gave rise to a very considerable theoretical interest in the problem of the rotatory dispersion of helical macromolecules,[18–22] but it has not been possible to take advantage of much of this because of the limited number of parameters that can be uniquely determined from the experimental data.

From the practical point of view, it is sufficient to say that the following equation was found adequate to express the observed data[23–25]:

$$[m'] = \frac{3}{n^2+2}\frac{M_0}{100}[\alpha] = (a_0^R + a_0^H)\left(\frac{\lambda_0^2}{\lambda^2-\lambda_0^2}\right) + b_0\left(\frac{\lambda_0^2}{\lambda^2-\lambda_0^2}\right)^2. \quad (1)$$

Here, the specific rotation is multiplied by two factors that eliminate the trivial effect of residue weight, $M_0$, and refractive index, $n$. The corrected specific rotation, called the effective residue rotation, $[m']$, is then assumed to be made up of two components. The first of these is a Drude term but with two parts to its coefficient, one depending on the intrinsic residue rotation (asymmetric carbon), $a_0^R$, and the second is owing to the effect of the helical configuration, $a_0^H$. The second term is the anomalous one, being the square of a Drude term, and its coefficient is wholly owing to the effects of the helical configuration on the rotatory dispersion.

This equation is of the same form as that proposed by Moffitt and shown to be applicable to the first dispersion data obtained on helical polypeptides,[8,9,20] but it now seems proper to consider it an empirical proposal for two reasons. The equation itself is of the form expected for coupled oscillators. Secondly, Moffitt concentrated his attention on the second term, recognizing the complexities that may be associated with the first since, in any rigorous view, the first term must instead be replaced by several Drude terms each having a different $\lambda_0$. But if this last point is admitted, then the analysis of any available dispersion data is impossible, because there is not enough precision to allow the independent assignment of values to more than the three constants contained in Eq. (1), i.e., $(a_0^R+a_0^H)$, $b_0$, and $\lambda_0$. Consequently, in order to proceed, it was necessary to assume that $\lambda_0$ had the same value in both terms. This compromise, added to the later finding that the absolute value of $b_0$ cannot be computed with enough accuracy to be useful, forces us to accept Eq. (1) as empirical. It is important to make this point because, when it is recognized that Eq. (1) is not firmly supported by theory, it is more likely that it will be used with the caution that an empirical relation deserves.

The Fitts-Kirkwood theory[18,19] cannot be expressed in simple functional form, but since it consists of four different contributions, it is clear that these cannot possibly be resolved from experimental dispersion data since again too many parameters would have to be independently evaluated. However, if one takes the computed results of Fitts and Kirkwood,[22] one finds that they can be quite well fitted by Eq. (1). This observation emphasizes that the procedure that we have adopted in using Eq. (1) is not in conflict with the Fitts-Kirkwood theory.

Examine the adequacy of Eq. (1) in representing the dispersion data of polypeptides in the helical form. By dividing Eq. (1) by $\lambda_0^2/(\lambda^2-\lambda_0^2)$, it is clear that dispersion data obeying this relation should yield linear plots when $[m'](\lambda^2-\lambda_0^2)/\lambda_0^2$ is plotted against $\lambda_0^2/(\lambda^2-\lambda_0^2)$, provided the appropriate value of $\lambda_0$ has been selected. In this plot, the intercept of the straight line would then provide the value of $(a_0^R+a_0^H)$, which may be denoted $a_0$ for convenience, and the slope would provide the value of $b_0$. Since in this type of plot an empirical relation is being tested, it will be necessary to have specific rotation measurements at a number of wavelengths, because the absence of curvature in the plot must be demonstrated in each application.

In selecting the appropriate value of $\lambda_0$, a process of trial and error must be followed until that particular value which linearizes the dispersion data is found. In the fitting of the first dispersion data, a value of 212 mμ was found for poly-γ-benzyl-L-glutamate and poly-L-glutamic acid. Since that time, the value has been confirmed on both of these polypeptides[10,26] and found to apply to at least four other helical polypeptides, as well as to several copolypeptides.[27] Dispersion data plotted in this manner are shown in Fig. 4 for three samples of poly-γ-benzyl-L-glutamate.[26] One sample had an average chain length of 4 units and gave a horizontal plot indicative of no helical content. Another had an average chain length of 10 residues and gave an intermediate slope and intercept, which, in comparison with the high molecular-weight sample, indicates about 60% helical content.

In all of the cases referred to in the foregoing (poly-γ-benzyl-L-glutamate, poly-L-glutamic acid, poly-carbobenzoxy-L-lysine, poly-L-lysine, and the copoly-L-lysine-L-glutamic acid), the values of $b_0$ obtained from the slopes of what one might call the "coupled oscillator" plot have clustered about $-630$. Thus, there is good reason to accept this value with an uncertainty of no more than $\pm 10\%$ as a constant characterizing in an empirical way the anomalous dispersion of helical polypeptides of L-residues. In some cases, we have found somewhat lower values, but these have in each case been with polypeptides that had not been proved to be in the completely helical form.

Turning finally to the coefficient $a_0$, we do not expect to find constant values for different polypeptides, because $a_0$ consists of two components, one of which $(a_0^R)$ depends on the intrinsic rotatory power of the individual amino-acid residue. It is sufficient to say that, in the

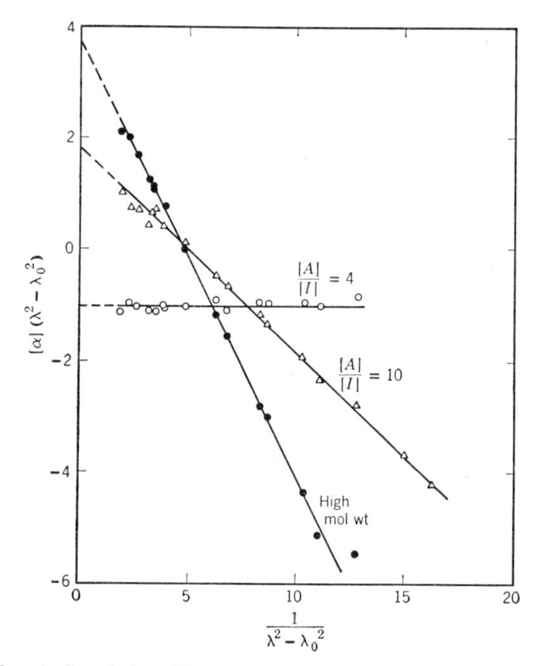


FIG. 4. Coupled oscillator plot of rotatory-dispersion data for poly-γ-benzyl-L-glutamate of various molecular weights. $[A]/[I]$ indicates degree of polymerization (number average).

cases of the five polypeptides mentioned in the previous paragraph, the values of $a_0$ gave an average of zero. Since in the helical form $a_0$ is found to equal 650, this must be approximately the value of $a_0^H$ itself.

One is now in a better position to examine the extent to which the work on polypeptides enables one to understand the optical activity and rotatory dispersion of proteins. Actually, one is forced to proceed in a very simple manner. The polypeptide studies have given quantitative information about the rotatory dispersion of two configurations of polypeptide chains. One can, therefore, only ask if the rotatory dispersion of proteins can be accounted for by a linear combination of these two characteristic rotatory dispersions. It is important to recognize that, if this is the case, the residues not occurring in the α-helical configuration need not be in a randomly coiled configuration. It is necessary only that they not be in any kind of periodic arrangement, because this is the condition that the rotatory dispersion be of the kind observed for the randomly coiled configuration, subject of course to "solvent effects" on the intrinsic residue rotation, i.e., $a_0^R$.

The data summarized in terms of $a_0^H$ and $b_0$ values can be translated directly into the contribution that the α-helical configuration would make to $[\alpha]_D$. It is $+117°$ for the values of $a_0^H=650$, and $b_0=-630$ if the mean residue weight is taken as 115. The value of $a_0^H$ contributes 84% to this result and, hence, it is the uncertainty of this term that affects most the predicted value. Taking the limits of $a_0^H$ to be 550 and 750 would make the helix contribution to $[\alpha]_D$ 17° higher or lower, respectively. Thus, the prediction is made that for proteins the purely helical configuration and the completely denatured form should be separated by 100° to 135° in specific rotation. The polypeptide studies, adjusted to mean residue weights on 115 and aqueous solutions, establish the lower end of this increment—that is, the denatured protein—at about 110° and the high end—the pure helix—at about +5°. This is nearly the same as that obtained earlier from the measurements on poly-L-glutamic acid.

In terms of the model proposed here, the rotatory dispersion of proteins in aqueous solution should be given by converting Eq. (1) into the following form[28]:

$$[\alpha]=1.39\left[\left(\sum a_0^R+fa_0^H\right)\left(\frac{\lambda^2}{\lambda^2-\lambda_0^2}\right)+fb_0\left(\frac{\lambda^2}{\lambda^2-\lambda_0^2}\right)\right],$$

where $f$ denotes the fraction of residues in helical form and $a_0^R$ the sum over the $a_0^R$ values characteristic of each residue in the protein. The model is not subject to a considerably more rigorous test than in the comparison dealing only with $[\alpha]_D$. This can be carried out in the following way. Rotatory-dispersion measurements are made on the protein both in aqueous solution and in the completely denatured state. For the former set of data, $[\alpha]1.39(\lambda-\lambda)$ is plotted against $(\lambda-\lambda)^{-1}$. The result should be a straight line yielding a slope equal to $fb_0$

TABLE I. Excess right-handed helical contents of ($f$) of various proteins in water.

| | $-b_0/630$ | $a_0^H/650$ |
|---|---|---|
| Tropomyosin | 0.88 | 0.87 |
| Insulin | 0.38 | 0.57 |
| Bovine serum albumin | 0.46 | 0.58 |
| Ovalbumin | 0.31 | 0.50 |
| Lysozyme | 0.29 | 0.39 |
| Pepsin | 0.31 | 0.26 |
| Histone | 0.20 | 0.30 |
| Ribonuclease | 0.16 | 0.17 |
| Globin (H) | 0.15 | 0.09 |

and an intercept equal to $(\sum a_0^R + f a_0^H)$. With the value of $b_0$ set at $-630$, the value of $f$, the fraction of residues in the helical configuration, is obtained at once.

The rotatory dispersion in the denaturated state should yield at once the value of $a_0^R$, since $f=0$ in this case. Indeed, only a measurement of the specific rotation at one wavelength is needed if it is known that $\lambda_0=212$ m$\mu$, as appears to be the case in 8$M$ urea. With $a_0^R$ known, the value of $f a_0^H$ can be obtained from the intercept evaluated in the plot described above. By taking the value of $a_0^H$ as $+650$, an independent estimate of $f$ can be made.

This kind of analysis has been applied to a number of proteins.[28] In each case, a linear plot was obtained. The test of the model consists then in seeing if the two different estimates of $f$ are reasonable, that is, lie between 0 and 1, and are in agreement. A selection of this data is shown in Table I. It is seen that the values of $f$ are indeed reasonable ones and that the agreement is fairly good.

The reasonably good performance of our model in this test is not, of course, proof that the model is a faithful description of the secondary structure of the proteins to which it has been applied. However, it does appear to represent an advance which deserves further and more-incisive testing. At present, it greatly increases our confidence in the existence of regions having the $\alpha$-helical structure in proteins, it sharpens our views on protein denaturation, and it provides a framework of reference in which studies relating protein structure and function can be at least tentatively interpreted.

Having observed the striking dependence of the configuration of polypeptides on solvent, we were led to wonder if the intermediate development of the helical structure of proteins suggested by the foregoing experiments could not be increased by altering the solvent. For this, a solvent that was miscible with water, of comparable polarity and cohesive energy density, but with less hydrogen-bonding capacity, was needed. Our search indicated that 2-chloroethanol was well suited, and it was found that the addition of this to aqueous solutions increased the helical content as measured by rotatory dispersion in nearly every case. Some examples are shown in Table II for proteins dissolved in chloroethanol.[28]

Thus, the expectation that the helical content of proteins can be increased by lowering the hydrogen-bonding capacity of the solvent is confirmed. This behavior is precisely the opposite of denaturation. In the few cases that we have studied in detail (e.g., ribonuclease and insulin), the configurational changes on going from water to chloroethanol and back to water have been completely reversible. Thus, it would appear that the helical content of proteins is a result of balance between the intramolecular hydrogen bonding of the protein and the hydrogen-bonding capacity of the solvent. When water is the solvent, the point of balance is in many cases near 50%, and can be shifted in either direction by proper alteration of the solvent.

It is of little interest to note that insulin was the only protein in the list whose helical content was not substantially increased in chloroethanol over water. Since the model of insulin proposed by Lindley has about 20% of the residues in the left-handed helical configuration, owing to restrictions imposed by the cystine bridges, it appears quite possible that insulin has approximately this configuration both in aqueous and in chloroethanol solutions. The residues in the left-handed helix cancel the effect of an equal number in the right and thus, even when the helical content is nearly 100%, only about half of this amount registers with the rotatory-dispersion method. This anomaly is removed when it is recognized that the method measures only the excess right-handed helical content, as indicated in the titles of the tables. Nevertheless, this is the sort of detail that emerges only on making additional measurements. Obviously, there is now a great need of a completely independent method to measure total helical content so that the rotatory-dispersion methods can be checked in regularly behaving cases, and the amount of left-handed helical configuration evaluated in those proteins where cystine bridges may prevent the normal development of right-handed helical configurations.

As a final illustration of the relation between rotatory dispersion and protein structure, no better example can be quoted than the current work of Kay and Bailey on *Pinna* tropomyosin.[29] Kay's study of this molecule by physical methods shows it to have the character of thin, rigid rods having the dimensions of an $\alpha$-helix of the proper molecular weight. Rotatory-dispersion measurements have now shown it to behave precisely like an $\alpha$-helix with values of $a_0^H$ and $b_0$ equal to those used

TABLE II. Excess right-handed helical contents of various proteins in chloroethanol.

| | |
|---|---|
| Tropomyosin | 110 |
| Insulin | 45 |
| Bovine serum albumin | 75 |
| Ovalbumin | 85 |
| Lysozyme | 63 |
| Pepsin | 44 |
| Histone | 72 |
| Ribonuclease | 67 |
| Globin (H) | 74 |

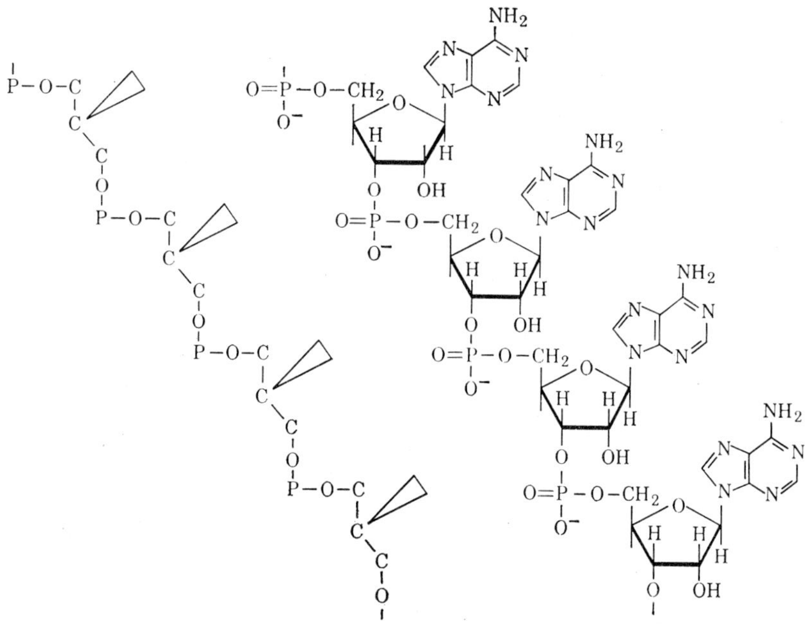

FIG. 5. The chemical structure of polyriboadenylic acid.

here.[30] Consequently, one has here the perfect example of the extreme of the scale of protein structure provided by our model. However, the work that is needed to establish this model as a framework of reference for protein structure in general remains to be done for the most part. It is certainly to be expected that numerous exceptions will be found, but there is even now reason to hope that the concept of describing the secondary structure of proteins by a partitioning of the residues between helical and nonhelical regions will play a useful role in the advance of knowledge of protein structure.

### POLYNUCLEOTIDES AND NUCLEIC ACIDS

From the viewpoint of chemical structure, nucleic acids are known to consist of chains of repeating units such as are shown in Fig. 5. The backbone is a phosphoester polymer with six chain atoms per repeating unit. The chain shown in Fig. 5 is of the ribonucleic-acid type. That of deoxyribonucleic acid differs in that the oxygen at the 2-position on the ribose ring is absent. Generally, four different monomeric units, differing in the heterocyclic rings attached to the ribose group, are found in a given nucleic acid. In ribonucleic acid (RNA), these four groups are adenine, uracil, guanine, and cytosine. In deoxyribonucleic acid (DNA), thymine (5-methyl uracil) replaces uracil. DNA is found in cell nuclei as the principal component of chromosomes, and plays the central role in carrying and passing on the genetic endowment of the cell. RNA occurs principally at the sites of protein synthesis in the cytoplasm, and is intimately involved with that process.

#### Deoxyribonucleic Acid

The nucleic-acid chain is a very flexible one and, in the absence of secondary structure, it would have the configuration of a random coil. About 10 years ago, it became evident that this could certainly not be the case for DNA. With increasing refinement in the study of DNA in solution, it became clear that this molecule was the most extended ever examined.[31] Light-scattering studies and, more recently, viscosity and sedimentation measurements, show that typical DNA samples have average molecular weights in the range of 5 to 10 million and occupy volumes in solution about one-half micron in diameter. Ordinarily, polyelectrolytes are characterized by their molecular size being very dependent on ionic strength. This was not the case for DNA. Thus, it appeared to be not only very greatly extended, but stiff rather than flexible, as well.

These observations, as well as a number of others, became understandable with the structural proposal made by Watson and Crick[32] in 1953 on the basis of the x-ray diffraction studies made by Wilkins and his collaborators.[33] This structure consisted of two antiparallel DNA chains united through hydrogen bonds connecting the heterocyclic rings (usually called bases). Only two types of pairings were permitted, those between adenine and thymine and those between guanine and cytosine as shown in Fig. 6. The resulting structure is a two-stranded helix about 20 A in diameter. With two residues every 3.4 A, a DNA molecule with a molecular weight of 10 million would have a length of 50 000 A (5 $\mu$). Such a long, thin structure would have a slight flexibility, presumably enough to account for several gentle folds that would reduce its maximum extent by about tenfold to agree with the molecular size found in solution. (Technically, it is correct to speak of this molecule as randomly coiled, but the degree of coiling is minute as compared with the single-chain coils to which this description is usually applied.)

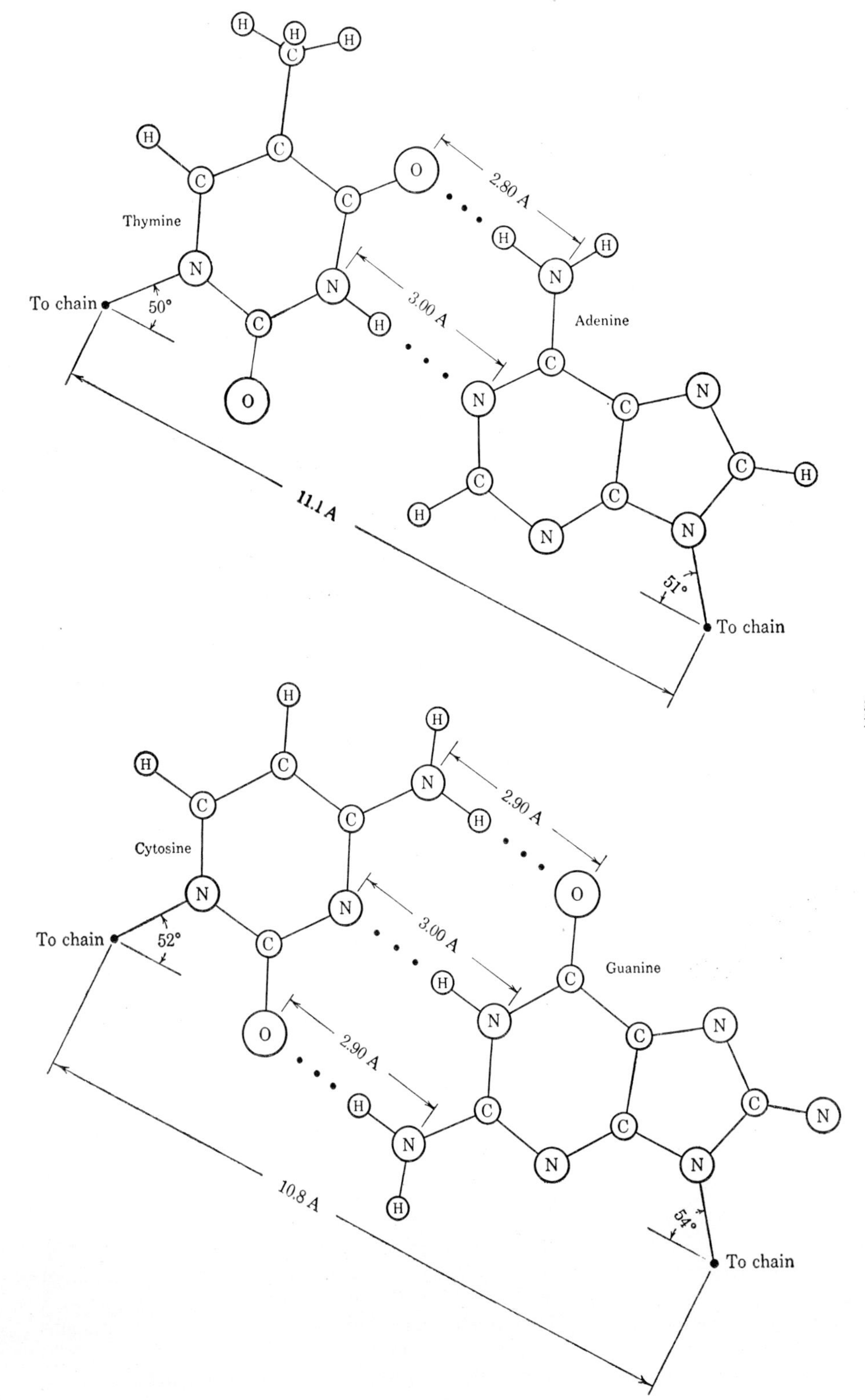


FIG. 6. Detail of the hydrogen-bonded base pairs occurring in DNA.

This periodic, secondary structure undergoes a cooperative melting out upon raising the temperature. The transition can be seen in terms of the viscosity, since the collapse of the structure is accompanied by a twenty-fold decrease in intrinsic viscosity. This is illustrated in Fig. 7.[34] The transition in this case is essentially irreversible. The reason for this is the low probability of the bases to find their original partners: without this the perfect arrangement in the original DNA cannot be recovered. This helix-coil transition can be induced by other means, for example by raising or lowering the *p*H, and it can be followed by other means, such as changes in optical rotation. However, the most sensitive indicator of structure has turned out to be the ultraviolet spectrum. DNA exhibits a broad maximum at about 2600 A, and it has been widely observed that the maximum of this absorption is substantially depressed relative to that found for the corresponding monomeric units (nucleotides). Thus, hydrolysis which brings about such a conversion from DNA in the helical form to individual nucleotides is accompanied by approximately a 40% increase in the optical density or extinction at 2600 A. This suppression of absorbance in the DNA is known as hypochromicity. Now, it is found also that, if the helical form is converted to the randomly coiled, denatured form by raising the temperature or by lowering the *p*H, a rise of nearly 40% in the extinction coefficient also occurs. Thus, electronic states of the base groups are substantially different, depending on whether or not they are held in a hydrogen-bonded, helically arranged configuration. It is not yet clear if the hypochromicity is a result of the hydrogen bonding or of the stacking of the base groups one on top of another.

### Polynucleotides

In 1955, Ochoa and Manago discovered a new enzyme which could bring about the polymerization of nucleoside diphosphates to form polymers having the chemical structure of ribonucleic acid. The study of pure polynucleotides which this has made possible has greatly increased knowledge of the basic properties which are combined in naturally occurring RNA itself. The problem that can be directly attacked with these new polyribonucleotides is the following. In 1956, Donohue[35] showed that the two specific base pairs used in the Watson-Crick structure of DNA are only two out of a large number of possible pairs. Thus, the availability of these polymers has made possible the investigation of which pairs are actually stable in aqueous solution. Warner[36] and Rich[37] quickly showed that polyadenylic acid (Poly A) and polyurdylic acid (Poly U) combined to form a double-stranded helix when their solutions were mixed. A number of other pairs have now been found, and in addition several triple-stranded helical complexes have been demonstrated. These are discussed elsewhere by Rich (p. 191).

Fresco, Klemperer, and the author have been concerned with self-pairing, as typically exhibited by Poly A.[38,39] We first investigated the sedimentation and intrinsic viscosity of a series of such polymers of different molecular weight at neutral *p*H. The molecular-weight dependence of these properties was found to be given by the molecular weight to the 0.45 and 0.65 power, respectively. These are self-consistent values for random coils. The absence of a significant birefringence of flow and the rise in viscosity that was observed when the ionic strength was lowered forced us to conclude that the configuration was that of randomly coiled, single chains. However, upon lowering the *p*H, we noticed a sharp transition from one type of ultraviolet adsorption to another. The titration curve showed a similar abruptness that could occur only if some cooperative transition were taking place. At *p*H's below this transition, the solutions showed very marked negative birefringence of flow. Moreover, the molecular weights were found to be greatly increased.

These observations clearly indicated that a cooperative association was taking place, upon lowering the *p*H through a critical value. The nature of this was further clarified in the following way. Neutral solutions of a single polymer were made up at a series of different concentrations. When these were acidified and the sedimentation and viscosities determined, the results showed the regular behavior that can be seen on the right-hand side of Fig. 8. It is seen that the molecular weights, or rather particle weights, span a twentyfold range and that the sedimentation constants and viscosities vary in the manner expected for homologous polymers. The weight and size increase with the concentration at which they were formed. The respective slopes are again self-consistent and their numerical values, 0.36 and 0.92, are indicative of a more-extended chain structure than was present in the neutral solutions. The results already mentioned at neutral *p*H are seen at the left of Fig. 8.

This evidence on the acid-stable form indicates fairly clearly that the polyadenylic-acid molecules have joined together in a regular structure. We were then able to show that the amino group of adenine is unavailable for reaction with formaldehyde in the acid-stable complex,

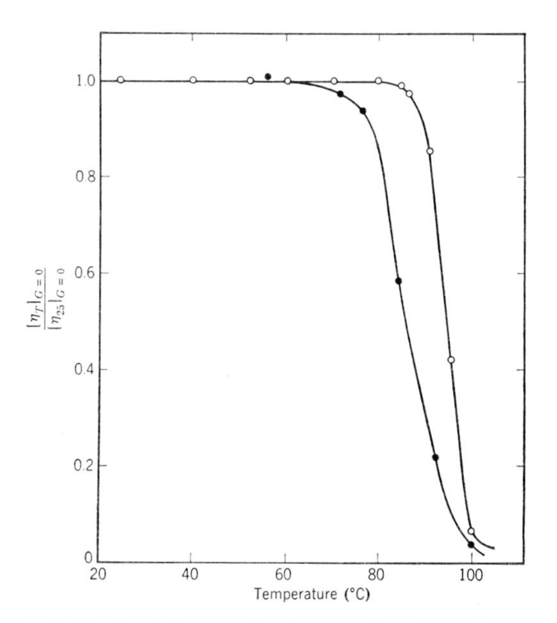


FIG. 7. The intrinsic viscosity (measured at 25°C) of DNA as a function of the temperature to which it has been heated for one hour.

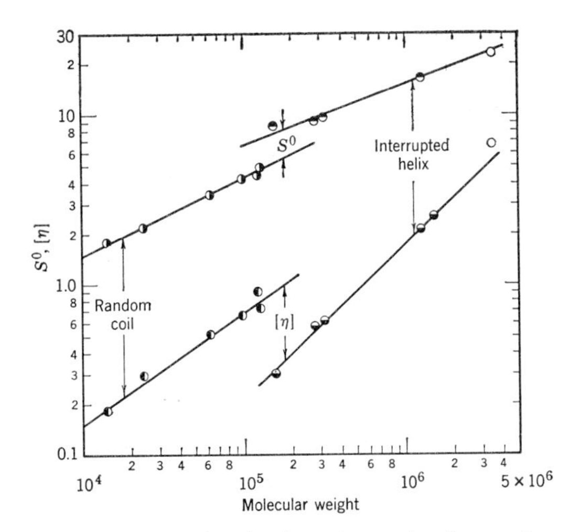


FIG. 8. The molecular-weight dependence of sedimentation constant and intrinsic viscosity for the two forms of polyadenylic acid.

but that it is highly reactive in the randomly coiled state. This showed that the pairing of the adenine groups was responsible for the association. When a model is built to satisfy this type of pairing, it is seen that it must resemble the double-helical configuration of DNA with the adenine bases nearly perpendicular to the axis. The structure requires that bases be 3.8 A apart. To test this point, J. Fresco[38] has taken x-ray photographs of solutions of Poly A (1 to 10%) above and below the transition. At low *p*H, he finds several very sharp rings, including one with a spacing of 3.8 A. At neutral *p*H, there are none. From all of this information, we conclude, therefore, that polyadenylic-acid molecules react as shown in Fig. 9. For the helix to form, the negative charge owing to the phosphate groups must be at least half-neutralized by the uptake of protons.

This behavior of Poly A is similar in many ways to other pairwise interactions that have been studied. Thus, we have nearly reached the point where all possible pairwise interactions have been cataloged. At present, it appears that, of the ten possible combinations between the four pure nucleotide polymers, 6 or 7 actually exist. The next problem is to list these pairwise interactions in order of their strength in analogy with the tables of bond energies one has in the case of covalent bonds. This information can be obtained by a careful study of the temperature at which the respective helix-coil transitions take place.

## Ribonucleic Acid

Although DNA and RNA are so very similar in their chemical structure, their configurations are strikingly different. This can be said even though the configuration of RNA is only now beginning to become clear. I conclude by summarizing some investigations that have been made in this laboratory on RNA from calf-liver microsomal particles by Hall[40,41] and on RNA from tobacco mosaic virus by Boedtker.[42]

All examinations of RNA solutions have indicated that the intrinsic viscosity of such solutions is suprisingly low when compared with the molecular weight. This is quite the opposite of the case of DNA. Furthermore, RNA shows only a small birefringence of flow, and this is of the opposite sign of DNA. Thus, it is certain that the configuration of RNA in solution is not that of the extended double-stranded helix found in DNA.

An examination of the dependence of sedimentation and intrinsic viscosity on molecular weight shows that it is very close to the 0.5 power in both cases. This can be unambiguously interpreted as showing that the molecules are highly coiled and that the coils are contracted somewhat in comparison with the usual polymer chain. This result is surprising because the RNA chain carries negative charges on each repeating unit and, as a consequence, at the ionic strength (0.01 *M*) where these observations were made, it would be expected that the electrostatic repulsions would result in a very highly swollen polymer coil. If this had been the case, the intrinsic viscosities would have been much larger and the molecular-weight dependences far different. These considerations clearly suggest that there are very substantial intramolecular attractions which hold the molecules in contracted configurations.

The next problem lies in the identification of these points of internal attraction. That they indeed could be broken, and reversibly formed again, was shown in two ways.[40,41] By removing the salt from the solution, the viscosity was found to rise indicting that, by increasing the electrostatic repulsion sufficiently, the intramolecular bonds could be broken, allowing the coil to expand in the manner expected for a polyelectrolyte.

Similarly, it was found that the viscosity would rise with the temperature of the RNA solution. This again indicated the breakup of the intramolecular bonding.

It is known that RNA is hypochromic relative to its hydrolysate. Consequently, it was of interest to see if the optical density increased upon removing the salt or upon heating. Such was indeed found to be the case. A quantitative study showed that the increase in extinction was parallel with the increase in viscosity. Thus,

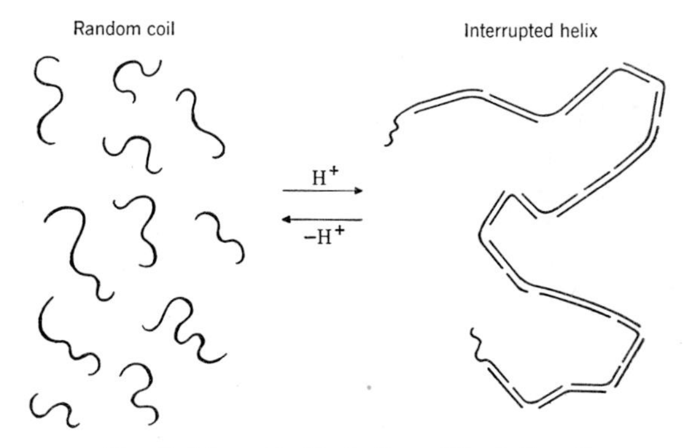


FIG. 9. Schematic illustration of the helix-coil transition in polyadenylic acid.

the base pairing through hydrogen bonds appears to be the cause of the contracted state of the RNA.

By referring to the extinction-coefficient changes accompanying helix-coil transitions in the study of polynucleotides, it is possible to estimate the fraction of base pairs that exist in RNA solutions at moderate ionic strength and at room temperature. This comes out to be the surprisingly large figure of about 50%. At present, this conclusion can only be tentative, but it does raise a very interesting question. The studies of polynucleotide interactions reveal the large number of different types of base pairing that are possible among the bases occurring in RNA. We now have the strong indication that such pairing does occur to a great extent. Obviously, the next problem to face is whether or not such extensive base pairing occurs in a random fashion, or whether there is sufficient organization in the RNA structure to justify describing it as secondary structure.

Finally, it is important to recognize that RNA occurs in the cytoplasm, not as a discrete substance, but in intimate association with protein in the microsomal particles. Unlike DNA, which can of itself be biologically active (as revealed in bacterial transformation), RNA performs its function as a complex with protein, and it is to the analysis of the configuration of both components in these particles that much future effort is certain to be directed.

## BIBLIOGRAPHY

[1] A more detailed exposition may be found in P. Doty, Proc. Natl. Acad. Sci. U. S. **42**, 791 (1956).

[2] L. Pauling, R. B. Corey, and H. R. Branson, Proc. Natl. Acad. Sci. U. S. **37**, 205 (1951).

[3] L. Pauling and R. B. Corey, Proc. Natl. Acad. Sci. U. S. **37**, 235 (1951).

[4] M. F. Perutz, Nature **167**, 1053 (1951).

[5] P. Doty, A. M. Holtzer, J. H. Bradbury, and E. R. Blout, J. Am. Chem. Soc. **76**, 4493 (1954).

[6] C. Cohen, Nature **175**, 129 (1955).

[7] P. Doty, J. H. Bradbury, and A. M. Holtzer, J. Am. Chem. Soc. **78**, 947 (1956).

[8] J. T. Yang and P. Doty, J. Am. Chem. Soc. **78**, 498 (1956).

[9] P. Doty, A. Wada, J. T. Yang, and E. R. Blout, J. Polymer Sci. **23**, 851 (1957).

[10] A. Wada and P. Doty, J. Am. Chem. Soc. (to be published).

[11] See, e.g., P. Doty and E. P. Geiduschek in *The Proteins*, H. Neurath and K. Bailey, editors (Academic Press, Inc., New York, 1953), Vol. I, p. 393.

[12] See, e.g., R. B. Simpson and W. Kauzmann, J. Am. Chem. Soc. **75**, 5139 (1953).

[13] J. T. Yang and P. Doty, J. Am. Chem. Soc. **79**, 761 (1957).

[14] J. Applequist and P. Doty, Abstracts, Am. Chem. Soc. Meeting, San Francisco (April 7–11, 1958).

[15] J. T. Yang and P. Doty (unpublished results).

[16] P. Doty, K. Imahori, and E. Klemperer, Proc. Natl. Acad. Sci. U. S. **44**, 424 (1958).

[17] W. Moffitt, J. Chem. Phys. **25**, 467 (1956).

[18] D. D. Fitts and J. G. Kirkwood, Proc. Natl. Acad. Sci. U. S. **42**, 33 (1956).

[19] D. D. Fitts and J. G. Kirkwood, J. Am. Chem. Soc. **78**, 2650 (1956).

[20] W. Moffitt and J. T. Yang, Proc. Natl. Acad. Sci. U. S. **42**, 596 (1956).

[21] W. Moffitt, D. D. Fitts, and J. G. Kirkwood, Proc. Natl. Acad. Sci. U. S. **43**, 723 (1957).

[22] D. D. Fitts and J. G. Kirkwood, Proc. Natl. Acad. Sci. U. S. **43**, 1046 (1957).

[23] K. Imahori, E. Klemperer, and P. Doty, Abstracts, Am. Chem. Soc. Meeting, Miami, Florida (April 1–7, 1957).

[24] K. Imahori, E. Klemperer, and P. Doty, "Estimate of $\alpha$-helical content of proteins," J. Am. Chem. Soc. (to be published).

[25] See W. Kauzmann [Ann. Rev. Phys. Chem. **8**, 413 (1957), reference 3], for a demonstration of the existence of a square term for coupled oscillators.

[26] J. C. Mitchell, A. E. Woodward, and P. Doty, J. Am. Chem. Soc. **79**, 3955 (1957).

[27] K. Imahori, P. Doty, and E. Blout (in preparation).

[28] K. Imahori, E. Klemperer, and P. Doty (in preparation).

[29] C. M. Kay and K. Bailey, Biochem. et. Biophys. Acta (to be published).

[30] C. M. Kay, Biochem. et Biophys. Acta **27**, 469 (1958).

[31] For a review, see for example: P. Doty, J. Cellular Comp. Physiol. **49**, Suppl. 1, 27 (1957).

[32] F. H. C. Crick and J. D. Watson, Proc. Roy. Soc. (London) **A233**, 80 (1954).

[33] M. H. F. Wilkins, W. E. Seeds, A. R. Stokes, and H. R. Wilson, Nature **172**, 759 (1953).

[34] S. A. Rice and P. Doty, J. Am. Chem. Soc. **79**, 3937 (1957)

[35] J. Donohue, Proc. Natl. Acad. Sci. U. S. **42**, 60 (1956).

[36] R. C. Warner, Ann. N. Y. Acad. Sci. **69**, 314 (1957).

[37] R. Rich, *The Chemical Basis of Heredity*, W. D. McElroy and B. Glass, editors (The Johns Hopkins Press, Baltimore, Maryland, 1957), p. 557.

[38] J. R. Fresco and P. Doty, J. Am. Chem. Soc. **79**, 3928 (1957).

[39] J. R. Fresco and P. Doty, "Polynucleotides II: The x-ray diffraction patterns of solutions of the coil and helical forms of polyriboadenylic acid," J. Molecular Biol. (to be published).

[40] B. Hall and P. Doty, "The configurational properties of ribonucleic acid isolated from microsomal particles of calf liver," *Symposium on Microsomal Particles*, American Biophysical Society (Pergamon Press, London, to be published).

[41] B. Hall and P. Doty, "The preparation and physical chemical properties of ribonucleic acid from microsomal particles," J. Molecular Biol. (to be published).

[42] H. Boedtker, "Some physical properties of infective ribonucleic acid isolated from tobacco mosaic virus," Biochem. et Biophys. Acta (to be published).

# 13

# Application of Infrared Spectroscopy to Biological Problems

G. B. B. M. Sutherland

*National Physical Laboratory, Teddington, Middlesex, England*

## INTRODUCTION

PROGRESS in biology at the molecular level is becoming increasingly dependent on physical and physicochemical techniques such as x-ray analysis, radioactive tracers, chromatography, light scattering, and spectroscopy. The purpose of this paper is to give a broad review of the actual and potential contributions of infrared spectroscopy to biology. Since several excellent articles and papers have already been written giving detailed accounts of the practical application of infrared methods to selected problems in biology, detail is avoided as much as possible in this presentation. In the writer's opinion, infrared methods have not been sufficiently exploited in the biological field and, since this may well be because many biologists find it hard to assess their value, it is hoped that a rather general discussion of the advantages and disadvantages of infrared analysis will be of value to biologists.

The infrared region of the spectrum is conventionally defined as extending from the limit of the visible (about 7500 A or 0.75 $\mu$) to the present lower wavelength limit of the microwave region (about 1000 $\mu$), but only a relatively small range of these wavelengths is of much interest to biologists, *viz.*, 2.5 to 15 $\mu$. There are two reasons for this: the first is that within these limits are found the vast majority of spectra which throw any light on molecular structure; the second is that this region can be conveniently scanned and recorded in about ten minutes by a commercially available instrument with a rocksalt prism as the dispersing element. The exclusion of infrared wavelengths beyond 15 $\mu$ does not seem so strange when expressed in frequency* rather than wavelength units (i.e., 670 to 10 $cm^{-1}$), since the majority of the fundamental vibration frequencies of molecules which can be readily assigned to highly localized motions of one or both of the atoms in a chemical bond lies between 4000 $cm^{-1}$ (2.5 $\mu$) and 600 $cm^{-1}$. The limits of 670 and 4000 $cm^{-1}$ given in the foregoing are set by the fact that beyond 670 $cm^{-1}$ absorption by the rocksalt prism rapidly becomes complete, while above 4000 $cm^{-1}$ the dispersion is rather low. However, it is becoming easy to replace prisms by gratings, and so these limits need no longer be set by the experimental technique. The first reason given is the dominant one, *viz.*, most of the interesting absorption bands which can be interpreted with certainty lie between 4000 and 600 $cm^{-1}$.

* The conventional frequency unit in infrared spectroscopy is the wave number ($cm^{-1}$), which expresses the number of waves in 1 cm.

The general interpretation of the infrared spectra of molecules is very straightforward; even the detailed interpretation is not difficult, provided the objective is, for instance, the identification of a compound in a mixture, or the establishment of the presence of a characteristic chemical grouping in a molecule. But, to determine the configuration of a biological polymer such as a protein is a difficult and complex problem which requires the same degree of expertise as that needed in the x-ray analysis of protein structure.

## GENERAL INTERPRETATION OF INFRARED SPECTRA OF MOLECULES

The average polyatomic molecule of biological interest has an absorption spectrum in the region between 600 and 4000 $cm^{-1}$ consisting of 20 to 30 discrete absorption bands of which about 5 to 10 will usually be much more intense than the remainder. This is because certain of the modes of vibration of such a molecule cause a periodic change in the dipole moment of the molecule. Thus, radiation of the appropriate frequency incident on the molecule can be absorbed and excite the corresponding mode of vibration. Some bands are more intense than others because the change in dipole moment is greater for some of the modes of vibration. Theoretically, a polyatomic molecule (of $n$ atoms) has an infinite number of modes of vibration, all of which can be built up from the $3n$-6 fundamental modes of vibration. However, the change in dipole moment associated with overtone and combination frequencies is usually an order of magnitude less than that associated with the fundamental modes, and in the region of the spectrum under discussion the majority of the observed bands can be assigned to fundamental modes.

How can these fundamental modes be vizualized? Although, strictly speaking, each fundamental mode involves some motion of every atom in the molecule and would, therefore, seem to be hard to predict and describe, it turns out that an appreciable fraction of the fundamental modes of any molecule can be assigned to easily vizualized vibrations of individual chemical bonds, or of small structural units in the molecule. This is especially true of the fundamental modes which are associated with the motions of hydrogen atoms. For instance, any molecule containing an NH group will be found to have a fundamental vibration near 3300 $cm^{-1}$ owing to the stretching and contraction of the NH bond and two other fundamentals between 1600 and 600 $cm^{-1}$ in which the motion of the hydrogen atoms is almost perpendicular to the NH bond. More generally, in the region between 3700 and 2500 $cm^{-1}$ are found fundamental modes owing to such "hydrogenic-stretching frequencies" in the following order:

OH 3650−3150 $cm^{-1}$,

NH 3500−3150 $cm^{-1}$,

CH 3150−2850 $cm^{-1}$,

SH 2650−2550 $cm^{-1}$,

PH 2450−2300 $cm^{-1}$.

No other fundamental modes appear in this region of the spectrum, and so identification of any of these bonds in a compound is generally easy. There are occasional limitations, e.g., difficulty of distinguishing OH from NH, and the weakness of the SH fundamental; these are not discussed here.

The corresponding "hydrogenic deformation frequencies" for OH, NH, and CH occur between 1650 and 600 $cm^{-1}$, but cannot be characterized so simply. Moreover, many other fundamental modes occur within this range whose assignment and identification present special problems. However, several groups (such as $CH_2$ and $CH_3$ and the peptide link) can readily be identified by characteristic fundamentals of this type.

Before leaving the hydrogenic frequencies, mention should be made of the ease with which hydrogen bonding can be detected through its effect on these frequencies. The stretching frequencies are lowered by a few hundred $cm^{-1}$, and the corresponding absorption bands become quite broad and diffuse. The deformation frequencies are increased to a much smaller degree and the broadening is not always so marked.

In the region between the hydrogenic-stretching and hydrogenic-deformation frequencies (*viz.*, 2300 to 1650 $cm^{-1}$) occur two other separable classes of fundamentals: firstly, those owing to the stretching of triple bonds such as C≡N and C≡C occurring between 2250 and 2100 $cm^{-1}$, and, secondly, those largely localized in the stretching vibrations of double bonds such as C=O, C=C, C=N. The latter start about 1800 $cm^{-1}$ and overlap a little the high-frequency end of the hydrogenic-deformation frequencies at 1650 $cm^{-1}$. This means that it is occasionally difficult to decide whether a band occurring between 1650 and 1500 $cm^{-1}$ is the result of the stretching of a double bond or of the deformation of a hydrogen atom. If the hydrogen atom can be substituted by deuterium, this uncertainty is often resolved, as a hydrogenic frequency will be reduced by almost a factor of $\sqrt{2}$, whereas the double-bond frequency will be unaffected. It should be added that occasionally a hydrogenic-deformation frequency interacts with a double-bond frequency, and the two fundamentals resulting can only be described as a superposition of these two motions.

A very large number of the chemical bonds in any polyatomic molecule are of the type C−C, C−O, C−N. Such bonds do not give rise to localized fundamental vibrations. The reason is that the masses of these atoms are not very different, nor are the restoring forces between them. This means that their characteristic stretching frequencies (which can be observed in small molecules such as ethane or methyl alcohol, where they occur in isolation) are all about the same magnitude (approximately 900 to 1100 $cm^{-1}$). Consequently, in a polyatomic molecule, strong coupling occurs and the resulting fundamental modes involve simultaneous motions of all of the bonds of this type in the molecule. Such fundamentals are usually referred to as "skeletal modes" and have a range between about 800 and 1250 $cm^{-1}$. The pattern of these skeletal frequencies is often the most characteristic physical property of a molecule. Two hydrocarbons or two steroids, which may be very hard to differentiate by chemical means or by other physical properties (e.g., refractive index or melting point), can usually be recognized instantly by comparing their infrared spectra between 700 and 1300 $cm^{-1}$.

## APPLICATION TO BIOLOGICAL PROBLEMS

There are three principal ways in which infrared spectroscopy can be of help in molecular biology, *viz.*:

(a) as an analytical tool, (b) as a means of establishing the structural formula of a biologically important compound, and (c) as a means of determining the spatial configuration of biological polymers.

### (a) Analysis

Infrared spectroscopy is now a well-established and widely used analytical tool in any up-to-date biochemical laboratory. The principal advantage of infrared over visible or ultraviolet spectroscopy is that the spectrum is much richer, since it is the result of vibrations in every part of the molecule. Spectra in the visible and ultraviolet arise from the excitation of an electron in one part of the molecule, usually in a double bond such as CO group. In analytical work, one therefore has to be certain that this group is not present in any of the other compounds in the mixture. On the other hand, infrared methods will usually reveal every one of the compounds in a mixture. The only serious disadvantage of infrared analysis is that the use of water as a solvent is generally impossible because of its intense absorption over a good deal of the working range. This can frequently be overcome either by the use of other solvents, or (in the case of a solid) by making a pressed disk of the compound ground very finely in an excess of KBr. One other point should be mentioned, since it is frequently very important in much biological work, *viz.*, the minimum quantity which is necessary in order to make an identification. It is impossible to give a precise figure here, because the intrinsic intensity of absorption in the infrared varies over such a wide range between compounds, and reference must be made to the literature on this subject. As a general guide, one may say that, by using specially designed cells, it is possible to get a spectrum from about a milligram of material, but that, by using a reflecting microscope attachment, this limit can be reduced to below one microgram.

### (b) Structural Formulas

One of the earliest applications of infrared spectroscopy to the structure of a biologically important molecule occurred in the case of penicillin. Here, the chemists were unable to agree on which of the three possible formulas was the correct one. Although x-ray analysis gave the first and most unequivocal proof that the β-lactam formula was the correct one, infrared methods gave an independent proof and demonstrated that no structural changes took place in going from the solid state (required for x-ray work) into solution, in which penicillin is known to be a rather labile molecule. This work depended on differences in the type and in the environment of the double bonds in the three possible structures. By investigating the spectra of model compounds containing double bonds of various types in appropriate environments, a decision could finally be made in favor of the lactam structure.

The biological molecules which have been most extensively investigated by this method are the steroids. In such compounds, it is now possible to answer quite a variety of important structural questions concerning the position of CO, OH, $CH_2$, and C=C groups. On the other hand, it is still not possible to recognize the class of steroids by any common feature running through all of their spectra, although certain closely related groups of steroids do show common features. We find the opposite extreme in the proteins which, as a class, are easily recognized and differentiated from nucleic acids and lipids, but show remarkably little variation among one another. The reason is that the spectrum of any protein is dominated by several intense bands arising from the identical peptide links. Only in the few cases where two or three of the constituent amino acids are in great excess is it possible to identify individual amino-acid residues (e.g., glycine and alanine in silk).

The general line of attack is to compare the spectrum of the compound whose structure is unknown with the spectra of compounds of known structure containing groups of atoms identical with and in a similar environment to those suspected of occurring in the unknown. Extensive collections of infrared spectra have now been compiled by various laboratories and organizations, and the comparisons can be made relatively rapidly by various mechanized sorting devices. At first, it is usually advisable to call in an experienced spectroscopist for final confirmation, but any laboratory which takes up infrared analysis in a particular chemical field is not long in developing its own expert. The logic behind the whole process is not unlike that behind the chemical attack on the same problem. The chemist identifies certain groups in an unknown structure by their reactive properties, and he degrades the molecules into simpler ones which he can identify from their well-known chemical or physical properties. The spectroscopist identifies individual chemical bonds in an unknown structure from their known spectroscopic earmarks and tries to supplement this by identifying groups or constituent units in a structure (e.g., peptide link, benzene ring) in a similar way. The most efficient approach is, of course, to combine the two methods.

### (c) Determination of Spatial Configuration of Biological Polymers

In order to understand many key biological phenomena, such as the mode of action of genes, protein synthesis, or the production of antibodies, at the molecular level, it will be necessary to establish in detail the molecular configuration of proteins and nucleic acids. This must be done by physical methods, of which the most successful so far has been x-ray analysis. However, in spite of the intensive efforts of several groups of very able workers in various countries over the past twenty years, the spatial configuration of any globular protein is still unknown. While it is true that one now seems to be very close to this goal in the case of myoglobin, the extension of the methods used for that protein to other proteins presents formidable problems. At present, infrared analysis seems to be the second most powerful physical method. Although it has not yielded such precise results, it has often given independent confirmation of the main features of a structure. More important is the fact that it can often give a lead at any early stage, either in laying down certain conditions, which the correct model structure must fulfill, or in ruling out certain models which might appear to fit a preliminary analysis of the x-ray data. Just as the x-ray method requires the polymers to be arranged in an orderly pattern (preferably in a single crystal), so the infrared method can be more precise the closer the arrangement of the polymers approaches that found in a single crystal. This is because the infrared method is based on observing the dichroism associated with key absorption bands when the spectrum is obtained using polarized radiation.

An oversimplified example will make this clearer. In the infrared spectrum of deoxyribonucleic acid is an absorption band which can be assigned to vibrations localized in the planes of the purine and pyrimidine bases. In the Watson-Crick model of DNA, these bases are nearly perpendicular to the axis of the double helix. If a highly oriented specimen of DNA is prepared in which the axes of the helices are roughly parallel to the direction of orientation, then this absorption band is very intense when the incident infrared radiation is polarized with the electric vector perpendicular to the direction of orientation, and very weak when the direction of polarization is rotated through a right angle (Fig. 1). This gives a general confirmation of one aspect of the Watson-Crick model. If it were possible to produce a single crystal of DNA of suitable dimensions in which all of the bases were parallel to each other, then the dichroism would become perfect (i.e., no absorption for

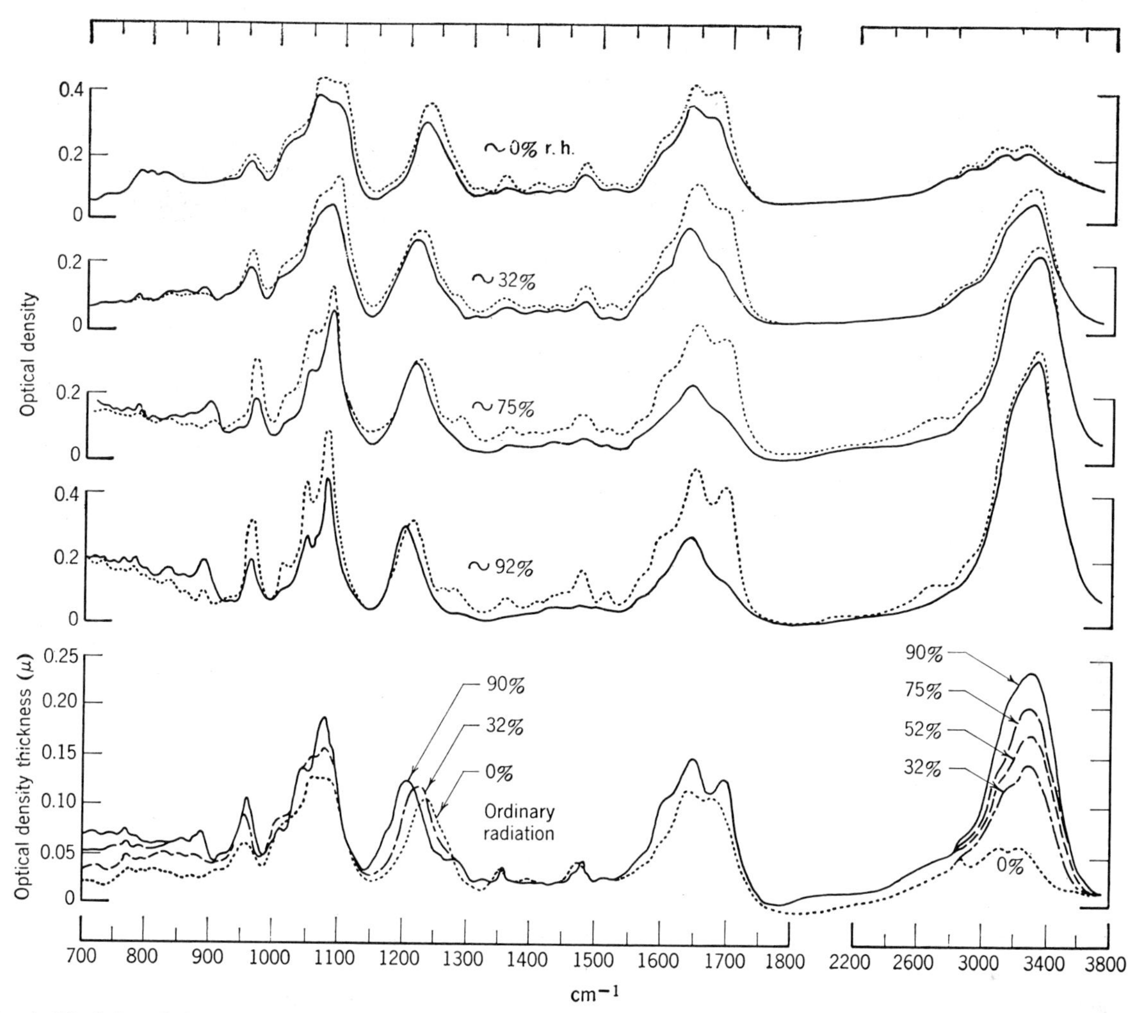


FIG. 1. The infrared absorption spectrum of NaDNA at various relative humidities. ——— indicates electric vector parallel to the orientation direction; ...... indicates perpendicular to the same direction. The bands referred to in the text are those between 1600 and 1700 $cm^{-1}$ [from G. B. B. M. Sutherland and M. Tsuboi, Proc. Roy. Soc. (London) **A239**, 446 (1957)].

one polarization direction) and a precise confirmation of this aspect of the model could be given.

From the foregoing, it might appear that the infrared methods should, therefore, be applied to single crystals of biological polymers. Unfortunately, there are severe practical difficulties about such an approach. The first is that the "suitable dimensions" referred to in the foregoing hardly ever can be realized, *viz.*, a flat crystal about 10 $\mu$ thick and 2×5 mm in area. A second difficulty is that many biological polymers crystallize only in association with a high proportion of water. This limits observation to the bands which are not obscured by the spectrum of $H_2O$ or $D_2O$. The third is that the molecular configuration is unknown and assumptions have, therefore, to be made about the orientations of recurring chemical groups or bonds in the polymer. The number of reasonable assumptions from which a choice has to be made can be quite large, and infrared analysis may not prove to be a sensitive enough discriminant between the various possibilities. In cases where the number of ways of fitting a long polymer into the unit cell is severely restricted, this method should give a precise answer, and it may be that the first experimental difficulty referred to can be overcome by investigating with a microspectrometer the reflection spectrum, instead of the absorption spectrum. Another line of attack is to measure the dichroism of the overtone and combination bands above 4000 $cm^{-1}$, since these are much weaker than the fundamentals and can be conveniently studied in a reasonably thick crystal. It should be added that there are still some theoretical problems to be overcome in the interpretation of such spectra.

Because of all of these difficulties, most of the infrared investigations to date have been made on thin sections of fibrous proteins such as porcupine quill, silk, etc. In such cases, the orientation is far from perfect and obviously the degree of disorientation must be estimated quantitatively before any reliable deductions can be made from the partial dichroism. A method of doing this has been recently worked out and applied to a number of proteins, *viz.*, porcupine quill, elephant hair, horse hair, horn, feather quill, silkworm gut, and collagen. For each of these proteins, seven different structures were tested, and it was possible to reject all but one or two of the model structures in most cases.

It must be emphasized that this method of attack depends on certain assumptions which may not always be correct. These assumptions are as follows:

(1) The direction of change of electric moment (associated with a particular absorption band) bears the same relation to the repeat unit of the polymer as that determined from infrared studies on single crystals of small molecules which contain the repeat unit; (2) the spatial distribution of the polymer chains has a certain degree of symmetry; (3) the specimen studied consists of one structural species.

The first assumption is necessary, because it is found from work on single crystals of small molecules containing a peptide link that the change of dipole moment in a bond stretching fundamental vibration (e.g., NH or CO) is not exactly along the bond in question, but may make an angle of as much as 20° with it. Thus, the directions of these bonds (and consequently the molecular configuration) cannot be determined unless the corresponding angles are known for a peptide link in a protein. Up to the present, single crystals of only two different molecules have been investigated. The angles in these two crystals did not differ by more than five degrees. However, more work on single crystals of a few other molecules is required in order to estimate accurately the degree of uncertainty arising from this assumption. The second assumption can be shown to be a reasonable one if the sample is prepared in a certain way. At first sight, the third assumption seems reasonable and is made by the x-ray analysts. The reason for doubting it is that the structure of many of the infrared bands of proteins is complex. Whether this arises from the presence of two different configurations, or from differences in frequency between a paracrystalline and less-ordered arrangements of the polymer molecules, cannot easily be determined. In simple synthetic polymers, such as polyethylene, differences in frequency are found between crystalline and amorphous forms of the polymer. Structure can also occur in the absorption band of a single crystal because of interaction between neighboring molecules.

It appeared at one time that infrared analysis might be able to distinguish between various configurations of a protein molecule (specifically between the $\alpha$-helix and the fully extended form of the polypeptide chain), even when the polypeptide chains were not arranged in any ordered pattern. This method depended on a change in the frequency of the CO absorption band in synthetic polypeptides which could be prepared in either of these two forms by precipitation from the appropriate solvent. However, there are so many possible reasons for a small change in the frequency of an absorption band that the application of any such rule to proteins (even on an empirical basis) leads to serious difficulties and inconsistencies. Until the reasons for these changes are well understood from the study of simpler model compounds, it seems very unwise to use them as a guide to protein structure.

The general situation with the infrared analysis of protein structure is similar to that which existed at an earlier stage in the x-ray analysis of the same problem. Before the interatomic distances in the peptide link were established by careful work on small molecules, it was impossible to establish the correctness of any protein structure by x-ray methods. One could only say that such and such a structure was generally consistent with the diffraction pattern. Now that these distances are known and that single crystals of proteins can be prepared containing a heavy atom, it is becoming feasible to locate the individual atoms. Similarly, before the infrared method can become precise, more work is required on single crystals of simple compounds containing the peptide link and on the experimental problems of obtaining satisfactory spectra from single crystals of proteins.

## BIBLIOGRAPHY

The following bibliography is not meant to be comprehensive. It gives references to most of the general articles on the subject within which references can be found to the original papers. A few references have been given to very recent papers which are not dealt with in the general articles.

M. Beer, G. B. B. M. Sutherland, K. N. Tanner, and D. L. Wood, "Infrared spectra and structure of proteins," Proc. Roy. Soc. (London) **A249**, 147 (1959).

L. J. Bellamy, *The Infrared Spectra of Complex Molecules* (Methuen and Company, Ltd., London, 1954).

E. R. Blout, M. Parrish, Jr., G. R. Bird, and M. J. Abbate, "Infrared microspectroscopy," J. Opt. Soc. Am. **42**, 966 (1952).

C. Clark, "Infrared spectrophotometry" in *Physical Techniques in Biological Research*, G. Oster and A. W. Pollister, editors (Academic Press, Inc., New York, 1955), p. 205. (This is the best general review of the subject.)

K. Dobriner, E. R. Katzenellenbogen, and R. N. Jones, *Infrared Absorption Spectra of Steroids* (Interscience Publishers, Inc., New York, 1953).

A. Elliott, "The structure of polypeptides and proteins" in *Proceedings of the Third International Congress of Biochemistry, Brussels, 1955* (Academic Press, Inc., New York, 1956), p. 106.

R. D. B. Fraser, "Infrared spectra of biologically important molecules" in *Progress in Biophysics*, J. A. V. Butler and J. T. Randall, editors (Pergamon Press, London, 1953), Vol. III, p. 47.

*Biological Applications of Infrared Spectroscopy*, Ann. New York Acad. Sci., **69**, 1–254 (1957).

H. M. Randall, R. G. Fowler, N. Fuson, and J. R. Dangl, *Infrared Determination of Organic Structures* (D. Van Nostrand Company, Inc., Princeton, New Jersey, 1949).

G. B. B. M. Sutherland, "Infrared analysis of the structure of proteins and nucleic acids" in *Convegno Antonio Baselli le Macromolecole Viventi*, Ist. Lombardo sci. lettere **89**, 67 (1955).

G. B. B. M. Sutherland and M. Tsuboi, "Infrared spectrum and molecular configuration of sodium deoxyribonucleate," Proc. Roy. Soc. (London) **A239**, 446 (1957).

D. L. Wood, "Infrared spectrophotometry" in *Methods in Enzymology* (Academic Press, Inc., New York, 1957), Vol. IV, p. **104**.

# 14
# Concentrated Macromolecular Solutions

BRUNO H. ZIMM

*Physical Chemistry Section, General Electric Research Laboratory, Schenectady, New York*

DILUTE solutions were covered in the papers of Rice (p. 69), Stockmayer (p. 103), and Doty (p. 61). This paper first discusses concentrated solutions in contrast to dilute solutions. Many properties in concentrated solutions are determined by the fact that the interactions between the molecules are of primary importance, whereas they are of secondary importance when the dilution is high.

In most macromolecular solutions, one may never neglect completely the interactions between the large molecules. The nearest exception to this is probably solutions of the globular proteins at high ionic strength and at reasonable concentrations—i.e., concentrations in which it is quite easy to do physical experiments, such as the determination of intrinsic viscosity, and so on. In any solution, interactions can be ignored fairly well, providing highly precise measurements are not desired. As soon as extended particles—like fibrinogen, or especially the nucleic acids, or the randomly coiled polymers such as the synthetic ones, or starch or other polysaccharides—become involved, it is practically impossible to make them sufficiently dilute to eliminate the interactions between the molecules, and, at the same time, to have enough material remaining in solution to measure by the standard techniques. In these cases, it might be said that one is dealing always with concentrated solutions.

For practical purposes, the interaction can be characterized by the following. The osmotic pressure *versus* the concentration is plotted. The ideal-solution laws predict that the osmotic pressure divided by the concentration should be practically constant. The van't Hoff law says that the osmotic pressure is proportional to the concentration. For an ideal solution, one obtains, then, the curve $a$ in Fig. 1. Most of the time, however, curve $b$ is obtained. As soon as the curve has deviated from the ideal law by about as much as the ideal law itself predicts, one has a concentrated solution.

This upward, or positive, deviation usually results from repulsion between the molecules, and the fact that the molecules are large means that there are unavoidably a lot of repulsions just from their space-filling properties. This is the common type of behavior in a macromolecular solution; in effect, the molecules elbow each other apart. There is the same type of behavior in the theory of imperfect gas where one must consider the "co-volume" of the molecules. The pressure is raised in consequence.

A similar effect is observed on the scattering of light. If there are these repulsive interactions, the fluctuations of concentrations can not take place as easily or as independently of each other, and with less fluctuation there is less scattering of light.

These are manifestations of the intermolecular forces. There is one trap into which the uninitiated easily fall, and that is to consider these intermolecular forces as acting as though the molecules were in a vacuum. The molecules are in a solvent, and the forces between the molecules and the solvent molecules and other solvent molecules are as important as the direct interaction between the macromolecules. What is seen here actually is a balance of forces.

Figure 2 shows some data obtained by Doty and Yang[1] on polybenzylglutamate. The fraction of internal hydrogen bonds in a molecule are shown. When they are all intact, the helical form results, and when none of them are intact, the random-coil form results. The transition region between these two forms depends upon the solvent composition, and this particular solvent is a mixture of dichloroacetic acid and dichloroethane. Taking the transition temperature as the reference point, it is seen that, when the temperature is reduced, one goes to the unbonded form. If the temperature is raised, one goes to the bonded form. It is incorrect to consider solely the interactions between parts of the macromolecule itself because, in that case, an increase in temperature should melt it— that is, break the bonds. In fact, just the opposite is done in raising the temperature; the bonds are formed. Obviously, there is competition between the bonding to the solvent and the internal bonding of the macromolecule. The solvent here actually is the determining factor. In fact, the entropy of this transition has the wrong sign. Instead of the entropy of the random coil being the greater and that of the helix the lesser, the reverse is true. The helical

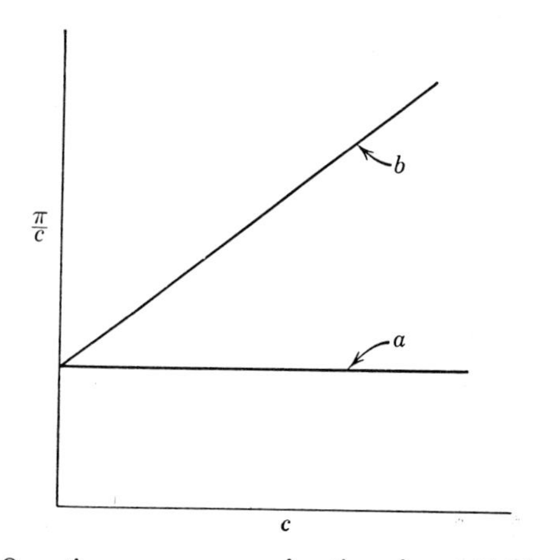


FIG. 1. Osmotic pressure $\pi$ as a function of concentration $c$ for real and ideal macromolecular solutions. See text.

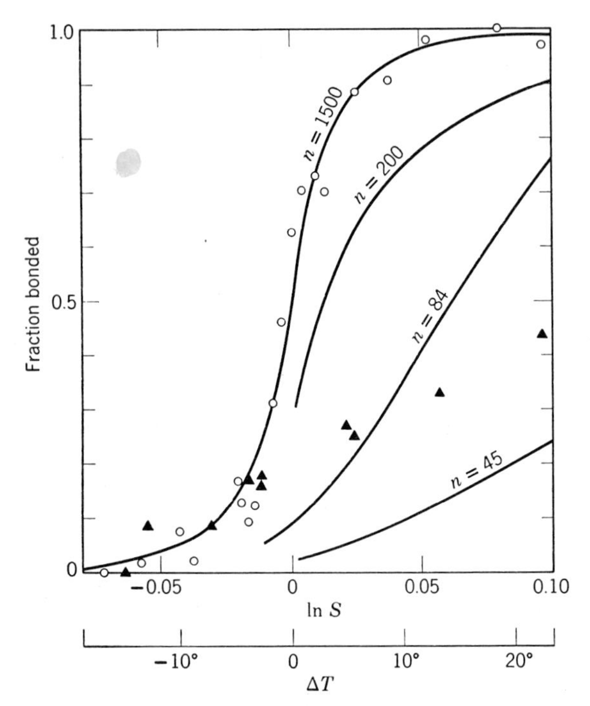


FIG. 2. Fraction of intramolecular hydrogen bonds in poly-$\gamma$-benzyl-L-glutamate; points calculated from optical-rotation data of Doty and Yang[1]; the curves represent an unpublished theory developed by J. K. Bragg and the author.

form is the one that is stable at high temperatures and the entropy of the helix is greater and that of the random coil is less. This means that the solvent is participating and that there is considerable entropy change in the solvent. This illustrates the pitfalls possible when the solvent is overlooked.

Consider highly concentrated solutions, such as 10 to 50% (or possibly more) dissolved material in solvent. What can be said about them? The kinetics of such systems are treated in the paper by Ferry (p. 130). The concern here is with statics. One of the first items of interest is, "What is the absorptive capacity of such solutions?" They can be called gels, for convenience. What then is the absorptive capacity of gels for foreign solvents, such as ions, etc.? For gels composed of long-chain macromolecules, the thermodynamic behavior with respect to the absorption of solvent is described by a theory advanced by Flory, Huggins, and Guggenheim.[2,3] The derivation contains numerous obviously wrong assumptions, over-simplifications, and so on; however, since it is a classical theory, one need say little about it here. If one simply wants to calculate an absorption isotherm, it is useful. It has an adjustable parameter that allows one to make it work fairly well in many different cases.

Another point of view should be mentioned. As shown in Fig. 3, one can plot the volume fraction of a solute $\phi_1$ as a function of the activity of the solvent $a_1$ ,defined in such a way that, in a pure solvent, $a_1$ has the value of unity. The activity $a_1$ is really little more than the relative vapor pressure—the vapor pressure of the solvent in the mixture over that of the solvent. To be very careful, gas-law corrections and such things must be included.

An ideal solution should give the perfectly straight diagonal of curve *c*. This shows the activity of the small molecule component proportional to the volume fraction. Raoult's law actually was stated in terms of mole fractions, but if it is modified slightly, it states that the ideal solution follows curve *c*. However, a real macromolecular solution usually gives something of the type of curve *d*. Sometimes the curve is more complicated, as is shown later.

Two characteristics are to be noted. At the upper right-hand corner, the slope of curve *d* is, as is usual, very much smaller than the ideal one. This is merely a manifestation of the repulsive interaction mentioned before, i.e., a positive deviation. The osmotic pressure is higher; the vapor pressure, therefore, also is higher in dilute solution than in the ideal case. At the lower left-hand corner, curve *d* usually rises much more steeply than the ideal. This slope, in cases where there are no specific interactions, is of the order of three to five times that of the ideal.

There is a very simple interpretation for this marked steepness in slope. Malcolm Dole probably was the first to apply it to any extent.[4] It is based upon the proposed mechanisms for rigid absorbing systems, like clays, inorganic catalysts, carbonblack, etc. It says that, in the absorbing material—in this case, the dry polymer—there are a number of sites which can take up the solvent, and that the rate with which the activity rises is an inverse function of the number of sites. If there is a large number of sites, then much absorbed material can be taken up without raising the activity very much. If there are a very few sites, however, they will fill up very fast, and the activity will rise much more rapidly than the volume fraction.

A mixture of two pure liquids, toluene and benzene, gives the straight line, curve *c* in Fig. 3. The number of

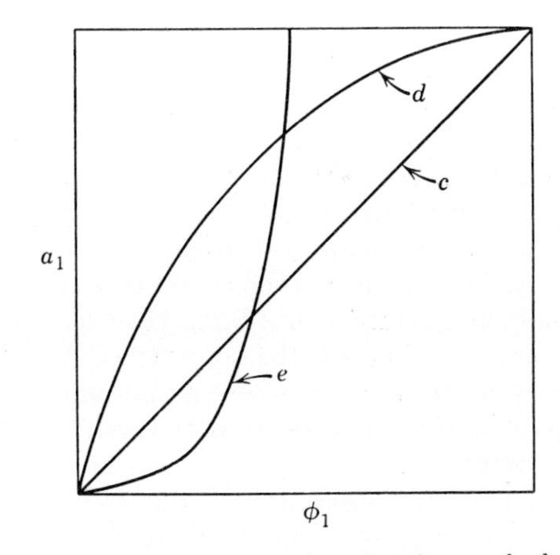


FIG. 3. Activity $a_1$ against volume fraction $\phi_1$ of solvent sorbed in a high polymer. See text.

sites in the mixture apparently increases in proportion to the amount of toluene put in, with one site available for every molecule added. The whole volume remains accessible to these molecules and curve $c$ results.

The macromolecular system is different. There are not nearly as many available sites in which to put solvent molecules; there appear to be about one-fourth as many. This suggests a very simple and reasonable hypothesis. Figure 4(a) shows the absorption of toluene in benzene. The benzene molecules can be elbowed out of the way, presenting little resistance to the added toluene, since the whole volume is accessible to its admission.

For macromolecules, the volume initially consists of intertwined chains [Fig. 4(b)]. If a toluene molecule is introduced at position $a$, a long cavity will have to be made with considerable wasted space at its ends, since the macromolecular chains are quite rigid. Therefore, the toluene molecule enters position $b$ preferentially where it can cause with greater ease the necessary rearrangements in the neighboring macromolecules. It appears obvious that material of this kind is relatively unreceptive to small molecules; it can fit them into certain specific places only.

One cannot predict the number of specific places, but the experiments quoted in the following indicate that about one-fourth of the volume is occupied by sites into which a small molecule easily can be introduced. This discussion refers to unactivated absorption where no strong specific forces are involved. An example is toluene in polystyrene where the forces are fairly well in balance.

In systems permitting hydrogen bonding, specific interactions are likely, with pronounced deviations resulting (curve $e$, Fig. 3). This curve, which is typical of water absorbed in protein, starts off with a very

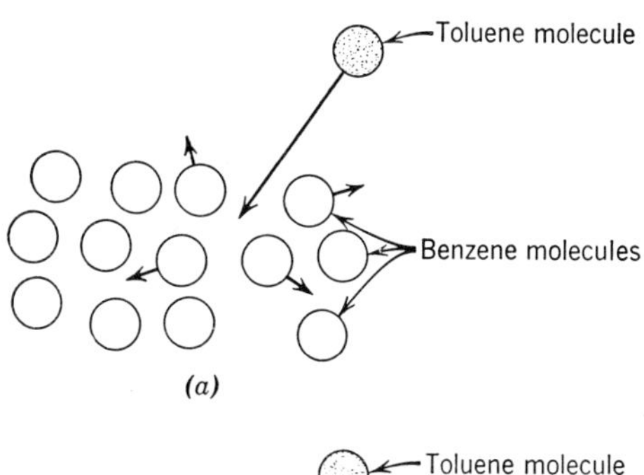


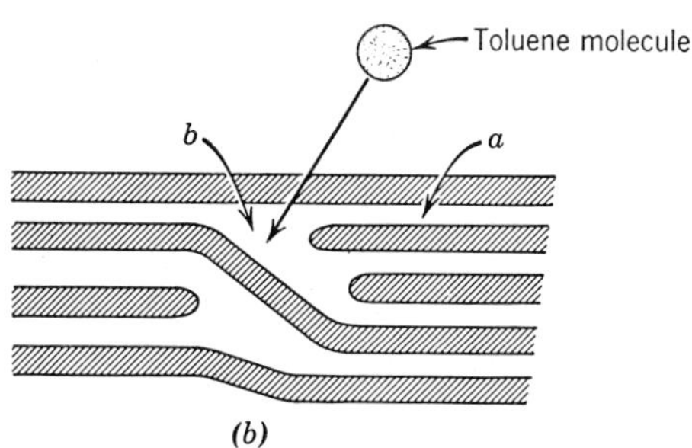


FIG. 4. Sorption of a small molecule, toluene, ($a$) in an ordinary liquid and ($b$) in a high polymer.

low, rather than a high, slope, indicating that some of the sites are very receptive to the water molecules. After these specific places have been filled up, the curve rises rapidly. It even may come up and hit the axis, which means that the protein is saturated and cannot take up any more water.

In this case, there are a few very specific active sites as compared with the previous case in which there are many inactive sites. But, when one attempts to apply the simple site hypothesis quantitatively, it is not very successful, because the manner in which the absorbed molecules affect the protein has been ignored. Apparently, the protein rearranges its state considerably, and it is difficult to describe the situation in simple terms. The whole concept of an inert absorptive site loses its applicability.

There is, however, another approach to this problem.[5] Define a function that measures the tendency of the absorbing molecules to cluster. This function, the "clustering function," has an exact molecular definition. It is given by an integral,

$$G_{11} = (1/V)\int\int [F_2(i,j)-1]d(i)d(j),$$

where $i$ and $j$ are molecules of component 1, the solvent; $V$ is the total volume; and $F_2(i,j)$, the distribution function, is defined by the statement that $(1/V^2)F_2(i,j)d(i)d(j)$ is the probability that the molecules $i$ and $j$ are each at the positions specified by the coordinates $(i,j)$ in the range of these coordinates $d(i)$ and $d(j)$. This distribution function is familiar in the case of spherical molecules where it is called the radial distribution function. It is simply the probability of finding one molecule in a given position with respect to another.

The quantity $\phi_1 G_{11}/v_1$, where $v_1$ is the molar volume of the solvent, is the mean number of solvent molecules *in excess of the random expectation* in the neighborhood of a given solvent molecule. Thus, it measures the clustering tendency of solvent molecules, and, for this reason, $G_{11}$ is called the clustering function. If $G_{11}$ is positive, the solvent molecules cluster; if it is negative, as frequently happens, they do not.

It should be noted that this clustering function has a simple and exact relation to the isotherm. There are no assumptions except for a very minor approximation of neglecting the compressibility of the system. The relation is as follows:

$$\frac{G_{11}}{v_1} = (\phi_1 - 1)\frac{\partial}{\partial a_1}\left(\frac{a_1}{\phi_1}\right) - 1.$$

Similar relations were known to Willard Gibbs,[6] but they were neglected until interest was revived recently by Mayer[7] and by Kirkwood[8] and their collaborators.

The clustering function can be determined from the isotherms, and they immediately reveal whether the

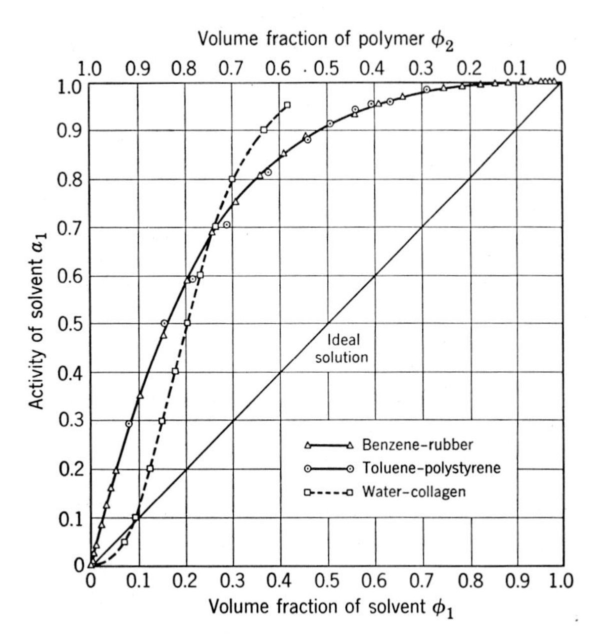


FIG. 5. Activities of solvents *vs* volume fractions of solvents and polymers for benzene-rubber, toluene-polystyrene, and water-collagen at 25° [from B. H. Zimm and J. L. Lundberg, J. Phys. Chem. **60**, 425 (1956)].

absorbing molecules tend to crowd into a few sites, or whether they tend to occupy isolated sites and prevent other molecules from entering. This would depend upon whether $G_{11}$ is positive or negative, and upon how much so.

Figure 5 shows the data for three systems, two of which are simple nonpolar systems. The heavy solid curve represents the activity of the solvent in the systems benzene-rubber and toluene-polystyrene, in which the intermolecular forces are neither specific nor strong. The two curves lie almost on top of one another. The initial slope is quite high, which indicates that the number of sites available for the absorption of the incoming solvent molecules is much smaller than it would be in a mixture of two ordinary liquids. A water-collagen system exhibits another type of behavior.[9] First, there is a region of very strong attraction for the first few molecules of water absorbed, with the water hardly showing any vapor pressure. Then, after the initial sites are filled, the activity rises very rapidly until it is seen to come up to that of liquid water. Dole and McLaren[9] found that the first region has a strong negative heat of absorption; that is, heat was evolved to the amount of about 6 kcal/mole of water in the region of $\phi_1$ less than 0.1.

Figure 6 shows the clustering function for these same systems. A clustering function of zero would mean no interactions at all between the molecules. The ideal solution has $G_{11}$ equal to $-1$. In the ideal solution, there are interactions; the molecules still have volume. One molecule cannot be placed on top of another, so the clustering function is $-1$; in other words, the first molecule excludes one molecular volume to the other molecules. For the hydrocarbons, a positive clustering tendency appears. $G_{11}$ is fairly uniformly $+1$ for all values of $\phi_1$. This means that the molecules of the liquid tend to cluster in the polymer, which is probably a manifestation of the same geometrical problems described in Fig. 4(b). It is not easy for the polymer to make room for the liquid molecules, but when one has been admitted and a cavity is opened, it is apparently easier to put in other liquid molecules by extending the same cavity than by opening up new cavities. Therefore, there is this mild clustering tendency which is not a manifestation of any special attractive properties between the benzene and rubber or between the toluene and polystyrene. Actually, in the former case, the heat of absorption is mildly negative, and, in the latter case, it is mildly positive. This contrast leads to the tentative conclusion that $G_{11}$ is affected more by the geometry of the packing than by the heat of sorption.

The Flory-Huggins-Guggenheim theory predicts values of $G_{11}$ of less than unity for the systems previously referred to. In general, it underestimates the amount of clustering. In other words, it does not take sufficiently into account the intrinsic heterogeneity of the mixture. This, in fact, is very difficult to do.

The water-collagen system presents a different picture. Initially, the clustering function is markedly negative. The first experimental point seems to be at about $-8$, and the curve is very steep. This means that the first molecule of water that is adsorbed excludes at least eight times its own volume to other molecules. One can interpret this as meaning that the first molecule occupies a special site, and that there are no other special sites

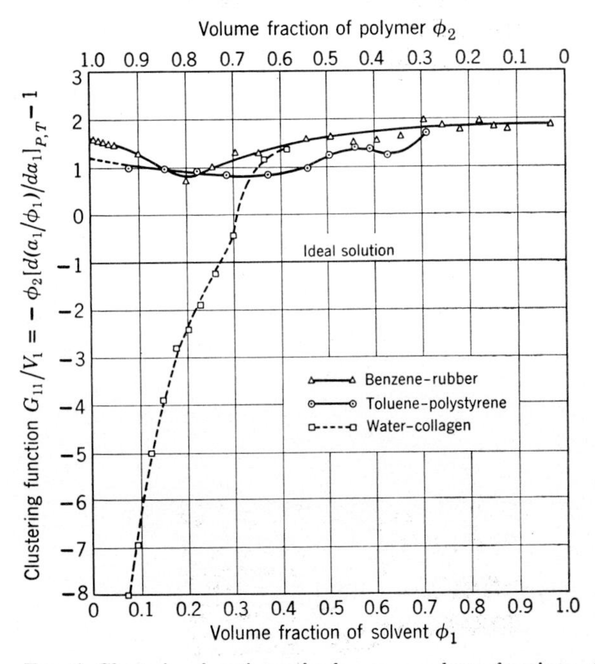


FIG. 6. Clustering functions of solvents *vs* volume fractions of solvents and polymers for benzene-rubber, toluene-polystyrene, and water-collagen at 25° [from B. H. Zimm and J. L. Lundberg, J. Phys. Chem. **60**, 425 (1956)].

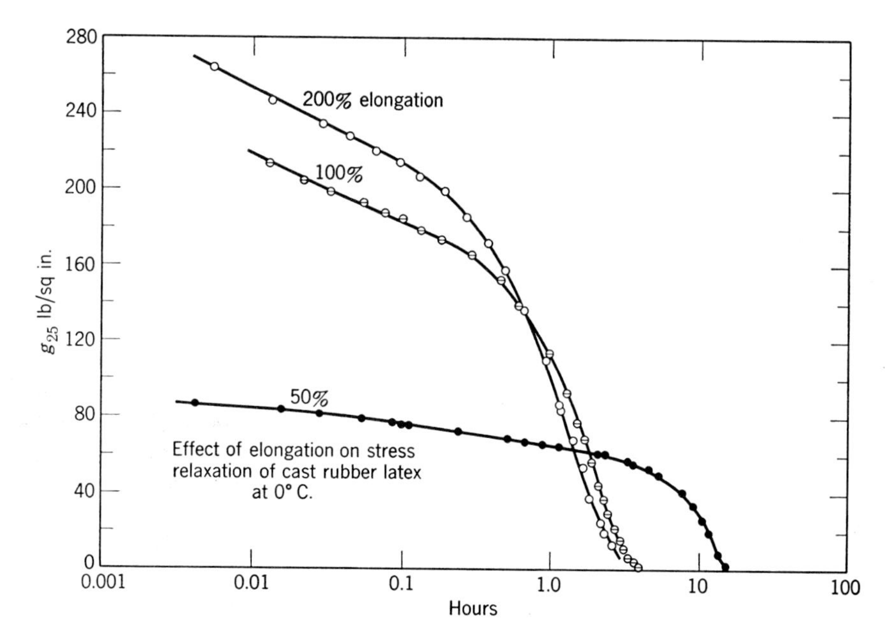


FIG. 7. Crystallization of stretched rubber [from A. V. Tobolsky and G. M. Brown, J. Polymer Sci. **17**, 547 (1955)].

within at least the cube root of eight molecular diameters around. Other molecules that enter will have to find a site elsewhere. After these special sites are filled, the system behaves more normally, and just before the collagen is saturated with water, it acts like an ordinary unspecific high polymer in which the water molecules show a mild tendency to cluster.

From the curve of the clustering function, it is apparent that the initial excluded volume for water on collagen is at least 10 water volumes, or 180 ml, which amounts to less than 0.8 mole of absorption sites per 100 g of collagen. By use of the Langmuir isotherm, Dole and McLaren estimated about 0.6 mole,[9] which seems to be reasonable agreement in view of the fact that the curve is still climbing steeply at the lowest point. If the Langmuir isotherm is used, one makes special assumptions, whereas, with the clustering function, none are made.

Consider now the elasticity of macromolecular systems. Most concentrated macromolecular solutions are not simple liquids. They show some kind of structural elasticity or structural viscosity. The following discussion is restricted to the equilibrium manifestations. [For further discussion, see Ferry (p. 130).] Rubber is classically the best-known example. When a rubber band is stretched, it warms up; the work of stretching is converted into heat. In contrast, when a steel spring is stretched, there is practically no thermal effect; in steel, the work of stretching is converted into internal energy. The obvious explanation is that in rubber the coiled macromolecular chains are straightened and their entropy is reduced. If this is done reversibly, an output of heat must result. Alternatively, the retractive force of the stretched rubber band is caused by the thermal vibration of the macromolecular chains which try to return to a state of increased entropy.

This is the entropic theory of rubber elasticity, and it works very well at elongations of less than 200 to 300%. At small elongations, the change of entropy accounts for nearly all of the retractive force. This is discussed by Flory.[2]

Beyond about 300% elongation, the material stiffens rapidly and the internal energy contributes heavily to the retractive force. The energy of extension becomes negative, which means that there is a decrease in internal energy on extension. With further extension, the entropy also becomes more negative than theory predicts. This marked decrease in entropy and energy is suggestive of the decreases that occur during crystallization where the molecules are ordered with the result that the entropy decreases; at the same time, they are placed in energetically favorable configurations so that the energy decreases also.

Direct evidence for this is shown in Fig. 7 from work by Tobolsky and Brown.[10] It is known that rubber does crystallize, since x-ray patterns are obtainable. If rubber is stretched at 0°C, which is well below the freezing point for the initially amorphous rubber, the retractive force decreases with time; the curve approaches the axis. (The apparatus is incapable of measuring a negative retractive force. In many cases, the rubber band extends spontaneously.) The macromolecular chains, which initially are lined up somewhat by the stretching, are crystallizing now into an even better linear array, and the band extends.

At present, polyethylene thread is available which, when stretched, oriented, and then cross-linked by irradiation, exhibits a striking phenomenon. If heated, it melts and contracts about 10 to 20%. When cooled and frozen at about 125°C, it extends again spontaneously. This phenomenon is the result of crystallization of oriented chains. A similar process occurs in biological

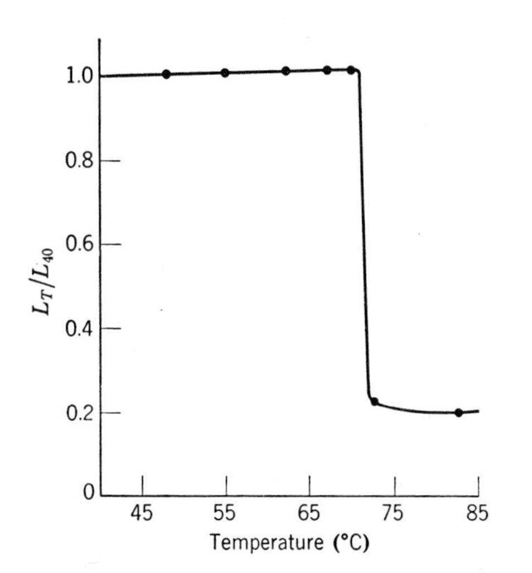


FIG. 8. Length as a function of temperature for formaldehyde-tanned rat-tail tendon under a constant small load [from P. J. Flory, J. Cellular Comp. Physiol. **49**, Suppl. 1, 175 (1957)].

materials, for example in collagen. Figure 8 illustrates some interesting data obtained by Dumitru, and discussed by Flory.[11] Collagen, normally, is in crystalline form, but, if heated to its melting point, it suddenly contracts. Unfortunately, when annealed below its melting temperature, collagen does not re-extend very markedly on cooling, although the density does go up somewhat. This striking change of length accompanying the melting of the oriented crystal is similar to the phenomenon observed in rubber.

A similar interesting calculation can be made for $\alpha$-helices. Instead of plotting the usual stress-strain diagram with length as a function of force, two convenient dimensionless variables can be introduced. In place of length $L$, substitute $L/n\delta$, where $n$ is the number of residues in the chain and $\delta$ is the length along the helix axis of one residue. Similarly, in place of force $f$, substitute $fb^2/3\delta kT$, where $b^2$ is the square of the effective length of a residue in the random form; that is, $b^2$ is the mean-square end-to-end length divided by $n$. These variables are plotted in Fig. 9.

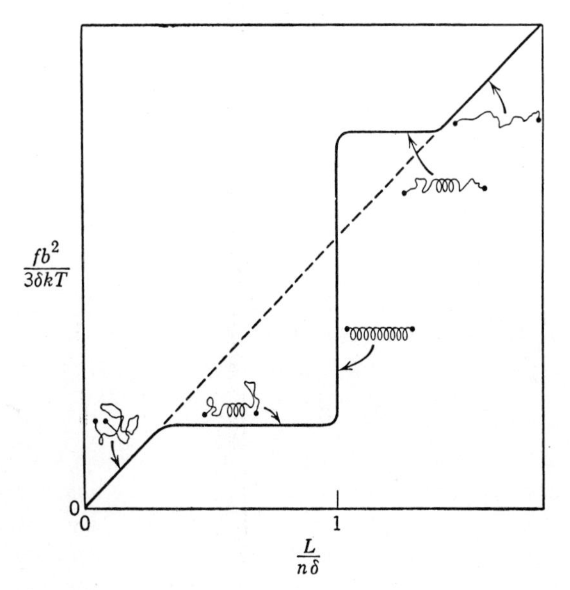


FIG. 9. Hypothetical stress-strain diagram for a polypeptide chain that can "crystallize" into a helix on stretching. See text for symbols.

Suppose that the material initially is in the random form. The stress-strain curve is linear and somewhat simpler than the curve for rubber, because only one molecule is considered rather than an array of molecules with different orientations. Initially, the slope of the curve is unity. Upon stretching, the helical form is favored because the helix is oriented in the direction of stretch. The coil-to-helix transition commences with an essentially two-phase situation. In this situation, considerable change in length results from a small change in force, since the proportion of helical to random form rapidly increases. Upon complete conversion into an $\alpha$-helix, the structure becomes very rigid, and large changes in the force produce little change in length. As illustrated in Fig. 9, the 45° line is crossed, and the random form represented by the dashed diagonal line would extend more readily than the helical form in the direction of stretching. Thus, the random form again becomes more favorable, and a second transition from the helical to the random coil occurs. As shown in the diagram, both of the horizontal regions of the curve (the transition regions) represent a mixture, with a segment of random coil followed by one of helix. There is probably always a random segment at the end, because of the disordering effect of the end.

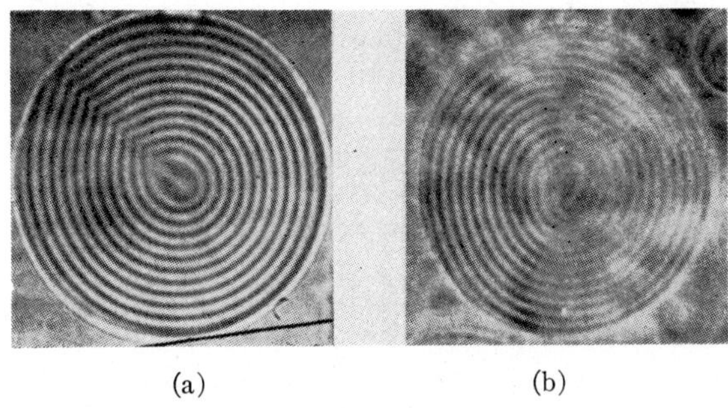


FIG. 10. Two spherulites of an anisotropic solution. (a) Large regular spherulite, showing radial fault. *R*1 in methylene chloride; spacing about 0.015 mm×154. (b) Large regular spherulite, fault not visible. Phase contrast; *R*1 in methylene chloride; spacing about 0.015 mm×131 [from C. Robinson, Trans. Faraday Soc. **52**, 571 (1956)].

An additional subject of interest is that of liquid crystals. Many macromolecular solutions contain a considerable amount of order. Tobacco mosaic virus particles are random in dilute solution, but, in a more concentrated form, the solution is intrinsically anisotropic and exhibits colors between crossed polaroids. Its x-ray diffraction pattern consists of arcs of circles. This pattern originally was investigated by Bernal and Fankuchen.[12] There are also many mixtures of soaps and water and of various lipids in water which have similar properties. There is, in fact, a large and rather neglected field of what are generally called liquid crystals, or crystalline liquids, or *mesophases*—different names for the same phenomenon.

Figure 10, a recent photograph taken by Conmar Robinson,[13] shows a liquid crystal, a drop of a solution of polybenzylglutamate in methylene chloride. The concentration is roughly 20% by weight. In this solvent, the molecules are in the $\alpha$-helix form, and the presence of the long extended structures promotes formation of liquid crystals. In a concentrated solution, the liquid crystals separate and form small drops of a separate phase. Figure 10 illustrates one of the drops surrounded by the ordinary isotropic solution. The diameter of the drop is approximately 300 $\mu$. Each illustrated layer is about 10 to 15 $\mu$ thick. This onion-like structure is remarkable. In Fig. 10(a), one sees a single spiral with something of the nature of a stem, or radial fault, on one side. Looked at from a different angle, Fig. 10(b), the stem radius is not apparent; instead, a double spiral is seen. The structure is very striking. For further reference, the reader is referred to Robinson[13] and to papers by Robinson and Frank.[14] Onsager[15] and Flory[16] have proposed quantitative theories which try to explain the formations of the liquid crystal as resulting from the anisotropic interaction between long rod-like molecules. These theories indicate very nicely the essential feature that, if too many long rods are packed in a given space, they pack very poorly unless they are aligned; if aligned, they can be packed more economically. Hence, a concentrated solution of rod-like molecules tends to form liquid crystal structures, but, if diluted, isotropic structures occur. In fact, phase boundaries can result between two solutions of the same materials at different concentrations.

## BIBLIOGRAPHY


1 P. Doty and J. T. Yang, J. Am. Chem. Soc. **78**, 498 (1956).

2 P. J. Flory, *Principles of Polymer Chemistry* (Cornell University Press, Ithaca, 1953).

3 E. A. Guggenheim, *Mixtures* (Oxford University Press, London, 1952).

4 M. Dole, Ann. N. Y. Acad. Sci. **51**, 705 (1949).

5 B. H. Zimm and J. L. Lundberg, J. Phys. Chem. **60**, 425 (1956).

6 J. W. Gibbs, *The Collected Works of J. Willard Gibbs* (Longmans, Green and Company, New York, 1931) II, 201.

7 W. G. McMillan and J. E. Mayer, J. Chem. Phys. **13**, 276 (1945).

8 J. G. Kirkwood and F. P. Buff, J. Chem. Phys. **19**, 774 (1951).

9 M. Dole and A. D. McLaren, J. Am. Chem. Soc. **69**, 651 (1947).

10 A. V. Tobolsky and G. M. Brown, J. Polymer Sci. **17**, 547 (1955).

11 P. J. Flory, J. Cellular Comp. Physiol. **49**, Suppl. 1, 175 (1957).

12 J. D. Bernal and I. Fankuchen, J. Gen. Physiol. **25**, 111, 147 (1941).

13 C. Robinson, Trans. Faraday Soc. **52**, 571 (1956).

14 C. Robinson and F. C. Frank, Discussions Faraday Soc. (to be published).

15 L. Onsager, Ann. N. Y. Acad. Sci. **51**, 627 (1949).

16 P. J. Flory, Proc. Roy. Soc. (London) **A234**, 73 (1956).

# 15
# Rheology of Macromolecular Systems

John D. Ferry
*Department of Chemistry, University of Wisconsin, Madison 6, Wisconsin*

## I. INTRODUCTION

As the paper by Zimm (p. 123) shows, mechanical energy can be stored in a flexible macromolecular network at elastic equilibrium because of the configurational restrictions imposed on it by an external deformation. The resulting rubber-like elasticity and its characteristic dependence on temperature and cross-linking density are familiar and intuitively reasonable phenomena. Less familiar is the fact that elastic energy can be stored in a macromolecular system *without* cross-links when it is subjected to steady-state flow; this elasticity is manifested in solutions of either flexible or rigid molecules and even at high dilution. In flow, of course, mechanical energy is also continuously dissipated as heat. In some cases, the energy storage is masked by the dissipation and is difficult to detect; in others, including numerous systems of biological interest, the stored energy is obvious as seen, for example, in elastic recoil after cessation of flow.

The presence of stored elastic energy is usually associated with non-Newtonian character of the flow; the apparent steady-flow viscosity falls with increasing shear rate. Moreover, the elastic strain (i.e., that part of the deformation which is recoverable after removal of stress) may not be directly proportional to shear stress, so that the elasticity may be non-Hookean. The province of rheology should include the description and explanation of these nonlinear relationships, as well as the simultaneous appearance of viscosity and elasticity. Nevertheless, for sufficiently small stresses, the deviations from Newtonian flow and Hookean elasticity can be made negligible, and most of the present discussion is restricted to such conditions, corresponding to so-called linear viscoelastic behavior. Although nonlinear phenomena are certainly important and potentially valuable sources of information, they are less well understood and are omitted for the sake of brevity.

With this simplification, a rheological description of the concomitant viscous and elastic behavior of a macromolecular system involves the following quantities:

(a) The steady-flow viscosity, $\eta=\mathfrak{T}/\dot{\gamma}$, where $\mathfrak{T}$ is the shear stress and $\dot{\gamma}$ the rate of shear. The energy dissipated per cc per sec in flow is $\eta\dot{\gamma}^2$.

(b) The steady-state compliance, $J=\gamma_e/\mathfrak{T}$, where $\gamma_e$ is the elastic strain. The energy stored per cc is $J\mathfrak{T}^2/2$.

(c) Various parameters representing characteristic times. For example, the ratio of energy stored to energy dissipated per second has the dimensions of time ($=J\eta/2$). The terminal relaxation time for a macromolecular solution, $\tau_1$, is usually also of the order of $J\eta$. The time required to attain steady-state flow under constant stress is approximately $\tau_1$; after cessation of flow, the stored energy is dissipated by relaxation processes within an interval which is also of the order of $\tau_1$. For a complete description of the approach to steady-state flow and of the course of relaxation, as well as of other time-dependent rheological processes, a spectrum of relaxation times is usually required.

From experimental measurements of stored energy and of the rate of approach to steady-state flow, or from equivalent information about response to oscillating stresses, conclusions can be drawn regarding the ease of motion of macromolecular segments through their surroundings and the nature of long-range coupling between molecules, as illustrated by the examples which follow.

## II. ENERGY STORAGE IN STEADY-STATE FLOW

The storage of energy in very dilute solutions during flow can be attributed qualitatively to the departure from random orientation, familiar from discussions of streaming birefringence, which leads to a decrease in the entropy content in the steady state. The complete rheological behavior can be predicted by considering the interactions between the frictional forces imposed on solute molecules by the flowing solvent and their own Brownian motions, as calculated by Kirkwood and Auer[1] for thin rigid rods and by Rouse[2] and Zimm[3] for flexible coils. The results correspond mathematically to the mechanical models in Fig. 1, where the solvent viscosity $\eta_s$ is shown added to various viscosity contributions from the solute. For rods, one of the solute contributions behaves as though in series with an elastic spring; for coils, there is an infinite series of solute contributions, each in series with a spring. The magnitudes of some of the model components are given in Table I in terms of the solution viscosity, $\eta$, the molecu-

Table I. Rheological characteristics of macromolecular solutions.

| | Thin rigid rods (Kirkwood and Auer[a]) | Flexible free-draining coils (Rouse[b]) |
|---|---|---|
| $G_1$ | $3cRT/5M$ | $cRT/M$ |
| $\eta_P$ | $(\eta-\eta_s)/4$ | ... |
| $\eta_1$ | $3(\eta-\eta_s)/4$ | $6(\eta-\eta_s)/\pi^2$ |
| $J$ (very dilute) | $3M[\eta]^2c/20RT$ | $2M[\eta]^2c/5RT$ |
| $J$ (concentrated) | [c] | $2M/5cRT$ |

[a] Reference 1.
[b] Reference 2.
[c] Rigid rods will not form a concentrated disordered solution, preferring thermodynamically to distribute themselves between a dilute disordered and a very concentrated ordered phase.[4,5]

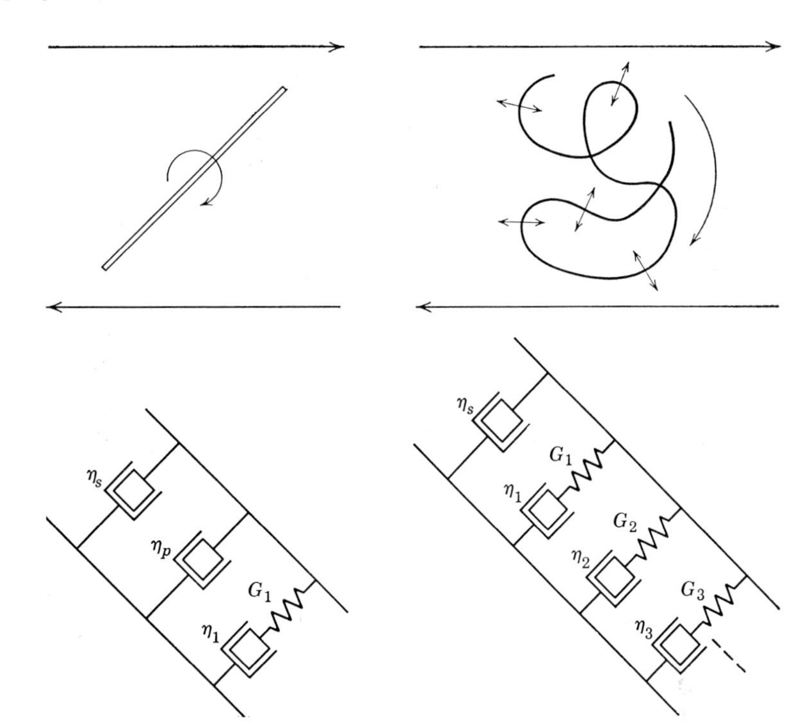


FIG. 1. Mechanical models corresponding to theoretical viscoelastic behavior of dilute rigid rods and flexible coils (below); schematic hydrodynamics (above).

lar weight, $M$, and the concentration, $c$ (in g/cc). The values for rigid ellipsoids[6,7] are quite similar to those for rigid rods, and those for coils with internal hydrodynamic interaction[3] are somewhat similar to those for the free-draining type. Of course, the portrayal of a mechanical model is not an essential feature, but it aids visualization. The terminal relaxation time (for rods, the only relaxation time) is $\tau_1=\eta_1/G_1$.

Table I also gives the steady-state compliance for these systems. In very dilute solutions, it is proportional to the square of the intrinsic viscosity $[\eta]$ (expressed here in cc/g) and to the molecular weight. It is interesting that the values for rods and coils differ only by a minor numerical factor. In concentrated solutions (defined by the condition that $\eta\gg\eta_s$), $J$ depends on the molecular weight only; if there are different macromolecular components, it is weighted strongly by those of highest molecular weight.[8]

Under the low stresses which would be necessary to avoid extreme non-Newtonian and non-Hookean effects, the magnitude of the stored elastic energy is quite small. For example, a dilute solution of tobacco mosaic virus[9] flowing under a shear stress of 10 dyne/cm$^2$ would store 25 cal/mole of solute; one of hyaluronic acid[10] (molecular weight 500 000), 0.25 cal/mole. At a concentration of 1%, the terminal relaxation times for these systems are calculated to be of the order of $10^{-3}$ and $10^{-4}$ sec, respectively, which are so short as to preclude direct experimental observation of a transient elasticity; macroscopic elastic recoil would be vitiated by the inertia of the solvent. The elasticity could be measured by dynamic experiments, however, as described in the following section. By contrast, the behavior of a concentrated solution of flexible coils may be illustrated by a 2% solution of the sodium salt of DNA, which[11] under a shear stress of 100 dyne/cm$^2$ would store about 200 cal/mole. (Sodium DNA molecules are of course not very flexible, but in concentrated solution exhibit some flexibility which may arise from hinges where the helix is unraveled for short distances.) The elastic shear would amount to about 50%, and the terminal relaxation time is calculated to be of the order of 1000 sec, so a substantial slow elastic recoil is readily observable following cessation of flow. It should be emphasized that to obtain such elastic effects, familiar in various biological systems (mucous and gelatinous secretions, slime molds), it is not necessary to have any linkages between the macromolecules or a netted gel structure.

Under higher stresses, substantial amounts of energy may be stored, but the complications of nonlinear effects prevent making numerical estimates. With a high degree of orientation, the structure of the system may be completely changed by the appearance of ordered phases,[4,5] and the changes may be irreversible as in the secretion of silk.[12] These phenomena may give some hint of the nature of rheological processes in protoplasm.[13]

### III. ENERGY STORAGE AND DISSIPATION IN OSCILLATING MOTIONS

When a stress is applied to a macromolecular system, a finite time of the order of $\tau_1$ is required for the stored energy to build up to its steady-state value. The changes during this transient period are determined by the characteristic relaxation times, or relaxation spectrum, mentioned in Sec. I. Equivalent information about the relaxation times can be obtained, often more easily, by subjecting the system to a sinusoidally oscillating shear

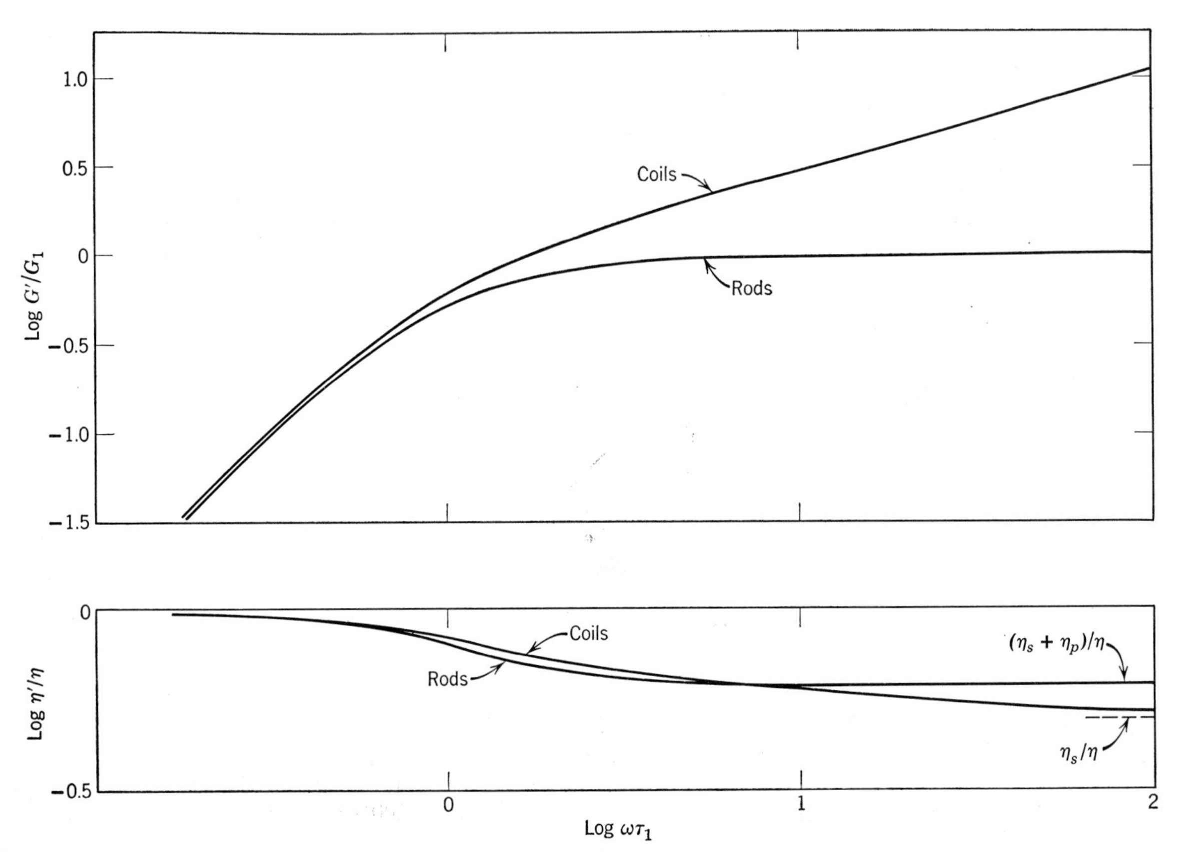


FIG. 2. Predicted frequency dependence of the dynamic storage modulus $G'$ and the real part of the complex dynamic viscosity $\eta'$, for dilute rigid rods and free-draining flexible coils, at concentrations such that $\eta = 2\eta_s$.

stress. Here some of the energy is stored and released each cycle, and some is dissipated. The respective values (for a given peak strain) are proportional to the storage modulus $G'$ and the real part of the dynamic viscosity $\eta'$, which are defined as the stress divided by those components of strain and rate of strain, respectively, in phase with the stress. Although it is doubtful whether any rhythmic biological processes can be approximated by this pattern, sinusoidal experiments are very useful for deducing the rates of molecular rearrangements.

At very low frequencies, $\eta'$ approaches the ordinary steady-flow viscosity $\eta$, and $G'$ approaches zero. However, the limiting value of $G'/(\omega\eta')^2$, where $\omega$ is the radian frequency, remains finite at vanishing $\omega$ and is in fact the steady-state compliance $J$. With increasing frequency, $G'$ increases and $\eta'$ decreases monotonically. These changes correspond to the obvious expectation in the mechanical models of Fig. 1 that with increasing frequency the motions will occur more in the springs and less in the dashpots. On a molecular scale, they mean that, as the period of oscillation is shortened, there is less opportunity for Brownian motion (rigid rotatory diffusion for rods, configurational rearrangements for coils) to erase the orientation imposed by the external deformations.

The characteristic frequency dependence expected for two simple cases is shown in Fig. 2, calculated for dilute solutions at a concentration such that $\eta = 2\eta_s$. For rigid rods,[1] there is a single relaxation time; $G'$ should approach a limiting high-frequency value of $G_1$, and $\eta'$ should approach $\eta_s + \eta_P$ (cf. Fig. 1 and Table I). Measurements of transverse wave propagation in solutions of rod-like intermediate polymers of fibrinogen are in rough agreement with this prediction.[14] For free-draining flexible coils,[2] there is a spectrum of relaxation times reflecting various degrees of cooperation between neighboring segments in configurational rearrangements; $\eta'$ should decrease and $G'$ increase more gradually with increasing frequency, following slopes of −0.5 and 0.5, respectively, on a doubly logarithmic plot, as the motions responding within the period of oscillation become restricted to shorter range cooperation. Eventually $\eta'$ should approach $\eta_s$, and $G'$ should attain a very high value of the order of the rigidity of a hard glass multiplied by the volume fraction of solute (this limit, which is not taken into account in the theories quoted,[2,3] corresponds to complete lack of configurational rearrangement within the period of oscillation, and is probably of no relevance to biological conditions). At present there are no data of this sort on polymers of biological origin; measurements on dilute solutions of synthetic polymers follow the general course shown in Fig. 2, although a somewhat different slope would be expected because of hydrodynamic interaction.[3]

Concentrated, rather than dilute, solutions are of primary interest in biological systems. Here the frequency dependence of dynamic mechanical behavior is more complicated, as shown in Fig. 3 for a 1.2% solution of sodium DNA.[11] Although the coils of this polymer have perhaps only moderate flexibility, the behavior follows a well-established pattern common to many flexibly coiling polymers of high molecular weight in concentrated solution.[15] There is an intermediate frequency range in which $\eta'$ falls steeply and $G'$ changes relatively little. This is associated with long-range coupling phenomena which cause the motions of one molecule to influence those of others separated by considerable distances but connected by some kind of chain of entanglements or coupling points.

The intermediate region—often called the plateau zone—lies between two extreme regions where the mechanical behavior is dominated by quite different physical processes. At high frequencies (corresponding to short times in transient experiments), the relaxation times are determined by the local friction of a short coil segment moving through its surroundings, oblivious of long-range coupling. The frequency dependence of $G'$ and $\eta'$ follows the Rouse theory for free-draining flexible coils, and from it the magnitude of the local friction can be calculated (Sec. IV). At low frequencies (corresponding to long transient times), the relaxation times are determined by the long-range coupling, which in turn is reflected in the steady-flow viscosity (Sec. V).

All the oscillatory phenomena described here relate to macroscopic deformation in shear. High-frequency measurements of $G'$ and $\eta'$ can be made by wave-propagation experiments provided shear waves are employed. In most sonic and ultrasonic wave-propagation experiments, however, the waves are longitudinal and the deformation is a combination of shear plus bulk compression. The energy storage is almost entirely due to the compression, in which, for a polymeric solution, the solute would play a minor role.

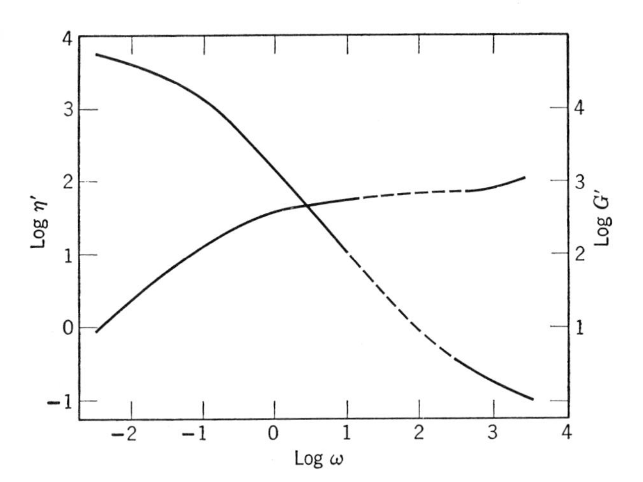


FIG. 3. Observed frequency dependence of $G'$ (ascending curve) and $\eta'$ (descending curve) in a 1.2% solution of sodium DNA.[11]

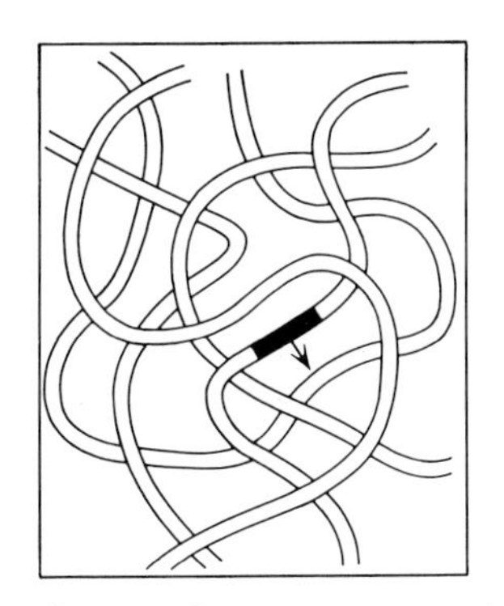
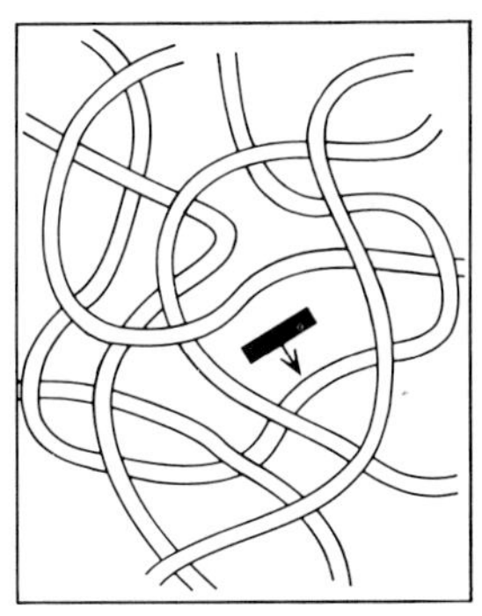

FIG. 4. Schematic representation of translational friction for a unit of a macromolecular chain (left) and a small foreign molecule in a macromolecular matrix (right).

## IV. LOCAL MOLECULAR FRICTION

The frictional forces encountered by a short segment of a flexible coil as it pushes its way through its surroundings in a concentrated solution are very complicated but can, on the average, be described by a friction coefficient per monomer unit, $\zeta_0$, which is force per unit velocity. This represents the same sort of translatory friction which a foreign molecule encounters in diffusion (Fig. 4). For certain soft synthetic polymers (with no diluent present), it has been possible to compare the friction coefficient per monomer unit of the wriggling chain, calculated from rheological measurements such as those in Fig. 3, with the friction coefficient of a foreign molecule similar in size to the monomer unit, obtained from diffusion measurements where only a trace of the foreign component is present.[16] They are actually of similar magnitude—for example, in polyisobutylene at 25°C, $4.5\times10^{-5}$ dyne sec/cm for the chain unit, and $3.8\times10^{-5}$ for a pentane molecule.

The friction coefficient reflects a sort of local viscosity which involves pushing aside solvent molecules together with short segments of other polymer coils in the immediate vicinity. A very crude estimate of the effective local viscosity $\eta_e$ can be obtained from Stokes' law ($\zeta_0=6\pi\eta_e r$), taking $r$ as the radius of a sphere with the volume of a monomer unit. When the solution is not highly concentrated, the limited information available indicates that $\eta_e$ is not far from the solvent viscosity,[17] and is thus much smaller than the macroscopic steady-flow viscosity. With increasing polymer concentration, $\zeta_0$ and $\eta_e$ increase rapidly, but not nearly as rapidly as does the macroscopic viscosity $\eta$. The most extreme divergence is reached in an undiluted soft polymer; for the polyisobutylene quoted above, for example, $\eta_e$ is about 70 poises and $\eta$ is $10^{10}$. For sufficiently high molecular weights, $\zeta_0$ and $\eta_e$ are independent of molecular weight, since the local motions are oblivious of the distant ends of the macromolecules. But $\eta$ increases very rapidly with molecular weight, often with the 3.4 power.[18]

It is evident that the effective local viscosity encountered in translatory friction of a foreign molecule

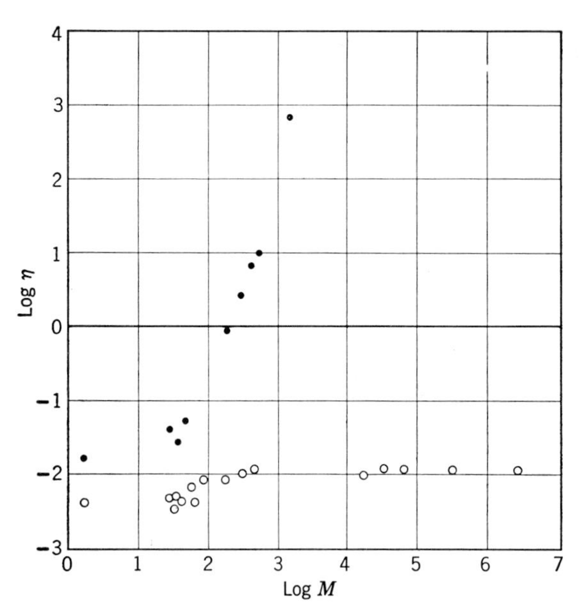


FIG. 5. Local effective viscosities of small organic molecules in soft rubber (black points) and in water (open circles), plotted against their molecular weights on a doubly logarithmic scale (after F. Grün[22]).

should increase continuously with molecular size as its motion involves increasingly long-range cooperation of the coil segments which rearrange around it. When the size reaches a magnitude such that long-range coupling affects its motion, there should be a large increase in $\eta_e$, and eventually for macroscopic dimensions of the moving particle $\eta_e$ must approach $\eta$.

There are some fragmentary examples of a striking dependence of local effective viscosity on the size of a moving particle, derived from measurements of centrifugation and diffusion in matrices of flexible coils. Bushy stunt virus molecules sedimented in an approximately 0.1% solution of sodium DNA encountered an effective viscosity only 20% higher than that of the solvent, in studies by Schachman and Harrington,[19] while much larger polystyrene latex particles encountered a viscosity higher by a factor of 30. A size dependence for the effective viscosity of latex particles of different diameters sedimenting through concentrated sodium-polyacrylate solutions has been found by Ferry and Morton.[20] Diffusion of sucrose in concentrated aqueous solutions of polyvinyl pyrrolidone, studied by Nishijima and Oster,[21] showed that the local effective viscosity encountered by the sucrose was independent of the molecular weight of the polymer and far smaller than the macroscopic viscosity. Finally, measurements of diffusion of various organic compounds, with molecular weights ranging up to 1000, in undiluted soft rubber enabled Grün[22] and Kuhn[23] to calculate effective local viscosities which increased very rapidly with molecular size, as shown in Fig. 5. (The figure also shows effective viscosities calculated from diffusion measurements in water, which, by contrast, simply correspond to the macroscopic viscosity of water over the whole molecular-weight range.)

The magnitude so fall relaxation times reflecting configurational changes of the sort described by the Rouse theory are proportional to $\zeta_0$. With increasing temperature, $\zeta_0$ falls and the relaxation spectrum shifts to shorter times without changing its shape; curves such as those in Fig. 3 shift to higher frequencies without changing their shapes. For biological systems, temperature changes are probably much less important than concentration changes. With a decrease in polymer concentration, which might arise from osmotic flow, $\zeta_0$ could drop sharply, speeding up all molecular rearrangements and conceivably allowing stored mechanical energy to be dissipated or to accomplish macroscopic movements.

## V. LONG-RANGE COUPLING PHENOMENA

The plateau zone in Fig. 3, observed in all high molecular-weight polymers in concentrated solution or the undiluted state, represents qualitatively the behavior to be expected if the macromolecules were coupled together quite strongly, though slipping to some extent, at widely separated points.[24] The same concept of long-range coupling explains another widely observed phenomenon in the dependence of the steady-flow viscosity on molecular weight.[25] A doubly logarithmic plot of viscosity against molecular weight or degree of polymerization shows a sharply increased slope above a critical value of the abscissa, which in typical cases corresponds to $M = 10\,000$ to $50\,000$ (Fig. 6). The long-range coupling cannot, in general, be attributed to secondary bonding at loci of attraction, because it appears in nonpolar polymers which are quite homogeneous in chemical composition. It is apparently re-

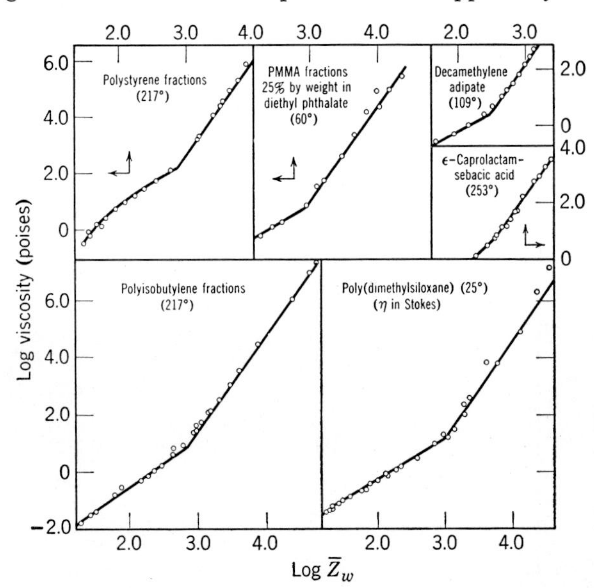


FIG. 6. Doubly logarithmic plots of viscosity against degree of polymerization for several types of synthetic polymers, showing proportionality to the 3.4 power of degree of polymerization or molecular weight above a critical value (after Fox and Loshaek[18]).

lated to the expectation that a molecule, in its long-range contour, will form a complete loop enclosing similar loops of other molecules,[26] but the topological requirements are not clear. Its influence increases with increasing molecular stiffness, as evidenced by comparing synthetic vinyl polymers, cellulose derivatives, and DNA.[11]

The terminal relaxation time in a concentrated solution is determined by the long-range coupling, and the extent to which motions of one molecule can influence others, through chains of mutual entanglement. Since the steady-flow viscosity is also dominated by the coupling, however, the order of magnitude of $\tau_1$ can still be estimated by the product $J\eta$.

Quite different from the above topological kind of coupling are the stronger linkages which cause some concentrated polymer solutions to gel. Here bonds of considerable permanence are formed, and the resulting network has no finite viscosity. It has no equilibrium elasticity either; under constant stress it continues to deform indefinitely at a very small and steadily decreasing rate. The spectrum of relaxation times extends indefinitely to long times, probably reflecting a mutual influence between molecular motions which extends throughout the system, communicated through the linkages. This is manifested, for low-frequency oscillating motions, by a very slight decrease in $G'$ with decreasing frequency together with an increase in $\eta'$ apparently without limit, as illustrated in Fig. 7 for a gel of polyvinyl chloride.[17] In some gels, the linkages are probably very small crystalline regions, tying together handfuls of flexible strands,[27] growth of the crystallites being limited by some kind of structural heterogeneity along the macromolecular chain. In others, there may be local secondary bonding of specific chemical groups.[28] As an extreme case, primary chemical bonds may join the macromolecules into a network, as proposed for gels of denatured serum albumin[29,30] and fibrin clots formed with fibrin-stabilizing factor.[31,32] Thus, while a netted structure is not essential to the storage of elastic energy

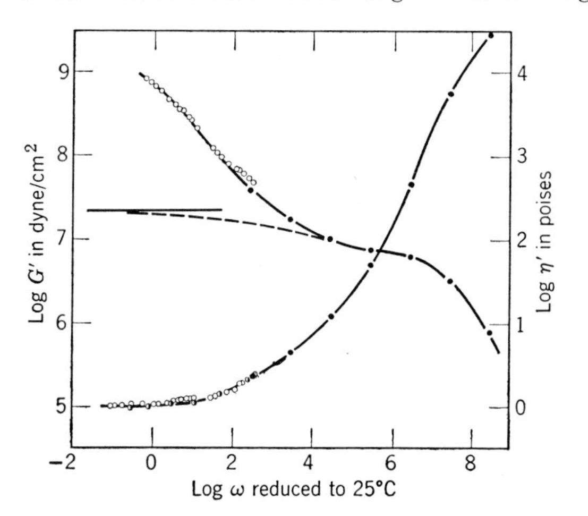


FIG. 7. Frequency dependence of $G'$ and $\eta'$ for a 10% gel of polyvinyl chloride in dimethyl thianthrene.[17]

under stress, it plays an important role when it is present.

The nature of molecular responses to very slow external motions in such networks, involving cooperation among widely separated network strands, is poorly understood. But if the strands between the linkage points are flexible macromolecular chains, the responses to rapid motions will be oblivious of the linkages and their rates will be governed by the local friction coefficient just as in the examples discussed in Sec. IV.

Changes in the consistency of protoplasm[13] suggest that within it network linkages are sometimes formed and sometimes broken. Such structural modifications should not affect responses to rapid motions but would enormously alter the relative storage and dissipation of energy in slow motions. Correspondingly, they should markedly influence frictional resistance to translation of large particles while leaving the resistance to small molecules essentially unchanged.

## BIBLIOGRAPHY

1 J. G. Kirkwood and P. L. Auer, J. Chem. Phys. **19**, 281 (1951).
2 P. E. Rouse, J. Chem. Phys. **21**, 1272 (1953).
3 B. H. Zimm, J. Chem. Phys. **24**, 269 (1956).
4 G. Oster, J. Gen. Physiol. **33**, 445 (1950).
5 P. J. Flory, Proc. Roy. Soc. (London) **A234**, 73 (1956).
6 R. Cerf, Compt. rend. **234**, 1549 (1952).
7 H. A. Scheraga, J. Chem. Phys. **23**, 1526 (1955).
8 J. D. Ferry, M. L. Williams, and D. M. Stern, J. Phys. Chem. **58**, 987 (1954).
9 M. A. Lauffer, J. Am. Chem. Soc. **66**, 1188 (1944).
10 T. C. Laurent, J. Biol. Chem. **216**, 263 (1955).
11 F. E. Helders, J. D. Ferry, H. Markovitz, and L. J. Zapas, J. Phys. Chem. **60**, 1575 (1956).
12 K. H. Meyer and J. Jeannerat, Helv. Chim. Acta **22**, 22 (1939).
13 L. V. Heilbrunn, *The Dynamics of Living Protoplasm* (Academic Press, Inc., New York, 1956).
14 J. D. Ferry and F. E. Helders, Biochim. et Biophys. Acta **23**, 569 (1957).
15 J. D. Ferry in *Die Physik der Hochpolymeren*, H. A. Stuart, editor (Springer-Verlag, Berlin, 1956), Vol. IV, Chap. VI.
16 J. D. Ferry and R. F. Landel, Kolloid-Z. **148**, 1 (1956).
17 J. D. Ferry, D. J. Plazek, and G. E. Heckler, J. chim. phys. **55**, 152 (1958).
18 T. G. Fox and S. Loshaek, J. Appl. Phys. **26**, 1080 (1955).
19 H. K. Schachman and W. F. Harrington, J. Am. Chem. Soc. **74**, 3965 (1952).
20 J. D. Ferry and S. D. Morton, unpublished experiments.
21 Y. Nishijima and G. Oster, J. Polymer Sci. **19**, 337 (1956).
22 F. Grün, Experientia **3**, 490 (1947).
23 W. Kuhn, Makromol. Chem. **6**, 224 (1951).
24 J. R. McLoughlin and A. V. Tobolsky, J. Colloid Sci. **7**, 555 (1952).
25 F. Bueche, J. Chem. Phys. **20**, 1959 (1952).
26 F. Bueche, J. Polymer Sci. **25**, 243 (1957).
27 J. T. Alfrey, Jr., N. Wiederhorn, R. Stein, and A. V. Tobolsky, J. Colloid Sci. **4**, 211 (1949).
28 G. Stainsby, editor, *Recent Advances in Gelatin and Glue[2] Research* (Pergamon Press, London, 1958).
29 C. Huggins, D. R. Tapley, and E. V. Jensen, Nature **167**, 592 (1951).
30 V. D. Hospelhorn, B. Cross, and E. V. Jensen, J. Am. Chem. Soc. **76**, 2827 (1954).
31 A. Loewy and J. T. Edsall, J. Biol. Chem. **211**, 829 (1954).
32 L. Lorand and A. Jacobsen, J. Biol. Chem. **230**, 421 (1958).

# 16
# Respiratory-Energy Transformation*

Albert L. Lehninger

*Department of Physiological Chemistry, The Johns Hopkins School of Medicine, Baltimore 5, Maryland*

## INTRODUCTION

THIS article summarizes the broad outlines of current knowledge on some enzymatic and biophysical aspects of the conversion of respiratory energy into phosphate-bond energy. There are many problems for the biophysicist in this area which not only are pertinent to electron transport and respiratory-chain phosphorylation but also have wide applicability to more-general problems of ultrastructure and arrangement of enzymes in biologically organized "solid-state" arrays. More-detailed treatments of the subject can be found in review articles.[1–5]

## ENERGY CYCLE IN AEROBIC CELLS

In all cells, the energy-requiring or *endergonic* functions are driven by chemical energy liberated during the energy-yielding or *exergonic* degradation of foodstuff molecules. The major energy-requiring functions or activities of cells include (a) *biosynthesis* of complex molecules from simple ones when such reactions proceed in an environment thermodynamically unfavorable for synthesis, as in the formation of glycogen from glucose in an aqueous system, (b) *mechanical work*, as in muscular contraction, (c) *active transport* or accumulation of substances against gradients of chemical potential, (d) *transmission* or conduction phenomena, and (e) *bioluminescence*. In aerobic cells, oxidative degradation of foodstuffs is the source of chemical energy. In strictly anaerobic cells, molecular oxygen does not intervene but internal oxido-reductions of fermentative or anaerobic metabolic cycles (such as alcoholic fermentation of glucose) are the primary energy source to drive endergonic functions.

The energy cycle in nature (as well as the carbon cycle) is then completed by the conversion of radiant energy of sunlight into chemical energy during photosynthesis, making possible reduction of carbon dioxide back to the oxidation level of carbohydrate, in reactions discussed in the companion paper by Calvin (p. 147). Respiratory-energy transformation and photosynthesis thus constitute the two complementary mainsprings of biological energy.

## CONCEPT OF PHOSPHATE-BOND ENERGY

The general medium of energy exchange between certain exergonic reactions of oxidative catabolism and the endergonic functions is the so-called "high-energy phosphate bond" of ATP.

Phosphate esters of cells can be classed into two major groups on the basis of the magnitude of the standard free energy of hydrolysis of the phosphate bond.[6–8] Simple phosphate esters, such as 3-phosphoglyceric acid, glucose-6-phosphoric acid, and $\alpha$-glycerophosphoric acid, etc., undergo hydrolysis enzymatically or by acid or base catalysis, often at widely different rates, with free energies of hydrolysis ($\Delta F°$) of $-0.8$–$-3.0$ kcal/mole at 25° and *p*H 7. These are the so-called "low-energy phosphate compounds." On the other hand, enol phosphate esters (phosphopyruvic acid), guanidine phosphates (phosphocreatine), acyl phosphates (1,3-diphosphoglyceric acid), and pyrophosphates (ATP) are in the high-energy class and have a $\Delta F°$ for hydrolysis of from 6 to 15 kcal/mole under standard conditions. This division is arbitrary and exact values of these esters are now in a state of flux because $\Delta F°$ values for hydrolysis of ATP are currently undergoing substantial revision.[9]

The *free energy of hydrolysis* represents the difference in energy content of reactants and hydrolysis products under standard conditions. It is *not* equivalent, of course, to the true bond energy or the zero-point energy; actually, input of energy is required to break chemical bonds. It must be made clear also that there is no particular biological magic involved in the difference between high- and low-energy compounds. Actually, the major reason for the relatively high free energy of hydrolysis of ATP and other high-energy compounds is that the immediate hydrolysis products of these compounds undergo secondary, spontaneous reactions with high equilibrium constants, leading to thermodynamically more-stable forms, either through resonance stabilization or tautomeric rearrangements. A second contribution to the high free energy of hydrolysis, in the case of ATP and phosphopyruvate, for example, is the separation and randomization of the neighboring like charges in these molecules at physiological *p*H. The structural, resonance, and charge contributions to the high-energy nature of these compounds have been discussed by Hill and Morales.[10] Classically, the free energy of hydrolysis of ATP has been assumed to have the value 12 kcal/mole, on the basis of a calorimetrically observed $\Delta H$ of approximately this figure and the assumption of only a small entropy change. However, recent work on the enzymatic equilibria of ATP and glutamate and the highly refined heat measurements by the important new method of Benzinger[11] have revealed that the standard free-energy change at

* Original experiments from the author's laboratory reported in this article were carried out with the aid of grants from the Public Health Service, U. S. Department of Health, Education, and Welfare, the National Science Foundation, The Nutrition Foundation, Inc., The Whitehall Foundation, and the Albert and Mary Lasker Foundation.

$1M$ reactants corrected for ionization changes at *p*H 7.5 is about $-8.4$ kcal, and that the true $\Delta H$ for hydrolysis is about $-4.8$ kcal (see Burton[9] for most recent evaluation). The value of the free energy of the hydrolysis of the ATP bond ($-8.4$ kcal) now seems to be a valid one; it is thus substantially lower than the classical value of 12 kcal. It must be pointed out, however, that, *under physiological conditions of* pH, *and intracellular concentration ranges of phosphate, ADP and ATP*, the recalculated $\Delta F'$ of ATP is about $-12.5$ kcal/mole or approximately the classical textbook value.[9] There is room for a great deal of careful work on the measurement of biochemically useful thermodynamic factors through refined heat measurements, enzymatic equilibria, and calculation of free energies of formation.[12,13] The values given above for ATP may be altered further when the hydrolysis is considered in a physiological milieu because of the lack of information on activity coefficients of the various ionic species of the reaction components—in particular, the Mg complexes of ATP and ADP in the presence of the high ionic strength background of intracellular fluid. Also, it must be pointed out that free-energy data for ATP obviously are reckoned on a macroscopic, statistical basis. In the intact cell, ATP-linked reactions occur with reactants bound to proteins (as in actomyosin) under nonequilibrium conditions where energy changes and transfers at the *microscopic* level dictate direction and rate of reactions.

There may be good reasons for the evolutionary choice of phosphoric acid, rather than sulfuric acid, HCl, or a carboxylic acid, as the vehicle for biological energy transfer through enzymatic group-transfer reactions. In the first place, phosphate anion has itself a rather high resonance energy; however, this property is not unique to phosphate. Perhaps it is more important that its anhydrides and amides are kinetically (as opposed to thermodynamically) much more stable than esters or amides of HCl, $H_2SO_4$, or carboxylic acids. Thus, acetic anhydride in $H_2O$ at *p*H 7.0 decomposes rapidly, whereas the anhydride pyrophosphoric acid at the same *p*H has a much greater half-life despite the fact that the free energy of hydrolysis of the two compounds is of the same order of magnitude. Phosphate anhydrides and amides may have been selected biologically not only for their resonance characteristics but also for their property of remaining relatively unreactive in an aqueous solution in the absence of suitable directing enzymes; less-stable derivatives would tend to react spontaneously and thus not be subject to enzymatic control.

## COMMON INTERMEDIATE PRINCIPLE IN UTILIZATION OF ATP ENERGY

Apart from certain photochemical and fluorescence phenomena, there is, in general, only one way in which energy can be transferred from one chemical reaction to another. The two reactions must be consecutive and *must have a common intermediate.*

The synthesis of sucrose provides a classical biochemical example.[14] The reaction

$$\text{glucose} + \text{fructose} \rightleftarrows \text{sucrose} + H_2O \quad (\Delta F = +5000) \qquad (1)$$

is strongly *endergonic*. Sugar-cane juice is $0.5M$ in sucrose. The concentrations of free glucose and fructose would have to be enormously greater than that of sucrose, and the water concentration low, in order for the reaction to proceed spontaneously in the sugar cane. This is not the case and sucrose is not made in this way. Enzyme studies show that it is formed at the expense of the hydrolysis of ATP in a coupled reaction. The driving reaction is the strongly *exergonic* hydrolysis of ATP:

$$\text{ATP} + H_2O \rightarrow \text{ADP} + \text{P} \quad (\Delta F = -12\,000). \qquad (2)$$

Reactions (1) and (2) are the partial thermodynamic reactions in sucrose synthesis. The actual enzymatic mechanism is the following:

$$\text{ATP} + \text{glucose} \longrightarrow \text{glucose}-1-P + \text{ADP} \quad (\Delta F = -7000 \text{ cal}) \qquad (3)$$

$$\text{glucose}-1-\text{P} + \text{fructose} \rightarrow \text{sucrose} + H_3PO_4 \quad (\Delta F = \sim 0 \text{ cal}). \qquad (4)$$

The over-all $\Delta F$ is $-7000$ cal and synthesis of sucrose proceeds spontaneously in the presence of the enzymes. In this sequence, glucose has been raised to the energy level of a glycoside by phosphorylation; glucose$-1-$P is now the common intermediate of two consecutive reactions which together have a negative net free-energy change, but in which a product of the first, *exergonic*, reaction has sufficient "group potential" in the form of the glycosyl phosphate to drive the second, *endergonic*, reaction to form sucrose.

In principle, this pattern underlies all endergonic biosynthetic reactions, such as formation of glycogen from glucose, proteins from amino acids, and complex lipids from fatty acids. ATP and an intervening phosphate-group transfer reaction represent the mode of energy transfer.

## COMMON INTERMEDIATE PRINCIPLE IN BIOLOGICAL CONVERSION OF OXIDATIVE ENERGY INTO PHOSPHATE-BOND ENERGY

From the foregoing, the breakdown of ATP to ADP and P is necessary to drive endergonic functions. The ATP then is regenerated continuously from ADP and P at the expense of catabolic reactions, primarily those of oxidation-reduction reactions, in order to complete the energy cycle.

The regeneration of ATP from ADP and P is very largely a matter of the transfer of the energy liberated in certain highly exergonic oxidations of metabolism to a reaction resynthesizing ATP from ADP and P. Such respiratory energy conversion also occurs by the principle of the common intermediate.

The best-understood example for this discussion is the enzymatic oxidation of glyceraldehyde—3—phosphate to 3—phosphoglyceric acid by DPN which occurs in the anaerobic breakdown of glucose to lactate, to which is coupled the synthesis of ATP from ADP and phosphate. The over-all reaction first can be broken down into two partial thermodynamic reactions. The first, the oxidation of an aldehyde to a carboxylic acid, is highly exergonic, and the second, formation of ATP, highly endergonic.

$$R-\underset{\underset{O}{\|}}{C}H + DPN_{ox} \rightleftarrows R-\underset{\underset{O}{\|}}{C}-O^{-} + DPN_{red} + H^{+} \qquad (\Delta F + -16 \text{ kcal}) \qquad (5)$$

$$P + ADP \rightarrow ATP + H_2O \ (\Delta F = +12 \text{ kcal}). \qquad (6)$$

The energy liberated in the oxidation of an aldehyde to a carboxylic acid is recovered as ATP, with energy left to spare, the net driving force being some 4 kcal, as is shown in the following over-all reaction:

$$\text{glyceraldehyde-3-phosphate} + P_i + DPN_{ox} \rightleftarrows \text{1,3-diphosphoglycerate} + DPN_{red} + H^{+}, \qquad (7)$$

$$\text{1,3-diphosphoglycerate} + ADP \rightleftarrows \text{3-phosphoglycerate} + ATP. \qquad (8)$$

It is seen that the oxido-reduction of reaction (7) results in formation of 1,3-diphosphoglycerate, a high-energy mixed anhydride of a carboxylic acid and phosphoric acid. This is the intermediate which is common in the two reactions and acts as a high-energy phosphate donor in the formation of ATP [reaction (8)]. Although this reaction is well understood, it accounts for only a very small fraction of ATP synthesis in aerobic cells, but it is an extremely important model of how the great bulk of ATP may be formed during respiration.

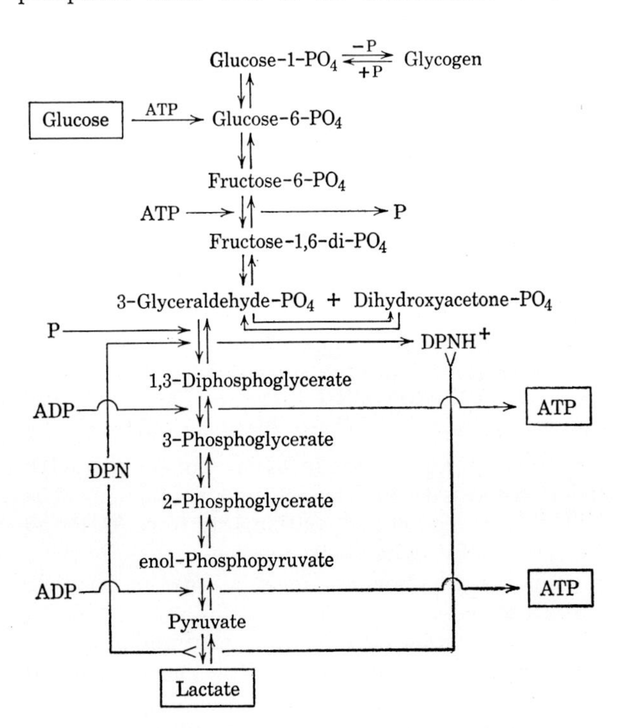


FIG. 1. The mechanism of anaerobic glycolysis.

### RELATIVE CONTRIBUTION OF FERMENTATIVE AND OXIDATIVE REACTIONS TO ATP SYNTHESIS

Figures 1 and 2 show the outlines of the mechanism of anaerobic glycolysis and the Krebs citric-acid cycle, the main aerobic phase of carbohydrate oxidation. However, the amount of energy liberated from a glucose molecule during glycolysis to lactate ($\Delta F = -49\,700$ cal) represents a few percent only of the total energy liberated by complete oxidation of glucose to carbon dioxide and water ($\Delta F = -686\,000$ cal/mole). However, this small yield of energy in the anaerobic phase is quite efficiently recovered by the two phosphorylations occurring in the cycle, and from the following equations it is seen that some 50% of the energy liberated when 2 moles of lactate are formed from 1 of glucose is recovered in the form of the phosphate-bond energy of two moles of ATP.

Over-all reaction:

$$\text{glucose} + 2ADP + 2P \rightarrow 2 \text{ lactate} + 2ATP \qquad (\Delta F = -25\,700 \text{ cal}).$$

Partial reactions:

$$\text{(a)} \quad \text{glucose} \rightarrow 2 \text{ lactate} \ (\Delta F = -49\,700 \text{ cal})$$

$$\text{(b)} \quad 2ADP + 2P \rightarrow 2ATP \ (\Delta F = +24\,000 \text{ cal}).$$

However, the real energetic mainspring of aerobic cells, regenerating 90% or more of the ATP required, is the oxidative phase of catabolism.

Despite the absence of phosphorylated intermediates in the Krebs cycle as it is depicted in Fig. 2, it is known now that a very large number of phosphorylations of ADP accompany such aerobic oxidations, approximately 36 per mole of glucose oxidized. In addition, the efficiency of energy recovery during this aerobic phase is even higher than in the anaerobic phase, approaching some 70%.

### OXIDATIVE PHOSPHORYLATION (SYNONYMS: AEROBIC PHOSPHORYLATION, RESPIRATORY-CHAIN PHOSPHORYLATION)

This term denotes the coupled phosphorylations of ADP which occur simultaneously with certain organized oxido-reduction reactions in the aerobic phase of metabolism. The biological significance of phosphorylations causally coupled to oxidations with molecular oxygen first was recognized by Kalckar in 1937, but the quantitative importance of these in cellular metabolism was not appreciated fully until the penetrating studies of Belitser in Russia in 1939–1941, which were

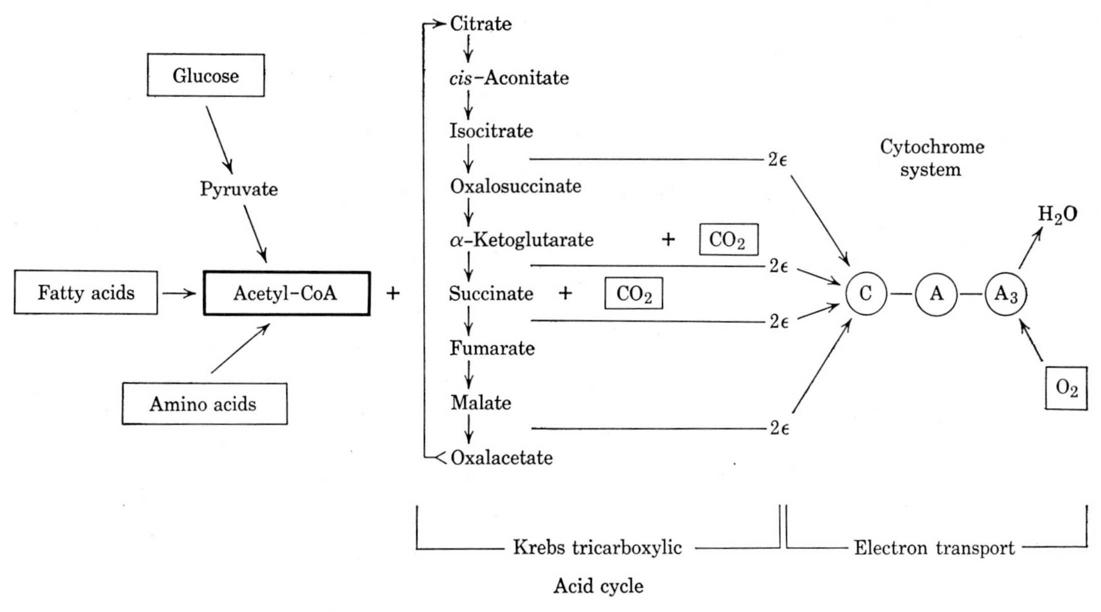


FIG. 2. The Krebs tricarboxylic-acid cycle.

followed during and after World War II by a number of independent and confirmatory studies by Ochoa, Cori, Hunter, and others (cf. references 4, 2, 3). These studies clearly demonstrated that, for the passage of every pair of electrons from a substrate of the Krebs citric-acid cycle to molecular oxygen during aerobic metabolism, an average of three moles of ATP were synthesized from ADP and $P_i$; i.e., the P:O ratio (moles $P_1$ taken up per atom oxygen used) was 3.0. Thus, the over-all equation for the oxidation of pyruvate to carbon dioxide and water as it occurs in the Krebs cycle was found to be

$$CH_3COCOOH+15P+15ADP+5O \rightarrow 3CO_2+15ATP+2H_2O. \quad (9)$$

This reaction may be broken down as follows:

$$CH_3COCOOH+5O \rightarrow 3CO_2+2H_2O \quad (\Delta F=-273 \text{ kcal}), \quad (10)$$

$$15P+15ADP \rightarrow 15ATP$$
$$(\Delta F=12\times 15)$$
$$(\Delta F=+12\times 15=180 \text{ kcal}), \quad (11)$$
$$\text{efficiency}=180\times 100/273=67\%.$$

## GENERAL PROPERTIES OF OXIDATIVE PHOSPHORYLATION

Oxidative phosphorylation classically has been recognized as an unstable and evanescent phenomenon which occurs only with relatively "native" and unfractionated extracts or homogenates of tissues. Aging of such extracts, or application of some of the common procedures for separation or purification of enzymes, quickly inactivated the coupled phosphorylation of ADP, whereas the purely oxidative activity of Krebs-cycle reactions remained rather stable. The findings thus suggested that coupled phosphorylation of ADP is not an obligatory accompaniment of oxidation, that it may be inactivated or uncoupled. The great lability of the coupling mechanisms to even relatively mild separation procedures has been a major technical obstacle to a fuller understanding of the mechanism of oxidative phosphorylation.

The great lability just described is explained in part by the fact that oxidative phosphorylation, along with the organized Krebs-cycle reactions and fatty-acid oxidation cycle, has been found to occur in rather highly organized subcellular structures, *the mitochondria*, as was first shown by Kennedy and Lehninger,[15] and independently by Schneider (cf. Ernster and Lindberg[16]), a finding which very quickly was confirmed and extended in many other laboratories.[17] Mitochondria isolated by the well-known sucrose procedure have become the standard study object in this field. It is important that mitochondria from many different cell types—plant, animal and microbial—show the respiratory-chain phosphorylation, thus providing a sound enzymatic basis for the suggestion by Claude that mitochondria are the power plants of the cell.

The mitochondria are almost completely self-contained units capable of carrying out both respiration and phosphorylation. It is necessary only to supplement a suspension of mitochondria with a substrate such as pyruvate, a trace amount of a 4-carbon catalyst such as fumarate or oxalacetate as substrates or fuel for the cycle, inorganic phosphate and ADP, and also $Mg^{++}$, which help to preserve mitochondrial structure. All of the necessary coenzymes and other co-factors are contained in the mitochondrial structure in such an organized manner that the complex cycle oxidations are smooth and complete with no obvious or conspicuous bottlenecks, indicating the high degree of enzymatic and morphological organization in the mitochondria.

The studies on oxidative phosphorylation have revealed, however, that relatively intact mitochondrial structure is necessary; simple procedures, such as freezing and thawing, treatment with detergents, exposure to hypotonic or strongly hypertonic agents, or solution, all cause uncoupling of the phosphorylation.[4]

Another important property of oxidative phosphorylation is that certain inhibitors in relatively minute concentrations are capable of uncoupling phosphorylation; that is, in their presence, no phosphorylation of ADP takes place, but oxidation proceeds nearly normally. These uncoupling agents include 2,4-dinitrophenol, as well as other nitro- and halo-phenols, the antibiotic gramicidin, and the anticoagulant Dicumarol, in addition to a number of oxidation-reduction dyes such as methylene blue. These agents have become extremely important tools for the elucidation of the character and sequence of the intermediate reactions of oxidative phosphorylation.[18]

### LOCALIZATION OF PHOSPHORYLATION SITES IN RESPIRATORY CHAIN[1–4]

To explain the occurrence of these oxidative phosphorylations, it was suggested at first that the intermediates of the Krebs cycle actually might occur in the form of phosphorylated derivatives rather than the free acids, analogs to the glycolytic reactions. Such phosphorylated intermediates never have been found. In 1940, however, Belitser postulated that the aerobic phosphorylations did not occur actually in the Krebs-cycle reactions proper, but rather during the phenomenon of electron transport from primary dehydrogenase to molecular oxygen via the known electron carriers making up the respiratory chain, such as diphosphopyridine nucleotide, flavoprotein, and the cytochromes. Belitser pointed out that the $\Delta F$ for the transfer of a pair of electrons from reduced diphosphopyridine nucleotide to molecular oxygen is approximately $-55\,000$ cal/mole. Since formation of one mole of ATP requires the input of 12 000 cal, it is evident that rather more than 4 moles of ATP theoretically could be generated during transport of each pair of electrons to oxygen, providing suitable enzymatic coupling mechanisms existed. As logical as the Belitser hypothesis appeared, experimental proof for this pattern was difficult to muster, and it was not until 1949–1951 that direct proof was adduced.[19,20] Highly purified, chemically reduced diphosphopyridine nucleotide was used as the actual substrate for the oxidation, in suspension of rat-liver mitochondria which had been pretreated in a hypotonic medium to produce sufficient alteration in permeability of the mitochondrial membrane so as to admit the reduced pyridine nucleotide to the oxidative centers within the mitochondria. Oxidation of reduced DPN via the known components of the respiratory chain, including the flavoprotein and cytochromes, culminated in oxygen uptake and coupled uptake of inorganic phosphate and the formation of ATP, with P:O ratios approaching the value 3.0. Oxidative phosphorylation thus occurs along the respiratory chain and may proceed in the absence of Krebs-cycle intermediates.

Similar experiments with pure cytochrome-*c*, used as either electron donor or acceptor, make possible further localization of the phosphorylations along the respiratory chain. Thus, passage of electrons from reduced DPN to cytochrome-*c* caused the formation of two moles of ATP.[4] Also, despite earlier experiments of Slater to the contrary, it finally was possible to demonstrate conclusively that the passage of a pair of electrons from cytochrome-*c* to oxygen yields a single phosphorylation.[21] More-recent spectroscopic investigations of Chance have confirmed these gross locations of phosphorylation sites.[1] Thermodynamic considerations of the known oxidation-reduction potentials of the carrier (Fig. 3), together with enzymatic and spectroscopic methods, suggest that one of the phosphorylations occurs between DPN and flavoprotein, the second between cytochrome-*b* and cytochrome-*c*. The site of the third phosphorylation is less certain: Chance suggests it lies between cytochrome-*c* and -*a*,[1] but other considerations indicate that it occurs between cytochrome-*a* and -$a_3$. These findings are summarized as follows:

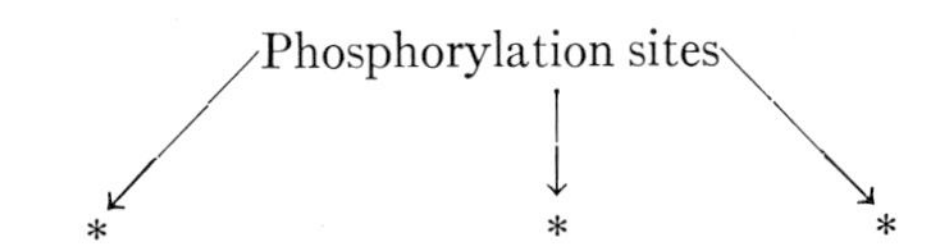


DPN→flavoprotein→cyt *b*→cyt *c*→cyt *a*→cyt $a_3$→O

The over-all reaction of the respiratory chain is thus

$$DPN_{red}+O+3P+3ADP \rightarrow DPN_{ox}+H_2O+3ATP, \quad (12)$$

and may be considered as the partial reaction

$$DPN_{red}+O \rightarrow DPN_{ox}+H_2O \quad (\Delta F=-55\,000 \text{ cal}) \quad (13)$$

$$3P+3ADP \rightarrow 3ATP+H_2O \quad (\Delta F=+36\,000 \text{ cal}). \quad (14)$$

From these results, it can be seen at once that the multimembered respiratory chain must represent a device for breaking up the large chunks of oxidative energy into smaller, biologically useful packets of approximately 12 kcal to accommodate the dimensions of the energetic currency of the cell—namely, the free energy of formation of ATP from ADP and phosphate. If all 55 kcal were released in one chemical-reaction step, even if coupled to a phosphorylation mechanism, only one ATP could be generated in this fashion.

### MECHANISM OF RESPIRATORY-CHAIN PHOSPHORYLATION

The mechanism of oxidative phosphorylation, in enzymatic and chemical terms, remains as one of the

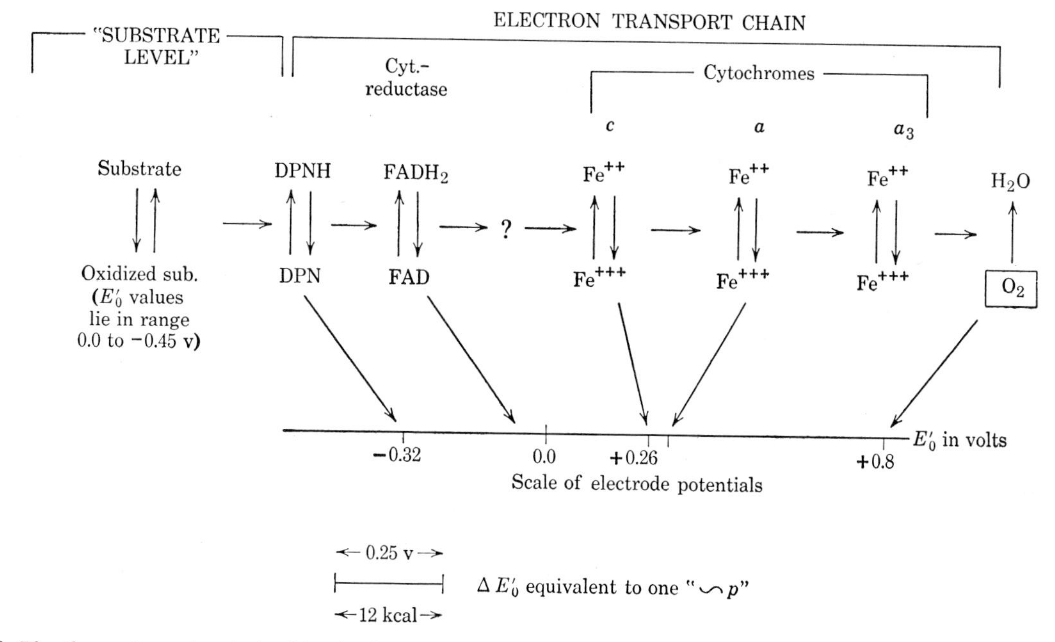


FIG. 3. The thermodynamic relationships in the respiratory chain. Electrons normally flow from substrates at the left to oxygen at the right. The approximate oxidation-reduction potentials ($p$H$=7.0$) of the carriers are shown on the scale. The free energy of hydrolysis of ATP of about 12 kcal is equivalent to the maximum free-energy change as a pair of electrons flow over a 0.25-v span, as is shown in the yardstick, from the relation $F=nf\Delta E$ [from A. L. Lehninger in *Harvey Lectures* (Academic Press, Inc., New York, 1953–1954), Vol. IL, p. 176].

most conspicuous and challenging unsolved problems of contemporary biochemistry. The immediate situation is that the classical approach of separation and isolation of enzymes making up a complex metabolic sequence, followed by *in vitro* reconstruction of the enzyme system, simply has not been possible to date, not only owing to the fact that the process is very labile but also because there is a requirement for geometrical organization of the many enzymes concerned in the characteristic morphology of the mitochondria. In addition, there is the fundamental dilemma of how a set of respiratory carriers of even moderate substrate specificity can catalyze both nonphosphorylating and phosphorylating electron transport, particularly in view of the fact that highly purified respiratory carriers, such as cytochrome-*c*, DPN, cytochrome-*c* reductase, and other respiratory carriers, show absolutely no ability to phosphorylate when tested in highly purified form in the test tube [cf. reference 3].

To facilitate discussion of the problem of the mechanism of oxidative phosphorylation and some recent successes in unraveling part of the coupling mechanism, it is best to consider first the general mechanisms which have been proposed to account for energy coupling in the respiratory chain, which again make use of the principle of the common intermediate described below.

The following mechanism is one we have proposed and for which considerable evidence has accumulated.[3,22,5] It incorporates ideas originally proposed by Lipmann.[8] Similar mechanisms have been discussed since by Lardy, Hunter, Slater, Chance, and others (see references 2, 1, 23). In skeleton form, it is

$$\text{carrier}_{\text{red}}+\text{X}+\text{oxidant}\rightarrow \text{carrier}_{\text{ox}}\sim\text{X}+\text{reductant}, \quad (15)$$

$$\text{carrier}_{\text{ox}}\sim\text{X}+\text{P}_i\rightleftarrows \text{carrier}+\text{P}\sim\text{X}, \quad (16)$$

$$\text{P}\sim\text{X}+\text{ADP}\rightleftarrows \text{ATP}+\text{X}. \quad (17)$$

In this scheme, the substance X is the vehicle of energy coupling and forms a high-energy compound with the carrier during the course of transfer of electrons from the reduced carrier to the next in the chain (designated as "oxidant"). Carrier$_{\text{ox}}\sim$X then undergoes phosphorolysis to form P$\sim$X (a high-energy compound, presumably a phosphoenzyme) which in turn can donate its phosphate to ADP. Thus, carrier $\sim$X is a common intermediate shared by the exergonic oxido-reduction of reaction (1) and by the endergonic reaction leading to formation of ATP in reactions (2) and (3). The energy liberated in the oxido-reduction thus is utilized to cause the formation of ATP. It is possible that phosphate interacts with the carrier directly to form a high-energy carrier$\sim$phosphate compound; however, for a number of reasons, it appears more likely that some compound other than phosphate is the first reactant with the carrier in the coupling mechanism. This hypothesis is written in skeleton form to indicate in the simplest way the principle of the common intermediate. It is possible and likely, however, that there are additional intermediate reactions between reactions (1) and (2) and also between (2) and (3).

This sequence now may be visualized as occurring

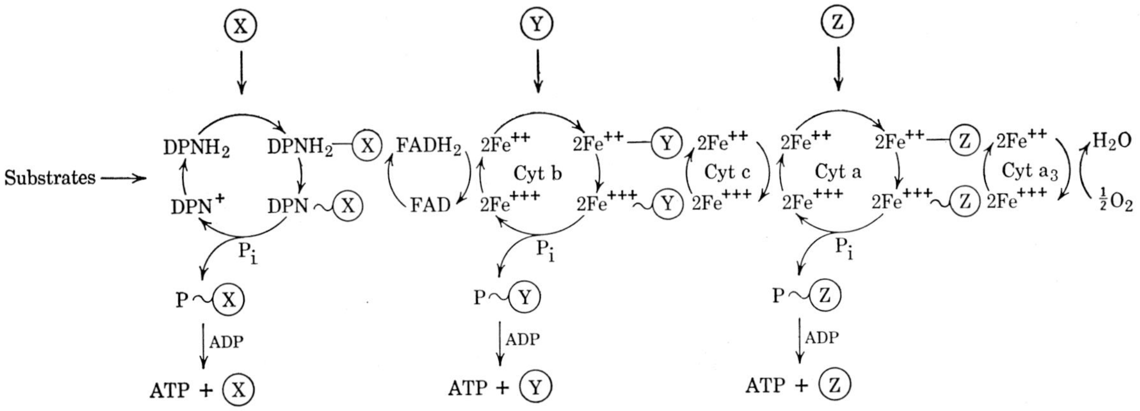

FIG. 4. A diagrammatic conception of the phosphorylation-coupled respiratory chain. This diagram is based on the coupling mechanism described in the text and assumes that the coupled carriers are DPN, cytochrome-*b*, and cytochrome-*a*. [from A. L. Lehninger, C. L. Wadkins, C. Cooper, T. M. Devlin, and J. L. Gamble, Jr., Science **128**, 450 (1958)].

with three different carrier pairs in the respiratory chain, as is shown in the diagram (Fig. 4) to account for the three different phosphorylations occurring along the chain.

The postulated skeleton-reaction mechanism indicates two possible approaches to unraveling the mechanism of energy coupling in oxidative phosphorylation. The first and seemingly more direct approach is to carry out investigations to isolate and to identify the molecular species of the respiratory carrier which exists in the energy-charged form—in short, the structure and enzymatic reactions of carrier $\sim$X or its analogs. However, not all of the known respiratory carriers have been obtained so far in purified soluble form suitable for examination of possible complexes of this kind and those which have been isolated to date have shown no propensity to phosphorylate in the test tube in simple systems. Furthermore, there now is increasing evidence that additional factors, such as vitamin $K_1$, $\alpha$-tocopherol, quinones, copper, nonheme iron, and other substances capable of undergoing reversible oxidation-reduction changes, may be members of the respiratory chain in addition to the classically accepted flavoprotein and cytochromes, but the site of these in the chain is still quite uncertain. Direct identification of the coupled carriers as a first step to determining the identity of X also is made difficult by the probability that a coupled form such as carrier $\sim$X reasonably could be expected to be rather labile. However, the spectroscopic studies of the kinetics of interaction of the carriers in the chain carried out by Chance in intact mitochondria provide some important landmarks for such efforts.[1]

There is another approach to the mechanism of oxidative phosphorylation, which sometimes is called the "back-door approach," and which we have chosen to take in our recent investigations.[5,18,24] The back-door approach starts with the end product of oxidative phosphorylation, namely ATP, and works back through the partial reactions to the carrier level. Our recent approaches to the mechanism of oxidative phosphorylation have been made possible by two developments. The first was our finding, in 1955, that phosphorylating submitochondrial fragments can be prepared from rat-liver mitochondria by the action of digitonin. These fragments are very much smaller than mitochondria but still retain phosphorylative activity. They are not simply miniature mitochondria since they no longer are capable of the Krebs-cycle reactions or of fatty-acid oxidation. However, these fragments, which are believed to be derived from the mitochondrial membrane, contain more or less intact respiratory chains together with the enzymatic factors responsible for energy coupling. Study of coupling in these so-called "digitonin fragments" is not complicated by the morphological compartmentation and permeability effects seen in intact mitochondria, which also catalyze many enzymatic reactions which are extraneous to oxidative phosphorylation.[25–31,22,5]

The second development is the finding that these particles catalyze two different isotopic exchange reactions of ATP which occur in the absence of electron transfers. The first is the ATP-$P_i^{32}$ exchange, in which labeled $P_i^{32}$ is rapidly incorporated into the terminal phosphate group of ATP.[22] The second is the so-called ATP-ADP exchange, in which $P^{32}$- or $C^{14}$-labeled ADP is incorporated bodily into ATP.[18] Both of the reactions in fresh fragments are completely inhibited by the classical uncoupling agent DNP, indicating their relationship to oxidative phosphorylation and distinguishing them from many other similar exchange reactions of ATP which have nothing to do with oxidative phosphorylation.

Close study of the interrelationship of these exchange reactions has revealed that the ATP-ADP exchange reaction is in reality the terminal reaction of oxidative phosphorylation

$$P\sim X + ADP \rightleftarrows ATP + X,$$

and is itself an intermediate step in the ATP-$P_i^{32}$ exchange. Since phosphate is not involved in the ADP exchange, but ADP is an obligatory component of the $P_i^{32}$ exchange, the sequence of the two terminal reactions of oxidative phosphorylation is established as

$$R{\sim}X + P_i \rightleftarrows P{\sim}X + R$$
$$P{\sim}X + ADP \rightleftarrows ATP + X$$

where R may be an electron-carrier molecule. Further, it has been found that the terminal reaction is not itself inherently sensitive to DNP but that sensitivity to DNP is conferred on it because it is in equilibrium with the preceding reaction, which is DNP-sensitive.[5,18]

These experiments have led very recently to the separation of the enzyme catalyzing the ATP-ADP exchange in soluble form from the mitochondrial fragments. It has been purified by chromatography on cellulose derivatives.[5] The nature of the active site participating in formation of the phospho-enzyme intermediate is now under investigation.

With this terminal enzyme in hand as a foundation for reconstruction approaches, we now propose to work nearer the carrier level and hope to soon obtain definitive information on the oxidation-reduction state of the energy-rich form of the carriers[1,5,32] which drives the phosphorylation, as well as the identification of the three coupled carriers in the chain.

## RESPIRATORY-ENZYME ASSEMBLIES AND THE STRUCTURE OF THE MITOCHONDRION

Consider now oxidative phosphorylation in the context of its natural habitat—namely, the mitochondrion.

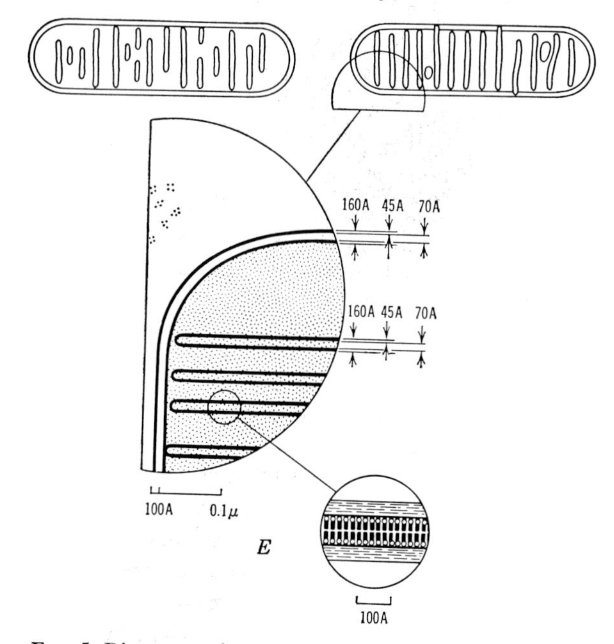


FIG. 5. Diagrammatic representation of the structure of mitochondrion according to Sjöstrand.[33] Note the double-layer character of outer membrane and the transverse "cristae." The magnified cross-section of the membrane suggests two protein monolayers separated by an oriented double layer of lipid molecules [from F. S. Sjöstrand in *Fine Structure of Cells—Symposium, Eighth Congress on Cellular Biology, Leyden, 1954* (Interscience Publishers, Inc., New York, 1955), pp. 16, 222].

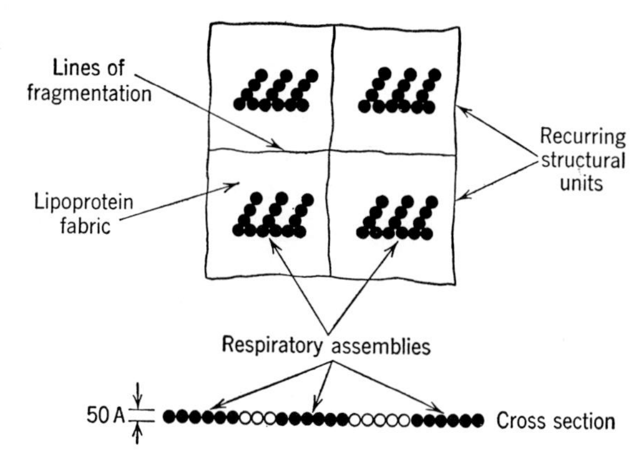


FIG. 6. Respiratory-enzyme assemblies in recurring units of mitochondrial membrane. The assembly consists of 6 electron-carrier protein molecules, supplemented by three sets of coupling enzymes, each consisting of three proteins as an approximation. This is a simplified representation since there may be several flavoprotein molecules per mole of cytochrome-*a* and several DPN-dehydrogenase molecules[1] [from A. L. Lehninger, C. L. Wadkins, C. Cooper, T. M. Devlin, and J. L. Gamble, Jr., Science **128**, 450 (1958)].

The structural details of mitochondria have been visualized by electron microscopy in thin sections of intact tissues in the laboratories of Sjöstrand, Palade, and others.[33] Although there are some differences in interpretation of electron micrographs of these organelles, the diagrammatic representation of Sjöstrand (Fig. 5) shows that the mitochondrion is surrounded by an outer double-layered membrane and also contains double-layered lamellar structures across the lumen called the "cristae" by Palade. These membranes, which are easily separable into fragmented form, are rich in lipid, containing some 30 to 40% which is mostly phospholipid. Sjöstrand has pointed out, as shown in his diagram, that the double-layered membranes correspond approximately to two monolayers of protein molecules separated by a double layer of oriented lipid molecules in the fashion suggested, and a similar arrangement in the cristae. These dimensions are constant in the mitochondria of all cell types examined.[33] The inner matrix of the mitochondrion contains soluble proteins and enzymes, electrolytes, and nucleotides which are released rather easily into the medium by mechanical or by osmotic rupture of the membranes. The results of many investigators suggest that most of the substrate-level enzymes of the Krebs cycle and the fatty-acid oxidation cycle are present either in the lumen or in the intercristal space. They are released easily in soluble form after damage of the membrane. On the other hand, it has been a universal finding that the enzymes involved in electron transport and coupled oxidative phosphorylation are present entirely in the insoluble membranes in organized "solid-state" systems; the separate enzymes of this system are not dissociated easily in truly soluble form.[5,16,34–36]

In phosphorylating subfragments of the mitochondrial membranes, the several electron carriers making up the respiratory chain occur in approximately equimolar ratios,[29] suggesting that each respiratory "assembly" consists of the six (or more) carrier molecules arranged in contiguous manner, presumably adjacent to each other. These are supplemented at three points along the chain by three sets of two or three (or more) auxiliary enzyme proteins which are necessary for the energy-coupling process, as in the coupling mechanism postulated. Such an arrangement thus would form a complete respiratory-phosphorylating multienzyme assembly, as is shown in Fig. 6.

The question arises as to the distribution of such respiratory assemblies along the membrane. We have subjected sonically-prepared fragments of the mitochondrial membrane to differential ultracentrifugation and have obtained a wide spectrum of particle sizes. The relative distribution of respiratory carriers in these particles of widely different sedimentation rate has been examined with the finding that, no matter what the size of the fragment, the ratio of the respiratory carriers to total protein was always constant, and the ratio of carriers to each other was relatively constant also. Phosphorylation activity is preserved in all fractions.[30] This finding indicates that the respiratory carriers are distributed more or less evenly along the membrane, and that fragmentation of the membrane produces pieces of different size made up of multiples of a basically recurring unit, as is indicated in the diagram, where each recurring unit contains a complete and organized respiratory assembly.[5]

Calculations involving the known extinction coefficients of some of the respiratory carriers such as cytochrome-*c* indicate that, in the membranes of a single rat-liver mitochondrion, there may be several thousand such respiratory assemblies. Still further analyses of this kind make it possible to estimate the total fraction of the mitochondrial membrane substance contributed by catalytically active proteins of these assemblies (assuming twelve protein molecules per assembly, each having a molecular weight of $10^6$) to be as high as 20% and possibly higher. From such considerations, it can be calculated that the average particle weight of a single respiratory assembly, together with its "filler" protein and lipid, may be $7 \times 10^{-7}$ approximately.[37]

From such considerations, it is clear that the mitochondrial membrane is not simply a "dead wall," but rather a highly complex fabric in which are embedded the respiratory assemblies in a highly oriented and purposeful manner. These respiratory assemblies thus represent the very warp and woof of the membrane fabric.

## MECHANICAL PROPERTIES OF MITOCHONDRIAL MEMBRANE

The mitochondrial membrane is not only the site of the highly organized respiratory assemblies, but it is also a structure capable of characteristic and selective permeability and active-transport phenomena. By optical measurements, it has been observed that the permeability may be altered by the ambient concentrations of ATP, magnesium, sulfhydryl reagents, and inorganic phosphate. In addition, we have found that the permeability of the mitochondrial membrane may be changed very drastically by addition of minute concentrations of the thyroid hormone, which causes rapid swelling and increased permeability to sucrose and other small molecules.[38–40] The mitochondrial membrane is capable of quite drastic extension and swelling under certain circumstances; it may increase in volume as much as several fold without actual rupture of the membranes (cf. Werkheiser and Bartley[41]). Phase contrast films of mitochondria in an intact cell show that they have remarkable plasticity of shape, and undergo rhythmic swelling and contraction as they wander through the canaliculi of the cytoplasm. The contractile properties of the mitochondrial membrane also can be observed *in vitro* when adenosine triphosphate is added to suspensions of heart mitochondria or when phosphorylating respiration is instituted. These agencies cause a rapid contraction of the mitochondrial membrane with the extrusion of water, and it has been postulated that the mitochondria, at least in specialized tissue such as the kidney, are active in water transport.[16,42] The contractility of the mitochondrial membrane thus may resemble the configurational changes in the myosin molecule similarly coupled to ATP energy. The number of water molecules extruded from mitochondria is out of all proportion to the number of molecules of ATP formed or utilized.

Very recently, we found that the permeability and contractility of the mitochondrial membrane are conditioned by the oxidation state of the respiratory carrier enzymes present in the membrane. Thyroxine and inorganic phosphate, which are potent swelling agents of mitochondria, increasing the permeability of the membrane, have this action only when the respiratory carriers are in the *oxidized* state. The action of these compounds results in a relaxation of the membrane and in increased permeability to certain test substances such as sucrose. On the other hand, when the respiratory carriers are maintained in the fully *reduced* state, the permeability of the mitochondrial membrane cannot be changed by addition of phosphate or thyroid hormone.[39,40]

The mitochondrial respiratory assemblies thus not only are the site of phosphorylating oxidation and major building blocks of the membrane, but also are controlling factors of rather dramatic mechanical and configurational changes which govern the rate of entry of essential metabolites such as substrates and inorganic phosphate, as well as the exit of respiration products such as $H_2O$, ATP, carbon dioxide, etc. The membrane-bound respiratory assembly thus becomes an example of a "mechanoenzyme" system, to use

Engelhardt's term for myosin. Indeed, the mitochondrial membrane has many characteristics resembling those of myosin, such as ATP-ase activity, DNP-stimulation, striking ionic strength relationships, as well as the configurational changes.

## SOME BIOPHYSICAL PROBLEMS ASSOCIATED WITH RESPIRATORY-ENERGY CONVERSIONS AND "SOLID-STATE ENZYMOLOGY"

Most well-known enzyme systems, such as the citric-acid cycle and glycolytic cycle, for example, involve low molecular-weight, rapidly diffusible intermediate molecules like di- and tri-carboxylic acids or phosphorylated sugars, as shuttles between relatively slowly diffusing enzyme proteins in an alternating manner. The respiratory-chain system, however, differs strikingly; it involves protein-protein interactions exclusively without low molecular-weight intermediates (Fig. 7). Since the rate of diffusion of protein molecules is extremely low as compared with that of simple sugars, amino acids, etc., a multienzyme sequence of a dozen consecutive reactions, each of which is brought about through protein-protein collisions alone, can be expected to occur at very low rates only if there were no means to direct collisions. In fact, it can be calculated readily that the known high rates of respiration of cells could not occur if all of the enzymes involved were distributed equally in free solution within the volume of the cell.[43] The assembly of the individual proteins making up this system into a solid-state array, as in the mitochondrial membrane, provides a molecular adaptation to overcome this kinetic problem. A variety of questions arises, however.

For one thing, do ordinary mass-law and ideal-collision considerations govern the behavior of protein molecules in such a "solid-state" array? The question is important and underlies all attempts to analyze reaction kinetics and reaction mechanisms in solid-state enzyme systems by the application of Michaelis-Menten principles and other kinetic parameters. Chance has commented on this problem.[1]

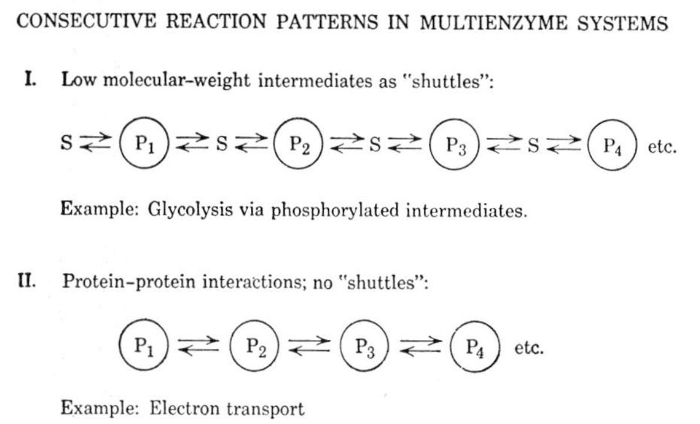

CONSECUTIVE REACTION PATTERNS IN MULTIENZYME SYSTEMS

I. Low molecular-weight intermediates as "shuttles":

$S \rightleftarrows (P_1) \rightleftarrows S \rightleftarrows (P_2) \rightleftarrows S \rightleftarrows (P_3) \rightleftarrows S \rightleftarrows (P_4)$ etc.

Example: Glycolysis via phosphorylated intermediates.

II. Protein-protein interactions; no "shuttles":

$(P_1) \rightleftarrows (P_2) \rightleftarrows (P_3) \rightleftarrows (P_4)$ etc.

Example: Electron transport

FIG. 7. Diffusion rates as factors in molecular interactions in multienzyme systems. Low molecular-weight substrate (product) molecules S, $S_z$, etc., have high-diffusion coefficients and may "shuttle" between high molecular-weight enzyme proteins; in the respiratory chain, slow-moving proteins must interact.

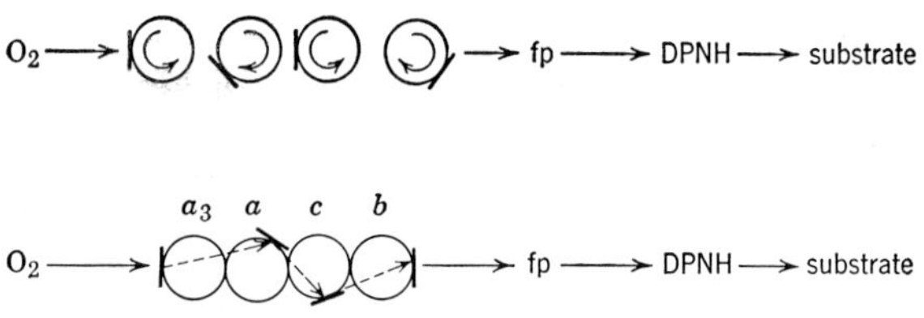

$O_2$ → fp → DPNH → substrate

$a_3$ $a$ $c$ $b$

$O_2$ → fp → DPNH → substrate

FIG. 8. Modes of electron transfer along chain of fixed electron-carrier molecules (after Chance and Williams[1]).

New methods thus will be required to examine the interaction of the individual catalyst molecules in such solid-state arrays, and to determine the translational, rotational, and vibrational degrees of freedom involved in these protein-protein interactions. Possibly, useful information will come from refined application of both electron-paramagnetic and nuclear-magnetic resonance spectroscopy, to supplement the already-extensive light spectroscopic studies of Chance and his colleagues.[1] Another question arises. How can the mechanism of electron transfer along the respiratory chain be accounted for, in view of the known dimensions of some of these proteins and their prosthetic groups, even assuming close juxtaposition of carriers? Chance has shown[1] two alternatives for the mechanism of electron transfer from one protein carrier to another along the chain (Fig. 8). In one, restricted rotation is postulated to permit collision of the prosthetic groups. In the second, the molecules are fixed, and electrons then must pass through the protein moieties to the prosthetic groups, possibly by $\pi$-electron interactions as in the so-called "charge-transfer" complexes. The passage of electrons or energy through the protein moiety of heme proteins has been considered frequently, since the classical experiments of Bücher on the constant quantum efficiency of 1.0 in the photodissociation of carbon monoxide myoglobin at all wavelengths tested.[44] Recently, Shore and Pardee have considered the physics of such energy transfers, which may take place through fluorescence and sensitization phenomena.[45,46] All of the respiratory carriers have quite characteristic fluorescence spectra. Szent-Györgyi, Weber, and others have suggested that fluorescence phenomena may conceivably play a physiological role in energy transfers along the carrier chain.[47]

The occurrence of such solid-state arrays of catalytically active proteins, the individual members of which are so difficult and as yet impossible to separate from each other with preservation of activity, presents a whole new area of protein chemistry and techniques. To date, it has been sufficiently difficult to examine the physical parameters of proteins in more or less ideal solutions; the interaction of two proteins with each other in dilute solution actually has been just barely touched upon in recent years. However, these far more complex solid-state arrays described above obviously present difficulties of a much greater order of magnitude. Similarly, the separation and identification of such proteins making up solid-state protein arrays will involve a whole new category of experimental approaches, as students of this area of enzyme chemistry already are painfully aware. If new techniques finally will permit separation of all of the carrier enzymes in soluble form, then reconstruction of these systems *in vitro* also can be expected to present great difficulty, because of the problem of restoring the spatial relationships among the members of such a complex in the test tube. Perhaps full *in vitro* reconstruction to yield soluble systems of high-catalytic activity never really will be attained.

These are but a few of the many biophysical and molecular problems posed by the solid-state character of the respiratory-energy transformers of mitochondria.

## BIBLIOGRAPHY

[1] B. Chance and G. R. Williams, Advances in Enzymol. **17**, 65 (1956).

[2] F. E. Hunter, Jr., in *Phosphorous Metabolism*, W. D. McElroy and B. Glass, editors (The Johns Hopkins Press, Baltimore, Maryland, 1951), Vol. I, p. 297.

[3] A. L. Lehninger in *Phosphorous Metabolism*, W. D. McElroy and B. Glass, editors (The Johns Hopkins Press, Baltimore, Maryland, 1951), Vol. 1, p. 344.

[4] A. L. Lehninger in *Harvey Lectures* (Academic Press, Inc., New York, 1953–1954), Vol. IL, p. 176.

[5] A. L. Lehninger, C. L. Wadkins, C. Cooper, T. M. Devlin, and J. L. Gamble, Jr., Science **128**, 450 (1958).

[6] H. M. Kalckar, Chem. Revs. **28**, 71 (1941).

[7] N. O. Kaplan in *The Enzymes*, J. B. Sumner and K. Myrbäck, editors (Academic Press, Inc., New York, 1951), Vol. II, p. 55.

[8] F. Lipmann, Advances in Enzymol. **1**, 99 (1941).

[9] K. Burton, Nature **181**, 1594 (1958).

[10] T. L. Hill and M. F. Morales, J. Am. Chem. Soc. **73**, 1656 (1951).

[11] T. Benzinger and C. Kitzinger, Z. Naturforsch. **10b**, 375 (1955).

[12] K. Burton and H. A. Krebs, Biochem. J. **54**, 94 (1953).

[13] K. Burton, Biochem. J. **59**, 44 (1955).

[14] M. J. Johnson in *Respiratory Enzymes*, H. A. Lardy, editor (Burgess Publishing Company, Minneapolis, 1949).

[15] E. P. Kennedy and A. L. Lehninger, J. Biol. Chem. **172**, 847 (1948).

[16] L. Ernster and O. Lindberg, Ann. Rev. Physiol. **20**, 13 (1958).

[17] D. E. Green, Symp. Soc. Exptl. Biol. **10**, 30 (1957).

[18] C. L. Wadkins and A. L. Lehninger, J. Biol. Chem. (to be published).

[19] M. Friedkin and A. L. Lehninger, J. Biol. Chem. **178**, 611 (1949).

[20] A. L. Lehninger, J. Biol. Chem. **190**, 345 (1951).

[21] S. O. Nielsen and A. L. Lehninger, J. Am. Chem. Soc. **76**, 3860 (1954).

[22] C. Cooper and A. L. Lehninger, J. Biol. Chem. **224**, 561 (1957).

[23] E. C. Slater, Nature **172**, 975 (1953).

[24] C. L. Wadkins and A. L. Lehninger, J. Biol. Chem. (to be published).

[25] C. Cooper, T. M. Devlin, and A. L. Lehninger, Biochim. et Biophys. Acta **18**, 159 (1955).

[26] C. Cooper and A. L. Lehninger, J. Biol. Chem. **219**, 489 (1956).

[27] See reference 26, p. 519.

[28] C. Cooper and A. L. Lehninger, J. Biol. Chem. **224**, 547 (1957).

[29] T. M. Devlin, J. Biol. Chem. (to be published).

[30] T. M. Devlin and A. L. Lehninger, J. Biol. Chem. **219**, 507 (1956).

[31] T. M. Devlin and A. L. Lehninger, J. Biol. Chem. (to be published).

[32] C. L. Wadkins and A. L. Lehninger, J. Am. Chem. Soc. **79**, 1010 (1957).

[33] F. S. Sjöstrand in *Fine Structure of Cells—Symposium, Eighth Congress on Cellular Biology, Leyden, 1954* (Interscience Publishers, Inc., New York, 1955), pp. 16, 222.

[34] E. G. Ball and R. J. Barrnett, J. Biophys. Biochem. Cytol. **3**, 1023 (1957).

[35] D. E. Green in *Harvey Lectures* (Academic Press, Inc., New York, 1956–1957), Vol. LII, p. 177.

[36] D. M. Ziegler, A. W. Linnane, D. E. Green, C. M. S. Dass, and H. Ris, Biochim. et Biophys. Acta **28**, 524 (1958).

[37] A. L. Lehninger in *Recent Advances in Molecular Biology* (to be published).

[38] A. L. Lehninger in *Enzymes: Units of Biological Structure and Function*, O. H. Gaebler, editor (Academic Press, Inc., New York, 1956), p. 217.

[39] A. L. Lehninger and B. L. Ray, Biochim. et Biophys. Acta **26**, 643 (1957).

[40] A. L. Lehninger, B. L. Ray, and M. Schneider, J. Biophys. Biochem. Cytol. (to be published).

[41] W. C. Werkheiser and W. Bartley, Biochem. J. **66**, 79 (1957).

[42] C. A. Price, A. Fonnesu, and R. E. Davies, Biochem. J. **64**, 754 (1956).

[43] A. G. Ogston and O. Smithies, Physiol. Rev. **28**, 283 (1948).

[44] T. S. Bücher, Advances in Enzymol. 14, 1 (1953).

[45] V. G. Shore and A. B. Pardee, Arch. Biochem. Biophys. **60**, 100 (1956).

[46] V. G. Shore and A. B. Pardee, Arch. Biochem. Biophys. **62**, 355 (1956).

[47] G. Weber, Biochem. J. **51**, 155 (1952).

# 17

# Energy Reception and Transfer in Photosynthesis*

Melvin Calvin

*Department of Chemistry, University of California, Berkeley 4, California*

The article by Lehninger (p. 136) presents a detailed and excellent description of how a cell can obtain energy by the combustion of carbohydrate. This article describes the reverse process, namely, how the green cells of plants are able to transform electromagnetic energy into chemical energy—by the absorption of carbon dioxide and water, which are the end products of the animal cell, and by the absorption of light—and how they produce the foodstuffs which are the beginning of the process of combustion. Figure 1 illustrates diagrammatically the content of this article.

The starting points in this case are carbon dioxide and water which contain the elements carbon, hydrogen, and oxygen in their lowest energy forms with respect to biological processes. The chemical energy which is accumulated is represented here (Fig. 1) in the form of oxygen (molecular oxygen), on the one hand, and a carbohydrate, on the other. The process itself has been divided, both theoretically and physically, into two rather easily separable stages. The first of these is the absorption of light by chlorophyll or by some related pigments and the subsequent separation of water into a reducing agent, here represented by [H], and some oxidizing fragment not specifically designated here but presumably one of the *A*, *B*, *C*, series. The oxidizing agent, or the primary oxidant, ultimately becomes molecular oxygen. In the second stage, the reducing agent is used to reduce carbon dioxide to the level of carbohydrate and other plant materials.

In order to see how the energy of light actually is accumulated in chemical form, it seems wise to describe what is known about the sequence from carbon dioxide to carbohydrate, and to determine at what point in that sequence the energy ultimately derived from the light enters, and from that point on to recognize and to define the problem of the primary quantum conversion into its first recognizable chemical form. Consider first in some detail what is known about the path of carbon so that one can define more precisely into what sort of energy the light must be converted in order to carry out that process.

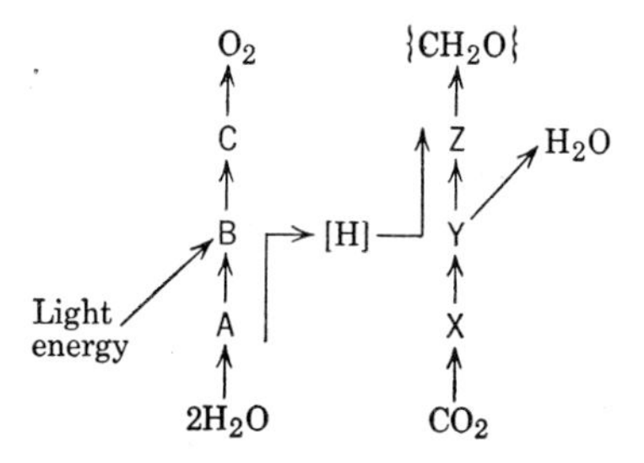


FIG. 1. Elementary photosynthesis scheme.

With the availability of radiocarbon (carbon-14) from the nuclear reactors some 15 years ago, it became possible for our laboratory to trace this sequence in some detail. The plant material used in most of the experiments was the unicellular green alga, *Chlorella*, and occasionally the alga, *Scenedesmus*; higher plants as well as separated photosynthetic material were used also. Figure 2 shows a photomicrograph of the algae cells commonly used; these are the *Chlorella* cells and the green stuff contained in a cup-shaped chloroplast can be seen. It is illustrated well by one of the cells in the upper right-hand corner.

The steps taken to trace the carbon sequence are as follows. The first operation constitutes a selection of cultures, which are grown in 200-cc flasks, and are transferred later into much larger continuous one-liter culture flasks. These are called shake-flask cultures in which algae can be maintained for years at a time.

The most recent type of culture device that is used in our laboratory is a continuous tube culture in which the density of the cells is monitored by a photoelectric cell which controls the automatic feeding of the medium, so that the cells are maintained in a steady state of growth.

The algal sample then is harvested and is used for the feeding of radiocarbon which is done in a special "hot box." In this box, the cells are placed in a little vessel (lollipop) between lights and are adapted with the concentration of normal carbon dioxide of interest. Radioactive carbon dioxide then is administered to the

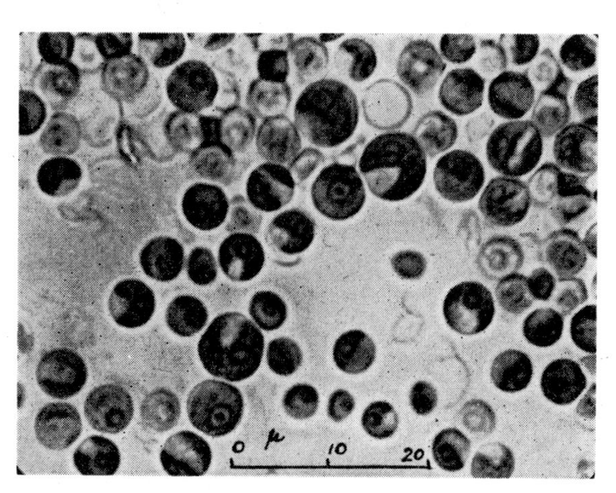


FIG. 2. Photomicrograph of *Chlorella* cells.

* The work described herein was sponsored by the U. S. Atomic Energy Commission.

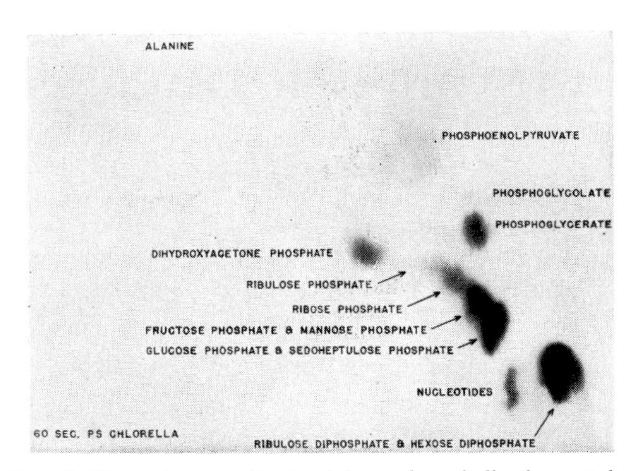


FIG. 3. Chromatogram of extract from algae, indicating uptake of radiocarbon during photosynthesis (60 sec).

adapted cells for a suitable length of time in order to trace the paths taken by the carbon atoms. The radioactive carbon usually is injected in the form of a solution of sodium bicarbonate. It is kept in contact with the cells for a specified period of time after which the cells are killed by a variety of methods; for example, the cells may be dropped into methanol at room temperature. The cell extract then is analyzed by the method of paper chromatography for the radioactive compounds which it may contain. In order to achieve this analysis, the extracts must be concentrated and a vacuum evaporator is used in a routine fashion to reduce the volume from 200 cc, or a liter, down to a cubic centimeter or so.

From this concentrated extract, an aliquot is taken and placed on the corner of a piece of filter paper for chromatographic separation. Prior to chromatography, the radioactivity of the origin is counted in a quantitative way. The filter paper then is hung in a box, in a trough in which a solvent is placed which passes over the filter paper and spreads the compounds down the side of the paper according to their relative solubilities in the solvent. The most soluble run the most rapidly. This procedure results in a set of spots along the side of the filter paper, depending upon the properties of the compound being analyzed. Some of these compounds overlap each other in one solvent system. The paper

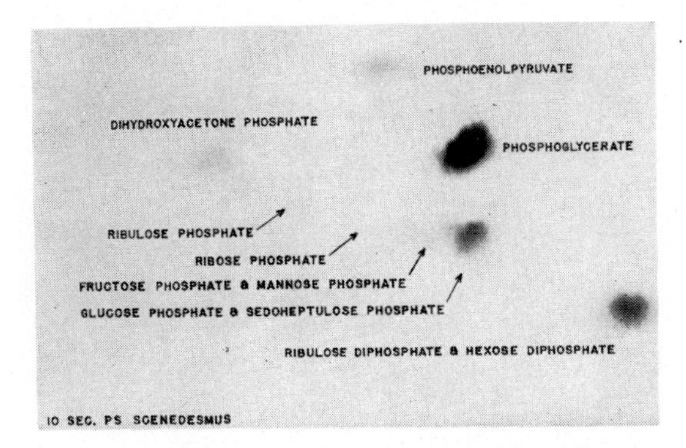


FIG. 4. Chromatogram of extract from algae, indicating uptake of radiocarbon during photosynthesis (10 sec).

then is removed from the box and dried overnight. The paper is rotated 90° and placed in another trough, in another box, with another solvent. The starting point is now a whole series of spots along the top edge of the paper. Another solvent is put in the trough and it spreads those spots out again in a similar operation. After this operation is completed, the paper is dried for a second time. The next problem is to locate the radioactive materials on the paper.

The coordinates of the material with respect to the origin constitute a physical property which is useful in the identification of the compound being analyzed. The compounds are not colored, and the only property that can be used to locate them is their radioactivity. This is done by placing the paper in contact with a sheet of photographic single-coat blue-sensitive x-ray film which becomes exposed by the radioactivity on the paper. Wherever there is a radioactive spot on the paper, there appears an exposed area on the film after a suitable period of time. For quantitative work, the film, covered by the paper, is placed on an x-ray viewer. The paper is explored with a Geiger counter. The compound can be identified then by its coordinates, and its amount can be determined by the amount of radioactivity found in the spot. A greater or lesser degree of resolution depends upon the nature of the solvent systems used and upon the time used for the chromatography.

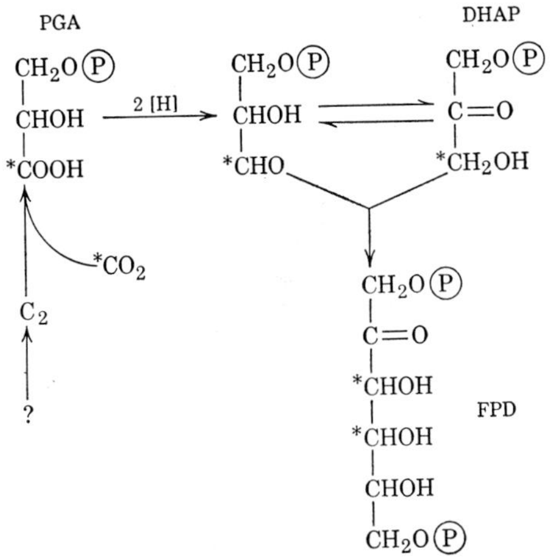

PGA: $CH_2O$Ⓟ — CHOH — *COOH $\xrightarrow{2\,[H]}$ $CH_2O$Ⓟ — CHOH — *CHO $\rightleftharpoons$ DHAP: $CH_2O$Ⓟ — C=O — $*CH_2OH$

$*CO_2$ + $C_2$ (← ?) → *COOH of PGA

*CHO + $*CH_2OH$ → FPD: $CH_2O$Ⓟ — C=O — *CHOH — *CHOH — CHOH — $CH_2O$Ⓟ

FIG. 5. Path of carbon from $CO_2$ to hexose during photosynthesis.

Figure 3 shows a chromatograph picture of the extract of a 60-sec illumination of *Chlorella*. One can see that a 60-sec illumination is much too long to find the earliest compounds into which the carbon enters in the course of its conversion from carbon dioxide to carbohydrate. Figure 4 shows a chromatogram of a shorter illumination (10 sec). Here, one compound, phosphoglyceric acid, dominates the scene. Our laboratory has been able to get the same type of sequence of events with isolated chloroplasts plus a number of cofactors. The phosphoglycerate appears no matter how the chloroplasts or the algae are killed: whether they are killed hot or cold, whether they are killed in alcohol, acetone, etc.

This, then, is an initial clue that the first isolable stable compound obtainable by these methods, or at least identifiable by these methods, is phosphoglyceric acid:

$$\underset{\displaystyle OPO_3H}{CH_2}\text{—}CHOH\text{—}CO_2H \quad ,$$

a three-carbon compound containing a low-energy phosphate group.

The next problem is to determine which of these carbon atoms is radioactive. This has been done by chemical-degradation methods. By taking the compound (phosphoglyceric acid) apart, one carbon atom at a time, it was found that the carboxyl group became radioactive first and the other two later. From this, together with the degradation of sugar molecules that came out in the same experiment, our laboratory was able to determine how the sugar molecule was constructed.

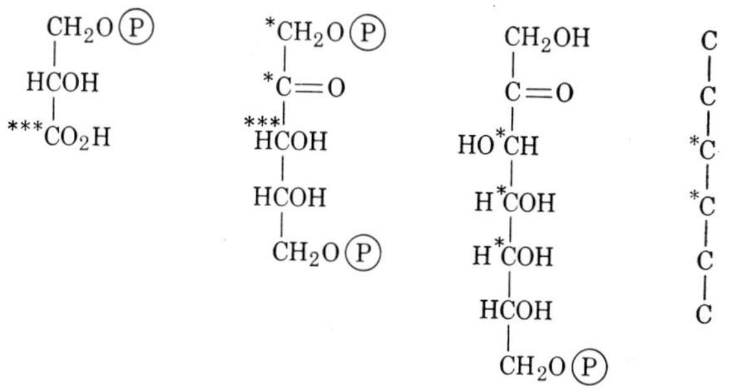

| PGA | RuDP | SMP | HMP |
|---|---|---|---|
| CH₂O(P) | *CH₂O(P) | CH₂OH | C |
| HCOH | *C=O | C=O | C |
| ***CO₂H | ***HCOH | HO*CH | *C |
| | HCOH | H*COH | *C |
| | CH₂O(P) | H*COH | C |
| | | HCOH | C |
| | | CH₂O(P) | |

Fig. 6. Distribution of radioactive carbon in certain sugars.

Figure 5 shows what was supposed to have occurred. The phosphoglyceric acid is shown as PGA. By reduction, this goes to triose phosphate. If the ketose phosphate is isomerized, and then combined with the isomer, a hexose diphosphate can be formed with the radioactive carbon atoms in the middle of the molecule. In this manner, the six-carbon molecule can be formed, but one does not know the origin of the three-carbon compound. Although two-carbon-containing molecules have not been found, Fig. 6 illustrates some findings of our laboratory.

In addition to the PGA, there is a five-carbon-atom compound, a sugar (ribulose diphosphate), a seven-carbon-atom sugar (sedoheptulose diphosphate), and, of course, the six-carbon-atom sugars. The stars on Fig. 6 indicate some idea of the way in which the radioactivity is distributed in these various sugar compounds. The heptose and the pentose can be made from the hexose, as shown in Figs. 7 and 8.

Figure 7 illustrates the method by which the heptose may be produced. From one molecule of hexose and one molecule of triose (taking off the top two carbon atoms of the hexose), a pentose and tetrose can be formed; the tetrose is labeled in the top two carbon atoms. The tetrose then can combine with a triose to make a heptose with the proper distribution of radioactive carbon.

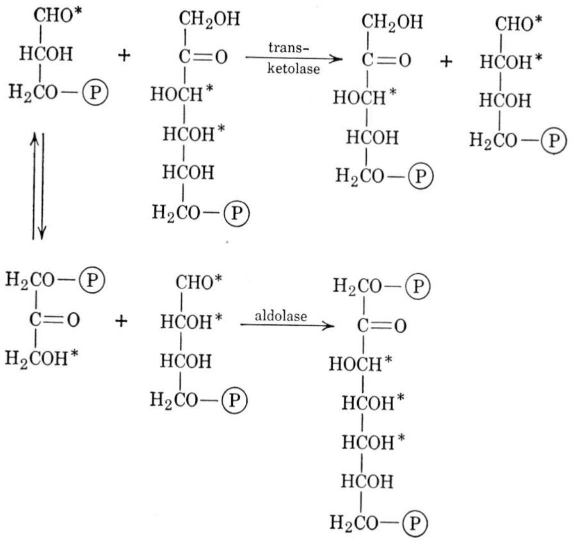

CHO* — HCOH — H₂CO—(P) + CH₂OH — C=O — HOCH* — HCOH* — HCOH — H₂CO—(P) →(transketolase) CH₂OH — C=O — HOCH* — HCOH — H₂CO—(P) + CHO* — HCOH* — HCOH — H₂CO—(P)

⇅

H₂CO—(P) — C=O — H₂COH* + CHO* — HCOH* — HCOH — H₂CO—(P) →(aldolase) H₂CO—(P) — C=O — HOCH* — HCOH* — HCOH* — HCOH — H₂CO—(P)

Fig. 7. Formation of a heptose from triose and hexose.

Figure 8 shows the way in which the pentose is put together, by combination of a heptose and a triose, in the same kind of reaction (the transketolase reaction) leading to two different pentoses which are in equilibrium with each other. This analysis does not distinguish between the two pentoses. All of these rearrangements are done at the sugar level; triose, tetrose, pentose, hexose, and heptose are all at the same redox level. They are all of them very nearly at the same energy level and there is thus practically no energy required for these rearrangements. However, no experiments of this type gave the desired information—namely, the origin of the three-carbon piece in the first place. This awaited a quite different kind of experiment, an experiment in which a steady state first was established in the organism, after which some environmental variable was changed suddenly. The transients that resulted from changing some of these variables were examined.

Figure 9 shows the results of such an experiment. A steady state is established by feeding the radiocarbon

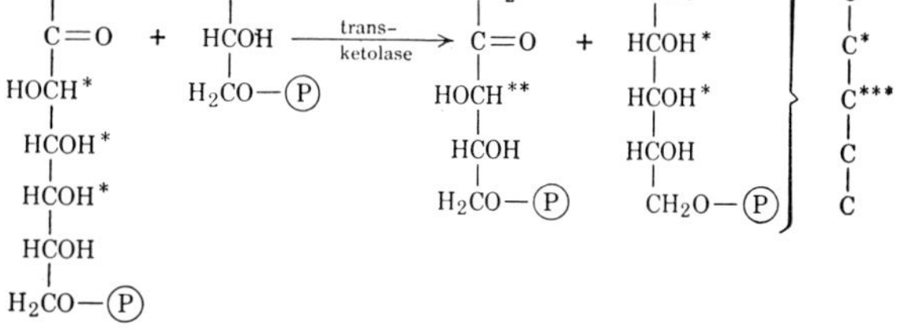

SMP: CH₂OH — C=O — HOCH* — HCOH* — HCOH* — HCOH — H₂CO—(P) + Phosphoglyceraldehyde: CHO** — HCOH — H₂CO—(P) →(transketolase) Xylulose monophosphate: CH₂OH — C=O — HOCH** — HCOH — H₂CO—(P) + Ribose monophosphate: CHO* — HCOH* — HCOH* — HCOH — CH₂O—(P) } C* — C* — C*** — C — C

Fig. 8. Proposed scheme for labeling of pentose.

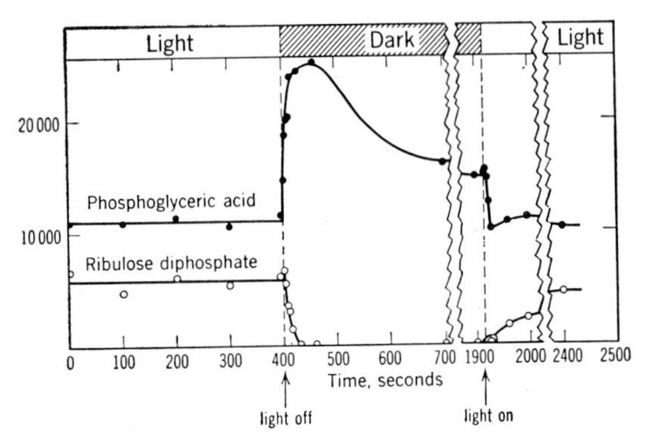


FIG. 9. Light-dark changes in concentrations of phosphoglyceric acid and ribulose diphosphate.

long enough to the plant to saturate the phosphoglyceric acid and the other compound mentioned. The lights are turned off suddenly, and the transient ensues. The phosphoglyceric acid rises suddenly and the ribulose diphosphate falls precipitously. This complementary behavior is the clue needed for the relationship between ribulose diphosphate and phosphoglyceric acid. It seemed as though the ribulose diphosphate was disappearing by combining with the carbon dioxide (that is, five carbons plus one, making a total of six carbon atoms) to produce two molecules of phosphoglyceric acid. If this is the case, then the relationships of the various compounds can be shown diagrammatically, as in Fig. 10.

In Fig. 10, the ribulose diphosphate combines with carbon dioxide to form phosphoglyceric acid which is reduced by light to triose, triose then going through this series of sugar rearrangements (shown in Figs. 6, 7, and 8) back again to the pentose. Turning the light off stops this reaction. When the reduction reaction is stopped, phosphoglyceric acid builds up and the ribulose diphosphate disappears. Figure 10 simply expresses in a scheme what the transient experiment revealed.

But this scheme (Fig. 10) predicts another type of transient. If the light is kept on and the $CO_2$ stopped, there should be a different kind of transient; namely, the ribulose diphosphate should build up suddenly and the amount of the phosphoglyceric acid should fall. This experiment has been accomplished with considerable difficulty. The results are shown in Fig. 11.

Figure 11 shows the steady state for ribulose diphosphate and the steady state for phosphoglyceric acid with the $CO_2$ at a concentration of 1%. At the vertical line, the $CO_2$ concentration is shifted from 1 to 0.003% by turning

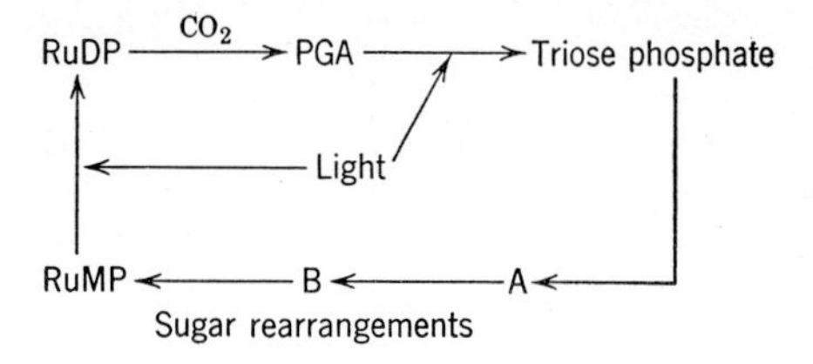


FIG. 10. Formation of PGA from RuDP.

the stopcocks. Under those circumstances, the predicted changes were observed, at least in the initial phase of the transient. The amount of phosphoglyceric acid fell and the amount of ribulose diphosphate rose. There is a number of kinetic oscillations here which are reminiscent of the kinds of oscillations one gets in circuitry, and possibly they are analogous. One or two attempts were made to reproduce these oscillations by putting first-order rate constants into the various reactions that are involved here and then running them through a digital computer. This kind of oscillation can be obtained but this work has not been pursued yet beyond the elementary stage of the first kind of transient. This kind of study will lead to much more-detailed knowledge of the mechanism of cellular response to changes in external or internal environments. It is a very simple system to use and one which is amenable to complete analysis, both experimentally in terms of the compounds

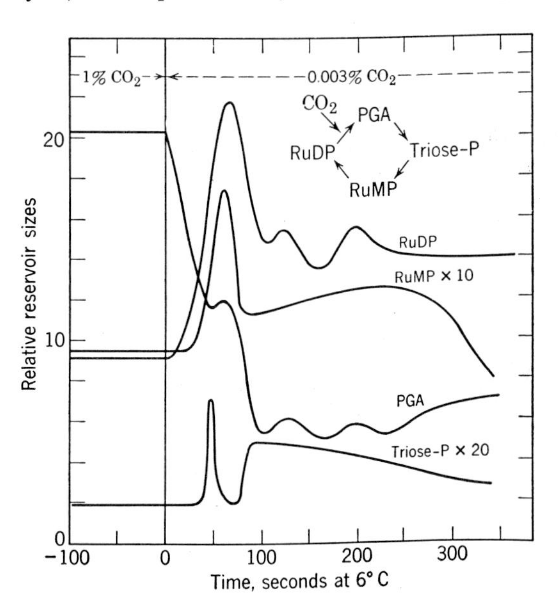


FIG. 11. Transients in the regenerative cycle.

involved and theoretically in terms of the simple kinetics involved.

Figure 12 shows the completed photosynthetic cycle in which there are put together all of the rearrangements of hexose and triose through heptose and pentose, back again to the ribulose diphosphate which then picks up carbon dioxide to make two molecules of phosphoglyceric acid.

In trying to visualize this particular step, a proposed mechanism for this reaction is shown in Fig. 13. Here, the ribulose diphosphate is written as the ene-diol, combining with bicarbonate ion to form an intermediate, hypothetical up to this point, an α-hydroxy-β-keto acid, which then is hydrolyzed to give two molecules of phosphoglyceric acid. This α-hydroxy-β-keto acid, according to our chemical knowledge, would be very unstable either to decarboxylation, in which case it would

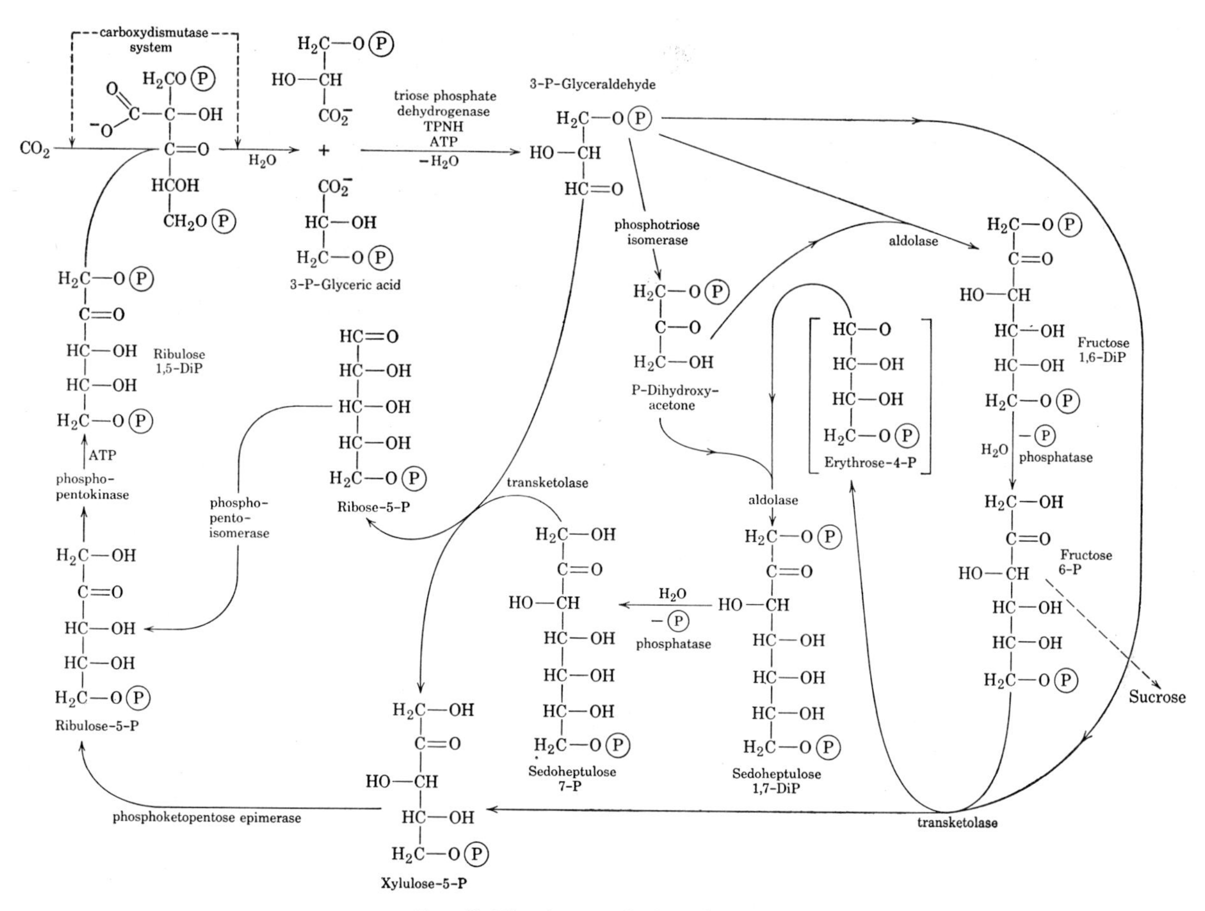


FIG. 12. The photosynthetic carbon cycle.

lead back to ribulose diphosphate, or to hydrolysis, in which case it would lead the other way. Fortunately, in our laboratory, a close relative of this intermediate, and probably some of the compound itself, has been found.[1] On theoretical grounds, it was decided that one might expect a compound of this sort to appear down in the diphosphate area of the chromatogram. Figure 14 shows that chromatogram which ran in a solvent for 48 hr in both coordinates. What was originally a single spot, which is dominantly ribulose diphosphate, now breaks up into at least three spots. The principal spot is the ribulose diphosphate; another one is hexose diphosphate and heptose diphosphate; and the last spot turned out to be a keto-acid diphosphate. It is not the $\beta$-keto-acid but rather the $\gamma$-keto-acid diphosphate

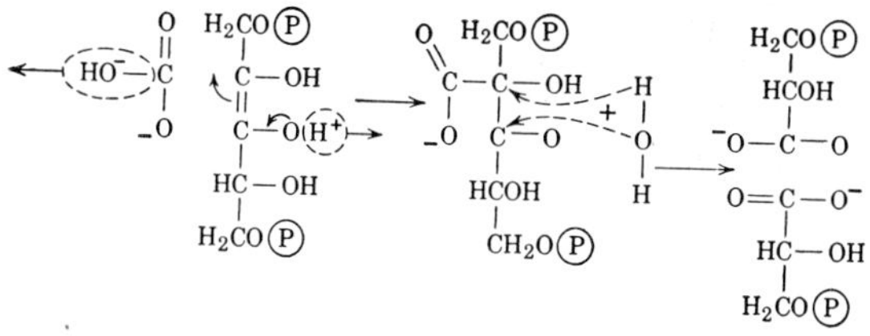

RuDP (ene-diol form) — Intermediate — 3-PGA

FIG. 13. Mechanism of carboxylation reaction.

which apparently is an artifact of the method of killing, but it does come from the $\beta$-keto-acid diphosphate which is still a labeled compound and does show its presence in small amounts.

In Fig. 15, one sees the diphosphate plus some dephosphorylated compounds, particularly the $\gamma$-keto-acid

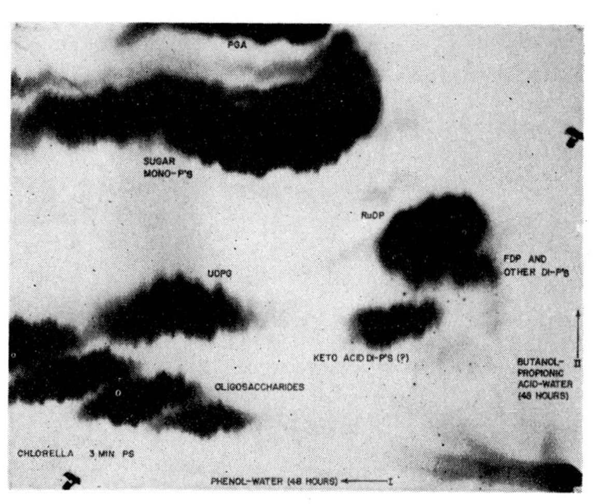


FIG. 14. Chromatogram of extract of *Chorella* after three minutes of photosynthesis in the presence of radiocarbon. The solvents were allowed to run for 48 hr in each dimension.

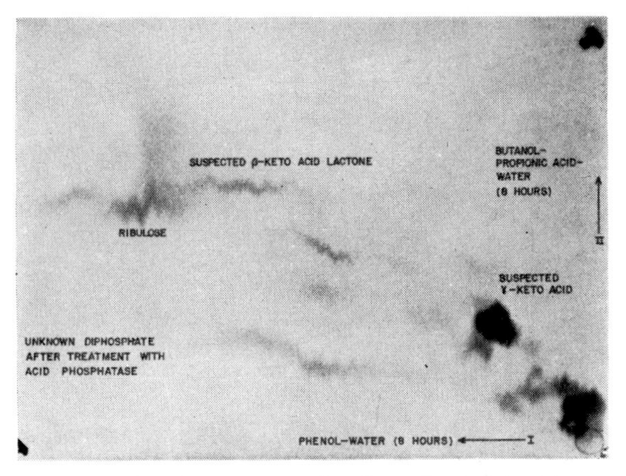


FIG. 15. Unknown sugar phosphate after treatment with acid phosphatase.

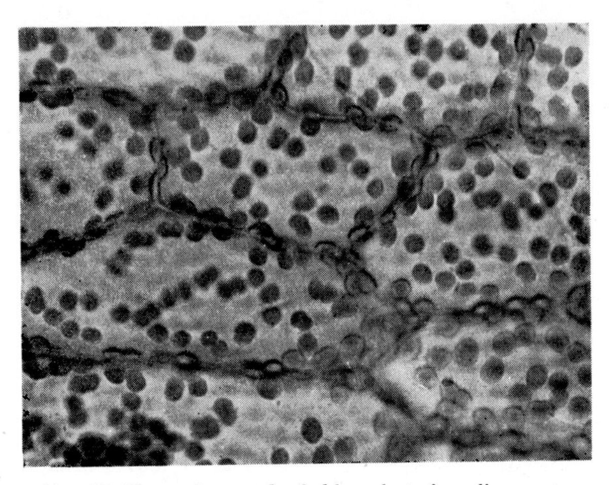

FIG. 17. Photomicrograph of chloroplasts from liverwort.

diphosphate. And here is seen a trace of the β-keto acid in its lactone form because it is stabilized as a lactone, enough to catch it on a chromatogram. Racker[2] has carried out this whole sequence (Fig. 12) by collecting all of the enzymes that were indicated in that figure. By putting in the suitable substrates, ribulose and carbon dioxide, he was able to pull out glucose phosphate from the $CO_2$. The immediate sources of energy are the two compounds adenosine triphosphate (ATP) and reduced pyridine nucleotide (TPNH). Figure 16 shows the needed relationships.

Figure 16 shows the photosynthetic carbon cycle in a simplified form. Carbon dioxide enters to make the β-keto acid which then goes to the phosphoglyceric acid. The only points of entry of energy into this system, as it is written now, are where there is a need for ATP and for reduced pyridine nucleotide. (These are the points of entry of energy into this wheel.) These points are the gears which drive the cycle in a forward direction.

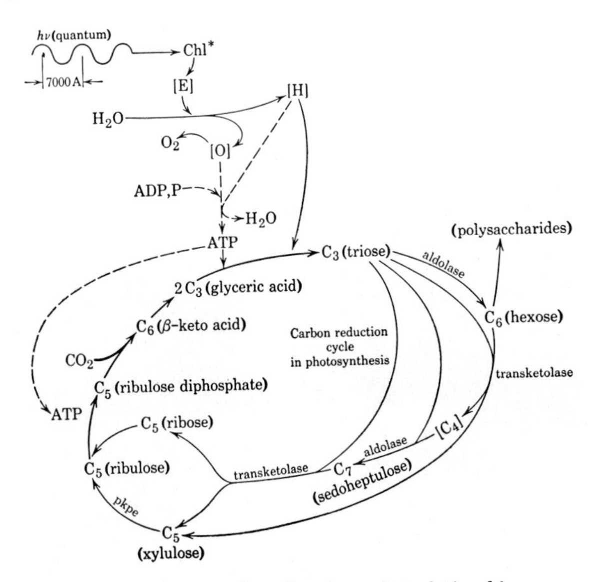


FIG. 16. Suggested cyclic scheme for relationships in photosynthesis.

Clearly the ATP and the TPNH (energy sources for the photosynthetic cycle) must come ultimately from the light. The remainder of this paper is devoted to this problem.

How does the light, which is absorbed by the chlorophyll, produce these two substances (ATP and TPNH) which are known to be required for carbon reduction? Before going into the details of a possible answer, consider some pictures of the apparatus which does it. Figure 17 shows a photomicrograph of liverwort tissue; one can see the cells, the cell walls, and the chloroplasts in which the chlorophyll is distributed very nicely inside the cells. Figure 18 shows isolated chloroplasts from spinach. (These are bigger chloroplasts and they have been isolated in sucrose solution.) All of this carbon reduction, oxygen evolution, phosphate production can be done with these chloroplasts removed from their natural habitat inside a cell. However, in order for that to be possible at anything approaching the rates at which it occurs in the living cell, one must add cofactors, some of which are heat stable, some of which are heat labile, and some of which are unknown but which are obtained out of the sap of the cells. In any case, this whole process can be done outside the cell.

Figure 19, from the work of Steinman and Sjöstrand,[3] shows an electron micrograph of a chloroplast. The picture on the right is shown at a higher magnification. The outstanding features of the chloroplast structures are lamellae, discussed in the following.

For some twenty years, it has been possible to carry out the photochemical evolution of oxygen by isolated chloroplasts using a suitable hydrogen acceptor such as ferrocyanide or quinone. This is called the Hill reaction. In the last five years, by preparing the chloroplasts in a manner which presumably does not destroy a chloroplast membrane or perhaps precipitates enzymes from the cytoplasm onto the chloroplast (i.e., preparing the chloroplasts in salt or sugar solutions), our laboratory has been able to carry out two other reactions with the chloroplasts. These reactions are the reduction of $CO_2$

as well as the evolution of oxygen, and, finally, the production of ATP by illumination of the chloroplasts. These three reactions, carbon-dioxide reduction (or, one step further back, the production of reduced pyridine nucleotide rather than $CO_2$ reduction), ATP production, and oxygen production are the three processes that one now can accomplish with the chloroplasts. The reduction of $CO_2$ requires two of the items and the evolution of oxygen may or may not require ATP.

How many of these things can be done simultaneously by the chloroplasts? In a recent conference,[4] it became evident that all of the pair combinations of these processes (i.e., $CO_2$ reduction, ATP production, and oxygen production) could be demonstrated.

It has been demonstrated that one could make one mole of pyridine nucleotide for every atom of oxygen produced. Simultaneously, one can demonstrate the production of one mole of ATP for every equivalent of reduced pyridine nucleotide produced. (One can demonstrate now that one mole of ATP is created for every equivalent of oxygen produced simultaneously.) This is something beyond the oxidative phosphorylation about which Lehninger writes (p. 136); that is, the oxidative phosphorylation would be the production of ATP by a recombination of TPNH and intermediate oxidant. It now appears that all three of these things can be reproduced equivalently at the same time.

The apparatus which does this in the plant has been shown in three magnifications—the whole chloroplasts in the cells, the chloroplasts outside the cells, and, finally, the lamellar structure of the chloroplasts as seen by electron microscopy. Studies of this lamellar structure have resulted in a particular conclusion which is sufficiently general to be stated; namely, the chloroplast lamellae seem to be (no matter what plant cell is investigated) disk-like in character; they seem to be connected at the edges to form a hollow disk—this is the lamella. The lamellae are quite long, about 2000 A in spinach chloroplasts. The lamellae do not appear in the chloroplast in the absence of chlorophyll or protochlorophyll. If, in some way, one prevents either the formation of protochlorophyll or of chlorophyll, one prevents the appearance of well-developed lamellae. Protochlorophyll alone will induce in cells which are normally capable of making them structures which look like these lamellae.

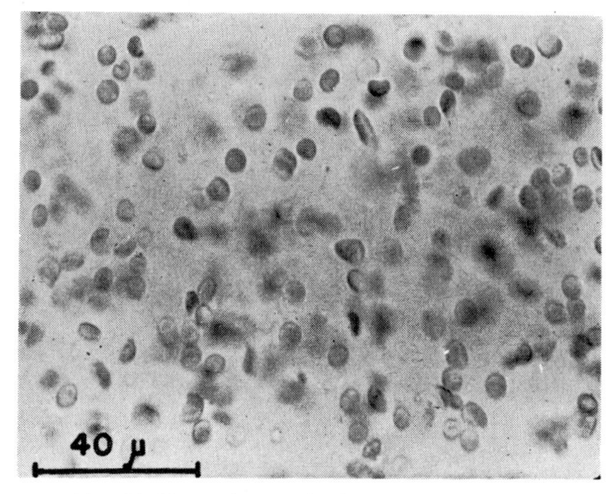


FIG. 18. Photomicrograph of spinach chloroplasts.

The possible function of this lamellar structure of the chloroplasts is discussed now. The basic problem of photosynthesis can be reduced to the problem of converting a 35- to 40-kcal quantum of energy into some chemical potential. In order to do this, one presumably has to find a reaction, which will take up 35 kcal at one time. The products of this reaction must not back-react. The 35 kcal are a great driving force for the back-reaction, so there must be some mechanism provided in the apparatus to prevent it.

There are a number of other difficult requirements which must be fulfilled in this quantum-conversion process with respect to the time constants involved. For

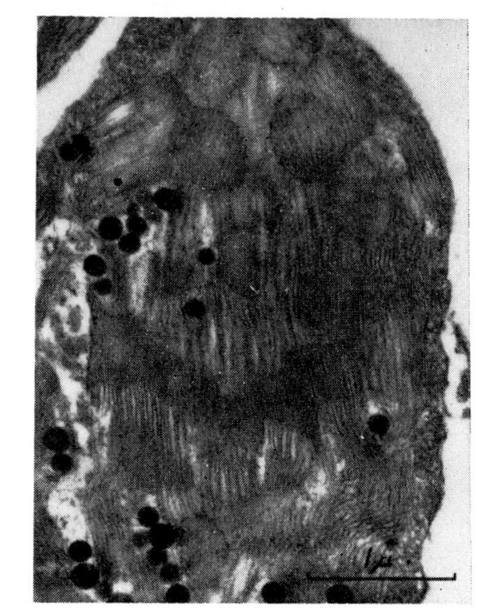


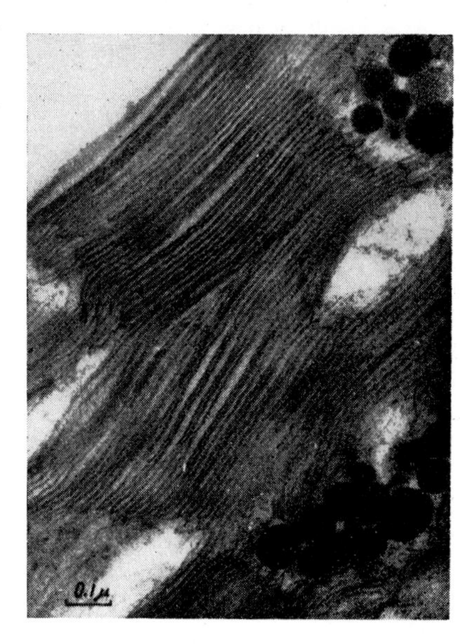


FIG. 19. Ultrastructure of chloroplasts [from E. Steinman and F. S. Sjöstrand, Exptl. Cell Research 8, 15 (1955)].

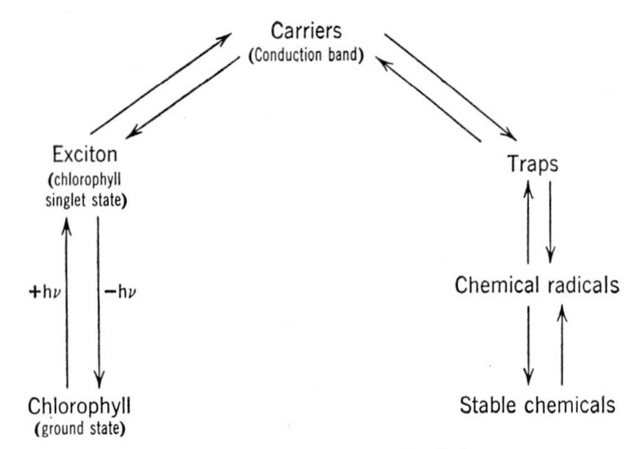


FIG. 20. Hypothetical scheme for light-energy utilization on chloroplasts.

example, following the absorption of the quantum, there must be a very efficient way in which the excited state of chlorophyll can be converted very quickly into a long-lived chemical potential because of the efficiency of the over-all process, regardless of whether one believes the maximum efficiency to be 30 or 60%.

There are a number of approaches to this problem which are based upon ordinary statistical solution photochemistry. In the past, I have looked for reactions unique to chlorophyll that might conceivably be used to store this 35 kcal of energy, such as reduction of chlorophyll to bacteriochlorophyll (that is, adding two more hydrogens to the chlorophyll molecule). Perhaps the process can be accomplished in the reverse manner, taking off two hydrogens from the chlorophyll molecule, making protochlorophyll, and hanging the hydrogen atoms onto something else. These are possible reactions of chlorophyll.

As a matter of fact, our laboratory demonstrated both of these reactions some years ago. More recently, and more elegantly, they have been demonstrated by Krasnovskii[5] in systems that are more nearly related to those which one finds in the living organism.

As a result of a variety of requirements, as a result of the recognition of this highly organized apparatus in which chlorophyll occurs in the chloroplast, and as a result of the failure to solve the problem with solution photochemistry, our group has turned to the notions of cooperative phenomena of organized systems such as those which are represented by barrier-layer cells in physics. Our group is trying to visualize how a lamellar structure such as this conceivably might be an unsymmetrical layer in which one could generate, by the absorption of light, an oxidant and a reductant, on opposite sides of the layer, so that they could not back-react easily and could persist for a long period of time. These substances (reductant and oxidant) should live long enough, by chemical standards, to be taken up efficiently, on the one hand, by electron acceptors to go on to make the reduced pyridine nucleotide, and, on the other hand, by electron donors to make molecular oxygen. I should like to present a proposal which fulfills all of the necessary requirements of the molecular interactions together with the need for conductivity (electrical conductivity) in certain parts of the lamellae and the consequent separation of charges.

The basic proposal is given in Fig. 20 which suggests how these lamellae achieve this energy conversion. Chlorophyll in the ground state absorbs light which brings it to its lowest singlet excited state. The excited state can move around among the chlorophyll molecules by resonance transfer (exciton migration ) until a point is reached where ionization occurs. Then charge separation can take place. The exciton can be visualized as a charge-pair which cannot move separately—a positive charge and a negative charge which must move together. When a suitable point in the chlorophyll lattice is reached where the charges can be uncoupled so that they can move separately, there is a conduction band. The electrons can move in one direction and the holes, or positive charges, in another. The electrons and holes move around until they find suitable places of lower potential energy into which they fall, and there sit for times sufficiently long so that suitable chemicals can come up and take off electrons, on the one hand, and the positive holes, on the other. This leads to chemical reactions which then produce stable chemicals such as a pyridine nucleotide and perhaps hydrogen peroxide, or something else of that sort, ultimately going on to the final products.

With this concept, consider how the structure of the lamella may be interpreted in terms of the molecular constitution. It is suggested that this layer is made up of at least four components [Fig. 21(b)]. The protein enzymes involved in carbon-dioxide reduction are on the outside of the disk. The protein enzymes on the inside of the disk are involved in oxygen evolution. The separation of the two processes (carbon-dioxide reduction and oxygen evolution) is achieved by a layer of chlorophyll packed in the characteristic aromatic way. This is a very characteristic pattern of packing. The aromatic rings do not pile flat on themselves; they lie at an angle, approximately 45° to the stacking axis. This type of packing is suggested for chlorophyll.

Figure 21(a) represents chlorophyll molecules tipped this way. Packed between them are carotenoids and the phospholipids. The proposal is that, after absorption, the exciton can migrate around among a few of these chlorophyll molecules to find a suitable point of ionization where the electrons may move in one direction and the positive holes in the other. Thus, one side leads to oxygen production, and the other to the reduction of carbon.

What kind of experimental evidence might detect this kind of system? Electrodes cannot be placed on these lamellae; they are too small. But one part of this scheme is susceptible to experimental observation—namely, the trapped holes and trapped electrons. It is expected that

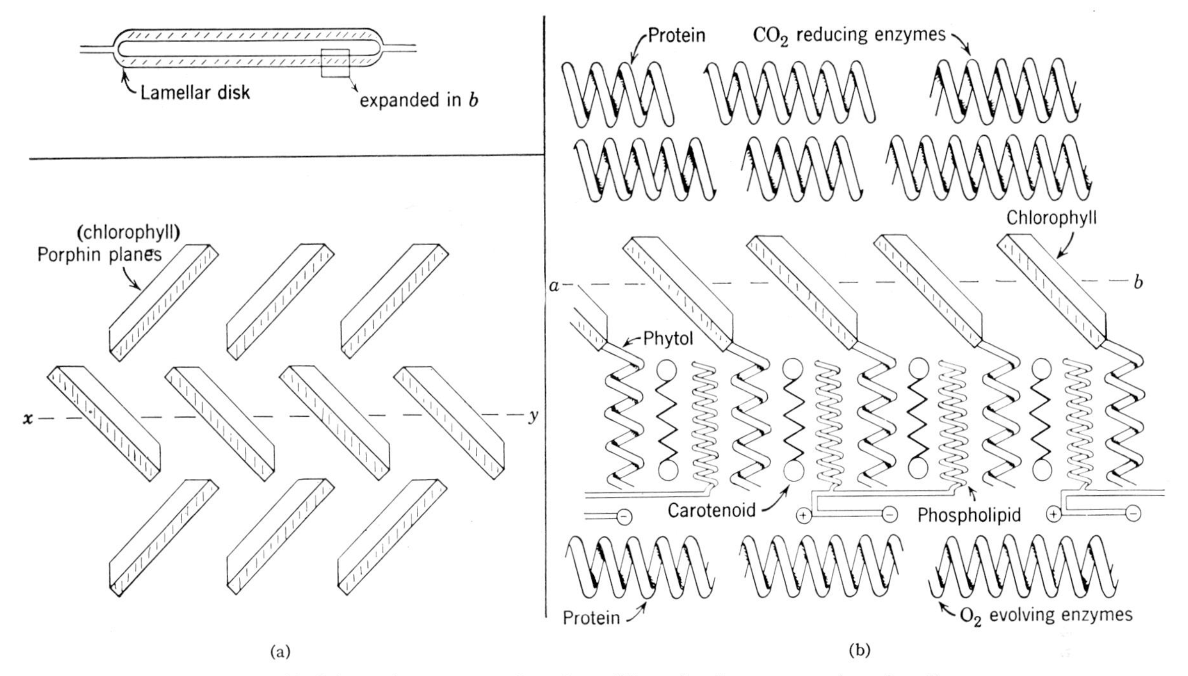


FIG. 21. Schematic representation of possible molecular structure for a lamella.

these trapped electrons are single, trapped electrons and, therefore, are detectable by paramagnetism. Our laboratory set out to search for photoinduced paramagnetism in the chloroplasts. Figure 22 shows the results of that search.[6–8] This is an illustration of electron spin-resonance signals for illuminated whole spinach chloroplasts at 25°C and at −150°C. (Similar signals, at least at room temperature, were reported from the St. Louis laboratory by Townsend.) The fact that one can get the signals at −150°C, either in chloroplasts or in algae, indicates that their production is not an enzymatic process.

The next question that may be asked is: "How fast can the signals be produced at −150°C as compared to 25°C? This is shown in Fig. 23. At 25°C, when the lights are turned on, the signals grow more rapidly than the instrument can follow them. At −150°C, the signals grow equally rapidly. The difference lies in the rate of decay of the signals. They have a complex decay—partly rapid and partly slow. At room temperature, the decay is rather rapid. At −150°C, there may be a rapid decay, but most of it is slow. This at least eliminates the possibility that the signals result from enzymatic formation. The questions remain, could the signal result from a triplet state—that is, a paramagnetic excited chlorophyll—or could the signal be the result of a

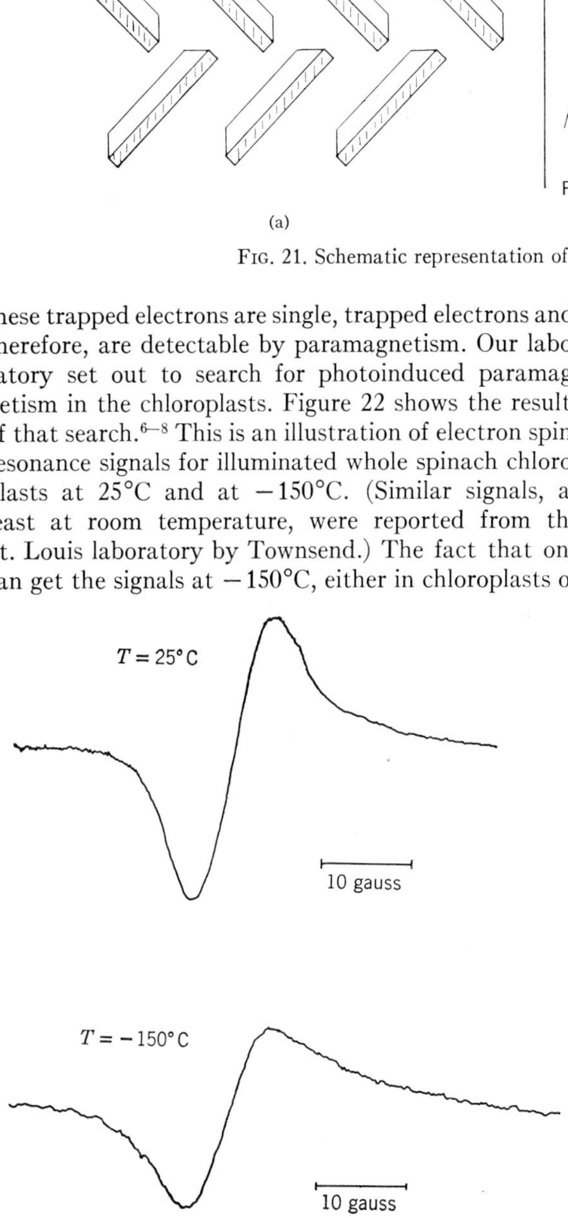

$T = 25°C$

10 gauss

$T = -150°C$

10 gauss

FIG. 22. Light signals from whole spinach chloroplasts.

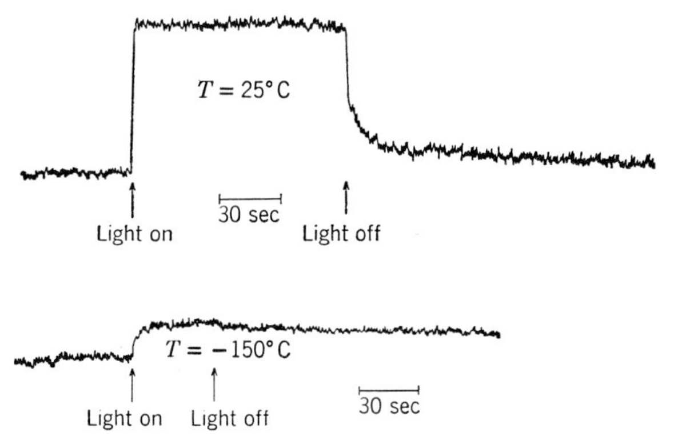


FIG. 23. Signal growth and decay time curves of whole spinach chloroplasts at 25°C and at −150°C.

photodissociation of chlorophyll, or something very closely associated with chlorophyll, to form chemical radicals, which process can take place at −150°C? These questions are discussed in the next article.

## BIBLIOGRAPHY

[1] V. Moses and M. Calvin, Proc. Natl. Acad. Sci. U. S. **44**, 260 (1958).

[2] E. Racker, Nature **175**, 249 (1955).

[3] E. Steinman and F. S. Sjöstrand, Exptl. Cell Research **8**, 15 (1955).

[4] Brookhaven National Laboratory Biology Conference No. 11 (June, 1958).

[5] A. A. Krasnovskii, J. chim. phys. (to be published).

[6] M. Calvin and B. P. Sogo, Science **125**, 499 (1958).

[7] P. B. Sogo, N. G. Pon, and M. Calvin, Proc. Natl. Acad. Sci. U. S. **43**, 387 (1957).

[8] P. B. Sogo, M. R. Jost, and M. Calvin, Radiation Research (to be published).

# 18
# Free Radicals in Photosynthetic Systems*

MELVIN CALVIN
*Department of Chemistry, University of California, Berkeley 4, California*

ONE bit of evidence introduced in the preceding paper concerned the possibility that the quantum-conversion act of photosynthesis might be a production of unpaired electrons which were trapped successively and then handed down to other acceptors (chemical acceptors) to do their job by ultimately reducing carbon dioxide.

That particular evidence was not explained. It simply was stated that there was evidence for unpaired electrons. The nature of this evidence now may be explained. These curves previously shown are known as electron spin-resonance absorption spectra. This article describes electron spin resonance, how it may be used for the detection of unpaired electrons in biological systems, how it has been used, and how it further might be used to identify the nature of the unpaired electrons that might and do occur in biological systems.

Electron spin resonance is another spectroscopic method. Zavoisky[1] was the first to make use of it to detect unpaired electrons in physical systems. It is based on the principle that an electron has a spin giving it a magnetic moment such that, when placed in a magnetic field, the electron orientates itself with respect to the magnetic field in certain specific directions. In the case of the electron, the spin is said to be one-half of a unit, and this leads to only two possible orientations of the electron in an external field—with and against the field.

Figure 1 shows the diagrammatic representation of this situation. On the left is the case in which the electrons are in between the pole faces of a magnet, but the electric current is not flowing yet and the electron magnetic moments are arranged randomly. When the field is turned on (as on the right), some of the electrons orientate themselves with the external field, and some against it. These two orientations do not have the same energy. The energies are designated as $E_1$ and $E_2$ in the figure, and the difference between them is equal to the product of the magnetic moment $\mu_0$, the gyromagnetic ratio $g_0$, and the value of the external magnetic field $H_0$. This difference in energy can be expressed in terms of the frequency $h\nu$, and for an electron in a field of about 3000 gauss, the wavelength corresponding to this transition is about 3 cm.

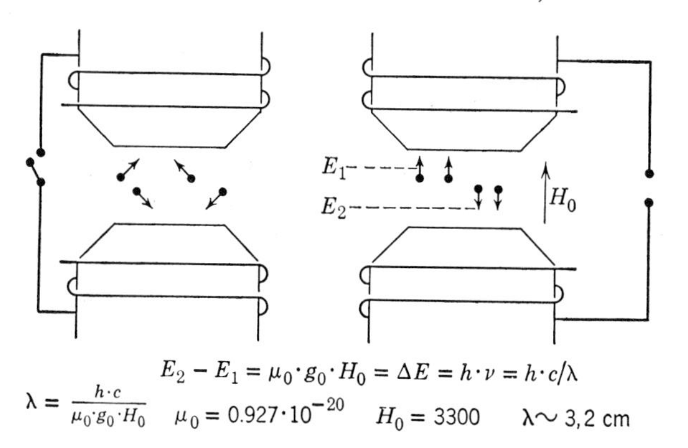


FIG. 1. Energy states of free electrons in an external magnetic field.

One may observe unpaired electrons by shining them with electromagnetic energy of a wavelength corresponding to the transition, and by watching for the absorption of this characteristic frequency, $\nu$, when the magnetic field, $H_0$, is varied. One then determines the magnetic field at which absorption occurs.

Figure 2 indicates the manner in which the experiment is done, showing the 3-cm generator and the external magnetic field. In the presence of the magnetic field, some of the electrons are oriented parallel with the field, some antiparallel. There will be more in the lower (parallel) state than in the upper (antiparallel), so there results a net energy absorption as transitions between states occur.

Most of the apparatus in use today does not record direct absorption, recording instead the derivative of this absorption. The two graphs in the preceding paper (Figs. 22 and 23) are these derivatives of absorption rather than the absorption itself.

This experiment thus provides a method for the specific detection of unpaired electrons (since paired electrons mutually each quench the magnetic moment of the other). It provides a highly sensitive device, more sensitive in general than that provided by any mass susceptibility measurement, because the latter always must correct for the diamagnetic material in which the electron is buried. In recent years, this device has been used more and more in chemistry and now is being used on biological materials to detect the presence of free radicals and to determine whether or not free radicals

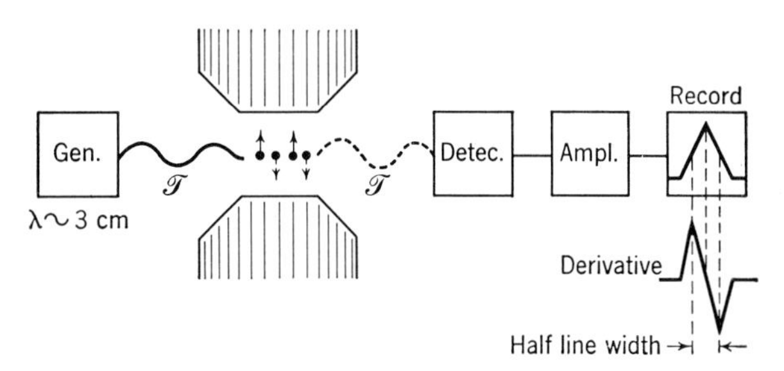


FIG. 2. Absorption of 3-cm waves resulting from transition of free electrons between energy states in an external magnetic field.

* The work described herein was sponsored by the U. S. Atomic Energy Commission.

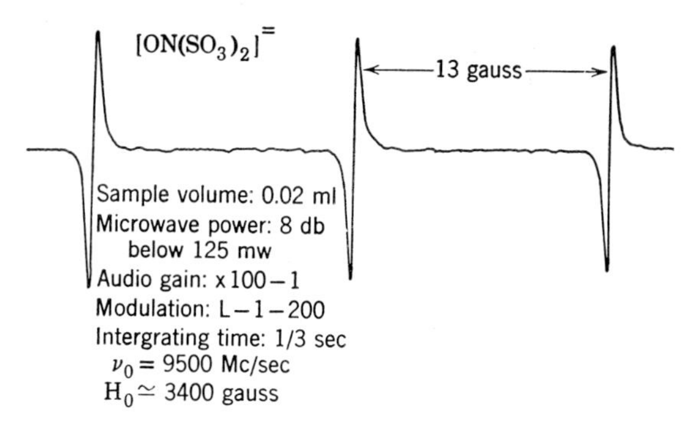


FIG. 3. ESR spectra of nitrosyl disulfonate [from Varian Associates Technical Information Bulletin, No. 2, 11 (1958)].

and unpaired electrons are participants in biochemical transformations.

The measurements obtained by this method not only can reveal how many electrons there are, but also, by the nature of the absorption record, may reveal something of the environment in which the electrons are situated. One recalls that the energy of the transition is dependent upon the field that the electron sees—that is, the external field due to the magnet, plus any magnetic field which the molecule may cause. The molecule itself is made up of nuclei and other electrons. Most of the other electrons, of course, are paired off, but the nuclei may have magnetic moments. Some have magnetic moments that produce magnetic fields with which the unpaired electrons may interact. If the odd electron sees not only the external magnetic field, but also that of the molecule, a spectrum similar to that shown in Fig. 3 from Varian Associates[2] may be observed. In this case, the molecule is one which has in it a nitrogen atom, which has a nuclear spin of one unit. This means that the nitrogen nucleus can take up one of three orientations in the external field—parallel, antiparallel, or normal. The electron near the nitrogen atom actually

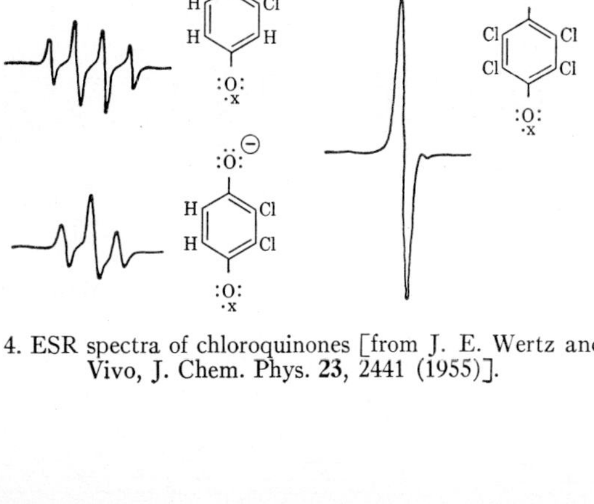

FIG. 4. ESR spectra of chloroquinones [from J. E. Wertz and J. Vivo, J. Chem. Phys. **23**, 2441 (1955)].

may be subjected to three different magnetic fields. In one configuration, the magnetic field of the nucleus is added to the external field. In another, it is subtracted from the external field. In the third case, there is no effect on the external field. So instead of one there should be three peaks. Figure 3 shows the three peaks attributable to the interaction with the spin of the nitrogen nucleus: one in the middle, and one on each side resulting from the three possible orientations of the nitrogen nucleus.

Nitrogen is not the only nucleus having a magnetic moment. One of the most common in organic substances is hydrogen. Hydrogen has a spin of a half, and it also

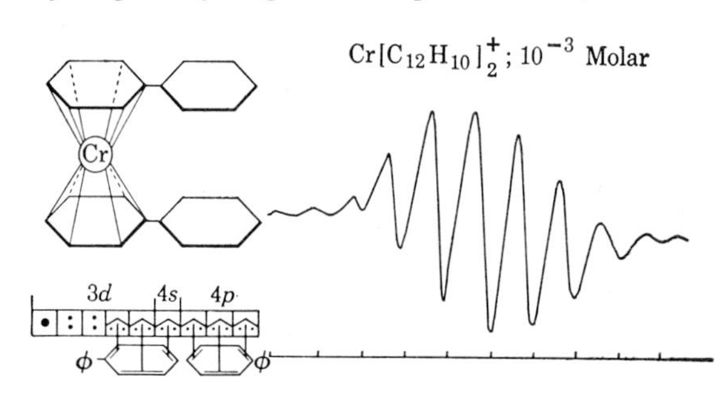


FIG. 5. Paramagnetic-resonance absorption spectrum of bisdibenzene chromium cation.

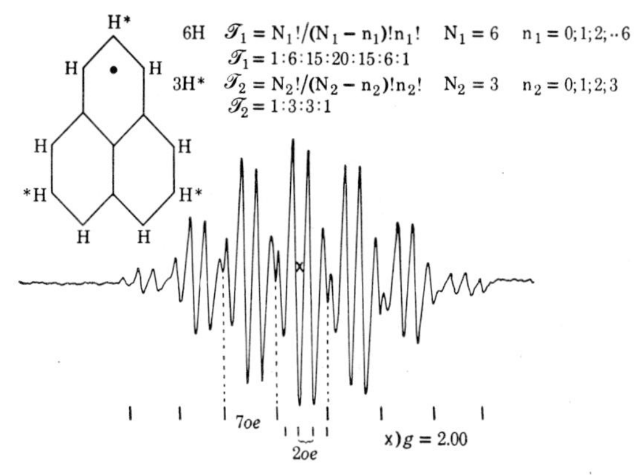


FIG. 6. Paramagnetic-resonance absorption spectrum of perinaphthenyl radical.

can influence the spin spectrum of an odd electron in the case where molecules are constructed such that the electron interacts with these protons.

Figure 4, from work by Wertz and Vivo,[3] shows the case for tetrachlorohydroquinone (lower right). The resonance field is not modified by any of the atoms in the molecule since their nuclei are all of spin zero (no magnetic moment), and thus only one line obtains. If, however, one removes one of the chlorines and replaces it by a proton, and if the electron can interact with that proton, the proton, having two possible orientations in the magnetic field, splits the resonance into two lines.

How many different arrangements can one obtain to

change the external magnetic field if two protons are present? Both of the magnetic moments may be had with or against the field, or they can be had one with and one against the field. There are, therefore, three different ways in which these two protons can be arranged with respect to the external magnetic field. The third situation is twice as probable as either one of the other two. Therefore, the middle peak should be twice as high as either of the two outside ones. Indeed, this is the case. In both cases, they produce no net modification of the external field. If three protons are present, there are four possible arrangements, and one gets four peaks; with four protons, there are five possible arrangements, and the amplitude ratios are the ordinary binomial coefficients, as seen in the figure.

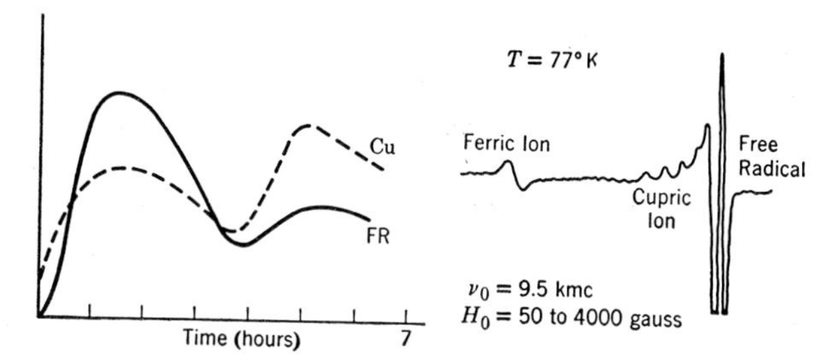


FIG. 7. ESR spectra from coenzyme-*A* dehydrogenase plus substrate [from H. Beinert, EPR Talk No. 10, Varian Associates (1958)].

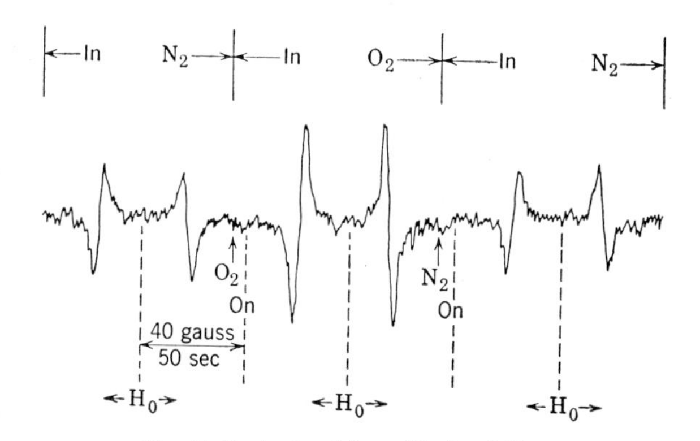


FIG. 8. Dark signal from *Rhodospirillum* methanol extract at room temperature

Not only may one determine, therefore, the number of odd electrons present in a system, but also one may ascertain something about the kind of an environment in which they are. Figure 5 shows the compound, dibenzene-chromium cation, which has one unpaired electron. The question of interest is the location of this unpaired electron. It was found, in our laboratory, that this unpaired electron of the chromium actually can see the ten protons of the two benzene rings of this compound associated with the chromium atom. Ten protons have eleven possible arrangements, and there should be eleven peaks in the curve. Those on the outside are very weak, because the probability of having all ten protons oriented in the same direction is very small as compared with the mixed configurations. One can determine, therefore, something about the environment of the transition elements by this method.

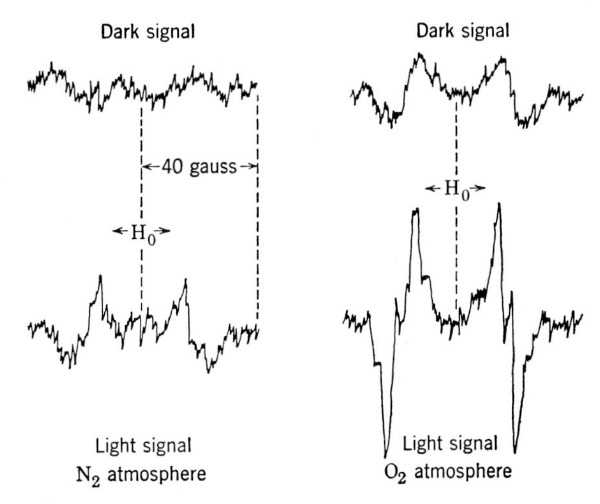


FIG. 9. Light and dark signals from *Chlorella* methanol extracts in two atmospheres at room temperature.

As a final chemical example, consider a perinaphthalene radical—a beautifully symmetrical radical (Fig. 6). During the preparation of the perinaphthalene, it accidentally became oxidized, and a very formidable absorption spectrum was obtained. The odd electron sees two different kinds of protons. On careful examination, one sees that there are seven groups of lines—each group a quadruplet. This means that there are six protons of one kind and three of another. The electron interacts more strongly with the six-proton group than it does with the three-proton group.

How much of this can be used for the investigation of biological free radicals? Beinert[4] has been working with a fatty acyl CoA which he found produced a transient, colored intermediate when mixed with its substrate. This, he proposed, was a free-radical intermediate, and when it was examined at Stanford University and at Varian Associates, an electron spin-

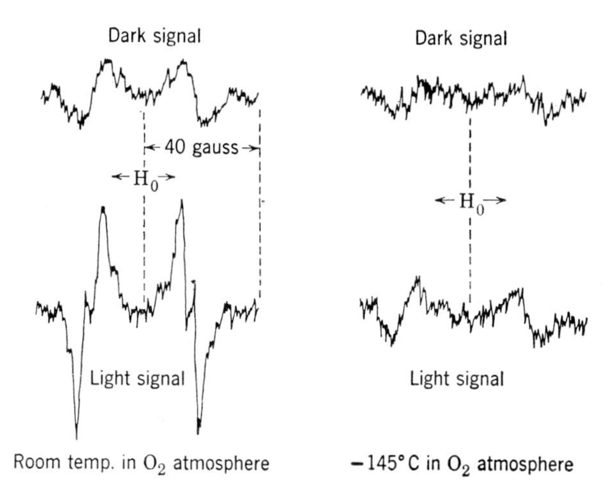


FIG. 10. Light and dark signals from *Chlorella* methanol extract at two temperatures.

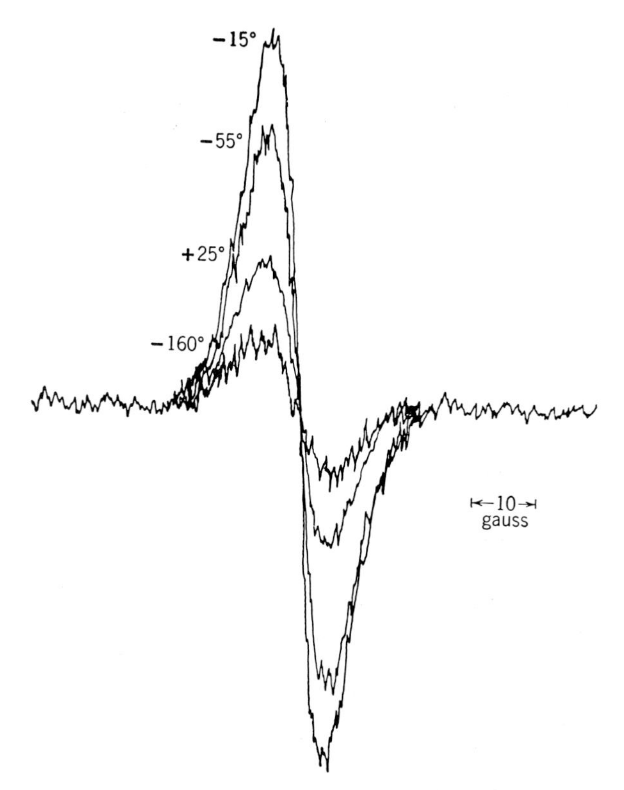


FIG. 11. ESR signals from *Rhodospirillum rubrum*, 5 min continuous illumination.

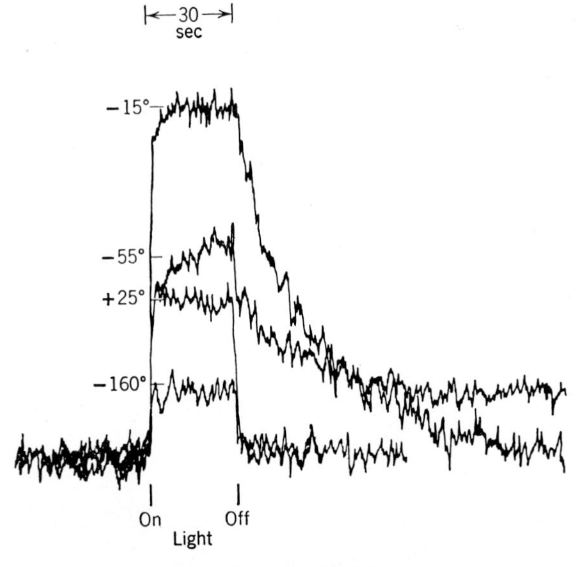


FIG. 12. Rise and decay of ESR signals from *Rhodospirillum rubrum*.

resonance signal was seen which could be distinguished from the resonance owing to the $Cu^{2+}$ in the enzyme (Fig. 7). Commoner[5] and co-workers presented evidence of free-radical intermediates in the reactions of alcohol dehydrogenase and of cytochrome oxidase when these enzymes are brought into contact with their substrates.

How is this kind of a signal of a free radical distinguished from that shown in the preceding paper (Fig. 22)? Firstly, radicals of this kind do not fade, at low temperatures. They are frozen in. That is exactly what happens upon cooling the system of chloroplasts as described earlier. The fact is that structured systems can be found wherein the signal can be induced at a low temperature, but wherein it also fades at a low temperature at a very high rate.

The occurrence of a signal owing to an oxidation mechanism alone is illustrated by Fig. 8, which shows the behavior of a methanolic extract of *Rhodospirillum* when exposed to oxygen and nitrogen, alternately. When a similar methanolic extract of *Chlorella* is illuminated first in an oxygen and then in a nitrogen atmosphere, the increased signal of the former (Fig. 9) is considered the sum of the contributions of an oxidative, a photooxidative, and an odd electron produced and trapped in a free radical. The relative contributions of these processes are apparent from the figure. Figure 10 shows how cooling, even in oxygen, reduces the signal amplitudes, and one sees that cooling to −145°C nearly has eliminated the dark spectrum.

Figure 11 shows the spin-resonance signal for *Rhodospirillum* after 5 min of illumination at the indicated temperatures. At 25°C, it is the next smallest signal;

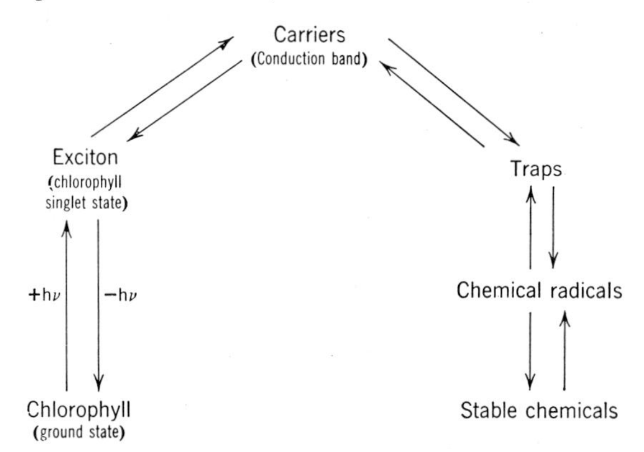


FIG. 13. Hypothetical scheme for light-energy utilization on chloroplasts.

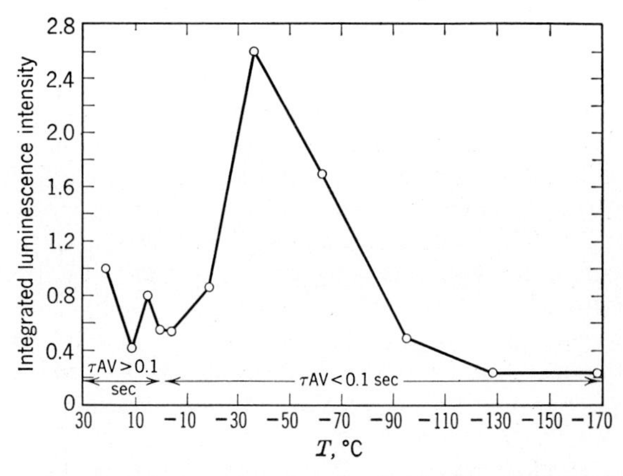


FIG. 14. Integrated intensity of spinach-chloroplast luminescence from 0.0015 to 5.0 sec after excitation, as a function of temperature.

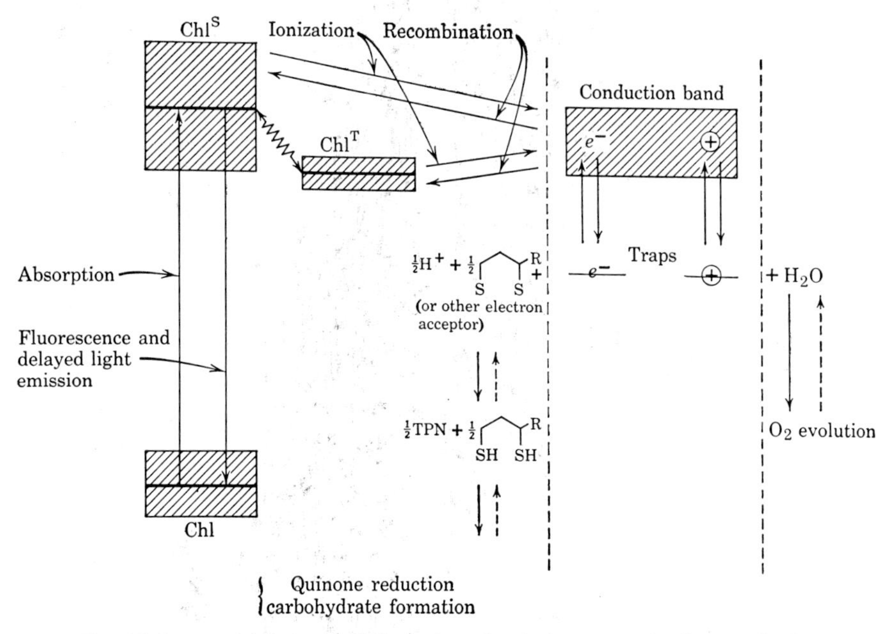


FIG. 15. Proposed scheme for various photochemical processes in photosynthesis.

at −55°C, it is larger; and at −150°C, it is larger yet. When the temperature is further reduced to −160°C, the intensity falls to the smallest shown. The time behavior is shown in Fig. 12. The spectrometer is set on the peak of the signal, and one may observe the rates of rise and of decay of each of these signals at each temperature as a function of time. The room-temperature signal is the next smallest shown, and it rises as rapidly as the instrument can respond. This signal decays immediately after illumination ceases, to within the time constant of the apparatus. There is, however, no rapid decay at −15°C; at −55°C, there is a very rapid rise, a further slow rise, followed by a small component of rapid decay, and a much longer slow decay. At −160°C, all that remains is very rapid rise and very rapid fall. Thus, I believe the possibility has been eliminated that these free radicals are being produced directly by illumination. This behavior must be the result either of untrapped carriers or of trapped carriers in well-shielded traps.

How can this whole sequence of events be accounted for? Figure 13 (discussed in the previous article) indicates the first act, the absorption of light to produce an exciton. The exciton then is converted into conductive carriers, these into chemical radicals, and the chemical radicals lead to stable chemicals. To account for the data shown in Fig. 12, it is necessary to presume that each one of these successive acts has a higher temperature coefficient. The first step has no temperature coefficient, the second may have a very slight one, the third a higher one, and so on. At room temperatures, the energy is transported into the chemical radicals, but these decay very rapidly into stable chemicals by the usual enzymatic process. The possibility that a triplet state exists is eliminated by the nature of the signal. A triplet signal would be a double band very characteristic of two unpaired electrons in the same molecule. A similar sequence of temperature effects is seen in the luminescence intensity curve (Fig. 14) and similarly is accountable, presuming the integrated intensity to be a measure of the back-reaction.

Another representation of the same scheme is shown in Fig. 15. The chlorophyll ground state is represented by a band instead of by a single line—as a result of the interaction of the chlorophyll molecules in these arrays previously shown in electron micrographs—and the chlorophyll excited singlet state actually would be a broader band. The triplet state is shown as a rather narrow band overlapping the excited sequence. The triplet band has not been put directly in line because triplet light emission is not observed.

The foregoing scheme gives a working hypothesis to account for some of the problems, outlined at the onset —that the initial 35 kcal of energy be converted into useful chemical potential.

## BIBLIOGRAPHY

[1] E. Zavoisky, J. Phys. U.S.S.R. **9**, 211 (1945).

[2] Varian Associates Technical Information Bulletin, No. 2, 11 (1958).

[3] J. E. Wertz and J. Vivo, J. Chem. Phys. **23**, 2441 (1955).

[4] H. Beinert, EPR Talk No. 10, Varian Associates (1958).

[5] B. Commoner, J. J. Heise, B. B. Lippincott, R. E. Norberg, J. V. Passonneau, and J. Townsend, Science **120**, 57 (1957).

# 19

# Relation between Exciton Bands and Conduction Bands in Molecular Lamellar Systems*

MICHAEL KASHA

*Department of Chemistry, Florida State University, Tallahassee, Florida*

## 1. INTRODUCTION

BIOPHYSICAL investigations on a molecular level turn naturally to the phenomena of energy transfer and electron transfer for the interpretation of the behavior of the ordered structures found in biological systems. The famous Korányi lecture[1,2] of Szent-Györgyi has done much to stimulate interest in these phenomena, although earlier recognition of their importance, e.g., in the photosynthetic system, had been made.

The general tendency in this field at present seems to be the adoption without caution of the ideas and terminology of the solid-state physics of atomic and ionic semiconductors, to the interpretation of biological systems. Since the latter are usually molecular aggregates bound weakly by van der Waals forces (molecular lamellae) or very strongly by overlap forces (intramolecular forces in proteins), it is worthwhile to examine critically the differences between the atomic and molecular array problems, to see how valid is the application of the ideas of solid-state physics to biological mechanisms.

Since most of the treatments of the exciton theory are presented in fairly intricate mathematical terms, an attempt is made here to give a simple but accurate account of excitation-energy transfer as well as of electronic conduction in atomic and molecular aggregates.

## 2. EXCITONS

The theory of excitons in atomic lattices was developed by Frenkel,[3,4] and elaborated later by other physicists.[5–10] The application of exciton theory to molecular crystals was first made by Davydov.[11,12] A semiclassical treatment of excitons in van der Waals pigment polymers was given by Förster.[13] Qualitatively the same idea was also used by Förster to interpret the spectra of pigment dimers, and has been treated quantum mechanically by Simpson's group.[14] The extension to excitons in pigment aggregates of various geometries has been made by McRae and Kasha.[15] The problem of energy transfer between pairs of randomly oriented molecules was treated quantum mechanically by Förster.[16] All of the foregoing studies make use of the same physical basis, which is described in the following. Several papers of a descriptive nature on the topic of energy transfer have been published,[17–2] mainly from the phenomenological viewpoint.

Three different definitions of the *exciton* have been given, illustrating different aspects of the idea, but not all of which seem equally suitable for a critical understanding:

(1) Exciton states involve excitation of an assembly of atoms (or molecules) in concert, instead of localized excitation of individual species of the assembly.

(2) An exciton can be described by a wave packet traveling through an assembly of atoms (or molecules), and arising from a superposition of exciton states (defined in the foregoing).

(3) An exciton is a neutral excitation "particle," consisting of an electron and a positive hole, traveling together through the lattice.

Each of these definitions is next considered in order to clarify the picture of an exciton.

For a discussion of the first definition of an exciton, consider a simple two-dimensional model of an atomic and a molecular lattice [Figs. 1(a) and 1(b)]. The atomic lattice is assumed to be an ionic one held together by Coulombic forces (for simplicity, the positive and negative ions are not distinguished, the ions are assumed to be isoenergetic in their electronic states throughout, and overlap is assumed to be negligible in the ground state of the lattice). All but one of the ions are shown to be in their ground state, indicated schematically by circles representing spherically symmetrical orbitals.† The ion (2,2),‡ however, is shown with an electron excited to a *p*-type orbital. This model of the excited lattice forms the basis of what is designated as a *zeroth-order* description, because it is not possible to localize excitation to one atom in such a system. Thus, atom (3,2) might equally well be excited, etc. Abstracting the second row of atoms for simplicity, the following zeroth-order descriptions could be given of singly excited states (where asterisk denotes excitation of atom $A_n$)

$$(\text{i})=A_1{}^*A_2A_3A_4,$$

$$(\text{ii})=A_1A_2{}^*A_3A_4,$$

$$(\text{iii})=A_1A_2A_3{}^*A_4,$$

$$(\text{iv})=A_1A_2A_3A_4{}^*.$$

---

* This study was made during the Study Program in Biophysical Science held in Boulder, Colorado, in the summer of 1958, under the sponsorship of the National Institutes of Health, Public Health Service, U. S. Department of Health, Education, and Welfare.

† The biologist unfamiliar with the characteristic three-dimension orbital wave functions of atoms and molecules will find it helpful to consult *The Chemical Aspects of Light* by E. J. Bowen [(Clarendon Press, Oxford, England, 1946), second edition], especially pages 68, 100–115, and 139.

‡ Row 2, column 2, from left upper corner.

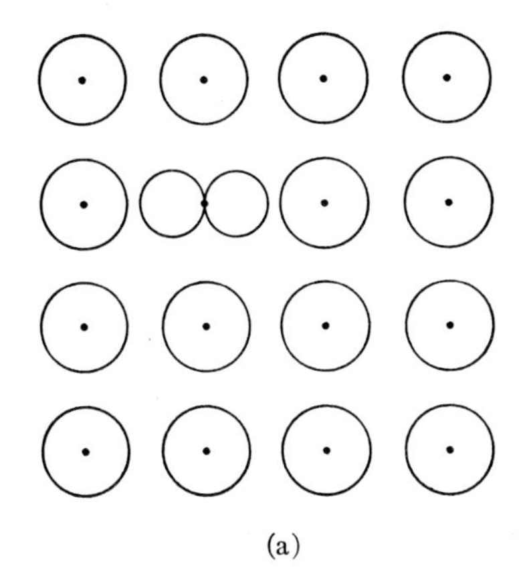


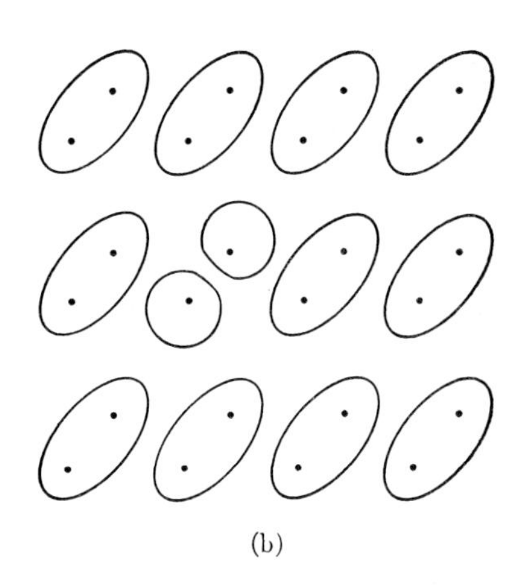


FIG. 1. (a) Zeroth-order description of a section of an atomic lattice having one atom with an electron in its lowest excited orbital, the remaining atoms with electrons in their ground orbitals. (b) Zeroth-order description of a section of a molecular lattice with one molecule with an electron excited to its next highest orbital, the remaining molecules with electrons in their ground orbitals.

None of these is by itself a suitable description of the excited array, since they could (according to the assumption) all have the same energy, and if they could *interact* with each other, new states which are true stationary states in the quantum-mechanical sense will arise. These new states are the *exciton states* of the first definition; if only the foregoing four atoms were considered, they would have the form

$$\mathrm{I}=\frac{1}{\sqrt{4}}[(\mathrm{i})+(\mathrm{ii})+(\mathrm{iii})+(\mathrm{iv})],$$

$$\mathrm{II}=\frac{1}{\sqrt{4}}[(\mathrm{i})+(\mathrm{ii})-(\mathrm{iii})-(\mathrm{iv})],$$

$$\mathrm{III}=\frac{1}{\sqrt{4}}[(\mathrm{i})-(\mathrm{ii})-(\mathrm{iii})+(\mathrm{iv})],$$

$$\mathrm{IV}=\frac{1}{\sqrt{4}}[(\mathrm{i})-(\mathrm{ii})+(\mathrm{iii})-(\mathrm{iv})],$$

where the signs (see later) result from the requirement (of orthogonality) that the stationary states (I) to (IV) be completely independent. This statement would be true only if it were appropriate to assign an equal weighting to each of the zeroth-order states, i.e., for an infinite array. The present argument is artificial in the sense that it attempts to simplify the picture by abstracting four molecular units from a very large aggregate. It should be noted that each of the stationary exciton states involves all of the zeroth-order descriptions; that is, *each of the stationary exciton states involves excitation of all of the atoms considered as locally excited.* Thus, in the exciton states the excitation is delocalized.

The second definition arises from the possibility, pointed out by Frenkel,[3,4] of construction of wave packets from the stationary exciton states. This is a useful aspect for discussions of electronic energy transfer in atomic and molecular systems. The properties of the wave packet can be calculated by quantum-mechanical methods.[7,9,10]

The third definition can be misleading, so its meaning is examined more fully. In Fig. 1(a), the atomic ion (2, 2) is schematically shown to have an electron in a *p*-type excited orbital. Both the excited atom and the ground-state atom are electrically neutral and electronically symmetrical. The excited atom is produced from a ground-state atom by the polarizing influence of the electric vector of the interacting light wave (the electric vector having been taken to be horizontal, and in plane). While the light wave impinges, a *transitory dipole moment* is induced in the atom, but no permanent charge separation, i.e., ionization, takes place. Thus, the definition of an exciton as an "electron-positive hole" combination can be meaningful largely in a figurative sense, since the exciton involves excited *bound-electron* states of the electron and its originating atom. To actually ionize the "pair" requires the additional energy between the exciton level and the ionization limit.

On the other hand, the transitory dipole induced by the light wave would be directly the cause of the interaction of the zeroth-order states, since all neighboring dipoles could interact with each other in a purely electrostatic way.§ Incidentally, the signs in the exciton states (I) to (IV) should not be taken to mean electron nodes (as in ordinary electronic wave functions), but nodes signifying the phase relation between the transitory dipole moments of the zeroth-order states (see Sec. 4).

The molecular lattice of Fig. 1(b) can be discussed analogously. It is assumed that here van der Waals forces bind the molecules into ordered layers (crystals, or lamellae), and that negligible electron overlap is present in the ground and lowest excited states. The molecular axis is shown inclined at 45° to the layer direction; for simplicity, a two-center molecular orbital is considered. The molecule shown excited (2, 2) has an electronic orbital node perpendicular to the molecular axis. The excited molecule is thus electrically neutral and symmetrical, and is produced from a ground-state molecule by a light wave of the right frequency with

§ In the formal treatment of the theory, the expansion of the intermolecular interaction potential results in a series of terms, of which the dipole-dipole terms are the predominant ones for formally allowed electronic excitation.

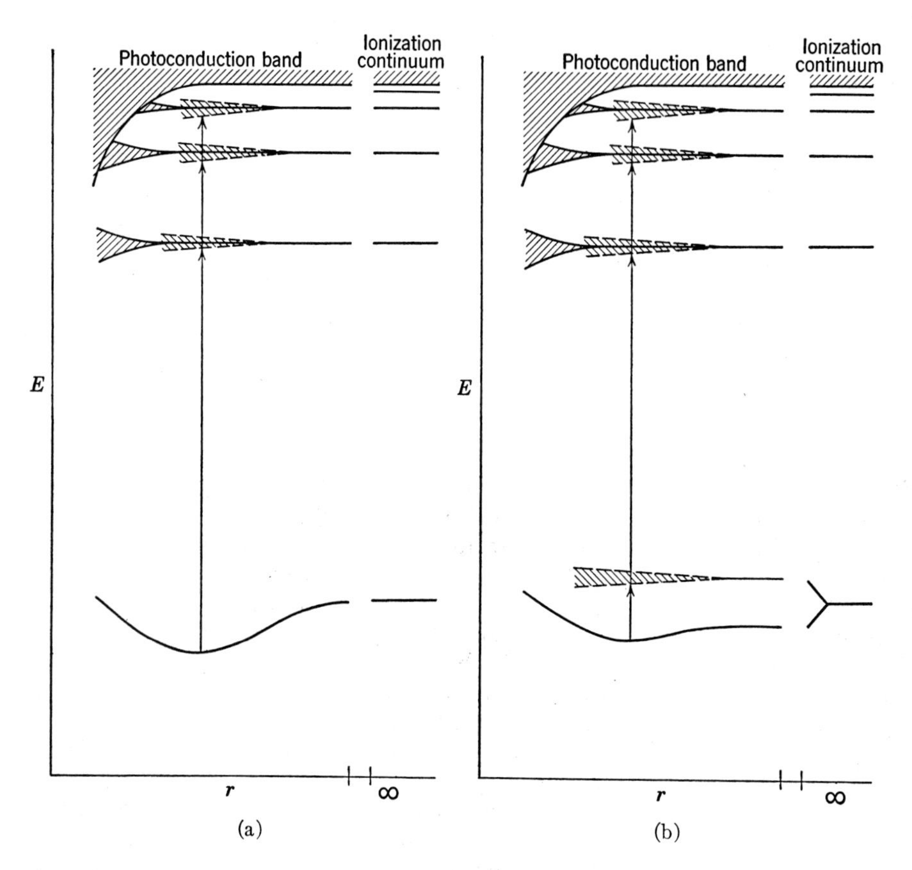


FIG. 2. Schematic energy *vs* interspecies-distance curves for a crystal or aggregate taken as a whole. (a) atomic lattice, (b) molecular lattice or lamellar structure, showing onset of exciton bands and conduction bands. Exciton-band splitting exaggerated, and overlap interaction shown independent of dipole-dipole interaction (cf. Fig. 3).

its electric vector oriented along the molecular axis. The identical symbolic description of the zeroth-order states and exciton states, used for the foregoing atomic lattice model, is applicable to this molecular model.

From the foregoing discussion, it should be clear that exciton states cannot impart electrical conductivity. It is found experimentally that excitation of atomic lattices to exciton states produces no photoconduction of electrons (see Sec. 4).

### 3. ORIGIN OF EXCITON BANDS AND CONDUCTION BANDS

The origin and relation of exciton bands and conduction bands may be clarified by a discussion of their dependence on interspecies distance. In Fig. 2 are given, schematically, energy *vs* interatomic (interionic) distance (a) and energy *vs* intermolecular distance (b) for the energy levels of a crystal or aggregate taken as a whole.

The atomic array is assumed, as before, to be made up of ions with isoenergetic electronic states, brought together into a Coulombic lattice (positive and negative ions not distinguished). Since the *Coulombic* potential varies as $1/r$, it would be effective at the greatest interatomic separations, and strong ground-state binding results [Fig. 2(a)]. Next in distance dependence would be the long-range $1/r^3$ *dipole-dipole* interaction between excited states. Depending upon the strength of the interaction, each state of the isolated ion would split into an $N$-fold exciton band (where $N$ equals the number of ions in the assembly). The exciton-band width[15] is proportional to the intensity of the transition in the unit species, and depends upon the orientation of the transition dipole moments (besides the $1/r^3$ dependence, and a rather insensitive function of the number of unit species in the array). Finally, there is the extremely short-range electron *overlap* interaction. This short-range characteristic of overlap forces arises from the exponential fall of electron orbital wave functions in the exterior of atoms (and molecules). In Fig. 2(a), it is assumed that negligible overlap is present in the ground state of the atoms, but that, in the excited state, overlap may occur yielding a conduction band, because of crbital expansion upon high quantum-number excitation. If there is overlap in the ground state, a *valence band* is said to arise, analogous to the conduction band for the excited states.

The point to be noted from the above is that, because of the long-range nature of the dipole-dipole interaction, nonphotoconducting exciton bands can exist discrete from the photoconduction bands. Nevertheless, atomic exciton bands usually lie not far below the photoconduction band of atomic lattices.

The molecular case has some qualitatively different features. However, it is well known that gaseous molecules exhibit converging series absorption in the vacuum ultraviolet (molecular Rydberg series) whose convergence limits represent the ionization potentials for various electrons in the molecule. One could antici-

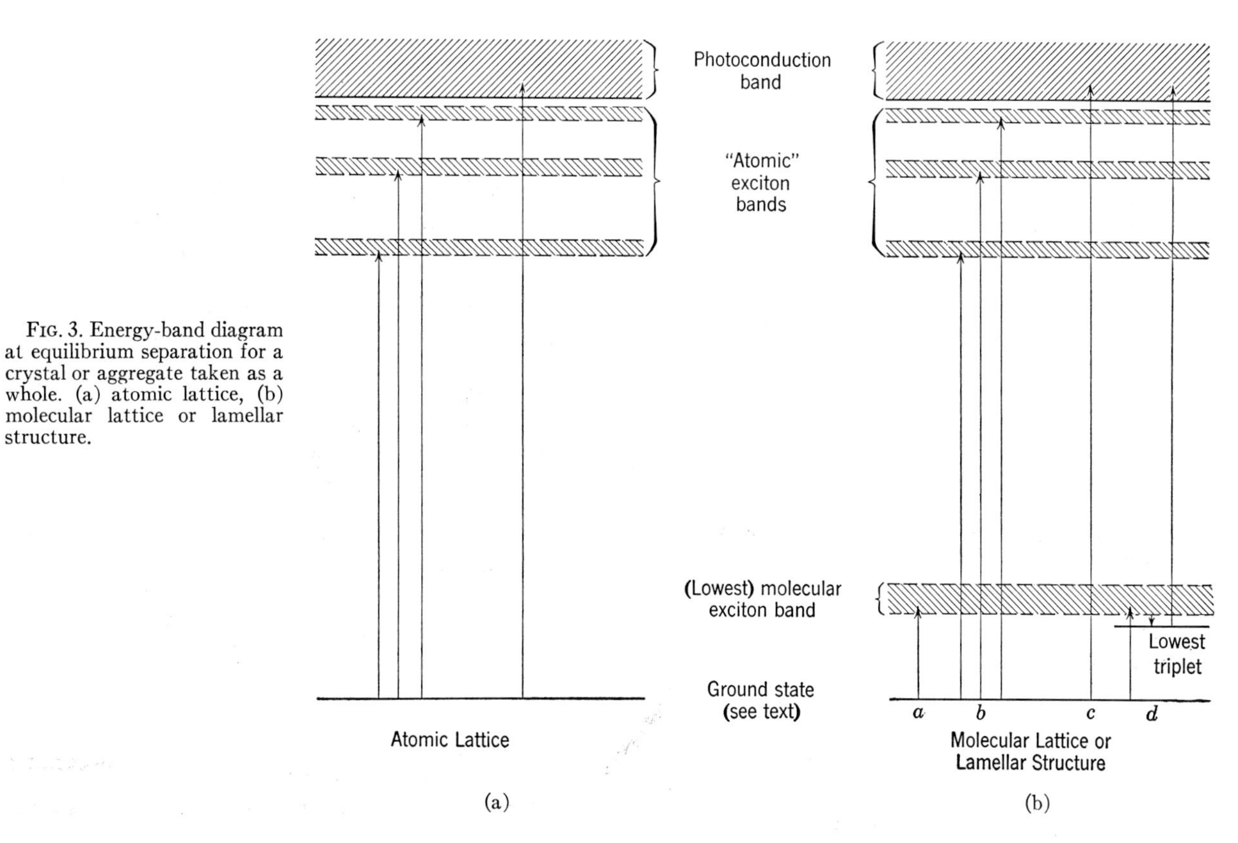


FIG. 3. Energy-band diagram at equilibrium separation for a crystal or aggregate taken as a whole. (a) atomic lattice, (b) molecular lattice or lamellar structure.

pate that dipole-dipole interactions between the electrons in Rydberg levels would give rise to "atomic" exciton bands. These are indicated in the upper part of Fig. 2(b), analogous to the exciton bands of Fig. 2(a).

In addition, a molecule during its formation from atoms [at $r=\infty$ for molecular lattice; lower right corner of Fig. 2(b)] gives rise to excited (unfilled) molecular orbitals. For simplicity, only one filled and one excited orbital are indicated. In the formation of the molecular lattice, therefore, while the ground state becomes slightly lowered in energy through a weak van der Waals interaction, the molecular excited states of the assembly spread by dipole-dipole interaction into an $N$-fold exciton band ($N$ equals the number of molecules in the assembly). Since molecular transitions are frequently very intense (oscillator strength $f\sim 1$), a large band splitting is to be expected (see later). Moreover, since highly conjugated molecules (i.e., pigment or dye molecules) have rather low-energy excited states, exciton bands far from the conduction band should be observed.

#### 4. PROPERTIES OF EXCITON BANDS

The electrically nonconducting characteristic of exciton bands has been referred to above, and the physical basis for this property has been discussed. Some additional characteristics of atomic exciton bands are now mentioned. It is typical of atomic lattices that exciton bands are usually observed just below the conduction band. The optical properties of, e.g., the alkali halide crystals conform roughly to a behavior described by Fig. 2(a), which for the equilibrium interatomic separation yields an energy-level diagram such as Fig. 3(a). The pure alkali halide crystals are properly described as electrical *insulators* rather than as semiconductors, as far as their electronic behavior is concerned. For example, the KCl crystal[8,10] exhibits its first exciton band as a broadened line at 7.6 ev ($\bar{\nu}\sim 61\,000$ cm$^{-1}$, $\lambda\sim 1620$ A). Excitation of this band produces no photoconductivity. At 9.44 ev ($\bar{\nu}\sim 76\,000$ cm$^{-1}$, $\lambda\sim 1300$ A) photoconduction is observed. Thus, both types of bands are observed in the vacuum ultraviolet not far from each other. No contribution to dark conductivity would be expected at ordinary temperatures from this high-energy conduction band.

Another characteristic of atomic exciton bands is their occurrence in sets which fall into a converging series.[10] This behavior could be anticipated from their atomic character; i.e., the orbitals which would be used in the zeroth-order description of an atomic-lattice exciton are essentially hydrogenic in character, although extending over many lattice points. The series formulas are, of course, modified by the dielectric constant of the lattice, lattice interactions, etc.[10]

The band width for molecular exciton cases has been calculated[15] to be approximately 0.2 ev ($\Delta\bar{\nu}\sim 1600$ cm$^{-1}$) in linear pigment assemblies for an oscillator strength of 1 for the isolated molecule transition, at intermolecular distances of 10 A, and in a suitable geometry. Experimental studies indicate that this is the order of

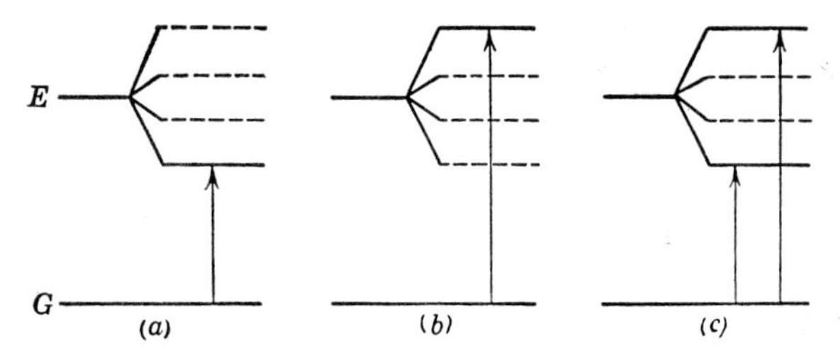


FIG. 4. Selection rules for exciton levels in linear arrays (cf. reference 15), for a hypothetical 4-unit molecular array. (*a*) head-to-tail, (*b*) card-pack, (*c*) alternate-translational-arrays of transition dipoles. Dashed line: forbidden exciton level; solid line: allowed exciton level. Diagram shows dipole-dipole splitting of zeroth-order excited states.

magnitude observed. For weaker optical transitions, or for greater intermolecular separations, smaller exciton-band widths can be expected.

The foregoing comments on exciton-band width are not meant to imply that optical transitions of isolated molecules are necessarily to be broadened in the lamellar aggregates. On the contrary, one of the characteristics[3,4,11,15] of exciton bands is the restriction of optical transitions to only certain components of an exciton band. For example, once more consider the abstracted linear array of four atoms of Fig. 1(a). The exciton states (I) to (IV) can be depicted by vector models representing the dipole moments induced by the light wave with electric vector in the direction of the array,

$$\begin{aligned} &\mathrm{I} \rightarrow \rightarrow \rightarrow \rightarrow, \\ &\mathrm{II} \rightarrow \rightarrow \leftarrow \leftarrow, \\ &\mathrm{III} \rightarrow \leftarrow \leftarrow \rightarrow, \\ &\mathrm{IV} \rightarrow \leftarrow \rightarrow \leftarrow. \end{aligned}$$

The dipole-dipole interaction would give (I) the lowest energy and (IV) the highest. Also, the resultant transition moment (vector sum of the individual moments) is finite in the case of (I), but zero for the remaining three states. Thus, the exciton band for this case would be observed to have the structure shown in Fig. 4(a), with only the lowest exciton state permitted in optical absorption. Thus, for this example, a long wavelength or red shift would be observed in the linear aggregate compared with the isolated atomic species, rather than a band broadening. Cases 4(b) and 4(c) correspond to other geometries for a linear array, the former for a card-pack orientation of dipoles, and the latter for an alternate translational orientation. The staggered arrangement of naphthalene molecules in the crystalline state also leads to this last result.[11] For an $N$-fold linear array, even though the exciton band will now have $N$ components, analogous selection rules to those illustrated in Fig. 4 still apply.[15] Obviously, the ground state is not affected by the exciton process.

Finally, in the special case of dye-molecule aggregates, one additional result of exciton-band formation is to be noted. This is the enhancement of triplet excitation accompanying molecular aggregation.[14,15] Referring to Fig. 3(b), four separate electronic processes in molecular aggregates can be delineated: (a) molecular exciton-band excitation, (b) "atomic" exciton-band excitation, (c) photoconduction excitation, and (d) triplet excitation following exciton-band excitation.[14,15] The recently reported, dramatic experiments by Albert Szent-Györgyi, demonstrating[21] great enhancement of phosphorescence in dyes frozen in water solutions, seem to be due to the foregoing phenomenon.[15]

Furthermore, since triplet states of molecules have lifetimes of a millisecond or longer, double quantum processes such as the triplet to conduction-band process shown in Fig. 3(b) may be expected, following exciton-band, then triplet excitation. The exciton-band-to-triplet process is radiationless. In most dyes, the radiationless excitation of triplets is very weak in isolated molecules.

The Förster theory,[16] with its phenomenological aspect, permits useful calculations to be made on distance of dipole-dipole transfer of excitation energy. Karreman and Steele[22] have shown that excitation energy transfer in proteins by dipole-dipole interaction through the aromatic residues is plausible according to the Förster formula.

In this section, the complications[23] of intramolecular vibrational-electronic coupling in exciton interactions have been omitted for simplicity.

## 5. MOLECULAR SYSTEMS AS PHOTOCONDUCTORS AND SEMICONDUCTORS

If the energy gap between the ground state (or valence band) and the conduction band of an atomic lattice is small, intrinsic electrical conductivity is observed. For example, in pure germanium and pure silicon, the energy gaps are 0.7 and 1.10 ev ($\bar{\nu}\sim5600$ and 9000 $\mathrm{cm}^{-1}$), respectively.[10] In these solids, enough electrons can be thermally excited at room temperature so that dark electrical conductivity can be observed. The term semiconductor is applied,[10] arbitrarily, to solid materials which have a resistivity in the range $10^{-2}$ to $10^{9}$ ohm cm. For pure diamond, with an energy gap of 6 ev ($\bar{\nu}\sim49\,000$), light of $\lambda\sim2100$ A would be required to raise an electron to a conduction band. Materials which exhibit only photoconductivity are properly classified as insulators. Since molecular ionization potentials (gas phase) are in the range[24] 8 to 10 ev, one would expect pure molecular crystals and aggregates to act as insulators, having photoconductivity perhaps for 6 to 8 ev quanta, not intrinsic electrical semiconductivity. Observations of a weak photoconductivity have been made in organic crystals absorbing near-ultraviolet light, but evidence has accumulated that, in pure single crystals, even this weak photoconductivity becomes vanishingly small as surface effects, impurities, etc., are eliminated.[25]

On the other hand, there are several reports of semiconductivity in certain organic substances in the class of molecular crystals, especially those of Eley *et al.*[26–28]

The general observation seems to be that, if a molecule has an optical transition at around 1 ev ($\bar{\nu}\sim 8067$ cm$^{-1}$, $\lambda\sim 12\,500$ A), semiconductive behavior may be observed. This seems to be a necessary condition for semiconductivity, but not a sufficient one (see the following). For example, metal-free phthalocyanine exhibits semiconductivity (temperatures up to 400° C are used), with an indicated energy gap of $\sim 1.2$ ev ($\bar{\nu}\sim 9900$ cm$^{-1}$, $\lambda\sim 10\,200$ A). However, the lowest allowed optical transition in phthalocyanine is near the infrared, but not in the infrared. The side question that arises is whether the energy gap determined by the temperature dependence of the semiconductivity is in error, or whether the lowest triplet state (which seems to be in the region of $\lambda\sim 10\,000$ A[29]) is involved. Contrariwise, those substances which absorb only in or near the ultraviolet, such as diphenylbutadiene and diphenyloctatetraene, exhibit no dark semiconductivity,[26–28] indicating an energy gap of greater than 2.2 ev ($\sim 50$ kcal/mole, $\bar{\nu}\sim 17\,000$ cm$^{-1}$, $\lambda\sim 5800$ A).

For intrinsic semiconductivity, not only must there exist a small gap between the filled levels and the unfilled levels, but there must exist sufficient intermolecular orbital overlap for a true conduction band to develop. Apparently, in the case of near ultraviolet-absorbing hydrocarbons, only a nonconducting exciton band may exist in the crystal. The stronger the van der Waals forces become, the greater is the chance that overlap interaction resulting in a conduction band will arise. Especially in the case of such dye-like molecules as pure phthalocyanines, chlorophylls, and cyanine dyes, one might anticipate significant development of a conduction band from overlap of the lowest excited-state orbitals. However, since resistivity measurements cover a wider decadic range[10] than almost any other physical property, and since conductivity measurements using powder samples are so subject to experimental artifacts, one cannot accept the observed results without caution as indicating that molecular crystals are good semiconductors at room temperature.

The semiconductivity reported[26–28] for proteins occupies a strange position. It is well known that proteins absorb in the ultraviolet, which would indicate that only photoconductivity could result, again if there were sufficient orbital overlap. Evans and Gergely,[30] in a widely misinterpreted paper, proposed that $\pi$-electron overlap *across* hydrogen bonds could give rise to sufficient overlap interaction to yield a "conduction" band. Using three different models, they obtained choices for the energy gap between the valence band (ground state) and conduction band of 3.5, 4.8, and 3.2 ev ($\lambda\sim 3500$, 2600, and 3800 A), respectively. [Eley *et al.*[26–28] quote an energy gap as given by Evans and Gergely to be $\sim 2$ ev ($\lambda\sim 6200$ A), but this is their value for the gap between two *filled* valence bands.] Thus, according to this discussion, and according to the pattern of results of Eley *et al.*,[26–28] *Evans and Gergely proved* (within the severe limits of their calculation) *that a pure protein should be an electrical insulator, exhibiting only photoconductivity and not semiconductivity.*

The semiconductivity observed[26–28] in proteins cannot be *intrinsic*, but if it is truly electronic, it may be *extrinsic*.[10] Extrinsic semiconductivity arises generally from the presence of lattice imperfections or impurity centers, which introduce new energy levels into the term scheme for a lattice structure. In biological systems, such perturbations of the perfect lattice probably play a very important role, and all of these discussions of the pure or perfect molecular lamellar structures are modified by their consideration. New energy levels appear in such systems, often greatly lowering the energy required for conductivity phenomena.[10]

The experiments of Arnold and Sherwood[31] demonstrate that, in the complex, heterogeneous structure of chloroplast, electron traps exist and are involved in luminescence and conductivity properties of the chloroplast. This type of investigation probably has valuable extension to other complex biological systems, and much valuable correlation with solid-state physics might be made in this connection. On the other hand, in the extremely small conductivity of weak semiconductors and insulators, it is difficult to unravel[8] ionic from electronic semiconductivity. Probably in proteins and chloroplasts, there is an appreciable ionic component of electrical conductivity.

### 6. PHOTOPHYSICAL STEPS IN THE PRIMARY PROCESS OF PHOTOSYNTHESIS

Since the work of Emerson and Arnold[32,33] and of Gaffron and Wohl,[34] it has been evident that absorption of energy by chlorophyll in chloroplast involves exciton behavior (called "resonance energy migration" in most of the literature). The review by Rabinowitch[20] summarized the recent status of this topic. However, since newer investigations have revealed that in pigment lamellar aggregates triplet excitation is facilitated,[14,15] the detailed primary photophysical steps require re-examination.

The lowest triplet states of chlorophyll-*a* and chlorophyll-*b* have been established spectroscopically by Becker and Kasha[29] to be at approx 1.42 ev ($\bar{\nu}\sim 11\,500$ cm$^{-1}$, $\lambda\sim 8700$ A). The quantum yield of triplet excitation (via excited singlet state) is so small in monomeric chlorophyll as to suggest that the lowest triplet state could not participate in the efficient process of photosynthesis. However, the finding[14,15] that triplet excitation is greatly enhanced by aggregation of pigment molecules suggests strongly that this mechanism may play an important role in the chloroplast. The process would involve excitation to the chlorophyll exciton band, followed by triplet excitation [Fig. 3(b), first steps of process *d*]. As in other pigments,[14,15] the quantum yield of fluorescence of aggregated chlorophyll (in the leaf) is diminished many-fold over that of isolated molecules in solution,[35] as would be required if enhancement of triplet excitation occurred.

The detection of photoproduced radicals in chloroplast by electron paramagnetic-resonance studies made by Commoner *et al.*[36–38] and investigated further by Calvin *et al.* (p. 157 and references 39–41) poses the question of the relation of the primary photophysical process of exciton-band excitation to the secondary photochemical process of radical production. The picture presented by Calvin and co-workers[41] proposes ionization at the exciton stage. It is possible that in the heterogeneous chloroplast structure some low-energy ionization center could exist, which traps the exciton energy, leading to the direct production of the radicals detected. But the energy of this exciton band is small ($\sim$1.8 ev, or $\sim$41 kcal/mole), and the existence of such low-energy ionizations is unknown in molecules. Moreover, the lowest exciton state is very short-lived, somewhat shorter in life than the lowest excited singlet state of isolated chlorophyll, so that it may not be a very efficient state for a photochemical act.

The lowest triplet state of chlorophyll has been observed to have a mean lifetime of approximately 1 msec in solution,[42,43] and would be expected to have a somewhat longer lifetime in the compact chloroplast structure. It is natural to favor the triplet state for participation in the photochemical act, especially in view of the previously discussed enhancement possibility. Moreover, it is quite possible that these favorable properties of the triplet state would permit a two-quantum process to be important in the photosynthetic system. For example, as shown in Fig. 2(b), even triplet-to-conduction-band excitation might be considered. The two-quantum hypothesis of Franck[44,45] has now mainly the objection[46] that the upper state he proposed for the second quantum was a triplet state, which must have a mean lifetime of less than $10^{-12}$ sec.

If the lowest triplet state of chlorophyll is involved in the primary process in photosynthesis, its steady-state concentration could be too low to detect by electron paramagnetic-resonance absorption. Moreover, special orientation problems are involved which have been only recently solved.[47] It seems fairly certain that a triplet state of chlorophyll in chloroplast could not be detected by electron paramagnetic resonance.

The phenomena of solid-state physics have much to contribute to the understanding of the physical behavior of biological systems, but the richest development of experimental approaches and theoretical understanding requires that detailed attention be paid to the typically molecular behavior of the materials involved.

## ADDENDUM

Since completing this manuscript, the writer has had the opportunity of studying the full text of Garrett's chapter[25] in *Semi-Conductor Chemistry*. His comprehensive review covers additional researches not included for consideration in the foregoing and emphasizes the difficulty of making significant semiconductivity measurements.

The quantitative treatment of excitons in molecular aggregates is in preparation by E. G. McRae and M. Kasha, and will be published shortly elsewhere. Reference to a preliminary communication on this work[15] has been made in the text.

## ACKNOWLEDGMENTS

The author gratefully acknowledges a very useful discussion with Professor R. E. Peierls, who was Visiting Professor of Physics at the University of Colorado concurrently with the Boulder conference.

The author is indebted to Dr. C. G. B. Garrett and to Professor C. A. Hutchison, Jr., for preliminary copies of their manuscripts, referred to in the text.

## BIBLIOGRAPHY

1 Albert Szent-Györgyi, Science **93**, 609 (1941).
2 Albert Szent-Györgyi, Nature **148**, 157 (1941).
3 J. Frenkel, Phys. Rev. **37**, 17, 1276 (1931).
4 J. Frenkel, Physik. Z. Sowjetunion **9**, 158 (1936) (in English).
5 R. Peierls, Ann. Physik **13**, 905 (1932).
6 G. H. Wannier, Phys. Rev. **52**, 191 (1937).
7 W. R. Heller and A. Marcus, Phys. Rev. **84**, 809 (1951).
8 Cf. F. Seitz, *The Modern Theory of Solids* (McGraw-Hill Book Company, Inc., New York, 1940).
9 Cf. R. E. Peierls, *Quantum Theory of Solids* (Clarendon Press, Oxford, England, 1955).
10 Cf. C. Kittel, *Introduction to Solid State Physics* (John Wiley and Sons, Inc., New York, 1956), second edition, p. 504.
11 A. S. Davydov, Zhur. Eksptl. i Teoret. Fiz. **18**, 210 (1948). (English translation available from the present author).
12 Cf. D. Fox and O. Schnepp, J. Chem. Phys. **23**, 767 (1955).
13 Th. Förster, Naturwissenschaften **33**, 166 (1946).
14 G. S. Levinson, W. T. Simpson, and W. Curtis, J. Am. Chem. Soc. **79**, 4314 (1957).
15 E. G. McRae and M. Kasha, J. Chem. Phys. **28**, 721 (1958).
16 Th. Förster, Ann. Physik **2**, 55 (1948).
17 J. Franck and E. Teller, J. Chem. Phys. **6**, 861 (1938).
18 J. Franck and R. Livingston, Revs. Modern Phys. **21**, 505 (1949).
19 R. Livingston, J. Phys. Chem. **61**, 860 (1957).
20 E. Rabinowitch, J. Phys. Chem. **61**, 870 (1957).
21 Albert Szent-Györgyi, Science **124**, 873 (1956).
22 G. Karreman and R. H. Steele, Biochim. et Biophys. Acta **25**, 280 (1957).
23 W. T. Simpson and D. L. Peterson, J. Chem. Phys. **26**, 588 (1957).
24 W. C. Price, Chem. Revs. **41**, 257 (1947).
25 C. G. B. Garrett in *Semi-Conductor Chemistry* (American Chemical Society Monograph), Chap. 15 (to be published).
26 D. D. Eley, Nature **162**, 819 (1948).
27 D. D. Eley, G. D. Parfitt, M. J. Perry, and D. H. Taysum, Trans. Faraday Soc. **49**, 79 (1953).
28 D. D. Eley and G. D. Parfitt, Trans. Faraday Soc. **51**, 1529 (1955).
29 R. S. Becker and M. Kasha, J. Am. Chem. Soc. **77**, 3669 (1955).
30 M. G. Evans and J. Gergely, Biochim. et Biophys. Acta **3**, 188 (1949).
31 W. Arnold and H. K. Sherwood, Proc. Natl. Acad. Sci. U. S. **43**, 105 (1957).

[32] R. Emerson and W. Arnold, J. Gen. Physiol. **15**, 391 (1932).

[33] R. Emerson and W. Arnold, J. Gen. Physiol. **16**, 191 (1932).

[34] H. Gaffron and K. Wohl, Naturwissenschaften **24**, 81, 103 (1936).

[35] P. Latimer, T. T. Bannister, and E. Rabinowitch, Science **124**, 585 (1956).

[36] B. Commoner, J. Townsend, and G. E. Pake, Nature **174**, 689 (1954).

[37] B. Commoner, J. J. Heise, and J. Townsend, Proc. Natl. Acad. Sci. U. S. **42**, 710 (1956).

[38] B. Commoner, J. J. Heise, B. B. Lippincott, R. E. Norberg, J. V. Passonneau, and J. Townsend, Science **126**, 57 (1957).

[39] M. Calvin and P. B. Sogo, Science **125**, 499 (1957).

[40] P. B. Sogo, N. G. Pon, and M. Calvin, Proc. Natl. Acad. Sci. U. S. **43**, 387 (1957).

[41] G. Tollin, P. B. Sogo, and M. Calvin, J. chim. phys. (to be published).

[42] R. Livingston and V. A. Ryan, J. Am. Chem. Soc. **75**, 2176 (1953).

[43] R. Livingston, J. Am. Chem. Soc. **77**, 2179 (1955).

[44] J. Franck, Daedalus **86**, 17 (1955).

[45] J. Franck in *Research in Photosynthesis*, H. Gaffron, editor (Interscience Publishers, Inc., New York, 1957), pp. 19–30, 142–146.

[46] G. Tollin, Arch. Biochem. Biophys. **75**, 539 (1958).

[47] C. A. Hutchison, Jr., and B. W. Mangum, J. Chem. Phys. **29**, 952 (L) (1958).

# 20
# General Patterns of Biochemical Synthesis

RICHARD BROOKE ROBERTS

*Department of Terrestrial Magnetism, Carnegie Institution of Washington, Washington 15, D. C.*

THE general pattern of biochemical synthesis is outlined in Fig. 1. The first block indicates the processes of photosynthesis which Calvin describes (p. 147). The second block represents the processes of intermediary metabolism wherein the raw materials of the medium are converted into the building blocks and energy needed for synthesis of the macromolecules. The third block represents the processes of ordering and polymerizing the intermediates. And lastly, the fourth block indicates the final folding and arranging of the macromolecules into products which may be excreted, used to maintain existing cells, or used to form new cells.

The division into these blocks is quite obvious from consideration of the growth requirements of various cells. A large variety will grow with $CO_2$ and light as the sole carbon and energy sources whereas others lack this capacity. Furthermore, even those cells which have this ability can hold it in abeyance and use chemical energy when the occasion demands. Accordingly, it seems quite reasonable to segregate the photosynthetic processes from the rest of the cell. Likewise, there are a number of cells which can synthesize all of their requirements from a single carbon source, such as glucose, whereas others require an exogenous supply of one or more of the more complicated molecules. In addition, most cells utilize preformed intermediates whenever they are available. Again, it seems reasonable to indicate a distinction between those processes and other operations of the cell. The division of the third and fourth stages is somewhat more arbitrary, but Kendrew's model of a protein molecule (p. 94) is quite convincing evidence that these are at least two distinct steps in the synthesis of protein.

Actual cells, in fact, do show some of the localizations of function indicated in the model cell. Calvin showed pictures of the chloroplasts which correspond to region 1, and Lehninger showed a similar localization of the Krebs-cycle reactions in the mitochondria (p. 136). Synthesis of protein and nucleic acid may occur in other special regions of the cell, such as the microsomes and the nucleus.

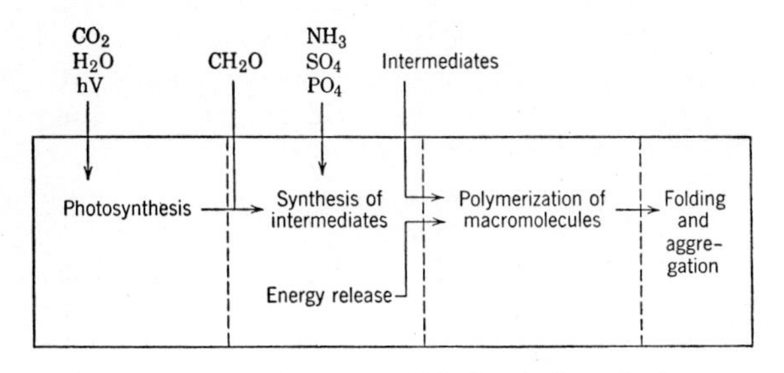


FIG. 1. General pattern of biochemical synthesis.

However, the main purpose of this diagram is not to indicate the workings of a cell, but to indicate the outline of this paper. After Calvin and Lehninger, nothing remains to be said about photosynthesis or the energy-yielding reactions. Macromolecular synthesis is covered in later papers. Thus, emphasis here is on the flow of material in the synthesis of the precursors of the macromolecules.

Before getting into the details of the synthetic pathways, there are two assigned topics which should be disposed of. The first is "least common denominators of life." One of the pleasures of biophysics is that such fascinating subjects come within the legitimate field of interest. How can life be defined? If one attempts to create life, how can success be recognized? Is the slimy precipitate in the bottom of an autoclaved mixture of organic chemicals alive or not? In part, this is a question of semantics, but the time spent in framing a definition is not wasted as it is spent in trying to select the essential from a multitude of nonessential properties.

In the model cell of the diagram, clearly photosynthesis is not essential; neither is the synthesis of intermediates, as many obviously living cells lack these capacities. Cells lacking an energy system and requiring ATP surely would be considered living. In a similar way, if some of the cells grown in tissue culture are shown to require one or more proteins, they are none the less alive! What then is essential? One obvious feature of living organisms is their capacity for growth and reproduction. Yet, these qualities alone do not define life. In framing a definition, it is necessary to choose words and concepts that include such creatures as mules which cannot reproduce and that exclude salt crystals which can grow. Furthermore, objects such as seeds and perhaps even viruses should be included. Both have the capability of catalyzing the synthesis of more of their kind even though they may not exhibit growth or other metabolic activity for long periods of time.

There presently seems to be no simple and universally acceptable answer as to how newly created life might be recognized. On the other hand, there are two particularly important properties. One is autocatalysis with its implications for growth and reproduction; another is a capacity for evolution. It would be difficult to recognize anything as alive which lacked either of these capabilities. Yet, any definition even framed in these terms is arbitrary, as the choice to include or exclude virus is largely a matter of taste.

Another assigned topic is "comparative biochemistry at the molecular level." At this level, the processes are amazingly similar in different organisms. The reactions of glycolysis originally were worked out in yeast and in mammalian muscle; the synthesis of certain amino acids is identical in a wide variety of creatures from bacteria to man. The ribonucleoprotein particles of the microsome fraction (ribosomes for short) appear identical whether isolated from bacteria, yeast, plants, or animals. Thus, to a first approximation, all cells are the same. One can consider *E. coli* as the problem to be solved and treat all other cells as perturbations of *coli*. Thus, *Chlorella* is a variant of *coli* which has gained some additional capacities, while man is a deficient mutant which has lost a number of important synthetic pathways.

The differences between cells are often quantitative and much less frequently qualitative. The most frequently observed difference is that a pathway is completely lacking or quantitatively unimportant. It is much less common to find that two cells have found quite different solutions to a synthetic problem. Perhaps the space biologists will find some radically different mechanisms on another planet.

Consider, now, some of the details of synthetic pathways. Suppose there exists in the cell a series of reactions catalyzed by the enzymes $E_1$, $E_2$, $E_3$; how can these be demonstrated?

$$A \xrightarrow{E_1} B \xrightarrow{E_2} C \xrightarrow{E_3} D \cdot$$

The most common approach is to disrupt the cells and then to isolate the enzymes and intermediates from the extract. The individual reactions can be studied then in cell-free systems. This is the classical method of biochemistry mentioned in Kornberg's introduction (p. 200). In spite of the facts that the intermediates are often present in extremely low concentrations and that the enzymes are frequently unstable, most of the important pathways of intermediary metabolism have been worked out in detail by this method. As Lehninger mentions, there are difficulties when the spatial arrangement of the enzymes in the cell is important. This method is amply illustrated in other papers (Kornberg, p. 200; Meister, p. 210; Lehninger, p. 136).

Another versatile tool is the deficient-mutant technique. Mutants which require a certain compound ($D$) for growth are isolated. Among these mutants, some are able to grow with one or another precursor ($B$ or $C$) depending upon which enzyme has been lost in the mutation. The growth requirement of these different classes of mutants often suggests the reaction sequence. Such evidence also can be strengthened by demonstrating the accumulation of intermediates or the absence of a particular enzyme.

Calvin describes the kinetic approach to the identification of intermediates and his experiments demonstrate its power (p. 147). It presents certain difficulties when applied to amino-acid synthesis, because the reactions are rapid and the quantities of intermediates small. The kinetic delays, consequently, are extremely short.

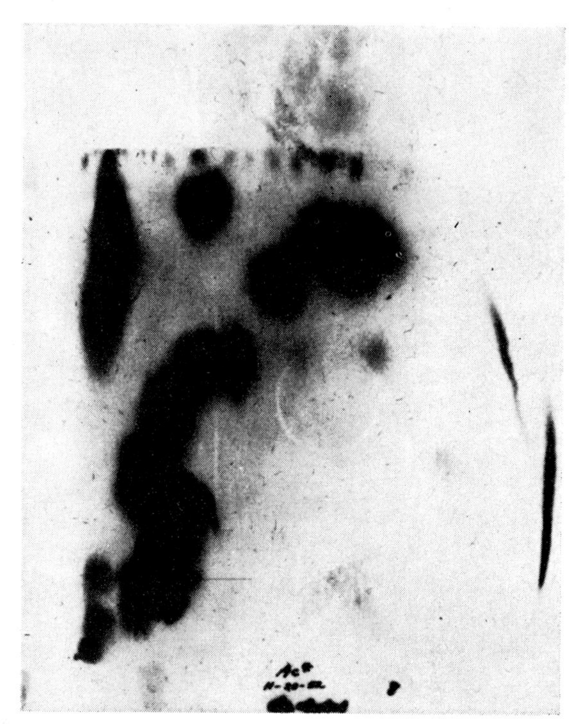

FIG. 2. Radioautograph of hydrolyzed protein of cells given with $C^{14}$ acetate as sole carbon source.

Another technique, one used extensively in our laboratory, is isotopic competition. Two different molecules, one having an isotopic label, are allowed to compete for a place in the final product. Thus, if the starting material ($A$) has a $C^{14}$ label, $C^{14}$ will appear in the final product ($D$). If, however, nonradioactive precursors ($B$, $C$) also are supplied, the radioactivity of the product ($D$) often is reduced strongly. Alternatively, radioactive forms of suspected precursors ($B$, $C$) may be added to compete with a nonradioactive carbon source ($A$).

The figures below show some examples of isotopic competition. Cells are grown with $C^{14}$ acetate as the sole carbon source; they then are harvested; the protein is isolated and hydrolyzed, and the amino acids are displayed on a paper chromatogram. A radioautogram of the paper (Fig. 2) shows all of the amino acids. No competition is involved in this case and no information is gained concerning the synthetic pathways. Measurement of the different spots gives only the amino-acid content of the protein.

If $C^{12}$ glucose also is present, the acetate enters only a limited group of amino acids (Fig. 3). Thus, a class of amino acids, which are derived more directly from acetate, is immediately segregated. If, in addition, $C^{12}$ aspartic acid is added (Fig. 4), little radioactivity

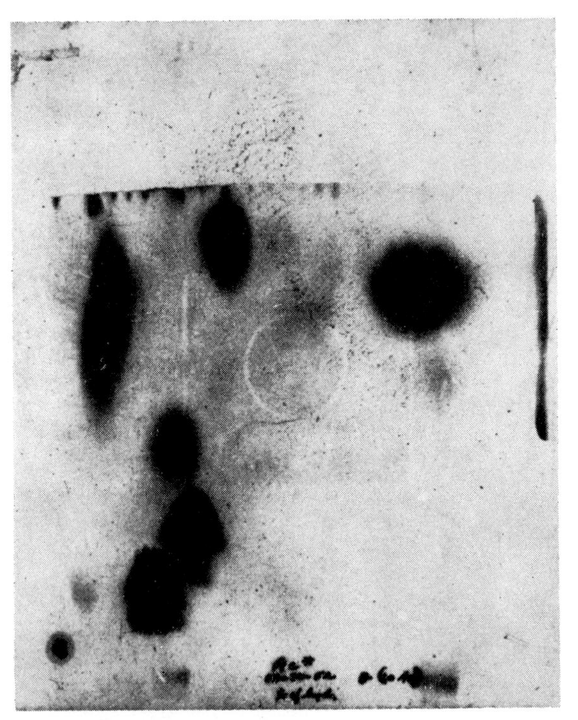

FIG. 3. Cells grown with $C^{12}$ glucose and $C^{14}$ acetate show radioactivity in the amino acids which incorporate acetate.

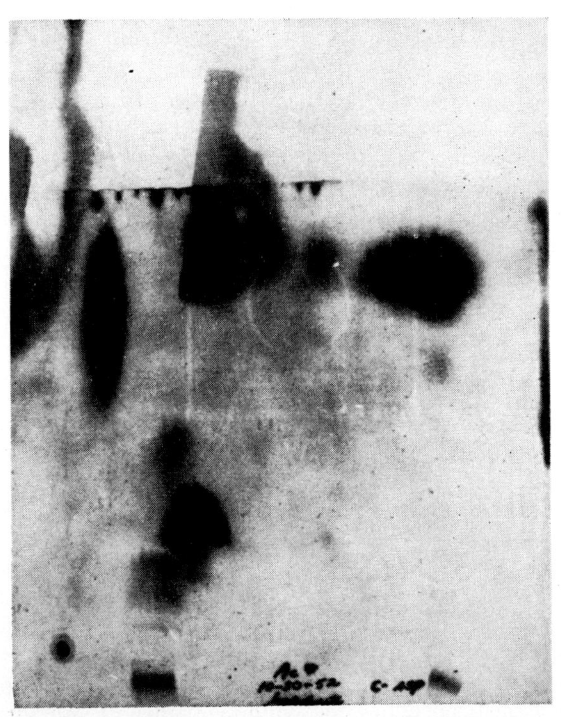

FIG. 4. $C^{12}$ aspartic acid (in addition to $C^{12}$ glucose and $C^{14}$ acetate) reduces the radioactivity of aspartic acid and the amino acids derived from it.

appears in aspartic acid, threonine, methionine, isoleucine, and lysine, as these amino acids are derived from the exogenous aspartic acid. Figure 5 shows the effect of adding glutamic acid. The incorporation is

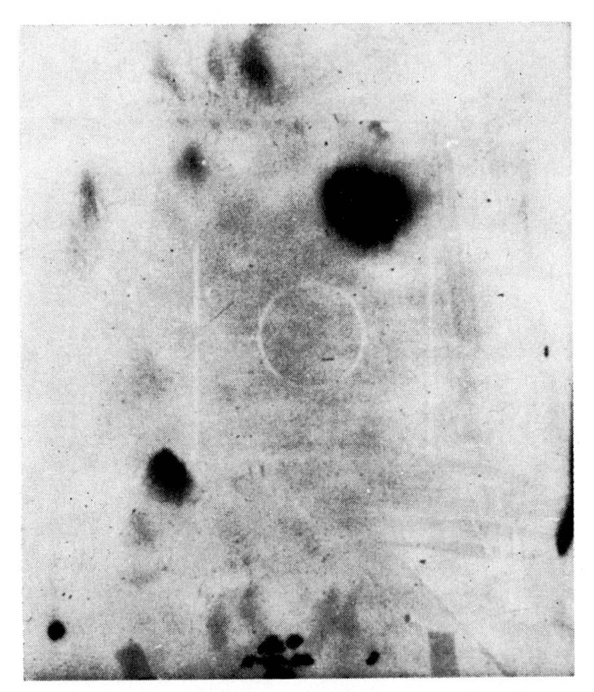

FIG. 5. $C^{12}$ glutamic acid prevents the incorporation of acetate into the amino acids derived from the Krebs cycle leaving only leucine strongly radioactive.

suppressed in all of the amino acids except leucine. $C^{12}$ leucine blocks the incorporation into leucine (Fig. 6) but has no effect on the other amino acids. Further details of pathways can be established by adding suspected intermediates as competitors.

This technique also gives information concerning the flow along the pathway. When radioproline is added, its sole end product is protein-bound proline. The rate of incorporation is equal to the proline content times the growth rate, as there is little if any incorporation by exchange in the growing *coli*. Glutamic acid, on the other hand, supplies carbon for proline and arginine.

A few additional experiments permit an analysis of the flow pattern in the Krebs cycle (Fig. 7). The flow of four carbon units into the cycle is equal to the flow out of the cycle into glutamic acid and its products, aspartic acid and its products, plus leakage products found in the medium. The circulation in the cycle can be calculated from the relative specific radioactivities of aspartic acid and glutamic acid when $CO_2$ or acetate is used as tracer. The resulting flow pattern shows that, when these cells grow on glucose, the Krebs cycle is not the main source of energy in the cell but is principally concerned with synthesis.

In contrast, when acetate is the sole carbon source, the cells must derive all of their energy from the reactions of the Krebs cycle. As a consequence, the circulation increases so that the input of acetate is roughly ten times the input of four carbon units. The reactions of the cycle must be considered quite flexible responding to changes in the environment.

This type of analysis can be extended to include the flow of carbon into nucleic acids, lipids, and the other families of amino acids. More than 80% of the carbon incorporated can be accounted for. Minor flow patterns can be discerned, and the changes in the flow patterns caused by adaptation to other carbon sources can be observed.

Several features are notable. First, the flows are well balanced. The amino acids, although synthesized in several different systems, are each supplied at the rate needed for protein synthesis. Usually, there is very little leakage of amino acids to the medium. On occasion, alanine and valine are synthesized in excess and pour out into the culture fluid.

One clue to this balance is the control of synthesis by the product. When proline is added to the medium, even at low concentrations, the synthesis of proline promptly stops. This is not simply a reversal of the usual reactions, as the exogenous proline only is converted back to glutamic acid when present at very high concentrations. Other amino acids which strongly inhibit their own synthesis are arginine, threonine, serine, and methionine. A less complete inhibition is observed with lysine, cystine, leucine, and isoleucine.

Glutamic acid, aspartic acid, glycine, alanine, and valine, however, continue to be synthesized from glucose even though they are present in the medium. In this case, the observed competition effects must come about by a simple isotopic dilution of the internal pools of amino acids. The balance of synthesis among these and the other amino acids must be the result of other controls.

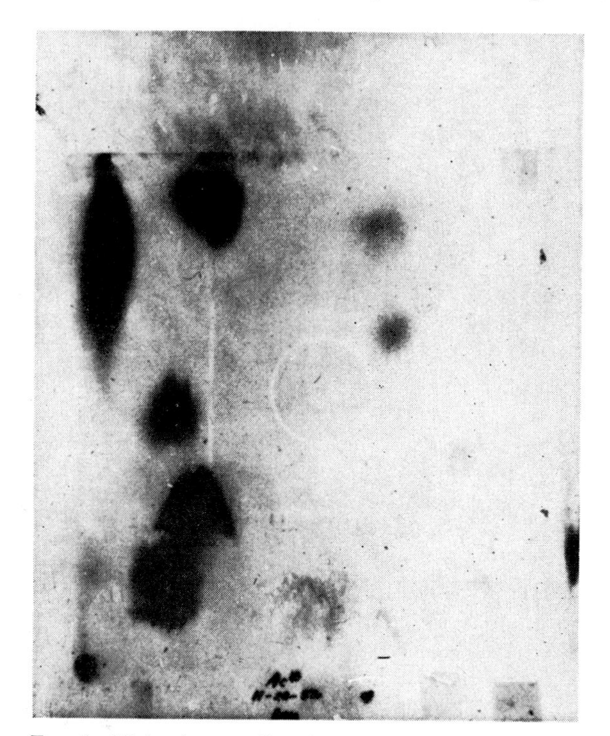

FIG. 6. $C^{12}$ leucine supplies the carbon for leucine without influence on the radioactivity of the Krebs-cycle group of amino acids.

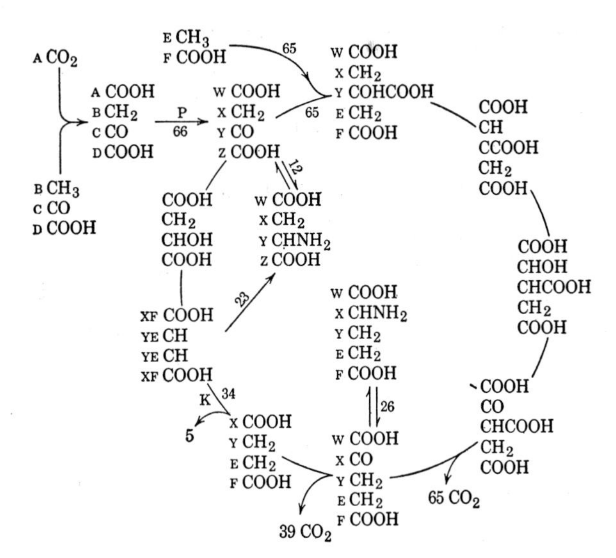


FIG. 7. Flow pattern of the Krebs cycle for cells growing on glucose as carbon and energy source. The flow numbers are expressed as micromoles per gram dry cells per 100 seconds.

Perhaps the enzymes are present in exactly the distribution required to achieve the balance, but there may also be other less obvious control mechanisms. For example, there is a coupling between phosphorus and nitrogen metabolism which is quite obscure; external phosphate does not exchange with internal phosphate unless a nitrogen source is present. This is an unexpected result, as glycolysis continues. Growth stops when the exogenous adenine is exhausted, but the pools of adenine ribotides are not depleted.

It is clear, however, that the cell has an immediate chemical response to changes in its environment, in addition to the biological response of enzyme induction. Also, the entire system is closely coupled so that seemingly unrelated parts interact strongly.

Competition studies provide some other clues to the degree of organization of the cell. Exogenous threonine is cleaved promptly to supply the carbon of glycine. Internally synthesized threonine contributes no carbon to glycine. The enzyme which splits threonine is clearly present, yet it does not have access to the endogenous threonine. As another example, the intermediates of glycolysis do not compete with glucose. Fructose-6-phosphate, in particular, contributes no carbon to cells growing on glucose even though it supplies all of the phosphorus and completely suppresses the incorporation of orthophosphate. It will, however, compete successfully with acetate as a carbon source. These anomalies suggest highly organized enzyme systems wherein substrate molecules proceed from one active site to another, perhaps by surface diffusion.

Isotopic competition in common with the mutant technique has one basic ambiguity. It fails to distinguish whether a compound is a true intermediate or

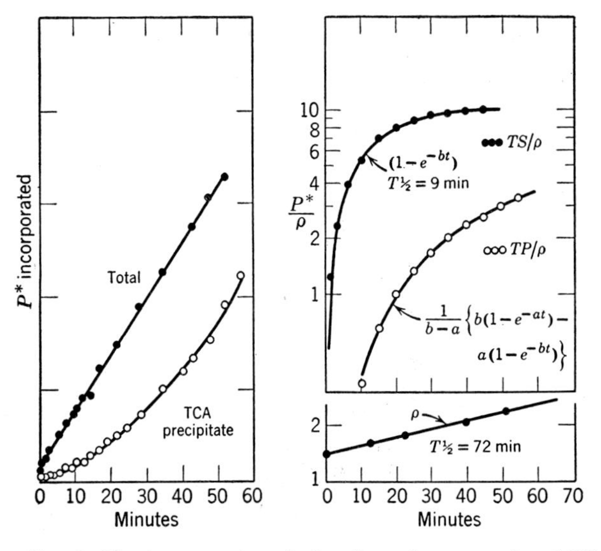


FIG. 8. The incorporation of phosphate shows transfer of $P^{32}$ from internal pool of small TCA-soluble molecules to TCA-insoluble material. A precursor-product relationship is indicated.

merely on a side line. For example, it is not possible to tell by competition studies with $C^{14}$ whether the keto analog of leucine is on the main line of the sequence from valine to leucine or whether it is merely a compound which is readily animated to give leucine. In this case, additional experiments using $N^{15}$ tracers are needed.*

As a next stage in intermediary metabolism, one might expect to find the synthesis of small peptides and small polynucleotides. Unfortunately, the same methods cannot be used. Peptides are seldom found in the cells; mutants requiring peptides have not been isolated; isotopic competition fails because the peptides used as competitors are rapidly split and incorporated as individual amino acids. To investigate the reactions of amino acids and nucleotides prior to their incorporation into macromolecules, it is necessary to use kinetic methods.

In our laboratory, we have used a technique of adding a tracer compound to a growing culture and then taking a rapid succession of samples with a syringe. The cells are separated from the culture medium by filtration which requires less than two seconds. The filter can then be counted directly or extracted for chromatography.

Alternatively, the samples can be squirted directly into 5% trichloracetic acid (TCA) which extracts the small molecules. Subsequent filtration provides a sample which measures incorporation into the large molecules.

Figure 8 shows the incorporation of phosphate and its subsequent transfer from small molecules to large ones. To the accuracy of this experiment, the TCA-soluble and the TCA-precipitable material follow the curves expected for a precursor and its product. Greater accuracy and more-detailed information can be obtained by using a "pulse" of radioactivity. Figure 9 shows such an experiment; the pulse was terminated at 63 sec by adding an excess of $P^{31}O_4$. Also, the time scale was stretched out by lowering the temperature to 15° C to increase the generation time to 3.5 hours. There is a noticeable delay in the incorporation into nucleic acid after which the maximal rate is maintained for nearly 8 min. Figure 10 shows radioactivity of two nucleotides of uracil separated out by chromatography. The radioactivity of the individual phosphorus atoms of adenosine triphosphate is shown in Fig. 11. The

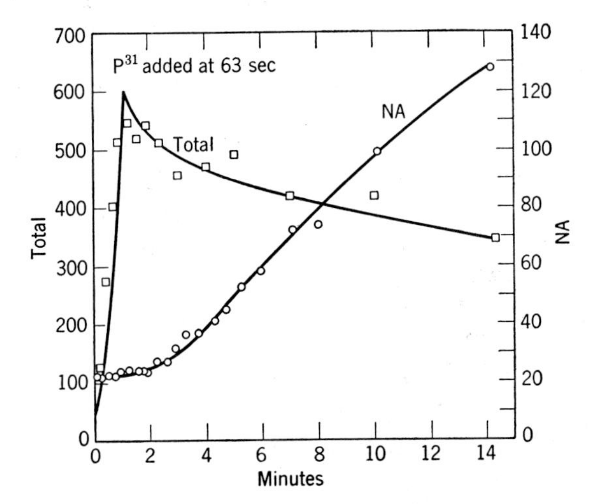


FIG. 9. Incorporation of $P^{32}$ into total cell and nucleic-acid fraction when the radioactivity is available for a short interval.

outermost phosphorus equilibrates rapidly as it enters into the energy-transfer reactions. The innermost phosphorus appears to be the precursor of nucleic acid, as its specific radioactivity corresponds to the slope of the nucleic-acid curve.

A large number of similar experiments have been performed to follow the incorporation of amino acids into protein. As a first step, the amino acids enter the cell to form a pool wherein the concentration is typically several thousand times the concentration of the medium. This pool has extremely complicated characteristics which cannot be interpreted either in terms of simple adsorption on sites or in terms of simple transport across an impermeable membrane. A model which includes transport by a limited quantity of carrier to an array of sites, however, seems adequate for *E. coli*. In yeast, the pools are considerably more complicated.

Fig. 12 shows a typical time course of incorporation. As the size of the pool depends on the concentration in

* The experimental work described in this paper was done by the Biophysics Section, Department of Terrestrial Magnetism, Carnegie Institution of Washington, which at present includes E. T. Bolton, R. J. Britten, D. B. Cowie, and R. B. Roberts. The material up to this point is described in detail in "Studies of Biosynthesis in *E. coli*," R. B. Roberts, P. H. Abelson, D. B. Cowie, E. T. Bolton, and R. J. Britten, Carnegie Institution of Washington Publication No. 607, 1955. A more complete account of the remainder of the experimental work can be found in the Carnegie Institution of Washington Yearbooks Nos. 54, 55, 56, and 57 (1954–1958) and in *Microsomal Particles and Protein Synthesis*, R. B. Roberts, editor (Pergamon Press, New York, 1958).

the medium, the cells first are given a $C^{12}$ amino acid to establish a pool of the desired size, and then the tracer is added. The pool exchanges rapidly with the external medium, as shown by the curvature of the line representing the total incorporation. Often the outflow is two-thirds the inflow. The rate of incorporation into the protein is proportional to the specific radioactivity of the pool as would be expected in a precursor-product relationship. Chromatography shows that the material extracted (either by TCA, alcohol, or water) is the unaltered amino acid and the nonextractable material is protein. Other tests showed that the amino acids incorporated during the first fifteen seconds are not concentrated at the ends of peptide chains and that they are released by acid hydrolysis at the usual rate.

These curves show that there is no appreciable quantity of partially formed proteins. If such were the case, the first protein to appear would be deficient in radioactivity. The only kinetic delay is accounted for by the time to build up radioactivity in the pool.

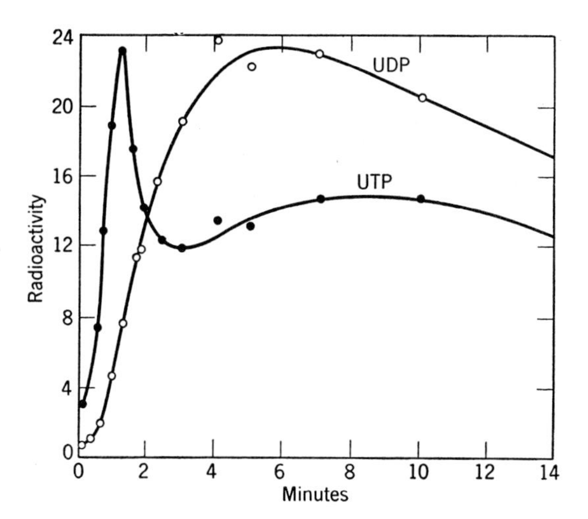


FIG. 10. Corresponding incorporation of $P^{32}$ into uridine diphosphate and uridine triphosphate.

Similar experiments using radiosulfate as a tracer showed that glutathione (the only peptide found in *E. coli*) is not an intermediate in protein synthesis although it can serve as a sulfur source in sulfur-deficient medium. No evidence of other peptides was found even when protein synthesis was blocked either by chloramphenical or by lack of a required amino acid.

Activated amino acids and amino acids attached to ribonucleic acid are suspected intermediates which are presently under intense study. These compounds and their possible role in protein synthesis are discussed in detail by Meister (p. 210). Unfortunately, they are present in *E. coli* in such small quantities that our methods have not given any conclusive evidence of whether or not they are intermediates of protein synthesis.

These various attempts to find intermediates beyond

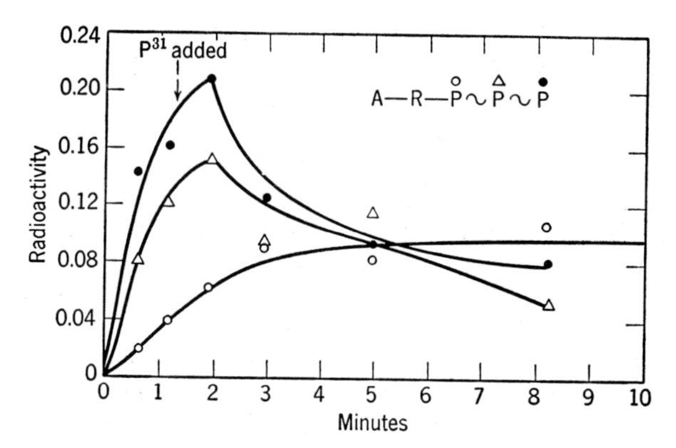


FIG. 11. Corresponding incorporation into individual phosphorus atoms of adenosine triphosphate.

the amino acids and nucleotides have failed. Looking at macromolecular synthesis from the point of view of the intermediates, one sees both amino acids and nucleotides held in the cell but exchangeable with the outside medium, until suddenly and without further preparation they appear linked into macromolecules. At this stage, it becomes necessary to determine in which macromolecules they appear. The ribosomes are of particular interest because it has been suspected for some time that RNA plays an important role in protein synthesis.

A column of Sober's modified cellulose (DEAE) provides an adequate resolution of the juice from disrupted cells. Using a salt-gradient elution, the proteins are well spread out and the protein of the ribosomes stands out as a distinct peak in the distribution. The ultraviolet absorption shows a corresponding peak with a secondary peak of nucleic acid.

Following a short exposure to radiosulfate or to radioactive amino acids, the specific radioactivity of the ribosome peak is somewhat lower than that of the other proteins. Neither is there any extra radioactivity in the proteins which spin down with the ribosomes.

Corresponding experiments with $P^{32}O_4$ show at first

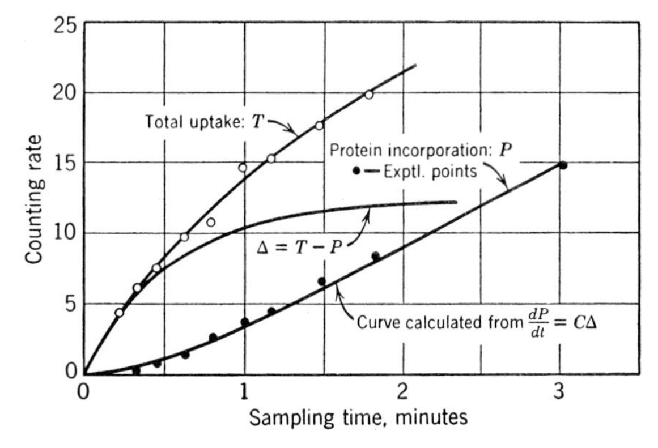


FIG. 12. Radioactive proline is rapidly incorporated by *E. coli*. The difference between the total and the TCA-insoluble (protein) measures the pool of extractable amino acid. The rate of incorporation into protein is proportional to the specific radioactivity of the pool, indicating a precursor-product relationship.

a new and distinct peak with its radioactivity transferring only after a considerable kinetic delay to the nucleic-acid peak and later still to the ribosome peak.

These experiments, which indicate that the ribosomes are a late product of synthesis, differ from the observation in several other laboratories that radioactive amino acids appear first in the ribosomes. Two explanations for this discrepancy seem plausible. One is that in *E. coli* incorporation of amino acids by exchange is negligible whereas it may be important in mammalian cells. The prompt incorporation into the ribosomes of mammalian tissues may be owing to exchange. Another possibility is that the ribosomes of *E. coli* work much more rapidly, and there is correspondingly much less newly formed protein adhering to them. It should be pointed out also that the recovery from the columns is not complete and an important part of the kinetics thereby may be missed. In particular, the cell membranes which are lost on the column are initially high in radiophosphate.

In the space allowed, it is difficult to present more than a brief glimpse of the operation of the synthetic machinery of a cell, of the way its macromolecules are bathed in an internal medium which is different from but rapidly responding to the external environment, of the way the reactions change to meet changing conditions in the environment. It has been necessary to discuss the various elements separately, but the intercouplings and interactions are equally important to maintain a balance among the different systems. I have attempted to emphasize those findings in the study of intermediary metabolism which throw some light on the more obscure processes of protein and nucleic-acid synthesis, but they provide little information, only a few boundary conditions which must be met by any model of macromolecular synthesis.

# 21
# Mechanisms of Enzyme Action*

ROBERT A. ALBERTY

*Department of Chemistry, University of Wisconsin, Madison 6, Wisconsin*

ALMOST all reactions in living things occur spontaneously at insufficient rates and have to be catalyzed. The biological catalysts, enzymes, have all been found to be proteins. With respect to their turnover numbers, specificity, and variety of reactions catalyzed, enzymes are the most remarkable catalysts known, especially since they provide the means by which the rates of various reactions are controlled so that the rates of the large number of necessary reactions have a proper relation to each other and to the requirements of the situation.

Thus, one is presented with the tremendous challenge of trying to understand these catatalytic actions. One can distinguish roughly between two types of approaches to this problem, the kinetic approach and the chemical approach. Of course, these distinctions are arbitrary and it is really not possible to make a clear separation. Through the kinetic approach, which is emphasized in this article, one studies the effect of various independent variables on the rate of reaction and the interpretations of these effects. Through the chemical approach, one studies the effect of chemical modification of the substrate and of chemical reactions of the enzyme.

The general features of the kinetic approach have been reviewed in the literature.[1–9] For any given mechanism, it is possible to write an equation for the rate of formation of each substance involved. These equations consist of summations of concentrations or products of concentrations, each multiplied by a rate constant. These rate constants are constants under a given set of conditions, but the values of the rate constants are affected in a number of ways by components of the solution. The effects of constituents of the reaction medium may be classified according to: (a) whether they have a significant effect on the activity coefficients of reactants and the transition state; or (b) whether they are directly involved in steps in the mechanism. Effects of the first type are encountered always when the electrolyte concentration is changed. Effects of the second type are produced by hydrogen ions, metal ions, inhibitors, coenzymes, and even by buffer ions.

The solution of the simultaneous differential equations for a mechanism yields the functional form for the dependencies of the various concentrations upon time. As a further aid in this analysis, there are two types of relations between the various rate constants in a mechanism. According to the first, provided by thermodynamics, the values of the rate constants for all of the forward and reverse steps must give the correct equilibrium constant for the over-all reaction. According to the second, provided by irreversible thermodynamics, the frequency of transitions from $A$ to $B$ must be equal to the frequency of transitions from $B$ to $A$ at equilibrium, even though there are other indirect paths by which this equilibrium may be maintained. This principle of detailed balancing[10–13] applies when there are cyclic paths in a mechanism.

The kinetic behavior of a given mechanism always can be represented by a theoretical rate equation, but in the laboratory one is faced with the inverse problem of having to devise a mechanism to explain certain kinetic observations. This process requires imagination and is not one that can be summarized by a series of rules. Kinetic data show the composition of the activated complexes for the various steps in the reaction but do not prove a particular mechanism. Many conceivable mechanisms, however, may be eliminated by the use of such data.

A minimum requirement for devising a mechanism to account for kinetic data is a complete understanding of the kinetic consequences of any given mechanism. As an illustration, the simplest type of enzymatic mechanism is considered first, in which an intermediate $X$ is formed by reaction of enzyme and substrate or of product and substrate. The concept of an intermediate complex dates back at least to 1902 when Henri and Brown independently sought to explain the fact that the velocity of inversion of sucrose by invertase is independent of the sucrose concentration at high concentrations and is directly proportional to sucrose concentration at low concentrations. In 1913, Michaelis and Menten[14] attracted considerable attention with their derivation of the initial-velocity equation assuming that the reaction $E+S \rightleftarrows X$ was in equilibrium, a rather restrictive assumption.

### ENZYMATIC REACTION WITH TWO SUBSTRATES

| | $E$ | $+$ | $S$ | $\underset{k_2}{\overset{k_1}{\rightleftarrows}} X \underset{k_4}{\overset{k_3}{\rightleftarrows}}$ | $E$ | $+$ $P$ |
|---|---|---|---|---|---|---|
| Initial conc. | $e_0$ | | $s_0$ | $0$ | $e_0$ | $0$ |
| Conc. at $t$ | $e=e_0-x$ | | $s=s_0-x-p$ | $x$ | $e=e_0-x$ | $p$ |

(1)

If there is more than one site per enzyme molecule, this

* This research was supported by grants from the National Science Foundation and from the Research Committee of the Graduate School of the University of Wisconsin from funds supplied by the Wisconsin Alumni Research Foundation.

derivation will be applicable if there is no site-site interaction. If the reactants initially present are $E$ and $S$, it is convenient to introduce the conservation equations ($e_0=e+x$ and $s_0=s+p+x$) and express the various concentrations at any time in terms of the instantaneous concentrations $x$ and $p$ of intermediate $X$ and product $P$ and the initial molar concentrations $e_0$ and $s_0$ of enzymatic sites and $S$. Owing to the conservation equations, only two of the four rate equations which may be written are independent. According to mechanism 1, the rates of change of concentration of $X$ and $P$ are given by

$$dx/dt=k_1se+k_4pe-(k_2+k_3)x \tag{2}$$

$$dp/dt=k_3x-k_4pe. \tag{3}$$

When these equations are written in terms of the two unknown concentrations $x$ and $p$ by use of the expressions under mechanism 1 and are rearranged, they become

$$dx/dt=k_1s_0e_0+(k_4-k_1)e_0p - [k_1e_0+k_1s_0+(k_4-k_1)p+k_2+k_3]x+k_1x^2 \tag{4}$$

$$dp/dt=(k_3+k_4p)x-k_4e_0p. \tag{5}$$

## Transient Phase

Apparently, these simultaneous nonlinear differential equations can be solved in complete generality only with the use of a differential analyzer. The general character of the solutions is that $x$ rises from zero and may or may not go through a maximum.[7,9,15] Families of plots of $x$ and $p$ for this mechanism have been given by Chance[16] for the case that $k_4=0$. Various approximate solutions have been discussed for the transient phase of the reaction when $k_4=0$.[17–20]

For the particular case that $k_1=k_4$, a condition which is met in practice at certain $p$H values, Eq. (4) is simplified so that it is immediately integrable to yield $x$ as a function of time.[21] Substitution of this relation into Eq. (5) and integration yields $p$ as a function of time. Rather than give the resulting equations, which are rather complicated, only the forms obtained when $s_0 \gg e_0$ are given; they are:

$$x=\frac{k_1e_0s_0}{k_1s_0+k_2+k_3}\{1-\exp[-(k_1s_0+k_2+k_3)t]\} \tag{6}$$

$$p=\frac{k_3s_0}{k_2+k_3}\left\{1-\exp\left[-\frac{k_1(k_2+k_3)e_0t}{k_1s_0+k_2+k_3}\right]\right\} - \frac{k_1k_3s_0e_0}{(k_1s_0+k_2+k_3)}\{1-\exp[-(k_1s_0+k_2+k_3)t]\}. \tag{7}$$

Thus, for this special case, the concentration of the intermediate $X$ rises in a first-order manner with a half-life of

$$\tau_1=0.693/(k_1s_0+k_2+k_3). \tag{8}$$

Equation (7), for the concentration of $p$, contains two first-order terms. The half-life for the first term is given by $\tau_1$ and that for the second term is given by $\tau_2$.

$$\tau_2=\frac{0.693(k_1s_0+k_2+k_3)}{k_1(k_2+k_3)e_0}. \tag{9}$$

Since $s_0 \gg e_0$, $\tau_2 \gg \tau_1$; that is, the half-life for the appearance of $P$ is very large as compared with that for the appearance of the intermediate $X$. The second term in Eq. (7) is identical with the solution obtained below with the steady-state approximation.

If $k_1 \neq k_4$ and $s_0 \gg e_0$, Miller and Alberty[21] have obtained a perturbation solution which is useful if $0.1k_1<k_4<10k_1$.

The study of the transient phase of enzymatic reactions requires very sensitive and rapidly responding analytical methods. When suitable methods can be developed as Chance has done for the peroxidase and catalase reactions and for the respiratory chain,[22] more information can be obtained about the intermediates than could be obtained in general from steady-state measurements.

## Steady State

If the substrate concentration is high as compared with the concentration of enzymatic sites, $E$ must be converted to $X$ and reconverted to $E$ many many times before equilibrium is reached. As a result, $dx/dt$ is very small as compared with other terms in the rate equation and, to a high degree of approximation, may be taken equal to zero. This is an approximation of considerable value for most enzyme kinetic experiments. Setting $dx/dt$ in Eq. (4) equal to zero for the case that $s_0 \gg e_0$ yields

$$x=\frac{k_4s_0+(k_1-k_4)(s_0-p)}{(k_2+k_3)+k_4s_0+(k_1-k_4)(s_0-p)}. \tag{10}$$

The concentration of the intermediate $X$ may either increase or decrease during the steady-state phase of the reaction,[15] depending upon the sign of $k_1-k_4$. If $k_4>k_1$, the concentration of $X$ increases; if $k_1>k_4$, the concentration of $X$ decreases; and if $k_1=k_4$, the concentration of $X$ remains sensibly constant for $t \gg \tau_1$. If the concentration of $X$ increases in the steady state, it does not pass through a maximum; if the concentration of $X$ decreases in the steady state, it does pass through a maximum before the steady state is established.

Substitution of Eq. (10) into Eq. (5) yields the steady-state rate equation[23] for mechanism 1.†

$$-\frac{ds}{dt}=\frac{dp}{dt}=\frac{(V_S/K_S)s-(V_P/K_P)p}{1+s/K_S+p/K_P}, \tag{11}$$

† Writing the equation in this form rather than in terms of the individual rate constants has the advantage that the constants are those directly determinable from experiment and that the steady-state rate equation for the $n$-intermediate case can be written in the same form.

where

$$V_S = k_3 e_0 \qquad V_P = k_2 e_0$$
$$K_S = (k_2 + k_3)/k_1 \quad K_P = (k_2 + k_3)/k_4. \tag{12}$$

The values of the kinetic constants usually are obtained by measuring the steady-state velocity $v_f$ of the forward reaction immediately after the transient phase of the reaction but before the second term in the numerator and the third term in the denominator of Eq. (11) become important. Under these conditions, the velocity of the over-all reaction is given by the Michaelis-Menten equation.

$$v_f = V_S/(1 + K_S/s_0). \tag{13}$$

The initial steady-state velocity of the reverse reaction may be measured similarly if the equilibrium constant is not too large.

$$v_r = V_P/(1 + K_P/p_0). \tag{14}$$

According to this mechanism, as $s_0$ (or $p_0$) is increased, the initial steady-state velocity approaches $V_S = k_3 e_0$ (or $V_P = k_2 e_0$) which is referred to as the maximum initial velocity. If $V$ is expressed in moles per liter per unit time and if $e_0$ is expressed in moles per liter, $k$ is the turnover number of the enzyme, the number of moles of product produced per enzyme molecule (or better, per enzymatic site) per unit time. Under optimal conditions, the turnover number for catalase is greater than $10^8$/min; for fumarase it is $10^5$/min; and for chymotrypsin it is 0.01–10/min, depending upon the substrate.

The constants $K_S$ and $K_P$ in Eqs. (13) and (14) are referred to as Michaelis constants. It is evident that Michaelis constants have the units of concentration and are equal to the substrate concentration required to produce an initial steady-state velocity half as large as the maximum initial velocity. At substrate concentrations well below the Michaelis constant, the initial velocity is directly proportional to the substrate concentration.

In order to determine $v_f$ or $v_r$, a very sensitive analytical method is very much needed since these velocities must generally be determined during the first several percent of reaction. Spectrophotometric, fluorometric, and titrimetric methods are those most often used, and there is a real need for better analytical methods for measuring small changes.

If the Michaelis constant for the product is small so that the third term in the denominator of Eq. (11) quickly becomes important[24] or the equilibrium constant is unfavorable, the steady-state velocity decreases rapidly during a kinetic experiment making it difficult to determine the initial velocity. To determine $v_f$, it is convenient to plot $p/t$ or $-(1/t)\ln(1-p/p_{eq})$ *versus* $t$ and extrapolate to $t=0$ to obtain $v_f$ or $v_f/p_{eq}$, respectively. Such an extrapolation is linear at sufficiently short times and further kinetic information may be obtained from the initial slope.[25]

The values of the four kinetic parameters are not independent of the equilibrium constant for the over-all reaction. At equilibrium, $ds/dt = 0$, therefore, the following equation[23] is obtained from Eq. (11):

$$K_{eq} = \frac{p_{eq}}{s_{eq}} = \frac{V_S K_P}{V_P K_S} = \frac{k_1 k_3}{k_2 k_4}. \tag{15}$$

This useful test of the validity of kinetic data applies even if there are further complications in the mechanism, such as hydrogen-ion equilibria, or metal-ion binding, provided only that the effects of substrate and product concentration under a given set of conditions are represented by Eq. (11). Also, when the reaction goes essentially to completion, $V_P$ may be calculated using this relation, provided that the other three kinetic constants have been determined and that $K_{eq}$ is known.

## Integrated Steady-State Rate Equation

Integration of Eq. (11) which is applicable if $s_0 \gg e_0$, and introduction of Eq. (15) yields[5]

$$\frac{V_S}{K_S}\left(1+\frac{1}{K_{eq}}\right)t = \left(\frac{1}{K_S} - \frac{1}{K_P}\right)p - \left[1 + \frac{s_0}{K_P} + \frac{(1/K_S)-(1/K_P)}{1+K_{eq}} s_0\right] \ln\left[1 - \frac{p}{p_{eq}}\right]. \tag{16}$$

If $K_{eq} \gg 1$ and $s_0 \ll K_p$, this equation becomes[26]

$$V_S t = p = K_S \ln(1 - p/s_0). \tag{17}$$

Thus, plots of $(1/t)\ln(1-p/p_{eq})$ *versus* $p/t$ are linear. However, since it is rather complicated to calculate the maximum initial velocities and Michaelis constants from the slopes and intercepts, this approach has not been used very often.[26–28]

Once the values of $K_S$, $V_S$, $K_P$, and $V_P$ have been determined, the values of the rate constants in mechanism 1 may be calculated from

$$k_1 = (V_S/e_0 + V_P/e_0)/K_S, \tag{18}$$

$$k_2 = V_P/e_0, \tag{19}$$

$$k_3 = V_S/e_0, \tag{20}$$

$$k_4 = (V_S/e_0 + V_P/e_0)/K_P. \tag{21}$$

Thus, from steady-state kinetic studies, it is possible to obtain the values of all four of the rate constants in mechanism 1. If the number $n$ of enzymatic sites per molecule is unknown, the calculated values of $k_1$ to $k_4$, using the molar concentration of the enzyme in Eqs. (18) to (21), will be larger than the true constants by an integer factor $n$.

Although mechanism 1 has certain features which are

common to many enzymatic reactions, it can be stated confidently that Eq. (2) *is too simple to represent the mechanism of any enzymatic reaction.* However, this mechanism is of great importance for two reasons. First, it can be extended in many ways to include further steps. Second, under a given set of conditions, reactions following more-complicated mechanisms may appear to follow mechanism 1. Thus, it is often useful to calculate the magnitudes of the four apparent rate constants. As an example of the extension of the simple mechanism, the mechanism of the fumarase reaction is considered.

### Effect of *p*H Changes on the Fumarase Reaction

Fumarase catalyzes the hydration of fumarate (F) to L-malate (M). The effect of *p*H changes on the

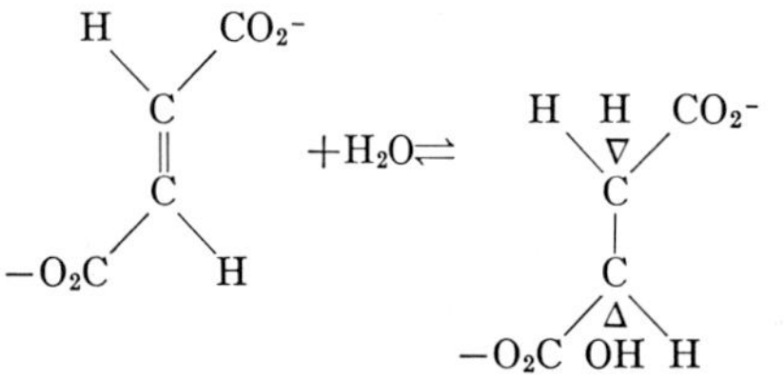

$$\mathrm{H(CO_2^-)C{=}C(H)(^-O_2C)} + H_2O \rightleftharpoons \mathrm{H\,H\,CO_2^-}\text{-}\overset{\nabla}{C}\text{-}\overset{\Delta}{C}\text{-}{}^-O_2C\ OH\ H \quad (22)$$

maximum initial velocities and Michaelis constants for the forward and reverse reactions can be accounted for in terms of the mechanism[29] shown in Fig. 1. Two intermediates are required because the maximum initial velocities for the forward and reverse reactions have different *p*H optima.

As expected for the steady-state treatment of a mechanism of this type, only the equilibrium constants for the acid dissociations (represented by $K$) are involved and not the individual rate constants which make up these equilibrium constants. The steady-state rate equations may be arranged in the form of Eqs. (13) and (14).[15] The Michaelis constants and maximum velocities are given by

$$V_F=\frac{V_F'e_0}{1+(\mathrm{H}^+)/K_{aEF}'+K_{bEF}'/(\mathrm{H}^+)} \quad (23)$$

$$K_F=K_F'\frac{1+(\mathrm{H}^+)/K_{aE}+K_{bE}/(\mathrm{H}^+)}{1+(\mathrm{H}^+)/K_{aEF}'+K_{bEF}'/(\mathrm{H}^+)} \quad (24)$$

$$V_M=\frac{V_M'e_0}{1+(\mathrm{H}^+)/K_{aEM}'+K_{bEM}'/(\mathrm{H}^+)} \quad (25)$$

$$K_M=K_M'\frac{1+(\mathrm{H}^+)/K_{aE}+K_{bE}/(\mathrm{H}^+)}{1+(\mathrm{H}^+)/K_{aEM}'+K_{bEM}'/(\mathrm{H}^+)}. \quad (26)$$

The ten kinetic parameters in these four equations represent the experimental data over the range *p*H 5 to 8.5, and they have been determined at five ionic strength values at 25°C using tris-(hydroxymethyl)-aminomethane acetate buffers. Only two of them, $K_{aE}$ and $K_{bE}$, can be identified with constants in the mechanism. The other eight kinetic parameters are related to the constants in the mechanisms by

$$V_F'=\frac{k_3k_5}{k_3+k_4+k_5} \quad (27)$$

$$K_F'=\frac{k_2k_5+k_2k_4+k_3k_5}{k_1(k_3+k_4+k_5)} \quad (28)$$

$$V_M'=\frac{k_2k_4}{k_2+k_3+k_4} \quad (29)$$

$$K_M'=\frac{k_2k_5+k_2k_4+k_3k_5}{k_6(k_2+k_3+k_4)} \quad (30)$$

$$K_{aEF}'=\frac{k_3+k_4+k_5}{(k_4+k_5)/K_{aEF}+k_3/K_{aEM}} \quad (31)$$

$$K_{bEF}'=\frac{(k_4+k_5)K_{bEF}+k_3K_{bEM}}{k_3+k_4+k_5} \quad (32)$$

$$K_{aEM}'=\frac{k_2+k_3+k_4}{(k_2+k_3)/K_{aEM}+k_4/K_{aEF}} \quad (33)$$

$$K_{bEM}'=\frac{(k_2+k_3)K_{bEM}+k_4K_{bEF}}{k_2+k_3+k_4}. \quad (34)$$

The $K_a'$ and $K_b'$ values are calculated from the plots of $V$ *versus* *p*H and $V/K$ *versus* *p*H.[30] The eight relations of Eqs. (27) to (34), however, are insufficient to determine the values of the other ten constants for the mechanism illustrated in Fig. 1. But, these relations do place considerable restrictions on the values of the various rate constants and acid-dissociation constants. A detailed analysis[15] of the data shows that $k_3$ and $k_4$ may be determined within their experimental uncertainties, and *minimum* values are obtained for $k_1$, $k_2$, $k_5$, and $k_6$.

The experimental data show that the minimum values of the rate constants for the bimolecular reaction of the substrate with the enzymatic site are of the

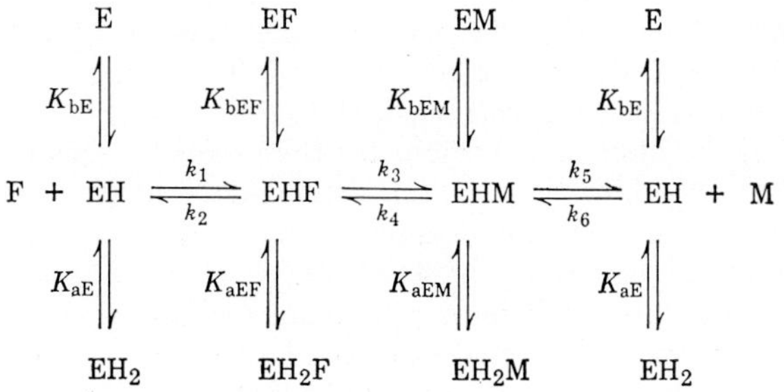

$$\begin{array}{ccccccc} \mathrm{E} & & \mathrm{EF} & & \mathrm{EM} & & \mathrm{E} \\ K_{bE}\updownarrow & & K_{bEF}\updownarrow & & K_{bEM}\updownarrow & & K_{bE}\updownarrow \\ \mathrm{F + EH} & \underset{k_2}{\overset{k_1}{\rightleftharpoons}} & \mathrm{EHF} & \underset{k_4}{\overset{k_3}{\rightleftharpoons}} & \mathrm{EHM} & \underset{k_6}{\overset{k_5}{\rightleftharpoons}} & \mathrm{EH + M} \\ K_{aE}\updownarrow & & K_{aEF}\updownarrow & & K_{aEM}\updownarrow & & K_{aE}\updownarrow \\ \mathrm{EH_2} & & \mathrm{EH_2F} & & \mathrm{EH_2M} & & \mathrm{EH_2} \end{array}$$

FIG. 1. Mechanism of the fumarase reaction required to explain the effect of *p*H on the kinetics of the forward and reverse reactions.

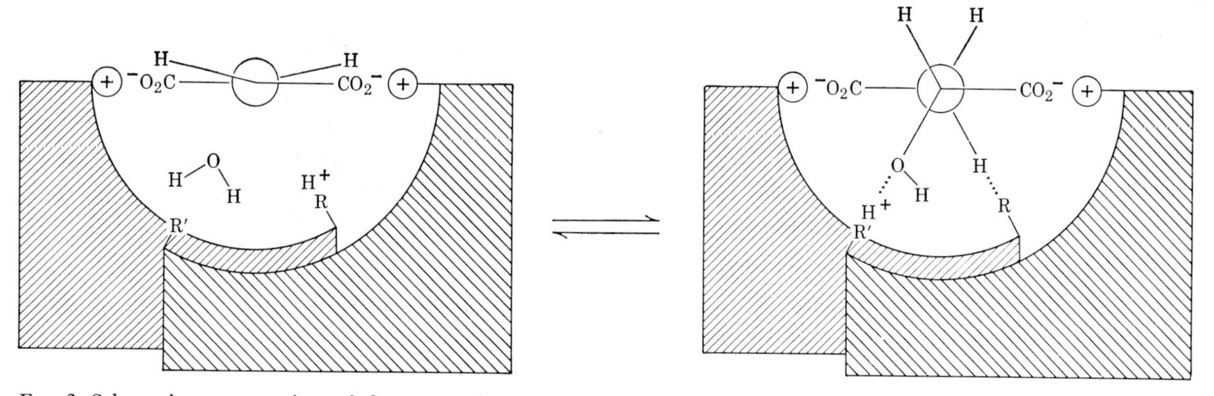


FIG. 2. Schematic representations of the enzyme-fumarate and enzyme-L-malate complexes indicating the two weak acidic groups R and R′ in the catalytic site and accounting for the *cis*-hydration [from T. C. Farrar, H. S. Gutowsky, R. A. Alberty, and W. G. Miller, J. Am. Chem. Soc. **79**, 3978 (1957)].

order of $10^9$ liter mole$^{-1}$ sec$^{-1}$ when extrapolated to zero ionic strength. This high value for the rate constant of a second-order reaction in solution indicates that the reaction rate is controlled by diffusion. The maximum rate with which the substrate could diffuse into a hemispherical sink on a flat surface of a protein molecule is expressed as a second-order rate constant (in liter mole$^{-1}$ sec$^{-1}$) by

$$k=\frac{2\pi N}{1000}R_{12}D_{12}f, \tag{35}$$

where $N$ is Avogadro's number, $R_{12}$ is the reaction radius (sum of the radii of site and substrate), $D_{12}$ is the sum of the diffusion coefficients of substrate and enzyme, and $f$ is an electrostatic factor which corrects for the slowing down or speeding up of the diffusion process owing to electrostatic interactions between the substrate and enzymatic site.[31] The theoretical values of the second-order rate constant calculated for fumarase using Eq. (35) are greater than the minimum value obtained from the kinetics by a factor of 3 at most.[32] Therefore, it is likely that the first step in the enzymatic reaction is simply the formation of an ion pair which subsequently undergoes an intermolecular reaction. The effect on the second-order rate constant of changing the ionic strength indicates that the enzymatic site has two positive charges.[32]

Because of the high value of this second-order rate constant, the half-time for the rise in concentration of intermediates in the fumarase reaction is of the order of $10^{-5}$ to $10^{-4}$ sec. Such rapid changes cannot be studied by mixing methods, but may be measured by relaxation methods such as those developed by Eigen.[33]

Our interpretation of the total effect of two acid groups in the enzymatic site is that these groups donate and accept protons in the catalytic process. This is illustrated in Fig. 2 which shows the ionizable groups on one side of the bound substrate ion as required by the stereochemistry of the deuteromalic acid obtained when the reaction is carried out in $D_2O$.[34] According to this mechanism, the enzyme is a clever acid-base catalyst which makes available a proton at exactly the required position in space and which has a site to accept a proton at precisely the right position. This sort of mechanism accounts for the absolute stereochemical specificity.

### Inhibition and Activation

Many substances combine with enzymes in such a way as to reduce or even to increase the steady-state velocity. Compounds which are structurally similar to the substrate often combine at the enzymatic site. This may be represented by adding the reaction

$$E+I=EI, \tag{36}$$

where $EI$ cannot combine with substrate and $I$ is the inhibitor. The steady-state rate equation for reactions (1) and (36) becomes

$$v_f=\frac{V_s}{1+K_S(1+i/K_I)/s}, \tag{37}$$

where $i$ is the concentration of the inhibitor $I$ and $K_I$ is the dissociation constant for the $EI$ complex. An inhibitor whose effect on the rate may be represented in this way is called a competitive inhibitor. Information on the competitive-inhibition constants for a number of inhibitors is useful in learning about the nature of the enzymatic site. However, since inhibition constants will, in general, depend upon the *p*H, what is needed in the study of inhibitors are kinetic data over a range of *p*H, so that the affinity of a *particular* ionized form of the enzymatic site for the inhibitor may be calculated.[35]

If the inhibitor combines only at a site where substrate does not combine, the effect on the kinetics is not competitive and the inhibitor concentration enters the rate equation in a different way. A mixture of competitive and noncompetitive effects may be accounted for by a mechanism of the type[5]

$$
\begin{aligned}
&E+S \underset{k_2}{\overset{k_1}{\rightleftharpoons}} ES \overset{k_1}{\rightarrow} E+P \\
&E+I \underset{k_5}{\overset{k_4}{\rightleftharpoons}} EI \\
&E+I \underset{k_7}{\overset{k_6}{\rightleftharpoons}} IE \\
&IE+S \underset{k_9}{\overset{k_8}{\rightleftharpoons}} IES \overset{k_{10}}{\rightarrow} IE+P \\
&EI+I \underset{k_{12}}{\overset{k_{11}}{\rightleftharpoons}} IEI,
\end{aligned}
\tag{38}
$$

where writing $I$ to the left of $E$ means that it is combined at a neighboring site. The combination of an ion at a neighboring site may increase the rate of reaction of the enzyme-substrate intermediate to yield product. The steady-state rate equation for this mechanism is the Michaelis equation with

$$V=k_3(E)_0\frac{1+k_{10}K_{ES}i/k_3K_{IES}K_{IE}}{1+K_{ES}i/K_{IES}K_{IE}} \tag{39}$$

$$K=K_{ES}\frac{1+i(1/K_{EI}+1/K_{IE})+i^2/K_{EI}K_{IEI}}{1+K_{ES}i/K_{IES}K_{IE}}, \tag{40}$$

where $K_{ES}=(k_2+k_3)/k_1$, $K_{IES}=(k_9+k_{10})/k_8$, $K_{EI}=k_5/k_4$, $K_{IE}=k_7/k_6$, and $K_{IEI}=k_{12}/k_{11}$. Accordingly, linear plots of $1/v$ *versus* $1/s$ are obtained, but both the slope and intercept vary with the concentration of inhibitor in this general case. Mechanisms with further steps have been discussed in detail in the literature.[36]

Buffer ions, and even substrate ions, may cause activation or inhibition as discussed above for substance $I$.[37] If substrate is bound at a neighboring site and this alters the rate constants in the catalytic mechanism, the steady-state kinetics will not follow the Michaelis-Menten rate equation except in the limit of low substrate concentrations. The binding of substrate at higher substrate concentrations may cause either inhibition or activation, effects which are not provided for by the simple Michaelis-Menten mechanism.

## ENZYMATIC REACTIONS WITH FOUR SUBSTRATES

Such a reaction may be represented in general by

$$A+B=C+D. \tag{41}$$

An example is the alcohol-dehydrogenase reaction

$$\text{alcohol}+\text{DPN}=\text{aldehyde}+\text{DPNH}+\text{H}^+, \tag{42}$$

where DPN is diphosphopyridine nucleotide (a coenzyme) and DPNH is the reduced form. This reaction is catalyzed by the enzyme alcohol dehydrogenase. The kinetics of this reaction have been extensively investigated by Theorell and co-workers.[38–41]

Steady-state treatments have been given for a number of simple mechanisms that can be conceived of for a reaction of this type.[42] The objective of the kinetic experiments is to determine which of the paths is followed and which are the important intermediates and rate-determining steps. The same type of rate law is obtained for several of these mechanisms, and, as shown by Dalziel,[43] it is convienient to write the steady-state rate equations for the initial velocity when $A$ and $B$ are mixed with enzyme in the general form

$$\frac{e_0}{v}=\phi_0+\frac{\phi_1}{a}+\frac{\phi_2}{b}+\frac{\phi_{12}}{ab} \tag{43}$$

wherever this is possible. Here $a$ and $b$ represent the initial concentrations of $A$ and $B$, and $\phi_0$, $\phi_1$, $\phi_2$, and $\phi_{12}$ are parameters which may be obtained readily from the experimental data by making plots of $1/v$ *versus* $1/a$ with $b$ held constant and $1/v$ *versus* $1/b$ with $a$ held constant. There are certain relationships between these parameters for certain mechanisms and various relations between the parameters for the forward and reverse reactions. This is illustrated for the simplest type of mechanism only. Both steady-state and transient-state kinetic data consistent with the following mechanism have been obtained[38–41] for the alcohol-dehydrogenase reaction

$$
\begin{aligned}
&E+A \underset{k_{-1}}{\overset{k_1}{\rightleftharpoons}} EA \\
&EA+B \underset{k_{-2}}{\overset{k_2}{\rightleftharpoons}} EC+D \\
&EC \underset{k_{-3}}{\overset{k_3}{\rightleftharpoons}} E+C.
\end{aligned}
\tag{44}
$$

The steady-state rate equations for the forward and reverse reaction are conveniently written in the forms

$$\frac{e_0}{v_f}=\left(\frac{1}{k_3}+\frac{1}{k_1a}\right)+\frac{1}{k_2b}\left(1+\frac{k_{-1}}{k_1a}\right), \tag{45}$$

$$\frac{e_0}{v_r}=\frac{1}{k_{-1}}+\frac{1}{k_{-3}c}+\frac{1}{k_{-2}d}+\frac{k_3}{k_{-2}k_{-3}cd}. \tag{46}$$

It can be seen that, from studies of the forward reaction, the values of $k_1$, $k_2$, $k_3$, and $k_{-1}$ may be obtained, and that, from the reverse reaction, the values of $k_{-1}$, $k_{-2}$, $k_{-3}$, and $k_3$. Thus, all six of the rate constants are obtained and check values are obtained for $k_3$ and $k_{-1}$. There is a further relation between these kinetic parameters which must be obeyed; namely,

$$K_{eq}=\frac{c_{eq}d_{eq}}{a_{eq}b_{eq}}=\frac{k_1k_2k_3}{k_{-1}k_{-2}k_{-3}}. \quad (47)$$

Further checks may be obtained by the use of experiments in which one of the products is added along with the two reactants.[44] For example, if $A$, $B$, and $C$ are added, the steady-state rate equation becomes

$$\frac{(E)_0}{v_f}=\frac{1}{k_3}+\frac{1}{k_1a}\left(1+\frac{k_{-3}c}{k_3}\right)+\frac{1}{k_2b}+\frac{k_{-1}}{k_1k_2ab}\left(1+\frac{k_{-3}c}{k_3}\right). \quad (48)$$

Thus, by determining the effect of $C$ on the initial steady-state velocity, it is possible to obtain $k_{-3}/k_3$ and, consequently, a value of $k_{-3}$ from the forward reaction, since $k_3$ is known. By adding $D$, it is possible to obtain $k_{-2}$. Thus, all six of the rate constants can be obtained from steady-state velocities *of the forward reaction only*.

If there is a ternary complex in the second step of mechanism 44, there are eight rate constants in the mechanism, and all eight may be determined by means such as those just described. Actually, there must be a ternary complex in this mechanism, since a proton is transferred stereochemically specifically from DPNH to aldehyde to form an optically active monodeuteroethanol, but apparently its steady-state concentration is very low under the usual experimental conditions.

Hearon[7] has given a generalized mathematical treatment of mechanisms of reactions of this type which provides for the possibility of ternary complexes and for further reactants and products. Some rather unusual kinetic features are encountered in the steady-state studies of coenzyme mechanisms.[24]

## USES OF ISOTOPES IN ELUCIDATING MECHANISMS

The use of isotopes often permits the determination of the cleavage point in an hydrolysis reaction, as represented by[45,46]

$$AOB+H_2O^*=AO^*H+BOH \quad \text{or} \quad AOH+BO^*H, \quad (49)$$

or of the position of the hydrogen atom transferred, as in

$$CH_3CD_2OH+DPN^+=CH_3CDO+DPND. \quad (50)$$

Isotope-exchange experiments can be used to distinguish between the following two mechanisms for a transfer reaction.

$$AB+D=A+BD \quad (51)$$

(I)
$$AB+E=EAB$$
$$EAB+D=EABD=EBD+A$$
$$EBD=E+BD$$

(II)
$$AB+E=BE+A$$
$$D+BE=BD+E$$

If the mechanism is given by (II), a small amount of free $A$ is formed when $AB$ is added to the enzyme, even when $D$ is absent. If labeled $A$ is placed in the medium, the label appears in $AB$. If the mechanism is given by (I), no incorporation of label into $AB$ occurs unless $D$ is present also.

When the fumarase reaction is carried out in $D_2O$, a specific monodeutero-L-malate is formed. The absence of an isotope effect in the enzymatic dehydration of this monodeutero-L-malate suggested that the deuterium might be incorporated in a rapid reversible step. In order to determine whether such an exchange precedes the rate-determining step in the fumarase reaction, an experiment was carried out with L-malate as substrate in $D_2O$, and L-malate was isolated at various times and analyzed for deuterium.[47] The experiment was carried out under conditions where the Michaelis constants for malate and fumarate were equal so that the over-all reaction was pseudo first order, and the incorporation expected in the absence of exchange could be calculated. The experimental data were in agreement with this calculation showing that there is no direct exchange of the specific proton of malate with the medium.

Investigations of the type discussed in the foregoing reveal the nature of the kinetically important steps in the reaction and the values of the rate constants, but they do not reveal much about the chemical nature of these intermediates. A basic difficulty in understanding the chemistry of the intermediate compounds at the present time is the lack of information about the chemical structure of the enzymatic site. Progress is being made, however, and recent gains in the understanding of the chemical action of chymotrypsin (Neurath, p. 185) is an example, where the chemical mechanism has been filled into these blank spaces in the formal mechanism.

## BIBLIOGRAPHY


1 B. Chance, Advances in Enzymol. **12**, 153 (1951).

2 J. Z. Hearon, Physiol. Rev. **32**, 499 (1952).

3 W. D. McElroy and B. Glass, editors, *The Mechanism of Enzyme Action* (The Johns Hopkins Press, Baltimore, Maryland, 1954).

4 *The Physical Chemistry of Enzymes*, Discussions Faraday Soc. **20** (1956).

5 R. A. Alberty, Advances in Enzymol. **17**, 1 (1956).

6 M. Dixon and E. C. Webb, *Enzymes* (Academic Press, Inc., New York, 1958), p. 62.

7 J. Z. Hearon, S. A. Bernhard, S. Freiss, D. J. Botts, and M. F. Morales in *The Enzymes*, P. D. Boyer, H. Lardy, and K. Myrbäck, editors (Academic Press, Inc., New York, 1958), second edition (to be published).

[8] R. Lumry in *The Enzymes,* P. D. Boyer, H. Lardy, and K. Myrbäck, editors (Academic Press, Inc., New York, 1958), second edition (to be published).
[9] R. A. Alberty in *The Enzymes,* P. D. Boyer, H. Lardy, and K. Myrbäck, editors (Academic Press, Inc., New York, 1958), second edition (to be published).
[10] R. H. Fowler, *Statistical Mechanics* (Cambridge University Press, Cambridge, England, 1929).
[11] G. N. Lewis, Proc. Natl. Acad. Sci. U. S. **11**, 179 (1925).
[12] L. Onsager, Phys. Rev. **37**, 405 (1931).
[13] J. S. Thomsen, Phys. Rev. **91**, 1263 (1953).
[14] L. Michaelis and M. L. Menten, Biochem. Z. **49**, 333 (1913).
[15] R. A. Alberty and W. H. Peirce, J. Am. Chem. Soc. **79**, 1526 (1957).
[16] B. Chance, J. Biol. Chem. **151**, 553 (1943).
[17] M. F. Morales and D. E. Goldman, J. Am. Chem. Soc. **77**, 6069 (1955).
[18] H. Gutfreund, Discussions Faraday Soc. **20**, 167 (1955).
[19] K. J. Laidler, Can. J. Chem. **33**, 1614 (1955).
[20] P. A. T. Swoboda, Biochim. et Biophys. Acta **23**, 70 (1957).
[21] W. G. Miller and R. A. Alberty, J. Am. Chem. Soc. **80**, 5146 (1958).
[22] B. Chance and G. R. Williams, Advances in Enzymol. **17**, 65 (1956).
[23] J. B. S. Haldane, *Enzymes* (Longmans, Green and Company, London, 1930).
[24] E. L. King, J. Phys. Chem. **60**, 1378 (1956).
[25] R. A. Alberty and B. M. Koerber, J. Am. Chem. Soc. **79**, 6379 (1957).
[26] A. C. Walker and C. L. A. Schmidt, Arch. Biochem. **5**, 445 (1944).
[27] T. H. Applewhite and C. Niemann, J. Am. Chem. Soc. **77**, 4923 (1955).
[28] K. A. Booman and C. Niemann, J. Am. Chem. Soc. **77**, 5733 (1955).
[29] C. Frieden and R. A. Alberty, J. Biol. Chem. **212**, 859 (1955).
[30] R. A. Alberty and V. Massey, Biochim. et Biophys. Acta **13**, 347 (1954).
[31] P. Debye, Trans. Am. Electrochem. Soc. **82**, 265 (1942).
[32] R. A. Alberty and G. G. Hammes, J. Phys. Chem. **62**, 154 (1958), and forthcoming publication.
[33] M. Eigen, Discussions Faraday Soc. **17**, 194 (1954).
[34] T. C. Farrar, H. S. Gutowsky, R. A. Alberty, and W. G. Miller, J. Am. Chem. Soc. **79**, 3978 (1957).
[35] R. A. Alberty in *Molecular Structure and Biological Specificity,* L. Pauling and H. A. Itano, editors (American Institute of Biological Sciences, New York, 1957).
[36] J. Botts and M. F. Morales, Trans. Faraday Soc. **49**, 696 (1953).
[37] R. A. Alberty and R. M. Bock, Proc. Natl. Acad. Sci. U. S. **39**, 895 (1953).
[38] H. Theorell and R. Bonnichsen, Acta Chem. Scand. **5**, 1105 (1951).
[39] H. Theorell and B. Chance, Acta Chem. Scand. **5**, 1127 (1951).
[40] H. Theorell, A. P. Nygaard, and R. Bonnichsen, Acta Chem. Scand. **8**, 1490 (1954).
[41] H. Theorell, A. P. Nygaard, and R. Bonnichsen, Acta Chem. Scand. **9**, 1148 (1955).
[42] R. A. Alberty, J. Am. Chem. Soc. **75**, 1928 (1953).
[43] K. Dalziel, Acta Chem. Scand. **11**, 1706 (1957).
[44] R. A. Alberty, J. Am. Chem. Soc. **80**, 1777 (1958).
[45] D. E. Koshland and S. S. Stein, J. Biol. Chem. **208**, 139 (1954).
[46] H. F. Fisher, E. E. Conn, B. Vennesland, and F. H. Westheimer, J. Biol. Chem. **202**, 687 (1953).
[47] R. A. Alberty, W. G. Miller, and H. F. Fisher, J. Am. Chem. Soc. **79**, 3973 (1957).

# 22
# Protein Structure and Enzyme Action

HANS NEURATH
*Department of Biochemistry, University of Washington, Seattle 5, Washington*

THE idea that proteins are enzymes is so firmly entrenched in our minds that it may be regarded as a "self-evident truth." However, only a few decades ago, this concept was the subject of a lively controversy, and the first report, by J. B. Sumner,[1] of the crystallization of a protein having all of the properties of the enzyme urease, was greeted with critical reservations. The subsequent crystallization of other enzymes, including several proteolytic enzymes, notably by Northrop, Kunitz, and Herriott,[2] has added support to the identification of enzymes as proteins, and today over 130 enzymes of known specificity and function have been obtained in crystalline form (see Dixon and Webb[3]).

Of course, crystallinity, *per se*, is neither a necessary nor a sufficient criterion for the protein nature of a compound. Suffice it to say that every criterion of protein chemistry has been found applicable to purified enzymes, and no enzyme has yet been found which would be an exception to this statement. Thus, it has been known for many years that processes leading to protein denaturation, in turn, are likely to cause enzyme inactivation, and, when denaturation could be reversed, enzyme activity would reappear. Typical examples for this statement are the work of Kunitz,[4] on the heat denaturation of the soy-bean trypsin inhibitor, and the work of Eisenberg and Schwert,[5] on the reversible denaturation of chymotrypsinogen. As denaturation may be described as an effect on the secondary and tertiary structure of proteins, it may be said, therefore, that enzymatic activity is related to a specific configuration of the protein molecule or of a part thereof. The enzymatic activity is evidently also related to the primary structure of the protein molecule; i.e., the sequence of amino acids along the polypeptide chain but not all amino-acid side chains are required for activity. Thus, in some cases, some of the side chains may be chemically modified without adversely affecting catalytic functions, and in others, some portions of the polypeptide chain may be split off without deleterious effects on activity. Furthermore, enzymes of like specificity but of different origin have been found to differ from one another in amino-acid composition. It is evident, therefore, from these sketchy comments, that not the entire molecular structure of the protein is necessarily involved in the enzymatic process, a concept which is fully consistent, on the molecular level, with the idea previously held by enzymologists that an enzyme has an "active site" or "active center" which is the primary locus of the catalytic process. Such a view, though probably an oversimplification, provides a convenient starting point for an inquiry into the relation of protein structure to enzymatic function.

It seems desirable at this point to differentiate between the catalytic activity as such and the specificity of the enzyme, because, unlike inorganic catalysts or catalysts of simpler molecular structure, enzymes are endowed with a high degree of substrate specificity. Though the same type of bond may be hydrolyzed by an enzyme in different substrates, the chemical environment of the bond being split will determine the specificity requirements of an enzyme. Consider, as an example, the proteolytic enzymes which, as the name implies, catalyze the hydrolysis of proteins. A protein may be considered a repeating pattern of peptide bonds, but each peptide bond will find itself in a different environment, depending on the nature of amino acids contributing the CO and NH groups. Table I indicates the specificity requirements of trypsin, chymotrypsin, and pepsin.

This specificity is related to the binding of substrate by the enzyme, the simplest postulate being that the enzyme has a structure complementary to that of the substrate. Conversely, if the enzyme is presented with a typical substrate, the bond being hydrolyzed can be varied within relatively wide limits provided that the various bonds do not differ greatly in bond strength.

TABLE I. Peptide bonds hydrolyzed by proteolytic enzymes.

| Enzyme | Substrate | Bonds hydrolyzed |
|---|---|---|
| Trypsin | Insulin (oxidized *B*) | −Arg−Gly−; −Lys−Ala− |
| | Ribonuclease (oxidized) | −Lys−Phe−; −Lys−Ser−; −Arg−Asp($NH_2$)−; −Lys−Asp−; −Lys−Asp($NH_2$)−; −Lys−His−. |
| Chymotrypsin | Insulin (oxidized *B*) | −Tyr−Leu−; −Phe−Tyr−; −Tyr−Thr−; (−Leu−Tyr−) |
| | Melanophore-stimulating hormone | −Tyr−Lys−; −Phe−Arg−; Try−Gly−. |
| Pepsin | β-Corticotropin | −Glu−Asp−; −Glu($NH_2$)−Leu−; −Leu−Ala−; −Phe−Pro−. |

TABLE II. Types of bonds hydrolyzed by chymotrypsin.

| | |
|---|---|
| Peptide | $R_1CONH-CHR_2-CO-\nearrow-NH-$ |
| Ester | $R_1CONH-CHR_2-CO-\nearrow-OR_3$ |
| Amide | $R_1CONH-CHR_2-CO-\nearrow-NH_2$ |
| Hydroxamide | $R_1CONH-CHR_2-CO-\nearrow-NHOH$ |
| Hydrazide | $R_1CONH-CHR_2-CO-\nearrow-NHNH_2$ |
| $-C-C-$ | $C_6H_5OH-CH_2-CH_2-CO-\nearrow-CH_2-COOC_2H_5$ |

Thus, chymotrypsin can hydrolyze not only peptide bonds but also amide, ester, hydroxamide, hydrazide, and certain labile C—C bonds (Table II). The feature common to all these substrates is that the surroundings of the bonds are similar and meet the specificity requirements of the enzyme.

Having thus proposed the existence of a catalytic site and a specificity site within the enzyme molecule, one may start the inquiry by posing the following questions:

(1) What is the chemical configuration of the active site responsible for catalysis?

(2) Is the same or a different site responsible for the specificity of the enzyme, and what is the chemical nature of that site?

(3) How many active sites (including both the catalytic and the specificity sites) exist in an enzyme molecule?

(4) What is the function of the remainder of the mass of the enzyme molecule, or, to put this question in an operational frame, how extensively can an enzyme be degraded before it loses its enzymatic activity?

At this point, one should differentiate clearly between the protein portion of the enzyme and the nonprotein portion, which may also be part of the active site. The nonprotein moiety, usually referred to as a prosthetic group, may be an organic molecule of a different kind such as the flavin group in a flavoprotein, it may be a metal ion, or it may be a combination of an organic molecule and a metal ion. Such a prosthetic group and the particular group of the protein involved in the specific interaction are operationally included in the definition of the active site. The important role of metals in these catalytic processes is discussed in recent reviews by Vallee[6] and by Smith *et al.*[7] which should be consulted.

Consider now enzymes which, as far as is known, do not contain a prosthetic group, and where the active site seems to be composed entirely and exclusively of specific amino-acid residues. Such enzymes include certain, but by no means all, proteolytic enzymes. Trypsin, chymotrypsin, and papain are proteolytic enzymes which are believed to be free of nonprotein prosthetic groups, whereas carboxypeptidase is definitely known to be a zinc-metalloenzyme[8] and is not to be considered here. The three proteolytic enzymes mentioned in the foregoing have each been fully characterized with respect to molecular weight and certain other physicochemical properties,[9,10] some of them with respect to their amino-acid composition and partial amino-acid sequences.*,[11] The physicochemical properties of an inactive precursor† of one of these enzymes, chymotrypsinogen, are summarized in Table III. It is clear that chymotrypsinogen has a molecular weight of about 25 000, a well-defined isoionic point, and is composed of a single polypeptide chain with a single half-cystine as the *N*-terminal residue and a single asparagine as the *C*-terminal residue.‡ The 25 000 molecular-weight unit contains 242 amino-acid residues.[12]

The relation of the primary structure of the enzyme to its catalytic function may be elucidated by a study

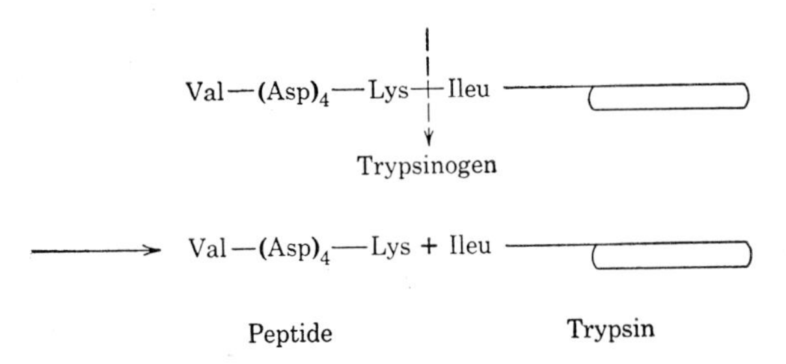


FIG. 1. Schematic representation of the tryptic activation of trypsinogen.

of reactions which involve the hydrolysis of a limited number of peptide bonds ("limited proteolysis"). Two types of reactions may be considered—namely, (1) those which involve the conversion of enzyme precursors to the active form (zymogen activation), and (2) those which lead to the partial degradation of active enzymes.

The controlled hydrolysis of one or more peptide bonds in trypsinogen or chymotrypsinogen causes these zymogens, having potential enzymatic activity, to become converted to active enzymes. As a result of this primary event, an intramolecular rearrangement occurs which presumably leads to the formation of the active site.[9] Figure 1 shows schematically the

TABLE III. Structural characteristics of chymotrypsinogen.

| | |
|---|---|
| Molecular Weight | |
| Chemical analysis | 25 081±336 |
| Sedimentation, diffusion, *p*H 3.0 | 24 400 |
| *p*H 7.5 | 24 000 |
| Light-scattering | 26 100±1000 |
| X-ray | 25 000±800 |
| Osmotic Pressure | 24 000±500 |
| Isoionic point | 9.4 |
| *N* terminal | CySS/2 |
| *C* terminal (Carboxypeptidase, hydrazinolysis) | None detectable in native protein |

* In the case of one enzyme, ribonuclease, the entire sequence of amino acids had been recently determined by Moore, Stein, and co-workers at the Rockefeller Institute.[11] Ribonuclease is not a proteolytic enzyme, but catalyzes the hydrolysis of ribonucleic acid.

† The suffix *-ogen* denotes a precursor of an enzyme which, after certain enzymatic changes, becomes converted into an active enzyme. Such an enzyme precursor is also known as a zymogen.

‡ The *C*-terminal asparagine residue appears to be inaccessible to carboxypeptidase in intact chymotrypsinogen, and it remains inaccessible in the chymotrypsins. It can be detected after rupture of the disulfide bonds, by reduction or by oxidation.

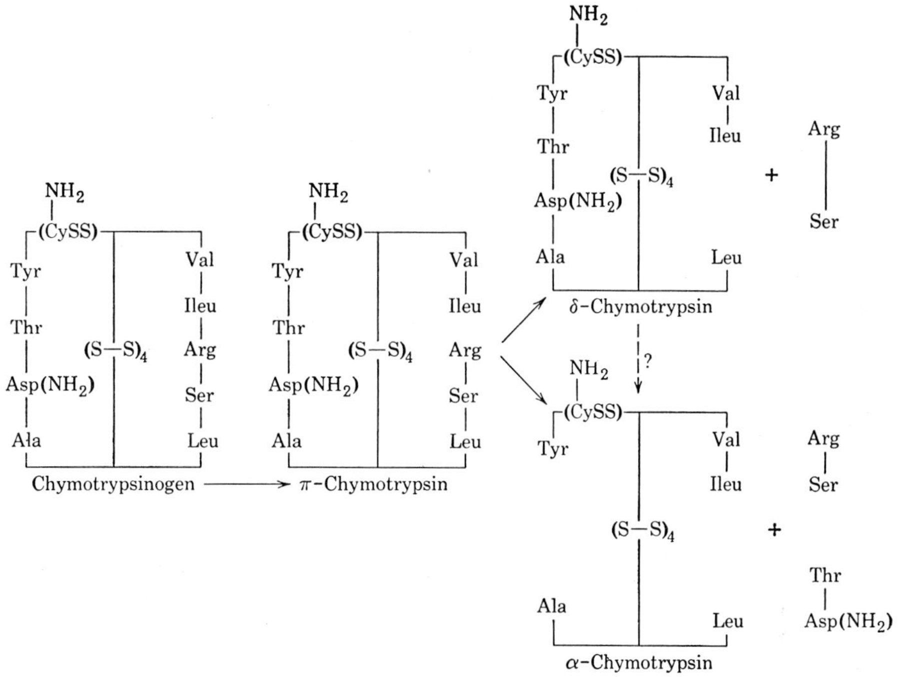

FIG. 2. Schematic representation of the activation of α-chymotrypsinogen. The first step, from left to right, is catalyzed by trypsin, all subsequent steps by chymotrypsin. The formation of δ- and α-chymotrypsins is believed to involve alternate rather than consecutive processes.

conversion of trypsinogen to trypsin. The trypsin-catalyzed hydrolysis of a specific, single peptide bond near the *N*-terminal region of the molecule, causes the hexapeptide, HVal—Asp—Asp—Asp—Asp—LysOH to be released and the *N*-terminal valine residue of the protein to be replaced by an isoleucine group. This seems to be the one and only chemical event which underlies the conversion of the inactive zymogen into the active trypsin enzyme.[13]

Figure 2 illustrates similarly the conversion of chymotrypsinogen to chymotrypsin. Here, only one of the seventeen bonds susceptible to hydrolysis by trypsin needs to be split in order to convert the zymogen into an active enzyme. These are fruitful processes in the sense that an active enzyme is formed from an inactive precursor.

Another example of peptide-bond hydrolysis relates to the partial degradation of an enzyme with or without concomitant loss of biological activity.[14] For instance, the removal from chymotrypsin of a few amino acids from the carboxyl end, with the aid of carboxypeptidase, yields a derivative which is still fully active. Similarly, in the case of ribonuclease, which consists of a single polypeptide chain, removal of as many as three amino acids from the carboxyl terminus yields an active derivative. If, however, four or more amino acids are removed from the same position, the product will be inactive. Another case is that of the ACTH, the adrenocorticotrophic hormone, which contains 39 amino acids. The removal of three amino acids from the *C*-terminal portion of the chain has no effect on hormonal activity, whereas the removal of one or two amino acids from the *N*-terminal portion, by use of leucine aminopeptidase, causes the hormone to become inactive. Perhaps the most dramatic case known today is that of papain, studied by Emil Smith and coworkers,[10] a proteolytic enzyme containing 180 amino acids. One hundred and twenty of these can be removed by stepwise hydrolysis with leucine aminopeptidase, and the derivative still maintains enzymatic activity, suggesting that the active center, including both the catalytic and specificity sites, is located in the carboxyl terminal region of the polypeptide chain and that the remainder of the molecule is not essential for over-all catalytic function.

In order to identify the chemical environment of the active site, it is necessary that it be labeled. In what follows, consider the reactions which are involved in the labeling of the active site of an enzyme and in the subsequent identification of the amino-acid composition of its environment.

The kinetics of the reaction of an enzyme—in this particular case, a proteolytic enzyme reacting with a substrate—indicate that at least two steps are involved. In the first step, the enzyme reacts with the substrate forming an acyl derivative; in the second step, this intermediate then reacts with water to form a free acid, and the enzyme is reconstituted.

$$\mathrm{RCOX+EH \rightarrow RCOE+HX,}$$

$$\mathrm{RCOE+HOH \rightleftharpoons RCOOH+EH.}$$

It is this type of reaction which may be thought to underline the processes which Alberty describes for other systems (p. 177). In the reaction of a proteolytic enzyme with esters, the acyl-enzyme is formed first with the liberation of the alcohol, subsequent hydrolysis

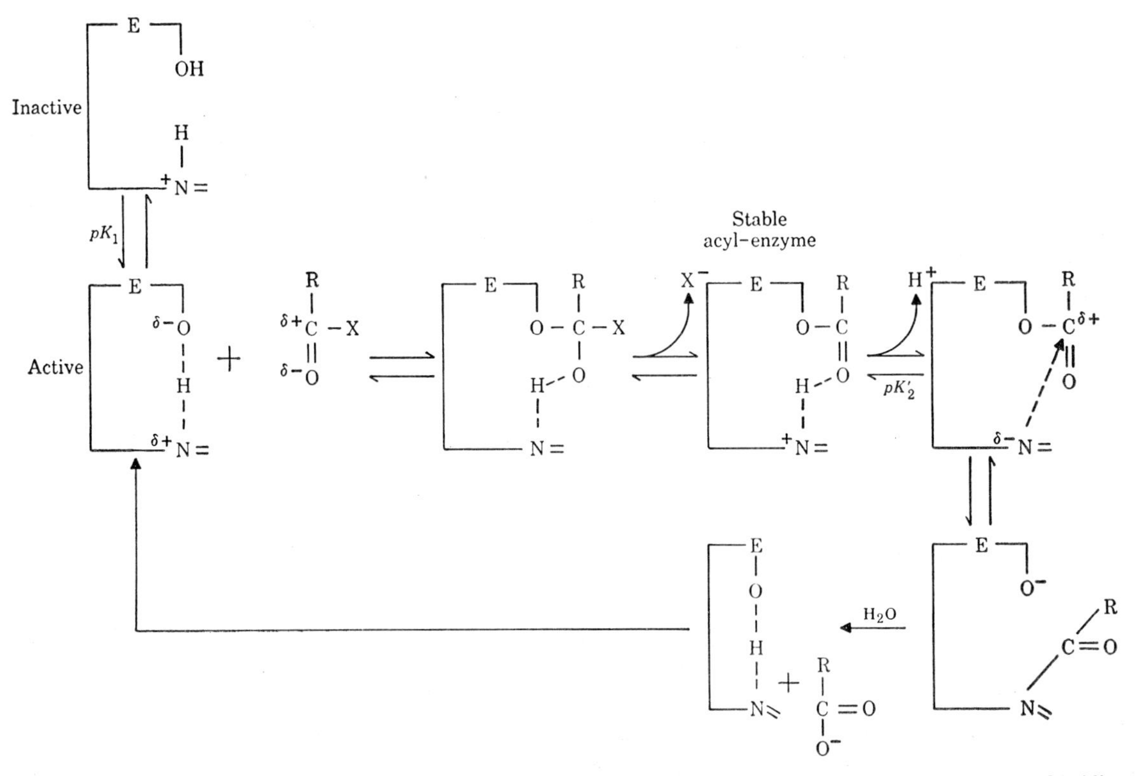


FIG. 3. Proposed mechanism of enzymatic ester hydrolysis, involving the interaction of the imidazole nitrogen of a histidine side chain and the hydroxyl group of a serine side chain.

of the acyl-enzyme leading to the free acid. This type of reaction has been studied in considerable detail, because substrates such as *p*-nitrophenylacetate, which contain a chromophoric group absorbing in the visible region of the spectrum, lend themselves excellently to precise kinetic studies.[15–17]

In the reaction of chymotrypsin or trypsin with organic phosphates such as diisopropyl phosphorofluoridate (DFP), which has been studied for a number of years, a phosphoryl-enzyme is formed instead of an acyl-enzyme. This is a convenient reaction for purposes of labeling, because the second step normally does not occur.

$$\mathrm{EH+F-\underset{\underset{O}{\|}}{P}(OC_3H_7)_2 \rightarrow E-\underset{\underset{O}{\|}}{P}(OC_3H_7)_2+HF} \quad \text{(stable).}$$

It requires the action of a more strongly nucleophilic reagent than water for the phosphoryl group to be removed from the enzyme. It is, therefore, possible to react an enzyme such as chymotrypsin with DFP$^{32}$, leading to the incorporation of the diisopropyl phosphate into the protein, and then to degrade the protein and see where the P$^{32}$ went.

This type of study has shown that, in the case of trypsin, chymotrypsin, cholinesterase and in the case of the enzyme thrombin which is involved in the blood coagulation process, only one mole of DFP was specifically taken up per mole of enzyme, suggesting that these enzymes contain only one active center. In each case, the organic phosphate was found to be esterified to the hydroxyl group of serine. The numerous studies designed to identify the peptide sequence around the reactive serine residue have been described in considerable detail in recent reviews and are not considered here.[9,13,18] Suffice it to summarize, as shown in Tables IV and V, the composition and structure of the peptides obtained from DIP-trypsin and chymotrypsin, respectively. The serine group which is involved here must be

TABLE IV. DIP-peptides from esterases.

| | α-Chymotrypsin | Trypsin | Liver aliesterase |
|---|---|---|---|
| DIP | 1 | 1 | 1 |
| Ser | 1 | 1 | 1 |
| Gly | 2–3 | 2–3 | 2 |
| Ala | ... | ... | 1 |
| Asp | 1 | 1 | ... |
| Glu | ... | ... | 1 |
| Pro | 1 | 1 | 1 |
| Leu or Ileu | 1 | ... | 1 |
| Val | ... | 1 | ... |

TABLE V. Sequence of DIP-peptides.

| | |
|---|---|
| α-Chymotrypsin: | −gly−asp−ser(DIP)−gly−gly−pro−leu |
| Trypsin: | −asp−ser−cyS−glu−gly−gly−asp−ser(DIP)−gly−pro−val−cyS−ser−gly−lys |

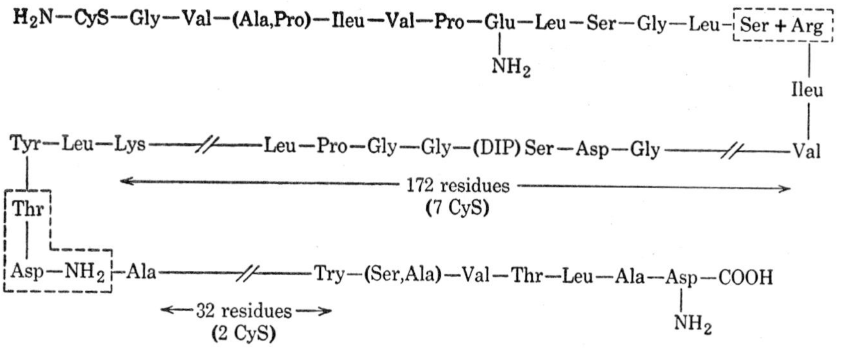

H₂N—CyS—Gly—Val—(Ala,Pro)—Ileu—Val—Pro—Glu—Leu—Ser—Gly—Leu—[Ser + Arg]
NH₂ (on Glu)
Ileu
Tyr—Leu—Lys—//—Leu—Pro—Gly—Gly—(DIP) Ser—Asp—Gly—//—Val
← 172 residues (7 CyS) →
[Thr]
[Asp—NH₂]—Ala—//—Try—(Ser,Ala)—Val—Thr—Leu—Ala—Asp—COOH
← 32 residues (2 CyS) →
NH₂ (on Asp)

FIG. 4. Fragmentary representation of the amino-acid sequence of the single polypeptide chain of α-chymotrypsinogen. The dipeptide sequences delineated by the dotted lines are split off during the activation (see Fig. 2).

of unusual reactivity since there are twenty to thirty serine groups in trypsin and chymotrypsin, respectively, but only one of these is involved in the binding of DFP. Incidentally, the same seryl group is involved in the reaction of these enzymes with acylating agents such as *p*-nitrophenylacetate. Simpler seryl peptides do not behave in this way, and, if the enzyme is first denatured, the reaction with DFP does not take place at all. Conversely, if chymotrypsin is allowed to become acylated at *p*H 5, the acetyl group can be removed at this *p*H by reaction with hydroxylamine; but, if the protein is denatured by 8 molar urea after acylation, the acetyl group cannot be removed until denaturation is reversed, suggesting that separation and uncoiling of polypeptide chains removes in space another group or grouping which confers the high reactivity on this particular serine.[18] A number of lines of reasoning and experimental data has led to the conclusion that the second group involved in the enzymatic reaction is the imidazole group of a histidine side chain. This subject has been considered in detail in a recent review[13] and no attempt is made, therefore, to evaluate the experimental evidence for this hypothesis. The presence of histidine in the active region could account for the special reactivity of an adjacent serine residue, and, in a scheme developed by Cunningham, such an interaction has been proposed to include a hydrogen bonding between serine and histidine, as shown in Fig. 3. According to this scheme, the hydroxyl group of the serine and the nitrogen of the imidazole group of the histidine interact by hydrogen bonding, forming the active configuration; the *p*H dependence of the reaction is related to the *p*H dependence of the formation of these bonds.

The exact position of the serine and histidine within the structure of trypsin or chymotrypsin is unknown. In order to describe the structure of the active site, it would be necessary to know the complete amino-acid sequence, whereas, in fact, only partial sequences can be established, as shown in Fig. 4 for α-chymotrypsinogen. The exact tertiary structure of the polypeptide chain likewise needs to be elucidated. While this goal is still a long way off, nevertheless certain interpretations can be advanced on the basis of the data at hand, as shown schematically in Fig. 5, for the formation and the structure of the active site of trypsin.

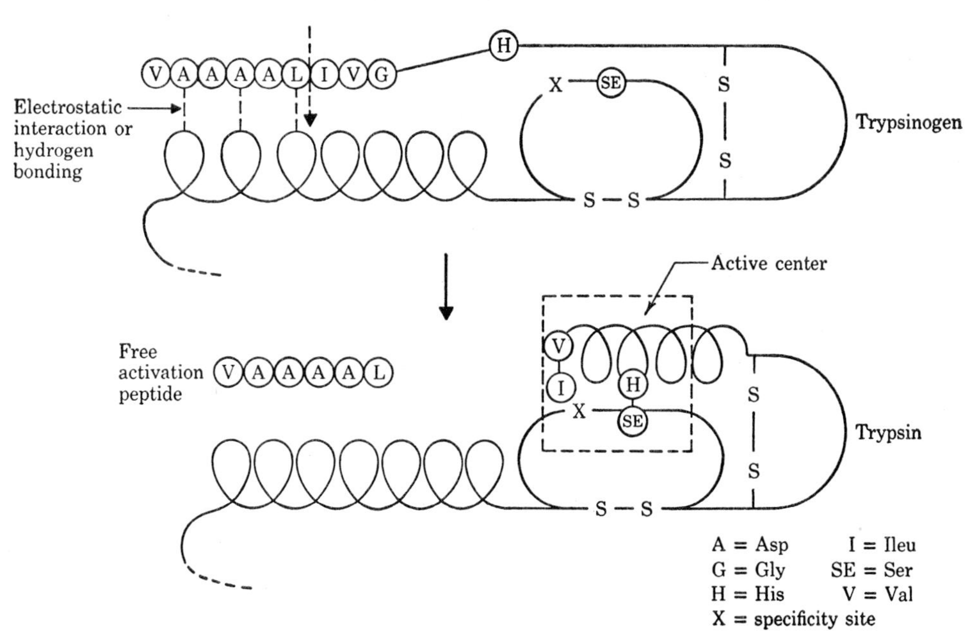


FIG. 5. Schematic representation of the structural changes involved in the tryptic activation of trypsinogen. Rupture of the lysyl-isoleucine bond in the *N*-terminal region (dotted arrow) leads to the liberation of the activation peptide, and causes the newly formed *N*-terminal region of the polypeptide chain to assume a more nearly helical configuration. This, in turn, permits a histidine and serine side chain to come into juxtaposition so as to form the esteratic site (see Fig. 3). The specificity side (X) is believed to be pre-existent in the zymogen molecule.

The active site does not exist prior to activation of trypsinogen (or chymotrypsinogen), although the ability to bind the substrate may be pre-existent in the zymogen molecule. Furthermore, although the serine-histidine interaction would account for a generalized process of ester or peptide hydrolysis, it would fail to account for the high degree of specificity characteristic of proteolytic enzymes. It becomes of paramount importance, therefore, to elucidate those aspects of structure which differentiate between trypsin and chymotrypsin, and which, at the same time, may be correlated with their respective enzymatic specificities.

It is not fruitful to speculate about the reasons why enzymes are as large as they are just as no one has given a physical interpretation for the role of the complex and large structures of cholesterol or porphyrin. The fact that one has just begun to identify the functional relations among a few amino-acid residues in a molecule containing 200 or more does not justify the conclusion that the remainder is without function. On the other hand, the demonstration that *in vitro* one-half of a molecule may have similar if not identical functions as the whole does not lead to the compelling conclusion that, in a complex system such as living cell, the other half is of no consequence. It remains for the future to clarify fully what today can at best be considered as a promising beginning.

## BIBLIOGRAPHY

[1] J. B. Sumner, J. Biol. Chem. **69**, 435 (1926).

[2] J. H. Northrop, M. Kunitz, and R. M. Herriott, *Crystalline Enzymes* (Columbia University Press, New York, 1948), second edition.

[3] M. Dixon and E. C. Webb, *Enzymes* (Academic Press, Inc., New York, 1958).

[4] M. Kunitz, J. Gen. Physiol. **32**, 241 (1948).

[5] M. A. Eisenberg and G. W. Schwert, J. Gen. Physiol. **34**, 583 (1951).

[6] B. L. Vallee, Advances in Protein Chem. **10**, 318 (1955).

[7] E. L. Smith, N. C. Davis, E. Adams, and D. H. Spackman in *The Mechanism of Enzyme Action*, W. D. McElroy and B. Glass, editors (The Johns Hopkins Press, Baltimore, Maryland, 1954), p. 291.

[8] B. L. Vallee and H. Neurath, J. Biol. Chem. **217**, 253 (1955).

[9] H. Neurath, Advances in Protein Chem. **12**, 320 (1957).

[10] J. R. Kimmel and E. L. Smith, Advances in Enzymol. **19**, 267 (1957).

[11] C. H. W. Hirs, W. H. Stein, and S. Moore in *Symposium on Protein Structure*, A. Neuberger, editor (Methuen and Company, London; John Wiley and Sons, Inc., New York, 1958), p. 211.

[12] P. E. Wilcox, E. Cohen, and W. Tan, J. Biol. Chem. **228**, 999 (1957).

[13] G. H. Dixon, H. Neurath, and J.-F. Pechère, Ann. Rev. Biochem. **27**, 489 (1958).

[14] C. B. Anfinsen and R. R. Redfield, Advances in Protein Chem. **11**, 1 (1956).

[15] B. S. Hartley and B. A. Kilby, Biochem. J. **50**, 672 (1952).

[16] H. Gutfreund and J. M. Sturtevant, Biochem. J. **63**, 657 (1956).

[17] G. H. Dixon and H. Neurath, J. Biol. Chem. **225**, 1049 (1957).

[18] H. Neurath and G. H. Dixon, Federation Proc. **16**, 791 (1957).

# 23
# Molecular Structure of the Nucleic Acids

ALEXANDER RICH

*Department of Biology, Massachusetts Institute of Technology, Cambridge 39, Massachusetts*

ONE of the most useful, although simplified, views of the role of the nucleic acids in cells is that they function in a manner analogous to a punched tape in a computer. The punched tape and the nucleic acids are very long elements which direct the larger unit, either computer or cell, in which they are located. They both carry information through the linear arrangement of a few fundamental repeating units along their length. The nucleic acids are very long polymeric molecules which are built up by the repeated connection of a very few small molecules. It is generally assumed that the specificity of nucleic-acid function arises from the particular sequence of its constituent residues, as well as from their geometrical configuration.

There are two classes of nucleic acids, deoxyribonucleic acid (DNA) and ribonucleic acid (RNA), and they differ in structure as well as in function. However, they are made up of similar though not identical chemical units. The fundamental building block of the nucleic acids is the nucleotide. It is a complex molecule consisting of a purine or pyrimidine base, a sugar residue, and a phosphate group. In DNA, the sugar is deoxyribose, while in RNA, it is ribose. These sugars differ by the presence of a hydroxyl group on $C_2'$. Both of the nucleic acids have four types of bases, two purines and two pyrimidines, and three of these are found both in the deoxyribose and in the ribose polymers. DNA and RNA contain the purines adenine (A) and guanine (G), as well as the pyrimidine, cytosine (C) (see Figs. 3 and 4). In addition, RNA has the pyrimidine uracil (U), whereas DNA has the closely related pyrimidine, 5 methyl-uracil (thymine, T). Thus, both of the nucleic acids have a similar chemical composition, and only differ by the presence of a systematic hydroxyl group on each nucleotide of RNA and by the absence of a methyl group on one of the bases. While this description of the chemical composition of the nucleic acids is roughly correct, it should be pointed out that some nucleic acids have modified purine or pyrimidine residues, such as a $-CH_2OH$ group or a glucose residue attached to cytosine in the bacteriophages.

Both ribose and deoxyribose are five-carbon sugars which are in the furanose form—i.e., in the form of a ring involving four of the carbons and one oxygen. The nucleotides of DNA and RNA are connected by the same linkage through the phosphate group which is attached to the $C_3'$ and the $C_5'$ atoms of successive sugar residues. In a schematic way, the polynucleotide chain for both RNA and DNA can be written as shown in Fig. 1. It should be pointed out that the chain in Fig. 1 is asymmetric in that it has a direction which is most easily seen by the sense of the $C_3'-C_5'$ linkage in the sugar residue.

### INFORMATION TRANSFER AND THE NUCLEIC ACIDS

It is generally believed that DNA alone functions as the carrier of genetic information. This understanding is based upon the classic experiments of Avery who discovered bacterial transformation—that is, the ability of purified DNA from one bacterial species to alter the metabolic characteristics of another bacterial species in an inheritable manner. This interpretation of DNA function was further strengthened by the demonstration by Hershey and Chase[1] that DNA alone is the infective component and hence the carrier of genetic information in the bacterial viruses.

All cells of a given organism have the same DNA content. The only exception to this statement is to be found in spermatozoa and ova, where the DNA content is one-half of the normal amount. Further, if one analyzes the chemical composition of the DNA in all tissues of a given animal, it is found to be the same. Thus, it is believed that all cells contain the same set of DNA molecules. The DNA is located in the chromosomal material of the nucleus, and during cell division the DNA is replicated in some manner so that an equal amount is found in the two daughter cells, and with the same chemical composition as that found in the parental cell. One of the goals of molecular structural work in the nucleic acids is to discover the fundamental interpretation for these phenomena.

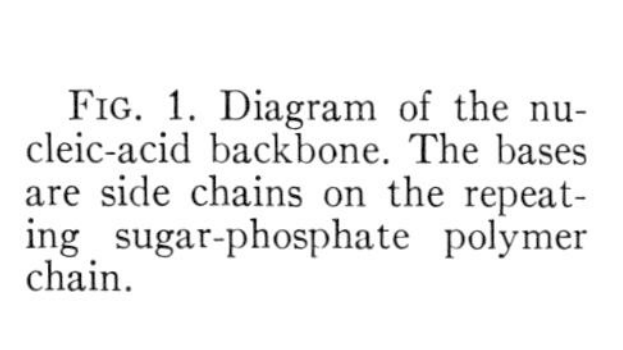

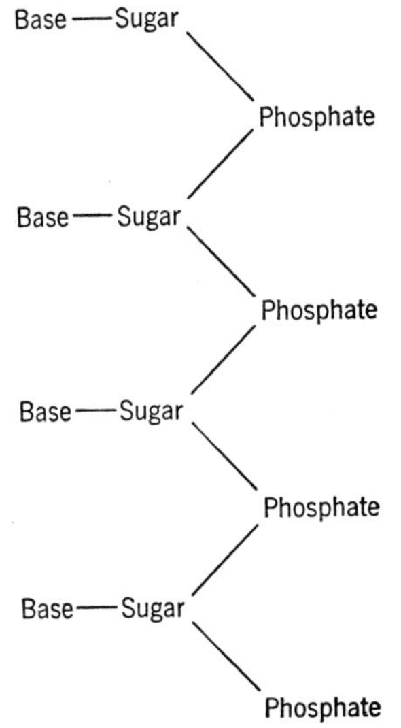


FIG. 1. Diagram of the nucleic-acid backbone. The bases are side chains on the repeating sugar-phosphate polymer chain.

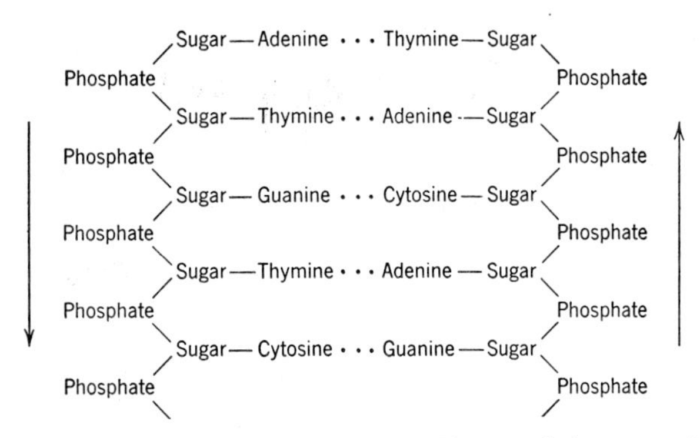


FIG. 2. Schematic diagram of DNA. The two chains are antiparallel, as shown by the arrows. The dotted lines between the bases represent hydrogen bonding. Although the chains are drawn as flat in the diagram, they are actually wound around each other in the molecule.

Although the amount of DNA is the same in each cell of a given organism, the amount varies from species to species. In general, the more complex species have more DNA. A bacterial cell has the order of $10^8$ nucleotides in its DNA, which would make a molecular strand about 2 cm in length. In mammalian species, there are the order of $10^{10}$ nucleotides which corresponds to a total molecular length of 1 to 2 m/cell. Thus, the actual length of the primary coding material in a living cell is in the range of macroscopic dimensions and is much longer than the metabolic machine (or cell) which it directs. The same is often true for the punched tapes which are fed into computers.

## MOLECULAR STRUCTURE OF DEOXYRIBONUCLEIC ACID

One of the most stimulating suggestions in molecular biology was a proposal made by Watson and Crick[2] that the molecular structure of DNA may consist of two polynucleotide chains helically wrapped around each other, with the sugar-phosphate chain on the outside and the purine and pyrimidine bases on the inside. They suggested that the purine and pyrimidine bases from the two chains are joined by hydrogen bonds to form specific pairs. Thus, the adenine residue hydrogen bonds with thymine, and guanine hydrogen bonds with cytosine. These hydrogen-bond pairs are specific in that only these combinations have the necessary stereochemistry to fit into the repeating lattice formed by the regular helical polynucleotide chains. In a schematic way, the DNA molecule is illustrated in Fig. 2 which shows the pairing relationship between the two polynucleotide chains. The arrows indicate the direction of the sugar-phosphate backbone. The two strands are organized in an antiparallel fashion so that the molecule looks the same even if it is turned about by 180°. If one ignores the varied base sequence, the backbone sugar-phosphate chains are organized about a diad axis perpendicular to the fiber axis and passing through the center of each base pair. The pairing of the bases is shown in Figs. 3 and 4. An important feature of the Watson-Crick hypothesis is the identity of the two types of base pairs. That is, the distance between the two sugar-phosphate chains must be the same both for the adenine-thymine pair and for the guanine-cytosine pair. In this way, both base pairs could fit into the helix interchangeably. Pauling and Corey[3] have made a critical survey of x-ray diffraction results obtained from crystals containing purines and pyrimidines. From this, they concluded that the cytosine and guanine residues are probably held together by three hydrogen bonds (Fig. 4), while the adenine-thymine residues are held together by two. The dimensions of the hydrogen-bond pairs suggested by Pauling and Corey are shown in Figs. 3 and 4. Within experimental error, the positions and angles of the two chains relative to the base pairs are identical.

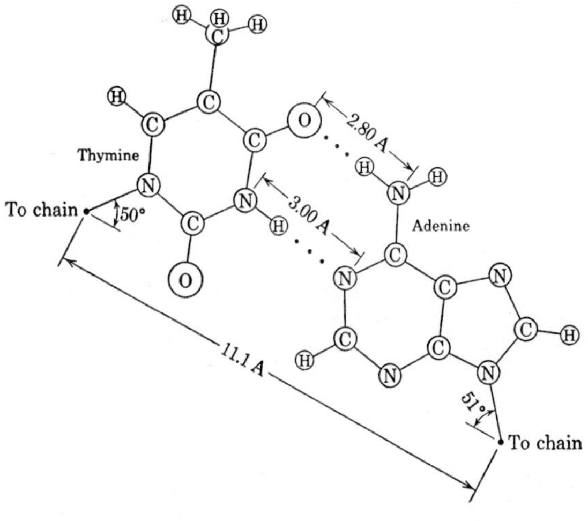

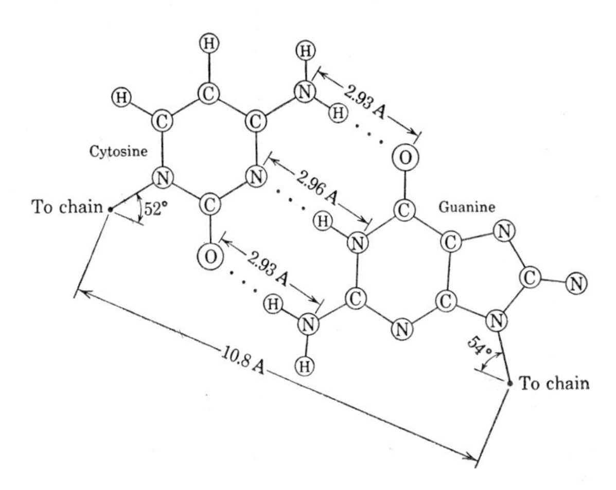


FIG. 4. Diagram showing the hydrogen bonding between guanine and cytosine in DNA. This pair is held together by three hydrogen bonds in contrast to the two found in the adenine-thymine pairing [from L. Pauling and R. B. Corey, Arch. Biochem. Biophys. **65**, 164 (1956)].

FIG. 3. Diagram showing the hydrogen bonding between adenine and thymine in DNA. The dimensions of the base pair are those discussed by Pauling and Corey.[3]

Early x-ray diffraction studies by Wilkins[4] and by Franklin,[5] and their collaborators showed that DNA is a helical molecule with 10 residues per turn. The gross features of the diffraction pattern were, at this early stage, shown to be compatible with the double-stranded form proposed by Watson and Crick. Wilkins and his collaborators have continued to carry out extensive studies on the diffraction patterns of DNA, and they are responsible for most of the knowledge concerning the detailed configuration of DNA.[6–8]

The DNA molecule exists in several forms. At lower relative humidities (about 70%), the molecule crystallizes in the *A*-form—that is, a face-centered monoclinic lattice with $a=22.2$ A, $b=40.0$ A, $c=28.1$ A, and $\beta=97.1$ A. This unit cell contains a repeat unit of two DNA molecules with the helical axis along $c$. The water content of this lattice is about 40%, and the bases in it are tilted about 25° from the fiber axis. In this form, the DNA is a true crystal and produces about 100 independent reflections. This implies that there is a high degree of regularity in the structure in all directions.

At higher relative humidities, the *B*-form which is paracrystalline appears with the molecules all parallel to each other but with random rotation about their molecular axes. The layer lines on these paracrystalline diffraction patterns show a continuous distribution of scattering intensity rather than the sharp spots characteristic of a crystalline lattice. In the *B*-form, the fiber-axis repeat is 34.6 A, and there is an extremely strong x-ray reflection on the meridian at 3.4 A. More recently, Wilkins and his collaborators[8] have been able to obtain truly crystalline diffraction patterns of the lithium salt of DNA in the *B*-configuration. In this form, the lithium DNA crystals are orthorhombic and have the dimensions $a=22.7$ A, $b=31.3$ A, and $c=33.6$ A. The axis of the helical molecules is along $c$, and two molecules pass through the unit cell.

The *B*-form of the DNA molecule is shown in Fig. 5 where atoms have been drawn with approximately their van der Waals radii. The base pairs are shown horizontally in the middle of the diagram, and the two sugar-phosphate chains are helically wrapped around the stacked bases. As can be seen in the diagram, there are two helical grooves which go round the DNA molecule. One of them is wider than the other because of the orientation of the sugar-base bonds shown in Figs. 3 and 4. The phosphates on the outside of the molecule are found at a radius of 9 A, and they are just over 7 A apart along a given chain. Wilkins and his co-workers have studied the organization of the polypeptide chain of protamine in a DNA-protamine combination.[7] The polypeptide chain is arranged as a third coaxial helix which fills the narrow groove in the DNA structure. In this position, the positively charged arginine side chains can interact with the negatively charged phosphate groups and stabilize the molecule.

Whenever a parent cell divides, the genetic information in that cell has to be replicated in order to insure continuity of inheritance to the daughter cell. The structural model of DNA suggested to Watson and Crick[9] a method for the replication of this molecule. They felt that the parent molecule could unwind as shown in Fig. 6 so that its two strands separated. These individual strands could then serve as a template for organizing the individual nucleotides which are necessary to make the second strand of the DNA daughter molecules. The template specificity is assured by the specificity of the hydrogen bonds between the purine-pyrimidine base pairs. In this way, a single molecule could twist about its axis, and two daughter helixes would form on the unraveled ends. Although this molecular model of genetic replication has not been established, recent experiments suggest that it is probably correct. This is a good example of the way in which a molecular structure can suggest a molecular mechanism.

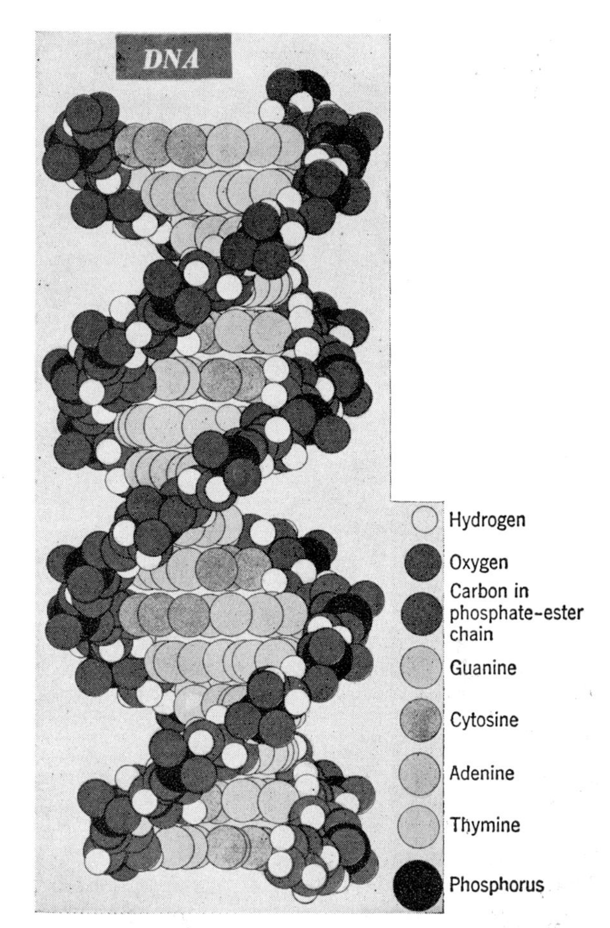


FIG. 5. A drawing of the DNA molecule using solid circles to illustrate atoms. It can be seen that there are two helical grooves of unequal size on the outside of the DNA molecule (after M. Feughelman *et al.*[7]).

## ROLE OF RIBONUCLEIC ACID

In addition to carrying out its replication activities, it is necessary for the DNA molecule, acting as a genetic material, to influence and guide the metabolism of the

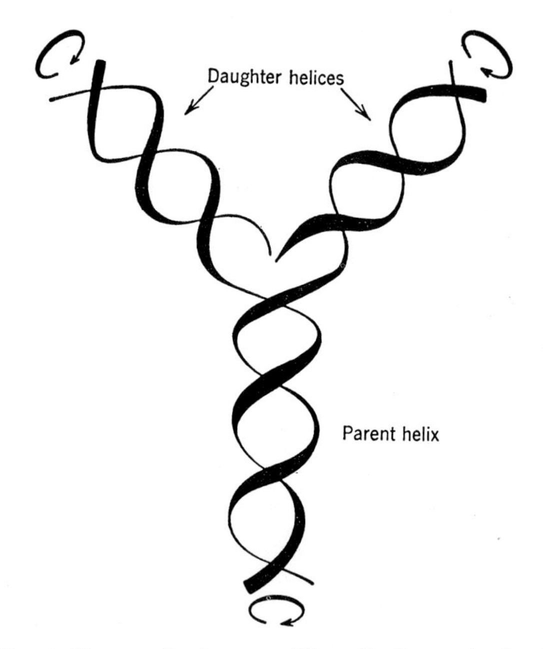


FIG. 6. Diagram showing a possible replication mechanism for DNA. The parent helix unwinds, and the two separated strands serve as sites for forming two daughter molecules. Both parent and daughter helices must wind simultaneously [from M. Delbrück and G. S. Stent in *The Chemical Basis of Heredity*, W. D. McElroy and B. Glass, editors (The Johns Hopkins Press, Baltimore, Maryland, 1957), p. 699].

cell. It is not known for certain how this is carried out. However, there is a large body of indirect information which suggests that this is carried out by using another nucleotide polymer which is somewhat similar to DNA—i.e., ribonucleic acid, which probably acts as an intermediary between DNA and the proteins which are synthesized and constitute the working chemical machinery of the cell.

The relationship between RNA and protein synthesis is not completely worked out at the present time. Proteins appear to be polymerized from their amino acids at the site of the small particulate bodies which are known as microsomal particles. These units are widely distributed in the cytoplasm, and appear as approximately spherical particles with a diameter of 180 A. They usually contain about half RNA and half protein, and it is likely that the RNA component plays a fundamental role in the protein-synthetic process, such as partly or wholly determining the sequence of amino acids. Thus, in order to control protein specificity, the information present in the DNA molecule in its nucleotide sequence may pass through the RNA molecule before ultimately emerging in a particular sequence of amino acids which define a particular protein. If these views are correct, one would like to know how it is that the DNA molecule "makes" RNA and how in turn this RNA molecule organizes amino acids. Unfortunately, at the present time, these questions cannot be answered.

The role of RNA in the metabolic cycle is not simple. In addition to being implicated in protein synthesis, it has also been demonstrated that RNA can function as a carrier of genetic information in a manner very similar to that of DNA. The pure RNA isolated from the tobacco mosaic virus is capable of infecting the tobacco leaf, and this infection ultimately produces a large number of new virus particles which carry the same genetic markers as those of the original virus.[10,11] Thus, RNA has some functions in common with DNA, but also appears to have unique activities as well.

Unfortunately, it is not possible to describe the molecular structure of RNA in a final manner, as is the case for DNA. RNA has been isolated in fibrous form and x-ray diffraction photographs taken of this material show a helical diffraction pattern with a fiber-axis repeat of approximately 26 to 28 A and strong meridional reflections at 3.3 and 4.0 A.[12] However, the diffraction photographs are not of sufficiently high resolution to allow one to make a unique structural interpretation. Experimental work on the structure of RNA offered little hope of achieving a solution to this molecular structural problem until the discovery of an enzyme capable of polymerizing synthetic polyribonucleotides.

## Synthetic Polyribonucleotides

Grunberg-Manago and Ochoa[13] discovered an enzyme which converts nucleotide diphosphates into polyribonucleotides. The enzyme removes the terminal phosphate groups from the diphosphates and assembles the resultant nucleotide residues to form polyribonucleotides. Polymers obtained in this fashion resemble naturally occurring ribonucleic acid in that they have the same covalent ribophosphate backbone, and they have been shown to undergo similar enzymatic hydrolysis.[14] Polymers have been made from all of the purine and pyrimidine bases which are found in RNA. In addition, the enzyme will also make polymers which contain other purine or pyrimidine bases; e.g., polyinosinic acid has been made which contains the purine hypoxanthene, and recently a polymer has been made which contains thymine—i.e., a normal constituent of DNA but not of RNA. These polynucleotide molecules can be made either as pure molecules involving only one residue, or as copolymers involving two or more of the purine-pyrimidine side chains. The similarity between RNA and the synthetic polyribonucleotides can be shown by an x-ray diffraction study of synthetic copolymers, since they produce an x-ray diffraction pattern identical with that of native RNA.[15] This suggests that it might be possible to study the molecular configuration of the synthetic polyribonucleotides and thereby learn something about the configuration of naturally occurring RNA.

The synthetic polyribonucleotides are very reactive molecules. Soon after they were polymerized, it was shown that a complex formed when polyadenylic acid was mixed with polyuridylic acid.[16] Using x-ray diffraction analysis, it was found that these two molecules wrap around each other in solution to form a two-stranded helical molecule very similar to naturally occurring DNA.[17] The discovery of this remarkable interaction has been followed by a variety of similar discoveries among the other polynucleotides. At present, we know of the existence of several of these elongate macromolecules which form two-stranded and three-stranded helical complexes.

### Formation of Synthetic Two-Stranded Helical Molecules

If a dilute salt solution at neutral *p*H contains both polyuridylic acid (Poly U) and polyadenylic acid (Poly A), these two molecules complex together. The reaction is shown schematically in Fig. 7, where the bands represent the polynucleotide chains. On meeting each other, the molecules wrap about to form a two-stranded helix. Evidence for this reaction can be obtained from an x-ray diffraction study of a fiber drawn from a lyophilized mixture of the two polymeric species. The fiber has strong negative birefringence, and produces an x-ray diffraction pattern which has many similarities to a diffraction pattern of DNA. The distribution of scattering intensity is that which is characteristic of a helix: it has a large area on the meridian which is clear, and the scattering intensity is distributed in the form of a "cross" through the origin. The layer-line spacing varies slightly with humidity.[18] However, both DNA and (Poly A+Poly U) have a layer-line spacing of 34 A. This spacing represents the helical pitch of the molecule. From the strong meridional reflections in the region of 3 to 4 A, it can be shown that there are 10 residues per turn of the helix in both DNA and the (Poly-A+Poly-U) molecules. The birefringence of both materials is identical when the (Poly-A+Poly-U) molecules crystallize in a hexagonal lattice with a distance between the molecules of 28.8 A. This is approximately 6 A greater than that observed for the DNA molecule.

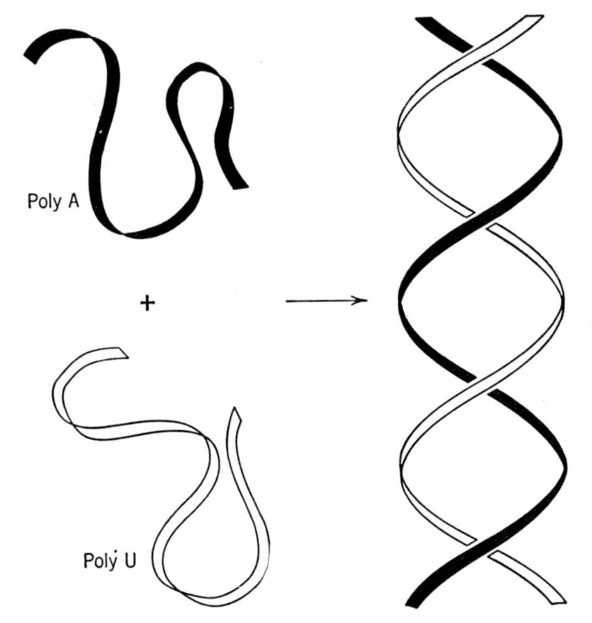


FIG. 7. Diagram illustrating the chemical reaction between the two polymers, polyadenylic acid (A) and polyuradylic acid (U). The irregular contours on the left represent the molecules in a random-coil configuration. After reacting on the right, the molecules are organized into a regular two-stranded helix. The bases connecting the two molecules are not shown in this diagram.

With the exception of the diameter of the molecule, the two diffraction patterns are similar enough to suggest that they arise from a similar helical structure. In the solution, the adenine residues of polyadenylic acid meet with and hydrogen bond onto the uracil residues of polyuridylic acid in a way which is identical to the kind of hydrogen bonding which occurs between adenine and thymine in DNA (Fig. 3). The only difference is that the uracil does not have the methyl group which is found on thymine. However, this does not affect the hydrogen bonding. Since the remainder of the molecule is similar to DNA, it forms the stablest structure possible—i.e., a DNA-like configuration. There is an additional hydroxyl group in the sugar residue of the polyribonucleotides relative to DNA, and this increases the diameter of the molecule slightly through its interaction with the other atoms of the sugar ring. This alters the hexagonal spacing mentioned above.

There are other methods which can be used to study the interaction between these two molecular species. When they react in solution, the optical density at 259 m$\mu$ decreases. This effect has been utilized in a quantitative study of the interaction.[19] A series of mixtures of polyadenylic acid and polyuridylic acid is made wherein the total concentration of phosphate groups remains constant, but the mole ratio of the two species varies continuously. The optical density is measured for this continuous series of solutions and the results are shown in Fig. 8. The dashed line shows the optical density of various mixtures of polyadenylic acid and polyuridylic acid at neutral *p*H in 0.1 molar sodium chloride, plotted as a function of mole ratio. It can be seen that the optical density falls quite sharply, and a minimum is reached at 50% mole ratio when the number of adenine residues in the solution just equals the number of uracil residues. This strongly suggests that a new species is being formed which is a 1:1 mixture of the two polymeric molecules. This interpretation is reinforced by studying this reaction in an ultracentrifuge, since there is an increase in molecular weight and sedimentation velocity when the two molecules combine.

Making careful measurements of the type shown in Fig. 8 for the 1:1 complex, it can be shown that over 95% of the residues have reacted, as judged by the sharpness of the drop in optical density. This is a measure of the high equilibrium constant for the reaction. One of the consequences of this figure is the inference that the reaction must be reversible in order to have all of these residues react.[20]

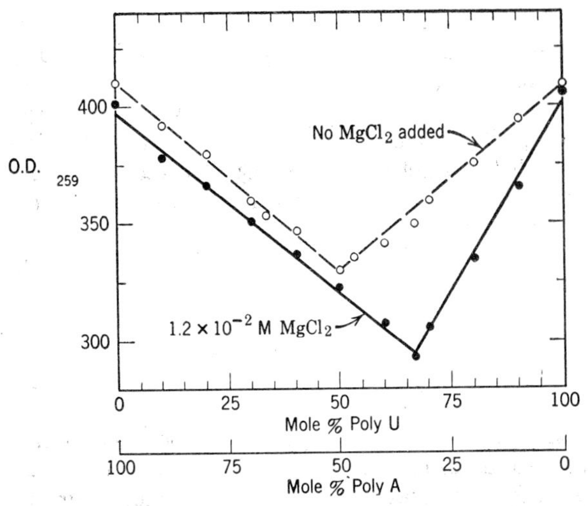


FIG. 8. The optical density of various mixtures of polyadenylic acid (A) and polyuradylic acid (U) at 259 mμ. The total number of moles of polymer is constant for all points, but the ratio of molecular species varies as indicated. All solutions are in 0.1 molar sodium chloride at neutral *p*H. The dashed line shows the formation of a 1:1 complex. The addition of a small number of divalent cations induces the formation of the three-stranded molecule [from G. Felsenfeld and A. Rich, Biochim. et Biophys. Acta **26**, 457 (1957)].

These experiments with mixtures of polyadenylic acid and polyuridylic acid clearly demonstrate the stability of the DNA configuration. In addition, they show that it is possible for the RNA covalent backbone to assume the form of a two-stranded complimentary duplex of the DNA type. This is significant because the work mentioned above on tobacco mosaic virus showed that the RNA molecule from the virus carries the genetic information residing in the virus. The molecule is also probably capable of carrying out the molecular replication necessary to virus multiplication in the leaf. In view of the fact that the RNA backbone can assume the DNA configuration, it seems quite reasonable to assume that the molecular replication of RNA may be carried out by a mechanism very similar to that involved in the molecular replication of DNA.

The reaction between polyadenylic acid and polyuridylic acid was perhaps not completely unexpected in view of the fact that DNA is composed of two strands held together by hydrogen-bonded purine-pyrimidine base pairs. Since uracil is so close to thymine, it is expected that the stability of (Poly A+Poly U) may be related to the stability of DNA itself. However, it is perhaps unexpected to find that it is possible to make a stable two-stranded helical molecule composed of polyadenylic acid and polyinosinic acid—i.e., a molecule very similar to the DNA molecule, except that it has purine-purine base pairs instead of purine-pyrimidine base pairs.

The evidence for this combination parallels that mentioned in the foregoing for polyadenylic acid and polyuridylic acid. When polyadenylic acid is mixed with polyinosinic acid under appropriate conditions, a lowering of the optical density occurs producing a minimum at the 1:1 mole ratio point, just as shown in Fig. 8.[21] Further evidence for the formation of this complex is also seen in ultracentrifuge experiments, since the complex has a larger molecular weight and sedimentation constant than either original molecule.

An x-ray diffraction pattern of a fiber of polyadenylic acid plus polyinosinic acid (Poly A+Poly I) is similar also to the *B*-form of deoxyribonucleic acid. The (Poly-A+Poly-I) molecules crystallize in a hexagonal array on the equator with $a = 24.4$ A. The fundamental screw operation for generating the (Poly-A+Poly-I) helix is a translation of 3.4 A and a rotation of 31.5°, just slightly less than the DNA and the (Poly-A+Poly-U) molecules.

The purine base in polyinosinic acid is hypoxanthine. It is closely related to guanine in that it has an oxygen on $C_6$ of the purine ring, even though it lacks the amino group present in guanine on $C_2$. It is likely that the hypoxanthine is in the keto tautomeric form and that it hydrogen bonds to the adenine residue as shown in Fig. 9. The keto oxygen of hypoxanthine is hydrogen bonded to the amino group of adenine, while the hydrogen on $N_3$ of hypoxanthine is bonded to the corresponding ring nitrogen in adenine. As can be seen when comparing this with Figs. 3 and 4, the hydrogen-bonding system has some similarities to what is observed in DNA. The major difference is the additional imidazole ring present in the hypoxanthine base. The hydrogen bonding shown in Fig. 9 could be used in the naturally occurring nucleic acids if the hypoxanthine base were replaced by guanine, since the additional amino group attached to $C_2$ of the purine ring would not introduce any steric interference.

## Three-Stranded Helical Molecules

It was found that small amounts of divalent salts had a profound effect on the optical density-composition diagram in the Poly-A and Poly-U system.[22] The solid line in Fig. 8 shows the change brought about by making the solution $10^{-2}$ molar in magnesium chloride. A new minimum appears at 67% polyuridylic acid and 33%

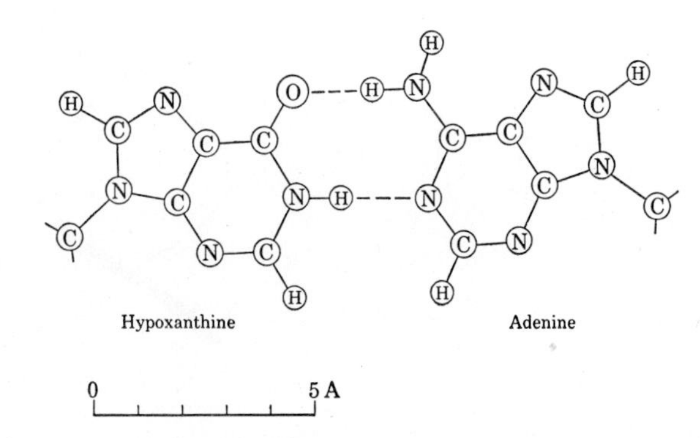


FIG. 9. Diagram showing the hydrogen bonding between the adenine base of polyadenylic acid and the hypoxanthine base of polyinosinic acid [from A. Rich, Nature **181**, 521 (1958)].

polyadenylic acid. The new minimum is quite sharp, indicating that the new complex is quite stable. This was the first indication that a three-stranded complex was forming from two strands of polyuridylic acid and one of polyadenylic acid.

Additional evidence for the formation of the three-stranded molecule is obtained by studying the sedimentation constant of the 1:1 complex as compared with the 2:1. The mean sedimentation constant for the three-stranded complex is about 50% greater than for the two-stranded complex. This increase would be expected if the two-stranded molecule were to take on a third strand. The third strand probably fills the deep helical groove in the two-stranded molecule which is similar to the deep groove seen in DNA. Since it would displace water molecules from that site and would not appreciably alter the frictional forces or shape factor of the molecule, the net density increment of the molecule over the solvent would result in approximately a 50% increment in sedimenting velocity.

It has been suggested that the second uracil residue is hydrogen bonding to the original adenine-uracil pair by forming two strong hydrogen bonds onto $N_7$ and $N_{10}$ of the adenine ring. Such an additional third strand would not involve an increase in radius or helical pitch of the molecule, but could account for the approximately 50% increase in sedimentation velocity. X-ray diffraction photographs have also been obtained from this three-stranded complex.

The kinetics of the formation of the three-stranded molecules from a mixture of two-stranded (Poly-A + Poly-U) molecules and single-stranded Poly-U molecules have been investigated by measuring the optical density at 259 m$\mu$ as a function of time. Study of these curves for various concentrations of magnesium chloride or manganese chloride has shown that the reaction is second order for divalent cations; i.e., two divalent cations are present for each triplet of bases.

It is important to note that divalent cations do not have a unique role in forming the three-stranded molecule, since this complex will fully form in sodium-chloride solutions which are 0.7 molar. Thus, it is likely that the cations are necessary to overcome the electrostatic repulsion between the negatively-charged phosphate groups in the three polynucleotide chains. Divalent cations are much more effective than monovalent cations, probably because they form stable complexes with phosphate groups. Nonetheless, they are not necessary, since monovalent cations alone are capable of carrying out this reaction.

In a completely analogous fashion, the two-stranded polyadenylic-acid plus polyinosinic-acid molecule will take on a third strand of polyinosinic acid to become three-stranded.[21] This has been shown both spectrophotometrically as well as in the ultracentrifuge.

Up to this point, the discussion has been concerned with two- and three-stranded molecules composed of different kinds of residues hydrogen bonded together. However, polyinosinic acid forms another kind of helical structure which involves only one type of molecule.[23]

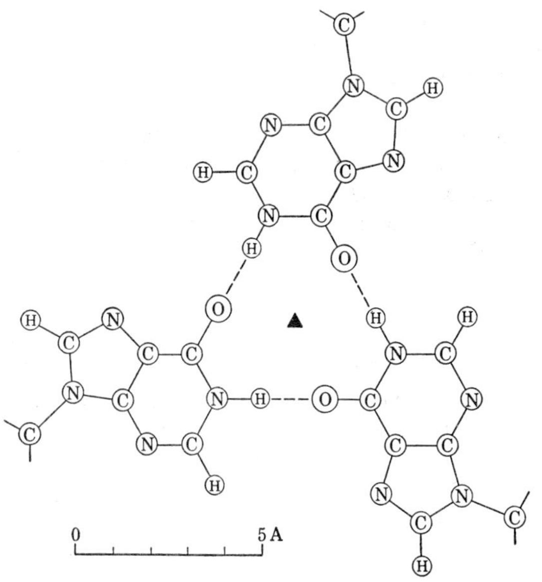

FIG. 10. Diagram showing the hydrogen bonding between three hypoxanthine bases in the three-stranded model of polyinosinic acid. The molecules are organized around a threefold rotation axis [from A. Rich, Biochim. et Biophys. Acta **29**, 502 (1958)].

If polyinosinic acid is prepared as a high molecular-weight polymer, it can be drawn into an oriented fiber which is negatively birefringent and which produces an unusual diffraction pattern when compared with the diffraction photograph of the mixtures described above. One unusual feature is that the first layer line is found at a spacing of 9.8 A, in contrast to the 30- to 40-A spacings discussed in the foregoing. Even more unusual is the appearance of the second layer line at 5.2 A and of the third intense meridional layer line at 3.4 A. These are nonintegral; i.e., they are not successive orders of one fundamental repeat distance. This feature and the fact that the first and second layer lines do not appear on the meridian of the diagram point uniquely to a helical configuration for the molecule. The meridional reflection at 3.4 A is undoubtedly attributable to the stacking of the purine residues at right angles to the fiber axis in agreement with the negative birefringence. The largest equatorial reflections occur at a spacing of 23.8 A.

This is an example of a helical diffraction pattern which points to a multiple-stranded structure, largely because it will not fit any single-stranded model. A three-stranded model will fit the diffraction data if all three strands are parallel to each other. The bases in this model are organized around a threefold rotation axis, utilizing the hydrogen-bonding system shown in Fig. 10. Here, the three hypoxanthine residues are hydrogen bonded in a cyclic manner, so that the keto oxygen attached to position 6 in the purine ring is hydrogen bonded to the nitrogen at position 1 on an adjoining ring. The threefold rotation axis has the effect of decreasing the repeat distance along the fiber axis from 29.4 to 9.8 A.

Polyinosinic acid in solution can be converted from an organized helix into a random coil, and vice versa. This alteration is characteristic of all the polynucleotide molecular complexes. There are certain conditions under which they form stable, organized aggregates, and other conditions under which they separate in solution and are no longer organized. It is only when polyinosinic acid is in a single-chain random coil that it is able to react with polyadenylic acid to form the two- and three-stranded molecules already described; if polyadenylic acid and polyinosinic acid are mixed together in high salt concentration, no reaction occurs at all.

Several other combinations of polynucleotides have been studied and are mentioned only. Polyadenylic-acid chains combine with themselves to form two-stranded helixes in which both chains are parallel to each other. Polyinosinic acid, in addition to combining with itself and with polyadenylic acid, will also combine with polycytidylic acid.[24] Recently, it has been possible to synthesize polyribothymidylic acid.[25] This is a polymer with an RNA backbone even though it contains a DNA base. This molecule also combines with polyadenylic acid to form two- and three-stranded molecules, in a manner analogous to the combination with polyuridylic acid.[26]

## DISCUSSION

The synthetic polyribonucleotides are extraordinarily reactive, and they can readily form a variety of multi-stranded helical structures, in some cases with molecules which are all alike, and in other cases with molecules which are different. There are, however, some generalities which emerge from this study which undoubtedly reflect some of the fundamental stabilizing features found in helical nucleic-acid molecules. For example, all of the helical molecules discovered so far have been either two- or three-stranded, but in no case has there been a stable single-stranded molecule. In the structures described in the foregoing, including DNA, the over-all architecture has been similar in that the purine or pyrimidine bases appear to be largely on the inside of the molecule, whereas the charged sugar-phosphate chain is on the outside of the molecule. The bases are hydrogen bonded together, usually with two or occasionally three strong hydrogen bonds holding each base. The molecules are stabilized by the van der Waals packing of the stacked planar purines or pyrimidines, in addition to the hydrogen bonding. It may be that a single-chain helix cannot form in a polyelectrolyte molecule such as the nucleic acids, because of the electrostatic repulsion. When two or more chains are present, the electrostatic repulsion in the helical molecule usually tends to pull the two chains closer to each other since they are coiled. In a single-stranded molecule, the effect of electrostatic repulsion would be only to elongate the molecule and break up any organized helical structure.

All of the synthetic organized helices are formed reversibly, and they can be made to assume a random coil in solution when conditions are appropriately modified. In most cases, a simple reduction of ionic strength is enough to drive the two chains apart; in other cases, altering the *p*H is sufficient. Usually, there is only a given *p*H range over which a multistranded molecule is stable. For example, the polyinosinic-acid helix breaks up when the *p*H is raised above 10 owing to the fact that a proton necessary in the hydrogen-bonding system is removed. Similarly, the other helices are stable only over *p*H ranges in which the necessary tautomeric forms exist. Studies such as these are useful in developing an understanding of the stability of the naturally occurring nucleic acids and, since several of these molecules are similar to DNA, they can be used as model systems for studying nucleic-acid reactivity.

In this regard, an especially attractive hypothesis may be made concerning the three-stranded molecules. It is possible that these may be analogs for the formation of a single-stranded RNA by a two-stranded DNA. DNA has a deep helical groove in it between the two strands. It is this groove which is filled in the (poly-adenylic plus polyuridylic) molecule by the oncoming strand of polyuridylic acid, and, as such, it may be an example of a physiologically important type of reaction. Since DNA itself has two kinds of base pairs, there are a total of four different sites on the DNA molecule. These four types of sites may serve as templates for the four kinds of ribonucleotides which must be polymerized together to make the RNA molecule. Schemes such as this have been worked on by several investigators for many years, since it is an attractive and simple method for transferring sequence information from DNA to RNA. However, there has been as yet no convincing demonstration of a detailed molecular mechanism which has the requisite specificity. Further work is necessary to fully evaluate mechanisms of this sort.

Despite the variety of polynucleotide structures which are now understood, the structure of naturally occurring RNA itself remains unknown. There is undoubtedly a difference in the configuration of the RNA which is found in the microsomal particle from that found in the small RNA molecules present in the soluble supernatant. In addition, it is possible that the nuclear RNA is in yet a different configuration. The most interesting configuration for RNA is perhaps that which is found in the microsomal particle. According to the current hypothesis, the nucleotide sequence in this molecule is translated by some means into the amino-acid sequence in the proteins which are synthesized at that site. The configuration of RNA in these particles is indeed an interesting problem, and there is no way of knowing whether or not the studies on the synthetic polyribonucleotides will yield an answer which is at all applicable to the problem. Nonetheless, these studies are producing a variety of structures and configurations and it can only be hoped that future work will yield promising results in this most interesting aspect of the problem.

## BIBLIOGRAPHY


[1] A. D. Hershey and M. Chase, J. Gen. Physiol. **36**, 39 (1952).

[2] J. D. Watson and F. H. C. Crick, Nature **171**, 737 (1953).

[3] L. Pauling and R. B. Corey, Arch. Biochem. Biophys. **65**, 164 (1956).

[4] M. H. F. Wilkins, A. R. Stokes, and H. R. Wilson, Nature **171**, 738 (1953).

[5] R. E. Franklin and R. G. Gosling, Nature **171**, 740 (1953).

[6] M. H. F. Wilkins, W. E. Seeds, A. R. Stokes, and H. R. Wilson, Nature **172**, 759 (1953).

[7] M. Feughelman, R. Langridge, W. E. Seeds, A. R. Stokes, H. R. Wilson, C. W. Hooper, M. H. F. Wilkins, R. K. Barclay, and L. D. Hamilton, Nature **175**, 834 (1955).

[8] R. Langridge, W. E. Seeds, H. R. Wilson, C. W. Hooper, M. H. F. Wilkins, and L. D. Hamilton, J. Biophys. Biochem. Cytol. **3**, 767 (1957).

[9] J. D. Watson and F. H. C. Crick, Nature **171**, 964 (1953).

[10] A. Gierer and G. Schramm, Nature **177**, 702 (1956).

[11] H. Fraenkel-Conrat, J. Am. Chem. Soc. **78**, 882 (1956).

[12] A. Rich and J. D. Watson, Proc. Natl. Acad. Sci. U. S. **40**, 759 (1954).

[13] M. Grunberg-Manago and S. Ochoa, J. Am. Chem. Soc. **77**, 3165 (1955).

[14] L. A. Heppel, P. J. Ortiz, and S. Ochoa, J. Biol. Chem. **229**, 679 (1957).

[15] A. Rich, Ann. N. Y. Acad. Sci. Special Publ. **5**, 186 (1957).

[16] R. C. Warner, Federation Proc. **15**, 379 (1956).

[17] A. Rich and D. R. Davies, J. Am. Chem. Soc. **78**, 3548 (1956).

[18] A. Rich in *The Chemical Basis of Heredity*, W. D. McElroy and B. Glass, editors (The Johns Hopkins Press, Baltimore, Maryland, 1957), p. 557.

[19] G. Felsenfeld and A. Rich, Biochim. et Biophys. Acta **26**, 457 (1957).

[20] G. Felsenfeld, Biochim. et Biophys. Acta **29**, 133 (1958).

[21] A. Rich, Nature **181**, 521 (1958).

[22] G. Felsenfeld, D. R. Davies, and A. Rich, J. Am. Chem. Soc. **79**, 2023 (1957).

[23] A. Rich, Biochim. et Biophys. Acta **29**, 502 (1958).

[24] D. R. Davies and A. Rich, J. Am. Chem. Soc. **80**, 1003 (1958).

[25] B. E. Griffin, A. R. Todd, and A. Rich, Proc. Natl. Acad. Sci. U. S. **44**, 1123 (1958).

[26] Author's unpublished observations.

# 24
# Biosynthesis of Nucleic Acids

ARTHUR KORNBERG

*Department of Microbiology, Washington University School of Medicine, St. Louis 10, Missouri*

INTEREST in the biosynthesis of nucleic acids stems from the crucial importance of these compounds in heredity and reproduction and in the development of the cellular machinery.

Every cell has DNA in its nucleus and, from a variety of considerations, DNA is regarded as the prime source of genetic information. Consider a cell which just has been pinched from its mother and which has little more than a faithful replica of the maternal DNA. In its growth to maturity, the young cell will have to make the enzymatic machinery to carry on energy metabolism on an expanding scale. At the same time, the cell must synthesize all of the special structures (be they lipid, carbohydrate, protein, or of any other special nature) that will characterize the shape, structure, appearance, and behavior of the cell. The sequence can be represented, oversimplified, as

$$\text{DNA} \rightarrow \text{enzymes} \rightarrow \text{everything else.}$$

From all indications, it can be said with some confidence that the structure of DNA is fixed, that it does not undergo turnover under the most extreme variations of cellular nutrition and physiology, and, facetiously, that it is immutable, except for mutations. Aside from having to serve as an original template for the synthesis of the enzymes, it has one other major function and that is to provide the template for the synthesis of a replica of itself in the reproductive act of forming a daughter cell. Discussion of the biosynthesis of DNA is limited here to its replication during reproduction.

## RNA BIOSYNTHESIS

What is known about the function and biosynthesis of RNA? In contrast with DNA, this problem appears more complex. RNA occurs both in the nucleus and in a variety of cytoplasmic structures. Its concentration in the cell may vary within wide limits depending upon the age, nutrition, and other environmental factors, and it obviously undergoes considerable turnover and replacement. While the DNA of the chromosome might be regarded as a single, exceedingly complex molecular unit, RNA is, by comparison, physically and metabolically heterogeneous and there is no reason to speak of RNA in the singular, except in the generic sense that one regards protein.

It would appear from current studies that RNA may provide a link between DNA and protein synthesis, translating the information in the DNA code to a form that is a proper template for the manufacture of distinctive proteins. What is known, in fact, from the work of Zamecnik *et al.*[1] with animal cells, is that RNA is essential for the assembly of amino acids into polypeptides, and from the work of Berg[2] with a bacterial system, is that only about 5 to 10% of the RNA seems to be active in the fixation of amino acids. This function of RNA as a link between DNA and enzyme synthesis is at present an educated and attractive speculation only.

Regarding the biosynthesis of RNA, the problem is complicated by its heterogeneity and by ignorance of its functions. There is a large and rapidly growing literature on the subject, but brief reference is made to only a few of the contributions.

(1) Ochoa[3] isolated from a variety of bacterial cells an enzyme which condenses various nucleoside diphosphates to high molecular-weight ribonucleic-acid-like polymers. These, discussed by Doty (p. 107) and by Rich (p. 191), are polymers of adenylic, cytidylic, or uridylic acid, or of other nucleotides or mixtures of them. As seen in Fig. 1, the reaction involves a condensation between the hydroxyl group in position 3 of one nucleotide with the inner phosphate group of the other nucleotide, with a consequent elimination of inorganic orthophosphate (Pi). There is little free-energy change in this reaction and, as indicated, the polymers are split by inorganic phosphate at physiological levels. For this enzyme to form the highly specific polymers required in protein synthesis, it must have controls in the cell from which it has escaped upon isolation. A reasonable possibility is that this enzyme provides for the conservation and storage of nucleotide units which can be called upon to form the adenylic, uridylic, cytidylic, and other coenzymes essential for energy metabolism, for carbohydrate, protein, and lipid biosynthesis, and for specific types of RNA.

(2) Several groups of investigators[4–9] have recognized, during the past year or so, the existence of enzyme systems which extend ribonucleic-acid chains by one or a few nucleotide units, and which employ nucleoside triphosphates rather than the diphosphates. There are suggestive indications of relationships of these reactions to amino-acid fixation leading to polypeptide formation.

(3) A reaction which has the intriguing possibility of relating RNA synthesis to DNA is the one recently described by Hurwitz.[10] An enzyme from *E. coli* was found which required DNA for the incorporation of ribonucleotides into a polymeric structure. With a mixture of radioactive ribonucleoside triphosphates and deoxynucleoside triphosphates as substrates, the prod-

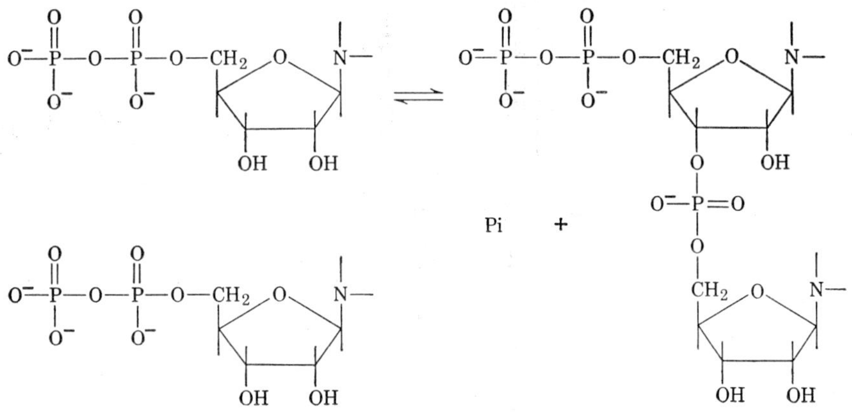

FIG. 1. Mechanism of RNA synthesis from ribonucleoside diphosphates.

uct formed was shown to contain the incorporated units in typical 3′-5′-phosphodiester linkage. Samples of RNA from several sources failed to replace the DNA requirement. Further purification of this enzyme system should clarify the specificities of the reaction and they are awaited with great interest.

(4) Turning now to studies with whole cells which attempt to relate nucleic-acid and protein synthesis, mention should be made first that, by the addition of chloromycetin to growing bacterial cultures, it is possible to suppress protein synthesis completely and still maintain the new formation of RNA and DNA. The synthesis of RNA, however, can be shown to be dependent upon the availability of all of the amino acids essential for protein synthesis.[11–13] Thus, a strain of *E. coli* requiring tryptophan for growth fails to make RNA in the presence of chloromycetin unless tryptophan is added to the medium. Yet this and other amino acids for which this effect has been demonstrated do not appear to be precursors of RNA. A dependence of RNA synthesis upon amino acids, despite the exclusion of protein formation, is implied, therefore, but an enzymatic basis for these observations as yet is not apparent.

(5) Finally, the studies of RNA biosynthesis in virus-infected bacterial cells are noteworthy. In a logarithmically growing population of bacterial cells, there is an exponential increase of protein, RNA, and DNA. At the time of virus infection, as with *T2* infection of *E. coli*, synthesis of bacterial components ceases abruptly and the cellular machinery is devoted completely to the synthesis of viral protein and viral DNA. Since the virus lacks RNA, there is—or rather, there was thought to be—no RNA synthesis. Recent tracer studies by Volkin and Astrachan[14] have shown a burst of RNA synthesis, very small in amount but distinctive in composition. One interpretation of these findings is that this RNA synthesis is in response to the unique instructions carried in the viral DNA, and is a necessary prerequisite for the formation of the proteins essential to virus formation. There is here a highly simplified situation in which to study a form of RNA biosynthesis that is relevant to the link between DNA and protein and that is inviting to biochemical study.

## DNA BIOSYNTHESIS

To review briefly the chemical structures which serve as the basic building units of the DNA molecule (Fig. 2): Deoxyadenylate (deoxyadenosine 5′-phosphate) is composed of a purine base linked by a glycosidic bond to the sugar that is unique to DNA, 2-deoxyribose. It is in the form of a furanose ring and lacks an oxygen at the 2-carbon position; esterified at carbon-5 is a phosphate residue. Thymidine 5′-phosphate, or thymidylate, is composed of a pyrimidine, thymine, linked as in the case of adenine to a 2-deoxyribose 5′-phosphate. The methyl group on the 5 carbon distinguishes thymine from uracil, also a pyrimidine, which is a component of

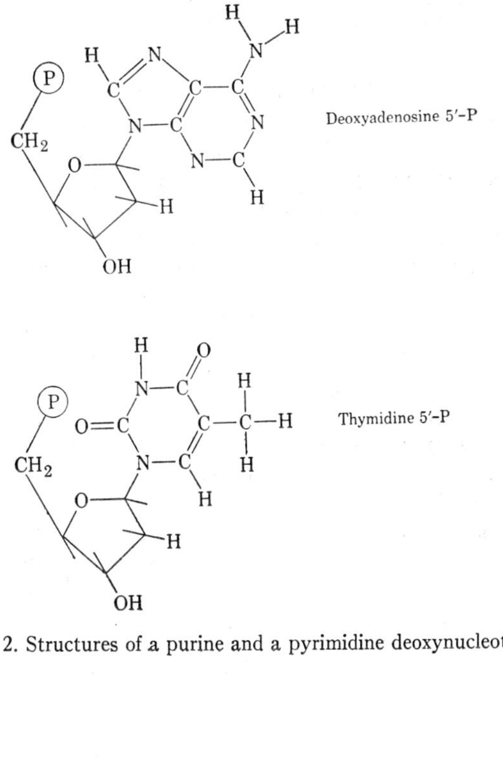

Deoxyadenosine 5′-P

Thymidine 5′-P

FIG. 2. Structures of a purine and a pyrimidine deoxynucleotide.

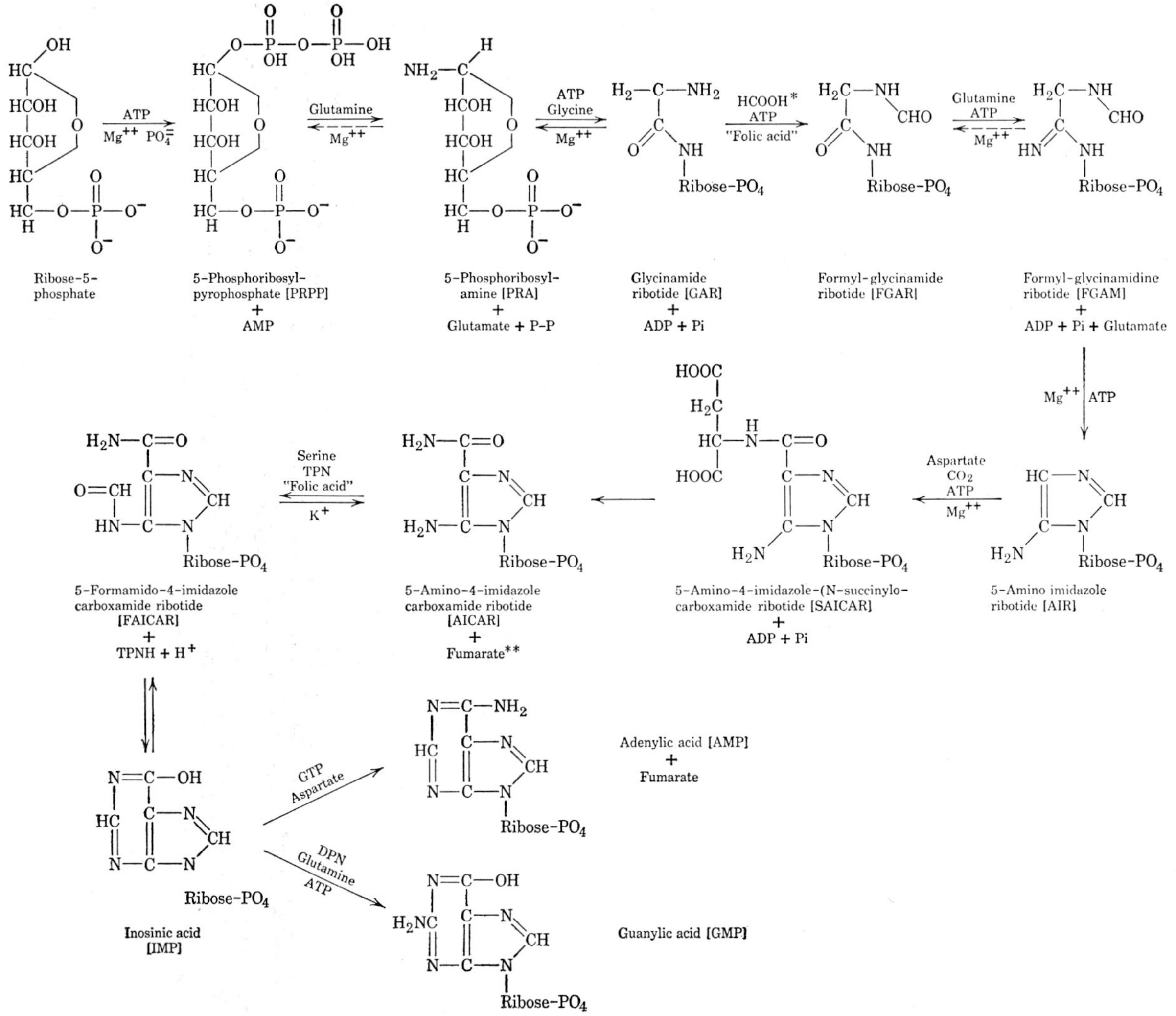

Ribose-5-phosphate → (ATP; $Mg^{++}$ $PO_4^{=}$) → 5-Phosphoribosyl-pyrophosphate [PRPP] + AMP ⇄ (Glutamine; $Mg^{++}$) → 5-Phosphoribosyl-amine [PRA] + Glutamate + P-P → (ATP, Glycine; $Mg^{++}$) → Glycinamide ribotide [GAR] + ADP + Pi → ($HCOOH^*$, ATP, "Folic acid") → Formyl-glycinamide ribotide [FGAR] ⇄ (Glutamine, ATP; $Mg^{++}$) → Formyl-glycinamidine ribotide [FGAM] + ADP + Pi + Glutamate → ($Mg^{++}$ ATP) → 5-Amino imidazole ribotide [AIR] → (Aspartate, $CO_2$, ATP, $Mg^{++}$) → 5-Amino-4-imidazole-(N-succinylo-carboxamide ribotide [SAICAR] + ADP + Pi → 5-Amino-4-imidazole carboxamide ribotide [AICAR] + Fumarate** ⇄ (Serine, TPN, "Folic acid", $K^+$) 5-Formamido-4-imidazole carboxamide ribotide [FAICAR] + TPNH + $H^+$ ⇄ Inosinic acid [IMP]

Inosinic acid [IMP] → (GTP, Aspartate) → Adenylic acid [AMP] + Fumarate

Inosinic acid [IMP] → (DPN, Glutamine, ATP) → Guanylic acid [GMP]

FIG. 3. Schema of enzymatic synthesis of purine ribonucleotides [from J. M. Buchanan, J. G. Flaks, S. C. Hartman, B. Levenberg, L. N. Lukens, and L. Warren, *Chemistry and Biology of Purines,* Ciba Foundation Symposium (J. and A. Churchill, Ltd., London, 1957), p. 233].

ribonucleic acid and of coenzymes prominent in carbohydrate metabolism. These and two other deoxynucleotides—a purine one, deoxyguanylate, and a pyrimidine one, deoxycytidylate—comprise the four deoxynucleotides which commonly occur in samples of DNA from bacterial, plant, and animal cells, and from some of the viruses. How are these units assembled?

This type of biosynthetic question has been approached in several ways. Earlier, Calvin (p. 147) and Roberts (p. 170) described the use of isotopic tracers to chart biosynthetic pathways. A suspected intermediate containing an isotopic marker is administered in the medium, and the cellular constituents which incorporate this marker are identified after brief or extended time exposures. A variation of this technique is to label the exclusive carbon or nitrogen source of the cell (e.g., $C^{14}$ glucose or $N^{15}$ ammonia), and then to determine whether or not certain (unlabeled) compounds in the medium can reduce the specific radioactivity of particular molecules in the cell under study. Quite another approach involves the use of a mutant which accumulates intermediates because of an enzymatic deficiency at some point in the biosynthetic process. The facility with which mutants can be selected for service in a particular situation has made this technique extremely versatile and effective. These and other quite diverse methods have been used with considerable success in mapping pathways in the intact cell. But no combination of these methods can establish unequivocally a biosynthetic sequence! The walls and membranes of the cell form an "iron curtain" which either prevents certain molecules from entering the cell or alters them upon their entry or exit. It is a curtain which obscures the details to the point where a spur from a pathway is indistinguishable from the main line. Hence, a study of the enzymatic apparatus of the disrupted cell is an indispensable approach to the problem of biosynthesis. The objective of such an enzymatic attack on the problem is the isolation from cells of separate enzymes, each of which effects a single, chemically rational step. These steps arranged in proper sequence should lead to the total synthesis of complex molecules. To assume real significance, the rates and conditions under which a reconstructed biosynthetic pathway operates ultimately must be reconciled with all of the isotopic, nutritional, and other observations made with the intact cells.

Turning now to what is known about the details of the enzymatic synthesis of the nucleotides (Fig. 3), one sees an outline of the discrete steps in the synthesis of adenosine 5′-phosphate starting with ribose 5-phosphate, glycine, glutamine, and, as an energy source, adenosine triphosphate (ATP). This knowledge is largely the result of the work of Buchanan[15] and of Greenberg[16] and their colleagues. A comparable scheme could be shown for the pyrimidine nucleotides. The origin of the deoxyribonucleotides would appear, from current indications, to derive from the ribonucleotides by a direct reduction step.

Several years ago, experimental work was begun by our laboratory to determine the biochemical mechanism for the replication of DNA chains. One might suggest many permutations for how such chains might be

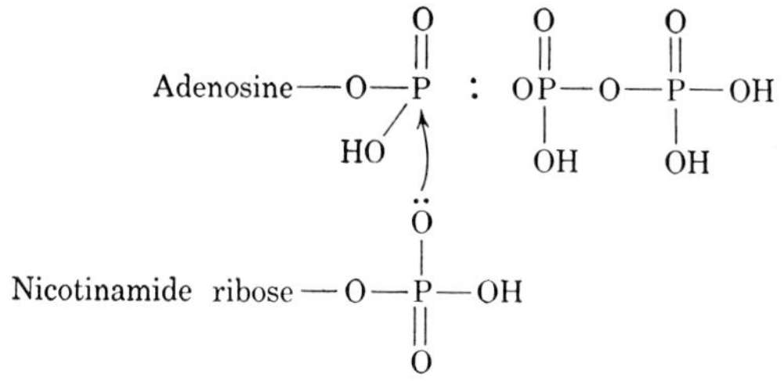

FIG. 4. Nucleophilic attack of nicotinamide mononucleotide on ATP in DPN synthesis.

assembled. We were guided by the knowledge of how the simplest and best known of the complex nucleotides, the coenzymes, are synthesized by the cell.[17–19] For example, the enzymatic synthesis of diphosphopyridine nucleotide (DPN) (Fig. 4) involves a condensation of ATP with nicotinamide mononucleotide and a resultant elimination of inorganic pyrophosphate (PP). Similarly, in the synthesis of flavin adenine dinucleotide (FAD), or coenzyme-*A*, there is a reaction of flavin mononucleotide or pantetheine phosphate, respectively, with ATP, and again inorganic pyrophosphate is eliminated (Fig. 5).[20] In each case, the adenyl coenzyme is produced by a reaction with an activated adenyl derivative such as ATP. More recently, the enzymatic synthesis of the adenyl derivatives of the fatty acids and amino acids have been shown to proceed by a similar mechanism.[21] What is shown in Fig. 5 to apply to the synthesis of adenyl derivatives applies also to the synthesis of uridyl, cytidyl, and guanosyl coenzymes.[22–24] Along lines proposed by Koshland,[25] one may visualize (Fig. 4) a nucleophilic attack on the activated adenyl derivative, in the case of DPN synthesis, by nicotinamide mononucleotide, to form an anhydride bridge with the adenyl group and to displace inorganic pyrophosphate. By analogy, it can be conjectured that chains of nucleotides might be formed by a reaction of the end of a chain with an activated nucleotidyl molecule, such as a nucleoside triphosphate. In other words, it is assumed that the basic building block is a 5′-phosphate ester of a deoxynucleoside, activated perhaps by linkage to a pyrophosphate group. As shown in Fig. 6, the end of a DNA chain might be

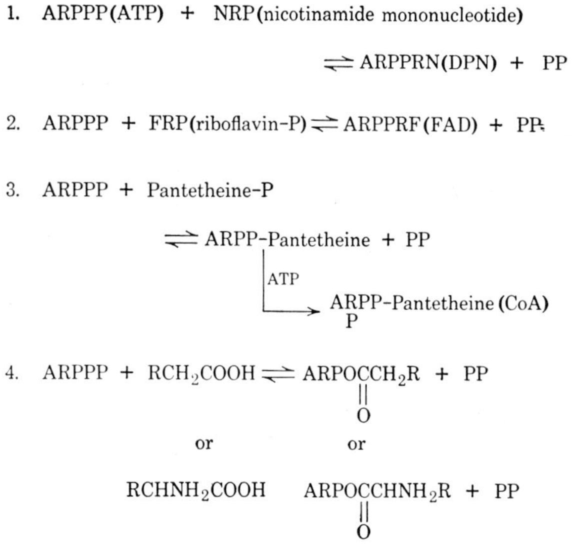

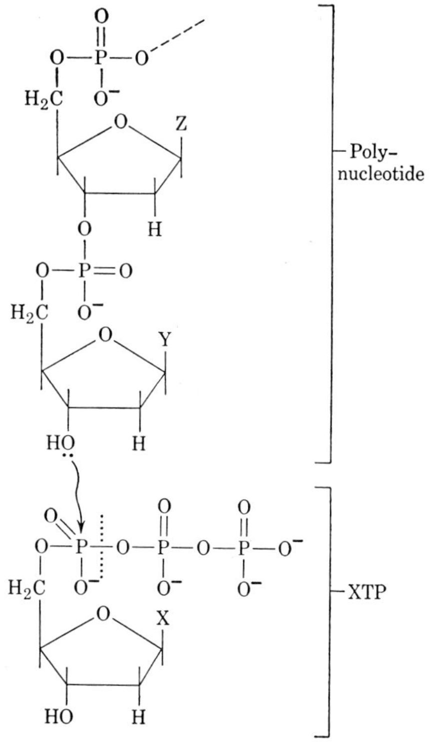

1. ARPPP(ATP) + NRP(nicotinamide mononucleotide) ⇌ ARPPRN(DPN) + PP

2. ARPPP + FRP(riboflavin-P) ⇌ ARPPRF(FAD) + PP

3. ARPPP + Pantetheine-P ⇌ ARPP-Pantetheine + PP
   —ATP→ ARPP(P)-Pantetheine (CoA)

4. ARPPP + $RCH_2COOH$ ⇌ $ARPOC(=O)CH_2R$ + PP
   or $RCHNH_2COOH$ or $ARPOC(=O)CHNH_2R$ + PP

FIG. 5. Synthesis of adenine-nucleotide coenzymes.



FIG. 6. Postulated mechanism for extending a DNA chain.

TABLE I. Enzyme purification.

| Fraction number | Step | Units[a] per ml | Units[a] Total | Protein mg per ml | Specific activity units per mg protein |
|---|---|---|---|---|---|
| I | Sonic extract | 2.0 | 16 800 | 20.0 | 0.1 |
| II | Streptomycin | 13.0 | 19 500 | 3.0 | 4.3 |
| III | DNase, dialysis | 12.1 | 18 100 | 1.80 | 6.7 |
| IV | Alumina gel | 15.4 | 12 300 | 0.78 | 20. |
| V | Concentration of gel eluate | 110. | 9 000 | 4.90 | 22. |
| VI | Ammonium sulfate | 670. | 6 030 | 8.40 | 80. |
| VII | DEAE resin | 120. | 3 600 | 0.60 | 200.–400. |

[a] A unit is defined as the amount of enzyme causing the incorporation of 10 mμmoles of thymidine triphosphate into an acid-insoluble fraction during the assay period of 30 min at 37°. See Table II for further details about the assay incubation mixture.

involved in a nucleophilic attack on a deoxynucleoside triphosphate, thus extending the chain by one unit and displacing pyrophosphate.

We approached the problem of nucleic-acid synthesis experimentally by mixing together and incubating the following: a deoxynucleoside labeled very intensely in one of the carbon atoms of its base, an extract made from an exponentially growing culture of *E. coli* (generation time 20 min), ATP as an energy source, and Mg ions. To determine whether any nucleic-acid synthesis had taken place, one relied upon the very simple and fortunate chemical property of DNA that it is precipitated at acid *p*H, while the deoxynucleotides which were the substrates are completely soluble under these conditions. It was ascertained many times that thorough washing of the DNA precipitate removed even the smallest traces of the deoxynucleoside substrates. Thus, in our earliest experiments, traces of radioactive thymidine (thymine deoxyriboside) were found incorporated into an acid-precipitable substance that could be rendered *non*precipitable by treatment with purified deoxyribonuclease (DNase).[18] In this crude system, it was possible to detect 50 counts in the precipitate out of 5 000 000 counts in 1 μmole of the starting substrate. This indicated an incorporation of only 10 μμmoles or about 1/10 000 of the amount detectable by the most sensitive microchemical methods.

Using the incorporation of $C^{14}$-thymidine into a DNase-sensitive acid-precipitable material as the assay, purification was begun of the components of the *E. coli* extract essential for this reaction. It soon was recognized that one of the enzymatic functions of the crude fraction was the conversion of the thymidine through thymidylate to the triphosphate level, and thymidine triphosphate was used thereafter as the substrate.[26,27]

The enzyme-purification procedure at present[28] involves first the preparation of an extract of *E. coli* by sonic treatment, then several fractionation steps (Table I) resulting in an enrichment with respect to protein of about 2000 to 4000 times over the crude extract. With this preparation, certain interesting features of the reaction have become apparent and are discussed presently.

Even at an intermediate stage of purification, it became clear that, in order for one of the triphosphates to be incorporated into DNA, the triphosphates of all four of the deoxynucleosides which commonly occur in DNA, and also DNA itself, must be present.[29] This is illustrated in Table II. The incubation mixture contained the triphosphates* of thymidine, deoxycytidine, deoxyguanosine, and deoxyadenosine, calf-thymus DNA, $Mg^{++}$, and enzyme (with the most purified of the enzyme preparations, about 0.1 μg of protein was used). The isotopic marker used has been either $C^{14}$ in carbon-2 of thymine or $P^{32}$ in the innermost phosphate group of one of the triphosphates. Under these conditions, 0.5 mμmole of the labeled nucleotide was incorporated. Omission of any one of the triphosphates, of the DNA, or of $Mg^{++}$, or pretreatment of the DNA with crystalline pancreatic DNase reduced the reaction to a level below the limits of detectability. Replacement of the triphosphates by the corresponding deoxynucleoside diphosphates reduced the reaction to or near the limits of detectability ($<5\%$). Nor was there any detectable reaction when ribose analogs such as ATP or CTP were used in place of dATP or dCTP, respectively. Also, there was no reaction when DNA was replaced by RNA or by DNA degraded by acid treatment or sonic irradiation. However, as is mentioned later, DNA from a variety of sources, plant, animal, and virus, served in the reaction.

The enzymatic incorporation of deoxynucleotide into a DNA fraction using incubation conditions similar to those in Table II also has been observed with the use of sonic extracts of other bacterial species (*Hemophilus influenzae*, *Aerobacter aerogenes*), and with extracts of several types of animal cells (HeLa cell cultures, lymph-gland and leukemic cells).[29] While assays for the DNA-

TABLE II. Requirements for deoxynucleotide incorporation into DNA.

| | mμmoles |
|---|---|
| Complete system[a] | 0.50 |
| Omit dCTP, dGTP, dATP | $<0.01$ |
| Omit dCTP | $<0.01$ |
| Omit dGTP | $<0.01$ |
| Omit dATP | $<0.01$ |
| Omit $Mg^{++}$ | $<0.01$ |
| Omit DNA | $<0.01$ |
| DNA pretreated with DNase | $<0.01$ |

[a] The complete system contained 5 mμmoles of $TP^{32}PP$ ($1.5\times10^6$ cpm per μmole), dATP, dCTP, and dGTP, 1 μmole of $MgCl_2$, 20 μmoles of glycine buffer (pH 9.2), 10 γ of calf-thymus DNA, and 3 γ of enzyme Fraction V, in a final volume of 0.30 ml. The incubation was carried out at 37° for 30 min.
DNase treatment of DNA was carried out in an incubation mixture containing 60 γ of thymus DNA, 50 μmoles of Tris buffer, *p*H 7.5, 5 μmoles of $MgCl_2$, and 5 γ of pancreatic DNase, in a final volume of 0.50 ml. After 30 min at 37°, 0.02 ml of 5% bovine serum albumin and 0.05 ml of 5*N* perchloric acid were added. The precipitate was centrifuged and dissolved in dilute alkali and neutralized. A control incubation from which DNase was omitted yielded fully active DNA.

* For the triphosphates of thymidine, deoxycytidine, deoxyguanosine, and deoxyadenosine, respectively, the following abbreviations are used: TTP, dCTP, dGTP, and dATP. The symbol $TP^{32}PP$ indicates that the triphosphate is labeled with $P^{32}$ in the innermost phosphate group.

synthesizing system cannot be regarded as accurate in the crude extracts, it is significant that the values for the bacterial extracts are of the order of 20 to 50 times greater than those for the animal-cell extracts. Bollum and Potter[31,32] have reported the incorporation of $H^3$-thymidine into DNA by extracts of regenerating rat liver.

With the most purified *E. coli* enzyme preparation, it has become possible to demonstrate net synthesis by the use of more-direct chemical methods. Several experiments are illustrated in Table III. In experiment 1, there was an increase in DNA of a little over twofold, measured by spectrophotometry, deoxypentose assay, or by incorporation of radioactive tracer. In the other experiments, increases of DNA by a factor of 10 to 20 were obtained, and 90 to 95%, therefore, of isolated DNA was derived from the deoxynucleotide substrates supplied. The factors responsible for cessation of the reaction are currently under study.

On the basis of the foregoing results, one can consider the following over-all equation for the enzymatic synthesis of DNA:

$$\begin{array}{l} n\ \mathrm{T\mathbf{P}PP} \\ n\mathrm{dG\mathbf{P}PP} \\ n\mathrm{dA\mathbf{P}PP} \\ n\mathrm{dC\mathbf{P}PP} \end{array} + \mathrm{DNA} \rightleftharpoons \mathrm{DNA} - \begin{bmatrix} \mathrm{T\mathbf{P}} \\ \mathrm{dG\mathbf{P}} \\ \mathrm{dA\mathbf{P}} \\ \mathrm{dC\mathbf{P}} \end{bmatrix}_n + 4(n)\mathrm{PP}$$

The four triphosphates+DNA yield a product which contains the four nucleotidyl residues linked in some covalent and hydrogen-bonded fashion with the DNA. It is apparent from the equation that inorganic pyro-

TABLE III. Net synthesis of DNA.

| Expt. No.[a] | Estimation | Control (no enzyme) | Complete | Δ |
|---|---|---|---|---|
| | | μmoles | μmoles | μmoles |
| 1 | $P^{32}$ incorporation | 0.00 | 0.28 | 0.28 |
| | Optical density | 0.19 | 0.46 | 0.27 |
| | Deoxypentose | 0.19 | 0.40 | 0.21 |
| 2 | Optical density | 0.06 | 0.63 | 0.57 |
| 3 | Optical density | 0.05 | 0.58 | 0.53 |
| 4 | Optical density | 0.05 | 0.64 | 0.59 |
| 5 | Optical density | 0.04 | 0.89 | 0.85 |

[a] In experiment 1, the incubation mixture (3.0 ml) contained 0.15 μmole of $dAP^{32}PP$ ($1.3\times10^6$ cpm per μmole), 0.3 μmole of dGTP, 0.15 μmole of dCTP, 0.15 μmole of dTTP, 200 μmoles of potassium-phosphate buffer (*p*H 7.4), 20 μmoles of $MgCl_2$, 0.1 mg of calf-thymus DNA, and 12 γ of enzyme Fraction VII. The mixture was incubated at 37° for 180 min. DNA was precipitated, washed, taken up in 1.2 ml of 0.5*N* perchloric acid, and heated for 15 min in a boiling water bath. Optical-density measurements were made at 260 mμ and converted to nucleotide equivalents using a molar extinction coefficient of 8960 (derived from the calculated values for an acid hydrolyzate of calf-thymus DNA). In the $P^{32}$ estimation of DNA synthesis, incorporation of deoxyadenylate was multiplied by a factor based on its percentage in calf-thymus DNA. The radioactivity actually observed for the controls did not exceed the background count. In experiments 2, 3, 4, and 5, the reaction mixture (1.0 ml) contained 0.32 μmole of each of the four triphosphates, 30 γ of calf-thymus DNA, 60 μmoles of potassium-phosphate buffer (*p*H 7.4), 6 μmoles of $MgCl_2$, and 8 γ of enzyme Fraction VII. The mixture was incubated for 250 min at 37°. 2*M* NaCl then was added to give a final concentration of 0.2*M* and the mixture was heated for 5 min at 70°. Unreacted triphosphates were removed by exhaustive dialysis against 0.2*M* NaCl. The product contained no acid-soluble material. Optical density at 260 mμ was determined and converted to nucleotide equivalents using a molar extinction for DNA of 6900.

TABLE IV. Physical properties of DNA.

| | | | Heated at 100°C, 15 min | |
|---|---|---|---|---|
| | Primer | Product | Primer | Product |
| Sedimentation coefficient | 25 | 20–25 | 20 | 14 |
| Intrinsic viscosity | 50 | 15–35 | <1 | <1 |
| Molecular weight | $8\times10^6$ | $4\text{–}6\times10^6$ | $<1\times10^6$ | $<0.7\times10^6$ |

phosphate (PP) must be split out during the reaction and in quantities equimolar to the amounts of deoxynucleotide incorporated into the DNA. It might be considered also, especially in the light of the Ochoa studies[33] of the reversibility of synthesis of ribonucleic-acid polymers, that there might be a pyrophosphorolytic reversal of the reaction. It has been shown[18] that, with the incorporation of $C^{14}$-thymidylate into DNA, equimolar amounts of PP were released; this was isolated and identified by ion-exchange chromatography.

Evidence for reversal of the reaction has been obtained.[29] When PP is present in concentrations of $2\times10^{-3}$ M—i.e., about 100 times the concentration of the deoxynucleoside triphosphates—the synthetic reaction is inhibited by about 50%. Under such conditions, when $PP^{32}$ is used, its incorporation into the terminal PP groups of the four triphosphates has been observed. The rate of the reaction is comparable to the synthetic rate. The reaction is absolutely dependent upon the presence of DNA. DNA degraded with DNase is inert. Inorganic orthophosphate fails to replace PP. A distinctive feature of the reaction is that triphosphates are required. When they are omitted, there is only a small reaction, which may be interpreted as a very limited pyrophosphorolysis of DNA. It is necessary to have only one triphosphate to augment the pyrophosphorolysis reaction considerably.

Up to this point, the discussion has been concerned with how a purified enzyme preparation can synthesize, in the presence of four deoxynucleoside triphosphates and of DNA, presumably acting as primer, a product that is insoluble at acid *p*H and which is nondialyzable and may be degraded to acid-soluble fragments by the action of pancreatic DNase. It has been established with more-sensitive techniques that even a single triphosphate may react, but to an extent so limited as to suggest the addition of one or very few residues to the ends of the DNA chains added as acceptor.[34]

A more detailed consideration is presented below of the physical and chemical nature of the enzymatic product, DNA, prepared under conditions where 90% or more of the polymerized material is newly synthesized.

In physicochemical studies, we have been able to show that the enzymatically synthesized product has essentially the same physical characteristics as DNA carefully prepared from calf thymus.[35] The sedimenta-

FIG. 7. Sedimentation of DNA primer and the enzymatically synthesized product.

tion behavior (Fig. 7) of the calf-thymus and enzymatically synthesized samples were quite similar, although the enzymatically synthesized sample showed a greater polydispersity. The latter may be the result of the action of contaminating DNases in the enzyme preparation. The sedimentation constants in this and in a great many other runs have been in the neighborhood of 25 to 30. Viscosimetric determinations also have yielded values of 15 to 35 deciliters/gram which are comparable to those obtained for calf-thymus DNA (40 to 50); from these, average molecular weights of 4 to 6 million for several samples of enzymatically synthesized DNA were calculated (Table IV). It may be inferred from the sedimentation and viscosity characteristics of the product that it consists of highly ordered rigid structures with effective volumes greater than would be expected from single polynucleotide chains with freedom of rotation at each link in the backbone. In support of this view was the collapse of the macromolecular structure when the DNA was heated for 15 min at 100°C; like thymus DNA, the viscosity decreased to immeasurably low levels, while the sedimentation rate decreased only slightly. Furthermore, there was found a typical hyperchromic effect[36] upon degradation of the product with DNase. The curves in Fig. 8 show that just as there is an increase in optical density upon digestion of calf-thymus DNA with pancreatic DNase, so is there a kinetically similar increase with the enzymatic product and to the same extent of about 30% above the starting values.

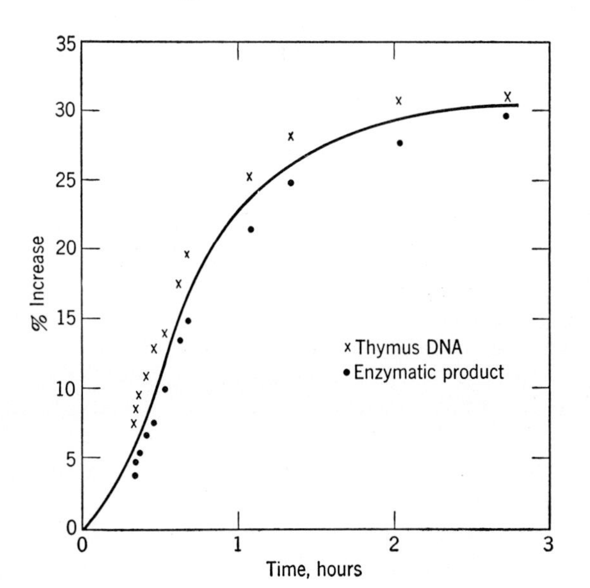


FIG. 8. Increase in ultraviolet absorption of DNA upon digestion with pancreatic DNase.

One now comes to a consideration of the chemical structure of the product. To begin with, it can be affirmed that the substrates are linked in the DNA product by typical 3′–5′ diester bonds.[29] What can be said of the base composition and the ratios of bases of the product? Do they bear any relation to the DNA primer added to the reaction?

In Table V are the base-composition data on enzymatically synthesized DNA. A close correspondence exists in the enzymatically synthesized DNA's between the contents of adenine and thymine on the one hand and of guanine and cytosine on the other, so that the ratio of purines to pyrimidines (A+G/C+T) is in each case nearly unity. Furthermore, good agreement exists for the ratio (A+T/G+C) between the enzymatic product and its corresponding primer, ranging from values of 0.59 for *Mycobacterium phlei* to greater than 40 for an enzymatically synthesized copolymer of deoxyadenylate and thymidylate. The latter is formed by the DNA-synthesizing enzyme in the absence of primer under rather specialized, but poorly understood, conditions: specifically, after lag periods of from 3 to 6 hr. Once formed, such a polymer then can be replicated without a lag and will consist solely of adenine and thymine, although all four of the deoxynucleoside triphosphates are provided in the reaction mixture. This polymer, therefore, represents an extreme case in which the base composition of the enzymatic product reflects that of the primer.

TABLE V. Purine and pyrimidine composition of enzymatically synthesized DNA.[a,b]

| DNA | | Number of analyses | A | T | G | C | $\frac{A+T}{G+C}$ | $\frac{A+G}{T+C}$ |
|---|---|---|---|---|---|---|---|---|
| *M. phlei* | primer | 3 | 0.65 | 0.66 | 1.35 | 1.34 | 0.49(0.48–0.49) | 1.01(0.98–1.04) |
| | product | 3 | 0.66 | 0.80 | 1.17 | 1.34 | 0.59(0.57–0.63) | 0.85(0.78–0.88) |
| *A. aerogenes* | primer | 1 | 0.90 | 0.90 | 1.10 | 1.10 | 0.82 | 1.00 |
| | product | 3 | 1.02 | 1.00 | 0.97 | 1.01 | 1.03(0.96–1.13) | 0.99(0.95–1.01) |
| *E. coli* | primer | 2 | 1.00 | 0.97 | 0.98 | 1.05 | 0.97(0.96–0.99) | 0.98(0.97–0.99) |
| | product | 2 | 1.04 | 1.00 | 0.97 | 0.98 | 1.02(0.96–1.07) | 1.01(0.96–1.06) |
| Calf thymus | primer | 2 | 1.14 | 1.05 | 0.90 | 0.85 | 1.25(1.24–1.26) | 1.05(1.03–1.08) |
| | product | 6 | 1.19 | 1.19 | 0.81 | 0.83 | 1.46(1.22–1.67) | 0.99(0.82–1.04) |
| *T*2 phage | primer | 2 | 1.31 | 1.32 | 0.67 | 0.70 | 1.92(1.86–1.97) | 0.98(0.95–1.01) |
| | product | 2 | 1.33 | 1.29 | 0.69 | 0.70 | 1.90(1.82–1.98) | 1.01(1.01–1.03) |
| "Synthetic A-T Copolymer" | ... | 1 | 1.99 | 1.93 | $<0.05$ | $<0.05$ | $>40.$ | 1.05 |

[a] A, T, G, and C refer, respectively, to adenine, thymine, guanine, and cytosine, except that C in the case of *T*2-phage primer refers to hydroxymethylcytosine.
[b] The figures in parentheses represent the range of values obtained.

The somewhat higher values of (A+T/G+C) observed for some of the products most probably can be attributed to contamination of these DNA's with traces of this deoxyadenylate-thymidylate copolymer. The obvious implications of these early results are that the added DNA is serving as a template for an enzymatic replication of DNA, but it is evident too that more-extensive documentation is necessary before this conclusion can be considered to be established.

In reviewing the specificity of this DNA-synthesizing system, the results have indicated that samples of DNA from a variety of origins can serve as primers. It has been mentioned also that only the *tri*phosphates of the *deoxy*nucleosides are reactive. What can be said of the specificity of the substrates with respect to the structure of the pyrimidine and purine bases? From the many interesting reports on the incorporation of bromouracil,[37–39] of azaguanine,[40] and of other analogs into bacterial and viral DNA, it might be surmised that some latitude in the structures of the bases can be tolerated provided there is no interference with their hydrogen bonding. It would be well to reiterate at once what Rich mentioned earlier (p. 191). Analysis of the composition of samples of DNA from a great variety of sources and by many investigators (reviewed by Chargaff[41]) reveals the remarkable fact that the purine content always equals the pyrimidine content. Among the purines, the adenine content may differ considerably from the guanine, and among the pyrimidines, the thymine from the cytosine. There is an invariable equivalence, however, between bases with an amino group in the 6 position of the ring, such as in adenine and cytosine, and the bases with a keto group in the 6 position of the ring, such as in guanine and thymine. These facts[41,42] were interpreted on the basis of hydrogen bonding by Watson and Crick[43] in their masterful hypothesis on the structure of DNA. In a given species, the DNA composition of all of the cells is characterized by a distinctive ratio of the number of adenine-thymine pairs to the number of guanine-cytosine pairs.

TABLE VI. Replacement of natural bases by analogs in enzymatic synthesis of DNA.

| Expt. No. | Control value[a] (mμmoles) | Base analog used | Natural base omitted: Thymine | Adenine | Guanine | Cytosine |
|---|---|---|---|---|---|---|
| | | | (Percent of control)[b] | | | |
| 1 | 0.50 | Uracil | *54* | 4 | 6 | |
| 1a | 0.88 | Uracil | | | | 3 |
| 2 | 0.43 | 5-bromouracil | *97* | 2 | 4 | |
| 2a | 0.42 | 5-bromouracil | | | | 4 |
| 3 | 0.51 | 5-bromocytosine | | 4 | 4 | *118* |
| 3a | 0.40 | 5-bromocytosine | 4 | | | |
| 4 | 0.58 | 5-methylcytosine | | 2 | 3 | *185* |
| 4a | 0.52 | 5-methylcytosine | 2 | | | |
| 5 | 0.37 | Hypoxanthine | | 3 | *25* | 5 |
| 5a | 0.27 | Hypoxanthine | 4 | | | |

[a] Control values are mμmoles of radioactive deoxynucleotide incorporated into DNA in the absence of analog. Incubation mixtures contained in 0.3 ml, 5 mμmoles each of TTP, dATP, dCTP, and dGTP; 2 μmoles of $MgCl_2$; 20 μmoles of potassium phosphate (*p*H 7.4); 10 μg of calf-thymus DNA; and 1μg of enzyme fraction VII-R. Experiments were performed at 37° for 30 min. Labeled substrates were: $dCP^{32}PP$ in Expts. 1, 2, 5a; $TP^{32}PP$ in Expts. 1a, 3,4,5; and $dGP^{32}PP$ in Expts. 2a, 3a, 4a.
[b] The percentage value represents the fraction of the labeled substrate incorporated when the analog (5 mμmoles) was used instead of a natural base. All bases, natural or analog, were supplied as the deoxynucleoside triphosphates. Values of 5% or below are near the limit of detectability and are of questionable significance.

Hydrogen Bonding of Adenine to Thymine

| R | |
|---|---|
| $-CH_3$ | Thymine |
| $-H$ | Uracil |
| $-Br$ | Bromuracil |

Hydrogen Bonding of Guanine to Cytosine

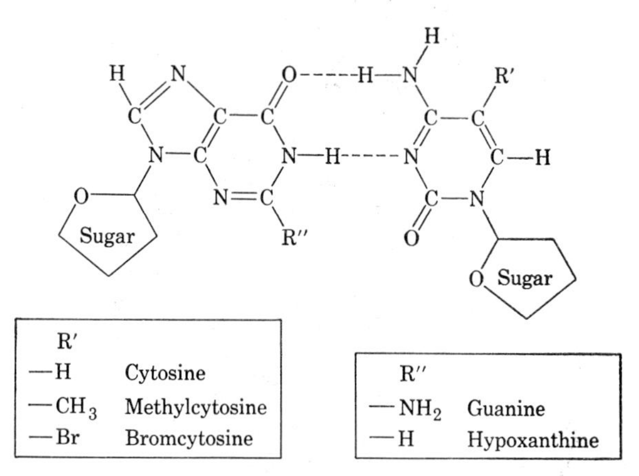


FIG. 9. Hydrogen bonding of the bases.

As can be seen from Table VI,[44] deoxyuridine triphosphate used in place of thymidine triphosphate supported DNA synthesis at 54% of the rate of the control value but failed to support synthesis when used in place of the triphosphates of deoxyadenosine, deoxyguanosine, or deoxycytidine. 5-bromodeoxyuridine triphosphate was more effective as a replacement for thymidine triphosphate but was unable to substitute for any of the other triphosphates. The 5-bromo- and 5-methyl-deoxycytidine triphosphates replaced only deoxycytidine triphosphate and, for unexplained reasons, were even more effective than deoxycytidine triphosphate itself. Deoxyinosine triphosphate permitted a reduced rate of DNA synthesis in the absence of deoxyguanosine triphosphate, but essentially no synthesis when any one of the other triphosphates was absent.

Thus, uracil and 5-bromouracil specifically replaced thymine; 5-methyl- and 5-bromo-cytosine replaced cytosine, and hypoxanthine substituted for guanine; xanthine was not incorporated into DNA. As seen from Fig. 9, the specific replacement of the natural bases by these analogs is consistent with and offers additional support for the base-pairing relationships in the double helix proposed by Watson and Crick for the structure of DNA.

The current status of knowledge of the biochemical aspects of RNA and DNA synthesis has been sketched in the foregoing. With respect to RNA, the enzymatic information available is inadequate to explain the metabolic behavior of cells and tissues. In the case of DNA, the enzymatic studies of replication can be reconciled with genetic phenomena, but much remains to be clarified in the mechanism of the reaction. While the biological implications of these problems are exciting and pressing, the most immediate obstacles are the limited resources available to separate and characterize these macromolecules. New techniques are desperately needed now in the nucleic-acid field to cope with such problems as were solved in the protein field over the last 50 years.

## BIBLIOGRAPHY

[1] P. C. Zamecnik, M. L. Stephenson, and L. I. Hecht, Proc. Natl. Acad. Sci. U. S. **44**, 73 (1958).

[2] P. Berg and E. J. Ofengand, Proc. Natl. Acad. Sci. U. S. **44**, 78 (1958).

[3] S. Ochoa, Federation Proc. **15**, 832 (1956).

[4] E. S. Canellakis, Biochim. et Biophys. Acta **23**, 217 (1957).

[5] E. S. Canellakis, Biochim. et Biophys. Acta **25**, 217 (1957).

[6] M. Edmonds and R. Abrams, Biochim. et Biophys. Acta **26**, 226 (1957).

[7] E. Herbert, V. R. Potter, and L. I. Hecht, J. Biol. Chem. **225**, 659 (1957).

[8] P. C. Zamecnik, M. L. Stephenson, J. F. Scott, and M. B. Hoagland, Federation Proc. **16**, 275 (1957).

[9] C. W. Chung and H. R. Mahler, J. Am. Chem. Soc. **80**, 3165 (1958).

[10] J. Hurwitz, Federation Proc. **17**, 247 (1958).

[11] A. B. Pardee and L. S. Prestidge, J. Bacteriol. **71**, 677 (1956).

[12] F. Gros and F. Gros, Biochim. et Biophys. Acta **22**, 200 (1956).

[13] A. I. Aronson and S. Spiegelman, Biochim. et Biophys. Acta **29**, 214 (1958).

[14] L. Astrachan, Federation Proc. **17**, 183 (1958).

[15] J. M. Buchanan, J. G. Flaks, S. C. Hartman, B. Levenberg, L. N. Lukens, and L. Warren, *Chemistry and Biology of Purines*, Ciba Foundation Symposium (J. and A. Churchill, Ltd., London, 1957), p. 233.

[16] G. R. Greenberg, Federation Proc. **12**, 651 (1953).

[17] A. Kornberg in *Phosphorus Metabolism*, W. D. McElroy and B. Glass, editors (The Johns Hopkins Press, Baltimore, Maryland, 1951), Vol. I, p. 392.

[18] A. Kornberg, Advances in Enzymol. **18**, 191 (1957).

[19] A. Kornberg in *The Chemical Basis of Heredity*, W. D. McElroy and B. Glass, editors (The Johns Hopkins Press, Baltimore, Maryland, 1957), p. 579.

[20] M. B. Hoagland and G. D. Novelli, J. Biol. Chem. **207**, 767 (1954).

[21] P. Berg, J. Biol. Chem. **222**, 991, 1015, 1025 (1956).

[22] A. Munch-Petersen, H. M. Kalckar, E. Cutolo, and E. E. B. Smith, Nature **172**, 1036 (1953).

[23] A. Munch-Petersen, Arch. Biochem. Biophys. **55**, 592 (1955).

[24] E. P. Kennedy and S. B. Weiss, J. Am. Chem. Soc. **77**, 250 (1955).

[25] D. E. Koshland in *The Mechanism of Enzyme Action*, W. D. McElroy and B. Glass, editors (The Johns Hopkins Press, Baltimore, Maryland, 1954), p. 608.

[26] A. Kornberg, I. R. Lehman, and E. S. Simms, Federation Proc. **15**, 291 (1956).
[27] A. Kornberg, I. R. Lehman, M. J. Bessman, and E. S. Simms, Biochim. et Biophys. Acta **21**, 197 (1956).
[28] I. R. Lehman, M. J. Bessman, E. S. Simms, and A. Kornberg, J. Biol. Chem. **233**, 163 (1958).
[29] M. J. Bessman, I. R. Lehman, E. S. Simms, and A. Kornberg, J. Biol. Chem. **233**, 171 (1958).
[30] C. G. Harford and A. Kornberg, Federation Proc. **17**, 515 (1958).
[31] F. J. Bollum and V. R. Potter, J. Am. Chem. Soc. **79**, 3603 (1957).
[32] F. J. Bollum, Federation Proc. **17**, 193 (1958).
[33] M. Grunberg-Manago, P. J. Ortiz, and S. Ochoa, Biochim. et Biophys. Acta **20**, 269 (1956).
[34] J. Adler, I. R. Lehman, M. J. Bessman, E. S. Simms, and A. Kornberg, Proc. Natl. Acad. Sci. U. S. **44**, 641 (1958).
[35] H. K. Schachman, I. R. Lehman, M. J. Bessman, J. Adler, E. S. Simms, and A. Kornberg, Federation Proc. **17**, 304 (1958).
[36] M. Kunitz, J. Gen. Physiol. **33**, 349 (1950).
[37] V. F. Weygand, A. Wacker, and H. Dellweg, Z. Naturforsch. **7b**, 19 (1952).
[38] D. B. Dunn and J. D. Smith, Nature **174**, 305 (1954).
[39] S. Zamenhof and G. Griboff, Nature **174**, 306 (1954).
[40] M. R. Heinrich, V. C. Dewey, R. E. Parks, Jr., and G. W. Kidder, J. Biol. Chem. **197**, 199 (1952).
[41] E. Chargaff in *Nucleic Acids*, E. Chargaff and J. N. Davidson, editors (Academic Press, Inc., New York, 1955), Vol. I, p. 307.
[42] G. R. Wyatt, Exptl. Cell. Research Suppl. 2, 201 (1952).
[43] J. D. Watson and F. H. C. Crick, Nature **171**, 737 (1953).
[44] J. M. Bessman, I. R. Lehman, J. Adler, S. B. Zimmerman, E. S. Simms, and A. Kornberg, Proc. Natl. Acad. Sci. U. S. **44**, 633 (1958).

# 25
# Approaches to the Biosynthesis of Proteins

ALTON MEISTER

*Department of Biochemistry, Tufts University School of Medicine, Boston 11, Massachusetts*

IT is evident that, in considering protein synthesis, one must think about amino acids. Figure 1 provides a very general summary of what is known about the metabolic transformation of the amino acids. The figure shows that several hundred separate metabolic steps that do not lead to protein synthesis have been recognized. Each of these steps probably requires at least one specific protein. These enzymes, many other enzymes, and still other proteins that do not exhibit specific catalytic properties, are formed by processes about which very little is known—these are represented by a single large upward arrow in Fig. 1. All of the other arrows represent "pathways of metabolism," a series of stepwise processes. It is known that protein molecules of enormous complexity are synthesized by living cells at relatively rapid rates. Partly for this reason, there is a tendency to regard protein synthesis as a rapid spontaneous condensation of amino acids on a template. It is reasonable to believe that some type of model or template is needed for the synthesis of a specifically organized macromolecule composed of about 20 different building blocks. It is also clear that the specificity of proteins must ultimately be determined by genetic information. Nevertheless, it is highly improbable that the amino acids are brought together simultaneously by a process in which there are no intermediates. Failure to detect intermediates (peptide or other) in protein synthesis—or in intracellular protein degradation (which is often thought to be a stepwise process)—may probably be ascribed to inadequacies in the experimental approaches employed.

A complete description of the process of protein synthesis must include the mechanism of peptide-bond synthesis, and also the manner in which the amino-acid building blocks are arranged in specific sequences. Such a description should also explain the processes responsible for the formation of the specific configurations and linkages of the peptide chains of protein and for the binding to proteins of a variety of low molecular-weight compounds. In short, it is necessary to understand how essentially the same building blocks are used by living cells to yield such widely different structures as the contractile protein of muscle and the proteins of the blood, milk, silk, or hair. An attempt is made here to review some approaches to the problem.

## STUDIES ON INTACT ORGANISMS, TISSUES, AND CELLS

Several general aspects of the mechanism of protein synthesis are apparent from studies on intact organisms. For example, nutritional studies on rats showed that growth was greater when all of the dietary essential amino acids were available to the animal at the same time.[1,2] If some of the essential amino acids were fed several hours after the others had been fed, no growth or less growth was observed, as compared with animals that received all of the essential amino acids at the same time. The necessity for the simultaneous availability of the amino acids for growth leads to at least two important conclusions concerning protein synthesis. First, it is evident that the amino acids are not stored in the body to an appreciable extent, but are removed by degradative reactions or by excretion. Second, it is clear that protein synthesis takes place relatively rapidly. If an adult animal is deprived of even one of the essential building blocks of proteins, there is prompt loss of appetite, decrease of dietary intake, and development of negative nitrogen balance, with eventual loss of weight. These effects are promptly reversed by administration of the missing amino acid.[3] These observations suggest that there is continual synthesis of protein in an animal that is not growing, and also that, in such an animal, there must be a balance between protein breakdown and synthesis.

Studies in which isotopic amino acids were administered to animals demonstrated that such amino acids were incorporated into the tissue proteins of growing and nongrowing animals.[4] These experiments proved the existence of an over-all dynamic equilibrium in which there was a continuous synthesis and breakdown of body protein, and made it possible to determine rates of turnover. Thus, it was observed that certain proteins (e.g., those of intestinal mucosa, liver, kidney, spleen) turn over rapidly, while in others (skin, muscle, brain) the turnover of protein amino acids is relatively slow. In a recent study,[5] Thompson and Ballou administered

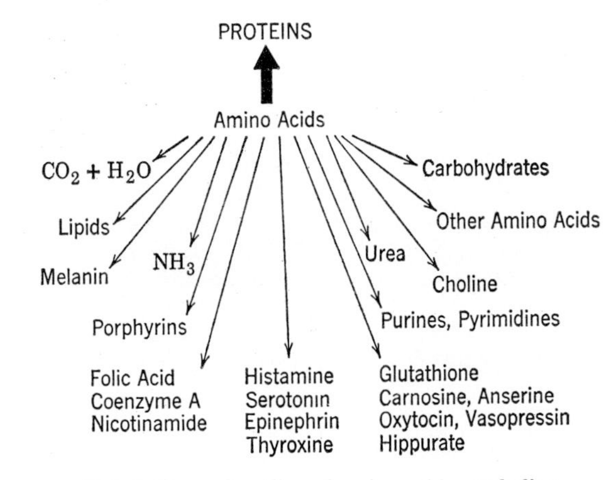


FIG. 1. General outline of amino-acid metabolism.

tritium oxide to rats daily from conception to 6 months of age; the animals were then sacrificed at various times over a 300-day period. Also, tritium oxide was administered for 124 days to a group of mature rats, which were then sacrificed at various times over a period of 360 days. It was found that components with half-lives of several days constituted a very small proportion of the total animal, and that about half of the body components exhibited half-lives of greater than 100 days. About three-quarters of the collagen fraction exhibited an apparent half-life of 1000 days, in general agreement with other studies.[6] Collagen and the muscle proteins represent a major fraction of the total body proteins; therefore, a relatively small fraction of the body protein is extensively involved in the dynamic state. However, this fraction includes proteins that are being degraded and synthesized at remarkably rapid rates. The observed turnover phenomena may be explained in part in terms of cell destruction and cellular protein secretion. Thus, the intracellular protein molecules may be stable until they are secreted by the cell or until the cell is destroyed. The latter interpretation applies to the hemoglobin of the red blood cells, which does not turn over until cellular destruction occurs.[7]

In *E. coli* cells synthesizing adaptive enzymes, the new proteins are formed almost entirely from free amino acid in the medium, and there is practically no utilization of the amino acids of the preformed proteins.[8–10] Synthesis of induced enzymes in *P. saccharophilia*, however, was accompanied by some utilization of preformed protein in growing cells, and a larger proportion was used in resting cells.[11] In nongrowing *E. coli* cells, considerable (4 to 5% per hour) degradation and synthesis of protein has been demonstrated.[12] It appears that, in growing *E. coli* cells, the rate of protein synthesis is very rapid as compared with the rate of degradation and, therefore, that the synthesis of new protein does not involve the utilization of significant quantities of amino acids derived from the degradation of other cellular proteins. However, in *E. coli* cells that are not carrying out net synthesis of protein, considerable degradation and synthesis of protein can be demonstrated.

In studies on Ehrlich's ascites carcinoma cells, Moldave[13,14] demonstrated an uptake of labeled amino acids into the cellular proteins. Incubation of labeled cells in a medium containing $C^{12}$-amino acids resulted in intracellular release of radioactive amino acids without concomitant net loss of protein. Similar results were obtained by Piez and Eagle[15] in experiments on HeLa cells grown in tissue culture. It was shown that there was no appreciable reutilization of degraded cells or of secreted proteins. The available data do not indicate whether all of the individual protein molecules of the cell turn over or whether they turn over at similar rates. It is also difficult to establish or to exclude the existence or exchange, i.e., replacement, of a single amino acid or a group of amino acids of a peptide chain without complete synthesis of the protein. Such exchange might occur at an intermediate stage of protein synthesis. Exchange of an amino acid with an amino-acid residue in the interior of a protein would require opening two peptide bonds and a mechanism that would bring the two portions of the peptide chain together after acceptance of the new amino-acid molecule. That some of the steps of protein synthesis may be reversible must be considered in connection with the turnover and exchange phenomena considered above. Studies by Simpson[16] showed that the release of amino acids from rat-liver slices is depressed by conditions (anaerobiosis, 2,4-dinitrophenol, cyanide) that inhibit utilization of energy. Amino-acid incorporation is also reduced under these conditions. The processes of degradation and synthesis may be closely related and could perhaps represent separate aspects of the same mechanism.

It is evident that information concerning the chemical reactions leading to protein synthesis will be needed to interpret the observed exchange phenomena. However, experimental work bearing on this question has arisen from studies designed to determine whether protein synthesis takes place by a stepwise process (in which free peptides or other peptide intermediates are formed), or by a mechanism in which the amino acids are arranged at one time on a template corresponding to the sequence of the protein to be synthesized and then dislodged from the template in a single step. The "template" idea is attractive; it provides a mechanism for a specific amino-acid sequence and is consistent with the "all-or-none" aspects of nutritional studies. Experimental studies relating to this problem fall in two categories: those which show that labeled amino acids are incorporated to the same extent into different positions of the peptide chains of proteins, and those which show that the labeling is unequal, i.e., that the specific activities of amino acids obtained from different portions of the protein molecule are not the same. The studies of Muir *et al.*[17] fall into the first category. They found that the *N*-terminal valine and the total of the nonterminal valine residues of the hemoglobin of rats previously injected with $C^{14}$-valine exhibited the same specific activity. Askonas *et al.*[18] injected a lactating goat with labeled amino acids and found that the specific activities of amino acids obtained from different portions of the casein molecule were, within experimental error, the same. On the other hand, Steinberg *et al.*[19] labeled the ovalbumin of hen's oviduct by incubating this tissue with $C^{14}O_2$, and found that the specific activity of the aspartate of the hexapeptide (cleaved by enzymatic conversion of the ovalbumin to plakalbumin) was significantly greater than that of the plakalbumin. Similar results were obtained in *in vivo* experiments, and with other proteins. The finding of unequal labeling is consistent with the existence of peptide intermediates, but other interpretations are not excluded. Uniform labeling would be the expected result if amino acids were added at one time to a template. Uniform labeling might result with peptide intermediates also, if these intermediates equilibrated rapidly with the available free amino acids. The recent discovery of separate amino-acid activating enzymes and ribonucleic-acid acceptors (see p. 217) makes it possible to conceive of mechanisms by which amino acids may be incorporated into different types of intermediates for protein synthesis. On the other hand, it is possible that templates exist for certain portions of protein molecules and that, in the final stages of protein synthesis, one or more large peptides prepared on different templates are combined. For example, the *A*- and *B*-peptide chains of insulin might require separate templates, and, if the two chains were not synthesized at the same rates, or if the templates were filled by amino acids from different pools, unequal labeling of the final molecule would result. Yet, in studies in which insulin was synthesized by calf-pancreas slices in the presence of labeled amino acids, unequal labeling within both of the *A*- and *B*-chains was observed; also, unequal labeling of ribonuclease was observed.

The finding of unequal labeling is clearly more significant than that of uniform labeling. Although unequal labeling might occur if a newly synthesized, uniformly labeled protein participated in subsequent "exchange" reactions, the evidence indicates that unequal labeling is observed in studies where the interval of time between introduction of free, labeled amino acid and isolation of the protein is relatively short. With longer time intervals, in the same system, labeling becomes uniform. Although the available data do not permit unequivocal conclusions, it is clear that the unequal labeling phenomenon, which has now been observed with several proteins, must be taken into account in any complete explanation of the process of protein synthesis.

As discussed above, the studies in which labeled amino acids were administered to intact animals revealed that relatively high concentrations of isotope were found in those organs known to be capable of very active protein synthesis, such as the liver and pancreas. When these organs were separated by differential centrifugation into subcellular fractions (nuclei, mitochondria, microsomes), the highest concentrations of isotope were found in the microsomal fraction. Thus, Hultin found that injected $N^{15}$-glycine was taken up most rapidly by the microsomal fraction of the liver of chicks.[20] Similar results were observed in experiments with mammalian tissues.[21,22] The concept has, therefore, developed that the microsomal particles of these tissues represent the most active fraction of the cell in protein synthesis. The microsome fraction (as usually obtained by differential centrifugation of liver homogenates) includes ribonucleoprotein particles which are attached to membranous material, and perhaps also to other cellular matter.[23,24] When the microsomal particles are treated with sodium deoxycholate, most of the lipid and lipoprotein portions of the microsome are removed, leaving insoluble ribonucleoprotein particles which contain about equal weights of protein and ribonucleic acid. These particles contain most of the ribonucleic acid of the microsomal fraction and about one-sixth of the total protein of this fraction.[25] When such particles were isolated from the livers of animals previously injected with labeled amino acids, they were found to have incorporated isotope more rapidly than the lipoprotein portion of the microsome and the other subcellular fractions.[26] Similar findings have been made in different laboratories, on several tissues and in various species. The findings, which are remarkably similar, lead to the conclusion that the most rapid uptake of amino acids takes place in the ribonucleoprotein particle of the microsomal fraction. However, as indicated below, mitochondria and nuclei also incorporate amino acids.

A close relationship between nucleic acids and protein synthesis has become increasingly evident.[27,28] It has been suggested that nucleic-acid synthesis must proceed concomitantly with the synthesis of protein. This conclusion is based on experiments which have shown a close correlation between the two synthetic processes (cf. Spiegelman *et al.*[29]). Nevertheless, it must be emphasized that it is not proven that the two syntheses are interdependent, and there is evidence inconsistent with this hypothesis. For example, in studies on phosphate-starved yeast, synthesis of protein was observed, whereas there was no synthesis of RNA.[30] Synthesis of amylase in pigeon-pancreas slices was not accompanied by net RNA synthesis.[31] In view of these findings, it is difficult to accept the view that protein and nucleic-acid synthesis are obligatorily linked, although there is much evidence for the belief that RNA plays a role in protein synthesis.

## CELL-FREE SYSTEMS

Although important concepts concerning protein synthesis have arisen from studies on intact organisms and tissue slices, a detailed explanation of protein synthesis requires isolation and study of the cellular catalytic components. Several approaches have been made to the study of protein synthesis in cell-free systems. One of these concerns enzyme systems, initially recognized because of their ability to catalyze hydrolysis of peptide bonds. Another line of investigation has been directed toward study of certain "model" systems that synthesize peptide or pseudo-peptide bonds. Finally, attempts have been made to study incorporation of isotopically labeled amino acids into proteins in cell-free systems. This work has been facilitated by the studies on model systems and by information obtained from cytological investigations.

### Formation of New Peptide Bonds Catalyzed by "Hydrolytic" Enzymes

The ability of proteolytic enzymes to catalyze the synthesis of peptide bonds was observed first more than

60 years ago. In these early experiments and in later ones, partial hydrolyzates of proteins were incubated with proteolytic enzymes; the formation of insoluble, high molecular-weight (2000 to 400 000) polymers (plasteins) occurred.[32,33] Although the free energy change associated with the formation of plasteins must be relatively small, the hydrolysis of a dipeptide to free amino acids proceeds spontaneously and virtually to completion. Significant reversal of such a reaction could probably not occur at the concentrations of amino acids usually present in the cell, unless a mechanism existed for continuous removal of the peptide from solution. The free energy change for the formation of glycylglycylglycylglycine from glycylglycine is about half of that for the formation of glycylglycine from glycine.[34] Thus, it appears that the free energy change for the formation of small peptides from amino acids is much greater than that for condensation of peptides to form larger peptides. If this type of reaction plays a role in the synthesis of protein, it would be expected to be significant later in the process when relatively large molecules have been formed by other reactions.

The recognition that proteolytic enzymes could catalyze replacement reactions and thus form new peptide bonds has led to a number of studies on transpeptidation reactions.[35–37] Fruton and collaborators[35,36] discovered that cathepsin-*C* (which hydrolyzes certain dipeptides at *p*H values near 5) catalyzes polymerization reactions at *p*H 7.5. Thus, glycyl-L-phenylalaninamide reacts to form an octapeptide amide as follows:

$$\text{gly-phe-NH}_2+\text{gly-phe-NH}_2$$
$$\downarrow$$
$$\text{gly-phe-gly-phe-NH}_2+\text{NH}_3$$
$$\downarrow \text{ gly-phe-NH}_2$$
$$\text{gly-phe-gly-phe-gly-phe-NH}_2+\text{NH}_3$$
$$\downarrow \text{ gly-phe-NH}_2$$
$$\text{gly-phe-gly-phe-gly-phe-gly-phe-NH}_2+\text{NH}_3$$

The reaction is similar in some respects to the formation of amylose from glucose-1-phosphate and also to the enzymatic synthesis of polynucleotides from nucleoside diphosphates.

$\gamma$-Glutamyl transpeptidation reactions may be involved in the synthesis of the polyglutamic acid produced by certain bacteria.[38–40] An enzyme preparation from *B. subtilis* catalyzes a reaction in which the $\gamma$-glutamyl group of glutamine is transferred to D-glutamic acid or to $\alpha$-D-glutamyl-D-glutamic acid to yield di- and tri-peptides. Continuation of this process could yield polyglutamic-acid molecules of considerable size. Accordingly, the glutamine-synthesis system and the transpeptidation enzyme would be the major catalytic components required for the synthesis of polyglutamic acid from glutamic acid.

It has recently been shown that transpeptidation reactions involving proteins can take place.[41] Thus, when insulin and $C^{14}$-glycyl-L-tyrosinamide were incubated with cathepsin-*C*, labeled insulin could be recovered, and it was found that the labeling occurred by substitution of an *N*-terminal amino-acid residue (probably mainly the $\alpha$-amino group of the *N*-terminal glycyl residue). Similarly, incubation of rat-liver mitochondria with $C^{14}$-tyrosinamide resulted in incorporation of isotope into the mitochondria.

### Incorporation of Amino Acids into Proteins

It was shown about ten years ago that homogenates of animal tissues could incorporate labeled L-amino acids into proteins.[42] The general procedure employed involved incubation of a tissue preparation with a labeled amino acid followed by precipitation of the protein with trichloroacetic acid. Subsequent treatment of the precipitate was designed to remove nonprotein components and labeled amino acid not bound to protein. It is obvious that this approach to the study of protein synthesis has a number of inherent difficulties. For example, types of binding other than those involving peptide linkage may occur. Furthermore, the relatively small quantities of isotope incorporated and the limited amounts of labeled material available are factors that have thus far prevented isolation of pure proteins. It was subsequently recognized that incorporation was inhibited by anaerobiosis and by inhibitors of oxidation and phosphorylation,[43] and that ATP could supply the energy for amino-acid incorporation.[42–45] As in the *in vivo* experiments, incorporation studies with liver homogenates led to greater labeling of the microsomes than of the other subcellular fractions.[45] Incorporation of labeled amino acids into protein was shown to take place in a system obtained from rat-liver homogenates consisting of a microsome-rich fraction, a soluble nondialyzable fraction, and an ATP-generating system.[46] The incorporated amino acids were not released from the microsomes by subsequent incubation with unlabeled amino acids. When the soluble nondialyzable fraction required for the incorporation of amino acids into microsomes was precipitated at *p*H 5, it was less active in incorporating amino acids, but activity was restored by the addition of either guanosine diphosphate or guanosine triphosphate.[47] These studies were extended to rat-hepatoma and mouse-ascites tumor cells.[48] It is of interest that the microsomes obtained from one source were labeled by incubation in a system containing soluble fraction obtained from another. The incorporation of amino acids was additive rather than competitive; the incorporation of one amino acid was not stimulated by the addition of 17 other amino acids. It seems probable that the preparations employed contain a pool of amino acids, perhaps complete, although this does not seem to have been specifically investigated.

Although there can be little doubt that the microsomes actively incorporate amino acids, it appears that other subcellular fractions can also carry out this activity. Thus, Mirsky and collaborators[49,50] observed incorporation of amino acids into isolated calf-thymus nuclei. In certain respects, incorporation was similar to that observed with microsomal preparations. Thus, anaerobiosis and dinitrophenol inhibited incorporation. L-Alanine was incorporated, whereas D-alanine was not. Nuclei labeled by incubation with a $C^{14}$-amino acid did not lose significant radioactivity on subsequent incubation in a medium containing the corresponding unlabeled amino acid. Incubation with a complement of L-amino acids did not stimulate $C^{14}$-alanine incorporation. Treatment of the nuclei with deoxyribonuclease led to reduced incorporation; the reduction in incorporation did, in fact, parallel the removal of DNA. The decrease in incorporation observed after removal of the DNA appears to be associated with a decrease in the ability of the nuclei to synthesize ATP; thus, recent work has shown that addition of polynucleotides restores ATP synthesis.[51] Separation of the nuclei into different fractions was carried out after incorporation, and the greatest specific activity was found in a nonhistone protein closely associated with the DNA; the incorporation into histones was relatively low. A ribonucleoprotein complex that was easily extractable in *p*H 7.1 buffer was also highly labeled.

Incorporation of amino acids into the ribonucleoprotein particles of plants has been observed.[52] The features of the particulate amino-acid incorporating system of plants are similar to those of the animal-tissue systems.

Gale and collaborators have carried out an extensive study of amino-acid incorporation in disrupted bacterial cells.[53–55] In the presence of the necessary amino acids, development of certain enzyme activities in disrupted cell preparations of *Staphylococcus aureus* has been reported. Gale has concluded that when disrupted preparations of *Staphylococcus aureus* are incubated with a single amino acid and a source of energy (ATP and hexose diphosphate), incorporation occurs by an exchange between the added amino acid and protein-bound amino acids. In these experiments, incubation of the labeled preparation with the corresponding unlabeled amino acid and a source of energy led to release of the radioactive amino acid from the disrupted cell preparation. Removal of nucleic acid from the disrupted cell preparations reduced the extent of incorporation. Reactivation of the system was accomplished by the addition of DNA or RNA. Digestion of staphylococcal RNA by ribonuclease was reported to increase the ability of the RNA preparation to stimulate amino-acid incorporation. A number of active components were isolated from ribonuclease digests of staphylococcal or yeast RNA. The nature of these factors is not yet entirely clear, but it appears that certain of them are of low molecular weight and can be separated from polynucleotide material.

Simpson *et al.*[56,57] found that the rate of amino-acid incorporation into the mitochondria of muscle was of the same order of magnitude as that into the microsomes, and that isolated muscle and liver mitochondria were able to incorporate amino acids into protein.[58] Incorporation into liver mitochondria required ATP, and was increased by addition of magnesium ions and a soluble fraction obtained from rat liver. The mitochondrial-incorporation system appears similar to those for incorporation of amino acids into microsomes and thymus nuclei. One major and somewhat surprising difference, however, is that the incorporation of amino acids is increased by treatment of the mitochondria with ribonuclease. Inhibition by the nucleases is a characteristic feature of the incorporation systems of nuclei and microsomes.

Ochoa and Beljanski[59] have reported that a particulate fraction of sonically disrupted *Alcaligenes faecalis* incorporated $C^{14}$-L-amino acids into protein. Incorporation was dependent upon oxidative phosphorylation for the generation of ATP; however, the system was reported not to contain amino-acid activating enzymes (see below) as measured by the absence of amino-acid stimulated, pyrophosphate incorporation into ATP. The incorporation of one $C^{14}$-amino acid was stimulated by the addition of 18 other amino acids. They also observed small net increases of protein-nitrogen. Incorporation was decreased by treatment of the particles with *M* sodium chloride; reactivation could be obtained by addition of RNA or of a soluble enzyme ("amino-acid incorporation enzyme") that was purified from the supernatant. Incorporation was not reversible, nor was it inhibited by chloramphenicol. It is of considerable interest that incorporation did not appear to require activating enzymes of the type found in other systems.

The studies on disrupted bacterial systems have led to results somewhat different from those observed with the animal microsomal, mitochondrial, and nuclear systems. The major differences include the reversibility of incorporation of amino acids, and the observation of net protein synthesis. These phenomena are, of course, characteristic of intact cell systems. That some of the incorporation observed in studies with bacteria may represent synthesis of cell-wall material is suggested by recent work. Mandelstam and Rogers[60] incubated *Staphylococcus aureus* cells with radioactive glycine, glutamic acid, or lysine in the presence of chloramphenicol. The total precipitate obtained after treatment with trichloroacetic acid or organic solvents, and (in separate experiments) the cell-wall material and protein were isolated. The data obtained indicate that the amino acids were incorporated into the cell-wall material, and that the "protein synthesis" observed in the total precipitate was, in fact, an increase in the mass of the cell wall. Incorporation into cell-wall material is not significantly affected by chloramphenicol; some incorporation into preparations of disrupted staphylococci can also occur in the presence of chloramphenicol. It is obviously important to determine how much of the observed incorporation by bacterial preparations represents cell-wall or cell-membrane synthesis.

### Possible Intermediates in Protein Synthesis

Clues to the mode of formation of the peptide bonds of proteins might be expected to come from studies of the synthesis of smaller and chemically characterized compounds possessing peptide or pseudo-peptide linkages, e.g., benzoylglycine (hippuric acid), glutathione, glutamine, pantothenic acid. The major outlines of the enzymatic reactions leading to the synthesis of these compounds are now known. Study of these and of similar systems has contributed significantly to the understanding of amino-acid activation and the intermediates in protein synthesis. It may, therefore, be valuable to review briefly several activation phenomena leading to formation of peptide or pseudopeptide bonds.

Studies on the enzymatic synthesis of glutathione demonstrated that the synthesis from the three amino-acid components takes place by a stepwise process:[61,62]

$$\text{L-glutamic acid} + \text{L-cysteine} + \text{ATP} \xrightleftharpoons{Mg^{++}} \text{L-}\gamma\text{-glutamylcysteine} + \text{ADP} + \text{Pi}, \quad (1)$$

$$\text{L-}\gamma\text{-glutamylcysteine} + \text{glycine} + \text{ATP} \xrightleftharpoons{Mg^{++}} \text{L-glutathione} + \text{ADP} + \text{Pi}. \quad (2)$$

Both reactions have been separately demonstrated. As yet, there is no evidence for the formation of a free intermediate in either reaction.

The first reaction in glutathione synthesis is similar to that catalyzed by the glutamine synthesis enzyme:[63–70]

$$\text{glutamic acid} + NH_3 + \text{ATP} \xrightleftharpoons{Mg^{++}} \text{glutamine} + \text{ADP} + \text{Pi}. \quad (3)$$

The purified glutamine synthesis enzyme also catalyzes synthesis of $\gamma$-glutamyl hydroxamic acid when hydroxylamine is substituted for ammonia, as well as a transfer reaction which requires $Mg^{++}$ (or $Mn^{++}$) and catalytic quantities of ADP and inorganic phosphate:

$$\text{glutamine} + NH_2OH \rightleftharpoons \gamma\text{-glutamylhydroxamic acid} + NH_3. \quad (4)$$

No free intermediates have been isolated. Incubation of $O^{18}$-labeled glutamic acid in this system leads to the formation of $O^{18}$-inorganic phosphate and glutamine in stoichiometric quantities.[71,72] It is possible that the transfer of $O^{18}$ occurs via $\gamma$-glutamylphosphate, or perhaps by a more complex mechanism. Experiments with synthetic $\gamma$-glutamylphosphate revealed that, although this compound reacted with very low concentrations of ammonia to form glutamine, there was no significant acceleration of the rate of glutamine synthesis in the presence of enzyme.[70] The possibility remains that enzyme-bound $\gamma$-glutamylphosphate is the active intermediate, according to the following scheme:

$$\text{enzyme} + \text{glutamate} + \text{ATP} \rightleftharpoons \text{enzyme-}\gamma\text{-glutamyl phosphate} + \text{ADP}, \quad (5)$$

$$\text{enzyme-}\gamma\text{-glutamyl phosphate} + NH_3 \rightleftharpoons \text{enzyme} + \text{glutamine} + \text{phosphate}. \quad (6)$$

Although the glutamine-synthesis enzyme catalyzes a transfer reaction [reaction (4)], there is thus far no evidence for the existence of a natural acceptor. Several other enzymes catalyze transfer reactions involving the $\gamma$-carboxyl group of glutamic acid, e.g., $\gamma$-glutamyl transpeptidases that act on glutathione,[37] various glutamine transferases.[73,74] With the possible exception of the $\gamma$-glutamyl transpeptidase of *B. subtilis*,[38–40] there is as yet no clue as to the physiological roles of these enzymes that act on the $\omega$-amide or $\omega$-carboxyl groups of the dicarboxylic amino acids. There is no evidence for $\gamma$-glutamyl or $\beta$-aspartyl linkages in proteins, although this type of linkage occurs in certain peptide antibiotics.[75] It is conceivable that activation of the dicarboxylic acids for protein synthesis occurs by primary attack on the $\omega$-carboxyl group followed by rearrangement to an $\alpha$-carboxyl-activated molecule. This type of rearrangement has been observed nonenzymatically with derivatives of glutamic acid and aspartic acid.[70] Another possible mechanism for dicarboxylic amino-acid activation is cyclic anhydride formation. Such derivatives might react enzymatically to give $\alpha$-carboxyl substituted products; similar reactions occur nonenzymatically.[76]

In contrast to the activation of glutamic acid and of $\gamma$-glutamylcysteine, which are associated with a cleavage of ATP to ADP, the synthesis of pantothenic acid[77] from pantoic acid and $\beta$-alanine and of benzoylglycine from benzoic acid and glycine involves formation of pyrophosphate from ATP. The enzyme system responsible for pantothenic-acid synthesis catalyzes an incorporation of PP into ATP in the presence of pantoic acid. A similar exchange is catalyzed by the acetate-activating enzyme:

$$\text{Acetate} + \text{ATP} + \text{CoA} \rightleftharpoons \text{acetyl CoA} + \text{AMP} + \text{PP}. \quad (7)$$

Evidence for the existence of an acyl-adenylate intermediate was first achieved by Berg[78,79] in studies on the acetate-activating system. Thus, it was shown that addition of synthetic acetyl adenylate to the enzyme system could yield either ATP or acetyl CoA:

$$\text{Acetate} + \text{ATP} \rightleftharpoons \text{acetyl AMP} + \text{PP}, \quad (8)$$

$$\text{Acetyl AMP} + \text{CoASH} \rightleftharpoons \text{acetyl CoA} + \text{AMP}. \quad (9)$$

Evidence suggesting participation of acyl adenylates in activation of fatty acids was subsequently obtained.[80] Earlier, it had been shown by Chantrenne[81] that CoA

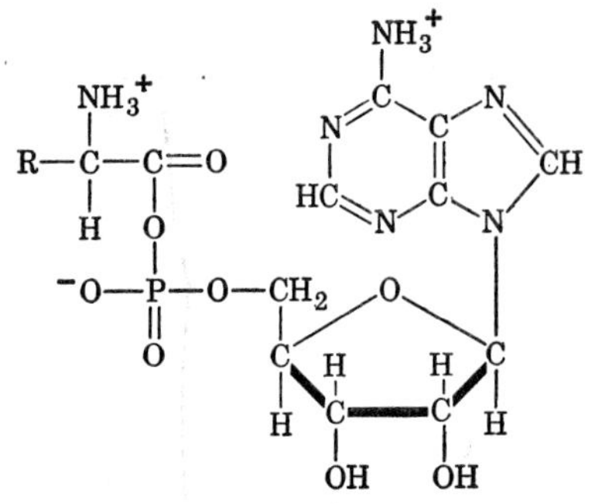

FIG. 2. General structure of aminoacyl adenylate.

was required for benzoylglycine formation. The synthesis of benzoylglycine appears to take place as follows[82–86]

$$\text{Benzoic acid} + \text{ATP} \rightleftharpoons \text{benzoyl AMP} + \text{PP}, \quad (10)$$

$$\text{Benzoyl AMP} + \text{CoA-SH} \rightleftharpoons \text{benzoyl CoA} + \text{AMP}, \quad (11)$$

$$\text{Benzoyl CoA} + \text{glycine} \rightleftharpoons \text{benzoylglycine} + \text{CoA-SH}. \quad (12)$$

Although formation of benzoyl AMP was not demonstrated, when added to the system it was active (in the presence but not in the absence of CoA-SH) in forming benzoylglycine in human liver and kidney preparations. The synthesis of phenylacetyl-L-glutamine by human tissues appears to occur by an analogous reaction.[85,86]

It thus appears that acyl adenylates are intermediates in the enzymatic activation of acetate and fatty acids, and in the synthesis of benzoylglycine and phenylacetylglutamine. The evidence for these intermediates is incomplete, however, for the enzymatic formation of these acyl adenylates has not been demonstrated. It is probable that the actual intermediates are enzyme-bound and, therefore, that the number of molecules of the intermediate present at a given time is no greater than the number of acceptor sites on the enzyme.

Amino-acid dependent, enzymatic PP-ATP exchange was independently observed by Hoagland *et al.*[87,88] and by Berg,[78,79,89] and several features of the reaction suggested that the activated amino-acid intermediate was an aminoacyl adenylate (Fig. 2). The phenomenon has now been observed and studied in a number of laboratories[90–93]; several specific amino-acid activating enzymes have been purified, and evidence has been obtained for the formation of an enzyme-bound, aminoacyl adenylate intermediate. The general reaction may be written as follows:

$$\text{Amino acid} + \text{enzyme} + \text{ATP} \xrightleftharpoons{Mg^{++}} \text{enzyme-aminoacyl adenylate} + \text{PP}. \quad (13)$$

In the presence of the specific amino acid, the enzyme catalyzes incorporation of PP into ATP. When the reaction is carried out in the presence of enzyme, ATP, $Mg^{++}$, amino acid, and high concentrations of hydroxylamine, the corresponding aminoacyl hydroxamate is formed. That the intermediate formed is an aminoacyl adenylate is suggested, in analogy with acetate activation, by the formation of pyrophosphate rather than orthophosphate. DeMoss *et al.*[94] found that leucyl-adenylate reacted enzymatically with pyrophosphate to give ATP (reversal of reaction[13]). Further evidence for formation of an anhydride linkage between the phosphoric-acid group of adenylic acid and the carboxyl group of amino acids was obtained in studies in which transfer of $O^{18}$ from the carboxyl group of an amino acid to adenylic acid was observed during enzymatic activation.[95] More-direct evidence for the aminoacyl-adenylate intermediate was obtained by Karasek *et al.*,[96] who extracted a compound that exhibited the properties of synthetic tryptophanyl adenylate from reaction mixtures containing large amounts of pancreatic tryptophan-activating enzyme.[97] Also, it was found that the tryptophan-activating enzyme could catalyze ATP synthesis from pyrophosphate and a wide variety of $\alpha$-aminoacyl adenylates including those of D-amino acids.[96,98] ($\beta$-Alanyl adenylate, acetyl adenylate, and carbobenzoxy $\alpha$-amino acyl adenylates were not active.) On the other hand, only L-tryptophan was active in the enzymatic hydroxamate-forming reaction and in the pyrophosphate-ATP exchange. Similar results were observed with a yeast methionine-activating enzyme.[99] The explanation for this phenomenon is not yet clear. The findings with the D- and L-amino-acid derivatives are reminiscent of the observation that D-glutamic acid is activated almost as rapidly as L-glutamic acid by the glutamine-synthesis enzyme, whereas only the L-isomer of glutamine is significantly active in the transfer reaction.[67]

It would appear that the aminoacyl adenylate is stabilized by its binding to the enzyme. Such stabilization may be accomplished by a linkage involving the $\alpha$-amino group; it has been found that *N*-substituted aminoacyl adenylates (e.g., carbobenzoxy-aminoacyl adenylates) are much more stable than are the corresponding free amino compounds.[100] The high reactivity of the aminoacyl adenylates is indicated by their rapid nonenzymatic reaction with hydroxylamine to form aminoacyl hydroxamates, with ammonia to yield amino-acid amides, and with amino acids to form peptides. Synthetic aminoacyl adenylates can react nonenzymatically with nucleic acid[100] and with microsomal[100] and mitochondrial[41] proteins. The incorporation of amino acids from synthetic aminoacyl adenylates into the particulate proteins and into soluble proteins represents mainly acylation of the available free amino groups of the protein. Heated mitochondrial and microsomal preparations are acylated to a much greater extent than are the corresponding unheated materials. Presumably, heat denaturation increases the number of available free amino groups. It is evident from the observations which demonstrate the high reactivity of the aminoacyl adenylates that they must be selectively transferred, and that their participation in other reactions must involve a mechanism which prevents the occurrence of nonspecific acylation reactions.

It appears that very little (if any) *free* aminoacyl adenylate is formed enzymatically. However, compounds other than hydroxylamine can react with enzymatically synthesized aminoacyl adenylates. Thus, incubation of tryptophan-$C^{14}$ with ATP, $Mg^{++}$, activating enzyme, and various protein preparations (rat-liver microsomes, serum albumin, ovalbumin) gave labeled protein preparations, and similar studies with soluble RNA preparations yielded RNA that was labeled with amino acids. The reaction of enzyme-bound aminoacyl adenylates with RNA appears to be more specific, since there is evidence for the existence on soluble RNA molecules of specific amino-acid acceptor sites (see following). Aminoacyl adenylates, whether produced enzymatically or by chemical synthesis, can react with nucleotides to yield products which appear to be bound by ester linkage to the 2′ (or 3′) hydroxyl groups of the ribose moieties.

A number of specific amino-acid activating enzymes have been found. Thus, enzymes specific for methionine,[89,99] tryptophan,[97] tyrosine,[92,101] valine,[102] and leucine[101] have been described. Novelli[98] has studied the activation of amino acids by preparations of microorganisms, plants, and animal tissues. With most of these preparations, only 8 to 10 amino acids stimulated pyrophosphate-ATP exchange, although it is possible that the background exchange (owing to endogenous amino acids) may have obscured the effects of the added amino acids. With guinea-pig tissue extracts, all of the amino acids appeared to catalyze at least a small amount of exchange.[93] It is obviously of importance to determine whether or not this type of activating mechanism exists for all of the amino acids. It is curious that the tryptophan-activating enzyme should be so prominent in beef pancreas and in other sources, since this amino acid is present in very low concentrations in proteins and is absent from some proteins. However, it is possible that improved methods of isolation will be needed to obtain all of the individual enzymes.

There are difficulties associated with the study of the $\alpha$-activation of glutamic acid and aspartic acid; thus, these amino acids (and their $\omega$-amides) may form $\omega$-carboxyl-linked intermediates. ATP exchange reactions and the hydroxamic-acid test system may, therefore, be unreliable guides as to the occurrence of $\alpha$-activation. It is conceivable that these amino acids are activated by a different mechanism as suggested above.

Evidence for the accumulation of labeled intermediates during the incorporation of $C^{14}$-leucine into rat-liver microsomes was reported by Hultin and Beskow.[103] Hoagland, Stephenson, and Zamecnik[104,105] provided evidence for the transfer of labeled amino acids to ribonucleic acid of the soluble enzyme fraction of liver homogenate (which also contains amino-acid-activating enzyme activity). They also obtained evidence for the further transfer of the amino-acid moiety to microsomal protein. The transfer to protein was dependent upon the presence of guanosine triphosphate. Other studies have provided evidence for the hypothesis that soluble ribonucleic acid serves as an acceptor for activated amino acids.[101,102,106–108] The reaction, which appears to be analogous to other activation reactions [Eqs. (7) to (12)], but with ribonucleic acid instead of CoA as the acceptor, may be formulated as follows:

$$\text{Amino acid} + \text{ATP} \rightleftharpoons \text{aminoacyl adenylate} + \text{PP} \quad (14)$$

$$\text{Aminoacyl adenylate} + \text{RNA} \rightleftharpoons \text{amino acid-RNA} + \text{AMP}. \quad (15)$$

It appears possible that both reactions are catalyzed by the same enzyme. When the total fraction of soluble RNA is incubated with a specific amino-acid activating enzyme and the corresponding amino acid, the RNA becomes labeled. A separate activating enzyme appears necessary for linking each amino acid to RNA. There may be specific RNA sites for each amino acid; thus, labeling by several amino acids is approximately the sum of maximum labeling with each. There are also preliminary indications[101,102] that there may be separate RNA molecules for each amino acid. Formation of the RNA amino-acid complex is reversible according to Eqs. (15) and (16); thus, incubation of the RNA amino-acid complex with AMP and pyrophosphate yields ATP.[101,102,109] The RNA amino-acid complex is alkali-labile, but considerably more stable than are aminoacyl adenylates. This suggests that the linkage of amino acid to RNA does not involve an anhydride linkage. Potter and Dounce[110] isolated alkali-stable polynucleotide fractions from calf pancreas that contained amino acids and peptides; they suggested that the amino acids were bound to the nucleotides by phosphoamide bonds. The stability of this linkage to alkali effectively excludes it as the binding mechanism in the enzymatically formed RNA/amino-acid complexes. An attractive possibility, consistent with the alkali-lability of the bond, is an ester linkage involving the 2′ or 3′ hydroxyl group of ribose. Recent work by Hecht, Stephenson, and Zamecnik,[111] has shown that the soluble RNA of ascites-cell tumors incorporates adenine nucleotide into the terminal position in the presence of a soluble enzyme and ATP. Such incorporation is increased by adding cytidine triphosphate. Their evidence was interpreted to mean that cytosine and adenine nucleotides are added to RNA in this sequence. It is of interest that the incorporation of a great many amino acids was enhanced by addition of these nucleotides, suggesting that these end groups participate in the attachment of activated amino acids to RNA.

The further transfer of amino-acid moieties from RNA to microsomal protein has been described by Hoagland *et al.*, as mentioned above. The work of Hultin and Beckow,[103] which demonstrated a two-step incorporation of leucine into rat-liver microsomal protein, now may be interpretated in terms of the formation of an intermediate RNA-leucine complex followed by transfer to protein. The mode of transfer and the relationship of the incorporated amino acids to specific characterized proteins are major problems that await solution.

### LATER STAGES OF PROTEIN SYNTHESIS

The incorporation of amino acids into the microsomal particles of animal cells has thus far not been shown to be associated with the formation of recognizable specific proteins. It is possible that the synthesis of protein molecules is completed in the ribonucleoprotein particle. Failure to detect it at this stage may be ascribed to the very small quantity of protein present, or to binding of the newly formed protein to the particle. If the latter interpretation is correct, existence of additional reactions that remove the protein and transport it to other parts of the cell must be postulated. Another explanation is that the newly formed peptide material incorporated by the microsomal particle undergoes additional transformation before becoming a recognizable protein molecule. The existence of such a protein precursor has been considered by several authors.[112,113] Such transformations might involve reactions that link relatively large peptide chains together, introduce disulfide bonds, and add smaller nonpeptide molecules. Transpeptidation reactions may function at this stage of protein synthesis. Experimental findings that appear to bear on this question have arisen from studies on the synthesis of pancreatic enzymes by Daly and Mirsky.[113] This work showed that there was no significant change in the total protein content of pancreas during secretion and synthesis; however, there were significant and marked changes in enzyme concentration. There appears, therefore, to be a distinction between rapidly formed precursor protein and specific enzyme protein. The transformation of precursor protein to specific proteins associated with secretion may be a slower process than the formation of the precursor protein. Similar conclusions are suggested by experiments of Campbell and Work[114] on incorporation of amino acids by the rabbit mammary gland, by Green and Anker[115] on the synthesis of $\alpha$-globulin, and by Putnam *et al.*[116] on Bence-Jones protein synthesis. Peters[117,118] has reported that labeled carbon appears more slowly in serum albumin than in the total protein fraction, when chicken-liver slices are incubated with radioactive carbon dioxide or glycine. The appearance of isotope in the total liver-protein fraction preceded the labeling of albumin by 15 or 20 min. Further studies indicate that serum albumin, isolated by an immunological procedure, accumulates in the cytoplasmic granules before its appearance in the soluble fraction. These findings suggest that an antigenically reactive serum-albumin molecule is formed on the ribonucleoprotein particle or the associated lipoprotein membrane.

The ribonucleoprotein particles obtained from various sources have been found to contain about equal weights of RNA and protein. Such particles have been obtained from microorganisms, higher plants and animal tissues. Recent work indicates that particles obtained from the same source are not homogeneous, structurally or biochemically. Furthermore, Simkin and Work[119] found that amino acids are incorporated *in vitro* into the proteins of different microsomal subfractions at different rates, and that the pattern of *in vitro* incorporation is not the same as that observed *in vivo*.

The amino acids incorporated into isolated thymus nuclei become linked to a specific protein closely associated with the DNA.[49,50] In the mitochondrial system, incorporation of labeled valine into cytochrome-*c* was observed.[58] These findings are consistent with the belief that specific proteins may be synthesized in nuclei and mitochondria.

### DISCUSSION

The available experimental data indicate that virtually all cells can incorporate amino acids, and that amino-acid incorporation (and perhaps also protein synthesis) can take place in the microsomal, nuclear, and mitochondrial fractions of certain animal cells. The general features of amino-acid incorporation into these structures are remarkably similar. The need for amino-acid activation is apparent in these systems and also in those obtained from plants and microorganisms. Additional data are required before it can be concluded that all of the amino acids are activated by formation of aminoacyl adenylates; for example, the evidence concerning the dicarboxylic amino acids and the corresponding $\omega$-amides is incomplete, as observed above. Although acyl adenylates appear to be the reactive intermediates in acetate, fatty acid, benzoate, and phenylacetate activation, another type of activation, perhaps involving carboxyl phosphate anhydrides, appears to be involved in both steps of glutathione synthesis and in glutamine synthesis. The studies of Ochoa and Beljanski[59] lend support to the possibility that other types of amino-acid activation may occur. That phosphorylated RNA may serve as a source of energy for protein synthesis has been considered and a phosphorylated template has also been suggested. The conversion of amino acids to aminoacyl adenylates and to RNA/amino-acid complexes may form part of the mechanism responsible for transport of amino acids into the cell. Amino-acid activation may play a role in both transport and protein synthesis.

There have been many published speculations concerning the existence and nature of templates for protein synthesis. The concept of a template or model for protein formation seems almost indispensable in order to explain the high degree of specificity apparently associated with the arrangement of amino acids in proteins. A system of separate and specific activating enzymes and RNA acceptor sites could provide some specificity in preparing amino acids for synthesis. Yet it appears that neither the activating systems nor the complete synthesizing system are absolutely specific, for such unnatural amino acids as ethionine and tryptazan can be activated and incorporated into protein. The incorporation of a number of "foreign" amino acids into "false" protein has now been observed. If ethionine can occasionally replace methionine in the synthesis of protein, and if fluorophenylalanine can sometimes be incorporated in place of phenylalanine or tyrosine, is it possible that the naturally occurring amino acids may themselves occasionally be "misplaced" on the peptide chains? Thus, valine might "accidentally" replace isoleucine or vice versa, and so forth. It seems possible that the specificity of amino-acid sequences may not be absolute.

The participation of soluble and particulate RNA in the process of amino-acid incorporation is consistent with the belief that RNA functions as a template. However, there is no evidence that unequivocally excludes the possibility that protein may serve as a template. Specific enzymes are apparently needed throughout the process of protein synthesis, and it appears very probable that both protein and RNA function in the formation of specific amino-acid sequences. The nature of the process by which genetic information is supplied to the protein-synthesizing mechanisms, and discussions of possible relationships between the base sequences of nucleic acids and the amino-acid sequences of proteins are considered elsewhere in this volume (see p. 227). Information relating to these important problems may arise from studies on the structure of the soluble RNA molecules that accept the amino-acid moieties of aminoacyl adenylates.

The observation that incorporation of amino acids into microsomes from one source could be carried out with soluble enzymes obtained from another is very interesting. If incorporation represents synthesis of a specific protein, then it would appear that the specificity must reside in the microsomal particle; alternatively, a "mixed" system of this type synthesizes a "hybrid" protein. It is possible that the microsomal-incorporation system represents only a portion of the cellular-protein synthetic system, and that additional components present in the intact cell must be added in order to obtain net protein synthesis. As yet, there is no unequivocal demonstration of net synthesis of a specific protein in a cell-free system. Such an achievement would be an important step toward the goal of understanding the process of protein synthesis.

## BIBLIOGRAPHY


[1] L. E. Frazier, R. W. Wissler, C. H. Steffee, R. L. Woolridge, and P. R. Cannon, J. Nutrition **33**, 65 (1947).

[2] E. Geiger, Science **111**, 594 (1950).

[3] W. C. Rose, Federation Proc. **8**, 546 (1949).

[4] R. Schoenheimer, *The Dynamic State of Body Constituents* (Harvard University Press, Cambridge, 1942).

[5] R. C. Thompson and J. E. Ballou, J. Biol. Chem. **223**, 795 (1956).

[6] A. Neuberger and H. G. B. Slack, Biochem. J. **53**, 47 (1953).

[7] D. Shemin and D. Rittenberg, J. Biol. Chem. **166**, 621, 627 (1946).

[8] B. Rotman and S. Spiegelman, J. Bact. **68**, 419 (1954).

[9] D. S. Hogness, M. Cohn, and J. Monod, Biochim. et Biophys. Acta **16**, 99 (1955).

[10] A. L. Koch and H. R. Levy, J. Biol. Chem. **217**, 947 (1955).

[11] A. Markovitz and H. P. Klein, J. Bact. **70**, 641, 649 (1955).

[12] J. Mandelstam, Biochem. J. **69**, 110 (1958).

[13] K. Moldave, J. Biol. Chem. **221**, 543 (1956).

[14] K. Moldave, J. Biol. Chem. **225**, 709 (1957).

[15] K. Piez and H. Eagle, Federation Proc. **17**, 289 (1958).

[16] M. V. Simpson, J. Biol. Chem. **201**, 143 (1953).

[17] H. M. Muir, A. Neuberger, and J. C. Perrone, Biochem. J. **52**, 87 (1952).

[18] B. A. Askonas, P. N. Campbell, and T. S. Work, Biochem. J. **58**, 326 (1954).

[19] D. Steinberg, M. Vaughan, and C. B. Anfinsen, Science **124**, 389 (1956).

[20] T. Hultin, Exptl. Cell Research **1**, 376 (1950).

[21] H. Borsook, C. L. Deasy, A. J. Haagen-Smit, G. Keighley, and P. H. Lowy, J. Biol. Chem. **187**, 839 (1950).

[22] V. G. Allfrey, M. M. Daly, and A. E. Mirsky, J. Gen. Physiol. **37**, 157 (1953).

[23] G. E. Palade, J. Biophys. Biochem. Cytol. **1**, 59 (1955).

[24] F. S. Sjöstrand and V. Hanzon, Exptl. Cell Research **7**, 393 (1954).

[25] R. B. Loftfield, Progr. in Biophys. and Biophys. Chem. **8**, 347 (1957).

[26] J. W. Littlefield, E. B. Keller, J. Gross, and P. C. Zamecnik, J. Biol. Chem. **217**, 111 (1955).

[27] T. O. Caspersson, *Cell Growth and Cell Function* (W. W. Norton and Company, Inc., New York, 1950).

[28] J. Brachet in *The Nucleic Acids*, E. Chargaff and J. N. Davidson, editors (Academic Press, Inc., New York, 1955), Vol. II, p. 475.

[29] S. Spiegelman, H. O. Halvorson, and R. Ben-Ishai, *A Symposium on Amino Acid Metabolism* (The Johns Hopkins Press, Baltimore, Maryland, 1955), p. 124.

[30] G. Schmidt, K. Seraidarian, L. M. Greenbaum, M. D. Hickey, and S. J. Thannhauser, Biochim. et Biophys. Acta **20**, 135 (1956).

[31] L. E. Hokin and M. R. Hokin, Biochim. et Biophys. Acta **13**, 236, 401 (1954).

[32] H. Borsook, Advances in Protein Chem. **8**, 127 (1953).

[33] H. Wasteneys and H. Borsook, Physiol. Rev. **10**, 110 (1930).

[34] K. U. Linderstrøm-Lang in *Lane Medical Lectures, 1951* (Stanford University Press, Stanford, 1952), Vol. VI.

[35] J. S. Fruton in *Essays in Biochemistry* (John Wiley and Sons, Inc., New York, 1956), p. 106.

[36] J. S. Fruton in *Harvey Lectures* (Academic Press, Inc., New York, 1955–1956), Vol. LI, p. 64.

[37] C. S. Hanes, F. J. R. Hird, and F. A. Isherwood, Biochem. J. **51**, 25 (1952).

[38] W. J. Williams and C. B. Thorne, J. Biol. Chem. **210**, 203 (1954).

[39] W. J. Williams and C. B. Thorne, J. Biol. Chem. **211**, 631 (1954).

[40] W. J. Williams and C. B. Thorne, J. Biol. Chem. **212**, 427 (1955).

[41] C. Zioudrou, S. Fujii, and J. S. Fruton, Proc. Natl. Acad. Sci. U. S. **44**, 439 (1958).

[42] H. Tarver in *The Proteins*, H. Neurath and K. Bailey, editors (Academic Press, Inc., New York, 1954), Vol. II, p. 1199.

[43] H. Borsook, C. L. Deasy, A. J. Haagen-Smit, G. Keighley, and P. H. Lowy, J. Biol. Chem. **186**, 297 (1950).

[44] H. Borsook, C. L. Deasy, A. J. Haagen-Smit, G. Keighley, and P. H. Lowy, J. Biol. Chem. **196**, 669 (1952).

[45] P. Siekevitz, J. Biol. Chem. **195**, 549 (1952).

[46] P. C. Zamecnik and E. B. Keller, J. Biol. Chem. **209**, 337 (1954).
[47] E. B. Keller and P. C. Zamecnik, J. Biol. Chem. **221**, 45 (1956).
[48] J. W. Littlefield and E. B. Keller, J. Biol. Chem. **224**, 13 (1957).
[49] A. E. Mirsky, S. Osawa, and V. G. Allfrey, Cold Spring Harbor Symposia Quant. Biol. **21**, 49 (1956).
[50] V. G. Allfrey, A. E. Mirsky, and S. Osawa, J. Gen. Physiol. **40**, 451 (1957).
[51] V. G. Allfrey and A. E. Mirsky, Proc. Natl. Acad. Sci. U. S. **43**, 589 (1957).
[52] G. C. Webster, J. Biol. Chem. **229**, 535 (1957).
[53] E. F. Gale and J. P. Folkes, Biochem. J. **55**, 721 (1953).
[54] E. F. Gale and J. P. Folkes, Biochem. J. **59**, 661, 675 (1955).
[55] E. F. Gale, Biochem. Soc. Symposia **14**, 47 (1957).
[56] M. V. Simpson and J. R. McLean, Biochim. et Biophys. Acta **18**, 573 (1955).
[57] J. R. McLean, G. L. Cohen, and M. V. Simpson, Federation Proc. **15**, 312 (1956).
[58] H. M. Bates, V. M. Craddock, and M. V. Simpson, J. Am. Chem. Soc. **80**, 1000 (1958).
[59] M. Beljanski and S. Ochoa, Proc. Natl. Acad. Sci. U. S. **44**, 494 (1958).
[60] J. Mandelstam and H. J. Rogers, Nature **181**, 956 (1958).
[61] J. E. Snoke and K. Bloch in *Glutathione*, S. P. Colowick, editor (Academic Press, Inc., New York, 1954), p. 129.
[62] J. E. Snoke and K. Bloch, J. Biol. Chem. **213**, 825 (1955).
[63] J. F. Speck, J. Biol. Chem. **168**, 403 (1947).
[64] J. F. Speck, J. Biol. Chem. **179**, 1387, 1405 (1949).
[65] W. H. Elliott, Biochem. J. **49**, 106 (1951).
[66] W. H. Elliott, J. Biol. Chem. **201**, 661 (1953).
[67] L. Levintow and A. Meister, J. Am. Chem. Soc. **75**, 3039 (1953).
[68] L. Levintow, A. Meister, G. H. Hogeboom, and E. L. Kuff, J. Am. Chem. Soc. **77**, 5304 (1955).
[69] L. Levintow and A. Meister, J. Biol. Chem. **209**, 265 (1954).
[70] A. Meister, Physiol. Rev. **36**, 103 (1956).
[71] P. D. Boyer, O. J. Koeppe, and W. W. Luchsinger, J. Am. Chem. Soc. **78**, 356 (1956).
[72] A. Kowalsky, C. Wyttenbach, L. Langer, and D. E. Koshland, Jr., J. Biol. Chem. **219**, 719 (1956).
[73] H. Waelsch, Advances in Enzymol. **13**, 237 (1952).
[74] H. Waelsch, Advances in Protein Chem. **6**, 299 (1951).
[75] E. P. Abraham, *Biochemistry of Some Peptide and Steroid Antibiotics* (John Wiley and Sons, Inc., New York, 1957).
[76] M. Bergmann, L. Zervas, and L. Salzmann, Ber. dent. Chem. Geo. **66**, 1288 (1933).
[77] W. K. Maas, 3rd Intern. Congress Biochem. Abstracts 4–36, Brussels (1955).
[78] P. Berg, J. Am. Chem. Soc. **77**, 3163 (1955).
[79] P. Berg, J. Biol. Chem. **222**, 991 (1956).
[80] W. P. Jencks and F. Lipmann, J. Biol. Chem. **225**, 207 (1957).
[81] H. Chantrenne, J. Biol. Chem. **189**, 227 (1951).
[82] D. Schachter and J. V. Taggart, J. Biol. Chem. **203**, 925 (1953).
[83] D. Schachter, and J. V. Taggart, J. Biol. Chem. **208**, 263 (1954).
[84] D. Schachter and J. V. Taggart, J. Biol. Chem. **211**, 271 (1954).
[85] K. Moldave and A. Meister, Biochim. et Biophys. Acta **25**, 434 (1957).
[86] K. Moldave and A. Meister, J. Biol. Chem. **229**, 463 (1957).
[87] M. B. Hoagland, Biochim. et Biophys. Acta **16**, 288 (1955).
[88] M. B. Hoagland, E. B. Keller, and P. C. Zamecnik, J. Biol. Chem. **218**, 345 (1956).
[89] P. Berg, J. Biol. Chem. **222**, 1025 (1956).
[90] J. A. DeMoss and G. D. Novelli, Biochim. et Biophys. Acta **18**, 592 (1955).
[91] J. A. DeMoss and G. D. Novelli, Biochim. et Biophys. Acta **22**, 49 (1956).
[92] R. S. Schweet, R. W. Holley, and E. Allen, Arch. Biochem. Biophys. **71**, 311 (1957).
[93] F. Lipmann, Proc. Natl. Acad. Sci. U. S. **44**, 67 (1958).
[94] J. A. DeMoss, S. M. Genuth, and G. D. Novelli, Proc. Natl. Acad. Sci. U. S. **42**, 325 (1956).
[95] M. B. Hoagland, P. C. Zamecnik, N. Sharon, F. Lipmann, M. P. Stulberg, and P. D. Boyer, Biochim. et Biophys. Acta **26**, 215 (1957).
[96] M. Karasek, P. Castelfranco, P. R. Krishnaswamy, and A. Meister, J. Am. Chem. Soc. **80**, 2335 (1958).
[97] E. W. Davie, V. V. Koningsberger, and F. Lipmann, Arch. Biochem. Biophys. **65**, 21 (1956).
[98] G. D. Novelli, Proc. Natl. Acad. Sci. U. S. **44**, 86 (1958).
[99] P. Berg, J. Biol. Chem. **233**, 601 (1958).
[100] P. Castelfranco, K. Moldave, and A. Meister, J. Am. Chem. Soc. **80**, 2335 (1958).
[101] R. S. Schweet, F. C. Bovard, E. Allen, and E. Glassman, Proc. Natl. Acad. Sci. U. S. **44**, 173 (1958).
[102] P. Berg and E. J. Ofengand, Proc. Natl. Acad. Sci. U. S. **44**, 78 (1958).
[103] T. Hultin and G. Beskow, Exptl. Cell. Research **11**, 664 (1956).
[104] M. B. Hoagland, P. C. Zamecnik, and M. L. Stephenson, Biochim. et Biophys. Acta **24**, 215 (1957).
[105] M. B. Hoagland, M. L. Stephenson, J. F. Scott, L. I. Hecht, and P. C. Zamecnik, J. Biol. Chem. **231**, 241 (1958).
[106] K. Ogata and H. Nohara, Biochim. et Biophys. Acta **25**, 659 (1957).
[107] S. B. Weiss, G. Acs, and F. Lipmann, Proc. Natl. Acad. Sci. U. S. **44**, 189 (1958).
[108] P. Zamecnik, M. L. Stephenson, and L. I. Hecht, Proc. Natl. Acad. Sci. U. S. **44**, 73 (1958).
[109] R. W. Holley, J. Am. Chem. Soc. **79**, 658 (1957).
[110] J. L. Potter and A. L. Dounce, J. Am. Chem. Soc. **78**, 3078 (1956).
[111] L. I. Hecht, M. L. Stephenson, and P. C. Zamecnik, Federation Proc. **17**, 239 (1958).
[112] J. H. Northrup, M. Kunitz, and R. M. Herriott, *Crystalline Enzymes* (Columbia University Press, New York, 1955), second edition, p. 232.
[113] M. M. Daly and A. E. Mirsky, J. Gen. Physiol. **36**, 243 (1952–1953).
[114] P. N. Campbell and T. S. Work, Biochem. J. **52**, 217 (1952).
[115] H. Green and H. S. Anker, J. Gen. Physiol. **38**, 283 (1955).
[116] F. W. Putnam, F. Meyer, and A. Miyake, J. Biol. Chem. **221**, 517 (1956).
[117] T. Peters, Jr., J. Biol. Chem. **200**, 461 (1953).
[118] T. Peters, Jr., Federation Proc. **16**, 369 (1957).
[119] J. L. Simkin and T. S. Work, Biochem. J. **67**, 617 (1957).

# 26
# Coding and Information Theory

PETER ELIAS

*Department of Electrical Engineering and Research Laboratory of Electronics,* Massachusetts Institute of Technology, Cambridge 39, Massachusetts*

## INTRODUCTION

"INFORMATION theory" is used in at least three senses. In the narrowest of these senses, it denotes a class of problems concerning the generation, storage, transmission, and processing of information, in which a particular measure of information is used. This area is also called "coding theory" and, especially in Britain, "the mathematical theory of communication," which is the title of Shannon's original paper[1] from which the field is derived. This is the usage in the titles of the books by Khinchin[2] and Feinstein,[3] which are rather abstract mathematical presentations.

In a broader sense, information theory has been taken to include any analysis of communications problems, including statistical problems of the detection of signals in the presence of noise, that make no use of an information measure. Woodward[4] has shown the relationship of information measure to some of these problems. A book by Wiener,[5] an article by Rice[6,7] (reprinted in a book edited by Wax[8]), and a recent textbook by Davenport and Root[9] discuss the use of statistical techniques in problems concerning analysis of signals and noise, and books by Blanc-LaPierre and Fortet,[10] Loeve,[11] and Doob[12] provide the (abstract) mathematical background. It is in this broader sense that the word is used, for example, in the title of the Professional Group on Information Theory of the Institute of Radio Engineers, whose *Transactions* contains articles both on coding and on signal-noise problems.

In a still broader sense, information theory is used as a synonym for the term "cybernetics" introduced by Wiener[13] to denote, in addition to the areas listed in the foregoing, the theory of servomechanisms, the theory of automata, and the application of these and related disciplines to the study of communication, control, and other kinds of behavior in organisms and machines. This is the usage in the titles of three meetings held in London (two of which have published proceedings[14,15]) and two held at the Massachusetts Institute of Technology.[16,17] *Information and Control* publishes articles in this broad area.

Only the area covered by the narrowest definition is discussed here—not because it is necessarily the most important for biological applications, but because of the limitations of space.

* The work of this Laboratory is supported in part by the U. S. Army (Signal Corps), the U. S. Air Force (Office of Scientific Research, Air Research and Development Command), and the U. S. Navy (Office of Naval Research).

## INFORMATION MEASURE. SOURCES

The first problem is to assign a measure to information. Figure 1 shows two representations of a message and the code book which connects them. At first one might suppose that a long message has more information than a short one, but, as the figure shows, the message length—or any other characteristic of the representation itself—can be changed drastically by coding at the transmitter. If the receiver decodes correctly, the message has been transmitted successfully by using a very brief form. In a communications system, any message has a variety of different representations in different places. One may want to say that it contains the same amount of information. An amount of information, then, cannot depend upon the form of representation. It seems reasonable, however, to make it depend upon how many messages there are. If the number of messages is large, longer code words will have to be used in order to distinguish between them. One starts, therefore, with the hypothesis that the information in a message is some function of the number of messages in the set. One would like to say that two successive, independent selections from the same set have twice the information value of a single selection, *if the two messages are equally probable.* This demands a function $f(m)$ of the number $m$ of messages in the set, for which $f(m^2)=2f(m)$, and the logarithm is the only respectable function with this property. For a selection between two messages, this gives [log 2] units of information. This can be rewritten as $[-\log(\frac{1}{2})]$ or, in general, as the negative of the logarithm of the message probability. This, in fact, is what one chooses to gener-

*CODED MESSAGES*

ABBA . . .

THE QUICK BROWN FOX JUMPED OVER THE LAZY DOG NOW IS THE TIME FOR ALL GOOD MEN TO COME TO THE AID OF THE PARTY NOW IS THE TIME FOR ALL GOOD MEN TO COME TO THE AID OF THE PARTY THE QUICK BROWN FOX JUMPED OVER THE LAZY DOG . . .

*CODE BOOK*

A↔THE QUICK BROWN FOX JUMPED OVER THE LAZY DOG
B↔NOW IS THE TIME FOR ALL GOOD MEN TO COME TO THE AID OF THE PARTY
I(A)=I(THE QUICK BROWN FOX JUMPED OVER THE LAZY DOG)=LOG 2=−LOG(1/2)=−LOG Pr[A]=1 BIT/SYMBOL

FIG. 1. Two representations of a message and the code book which connects them.

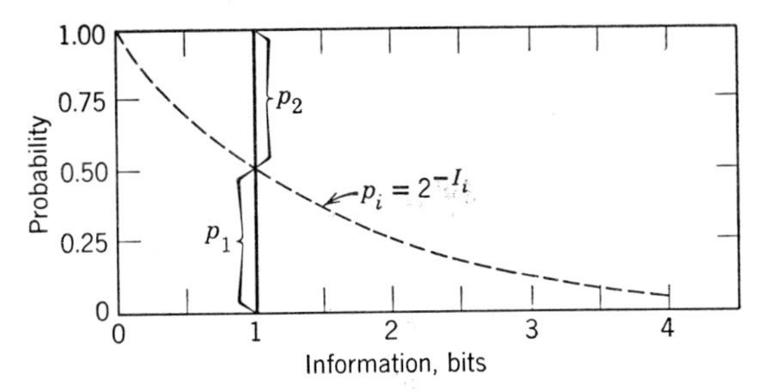


FIG. 2. Distribution of information for a two-symbol source whose symbols have equal probabilities.

$$p_1 = p_2 = \tfrac{1}{2}, \quad I_1 = I_2 = -\log \tfrac{1}{2} = 1 \text{ bit}$$
$$\bar{I} = H = \tfrac{1}{2} + \tfrac{1}{2} = 1 \text{ bit}$$

alize as the information associated with a message. The choice of a logarithmic base determines the unit of information, and the base 2 is chosen, the unit being the "bit." Thus, a selection between two equiprobable alternatives requires one bit of information.

This definition is plausible, but it does not justify the information measure. The adopted measure is justified, in a practical sense, on the average, because a source that generates information more rapidly than another, also requires more communications facilities—more band width, time, signal-to-noise ratio—to transmit its output successfully.

The information source, Fig. 1, is characterized by the distribution of information, Fig. 2. Since this source generates one bit of information per selection regardless of which message is selected, its information distribution is degenerate: all of the probability is piled up at one bit. Each symbol has a probability given by the value of the exponential at the information value of the symbol, as the plot is just probability *vs* minus log probability. Figure 3 shows the information distribution for another two-symbol source whose symbols have the unequal probabilities $\frac{3}{4}$ and $\frac{1}{4}$. This source generates only 0.42 bits when it selects an $A$ and 2.0 bits when it selects a $B$. Its average rate is 0.81 per symbol, which is smaller than in Fig. 2 with equiprobable symbols. This is a general characteristic of sources. Any con-

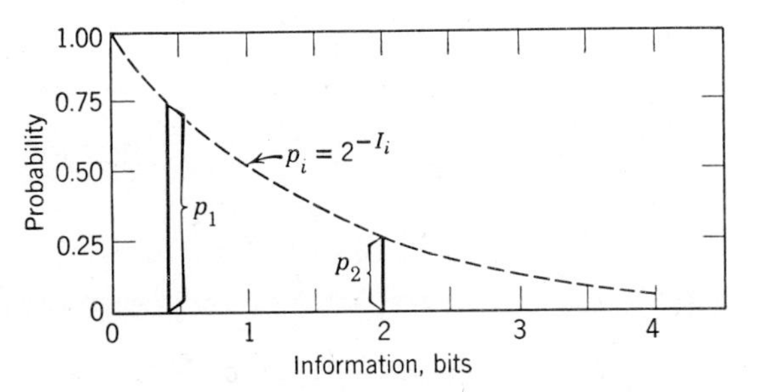


FIG. 3. Distribution of information for a two-symbol source whose symbols have the unequal probabilities $\frac{3}{4}$ and $\frac{1}{4}$.

$$p_1 = \tfrac{3}{4}, \quad I_1 = -\log \tfrac{3}{4} = 0.42 \text{ bit}$$
$$p_2 = \tfrac{1}{4}, \quad I_2 = \log \tfrac{1}{4} = 2.00 \text{ bits}$$
$$\bar{I} = \sum_{i=1}^{2} p_i I_i = H = \sum_{i=1}^{2} -p_i \log p_i = 0.31 + 0.50 = 0.81 \text{ bits}$$

straint on the number of symbols or sequences that a source may generate and any shift away from equal probabilities will reduce the average source rate. A source constrained to generate sequences of letters which spell out sentences in English has a lower average rate than a source which selects successive letters of the alphabet with statistical independence.

There are, of course, many questions which arise in connection with coding and with the description of sources having sequential constraints. What has been covered here is a mathematical theory, and problems arise in its application. One of these is illustrated in Fig. 4. The problem consists in identifying the alphabet which is relevant in a given situation. Suppose that, in observing the output of a neuron, the eight wave forms shown in the figure are seen with equal frequency. This would give three bits of information per wave form. However, these signals may not all be distinguishable by the system being observed. The system may act only as a pulse counter over this time interval, and may recognize only four different signals—no pulse,

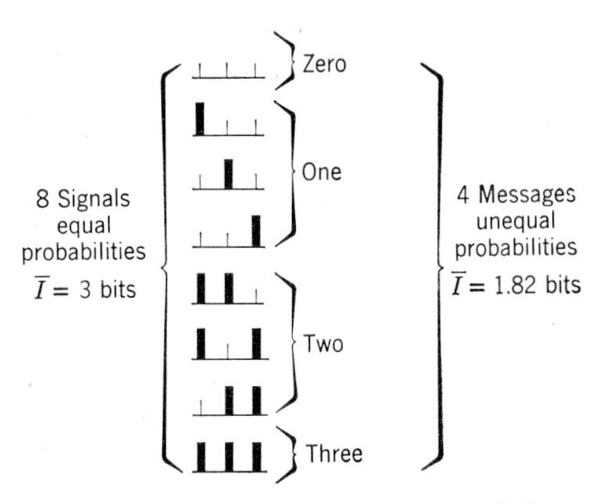


FIG. 4. Dependence of source rate on alphabet.

one pulse, two pulses, and three pulses. Then the average rate is only 1.82 bits per signal. There is no way of telling from the outside which of these alphabets is actually in use—if, in fact, either is. It is necessary to determine that the system does have different responses to two signals before they can be defined as being distinct letters of an alphabet. The problem of recognizing the relevant alphabet always is present and shows up in many ways, including the selection of scales of resolution to be used in amplitude and time to distinguish different signals. In Fig. 4, the average rates for the two alphabets are not too far apart, but in a train of 100 pulses, if all observed wave forms are equiprobable, there is a factor of about 20 between the rates obtained for the two corresponding alphabets.

## NOISY CHANNELS

There are, then, sources to select symbols, and channels are needed to transmit them. Figure 5 shows one—a noisy channel. In communications, a noisy channel usually is a medium which separates transmitter and receiver. In information storage, a noisy channel may model the action of the environment which, through thermal agitation or other forces, may cause changes in the stored information. In Fig. 5, the channel eliminates some of the distinctions present in its input. The source selects from among four equiprobable symbols $a$, $b$, $c$, and $d$. The receiver on reception of $A$ or $B$ can eliminate two input possibilities, but it cannot choose between the other two.

To analyze informationally what happens in the channel, consider a particular transmission event: the selection of $a$ at the transmitter and the reception of $A$ at the receiver. It is assumed that the receiver knows the input-letter probabilities and the channel. Thus, before reception of $A$, the estimate by the receiver of the probability that $a$ will be transmitted is $\frac{1}{4}$. After reception of $A$, the receiver knows that only either $a$ or $b$ could have been transmitted, and both are equally likely. Thus, *a posteriori*, the probability of $a$ (on the evidence $A$) is $\frac{1}{2}$.

The receiver *a priori* needs log 4=2 bits of information to select $a$ as the transmitted letter.

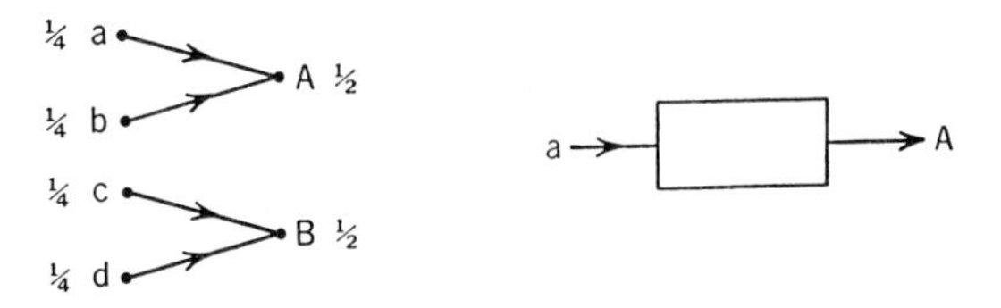


FIG. 5. The mutual information between a transmitted symbol and a received symbol.

$$Pr(a)=\tfrac{1}{4} \qquad Pr(a|A)=\tfrac{1}{2}$$
$$I(a)=2 \text{ bits} \qquad I(a|A)=1 \text{ bit}$$
$$\Delta I=\log Pr(a)-[-\log Pr(a|A)]=1 \text{ bit}$$
$$=\log\frac{Pr(a|A)}{\Pr(a)}=\log\frac{Pr(a,A)}{Pr(a)Pr(A)}=I(a;A)$$

*A posteriori*, it still needs log 2=1 bit to select $a$ after receiving $A$. The channel is credited with the 1-bit difference, which is defined as the amount of information which the receipt of $A$ gives about the transmission of $a$. The quantity expressed in Fig. 5 as $I(a;A)$ is called the *mutual information* of $A$ about $a$, or the *transmitted information*.

In general, for each possible pair of transmitted and received symbols, $x_i$ and $y_j$, there is a probability of occurrence $\Pr(x_i, y_j)$ [which may be zero as in Fig. 5 for $\Pr(a, B)$] and an information value $I(x_i, y_j) = \log \Pr(x_i, y_j)/\Pr(x_i)\Pr(y_j)$, which measures the change in the logarithm of the probability of $x_i$ owing to knowledge of $y_j$. This quantity is positive if $y_j$ makes $x_i$ more probable than it was, and is negative if $y_j$ makes $x_i$ less probable than it was. A plot can be made of the *mutual-information distribution* of a channel-source combination. For the channel of Fig. 5, this plot is exactly like the plot of Fig. 2: one bit of information is always

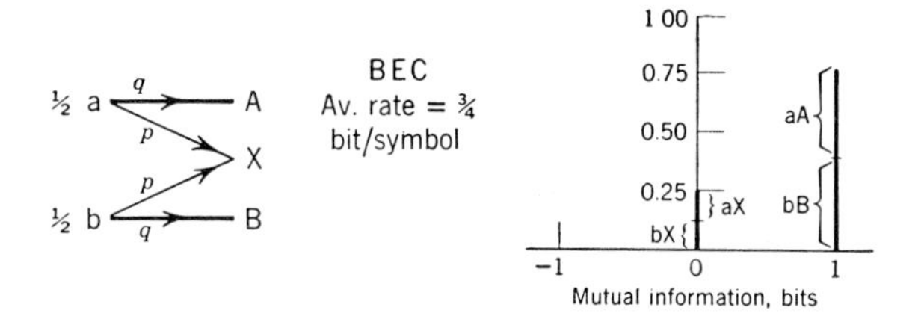


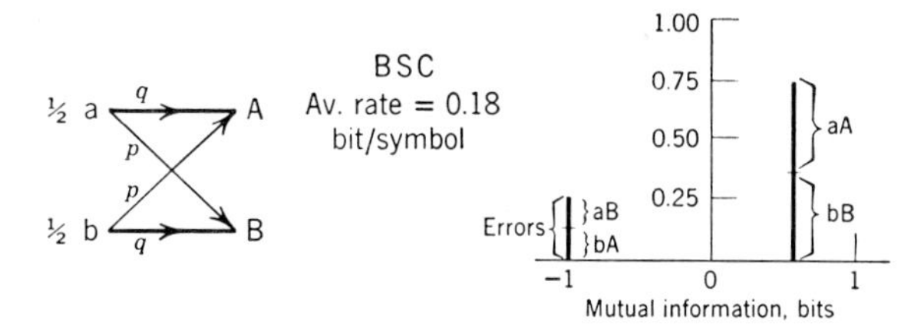


FIG. 6. The binary erasure channel, the binary symmetric channel, and their mutual information distribution.

transmitted regardless of which input-output pair happens.

Two more-complicated channels and their mutual-information distributions are shown in Fig. 6. In the binary erasure channel (BEC), one of two symbols is selected at the transmitter. The input symbol may be received correctly, with probability $q$ taken here as $\frac{3}{4}$, or it may be erased with probability $p=\frac{1}{4}$: The receiver then receives an $X$. Evaluating mutual information gives the distribution shown. The channel sends one bit per symbol three-quarters of the time, when no erasure occurs; one-quarter of the time, when the transmitted symbol is erased, no information is transmitted. The average rate is just $\frac{3}{4}$ bit per symbol, if the input symbols are equiprobable.

In the binary symmetric channel (BSC), there are only two input and two output symbols, and true errors occur. Here less than one bit is transmitted when there is no error, and a negative amount of mutual information is transmitted if there is an error. One takes error probability $p=\frac{1}{4}$. The average rate of transmission here is only 0.18 bits per symbol with equiprobable inputs.

Returning to Fig. 5, one interprets the average rate of transmission of 1 bit per symbol as the average rate at which the receiver's ignorance is reduced—from 2 bits to 1 bit for each letter transmitted. However, the receiver's ignorance is not reduced to zero. There is an apparent difference between this channel and another with the same rate which transmits two equiprobable symbols without error. In the latter case, the ignorance of the receiver is reduced from 1 bit to zero bits for each transmission. One would like to be able to say that the average rate over a channel is the significant parameter. This requires that, for the channel of Fig. 5, one finds some method for which the rate of putting information in is reduced in such a way as to reduce both the initial and final ignorance of the receiver, keep-

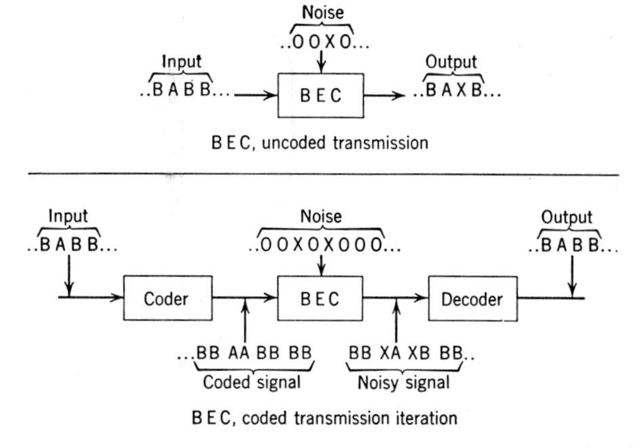


FIG. 7. Transmission over the binary erasure channel uncoded and coded by iteration.

ing their difference fixed. In this case, it is easy to do. The transmitter agrees to send only two symbols, $a$ and $c$, with equal probabilities. The receiver then can decode unambiguously, receiving one bit of mutual information per symbol with no residual ignorance.

## CODING FOR NOISY CHANNELS

The foregoing result is, in fact, general for a broad class of noisy channels, although the way in which the input information rate is reduced is usually more complicated. It is not possible in the BEC or BSC to find a set of input symbols which will leave the receiver with no ambiguity. All input symbols are used, therefore, but they are used in sequences, and only a fraction of the possible sequences occur.

Figure 7 shows the BEC, first used in ordinary, uncoded transmission and then used with a simple coding and decoding scheme. In the upper picture, one bit per symbol is put into the channel, $\frac{1}{4}$ of the output symbols are erased, and an average of $\frac{3}{4}$ bit per symbol comes out. Below, a coder is added, which duplicates each input symbol. For a fixed channel, input symbols can be accepted now only half as often, since two symbols go into the channel for each input symbol to the coder. The input rate is then $\frac{1}{2}$ bit per channel symbol. Again $\frac{1}{4}$ of the output symbols are erased, but both of the copies of an input symbol are erased $\frac{1}{16}$ of the time only, so reliability has improved.

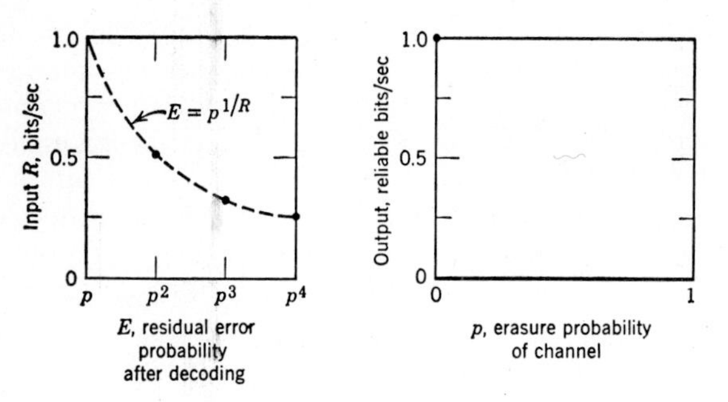


FIG. 8. Rate and error probability for iteration coding of the BEC.

Of course, each digit could be sent three or more times instead of twice. This would reduce further both the rate of transmission and the probability of total erasure, or of ambiguity after decoding. The relationship is shown in Fig. 8. On the left plot, it can be seen that sufficient reduction of rate reduces residual probability of total erasure as low as is desired, but, to get arbitrarily low probability, an arbitrarily low rate of transmission is necessary. If one looks at the channel in a different way, demanding transmission with *arbitrarily low* total-erasure probability, and asking how average rate varies as the channel-erasure probability $p$ is varied, one sees the plot on the right in Fig. 8. If $p=0$, one can send one reliable bit per symbol, but if $p\neq 0$, none can be sent. Time must be spent on repeating the first input symbol to make its reliability arbitrarily good, and one never can get around to sending anything else.

Certainly, one way of avoiding misunderstanding is never to say anything new, but it would be discouraging

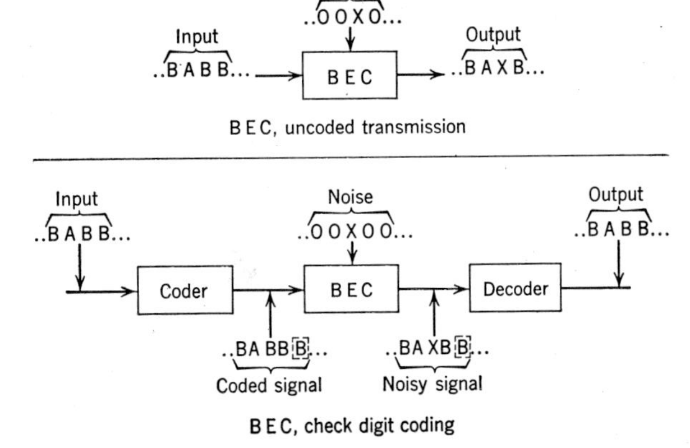


FIG. 9. The transmission erasure channel, the binary symmetric channel, and their mutual information distribution with check digits.

if this were the only way. An alternative is shown in Fig. 9.

Here one starts with uncoded transmission again. Now, however, instead of duplicating each input symbol, a general purpose replacement, called a check digit or a parity check, is inserted so that a replacement will be available in case it it erased. The digit enclosed in the dotted box is selected to make the total number of $B$'s in the sequence of 5 symbols an *even* number. At the receiver, if only one symbol has been erased, the knowledge that an even number of $B$'s should be present makes the decoding unique. The rate of transmission has been reduced to $\frac{4}{5}$ bit per symbol and increased reliability is obtained. There is still a danger, however, that two or more erasures may occur in the same block of five digits, in which case decoding still would be ambiguous.

The situation can be improved by adding further

check digits as shown in Fig. 10. Here the sequence of symbols is shown above. Added check digits have bars over them. The check digits are computed as shown below: each symbol to the right of a row is selected to make the total number of *B*'s in the row even; each symbol at the foot of a column is selected to make the total number of *B*'s in the column even. First, those rows with single erasures are decoded, and then the columns having only single erasures remaining after row correction. In the case illustrated, this corrects all erasures. Further high-order check digits can be added indefinitely to give a total-erasure probability which is arbitrarily small without reducing the transmission rate to zero. Figure 11 shows the kind of relationship between rate and residual-erasure probability which results. Here an erasure probability $p$ of 1/20 and an initial check group of ten, rather than of five, symbols have been used. Rate still goes down as reliability increases, but it now has a positive asymptote at 0.80 bits per symbol, at which rate it is possible to get arbitrary reliability. The plot of reliable rate *vs* the channel-erasure probability $p$ is now continuous, as shown by

..B A B B $\bar{\text{B}}$ A B A B $\bar{\text{A}}$ A A B A $\bar{\text{B}}$ B B A B $\bar{\text{B}}$ B A B B $\bar{\text{B}}$

A B A A $\bar{\text{B}}$ B A A B $\bar{\text{A}}$ A B B A $\bar{\text{A}}$ A B B B $\bar{\text{B}}$ $\overline{\text{A B B A A}}$...

Transmitted

| | | | | |
|---|---|---|---|---|
| B | A | B | B | B |
| A | B | A | B | A |
| A | A | B | A | B |
| B | B | A | B | B |
| B | A | B | B | B |
| A | B | A | A | B |
| B | A | A | B | A |
| A | B | B | A | A |
| A | B | B | B | B |
| A | B | B | A | A |

Received

| | | | | |
|---|---|---|---|---|
| B | A | B | B | B |
| A | X | A | B | A |
| A | A | X | A | B |
| B | B | A | B | X |
| B | A | B | B | B |
| X | B | X | A | B |
| B | A | A | B | A |
| A | B | B | X | A |
| A | B | B | B | B |
| A | B | B | X | A |

Decoded

| | | | | |
|---|---|---|---|---|
| B | A | B | B | B |
| A | B | A | B | A |
| A | A | B | A | B |
| B | B | A | B | B |
| B | A | B | B | B |
| A | B | A | A | B |
| B | A | A | B | A |
| A | B | B | A | A |
| A | B | B | B | B |
| A | B | B | A | A |

FIG. 10. Correction of erasures by iteration of parity-check digits.

the solid curve. This kind of iterated check-symbol coding can be used for the BSC[18] as well as for the BEC.[19]

Although a very special case of noisy-channel coding has been shown, it illustrates a general result. Given any noisy channel and source, with an average rate of transmission, it is possible to code the input in long blocks and to reduce both the input rate and the receiver's residual uncertainty, while keeping the average transmission rate fixed. The ideal behavior is illustrated in the dotted line on Fig. 11 for the BEC; to approach it, coding in large blocks is required.

One other point should be noted. The code we constructed was engineered carefully, and its reliability may appear to be atypical. In fact, by random selection of long sequences of symbols, it is possible to get the same results—the "ideal behavior" of Fig. 11 is obtained by just such random coding. It is necessary only to *not* select too many possible input sequences. Thus, a code suitable for reliable transmission over a noisy channel might occur quite accidentally. However, no

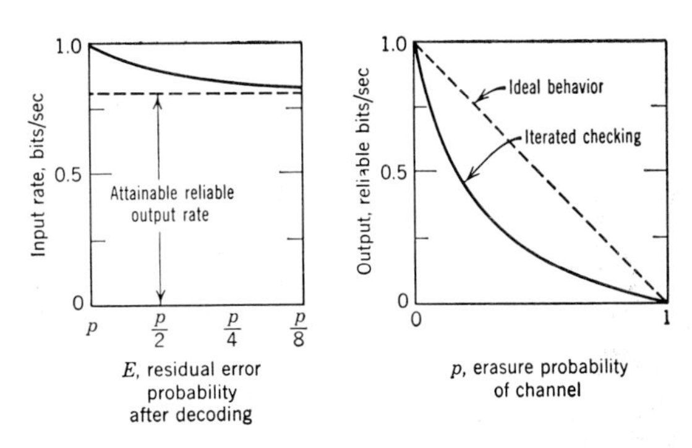


FIG. 11. Rate and error probability for iterated parity checking.

accident seems likely to account for the necessary decoding equipment and its organization.

## CONCLUSION

In conclusion, it might be appropriate to point out what applications have been made to biological problems. The most successful ones, I think, have been to experiments in human communication where the channel capacity of a human being for handling information has proved a useful concept in a number of experimental situations [see e.g., Rosenblith (p. 485) and reference 20]. There also have been applications at the neurophysiological level, but in terms of statistical signal analysis rather than of information theory *per se;* some of these are referred to in the second paper by Rosenblith (p. 532). Computations of neuron channel capacity have been made, but they are of dubious value in view of the alphabet problem illustrated in Fig. 4. It seems likely that real applications are forthcoming at this level.

There have been applications to chemical specificity, etc., in biological systems (see, e.g., references 21 and 22). My feeling is that these use information measure either as a language for the discussion of purely combinatorial problems or as a useful statistic, but they do not use it in any coding sense which would imply that the informational treatment was at all necessary or unique. I think that the only other immediate application might arise in connection with the genetic-coding problem. Here, the most urgent need obviously is more data about nucleotide-amino group correspondences and the statistics of series of each. Although informational ideas may be useful here, it seems unlikely that they are essential. That is, it seems unlikely that high orders of redundancy and error-correction are being used. Some data on how local the coding is would be useful in reaching a decision on this point.

## BIBLIOGRAPHY

[1] C. E. Shannon, Bell System Tech. J. **27**, 379, 623 (1948); reprinted in C. E. Shannon and W. Weaver's *The Mathematical Theory of Communication* (The University of Illinois Press, Urbana, Illinois, 1949).

[2] A. I. Khinchin, *Mathematical Foundations of Information Theory* (Dover Publications, New York, 1957).

[3] A. Feinstein, *Foundations of Information Theory* (McGraw-Hill Book Company, Inc., New York, 1958).

[4] P. M. Woodward, *Probability and Information Theory* (McGraw-Hill Book Company, Inc., New York, 1953).

[5] N. Wiener, *The Extrapolation, Interpolation and Smoothing of Time Series* (Technology Press, Cambridge, Massachusetts and John Wiley and Sons, Inc., New York, 1949).

[6] S. O. Rice, Bell System Tech. J. **23**, 282 (1944).

[7] S. O. Rice Bell System Tech J. **24**, 46 (1945).

[8] N. Wax, editor, *Noise and Stochastic Processes* (Dover Publications, New York, 1954).

[9] W. B. Davenport, Jr., and W. J. Root, *An Introduction to the Theory of Random Signals and Noise* (McGraw-Hill Book Company, Inc., New York, 1958).

[10] A. Blanc-LaPierre and R. Fortet, *Théorie des fonctions aléatoires* (Masson et Cie., Paris, 1953).

[11] M. Loeve, *Probability Theory* (D. Van Nostrand Company, Inc., New York, 1955).

[12] J. L. Doob, *Stochastic Processes* (John Wiley and Sons, Inc., New York, 1953).

[13] N. Wiener, *Cybernetics* (Technology Press, Cambridge, Massachusetts and John Wiley and Sons, Inc., New York, 1948).

[14] W. Jackson, editor, *Proceedings of a Symposium on Application of Information Theory, London, 1952* (Butterworth Scientific Publications, London, 1953).

[15] C. Cherry, editor, *Information Theory, Proceedings of a Symposium* (Butterworth Scientific Publications, London, 1956).

[16] IRE Trans. Professional Group on Information Theory **PGIT-4** (1954).

[17] IRE Trans. on Information Theory **IT-2** No. 3 (1956).

[18] P. Elias, IRE Trans. Professional Group on Information Theory **PGIT-4**, 29 (1954).

[19] P. Elias, in *Handbook of Automation, Computation and Control*, E. Grabbe, editor (John Wiley and Sons, Inc., New York, 1958), Vol. I, p. 16–01.

[20] H. Quastler, editor, *Information Theory and Psychology* (Free Press, Glencoe, Illinois, 1955).

[21] H. Quastler, editor, *Essays on the Use of Information Theory and Biology* (The University of Illinois Press, Urbana, Illinois, 1953).

[22] H. P. Yockey, R. L. Platzman, and H. Quastler, editors, *Symposium on Information Theory in Biology* (Pergamon Press, New York, 1958).

# 27
# Genetics and the Role of Nucleic Acids

CYRUS LEVINTHAL

*Department of Biology, Massachusetts Institute of Technology, Cambridge 39, Massachusetts*

THE purpose of this article is to review some of the basic ideas in genetics and to set the stage for a discussion of certain specialized problems. The first question to be considered is the nature of the genetic map: its construction and meaning. The genetic map is an abstract entity defined in terms of the probabilities with which certain types of offspring are produced in matings. In a sense, it is related to a physical structure of the cell; however, this relationship must be established experimentally, and it is not a logical consequence of the definition of the map.

For any organism which can duplicate without a sexual process, one finds that progeny organisms are, in general, identical with the parental. Occasionally, an alteration occurs. If this alteration persists—that is, if all of the descendants of the altered organism have the same alteration—then, the change is called a mutation. In an operational sense, mutations form the basic entities of genetics. Mutations may occur which lead to red hair, to six toes, or to any of a number of other attributes in a higher organism. Or for bacteria, a mutation may result in a small colony or in a colony which is red in a certain medium. For the moment, assume that one can deal with several such mutations which are recognizable and recognizably different from one another. Furthermore, assume that an organism which contains several such mutations can be detected, and that the total number and types of mutations it contains can be enumerated. The discussion of the possible physicochemical bases of such mutations is undertaken in a later article (p. 249). For the moment, it is only necessary that the mutations give a recognizable result. The analysis begins with the definition of some organism as the standard or "wild" type. Suppose that one has three different, recognizable mutations which arise from the wild type; call them $A$, $B$, and $C$. The standard wild-type organism is called, by definition, "plus" with respect to each of these characters. The organism with the first mutation having lost the property $A$, is designated $A\ B+\ C+$; that is, it has lost the property $A$, but maintains the properties $B+$ and $C+$. Organisms of this type also may be designated simply by the letter "$A$".

A group of organisms may be selected, one of which has the mutation $A$, one $B$, and one $C$. If one works with microorganisms, then from a single individual of each of these types, stocks containing many millions of microorganisms can be grown. Except for secondary mutations, each organism within a stock will have lost a specific property.

Any two of these mutant types may be allowed to mate with one another. The mechanism of the mating is not of concern at the moment. It will suffice to say that, in general, the experiments with microorganisms involve mixing of the two types under suitable conditions and allowing the interactions to occur at random.[1] If a stock of organisms of one type $A$ is allowed to mate with a stock of another type $B$, the experiment is called a cross between $A$ and $B$ and is described by the symbol $A \times B$.

The organisms discussed here are called haploid, meaning that they have only one set of genetic determinants. In many higher organisms, complications can arise which result from the double set of genetic determinants which they contain. With regard to the discussion of a genetic map, none of these complications represents an alteration to the general conceptual scheme. Thus, for the purpose of simplification, they are left out entirely and only haploid organisms are considered.

Among the progeny arising from the cross $A \times B$ are found organisms of the type $A$ and organisms of the type $B$ which are identical with either of the parents. Progeny organisms also are found which lack both the character $A$ and the character $B$. Still others are found which are like the wild type in having both $A+$ and $B+$. $A$ and $B$ are parental in their genetic composition and the two types $A\ B$ have the property $B$ (i.e., the loss of $B+$) and the property $A$ (i.e., the loss of $A+$) which it must have received, one from one parent and one from the other parent. And the organism $A+B+$ must have received the property $A+$ from one parent and the property $B+$ from the other parent. The new types, $AB$ and $A+B+$, are called *recombinants*.

The fraction of the descendants which are of the recombinant types is called the recombinant frequency and is the only number which is needed to describe the results of the cross. This is true, however, only if there has been no selection of one of the types in preference to the others. If a cross has been carried out involving equal numbers of $A$ and $B$ parents and if the analysis takes into account any possible difference in the mating type (maleness or femaleness), then one finds that the two recombinant types $A\ B$ and $A+B+$ occur with equal frequency. Again, this is true *only* if the presence of the property $A+$ or the property $B+$ in the organism does not give it a selective advantage in the growth which occurs before the final progeny are selected and scored. If there were any such selective advantage in the growth, it would constitute a physiological effect which is not relevant to the genetic phenomenon being discussed. Therefore, only those cases are considered in which the reciprocal recombinants arising from a cross are, on the average, equal to each other. In practice, this is not a serious limitation, since in most cases such selection easily is eliminated.

With a given pair of mutations, two crosses can be carried out. The two mutant types can be crossed with each other; that is, $A\ B+ \times A+B$, or the double mutant $A\ B$ can be crossed with the wild type $A+B+$. In the first case, the recombinants are $A\ B$ and $A+B+$ while in the second they are $A\ B+$ and $A+B$. For the systems discussed here in which none of the possible types has any growth advantage over the others, the resulting recombinant frequency found in these two different crosses is the same. Thus, one may conclude that there is only one number needed to describe the results of crosses between any pair of mutants—namely, the recombinant frequency in the progeny.

In dealing with three different mutants, three different crosses can be carried out between pairs of them: $A$ with $B$, $A$ with $C$, and $B$ with $C$. The recombinant frequency obtained in $A \times B$ is designated $A \cdot B$, and the same notation is used for the recombinant frequency in $B \times C$ and $A \times C$. If the markers are neither too close nor too distant from each other, it is found that the following relationship holds as a good first approximation:

$$A \cdot B + B \cdot C = A \cdot C.$$

That is, two of the three recombination frequencies, when added together, equal a third. This additivity obtains if the recombination frequencies are not too large. If the recombination frequency is large, since recombination can be assumed to result from some kind of switching event, then there is a high probability of more than one switch taking place in the same mating. If an even number of switches occur between two genetic markers, then no recombinant is observed. Correcting for these multiple events by calculating the total probability of switches, under the assumption that they occur at random, one finds that the results of most crosses are now consistent with the additivity relationship except when extremely closely linked markers are involved. To the extent that this relationship holds, mutants can be arranged as points along a line, the distances between the points being proportional to the recombination frequencies. This line with the mutants arranged along it is called the *genetic map*.

The genetic map defined in this way has two important characteristics which must be distinguished. The first is that the mutants can be arranged in unique order along a line without requiring any branch points or any other configuration which implies a two-dimensional surface. The second is that the distances along the line are proportional to the recombination frequencies. For the latter to be true, strict additivity must hold. If it does hold, then it is true also that a one-dimensional map will be demonstrated.

However, the one-dimensional character of the map can be demonstrated even if additivity does not apply. Three mutations can be ordered with respect to each other in two different ways. As in the foregoing, one can measure the recombination frequencies when each of the three pair-wise crosses are performed. The pair which shows the highest recombination frequency between them can be taken as the outside pair, while the one which shows a lower recombination frequency with each of the other two can be placed between them. A more precise method of ordering three mutations is by means of a three-factor cross. This method utilizes the fact that if the probability is low for a switching event which produces a recombination, the occurrence of two such events is much less likely than the occurrence of a single one. Three-factor crosses are done if an organism has, for example, both mutations $A$ and $B$ crossed with an organism which has mutation $C$. The other two crosses of this type also can be done—namely, $A\ C \times B$ and $B\ C \times A$. In these crosses, one can examine the progeny for recombinants which are like the wild type in that they contain all three properties, $A+$, $B+$, and $C+$. One of the three crosses will yield, in general, fewer recombinants of this type than the other two. Suppose, for example, the cross $A\ C \times B$ produces the smallest number of recombinants. This would imply that $A$ and $C$ are on the outside and $B$ between them.

This ordering of three elements in an operationally unique way still implies nothing about the nature of the genetic map, since the ordering, by definition, is always possible. But suppose that there is a large number of mutations and all possible triplets are ordered in the same way. Then, it is possible to test whether or not the data are consistent with a one-dimensional map. In all cases so far studied, it has been found that the mutants can be arranged uniquely along one or more linear elements. In some cases, one set of mutants will assort randomly with respect to another set of mutants in crosses. For extremely close markers (those which give very little recombination in crosses between them), one finds a marked deviation from additivity. This is related to the phenomenon of interference in which the occurrence of one switch has an influence on the probability of another occurring immediately adjacent to it. In some systems, the probability of the second switch is reduced, in which case the effect is called interference. In other systems, the probability of multiple switches in the same region is very high, in which case the effect is called negative interference. In the instance of bacteriophage, strong negative interference is found,[2] only for sets of mutants which are very closely linked. Benzer[3] has developed an extremely elegant and essentially new method of establishing the one-dimensionality of the genetic map by making use of mutants which act as though they extend over a finite length of the genetic map. Thus, the conclusion from many types of genetic analyses is that independently arising mutations can be ordered along one-dimensional structures whether strict additivity applies or not.

So far, the recombinational event itself has been discussed in very generalized terms. It is an event which occurs with constant probability per unit length along the genetic map and which results in the formation of recombinants. For higher organisms, there is good reason to believe that recombinants are formed by *crossing over*. In this process, the two homologous structures which correspond physically to the genetic map pair each with the other, with a break occurring at the corresponding points. The broken ends reunite in such a way as to form two new structures, each of which contains parts of the genetic structure of the parents. One of the characteristics of the crossing-over model is that recombinants always are formed in reciprocal pairs; that is, if the cross $A \times B$ is done, the wild-type and the type $A\ B$ recombinants always are formed in the same event. This is a much stronger statement than that made earlier about the equality of these reciprocal recombinants on the average.

Only by special tricks can one determine whether or not reciprocal recombinants are formed in the same act. However, in those higher organisms where the phenomenon can be studied—and this is primarily with the molds, which, for present purposes, are considered as "higher"—one can show that the predictions of the crossing-over model are borne out. Where the corresponding experiments are possible with viruses and bacteria, it is generally found that reciprocal recombinants are not formed in the same event. In those cases where recombinants are formed without associated formation at their reciprocals, there is also evidence which suggests that the parental structures are not broken during the recombination act. One is then led to a model of recombinant formation which involves the copying of the genetic information along one structure for part of its length, followed by a shift after which the copying continues along another parent. This mechanism has been called *copy-choice* (Fig. 1).

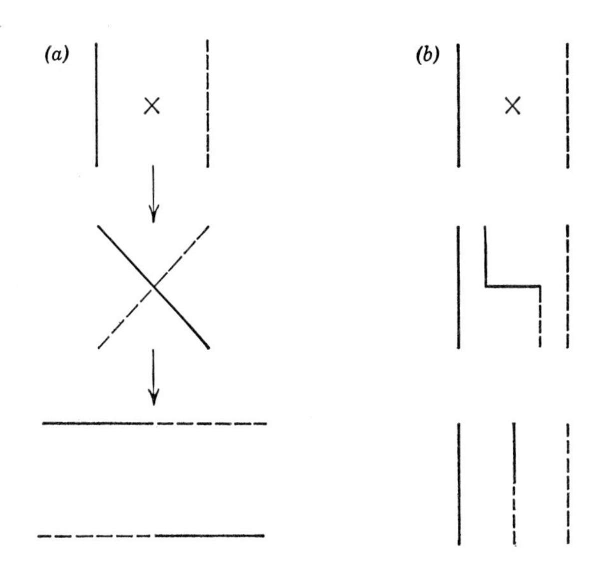


FIG. 1. Two models by which recombinants could be formed considering the genetic map only; (a) is the crossing-over model and (b) is copy-choice.

It should be kept in mind that the generalizations about the genetic map are independent of the mechanism of recombination. Whether recombinants arise by crossing-over or by copy-choice, the one dimensionality of the map and the roughly additive distances along it can be established. The event which produces the recombinants, regardless of its nature, must, by definition, occur with a constant probability per unit length of the map.

The next question concerns the relationship between the genetic map and the physical constituents of the cell. In the case of a few higher organisms, it has been possible to show by very direct methods that there is a one-to-one correspondence between the position of a genetic marker along the map and the position of an observable alteration on the chromosome of the organism.[1] The chromosomes are cellular components originally characterized by the color they demonstrate with particular stains. The chromosomes of the *Drosophila* salivary gland are abnormally large and, because of characteristic banding, individual portions can be identified by direct microscopic observation. Inverted or missing segments of a chromosome arise in some mutant organisms and can be recognized morphologically. Corresponding to the inversion in the visible chromosome is an inversion in the mapping of a short region of the genetic map. The same holds true when deletions are observed cytologically. In this case, a small region of the genetic map acts as though it had been deleted. By experiments with mutants of this type, it has been possible to establish a one-to-one correspondence between the position along the genetic map and the position along a chromosome (Fig. 2). This correspondence, however, has been made in detail only with the giant chromosomes of *Drosophila* which contain approximately a thousandfold more material than normal chromosomes of *Drosophila* and are in cells which cannot undergo further division.

In all of the organisms where it is technically possible to do the combined cytological and genetic studies, the correspondence between the genetic map and the chromosomes can be demonstrated. The chromosome, however, is a complicated structure which contains several chemical components. Of these, only deoxyribonucleic acid (DNA) is constant in amount in all normal cells of a given organism. Those cells which have only one set of genetic determinants contain only half as much DNA as the amount contained in the majority of cells known to possess a double set of chromosomes.

There is a considerable body of evidence indicating that the chemical structure which contains the genetic information and which, in fact, corresponds to the

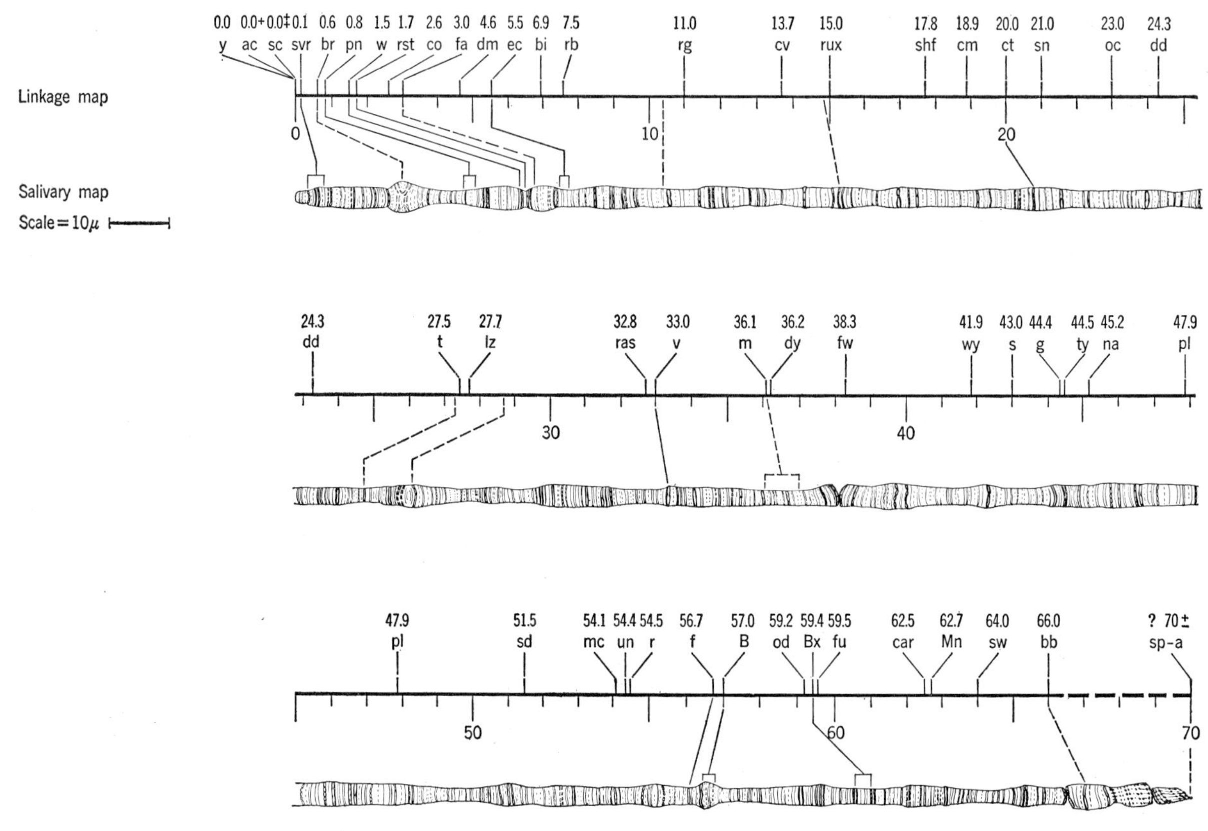


FIG. 2. The genetic map (upper line) and the *X*-chromosome of the salivary gland of *Drosophila Melanogaster* [after J. L. Bridges, J. Heredity **29**, Suppl. 1 (1938)].

genetic map is nucleic acid. Since this conclusion has come to be very widely accepted, it is important to examine in some detail the experimental basis for it. Four types of experiments are reviewed which lend strong support to the view that nucleic acid can carry genetic information. The acceptance of the generalization, however, that nucleic acids do carry the genetic information in all organisms is still very much a matter of personal taste. Most of the direct experimental demonstrations have come from studies with microorganisms, and, in general, these demonstrations have consisted of showing that genetic information, either of bacteria or viruses, could be transferred to a recipient cell by the sole means of nucleic acid.

The most complete analyses have been done in experiments with bacterial transformation.[5] Pure DNA is extracted from a bacterial suspension of one genetic type and added to bacteria having different genetic properties. Genetic characters are transferred in this way from the original cells, which donated the pure DNA, to the recipient cells. The material which produces the transformation has been purified to the point where no bacterial substances other than DNA are detectable, and the amount of protein which might be in these preparations and escape detection is less than 0.02%. That any small undetected component might have biological activity is made unlikely by the fact that, as the purification proceeds, the amount of transforming activity in the preparation is proportional to the quantity of DNA and independent of the quantity of any other components. Furthermore, it has been shown that the number of transferred cells is proportional to the amount of DNA actually absorbed by the recipient cells. Additional evidence has been provided by experiments in which enzymatic degradation is used specifically to degrade various components. Only the enzymes which degrade DNA eliminate the ability of the purified material to transform cells genetically.

The results of the experiments, which seem to be quite unambiguous, show that one can transfer genetic information from one cell to another by means of purified DNA. Thus, it can be concluded that, at least at some stage in the life cycle of the bacteria, the genetic information must be contained in DNA. However, the next question arises when the duplication of the cells is considered. Under these conditions, it is clear that the genetic material also must duplicate; the simplest hypothesis for this would seem to be that DNA of a specific type makes more of its own kind. One cannot eliminate the possibility that, in the process of duplication, the information must be passed to some other substance before it is passed back to new DNA. Although there has been a good deal of speculation along these lines,[6] there are no experiments which definitively show that the information is transferred to anything but DNA.

Other experimental results which indicate that the genetic information is contained in nucleic acids have been obtained with viruses and with genetic recombination in bacteria. The first evidence that only the DNA of viruses carries genetic information came from the experiments by Hershey and Chase[7] using the bacteriophage *T*2. This is a virus which can duplicate only after it has attached to a bacterial cell. The virus particle consists of a head and a tail. The tail is composed of proteins exclusively and the head of a protein membrane surrounding DNA. When the virus attaches to a bacterial cell, it uses the tip of its tail, and after attaching, the DNA which was contained in the head of the virus is transferred into the bacterial cell, leaving the head membrane and most of the tail on the outside of the bacterium. This residual protein which remains on the outside can be sheared off by stirring the infected cell in a Waring Blendor and the bacterium, with the DNA of the virus inside it, will develop then and produce a large number of virus particles identical with the one used for infection. Thus, most of the viral protein can be eliminated, indicating that it plays no role in the formation of the new virus particle.

The most obvious interpretation of these experiments is that the protein acts as a microsyringe injecting the DNA into the host cell, and that this DNA carries the genetic information necessary for the formation of the new virus particles. On the other hand, there is a measurable amount of protein which cannot be eliminated and which does enter into the cell with the injected nucleic acid. The fact that the atoms of the associated protein do not appear in the progeny virus, while more than 50% of the atoms of the injected DNA do, also suggests here that it is solely the DNA which is the genetic material. It must be kept in mind, however, that the associated protein may play an essential role in initiating the process of growth in the invaded cell.

In the case of tobacco mosaic virus (TMV), similar results have been obtained, but a more definitive analysis has been possible. In this case, one can completely separate the protein and the nucleic acid of the virus by chemical means. It is found, here, that the nucleic acid is infective and that the protein is not.[8] If one reconstitutes the nucleic acid from one strain of TMV and the protein from another into a single particle, then the reconstituted particles have the genetic proper ties of the strain from which the nucleic acid was taken but not of the strain from which the protein was taken.[9] It is important to note, however, that, in the case of TMV, the nucleic acid involved is ribonucleic acid (RNA). The same kinds of analytical and enzymatic demonstration of the purity of the RNA have been carried out as were discussed in the foregoing for the transforming factor. Here, the results indicate that the genetic information is carried in the RNA exclusively. Again, as in the case of DNA, one can argue about the lower limits of detectibility of some odd component, but this is also a case where one has to push the evidence rather far to escape the conclusion that it is nucleic acid which contains the genetic information.

Genetic recombination takes place between fertile strains of bacteria by direct contact of the donor and recipient cells.[10] It has been shown that a cytoplasmic bridge is formed between these two cells and that DNA passes from the donor to the recipient. It has been demonstrated that the amount of DNA transferred is proportional to the amount of the genetic material as measured along the genetic map. These experiments have not yet been done in such a way as to eliminate the possibility of small amounts of other components also being transferred, but the apparent proportionality between the amount of DNA and the length of the genetic map again suggests that the genetic information is contained in DNA.[10,11]

Since the structure of RNA has not been worked out yet, it is difficult to say very much about the significance of those cases where RNA carries genetic information. In the case of the genetic information contained by the DNA, however, it is clear that the information must reside in the order of the four bases along the DNA chain. This is evident from the fact that, as far as the sugars and the phosphate-ester linkages are concerned, the DNA molecule is extremely monotonous with every part of it identical to every other part. One can look, therefore, at the three essential properties of any genetic material in terms of the base sequence of the DNA molecule containing the genetic information. The first activity that a genetic entity must be able to perform is reproduction and, in this regard, the suggestion made by Watson and Crick[12] as to the possible mechanism for the duplication of DNA is extremely hopeful. At present, there are no experiments to contradict this hypothesis and there are a large number which seem to be consistent with it. In addition to duplicating its own kind, genetic material must be such that recombinants can be formed. Although a great deal already is known about the way in which recombinants are formed, it will probably be necessary to know much more about the physicochemical properties of DNA in solution before a self-consistant model can be constructed with any confidence. Specifically, one must find out whether the DNA in cells has all of its possible hydrogen bonds intact or whether some regions of the molecule have two unpaired chains available for pairing with homologous parts of other DNA molecules. If this is so, then it is possible[1] to construct a model for the recombinational event consistent with all of the available genetic data. Finally, the genetic material must be able to exercise its control; that is, it must be capable of controlling the formation of proteins and perhaps other cell components with specific composition and configuration. This problem is taken up in a later article (p. 249).

## BIBLIOGRAPHY

[1] C. Levinthal, "Bacteriophage genetics," in *The Viruses*, F. M. Burnet and W. M. Stanley, editors [Academic Press, Inc., New York (to be published)], Vol. II, Chap. 8.

[2] M. Chase and A. H. Doermann, "High negative interference over short segments of the genetic structure of bacteriophage *T*4," Genetics (to be published).

[3] S. Benzer in *The Chemical Basis of Heredity*, W. D. McElroy and B. Glass, editors (The Johns Hopkins Press, Baltimore, Maryland, 1957), p. 70.

[4] E. W. Sinnott and L. C. Dunn, *Principles of Genetics* (McGraw-Hill Book Company, Inc., New York, 1958), fifth edition.

[5] R. D. Hotchkiss in *Phosphorus Metabolism*, W. D. McElroy and B. Glass, editors (The Johns Hopkins Press, Baltimore, Maryland, 1952), p. 426.

[6] G. S. Stent in *Advances in Virus Research* (Academic Press, Inc., New York, 1958), Vol. V, p. 95.

[7] A. D. Hershey and M. Chase, J. Gen. Physiol. **36**, 39 (1952).

[8] A. Gierer and G. Schramm, Nature **177**, 702 (1956).

[9] H. Fraenkel-Conrat, B. A. Singer, and R. C. Williams in *The Chemical Basis of Heredity*, W. D. McElroy and B. Glass, editors (The Johns Hopkins Press, Baltimore, Maryland, 1957), p. 501.

[10] F. Jacob and E. L. Wollman in *Biological Replication of Macromolecules*, Symp. Soc. Exptl. Biol. **12**, 75 (1958).

[11] A. Garen and P. D. Skaar, Biochim. et Biophys. Acta **27**, 457 (1958).

[12] J. D. Watson and F. H. C. Crick, Cold Spring Harbor Symposia Quant. Biol. **18**, 123 (1953).

# 28
# Replication of Nucleic Acids

Robley C. Williams

*Virus Laboratory, University of California, Berkeley 4, California*

SOME of the evidence for the belief that genetic information resides in the structure of nucleic acid has been presented by Levinthal (p. 227). A key question in molecular biology is the mechanism by which the replication of such genetic information takes place. It is certainly agreed that nucleic-acid replication takes place within living cells, and, in a few cases mentioned later, some information bearing on the mechanism of replication has come directly from observations on growing, normal cells. But it is also possible to observe phenomena related to nucleic-acid replication in cases where the nucleic acid of interest has been allowed to enter the cell from the outside. Three distinct systems of this nature are known: (1) Some bacterial cells release a product, called the "transforming principle" and now known to be pure DNA, that is capable of changing in an hereditary manner certain characteristics of other cells which have been exposed to the transforming material.[1] (2) Infection of bacterial cells by DNA-containing viruses (bacteriophages) causes the viral DNA to be replicated. (3) Purified RNA, notably from tobacco mosaic virus (TMV), can be mechanically inoculated into plant cells, thereby initiating the replication of more TMV-RNA encased within whole virus particles.[2] The greater part of this paper deals with experiments with bacteriophages, since it is with these systems that the greatest amount of quantitative and precise information is available.

## GROWTH OF BACTERIOPHAGE WITHIN INFECTED CELLS

Although dozens of different bacteriophages are known, the group of the "*T*" phages that infect and lyse cells of *E. coli* has been worked with by far the most extensively. Within the *T*-group itself, there are three closely related strains, *T*2, *T*4, and *T*6 (the *T*-even phages), which have furnished much of the biological and biochemical information that bears on nucleic-acid replication.

There are several important observed facts about the intracellular growth of the *T*-even phages.[3–5] In the extracellular form, the infective phage particle consists of a head and tail. The head consists of a thin, protein membrane, within which is stuffed all the phage DNA, amounting to about 300 000 nucleotides. The tail is solely protein and serves as the attachment organ, having a high degree of host-range specificity, and through which the DNA is injected into the cell upon infection. Bacteria usually can be infected with several phage particles, with the likelihood of success diminishing with the length of time elapsing since infection by the first particle. Except for a small amount of soluble protein, polypeptides, and polyamines, the only phage material transferred to the interior of the cell is the DNA.[5–7]

Shortly after infection the host cell knows that it is in for trouble. The synthesis of host RNA drops almost immediately to about 3% of the normal rate. Although the synthesis of total protein is unaffected, it is not long before most of it is directed toward making phage-specific structures. Total DNA synthesis is held up at first, and after about 10 min at 37°C the pre-existing host DNA begins to be transformed into phage DNA. Just before lysis, most of it has disappeared in this manner. In the case of the *T*-even phages, this depletion is made evident by the complete loss early in infection of the structural integrity of the cell nucleus.

In the meantime, there is activity in the synthesis of phage-specific materials. For about 15 min, no infectivity can be demonstrated in the contents of artificially disrupted cells,[8] this observation defining the so-called eclipse period. At the end of only about 6 min, some RNA, related to phage infection, begins to be synthesized. Since it does not appear in the mature phage, it is not a precursor material, but it does differ in its base composition from normal host RNA.[9] At about 8 min, phage-precursor DNA begins to be synthesized (and also to be converted from host DNA) so rapidly that 50 phage units of it have accumulated at the end of the eclipse period. At about the same time, various protein fractions that are phage-specific antigens begin to be made. Some of this material is not incorporated into infective phage and remains as surplus antigen. After about 15 min, it is possible to detect infective phage units in the contents of disrupted bacteria. Their number increases rapidly at first and then more slowly as the average lysis time is approached.[10] The average burst size is something like 100 phage particles, although this number varies greatly from cell to cell. Even at the time of lysis there exist within the cell about 20 phage units of excess phage DNA, the same amount of surplus phage antigen, and about 10 units of nonhost RNA. Evidently lysis does not result from a completion of phage maturation, but represents rather a catastrophic collapse of the integrity of the cell wall (see Fig. 1 for a summary of this paragraph).

There are two distinguishable periods during the infective cycle. During the eclipse period, the components of the mature phage are being synthesized and are formed into a precursor pool. At this stage, DNA replication and interchange begin to take place. Subsequently, preformed structural units begin to be as-

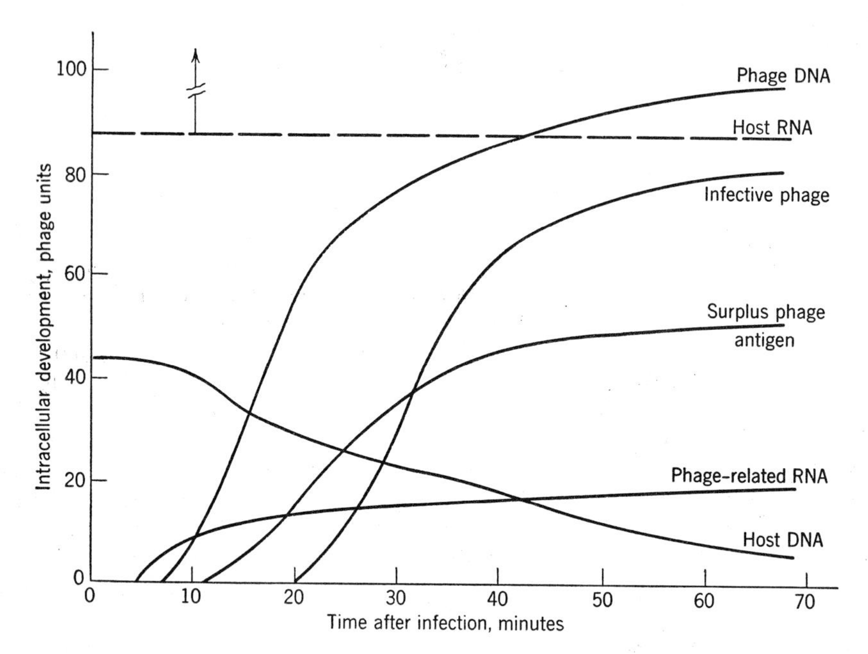


FIG. 1. A semischematic representation of the time course of the intracellular development of certain protein and nucleic-acid components following the infection of *E. coli* with *T*2 bacteriophage in a one-step growth experiment.

sembled into whole particles, although during this time some synthetic activity still persists. For our purposes, this period of maturation is of less interest than the one that precedes it, the so-called vegetative stage.

Inquiry into the origin of the DNA which ends up in the progeny phage is greatly aided by isotopic labeling, and by the fact that the DNA of the *T*-even phages can be distinguished from host DNA by the presence of the pyrimidine, 5-hydroxymethylcytosine (5-HMC).[11] It is found that *T*2-phage DNA comes mostly from material that is present in the growth medium at the time of infection; pre-existing host DNA contributes less than one-third of the total finally assembled into phages. This host DNA is evidently broken down into low molecular-weight fragments since it contains 5-HMC when repolymerized into phage DNA. As might be expected, the earliest phage to be matured are relatively rich in DNA from the host. The last phage to be matured contain some host DNA, further evidence that a large pool is formed, consisting of host DNA, newly synthesized DNA, and DNA contributed by the parent particles.

In contrast to the DNA, the phage protein is found to be synthesized solely from the growth medium. The protein that is made within 5 min of infection is found very sparsely in the progeny phage, while about half that made later is found in them. It is important to note these two kinds of protein: the nonprecursor and the precursor variety. When the formation of phage precursor protein is blocked by chloramphenicol, with the result that no progeny phage are formed, the synthesis of phage DNA goes on unabated.[12]

Although it might seem more profitable to concern ourselves with phage-precursor material only, it should not be forgotten that the kinetics of the synthesis of nonprecursor material may have a bearing on the understanding of phage DNA replication. It might be expected that the formation and fate of RNA within a cell would be unaffected by infection, since no RNA is in mature phages. But it is found that the RNA assembled after infection is unlike normal host RNA in two respects:[9] (1) its base composition is different, and (2) it suffers a rapid metabolic turnover, in contrast to the RNA of uninfected cells. This observation may be important in later considering DNA replication.

As already mentioned, two kinds of protein appear to be synthesized in infected cells: a precursor protein, and a nonprecursor one that is formed preferentially immediately after infection. Its synthesis is closely related to the formation of phage DNA, for, if protein synthesis in general is blocked by chloramphenicol during the first 5 min, no phage DNA is formed. But if blocking is delayed, phage DNA can be formed in amounts that are greater, the longer the delay. There is apparently a stage at which the assembly of protein and DNA become independent of each other, although they are severely dependent at the start. Experiments also have shown the inverse relation: that, if the assembly of phage DNA has started, its subsequent blocking by ultraviolet irradiation will not halt the synthesis of phage-precursor protein.

## FATE OF PARENTAL DNA

After attachment of phage particle to bacterium, almost all of the phage DNA enters the cell. Experiments using $P^{32}$ and $C^{14}$ have been used to determine how much of the parentally injected DNA is found in the first and subsequent progeny generations. The experimental results with *T*-even phages are these: at each generation about one-half of the $P^{32}$-labeled nucleotides find their way into the progeny particles.[13]

(Whether or not the transfer is as inefficient as this is somewhat in doubt because of technical limitations in the recovery of all of the labeled material.) It is unlikely that the loss of the parental $P^{32}$ atoms is due to a drastic breakdown of the phage DNA resulting in a redistribution of the $P^{32}$ atoms among both the phage-precursor and the host DNA. Experiments of both biochemical and genetic nature strongly indicate that fairly large nucleotide sequences are transferred intact from parent to progeny. It also appears that phage DNA is not made up of a portion that is transferred with 100% efficiency, and a portion transferred not at all. If this were so, the efficiency of transfer from first-generation progeny to second-, and from second- to third-, would approach 100%, as distinct from the observed constancy of the efficiency factor. The most likely explanation is that there is a combination of apparent losses owing to technical reasons, and of random losses of fairly large phage nucleotide sequences during the process of infection, replication, and maturation.

The kinetics of DNA transfer can be followed by breaking open the infected cell at various times and examining the state of aggregation of the parental DNA.[14] During the eclipse period, such DNA is found partly in free polymers and partly in lower molecular-weight material. After the eclipse period is over, the earlier phages to mature contain more of the parental DNA, with the latest to mature containing practically none. These results are in agreement with all others which suggest that all phage-precursor substances lie around in a pool until the onset of maturation begins to withdraw them from it, apparently at random.

So far, only the content of parental DNA among the average of the progeny particles has been considered. It might be expected, however, that the distribution of the parental DNA from one individual progeny particle to another would throw more light on replication mechanisms. Two extremes are possible: (1) all of the transferred DNA winds up in one particle; (2) it is equally divided among all progeny particles. The techniques used to determine this distribution are the counting method of Levinthal and Thomas,[15,16] in which the $P^{32}$ content of the phage particles is assayed by counting $\beta$-decay "stars" on nuclear emulsions, and the "suicide" method of Stent,[17] in which the content of $P^{32}$ is assayed by its inactivation of those phage particles that harbor it. The former method has high precision in assessing the *size* of stars (the $P^{32}$ content of the phages), but a low precision in assessing the *number* of phages per infected bacterium that produce detectable stars. The relative sensitivities are in the inverse order in the suicide experiments—the $P^{32}$ content of highly labeled phage is poorly determined, while the number of phages containing at least some minimal amount of $P^{32}$ is precisely determined.

Both methods agree in showing that the distribution of parental DNA is neither all-or-none, nor is it uniform from particle to particle. Rather, the experiments show that a very few of the first-generation progeny particles, per bacterium, individually contain as much as 20 to 25% of the parental $P^{32}$. The remainder of the progeny particles contain far less, with some containing no detectable amount. About half of the parental DNA goes into one first-generation progeny particle. Since Levinthal and Thomas have found that phage particles, uniformly labeled in their $P^{32}$, when osmotically shocked, disclose a nondispersed DNA piece containing about 40% of the original label, they have been inclined to believe that just half of this initial "big piece" stays intact and goes into the highly labeled phage particle of the first-generation progeny—i.e., it is one-half of the phage chromosome, presumably a DNA duplex. When the first-generation progeny are caused to infect, the $P^{32}$ label continues to be transferred to the second generation, but the quantitative measurement of the results is less reliable. Levinthal finds that the star size is undiminished—i.e., that the big piece stays intact—but he cannot be certain as to the number of such stars per bacterium. Stent finds that the *number* of highly labeled phage per bacterium is the same as with the first generation, but he cannot be sure that the $P^{32}$ content of these has remained the same. However, it appears certain that there is a highly nonuniform transfer of parental DNA; whether or not it is a single piece that could be called the phage chromosome awaits extensive and convincing genetic investigation. The idea that the big piece is the genetic template does have the merit of simplicity. It can be visualized that upon infection the big piece of the parental phage remains intact until at some stage it comes apart exactly into two halves. One half may be broken down into smaller units like the DNA units of the parental phage outside the big piece. The other half serves as a template for the assembly of other "half big pieces," two of which enter every newly matured phage. The original half big piece is preserved intact to enter one of the progeny phage (and to result in a star, if the parental phage is $P^{32}$ labeled). About one-half of the DNA in each progeny phage will be in the form of smaller units which had as their templates either a portion of the original half big piece or the smaller DNA units from the parental phage.

### DIFFERENCES AMONG PHAGE TYPES

So far, the biological and biochemical events associated with infection by only the *T*-even phages have been discussed, but it would be incorrect to conclude that they are examples of all phage types that have been studied. No comparable amount of information respecting DNA synthesis is known for the other phages for the reason that they do not contain a DNA species distinguishable from the host DNA. Only the *T*-even phage have been found to have 5-hydroxymethylcytosine (5-HMC).

Distinct differences do exist between infection with

TABLE I. Characteristics associated with infection by phages of different types. Most phages are found to fall into one of two classes, here called class *A* and class *B*.

| Class | uv Sens. | Mult. React. | Capacity of uv cells Normal phage | Capacity of uv cells uv phage | Cell nucleus | Vegetative radiosens. | Recomb. | Lysog. |
|---|---|---|---|---|---|---|---|---|
| *A*<br>*T*-even<br>*T*5 | High | High | Resistant | Resistant | Lost | Decreases with time | Frequent | Never |
| *B*<br>λ, *P*22<br>*p*8, (*T*1,*T*3) | Low | Low | Sensitive | Less resistant | Intact | Constant with time | Rare | Sometimes |

various phages, at any rate in a biological sense, and they are worth recounting cursorily for the insight they lend toward a proper feeling of uncertainty about phage development. In a general way, phages may be put into two categories, which can be termed *A* and *B*. The *T*-even phages are examples of *A*; phages *T*1, *T*7, λ, *p*8, and P22 are among the *B* list. These may be contrasted in several different ways[3] (Table I):

(1) The host-cell nucleus is destroyed upon infection with *A*; preserved with *B*.

(2) Upon multiple infection with homologous phages with different genetic markers, it is found that genetic exchange (recombination) is frequent among the *A* types and infrequent among *B*.

(3) Phages in the *A* category are found never to enter into a lysogenic[18] relation with the cell—i.e., to enter into a nonvirulent stage within the host cell—conferring upon it the *hereditary* property of later producing phage either spontaneously or by induction. *B* phages sometimes do become lysogenic.

(4) In experiments with ultraviolet-light inactivation of virulent phages, it is found that the phages of *A* are relatively more sensitive than those of *B*.

(5) When phages of category *A* are inactivated with ultraviolet light and then are allowed to infect cells multiply, much of their infectious activity can be restored, a phenomenon called "multiplicity reactivation." Phages of *B* are not nearly as capable of multiplicity reactivation.

(6) Phages *A* and *B* also differ with respect to the "capacity" of their host cells, defined as the ability of cells to reproduce infective phage, particularly after the cells have been damaged with agents as ultraviolet light and x-rays. For phages of class *A*, the capacity of cells is much more resistant to ultraviolet irradiation than it is for class *B*.

(7) Cells may be irradiated also while the phage is in the vegetative state, allowing measurements to be made of the radiosensitivity of the infected host to the production of infective virus. The phages of class *A* (and their host cells) have a radiosensitivity that rapidly decreases with time after infection, while those of class *B* very nearly retain their initial sensitivity.

The simplest generalized conclusion that can be drawn from the foregoing set of experimental facts is that the two classes of phages differ in the degree to which, during the vegetative stage, the genome of the infecting phage is associated with that of the host cell. It appears likely that the genome of phages *T*2 and *T*4, as examples of class *A*, replicates with considerable independence of the host, while the phages of class *B* multiply only as a consequence of genomic interaction with the host cell. A virtue of this unitary explanation of what appear to be different phenomena is that it allows the preservation of a concept that has proven fruitful in the cases of the *T*-even phages. This is the notion, given quantitative elucidation by Visconti and Delbrück,[19] that *mating* is a frequent activity in the DNA pool of the vegetative state. Mating is operationally defined as having occurred if two vegetative phages have interacted in a manner such that the probability is 0.5 that they have interchanged two distantly linked genetic loci, an occurrence detected by the production of recombinant phages. Conceptually, mating may be thought of as the pairwise physical interaction of two (or more) homologous genomic structures. The arithmetical conclusion of the Visconti-Delbrück theory is that each act of genomic *replication* is associated with a mating act, the number of such double acts being about 5 for the *T*2 phages during the vegetative phase. Strength is accorded this notion by the observation that, if lysis is inhibited and the infected bacteria opened at various times, the proportion of recombinants in the matured phages increases with time after infection—i.e., the longer the time spent in the mating pool, the greater the chances of recombination. But the experiments showing a low frequency of recombinants with such phages as λ (class *B*) have cast doubt on the correctness of the Visconti-Delbrück theory. The theory can be rescued qualitatively by assuming that mating still accompanies replication, but in the case of the class *B* phages the mating is frequently between the phage and the host genomes.

## MECHANISMS OF DNA REPLICATION

The paper by Levinthal (p. 227) shows that there is good evidence that DNA can be the bearer of genetic information; that the detailed structure of one chain determines that of the other; and that presumably the structure of newly formed DNA is determined by that of the parental DNA. (It should not be inferred, however, that only DNA is capable of transmitting such information, since these are examples of the apparent genetic stability of the RNA-containing viruses.) The problem to consider now is the mechanical detail of how the DNA that is in the process of being formed is so controlled in its sequence of nucleotides that it usually will be genetically identical (and presumably structurally identical) with the DNA that is acting as a template. A correlative problem is how the template DNA, either directly or through intermediaries, controls the specific assembly of amino acids into proteins, to confer upon progeny structures some of their phenotypic characteristics.

It is an attractive consequence of the two-chain structure of DNA to assume that the obvious is true: that replication is preceded or accompanied by a separation of the two chains, each of which then acts as a template for the production of its mate. This could be done by allowing the chains initially to separate at an end of the helix, forming a short-pronged Y.[13] New nucleotides then would start assembling in proper sequence along the branches of the Y, the residual helix untwisting as the polymerization of the two new chains proceeded. A difficulty with any scheme of this or similar nature is the unwinding process; how can two chains involving several hundred turns separate by untwisting, both in consideration of the energy required and the hazards of hopeless tangling? To avoid this dilemma, several schemes involving rather *ad hoc* notions of mechanical processes have been proposed, such as backbone breakages occurring at every turn of the helix during replication, but all of the schemes seem to introduce more difficulties than they remove. Actually, the energy requirements of unwinding the helix are not as great as might be imagined, if the helix is restricted to turning on its own axis. If it always turns on that axis, there is also a minimal chance of tangling. Levinthal and Crane[20] showed that the energy required per turn of the helix is very small as compared with the energy of formation of the phosphate-diester linkages per turn. It seems inevitable that the transfer of information from parental DNA to new DNA and to new, specific structures of other kinds, such as RNA and protein, involves unwinding at one or more stages.

As there seems to be no likelihood in the near future of directly determining the mechanics of replication, it is profitable to examine phenomena that indirectly might indicate what kind of mechanism is the most probable. Here some of the phage experiments are relevant. Unfortunately, the transfer experiments shed very little light on the problem; they indicate that there is some conservation of large pieces of $P^{32}$-containing material and some dispersal of smaller pieces among several progeny particles. A suggestion by Stent,[3] which appears attractive, formulates a sequence of events during the vegetative stage whose interpretation can point toward a mechanism of replication not mentioned above. These events are characterized chiefly by three facts: (1) phage-directed RNA synthesis begins immediately after replication; (2) the blockage of protein synthesis stops phage-DNA synthesis if it occurs soon enough, otherwise DNA synthesis proceeds in any event; (3) with $T$-even phage infection, the ultraviolet light and $P^{32}$ sensitivity decreases with time during the vegetative phase. These facts can fit a replication scheme wherein the parental DNA structure is conserved initially, at any rate, and serves as template in the *grooves* of which is assembled a single chain of *ribonucleoprotein*. This chain subsequently unwinds off the template, leaving the latter to serve in this capacity again. The RNA-protein chain wanders off in search of a mate, which may be a chain copied from another phage-parental DNA helix or from a DNA helix of the host itself (class $A$ *vs* class $B$). Mating consists in forming a duplex helix of two reasonably identical or homologous RNA-protein chains. The new DNA is then assembled by a process the inverse of the formation of the RNA-protein chains, wherein the two chains of the DNA are helically assembled around one of the chains of the RNA-protein duplex, the other chain disengaging during the DNA assembly. The final, RNA-free DNA is then disengaged from the template RNA-protein chain by unwinding. The RNA-protein intermediaries may be used also to serve as templates for the assembly of phage-specific protein. The handing on of the guidance of DNA synthesis to a ribonucleoprotein template would account for the production of RNA shortly after infection, and for the blocking phenomena respecting protein and new DNA, since DNA synthesis could readily proceed after a few RNA-protein templates became assembled. Further, since the secondary template is now made of all-new RNA and *protein*, one might expect it to be relatively resistant to damage by ultraviolet light and to $P^{32}$ decay. The scheme has the further merit of predicting that the effects of slightly damaged parental (or host) DNA, such as single-chain breakage by ultraviolet light or x-rays, could be effectively overcome since all that is required is that the DNA helix hold together as a duplex. It is hard to see how single-chain breaks, spotted here and there along both chains of the helix, could lead to anything other than abortion if the original Watson-Crick replicating scheme is correct. The Stent hypothesis also provides for mating at each replication, although other schemes arrange for this too.

It does not seem that the observed facts of genetic recombination can point unequivocally to either a replicating scheme that involves DNA assembly directly upon existing DNA templates or to one involving an intermediary helical structure. It is grossly inferred by light-microscopic observations of *cellular* organisms that recombination is due to the existence of breaks and subsequent crossing-over in the dividing chromosomes. With phage, however, the genetic and biological evidence is contrary to the expectation that crossing-over among genetic structures exists. The alternate possibility is that recombinants are produced by a mechanism of "copy choice,"[21] in which a daughter genetic structure switches from one template to a closely-adjoining but slightly differing one during its stepwise assembly. Such switches may be multiple, incorporating several genetic loci from the two mated structures serving as templates. But a hypothesis of copy-choice used in interpreting genetic recombination does not result in any unequivocal answers with respect to the *mechanism* of replication, except to enhance the notion that replication is associated with mating.

Replication schemes can be categorically distinguished (on paper) with respect to the resulting distribution of the parental DNA. If the parental DNA remains intact during all replication processes, resulting in only one parental structure and numerous other DNA's containing no parental material, a *conservative* system is said to exist. A *semiconservative* system is one in which the single DNA chains of the parental duplex retain their integrity but are physically and permanently separated during replication. The original Watson-Crick replicating scheme is a semiconservative one. A *dispersive* mechanism is one in which the atoms of the parental DNA are dispersed by some kind of fragmentation among all the daughter DNA structures. It might be hoped that experimental determinations of the kind of mechanisms so categorized would lead at least to the *elimination* of some replication schemes.

## REPLICATION MECHANISMS AT THE CELLULAR LEVEL

Some autoradiographic work done by Taylor and colleagues[22] on the dividing chromosomes of root cells of the English broad bean is next to be considered. The growing cells are first fed tritium-labeled thymidine, which is known to be incorporated only in the cellular DNA, and are then observed after one or two divisions in a normal medium. Colchicine is used to preserve both generations of chromosome division within one nuclear structure. The upshot of the experiment is that the chromatid structures of these cells multiply in a way that can be interpreted as semiconservative at the level of the DNA duplex. That is, if each chromatid is thought of as consisting of numerous DNA helices (arranged in presumably some regular fashion), the helices associated with any one chromatid separate into two strands with each strand preserving its identity. At the half-chromatid level, the mechanism is conservative: the DNA chains that are associated together in any one half-chromatid stay together indefinitely.

In experiments with growing *E. coli*, Meselson and Stahl[23] have been able to investigate the type of replicating mechanism at the level of the DNA polymers. *E. coli* that have been grown on $N^{15}$ are subsequently allowed to grow for a few divisions on $N^{14}$, and their DNA is examined by density equilibrium centrifugation, using a caesium-chloride gradient. A gradient can be established with a sufficiently gentle density change to distinguish structures such as DNA by their relative content of $N^{15}$ and $N^{14}$. DNA containing solely $N^{15}$ will form a band in the gradient cell that is distinctly more centrifugal than will a DNA containing only $N^{14}$. It is found that at first, of course, all of the *E. coli* DNA is $N^{15}$-containing. As growth goes on, a band midway between $N^{15}$ and $N^{14}$ appears, followed later by a band at $N^{14}$. At no time does material appear at the interband region, thus establishing that dispersive mechanisms do not operate in this system. The $N^{14}$:$N^{15}$ band shows that the initial $N^{15}$-containing DNA helices are diluted just one-half by $N^{14}$ during the growth cycle. The most obvious interpretation of the results is that DNA replication is semiconservative at the level of the DNA duplex. This is an elegant technique indeed and it might be hoped that it would give straight answers to the mechanism of replication of *phage* DNA. Although only a preliminary report of such investigation is available at present, it appears that the answer will not come easily. When phage previously grown on $N^{14}$ are allowed to infect $N^{15}$-containing cells, in an $N^{15}$ medium, the progeny phages can be purified and the DNA examined for $N^{14}$:$N^{15}$ content. What is then seen in the gradient centrifuge is a single $N^{15}$ band. If the cells are disrupted prematurely to expose the DNA vegetative pool, a very small amount of what appears to be an $N^{14}$:$N^{15}$ hybrid DNA is found. The experiments are very likely inconclusive. If one takes the results with the purified progeny phage as they come, one would conclude that there is conservative replication. But it is easy to calculate that any $N^{14}$:$N^{15}$ hybrid band which might exist would be just on the limit of detectability. So the experiments prove nothing, so far.

## CRITERIA FOR ACCEPTABLE REPLICATION SCHEMES

Let us return again to the problem raised by the existence of the big piece of phage DNA, along with the implied existence of small pieces. All that is known about these structures is that a $P^{32}$-containing big piece is held intact during osmotic shock, and that very likely half of this big piece is passed on to individual particles of progeny phage. But it is not known how the halving of the big piece comes about during replication. It could conceivably (although improbably) be two separable half-pieces, and one of these could remain conservatively integrated as a DNA duplex during replication. And what is the role of the small pieces? Can it be that the big piece and the small pieces have different roles? Until the significance of the apparent bipartite character of phage DNA is known, it is unlikely that any replication scheme will win complete adherence.

It is tantalizing to find that something as apparently straightforward as the problem of synthesizing more DNA on a template of pre-existing DNA should present so many puzzles. Let us recapitulate some of the experimental conclusions which any scheme of DNA replication should either explain or, at any rate, be consistent with. "Experimental conclusions" include presently available facts, quite reasonable inferences, and conclusions from experiments yet to be conclusively performed.

(1) Replication schemes must agree with any observations on the densities of $N^{14}$ and $N^{15}$ DNA, since these may well indicate whether or not replication is conservative, semiconservative, or dispersive. A conclusion based on experiments with a certain type of replicating system, such as bacterial cells, is not necessarily subject to generalization to all systems, such as phages.

(2) Provision must be made for information transfer from DNA to the synthesis of non-DNA structures, such as specific proteins.

(3) With respect to phage replication there are these specific points:

(a) The numerical equivalence of rounds of mating with rounds of replication seems to be established, especially with the *T*-even phages. It is an attractive corollary that mating and replication are not independent events, and a replicating scheme might be expected to provide for this dependence.

(b) Genetic recombination is experimentally inferred to take place by some sort of copy-choice mechanism, rather than by the occurrence of breaks and crossing-over.[21] Copy-choice recombination is reasonably believed to take place during a replicating act that occurs under the influence of mated structures. Provision must be made to have homologous templates arranged in one-to-one structural opposition, thus implying the existence of highly specific medium-range forces of interaction.

(c) If copy-choice recombination takes place in the presence of mated structures, the latter must be so mutually arranged that the newly assembled DNA can become disengaged from both mated structures without tangling and without a high degree of fragmentation of the templates, *if* these are the parental DNA duplexes.

(d) The phenomena of "marker rescue" by multiplicity reactivation[24] and cross-reactivation should be accounted for. If it turns out that parental DNA may be rescued (in its ability to be replicated) even after scission of one of the two chains, this phenomenon is to be accounted for by any replicating scheme that envisions the unwinding of parental DNA.

(e) The observed heterogeneous, but nondispersive, distribution of parental DNA in the progeny must be accounted for.

(f) Biochemical events during the vegetative phase are to be made consistent with the replication scheme. In particular, the observation is important that there is an early synthesis of phage-specific RNA and that only early blocking of protein synthesis will prevent DNA synthesis.

No present scheme of replication of phage DNA satisfies all of these requirements, and only for phages is enough known to allow schemes to be anything other than exercises in model building. Any scheme involving the disengaging of DNA chains by breaking and rejoining them is unlikely on the grounds that it implies a large degree of dispersion of the parental DNA, and it meets with difficulty in mating and recombination. The obvious semiconservative scheme based on the Watson-Crick model satisfies many of the criteria but has difficulty in handling single-chain scissions, and, in its simplified form, does not account for synthesis of phage-specific proteins. It also seems not to be sensitive to the effect of early blockage of protein synthesis. Copy-choice replication in mated DNA structures also gives difficulty, for the reasons that there seems to be no reason why complete DNA duplexes should mate and that the stereo problems of disengagement are large. The scheme of invoking a ribonucleoprotein intermediary template is consistent with many of the experimental conclusions mentioned in the foregoing. It suffers from the uncertainty as to whether it is stereochemically possible in atomic detail, nor does it make any specific predictions as to the fate of the integrated parental DNA duplex. If this remains intact throughout phage replication and maturation, a conservative type of replication is demanded. But if experiments show that phage replication is precisely semiconservative, only *ad hoc* disposition of the parental DNA structure would save this model at the moment.

## REPLICATION IN RNA VIRUSES

We have concentrated so far on the problems of replication of DNA that has been introduced into cells, the best examples of which are the DNA of the transforming factor and of bacteriophage. But there exist many cases in which the genetic message must have been introduced through the medium of viruses that contain no demonstrable DNA. Actually, most known viruses, both plant and animal, contain no DNA. Although none of these RNA viruses can be observed with the quantitative precision inherent in phage work, it is known that the general facts of replication are similar. They possess genetic continuity, are found occasionally to mutate, and in one instance it appears that the character of the virus genome is controlled by the type of host in which the virus is multiplied. Their general structure is similar to that of phage, with a protein exterior and some kind of core containing the RNA.

Interest in the replicating potentialities of RNA has been enhanced by the discovery that a highly purified RNA obtained from tobacco mosaic virus (TMV) is infectious.[2,25] If a solution of this material is rubbed upon tobacco leaves in the same way as TMV is applied for assay purposes, local lesions are developed, or systemic infection is initiated. At present, pure RNA can be prepared whose infectivity per unit weight is about one percent of the same weight of RNA as it exists in native virus[26] particles. The infectivity can be greatly increased by reconstituting the virus—i.e., combining the RNA with protein fragments of the virus in a way such as to re-form virus particles indistinguishable from native ones. The infectivity may then rise to a figure as high as 60%. Although reconstitution has not been achieved for any other virus, it has so far been found that the RNA from another plant virus, turnip yellow mosaic,[27] is infectious, as is the RNA from polio virus.[28]

The findings just described allow RNA to be studied in a way that has some aspects of both phage and transformation experimentation. Like the transforming factor, RNA can be biologically assayed after removal from its natural host, and after purification and laboratory manipulation. Like phage DNA, its activity is demonstrated by the production of copies of the virus originally containing the nucleic acid. It is frustrating, however, that plant viruses cannot be studied with a quantitative precision even approaching that enjoyed by the phage experiments.

The work with infectious RNA has largely taken the lines of finding out what happens when artificial "hybrid" viruses are reconstituted, and of deducing the physicochemical characteristics of the infectious unit. As might be expected, when RNA from a virus of one strain is reconstituted with protein from another strain, the progeny particles are copies of the virus that furnished the nucleic acid. The protein evidently serves primarily as a protecting coat to preserve the integrity of the RNA during inoculation. It is also possible to make what might be called artificial recombinants: reconstituted TMV particles that are believed to contain nucleic acid from two parent strains.[2] When such virus particles are assayed on local-lesion hosts, most of the lesions found resemble either those normally caused by one or the other nucleic acids, but some have a mixed character. When the latter type of lesion is subinoculated by using it as inoculum for systemic infection, the symptoms developed and the character of the progeny virus from the systemic infection are wholly those normally associated with either one strain of RNA or the other. If the reconstituted particles have two kinds of RNA packaged within them, the process of infection evidently allows each kind of RNA to be independently replicated, a not surprising result.

The experiments designed to find the minimum weight of RNA that is infectious are obviously important in their bearing on the question as to whether or not there is redundancy in the amount of information-carrying material in a viral genome. Great interest would attach to the finding, for example, that a structural unit of considerably less mass than all of the RNA of the virus is sufficient to act as the replicating template. Is there a TMV-RNA big piece and are there little pieces that are infectious? The answers to these questions, if they are positive, should come more directly with the RNA from TMV than they are likely to come with the phages. So far, the results are equivocal; RNA is not easy to fractionate into differing sizes nor is it easy to characterize physically. At present, it appears that RNA pieces smaller than the RNA content of one TMV particle are infectious.

It seems that, at present, the importance of the finding that purified RNA is virogenic is a conceptual one. Even though RNA viruses have been recognized for a long time, there has been no compelling reason to believe that their infectious proclivity is localized in their RNA, in analogy with the phage-injected DNA. Now it appears that the biochemical aspects of the earliest stage of infection are similar—that in both cases reasonably pure nucleic acid can act in the initiation of infection and the messenger of genetic information. Any master plan of replication, therefore, must show how *RNA* can direct the formation of its copies, as well as how *DNA* can do this. It certainly will not be surprising to find that some kind of protein synthesis, acting in specific and intimate contact with the nucleic acids, is a necessary correlate in the whole replication process. Nor will it be surprising to find that RNA is involved in DNA replication, and that DNA is involved in RNA replication.

Where have these extended but inconclusive remarks on the replication of nucleic acids as evidenced by the viruses led us? My own feeling is that it is currently premature and naive to expect that the geometrical nicety implied in the two-strand helix of DNA will allow us to get out our poppet beads and come up with a jigsaw puzzle answer to the way such structures are assembled within cells. It is not known even that DNA and RNA *are* replicated, in the sense, for example, that the newly formed nucleic acids of genetically identical progeny viruses are identical on atomic dimensions with the parental nucleic acids (among which, of course, there may be diversity from parent to parent). Model building is fun, and its results are provocative and suggestive of what experiments to do next, but we are a long way from being able to identify the model with reality. We do not know how nucleic acids replicate, even in terms of quite nebulous mechanisms. The danger of premature model building is that it tends to direct our experimentation along subjectively controlled lines, perhaps to blind us to the key experiments that do not currently fit our preconceptions.

## BIBLIOGRAPHY

[1] H. Ephrussi-Taylor in *The Chemical Basis of Heredity*, W. D. McElroy and B. Glass, editors (The Johns Hopkins Press, Baltimore, Maryland, 1957), p. 299.

[2] H. Fraenkel-Conrat, B. A. Singer, and R. C. Williams in *The Chemical Basis of Heredity*, W. D. McElroy and B. Glass, editors (The Johns Hopkins Press, Baltimore, Maryland, 1957), p. 501.

[3] G. S. Stent, Advances in Virus Research **5**, 95 (1958).

[4] G. S. Stent in *The Viruses: Biochemical, Biological and Biophysical Properties*, F. M. Burnet and W. M. Stanley, editors [Academic Press, Inc., New York, (to be published)], Vol. I, Chap. 3.

[5] S. E. Luria in *The Viruses: Biochemical, Biological and Biophysical Properties*, F. M. Burnet and W. M. Stanley, editors [Academic Press, Inc., New York (to be published)]. Vol. I, Chap. 8.
[6] A. D. Hershey and M. Chase, J. Am. Physiol. **36**, 39 (1952).
[7] A. D. Hershey, Advances in Virus Research **4**, 25 (1957).
[8] A. H. Doermann, J. Gen. Physiol. **35**, 645 (1952).
[9] E. Volkin and L. Astrachan, Virology **2**, 149 (1956).
[10] C. Levinthal and H. Fisher, Biochim. et Biophys. Acta **9**, 419 (1952).
[11] G. R. Wyatt and S. S. Cohen, Nature **170**, 1072 (1952).
[12] A. D. Hershey and N. E. Melechen, Virology **3**, 207 (1957).
[13] M. Delbrück and G. S. Stent in *The Chemical Basis of Heredity*, W. D. McElroy and B. Glass, editors (The Johns Hopkins Press, Baltimore, Maryland, 1957), p. 699.
[14] G. S. Stent and N. K. Jerne, Proc. Natl. Acad. Sci. U.S. **41**, 704 (1955).
[15] C. Levinthal and C. A. Thomas, Biochim et Biophys. Acta. 2?, 453 (1957).
[16] C. Levinthal, Proc. Natl. Acad. Sci. U.S. **42**, 394 (1956).
[17] G. S. Stent, J. Gen. Physiol. **38**, 853 (1955).
[18] F. Jacob, *Les Bactéries lysogènes et la notion de provirus* (Masson et Cie., Paris, 1954).
[19] N. Visconti and M. Delbrück, Genetics **38**, 5 (1953).
[20] C. Levinthal and H. R. Crane, Proc. Natl. Acad. Sci. U. S. **42**, 436 (1956).
[21] J. Lederberg, J. Cellular Comp. Physiol. **45**, Suppl. 2, 75 (1955).
[22] J. H. Taylor, P. S. Woods, and W. L. Hughes, Proc. Natl. Acad. Sci. U. S. **43**, 122 (1957).
[23] M. Meselson and F. W. Stahl, Proc. Natl. Acad. Sci. U. S. **44**, 671 (1958).
[24] S. E. Luria, Proc. Natl. Acad. Sci. U. S. **33**, 253 (1947).
[25] G. Schramm and A. Gierer, *Cellular Biology, Nucleic Acids, and Viruses* (Special Publications of the New York Academy of Sciences, 1957), p. 229.
[26] H. Fraenkel-Conrat and R. C. Williams, Proc. Natl. Acad. Sci. U. S. **41**, 690 (1955).
[27] R. L. Steere (unpublished observations).
[28] J. S. Colter, H. H. Bird, A. W. Mayer, and R. A. Brown, Virology **4**, 522 (1957).

# 29
# Genetic Fine-Structure Analysis

E. S. Lennox

*Department of Chemistry and Chemical Engineering, University of Illinois, Urbana, Illinois*

The question of concern here is: To what extent can properties of the genetic map be made to correspond to some physical structure in the organism? Since the bulk of the evidence points to deoxyribonucleic acid (DNA) as the molecule involved, this becomes a question of refining genetic experiments to the point where they can critically examine properties of DNA which are involved in its acting as genetic material.

The experiments discussed herein are those involving genetic measurements only—essentially, recombination frequencies in certain standard kinds of crosses (see Levinthal, p. 227). Experiments attempting to correlate genetic properties with synthetic capacities of cells are discussed in the second paper by Levinthal (p. 249).

One of the most fruitful areas of investigation regarding this problem has been that of detailed fine-structure analysis. In essence, the method has been that of pushing recombination frequencies to such a limit that sizes of genetic units may be compared with the molecular subunits of nucleic acid, the nucleotides. It is a comparison between the fine-grain structure of the genetic map and the fine-grain structure of DNA.

Much of the following discussion of fine-structure analysis is based upon experiments by Benzer on the bacteriophage *T*4.[1] Discussed also are experiments with bacteria bearing on the question posed above.

The reasons for choosing microbial systems for these kinds of studies are several. Mutants are available in large numbers and varieties, and selective techniques make it possible to screen large numbers of organisms in the search for a desired type. For example, in the *T*4 system (discussed in detail in the following), one wild-type bacteriophage particle can be detected among $10^6$ to $10^7$ mutant types, if the proper bacterial strain is used as an indicator. This allows examining recombination frequencies of the order 1 in $10^6$ to $10^7$. Another reason for the choice of microbial systems is that homogeneous preparations can be made with relative ease for chemical studies.

The work of Benzer made it necessary to define carefully the genetic units in terms of genetic measurements. The classical unit of heredity, the gene, has several operational definitions which had, in general, not been distinguishable experimentally in *Zea mays* and in *Drosophila*, the principal organisms of research in genetics. The definitions are as follows: (1) The unit of recombination—the *recon*—is defined as the smallest element interchangeable in recombination experiments. In such experiments, one uses easily-recognizable characters which enter the crosses in the genetic material of the parents and which sort among parental-type or recombination-type progeny. This element is not necessarily the same as that defined by another set of operations—isolation of mutant types (either spontaneous or induced) and mapping of their sizes. This yields another unit. (2) The unit of mutation—the *muton*—is the smallest element, alteration of which gives a type recognizably different from the standard form chosen as wild type. What is recognized as a mutant depends on the method of observation. It is, of course, conceivable that different mutant characters will be of different sizes. (3) The last unit that needs definition is the unit of function—the *cistron*. This unit is more difficult to define because what is recognized as the function of the genetic material depends upon the level at which observation is made between the primary synthetic act of the genetic material and an observable result of this act. If, for example, wing shape in *Drosophila* is the character being scored, then a large number of steps leading to the mutant form may be involved. One needs a way of deciding that two mutants are blocked in the same function, in the absence of knowing what the precise function is in terms of primary genetic product. A definition of a functional unit by genetic experiments is provided by the (*cis*-)(*trans*-) test devised originally for *Drosophila* by Lewis,[2] and used so elegantly by Benzer in his study of *T*4 bacteriophage. One asks the question: Are two given mutants with the same outward appearance affected in the same functional unit? This is answered by allowing the genetic material of both mutants simultaneous expression in the same cell. In *Drosophila*, this is accomplished by introducing one mutant into the egg and the other into the sperm. The resulting cells then have the two mutants' markers in homologous but not identical sets of genetic material. In the bacteriophage system, the test is performed by simultaneous infection of the host bacterium with two mutant particles. In each case, one wants to know whether the resulting combination (the *trans*-configuration) is wild type or mutant in its functioning. The control experiment is the similar situation in which both mutants (as a double mutant) are in one complement of genetic material, the other complement being wild type (*cis*-configuration). What are the expected results of such experiments? First the control: this must function as wild type if the other tests are to be revealing. If, in the *trans*-test, both mutants are in the same functional unit, the combination is not expected to be competent. If, on the other hand, they are mutant in different functional regions, then the combination should be competent, since each can make up for the deficiency of the other. Note that the competence does not depend upon recombination to give wild type, for the genetic material would in subsequent tests still be revealed as mutant. Two mutants then are said to belong to the same functional unit—i.e., the same *cistron*—when in the *trans*-configuration they function as mutants.

Consider in some detail the system exploited by Benzer. The bacteriophage mutants used were many independent isolates from wild-type parents which show different plaque morphology (here designated *r*-type) on the host strain *E. coli B*. On other bacterial strains, these mutants have properties which are compared with wild type in Fig. 1.

The classification of mutants into groups *rI*, *rII*, and *rIII* is done on the basis of the characteristics indicated in Fig. 1. The group of mutants *rII* is especially interesting in not giving plaques on strain *K*. This allows in a cross selecting against two parental *rII* mutants and selecting for rare wild-type recombinant.

Figure 2 shows how the appearance of the genetic map depends upon the choice of the bacterial host for examining the bacteriophage mutants. The designation of *rII* on *K* as *m* comes from the fact that an infrequent

| Phage strain | Bacterial host strain *B* | *S* | *K* |
|---|---|---|---|
| wild | wild | wild | wild |
| *rI* | *r* | *r* | *r* |
| *rII* | *r* | wild | . . . |
| *rIII* | *r* | wild | wild |

Fig. 1. Plaque morphology of *T*4 strains (isolated in *B*) plated on various hosts [from S. Benzer in *The Chemical Basis of Heredity*, W. D. McElroy and B. Glass, editors (The Johns Hopkins Press, Baltimore, Maryland, 1957), p. 70].

*rII* mutant gives a minute turbid plaque on this host—most of them, however, give no plaques. In the physiological effect of the mutational event, these mutants vary, then, from lethal to unnoticeable, depending on the host used for phage assay.

Turn now to the problems of measuring, in terms of recombination frequencies, the sizes of the genetic units defined in the foregoing.

To measure the unit of recombination, the appropriate bacterial cell (here, *E. coli B*) is infected with two parental *rII* mutants. One measures the frequency with which this pair of mutants recombine to give wild type (assayed on *K*). By searching for many *rII* mutants and by mapping each against the others, one sees if there is a limiting distance between positions at which two mutational events occurred.

To measure the muton length, one observes whether, with three closely linked markers, one can find aberrations in recombinant frequencies owing to the size of the mutation (Fig. 3). This size is reflected by a nonadditivity of distance 1 to 2 and 2 to 3 as compared with 1 to 3.

To determine a cistron length, many isolates of *rII*-type mutants are tested in pairs in a *trans*-

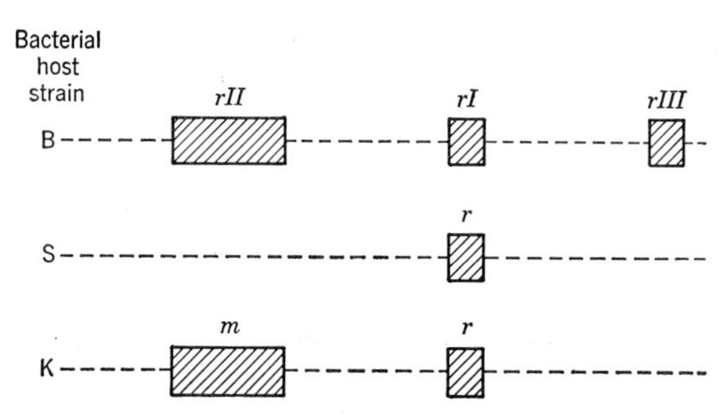


Fig. 2. Dependence of the genetic map of *T*4 upon the choice of host. Three regions of the map are shown as they probably would appear if *E. coli* strains of *B*, *S*, or *K* were used as the host [from S. Benzer in *The Chemical Basis of Heredity*, W. D. McElroy and B. Glass, editors (The Johns Hopkins Press, Baltimore, Maryland, 1957), p. 70].

configuration; e.g., *rII* mutant *m* and *rII* mutant *n* are put into the same cell, and this combination scored as wild type—i.e., yields phage, or as mutant, i.e., no phage yielded. This establishes whether the two mutants are in the same cistron. Two mutants in the same cistron are then mapped relative to each other and to other members of the cistron. The largest distance between two mutational markers within the same functional unit gives a lower bound to the size of this unit.

The amount of labor involved in mapping the numerous mutants necessary for such studies is considerably reduced by mapping each newly discovered mutant against a set of deletions that span the regions to be mapped. Mutants labeled as deletions behave as if a portion of the genetic map were missing. They do not revert to wild type at a measurable frequency. One can find for any deletion a group of reverting mutants which among themselves yield recombinant types, but none of which do so with the given deletion. The extent of the deletion covers this group of reverting mutants. The reverting mutants themselves are mapped as points on the map.

The enormous advantage of using these deletions is that they span blocks of the map, allowing any new mutant to be located quickly without an excessive number of crosses. Detailed mapping of a newly found mutant is then done only with regard to those pre-

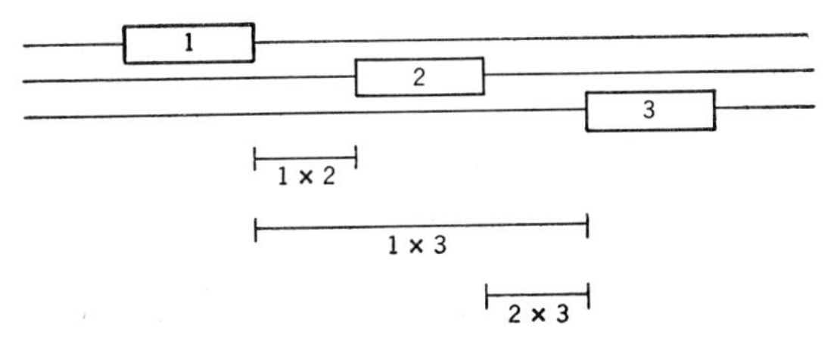


Fig. 3. Method for determining the "length" of a mutation. The discrepancy between the long distance and the sum of the two short distances measures the length of the central mutation [from S. Benzer in *The Chemical Basis of Heredity*, W. D. McElroy and B. Glass, editors (The Johns Hopkins Press, Baltimore, Maryland, 1957), p. 70].

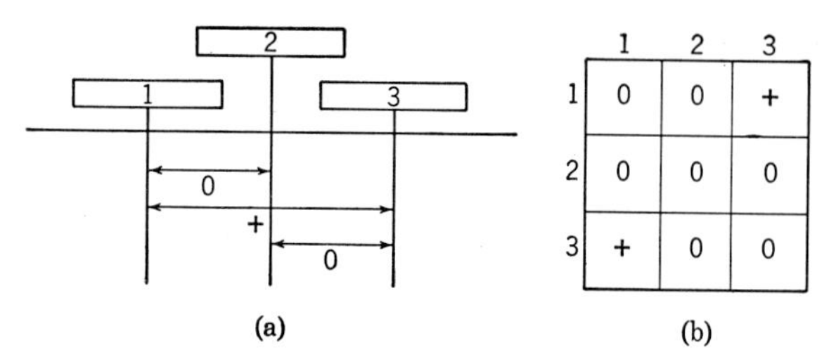


FIG. 4. The method of overlapping deletions. Three mutants are shown, each differing from wild type in the deletion of a portion of the genetic material. Mutants No. 1 and No. 3 can recombine with each other to produce wild type, but neither of them can produce wild recombinants when crossed to mutant No. 2. The matrix (b) represents the results obtained by crossing three such mutants in pairs and testing for wild recombinants; the results uniquely determine the order of the mutations on the map (a) [from S. Benzer in *The Chemical Basis of Heredity*, W. D. McElroy and B. Glass, editors (The Johns Hopkins Press, Baltimore, Maryland, 1957), p. 70].

viously discovered ones that lay within the same deletion.

Mapping of the deletions with regard to each other is also done relatively simply (Fig. 4). In the figure, mutants 1 and 2, for example, will not give wild type in a mixed infection and are defined as overlapping deletions. On the other hand, mutants 1 and 3 will give wild-type recombinants and are nonoverlapping deletions.

By these methods, Benzer was able to focus his attention on a relative small subregion of the ***rII*** region of the *T*4 map. By mapping each new mutant against a fixed standard deletion, it was screened either as one to be ignored for the moment (outside the deletion) or as one to be mapped further—first, against smaller subdeletions, and, finally, with the mutants in each subblock defined by the smallest deletions. In this way, refined genetic analyses were performed making possible measurements of the genetic units defined earlier.

One of the first results to come out of such a program of mapping is that the ***rII*** region defined originally in terms of plaque-forming characteristics on the bacterial strains *B*, *K*, and *S* consists of two functional regions, the *A*-cistron and the *B*-cistron.

Restricting attention further to one cistron allows more-refined mapping. Figure 5 indicates the results of

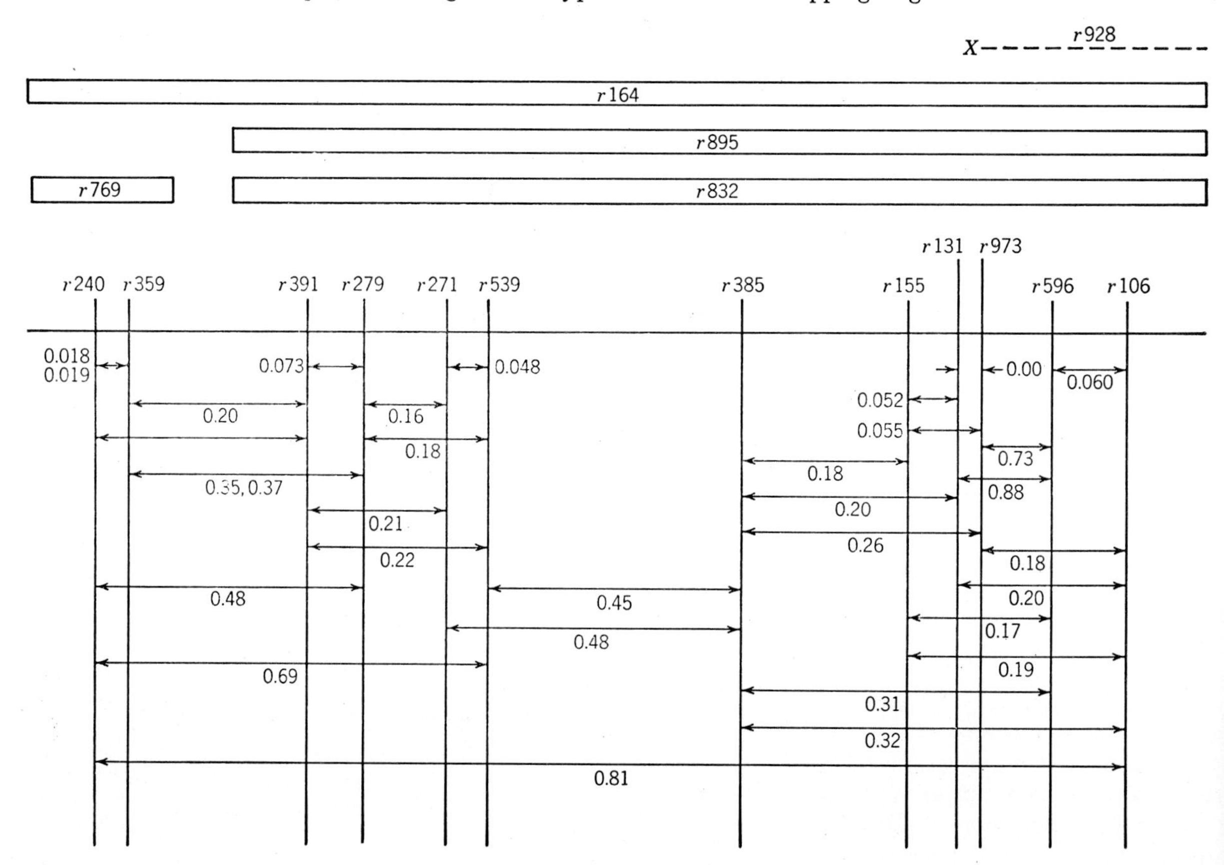


FIG. 5. Map of the mutants in the *r*164 segment. The numbers give the percentage of recombination observed in standard crosses between pairs of mutants. The arrangement on this map is that suggested by these recombination values; it has not yet been verified by three-point tests. Stable mutants are represented as bars above the axis; the span of the bar covers those mutants with which the stable mutant produces no detectable wild recombinants. The stable mutant ***r*928** appears to be a double mutant having one mutation at the highly mutable ***r*131** location and a second mutation at a point in the *B*-cistron. Mutants ***r*131** and ***r*973** are separated on the map so that the data for each can be indicated. Some of the data here given differ from (and supersede) previously published data based upon unconventional crosses which turned out to be incorrect [from S. Benzer in *The Chemical Basis of Heredity*, W. D. McElroy and B. Glass, editors (The Johns Hopkins Press, Baltimore, Maryland, 1957), p. 70].

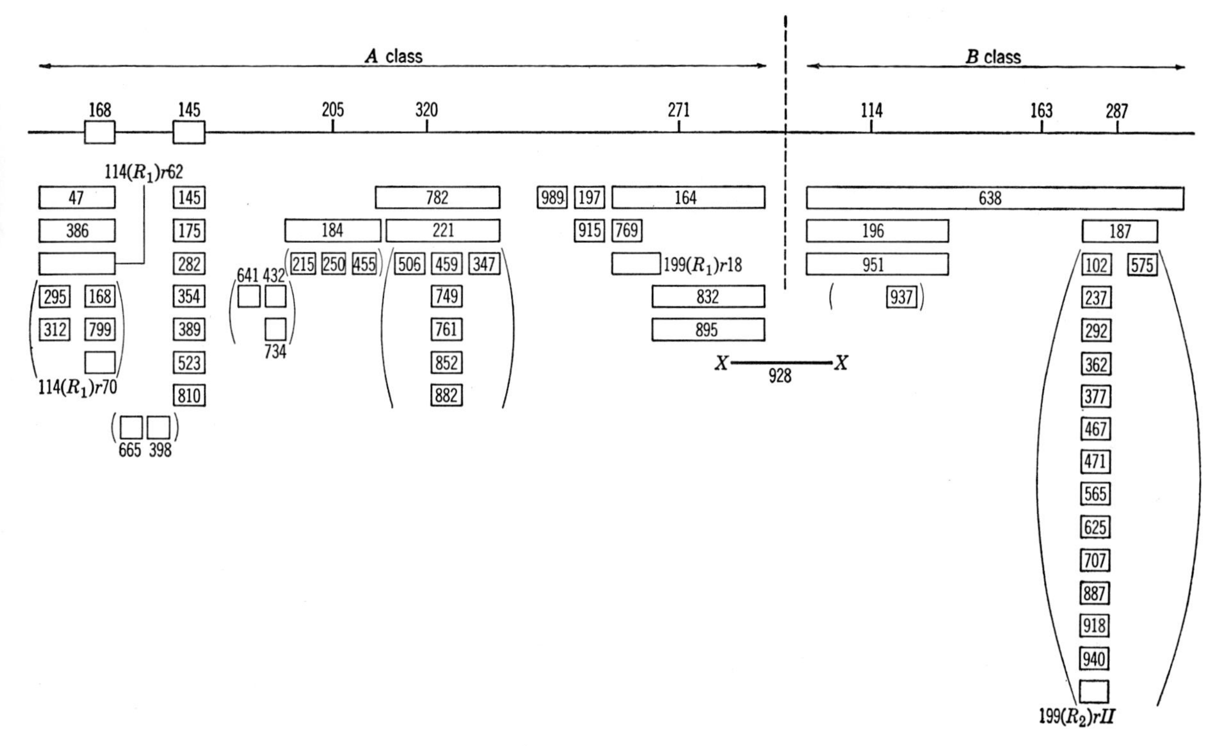


FIG. 6. Preliminary locations of stable *rII* mutants. Mutants producing no wild recombinants with each other are drawn in overlapping configuration. Pairs which produce small amounts are placed near each other. Since there remain some gaps, the order shown depends upon that established by Doermann (in personal communication with S. Benzer) for the mutants shown on the axis. The scale is somewhat distorted in order to show the overlap relationships clearly. Brackets indicate groups, the internal order of which is not established. Ten stable mutants of the *A* class and six of the *B* class were not sufficiently close to any others to permit them to be placed on the map. (*A* class equals *A*-cistron; *B* class equals *B*-cistron) [from S. Benzer in *The Chemical Basis of Heredity*, W. D. McElroy and B. Glass, editors (The Johns Hopkins Press, Baltimore, Maryland, 1957), p. 70].

such detailed mapping in a restricted region of the *A*-cistron. The number indicative of the size of the recon is about 0.02%, the closest distance between any two mutants in this region. It is seen in the following how this recombination frequency can indicate a size of the recon in molecular terms.

The recombination map as shown in Fig. 5 is far from filled, in the sense that very little of the total span has mutants crowded together as close as the above-indicated apparent limiting distance. Verification of this as a limiting distance will depend upon finding many more mutants to fill up the map.

The next step was to see if the map could be saturated using the deletions already discussed. Can the whole region of the *A*-cistron, concerned apparently with a single function, be spanned by the deletions? Also, can these deletions be ordered into a linear array that does cover region *A*? The overlapping of two deletions is indicated by the fact that each may yield recombinants with one-third (nonoverlapping) deletion, but not with each other. The ordering of a triplet of deletions is done in the standard way by comparing the frequencies of wild-type recombinants of each pair of the triplet. Figure 6 shows the result of the ordering of a large number of such deletions in the *A*- and *B*-cistrons. Two facts stand out. Firstly, each deletion, with one exception which is yet to be explained, falls clearly into either the *A*- or the *B*-cistron; i.e., control of the same function falls into a well-bounded map region. There were no *a priori* reasons why *A*- and *B*-mutants should not have been scattered among each other in the same region. Secondly, the array of mutants can be arranged in a unique sequence along a line.

Both of these facts and the one referred to in the foregoing, that there seems to be a smallest separation found between two mutants, are the heart of the interpretation which correlates the functioning of the genetic material with a molecular structure.

The lengths (in terms of recombination frequencies) of genetic units were found to be (1) the recon ~0.01%, (2) the muton ~0.05%, and (3) the cistron ~4%. These are not precise figures, for clearly the map has not been saturated sufficiently with markers to be confident of them. Probably the one in largest error is the muton.

In a rough calculation, Benzer translated these units into numbers of nucleotide pairs. The total amount of DNA in a phage particle has been well measured. Knowing roughly the total recombination frequency from genetic experiments and with the following assumptions, one can make the appropriate calculations:

(1) All of the DNA is genetic and corresponds to one copy of the genetic information.

(2) The map is uniform; i.e., a given recombination frequency anywhere on the map corresponds to the same amount of DNA.

With these assumptions, the ratio of total nucleotide pairs to recon size gives the order of magnitude 1. The limitation and expected variation in this calculation are discussed in the paper by Benzer.[1]

Consider next experiments with bacteria, investigating the correspondence between molecular properties of DNA and the genetic map. The question here is the same as for the bacteriophage system: Is there a physical structure in bacteria whose properties correspond to the genetic structure?

Bacterial strains *E. coli* exhibit two primary mating types which contribute asymmetrically to the formation of zygotes from which recombinants are derived.[3] One type, *Hfr*, acts as a donor of material and the other type, $F^-$, acts as the acceptor. In general, the characteristics of the recombinants formed by a mating event are derived principally from the $F^-$ parent. The *Hfr* parent contributes only a part of the total genetic map.

Conjugation of these two mating types is accompanied by formation of a tubule between the bacteria.[4] The sequential character of genetic transfer is revealed by experiments in which mating bacteria are subjected to high shearing forces in a Waring Blendor to interrupt the mating process. This interruption, at different intervals after initiation of mating, allows the formation of zygotes with varying amounts of the genetic map from the *Hfr* parent. This process reveals a sequence of genetic markers and distances among them closely corresponding to those found by the more usual genetic crosses. Other experiments indicate that the effect of the Blendor is on the genetic-transfer process[3] rather than at later stages of incorporation of the genetic material.

To study how this genetic transfer is correlated with a material transfer, Fuerst, Jacob, and Wollman (referred to in reference 5) used $P^{32}$ labeling in the *Hfr* parent. Earlier experiments[6] had shown that bacteria having high specific-activity $P^{32}$ lose with time their ability to multiply. This loss varies with the number of $P^{32}$ decays that have occurred in such a way as to indicate that each decay has a constant probability of rendering the cell incapable of survival. Two related experiments were done by Fuerst, Jacob, and Wollman. (1) *Hfr* bacteria were grown with high levels of $P^{32}$, washed free of excess $P^{32}$, frozen to stop metabolic activity, and stored frozen. At regular intervals, samples were thawed, mixed under standard conditions with an $F^-$ strain, and mating was allowed to occur. They measured the probability that a given genetic length of chromosome be incorporated in the recombinants as a function of the total amount of $P^{32}$ decay in the *Hfr* parent. (2) *Hfr* bacteria were grown with high levels of $P^{32}$, allowed to conjugate with the $F^-$ parent to form zygotes which are then frozen, and stored as above. These are thawed at regular intervals, and the number of recombinants determined.

The two types of experiments differ in that, in the latter, injection occurs before $P^{32}$ damage, whereas, in the former, $P^{32}$ decay acts in destroying the capacity of a given map length to be transferred and to be incorporated in the recombinants. In the second type of experiment, the effect on recombination frequency is only on the probability of incorporation of a given segment of map.

In these experiments, two results were clear:

(1) The rate of inactivation of the genetic map varies with the length of map selected.

(2) The transfer probability varies according to the length of map.

In both of these results, the kinetics of decay indicate a constant probability of destruction of both the transfer probability and the incorporation per given length of genetic segment—the length being measured by the previous timed-injection Blendor experiments. This correlation of rate of inactivation with genetic length fits very well with the assignment of the phosphorus containing DNA as the genetic molecule.

There is still another kind of genetic experiment performed with bacteria that bears on the correlation of genetic proximity and functional relatedness. This is concerned with the mapping of genetic units controlling related biochemical functions.

The bacterial systems investigated have the advantage over the bacteriophage system in that it is easy to select for mutations controlling production of a specific enzyme catalyzing a single chemical step. With the bacteriophage system, it is easy to ascertain whether two mutants belong to the same functional region [by the (*cis-*)(*trans-*) test], but it has, so far, not been possible to ascribe synthesis of a specific protein as the function of such a region. With bacteria, finding the function is easy, but confirming that this function corresponds to a cistron as defined by Benzer has been more difficult.

In these experiments, a group of biochemically deficient mutants is isolated from a standard wild type, characterized with regard to the specific nature of the biochemical block, and each mutant is mapped.

Consider as a specific example the mapping of histidine-deficient mutants in *Salmonella typhimurium*.[7] The wild-type organism can grow on a synthetic medium using glucose as a source of carbon and ammonia as a source of nitrogen. From this wild-type parent, Hartman[7] isolated a large number of mutants having in common a requirement for histidine, in addition to the medium supporting growth of the wild-type organism. The synthesis of histidine consists of many known steps.[8] These mutants were characterized as to which particular steps were blocked—presumably through failure to make the necessary protein for catalysis of this step, or through making a

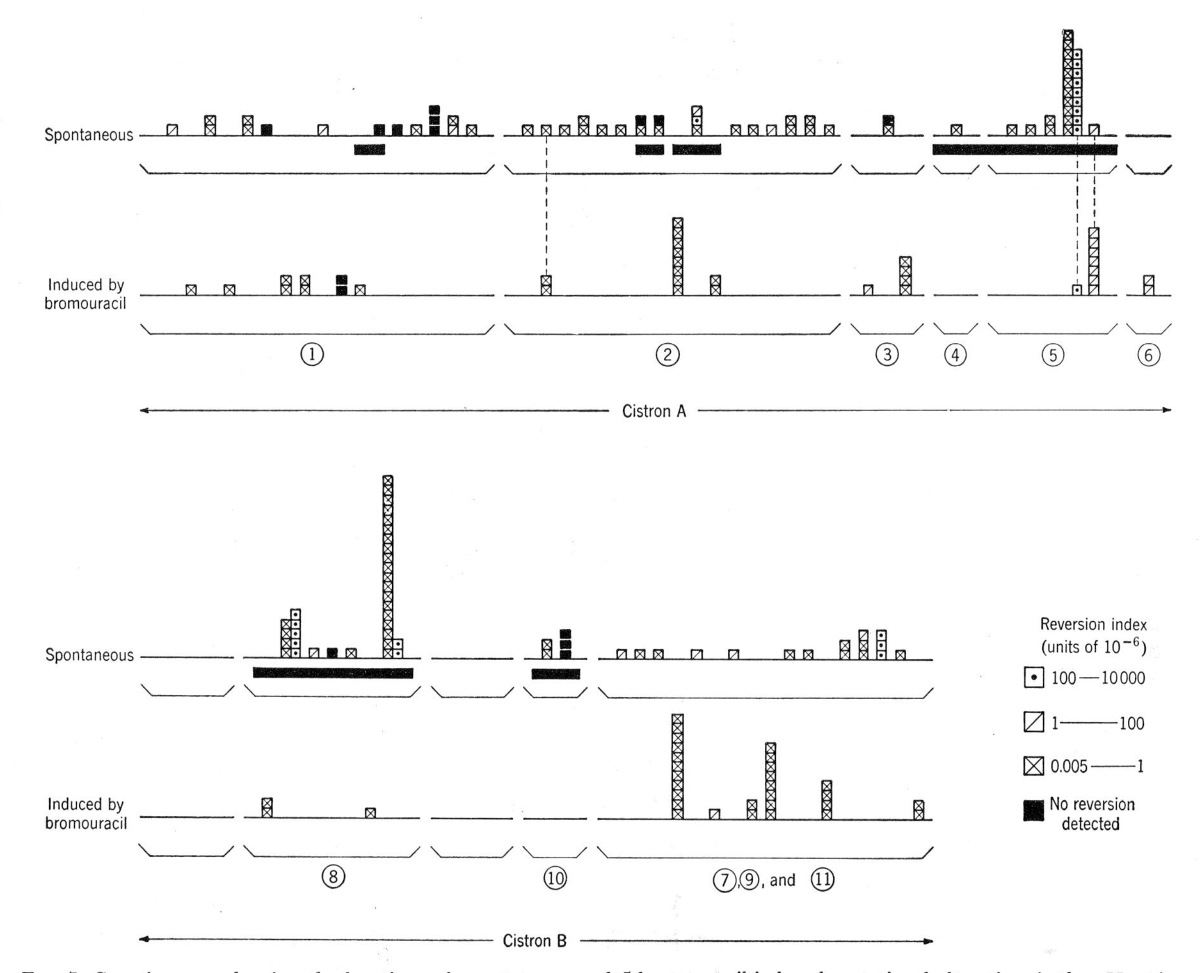


FIG. 7. Genetic maps showing the locations of spontaneous and 5-bromouracil-induced mutational alterations in the *rII* region of phage *T*4 divided into cistron *A* and *B*. Each mutant is represented by a box, placed in the proper segment of the map and shaded to indicate its reversion index. The relative order (but not the lengths) of segments 1, 2, 3, 4, 5, 6, 8, and 10 have been established. Segments 7, 9, and 11 are lumped together and considered as one segment. The order and position of different mutations within one segment have not been determined [from S. Benzer and E. Freese, Proc. Natl. Acad. Sci. U. S. 44, 112 (1958)].

protein with a defect, rendering it incapable of catalyzing the reaction step.

Comparing two histidine-requiring mutants, biochemically, reveals that they are blocked in the same or different biochemical step. Mapping of these mutants revealed two facts:

(1) All of the biochemical blocks in histidine synthesis are close to each other on the map.

(2) Those markers involved in blocking at the same step are in clusters; the markers from different steps seem to belong to separate but nearby clusters.

It is, of course, natural to speculate that such clusters are the cistrons defined by Benzer, though his defining (*cis-*)(*trans-*) test is lacking for this system.

There are many other cases in the mapping of biochemical blocks in bacteria where similar evidence has accumulated.[9–11] In all of these, the clear definition of a cistron is lacking. The experiments of Jacob and Wollman on the formation of partial zygotes by interrupted mating make such tests possible.

To summarize, experiments mapping functional blocks in bacteria have strengthened the idea that a unit function is under the control of a small region of the genetic material.

The last experiments discussed here return attention to *T*4 bacteriophage where mapping had been carried to a degree of fineness that the recombination and mutation units could be defined in terms of relatively few nucleotides.[1] This opened a new avenue to the study of mutation and to bringing into correspondence the properties of DNA with those of the genetic map.

Benzer had made a profile of the mutation frequency in various portions of the *A*- and *B*-cistrons. The spontaneous mutation frequency is not uniform throughout the map, but does have local "hot spots" of high mutation frequency. Benzer and Freese[12] then investigated whether a mutagenic agent which could be expected to interfere with the incorporation of a particular nucleotide into DNA would alter the profile of mutation. For this purpose, they used 5-bromouracil, which had been shown to be incorporated into the DNA of phage replacing thymine[13,14] and to be mutagenic in bacteriophage.[15]

Since the spontaneous mutational events probably arise from alterations at more than one kind of base site, and since the mutagenic agent 5-bromouracil would be expected to interfere principally by replacing thymine, then the profile of induced mutation would be expected to be very different from the spontaneous one.

A comparison is shown in Fig. 7, which plots the mutational frequency in various segments of the *A*- and *B*-cistrons. The profile of spontaneous mutants is markedly different from the 5-bromouracil induced ones.

Such studies, extended to other mutagenic agents, may form a basis for studies of mutagenesis on a molecular level.

The experiments discussed here—on bacteriophage and on bacteria—have yielded the following observations:

(1) The genetic map is linear down to its smallest dimensions.

(2) Portions of the map controlling the same function are compact and not intermixed with portions controlling other functions.

(3) The units of recombination and mutation have sizes of the order magnitude of a few nucleotide pairs; the size of the unit of function is the order of magnitude of $10^2$ to $10^3$ pairs.

(4) There is a correspondence between map distance and phosphorus content of the genetic material.

This summary is consistent with the identity of DNA and the genetic material. The real problems start here—to specify at a molecular level the detailed structure of DNA in its acting as genetic memory and in its role of making genetic products.

## BIBLIOGRAPHY

[1] S. Benzer in *The Chemical Basis of Heredity,* W. D. McElroy and B. Glass, editors (The Johns Hopkins Press, Baltimore, Maryland, 1957), p. 70.

[2] E. B. Lewis, Cold Spring Harbor Symposia Quant. Biol. **16**, 159 (1951).

[3] E. L. Wollman, F. Jacob, and W. Hayes, Cold Spring Harbor Symposia Quant. Biol. **21**, 141 (1956).

[4] T. F. Anderson, E. L. Wollman, and F. Jacob, "Sur les processus de conjugaison et de recombinaison génétique chez *E. coli III*. Aspects morphologiques en microscopie électronique," Ann. Inst. Pasteur (to be published).

[5] F. Jacob and E. L. Wollman, Symp. Soc. Exptl. Biol. **12**, 75 (1957).

[6] C. R. Fuerst and G. S. Stent, J. Gen. Physiol. **40**, 73 (1956).

[7] P. E. Hartman in *Genetic Studies with Bacteria,* Carnegie Institution of Washington Publication No. 612 (1956), p. 35.

[8] B. N. Ames in *Amino Acid Metabolism,* W. D. McElroy and B. Glass, editors (The Johns Hopkins Press, Baltimore, Maryland, 1955), p. 357.

[9] P. E. Hartman in *The Chemical Basis of Heredity,* W. D. McElroy and B. Glass, editors (The Johns Hopkins Press, Baltimore, Maryland, 1957), p. 408.

[10] C. Yanofsky, E. Lennox, and C. Richards, "Linkage relationships of the genes controlling tryptophan synthesis in *Escherichia coli*" (manuscript in preparation).

[11] M. Demerec and Z. E. Demerec, Brookhaven Symposia, Biol. No. 8 (1956), p. 75.

[12] S. Benzer and E. Freese, Proc. Natl. Acad. Sci. U. S. **44**, 112 (1958).

[13] D. B. Dunn and J. D. Smith, Nature **174**, 305 (1954).

[14] M. J. Bessman, I. R. Lehman, J. Adler, S. B. Zimmerman, E. S. Simms, and A. Kornberg, Proc. Natl. Acad. Sci. U. S. **44**, 633 (1958).

[15] R. M. Litman and A. B. Pardee, Nature **178**, 529 (1956).

# 30
# Coding Aspects of Protein Synthesis

Cyrus Levinthal

*Department of Biology, Massachusetts Institute of Technology, Cambridge 39, Massachusetts*

TWO of the foregoing articles have dealt with certain aspects of formal genetics and others have dealt with the physicochemical studies of the structure of nucleic acids and protein. This article presents some of the current ideas about the relationship between genetics and the structure of these molecules. Many of these ideas are highly speculative and should be thought of only as current working hypotheses (or dogmas) which are being subjected to experimental tests.

Evidence indicating that genetic information can be contained in DNA was presented in previous articles and it is assumed then that there is a one-to-one correspondence between the position on the genetic map and the position along a DNA molecule. If DNA itself contains genetic information, it must be contained in the order of the bases along the two chains. But since only the base pairs adenine-thymine and guanine-cytosine are possible in the DNA molecule, the information is contained twice, once along each of the two chains. If the sequence of the bases along one chain is known, then that along the other chain is determined. This fact is, of course, the basis of the Watson and Crick hypothesis of the self-complementary nature of the duplication of the DNA molecule. However, for the moment, only the base sequence along one of the two chains is considered as being the genetic determinant.

From the fine-structure genetic analysis of Benzer,[1] Demerec,[2] and Pontecorvo,[3] it is known that functionally related mutations are near each other on the genetic map. As discussed by Lennox (p. 242), the cistron is a well-defined region of the map which is responsible for the control of a particular function. In interpreting this in chemical terms, Benzer made the hypothesis that a cistron corresponds to a length of DNA which controls the formation of a single polypeptide chain in a protein molecule. Both the cistron and the single polypeptide chain are operationally well-defined entities, and the definition of one does not depend upon the definition of the other. Thus, although this hypothesis is yet to be tested, it is subject to direct experimental verification.

There is another hypothesis[4] which is central to any discussion of the genetic control of protein structure. It states that the genetic information determines only the sequence of the amino acids within a protein and that the three-dimensional configuration of the formed protein molecule is a direct consequence of the amino-acid sequence. The main reason for accepting such a hypothesis is that the genetic map and the base sequence along the DNA molecule form one-dimensional structures, and apparently the only corresponding linear order in a protein molecule is the order of the amino acids along the polypeptide chains. It is evident that, in a formed protein molecule, the three-dimensional configuration is maintained by secondary bonds which exist between amino acids, linking either different polypeptide chains or widely spaced amino acids on the same chain. It is assumed, however, that the information specifying how those secondary bonds are formed is itself contained in the order of the amino acids. For example, covalent sulfur—sulfur bonds play an important role in maintaining the structure of many proteins, but it is assumed that their position in the molecule is determined primarily by the position along the peptide chain of the two cysteine molecules required for S—S bond formation. It should be stressed, however, that there is no direct evidence in support of this assumption, nor can a precise mechanism be formulated at present which might explain how the required bonds would be made. One only can speculate that the folding of a polypeptide chain which was formed along a linear template may require the sequential unpeeling of the chain from its template beginning from some specified place. One can imagine, too, that the correct folding could be influenced by some small molecule with which the finished protein would interact; as, for example, the inducer molecule in the case of induced enzyme synthesis. From the point of view of the over-all information transfer, such speculation as to the details of the mechanism is, of course, irrelevant, and is mentioned, at this point, only to indicate that the hypothesis is not entirely unreasonable.

In most of the cases where they have been studied, the active sites, or the business regions, of enzyme molecules are found to contain only a few amino acids. However, in order to carry out its enzymatic function, the amino acids in the active region of the enzyme must be arranged in a very precise, three-dimensional configuration. One can suppose that the information contained in the order of the *other* amino acids in the protein is necessary to specify the three-dimensional position of each of the amino acids in the active region relative to each other. Many of the *aa*'s coming from the active site may be necessary for the correct folding to occur. Once this has been accomplished, some of them could be eliminated without altering the active site. If this interpretation is correct, it follows that, in some portions of an enzyme molecule, one amino acid can replace another without producing any observable change in the enzymatic function of the molecule. For example, in some regions, the polypeptide chain may be folded in such a way that it forms an α-helix for a short region. Since most amino acids will do equally well as the components of an α-helix, there could be many amino-acid replacements in such a region which would not affect the enzymatic function unless the particular amino acids were involved with secondary bonds to other regions of the molecule. On the other hand, it is likely that any amino-acid replacement in the active site of the molecule would cause a change in the enzymatic activity.

If it is assumed that the genetic information controls only the sequence of the amino acids along a polypeptide chain, then certain ideas suggest themselves which are independent of the detailed chemical mechanism by which proteins are synthesized. These arise from the fact that, although there are only four different kinds of bases in the DNA molecule, their sequence must uniquely determine the sequence of approximately twenty amino acids along the polypeptide chain. Thus, it would require at least three bases to select uniquely from among the twenty amino acids ($4^2$ is only 16 whereas $4^3$ is 64). But the distance required for three nucleotide pairs along the DNA molecule is about 10 A whereas the distance between the amino acids on an extended polypeptide chain is only about 3.7 A. The first solution for getting around this apparent difficulty was suggested by Gamow.[5] He imagined a coding scheme in which $n$ amino acids would be determined by a sequence of $n+2$ bases. Each amino acid would be determined by a sequence of three bases, but two adjacent amino acids would have two bases in common. Thus, two amino acids would be determined by a sequence of four bases; the first three members of the sequence, determining the first amino acid and the last three, determining the second amino acid. This type of overlapping code has two immediate consequences: First, there will be many mutations affecting two or even three amino acids; to date, no mutations of this type have been observed. Second—and this leads to a more serious objection to an overlapping code—not all pairs of amino acids can lie next to each other. In fact, the restrictions on the possible sequences of amino acids are so severe in the Gamow code that it seems to be impossible to use it to code some of the known amino-acid sequences. Brenner[6] has analyzed the known sequences of amino acids in proteins and has found that they are not consistent with *any* type of overlapping code in which two letters are used in common for adjacent amino acids. Brenner's analysis does rest on one further assumption, that the same code is used in all biological systems and for all proteins. This was a necessary assumption because, so far, only relatively small proteins have been analyzed as to the sequence of their amino acids and because it is not yet possible to obtain a sufficient number of adjacent pairs from any one protein. These two arguments, however, make it seem very unlikely that any variant of a double overlapping code is used in the synthesis of proteins.

Crick *et al.*[7] have avoided the spatial difficulty by making the chemically more plausible assumption that, when either DNA or RNA acts as a template for the formation of a protein, it is not the amino acid itself which fits on the template. They assumed that there could be some type of small adapter molecule which would attach to a specific amino acid on one end and to three nucleotides on the other. From what now is known of the chemistry of protein synthesis, one might suppose that the genetic information in the DNA is transferred first to a large RNA molecule which would act as the template for protein synthesis, and that the soluble RNA to which specific amino acids are attached could then be the adapter molecules. There is, as yet, no direct evidence that this model is correct, but it is interesting to note that the adapter hypothesis was made two years before there was any chemical information on the attachment of amino acids to soluble RNA. Having made this assumption, one is freed from the spatial argument and only need consider the problem from the point of view of coding. Crick *et al.*[7] have pointed out that the synthesis of proteins would be very much more efficient or, at least, more rapid, if each amino acid, presumably attached to its adapter, could place itself along the template independent of the other amino acids. Thus, a given position on the template should be such that it would accept a particular amino acid whether or not the adjacent spots were filled. In order to satisfy this condition and to use every three nucleotides to specify an amino acid, it is necessary to impose the condition that certain triplets (called *sense* triplets) of the nucleotides correspond to amino acids and that others (called *nonsense* triplets) do not. The code must be such that the six nucleotides required for two adjacent sense triplets should not contain any additional sense triplets among the overlap triplets. Thus, if both $xyz$ and $uvw$ correspond to amino acids, then $yzu$, $zuv$, $wxy$, and $vwx$ must not correspond to an amino acid. Such a code is called commaless, since no additional information is required to indicate where amino acids are to be placed along the nucleotide sequence. In the language used by Elias (p. 221), this is a "jitter-free" code because it is independent of the ability to read correctly from one end as to the distance from that end.

Most of the sixty-four words which can be written with four letters taken three at a time can be eliminated by the comma-free condition. This set of three-letter words which can be used in a comma-free message is called a dictionary. The word "AAA" cannot be included in such a dictionary, because, if the word were repeated twice, the sequence "AAA" would appear in the overlap and the position of the amino acid would not be well defined. This condition eliminates the four possible words which have three identical letters and reduces the number of possible words in the dictionary from sixty-four to sixty. Furthermore, no cyclic permutation of a word can be in the dictionary. If "ABC" is included, then neither "BCA" nor "CAB" can be, because if either one of them were included and repeated twice, "ABC" would occur in the overlap. Since, from any word, two others can be derived by cyclic permutation of the letters, the maximum size of the dictionary is reduced further from sixty to twenty.

Two twenty-word dictionaries which are essentially different from each other were presented by Crick *et al.*, and since no comma-free code can contain more than twenty words, this completes the proof that one is dealing with a maximal comma-free dictionary. Freudenthal[8] and Golomb *et al.*[9] have shown that a total of five, and only five, basic types of code can be constructed which satisfy the comma-free condition and contain twenty words (Fig. 1).

There is an additional difficulty with this type of code which is related to the fact that the information is contained twice in the DNA chain. Chemically, the two chains are antiparallel so that the over-all features of the molecule are invariant with respect to an end-to-end inversion. However, the informational content or the sequence of the nucleotide pairs clearly is not invariant with respect to this inversion. Or stating the problem differently, the information as to the protein to be formed by a given region of DNA will be different depending on which of the two chains is read. Delbrück[9] proposed that an additional requirement be added to the code such that, if the message were to read correctly along one chain, the complementary sequence which existed on the opposite chain would not contain any sense triplets. With this additional restriction, only ten words can be written with four letters taken three at a time, and it requires a sequence of four letters to supply enough information to specify twenty amino acids subject to this restriction. However, this additional requirement imposed by Delbrück is not absolutely necessary. One can, for instance, suppose that the information is

| I | | | II | | | III | | | IV | | | V | | |
|---|---|---|---|---|---|---|---|---|---|---|---|---|---|---|
| A | | A | A | | A | A | | A | A | | A | A | | A |
| B | | B | B | | B | B | | B | B | D | B | B | D | B |
| | D | | | D | | | D | | | | | | | |
| C | | C | C | | C | C | | C | C | | D | C | | D |
| | | D | | | D | | | D | | | | | | |
| A | | A | A | | A | A | | A | A | | A | A | | A |
| | C | | | | | | C | | | | | | | |
| B | | B | B | C | B | B | | C | B | C | B | B | C | B |
| | | C | | | C | | | | D | | C | D | | C |
| A | | A | A | B | B | A | | A | A | B | A | A | B | A |
| | B | | | | | | B | | | | | | | |
| | | B | | | | C | | B | | | B | | | |
| | | | B | A | A | | | | | | | B | A | A |

FIG. 1. The five possible types of commaless codes using four different letters taken three at a time (after Freudenthal[8]). In each group with a given middle letter, all combinations of the listed first and third letters are sense words.

transferred from a double-chain DNA molecule to a single-chain RNA molecule and that the information, specifying which of the two DNA chains will be used in forming the RNA molecule, is not present everywhere along the DNA, but only in short regions (for example, between cistrons). Thus, in the making of the RNA, one need not impose the commaless condition, whereas it still is required in the making of the protein from the RNA.

In the absence of experimental data, it is clear that such speculation can go on more or less indefinitely, and that very elaborate codes can be imagined. For example, there is no particular reason to assume that the information specifying a given amino acid necessarily comes from contiguous nucleotides. One could imagine a mechanism in which every tenth nucleotide (that is, one every turn of the DNA molecule) would specify an amino acid so that one might be specified by nucleotides 1, 11, and 21 and the next by 2, 12, and 22, etc. At the moment, there is very little experimental evidence relevant to these questions. However, a great deal of work is being done in an attempt to obtain some data. In the hope of providing some further understanding of the current status of the field, the experiments are discussed in terms of the kind of question which might be asked and of the kind of information which, in principle at least, might be obtained.

It seems unlikely that, in the near future, it will be possible to analyze the sequence of the nucleotides in a piece of DNA which is responsible for a particular protein molecule. In addition to the difficulty of carrying out the sequence analysis of the bases along the DNA, there is another problem which seems even more formidable. Except in the case of some of the small viruses, the method is not clear for obtaining a preparation of DNA wherein each molecule of the material controls the same protein. And even in the smallest viruses, the amount of nucleic acid is very large for an analysis of this type. The work that is being done involves the analysis of how protein molecules are altered as a result of specific mutations. Most of the currently available experimental results relevant to this problem have come from the work on human hemoglobin done by Ingram.[10] He analyzed the hemoglobin molecule by treating it with the proteolytic enzyme, trypsin, and then subjecting the resultant digest to a two-dimensional paper electrophoresis and chromatography. Trypsin splits the polypeptide chain at every arginine and lysine residue. In the case of hemoglobin, this digestion yielded 28 different peptides. Hemoglobin made by different individuals can be compared by noting the various movements of different peptides on the paper. Sickle-cell anemia, a disease which is presumably due to a single mutation, results in an altered, but not totally inactive, hemoglobin. This alteration was shown to correspond to the change of a single amino acid. A total of six altered hemoglobins have been examined by Ingram and, in three of these, it has been shown that only one amino acid is changed.[10] Two of the changes involve the same amino acid and one involves a change in a different region at the molecule. The other three alterations differ from these, but it has not been proved that only one amino acid is altered. In these studies with human hemoglobin, one selects for those mutations which produce a minor alteration in the functioning of the molecule. It is much less likely that an alteration which produces no physiological change would be examined. This is true also for any mutation resulting in the absence of a hemoglobin molecule which would appear as a recessive, lethal mutation, and, thus, would probably not be identified as affecting hemoglobin. However, it is important to note that no one of the cases so far examined represents changes in more than one of the peptides and of the three which have been totally characterized, none results in a change of more than one amino acid.

In several laboratories, work is directed toward obtaining a system with which fine-structure genetic analysis can be carried out and combined with chemical analysis of the protein controlled by the genetic region. In order to make such studies possible, one must work with a conditionally necessary protein whose absence prevents the organism from growing in some conditions but not in all. The substance made by the *rII* genetic region studied by Benzer[1] obviously satisfies this requirement. Although an intact protein is necessary if the virus is to grow in some bacterial hosts, it is not necessary for it to grow in other hosts. Likewise, there are many enzymes made by bacteria which are necessary for the growth of the bacteria only under certain environmental conditions. For example, fine-structure analysis has been done in bacteria by using the enzymes responsible for the synthesis of a certain amino acid. If the amino acid is supplied in the medium, the absence of the enzyme does not affect the growth of the organism. On the other hand, by placing the progeny organisms of crosses in a medium lacking the required amino acid, a very small number of recombinants can be selected which have the enzyme.

Under such conditions one can consider several classes of mutation. If the general concept of a commaless code is correct, then there will be many sequences of nucleotides which do not correspond to an amino acid. A mutation could occur which would result in such a nonsense sequence in the nucleotide chain. If "ABA" corresponded to a certain amino acid, then a mutation of the letter "B" to the letter "A" would result in the sequence "AAA" which is not in the dictionary. If this were to happen, it is possible that no protein molecule could be formed, since, presumably, the peptide chain no longer would be continuous at this point. This type of sense-to-nonsense mutation could result plausibly in the absence of an active protein, regardless of where it occurred in the molecule. In addition, there would be a class of sense-to-mis-sense mutations which could result in a totally inactive molecule. A replacement of valine by proline in an active site probably would result in a protein with no enzymatic activity. On the other hand, such a protein might still be formed and be recognized by its antigenic similarity to the normal enzyme. Also, the fact that the protein may be induced by the same inducing agent as caused the formation of the normal enzyme would be another factor in recognizing an alteration. Sense-to-mis-sense mutations are those which, in principle, may lead to a protein which could be analyzed chemically and whose difference from the normal enzyme could be determined. However, the sense-to-nonsense mutation would lead to the absence of a protein and, therefore, no information about the alteration of the amino-acid sequence. It may be possible, however, to obtain considerable information by studying the reversions to activity from the nonsense mutations. This is illustrated in Fig. 2(b) where the sequence equivalent to amino acid $\alpha$ first mutates so that it corresponds to a nonsense word. Then, reverse mutations are selected, some of which restore the amino acid $\alpha$ and others of which insert the amino acid $\beta$ or $\gamma$. In this sort of mutation-reversion study, one expects to find, in some instances, that several different reversions could be obtained which produce an enzyme differing from the normal. However, each of the reversions would be expected to differ from the normal at the same position in the molecule. It would be possible to test this conclusion despite an inability to carry out fine-structure genetic analysis as long as the system is such that the reversions with some activity have a selective advan-

(a)

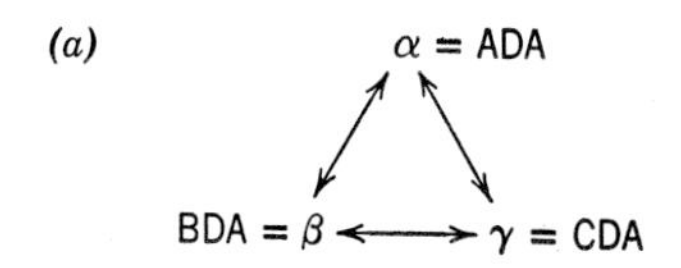


(b)

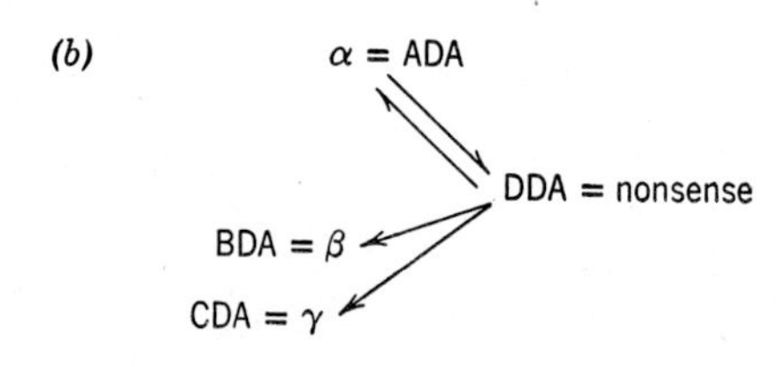


FIG. 2. (*a*) The conversion group $\alpha$, $\beta$, and $\gamma$ corresponds in all of the codes to changes in the first letter of the word, ADA. The amino acid $\alpha$=ADA would belong to two other conversion groups also whose size depend on which code was used. (*b*) ADA could mutate to the nonsense word, DDA, also, and in this special case all of the possible reversions to sense words would be in the same conversion group. However, it is not generally true that the reversions will be in the same conversion group.

tage. It has been shown[11] that, for all of the three-letter commaless codes, the number of different sense reversions from a nonsense mutation is at most four. This prediction could also be subject to experimental tests.

There are other classes of mutation which, although extremely useful in the genetic analysis, are not likely to be very helpful in this problem. These are the deletions or inversions which have been discussed earlier. In this case, one expects to find mutations which never revert to produce an active protein, and, as such, there could never be any analysis of an altered protein.

Another question which can be studied experimentally was mentioned earlier with reference to the work of Ingram.[10] It is concerned with whether or not a single mutation can affect more than one amino acid. In connection with this, mention should be made of one assumption which is implicit to all of these discussions. It is assumed that the probability of mutation is sufficiently low so that two letters in the nucleotide code will not be altered simultaneously. This assumption will be justified, if it is found that all mutations *do* affect a single amino acid only.

Another kind of analysis would become possible, if a sufficiently large number of transitions from one amino acid to another could be accumulated. Suppose [Fig. 2(a)] that a mutation occurs which carries amino acid $\alpha$ into $\beta$ and another also occurs which carries $\alpha$ into $\gamma$. One can ask if $\beta$ can mutate to $\gamma$. If it does, then the set of three amino acids is called a *conversion group*. A conversion set is defined as a set of amino acids, each member of which can be changed to every other member of the group by a single mutation. With the assumption that mutations change only a single nucleotide at a time, a conversion group must be produced by changing only one letter of the word which corresponds to an amino acid. Under these assumptions, the first prediction which would apply to any code is that no conversion group could contain more than four elements, corresponding to the four possible bases which could be inserted in a given position. If there are limitations on what alterations can occur in the nucleotide chain, the number of elements in a conversion group would be less. For example, if purines could be changed only to other purines and pyrimadines to other pyrimadines, no conversion group could contain more than two elements. A single amino acid could, of course, belong to several different conversion groups. The number of conversion groups to which a given amino acid belonged would be equal to the number of nucleotide letters which correspond to a single amino acid. Consequently, in any of the triplet codes, each amino acid would belong to three conversion groups. As can be seen from the five dictionaries indicated in Fig. 1, for the triplet commaless codes, some of the conversion groups would contain four elements, and others, three, two, or one. If a complete analysis of the conversion groups were possible, one could be led, unambiguously, to one of the five triplet codes. Conversely, if any code of more than three letters is written, the set of these conversion groups could be numerated and, therefore, in principle at least, any particular code could be tested with this type of analysis. However, in order to determine the conversion groups, a great deal of experimental data would be necessary.

All of the experiments mentioned so far require detailed chemical analysis of altered proteins but do not necessarily require fine-structure genetic analysis. However, if fine-structure genetic analysis can be combined with information as to the chemical alterations, two of the most fundamental questions can be investigated. The first involves the basic premise on which much of the speculation depends; that is, does the order and relative position of mutations along the genetic map correspond to the order and relative position of the corresponding altered amino acids along the polypeptide chain. If several different mutations can be mapped relative to each other and the corresponding altered proteins examined, one can investigate whether or not the relative positions are the same on the genetic map and on the polypeptide chain. One way of accomplishing this analysis would be to carry out the fine-structure mapping on the nonsense mutations which, since they produced no active protein, could be mapped with high resolution, and the amino-acid analysis on the reversions which have some activity and whose protein alteration could, in principle, be determined. Another problem which can be investigated, if the genetics could be combined with the chemical analysis, is to find whether or not the genetic determinants of a particular amino acid are adjacent to each other. If, for example, the information specifying one amino acid is interspersed along the nucleotide chain with the information controlling another amino acid, it would be evident from the finding that the mutants affecting one were interspersed among the mutants affecting another. In this way, the question as to whether or not the genetic code is truly a local code could be determined experimentally.

So far, this paper has dealt with much speculation, many questions, and a few experimental facts. The remainder of it is a brief review of the work going on in several laboratories which is directed toward obtaining some answers to questions raised. Ingram's studies with human hemoglobin are continuing, and the list of possible transitions from one amino acid to another is being extended. Even if the combined genetic and chemical analysis should become possible with proteins made by microorganisms, the work with hemoglobin will remain of prime importance because it can help to answer the question of whether or not the code is the same for different organisms.

Several systems are being investigated with microorganisms. The first is the enzyme trytophane synthetase which is being studied by Yanofsky. He has found[12] that this enzyme, under some conditions, can catalyze two sequential reactions which ultimately lead to the formation of tryptophane. He has found that this enzyme can be altered by mutation so that either one of the two enzymatic activities can be eliminated, leaving the other intact. Those mutations affecting one of the activities are localized and are in a region of the genetic map which is adjacent to the region in which the second set of mutations can occur. In many of these cases, the bacteria form a material which is antigenically related to the enzyme produced by the normal cell. The effects of many of the mutations affecting this enzyme can be reversed by suppressor mutations which occur elsewhere on the genetic map in regions not adjacent to the original mutations.[13] This finding of unlinked suppressors is, at first sight at least, in complete contradiction to many of the assumptions used in this analysis. However, the detailed investigation of the chemical alterations associated with these mutations and suppressors has been undertaken only recently, and, until it is completed, several interpretations of the results are possible. Some of the mutations leading to a loss of enzymatic activity possibly may be owing to a certain alteration of the enzyme so that it can be inactivated by some normal cellular component. The suppressor mutations could then affect the amount of this component present in the cells. At least one such case, involving tryptophane synthetase, has been found by Suskind.[14] However, the resolution of this question must await further chemical analysis of the altered proteins.

The proteins which make up the *T*-even bacterial viruses have been investigated also,[15] with regard to finding a system for doing combined genetic and chemical studies. A great deal of genetic analysis has been carried out with these phages, and several proteins have been isolated from them. Unfortunately, the *rII* region, which Benzer has studied so extensively, does not control a protein which is contained in the phage itself. Rather, the control seems to be over one of the several proteins required in the infected cell for the formation of the phage. Two of the proteins which are in the phage particle have been purified and characterized, and, although differences can be observed between the strains, *T*2 and *T*4, it has not yet been possible to obtain different proteins by mutation of any *one* of the strains. There are, however, several types of mutations of these phages and it is very likely that some of them will be found to control these proteins.

In addition to the coding problem already discussed, another type of question can be approached by using the phage proteins. The membrane around the DNA of these viruses is composed of a large number of identical protein molecules. Mutants have been found which result in many of the formed phages having defective heads, so that it is possible that information can be obtained as to the organization of these small molecules into a large and relatively complex structure. Whether or not any progress toward the solution of this question is possible using the phage, it is clear that, in order to understand genetic control of an organism, it is necessary to know how organized structures, as well as protein molecules, are controlled.

Another attempt to obtain a system in which a conditionally necessary protein can be studied chemically and genetically is being undertaken by Garen and Levinthal using the bacterial enzyme, alkaline phosphatase. This enzyme splits the phosphate-ester linkage of organic phosphates and is necessary for the bacterial growth only if there is organic phosphate, but no inorganic phosphate, supplied in the medium. Alkaline phosphatase is made in high concentration by bacteria only if they are starved for phosphorus. Under these conditions, it is made in very large amounts. The ability to make a single enzyme in this high concentration seems to be a characteristic of bacteria, which, although they have the genetic capability of making thousands of different enzymes, also have additional mechanisms which govern the amount of each of these enzymes actually made under given culture conditions. Approximately 5% of the total cell protein occurs as this one enzyme. When the bacteria are producing alkaline phosphatase at the 5% level, it is comparatively simple to purify the protein. Mutants have been isolated after irradiation with ultraviolet light which showed no enzymatic activity, and others have been isolated which have reduced activity. Several of the latter type have been analyzed and found to produce an altered protein in the same high concentration which was observed in the "wild" type. From the mutants which lack any enzymatic activity, reversions have been obtained, some of which are like the "wild" type, and others, like the mutants with reduced activity. Fine-structure genetics can be done with these mutants, but as yet no information has been obtained concerning the alterations in the amino-acid sequence in the enzymes made by the mutant organisms.

It can be seen from this discussion that, although there is some evidence that a code exists which translates genetic information into amino-acid sequence, there is no information as to the *nature* of this code. On the other hand, a number of well-defined questions which can be answered experimentally have been formulated, and several different systems which are likely to yield some answers are being studied.

## BIBLIOGRAPHY

[1] S. Benzer in *The Chemical Basis of Heredity*, W. D. McElroy and B. Glass, editors (The Johns Hopkins Press, Baltimore, Maryland, 1957), p. 70.

[2] M. Demerec, Z. Hartman, P. E. Hartman, T. Yura, J. S. Gots, H. Ozeki, and S. W. Glover, "Genetic studies with bacteria," Carnegie Institution of Washington, Publication No. 612 (1956).

[3] G. Pontecorvo, Cold Spring Harbor Symposia Quant. Biol. **21**, 171 (1956).

[4] F. H. C. Crick, Symp. Soc. Exptl. Biol. **12**, 138 (1958).

[5] G. Gamow, A. Rich, and M. Ycas, Advances in Biol. and Med. Phys. **4**, 23 (1956).
[6] S. Brenner, Proc. Natl. Acad. Sci. U. S. **43**, 687 (1957).
[7] F. H. C. Crick, J. S. Griffith, and L. E. Orgel, Proc. Natl. Acad. Sci. U. S. **43**, 416 (1957).
[8] H. Freudenthal, Koninkl. Ned. Akad. Wetenschap. Proc. **A61**, 253 (1958).
[9] S. W. Golomb, L. R. Welch, and M. Delbrück, Biol. medd. Dan. Vid. Selsk. **23**, 9 (1958).
[10] V. M. Ingram, Nature **180**, 326 (1957), and personal communication.
[11] I. Tessman (personal communication, 1958).
[12] C. Yanofsky and J. Stadler, Proc. Natl. Acad. Sci. U. S. **44**, 245 (1958).
[13] C. Yanofsky, Science **128**, 843 (1958).
[14] S. R. Suskind and L. I. Kurek, Science **126**, 1068 (1957).
[15] S. Brenner, S. Benzer, S. Champe, and G. Streissinger (personal communication, 1958).

# 31
# Induced Enzyme Synthesis

David S. Hogness

*Department of Microbiology, Washington University School of Medicine, St. Louis 10, Missouri*

## I. INTRODUCTION

A USEFUL hypothesis of modern genetics and that forming the basis of the preceding papers is that the genes (specifically, the cistrons) within a cell determine which proteins among all possible proteins a given cell can conceivably synthesize. Such genes are thus considered as the primary determinants of a given cell's allowance of proteins.

There are, however, other determinants of this allowance which, although secondary to genetic control, are of interest in any consideration of the mechanism of protein synthesis in that they specifically affect the synthesis of a single protein. Three phenomena have been sufficiently delineated to exemplify such secondary specific determinants. These are:

(1) Specific antibody synthesis resulting from exposure of cells to a given antigen.
(2) Induced enzyme synthesis.
(3) Repression of enzyme synthesis.

This paper is concerned solely with the last two phenomena, which quite possibly represent two forms of the same basic event.

Induced enzyme synthesis can formally be defined as the increase in the ratio of the rate of synthesis of a given enzyme to the rate of synthesis of total cell protein resulting from exposure of cells to compounds (inducers) which are identical or structurally related to the substrates of the given enzyme.[1] The vast majority of known instances of induced enzyme synthesis is derived from microorganisms, particularly the bacteria. This disproportionate representation is probably not real, for we have some reason to suspect that enzyme induction is operative in most cell types. Rather, it would seem that our observations are selected by the experimental techniques available for the manipulation of bacterial and fungal cell populations which have not, until very recently, been available for the culture of other cell types. A second consequence of this precision of bacterial manipulation is that the best-characterized and interpretable induced enzyme systems are found in the bacteria.

Induced enzyme synthesis in bacterial populations has been known for approximately seventy years, although, during a majority of its observed lifetime, this phenomenon has been shrouded in a teleological disguise by being named "enzymatic adaptation." It is only in the past decade that quantitative studies have allowed one to define this phenomenon as an induced enzyme synthesis not necessarily related to any increase in fitness of the cell in which it occurs.

Even in bacterial populations in which there are the most numerous examples of enzyme induction, it is quite clear that the majority of known enzyme-forming systems cannot be classified as inducible in that the enzymes are formed at considerable rates in the absence of exogenous inducers, and this rate of synthesis cannot be specifically increased by exposure of cells to their substrates or such structural analogs of these substrates as have been tested. Such enzymes are often referred to as constitutive enzymes in order to differentiate them from the induced variety. It must be stressed that the terms induced and constitutive do not describe the properties of an enzyme *per se*, but rather describe the properties of an enzyme-forming system. Thus, the same protein molecule can result from an induced enzyme synthesis in one bacterial population while being the resultant of constitutive enzyme synthesis in another, genetically distinct, bacterial population.

There is then an apparent dichotomy among enzyme-forming systems, being either inducible or constitutive. While at present the data are not sufficient for a unique explanation of this dichotomy, an attempt is made here to correlate these two systems to a common working hypothesis, thereby making the questions to be asked more specific and, it is hoped, experimentally answerable. It would not be profitable for the purpose of this paper to review the many known systems of enzyme induction or, in fact, even one of these systems in all of its detail. (For this purpose, see references 2–11). Rather, emphasis is placed on the salient features of one system that are relevant to a precise experimental illustration of the definition of induced enzyme synthesis given above and to the apparent dichotomy between induced and constitutive synthesis. The system of choice for this purpose is the β-galactosidase of *E. coli*, since it has been analyzed in perhaps the most detail as regards the enzyme protein itself, its induced synthesis, and the genetic and physiological relationship between its induced and constitutive synthesis.

## II. β-GALACTOSIDASE INDUCTION

### (A) Characteristics of the Enzyme Protein[8]

Since studies of the induction process depend upon the measurement of enzyme activity as a measure of the amount of enzyme protein present at any given time, it is obviously essential that a direct correlation between these two quantities can be made for the variety of conditions employed. This, in turn, demands that the catalytic and structural parameters of the enzyme in question be sufficiently determined. The β-galactosidase of *E. coli* has been subjected to such determinations and it is useful in the discussion of its induced synthesis if some of its basic properties are briefly described.

(a) The characteristic reaction catalyzed by β-galactosidase is represented in Fig. 1, namely, the hydrolysis of β-D-galactosides. Many glycosides have been tested as substrates or competitive inhibitors of this enzyme yielding the conclusion that the minimum requirement for affinity for the active site on the enzyme is the existence of the β-D-galactopyranosidic ring.[8] Whereas all of the β-D-galactosides tested are substrates, an interesting class of compounds, the β-D-thiogalactosides in which sulfur replaces the oxygen atom of the galactosidic linkage, function only as competitive inhibitors. The interest in these compounds lies in their capacity to function as inducers of β-galactosidase synthesis, without at the same time being hydrolyzed by the enzyme whose synthesis they induce.

(b) Purified preparations of β-galactosidase have been obtained which contain at most 1 to 2% contaminating protein.[8] This allows a direct correlation of the unit of catalytic activity* with the mass of enzyme, yielding the value of 1 catalytic unit per $3.0\times10^{-9}$ g of protein. Since Cohn[8] has determined by equilibrium-dialysis methods that there is one active site per molecular weight $1.3\times10^{5}$, activity measurements can yield a determinination of the number of active sites. Thus, one catalytic unit equals $1.4\times10^{10}$ such sites.

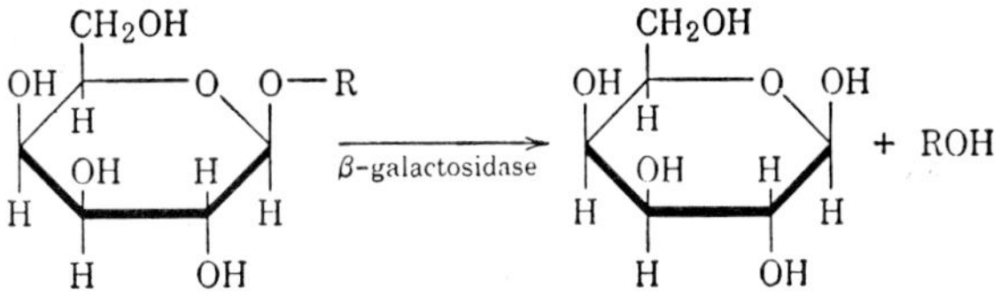

FIG. 1. Hydrolysis of a β-D-galactoside catalyzed by β-galactosidase.

### (B) Induction Phenomenon

A culture of *E. coli* growing in a medium of inorganic salts with a nongalactosidic carbon source, such as succinic acid, produces only trace amounts of β-galactosidase. The addition of a suitable galactoside to such a growing culture is immediately followed by a sharp increase of over 1000-fold in the rate of synthesis of this enzyme. This high rate of synthesis is maintained as long as the bacteria grow in the presence of the inducing galactoside (inducer). However, if the inducer is removed, the rate of synthesis falls directly to the original small value. The quantitative aspects of this situation are presented in Fig. 2. As shown in the left-hand graph of this figure, from 0 to 120 min the bacteria are growing in the absence of the inducer and the amount of enzyme per unit weight of bacteria is quite low, approximately 7 units per mg dry weight of bacteria, which is equivalent to 20 active sites per bacterium. However, immediately upon adding the inducer, one sees that the amount of enzyme in the culture increases rapidly against a background of constant growth rate. Thus, while the bacteria have only doubled in amount, the β-galactosidase activity has increased by a factor of approximately 1200. Plotted in this way, one observes the transition of the culture from one steady-state condition in which the inducer is absent (noninduced state) to a second steady-state condition with inducer present (fully induced state). The bacteria in the fully induced state contain $8.5\times10^{3}$ catalytic units per mg dry weight of bacteria or approximately $24\times10^{3}$ active sites per bacterium. This increase of about 1200-fold in the amount of enzyme per bacterium in the fully induced state over that in the noninduced state means that the rate of synthesis of β-galactosidase relative to the rate of synthesis of bacterial mass (i.e., the differential rate of β-galactosidase synthesis[12]) has increased via the induction process by the same factor of 1200.

It is this differential rate of enzyme synthesis which is of particular interest, since it is a more direct measure of the specific effect of the inducer on the rate of enzyme synthesis. One, therefore, wants to ask the question: how does this differential rate of β-galactosidase synthesis change during the induction process? This question is best answered by plotting the β-galactosidase activity of the culture *vs* the bacterial mass, a procedure which yields the right-hand graph of Fig. 2. Here, it is seen that the differential rate of β-galactosidase synthesis (i.e., the slope of the curve, *P*) changes from that of the noninduced state (7 units per mg dry weight) to that of the induced state ($8.5\times10^{3}$ units per mg dry weight) almost immediately after the introduction of the inducer, the transition taking place in less than 2 min, which is the minimum time detectable by the experimental techniques employed.

This high differential rate of synthesis remains constant as long as the inducer is present in the medium. However, as is shown in both graphs of Fig. 2, the inducer effect is readily reversible, since removal of the inducer immediately restores the differential rate of synthesis characteristic of the noninduced state, again with no appreciable change in the growth rate. Thus, in the case of β-galactosidase induction, it would appear that one has a system in which the synthesis of a given protein can be initiated or stopped by the simple addition or removal of a compound which, because it cannot function as a carbon or energy source, apparently does not influence the rate of synthesis of the vast majority of other cell constituents.

### (C) Enzyme Induction as *de novo* Synthesis

Consider what is meant by synthesis as used in the preceding paragraphs. While it is true that the physical

* One unit of catalytic activity is defined as that amount of enzyme which will cause the hydrolysis of *o*-nitrophenyl-β-D-galactoside to occur at the rate of 1 mμmole/min at 28°C and *p*H 7.1 in $1.0\times10^{-1}M$ sodium phosphate and $2.7\times10^{-3}M$ *o*-nitrophenyl-β-D-galactoside.

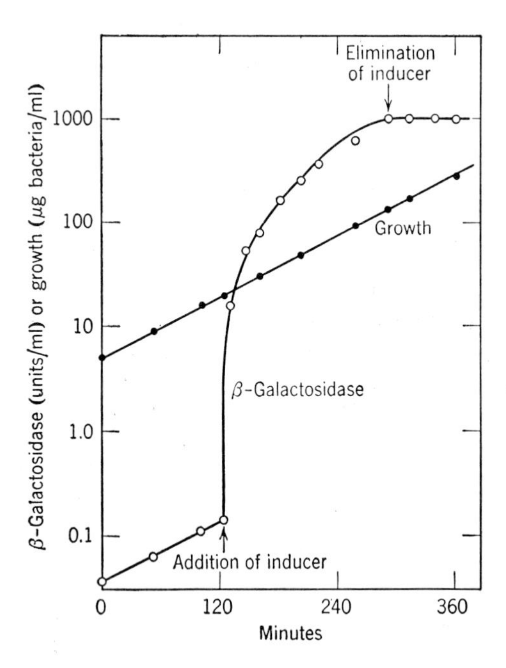


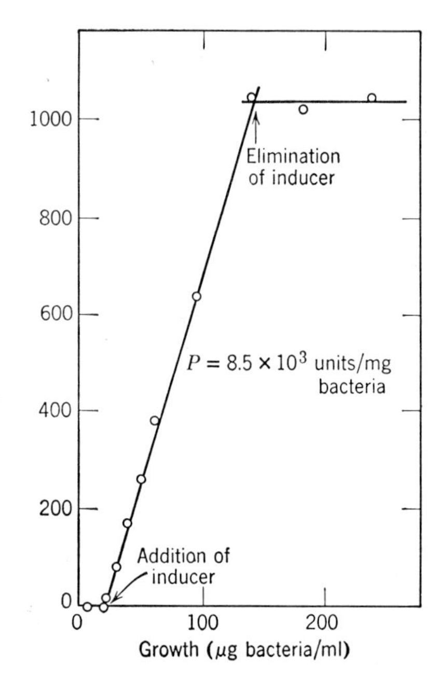


FIG. 2. Kinetics of β-galactosidase induction in *E. coli* ML 3. Inducer is isopropyl-β-D-thiogalactoside at $1.0 \times 10^{-3}$ *M*.

and catalytic parameters of the active enzyme are sufficiently well known that one can calculate the mass of the active enzyme from the measurements of the catalytic activity present in the culture, it does not necessarily follow from this knowledge alone that an increase in such activity (and, therefore, of the mass of active enzyme) results from the *de novo* synthesis of the active enzyme. Within the context of the information so far presented, it is conceivable that the increase in active enzyme involved in the induction process results from the formation of a new protein structure (active enzyme) derived from inactive proteins present in the noninduced state. This transformation from inactive to active material could result (a) from a zymogen- to enzyme-type reaction typical of protease activations, (b) from the formation of a complex active protein made up of inactive subunits, or (c) from the addition of a few amino acids in peptide linkage to an inactive protein. In all of these cases, the induction process would have to be considered as independent of *de novo* synthesis in the sense that protein precursors of active β-galactosidase would be present in the noninduced state, and, consequently, that the total protein potentially or actually available as active enzyme would not necessarily be changed as a result of induction. It is, therefore, critical that one determine whether the induction phenomenon corresponds to the activation of protein precursors or to *de novo* synthesis, before it is possible to assess the value of this phenomenon as a tool for the study of general mechanisms of protein synthesis.

Although, by 1954, there were several lines of evidence which led, rather indirectly, to the tentative conclusion that β-galactosidase induction did indeed correspond to the *de novo* synthesis of this enzyme,[12] it was not until that year that conclusive evidence was available for the proof of this correspondence.[13,14] The direct proof of this correspondence came from experiments carried out in collaboration with Cohn and Monod[13], and simultaneously from the work of Rotman and Speigelman.[14]

The experiments consist of determining whether or not proteins present in noninduced cells are ever incorporated in or associated with the induced β-galactosidase protein by simply labeling the proteins in the noninduced cells with a suitable radioactive isotope and measuring the amount of radioactivity present in the enzyme when induced in nonradioactive media. Thus, in our experiments, *E. coli* were first grown in a simple salts medium in which the sole source of sulfur was $S^{35}$ labeled sulfate. The amount of sulfate was so adjusted relative to the other components of the media that the bacteria stopped growing for lack of sulfate. In this starved condition, all of the radioactivity associated with the cells is contained within the trichloracetic-acid precipitable protein, and since no inducer was added to the medium, we have the condition of a unique radioactive labeling of the proteins of noninduced cells. This constitutes Step I (Table I) of the experiment.

In Step II, nonradioactive sulfate is added to the medium simultaneously with the inducer, methyl-β-D-thiogalactoside. This allows growth and induction to occur simultaneously. When the mass of the cells had increased so that 16% (aliquot *A*), 28% (aliquot *B*), and 43% (aliquot *C*) of the total growth had occurred in the nonradioactive inducing medium, aliquots of the bacterial population were removed, extracted, and, after purification of the β-galactosidase in these extracts, the amount of radioactivity associated with the enzyme determined.

For this experiment to succeed, it was necessary to develop purification techniques for the isolation of the enzyme from less than one gram of cells of which only 0.16% was β-galactosidase. That this was possible is demonstrated by the data in Table I relevant to the isolation control in which unlabeled enzyme was mixed with a labeled extract of noninduced cells, such that the ratio of radioactivity to enzyme activity ($K$) in the mixed extract was the same as that for the crude extract of aliquot $A$. By comparing the $K$-ratio of the enzyme isolated from this reconstruction mixture with that isolated from bacteria grown and induced in the $S^{35}$ labeled medium (fully labeled enzyme control), it is clear that the purification procedure yields an enzyme preparation which contains less than one percent bacterial protein contamination (0.4% in this case).

The enzyme purified from aliquots $A$, $B$, and $C$ contained, respectively, 0.1, 0.8, and 0.1% of the radioactivity associated with the fully labeled enzyme. Since this is within the limits of the purification procedure as indicated by the isolation control data, and since these values do not progressively decrease with increase in the fraction of total growth occurring in Step II, it was concluded that less than 0.8% of the sulfur of the enzyme formed in Step II was derived from proteins present in the noninduced state (Step I). This result, taken in conjunction with the fact that, in aliquot $A$, the enzyme level was only 5% of that found in fully induced bacteria, indicates that if any protein precursor of β-galactosidase exists in the noninduced bacteria, its level must be less than 0.04% of that for β-galactosidase in fully induced bacteria or less than 10 precursor molecules of molecular weight $1.3\times10^5$ (i.e., the molecular weight per active site of β-galactosidase) per bacterium, assuming of course that such precursors contain the same percent sulfur as does β-galactosidase. This conclusion is enforced by the fact that Rotman and Spiegelman[14] found essentially the same results using $C^{14}$ as the labeling material.

The results of these experiments allow one to conclude that, if there are any protein precursors existent in noninduced cells, they cannot form an appreciable contribution to the increase in active enzyme found upon

TABLE I. Incorporation of sulfur into β-galactosidase synthesized by labeled *E. coli* cells in nonradioactive medium.

| Experiment | Percent maximal enzyme level | $K = \frac{\text{radioactivity}}{\text{enzyme activity}}$ | Percent of $K$ for fully labeled enzyme[a] |
|---|---|---|---|
| Step I | 0.06 | (0.45)[b] | (100) |
| Step II—$A$ | 4.8 | 0.0050 | 0.1 |
| $B$ | 32 | 0.0043 | 0.8 |
| $C$ | 58 | 0.00072 | 0.1 |
| Controls | | | |
| Fully labeled enzyme | 100 | 0.45 | 100 |
| Isolation control | (4.8)[c] | 0.0018 | 0.4 |

[a] Corrected for basal activity.
[b] ( ) Basal level assumed to be equal to fully labeled enzyme.
[c] Reconstruction of Extract $A$ by mixing unlabeled enzyme with a labeled extract of noninduced cells.

the addition of the inducer, since it would take only a few seconds to form 10 new active sites per bacterium under the conditions of induction indicated in Fig. 2. Thus, when cells are exposed to inducer molecules the over-all rate of conversion of amino acids into a specific protein, β-galactosidase, is drastically and immediately increased, and consequently induced enzyme formation is equivalent to induced *de novo* synthesis. It would appear then, that the induction phenomenon should be accounted for in any working hypothesis that attempts an explanation of the general mechanism of protein synthesis.

Two further conclusions resulted from our labeling experiments. The first stems from the preceding experiment and states that, in exponentially growing *E. coli* cells, the vast majority of non-β-galactosidase proteins are stable in the sense that they are not broken down to their constituent amino acids at rates that are appreciable in relation to the rates of synthesis. If the state of proteins within these cells consists of a continual synthesis from and breakdown to their constituent amino acids (i.e., state of "dynamic equilibrium"), then one would expect the β-galactosidase synthesized in Step II of the above experiment to be labeled with $S^{35}$ as a result of the breakdown of the radioactive proteins. A simple calculation from the data of this experiment yields the conclusion that the average rate of breakdown of the non-β-galactosidase proteins must be less than one percent of their average rate of synthesis.

It is, therefore, not surprising to recall that, when the inducer is removed from an exponentially growing culture of *E. coli* cells (Fig. 2), the β-galactosidase formed while inducer was present remains constant, simply being diluted out in the exponentially increasing population. This would indicate that β-galactosidase is also stable once formed. Because of the possibility that this constancy of enzyme activity results from an equal rate of synthesis and degradation of β-galactosidase, and to test whether or not β-galactosidase is also stable in the presence of the inducer, we carried out further $S^{35}$ labeling experiments.[13] These experiments yielded the expected conclusion that the β-galactosidase molecules in exponentially growing cells are stable in both the absence and the presence of inducers. Thus, it can be concluded that the induced *de novo* synthesis of β-galactosidase is, like the synthesis of other proteins in *E. coli*, essentially an irreversible process, and the so-called "dynamic state" is not a concept that need be involved in an explanation of the mechanism of such synthesis.

### (D) Induction at the Cellular Level

While it is clear from the foregoing discussion that one may interpret the kinetics of formation of β-galactosidase activity during the induction of a bacterial population as the kinetics of *de novo* synthesis of this enzyme in that population, one can only extrapolate these kinetics to the cellular level if one makes the assumption that all cells participate equally and simultaneously in this induced synthesis. Even though one can demonstrate a high degree of genetic homogeneity for the bacterial population being induced, there is no *a priori* promise that this assumption of equal and simultaneous participation of all cells is justified. It is, therefore, critical to the correct interpretation of the kinetics of induced enzyme synthesis in cultures that one have a knowledge of the distribution of rates of induced synthesis among the cells of the bacterial population.

There are two known types of conditions that can cause a heterogeneous response of a bacterial population when exposed to inducer molecules. The simplest of these types is a condition whereby the catalytic activity of the first enzyme molecules synthesized by a given cell increases the probability of synthesis of future enzyme molecules by that cell. As an example of this condition, consider a culture of *E. coli* whose growth has stopped because of exhaustion of glucose in the medium. If lactose (4-glucose-β-D-galactoside) is added to such a culture, it will have two functions: to induce the synthesis of β-galactosidase and, as a substrate of this enzyme, to provide the only available carbon and energy source necessary for cell synthesis. Under these circumstances, the induced synthesis of the first enzyme molecules must depend upon traces of enzyme already present or upon internal bacterial reserves. Since synthesis of the first enzyme molecules by any cell will increase the availability of carbon and energy for further synthesis by that cell, the cells which have a head start will increase their advantage. Thus, any initial heterogeneity in the population with regard to internal reserves or initial amount of β-galactosidase would be expected to result in an exaggerated heterogeneity in respect to enzyme content during the initial stages of growth on lactose. This expectation has been verified experimentally by Benzer.[15]

Elimination of this factor favoring heterogeneous response is quite simple and consists of using an inducer which is not a substrate of β-galactosidase (e.g., the β-D-thiogalactosides), or alternatively of employing mutant bacteria which cannot further metabolize the products of hydrolysis of the inducer, while at the same time providing a nongalactosidic carbon and energy source (e.g., succinic or lactic acid) for cell synthesis. These conditions have been defined as "conditions of gratuity" by Monod and Cohn[3] and are the conditions employed in the experiment illustrated in Fig. 2. Benzer[15] found that, in the induced synthesis of β-galactosidase under conditions of gratuity and at saturating levels of inducer concentration, the bacterial population exhibits a homogeneous reaction in that essentially all cells participate to the same degree. Hence, under these conditions, the kinetics of induction of a culture represent the kinetics of induction of single cells.

Although most readers of Benzer's work concluded that conditions of gratuity were a sufficient guarantee for the assumption of population homogeneity during induction, Benzer himself cautioned against applying this assumption to conditions of less than saturating concentrations of inducer even though the induction was "gratuitous." Indeed, a second more subtle factor causing heterogeneous response at less than saturating concentrations of inducer has been discovered by Rickenberg, Cohen, Buttin, and Monod[16] (also referred to in reference 9). Thus, these workers found that, in wild-type *E. coli*, the β-galactosidase induction can be divided into two separate processes. The first of these involves an active transport of the inducer into the cell such that inducer concentrations inside the cell are much greater than those in the medium. The second is the actual induction of β-galactosidase synthesis inside the cell at a rate determined by the internal concentration of inducer. The active transport of inducer is accomplished by a unit, called galactoside-permease, which has been demonstrated to have most of the characteristics usually associated with enzymes and whose synthesis is induced by many of the same compounds which induce β-galactosidase synthesis, again at a rate determined by the internal concentration of inducer.

This transport mechanism obviously introduces another factor that would be predicted to encourage a heterogeneous response of a population upon exposure to inducer, providing the probability that a given cell will produce its first permease molecule per smallest detectable time unit is small. Thus, those cells which do synthesize their first permease unit will have an increased internal concentration of inducer which will, in turn, increase the probability of synthesizing the second permease unit above that for the first. Thus, those cells which by chance are the first to synthesize permease units will increase their advantage and rapidly pass to the stage of saturating internal concentrations of inducer and maximum rate of synthesis of both the permease and β-galactosidase. The distribution of rate of enzyme synthesis in the bacterial population at any given time after addition of the inducer will thus depend upon the probability that a given cell will synthesize its first permease unit and upon the time interval between this event and when maximum rate of synthesis is reached.

Novick and Weiner[17] and Cohn[8] have analyzed the kinetics of β-galactosidase induction in the wild-type *E. coli*, which is inducible for both the galactoside-permease and β-galactosidase. They have found that these kinetics are consistent with the foregoing theory with the addition that the probability of synthesis of the first permease units is determined by the inducer concentration, and that the time interval between this event and when maximum synthetic rate is achieved is very small, even at the lowest inducer concentrations studied. This means that, at inducer concentrations well below saturation, the distribution of rates of enzyme synthesis at any given time is essentially an all-or-none distribution; i.e., cells are synthesizing enzyme either at the maximum rate or at the minimum noninduced rate.

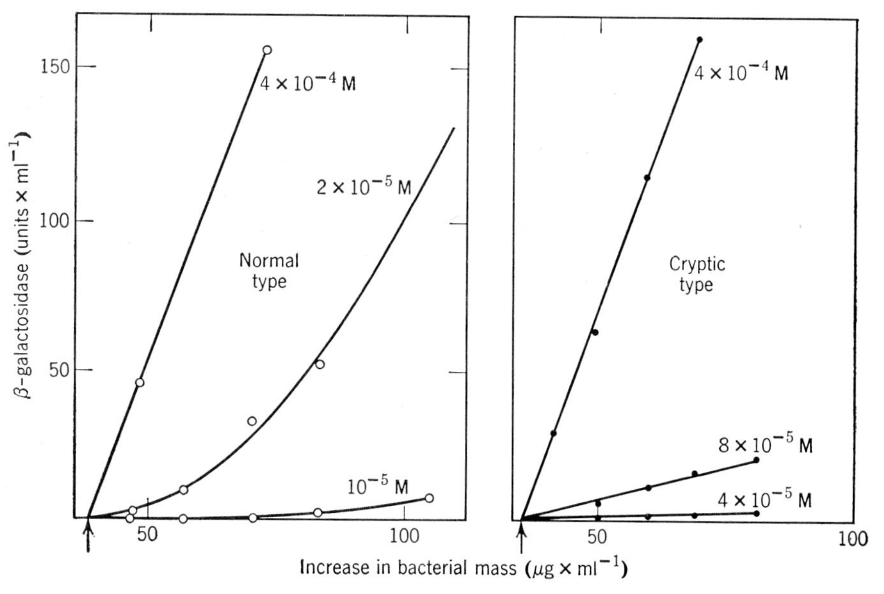


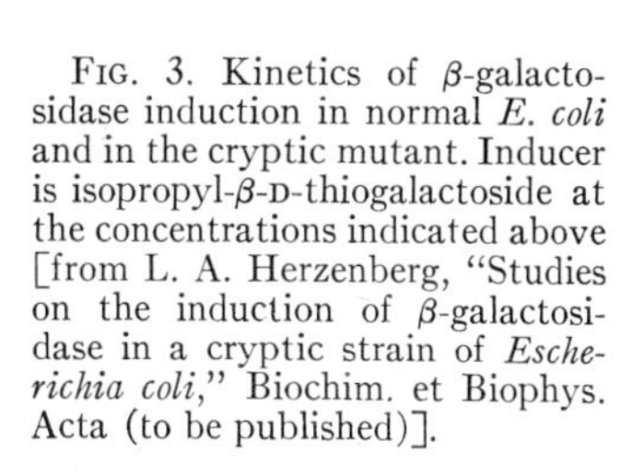
FIG. 3. Kinetics of β-galactosidase induction in normal *E. coli* and in the cryptic mutant. Inducer is isopropyl-β-D-thiogalactoside at the concentrations indicated above [from L. A. Herzenberg, "Studies on the induction of β-galactosidase in a cryptic strain of *Escherichia coli*," Biochim. et Biophys. Acta (to be published)].

As this induction at low inducer concentrations continues, a larger and larger fraction of the bacterial population is synthesizing enzyme at the maximum rate, and, consequently, a plot of enzyme activity *vs* increase in bacterial mass during the induction of such a culture will yield a curve with a slope (differential rate of synthesis) that increases with increase in growth to some final constant value (left-hand graph of Fig. 3). As the external inducer concentration is raised, the probability of synthesis of the first permease unit increases, and, consequently, the time duration of the phase of increase in differential rate of synthesis, or what may be termed the heterogeneous phase, decreases until it is no longer detectable. Under these conditions of saturating inducer concentration, for all intents and purposes, the distribution of rates of β-galactosidase synthesis in the population at any given time after addition of inducer is uniform, and, consequently, the kinetics of induction of the culture represents the kinetics of the individual cells.

However, it is quite clear that wild-type *E. coli*, in which both the galactoside-permease and the β-galactosidase are inducible, cannot be used to define the kinetics of induction of single cells at less than saturating concentrations of inducer. This could be accomplished if the galactoside-permease could be eliminated, thereby removing the last contributing factor toward heterogeneous response. The most convenient and complete method of such elimination is by the isolation and use of *E. coli* mutants which have lost the ability to form galactoside-permease, but retain the property of being inducible for β-galactosidase. Such mutants have been isolated and are commonly called inducible cryptics, in that, while β-galactosidase can be induced in these strains, because they lack the permease, the induced enzyme is essentially hidden from external substrate (e.g., lactose), the rate of hydrolysis in whole cells being limited by passive diffusion of the substrate.

One of these inducible cryptics (*E. coli* ML3) has been used extensively by Herzenberg[18] to define the kinetics of induction throughout the range of inducer concentrations that produce measurable increases in the rate of β-galactosidase synthesis. It is also the strain used in the experiment depicted in Fig. 2. An interesting result of the experiments of Herzenberg is that, for each inducer concentration tested, the differential rate of synthesis remains constant from the time of addition of the inducer, although the actual value of this rate is determined by the inducer concentration (Fig. 3). Under the assumption that, by employing conditions of gratuity and the inducible cryptic strains of *E. coli*, all factors encouraging a heterogeneous response during induction have been eliminated, one can interpret these kinetics of constant differential rate of synthesis for the culture as representing the kinetics of induction of individual cells. It must, however, be noted that, while this somewhat tortuous path of discovery and elimination of the sources of heterogeneous response during the induction process in *E. coli* wild-type makes this assumption quite reasonable, there is at present no direct experimental evidence available that completely justifies it.

If one admits this assumption, then the kinetics of constant differential rate of β-galactosidase synthesis during induction of the inducible cryptic strains (Fig. 3) allows one to conclude that the number of enzyme-forming units per cell that are activable by the inducer remains constant during induction. For, if the number of these units per cell should change during induction, then, in the presence of a constant subsaturating internal concentration of inducer, the differential rate of synthesis should change proportionately. The precision of the conclusion as to the constancy of β-galactosidase

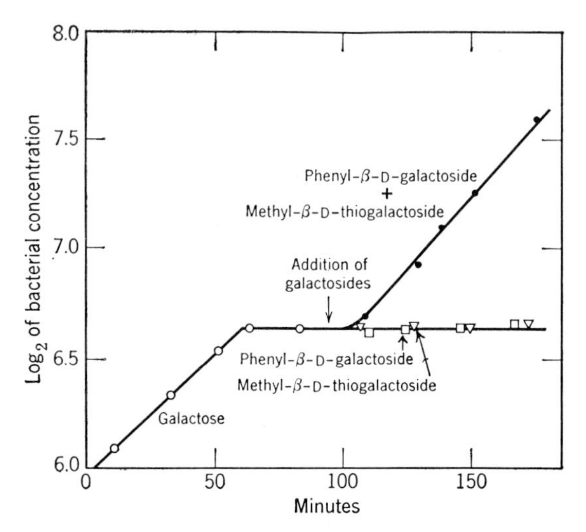


FIG. 4. Growth of *E. coli* ML 30 on galactose and the indicated galactosides demonstrating the independence of inducer and substrate functions.

forming units per cell is equivalent to the precision of the experimental observation of constant differential rate of synthesis. Consequently, if any change in the number of these units occurs during induction, it must occur in every cell of the population within two minutes of the moment the inducer is added. Since this seems extremely unlikely, it can be inferred that the same number of enzyme-forming units exist in noninduced cells as in induced cells, these units being active only in the presence of the inducer.

### (E) Independence of Substrate and Inducer Functions

A corollary to this conclusion is that the catalytic activity of β-galactosidase is not functional in its own synthesis; i.e., the process of enzyme induction is independent of enzyme action. The kinetic derivation of this postulate is confirmed by the observation that the property of being a substrate for β-galactosidase is neither a sufficient nor necessary condition for a compound to function as an inducer. Thus, phenyl-β-D-galactoside is an excellent substrate of β-galactosidase, but is not functional as an inducer. Methyl-β-D-thiogalactoside, on the other hand, is not a substrate of the enzyme, but is an excellent inducer. The inverse relationship of these two compounds is demonstrated quite clearly by a simple growth experiment shown in Fig. 4. An inducible strain of *E. coli* was grown in a medium containing galactose as the sole carbon source until growth stopped (*ca* 60 min) as a result of depletion of galactose. Since galactose does not induce β-galactosidase synthesis, these starved cells contain only trace amounts of the enzyme. After waiting a short time to insure complete starvation, the starved cells were divided into three parts, and the galactosides indicated in Fig. 4 were added as the sole carbon source. The cells did not grow in the presence of methyl-β-D-thiogalactoside because, although it can induce β-galactosidase synthesis, it is not a substrate and, consequently, cannot function as the necessary carbon source. No growth is observed in the presence of phenyl-β-D-galactoside, because it is not an inducer of the enzyme necessary for it to function as a carbon source. However, when the two galactosides are added together, growth occurs since methyl-β-D-thiogalactoside functions as the inducer and phenyl-β-D-galactoside serves as the substrate, yielding galactose, the necessary carbon source.

In addition to the qualitative data as to independence of substrate and inducer functions, quantitative measurements of the ability of many thiogalactosides to induce β-galactosidase synthesis and the ability to complex with the active site of this enzyme show no correlation. Thus, phenyl ethyl-β-D-thiogalactoside is an extremely good competitive inhibitor of β-galactosidase action ($K_I = 1.5 \times 10^{-5} M$), but can induce the synthesis of β-galactosidase to a rate that is only one-eightieth that obtainable with methyl-β-D-thiogalactoside, although this latter compound is about 800-fold less efficient as a competitive inhibitor ($K_I = 1.2 \times 10^{-2} M$). On the other hand, melibiose, an α-galactoside which is neither a substrate nor an effective competitive inhibitor of β-galactosidase, is quite effective as an inducer of this enzyme. Thus, it would appear that, in activating the enzyme forming units within the cell, the inducer reacts with some material having a different specificity than that associated with the active site of β-galactosidase.

There remains, however, one minimum structural requirement common both to inducers and to substrates or competitive inhibitors—namely, that they contain the galactopyranosidic ring structure.† The inference, then, is that the site at which the inducer reacts to activate the enzyme-forming unit is structurally similar but not identical to the active site of β-galactosidase.

In concluding this experimental definition of induced enzyme synthesis, the main conclusions that have been obtained are listed below.

1. The induction process involves the complete *de novo* synthesis of the enzyme from its constituent amino acids.
2. Induced enzyme synthesis is a virtually irreversible process, the enzyme being stable in the presence or absence of inducer.
3. The number of enzyme-forming units per cell remains constant during the induction process; that is, the inducer does not change the rate of synthesis of enzyme-forming units, but simply activates such units.
4. The site at which the inducer reacts to activate the enzyme-forming unit is structually similar but not identical to the active site on the enzyme.

† This is strictly true only if the *C6* carbon is not considered part of the ring, since the α-L-arabinosides (which are derivatives of β-D-galactosides lacking *C6*) exhibit the property of being weak substrates for β-galactosidase.

In addition to the above conclusions, one should mention another of equal importance but which, at present, cannot be derived from data on $\beta$-galactosidase induction. It is that the inducer acts as a catalyst in the sense that one molecule of inducer may cause the formation of more than one molecule of enzyme. This observation comes from experiments concerning the penicillin-induced synthesis of the enzyme penicillinase in *Bacillus cereus*.[11] It should be noted that each of the foregoing conclusions has been experimentally confirmed in only one, or, at the most, a very few systems and that, consequently, their generality remains to be shown.

### III. CONSTITUTIVE SYNTHESIS OF β-GALACTOSIDASE

Before analyzing some of the hypotheses attempting to explain the mechanism of induction, it is useful to first describe the phenomenon of the constitutive synthesis of $\beta$-galactosidase in *E. coli* and the origin of the bacteria responsible for it.

If one examines a $\beta$-galactosidase constitutive strain of *E. coli* that is growing in the absence of an inducer, one will find that $\beta$-galactosidase synthesis occurs at a very high rate, 1000-fold or more greater than is found for the inducible strains growing under the same conditions. Furthermore, the addition of an inducer to such a constitutive culture does not augment this already high rate of synthesis. Thus, the constitutive strains synthesize $\beta$-galactosidase in the absence of inducer at rates which are apparently maximal for these cells and which are approximately the same as the maximum rates of induced $\beta$-galactosidase synthesis found in the inducible strains.

The question that is immediately raised in such a comparison of constitutive and induced $\beta$-galactosidase synthesis is whether one is concerned with the synthesis of the same protein molecule in each case or whether two different proteins are included in the above use of the term, $\beta$-galactosidase, one being synthesized in inducible strains and the other in constitutive strains. This question has been answered with some certainty in favor of identity.[3,19] Thus, the titration of catalytic units precipitable by a given quantity of specific anti-$\beta$-galactosidase sera yielded the same results with enzyme preparations derived from constitutive bacteria as with those derived from induced cells, whether the antibody was formed in response to the induced or to the constitutive enzyme. This not only demonstrates the antigenic identity of the two enzymes, but also indicates that their turnover numbers (catalytic units per molecule) are the same. Similarly, an extensive comparison of the kinetic constants for several substrates, for sodium- and potassium-ion activation, and for thermal inactivation of the $\beta$-galactosidase in these two preparations did not reveal any measurable difference.

Taking these observations in good grace as sufficient evidence for the identity of the $\beta$-galactosidase in constitutive and induced cells, one should expect the basic genetic unit or units necessary for the synthesis of this protein to be common to both cell types, assuming, of course, that one accept the tenets of the DNA-protein doctrine presented by Levinthal (p. 227). What then is the genetic difference between inducible and constitutive *E. coli*?

The following series of mutation sequences indicates the relationship between inducible and constitutive strains[16,20,21]:

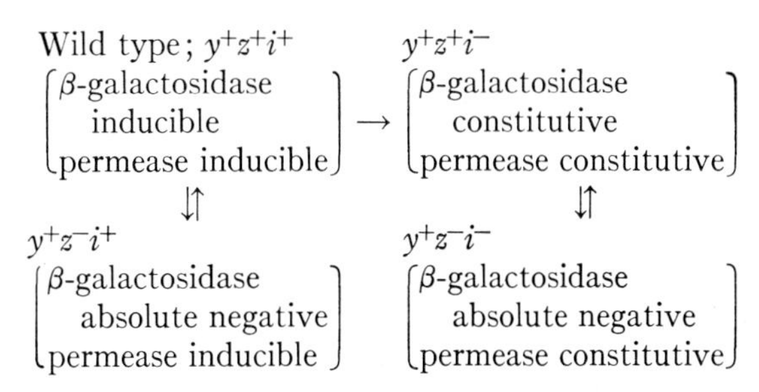


In this scheme, the genetic units have the following meaning:

1. $i^+ \rightarrow i^-$ designates the mutation involving a change of state from inducible ($i^+$) to constitutive ($i^-$) for both permease and $\beta$-galactosidase.
2. $z^+ \rightarrow z^-$ designates the mutation involving damage to the $z^+$ unit necessary for $\beta$-galactosidase synthesis such that the enzyme cannot be made.
3. $y^+$ indicates the unit necessary for the synthesis of the galactoside-permease, a $y^+ \rightarrow y^-$ mutation (not shown) being that involved in formation of the cryptic, permease negative mutants previously mentioned.

This genotypic interpretation of the mutation data is consistent with data from recombination studies[22] and, as with a unique type of *cis-trans* test demonstrating that $y^+$ and $z^+$ involve different cistrons.[23]

From the fact that the $\beta$-galactosidase-proteins synthesized by constitutive and induced bacteria are identical, and from the genetic evidence demonstrating that the change resultant from mutation of inducibles ($i^+$) to constitutives ($i^-$) involves genetic units (cistrons) different from those involved in the formation of absolute negatives ($z^-$), one comes to the conclusion that the mechanism of synthesis of $\beta$-galactosidase in induced and constitutive cells does not differ in its essentials—that is, the same amino acids and the same mechanisms of ordering such amino acids are employed in each case. Though slightly more specific, this conclusion is equivalent to the general unitary hypothesis for enzyme synthesis emphasized by Cohn and Monod.[4]

Thus, in addition to conforming to the conclusions drawn directly from the induction phenomenon (Sec. II), any useful hypothesis for the induced enzyme synthesis must also explain the inducible to constitutive transition via the agency of a single-step mutation not involving a change in the basic genetic units determining the structure of the enzyme.

## IV. HYPOTHESES FOR INDUCED ENZYME SYNTHESIS

The point of departure for all hypotheses purporting to explain enzyme induction within the context of the unitary principle is the choice between two possibilities: (1) an inducer is an essential part of the minimum requirements for rapid enzyme synthesis (i.e., the "generalized induction principle"[4]); or (2) the converse, inducers are *not* part of these minimum requirements.

Since there exists no direct experimental evidence which determines this choice, it is necessary to analyze the consequences of both possibilities. A number of schemes have been developed that invoke the inducer as an essential element of the minimum requirements for enzyme synthesis.[4–6] It is not proposed to analyze these here, since they either do not attempt to explain at what point and in what manner the inducers act in the over-all conversion of amino acids to enzyme molecules,‡ or they are not consistent with the conclusions formed in Sec. II from data on $\beta$-galactosidase induction.§ Rather, a model of enzyme synthesis is presented here in which the necessity of inducer participation can easily be visualized and which is consistent with the conclusions drawn thus far. The initial assumption in this construction is that the inducer is neither necessary nor functional in determining the amino-acid order of the induced enzyme, but instead acts to increase the rate of formation of its secondary or tertiary structure. This assumption has its basis in the coding theory of DNA function in which the amino-acid order in a given protein is uniquely determined by the nucleotide order in the DNA of some functional gene (cf. Levinthal, p. 249).

A diagrammatic representation of how an inducer might exert a catalytic function in tertiary structure formation is given in Fig. 5. In this diagram, the semicircle of capital letters represents a template of unspecified chemical composition (DNA, RNA, or other) whose structure is determined by a functional gene and which in turn determines the order of amino acids (lower-case letters) making up a specific polypeptide. Each capital letter then represents a code symbol for one amino acid, and it is assumed that the physical structure which determines a given code symbol specifically binds only one amino acid. Peptide-bond formation between amino acids is imagined to occur on this template with the condition that the binding force between a given amino acid and its code structure is not lost by the formation of the polypeptide. Consequently, polypeptide and template form a relatively stable complex, which momentarily may assume any one of many partially dissociated states. Certain of these transitory, partially dissociated configurations will be favorable for the formation of the tertiary structure necessary for enzyme formation (represented here by the formation of a disulfide bond via the oxidation of two sulfhydryl residues in the polypeptide chain). The formation of such tertiary structure is assumed to be concomitant with dissociation of polypeptide and template, thus allowing the formation of free enzyme.

With no inducer present, such favorable states are considered to be extremely short-lived, so that the probability of formation of the disulfide bond pictured in Fig. 5 is quite small. As a consequence, the rate of enzyme synthesis is very low.

The assumed function of the inducer, or some product derived from it, (X), is to stabilize the favorable configurations by interacting with them to form a complex of the type represented by the middle left-hand structure in Fig. 5. Such a stabilization would result in an increase in the probability of the critical disulfide-bond formation and, therefore, in the rate of enzyme formation.

This mechanism is offered more as an aid in visualizing how one can imagine the necessity of inducer participation in protein synthesis, rather than as a unique representation of this possibility. However, it is consistent with the conclusions developed concerning $\beta$-galactosidase induction. Thus, it accounts for the *de novo* synthesis of $\beta$-galactosidase during induction, since

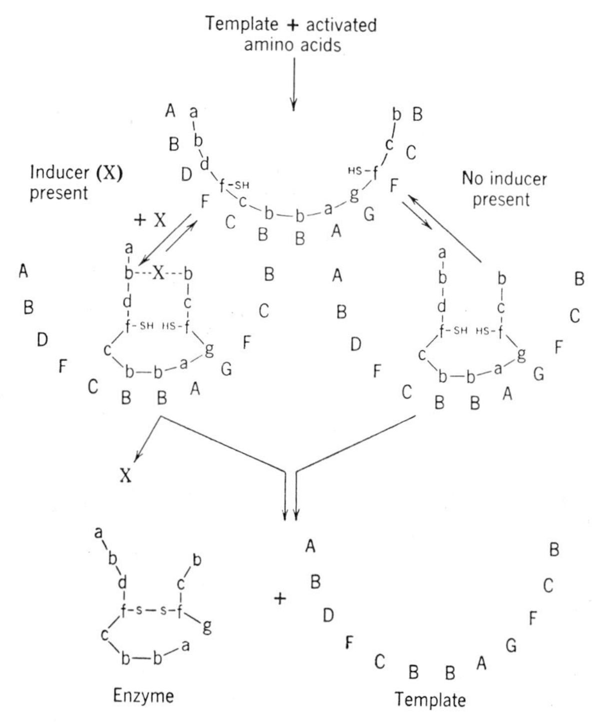


FIG. 5. Model for inducer function under the generalized induction hypothesis.

‡ The "organizer" hypotheses[4,5] insist that a derivative of the inducer (the "organizer") is an essential in enzyme synthesis but leave quite unspecified the mechanism by which such a derivative catalyzes enzyme synthesis.

§ The mass-action hypothesis by Yudkin[24] and its extensions[6] demand that enzyme induction be a *reversible* process in which the specificity of inducers of enzyme synthesis parallel the specificity of substrates and competitive inhibitors of this enzyme, conditions which are not consistent with the data derived from $\beta$-galactosidase induction (Sec. II).

the level of polypeptide precursors of the enzyme in noninduced cells could be no greater than the number of templates per cell. Indeed, the calculated limit of ten or less such precursor molecules per noninduced cell is not an unreasonable limit for the number of templates per enzyme per cell. The model also allows for an essentially irreversible synthesis of β-galactosidase during the induction of growing populations; for the activation of enzyme-forming sites (templates) by inducers and the constancy of such sites during induction; and, finally, for the similarity but nonidentity of the specificity of inducers and substrates (or competitive inhibitors) of the enzyme. It is perhaps of interest to note that, insofar as reversal of the synthetic reaction in the model is limited by competition of activated amino acids and enzyme for the template (the enzyme having a very small affinity for the template as compared to the precursor polypeptide), the rate of any possible reverse reaction would be predicted to be greater under conditions of low levels of activated amino acids (e.g., in nitrogen-starved cells) than when these compounds are plentiful (e.g., in fast-growing cells.)

With respect to this model, or any model that invokes an inducer as an essential in the minimum requirements for the synthesis of a given enzyme, the two genetic states of inducible and constitutive β-galactosidase synthesis can be interpreted by assuming that constitutive cells contain an endogenous inducer, and that inducible cells are unable to synthesize such an inducer for lack of a necessary enzyme. On this basis, the mutation of an inducible strain to a constitutive can be viewed as a repair of a damaged genetic unit responsible for the synthesis of the postulated enzyme.

Turning now to the second possibility—namely, that an inducer is not a necessary component of the minimum system for enzyme synthesis—one must accept the constitutive condition as representing this minimum system. A model built for such a system could be like that given in Fig. 5, with two important differences: (1) The favorable state leading to tertiary structure could be stabilized by the form of the template itself, thus eliminating the need for inducer; and (2) the affinity of the activated amino acid for its code structure on the template would have to be lost as a result of peptide-bond formation with its nearest neighbors. The latter condition could be allowed for if one supposes that the activated amino acid consists of the amino acid covalently linked to a residue, *R*, which is unique for each amino acid, and that the code structure on the template has an affinity for *R* rather than the amino acid itself. If peptide-bond formation entails the breaking of the *R*-amino acid bond, then the polypeptide thus formed is quite free to separate from the template.

However, regardless of the model one builds for the synthesis of an enzyme without inducer (endogenous or exogenous) participation, the induced enzyme synthesis must differ from constitutive synthesis, either by a difference in type of template or by the existence of an inhibitor of normal synthesis (i.e., constitutive type synthesis) in the inducible strain, an inhibition which is relieved by the addition of an exogenous inducer. In the first of these possibilities, one must imagine that mutation from inducible to constitutive invokes a change in template type from one requiring an inducer for synthesis (i.e., as in Fig. 5) to one in which no inducer is required. Furthermore, since the β-galactosidase molecules synthesized constitutively and by induction are identical, one must suppose that this change in template type does not involve a change in the order of the amino-acid code. This seems to be asking a lot from a single mutation, but since almost nothing is known of the template structure, one cannot say, *a priori*, that such a change is impossible.

The second possibility—namely, that induction is the release of inhibition of enzyme synthesis—is actually quite an old idea, but in the past it was generally offered facetiously as the *bête noire* of generalized induction theories, since it could not be eliminated. However, recent developments demand more positive consideration of this idea. The source of these developments lies in the observation of what Vogel[25,26] has termed "enzyme repression." Repression is the inhibition of the differential rate of synthesis of an enzyme resulting from exposure of cells to a given substance ("repressor"). The word repression was adopted simply to avoid confusion with the term "enzyme inhibition," which by long usage implies the inhibition of the catalytic function of a given enzyme and not of its synthesis. One of the best examples of enzyme repression is observed among the constitutive enzymes included in the biosynthetic pathway leading to the synthesis of ornithine, citrulline, and finally of arginine in *E. coli*. Thus, Vogel[25] found that arginine is a specific and very effective repressor of the synthesis of acetylornithinase. Similarly, Gorini and Maas[27] have shown that arginine also represses the synthesis of ornithine transcarbamylase. Perhaps the most significant implication of these observations is that there exists within the cell a mechanism of feedback control between the result of the catalytic action of the group of enzymes in a biosynthetic pathway (in this case, the synthesis of arginine) and the synthesis of these enzymes. For present purposes, however, they give plausibility to the supposition that the inducible strains could represent a case of repression of the constitutive synthesis of β-galactosidase by endogenous repressors present in such strains, but not in the constitutive strains. Indeed, it is known that the constitutive synthesis of β-galactosidase in *E. coli* can be inhibited by exposure of cells to galactose or one of a variety of β-D-galactosides[4]; i.e., in the constitutive strains, these substances act as exogenous repressors of enzyme synthesis.

One may, therefore, term this second possibility the repressor hypothesis. It states that in constitutive strains enzyme synthesis occurs without the aid of any inducer; that such synthesis is inhibited in the inducible cells because of repressor substances synthesized by these cells and not by constitutive cells; and that an inducer destroys this inhibition of enzyme synthesis either by combining with the repressor to yield a non-inhibiting complex, by competing with the repressor for some site on the template (or template-polypeptide complex)—which, if occupied by the inducer, does not inactivate the template, but if occupied by a repressor does so inactivate—or by inhibiting the synthesis of unstable repressors.

The three alternative explanations of induced and constitutive enzyme synthesis can be summarized as follows:

1. Generalized-induction hypothesis. Inducers are necessary components for enzyme synthesis both in constitutive and in inducible cells. Endogenous inducers are synthesized by constitutive cells but not by inducible cells, which must, therefore, receive exogenous inducers for enzyme synthesis.
2. Different-template hypothesis. Two different types of template exist for enzyme synthesis. In constitutive cells, the templates function without the aid of inducers. In inducible cells, the inducers are necessary to activate the template (as in Fig. 5). No endogenous inducer or repressor is assumed. It is assumed that the amino-acid ordering function of each template is the same.
3. Repressor hypothesis. The templates in inducible and constitutive cells are the same and do not require activation by inducers. Substances specifically inhibiting template function (repressors) are synthesized by inducible cells but not by constitutive cells. Exogenous inducers function by destroying this inhibition of enzyme synthesis.

## V. EVALUATION OF THE INDUCTION HYPOTHESES

The data presented thus far do not allow much more than an intuitive preference for one of the three hypotheses over the others. Thus, the enzyme repression observed in the enzymes involved in arginine biosynthesis can equally well be explained by the inhibition of endogenous induction as by direct inhibition of template catalysis not involving an inducer. However, experiments recently reported by Pardee *et al.*[23] have done much toward clarifying the relative weight that one can place on the three alternative explanations of enzyme induction. These experiments were designed to observe the behavior of $\beta$-galactosidase synthesis in *E. coli* zygotes having a heterogenotic structure for both the $z^+$, $z^-$ and the $i^+$, $i^-$ pairs in an effort to determine dominance in the $i^+$, $i^-$ pair. This determination is important in the evaluation of the hypotheses, since, in the generalized induction hypothesis as it has been presented, the positive and, therefore, dominant function of synthesizing an endogenous inducer is given only to the constitutive ($i^-$) strain, whereas in the repressor hypothesis the positive and, therefore, dominant function of synthesizing a repressor is given only to the inducible type ($i^+$). In the hypothesis of two template types, both inducible ($i^+$) and constitutive ($i^-$) units are given positive functions and, therefore, no dominance in the $i^+$, $i^-$ pair should necessarily exist.

The characteristics of *E. coli* conjugation allow the determination of dominance in a very unique manner. The male (*Hfr*) member of the mating pair injects its "chromosome" into the female ($F^-$) member through a small tubule connecting the two cells. The order in time of entrance into the female of the genetic units from a given strain of male cells is unique, and apparently no appreciable amount of cytoplasm enters with these genetic units (cf. Wollman *et al.*[28] and the chapter by Lennox p. 242). Thus, the initial cytoplasmic state of the zygote is determined by the cytoplasmic state of the female before conjugation. As a consequence, it is possible to form the $i^+z^+/i^-z^-$ zygote either by the injection of the $i^+z^+$ genetic units from an inducible male into an $i^-z^-$ formed cytoplasm of a constitutive absolute negative female or vice versa—to inject the $i^-z^-$ genetic units into an $i^+z^+$ formed cytoplasm. The question then asked is what is the behavior of $\beta$-galactosidase synthesis in the $i^+z^+/i^-z^-$ zygote in each case and to what hypothesis does this behavior correspond? In answer to this question, it is convenient to first describe the behavior predicted by each hypothesis.

1. Under the form of the generalized-induction hypothesis presented here, the mating of a male $i^+z^+$ cell to a $i^-z^-$ female in the absence of any exogenous inducer should yield the synthesis of $\beta$-galactosidase soon after the $i^+z^+$ units have been injected into the $i^-z^-$ cytoplasm, since this cytoplasm should contain endogenous inducers which can activate the template resulting from the entering $z^+$ unit. It should be noted here that the $z$ and $i$ units are very closely linked[23,29] so that, within the minimum time units employable in conjugation experiments, the two units enter simultaneously. Furthermore, the synthesis of $\beta$-galactosidase should continue in the $i^+z^+/i^-z^-$ zygote as long as this heterogenotic state remains intact.

In the reciprocal conjugation, the prediction would be the same, except that there might be expected to be a longer lag between the entrance of the $i^-z^-$ units into the $i^+z^+$ cytoplasm and the first appearance of enzyme synthesis owing to the necessity of the entering $i^-$ unit to catalyze the synthesis of the endogenous inducer in the zygote cytoplasm.

2. Under the hypothesis involving two types of templates for constitutive ($i^-$) and inducible ($i^+$) synthesis, the $i^+z^+/i^-z^-$ zygote should yield no $\beta$-galactosidase synthesis in the absence of exogenous inducer, unless the $z^+$ in one "chromosome" and the $i^-$ in the other can cooperate to yield the type of template synthesized in the constitutive $\beta$-galactosidase positive strain ($i^-z^+$). If this is possible, then the $\beta$-galactosidase synthesis should occur in the zygote with the same behavior, whether the $i^+z^+$ units are injected into a $i^-z^-$ female or

vice versa. It should be noted that the possibility of recombination leading to $i^-z^+$ on the same "chromosome" is negligible owing to the closeness of the $z$ and $i$ units.

3. Under the repressor hypothesis for induction, the injection of a $i^+z^+$ unit into a $i^-z^-$ cytoplasm in the absence of inducer should yield the synthesis of β-galactosidase soon after the injection, since initially there should be no repressor in the zygote cytoplasm, it being of the constitutive ($i^-$) type. However, different from the prediction of the generalized induction hypothesis, in this case one should expect the synthesis of β-galactosidase in the zygote to cease as soon as the injected $i^+$ unit has caused the synthesis of sufficient repressor in the zygote cytoplasm. At this time, the zygote should be phenotypically inducible rather than constitutive as it was immediately after the entrance of the $i^+z^+$ unit.

In the reciprocal mating, the cytoplasm of the zygote, being $i^+z^+$, should initially contain the repressor and should continue to synthesize repressor after the $i^+z^+/i^-z^-$ state is established. Thus, no synthesis of β-galactosidase in absence of exogenous inducer would be expected at any time after conjugation.

Each hypothesis yields a different prediction, and, consequently, the experiment should uniquely determine which, if any, is valid. The results are entirely consistent with the repressor hypothesis. Thus, in the conjugation of $i^+z^+$ males with $i^-z^-$ females in the absence of exogenous inducer, β-galactosidase synthesis can be detected within a few minutes after the entrance of the $i^+z^+$ genetic units. However, in the reciprocal conjugation, no enzyme synthesis could de detected in the absence of inducer, even after several hours. Furthermore, in the $i^+z^+$ male-to-$i^-z^-$ female conjugation the synthesis of β-galactosidase in absence of inducer ceases about two hours after zygote formation. This is shown in Fig. 6. This cessation of synthesis is not the result of segregation of the zygote, since this event cannot be detected until two hours after cessation of synthesis. As is shown in Fig. 6, the zygotes have become inducible by the time the constitutive synthesis stops. These results are exactly those predicted under the repressor hypothesis in which $i^+$ was predicted to be dominant over $i^-$, and they are inconsistent with the two other hypotheses. These results, furthermore, offer the evidence promised earlier that the $i$ and $z$ units involve different cistrons—i.e., different functional genetic units.

Does this experiment then offer the death knell to the hypothesis of different templates and to the generalized induction theory? It quite effectively eliminates the former of these possibilities, but, unfortunately, the generalized induction hypothesis can be made viable again by a very slight alteration. The supposition made under this hypothesis was that the $i^+$ to $i^-$ mutation involved the repair of a genetic unit necessary for the synthesis of endogenous inducer. This supposition is clearly eliminated by the experiments of Pardee, Jacob, and Monod. However, it can be assumed that, in both

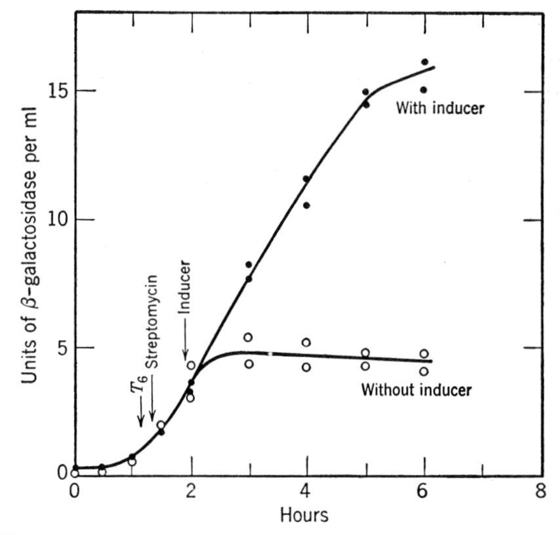


FIG. 6. Formation of β-galactosidase in zygotes from a ♂$i^+z^+$ × ♀$i^-z^-$ conjugation of *E. coli*. The abscissa indicates the time after mixing the parental populations. The experiment was performed in quadruplicate. Bacteriophage $T_6$ and streptomycin were added at the indicated times to stop conjugation and to eliminate the ♂. The inducer, methyl-β-D-thiogalactoside ($10^{-3}M$), was added at 115 min to two of the four cultures (filled circles), whereas the other two cultures (open circles) contained no inducer [from A. B. Pardee, F. Jacob, and J. Monod, Compt. rend. **246**, 3125 (1958)].

$i^+$ and $i^-$ cells, an endogenous inducer is synthesized, but that $i^+$ cells differ from $i^-$ cells in possessing an enzyme capable of destruction of this endogenous inducer. The $i^+$-to-$i^-$ mutation would then result in the loss of ability to synthesize this enzyme, and $i^+$ would be expected to be dominant to $i^-$. The results of the above experiment are equally well in accord with this extension of the generalized induction hypothesis as with the repressor hypothesis. Thus, this ingenious experiment has indeed limited the choice of hypotheses drastically; it has pushed the repressor hypothesis fully into the limelight, but it unfortunately does not offer a final unambiguous solution.

Nor should one expect such an unambiguous solution as long as one is forced to deal with the complexity of an entire cell. At this level, a solution will be unique only in that it unifies and poses the answer to a larger number of problems with a lesser number of assumptions than does its alternatives. In this sense, the repressor hypothesis is the more satisfying. It yields a mechanism for feedback control of enzyme synthesis that would seem to be necessary if the cell is not to run amuck. If the phenomenon of enzyme repression becomes a general observation, particularly in biosynthetic pathways such as that found in arginine synthesis, one can account for feedback control with the use of one or less repressor substances per enzyme. However, in choosing the generalized induction hypothesis, if one wishes to maintain the explanation of feedback control, one must involve inducers and substances inhibiting induction (i.e., repressors) for each enzyme. Thus, if both hypotheses are generalized, it would seem that feedback control is explained with a lesser number of components and consequently, assumptions, by the repressor hypothesis. On this basis, the inducible systems are seen as a peculiar form of a general repressible condition, peculiar in that they are relatively rare events of exceptionally strong endogenous repression. It is of interest to note that, should no exogenous inducer be known for such a system, it would appear as a typically negative mutant for the synthesis of a given enzyme.

If one wishes to find the unambiguous solution of enzyme induction and repression, it is the author's conviction that it will first be necessary to find a system of enzyme synthesis less complex than the whole cell. It is true that the two hypotheses pose clearly distinguishable alternatives in that one predicts the presence of an inducer in constitutive cells and the other of a repressor in inducible cells. However, it is also the strong suspicion of the author that the assay system necessary to test such exogenous inducers or repressors will require something much less complex than the whole cell. For this reason—and because resolvable subcellular systems of enzyme synthesis appear to be close at hand—the next step forward in the understanding of enzyme induction and repression will most probably attend the removal of the cell boundary from the assay system.

## BIBLIOGRAPHY

[1] M. Cohn, J. Monod, M. R. Pollock, S. Spiegelman, and R. Y. Stanier, Nature **172**, 1096 (1953).

[2] R. Y. Stanier, Ann. Rev. Microbiol. **5**, 35 (1951).

[3] J. Monod and M. Cohn, Advances in Enzymol. **13**, 67 (1952).

[4] M. Cohn and J. Monod in *Adaptation in Micro-organisms*, R. Davies and E. F. Gale, editors (Cambridge University Press, Cambridge, England, 1953), p. 132.

[5] M. R. Pollock in *Adaptation in Micro-organisms*, R. Davies and E. F. Gale, editors (Cambridge University Press, Cambridge, England, 1953), p. 150.

[6] J. Mandelstam, Intern. Rev. Cytol. **5**, 51 (1956).

[7] S. Spiegelman and A. M. Campbell in *Currents in Biochemical Research, 1956*, D. E. Green, editor (Interscience Publishers, Inc., New York, 1956), p. 115.

[8] M. Cohn, Bacteriol. Rev. **21**, 140 (1957).

[9] G. N. Cohen and J. Monod, Bacteriol. Rev. **21**, 169 (1957).

[10] M. R. Pollock and J. Mandelstam in *The Biological Replication of Macromolecules*, Society for Experimental Biology Symposia No. 12, (Cambridge University Press, Cambridge, England, 1958), p. 195.

[11] M. R. Pollock, "Inductive control of enzyme formation," in *The Enzymes*, P. D. Boyer, H. A. Lardy, and K. Myrbäck, editors (Academic Press, Inc., New York, to be published).

[12] J. Monod, A. M. Pappenheimer, Jr., and G. Cohen-Bazire, Biochim. et Biophys. Acta **9**, 648 (1952).

[13] D. S. Hogness, M. Cohn, and J. Monod, Biochim. et Biophys. Acta **16**, 99 (1955).

[14] B. Rotman and S. Spiegelman, J. Bact. **68**, 419 (1954).

[15] S. Benzer, Biochim. et Biophys. Acta **11**, 383 (1953).

[16] H. V. Rickenberg, G. N. Cohen, G. Buttin, and J. Monod, Ann. Inst. Pasteur **91**, 829 (1956).

[17] A. Novick and M. Weiner, Proc. Natl. Acad. Sci. U. S. **43**, 553 (1957).

[18] L. A. Herzenberg, "Studies on the induction of $\beta$-galactosidase in a cryptic strain of *Escherichia coli*," Biochim. et Biophys. Acta (to be published).

[19] M. Cohn (unpublished observations).

[20] G. Cohen-Bazire, and M. Jolit, Ann. Inst. Pasteur **84**, 937 (1953).

[21] J. Lederberg, E. M. Lederberg, N. D. Zinder, and E. R. Lively, Cold Spring Harbor Symposia Quant. Biol. **16**, 413 (1951).

[22] F. Jacob and J. Monod (unpublished observations).

[23] A. B. Pardee, F. Jacob, and J. Monod, Compt. rend. **246**, 3125 (1958).

[24] J. Yudkin, Biol. Revs. Cambridge Phil. Soc. **13**, 93 (1938).

[25] H. J. Vogel in *The Chemical Basis of Heredity*, W. D. McElroy and B. Glass, editors (The Johns Hopkins Press, Baltimore, Maryland, 1957), p. 276.

[26] H. J. Vogel, Proc. Natl. Acad. Sci. U. S. **43**, 491 (1957).

[27] L. Gorini and W. U. Maas, Biochim. et Biophys. Acta **25**, 208 (1957).

[28] E. L. Wollman, F. Jacob, and W. Hayes, Cold Spring Harbor Symposia Quant. Biol. **21**, 141 (1956).

[29] J. Monod (personal communication, 1958).

# 32

# General Features of Radiobiological Actions*

RAYMOND E. ZIRKLE

*Committee on Biophysics, The University of Chicago, Chicago 37, Illinois*

THIS paper and the three that follow (Pollard, p. 273; Wood, p. 282; Tobias, p. 289) are concerned with radiation biology. The potential scope of this field—effects of all types of radiations on all types of biological systems—and its relation to other areas of radiation research are indicated in Fig. 1, where an energy spectrum of radiations is plotted as abscissae and various inanimate and biological systems are "plotted" as ordinates in ascending order of presumed complexity. At the level of macromolecules, radiation biology shades imperceptibly into radiation chemistry and physics, the basic sciences on which it draws for necessary facts and concepts. At the level of groups of multicellular organisms, it approaches "radiation sociology," which is needed to deal with the problems of group behavior in irradiated populations.

Radiations below about 1 ev produce biological effects, if any, through heating. These effects usually are regarded as cognate to, but not included in, radiation biology. The region from 1 to 6 ev, comprising the near infrared, the visible, and the familiar ultraviolet wavelengths, is readily available for investigations as is the region from about 1000 ev upward. These two accessible portions of the spectrum conveniently are termed low-energy and high-energy, respectively. The intervening "transition" region is absorbed in air and biological material so readily that it can be used only on very small objects *in vacuo*. It is of great theoretical interest, in view of the great differences in basic response of biological systems to the low- and high-energy spectra.

Figure 1 shows that radiation biology potentially ramifies throughout biology as a whole. So does biophysics. What is the relationship, if any, between the two? In my personal view, each contains some of the other. The biophysical content of radiation biology is probably about as great, both potentially and actually, as that of general biology.

It is clearly impracticable, in four short papers, to cover all of radiation biology. Accordingly, the coverage is narrowed as follows. First, attention is concentrated on the high-energy radiations, with occasional comparative references to the ultraviolet. Second, specific samples are selected from the spectrum of biological systems (Fig. 1); after these brief general remarks, Pollard describes experiments on certain macromolecules and viruses (p. 273), Wood speaks on certain cellular effects that have been investigated intensively (p. 282), and Tobias discusses selected studies on cell populations and multicellular organisms (p. 289).

By and large, the effects of high-energy radiations, as well as those of the ultraviolet, are injurious to all or part of the irradiated system. Nevertheless, considerable effort goes into the study of these effects, for various reasons. Some persons find a fascinating field of research in the general problem of how very small amounts of radiation energy produce such drastic effects. I believe that the real attraction here is the peculiar combination of various portions of physics and chemistry and of many aspects of biology that must be brought to bear on any serious investigation of radiobiological mechanism. This peculiar versatility demands so much of a human life span that a radiation biologist is necessarily a specialist, much as he may strive to not become too differentiated.

However, most of the interest in radiation effects stems from their applications. In many branches of basic biology, they have long been used as powerful research tools. For example, the development of genetics has been accelerated immensely by the use of radiations as mutagenic agents; partial body irradiation has found fruitful application in embryology; partial cell irradiation is used to get information about the properties and functions of various cell parts. As basic biological tools, the various radiations have two properties that, in many situations, are of critical advantage: they do not disrupt membranes and other structures grossly, and the dosage usually is reproducible.

Radiation biology also has some important practical applications. The oldest, and still one of the most important, is radiation therapy. Another is its use in dealing with the widespread and multifarious problem of radiation hazards that, within the last decade and a half, have increased so explosively that they even figure in national and international politics.

Regardless of one's motivation to study or use radiation effects, there is obvious need to know as much as possible about their mechanisms. There is nothing basically unique about current methods of investigating these mechanisms; they are essentially those of the physical sciences and of the analytical biological sciences, with some special variations that stem from the unique physical properties of the high-energy radiations.

Like any story, a radiation action has a beginning, an end, and a middle. The beginning is the act of irradiation, and the end is the effect observed; there is considerable information about these, and the prospects of

* The following data are used in this paper: The usual energy unit is the electron volt (ev); 1 ev $=1.6\times10^{-12}$ erg $=3.8\times10^{-20}$ cal. For any radiation, energy times wavelength is $1.235\times10^{4}$ ev A. "Dose" is radiation energy transferred to unit mass of irradiated object; 1 rad $=100$ ergs/gram $=66$ ev/$\mu^3$, assuming density of object to be 1.05; 1 roentgen (r) *usually* is equivalent to 0.93 rad.

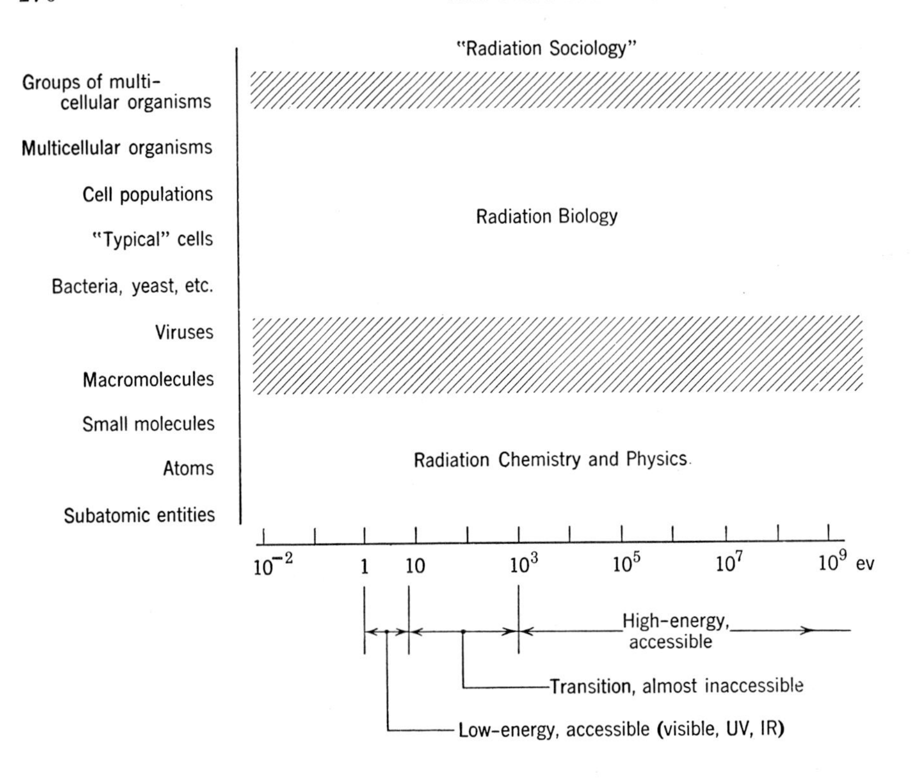


FIG. 1. Scope of radiation biology and its relations to other areas of radiation research. Abscissae, partial radiation energy spectrum in electron volts; ordinates, "spectrum" of inanimate and biological objects in order of presumed complexity.

getting more are good. The middle, frequently miscalled the "latent period," is essentially a domain of ignorance wherein most of the problems lie. The remainder of this article is devoted to an attempt to indicate some of the ways in which current research is directed to reduction of this ignorance.

Clearly, the end effect must be defined as accurately as possible. This implies information about the state of the investigated system *before* it is irradiated. The incompleteness of such information is demonstrated well by the need for a large conference such as the Study Program in Biophysical Science and by the remedial attempts so vividly described by earlier speakers. High-energy radiation in sufficient amount is capable of producing significant changes in practically any biological structure or function. In view of the great diversity and complexity of these structures and functions, it is not surprising that the list of known end effects is long. Some vigorously investigated examples are aberrations in chromosome structure, inhibition of cell division, inactivation of enzymes and other macromolecules, gene mutations, and induction of tumors in animals.

It is also essential to know as much as possible about the process of irradiation and the immediately consequent events. Here, one encounters facts and concepts that are not basic to other areas of biological research. The high-energy radiations differ in two important features from all other physical and chemical agents, including the low-energy radiations. First, a large fraction of the molecules that receive energy from the radiation are activated to very high states, even to ionization. As a result, in most biological material, radiation energy is transferred to the individual molecules of the many species usually present in a nonselective fashion that is chiefly dependent only on the molecular mass. Chromophores do not enter into consideration, as they do in work with low-energy radiations. This circumstance presents both advantages and disadvantages to the investigator.

Second, the energy transfers to individual molecules do not occur singly in space and time but in more or less linear groups. All of the high-energy radiations are either charged particles in motion (e.g., electrons or $\alpha$-particles) or are agents (neutrons; x-rays and $\gamma$-rays) that set such particles in motion. These charged particles typically have kinetic energies of thousands or millions of electron volts, whereas the average energy that can be accepted by an individual molecule is some tens of electron volts. Thus, successive energy transfers occur along the path of the high-energy particle. Some of these transfers are capable merely of exciting the molecules to higher energy states. Others involve enough energy to produce ionization, e.g., to eject electrons, some of which have kinetic energies sufficient to eject electrons from other molecules (secondary ionization). The ions produced in gases are well demonstrated by means of the Wilson cloud chamber. The trail of ions produced by a single primary charged particle is termed an ionization track. Similar tracks are recorded on photographic plates (linear sequences of blackened grains) and in suitable liquids at low temperatures (trails of bubbles). This, and other physical evidence, makes it highly probable that the distribution of energy transfer in a biological object is essentially the same as that displayed in the cloud chamber, except that the dimensions should be reduced by a factor of about eight hundred because of the difference in density.

These two properties—the high average value of the individual energy transfers and the grouping of these transfers into tracks—make the high-energy radiations unique in their mechanisms of action, even though these mechanisms, in some cases, produce end effects that are superficially indistinguishable from those of other agents.

To students of mechanism, probably the most significant single feature of high-energy radiation actions, especially on cells and on even more-highly organized systems, is the small amount of energy required. Of the many end effects that are known, most are produced to a significant degree by 10 000 rad (or less). This dose equals about 0.02 cal/g of biological material.

For some actions, the dose-effect relations ("kinetics") are suggestive. When samples of certain small cells (bacteria, haploid yeast, etc.) are given graded doses, the resulting survival curves are exponential; for more-complicated cells and organisms, the curves are sigmoid. The exponential curves suggest that the action on each cell is owing to some single primary event (formation of an ion pair, passage of a single ionizing particle, etc.). Sigmoid curves are correspondingly ascribed to multi-event types of action. In some cases, as Wood points out, the curve is exponential or sigmoid, depending on known properties of the cells. This is useful in devising models for the action.

Macromolecules and viruses typically exhibit exponential curves, although their shapes may be modified by various factors. By making simple assumptions concerning the nature of the hypothetical single event, the size of the "target" relevant to the radiation action can be calculated[1,2]; Pollard describes this procedure in detail (p. 273).

In dilute inorganic aqueous solutions, the alteration of the solute (e.g., oxidation of ferrous ion) typically follows zero-order kinetics, which indicates that the energy is absorbed principally by the solvent and that chemical intermediates, formed from solvent molecules, react with the solute. If two or more solutes are present, they compete for the intermediates and thus "protect" each other. Actions involving intermediates are termed "indirect."[3] "Direct" actions are those in which the radiation energy must be transferred to the solute molecules themselves.[4]† Current radiation chemistry identifies some of the intermediates[4,5]; they are OH and H radicals, $H_2O_2$ molecule, and, in the presence of molecular oxygen, $HO_2$ radical.

All of these concepts, derived from work on simple solutions, ramify throughout radiation biology. Indirect action has been demonstrated in many radiation effects on macromolecules in solution.[3] If a preponderant concentration of protective substances is present, a residual effect is observed that is ascribed to direct action. Indirect action has been invoked in analysis of various cellular effects[6]; the concept has been extremely fruitful in suggesting experiments, although in no case has its correctness been established for systems as complex as cells.

Many physical, chemical, and biological factors are known to modify the amount of absorbed radiation energy necessary to produce a given degree of effect. A few that have been found to operate in a wide variety of systems are mentioned here.

One is the physical parameter called linear energy transfer (LET), which is the amount of energy transferred per unit length of track of an ionizing particle to the molecules it traverses. This parameter varies with the square of the charge on the particle and increases as the velocity decreases. With the various kinds of radiation currently available, one can obtain LET values ranging from 0.025 to 25 ev/A, corresponding roughly to average spacings between primary ionizations that range from 4000 down to 4 A. Thus, it is possible to give the same dose to a biological system in various ways: a few ionizing particles with high LET, many particles with low LET, and intermediate values. Results are detailed elsewhere.[7] Not only does LET significantly influence the dose-effect relations in practically all radiobiological actions that have been investigated thoroughly in this respect, but it also operates in "simple" chemical actions, such as the formation of $H_2O_2$ in pure water. Thus, LET not only gives some geometrical notions about mechanism, but also gives some encouragement to use radiation chemistry as a basis for interpretation of radiobiological actions.

Another factor quantitatively influencing a wide variety of actions is molecular oxygen.[8] In all but a few of these actions, radiosensitivity increases with concentration of $O_2$ until a plateau is reached, usually at a value two or three times that observed when the $O_2$ concentration is zero. The basis of the "oxygen effect" is still controversial (cf. Wood, p. 282), but I think it significant that, like LET, it is encountered in radiation effects on simple aqueous solutions. There is also an interrelationship between $O_2$ and LET: the greater the LET produced by the radiation, the less is the influence of $O_2$.

In the foregoing, I have tried to communicate a concept of the present scientific state of basic high-energy radiation biology. Many effects on many diverse biological systems are known and cataloged; however, there is encouraging evidence that the mechanisms leading to these diverse effects have some strong resemblances. No one mechanism has been elucidated yet. On the other hand, several have been investigated intensively by means of the general approaches indicated, and, in a few cases, observed facts have been used as bases for theories which have been successful; i.e., they

† If no solvent is present, the action must be direct. Target theory, in its strict sense, presupposes direct action.

have suggested experiments which in turn have yielded new, significant facts. Good examples of such investigations are given in the three papers which follow.

## BIBLIOGRAPHY

[1] D. E. Lea, *Actions of Radiations on Living Cells* (Cambridge University Press, London, 1955), revised edition.

[2] E. C. Pollard, W. R. Guild, F. Hutchinson, and R. B. Setlow, Progr. in Biophys. and Biophys. Chem. **5**, 72 (1955).

[3] W. M. Dale in *Radiation Biology*, A. Hollaender, editor (McGraw-Hill Book Company, Inc., New York, 1954), Vol. I, p. 255.

[4] A. O. Allen, J. Phys. & Colloid Chem. **52**, 479 (1948).

[5] J. L. Magee, Ann. Rev. Nuclear Sci. **3**, 171 (1953).

[6] R. E. Zirkle and C. A. Tobias, Arch. Biochem. Biophys. **47**, 282 (1953).

[7] R. E. Zirkle in *Radiation Biology*, A. Hollaender, editor (McGraw-Hill Book Company, Inc., New York, 1954), Vol. I, p. 315.

[8] Z. M. Bacq and P. Alexander, *Fundamentals of Radiobiology* (Academic Press, Inc., New York, 1955).

# 33

# Radiation Inactivation of Enzymes, Nucleic Acids, and Phage Particles

ERNEST POLLARD

*Department of Biophysics, Yale University, New Haven, Connecticut*

## INTRODUCTION

IT is worthwhile, before the description of experimental work, to say a word about the motivation of the researches to be described and also something of the people engaged in them. The purpose has been to use ionizing radiation as a powerful, localized, and penetrating agent to study cell structure in relation to cellular function. This purpose was clearly in the mind of the late D. E. Lea, and in many ways we have been continuing lines which he began. The purpose can be usefully directed only if there is some knowledge of the actual character of ionizing radiation action on the three key elements of cellular systems: proteins, nucleic acids, and polysaccharides. Such knowledge is still imperfect, and what is here described is a series of studies, which enable one to make some preliminary hypotheses as to the action of ionizing radiation. The promise of immediate future progress is excellent, and, if research effort on the right scale were forthcoming, a year or two more would see the major features of the pattern properly exposed.

The people in the Yale Biophysics Department who have contributed are Hutchinson, Setlow, Guild, Preiss, Woese, Powell, Forro, Fluke, Jagger, Wilson, Till, and Whitmore. The author writes as representing this group and the work is theirs as much as his in every respect.

In the process of studying the inactivation of these key molecules, the basic problems of radiobiology are also being studied; and since radiobiology cannot be separated from biology, the problems of biology are also involved. This will become apparent, as has been stressed already by Zirkle (p. 269).

## GENERAL CHARACTER OF RADIATION INACTIVATION

The physical action of radiation is complex. All optically allowed molecular transitions presumably are capable of occurring, together with many forbidden transitions. The gradual dispersal of the intense local energy releases, and the accompanying "pre-equipartition" local high-temperature regions, plus the probable effect of local high temperatures, conspire to give the molecular physicist a choice of almost any personally favored mechanism of action. Such choices have been made, and it would be foolish to exempt this paper. In view of this great complexity, the traditional physical introduction is foregone and the experimental results on biological macromolecules are discussed directly.

The sort of experiment which is readily performed requires a set of assay tubes for some enzymes and a color agent that is dark for active enzymes. Since the reaction time is kept constant, the activity of the enzyme shows in the relative color density, and, for an enzyme heavily irradiated with $\gamma$-rays, the activity is clearly low. Such experiments were carried out first by Northrop[1] and by Hussey and Thompson.[2] Similar experiments on viruses are described later. Interestingly enough, the earliest recorded quantitative measurements on bacteriophage were made by the Sinclair Lewis hero, Martin Arrowsmith, which probably reflects the active mind of Paul de Kruif.

The dose-response curve found in these experiments generally obeys the relation

$$\ln(n/n_0) = \text{const} \times \text{dose}. \quad (1)$$

It can be explained statistically very simply, as suggested by Dessauer,[3] Crowther,[4] and Condon and Terrill[5] many years ago, by supposing that $I$ inactivating events per unit volume are distributed randomly and that there is a critical sensitive volume $V$ which may intercept one of these events. Since the average number of events in a volume $V$ is $IV$, then by the Poisson formula the probabilities of 0, 1, 2, 3, etc., events taking place in the volume are $P(0)$, $P(1)$, $P(2)$, $P(3)$, where

$$P(0) = e^{-IV} \qquad P(m) = e^{-IV}(IV)^m/m!$$

$$P(1) = e^{-IV}IV$$

$$P(2) = \frac{e^{-IV}(IV)^2}{2!}$$

These are mutually exclusive events so that $\sum P(m) = 1$, and, accordingly, one can reason thus: If the probability of escape is measured by the ratio of the number left active, $n$, to the number at the start, $n_0$, then for complete escape, $P(0)$, one has

$$n/n_0 = e^{-IV}. \quad (2)$$

If one "hit" can be withstood,

$$n/n_0 = e^{-IV} + e^{-IV}IV. \quad (3)$$

If two hits can be withstood,

$$n/n_0 = e^{-IV} + e^{-IV}IV + \frac{e^{-IV}(IV)^2}{2!}, \quad (4)$$

and so on.

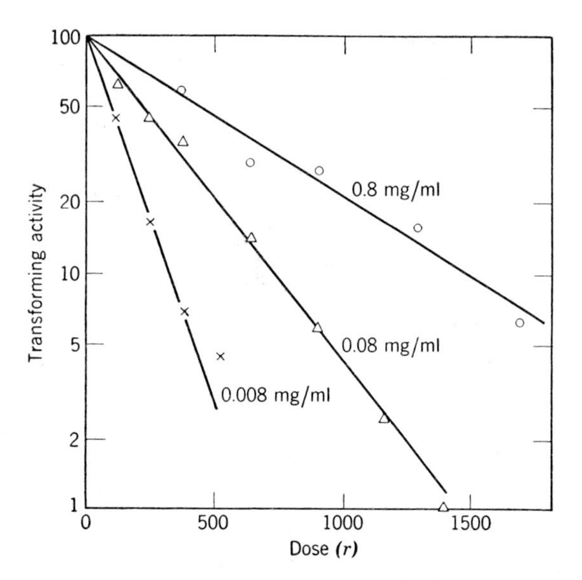


FIG. 1. The loss of transforming activity in pneumococcus DNA preparations in dilute solution. This is typical of irradiation in dilute solution where the majority of the effect is owing to the migration of radicals. That this is so can be seen from the increased inactivation in more dilute solution. Data due to DeFilippes and Guild.[6]

Usually, for enzymes and viruses, one finds the simplest, complete escape expression holding, or

$$\ln(n/n_0) = -IV. \tag{5}$$

Thus, $I$ can be identified somehow with the dose and $V$ with the constant in Eq. (1). The data of Figs. 1 and 2 provide examples of two inactivations which obey this relation. Figure 1, taken from work by DeFilippes and Guild,[6] shows the effect of x-rays on the transforming principle of pneumococcus, which Hotchkiss has shown to be pure DNA, and which was referred to by Rich (p. 191). The ordinate, the activity, is plotted logarithmically *vs* the dose in roentgens, and the relation for a one-hit effect holds. If the inactivation is ascribed to primary ionizations per unit volume, using a conversion factor of $5 \times 10''$ primary ionizations per cubic centimeter in water, an absurd value for $V$ is found, moreover one which depends upon the concentration of transforming principle (TP). The inactivation is due clearly to "activated water," probably free radicals which can migrate and so make $V$ many times larger than the volume of the molecule. In principle, if the radicals in activated water have infinite lifetime, and if the water and TP are pure, there always should be one molecule of TP inactivated per radical. In fact, radicals recombine; Hutchinson and Ross[7] and Smith[8] have estimated their half-life. In very pure water, it is as long as $10^{-4}$ sec. In yeast, it is $10^{-9}$ sec.

As a contrast, Fig. 2 shows some new data, taken by the author and Nancy Barrett, for inactivation of $\beta$-galactosidase. With one exception, the irradiations were in the dry state. All were done with a cobalt $\gamma$-ray source. The black dots show the effect of radiation on enzyme extracted from lactose-adapted bacteria. The relation $\ln(n/n_0) = -IV$ is obeyed. If one uses for $I$ the number of primary ionizations, or clusters of ions, one finds $V = 4.7 \times 10^{-19}$cc. Since one usually is not familiar with the volumes of molecules, this figure can be converted into a molecular weight by assuming a protein density of 1.3 and multiplying the mass of one molecule by Avogadro's number to get the molecular weight. The figure found is 370 000. Estimates for the molecular weight of $\beta$-galactosidase are mostly guesses, but, if one takes the molecule to be spherical, a sedimentation constant of 18, which has been quoted, and which is approximately checked by work in our laboratory by Langridge, gives a molecular weight of 390 000. So, in this case, the dry measurements lead to a statistical analysis in which the volume $V$ may well be the volume of the molecule.

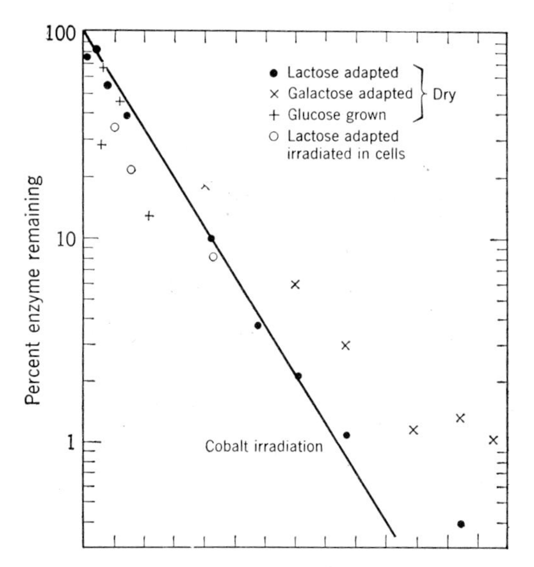


FIG. 2. The irradiation in the dry state of $\beta$-galactosidase grown under various conditions. The characteristic behavior is that an essentially uniform method of inactivation is observed dependent only on the direct hitting of some sensitive region. Irradiation in the wet state is put in for comparison. It is clear that in the wet state inside the bacterium the concentration is so high that inactivation is similar to the dry inactivation.

The "constitutive" enzyme, that small fraction which is present in unadapted cells, behaves about the same as does enzyme from galactose-adapted cells. The sensitivity of the enzyme in cells irradiated in C-minimal medium is not noticeably different.

Thus, one may draw radiation action on a molecule in two ways, as shown in Fig. 3. The shaded part in the center is possibly identifiable with the macromolecule itself, while the dotted lines are owing to migration of active chemical agents. For the purpose of the study of biological systems, the system can be forced very often into conditions where only the shaded part is operative. Sometimes this cannot be done, a familiar restriction to biologists.

**QUANTITATIVE STUDIES ON PROTEINS**

The result that the "inactivation volume" $V$ is closely related to the volume of the molecule first was suggested in work by Lea *et al.* in 1944.[9] In our laboratory, their work has been extended greatly to include many enzymes, albumin, and hormones; contributions by others, notably Fluke,[10] also have been made. A graph summarizing these, owing to Guild, is shown as Fig. 4. A log plot of the "radiation molecular weight" *vs* the accepted molecular weight is shown. There are some clear deviations, but the relation hardly can be overlooked. It is most remarkable and unexpected, and for some time no reasonable explanation could be given. In 1955, the suggestion was put forward[11] that, in a covalently bonded structure, there would be no reason why an initially positive region produced by ionization should be confined to one atom, but rather it should migrate and settle possibly in some point of weakness. Recent work by Gordy[12] using paramagnetic-resonance hyperfine structure has shown that, in irradiated material as extensive as proteins, there can be excited two characteristic patterns, one of which is clearly identified as owing to cysteine. Figure 5 shows some of his results.

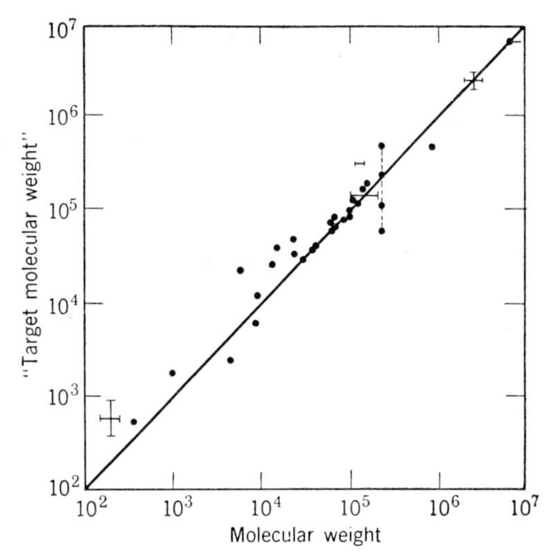


FIG. 4. A plot comparing the observed molecular weight with the target molecular weight, for a wide variety of bombarded substances. This graph has been prepared by Guild and shows the plausibility of the idea that a primary ionization anywhere within the molecule will cause the loss of biological activity. There are exceptions and quite clearly there can be means of influence which will alter the sensitivity of the target, but by and large there is a good relationship.

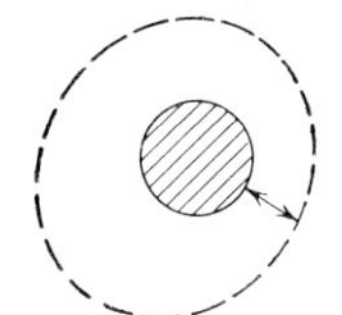


FIG. 3. Indication of the distance of migration of radicals in a cell. It has been estimated by Hutchinson that an increase in distance of 10 to 30 A corresponds to the distance of migration of a radical through a sensitive target. In dilute solution the distance of migration is much greater.

That irradiation, which hardly can have been so intense as to excite the whole set of amino acids, should consistently give such patterns, strongly suggests that migration of the positive charge indeed does occur, resulting in its statistically settling in one or two favored spots. We propose to call this the *primary lesion*.

Subsequently, the protein is exposed to water or to water and oxygen, and the primary lesion now exhibits a reactivity which is called the *chemical action*. This chemical action, by breaking an $-S-S$ bond, or by removing a side chain, gives an altered molecule. We thus feel that, as of 1958, the process is to be regarded as in Fig. 6. On this view, there should be a possibility of modifying the action at two steps, $A$ and $B$, and a lesser possibility at $a$. It has been shown that invertase and catalase are markedly more sensitive at temperatures just below the thermal inactivation region. This is possibly action at $a$ or $A$. Braams, Hutchinson, and Ray[13] have shown that ribonuclease dried in acetic acid is four times as sensitive, and in glucose, twice as sensitive as normally found. A variety of additives have no effect, among them salt and glycine, while glutathione reduces the sensitivity by a factor of two. A sensitivity modification which is eightfold thus can be achieved. Hutchinson[14] also has confirmed the finding of Alexander[15] that dry trypsin irradiated in air is 2.3 times as sensitive as in nitrogen, for $\gamma$-radiation. For deuteron or $\alpha$-particle irradiation, the effect is far less.

The fact that the relation shown in the log plot of Fig. 4 holds, means that most additives which occur in ordinary separation procedures are fairly uniform and not able to produce the effects described above.

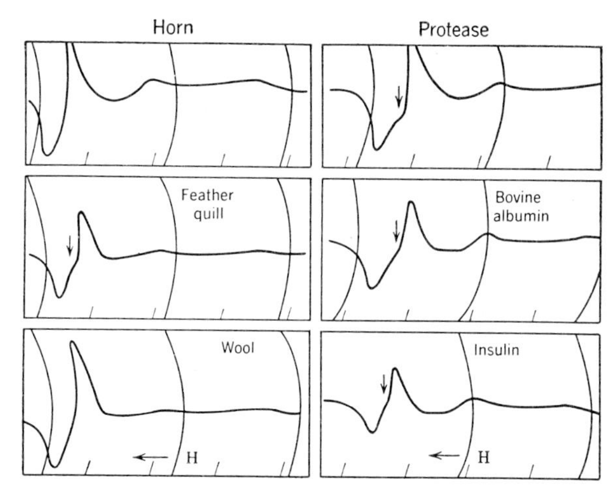


FIG. 5. Reproduction of data due to Gordy showing the paramagnetic resonance in a number of irradiated proteins. It is significant that while a variety of substances are shown quite similar resonance is observed in many of them. This means that there are probably regions in the molecule at which the formation of an unpaired electron is preferred and these can be regions of weakness for chemical action later on [from W. Gordy, W. B. Ard, and H. Shields, Proc. Natl. Acad. Sci. U. S. 41, 983 (1955)].

Ionization ⟶ primary lesion ⟶ chemical action ⟶ altered molecule
$a$ $A$ $B$

FIG. 6.

## QUANTITATIVE STUDIES ON NUCLEIC ACIDS

The transforming principle is pure DNA, and so affords a possible (though not very convenient) measure of the biological activity of DNA. Studies of radiation action on this DNA have been made by Fluke *et al.*,[16] by Marmur and Fluke,[17] by Ephrussi-Taylor and Latarjet,[18] and by Guild and DeFilippes.[19] A graph from the work of these last is shown as Fig. 7. The DNA was irradiated in the dry state and fast protons were used to bombard. The logarithmic relation is not simply obeyed, except at high doses. Two components are found, one with a very large inactivation volume which, when re-expressed as a molecular weight, is between 5 and 15 million, and a homogenous component of molecular weight equivalent of 300 000. The reason for the components is not clear. It may be that a fraction of the long DNA chains are so placed that they can become crosslinked by radiation action, and so cannot enter the pneumococcus to cause transformation. The smaller fraction cannot be crosslinked and is organized as polymers of an essential unit of 300 000 molecular weight, which must intercept an ionization to lose activity.

These studies again show the great sensitivity of DNA. More physical measurements have been made, notably early work by Hollaender and co-workers. More recently, Butler[20] has observed the change in viscosity of thymus DNA when irradiated by x-rays (Fig. 8). It is not plotted logarithmically, but the inactivation, as expressed by the fall in viscosity, is approximately logarithmic, and in the dry state the inactivation corresponds to an equivalent molecular weight of one million. Thus, both physical and biological measures show a strong sensitivity of DNA to ionizing radiation.

The sensitivity in aqueous solution is also high. Guild estimates that a unit of $1\frac{1}{2}$ million molecular weight can be inactivated by 100 ev of energy employed to make radicals in water.[6]

No clear proof of the precise nature of radiation action on DNA is available. Since it is so highly sensitive, and since it is a long thin molecule, the attractive hypothesis, which we adopt, is that it may be either broken, or crosslinked by ionizing radiation, whether dry or in aqueous solution. When this occurs, the result is biologically measurable. It is the biological sensitivity which is really responsible for the radiation sensitivities observed. If one chooses as the "measure" of DNA activity its ability to be digested by DNase, the radiation "molecular weight" is 4000, as was shown by Smith.[21] This is in contrast to the figure of 300 000 for transformation.

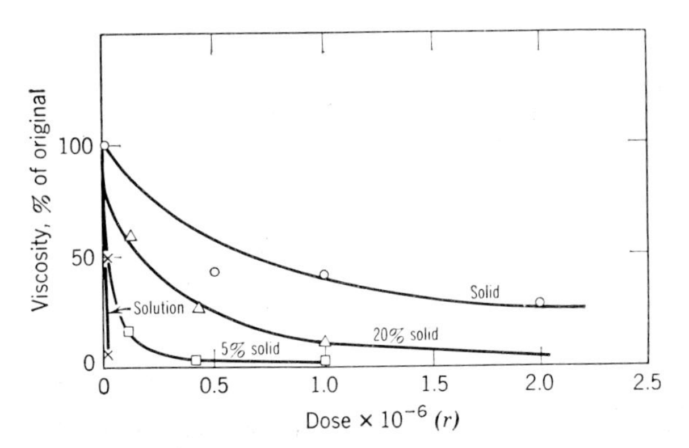


FIG. 8. A reproduction of data quoted by Butler showing the effect of 15 000 000-v electrons on solid and aqueous DNA. The sensitivity of the molecule DNA can be inferred from these data. [from J. A. V. Butler in *Ionizing Radiations and Cell Metabolism* (J. and A. Churchill, Ltd., London, 1956, and Little, Brown and Company, Boston, 1957), p. 59].

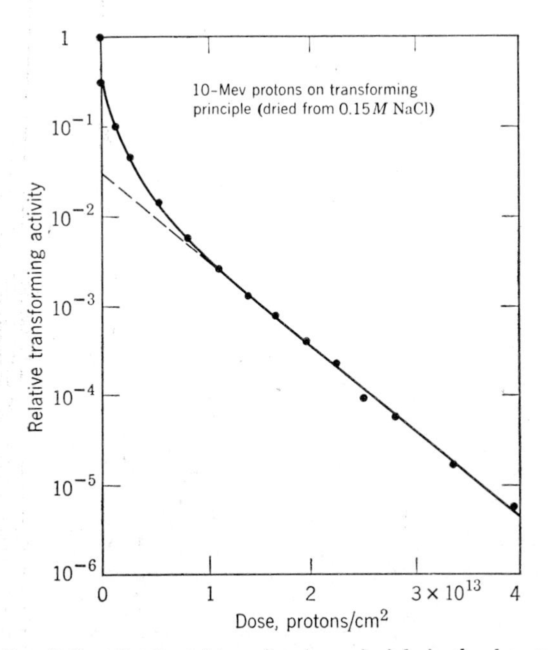


FIG. 7. Inactivation of transforming principle in the dry state as observed by Guild and DeFilippes. There is one very clear single component which corresponds to a molecular weight in the neighborhood of 300 000. For low doses, the behavior is more complex and possibly corresponds to the crosslinking of material which can then not enter the cell [from W. R. Guild and F. M. DeFilippes, Biochim. et Biophys. Acta **26**, 241 (1957)].

## POLYSACCHARIDES

Almost no work has been done on polysaccharides. A dysentery toxin was studied by Caspar and shown to have an inactivation volume corresponding to 11 000 molecular weight. Further work is clearly needed.

## EXAMPLE OF A RADIATION STUDY ON DRY PROTEIN

The statistical analysis of radiation action on an enzyme, antigen, or antibody can give more than a simple estimate of a volume. As an example, it is shown in outline how an enzyme, not purified particularly, can be characterized. Since the purification and measurement by physiochemical means has to be done yet, this work offers some chance to test the predictions of such studies. The enzyme is $\beta$-galactosidase, and one aspect of the study was shown in Fig. 2 where cobalt irradiation was employed. Such radiation is random in volume, and the product $IV$ can be found from the data, yielding $V$, the radiation sensitive volume. Using heavy charged particles, such as deuterons or $\alpha$-particles, a fraction of their ionization, generally about 75%, is confined to narrow tracks. A fraction is spread more widely, as seen in Fig. 9 where a crude representation of the ionization produced by a 4-Mev deuteron is shown. If that fraction of the inactivation of the enzyme due to the on-track ionization can be found, then two quantities can be measured: the area $S$ of the molecule exposed to radiation action, and the thickness $E$. Estimating the fractional inactivation due to the track is subject to some uncertainty, but can be done by using the Bohr

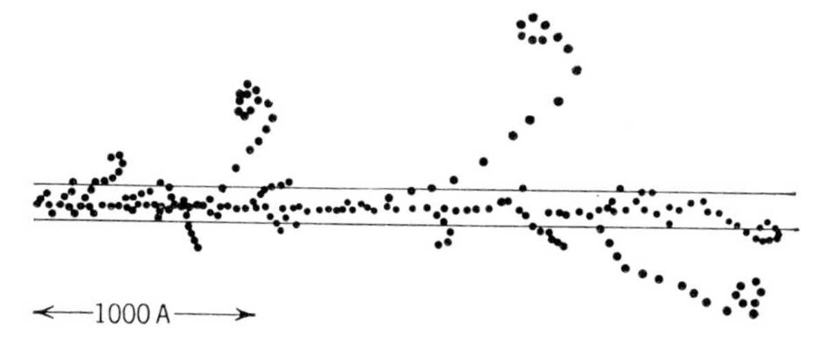


FIG. 9. Representation of the detail of a track of a deuteron. The black dots represent the ionization and the long spurs are $\delta$-rays. In considering the part of the ionization which is along a track, allowance has to be made for the $\delta$-rays which are off the track. The method of doing this has been worked out and permits consideration of just the part which is on the track.

theory of $\delta$-ray production, the Bethe-Bloch energy loss expression, and some modern range measurements for slow electrons in protein.

The effect of 4-Mev deuterons on $\beta$-galactosidase is shown in Fig. 10. The data can be analyzed by a relation

$$\ln(n/n_0) = -SD, \qquad (6)$$

where $D$ is a measure of the number of deuterons per square centimeter and $S$ is a cross section, which after correction, should bear some relation to the area of the molecule. In order to exploit the linear ionization process effectively, such cross sections can be found for different absorbers in the path of the deuteron beam. There results what is known in nuclear physics as a "Bragg curve," as seen in Fig. 11. From such a curve, the variation in cross section with the rate of energy loss of the deuteron or $\alpha$-particle can be found. This is shown in Fig. 12, which also shows the result of applying the correction for the off-track ionization.

From Fig. 12, a molecular area of $8.2\times10^{-13}$ cm$^2$ is deduced, and it is noted, in addition, that, for a rate of

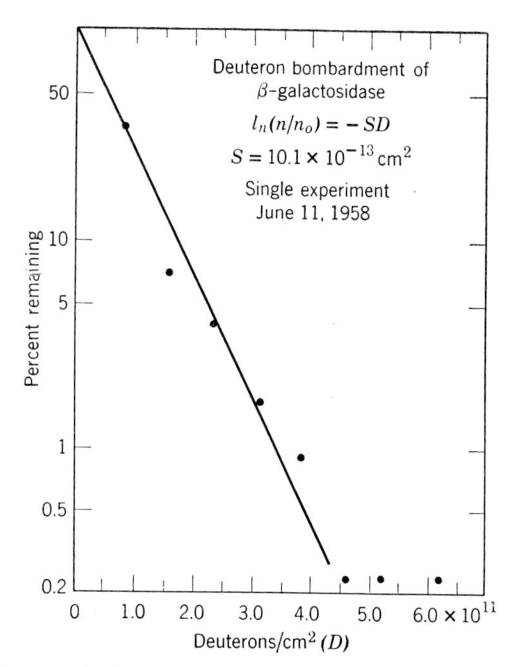


FIG. 10. The inactivation of $\beta$-galactosidase by deuteron bombardment. The relationship as indicated is obeyed and from it a target area can be inferred.

energy loss of 300 ev/100 A, the particles are 63% efficient. If this loss of efficiency is because of the fact that the ionizations are not dense enough along the track and that some "straddling" is occurring, then one should expect a relation

$$S = S_0(1 - e^{-it}), \qquad (7)$$

where $S_0$ is the maximum cross section, $i$ is the ionization per unit track length, and $t$ is the effective thickness. If the primary ionizations require on an average 110 ev one has, for 300 ev/100 A, $it=1$ and $i=2.7$ per 100 A, so $t=37$ A. Thus, one has the following data about

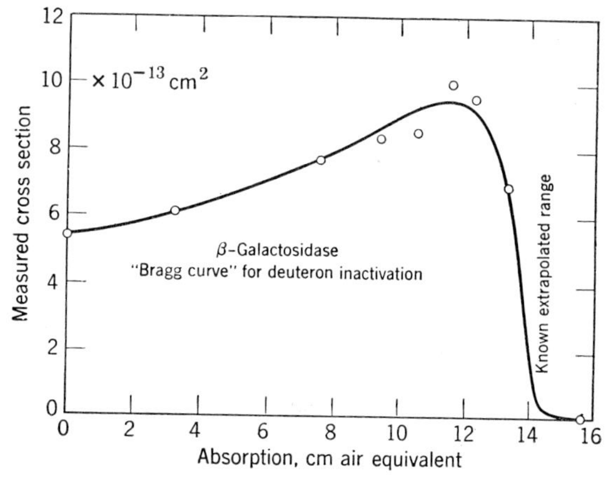


FIG. 11. A "Bragg curve" for the deuteron inactivation of $\beta$-galactosidase. The deuterons were systematically covered with foils of different thickness and their effectiveness measured. It can be seen that there is a rise as the ionization per unit path increases and then an abrupt fall as the end of the range is reached.

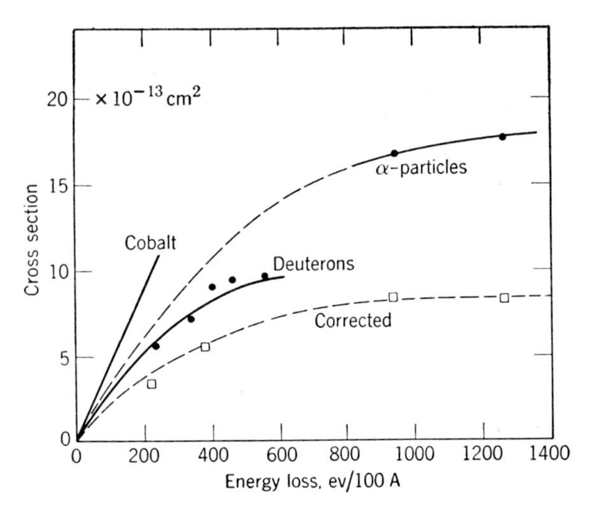


FIG. 12. The plot of the cross section *vs* rate of energy loss for deuteron and α-particle bombardment. A corrected line has been drawn in which allowance has been made for the δ-rays.

β-galactosidase:

$$\text{volume}=4.7\times 10^{-19}\ \text{cc}$$

$$\text{area}=8.2\times 10^{-13}\ \text{cm}^2$$

$$\text{effective thickness}=3.7\times 10^{-7}\ \text{cm}.$$

If it is assumed that the molecule is a cylinder, radius $r$ and length $l$, these can be solved for, from the volume and area, and $r=36$ A and $l=114$ A are obtained. The effective thickness should be $4r/\pi=47$ A, which has to be compared with 37 A, as deduced from Fig. 12. The agreement is fairly satisfactory. Thus, the enzyme should have a molecular weight 370 000, a length of 114 A, and a diameter of 72 A, and so is roughly spherical. Such a prediction should be subject to verification.

TABLE I. Virus properties available for cross-section measurement.

| Type of virus | Property | Rough results |
|---|---|---|
| Plant<br>Animal<br>Bacterial | Infectivity | Close to whole virus |
| Bacterial | Attachment | Very small part |
| Bacterial<br>Animal | Interference | About 1/20 of virus |
| Bacterial | Host killing | About 1/6 of virus |
| Plant<br>Animal<br>Bacterial | Serological affinity | Very small indeed |
| Plant<br>Animal<br>Bacterial | Ability to make antibodies | Not tested |
| Animal<br>Bacterial | Genetic recombination | Not tested |
| Some animal | Hemagglutination | Small |
| Bacterial<br>Animal | Complement fixation | Not tested |
| Bacterial | Latent period extension | About 1/3 of virus |
| Some animal | Enzyme action | Small |

## INACTIVATION OF VIRUSES AS RELATED TO THEIR STRUCTURE

Inactivation of a whole biological system, having a structure and a number of functioning parts, presents a new problem. It is reasonably certain that radiation inactivates at least one of the molecular units of the system, but now inquiry must be made as to whether or not it has any effect on the biological functioning, or rather, any measurable effect. Viruses offer perhaps the the simplest system, yet a virus *is* a system and this must always be remembered. The problem can be rendered schematic as follows:

radiation → molecular inactivation →
reduction of function
↓
biological consequence.

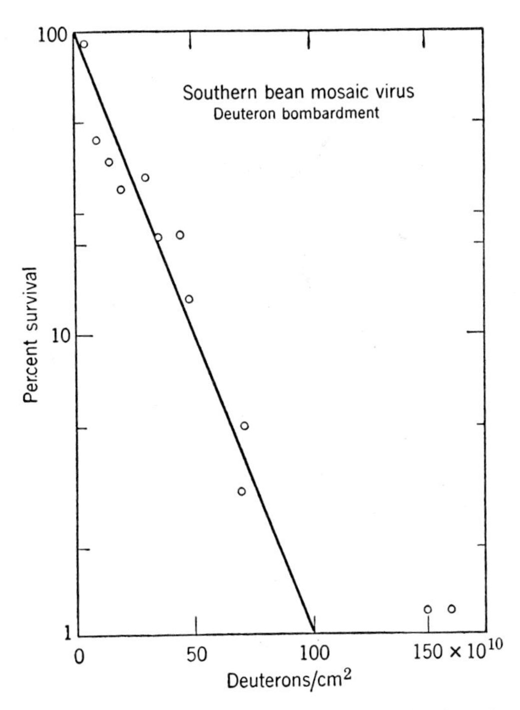


FIG. 13. Inactivation of Southern bean mosaic virus by deuterons as observed by Dimond and the author. From such data the cross section of the virus can be inferred and comes rather close to the observed cross section in electron micrographs.

Thus, when a virus is irradiated and its ability to multiply in the host is studied, one investigates the sum total of: (1) ability to survive in outside world, (2) ability to attach to host cell, (3) ability to invade the cell, (4) ability to multiply in the cell, and (5) ability to return to outside world. Thus, bland statements about "virus inactivation" should be regarded with suspicion.

Before treating separate inactivation studies, brief mention is made of two empirical correlates. The first is that the virus-inactivation cross section measured with heavy particles as bombarding agents (deuterons, $\alpha$-particles) is close to the electron-micrograph cross section. This correlation fails for the large viruses but works for influenza and below. The second is that the inactivation volume measured with $\gamma$-rays is close to the volume of the nucleic-acid content. This is less well established and does not work for polio or influenza, for example, but has been rather widely adopted as a simple interpretation of radiation inactivation. It has some validity, but must be used with caution.

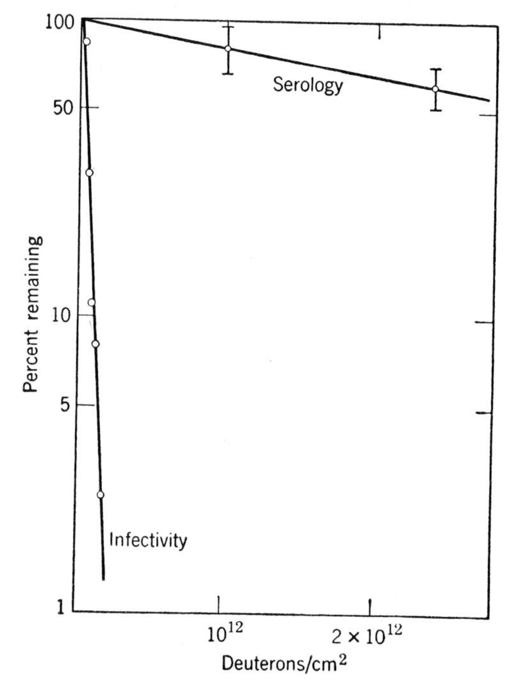


FIG. 14. Differential effect of deuterons on the infectivity and the serological combining power of tobacco mosaic virus. It can be seen that very much more radiation is needed to have any effect on the surface whereas the infectivity is very readily affected. This shows that there is a clear cut difference in the two functions of the virus and enables some estimate of the relative sizes to be made.

A chart is given, in Table I, of the various properties of viruses which are susceptible of study. It is clear to anyone who makes such differential radiation studies that a virus is *not* a simple molecule. In what follows, a few examples of virus studies are given, and in concluding consideration is given to what can be deduced about one virus from this kind of work. Figure 13, shows the inactivation of Southern bean mosaic virus by deuterons as measured by the ability to form local lesions on beans. The familiar relationship is found and the cross section, properly analyzed, corresponds to a spherical particle of radius 150 A, which agrees with electron microscopy.

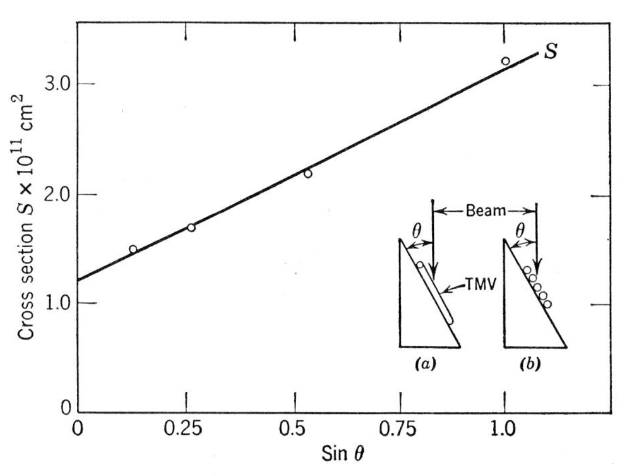


FIG. 15. Proof of the reality of irradiation target. Tobacco mosaic virus which is in long thin rods is irradiated in the manner shown in the insert of the diagram. In the first case, as the angle is varied, the target presents different aspects to the beam. This shows in the measurements as an increase in the cross section [from E. C. Pollard and G. F. Whitmore, Science **122**, 335 (1955)].

Figure 14 shows the effect of deuterons on dry tobacco mosaic virus (TMV) measured in terms of (a) its ability to produce local lesions, and (b) its ability to precipitate antibody to TMV. The loss of precipitating ability is

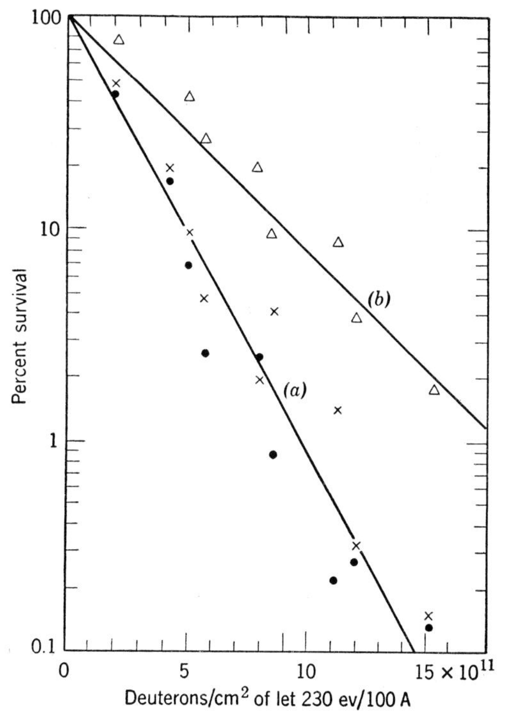


FIG. 16. Effect of deuterons on the cross reactivation of *T*1 bacterial phage as observed by Till and the author. In case (*b*), the bacterium was superinfected with undamaged virus and the fact that part of the virus material could be given from the undamaged virus to the damaged virus is shown in the lower sensitivity. The part which must be intact for cross reactivation to occur is approximately 60% of the whole virus [from J. E. Till and E. C. Pollard, Radiation Research **8**, 344 (1958)].

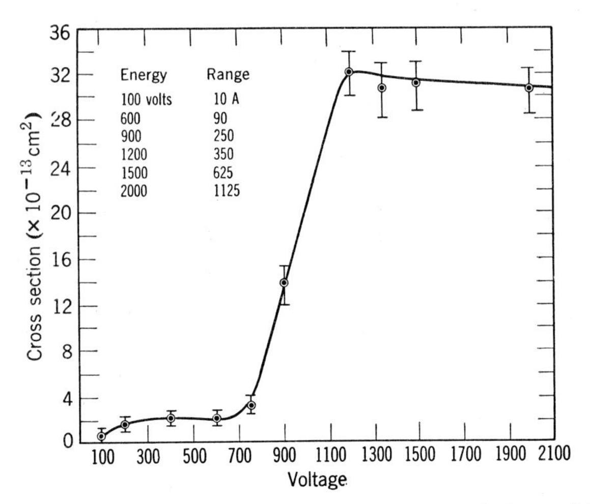


FIG. 17. The inactivation cross section of $T1$ bacteriophage with electrons of low voltage and penetration as indicated. It can be seen that, apart from a small inactivation owing presumably to some effect on the surface of the virus, the majority of the effect only occurs when the energy of the electron is sufficient to penetrate something like 250 A. This means that the sensitive part is inside the virus and also that none of it resides in the tail [from M. Davis, Arch. Biochem. Biophys. **49**, 417 (1954)].

very slight. In fact, it proved that most of the observed effect was on the virus and not on the combination with antibody. If one calculates the sensitive volume for the removal of ability to precipitate with antibody, it corresponds to a molecular weight of less than 20 000. Similar experiments on Southern bean mosaic viruses indicate that there are possibly two active sites of molecular weights 30 000 and 6000, respectively.

In view of all of this statistical reasoning, one sometimes needs more real evidence to bolster one's faith in numbers and deduction. An experiment, by Whitmore and the author,[22] on oriented samples of TMV is shown in Fig. 15. For samples of TMV held pointed at the beam, with varying angles, the cross section is least when the virus particles, which are long and thin, are held pointing at the beam and greatest when perpendicular to the beam. Controls held as at (*b*) showed no orientation effect. This experiment confirms the true physical character of the target.

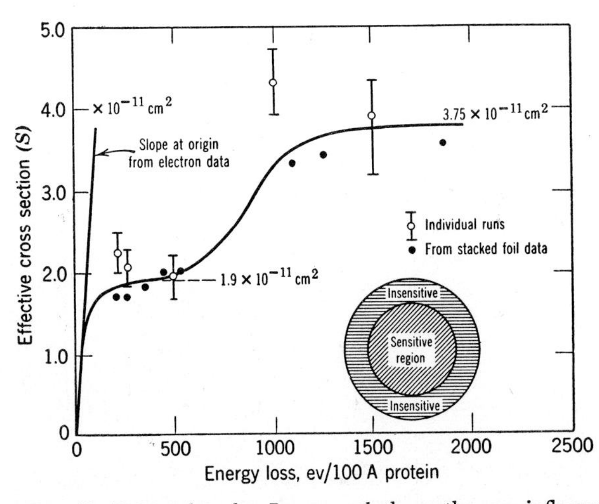


FIG. 18. Data taken by Jagger and the author on influenza virus. As the energy loss in the bombarding particle is increased, the effective cross section varies in the two-step manner. This would seem to fit with the idea that there is an inner sensitive region and an outer insensitive region and so that the virus as an internal structure [from J. Jagger and E. C. Pollard, Radiation Research **4**, 1 (1956)].

## BACTERIOPHAGE EXPERIMENTS

A series of phage properties has been studied. Two are selected as examples. The first is by Till and the author[23] and shows that $T1$ phage, when irradiated and allowed to enter a cell which has a second infection by a genetically different phage, can still "cross reactivate" the second phage, by which is meant that it can donate markers necessary for the progeny of both phages to function on a new host. The cross section for the loss of this ability is 60% of the single infection cross section, and it implies that 60% of the phage DNA is genetic in character while 40% is not. By quite different methods, for $T2$, Levinthal[24] has concluded somewhat similarly. The data are shown in Fig. 16.

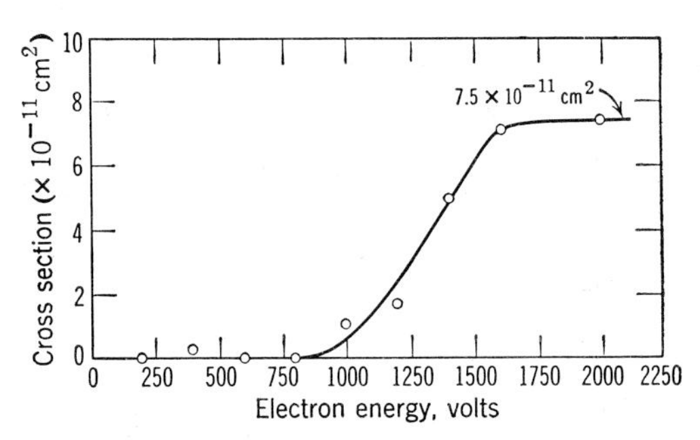


FIG. 19. Effect of low-voltage electrons on the infectivity of Newcastle disease as measured by Wilson and the author. It can be seen that there is no affect on the Newcastle disease until depth of penetration of about 200 A is observed. This means that the outer part of the virus is not concerned with the infectivity but there is a region inside which is actively concerned [from D. Wilson and E. C. Pollard, Radiation Research **8**, 131 (1958)].

A technique of low-voltage irradiation, using electrons of known penetrating power, has been developed by Hutchinson *et al.*[25] Applied to $T1$ (Fig. 17), only when electrons penetrate 125 A can they produce appreciable effect on $T1$. That is, the sensitive material is imbedded in the head, and, since the tail has a thickness of only 150 A, cannot be in the tail.

## NEWCASTLE DISEASE AND INFLUENZA VIRUS

Figure 18 shows an early experiment by Jagger and the author[26] on influenza virus. The infectivity in eggs was measured, and the cross section *vs* energy loss is shown. One way to analyze the data is indicated, in terms of an insensitive region surrounding one which is more sensitive. Figure 19 shows data by Wilson and the author[27] on Newcastle disease virus (NDV) using the low-voltage electron technique. Quite independent evidence for the existence of a protective coat is apparent.

Figure 20 shows the assembly of conclusions about inactivation experiments on infectivity, hemagglutination, and hemolysin ability for Newcastle disease virus. Such a semispeculative figure is representative of what can be done on one virus. So far, no very contradictory evidence has come to the attention of the author which invalidates this model of the virus.

## FURTHER APPLICATION AND OUTLOOK

The study of cellular structure in relation to its function by ionization inactivation can be carried beyond viruses. The effect of radiation on amino-acid uptake in *E. coli* can be analyzed to indicate that ribonucleoprotein particles, or ribosomes, are involved in

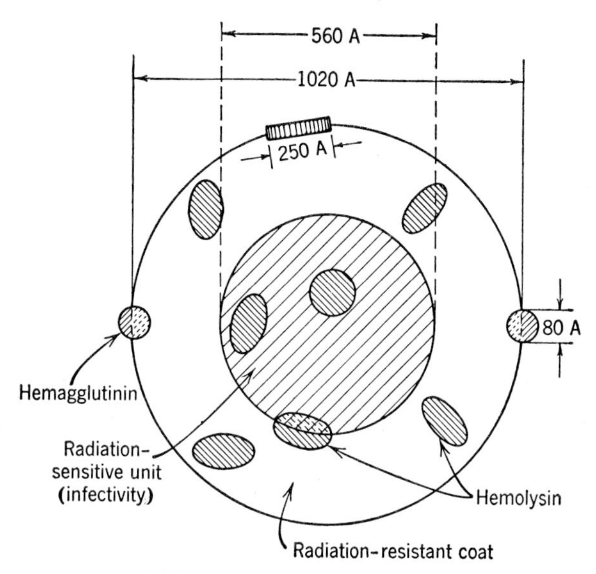


FIG. 20. Sketch of Newcastle disease virus, showing the relative size of several components. Such a drawing is schematic and serves primarily to indicate relative sizes [from D. Wilson and E. Pollard, Radiation Research 8, 131 (1958)].

the uptake. The low-voltage electron technique has been used by Preiss[28] to determine the location of the invertase in yeast cells. Currently, β-galactosidase in *E. coli* is under study and seems to be in the protoplast membrane. The technique is not harder than many others now in use and in the next few years should contribute appreciably to our knowledge of the cell.

## BIBLIOGRAPHY

[1] J. H. Northrop, J. Gen. Physiol. **17**, 359 (1934).
[2] R. G. Hussey and W. R. Thompson, J. Gen. Physiol. **5**, 647 (1923).
[3] F. Dessauer, Z. Physik **12**, 38 (1923).
[4] J. A. Crowther, Proc. Roy. Soc. (London) **B96**, 207 (1924).
[5] E. U. Condon and H. M. Terrill, Cancer Research **11**, 324 (1927).
[6] F. M. DeFilippes and W. R. Guild, Radiation Research (to be published).
[7] F. Hutchinson and D. A. Ross, Radiation Research (to be published).
[8] C. L. Smith, Arch. Biochem. Biophys. **50**, 322 (1954).
[9] D. Lea, K. M. Smith, B. Holmes, and R. Markham, Parasitology **36**, 110 (1944).
[10] D. J. Fluke and E. C. Pollard, Ann. N. Y. Acad. Sci. **59**, 484 (1955).
[11] E. C. Pollard, National Research Council Nuclear Science Series, Report No. 17, 1 (1954).
[12] W. Gordy, W. B. Ard, and H. Shields, Proc. Natl. Acad. Sci. U. S. **41**, 983 (1955).
[13] R. Braams, F. Hutchinson, and D. Ray, Nature (to be published).
[14] F. Hutchinson, Radiation Research **7**, 473 (1957).
[15] P. Alexander, Radiation Research **6**, 653 (1957).
[16] D. Fluke, R. Drew, and E. Pollard, Proc. Natl. Acad. Sci. U. S. **38**, 180 (1952).
[17] J. Marmur and D. J. Fluke, Arch. Biochem. Biophys. **57**, 506 (1955).
[18] H. Ephrussi-Taylor and R. Latarjet, Biochim. et Biophys. Acta **16**, 183 (1955).
[19] W. R. Guild and F. M. DeFilippes, Biochim. et Biophys. Acta **26**, 241 (1957).
[20] J. A. V. Butler in *Ionizing Radiations and Cell Metabolism* (J. and A. Churchill, Ltd., London, 1956, and Little, Brown and Company, Boston, 1957), p. 59.
[21] C. L. Smith, Arch. Biochem. Biophys. **45**, 83 (1953).
[22] E. C. Pollard and G. F. Whitmore, Science **122**, 335 (1955).
[23] J. E. Till and E. C. Pollard, Radiation Research **8**, 344 (1958).
[24] C. Levinthal, Proc. Natl. Acad. Sci. U. S. **42**, 394 (1956).
[25] F. Hutchinson, Ann. N. Y. Acad. Sci. **59**, 494 (1955).
[26] J. Jagger and E. C. Pollard, Radiation Research **4**, 1 (1956).
[27] D. Wilson and E. C. Pollard, Radiation Research **8**, 131 (1958).
[28] J. W. Preiss, Arch. Biochem. Biophys. (to be published).

# 34
# Some Aspects of Cellular Radiobiology*

T. H. Wood

*Department of Physics, The University of Pennsylvania, Philadelphia, Pennsylvania*

IN a preceding paper, Bloom (p. 21) describes experiments in which a stream of protons was collimated into a microbeam a few microns in diameter which could be used selectively to irradiate and inactivate chosen regions of living cells. Such studies should become increasingly important in modern biology as they allow correlations to be made between structure and function. However, before radiation probes can be used to give meaningful answers to such questions, it is necessary that one learn more about those events intervening between the original radiation insult and the observable end effect being assayed. In the complicated cellular systems discussed in the following, lack of information about these intervening processes is almost complete and few unambiguous answers are available.

Considered are two of the primary problems of cellular radiobiology. The first of these—With over-all cellular irradiation, what are the sites of damage as scored by various end-effects?—is a question which, in many cases at least, is answered. The second problem depends upon the answer to the first and is: Must the energy initiating the processes leading to the observable effect be laid down by the ionizing radiation particles directly within the target sites or can a portion of this activation energy migrate from the environment to the sensitive sites?

With the task at hand defined in rather broad terms, I should like to limit it by considering almost exclusively a particular microorganism (the yeast *Saccharomyces cerevisiae*) and by dealing with only one end point of radiation damage, namely, the loss of ability of an irradiated cell to produce a visible colony. Although one could use other end points such as particular genetic mutations, division delay, biochemical abnormalities, changes in cellular permeability, etc., the loss of ability to reproduce is, from a teleological point of view, of greatest importance.

Bloom, in his description of the microbeam experiments, has already given the clue that, in his test system, the nucleus is much more sensitive to irradiation than is the cytoplasm. So, the first question can be rephrased: What is the relative radiosensitivity of the nucleus and the cytoplasm in yeast as scored by colony formation?

It is possible to derive from a single strain of yeast related strains in which the chromosomal material is present in quantities which are multiples of a basic unit; such strains are designated as haploid, diploid, triploid, etc., up to hexaploid.[1] It has been most profitable to compare the radiosensitivities of related yeasts of various ploidies. In a typical experiment, yeast cells are suspended in an aqueous environment under various conditions and irradiated. Aliquots are then plated on nutrient agar on which a surviving cell, by our criterion, will produce a visible colony (about a million cells). The survival of such an irradiated population can then be compared with that of an unirradiated one. A semilogarithmic plot of the surviving fraction of haploid yeast as a function of dose is shown in Fig. 1. Exponential survival is obtained, the slope of the survival curve being a useful index of the radiosensitivity. The inactivation kinetics are first order, indicating that the inactivation process is caused by a single event, presumably an energy transfer to a single molecule. One might now ask the question: What single molecule is necessary for the continued functioning of the cell? One is led almost inevitably to the conclusion that the pertinent molecules must be molecules of the genetic apparatus of the cell.

The x-ray survival curve of diploid yeast is shown in Fig. 2.[2] Two points are noteworthy: (1) these cells are much more radioresistant than are haploid cells, and (2) the inactivation kinetics are of higher order than first. The first attempt to explain mechanistically the comparative response of haploid and diploid strains was that of Zirkle and Tobias.[3] They assumed that radiation damage in this system is owing to the induction of recessive lethal mutations. This is diagram-

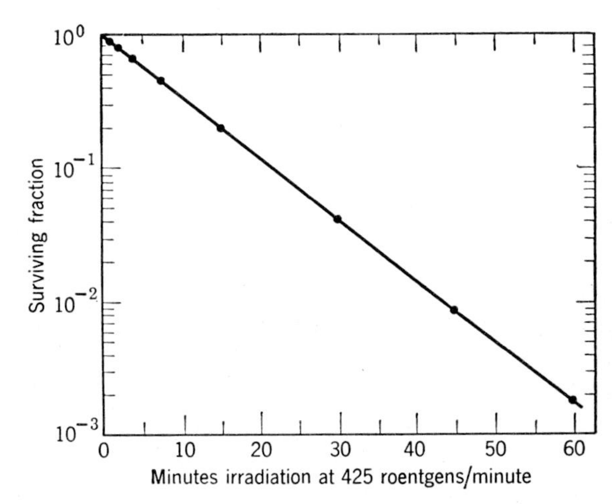


FIG. 1. X-ray survival curve of haploid yeast. Experimental points have errors of 2% or less (approximately the size of the points).

* A portion of the work reported herein was supported by a research grant (C–2390) from the National Cancer Institute, National Institutes of Health, Public Health Service, U. S. Department of Health, Education, and Welfare.

matically represented in Fig. 3 in which all of the chromosomal material of the haploid is shown in a continuous array; this will be duplicated in the diploid and will be present $m$-fold in yeast of ploidy $m$. To successfully inactivate a multiploid cell, corresponding sites on homologous chromosomes such as $c$, $c'$, $c''$, etc., must be simultaneously inactivated. Such a model can successfully correlate haploid and diploid survival and predicts increasing radioresistance for yeast of higher ploidy.

To test this hypothesis, Mortimer has extended these studies to yeast of ploidies up to six and finds that the triploid, tetraploid, pentaploid, and hexaploid are progressively *less* radioresistant, in direct contradiction to the predictions of the recessive lethal model.[1] Figure 4 shows that the doses required to produce the same

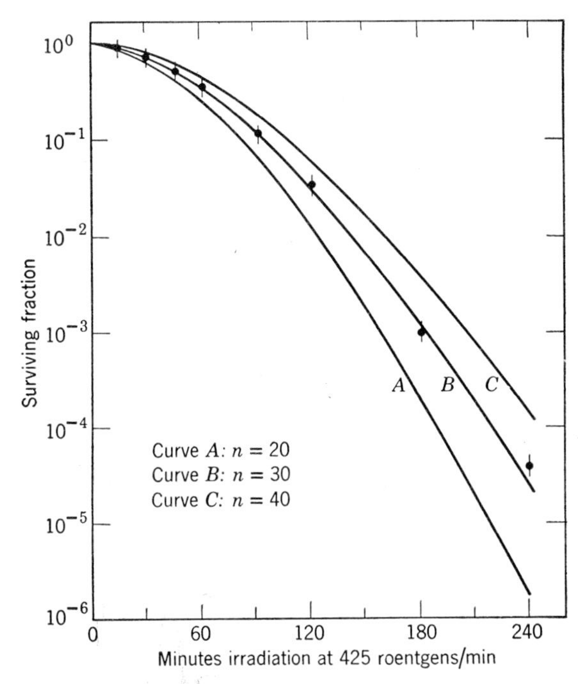


FIG. 2. X-ray survival curve of diploid yeast.

percentage inactivation decrease inversely with the ploidy for ploidies above haploid.[4]

In another series of elegant experiments, Mortimer has shown that x-ray inactivation of haploid yeast may be owing to the induction of either recessive or of dominant lethal mutations.[5] By micromanipulation, individual irradiated cells of various ploidies are mated with unirradiated cells. A diagram of such an experiment with haploid cells is shown in Fig. 5. Large numbers of cells have to be mated as such interpretations must be based on statistical evidence. Operationally, recessive lethal mutations in haploid cells are defined as those not causing loss of ability to divide when mating with an undamaged haploid cell; dominant lethals are defined as those causing loss of ability to divide.

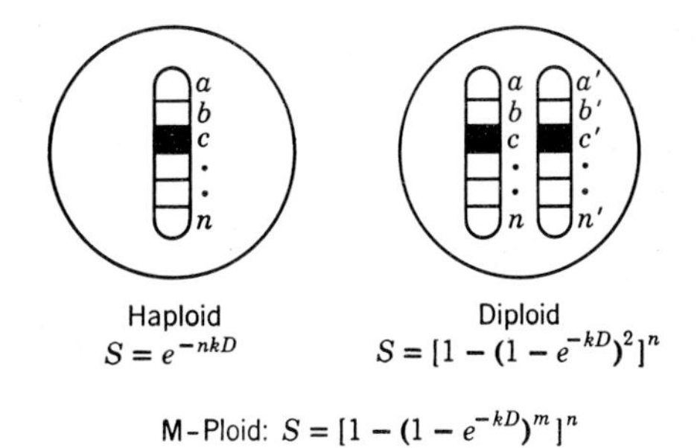


FIG. 3. Diagram for recessive lethal model.

Dominant lethals would be expected to be proportional to the total chromosome length, that is, to the ploidy. The data of Fig. 4 indicate that with yeast cells of higher ploidy the radioresistance is inversely related to the ploidy, or, the radiosensitivity is directly proportional to the ploidy. Thus, with yeasts of ploidy higher than diploid, the predominant path of inactivation as scored by colony formation is interpreted by Mortimer as owing to the induction of dominant lethal mutations.

It is very difficult to relate this dependence of radiosensitivity on ploidy to cytoplasmic damage resulting from irradiation. Cell volume is proportional to ploidy[1] and the diploid is more radioresistant than the haploid. Thus, if radiation damage were the result of cytoplasmic effects, the radioresistance should continue to increase with ploidies above two in contradiction to the observed results.

Additional evidence that the predominant path of radiation damage is by means of genetic damage is afforded by Tobias' experiments with defective diploid cells in which recessive radiation damage is propagated to the daughters of irradiated cells (p. 289).

In general, then, with yeast loss of ability to form

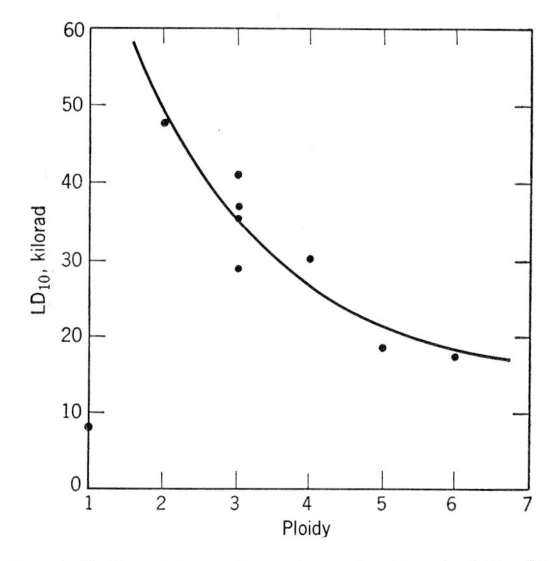


FIG. 4. Radioresistance of yeast as a function of ploidy [from R. K. Mortimer, University of California Radiation Laboratory Rept. 3902 (August, 1957)].

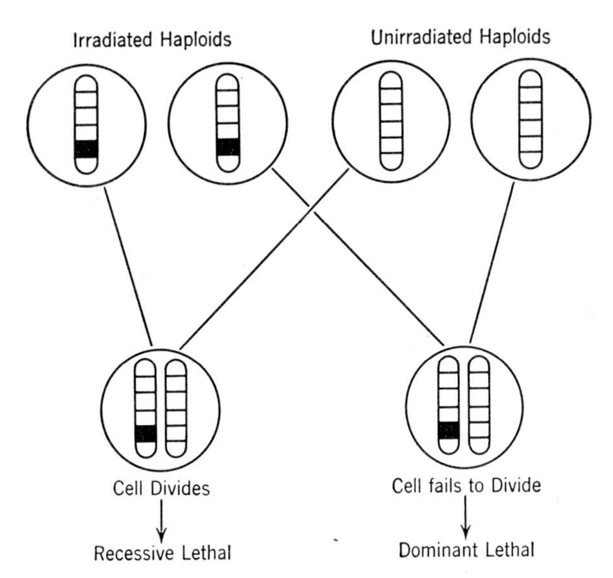


FIG. 5. Operational scheme for differentiating between recessive and dominant lethal mutations induced by x-rays in yeast.

colonies following x-irradiation can be explained in terms of genetic effects. The microbeam experiments of Zirkle and Bloom,[6] experiments with eccentrically located nuclei,[7,8] and many other experiments all indicate that nuclear inactivation is of primary importance in many types of cells. The relatively small number of cases to the contrary that have been reported (e.g., Duryee with frog eggs,[9] Bacq *et al.* with alga[10]) are interesting exceptions to this rule.

Turn now to the question of whether the radiation energy causing cellular inactivation is laid down directly in the molecules whose modification causes cellular inactivation or whether a portion of this energy can migrate from the molecular environment to the affected sites. Pollard (p. 273) has outlined in some detail the processes of direct-action inactivation in dried materials. There can be no doubt that such inactivation plays a role in cellular radiobiology. However, classical direct-action theory in its simplest form is unable to account for various experimental results, in particular those concerned with modification of radiation effects by various environmental factors such as oxygen concentration and phase state.[11] Furthermore, with the development of radiation chemistry and the demonstration of *in vitro* inactivation of various cellular constituents by free radicals produced by the irradiation of water, it became necessary to consider the importance of such processes in cellular systems. There thus evolved the indirect action hypothesis in which free radicals produced by the irradiation of water were postulated to bring about at least some of the initial radiation damage.

The most successful formulation of these concepts was in the migration model of Zirkle and Tobias.[3] In this model, a proper averaging was made of the contributions to inactivation by various species of free radicals, produced at various distances from the target molecules, under various experimental conditions. A single effective collision between an energy carrier and a critical molecule was postulated to be effective in bringing about molecular change in conformity with the first-order inactivation kinetics experimentally observed in many of the simple unicellular systems. This model also includes as a part of itself the direct-action type of inactivation.

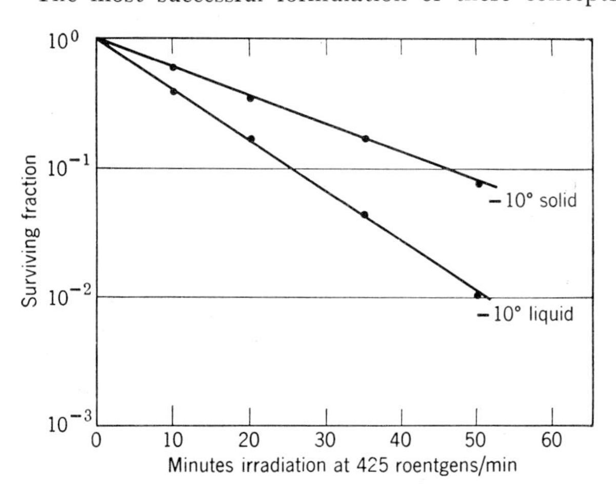


FIG. 7. X-ray survival curves of haploid yeast irradiated at $-10°C$, liquid and solid phases.

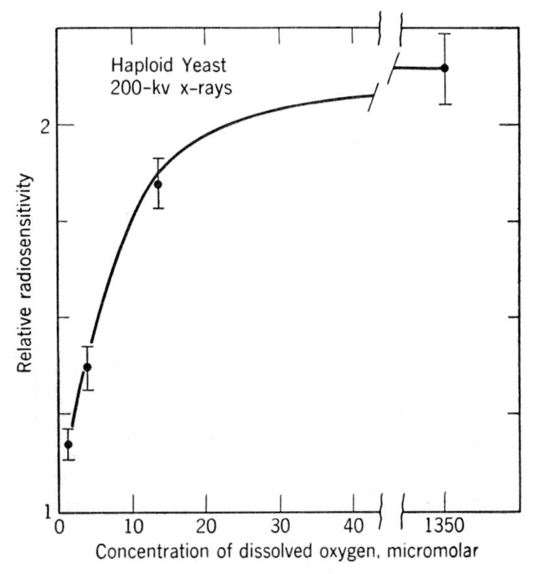


FIG. 6. Variation with oxygen concentration of x-ray sensitivity of haploid yeast [from P. Howard-Flanders and T. Alper, Radiation Research 7, 518 (1957)].

One of the most studied of the modifying parameters is the oxygen concentration in the cellular suspension during irradiation. Haploid yeast cells irradiated in the absence of oxygen are approximately twice as resistant as those irradiated under aerobic conditions (Fig. 6[12]). The oxygen effect is operative at oxygen tensions as low as 10 micromolar. A comparable oxygen effect has been found for a tremendous variety of materials using a considerable number of end results. Many chemicals have been found to decrease the radiosensitivity, but it is not yet certain that such protection acts other than by modifying the existing oxygen concentration. The phase state of the test system, whether frozen or liquid, greatly modifies radiosensitivity.[13] In Fig. 7 are shown survival curves for haploid yeast irradiated at −10°C in the liquid phase (supercooled) and in the frozen state.

Until quite recently, the modification of radiation response by such environmental factors as oxygen concentration, various chemicals, or phase-state change was considered as evidence that at least the modifiable portion of the radiation effect was mediated through indirect-action mechanisms. A simple formulation of indirect action is diagrammed in Fig. 8. Oxygen modification can be explained easily in this model by the elimination of those free radicals depending on the presence of oxygen for their formation. Modification by chemicals could be owing to competition for the free

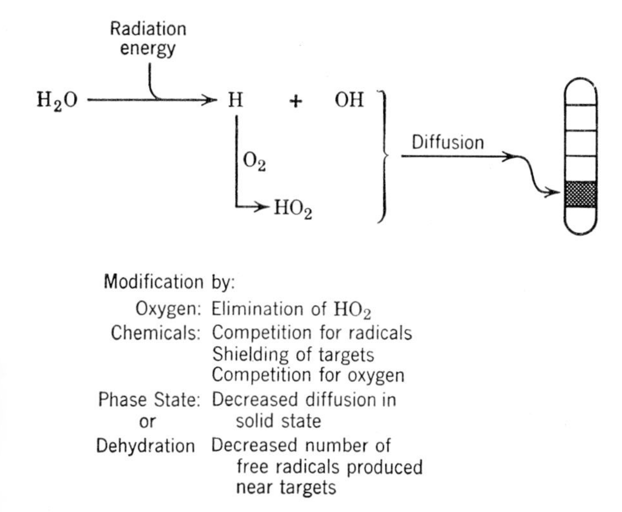


FIG. 8. Diagram of simplified indirect-action mechanisms.

radicals between the added chemicals and the target sites, to diminution of the oxygen effect, or to masking of sensitive molecular side chains (e.g., sulfhydryl groups). The phase-state change would operate by either removing water from availability for these reactions or by greatly decreasing the diffusion of free radicals. On the other hand, such modifying agents would not be expected to greatly modify the ionization levels of the target molecules and modification is difficult to explain using the concepts of classical direct action.

Alper, Howard-Flanders, and Alexander have questioned the assumption that modification is *a priori* evidence that an indirect action mechanism is operative.[12,14,15] For example, Alexander has shown that crystalline trypsin also shows an oxygen effect.[14] However, this evidence is not conclusive as this system still contains about 5% water which might play an important role in the oxygen effect. They suggest that the

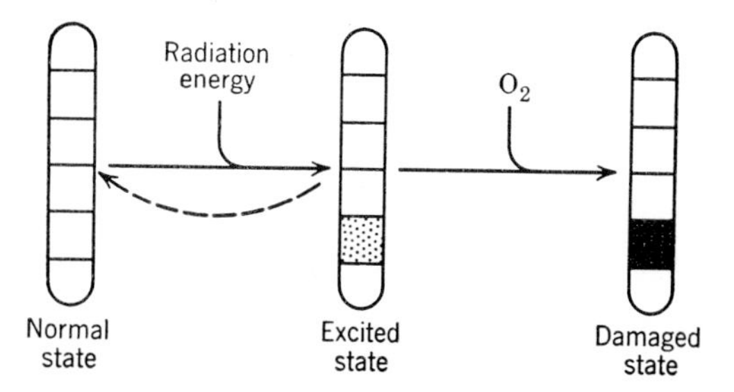


FIG. 9. Diagram of oxygen modification of radiosensitivity

oxygen effect is due to reaction of an excited state with oxygen to produce an irreversibly damaged site (Fig. 9). Recently, Hutchinson has interpreted his results on the *in vivo* inactivation of yeast enzymes by x-rays in terms of the migration model and has found that a diffusion distance for the pertinent inactivating free radicals of about 30 A is involved, a distance an order of magnitude less than had previously been estimated.[16] At the lowest oxygen concentrations at which the oxygen effect is important, the distance between oxygen molecules is about 700A. It is, therefore, unlikely in Hutchinson's system that there would be opportunity of reaction between oxygen and hydrogen free radicals produced by irradiation to give rise to the hydroperoxyl free radical ($H+O_2=HO_2$), the generally assumed mechanism for the oxygen effect. Other objections to the hydroperoxyl free radical as an important factor in cellular radiobiology have been discussed elsewhere in detail.[12] Thus, while modification of radiation damage in the truly dried state has not been demonstrated, the generally accepted mechanism for the oxygen effect via the hydroperoxyl free radical is subject to question. However, the mechanism diagramed in Fig. 9 could also be operative for indirect action events, and at the present time this seems to be the best operational hypothesis for the oxygen effect. Recently, estimates have been made on the lifetime of the excited states that may be involved in the oxygen effect. At the low oxygen concentrations necessary for the oxygen effect, there are approximately $10^6$ collisions/sec between oxygen molecules (assuming that they are free to diffuse within the cell) and a hypothetical target having a molecular weight of 100. Thus, it is necessary to assume that the excited state produced by the radiation energy must have a lifetime of at least a microsecond in length. Work by Howard-Flanders and Moore with bacteria in which cells irradiated under either aerobic or anaerobic conditions were quickly switched to the opposite condition indicates that the lifetime of the oxygen-sensitive excited state is less than 20 msec.[17]

Howard-Flanders has also found that nitric oxide can be used to mimic oxygen.[18] He suggests that the action of oxygen and nitric oxide in increasing radiosensitivity may be associated with the fact that both have unpaired electrons in $\pi$ orbitals.

The change in radiosensitivity associated with the

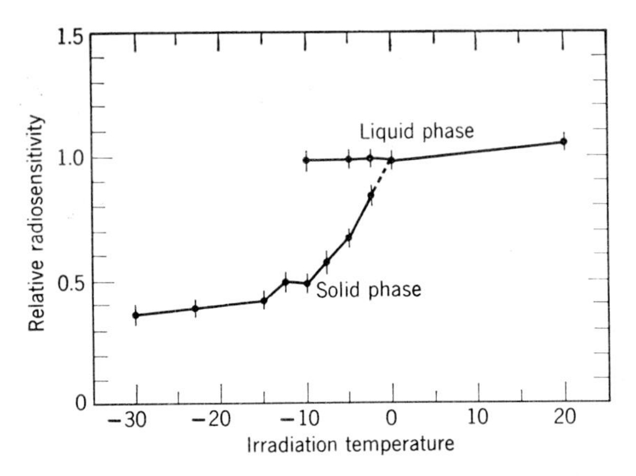


FIG. 10. Dependence of radiosensitivity on irradiation temperature: −30°C to 0°C, liquid and solid phases.

phase-state change from liquid to frozen cells has been mentioned already. Figure 10 shows the radiosensitivity over a temperature region down to −35°C in both the liquid and solid phase.[13] The progressive decrease in radiosensitivity with irradiation temperatures below 0°C suggested to us that the decrease might be associated with a progressive freezing of cellular water. To test this hypothesis, microcalorimetric studies were performed on frozen yeast suspensions. Figure 11 shows that cellular water is progressively frozen down to about −20°C, this region of constant freezing correlating nicely with the region of constant radiosensitivity. A certain fraction of the cellular water, 9%, is nonfreezable; this is one definition of bound water.[19]

A. M. Rosenberg, in our laboratory, has developed an interesting way to control effectively the amount of free water within yeast.[20] Very high concentrations of various chemicals (ethanol, glycerol, glucose, etc.) can be used to either dehydrate the cell or to tie up cellular water which may play a role in radiobiological processes. Figure 12 shows the progressive decrease in radiosensitivity with increase in glycerol concentration. The radioresistance increases by about a factor of four with 6.9 M glycerol. Additional protection is afforded by

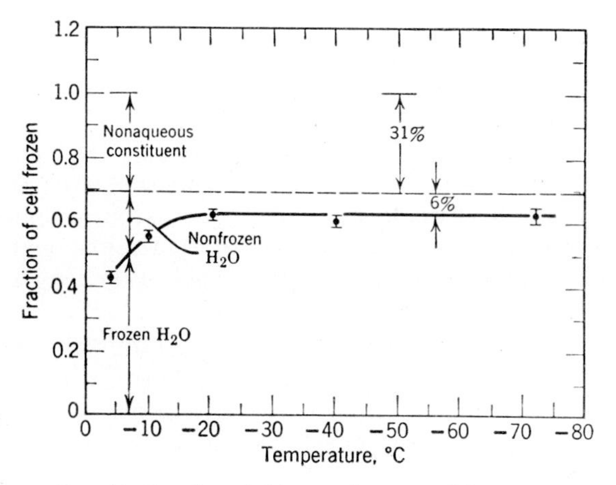


FIG. 11. Fraction of the normal yeast cell frozen as a function of temperature.

anoxic conditions, the maximum radioresistance being about 5 times that typical of aerobic cells in standard suspensions. We have been able to verify by several methods that glucose, for example, dehydrates the cell approximately to the degree expected from simple theory. Figure 13 summarizes the points mentioned in the foregoing—namely, that there is a correlation between the residual fractions of radiosensitivity and of unfrozen cellular water with respect to freezing temperature, and between the residual fractions of radiosensitivity and of cellular volume with respect to osmotic concentration. Furthermore, these two types of effective water removal are apparently closely related.

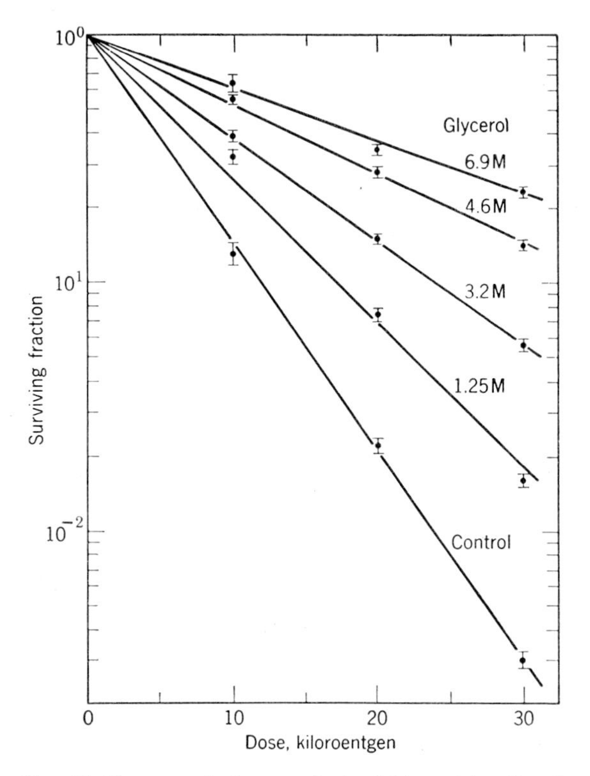


FIG. 12. X-ray survival curves for haploid yeast irradiated in different concentrations of glycerol.

From the results in Figs. 10 and 11, one might expect that the temperature dependence of yeast would not greatly change for irradiation temperatures below −35°C. However, when yeast suspensions are frozen at a rate characteristic of the irradiation temperature, an apparently anomalous behavior is observed below −35°C, the radiosensitivity progressively increasing in the region from −35°C to −72°C where a sensitivity is observed approximately equal to that typical of the liquid phase (Fig. 14[21]). This behavior can be explained by an examination of the physical processes that occur within the cell during freezing. Meryman and Platt[22] have shown that, in mammalian liver cells, the rate of freezing (the speed at which the ice-water boundary advances) is of great importance in determining whether

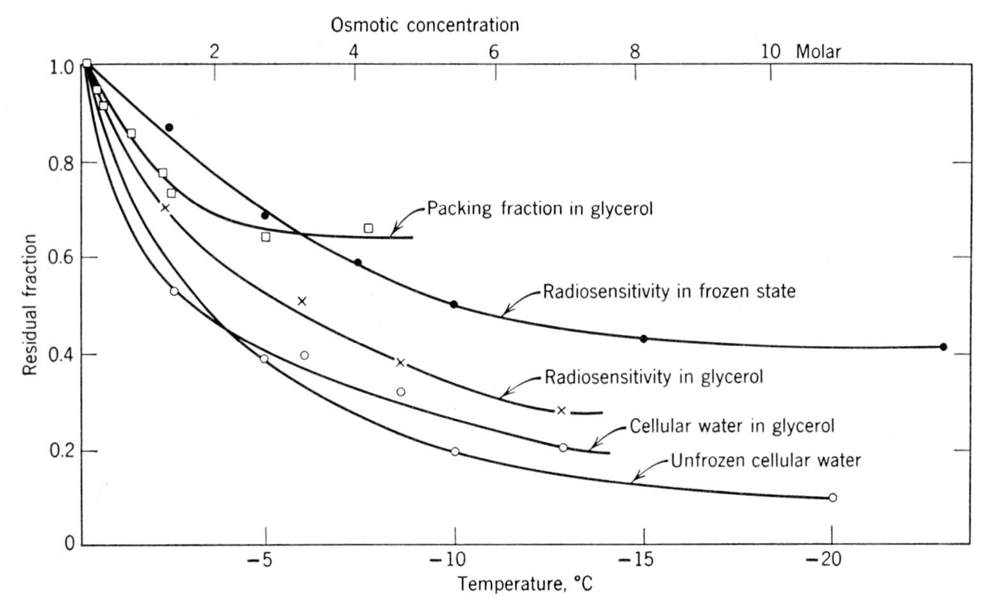


FIG. 13. Relationship of the radiosensitivity of yeast to the loss of cellular water. Glycerol concentrations are to be read at the uppermost abscissa, freezing temperatures at the bottom. The curve labeled "packing fraction in glycerol" shows the relative volume occupied by a standard amount of yeast after centrifugation in various concentrations of glycerol.

the cellular water is frozen within or without the cell. At low rates of freezing (less than 1.5 mm/min), the increase in external osmotic pressure owing to the freezing of the external medium causes water to move outward across the cell membrane where it is then frozen. Thus, the cell is effectively dehydrated. At high rates of freezing (greater than 1.5 mm/min), however, the cellular water freezes within the cell before the increase in external osmotic pressure is effective in dehydrating the cell. In the yeast system used here, this critical velocity of freezing (1.5 mm/min) occurs at −35°C; thus, for cells frozen at temperatures above −35°C, dehydration would occur. For freezing at temperatures below −35°C, some internal freezing of water would occur before dehydration would be complete. These mechanisms are diagramed in Fig. 15.

The results in Fig. 14 are interpretable in light of the migration model using the foregoing concepts of cellular freezing. On irradiation, free radicals are formed in all of the cellular water, whether liquid or frozen, whether within or without the cell, approximately in equal number per unit volume of water. We further assume that free radicals (or their products) produced by the irradiation of the ice are largely trapped within the ice crystal. The investigations of Stewart and Ghormley[23] give credence to this assumption. Now, if the frozen cellular water is largely exterior to the cell, the free radicals liberated on thawing will be effectively diluted by the external medium and the probability of an effective interaction between them and the sensitive regions of the cell will materially decrease. On the other hand, if a significant amount of the cellular water is frozen within the cell, the free radicals released on thawing will be in much the same positions they would have been in on irradiation of the unfrozen cell. Thus, if the radiobiologically significant cellular water is frozen within the cell, the radiosensitivity under these conditions would be not dissimilar to that typical of the liquid phase.

There are three critical tests for the foregoing hypothesis. If radiation sensitivity is primarily dependent on the degree of external dehydration rather than on the absolute amount of cellular water frozen (i.e., the temperature), then slow freezing to temperatures lower than −35°C by stepwise freezing at −5°C, −10°C, −20°C, etc. should dehydrate the cells to the same degree as freezing at −35°C. The resulting radiosensitivity should be the same as that obtained with cells frozen at −35°C; from Fig. 14 (slow $O_2$ curve) this can be seen to be the case. Secondly, if fast freezing at

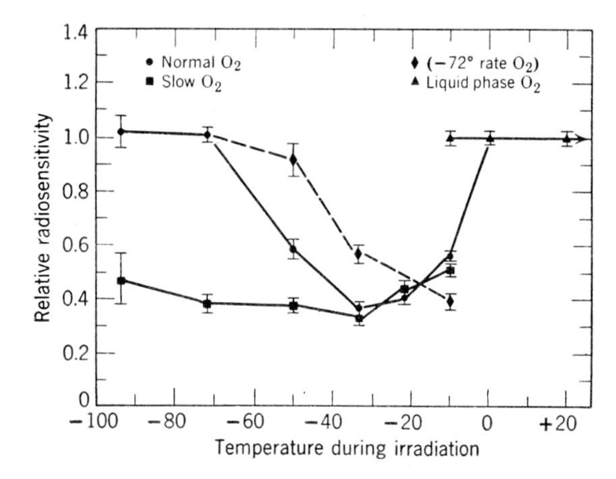


FIG. 14. Relative radiosensitivity of haploid yeast as a function of temperature, phase state, and rate of freezing.

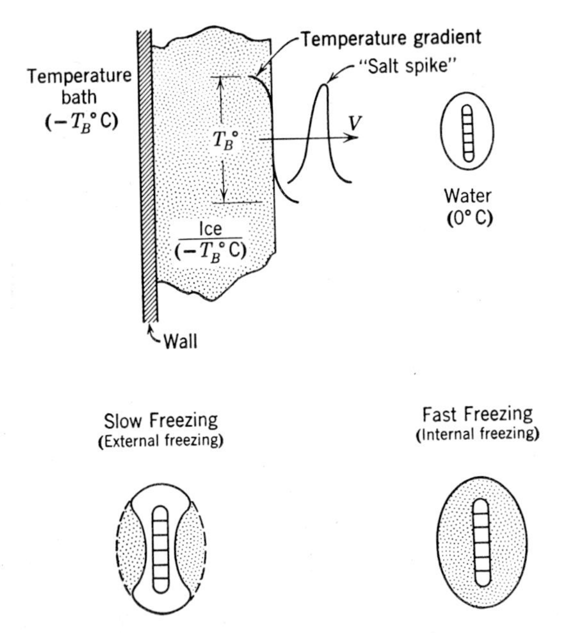


FIG. 15. Mechanisms for fast and slow freezing in yeast.

−72°C internally freezes the radiobiologically significant water, then freezing at even lower temperatures, say, −95°C, should produce the same radiosensitivity as that obtained at −72°C and in the liquid phase; this was also found to be the case. Thirdly, if fast freezing at −72°C freezes the radiobiologically significant water within the cell, then subsequent x-irradiation of such cells at temperatures higher than −72°C should produce the same effect as that obtained at −72°C. This is true at −50°C, but at higher temperatures there is a progressive loss of radiosensitivity. This decrease in sensitivity with increasing temperature for these prefrozen cells is qualitatively similar to the rate of disappearance of hydrogen peroxide produced by irradiation of ice that has been observed by Stewart and Ghormley and is probably the result of back reactions among free radicals not firmly trapped in the ice.[23]

No successful explanation of the phase effect in radiobiology based on the direct action model has been advanced (see Wood[24] for review). Probably the best way to differentiate between direct-action mechanisms and indirect-action mechanisms is by use of the phase effect. In the haploid-yeast system, 60% of the radiation damage under either aerobic or anaerobic conditions is capable of modification by the change in phase from liquid to solid state. Thus, if the above interpretations are correct, no more than 40% of the over-all radiation damage is owing to direct action. If oxygen modification operates through an indirect-action mechanism, then our results both with frozen, anoxic cells and with glycerol-dehydrated, anoxic cells indicate that at least 80% of the x-ray inactivation of this system is owing to indirect action.

In summarizing, it is difficult to make any broad statements that have general applicability for cellular radiobiology. The modification by oxygen of almost all radiobiological responses studied indicates the necessity of understanding the associated mechanisms. The generally accepted hydroperoxyl-radical mechanism is certainly questionable. At the present, the best operational tool to differentiate between direct and indirect types of inactivation is the change in radiosensitivity with phase state. The success of the migration model in dealing with modification of radiosensitivity by various physical and biological parameters is most encouraging and this model, even though its details may need revision, has been most valuable in suggesting new experiments.

In studying a biological unit as complicated as a cell, it is desirable to approach the problem from both the molecular level and from the gross phenomenological level. Phenomenological studies similar to those described here have greatly profited from work done with simpler systems such as those described by Pollard (p. 273). On the other hand, work with more complicated cellular systems has given a direction to molecular studies and has focused attention on the necessity for working with materials in an aqueous environment It is to be hoped that it will not be too long before these two types of studies have overlapped.

## BIBLIOGRAPHY

[1] R. K. Mortimer, Radiation Research **9**, 312 (1958).
[2] T. H. Wood, Proc. Soc. Exptl. Biol. Med. **84**, 446 (1953).
[3] R. E. Zirkle and C. A. Tobias, Arch. Biochem. Biophys. **47**, 282 (1953).
[4] R. K. Mortimer, University of California Radiation Laboratory Rept. 3902 (August, 1957).
[5] R. K. Mortimer, Radiation Research **2**, 361 (1955).
[6] R. E. Zirkle and W. Bloom, Science **117**, 487 (1953).
[7] R. C. von Borstel and H. Moser in *Progress in Radiobiology*, J. S. Mitchell, B. E. Holmes, and C. L. Smith, editors (Oliver and Boyd, London, 1956), p. 211.
[8] R. E. Zirkle, J. Cellular Comp. Physiol. **2**, 251 (1932).
[9] W. R. Duryee, J. Natl. Cancer Inst. **10**, 735 (1949).
[10] Z. M. Bacq, F. Vanderhaeghe, J. Damblon, M. Errera, and A. Herve, Exptl. Cell Research **12**, 639 (1957).
[11] D. E. Lea, *Actions on Radiations on Living Cells* (Cambridge University Press, London, 1955), revised edition.
[12] P. Howard-Flanders and T. Alper, Radiation Research **7**, 518 (1957).
[13] T. H. Wood, Arch. Biochem. Biophys. **52**, 157 (1954).
[14] P. Alexander, Radiation Research **6**, 653 (1957).
[15] T. Alper, Radiation Research **5**, 573 (1956).
[16] F. Hutchinson, Radiation Research **7**, 473 (1957).
[17] P. Howard-Flanders and D. Moore, Radiation Research **9**, 422 (1958).
[18] P. Howard-Flanders, Nature **180**, 1191 (1957).
[19] T. H. Wood and A. M. Rosenberg, Biochim. et Biophys. Acta **25**, 78 (1957).
[20] A. M. Rosenberg, Ph.D. thesis, University of Pennsylvania (June, 1958).
[21] T. H. Wood and A. L. Taylor, Radiation Research **7**, 99 (1957).
[22] H. T. Meryman and W. T. Platt, Naval Medical Research Institute Project NM 000 018.01.08 (January 3, 1955).
[23] A. C. Stewart and J. A. Ghormley, Radiation Research **1**, 230 (1954).
[24] T. H. Wood, Ann. Rev. Nuclear Sci. **8**, 343, (1958).

35

# Effects of Radiations on Populations of Cells and Multicellular Organisms

CORNELIUS A. TOBIAS

*Donner Laboratory of Biophysics and Medical Physics,* University of California, Berkeley 4, California*

THE animal body is an assembly of cells. In considering the effects of radiation on such an assembly, the relationships between the individual cells and their assembly must be defined and, moreover, performance criteria must be found by which to characterize both the radiation effects on individual cells and the relation of these effects to the performance of the cell population.

There are three important principles which characterize the biological effects of penetrating radiations:

A. Radiation effects at the subcellular level are of a statistical nature. A single ionizing particle may lead to a decisive single radiobiological event in one of the many DNA nucleoprotein molecules or in another essential molecule. As a result, many different radiobiological events occur in the cells of an irradiated population.

B. In establishing the effect of irradiation on a cell population, the measured results are determined by the defined criterion for an experimental test. Because of the replication of genetic alterations in mitotic cell division, the criterion used for measurement frequently involves the selection of certain more-fit cells and the rejection of others, leading to end results different from what might be expected from statistical averaging of the entire population.

C. Radiobiology is an important basic tool because its fundamental interactions occur at the atomic-molecular level in essential sites of the cell, sometimes in locations not accessible to the ordinary tools of physics and chemistry. These interactions are then amplified to a level involving the entire cell—i.e., by a factor of about $10^{12}$. Similarly, radiobiologic events that happened to a single cell or to a few cells may be amplified to the entire animal or human organism affecting the fate of as many as some $10^{14}$ cells.

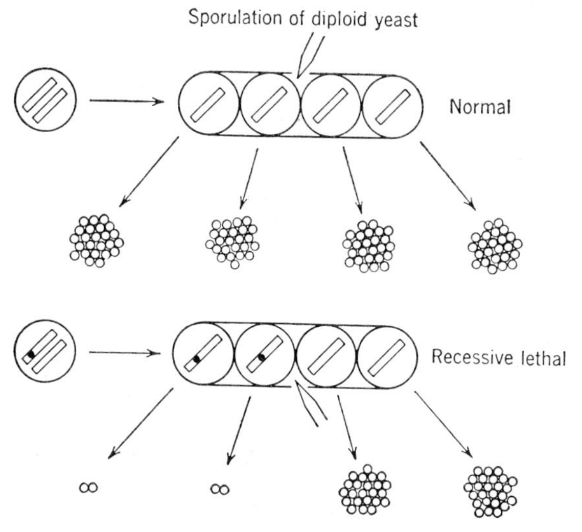


FIG. 1. Sporulation of normal and recessive lethal diploid yeast cells.

### CELL-DIVISION DELAY AND INHIBITION

Zirkle (p. 269) and Wood (p. 282) have admirably outlined the major causes of cell-division delay and inhibition in irradiated yeast cells. In order to consider population effects in detail, the discussion of the same cells is continued with the understanding that in other species many additional interesting and diverse radiobiological phenomena occur. One cannot describe in a single chapter the wealth of phenomena that have been observed.†

The following classes of genetic damage are of interest:

A. Dominant lethal damage. This leads to inhibition of colony formation in one, or in a few, cell divisions.[1]

B. Recessive lethal damage[2,3] can kill the haploid cell, but will allow survival of the diploid. In some forms of this damage, the genetic material appears to be rearranged so that progeny of such cells, for many generations, divide slower than normal cells. When the diploid which bears recessive lethal damage is sporulated, it produces two dead spores for each recessive damage. The presence of recessive lethal damage in a diploid "softens" the cell for reirradiation, already suggested by Latarjet and Ephrussi,[4] and the survival of such cells indicates greater sensitivity than the normal diploid, as shown by Tobias and Stepka[5] and by Beam.[6] Recessive lethal radiation damage is illustrated schematically in Figs. 1 and 2.

### EXTENT OF DAMAGE IN NUCLEAR MATERIAL

The existence of radiation-induced auxotropic mutants of yeast cells in haploid mating types, requiring specific nutrients, makes it feasible to carry out genetic analysis of the damage induced. By mating individual cells under the microscope, one can breed diploid cells

* Work at the Donner Laboratory reported in this paper was performed under the auspices of the U. S. Atomic Energy Commission.

† A good idea of the complexity of radiobiological research may be obtained by consulting the journal, Radiation Research, and the book, *Proceedings of the International Congress of Radiation Research, Vermont, 1958*, to be published by the Academic Press, Inc.

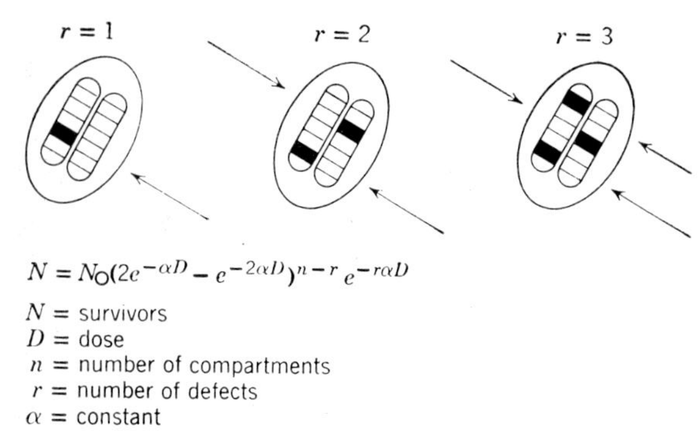


FIG. 2. Recessive model of radiation survival of diploid yeast cell with defects.

with recessive requirements for specific nutrients. Irradiation of such cells often makes the requirement dominant due to production of damage in allelic sites. In this fashion, it is found that damage often extends to regions greater than a single biochemical locus; moreover, the probability of damage occurring from site to site is very different. The situation reminds one of chromosome breaks, which are seen to be produced copiously in other organisms. The damage extends most of the time to chromosome parts which include more than one gene. In relatively rare events, the damage extends to a single genetic locus only or to some part of it.[7]

## POSTIRRADIATION CHANGES IN NUCLEAR MATERIAL

There are gene mutations that are extremely stable and persist through very many generations. It is believed generally that these mutations persist, and do not recover, and that their production varies linearly with dose. Except perhaps in some mutations of haploid cells, this statement is not exactly true, when one regards all kinds of nuclear damage from radiation.‡ In cells of higher ploidy, the presence of genetic material in two or more sets allows postirradiation rearrangements in the course of several subsequent cell divisions. The fate of a typical diploid survivor in the postirradiation period is shown in Fig. 3, as reported by Tobias.[7] Here, the irradiated cell was placed on nutrient agar under a microscope and observed continuously. As cell divisions occurred, the mother and daughter cells were separated. It is seen from the graph that, in the course of seven subsequent cell divisions, several cells were produced that failed to divide again. The radiated cells also exhibited long cell-division delays. A typical chart showing how a normal cell gives rise to progeny is given also. The process of generation of cells lacking viability can go on for many generations so that some colonies arise where cell replication is below optimum value. For instance, by preirradiation it is easy to obtain colonies which produce one dead cell for every four normal ones.

One of the suggested mechanisms for such a postirradiation recovery effect is shown in Fig. 4. As the cell with recessive radiation damage in its nucleus divides, occasionally "chromosome crossover" occurs; i.e., segments of the chromosome exchange places with each other. The two progeny are not identical in chromosome content. In the instance shown, one of them has homozygous lethals, whereas the other will survive, having gotten rid of the damaged part. A method of experimentally demonstrating chromosome crossover uses recessive genetic markers. In the example shown, an adenineless locus was carried in the heterozygous state. When homozygous, the cells needing adenine become pink; otherwise they are ivory colored. Thus, a clone established from one irradiated diploid cell may appear color segmented, with approximately $\frac{1}{2}$, $\frac{1}{4}$, or $\frac{1}{8}$ of the cells colored, depending upon whether crossover occurred in the first, second, or third postirradiation cell division. This sort of delayed phenotypic appearance of induced chromosome changes has been known since the work with bacteria by Witkin, Demerec, and Latarjet.[8–10] Taking advantage of the detailed genetic knowledge available in yeast cells, James demonstrated ultraviolet-induced homozygosis[11] and Mortimer demonstrated the x-ray-induced phenomenon.[12]

Appearance of homozygosis, following damage, of recessive genetic traits is a defense mechanism, but it also serves to make recessive lethals existing in the population homozygous.

There are other nuclear changes suspected of occurring in the postirradiation period. In some large plant and animal cells, these are the consequences of chromosome breaks and their rejoinings. The result may be wholly- or partially-increased ploidy, euploidy, or aneuploidy.

## CHANGES IN RADIORESISTANCE

In a mixed population of cells, various stages of cell division are present. It has been shown in haploid yeast cells that, in the early phase of budding, the cells are more than ten times more resistant than in inter-

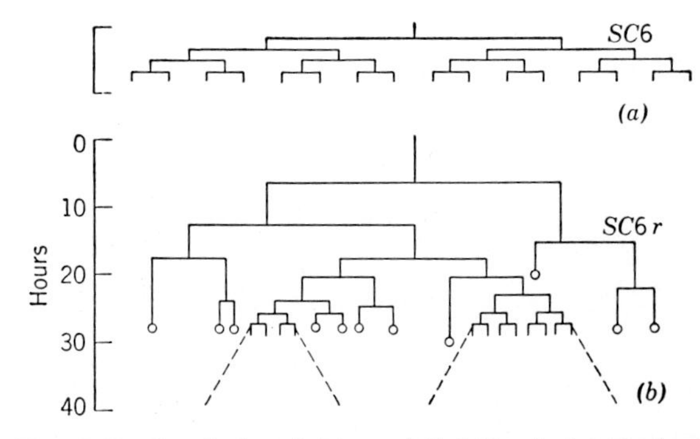


FIG. 3. Postirradiation divisions of diploid cell. (a) Nonirradiated. (b) Irradiated. Open circles indicate cells which have stopped dividing.

‡ *Note added in proof.*—Most of the evidence for linearity comes from radiation studies on the sperm of the fruit fly. Very recently, Russell *et al.* [Science **128**, 1546 (1958)] have furnished experimental proof of the dose-rate dependence of some mutations by irradiating male and female mice with x-rays and $\gamma$-rays.

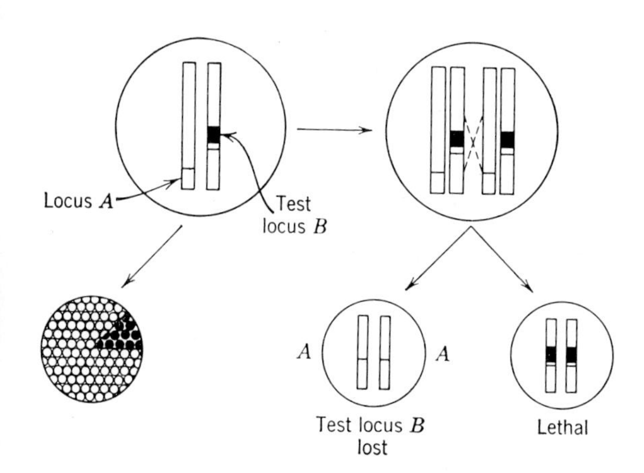


FIG. 4. Mechanism of postirradiation effect—chromosome crossover. If crossover occurs in delayed fashion, one may observe the appearance of segmented clones of cells, indicated on the left.

phase.[6,13] Moreover, kinetics of the sensitivity to radiation are different; while a single ionizing event in the proper location is sufficient for inhibition during interphase when most new DNA synthesis occurs, dividing cells require some 100 such events. Accepting DNA nucleoprotein as the site of action,§ this phenomenon is a clear indication that physical and perhaps chemical states of this substance must change radically in the course of cell division. DNA molecules undergoing synthesis are highly sensitive; existing DNA molecules exhibit decreased sensitivity. It would appear that, during mitotic division of yeast, the building blocks of the new chromosome are present in great multiplicity, after which they condense to make just two new nuclei.

## RECOVERY FROM MUTATIONAL DEFICIENCIES

Point mutations induced by ultraviolet- or x-rays generally have a low but definite spontaneous reversion rate in the haploid cell. It appears that within the same biochemical locus many kinds of damage may be produced and that each of these has a characteristic reversion rate. Since many kinds of mutations may be prepared by radiation at the same locus, these eventually may serve to help in the understanding of the chemical nature of the genes and their ability to transmit information to other cell constituents.||

When a diploid cell is made up of two haploids with heteroallelic auxotropic deficiency at the same locus, the cells cannot grow into colonies in the absence of the required nutrient. However, as Roman and Jacob have shown, ultraviolet irradiation of such a cell can bring about a striking reversion to independence of the same nutrient.[14] This occurs at unusually low doses and may be regarded as a beneficial effect of radiation, since it induces repair. The mechanism is not known. It is as though the irradiation helped the cell to make faultless replicas from the imperfect parts of the two allelic genes, a mechanism sometimes called the *copy-choice* method of replication.

## CYTOPLASMIC EFFECTS

While the cell nucleus undergoes injury, cytoplasmic deficiencies also take place. There is every reason to believe that RNA enzyme molecules and proteins are equally as vulnerable as DNA. Even if such injury occurs, however, the situation for the cell usually is not assumed to be too serious, since there are many identical RNA molecules and since it is assumed that, with the aid of DNA, they may be resynthesized; on the other hand, the loss of function of an essential DNA molecule may prove fatal to the cell.

There are some autonomous cytoplasmic particles in various cells. According to the work of Ephrussi,[15] these particles in yeast are known to be able to duplicate themselves and the rate of their duplication may occur somewhat independently from that of DNA.[16] When the granules carrying the cytochromes are lost from the cells, they cannot be resynthesized. Radiation produces these small-colony mutants, with ultraviolet- being much more efficient than x-rays (as first shown by Raut[17]).

Almost any other cytoplasmic variable tested has been shown to be effected by radiation, though generally requiring higher doses than do the nuclear effects. Examples consist of disturbances in ionic balance, release of ATP and other substances from the cells, changes in activity of various enzymes, etc. These effects in yeast cells are reviewed by Rothstein.[18]

## EXTRACELLULAR EFFECTS

Extracellular effects also are of importance in studying cell populations. These include the following factors:

A. Radiation effects on the nutrient medium.

B. Decrease in the autocatalytic action of cells which may synthesize important substances for other cells.

C. Release of growth-promoting and radiation-protective substances by the irradiated cells into the medium.

Interestingly enough, the last item of these seems to be of greatest importance. It first was detected in yeast cells under ultraviolet light by Loofbourow some 20 years ago, and more recently has been demonstrated for x-rays by Gunter and Kohn.[19] In mammalian tissue culture, Puck and Marcus[20] have shown such effects, while Révész[21] has worked on the problem in detail for ascites cells.

## CYTOLOGICAL EFFECTS OF PENETRATING RADIATIONS IN TERMS OF INFORMATION THEORY

Radiations generally decrease the information content of the genetic apparatus. They also increase the

§ Although one generally assumes that chromosome damage occurs as a result of a single ionizing particle, it has never been proved that this primary interaction is with the DNA molecule. Release of an enzyme capable of acting on DNA may have the same effect, and such an assumption indeed would explain the chain reaction which results in the breaking of whole chromosomes.

|| See chapters by Levinthal (p. 249) and by Lennox (p. 242).

noise of the information channels—namely, extragenic components—which are instrumental in transmitting instructions for cellular functions, particularly duplication. This necessitates a lag in cell division which has a definite limit, however, and, if complete information is not transmitted in due time, the new cell is lost. Systems with redundancy in the genetic apparatus can transmit the information faster and more reliably.

Radiation damage to the coding system not only consists of the knocking out of individual symbols, but also consists of deletions of extensive parts of the code or of their rearrangement in spatial order.

The cell has mechanisms to repair the damaged genetic code to some extent. Given time, it can develop a method of more-reliable information transmission even in the presence of continuous radiation which is producing new damage at a steady rate. Recovery may occur by increasing the channel capacity and, in some instances, by correcting the code or increasing its redundance. In the course of such events, groups of cells with maximum equivocation or greatest information-transfer efficiency receive preference and may take over the colony.

## POPULATIONS OF CELLS IN THE ANIMAL BODY

Radiation effects in the animal body are consequences not only of irradiation of its cells, but also of changes in functions of its organs and in their interactions with each other. For purposes of this discussion, an organ may be regarded as a heterogenous population of cells in a continuous state of activity—growth and anabolism for some cells, decay and catabolism for others. The cells are bathed in body fluids, which bring fresh nutrients and eliminate the products of cell metabolism and, at times, some of the cells themselves.

## YEAST-CELL POPULATIONS IN THE STEADY STATE

To approximate conditions in an organ, Welch in our laboratory[22] worked with a population of diploid yeast cells, propagated continuously in a device resembling the chemostat of Novick and Szilard.[23] A schematic view of this device is seen in Fig. 5. When a nutrient medium flows into this system at a given rate and when a medium with cells is eliminated from it at the same rate, steady-state populations of cells are established with a constant rate of cell division. Welch has irradiated such cultures continuously with x-rays for more than one hundred consecutive cell divisions. As the radiation delayed cell division, the flow rate of the chemostat was adjusted so that a new steady state of cell population was established at the same cell density as in the controls. In a population of this kind, radiation can cause killing by the dominant lethal mechanism. The recessive lethal mechanism can also kill, either by the change of heterozygosis into homozygosis, or by the accumulation of homozygous recessive lethals. If one assumes the view that no recovery of gene-radiation damage takes place, then it follows that in each generation the population can tolerate a very small fraction of the dose required for acute killing.

Actually, the population has maintained continuous proliferation up to the highest dose rate tested, about 6000 r/generation time. The main effects of radiation were the prolongation of the time required for cell division, and the generation of some cells incapable of further cell division.

From the number of viable and nonviable cells present, it is possible to evaluate the mean lifetime of cells in the population and the average number of daughter cells a mother cell is capable of producing before her ability to undergo mitotic division is lost.

Figure 6 indicates how the number of daughter cells from each mother cell decreased rapidly with increasing dose rate. Each control cell produced about 66 daughters; this number decreased to about seven at the highest dose rate. At the same time, however, the cells appeared to acquire greater radioresistance by a factor of two than they had had prior to irradiation. The increased radioresistance appeared to develop by adaptation, rather than by mutation. A few generation times after radiation was stopped, the cells regained their normal rate of cell division. Cells taken out of the steady-state culture and irradiated in conditions of starvation showed normal radioresistance.

Although a great deal more needs to be done with continuously proliferating cellular systems, Welch has demonstrated that diploid populations can live in the presence of surprisingly large radiation levels, and that part of the radiation-induced damage recovers in the course of successive cell divisions.

## RADIATION EFFECTS ON MAMMALS AND MAN

It is impossible to do justice to the intricate and detailed research work that has been going on in radia-

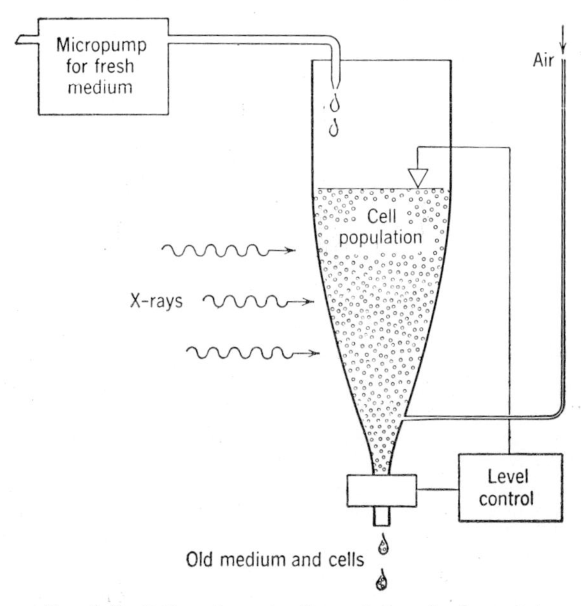


FIG. 5. Radiation of yeast-cell populations in chemostat.

tion physiology during the past decades. The reader si referred to the excellent review articles and books on the subject.[24–26] It is important to note that bacteriological techniques of propagation recently have been applied successfully to human cells by Puck¶ and that, as a result, some information is already available on the radiosensitivity of various diploid and tetraploid human-cell strains. The mean lethal doses range from about 90–300 r, and the mechanisms of inhibition of colony formation are perhaps not too different from those for yeast cells. It is not known, however, what the radiation resistance of differentiated human cells in the body is, where it is possible that the very rapid humoral and metabolic exchange and the changes of radiosensitivity due to cell division might lead to an increase of radioresistance, as noted in the steady-state yeast-cell cultures.

### RADIOSENSITIVE TISSUES

There is a 60-year-old law in radiobiology, originally proposed by Bergonie and Tribondeau; according to this law, the most radiosensitive tissues are those which have the highest mitotic index—i.e., the highest rate of cell division. One now understands that this law is only approximately true because radiation injury manifests itself mainly in the cell-division process. Radiated cells usually die when they attempt cell division; if their functions in body tissues do not require frequent cell division, the cells usually can perform ordinary metabolic and enzymatic functions for a long time post irradiation, since these functions are much less radiosensitive than cell division itself.

Correspondingly, in the acute phase, a single dose of radiation usually affects the most rapidly proliferating tissues; the bone marrow and its varied cell population and certain epithelial tissues are affected first. Early radiation deaths frequently are due to injury to the intestinal mucosa and to white-cell proliferation, while failure of the blood clotting mechanism and anemia make their appearance later. The neurons of the central nervous system of the adult are quite radioresistant, since almost no cell division occurs in them. In the developing mammalian embryo, however, as shown by Hicks,[27] nerve tissue is rapidly proliferating and has the greatest radiosensitivity.

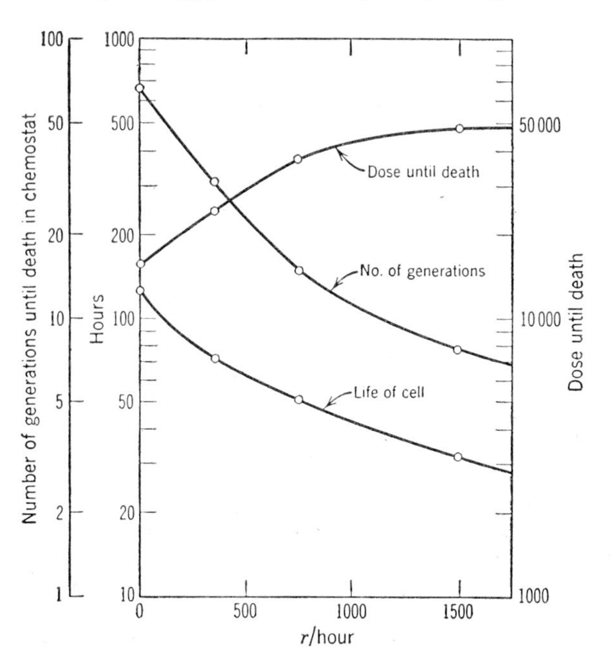


FIG. 6. Progeny decrease with dose rate.

### INDIRECT RADIATION EFFECTS

There are very intricate relationships between various organs of the body, and it is not surprising that remote effects of radiations have been observed. For example, if an organ receives a dose of radiation, depressive effects due to this injury manifest themselves elsewhere in the body. Starting with the work of Hevesy, it was shown that a small depression of cell division and of DNA synthesis occurs everywhere in the body if an organ, such as the spleen, is irradiated.[28] To illustrate remote effects of radiation, one may mention work done at the Berkeley cyclotron in irradiating pituitary and various regions of brain tissue by high-energy deuterons.[29] Because of their small scatter, these particles are well suited to make small radiolesions in various parts of the body. It was found that a large localized dose to the rat pituitary (size 1×2 mm) will cause progressive development of the hypophysectomized state.[30] The growth of the animal, calcification of its bones, and functions of adrenals, thyroid, and gonads are all affected; development of the remote effects has different dose-response curves for each kind of effect.

An organ that is shielded from radiation confers a degree of protection against effects from whole-body radiation to the animal. Shielding of spleen and bone marrow is of greatest benefit, sometimes increasing radiation tolerance by about a factor of two. Jacobson and his associates[31] have shown that transplantation of bone marrow, spleen, or embryo homogenate to mice which have received a dose of x-rays increases radiation tolerance. The greatest part of this effect is owing to transplantation of healthy; unirradiated bone-marrow cells, which prosper on the tissue bed of radiation-killed bone marrow. Apparently, one may allow death of all cells of the marrow by whole-body x-irradiation, and the animal may be resurrected by the administration of a sufficient quantity of serologically compatible healthy cells. This technique has given rise to valiant attempts now being made in various medical installations to save terminal leukemia patients. Patients receive a lethal dose of whole-body radiation, followed by a bone-marrow transplant obtained from healthy relatives. The radiation is supposed to kill all bone-marrow cells, including the leukemic cells. If the transplanted marrow saves the patient from the radiation effects, it

¶ See chapter by Puck (p. 433) for detailed discussion.

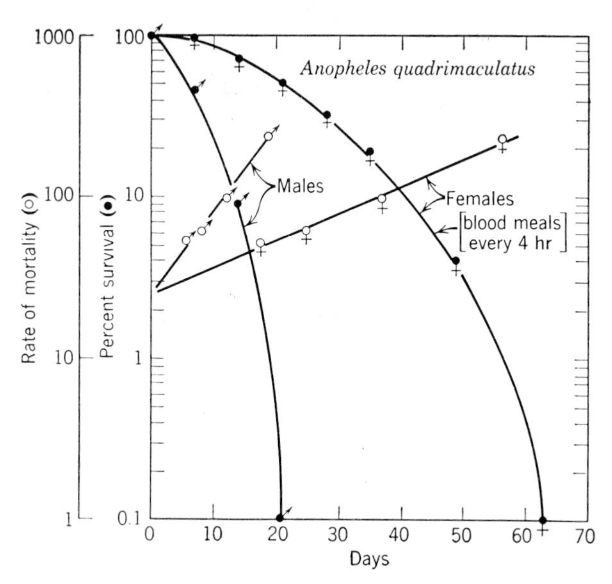


FIG. 7. Population-survival curve—mosquitoes [from W. E. Kershaw, T. A. Chalmers, and M. M. J. Lavoipierre, Ann. Trop. Med. Parasitol. **48**, 442 (1954)].

is hoped that leukemia will not recur. Because of immunological reasons, much research is needed before bone-marrow transplants are completely safe for humans.[32]

## LONGEVITY

In normal populations of humans, animals, invertebrates, and even unicellular organisms, there appear to be identical laws predicting the rate of death of the population. About a hundred years ago, Gomperz worked out details of this function, and at the present time Jones *et al.*[33] have applied it in detail to many problems of population survival.

The relative rate of death of a population of $N$ individuals as function of time $t$ is

$$\frac{d^2 \log N}{dt^2} = \alpha t,$$

where $\alpha$ is a characteristic constant of the population. To illustrate this principle, a normal survival curve of mosquitoes is reproduced in Fig. 7.[34] The relative rate of death is an exponentially increasing function. In adult humans, the death rate is described by a similar exponentially increasing function, with the rate doubling itself about every eight years. According to the theory by Jones, a single dose of radiation will modify the relative rate of death by shifting it upwards, parallel to itself. Continuous irradiation is supposed to increase the rate constant, $\alpha$. Although many longevity estimations were made on the basis of this theory which can be used for daily permissible-dose determinations, rigorous experimental proof for radiation effects is not as yet available. Several other theories have been proposed.[35]

## CARCINOGENESIS

It is of great interest to know something of radiation carcinogenesis. Cancer cells represent a population which is abnormal in the sense that (a) it has become independent of some of the growth-controlling factors characteristic of the tissue from which its cells originate; (b) it is frequently able to metastasize; and (c) it may have metabolic products which lead to death of the host. The rate of incidence of "spontaneous" cancer, chemical-, virus-, or radiation-induced neoplasms, usually, at least in some definite time interval, follows a law similar to that propounded for the rate of death of populations: The rate of onset increases as an exponential function of time. As an illustration, the rate of onset of bone tumors resulting from radioactive $Sr^{89}$ incorporated in bone is shown in Fig. 8; this work was done at the Argonne National Laboratory.[36]** The rate constant of tumor incidence increases with increased radiation dose and is a more important index of carcinogenesis than is the over-all increase of tumors in a lifetime, which, in addition to the cancer-inducing factor, depends upon factors determining longevity. In practice, such an exponential rate curve means a "time lag" of varying length before the demonstrable onset of neoplastic growth.

The detailed biochemical steps that lead to cancer are not known, but it is useful to draw an analogy between diploid cells of animal tissue that have received a dose of radiation and diploid yeast cells, discussed earlier in this paper. An individual, sublethally irradiated yeast cell frequently develops an abnormal colony distinguished by its slow rate of growth in the process of "recovery." Some of the cells regain their full ability to proliferate by processes outlined in the section in which the recovery from radiation damage was dis-

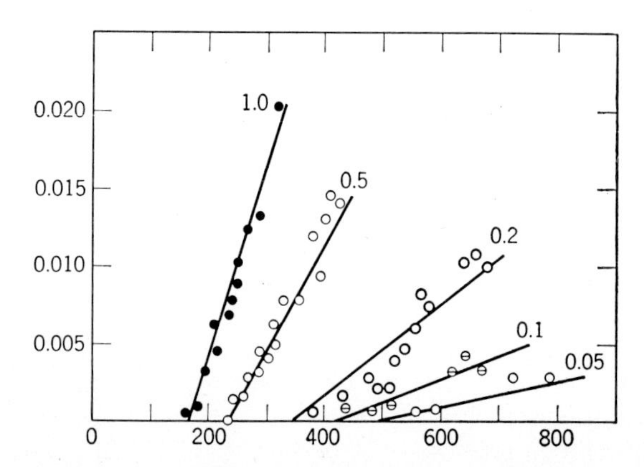

FIG. 8. Daily probability of bone-tumor development in the mouse following monthly injections of $Sr^{89}$ (after Brues[36]). Abscissas: time after first injection (days). Ordinates: daily probability of bone-tumor development. Curves show effect of monthly doses of $Sr^{89}$, of 1.0, 0.5, 0.2, 0.1, and 0.05 $\mu$C/g.

** *Note added in proof.*—A. Brues and M. Finkel have published recent interpretations of their experimental work in carcinogenesis [Science **128**, 637, 693 (1958)]. Examples of other views on this subject are found in the work of E. B. Lewis [Science **125**, 965 (1957)] and in references 36, 40, and 42.

cussed. This process often takes many cell generations, and in the course of it a number of changes can occur that will modify the genetic constitution of the cell. Changes resembling carcinogenesis in yeast cells also have been discussed by Lacassagne *et al.*,[37] Maisin *et al.*,[38] and Warburg.[39]

Most human cells normally are under hormonal and possibly neural control of the homeostatic apparatus and, as a result, they usually are in a state of proliferation which is below their maximum proliferative capacity. As an illustration, consider the human breast. This organ develops under elaborate hormonal influence, responding to secondary sex hormones elaborated by the ovaries and adrenals and to lactogenic hormone. In addition, the cells of this gland are influenced by growth hormone and posterior pituitary hormones. The state of this tissue and its metabolic activities are conveyed in some way to the hypothalamus and the pituitary, and as part of homeostatic control the gland may receive increased hormonal stimulation when its performance is below par.

There is increasing evidence available to show that hormonal balance is an important factor in radiation carcinogenesis. The intricate pattern of the hormonal factors has been studied for more than 80 years; here a few recent experiments only are noted, with mammary cancer as a restrictive example. Lacassagne has shown clearly that mammary cancer in mice may be induced by estrogen in females, as well as in castrated males. More recently, it appears that the joint application of estrogen and lactogenic hormone accelerates the onset of cancer. Bond *et al.*[40] found that a sublethal dose of whole-body irradiation in a strain of rats induced mammary cancer. However, the frequency of mammary cancer was decreased greatly if the ovaries of the animals were removed, thus eliminating most of the endogenous, secondary sex hormones. It is known that, following subtotal $I^{131}$ thyroidectomy in mice, thyroid tumors as well as pituitary tumors result. Furth *et al.*[41] found that some rats recovering from whole-body radiation developed pituitary tumors, apparently in response to hormonal demands from endocrine target organs. When such tumors are transplanted into other rats of the same inbred strain, hormonal stimulation from the graft includes adrenal, thyroid, and mammary tumors in the new hosts. Apparently, whenever some tissue in the hypothalamic-pituitary-endocrine organ system develops sublethal tissue damage involving chromosome derangements, so that the tissue is unable to perform its functions satisfactorily, there is strong hormonal and/or nervous stimulation of these tissues causing increased rate of proliferation. Under such conditions, genetic rearrangements may lead to neoplastic tissues. An interesting illustration is furnished by localized deuteron irradiation of the rat pituitary at Berkeley.[29] Here 945 rep to the rat pituitary led within one year to the development in every irradiated animal of pituitary tumors, with 3.5 tumors per gland. Higher doses led to lessened incidences of tumors, and when most pituitary cells were killed, none of the animals developed pituitary tumors.

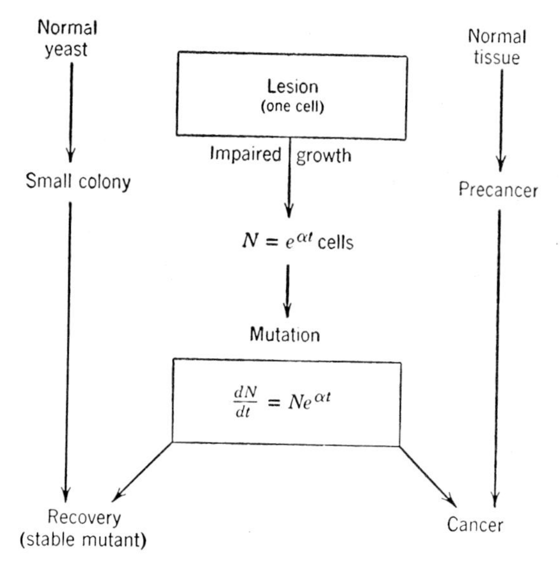


FIG. 9. Radiation carcinogenesis.

There are many other instances of hormonal influence over carcinogenesis. Perhaps the most complete investigations of this kind were made by Kaplan and his group.[42]

It would appear that a better understanding of homeostatic control mechanisms may lead to an understanding of the etiology of cancer and possibly to some methods of prevention or delay of its onset. In the meanwhile, it has been apparent that cancer cells that do not become completely undifferentiated in the carcinogenic process still retain some measure of dependence on hormonal control. Work is in progress in many countries to hypophysectomize patients with advanced and progressing metastatic carcinoma, in the hope that the cancer cells will stop growing and the tumors will regress.

The proton- and α-particle beams of the Berkeley 184-in cyclotron have proved to be promising tools for research on hypothalamic functions and for therapeutic investigations involving radiation hypophysectomy. The techniques and initial results have been described.[43] While hypophysectomy is not the final answer to cancer therapy, there are some patients who appear to have benefited from the pituitary irradiation. Detailed biochemical study of induced tumor regressions may bring additional clues with respect to the nature of the hormones which control normal or abnormal proliferation.

In conclusion, radiobiology has been useful as an aid to understanding not only the complex physiological changes following irradiation exposure, but also, the mechanism of carcinogenesis. Basic research on molecules and cells has helped in the understanding of some phenomena of cell division and of the nature of some genetic transformations. In Fig. 9, the present status of understanding of radiation carcinogenesis is shown in simplified fashion. It is apparent from the figure that there is a qualitative similarity between postirradiation events in diploid yeast cells and in somatic tissue cells. Many of the diploid yeast survivors develop small-colony mutants, some of them cytochrome deficient, that proliferate slowly and revert to rapid cell division many generations later. This process may be analogous to the slow development of benign precancerous lesions and the subsequent somatic mutation to cancerous cells. The time rate of occurrence of both sets of phenomena is quite similar. It is clear, then, that more work is needed to ascertain the genetic and biochemical features of delayed recovery in irradiated unicellular organisms and changes in the aerobic-anaerobic metabolism which, according to Warburg,[39] are known to occur in carcinogenesis. Perhaps such work may furnish some day direct clues to the detailed etiology and inhibition of cancer.

## BIBLIOGRAPHY


[1] R. K. Mortimer, Radiation Research **2**, 361 (1955).

[2] R. E. Zirkle and C. A. Tobias, Arch. Biochem. Biophys. **47**, 282 (1953).

[3] R. K. Mortimer and C. A. Tobias, Science **118**, 517 (1953).

[4] R. Latarjet and B. Ephrussi, Compt. rend. **229**, 306 (1949).

[5] C. A. Tobias and B. Stepka, "Mutations to increased radiosensitivity in yeast cells" [University of California Radiation Laboratory Report (UCRL-1922), 1952], p. 40.

[6] C. A. Beam, "Influence of metabolism on some radiation effects," in *Proceedings of International Congress of Radiation Research, Vermont, 1958* (to be published).

[7] C. A. Tobias, R. K. Mortimer, R. L. Gunther, and G. P. Welch, "The action of penetrating radiations on yeast cells" [University of California Radiation Laboratory Report (UCRL-8268), 1958]. To be published in *Proceedings of Second United Nations International Conference on the Peaceful Uses of Atomic Energy, Geneva, 1958*.

[8] E. M. Witkin, Proc. Natl. Acad. Sci. U. S. **32**, 59 (1946).

[9] M. Demerec and R. Latarjet, Cold Spring Harbor Symposia Quant. Biol. **11**, 38 (1946).

[10] M. Demerec, Proc. Natl. Acad. Sci. U. S. **32**, 36 (1946).

[11] A. P. James, Genetics **40**, 204 (1955).

[12] R. K. Mortimer, "Radiobiological studies on yeast (discussion)," in *Proceedings of International Congress of Radiation Research, Vermont, 1958* (to be published).

[13] C. A. Beam, R. K. Mortimer, R. G. Wolfe, and C. A. Tobias, "The relation of radio-resistance to budding in saccharomyces cerevisiae" [University of California Radiation Laboratory Report (UCRL-2345), 1953]; Arch. Biochem. Biophys. **49**, 110 (1954).

[14] H. Roman and F. Jacob, Compt. rend. **245**, 1032 (1957).

[15] B. Ephrussi, *Nucleo-Cytoplasmic Relations in Micro-Organisms, Their Bearing on Cell Heredity and Differentiation* (Clarendon Press, Oxford, 1953).

[16] F. Sherman, "A study of the effects of elevated temperatures on the growth and inheritance of Saccharomyces cerevisiae," thesis, University of California, Berkeley (1958) [University of California Radiation Laboratory Report (UCRL-8573), 1958].

[17] C. Raut and W. L. Simpson, Arch. Biochem. Biophys. **57**, 218 (1955).

[18] A. Rothstein, "Biochemical and physiological changes (in yeast cells)," in *Proceedings of International Congress of Radiation Research, Vermont, 1958* (to be published).

[19] S. E. Gunter and H. I. Kohn, J. Bacteriol. **72**, 422 (1956).

[20] T. T. Puck and P. I. Marcus, J. Exptl. Med. **103**, 653 (1956).

[21] L. Révész, J. U. S. Natl. Cancer Institute **20**, 1157 (1958).

[22] G. P. Welch, "Effects of chronic exposure to x-rays on a steady rate population (of Saccharomyces cerevisiae)." Thesis, University of California, Berkeley, (1957); University of California Radiation Laboratory Report (UCRL-3763), 1957.

[23] A. Novick and L. Szilard, Science **112**, 715 (1950).

[24] A. Hollaender, editor, *Radiation Biology* (McGraw-Hill Book Company, Inc., New York, 1954–1956), Vols. I–III.

[25] J. J. Nickson, editor, *Symposium on Radiobiology, Oberlin College, 1950. The Basic Aspects of Radiation Effects on Living Systems* (John Wiley and Sons, Inc., New York, 1952).

[26] Z. M. Bacq and P. Alexander, *Fundamentals of Radiobiology* (Academic Press, Inc., New York, 1955).

[27] S. P. Hicks, Am. J. Roentgenol. Radium Therapy Nuclear Med. **69** (2), 272 (1953).

[28] L. S. Kelly and H. B. Jones, Am. J. Physiol. **172**, 575 (1953).

[29] J. L. Born, A. O. Anderson, H. O. Anger, A. C. Birge, P. Blanquet, T. Brustad, R. A. Carlson, D. C. Van Dyke, D. J. Fluke, J. Garcia, J. P. Henry, R. M. Knisely, J. H. Lawrence, C. W. Riggs, B. Thorell, C. A. Tobias, P. Toch, and G. P. Welch, "Biological and medical studies with high energy particle accelerators" [University of California Radiation Laboratory Report (UCRL-8242), 1958]; in *Proceedings of the Second United Nations International Conference on the Peaceful Uses of Atomic Energy, Geneva, 1958* (to be published).

[30] D. C. Van Dyke, M. E. Simpson, A. A. Koneff, and C. A. Tobias, "Long term effects of deuteron irradiation of the rat pituitary" [University of California Radiation Laboratory Report (UCRL-8238), 1958]; Endocrinology (to be published).

[31] L. O. Jacobson in *Radiation Biology*, A. Hollaender, editor (McGraw-Hill Book Company, Inc., New York, 1954), Vol. I, p. 1029.

[32] A. Hollaender, C. C. Congdon, D. G. Doherty, R. Shapira, and A. C. Upton, "New developments in radiation protection and recovery," Oak Ridge National Laboratory Report in *Proceedings of the Second United Nations International Conference on the Peaceful Uses of Atomic Energy, Geneva, 1958* (to be published).

[33] H. B. Jones in *Advances in Biological and Medical Physics* (Academic Press, Inc., New York, 1956), Vol. IV, p. 281.

[34] W. E. Kershaw, T. A. Chalmers, and M. M. J. Lavoipierre, Ann. Trop. Med. Parasitol. **48**, 442 (1954).

[35] N. Berlin and F. L. DiMaggio, "A Survey of Theories and Experiments on the Shortening of Life Span by Ionizing Radiations" [Armed Forces Special Weapons Project Report (AFSWP-608), 1956].

[36] A. M. Brues and M. P. Finkel, quoted in R. H. Mole, Brit. Med. Bull. **14**, 184 (1958).

[37] A. Lacassagne, M. Schoen, and P. Beraud, Ann. Fermentations **5**, 129 (1939).

[38] J. Maisin, G. Lambert, and E. Van Duyze, Rev. belge Pathol. et méd. exptl. **21** (1), 52 (1951).

[39] O. Warburg, Science **123**, 309 (1956).

[40] V. P. Bond, E. P. Cronkite, C. J. Shellabarger, S. W. Lippincott, R. A. Conard, and J. Furth, "Mechanisms of radiation-induced mammary gland neoplasia related to dose dependence and hormonal status in rats," in *Proceedings of Second United Nations International Conference on the Peaceful Uses of Atomic Energy, Geneva, 1958* (to be published).

[41] J. Furth, E. I. Hirsch, H. J. Curtis, E. L. Gadsden, and R. F. Buffett, "Pathogenesis and character of radiation-induced pituitary tumors," presented before the Radiation Research Society, Rochester, New York, 1957; abstract: Radiation Research **7** (3), 317 (1957).

[42] H. S. Kaplan, C. S. Nagareda, and M. B. Brown in *Proceedings of the Laurentian Hormone Conference*, G. Pincus, editor (Academic Press, Inc., New York, 1954), Vol. X, p. 293.

[43] C. A. Tobias, J. H. Lawrence, J. L. Born, R. K. McCombs, J. E. Roberts, H. O. Anger, B. V. A. Low-Beer, and C. B. Huggins, Cancer Research **18**, 121 (1958).

# 36
# Fine Structure of Cell Nucleus, Chromosomes, Nucleoli, and Membrane

H. Stanley Bennett

*Department of Anatomy, University of Washington School of Medicine, Seattle 5, Washington*

THE most important concepts of the structure of the nucleus are based on information drawn from three rather separate approaches which are now converging. These approaches had their origins in the nineteenth century. The first of these—the optical approach—had its origin with Robert Brown's first recognition of the nucleus by the use of the light microscope in 1833. This distinguished botanist was also the discoverer of what is now called Brownian movement. The amplification of his original observation, as carried out with various optical tools, forms an exceedingly important part of present knowledge of the nucleus.

The second stream of knowledge is that of genetics. It can be regarded as beginning with Mendel's observation that inherited characteristics are transmitted from generation to generation in ratios of simple whole numbers. During the early part of the present century, derivatives of this discovery, correlated with optical observations on the nucleus, led to the field of cytogenetics. Subsequently, the information derived from genetic sources has played an essential part in concepts of nuclear structure. Indeed, the genetic data provide the only evidence regarding certain important structural features of the nucleus.

One can trace the third stream, the chemical one, to the work of Miescher, who published an important compendium in 1897 on the chemical characteristics of extracts derived from pus. Lamentably, this raw material was abundantly available in those days, before aseptic surgery had become popular. From this convenient product of human suffering, Miescher extracted substances which he named nucleic acids.

One can ask if the nucleus is an essential part of the cell. One can recall the red cell of the mammal, which is active for about 120 days without a nucleus. But how do other cells fare if the nucleus is removed? This has been investigated in a relatively small number of forms. It is possible to remove the nucleus from an amoeba by surgical means, as Mazia, for example, has done. It is possible also to amputate and study non-nucleated fragments of certain large algae, such as *Acetabularia.* One can summarize much by saying that a non-nucleated cell or cell fragment can survive for a certain period. An amoeba can eat, it can move about after enucleation, but in due time its synthetic capacities appear to degenerate, and, like the red cell, it perishes without reproducing. The red cell is perhaps the most successful of cells in functioning for a long period without a nucleus.

It is possible to isolate nuclei and to separate them from cytoplasmic components. If such isolated nuclei are studied biochemically, their metabolic behavior can be followed. The most elegant work along these lines has been carried out by Mirsky and Allfrey, based on earlier work by Dounce. The nuclei contain protein, DNA, RNA, and small amounts of lipid. Allfrey and Mirsky have found that isolated nuclei are capable of carrying out amino-acid incorporation into proteins and that certain other enzymatic activities are associated with the nucleus.

If one examines the unstained nucleus with a light microscope, either in the living cell or after fixation with ordinary reagents, one sees regions of varying density. This inhomogeneity of structure becomes even more striking following staining, since the nuclear components vary in their affinity for dyes. The components which bind the dyes have been called chromatin. As a rule, the term "chromatin" is used primarily to designate strongly-staining materials in the nucleus of the cell. It is a general term without any specific chemical significance. The term "nuclear chromatin" is applied to two distinct structures within nuclei. One comprises the chromosomes themselves; the second consists of nucleoli which are accumulations containing considerable quantities of RNA. Many of the staining reactions of chromosomes and of nucleoli are very similar or identical, but there are a few methods for distinguishing them from each other. These depend on chemical differences between DNA on the one hand, which dominates many of the properties of chromosomes, and RNA on the other, which has an important role in the nucleoli.

The appearance of nuclear chromatin can change greatly under different functional states of the cell. As the cell approaches a mitotic division, the chromatin material rearranges itself. At that time, one easily can see well-defined bodies, the chromosomes, so called because they can be stained so as to assume very vivid colors. It is well established that the chromosomes are present in the intermitotic nucleus, where they usually occur in a rather tenuous form which makes it difficult to recognize them morphologically. Yet chromosomal activity can be detected by a number of different means. The genetic method is perhaps the most powerful.

In certain specialized cases, the chromosomes may be present in a form quite different from the dispersed state which they assume in the normal intermitotic cell. For example, in the salivary gland cells of the fruit fly, *Drosophila,* the intermitotic chromosomes occur as discrete bodies known as "giant" chromosomes, which may be over 200 $\mu$ long. Under the light microscope, they show a longitudinal fibrillar structure and a distinct series of aperiodic crossbands. These crossbands are unique in the sense that they exist in distinct and recognizable patterns in each chromosome. These patterns are important for correlating the genetic influence of the chromosomes with its structure.

Consider now the question: How can one exploit the properties of the various chemical components of the nucleus in such a way as to gain information about their disposition, structure, interactions, and relationships? Both acid and basic dyes are bound by many of the components of the nucleus. From this, it is inferred that nucleoproteins behave as a mixture of anionic and cationic ion-exchange resins. These properties can be attributed to their negatively charged phosphate groups and positively charged protein groups. One can exploit these characteristics by setting up a suitable competitive system which introduces colored cationic molecules to bind to the nucleic-acid phosphate groups, replacing positively charged protein groups. This procedure tags phosphate groups of the nucleic acids with colored tracers or indicators. The capacity of anionic polymer groups, such as nucleic-acid phosphate groups, to bind basic dyes is spoken of as the property of "basophilia." Thus, nucleochromatin is often spoken of as being strongly basophilic. In this way, the nucleic acid can be tagged rather nonspecifically. Analogously, one can use anionic dyes such as azosulfonic acids to bind to polymers containing basic groups in tissue components. Such polymers are found in the nucleus in the form of the basic proteins which are associated with the nucleic acids to form nucleoproteins.

Most of these dye-binding methods do not distinguish between RNA and DNA. But, from the analytical data on isolated nuclei, there is reason to believe that both are present in the nucleus. If one turns to another property of nucleic acids—namely, the absorption spectrum in the ultraviolet of the purine and pyrimidine residues—it can be shown that the same structures which take up basic dyes also absorb strongly in the ultraviolet. This is consistent with the view that the nucleoprotein complex is responsible for the staining reaction and for the absorption in the ultraviolet. But the latter method does not permit a distinction between DNA and RNA components.

There is, however, a reaction which does distinguish between these two types of nucleic acids. It has proved to be exceedingly useful in characterizing the nucleic-acid distribution in cells. This reaction is known as the Feulgen reaction. It depends upon the presence in nucleic acids of linkages which are differentially susceptible to hydrolysis. Using appropriate conditions, dilute HCl will hydrolyze certain sugar-base linkages in DNA but not in RNA. The hydrolyzed sugars of DNA are thus converted to reducing groups resembling aldehydes. These can be labeled by coupling them to a suitable colored reagent which reacts with aldehydes. Each single aldehyde group formed by hydrolysis is then tagged with a colored chromophore. This permits the DNA to be distinguished sharply from RNA. This useful reaction can be used quantitatively.

There is another important approach to the problem of distinguishing the localization in cells of RNA and DNA. This involves the use of specific enzymes. The enzyme, ribonuclease, will depolymerize RNA but not DNA. If RNA is removed from the nucleus by action of this enzyme, a loss of dye-binding capacity may appear in certain structures. The material originally present which was removed by the ribonuclease is thus shown to be RNA. It is also possible to remove DNA by deoxyribonuclease, and thus to ascertain which dye-binding component contains DNA. Still another method, often less reliable and less rigorously specific than the others, can be used to stain DNA and RNA in contrasting colors in the same cell. These three methods—the Fuelgen reaction, the use of specific enzymes, and the use of differential staining—permit one to distinguish cytochemically between DNA and RNA and provide information about the localization of these substances within the cell.

It turns out that DNA is found characteristically in the chromosomes and frequently as a jacket surrounding the nucleolar material. RNA is abundant in the central portions of the nucleoli and small amounts can be found within the chromosomes or scattered about in the nucleus.

If a large number of nuclei from a given species are examined by the Feulgen or some other suitable method and the amount of DNA in the individual nuclei is measured by absorption microspectrophotometry, as carried out by Alfert and by Swift, a small number of cells (for example, sperm cells) display a certain unit quantity of DNA. However, the majority of cells contains approximately twice this unit amount, while a third group of cells yields values clustering around four times the unit amount. In certain organisms—in the human liver, for example—one may find nuclei with up to eight times the unit value. From this, it derives that the amount of DNA in individual nuclei tends to occur in integral multiples of some unit quantity.

The chromosomes in dividing cells can often be counted with precision. In most higher animals, spermatozoa, the spermatids, and fully mature ova are found to contain a certain number of chromosomes. This number is called the "haploid number," and the chromosomes making up this number comprise a single set of chromosomes. These cells contain a single unit quantity of DNA, as mentioned earlier. The great majority of cells in most multicellular organisms contains twice the number of chromosomes characterizing a single set, and contains twice the unit quantity of DNA. Such cells are said to contain the diploid number of chromosomes. In diploid cells, most of the chromosomes occur in homologous pairs. Each member of a pair resembles its mate closely, but morphological and genetic differences between the members of a homologous pair are often found. One may also find cells with four, six, eight, or some other small integral multiple of the haploid number of chromosomes, and with the same multiple of the unit amount of DNA found in haploid cells. Such cells are called tetraploid, hexaploid, or octaploid, respectively. In mammalian liver, one can find many tetraploid and some octaploid cells in a population which is predominantly diploid.

Each set of chromosomes contains a single set of genes. Thus, in diploid cells, many genes are represented twice, once in each member of a homologous pair. Such cells are said to be homozygous for the characteristics represented by genes which occur in identical form twice in each cell. But some gene loci are not identical in each member of a homologous pair of chromosomes. Such genes are represented only once in each cell, which is then said to be heterozygous with respect to the characteristic represented in those dissimilar loci. These data have contributed to the concept that there is a definite amount of DNA characteristic of each chromosome set, each chromosome, and each gene locus.

Some organisms live very well through most of their life cycle with haploid numbers of chromosomes in each nucleus. The fungus, *Neurospora*, for example, which has figured extensively in genetic studies, has a single set of chromosomes in each nucleus throughout long phases of the life cycle. The diploid phase may be very brief. In most metazoa, however, the diploid phase predominates.

One occasionally finds numbers of chromosomes that are *not* even multiples of the haploid number. Such cells are called "aneuploid" and contain amounts of DNA which are not integral multiples of the amount of DNA in haploid cells. Relevant examples are provided by certain cancers and leukemias. As a rule, the presence of aneuploidy and of morphologically abnormal chromosomes is interpreted as evidence for some genetic abnormality.

Levinthal (p. 227) remarks that information from genetics has led to the concept that the genetic carriers are arranged in linear sequence along some structure. The identification of the chromosome as this structure marked a major advance in the history of cytogenetics. Correlations between genetic and morphological data were most successfully worked out initially in *Drosophila*, where the chromosomes of the salivary glands are large and where genetic studies are readily carried out because of the short life cycle of the fly. Similar studies correlating genetic behavior with morphological abnormalities of chromosomes have now been carried out in a number of different forms. However, much of the corresponding work in virus and bacterial genetics is carried out conceptually, without direct visualization of chromosomes. The genetic data are used to construct a linear sequence, but no attempt is made ordinarily to correlate this sequence with morphological features of chromosomes—one important reason being that chromosomes have not been incontestably visualized in bacteria or in viruses. In these cases, the structural concept of linear sequence of units depends entirely upon genetic evidence.

At present, most of the important concepts of nuclear structure have been derived from genetic studies, from chemical analyses, and from examination with the light microscope. When combined ingeniously, these approaches have proved to be very powerful.

The electron microscope has yielded some additional information. It is evident that the intermitotic nucleus is surrounded by an envelope which consists of two unit membranes, an inner one and an outer one. These two membranes are joined to each other at certain sites. The lines of junctions surround and define pores several hundred Ångströms in diameter, perforating the nuclear membrane. Through these pores, the nuclear and cytoplasmic matrices communicate directly. The outer membrane of the nuclear envelope may also be continuous with membranes in the cytoplasm. The interior of the nucleus, however, is free from membranes, and is thus in sharp contrast to the cytoplasm, where membrane structures are frequent and often densely packed.

Electron micrographs usually show a rather irregular accumulation of granular material within the intermitotic nuclei. The granules presumably represent DNA and RNA, combined with protein. Very fine intranuclear helical threads have been described by some authors, but these are not seen with clarity and their significance is difficult to assess. No one has yet recognized structures in the nucleus corresponding to the nucleic-acid helices as studied by Hall and Doty, with the electron microscope, and pictured elsewhere in this volume (p. 107). One can find nucleoli in electron micrographs. They appear as rather irregular accumulations of granules more densely crowded together than elsewhere in the nucleus.

Mitotic cells show no nuclear envelope. The chromosomes are recognizable as dense accumulations of granular material without any membranous investment.

Thus, it appears that direct electron microscopy of the nucleus of ordinary cells has provided little information which is satisfying in the light of the physiological importance of the nucleus and its contents.

In certain specialized types of cells, organized structures have been detected with the electron microscope which are not characteristic of cells in general. Thus, in some sperm cells, dense striations running the length of the nucleus have been seen. Moses has detected more-delicate organized structures in certain spermatocytes. These appear as long thread-like elements with delicate side chains extending laterally. Efforts have been made to associate the filaments with some phase of chromosome structure. But, at the present time, it would be hazardous to do more than point out that the dimensions of the threads of the side loops are not inconsistent with those of a nucleic-acid helical chain.

Although the electron microscope shows that the nucleus contains particles which closely resemble the RNP particles of the cytoplasm, it reveals relatively little concerning the mechanism of transfer of nucleic acid from nucleus to cytoplasm. Such a transfer would provide a means whereby the nucleus could transmit information to the cytoplasm. Yet one can see images which suggest transition stages in the course of outward movement of RNA from the nucleus. One may occasionally see accumulations of particles in the cytoplasm which very closely resemble the nucleoli. Similar appearances observed with the light microscope have led to the view that in some cells whole nucleoli may be discharged from the nucleus carrying large packets of RNA to the cytoplasm. On the other hand, it may be that much of the RNA escapes into the cytoplasm through the pores in the nuclear envelope. Another mechanism whereby RNA could pass from nucleus to cytoplasm involves a binding of the newly found RNA to the inner nuclear membrane. The latter then would flow into the cytoplasm, carrying the bound particles with it. Such a mechanism, however, is not well documented.

We are forced, then, to conclude that the task of correlating the fine structure of the nucleus with its function and with the chemical and genetic evidence at our disposal is largely before us.

# 37

# Fine Structure of Cytoplasm: The Organization of Membranous Layers

Fritiof S. Sjöstrand

*Karolinska Institutet, Stockholm 60, Sweden*

When climbing the ladder of levels of organization of living matter, one finds nowadays that at least indications of steps are present along the whole ladder. Some steps may be weak and might break when exposed to too critical attention. Light microscopy has furnished information regarding a great number of subcellular structural elements, the so-called cell organelles, as well as of cytoplasmic differentiations of a more diffuse character. The most impressive of the cell organelles which are localized in the cytoplasm are the mitochondria and the Golgi apparatus. Basophilic regions of the cytoplasm represent more-diffusely distributed parts which long ago were called the ergastoplasm by Garnier. The reason for this term was the idea that these regions were especially active parts of the cytoplasm. In the exocrine pancreas cell, a filamentous or lamellar structure was observed within the basophilic regions of the cytoplasm, "die Basalfilamenten" of Heidenhain, or "die Basallamellen" of Zimmermann. These various components of the cytoplasm were considered immersed in the ground substance of the cytoplasm, a component which was assumed to lack any higher degree of specialized organization.

The submicroscopic organization of the regularly occurring cell organelles was completely unknown, however. A few cell components, such as the myelin sheath of peripheral nerve fibers and the outer segments of the retinal receptors of the vertebrate eye, were assumed to consist of alternating layers of lipid and protein molecules. This assumption was based upon polarization-optical data of these strongly birefringent components.

Electron microscopy has made possible the fairly close analysis of the submicroscopical organization of these various components. In some cases, the combination of electron-microscope data with polarization-optical and x-ray diffraction data has made it possible to propose models containing some features of the molecular architecture of these components. In this latter respect, the results can be considered rather crude.

On the other hand, electron microscopy has not revealed any definite new structural component of the cytoplasm. It has helped in bridging the gap between the molecular and the organizational levels of the cell organelles by making supramolecular elements available for direct observation.

The most striking feature of the ultrastructural organization of the cytoplasm is the frequent occurrence of membranous components in various cell organelles, as well as in the ground substance of the cytoplasm. Membranes of various dimensions and different organization appear to represent a common and basic principle of organization in the cytoplasm.

Some examples of such membranous elements are presented here, together with some arguments which justify an interpretation of some of the structural patterns as reflecting a certain molecular structure of the membranes. Also, the probability that the observed structural patterns represent preformed patterns existing in the intact living cell is discussed. Finally, some ideas are presented regarding the functional significance of the various membranous elements.

When evaluating the structural patterns observed by means of the electron microscope, it is wise to account for a certain degree of deformation introduced by preparatory techniques. When comparing two patterns, sometimes one pattern easily might be imagined as derived from the other as a result of disorganization or deformation. In such a case, the second pattern is selected as probably more directly related to the *in vivo* pattern. When a certain variation in the patterns is observed, the most frequently occurring pattern is selected as the most representative.

The chapter on the biochemistry of mitochondria by Lehninger (p. 136) makes it quite natural to start this survey with some electron micrographs of mitochondria. For this purpose, mitochondria may be chosen from almost any type of cell. Let us choose those of the retinal receptors of the eye. The first figure shows a schematic drawing of the rod type of receptor in the guinea pig retina. These receptors are segmented cells with each segment structurally organized in a characteristic and different way (Fig. 1). The outer segments which are most remote from the pupil of the eye contain the photochemically active molecules, such as rhodopsin in the rod cells. Here, the primary reactions in converting electromagnetic energy into chemical energy take place. The inner segment contains all of the mitochondria of the receptor and can be looked upon, therefore, as the energy-generating center of the cell.

The rod and cone fibers connecting the inner segment with the synaptic bodies are organized like unmyelinated nerve fibers. The vitreous end of the receptor cell forms the synaptic body where the synaptic contacts between receptors and nerve cells are located.

Figures 2 and 3 picture the inner segment, *IS*, of a cone cell in the perch retina. Most of the inner segment consists of a dense aggregation of closely packed mitochondria which together form the so-called ellipsoid.

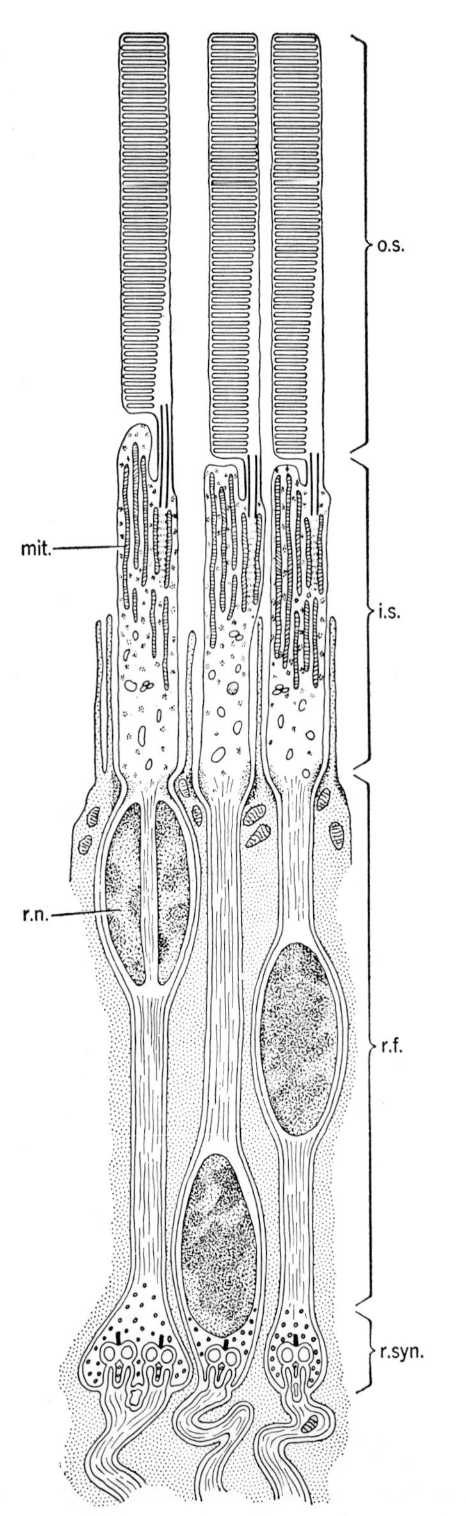


FIG. 1. Schematic drawing of retinal rods in the guinea pig eye. *o.s.*, outer segment; *i.s.*, inner segment; *r.f.*, rod fiber; *r.syn.*, rod synaptic body; *mit.*, mitochondria; *r.n.*, rod nucleus [from F. S. Sjöstrand, Intern. Rev. Cytol. **5**, 455 (1956)].

Each mitochondrion is bounded by a surface membrane, and a great number of inner membranes or platelets are oriented in a fairly parallel fashion in the interior. All of these membranes appear to be triple layered.[1–4] It is striking to note the rather constant spacing of the layers of these mitochondria membranes.

In 1952, when we first presented our structural model of mitochondria, Palade[5] proposed a different model consisting of a single-layered surface membrane which showed local infoldings which he called "cristae mitochondriales." These cristae left a central space in the mitochondrion free. After confirming our observations on the triple-layered character of the surface membrane, Palade changed his model by enveloping his original model in a peripheral opaque layer. He assumes[6] that this layer represents the surface membrane of the mitochondrion and that an outer space or chamber extends between the two opaque layers located at the surface of the mitochondrion. This space is continuous with spaces which extend in the cristae. An inner mitochondrion chamber forms a continuous central space. This interpretation differs from our own.

Our interpretation of the mitochondrial pattern is based upon certain observations that we had made on some typical lipoprotein systems—namely, the outer segments of retinal receptors[7,8] and the myelin sheath of peripheral nerves.[9,10] In both cases, polarization-optical data, and, in the myelin sheath, x-ray diffraction data had permitted the proposal of models showing alternating layers of lipid and protein molecules. X-ray diffraction data revealed some dimensions for the thickness of the layers in the myelin sheath and of the double layer of dried, mixed nerve lipids. For the latter, a value of 67.4 A was obtained.[11]

A discussion of the observations made on the outer segments of retinal receptors gives a necessary background for our interpretation of the mitochondrial pattern. In electron micrographs of sections (Figs. 4–6) oriented parallel to the long axis of the outer segments of rods and cones, alternating light layers and pairs of opaque layers are seen, the two opaque layers of a pair being fused along their rims. They thus bound a light interspace which, in guinea pig rods, has a thickness of 70 to 80 A. These pairs of layers form triple-layered disks, the total thickness of which varies from 100 to 250 A with the species and with the type of receptor, but remains constant for any one type of receptor. These disks are the unit structure of the outer segments.

The outer segments are 40% lipid in content.[12] More than 80% of these are phospholipids. The polarization-optical data have revealed that presumably the lipid molecules are oriented parallel to the long axis of the outer segment, and that the protein molecules form transversely oriented layers.[13]

When fragmenting isolated outer segments, it is possible to obtain round disks of varying thicknesses and also fragments of such disks.[7] The thicknesses of the disks isolated from the guinea pig retina are multiples of 140 A—that is, multiples of the thickness of the triple-layered unit disk which can be observed in sections and which measures 140 A in thickness in the

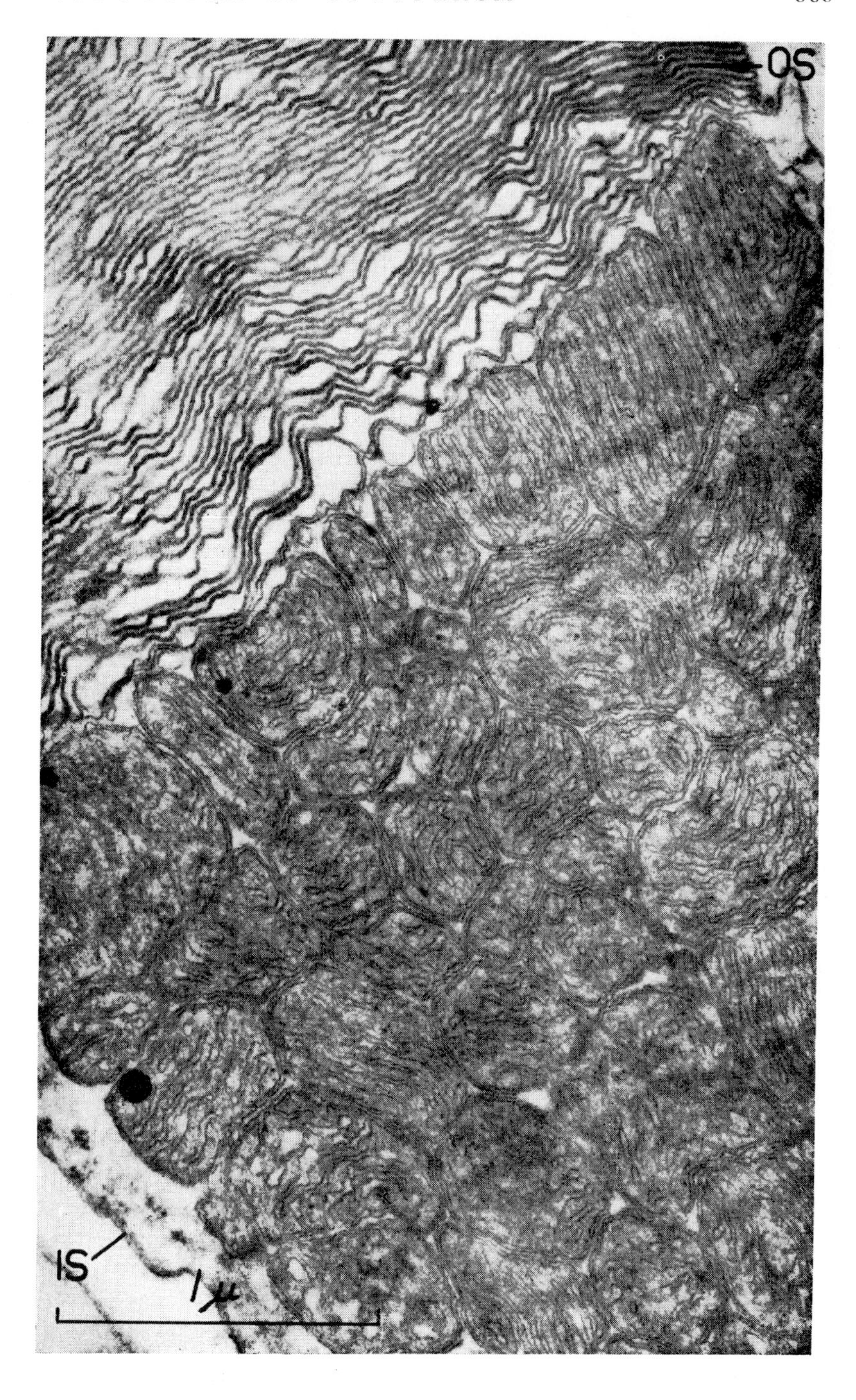


FIG. 2. Survey picture of the boundary between the outer, *OS*, and inner, *IS*, segments of a cone cell in the perch retina. Most of the inner segment consists of a dense aggregation of mitochondria [from F. S. Sjöstrand, *Ergebnisse der Biologie* (Springer-Verlag, Berlin, 1958), Vol. XXI, p. 128]. ×44 000.

guinea pig retina. The thickness of the thinnest membrane fragments is about 30 A. These fragments are osmiophilic. The thickness of the osmiophilic layers of the unit disks as observed in sections through guinea pig retinas is 30 to 40 A. There seem to be no doubt that these components in fragmented and in sectioned material are identical. The 140-A thick disks which were obtained by fragmentation could be demonstrated to consist clearly of two thin membranes, and under favorable conditions they could be split completely into two thinner components with a thickness of 70 A at the edge.

The fact that the thicknesses of the various disks obtained by fragmentation represent multiples of 140 A makes it justifiable to conclude that, when the outer segments are fragmented in piles of unit disks of different thicknesses and when these piles are dried on a supporting film, the interspaces between the triple-layered disks contribute insignificantly to the thicknesses of the piles. This means that these interspaces

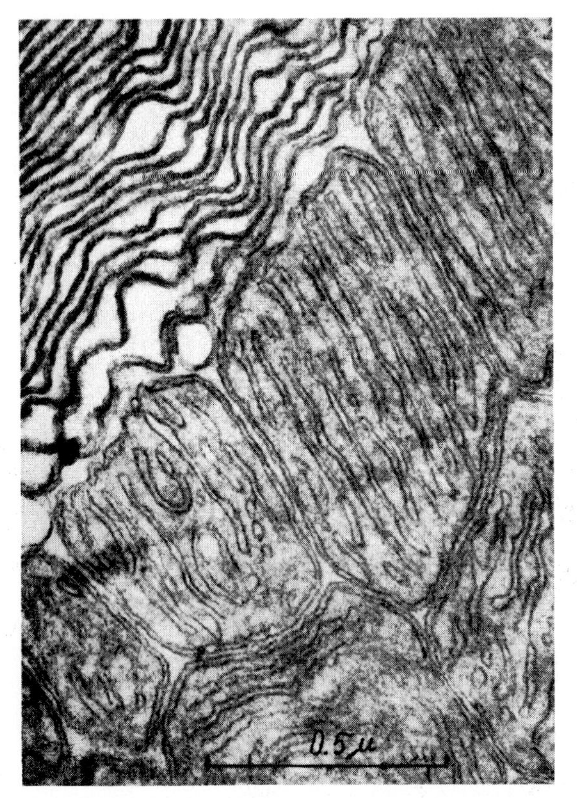


FIG. 3. Higher magnification of a region at the border between the inner and outer segments of a cone cell in the perch retina. Several mitochondria are observed densely aggregated. The triple-layered surface membrane and inner membranes show certain similarities to the triple-layered disks of the outer segment [from F. S. Sjöstrand, *Ergebnisse der Biologie* (Springer-Verlag, Berlin, 1958), Vol. XXI, p. 128]. ×65 000.

are filled mainly with an aqueous, ionic medium containing little or no lipids. The lipid molecules, therefore, are located presumably in the triple-layered disks. Their localization (Fig. 7) in the less-opaque space bounded by the 30- to 40-A thick osmiophilic layers was assumed because the dimension of this space, 70 to 80 A, could well accommodate a double layer of lipid molecules. Furthermore, the osmiophilic layers had shown a rather remarkable tensile strength in the fragmentation experiments which appeared to point to a protein nature. The total volumes of the opaque layers and of the spaces surrounded by these layers are estimated to be about equal. This observation is in accordance with the determined figure of 40% for the concentration of lipids in percent of dry weight.

The similarity of the mitochondrial pattern and that of the outer segments of retinal receptors is rather striking. Values as high as 40% were reported for the lipid content of mitochondria, and they showed positive birefringence after freeze-drying,[14] which could be assumed to be due to oriented lipid molecules. The triple-layered membranes of the mitochondria were interpreted as representing two protein layers sandwiching a double layer of lipid molecules (Fig. 8). This interpretation is based upon a series of evaluations of indirect evidence and upon a generalization which well might be erroneous.

The myelin sheaths of peripheral nerves also show a characteristic layered structure. Furthermore, peripheral nerves can be fragmented into thin membranous fragments.[15,16] In this instance, polarization-optical data, as interpreted by Schmidt,[13] revealed that the lipid molecules were oriented radially and that protein molecules formed concentric layers around the axis cylinder. X-ray diffraction data by Bear, Palmer, and Schmitt[11] were interpreted as showing a repeat period of 171 A in amphibian nerves, and 185 A in mammalian nerves. The repeat period was interpreted as containing two bimolecular leaflets of lipids, each about 67 A thick, and a layer of proteins about 25 A thick. A concentrically layered structure could be observed in the electron micrographs, obtained in 1952,[9,10] of osmium-fixed, sectioned peripheral nerves. Opaque layers about 25 A thick are arranged concentrically in the myelin sheath with a definite periodicity. In each period, a fainter opaque line is seen to halve the main period. The mean periodicity was estimated at 120 A in the electron micrographs of 20 nerve fibers.

Although the x-ray diffraction and the electron-microscope data did not coincide as to the length of the period, the over-all scheme predicted from the former data appears to be pretty well confirmed by the latter. From the dimensions, it was concluded that the opaque layers correspond to the protein layers, and the less opaque layers to bimolecular leaflets of lipids. The recent extensive work by Fernández-Morán and Finean[17] on this problem seems to support this interpretation.

A membranous component which is frequently present in the basophilic regions of the cytoplasm is discussed next. These membranes are particularly abundant in the excretory pancreas cells, the salivary gland cells, the plasma cells, and other types of cells in which a somewhat intense protein synthesis takes place.

In the exocrine pancreas cells (Fig. 9), most of the ground substance of the cytoplasm is filled with this type of membrane.[1,18] In osmium-fixed material, the membrane is easily identified by the numerous opaque particles that are attached to one surface. These particles have an average diameter of 150 A. Their form varies; the particles are more or less irregularly, angularly shaped. The membrane to which these particles are attached consists of an osmiophilic component which is about 40 A thick. These membranes are arranged in pairs and a pair bounds a narrow space or, as in the thyroid epithelium, rather large, irregularly shaped spaces. They thus appear to divide the cytoplasm into two different parts or compartments.[19]

Opaque particles of identical appearance, as those attached to the membranes, frequently occur free in the cytoplasm. Palade and Siekevitz[20] clearly demonstrated that the opaque particles are responsible for the RNA content of the microsome fraction from pancreas tissue. The terms microsomes, ribosomes, or microsomal

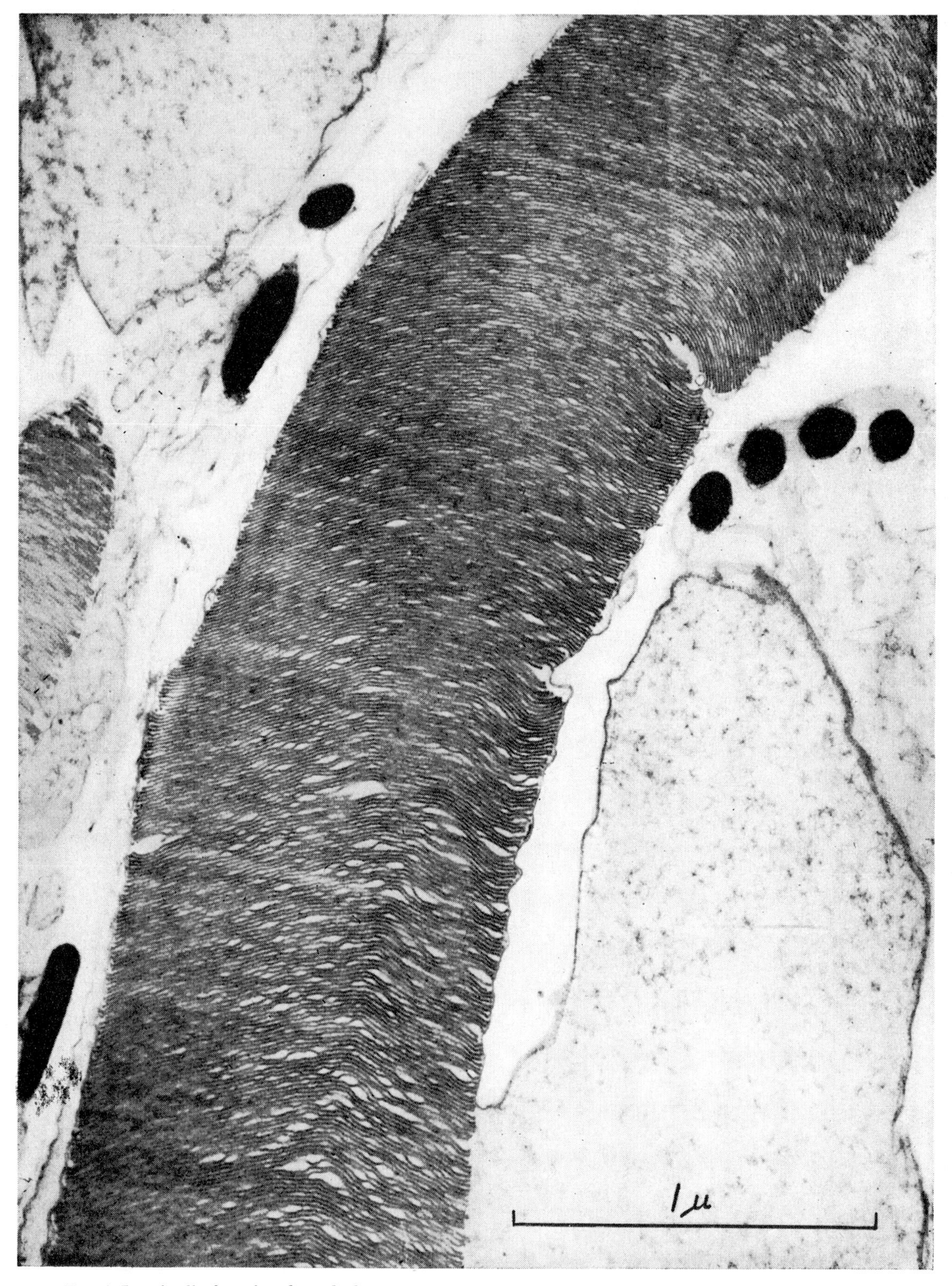


FIG. 4. Longitudinal section through the outer segment of a retinal rod cell from the perch eye showing the uniformly layered structure of the outer segment. ×59 000.

particles now seem to be used to classify this component which represents only a minor part of the classical microsome fraction.

At this point, it seems justifiable to make some comments on what is known about the fixation and embedding artifacts that are introduced when fixing

FIG. 5. Longitudinal section through a retinal-rod cell of the perch eye showing the triple-layered disks, which represent the elementary component of the outer segment [from F. S. Sjöstrand, *Ergebnisse der Biologie* (Springer-Verlag, Berlin, 1958), Vol. XXI, p. 128]. ×82 000.

and embedding labile living matter. The tissue is fixed in a solution of osmium tetroxide, dehydrated in a graded series of concentrations of ethyl alcohol, and then transferred to methacrylate which is polymerized before the tissue can be sectioned.

At the beginning, we were very much concerned about this problem and, therefore, we intentionally chose to analyze such structural components about which a little was known from an ultrastructural point of view through polarization-optical and x-ray diffraction analysis. In the outer segments of retinal receptors and in the myelin sheath, patterns in complete harmony with polarization-optical data obtained from fresh, unfixed material were observed by means of electron microscopy, as mentioned in the foregoing. Regarding the exocrine pancreas cells, there were no data available regarding birefringence of the cytoplasm *in vivo*. Living exocrine pancreas cells in mice were analyzed in polarized light and we found that the cytoplasm was birefringent.[1] When analyzed, this birefringence proved to be negative with the axis oriented perpendicularly to the cell membrane, a result that could fit in very well with the electron-microscope picture.

When the direct study of living material is excluded, the most rational way of checking the reliability of such a preparatory technique is to change the technique so radically that it appears unlikely that the two techniques could produce identical distortions. The second preferred preparatory technique is then preservation by means of freeze-drying. In fact, we proceeded in the opposite direction and studied frozen-dried material first[1] and later checked the observations made on exocrine pancreas cells by means of osmium fixation and birefringence analysis.

The freeze-drying technique is a rather delicate procedure with many pitfalls. The observations made on frozen-dried, sectioned material, stained with phosphotungstic acid, were rather surprising.[21] Similar structural patterns could be observed in the ground substance of the cytoplasm as well as in the mitochondria, with the remarkable exception that the patterns appeared as negative patterns, as compared to those obtained after osmium fixation (Figs. 10 and 11). These results mean that these patterns probably are preformed and exist in the living cell. At the same time, however, they may make it necessary to re-evaluate the interpretation of the molecular structure assumed to be responsible for these patterns.

One feature of the frozen-dried specimen must be noted. The so-called RNA particles do not show up.

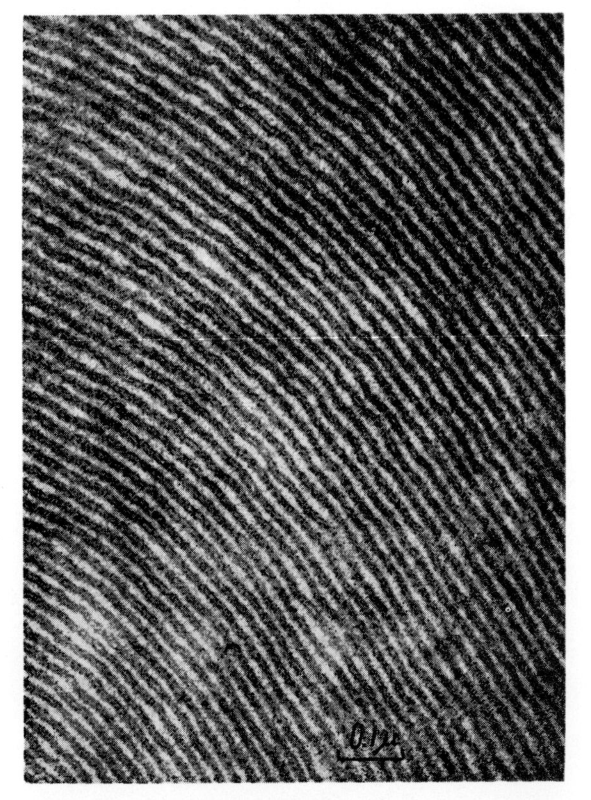


FIG. 6. Section oriented perpendicularly to the triple-layered disks of the outer segment of a cone cell in the perch eye. Notice the tendency to a uniform blackening of the whole triple-layered disk due to a staining of the middle layer [from F. S. Sjöstrand, *Ergebnisse der Biologie* (Springer-Verlag, Berlin, 1958), Vol. XXI, p. 128]. ×87 000.

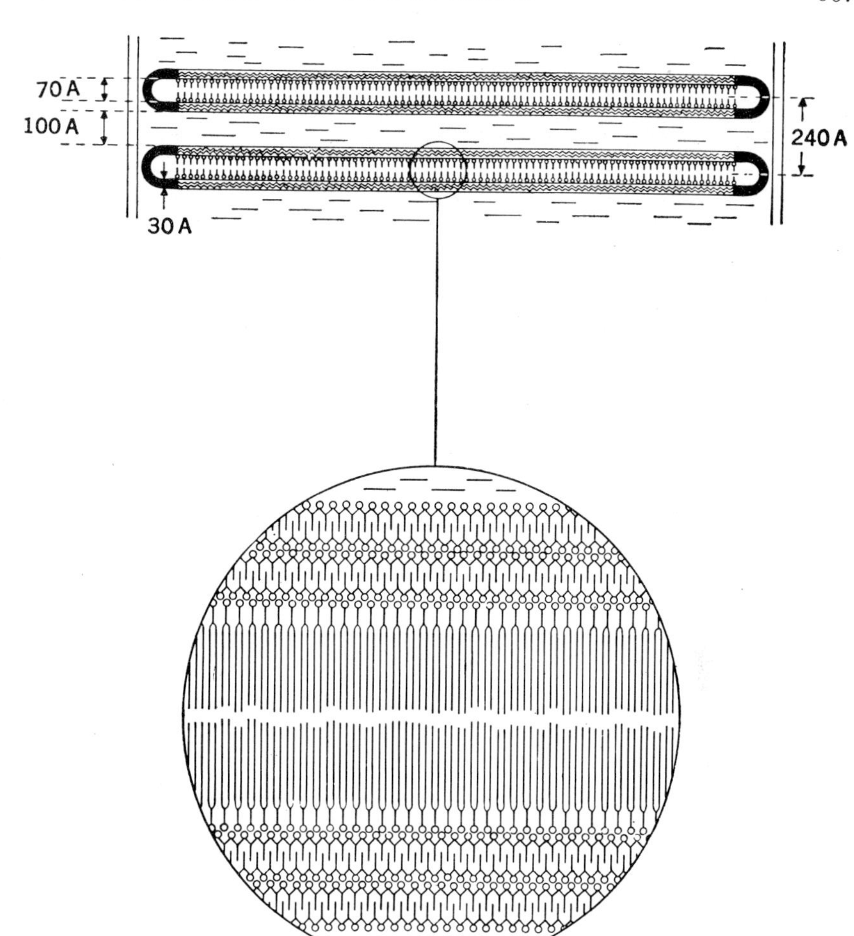


Fig. 7. Schematic drawing to illustrate a proposed interpretation of the localization and orientation of lipid and protein molecules in the triple-layered elementary disks of the outer segments of retinal receptors [from F. S. Sjöstrand, *Ergebnisse der Biologie* (Springer-Verlag, Berlin, 1958), Vol. XXI, p. 128].

The cytoplasm between the cytoplasmic membranes appears homogeneous even after long periods of staining. One cannot exclude, therefore, the possibility that these particles are artifacts formed through an aggregation or precipitation in connection with the preparatory procedure. This possibility has to be tested more extensively than has been done to date.

Even if some doubts can be raised regarding the particulate form of the RNA-protein components in the cytoplasm of intact living cells, there is little or no doubt that these particles indicate the approximate location of RNA in the cytoplasm. The firm connection of the particles with the cytoplasmic membranes, on the other hand, can be questioned as due to secondary absorption. This would explain why such particles are attached to the outer opaque layer of the nuclear membrane in cells where the RNA is located close to the nucleus. Furthermore, it explains why the particles are absent when the cytoplasmic membrane is located very closely to mitochondria where the amount of RNA easily can be assumed to be inadequate for the formation of such RNA particles.

Another structural component of the cytoplasm is the so-called *Golgi apparatus*. This component has been interpreted by light microscopists as an apparatus involved in secretory mechanisms, and it has been extensively studied in glandular cells. The secretory products were assumed to be either produced or accumulated in this region.

Electron microscopy has revealed another type of cytoplasmic membrane as the main component of the Golgi apparatus.[22–24] These membranes are arranged in pairs also and bound a space that can be either very narrow, about 100 A or less in width, or very wide appearing as large vacuolar, irregularly shaped spaces (Figs. 12 and 13). The membranes consist of one osmiophilic component, 60 A thick. Presumably, a less osmiophilic component separates the membrane pairs, because the space between the pairs frequently is very constant, measuring about 60 A in width.

It is rather striking that the zymogen granules which contain the secretory products of the pancreatic cells show a definite topographic relationship to the Golgi apparatus. The zymogen granules can be divided into two types—the fully-developed granules and the precursor granules. A complete series of intermediate-type granules bridges over from the one type to the other. The fully developed zymogen granules are spherical

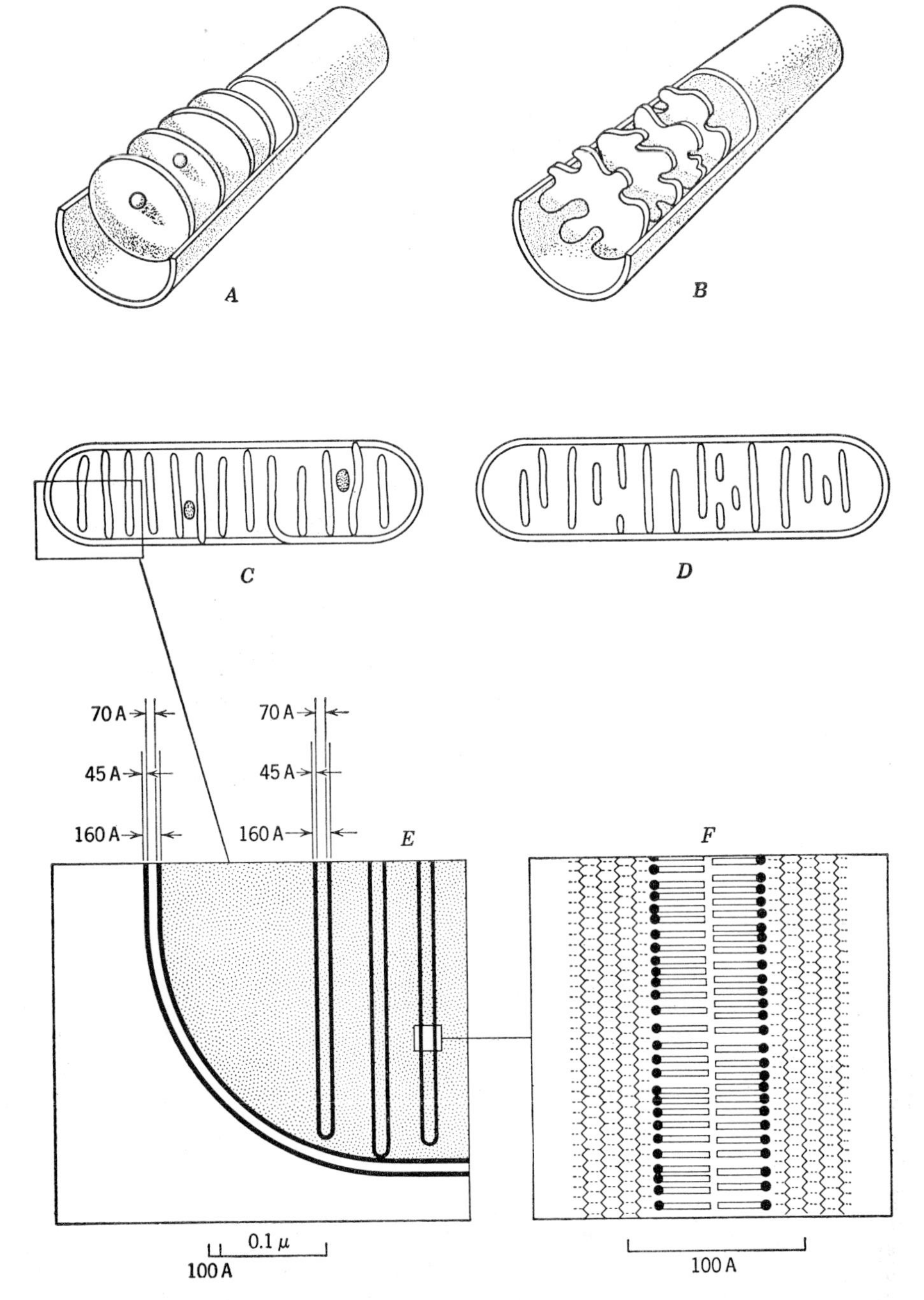


FIG. 8. Schematic drawing which illustrates an interpretation of the three-dimensional appearance of the mitochondria and of the molecular architecture of the mitochondria membranes. The three-dimensional representation *A* and *B* is based on an interpretation of the patterns observed in sections which can be, for instance, like the ones shown in *C* and *D*. *E* gives some commonly observed dimensions of the mitochondrial membranes, and *F* a scheme of the proposed arrangement of lipid and protein molecules in the membranes [from F. S. Sjöstrand, in *Fine Structure of Cells*, Proc. Symp. VIII Cong. Cell. Biol., Leiden, 1954 (Noordhoff, Ltd., Leiden, 1955), p. 16].

and of high density, and are bounded by a surface membrane about 50 A thick. The precursor granules are irregularly shaped and of lower opacity, but are bounded by a surface membrane which shows outpocketings and bulgings (Fig. 13). The precursor granules are found only in connection with the Golgi apparatus where the fully developed zymogen granules are either rare or completely absent. Furthermore, there is a definite polarity of the Golgi apparatus with the precursor granules arranged on the apical side of the Golgi membranes. On the basal side, the type of cytoplasmic membranes described earlier reaches close to the Golgi membranes. A zone of granulated cytoplasm, however, separates the territories of these two types of membranes.

Turning to the problem of the functional significance of these various kinds of membranes, the mitochondrial membrane is considered first. The article by Lehninger (p. 136) introduces the present concept regarding the importance of the structural organization of the mitochondria. The biochemical data point to a structural factor as important in connection with coupled phosphorylation. The electron transport seems to depend upon an organization of macromolecular particles. The

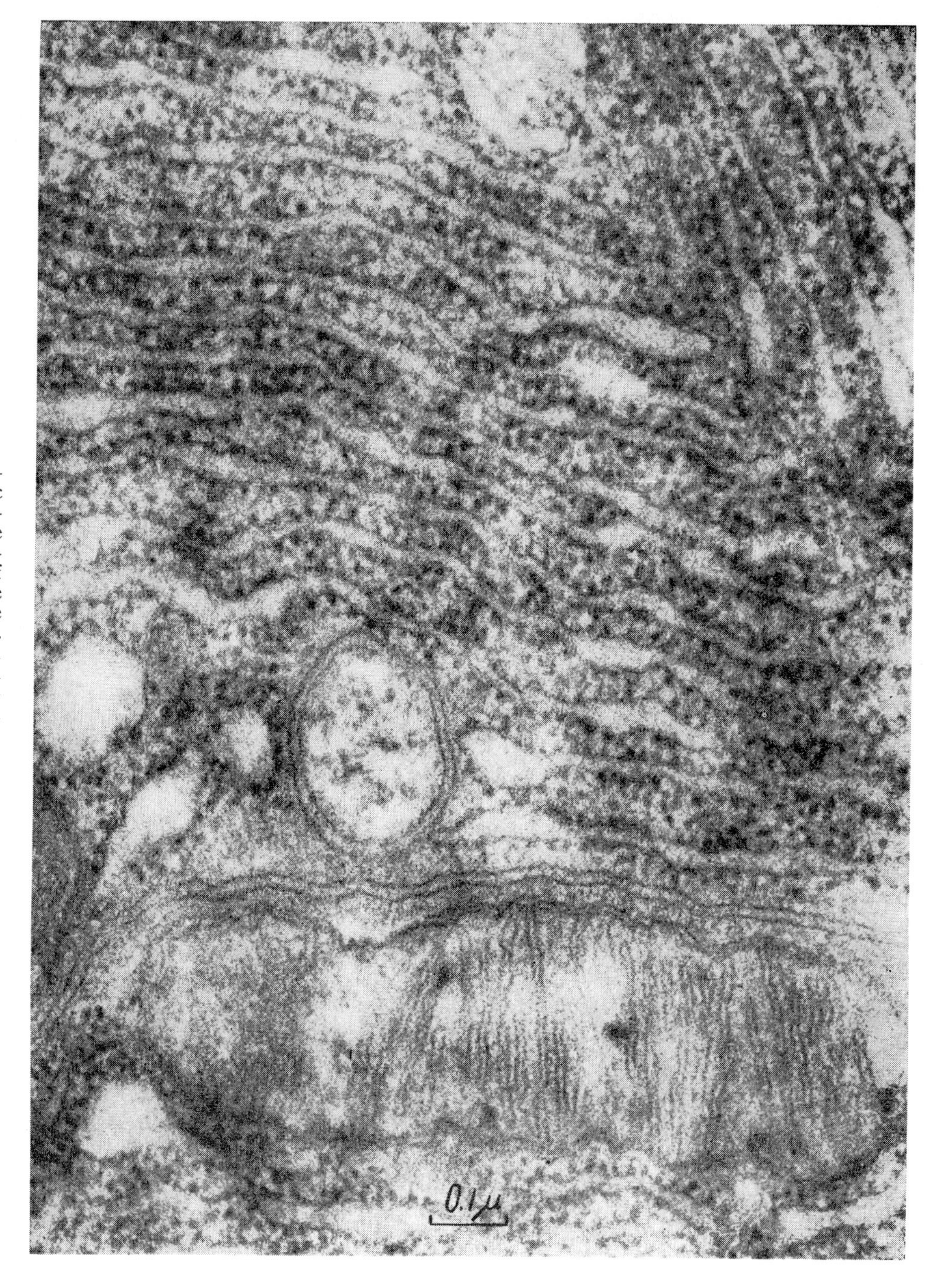


FIG. 9. Cytoplasmic membranes in the basophilic cytoplasm of exocrine pancreas cells. Notice the opaque 150-A particles attached to the one surface of the membranes which are arranged in pairs. In the lower part of the picture a cell boundary represented by two mutually parallel lines, and below the boundary a mitochondrion [from F. S. Sjöstrand and V. Hanzon, Exptl. Cell Research 7, 393 (1954)]. ×108 000.

smooth coordination of various enzymatic reactions was imagined as being the result of a topographic arrangement of the enzyme molecules according to certain patterns. The membranes of the mitochondria were assumed to be composed to a great extent, or altogether, of mitochondrial enzyme molecules oriented and spread out in layers in combination with layers of oriented lipid molecules.

The experiments done so far to isolate the membranous components of the mitochondria from the ground substance have been rather crude from a morphological point of view, but are interpreted as supporting this idea. The mitochondrial membranes are considered to be enzymatically active where the organization of the enzyme molecules in a plane increases the chances for a well-coordinated sequence of enzyme-substrate interactions. This organization is thought to be important for a rapid step-wise degradation of substrate molecules, and possibly for their step-wise synthesis.

One may assume that the synthesis of the secretory products in the exocrine pancreas cells takes place

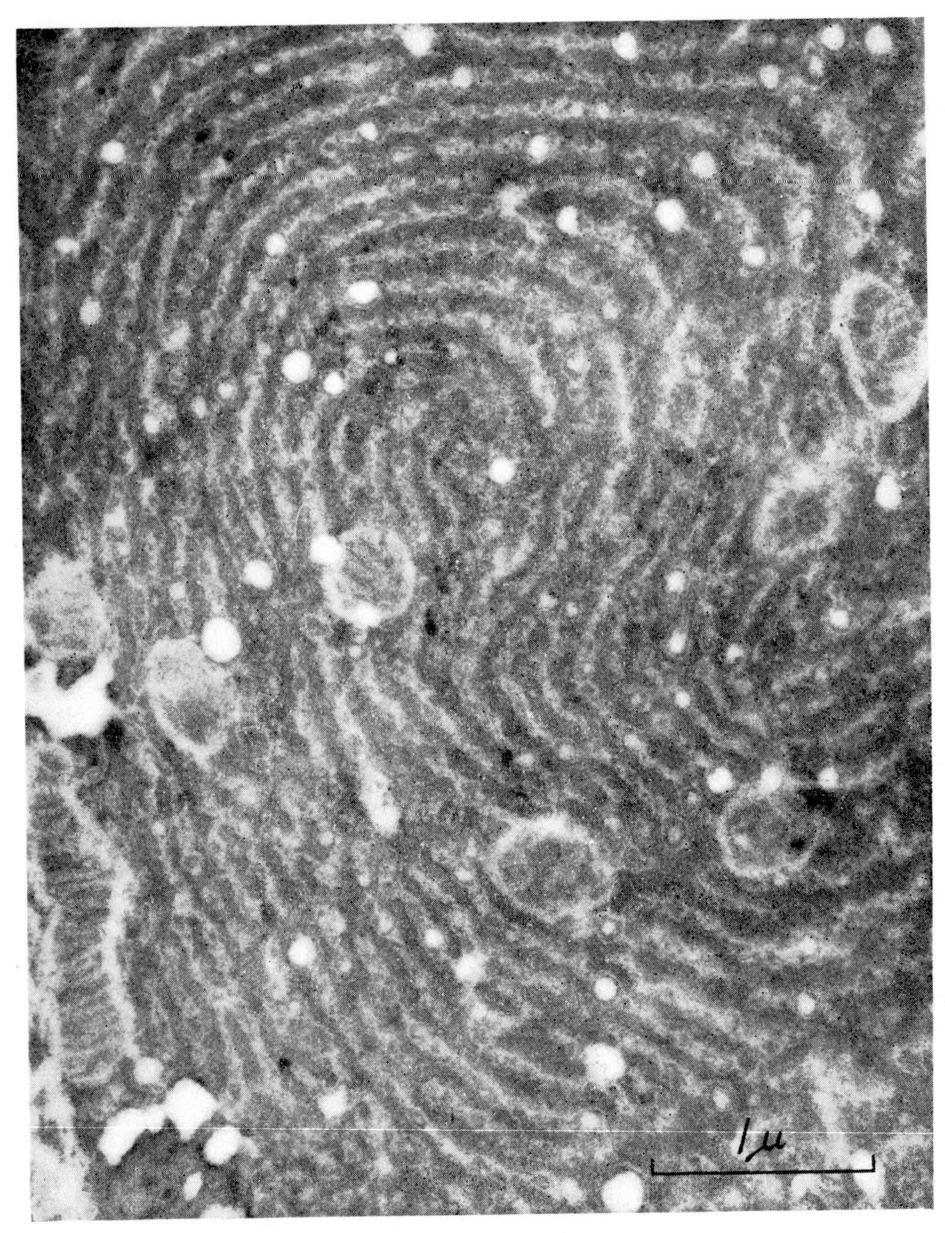


Fig. 10

Figs. 10–11. Sections through exocrine pancreas cells preserved by means of freeze-drying. The methacrylate section was stained with phosphotungstic acid before the examination in the electron microscope. The cytoplasmic membranes are visible as well as the mitochondria but the pattern is a negative one as compared to that shown in Fig. 2. The spaces between the membranes (Fig. 11) are uniformly filled with material in contrast to the lack of material, except the opaque 150-A particles, after osmium fixation. No such particles can be observed in these pictures. *Mi*, mitochondria; *CM*, cytoplasmic membranes; *V*, ice crystal vacuole [from F. S. Sjöstrand and R. F. Baker, J. Ultrastructure Research 1, 239 (1958)]. ×30 000 (Fig. 10); ×56 000 (Fig. 11) (Fig. 11 is on opposite page).

in steps. First, an enzyme-containing membrane is formed which is able to synthesize the secretory products (Fig. 14). This membrane is manufactured in the Golgi apparatus from material delivered by the cytoplasm as represented by the RNA particles. These particles swell, fuse, and mix with lipids, and extensive hydration results in the formation of vacuoles bounded by a thin membrane consisting of lipoproteins. The vacuoles collapse and pairs of membranes are formed which disintegrate into fragments forming the surface membrane of the precursor granules. The surface membrane then synthesizes the secretory products which are successively accumulated in the granule. Raw material can be delivered all of the time by the RNA-rich cytoplasm.

The foregoing is a wild hypothesis that could explain the morphological observations made so far. It is derived, however, from still pictures; the various stages

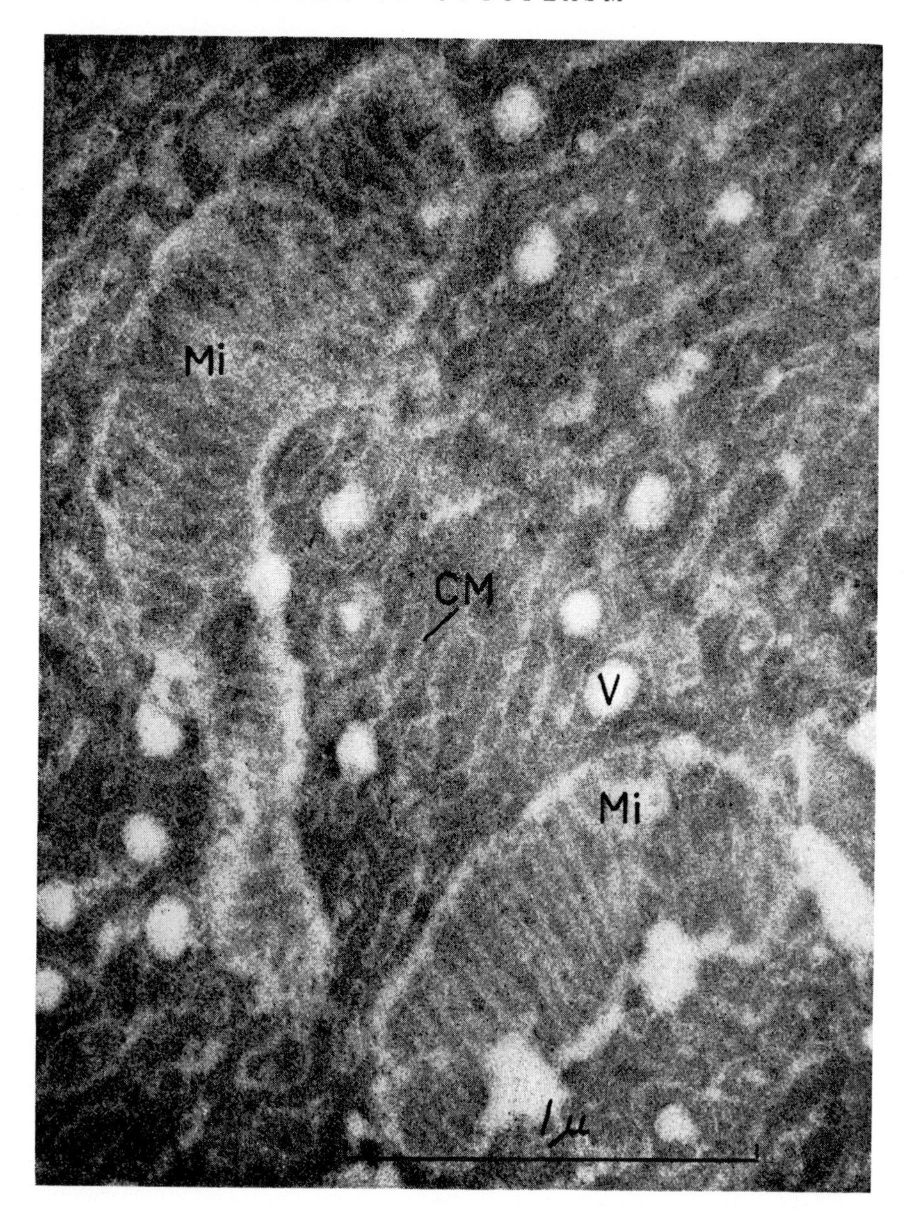


FIG. 11 (for legend, see Fig. 10).

have been selected subjectively from these pictures. Experimental evidence remains to be presented. One thing seems to be quite clear—namely, that the Golgi membranes can form vesicles or the membranes of secretory granules. This latter process is demonstrated better by the goblet cells in the intestinal epithelium. Holmberg,[25] in our laboratory, found that, after administration of diamox, an inhibitor for carbonic anhydrase, the Golgi apparatus in the ciliary epithelium of the eye partially disintegrated into vesicles and the whole cytoplasm became loaded with vesicles. These reversible changes could be described in a quantitative way and be correlated with a temporary 60% inhibition of the secretion of aqueous humor.

In certain types of cells, the plasma membrane appears extremely folded. It is not quite clear whether these folds are tight infoldings of the plasma membrane or whether they represent interdigitating ridges of the cell surfaces of adjacent cells. Zetterqvist,[26] in our laboratory, demonstrated that the plasma membrane can appear rather different at various parts of the surface of the same cell. In the intestinal epithelium (Fig. 15) the free surface of the cells forms a number of cylindrical processes, the so-called brush border. The plasma membrane bounding these processes appears as a 105-A thick triple-layered component (Fig. 16). Where two epithelial cells are in close contact, the plasma membrane appears as an osmiophilic layer 60 to 70 A thick. Between the two osmiophilic layers of the two adjacent epithelium cells, there is a light space with a very uniform thickness of about 100 A.

On the other hand, the plasma membrane which

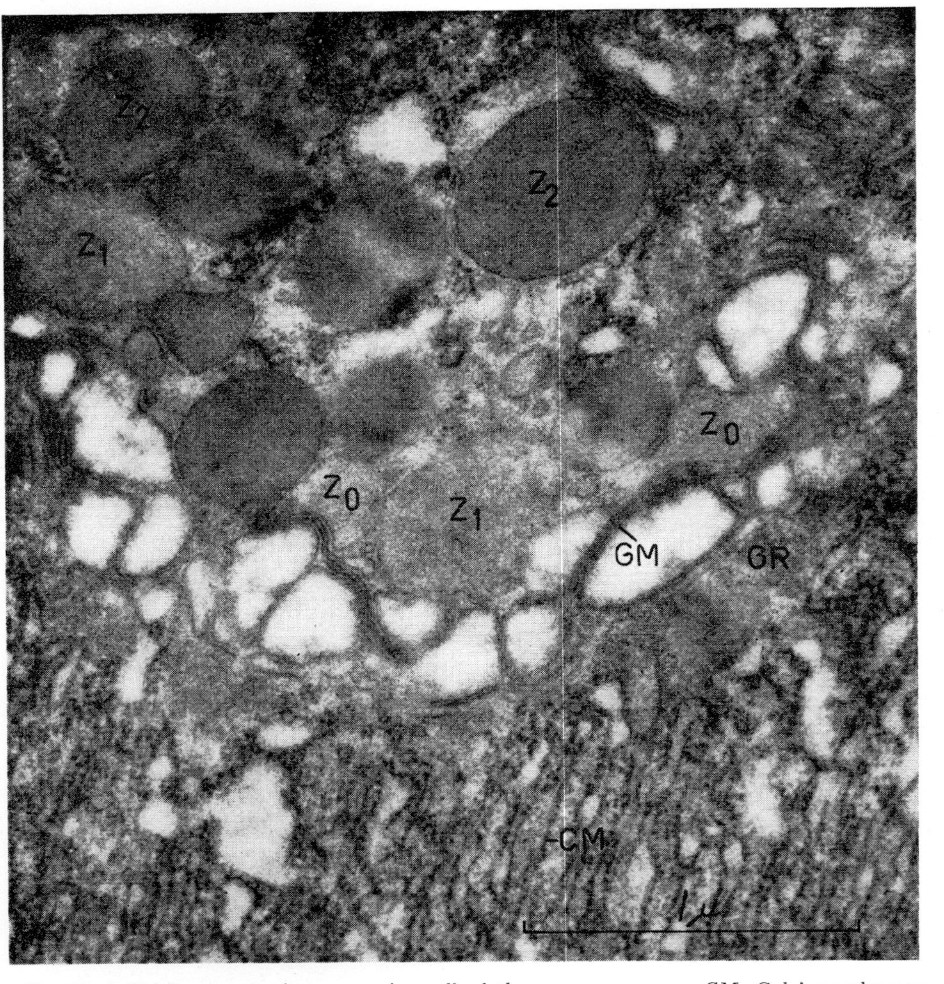


FIG. 12. A Golgi apparatus in an exocrine cell of the mouse pancreas. *GM*, Golgi membranes; *GR*, granules on the basal side of the Golgi apparatus; *CM*, cytoplasmic membranes; $Z_0$, prozymogen granules that appear to develop from the Golgi membranes; $Z_1$, prozymogen granules which can be interpreted as intermediate stages between $Z_0$ and $Z_2$, which are fully developed zymogen granules [from F. S. Sjöstrand and V. Hanzon. Exptl. Cell Research **7**, 415 (1954)]. ×44 000.

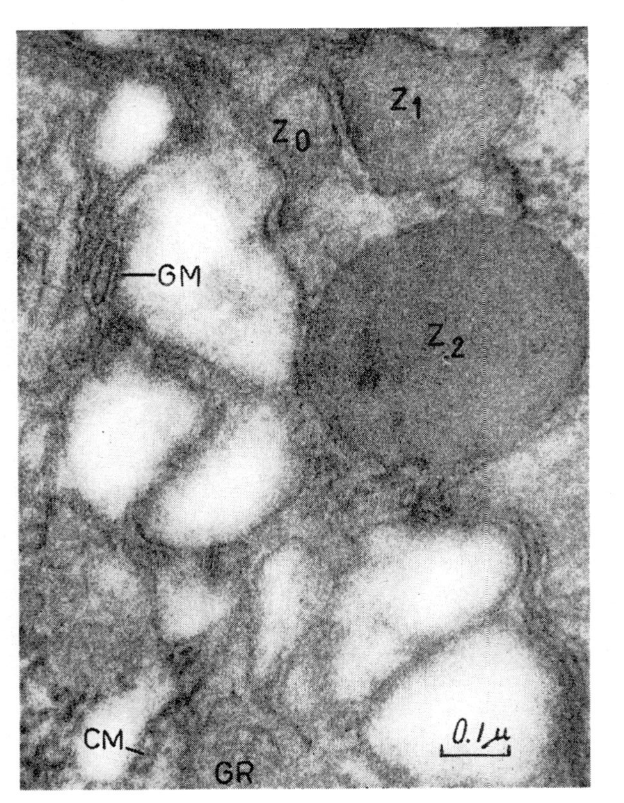


FIG. 13. Higher magnification of a part of Fig. 12. For explanation of indications, see Fig. 12. ×83 000.

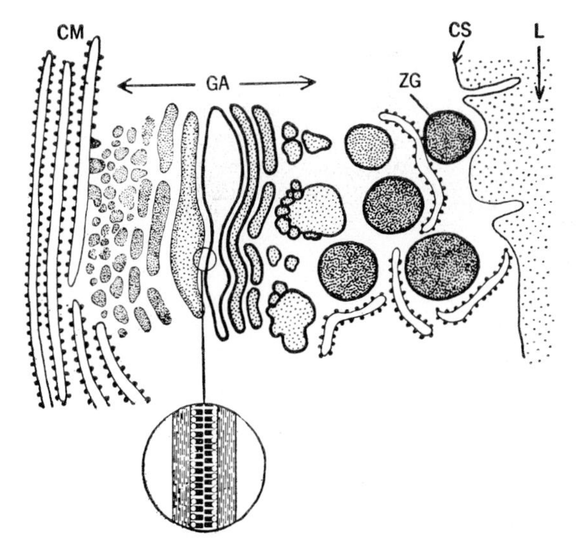


FIG. 14. Schematic drawing illustrating a hypothesis which aims at interpreting the functional significance of the Golgi apparatus of the excretory cells of the pancreas as an assembly line for membrane material with the power to synthesize the secretory products of the exocrine cells from raw material in the cytoplasm. *CM*, cytoplasmic membranes; *GA*, Golgi apparatus with formation of membranes from material delivered by the cytoplasmic membranes; *ZG*, zymogen granules formed from precursor granules in the Golgi apparatus; *CS*, cell surface facing the lumen of the excretory duct, *L* [from F. S. Sjöstrand, UNESCO Symp. Patterns of Cellular and Sub-Cellular Organisation, Edinburgh, 1957 (to be published.)]

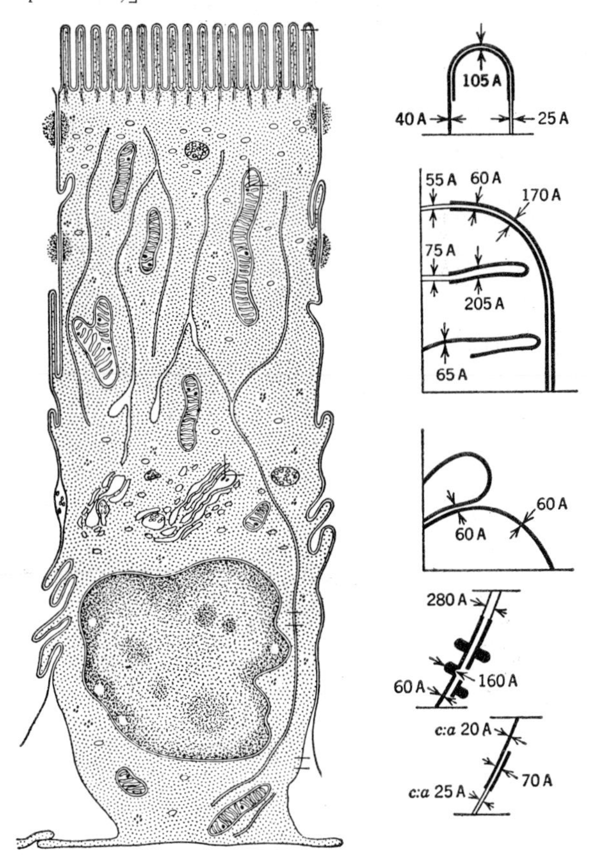


FIG. 15. Schematic drawing of a cylindrical epithelial cell lining the small intestine in the mouse. Notice the so-called brush border at the upper surface of the cell which faces the lumen of the gut [from H. Zetterqvist, *The Ultrastructural Organization of the Columnar Absorbing Cells of the Mouse Jejunum*. Thesis, Stockholm (1956)].

covers the surface of the cell facing the intercellular spaces and the basement membrane appears as a triple-layered structure which measures only 70 A in thickness. This structure consists of two opaque layers, each about 25 A thick, separated by a 20-A thick light interspace. A similar appearance has been reported recently by Robertson[27] for the plasma membranes of the Schwann cells. Robertson has interpreted this approximately 70-A thick triple-layered structure as representative of the whole plasma membrane (unit membrane) and which consists of two protein monolayers sandwiching a double layer of lipid molecules.

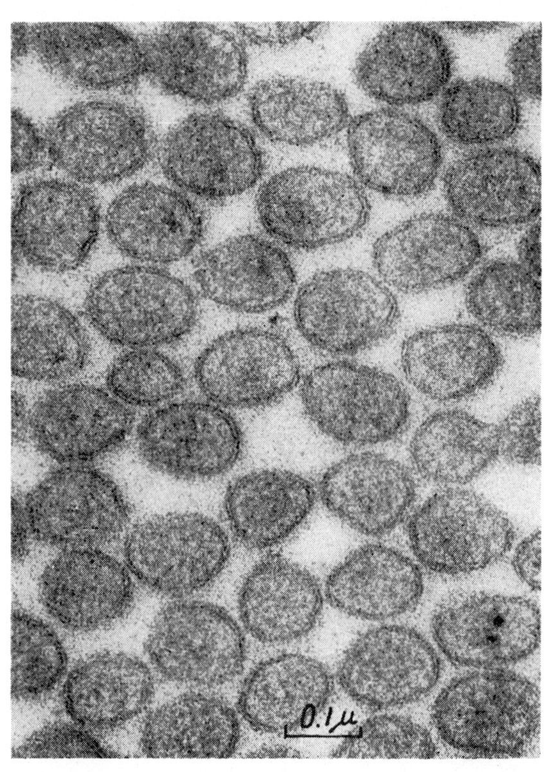


FIG. 16. Transversal section through the cylindrical processes of the brush border zone in an intestinal epithelial cell. The processes are bounded by a triple-layered surface membrane with an average total thickness of 105 A [from H. Zetterqvist, *The Ultrastructural Organization of the Columnar Absorbing Cells of the Mouse Jejunum*. Thesis, Stockholm (1956)]. ×103 000.

This interpretation is based on a great deal of evidence and it is discussed further by Schmitt (p. 455). It is assumed that the consistency of the width, about 100 A, of the light interspaces between two cells in close contact indicates the presence of some binding material, such as lipids or mucopolysaccharides.

Consider now the formation of membranous material in the outer segments of retinal receptors. It is seen, in the double cones of the perch retina, that the disks of

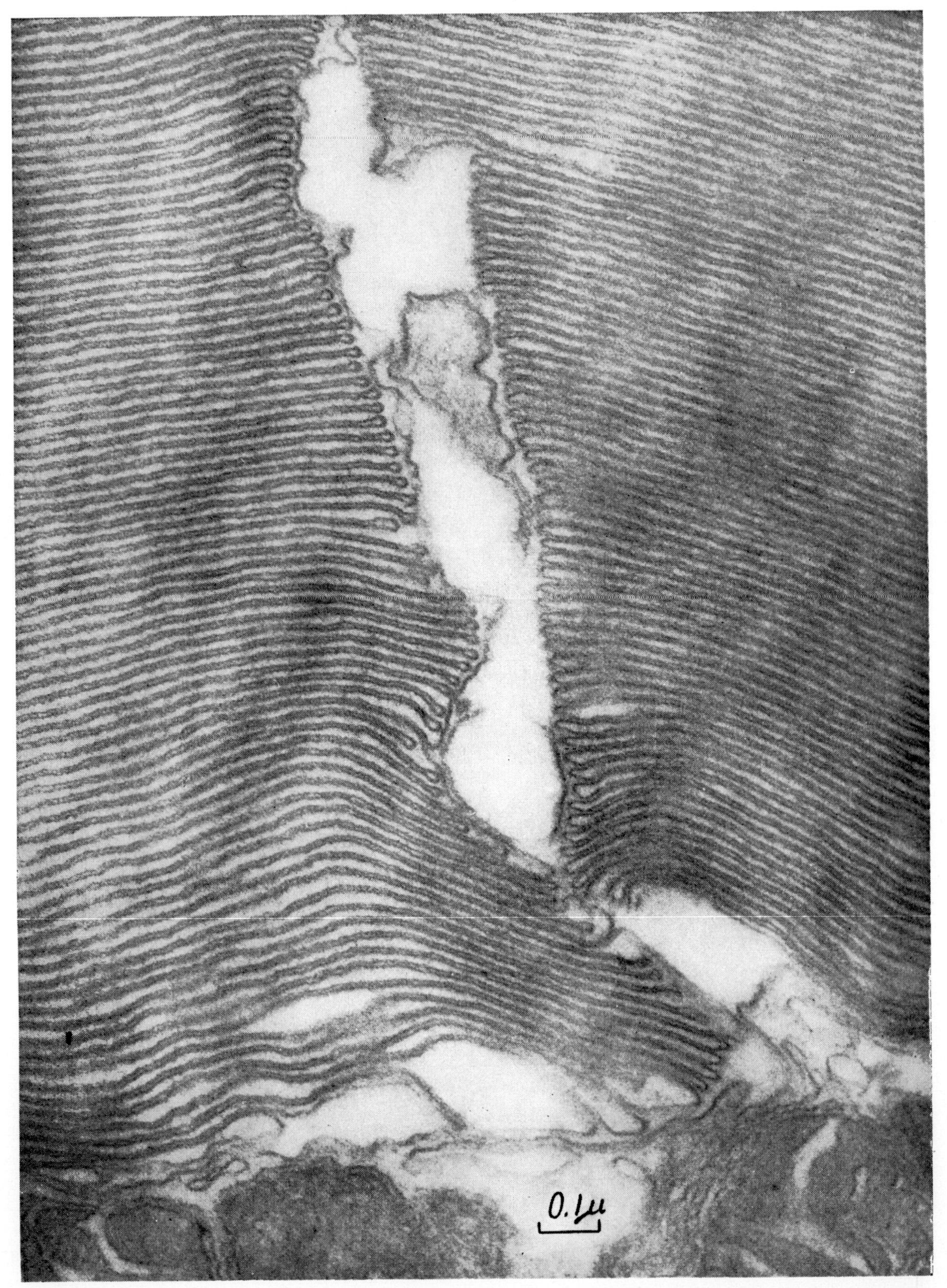


FIG. 17. The peripheral parts of two outer segments belonging to two twin cones of the perch retina. The outer segment shown in the lower half of the picture demonstrates that the less opaque interspaces in the triple-layered disks are open, and that the opaque layers of adjacent disks are continuous. At the opposite side of this segment, the conditions are those observed in the outer segment shown in the upper half of the picture. The less opaque interspaces are closed. This shows that the pile of disks is formed by a continuous membrane that is repeatedly folded. In other types of receptors, no such obvious folding can be observed but the disks are mutually connected through tube-like stalks which are continuous with the opaque layers of the disks. ×100 000.

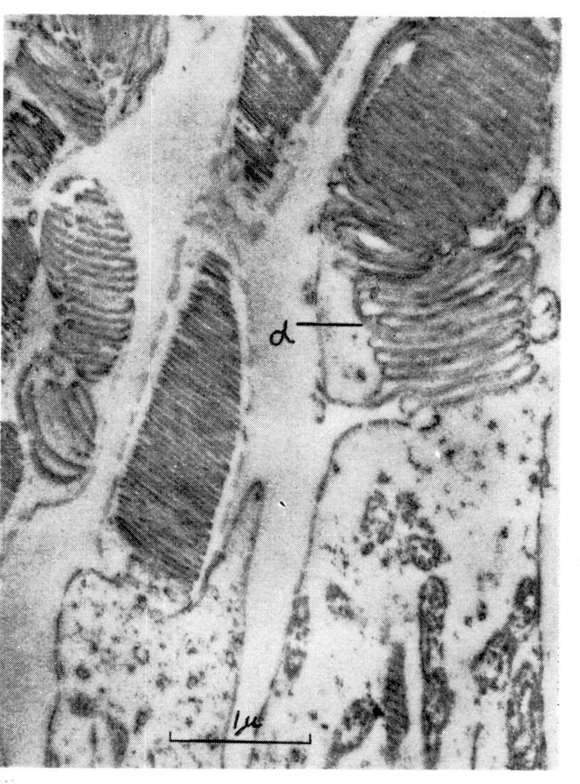


FIG. 18. Retinal receptors from an eye of a four-day-old kitten. At $\alpha$, a repeatedly folded membrane is observed which appears to represent a stage of the development of the outer segment. In the cat, the retinal receptors are differentiated after the birth. ×18 000.

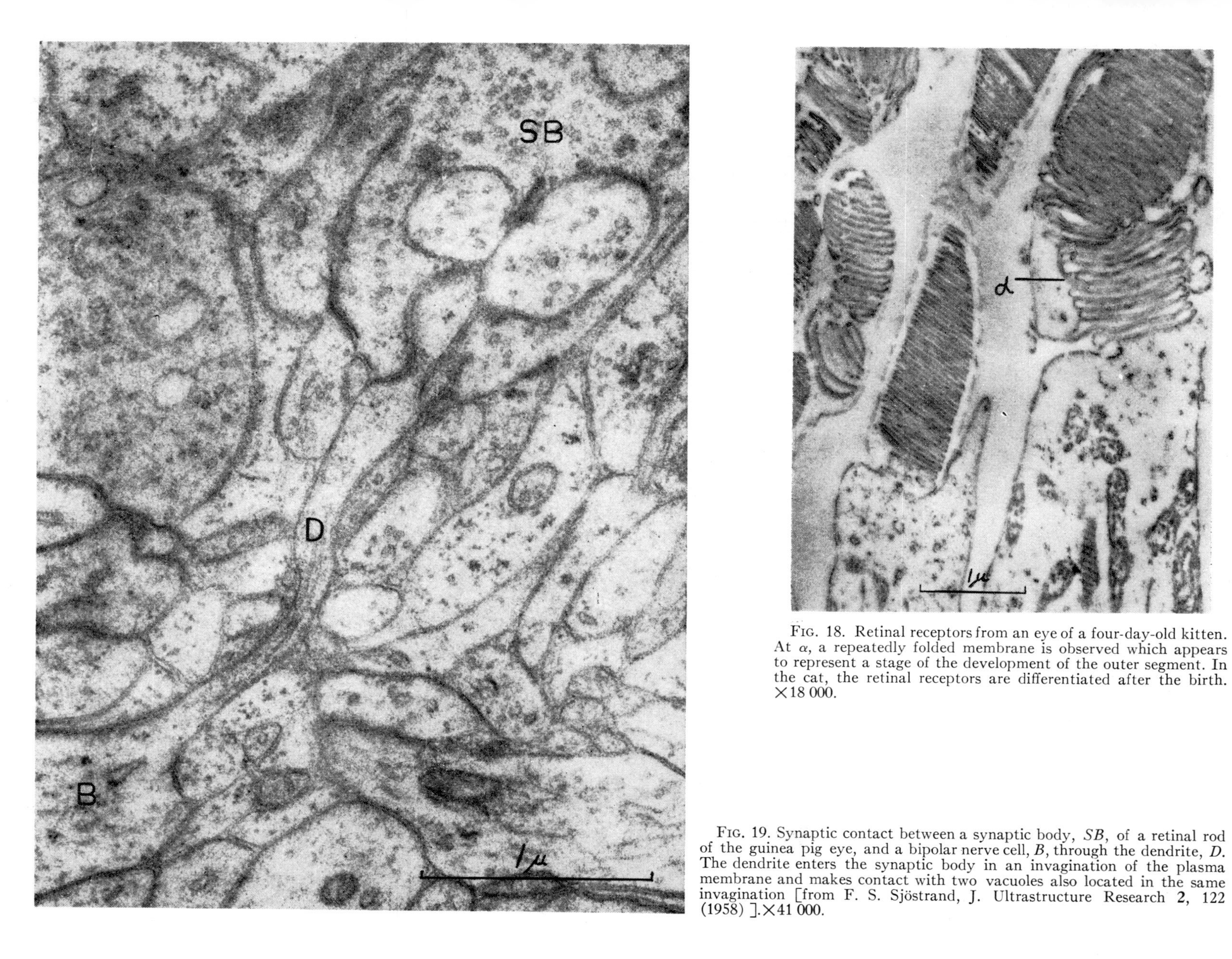


FIG. 19. Synaptic contact between a synaptic body, *SB*, of a retinal rod of the guinea pig eye, and a bipolar nerve cell, *B*, through the dendrite, *D*. The dendrite enters the synaptic body in an invagination of the plasma membrane and makes contact with two vacuoles also located in the same invagination [from F. S. Sjöstrand, J. Ultrastructure Research **2**, 122 (1958)].×41 000.

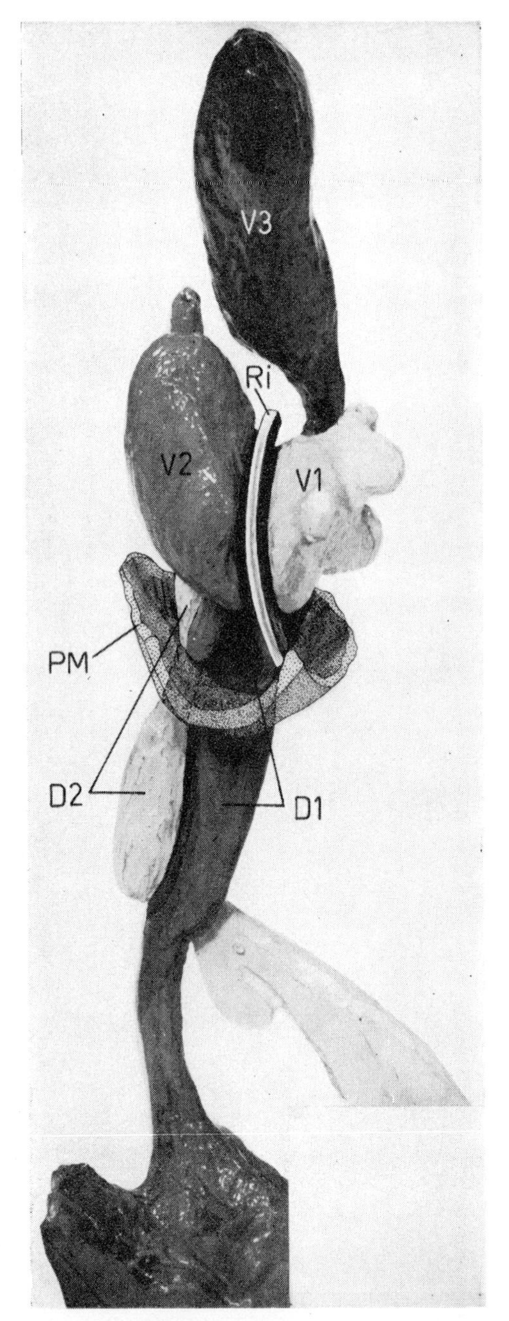


Fig. 20

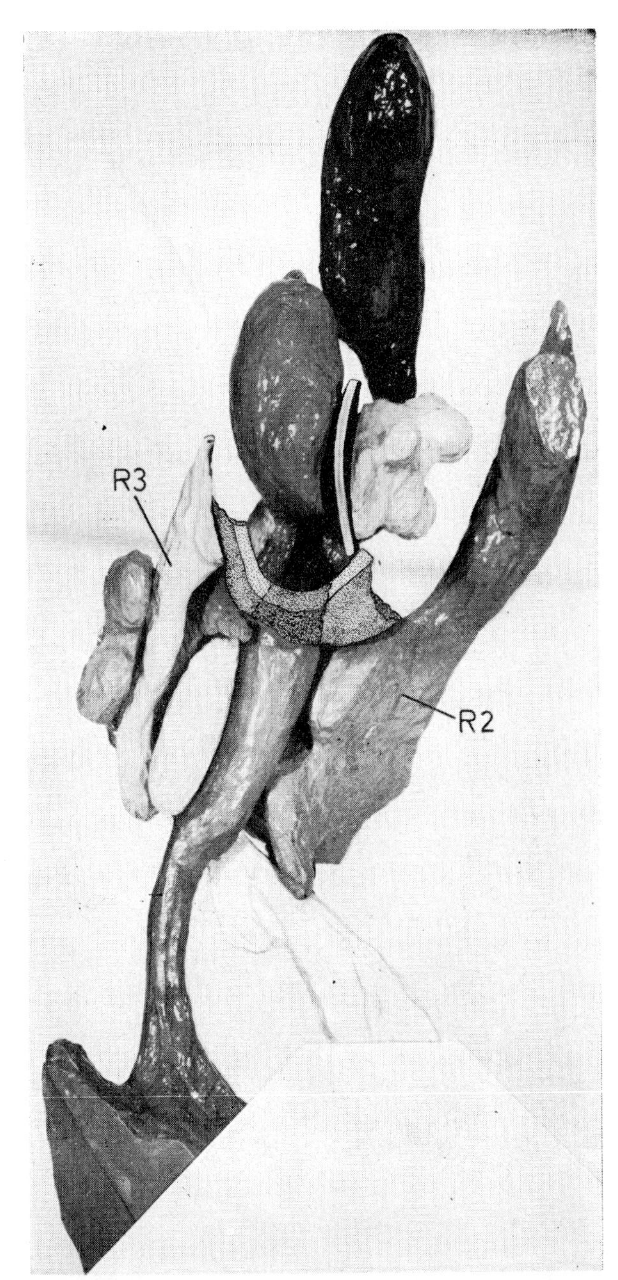


Fig. 21

Figs. 20–21. Three-dimensional reconstruction of the synaptic body of a retinal rod of the guinea pig eye made from a series of 40 sections according to the classical technique of the three-dimensional reconstructions for series of sections. Two dendrites, D1, and D2, from nerve cells are shown entering into synaptic contact with the synaptic body of the receptor. The plasma membrane bounding the synaptic body has been removed with the exception of a fragment, PM, in order to show the large vacuoles and the synaptic ribbon which, together with synaptic vesicles and granules, constitute the elementary components of the retinal synaptic bodies. In Fig. 21, some of the extensions from adjacent synaptic bodies, R2 and R3, of receptor cells are shown making contact with the surface membrane of the reconstructed synaptic body [from F. S. Sjöstrand, J. Ultrastructure Research 2, 122 (1958)].

the outer segments appear as folds of one continuous membrane (Fig. 17). If one looks at the outer segments of the kitten retina a few days after birth, one can see the outer segments in the process of being formed. They seem to be formed through a folding of a membrane (Fig. 18), each fold being transformed into a disk later on, each disk remaining associated with its neighbors through a tube-like connection located close to the center of the disk at the end of a deep incision.

It well might be that the mitochondrial membranes

are formed in the same way, which explains the fact that, in 1 to 10%, the light interspace of the inner mitochondrial membranes is continuous with the light interspace of the outer mitochondrial membrane. This would give one explanation of why this kind of contact between outer and inner membranes is so rare.

Recently, we succeeded in obtaining quite extensive series of serial sections. From one series of 40 sections with an average thickness of 250 A, the author has made a three-dimensional reconstruction of the synaptic connections and the inner structure of the synaptic bodies of the retinal rods in the guinea pig eye.[28] This reconstruction has made it possible to reveal the very complicated arrangement of various membranous components of this part of the cell (Figs. 19–21) and to draw the first primitive circuit diagram (Fig. 22) of the synaptic contacts of a receptor cell. This technique is, of course, of interest when analyzing the central nervous system, but it is necessary also for a detailed description of the topographical relations between different structural components of the cytoplasm.

One now may survey some of the types of membranes as observed in the cytoplasm (Fig. 23). Distinction can be made between the cytoplasmic membranes which, after osmium fixation, are associated with the RNA particles, the infolding of the plasma membrane, and the Golgi membranes. These various types of membranes easily can be distinguished morphologically and a neutral terminology is proposed for them: $\alpha$-, $\beta$-, and $\gamma$-cytomembranes.[29]

As assumed by Porter and Palade and accepted by most American electron microscopists, all of these types of membranes, as well as the nuclear membrane, the centrosphere, and various types of vesicles in the cytoplasm, have been considered to represent different aspects of one continuous canalicular system, the so-called endoplasmic reticulum, extending throughout the whole cytoplasm. There are no evidences that we have been able to confirm that might allow the conclusion that these various components form a continuous system. We represent an opposite standpoint in accepting morphological differences as indicating differences regarding function, and we stress that the cytoplasm is differentiated into a limited number of structurally and presumably functionally different components.

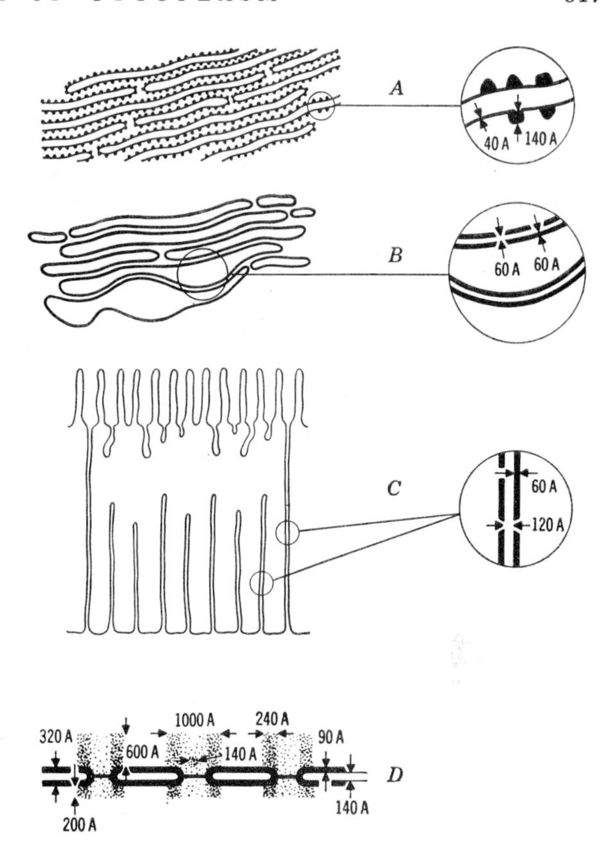


FIG. 23. Schematic presentation of some of the types of membranes that appear in cells. A. $\alpha$-cytomembranes; B. Golgi membranes ($\gamma$-cytomembranes); C. $\beta$-cytomembranes; D. nuclear membrane [according to B. A. Afxelius, Exptl. Cell Research **8**, 147 (1955)]. Some characteristic dimensions of the membranes are presented [from F. S. Sjöstrand, *Methods in Enzymology* (Academic Press, Inc., New York, 1957), Vol. IV, p. 391)].

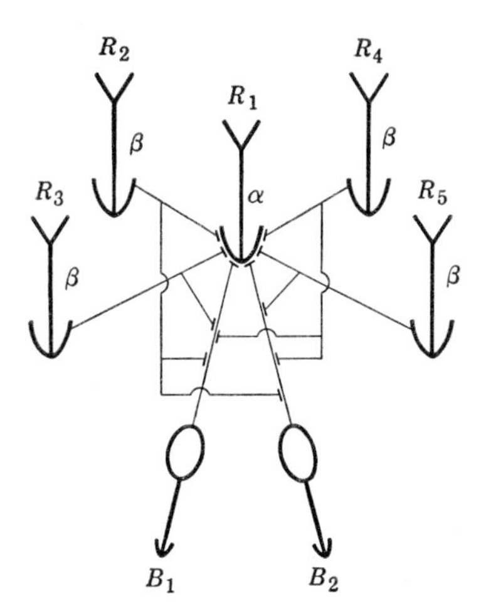


FIG. 22. A trial to make a circuit diagram of the synaptic contacts of a retinal receptor in the guinea pig eye showing the extensive inter-receptor contacts. $R_1$–$R_5$, receptor cells; $B_1$ and $B_2$, bipolar nerve cells [from F. S. Sjöstrand, J. Ultrastructure Research **2**, 122 (1958)].

For the future, it is probably less important to extend one's knowledge of the structure of cells to more and more types of cells. The structural patterns are repeated in a rather monotonous way. It is more important by far to try to analyze further the molecular architecture of the various components of the cytoplasm by using a variety of techniques. What appears most important, however, is to try to supplement the collected geometrical data with biochemical data. It is necessary then to work on fractions of homogenized cells, to improve the techniques of fractionation, to work out methods that make it possible to identify in the fractions the various cell components, and to estimate their relative concentrations in a quantitative way.

Membranes appear as the most common structural principle so far. When dealing with the cytoplasm, one gets the impression that the construction of a surface membrane with certain properties represents one of the first steps in the creation of life, and that this principle has been applied in a rather monotonous way by nature.

## BIBLIOGRAPHY


[1] F. S. Sjöstrand, Nature **171**, 30 (1953).

[2] F. S. Sjöstrand, J. Appl. Phys. **24**, 117 (1953).

[3] F. S. Sjöstrand and J. Rhodin, Exptl. Cell Research **4**, 426 (1953).

[4] F. S. Sjöstrand and J. Rhodin, J. Appl. Phys. **24**, 116 (1953).

[5] G. E. Palade, Anat. Record **114**, 427 (1952).

[6] G. E. Palade in *Enzymes: Units of Biological Structure and Function*, O. H. Gaebler, editor (Academic Press, Inc., New York, 1956), p. 185.

[7] F. S. Sjöstrand, J. Cellular Comp. Physiol. **33**, 383 (1949).

[8] F. S. Sjöstrand, Intern. Rev. Cytol. **5**, 455 (1956).

[9] F. S. Sjöstrand, J. Appl. Phys. **24**, 117 (1953).

[10] F. S. Sjöstrand, Experientia **9**, 68 (1953).

[11] R. S. Bear, K. J. Palmer, and F. O. Schmitt, J. Cellular Comp. Physiol. **17**, 335 (1941).

[12] F. S. Sjöstrand and T. Gierer, unpublished.

[13] W. J. Schmidt, Kolloid-Z. **85**, 137 (1938).

[14] F. S. Sjöstrand, Acta Anatomica, Suppl. 1 (1944).

[15] F. S. Sjöstrand in *Proceedings of the Conference on Electron Microscopy*, Delft, 1949, p. 144.

[16] F. S. Sjöstrand, Nature **165**, 482 (1950).

[17] H. Fernández-Morán and J. B. Finean, J. Biophys. Biochem. Cytol. **3**, 725 (1957).

[18] F. S. Sjöstrand and V. Hanzon, Exptl. Cell Research **7**, 393 (1954).

[19] K. R. Porter, J. Exptl. Med. **97**, 727 (1953).

[20] G. E. Palade and P. Siekevitz, Federation Proc. **14**, 262 (1955).

[21] F. S. Sjöstrand and R. F. Baker, J. Ultrastructure Research **1**, 239 (1958).

[22] A. J. Dalton and M. D. Felix, Am. J. Anat. **94**, 171 (1954).

[23] F. S. Sjöstrand and V. Hanzon, Experientia **10**, 367 (1954).

[24] F. S. Sjöstrand and V. Hanzon, Exptl. Cell Research **7**, 415 (1954).

[25] Å. Holmberg, *Ultrastructural changes in the ciliary epithelium following inhibition of secretion of aqueous humour in the rabbit eye.* Thesis, Stockholm (1957).

[26] H. Zetterqvist, *The ultrastructural organization of the columnar absorbing cells of the mouse jejunum.* Thesis, Stockholm (1956).

[27] J. D. Robertson, J. Biophys. Biochem. Cytol. **3**, 1043 (1957).

[28] F. S. Sjöstrand, J. Ultrastructure Research **2**, 122 (1958).

[29] F. S. Sjöstrand, *Physical Techniques in Biological Research* (Academic Press, Inc., New York, 1956), Vol. III, p. 241.

# 38
# Fine Structure of Biological Lamellar Systems*

H. Fernández-Morán

*Mixter Laboratories for Electron Microscopy, Neurosurgical Service, Massachusetts General Hospital, Boston 14, Massachusetts, and Department of Biology, Massachusetts Institute of Technology, Cambridge 39, Massachusetts*

## INTRODUCTION

LAMELLAR structures represent one of the most important and widespread components of biological systems.[1,2] Ranging from the individual membranes of cells to the complex multilayered structures found in the nerve myelin sheath, photoreceptors, and chloroplasts, they embody similar patterns of molecular organization. As derivatives of cell membranes, the lamellar structures exhibit the characteristic permeability properties, electrical activity, and other functional features essential to life. Moreover, by virtue of their repetitive and orderly arrangement, the multilayered systems are uniquely endowed for performing the functions of conversion, transfer, and storage of energy in organisms.

Investigation of the morphological substrate of these processes has been handicapped by the marked lability of membranous systems when subjected to the effects of analytical techniques. However, with recent advances, a general picture of the organization of these systems at the macromolecular level is now gradually emerging. The successful application of high-resolution electron microscopy in combination with polarization-optical and x-ray diffraction studies has revealed the extraordinary degree of regularity and structural differentiation which is present throughout the various hierarchies of organization down to the molecular level.

This review describes the salient features of the fine structure of the nerve myelin sheath, photoreceptors, and other representative lamellar systems. It is becoming increasingly evident that all lamellar systems appear to have certain common structural parameters at the molecular level, possibly reflecting an underlying functional analogy. Therefore, as more information becomes available on any one of these specialized systems, new approaches are suggested in the correlation of structure and function. Thus, many of the biochemical studies and recent concepts derived from solid-state physics in investigating the mechanisms of photosynthesis,[3–7] may eventually prove to be of operational value in elucidating the primary events of nerve function and sensory reception.

## FINE STRUCTURE OF THE NERVE MYELIN SHEATH

The myelin sheath of nerve fibers is one of the most highly ordered biological systems, and exhibits all of the properties of the smectic fluid-crystalline state. Despite the high water content and marked lability of the myelin sheath, its ultrastructure can be studied by combined application of different techniques under various conditions. The analysis of the fine structure of the myelin sheath can be regarded, in fact, as one of the best examples of the systematic application of complementary biophysical and biochemical methods. Since the myelin sheath derives from a multiply-folded Schwann-cell surface,[8,9] it can also be considered a model system for the study of cell-membrane structure in general.

Polarization-optical studies of the myelin sheath[10–12] indicated that it is composed of concentrically arranged protein or lipoprotein lamellae which alternate with layers of lipid molecules oriented with their long axes in the radial direction. This concept was confirmed and extended by x-ray diffraction studies of fresh nerve.[13–18] It was assumed that the thickness of the concentric lipid-protein unit layers corresponded to the fundamental small-angle x-ray diffraction spacings of 170 A recorded in amphibian, and of 180 to 185 A found in mammalian peripheral nerves.

Subsequently, the postulated layers and the concentric laminated structure of the sheath were observed directly in electron micrographs of thin nerve sections fixed with osmium tetroxide.[19–23] However, since the preparation of these specimens required fixation, dehydration and embedding in a plastic medium, followed by slicing into ultrathin sections, the possibility of artifacts had to be seriously considered. Moreover, the regular patterns of selective deposition of osmium could not be interpreted in terms of specific regions containing the lipids, lipoproteins, water, and other components of the normal myelin structure.

One is obviously dealing here with a case in which combined application of low-angle x-ray diffraction techniques and high-resolution electron microscopy can supplement each other in many ways. It is instructive to consider the merits and drawbacks of each technique in order to appreciate the type of approach required in analyzing biological systems.

The x-ray diffraction method offers the great advantage of permitting examination of the intact nerve trunk in the living animal.[13,15] The information derived from the x-ray diffraction pattern of normal active nerve can, therefore, be considered a reliable basis for all structural analyses of the myelin sheath.[13,18,24] From the low-angle x-ray diffraction pattern, the dimensions and

* Part of these studies were aided by a research grant (C-3174) from the National Institute of Neurological Diseases and Blindness of the National Institutes of Health, Public Health Service, U. S. Department of Health, Education, and Welfare.

approximate distribution of scattering groups in the radial direction of the sheath can be deduced. Reliable reference parameters are, therefore, provided for checking the results of electron microscopy in studies of the normal and modified myelin sheaths.[18,24]

Finally, the x-ray diffraction pattern recorded in a few hours represents an average of the main structural parameters of all of the nerve fibers contained in the exposed nerve trunk which contribute to the diffraction. The practical significance of this is immediately apparent when comparing it with the far greater time factor involved in the corresponding electron-microscope study. Examination of nearly 100 000 serial ultrathin cross sections would be required for a complete electron-microscope study of a 1-mm long segment of a few fibers only.

Conversely, electron microscopy offers the unique advantage of revealing directly the complex patterns of macromolecular organization in selected areas of the specimen.

### Structure of the Normal Myelin Sheath

X-ray diffraction patterns recorded from fresh peripheral nerve in a direction perpendicular to the fiber axes feature a series of very well-defined reflections at low angles. The wide-angle reflections at about 4.7 A exhibit meridional intensifications, while the low-angle reflections are precisely oriented in the equatorial direction. The low-angle reflections can be interpreted in terms of a fundamental radial repeating unit varying from 170 A in amphibians to 184 A in mammalian peripheral nerve. The low-angle reflections also show a characteristic alternation of intensities in the even and odd orders [Fig. 1(b)]. It is assumed that the fundamental repeating unit consists of two parts having very similar distributions of x-ray scattering power; from the intensities of the odd-order reflections, the magnitude of the difference between the two parts can be estimated.[18] Finean[15–18] has determined that this "difference factor" is appreciable in peripheral nerve myelin, but appears to be negligible in the optic nerves.

High-resolution electron micrographs of ultrathin sections of osmium-fixed peripheral nerve (Fig. 1) show a series of concentrically arranged, dense lines separated by lighter spaces with an average period of 130 to 140 A. There is, thus, a correspondence of the periodic layers observed in the electron micrographs of myelin preparations and the fundamental radial unit derived from low-angle x-ray diffraction patterns.

The discrepancy of 20 to 30 A between the layer thickness and the x-ray fundamental spacing is owing to the shrinkage effects introduced by osmium fixation and other preparative procedures connected with the examination of the thin sections in the electron microscope.[24] In addition to the periodic dense bands about 30 A wide, the electron micrographs feature light intermediate bands which are much narrower (10 to 15 A). This variation in the densities of the two principal bands of osmium deposition is related to the "difference factor" which manifests itself in the low-angle pattern as an increase in the intensity of the first-order reflection.

The present lack of precise information on the chemistry and localization of the sheath proteins, the lipids, the water layers, and other components of myelin has not as yet permitted a direct correlation with the structural data.[25–27] However, by combined electron-microscope and x-ray diffraction studies of controlled physical and chemical modifications of the myelin sheath, a general picture of the molecular arrangements in the radial units can be obtained.[15,18,21,24,28] Satisfactory correlation and interpretation of structural relationships at the macromolecular level requires the preparation of a large number of ultrathin (100 to 200 A) undistorted serial sections of the same specimens utilized for the parallel x-ray diffraction studies. This has been greatly facilitated by improved thin-sectioning techniques using a diamond knife.[24,29]

Such a combined approach has led to a detailed analysis of the preparation procedures, thus validating important findings and defining artefact sources. The extraction and enzymatic-digestion experiments performed on fresh myelin proved to be particularly revealing.

### Lipid-Extraction Experiments

Extraction of fresh nerve with acetone at 0°C removes about 30% of the cholesterol,[18] leaving the other lipid components essentially intact within the framework of a still highly organized residual myelin sheath.[24] The main modifications revealed by electron microscopy and x-ray diffraction [Figs. 2(d) and 2(e)] of the residual sheath comprise expansion of the layered structure with internal rearrangements and formation of collapsed layer systems [Figs. 2(a) and 2(b)]. The dense lines are relatively resistant to lipid solvents, and their persistence in the collapsed layered structure may indicate that the heaviest osmium deposition is in the region of the protein or stable lipoprotein constituents. These modifications emphasize the importance of cholesterol in the myelin structure, and lend further support to the existence of the phospholipid-cholesterol complex postulated in the unit cell by Finean.[18]

Low-angle diffraction patterns of the lipid extract show a strong 34.2-A reflection [Fig. 2(c)] characteristic of cholesterol. When the dried lipid extract is examined by electron microscopy, thin crystalline lamellae are found embedded between condensed multi-layered aggregates. These crystalline lamellae give electron-diffraction patterns which are similar to those recorded from pure cholesterol.

A more extensive breakdown of the sheath is observed after alcohol extraction. The fine-layered structures observed in certain areas may represent the refractory protein framework in addition to recrystallized lipid components. The extraction experiments suggest, therefore, that the dense bands observed in electron micro-

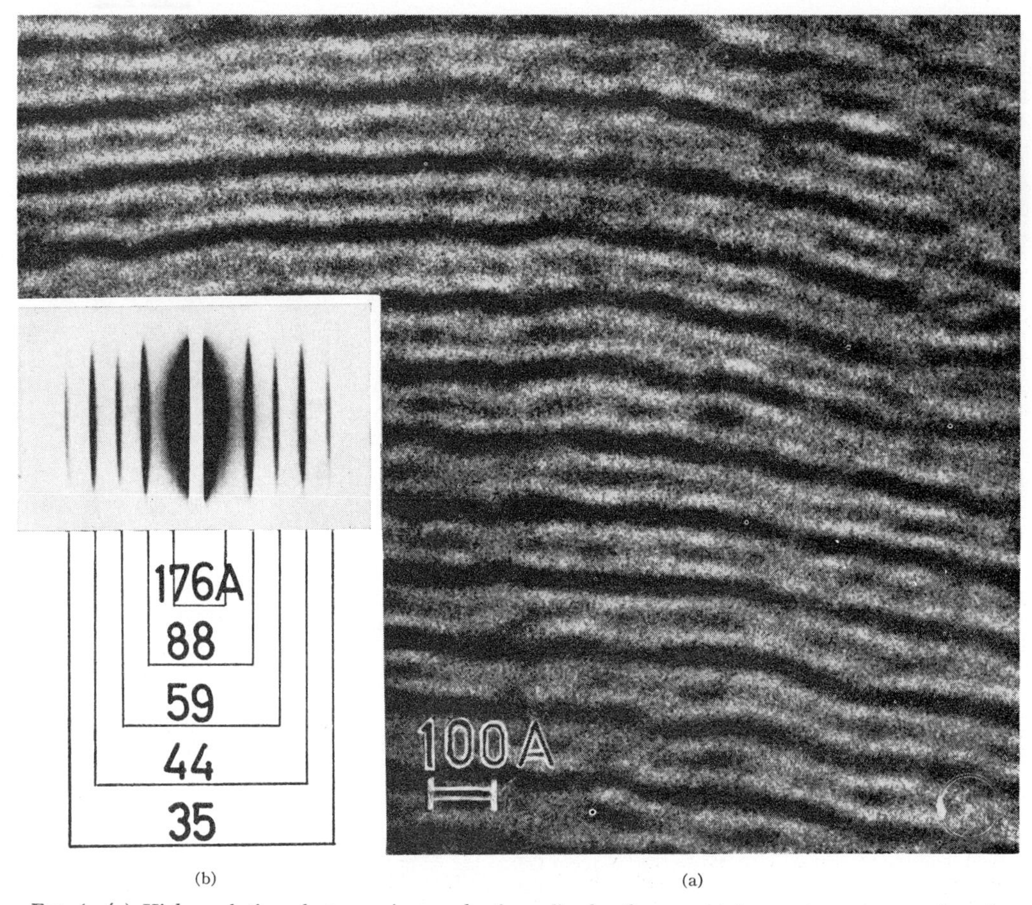


FIG. 1. (a) High-resolution electron micrograph of myelin-sheath segment from a transverse section of an osmium-fixed rat sciatic nerve embedded in gelatin. The average layer spacing is 110 A. Note compact and well-preserved dense lines with moderate enhancement of the intermediate lines. ×850 000. (b) Low-angle x-ray diffraction pattern of rat sciatic nerve recorded with Finean camera. This pattern features a fundamental period of 176 A, with characteristic alternation of the intensities of the even and odd orders.

graphs of the normal myelin sheath represent areas of selective osmium deposition at lipoprotein interfaces, whereas the light bands are regions occupied by lipid chains and associated myelin components which do not react primarily with osmium tetroxide.

### Enzymatic-Digestion Experiments

Although digestion with trypsin is not specific, its application to fresh nerve fibers produces characteristic modifications of the myelin fine structure.[19,24] In addition to slight expansion of the concentric layers, the uniform dense lines appear to dissociate into rod-shaped granules, 30 to 50 A wide and 40 to 60 A long (Fig. 3). The unit myelin lamellae isolated from the sheath after trypsin digestion likewise exhibit extensive dissociation into composite elongated granules of similar dimensions.[19,20] An analogous granular fine structure of the dense layers is commonly found after freezing and thawing of fresh nerve, $KMnO_4$ fixation, and other types of preparations.[24] Moreover, there are also numerous indications of a compact granular fine structure in the dense layers of the normal myelin sheath.

X-ray diffraction data furnish supporting evidence for the presence of a regular organization within the plane of the layers. A strong 60- to 70-A vector, which appears necessary to account for the relative intensities of the low-angle reflections, can be related to this type of organization within the planes of the lipoprotein layers.[24]

### Nerve-Degeneration Studies

The breakdown of myelin known to occur during *in vitro* degeneration of nerve affords an opportunity to study the fine structure of the disintegrating sheath without introducing extraneous reagents. Comprehensive studies of nerve-fiber degeneration, which are now being carried out by the author with combined application of high-resolution electron microscopy and low-angle x-ray diffraction techniques, reveal interesting details of the layer structure.

When fresh peripheral nerve is enclosed in a glass capillary under aseptic conditions and left to degenerate at 20°C, low-angle x-ray diffraction patterns recorded at periodic intervals show characteristic changes

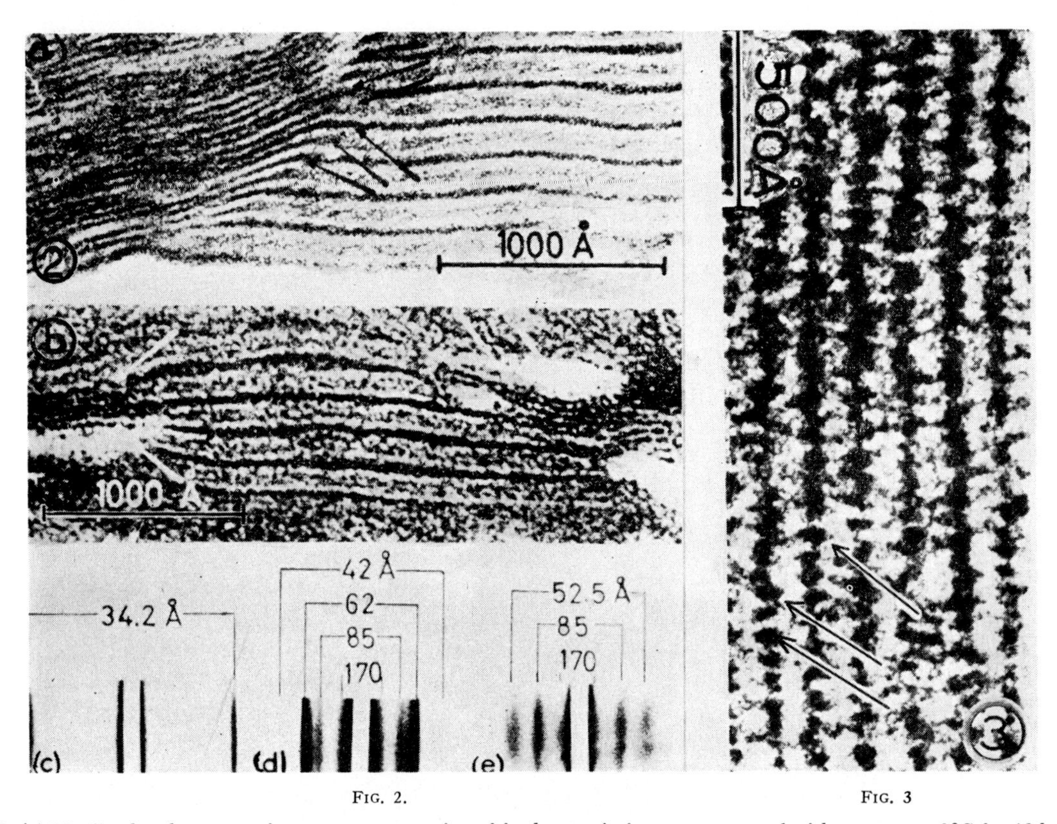


FIG. 2. FIG. 3

FIG. 2. (a) Myelin-sheath segment from transverse section of fresh rat sciatic nerve extracted with acetone at 0°C for 12 hours prior to fixation with 2% osmium tetroxide and embedding in butyl methacrylate. Notice the expanded periods (160 A) of the modified layers at the right, and the transitions (arrows) to the collapsed period (43 A) at the left. ×300 000. (b) Myelin-sheath segment from longitudinal section of fresh rat sciatic nerve extracted with acetone at 0°C for 12 hr prior to fixation with osmium tetroxide and embedding in butyl methacrylate. Notice the transitions (arrows) from the expanded layer system to the collapsed layer system ×280 000. (c) Low-angle x-ray diffraction pattern of the lipid material extracted from fresh sciatic nerve by immersion in acetone at 0°C for 12 hours, showing the characteristic cholesterol spacing at 34.2 A. (d) Low-angle x-ray diffraction pattern of residual dried rat sciatic nerve after extraction with acetone at 0°C for 12 hours. (e) Low-angle x-ray diffraction pattern of acetone extracted nerve fixed with buffered 2% osmium tetroxide and embedded in methacrylate.

FIG. 3. Myelin-sheath segment from a rat sciatic nerve which was incubated with 1% crystalline trypsin for 12 hours prior to fixation in buffered 1% isotonic osmium-tetroxide solution and embedding in butyl methacrylate. Notice the marked dissociation of the dense lines (arrows) into elongate granules. ×520 000.

(Fig. 4). Compared with the normal pattern, marked modifications of the distribution of x-ray scattering power within the unit are noted as degeneration progresses. There is marked intensification of the second-order diffraction, followed by the appearance of a 70-A reflection, and gradual extinction of the lower orders in later stages of degeneration. The corresponding electron micrographs (Fig. 5) show various forms of granular dissociation of the dense layers, resembling the structures observed after trypsin digestion. In addition to this supporting evidence of granular structure within the layers, numerous other features of myelin organization are being uncovered as the complex dissolution process is followed at the submicroscopic level.

### Suggested Molecular Organization of Myelin

Based largely on their pioneering x-ray diffraction studies, Schmitt and co-workers[12–14,30] concluded that the fundamental radial unit of the myelin sheath in the internodal portion of peripheral nerve comprised two lipoprotein layers, each of which consisted of bimolecular leaflets of 67 A sandwiched between protein layers of 25 A, with interposed water layers contributing a further 25 A. Taking into consideration the contraction of the lipid layers during drying, and the results of other experimental modifications, Finean[15–18] arrived at a more detailed picture of the molecular arrangement in the myelin unit. According to this conception, each lipoprotein layer would consist of two phospholipid-cholesterol complexes associated with a cerebroside molecule, intercalated between monolayers of protein. The two units are distinguished by a "difference factor" which has not been accurately determined, but can nevertheless be explained by the mechanism of myelin formation. As Geren[8,9] showed, the myelin sheath is formed by an infolding and multiple wrapping of the Schwann-cell membrane around the axon in embryonic fibers. Assuming that the Schwann-cell lipoprotein membrane is asymmetric, then this process of rolling

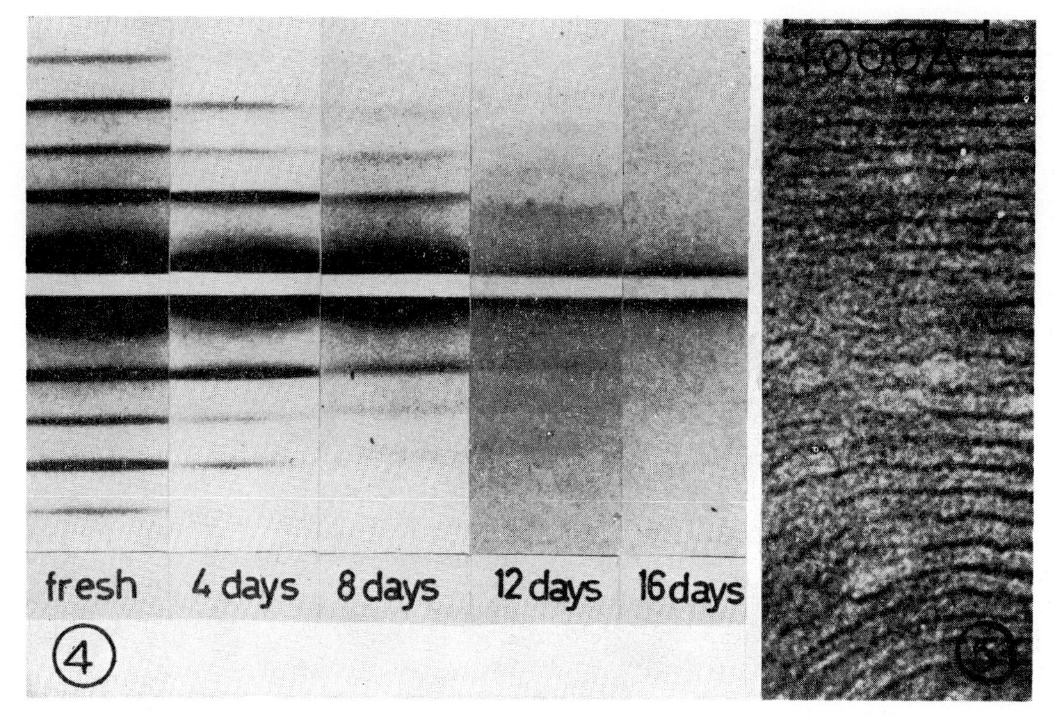


FIG. 4. FIG. 5.

FIG. 4. Low-angle x-ray diffraction patterns of rat sciatic nerve recorded during *in vitro* degeneration at 20°C. Notice marked modifications of the distribution of x-ray scattering power within the radial unit as degeneration progresses.

FIG. 5. Electron micrograph of nerve myelin sheath during *in vitro* degeneration (4 days), showing dissociation of the dense lines into granular structures, disappearance of the intermediate line, and other modifications of the layered structure. ×280 000.

onto the axon will produce the observed symmetry difference in successive layers.[18]

The lipids are considered to be oriented with their long axes radially in the sheath. The 4.7-A, meridionally intensified ring recorded from fresh nerve has been related to the cross sections of the lipid molecules, indicating the average interchain separation of their hydrocarbon chains. However, in order to account for the permeability properties of the membranes, various possibilities including the existence of real or potential "holes" have been discussed. One of the suggested configurations envisages a continuous layer of lipids sandwiched between monolayers of protein with a fenestrated or open-network type of structure. The filtering action of the protein layer would combine with the selective solubility of the penetrating ions or molecules in the lipid layer to account for the sieve mechanism of permeability. The formation of submicroscopic pores by radial extensions of the protein layers across the lipid layers has also been considered, but no direct evidence for the existence of these discontinuities is yet available. Nevertheless, the x-ray diffraction and electron-microscope data clearly indicate that there must be a considerable degree of organization within the plane of the lipoprotein layers which remains to be investigated.

Water is one of the most important components of fresh myelin, constituting at least 35% of the myelin sheath.[12,30] Information on the precise localization of water within the myelin layers is also of great importance when considering the possible pathways of diffusion of ions and other solutes between the axon and the extracellular fluids.[28,31] The water layers at the aqueous interfaces of the fundamental repeating unit in the compact myelin are about 12 to 15 A thick.[31] Under certain experimental conditions,[17] larger amounts of water can be incorporated between the aqueous interfaces in the myelin unit. The water is presumably coordinated on the protein layer located at the aqueous surface of each bimolecular leaflet. The thickness of the water layers is largely determined by the electrical charge density at the aqueous interfaces,[31] and by the ionic strength, since water is expelled from lipid layers in the presence of ions.[14] These changes in thickness of the aqueous interfaces may be of significance in connection with ion movements in the intermembrane spaces. However, as Schmitt[31] has pointed out, the physicochemical properties, such as *p*H and ionic strength in capillary spaces of these macromolecular dimensions, would be quite different from those in bulk solutions. The effects of the electrically charged lipid groups on the structure of the hydration water in these aqueous interfaces may also induce the formation of a crystalline arrangement resembling "frozen" hydration sheaths postulated in protein molecules.[32]

Through the application of high-resolution nuclear

FIG. 6. Single unit disk isolated from the outer segment of frog retinal rod, showing the granular surface structure and the margina cord. Latex particles (2800 A diam) have been added for calibration in this shadowed preparation. ×25 000.

magnetic-resonance spectrometry, it now appears possible to obtain important supplementary information on the water content of fresh whole nerve and the hydration state of its components, in a rapid and nondestructive way.[33] The size of the proton magnetic-resonance signal recorded from biological materials is proportional to the water content.[34] Since the mobility of the protons in the solid, nonaqueous constituents is much less than in the aqueous phase, the resonance spectrum consists of a narrow line owing to the water, superimposed on a broad line owing to protons in the supporting solid. After suitable calibration, the water content can be accurately determined by evaluating the line width of the proton resonance, or by measuring the amplitude of the derivative of the narrow absorption curve. Both methods have been applied in preliminary

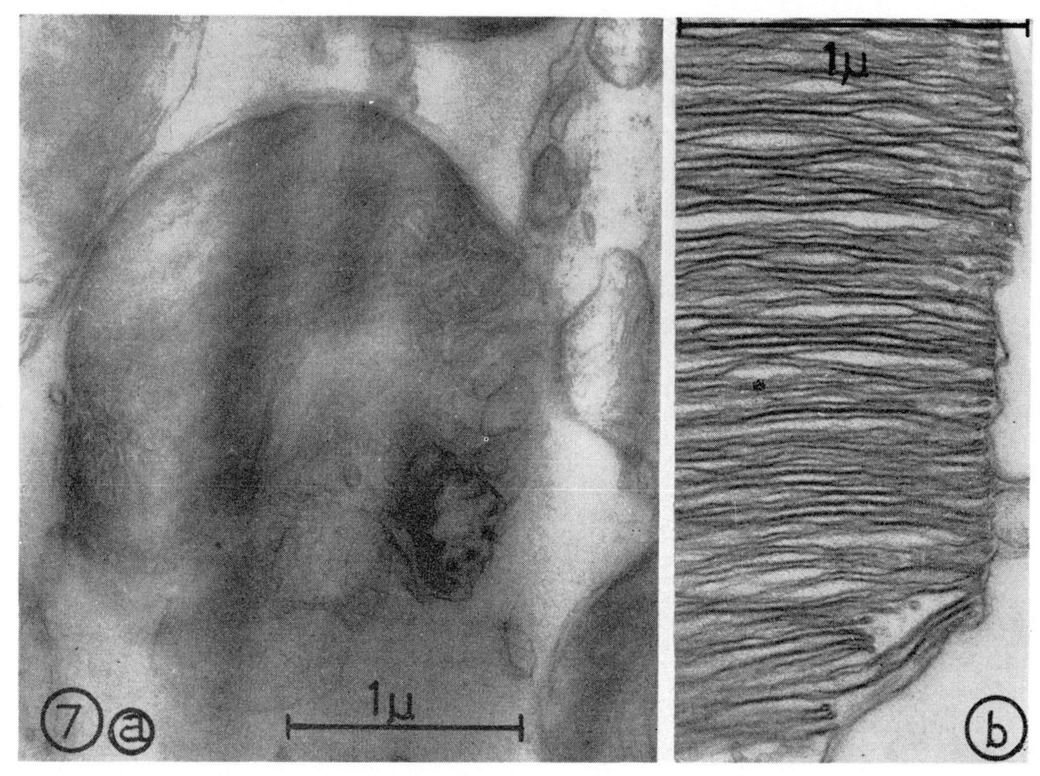


FIG. 7. (a) Transverse section through the basal region of the outer segment of a retinal rod of the guinea pig, showing the characteristic tubular structures and vesicular formations. Notice cross-sectioned filaments which establish connection with the inner rod segment. ×32 000. (b) Longitudinal ultrathin section through the basal part of the outer segment of retinal rod from guinea-pig eye showing the regular arrangement of the double membrane disks. ×48 000.

determinations of the water content of fresh nerves under various conditions, using commercially available equipment and a special transistorized NMR spectrometer.[33,35]

The protons in the hydration shells postulated around the macromolecules lack the high mobility of the water in the bulk aqueous environment, and give a broad and weak resonance line. By comparison with the strong narrow line owing to the more-mobile water protons, the proportion of water forming the hydration shells of nucleic acids has been estimated.[36] Similar experiments can be carried out on living nervous tissues, and should yield valuable information on the hydration state of the bound water in the myelin sheath.

The present concepts of the myelin's molecular architecture refer almost exclusively to the lipoprotein framework which stands up to various analytical techniques. Within this framework, no detailed localization is yet possible of the numerous enzymes, electrolytes, trace metals, and other essential components of the myelin sheath. Although deuterium and radiophosphate studies[37] indicate that myelin has a relatively low metabolic turnover, more information is required on the structural relationships with the phospholipases, phosphatases, and other enzymes involved in lipid metabolism.[38,39] Likewise, the molecular models will be incomplete until specific localizations can be assigned to the cholinesterases[40] and related enzymes which play an important role in nerve conduction. Recent investigations which link the metabolism of copper with certain demyelinating diseases[38] also focus attention on the possible relationships of trace metals with the lipoprotein layers. In view of the close association of the porphyrins[41] and related photodynamic agents with the central myelin, it may prove of interest to establish a more specific correlation with the ultrastructure of the myelin sheath.

## FINE STRUCTURE OF PHOTORECEPTORS

Electron microscope studies have revealed that most of the photoreceptors in biological systems, where light is converted into chemical or electrical energy, feature a regular lamellar fine structure. In the vertebrate eye, the photoreceptor elements are composed of submicroscopic plates orderly disposed in a pile; whereas in the invertebrates, closely packed thin-walled tubules build up the light percipient constituents. The process of visual excitation is initiated by the absorption of light in the characteristic photopigments which form an integral part of all photoreceptors. In the vertebrates, the visual pigments, rhodopsin and iodopsin, are located in the outer segments of the rods and cones, respectively. Polarized-light studies[42] had already indicated that the rod outer segments consist of transversally oriented protein layers alternating with layers of longi-

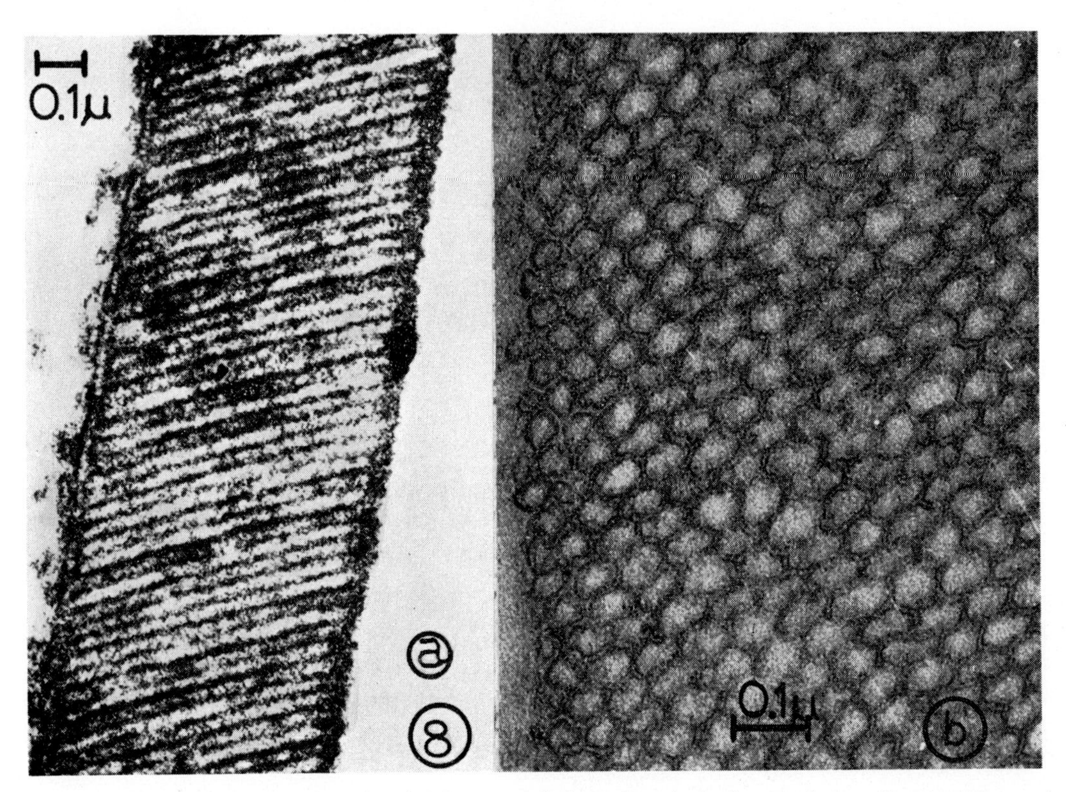


FIG. 8. (a) Oblique longitudinal section through a rhabdomere of the housefly retinula showing the lamellar type of internal structure. The highly regular submicroscopic organization is illustrated by the nearly perfect parallel array of several score dense bands occupying the entire rhabdomere. These parallel bands correspond to the dense walls of the tubular compartments sectioned longitudinally. ×60 000. (b) High-resolution electron micrograph of longitudinal ultrathin section through a rhabdomere of the housefly retinula, demonstrating the fine structure of the tubular compartments. The regular array of sharply-outlined annular profiles (400 to 500 A in diameter) bounded by a dense osmiophilic substance is interpreted as representing cross sections of the closely packed tubular compartments. This differentiation of the dense "walls" of the compartments into a thin, osmiophilic boundary line, 20 to 30 A wide, associated with a dense layer of approximately 60 A, may be of interest in relation to the postulated models of the macromolecular photoreceptor complexes. ×100 000.

tudinally oriented lipid molecules, and could, therefore, be compared with a radial cylinder of the myelin sheath. Electron microscopy confirmed that the entire rod outer segment consists of several hundred unit disks, about 150 A thick, stacked up in regular array.[21,43,44]

The unit disks of the frog retinal rods have a diameter of about 6μ and exhibit a peculiar lobulated shape with numerous incisures outlined by a dense marginal cord (Fig. 6). Thin sections disclose the remarkably regular layering of the unit disks [Fig. 7(b)], which are arranged with their planes perpendicular to the rod axis. Each disk consists of dense double membranes which stain intensely with osmium and enclose a light space of about 70 to 80 A. The repeating unit along the axis of the rod in osmium-fixed preparations corresponds closely to the spacing of about 320 A recorded in the low-angle x-ray diffraction patterns.[45] Recent studies demonstrate tubular processes at the incisures which are closely associated in the basal portion with the bundle of thin fibrils connecting the outer and inner rod segments [Fig. 7(a)]. The inner segments contain numerous densely-aggregated long mitochondria,[44,46] which are regularly encountered in all types of photoreceptors.[47,48] The retinal cones show a similar layered structure.[44]

It has been tentatively assumed that the protein membranes stain densely with osmium, while the lipid molecules may be predominantly localized in the intermembrane compartments of the disk.[44] However, knowledge of the detailed chemical composition of the photoreceptors is still too meagre, and controlled extraction and digestion studies of the type described earlier remain to be done. The exact location of the visual pigments is also unknown, but Wolken and co-workers[49] have suggested that the pigment molecules are oriented as monolayers at the aqueous protein and lipoprotein interfaces. The chloroplasts exhibit a very similar layered structure, and the chloroplastin-pigment complex appears likewise to reside in the aqueous protein-lipoprotein interfaces.

### Light Receptors in Insect Compound Eyes

In the insect compound eye, each separate visual element or ommatidium consists of eight sensory retinula cells.[50] The differentiated medial part of each retinula cell, known as a rhabdomere, is built up of approximately 20 000 to 30 000 closely packed tubular

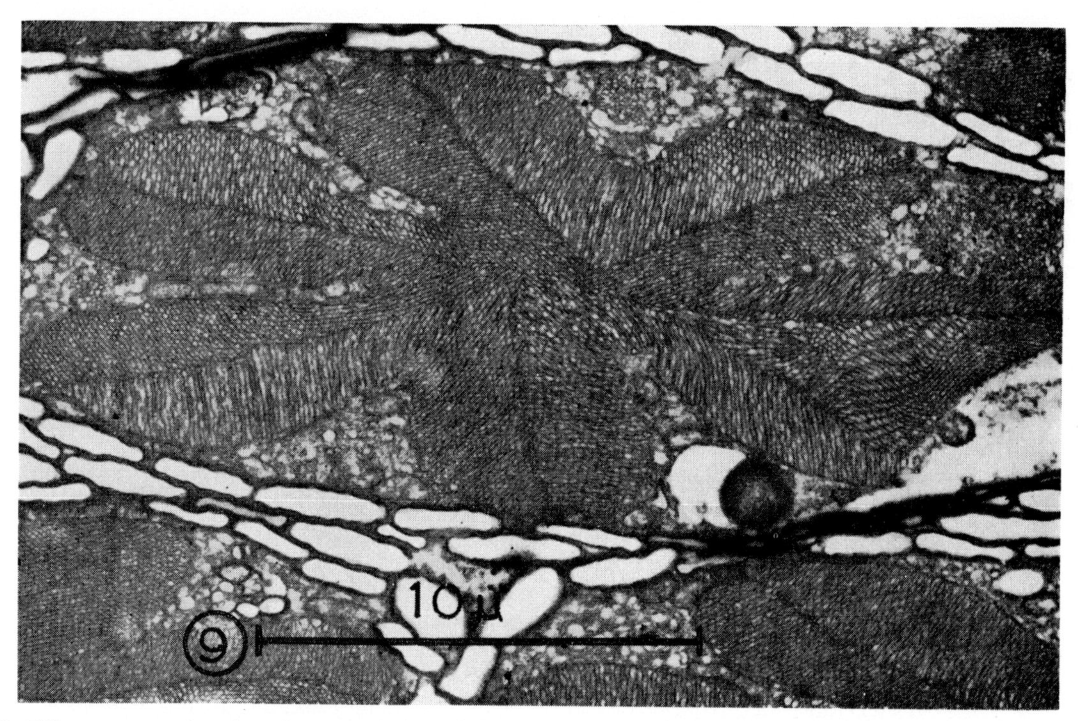


FIG. 9. Oblique cross section through a retinula from the eye of the giant tropical moth, *Erebus*, demonstrating the symmetrical arrangement of the fused rhabdomeres. Note that the general configuration and orientation of the tubular compartments of any one of the rhabdomeres is matched by the corresponding rhabdomere pattern directly or diagonally opposite to it. The equivalent rhabdomeres of contiguous ommatidia are also oriented with their main axes in the same direction, thus introducing a degree of "structural polarization" in the over-all pattern which must bear relation to the analysis of polarized light in the insect eye. ×6000.

compartments which vary from 400 to 1200 A in diameter, and are oriented in highly regular array [Fig. 8(a)] with their long axes normal to the longitudinal rhabdomere axis.[47–49,51] These tubular rhabdomere compartments are regarded as differentiated microvilli of the retinula cell membrane. The walls of the tubular compartments exhibit a dense boundary line of about 30 A in which the oriented visual pigments may possibly be located[47] [Fig. 8(b)]. Within each ommatidium, the rhabdomeres are radially arranged in a symmetrical pattern formed by matched pairs of opposite rhabdomeres displaying similar orientation of their layered structure (Fig. 9).

This differentiated submicroscopic organization of the visual elements—and their structural coupling in coordinated groups[47]—suggests a correlation with the remarkable ability of insects to recognize the regional patterns of polarization of the sky as a basis for light-compass orientation.[52–54] It is assumed that the radially arranged rhabdomere pairs containing dichroic visual pigments similarly oriented in their periodic tubular compartments may correspond to the functional units of the analyzer for polarized light postulated in the compound eye.[47–49,51,55]

The visual system of the insects and other arthropods features an intimate association with a complex three-dimensional network of air-filled tubules or tracheoles and with pigment granules, which is probably of considerable functional significance.[47]

The structural analysis of the insect compound eye is proving to be one of the most revealing examples of how the highly differentiated tubular and lamellar textures can provide a periodic array of gas-liquid-solid interfaces at the supramolecular level, which are ideally suited for selective interaction (absorption, reflection, diffraction, scattering, etc.) with incoming light signals.

## RELATION OF LAMELLAR STRUCTURES TO ENERGY-TRANSFER PROCESSES

Essentially the same type of macromolecular architecture described in the unit layers of the myelin sheath and the photoreceptors is encountered with only minor modifications in the organization of the mitochondria, intracytoplasmic lamellar complexes, Golgi apparatus, and other specialized lamellar systems.[21,31,33] If this striking structural similarity is actually the expression of an underlying functional analogy, then interesting general correlations may be derived from critical collation of the available information.

Correlation of fine structure with energy-conversion processes has been most fruitful in those systems—like mitochondria and chloroplasts—which can be isolated in quantity and studied outside of the cell without appreciably impairing their functional activity. Biochemical studies[56–58] have shown that certain fragments of mitochondria are still capable of carrying out the essential functions of electron transport and oxidative phosphorylation. Based on this evidence, the mitochondrion has been compared with a giant polymer made up of certain repeating units incorporated in the internal and external membranes.[57] Recently, it has been possible by digitonin treatment to obtain fragments of the mitochondrial membrane which contain relatively intact "respiratory chain assemblies" including the enzymes necessary for coupling phosphorylation to electron transport.[58] These fragments contain protein, phospholipid, and a small amount of cholesterol;[58] they may possibly correspond in many respects to the double membrane components obtained by sonic fragmentation of mitochondria.[56,57] The available evidence indicates that these enzymes and related components of the electron-transfer chain are precisely arranged in specific patterns built into the matrix of the lipoprotein layers, which provide the highly ordered "floor space" where these assemblies are anchored.[57,58]

It has been suggested that the electrons are transported in this comparatively rigid structure, approximating the solid state, by being relayed through small, mobile molecules within the lipoprotein matrix.[56,57] Considering the important participation of water in all lamellar systems, it is tempting to associate the water molecules or hydrated small molecules with this transfer function.

The general problem of structural interaction between an ordered substrate and its aqueous environment appears to be of particular significance in the fluid-crystalline lamellar systems with their high water content. Many unique features of proteins in aqueous solutions suggest that the macromolecules may be surrounded by a hydration sheath of icelike character.[32,36,59] The induction of a crystalline lattice of hydration water is ascribed to the local and long-range cooperative electric-field effects of the nonpolar side chains in proteins.[32] In the case of the 10- to 15-A thick, aqueous layers intercalated between the protein-lipoprotein layers, these effects are expected to be even more pronounced in determining the formation of crystalline lamellae of "frozen" water.

On first consideration, this concept of "frozen" hydration lamellae permeating in periodic array such a multilayered complex. would seem to be incompatible with the smectic fluid-crystalline character of biological systems. However, it should be borne in mind that the "icelike" hydration sheaths lodged in ultracapillary spaces of these molecular dimensions may exhibit physical and physicochemical properties which are quite different from those encountered in bulk solutions. Also, only that part of the hydration water which is in immediate contact with the protein or lipoprotein layers may be ordered in a crystalline lattice; while the remaining "liquid" aqueous strata, and the additional interstitial water channels would still allow sufficient freedom of movement of the flexible layered substrate. In this connection, it is interesting to note that the adsorbed hydration water in graphite facilitates slippage of the crystalline lamellae, and is largely responsible for the lubrication properties of this material.

It is, therefore, conceivable that the postulated crystalline hydration sheaths could provide an organized framework interconnecting the lipoprotein layers and serving primarily as a transfer medium for charged carriers like electrons, holes, protons, and ions. By virtue of the extensive hydrogen bonding in the icelike lattice, the effects of ion movements could be produced over comparatively long distances without involving bodily displacement of the ions, which would be an important factor in the relatively compact lamellar systems.

Regardless of these hypothetical assumptions, it is obvious that much more must be known about the hydration state and properties of water in lamellar systems before detailed functional correlations can be attempted. Nuclear magnetic-resonance spectrometry, supplemented by spin-echo techniques, now offers the possibility of obtaining essential quantitative data on relaxation times and correlation times of the water components in living biological systems under various physiological conditions. It is also suggested that a structural determination of the hydration shells of lattice-ordered water may eventually be possible by extending the neutron-diffraction studies already successfully carried out on ice[60] to suitable models of multilayered systems, or to liquid crystals of adequate size.

Important additional evidence that the processes of energy transfer and conversion in the chloroplasts are dependent on the highly ordered molecular arrangement in their layered structure has been furnished by comprehensive recent studies of the mechanisms of photosynthesis.[3–7,61] Calvin and co-workers[3,4,6] have strongly indicated that the initial process of energy transfer is a physical one, dependent on the specific physical organization of the chlorophyll and its associated pigments in the ordered, "quasicrystalline" layered structures. Absorption of a light quantum by chlorophyll raises an electron to a conduction band, leaving behind a "hole," both of which migrate together as an "exciton" through the oriented array of chlorophyll molecules. The conjugated carotenoid molecules act as conductors of electrons, while the lipids in the chloroplasts would function as an insulator, permitting a separation of charges when the electrons are "trapped" on one side of the layer and the "holes" on the other. This polar character is similar to that which exists at a *p-n* semiconductor junction. The electrons participate in the formation of a reducing agent leading to the reduction of carbon dioxide; the positive holes react with water to form oxygen, thus initiating the two basic reaction sequences of the photosynthetic mechanism. Direct experimental evidence has also been presented for the existence of trapped electrons and holes by recording electron spin-resonance signals from illuminated whole chloroplasts[3–6,61] at low temperatures

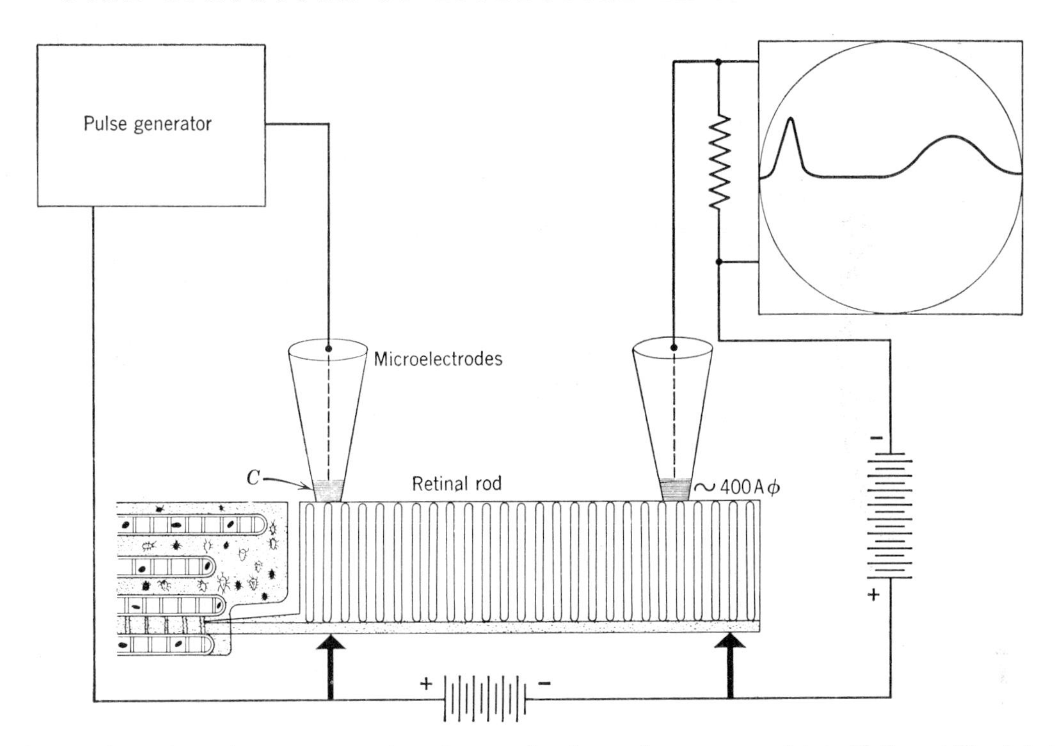


FIG. 10. Suggested experimental arrangement to investigate semiconductor phenomena associated with the mobility of electrons and holes in active biological lamellar systems. By applying microelectrodes (filled with a suitable solid free radical, "*C*") directly to the outer segments of retinal rods, it might be possible to determine the drift velocity and density of injected current carriers. Addition of a transverse magnetic field would then permit determination of the Hall effect and other magnetic concentration effects. Experiments of this type, carried out on photoreceptors and chloroplasts or on artificial multilayered structures containing specific photopigments, are needed to furnish direct experimental proof of the postulated semiconductor properties in biological systems.

(−150°C), and, therefore, excluding the possibility of any ordinary enzymatic process.

By tentatively extending these concepts to other lamellar systems, one arrives at a general picture of paracrystalline lipoprotein layers containing assemblies of specific enzymes associated with photopigments or specialized electron-transfer systems, all of which are organized in highly ordered patterns. These biological lamellar systems, permeated with hydration water of possible icelike character, might have many properties in common with semiconductors, although, as pointed out earlier,[21,62] the concepts of solid-state physics are not directly applicable to the fluid crystalline state. However, merely as a working hypothesis, the semiconductor analogy should prove useful as a possible basis for the physicochemical amplifying devices suggested in connection with the events leading from the primary process of sensory excitation to the complex bioelectric phenomena associated with nerve-impulse propagation.[63,64]

Although these ideas are admittedly speculative, they may, nevertheless, serve to encourage new experimental approaches. Of particular interest would be systematic investigations designed to furnish direct experimental proof in biological lamellar systems of the two processes of electronic conduction in semiconductors, corresponding to positive and negative mobile charges, which are characteristic of transistor action.[65] In the suggested experiment (Fig. 10), two capillary microelectrodes filled at the tip with a suitable solid free radical are applied directly to the outer segments of freshly isolated retinal rods, or, alternatively, to large chloroplasts. With the outlined arrangement, it might be possible to determine the drift velocity and the density of injected current carriers, in analogy to the classical experiments[65] carried out on small samples of germanium using micromanipulator techniques. By introducing a transverse magnetic field, determination of the Hall effect and other magnetic-concentration effects[65] should also be possible. Application of these techniques to simple model systems containing multilayers of organic semiconductors would probably be the first step in this type of study. The study of organic semiconductors[66,67] is just beginning now, and further developments in this field are bound to have far-reaching implications in the interpretation of biological phenomena.

Ultrastructural studies are confronted with more immediate tasks as a result of recent improvements in the preparation techniques and the useful resolving power of electron microscopy. Thus, continuation of earlier experiments based on rapid freezing of glycerine-treated tissues with liquid helium,[33] followed by special fixation and embedding procedures at low temperatures, may ultimately permit a more reliable correlation of organization at the molecular level with sequentially arrested states of activity in biological systems. Application of moire effects[68] to directly visualize equivalent configurations of atomic dimensions may also prove to be of considerable operational value in the study of paracrystalline components.

From the foregoing cursory survey, it is evident that the correlative investigation of fine structure and function of lamellar systems poses some of the most challenging and rewarding problems in modern biology, which will require an integrated approach enlisting the best efforts of biophysicists, biochemists, and physicists.

## ACKNOWLEDGMENTS

The author gratefully acknowledges stimulating discussions with M. Calvin, F. O. Schmitt, and J. Townsend, and their generous assistance in furnishing important bibliographical references.

## BIBLIOGRAPHY

[1] A. Engström and J. B. Finean, *Biological Ultrastructure* (Academic Press, Inc., New York, 1958).

[2] A. Frey-Wyssling, *Submicroscopic Morphology of Protoplasm* (Elsevier Publishing Company, Inc., Amsterdam, 1953), second edition.

[3] M. Calvin and P. B. Sogo, Science **125**, 499 (1957).

[4] G. Tollin, P. B. Sogo, and M. Calvin, "Energy transfer in ordered and unordered photochemical systems," University of California Radiation Laboratory Rept. (1957).

[5] P. B. Sogo and B. M. Tolbert, Advances in Biol. and Med. Phys. **5**, 1 (1957).

[6] J. A. Bassham and M. Calvin, *Currents in Biochemical Research*, D. E. Green, editor (Interscience Publishers, Inc., New York, 1956), p. 29.

[7] C. P. Whittingham, Progr. in Biophys. and Biophys. Chem. 320 (1957).

[8] B. B. Geren and J. Raskind, Proc. Natl. Acad. Sci. U. S. **59**, 880 (1953).

[9] B. B. Geren, Exptl. Cell Research **7**, 558 (1954).

[10] W. J. Schmidt, *Die Doppelbrechung von Karyoplasma, Zytoplasma und Metaplasma* (Gebrüder Borntraeger, Berlin, 1937).

[11] F. O. Schmitt and R. S. Bear, J. Cellular Comp. Physiol. **9**, 261 (1937).

[12] F. O. Schmitt and R. S. Bear, Biol. Revs. Cambridge Phil. Soc. **14**, 27 (1939).

[13] F. O. Schmitt, R. S. Bear, and G. L. Clark, Radiology **25**, 131 (1935).

[14] F. O. Schmitt, R. S. Bear, and K. J. Palmer, J. Cellular Comp. Physiol. **18**, 31 (1941).

[15] J. B. Finean, Exptl. Cell Research **5**, 202 (1953).

[16] J. B. Finean, Expt. Cell Research **6**, 283 (1954).

[17] J. B. Finean and P. F. Millington, J. Biophys. Biochem. Cytol. **3**, 89 (1957).

[18] J. B. Finean, Exptl. Cell Research **5**, 18 (1958).

[19] H. Fernández-Morán, Exptl. Cell Research **1**, 143 (1950).

[20] H. Fernández-Morán, Exptl. Cell Research **3**, 282 (1952).

[21] H. Fernández-Morán, Progr. in Biophys. and Biophys. Chem. 4, 112 (1954).

[22] F. O. Schmitt, Proc. Assoc. Research Nervous Mental Diseases **28**, 247 (1950).

[23] J. D. Robertson, J. Biophys. Biochem. Cytol. **1**, 271 (1955).

[24] H. Fernández-Morán and J. B. Finean, J. Biophys. Biochem. Cytol. **3**, 725 (1957).

[25] I. H. Page, *Chemistry of the Brain* (Charles C. Thomas, Springfield, 1937).

[26] J. Folch-Pi and W. M. Sperry, Ann. Rev. Biochem. **17**, 147 (1948).

[27] H. J. Deuel, Jr., *The Lipids* (Interscience Publishers, Inc., New York, 1951), Vol. I.

[28] F. O. Schmitt and N. Geschwind, Progr. in Biophys. and Biophys. Chem. **8**, 166 (1957).

[29] H. Fernández-Morán, Ind. Diamond Rev. **16**, 128 (1956).

[30] F. O. Schmitt in *Genetic Neurology*, P. Weiss, editor (The University of Chicago Press, Chicago, 1950), p. 40.

[31] F. O. Schmitt, Exptl. Cell Research, **5**, 33 (1958).

[32] I. M. Klotz, Science **128**, 815 (1958).

[33] H. Fernández-Morán in *Metabolism of the Nervous System* Second International Neurochemical Symposium, E. Richter editor (Pergamon Press, London, 1957), p. 1.

[34] T. M. Shaw and R. H. Elsken, J. Chem. Phys. **26**, 565 (1953).

[35] P. Denis, A. Csaki, M. Delco, J. Sprenger, H. Fernández-Morán, and W. Rawyler, Arch. sci. (Geneva) **10**, 223 (1957).

[36] B. Jacobson, W. A. Anderson, and J. T. Arnold, Nature **173**, 772 (1954).

[37] W. M. Sperry and H. Waelsch, Proc. Assoc. Research Nervous Mental Diseases **28**, 255 (1950).

[38] C. E. Lumsden, *Modern Trends in Neurology* (Butterworths Scientific Publications, London, 1957), Ser. II, p. 130.

[39] G. H. Sloane-Stanley, Biochem. Soc. Symposia **8**, 44 (1952).

[40] D. Nachmansohn in *Modern Trends of Physiology and Biochemistry*, E. S. G. Barron, editor (Academic Press, Inc., New York, 1952), p. 229.

[41] H. Klüver in *Biochemistry of the Developing Nervous System*, International Neurochemical Symposium, H. Waelsch, editor (Academic Press, Inc., New York, 1955), p. 137.

[42] W. J. Schmidt, Z. wiss. Mikroskop. **52**, 158 (1935).

[43] F. S. Sjöstrand, J. Cellular Comp. Physiol. **33**, 383 (1949).

[44] F. S. Sjöstrand, J. Cellular Comp. Physiol. **42**, 15 (1953).

[45] J. B. Finean, F. S. Sjöstrand, and E. Steinmann, Exptl. Cell Research **5**, 557 (1958).

[46] F. S. Sjöstrand, Intern. Rev. Cytol. **5**, 455 (1956).

[47] H. Fernández-Morán, Exptl. Cell Research **5**, 586 (1958).

[48] T. H. Goldsmith and D. E. Philpott, J. Biophys. Biochem. Cytol. **3**, 429 (1957).

[49] J. J. Wolken, J. Capenos, and A. Turano, J. Biophys. Biochem. Cytol. **3**, 441 (1957).

[50] V. B. Wigglesworth, *The Principles of Insect Physiology*, (Methuen and Company, Ltd., London, 1950), fourth edition.

[51] W. H. Miller, J. Biophys. Biochem. Cytol. **5**, 421 (1957).

[52] K. von Frisch, *Bees: Their Vision, Chemical Senses and Language* (Cornell University Press, Ithaca, New York, 1950).

[53] H. Autrum and H. Stumpf, Z. Naturforsch. **5b**, 116 (1950).

[54] H. Autrum, Exptl. Cell Research **5**, 426 (1958).

[55] H. Fernández-Morán, Nature **177**, 742 (1956).

[56] B. Chance and G. R. Williams, Advances in Enzymol. **17**, 65 (1956).

[57] D. E. Green in *Harvey Lectures* (Academic Press, Inc., New York, 1958), Vol. LII, p. 177.

[58] A. L. Lehninger, Ch. L. Wadkins, C. Cooper, Th. M. Devlin, and J. L. Gamble, Science **128**, 450 (1958).

[59] A. Szent-Györgyi, *Bioenergetics* (Academic Press, Inc., New York, 1957).

[60] S. W. Peterson and H. A. Levy, Phys. Rev. **92**, 1082 (1953).

[61] B. Commoner, J. J. Heise, B. Lippincott, R. E. Norberg, J. Passoneau, and J. Townsend, Science **126**, 57 (1957).

[62] H. Fernández-Morán, Bol. acad. cienc. (Caracas) **51**, 1 (1953).

[63] G. Wald, Exptl. Cell Research, **5**, 389 (1958).

[64] T. H. Bullock, Federation Proc. **12**, 666 (1953).

[65] W. Shockley, *Electrons and Holes in Semiconductors*, Bell Laboratories Series (D. Van Nostrand Company, Inc., New York, 1953).

[66] D. D. Eley, G. D. Parfitt, M. J. Perry, and D. H. Taysum, Trans. Faraday Soc. **49**, 79 (1953).

[67] D. D. Eley and G. D. Parfitt, Trans. Faraday Soc. **51**, 1529 (1955).

[68] D. W. Pashley, J. W. Menter, and G. A. Bassett, Nature **179**, 752 (1957).

# 39

# Fine Structure of Lamellar Systems as Illustrated by Chloroplasts*

ALAN J. HODGE

*Department of Biology, Massachusetts Institute of Technology, Cambridge 39, Massachusetts*

THE fine structure of specialized organelles such as mitochondria, which are known to be concerned with the metabolic side of the carbon cycle, has been discussed in the three foregoing chapters (Bennett, p. 297; Sjöstrand, p. 301; Fernández-Morán, p. 319). The structures concerned with the other half of this cycle—namely, the photosynthetic aspect—are considered here.

It is of interest, in terms of the evolution of specialized photosynthetic organelles, to examine the fine structure of certain primitive types of cells, since these are presumably representative of organisms arising very early in the evolutionary process. Such a group of organisms are the blue-green algae. Examination in the electron microscope shows that, in certain types such as *Nostoc* (Fig. 1), there are no specialized membrane-bounded organelles such as mitochondria, nuclei, and chloroplasts; rather there exists a generalized membrane structure ramifying throughout the cytoplasm. It seems reasonable to conclude that, in such cells, the various metabolic functions such as oxidative phosphorylation and photosynthesis are carried on in relation to an apparently undifferentiated membrane system (or perhaps in specialized patches of the membrane), rather than in well-defined organelles. In other blue-greens (e.g., *Anabena*), the photosynthetic material is present in a more specialized and segregated state, but the primitive organelles lack limiting membranes. It is of interest to note that, in higher organisms, an intracellular membrane system (the endoplasmic reticulum)[1] appears to be universally present, and to have clear continuity with the membranous envelope of the nucleus (and possibly with the limiting membranes of other organelles). It seems likely, therefore, that comparative study of such primitive cells might yield important clues concerning the origin and interrelationship of the endoplasmic reticulum and specialized organelles such as mitochondria, chloroplasts, and nuclei in the cells of higher organisms.

Proceeding a little higher in the evolutionary scale to a typical green alga, such as *Nitella* (Fig. 2), one observes typical membranous elements of the endoplasmic reticulum, the characteristically structured mitochondria found in all higher organisms, and chloroplasts, bounded by a well-defined, double limiting membrane and containing a number of relatively well-ordered dense lamella within a finely granular matrix material.[2] Chloroplasts of this structural type are highly characteristic of the lower forms of plant cells.[3–6]

In the higher plants, of which corn (*Zea mays L.*) is chosen as an example, the structure of the chloroplasts[7–10] usually differs from the more primitive pattern already described. Figure 3 shows in transverse section the leaf of a three-week-old corn plant. Around each vascular bundle are a number of parenchyma sheath cells, the chloroplasts of which are specialized for the formation and storage of starch. Most of the photosynthetic activity, however, is carried out in the chloroplasts of the mesophyll cells. Even at the light-microscopical level, there are obvious differences in appearance between the chloroplasts of the parenchyma-sheath cells and those of the mesophyll cells. In the electron microscope (Fig. 4), the parenchyma-sheath chloroplasts of *Zea*[7] resemble algal chloroplasts (cf. *Nitella*, Fig. 2) in their over-all plan of organization. Within a double external limiting membrane, a number of densely staining lamellae (each about 130 A thick) are set in a finely granular matrix which presumably contains most of the soluble enzymes of the chloroplast.

The mesophyll chloroplasts are similarly lamellated but, in addition, possess well-defined regions (the grana) in which the lamellae are more densely and regularly packed (Fig. 4). If the section is in the plane of the lamellae, the grana appear as circular profiles. If, on the other hand, the plane of the section is normal to the lamellar plane, the grana appear as regular rectangular-shaped regions. The concentration of lamellar surface within the grana is about twice as high as in the intervening (intergrana) regions, and it can be seen (Fig. 5) that this comes about because of a pairing or bifurcation of the lamellae at the edges of the grana. This type of chloroplast with well-defined grana is characteristic of higher plants, most of which do not possess plastids of the type found in the parenchyma-sheath cells of *Zea*.

In both types of chloroplasts, the individual lamellae (each 130 A thick) of both the grana and intergrana regions exhibit a compound-layer structure (Figs. 6 and 7), consisting of a central dense line (the *P* zone, about 40 A thick and often resolvable as a doublet) on both sides of which are less dense layers (the *L* zones). Finally, the entire compound layer structure is edged by very thin dense lines, the *C* zones. Such differences in density arise in part from differences in reactivity

* Most of the work described herein was carried out in the Chemical Physics Section, C. S. I. R. O., Melbourne, Australia, in collaboration with Dr. F. V. Mercer and Dr. J. D. McLean of the Botany Department, University of Sydney, N. S. W.

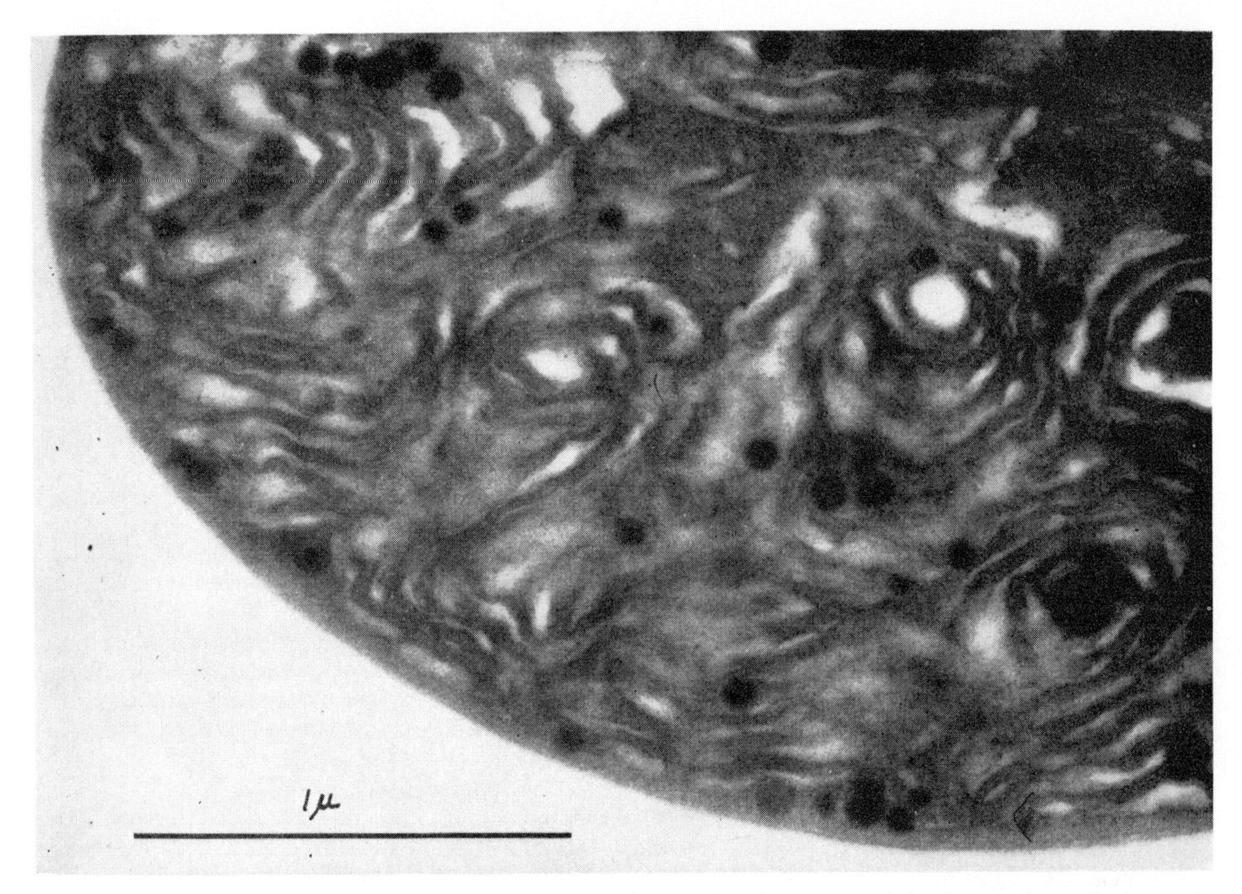


FIG. 1. Vegetative cell of a *Nostoc* spp., showing the generalized whorl-like lamellar system and the absence of specialized organelles such as chloroplasts and mitochondria. ×60 000. (Unless otherwise indicated, all illustrations are electron micrographs of thin sections of plant material fixed in osmium-tetroxide solutions of appropriate *p*H and tonicity, and embedded in methacrylate. Details are given in references 2, 7, 13, and 14.)

with osmium tetroxide as well as from intrinsic differences in electron density within the various components of the structure. Within the grana, the close packing of such compound lamellae results in close apposition of *C* zones and, thus, gives rise to a layer structure (Figs. 8 and 9) with a repeat period of 130 A, bearing a remarkable resemblance to that found in the myelin sheath of nerve fibers. The central dense lines of the compound lamellae correspond to the major dense lines of the periodic structure within the grana and are often observed as doublets (Fig. 9), each line of the pair being about 15 A thick. It is clear, therefore, that the symmetrical structure of the individual compound lamellae arises from the close apposition of two structurally asymmetric "unit membranes" (each about 70 A thick) as shown schematically in Fig. 11. The *I* zones, which occur midway between *P* zones in the grana, arise by apposition of the *C* zones of the unit membranes. The work of a number of people, notably Robertson,[11] indicates that the cell membranes and intracellular membrane systems of most, if not all, cells consist of such unit membranes in single, double, or compound array. This also appears to be true of plant cells. In Fig. 10, the tonoplast is seen as a single membrane with a pair of dense edges, and the double limiting membrane of the chloroplast consists of two such unit membranes. The intrinsic asymmetry of these unit membranes is not usually evident in the electron microscope except where they are stacked in double or multiple array, as in the lamellae and grana of chloroplasts and in myelin sheath. In both cases, this stacking results in a set of major dense lines with less-dense intermediate lines in between. As already indicated by Fernández-Morán (p. 319), there is good reason to believe that each 70-A membrane comprises a double layer of mixed lipids (the *L* zones) sandwiched between two thin monolayers of protein which stain densely with osmium tetroxide, with a probable contribution to this latter density arising from reaction of the hydrophilic groups of the phospholipids with osmium tetroxide. In the case of myelin sheath, a very satisfactory correlation between electron-microscopic and x-ray diffraction data has been achieved[12] and the chemical composition is reasonably well known. It seems fairly certain, therefore, that this type of structure is essentially correct for most cellular membranes, with only such minor differences in detail arising as are demanded by differences in function.

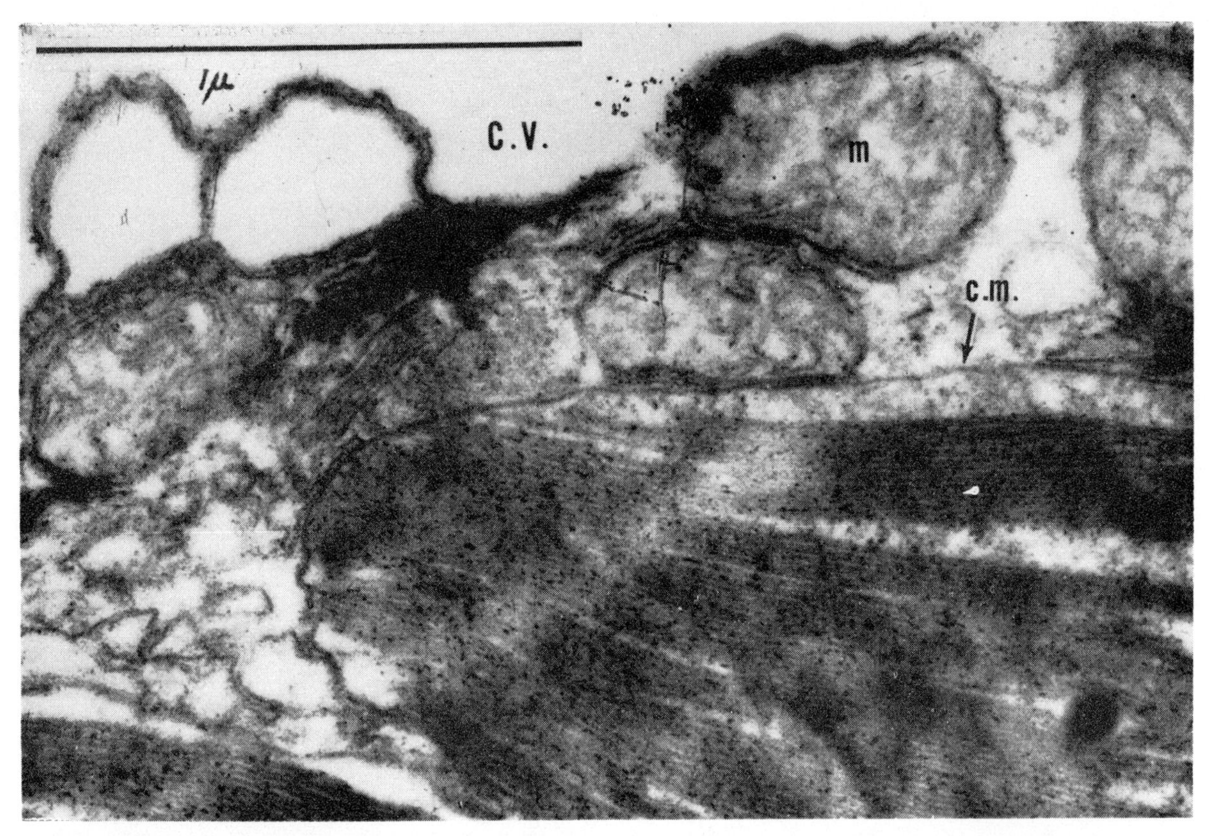


FIG. 2. Transverse section through the cytoplasm of a *Nitella* cell, showing the lamellated structure of a chloroplast surrounded by its limiting membrane *c. m.*, mitochondria *m*, various membranous elements within the cytoplasm, and the tonoplast lining the lumen of the central vacuole *C. V.* ×75 000.

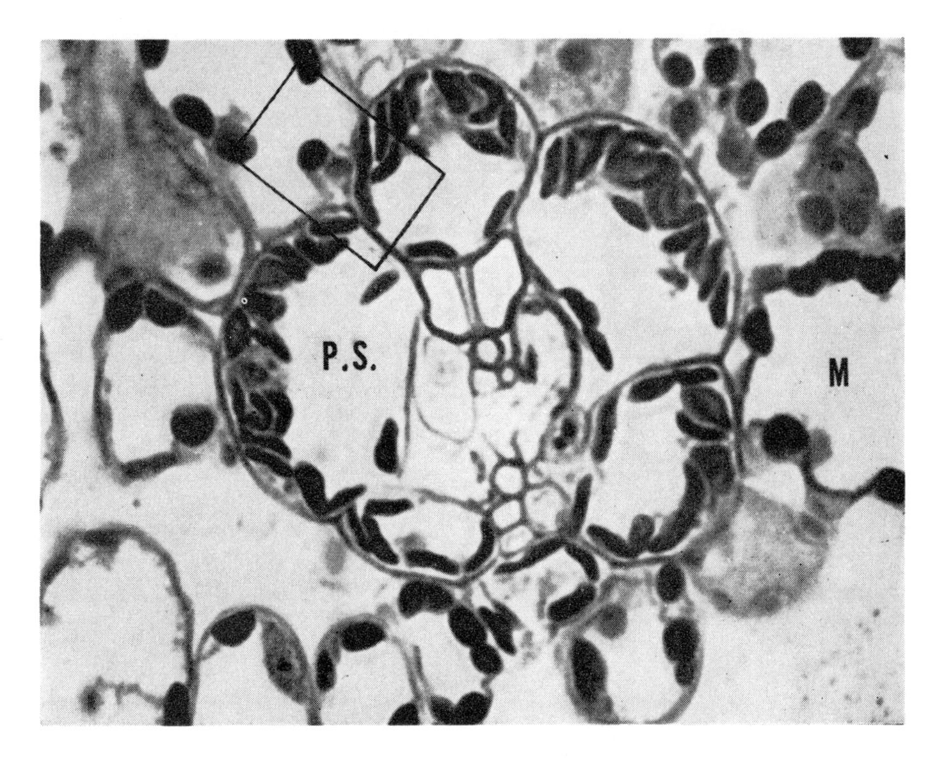


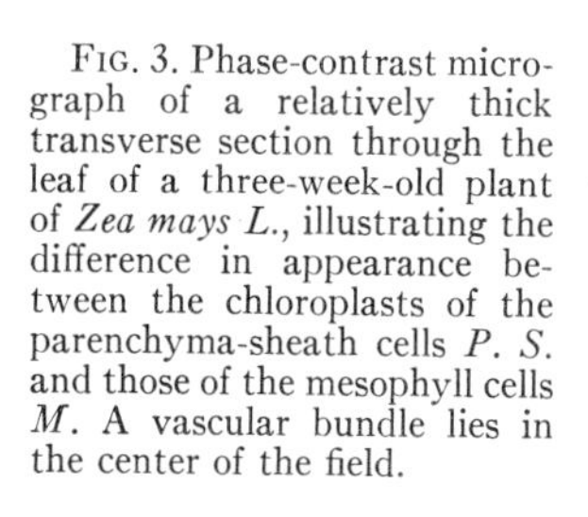
FIG. 3. Phase-contrast micrograph of a relatively thick transverse section through the leaf of a three-week-old plant of *Zea mays L.*, illustrating the difference in appearance between the chloroplasts of the parenchyma-sheath cells *P. S.* and those of the mesophyll cells *M*. A vascular bundle lies in the center of the field.

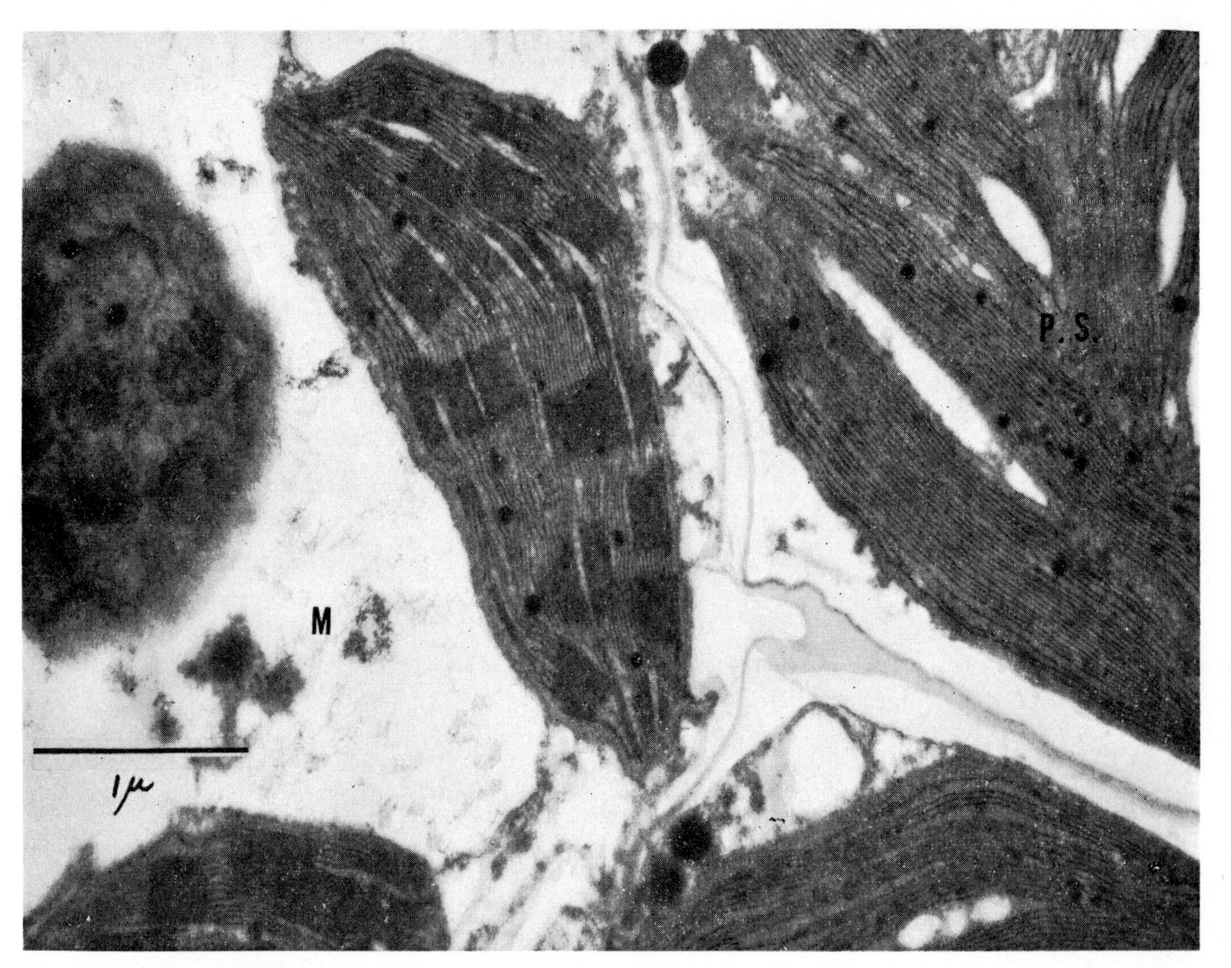


FIG. 4. Electron micrograph of a region similar to that outlined in Fig. 3, illustrating the lamellar structure both of parenchyma-sheath cells *P.S.* and mesophyll *M* chloroplasts of *Zea*. In the mesophyll chloroplasts, the grana appear as dense rectangular regions when the section is normal to the lamellar plane, and as circular regions when the plane of section is parallel to the lamellar plane. ×30 000.

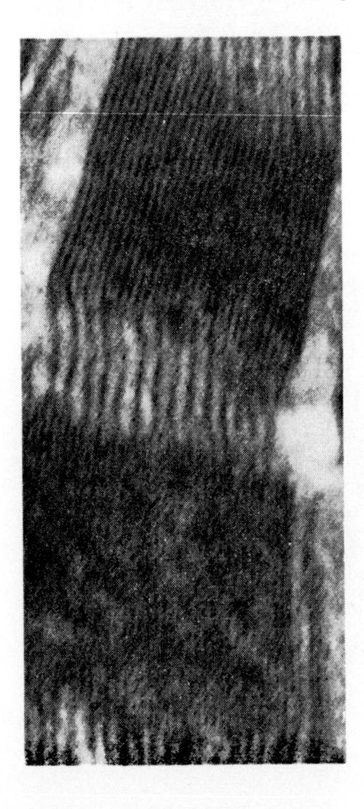

FIG. 5. Two grana in a mesophyll chloroplast of *Zea*, showing the type of connection between the grana lamellae and those in the intervening (intergrana) regions. (Cf. Fig. 11.) ×85 000.

In the case of the chloroplast, the situation is less well documented, but certain conclusions can be drawn from the available structural and chemical data, and by extrapolation from the better characterized myelin system. It can be seen from Fig. 10 that the membranes making up the lamellae and grana of the chloroplast are considerably more dense than the external limiting membranes of the chloroplast and the tonoplast, which exhibit densities more characteristic of the usual cellular membrane systems. In particular, the *L*-zones of the chloroplast lamellae exhibit higher densities than are found in cell membranes, endoplasmic reticulum, etc. A possible explanation of this may lie in differences of chemical composition. It is known that the chloroplast is deficient in phospholipids as compared with myelin sheath.[7] This deficiency is counterbalanced by the presence of considerable amounts of carotenoids and chlorophyll. Furthermore, as is shown later, there is good evidence to suggest that chlorophyll is an integral component of the chloroplast lamellae and that sufficient is present for its incorporation as a monolayer over the entire lamellar area of the chloroplast as estimated from electron micrographs. Thus, the general conclusion

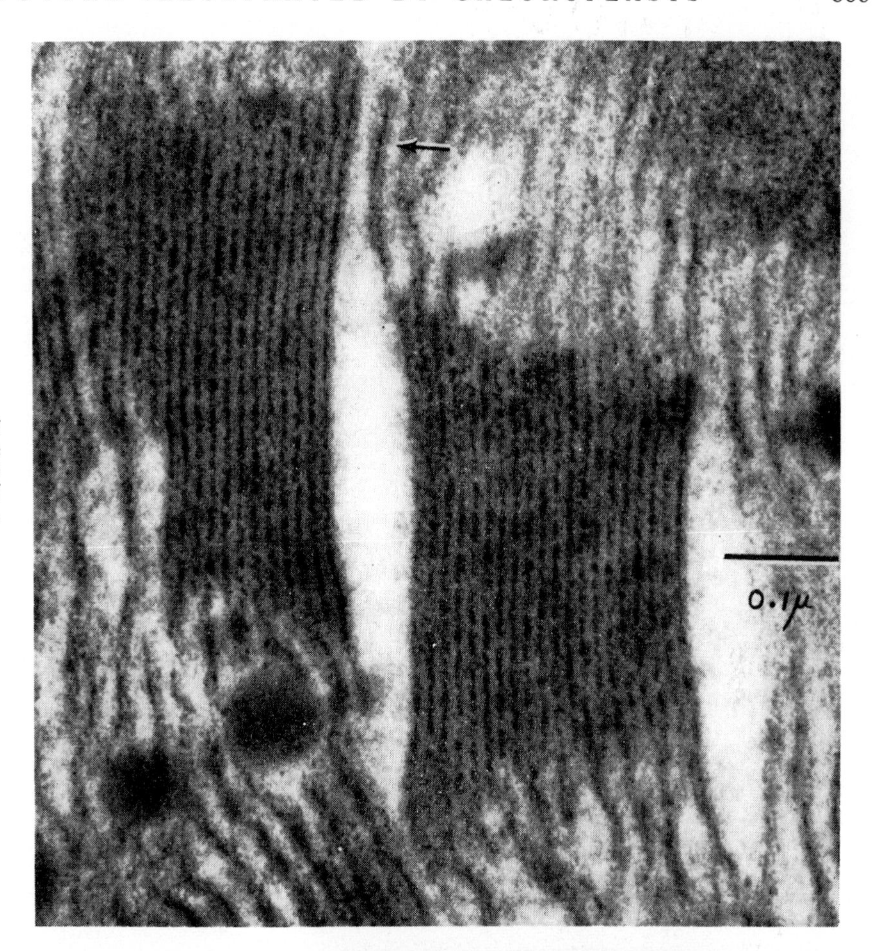


FIG. 6. Mesophyll chloroplast of *Zea*, illustrating the high degree of order within the grana, and the compound-layer structure of the intergrana lamellae (arrow). ×160 000.

seems justified that the photosynthetic lamellar elements of the chloroplast consist of structurally asymmetric unit membranes (Fig. 11) in which chlorophyll and carotenoids are incorporated in orderly array. It is of interest to note that the exciton-migration theory of energy transfer proposed by Calvin appears to require the presence of such a structural asymmetry. The type of model suggested by Calvin (p. 157) has the necessary asymmetry (Fig. 12) and appears to be consistent with the presently available chemical and structural evidence. Such a staggered, partially overlapping configuration of the chlorophyll molecules has the further advantage over other models (which usually involve coplanarity of the porphyrin "heads" of the chlorophyll) in that it offers a plausible explanation for the low values of dichroic effects so far observed in chloroplasts. Furthermore, it would allow a greater degree of $\pi$-electron interaction than allowed by the coplanar type of model, thus facilitating energy transfer, and is consistent with the type of packing found in crystals of polycyclic organic compounds by x-ray diffraction. As has been seen the chloroplasts of higher plants characteristically contain grana in which the unit membranes are arranged in a highly ordered and close-packed array. The significance of this specialization is unknown at the present time, but it is tempting to speculate that the high degree of order might lead to more efficient trap-

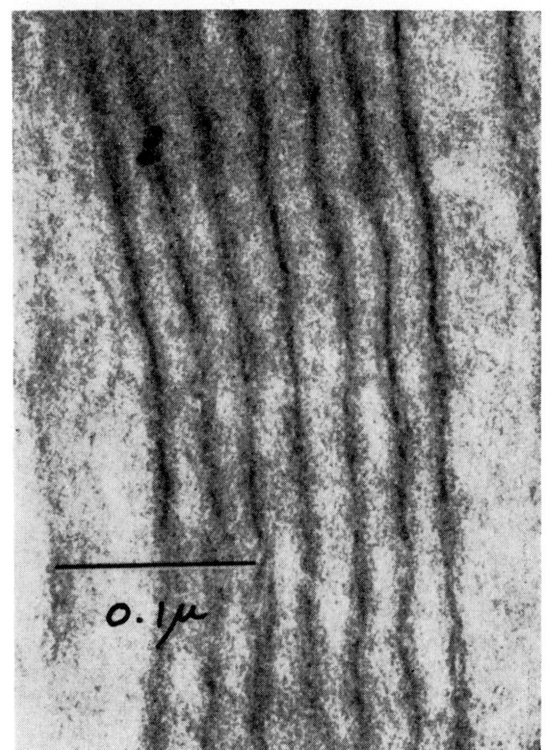


FIG. 7. Compound layer structure of the lamellae in a parenchyma-sheath chloroplast of *Zea*. ×270 000.

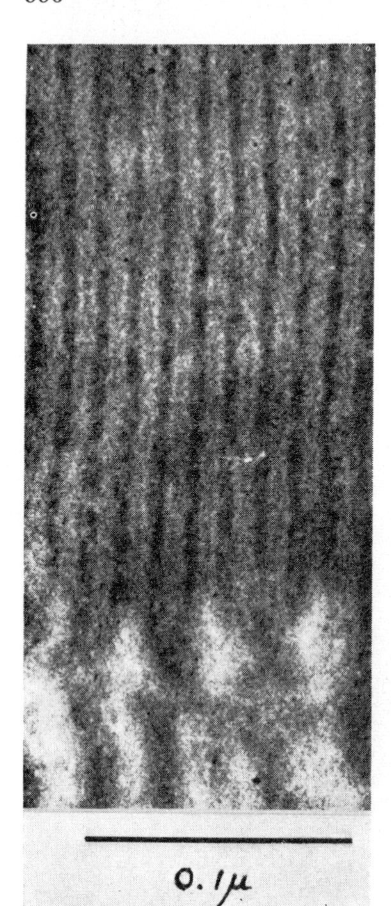


FIG. 8. Mesophyll chloroplast of *Zea* to illustrate the myelin-like structure within a granum. Note the main periodicity of about 130 A and the presence of fainter intermediate lines. ×370 000.

ping of photons, or perhaps result in facilitation of exciton migration as a result of a stabilization of the layer structures which possess the conduction bands necessary for energy transfer.

## SWELLING PROPERTIES OF THE CHLOROPLAST

The swelling characteristics of chloroplasts offer a striking demonstration of the "plasticity" of such lipoprotein layer systems. As already mentioned the swelling of mitochondria can be largely prevented by active expenditure of energy from ATP hydrolysis. A similar phenomenon exists in the case of chloroplasts. Klein (unpublished work) has shown that the swelling response of isolated chloroplasts depends on whether the suspension of plastids is illuminated or is kept in the dark.

When isolated *Nitella* chloroplasts are placed in a hypotonic medium, swelling results first in a separation of the lamellae to distances many times that characteristic of the intact chloroplast.[2] In highly hypotonic media, the lamellae break up. The hydrophobic hydrocarbon chains of the lipids thus are exposed at the free edges of the membranes, giving rise to an unstable situation, with the result that such free edges "zip" together to form membrane-bounded vesicles. Similar results have been obtained for isolated *Zea* chloroplasts (Fig. 13), and the characteristically vesicular structures found in so-called microsome fractions isolated by conventional procedures arise in similar fashion (e.g., Hodge[13]) from the endoplasmic reticulum. This plastic behavior is understandable in terms of what is known of membrane structure. In the type of layer structure under consideration here, it seems likely that the oriented lipid molecules possess a high degree of rotational freedom and considerable translational freedom within the plane of the membrane. Both properties are characteristic of the smectic fluid-crystal state and confer a remarkable capacity for changes in the topographical configuration of such membrane systems.

## CHLOROPLAST DEVELOPMENT

The development of the chloroplast is of interest in that it appears to take place by a process which is essentially the reverse of that involved in the swelling phenomena just discussed. In brief, the compound- and extended-layer structures of the lamellae and grana appear to be formed by a process involving the fusion of small vesicles or micelles one with another to form extended cisternae or "double-membrane structures."

The development of a leaf or of a plant is a complicated sequence of events, the time and spatial relationships of which lead to difficulty in following the process of chloroplast development. The etiolated plant (one grown from seed in total darkness) is a much more

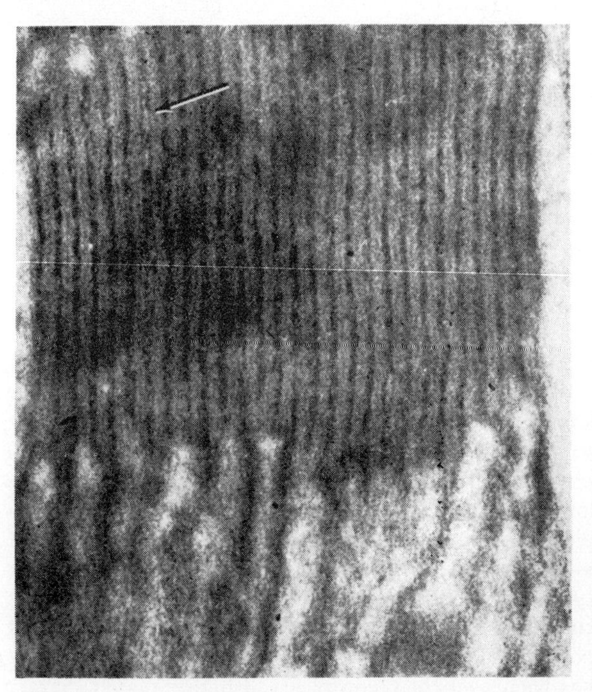

FIG. 9. A granum in a mesophyll chloroplast of *Zea*, illustrating the splitting of the main dense lines (arrow). The central dense lines (*P* zones) of the intergrana lamellae are also occasionally resolvable as doublets, thus showing that the compound lamellae consist of two closely apposed asymmetric unit membranes. ×220 000.

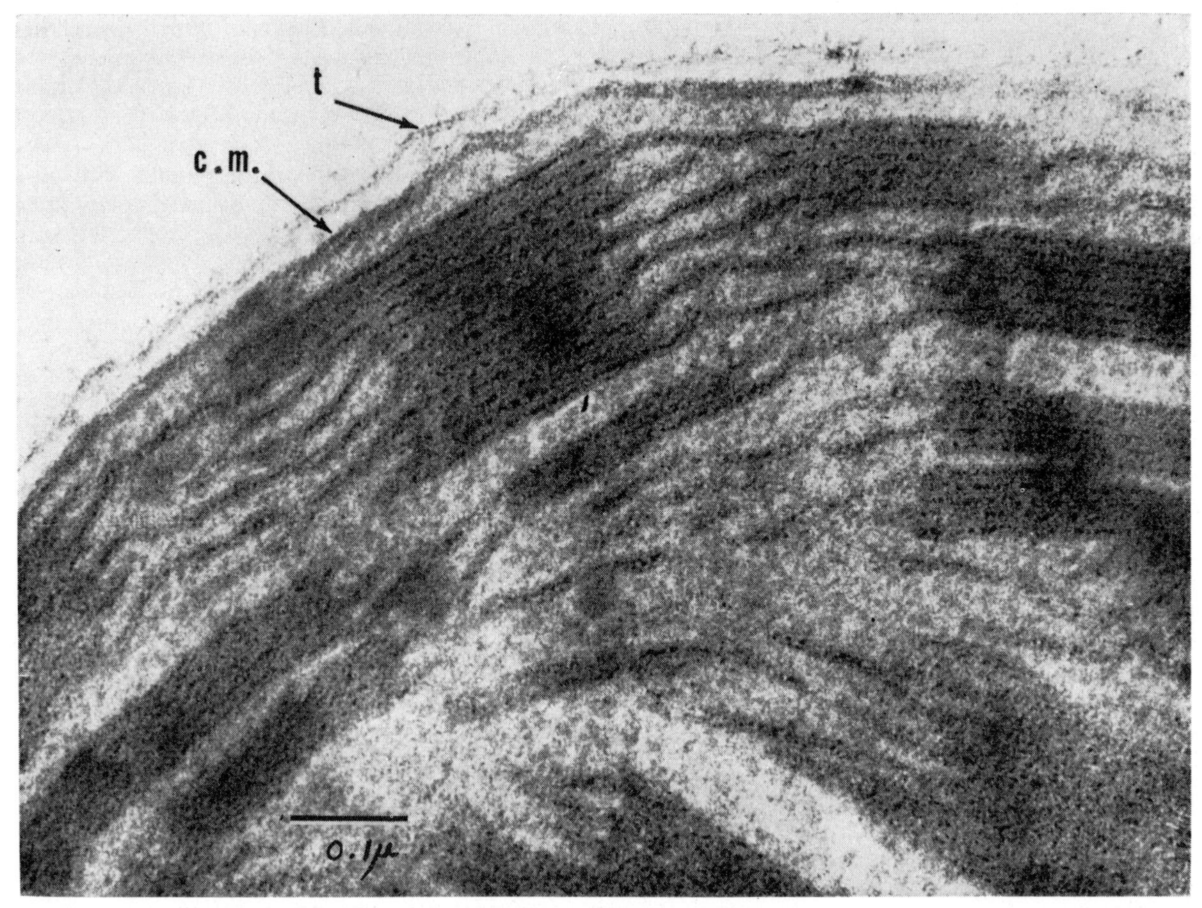


FIG. 10. Chloroplast in a three-week-old wheat plant, showing lamellar structure very similar to that in *Zea mays*. The tonoplast *t* appears as a single membrane, and where suitably oriented with respect to the plane of the section appears as two fine dense lines with a less dense layer between them, the over-all thickness being about 70 A. The chloroplast envelope *c. m.* comprises two unit membranes spaced about 100 A apart. ×160 000.

favorable system for such a study, since such seedlings can be exposed to light for various periods and the effects on chloroplast structure noted at definite time intervals following such exposure. This system depends on the fact that light is essential for chlorophyll synthesis in higher plants.

Figure 14 shows typical plastids in an etiolated leaf of *Zea*.[14] In the absence of chlorophyll, the plastids, although recognizable as well-defined organelles with external limiting membranes, fail to develop lamellae. Instead, the interior of each plastid is partially filled with a mass of small vesicles (the prolamellar body).

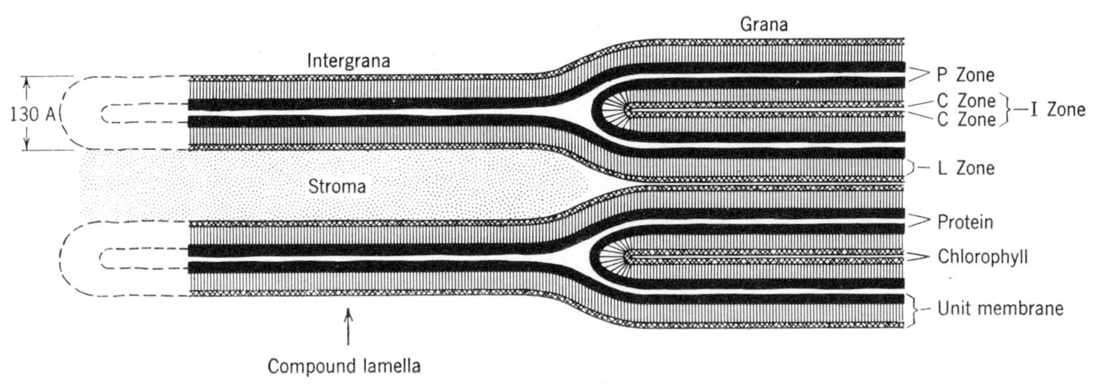


FIG. 11. Diagram to illustrate the densities observed in electron micrographs of osmium-fixed chloroplast lamellae and grana, and the structureal relationships involved in the formation of the compound lamellae (and grana) from structurally asymmetric unit membranes.

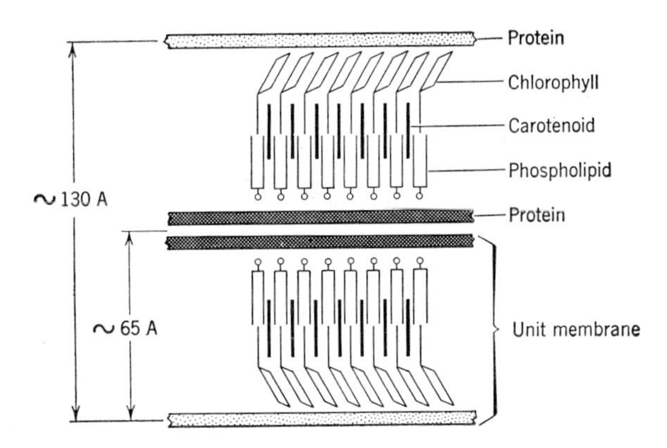


FIG. 12. Diagram to illustrate one way in which the compound lamellae observed in electron micrographs of chloroplasts could be built up from structurally asymmetric unit membranes having a molecular structure of the type proposed by Calvin (p. 157).

On placing the plant in daylight, chlorophyll is synthesized (as evidenced by a progressive greening of the leaves), and definite structural changes are observed in the plastids. Double membranes are observed (Fig. 15) apparently emerging from the prolamellar body.[14] At a later stage, typical chloroplast lamellae are seen, usually in a pattern radiating from a progressively decreasing prolamellar body (Fig. 16). It is of interest to note that in certain *Zea* mutants which are unable to synthesize chlorophyll, the plastids are similar to those of etiolated normal *Zea* in that they lack formed lamellar elements and contain only small vesicular structures. The evidence thus strongly suggests that chlorophyll is at least essential for the formation of typical chloroplast lamellae and is probably an integral component of these membrane structures.

At a later stage, rudimentary grana with a fine structure indistinguishable from that of the more mature grana already described may be observed (Fig. 17) within the mesophyll chloroplasts. After two days, the developing plastids closely resemble normal chloroplasts and no prolamellar bodies can be discerned, presumably as a result of complete conversion of the small vesicles into formed lamellar elements. At this stage and in earlier stages of recovery from etiolation, however, and in young chloroplasts of normal plants, one commonly observes immediately under the chloroplast limiting membrane a number of vesicles and sacs (Fig. 18) with membrane densities considerably lower than those characteristic of the already differentiated lamellae and grana. These observations suggest the possibility that this region, or the limiting membrane of the chloroplast itself, may be responsible for the formation of the small vesicles, perhaps by outpocketing and pinching-off from the limiting membrane, or by some direct synthetic process.

The concept of vesicle fusion either with other vesicles or with already existing extended membranes is attractive in relation to (1) the formation and maintenance of structures of higher order, such as the

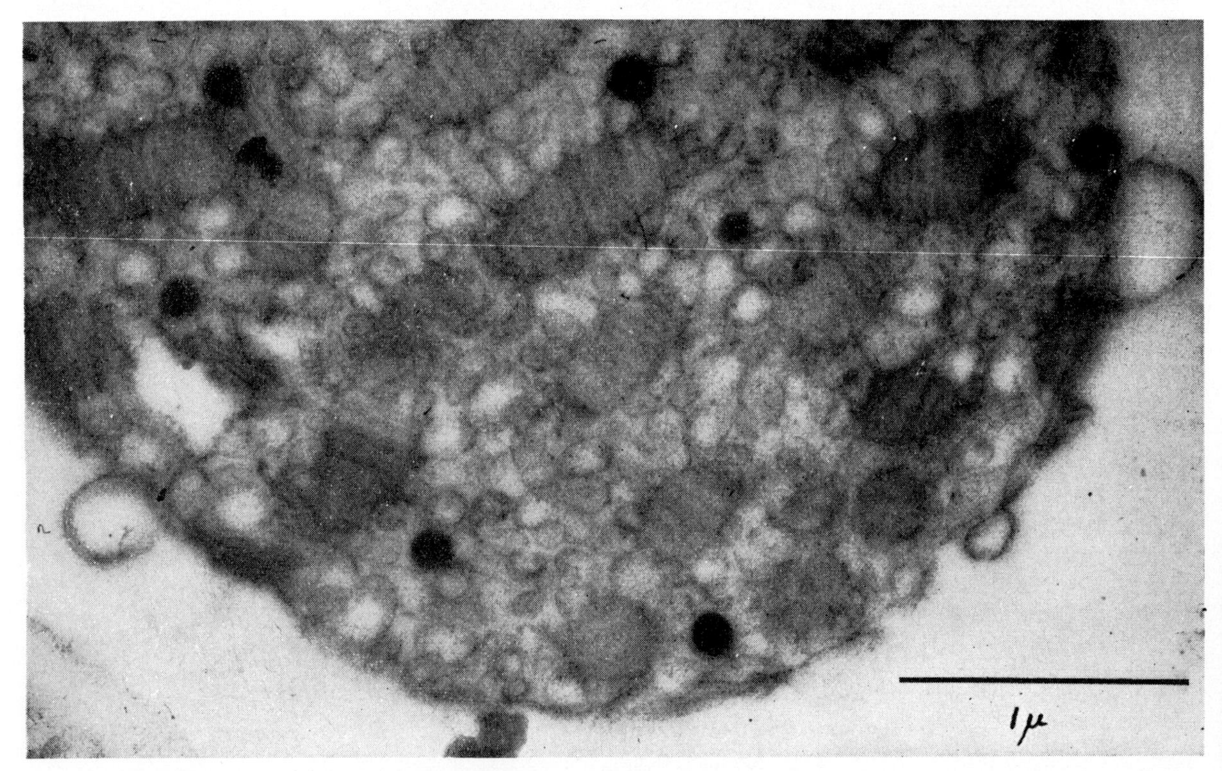


FIG. 13. Mesophyll chloroplast of *Zea* isolated in 0.5 M glucose in neutral phosphate buffer, showing a marked swelling reaction. (Compare with Figs. 4 and 6.) Note that the intergrana lamellae have broken up and formed large numbers of vesicles. × 40 000.

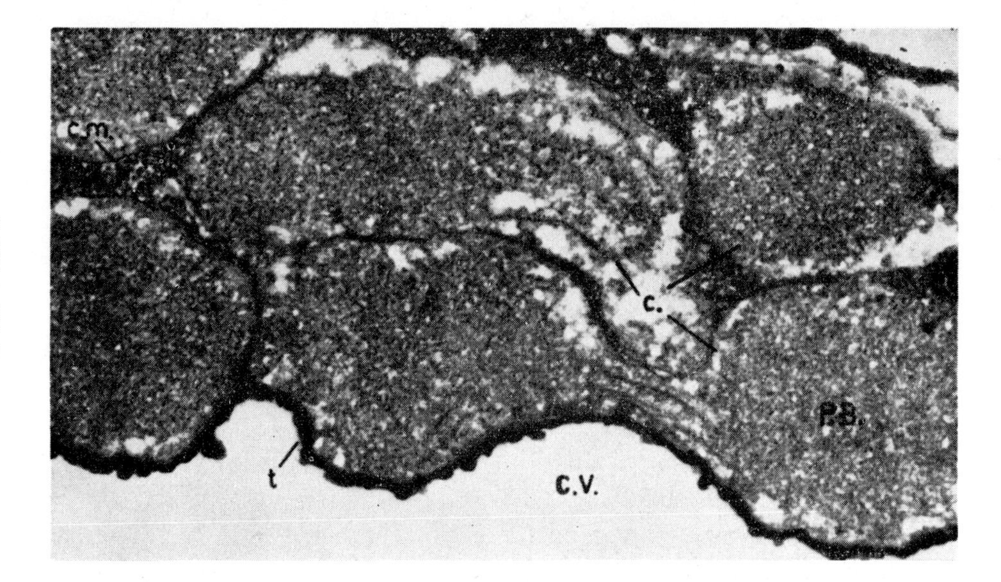


FIG. 14. Plastids in an etiolated leaf of *Zea*. The plastids contain masses of minute vesicles, but are devoid of lamellae, presumably because the absence of chlorophyll prevents their formation.

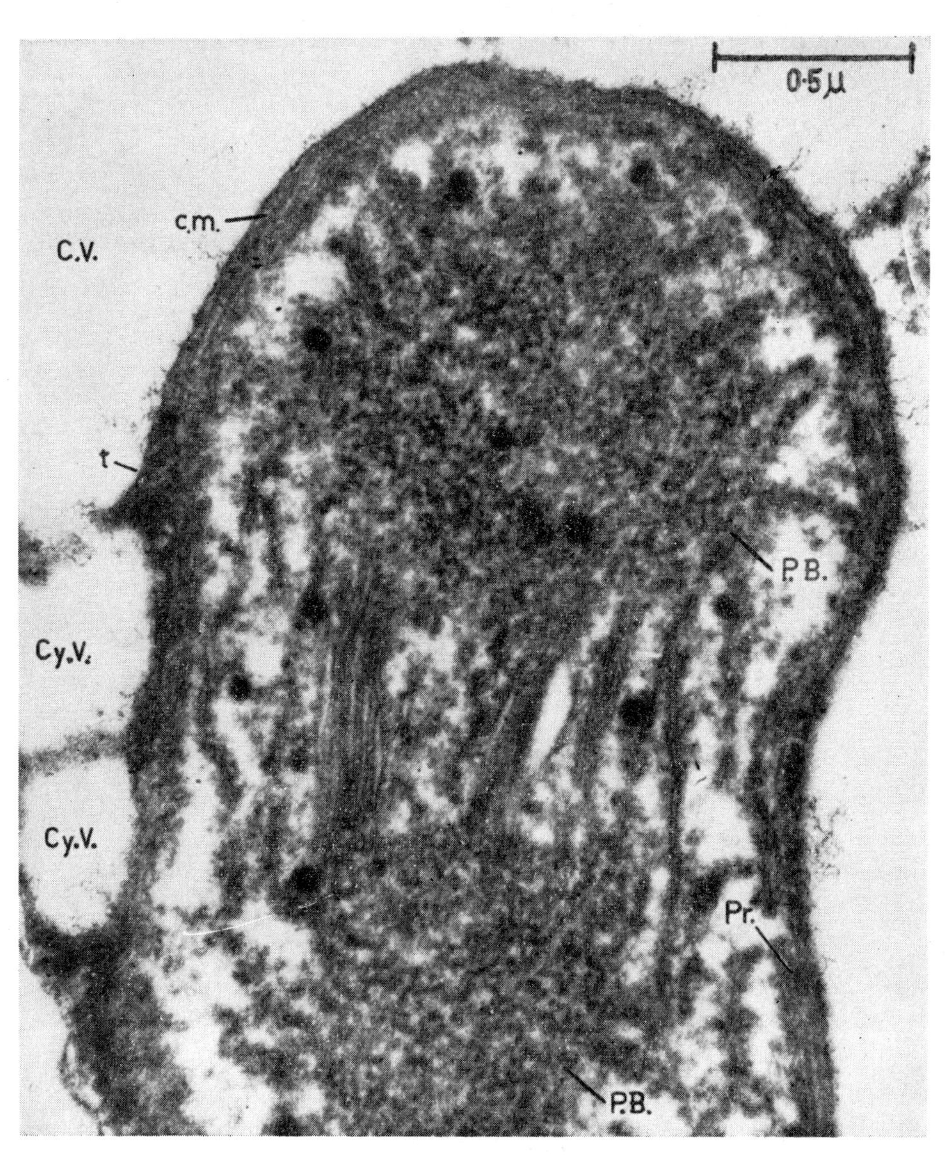


FIG. 15. Plastid from an etiolated leaf of *Zea* after several hours exposure to daylight showing an early stage in the formation of lamellae. Note that the lamellae appear to arise initially as double membranes, apparently by a process involving an orderly fusion of the small vesicles comprising the prolamellar body *P. B.* ×55 000.

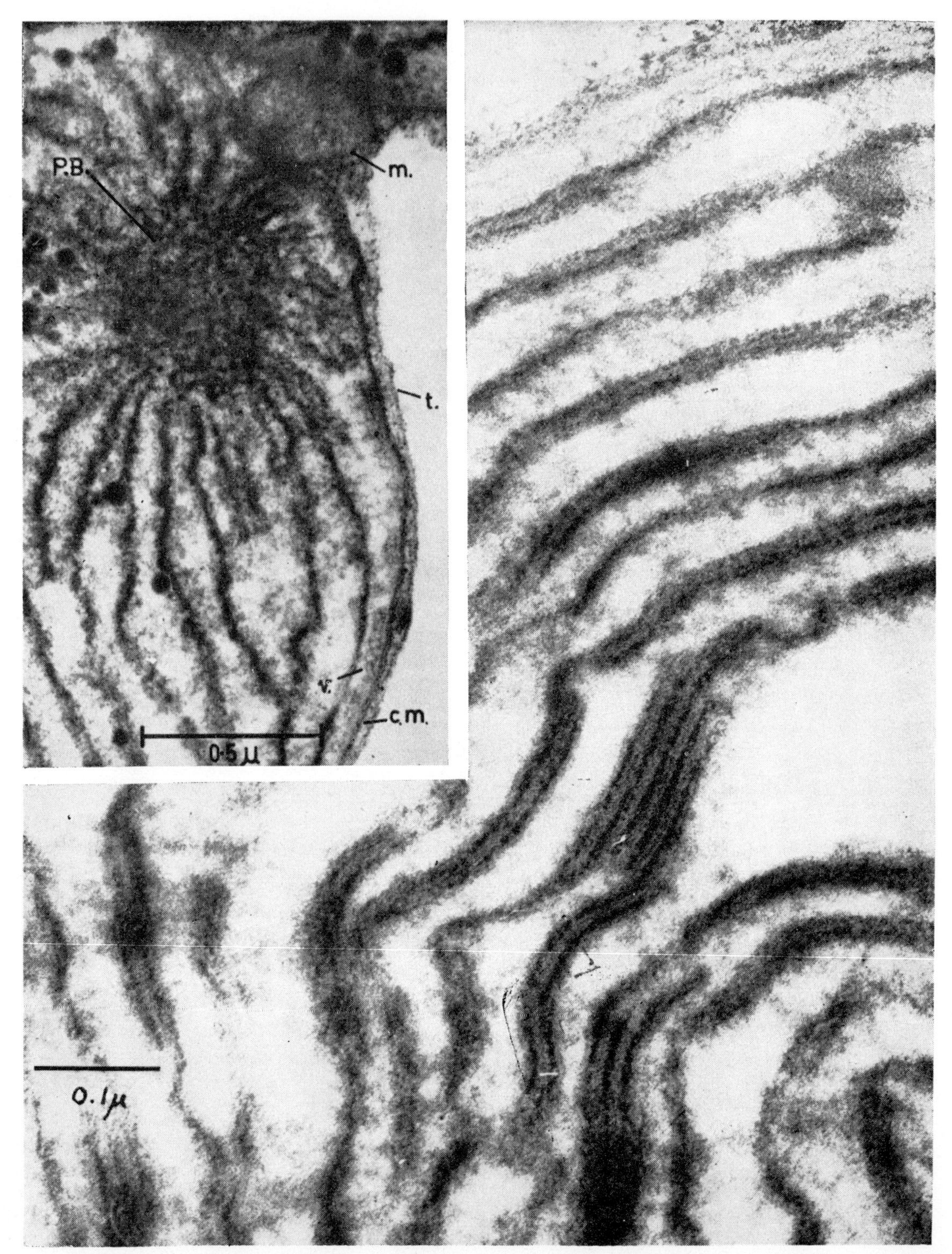


FIG. 16 (upper left). A somewhat later stage in recovery from etiolation than that shown in Fig. 15, illustrating the progressive reduction in size of the prolamellar body *P. B.* and formation of typically dense chloroplast lamellae.

FIG. 17 (upper right and lower area). Mesophyll plastid of etiolated *Zea* after 20-hr exposure to daylight showing the beginning of grana formation. ×210 000.

endoplasmic reticulum and the various organelles of the cell (the vast increase in membrane area required for the formation of the myelin sheath of nerve fibers could be accounted for by fusion of vesicles with the Schwann-cell membrane); and (2) the transport both into and out of the cell of various substances. Such a

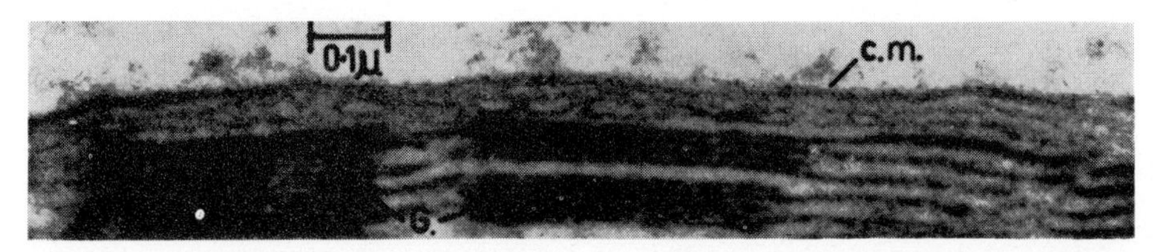


FIG. 18. Mesophyll plastid of etiolated *Zea* after 48-hr exposure to daylight. Typically dense chloroplast lamellae and grana are present. Note the presence of less-dense vesicles and cisternae in the region immediately beneath the chloroplast envelope *c. m.* ×85 000.

mechanism has been proposed for the extrusion of acetylcholine in "quantized" amounts from the nerve endings in myoneural junctions, and for the passage of zymogen granules from the apical regions of acinar pancreas cells through the limiting membrane into the lumen of the acinus. Similarly, the occurrence of the reverse process (i.e., pinocytosis) seems to be well documented. While mechanisms of this general type are widely supported by indirect experimental evidence, and appear to be thermodynamically feasible in terms of current knowledge concerning the structure of lipoprotein layer systems, the crucial problem of energy coupling in the physical control of such membrane systems remains unelucidated.

## BIBLIOGRAPHY

[1] K. R. Porter in *Harvey Lectures* (Academic Press, Inc., New York, 1957), Vol. LI.

[2] F. V. Mercer, A. J. Hodge, A. B. Hope, and J. D. McLean, Australian J. Biol. Sci. **8**, 1 (1955).

[3] R. Sager and G. E. Palade, Exptl. Cell Research **7**, 584 (1954).

[4] J. J. Wolken and G. E. Palade, Ann. N. Y. Acad. Sci. **56**, 873 (1953).

[5] P. A. Albertsson and H. Leyon, Exptl. Cell Research **7**, 288 (1954).

[6] H. Leyon, Exptl. Cell Research **6**, 497 (1954).

[7] A. J. Hodge, J. D. McLean, and F. V. Mercer, J. Biophys. Biochem. Cytol. **1**, 605 (1955).

[8] H. Leyon, Exptl. Cell Research **7**, 265 (1954).

[9] E. Steinmann and F. S. Sjöstrand, Exptl. Cell Research **8**, 15 (1955).

[10] M. Cohen and E. Bowler, Protoplasm **42**, 4, 414 (1953).

[11] J. D. Robertson, J. Biophys. Biochem. Cytol. **4**, 39 (1958).

[12] H. Fernández-Morán and J. B. Finean, J. Biophys. Biochem. Cytol. **3**, 725 (1957).

[13] A. J. Hodge, J. Biophys. Biochem. Cytol. Suppl. 2, 221 (1956).

[14] A. J. Hodge, J. D. McLean, and F. V. Mercer, J. Biophys. Biochem. Cytol. **2**, 597 (1956).

# 40
# Membrane Transport

Robert W. Berliner

*National Heart Institute, National Institutes of Health, Public Health Service, U. S. Department of Health, Education, and Welfare, Bethesda 14, Maryland*

IN discussing biological membranes one deals with a considerably higher level of organization than has characterized most of the material covered in the foregoing papers, even in the consideration of cellular biology, since one must deal with everything from the presumed limiting membranes of individual cells to entire layers of cells, each of considerable complexity and, in some cases, even consisting of several different cell types. However, in general, the problem is oversimplified here by disregarding whenever possible all details of structure which are not obviously concerned directly with the processes under immediate examination.

The animal body can be considered to be made up of a number of compartments separated one from another and from the external environment by a series of membranes. Each compartment has a more or less characteristic composition, often differing markedly from that of the adjacent compartment so that across these membranes there may be rather steep gradients of chemical concentration and often of electrical potential as well. The problem is to clarify the means by which various materials cross these membranes and by which the striking gradients are maintained. To some extent the penetration of various materials can be defined in terms of relatively simple forces; in other instances it is apparent that the movements of certain substances require the utilization of energy, a process which has been designated as active transport and which is currently the subject of numerous studies. In no case, however, can it be said that the mechanisms involved have been identified.

Consider, in a general fashion, the possible ways in which materials might cross biological membranes and the forces which might effect these movements. These considerations are based heavily upon the treatment of the subject of membrane transport as developed by Ussing. In general, the definable forces which produce movement across membranes can be classified as: (1) those due to gradients of chemical activity including differences in concentration or activity coefficients; (2) those due to gradients of electrical potential; and (3) those exerted upon solutes by the flow of solvent through solvent-filled channels. In addition, one finds in practically all biological systems movements of solute which are explained by none of these definable forces and which, therefore, are designated as "active transport." The definition of active transport is a rather arbitrary affair. As defined, it has the advantage of singling out those solutes which must be considered directly involved in the process by which metabolic energy is utilized to perform transport work. However, clearly all concentration or electrical gradients and flows in response to pressure or osmotic gradients ultimately depend upon metabolic work even though this may be remote from the particular membrane under consideration. So the requirement of metabolic work is not sufficiently exclusive for an adequate definition of active transport.

However, the definition does not include all processes which have specificities beyond simple diffusion and flow and does not include the process by which certain substances can cross membranes only, say, by combining with a specific carrier even though the movement is entirely downhill with respect to gradients of electrochemical potential. One must recognize, therefore, the arbitrariness of this definition. Presumably, when, and if, a better understanding of the discrete processes involved is attained, the need for the term active transport will disappear.

Of course, the manner in which solutes may cross membranes depends upon the nature of the membrane. If it is a continuous lipid layer, substances can pass this layer only by dissolving in it, although their passage through such a layer might be facilitated by combination with some component of the lipid layer which would enhance the lipid solubility of the solute. For such membranes, the permeability to particular solutes should, in general, be a function of lipid solubility. Indeed, observations going back many years have shown that in several chemical series the permeation of cells can be related to lipid solubility.[1] However, there are many features of cellular membranes which cannot be explained if they are assumed to consist of a solid lipid layer and it is generally believed that the lipid layer contains pores, through which water forms a continuous phase and through which the water soluble solutes pass. Indeed, it is only through such channels that forces produced by the flow of solvent can act. Since, as is pointed out later, there is evidence that solvent flow does produce movement of solute, it is apparent that such solvent-filled pores (and in this case, of course, the solvent referred to is water) are present in some membranes. For present purposes, it is convenient to accept the view that cell membranes generally consist of lipid layers penetrated by aqueous pores, the walls of which may bear an excess of fixed charged groups which will impede the passage of ionic species of the same sign. In some cases where membranes consist of more complicated structures than the plasma membrane of single cells and may consist of an entire layer of cells attached to a common basement membrane, there may be reason to believe that the passage of water soluble substances occurs between the cells rather than through them. This appears to be the case in blood capillaries,[2] but probably not in others such as the mucosal lining of the alimentary tract and the skin of frogs and toads.

Perhaps the best-studied biological membrane system and the one in which theoretical considerations have been applied most explicitly is the frog skin which, in particular, Ussing has used as a model system for extensive studies.[3–5] Unfortunately, from the structural point of view, the frog skin is rather complicated, but perhaps it may be simplified for present purposes by making the probably valid assumption that only the basal layer of cells is involved in the transport processes of concern here. The skin can be stripped rather easily from the abdomen of the frog and stretched *in vitro* as a diaphragm between two chambers. It will then survive in this isolated state, perform its metabolic operations, and retain a well-maintained capacity to perform active transport for some hours. The sheet of tissue thus obtained consists of a loose layer of connective tissue, a continuous basement membrane, and several layers of epithelial cells. As one proceeds from the basement membrane toward the outer skin surface, the cells progressively become more flattened and keratinized and, presumably, lose their metabolic activities. In any case, it may be assumed for present purposes that all of the observed phenomena are attributable to the basal layer of cells—an assumption to some extent supported by the fact that toad bladder has been shown to exhibit many similar properties but to have only the single layer of cells corresponding to the basal layer of cells in the skin.[6]

Now, it has been known for many years that the frog skin generates an electrical potential, the solution in contact with the inner surface being positive with respect to the outside. It has also been known that the frog in the intact state is able to extract salt from its surroundings even though the outside concentration of salt is far lower than in the tissue fluids of the animal. For instance, Krogh[7] found that the frog which has a concentration of sodium in its extracellular fluid of something over 100 mEq/L (milliequivalents per liter) was able to take up salt from its surroundings when the outside concentration was as low as one hundredth of a milliequivalent per liter—a concentration ratio of 10000:1. Furthermore, the uptake was completely specific for sodium—no potassium or calcium was taken up even when the concentration was many times higher. One other unusual observation was made by Hevesy *et al.*[8] when the technique of using isotopic tracers was introduced. They found that the frog sitting in water had a net uptake of water several times greater than predicted from the measured influx of tagged water and the assumption that the flux in each direction across the skin should be proportional to the water activity in the solution of origin—in this case, the lymph or extracellular fluid of the frog and the water in which the frog was immersed. Observations such as this one led to the hypothesis that water was actively transported across such membranes (similar observations were made by Visscher and his associates using the intestine of the dog[9])—however, a simpler physical explanation is shown to make the assumption of active water transport unnecessary. One further observation on the intact frog might be mentioned before going on to the more detailed observations—this is the fact that the rate of uptake of water from the surroundings can be enhanced greatly by injection of the frog with posterior pituitary extracts.[10] Hence, the term amphibian water-balance principle was introduced to describe this activity, which, however, has been since shown to be at least qualitatively reproduced by the use of the purified polypeptide hormones of the posterior pituitary.

These, than, are the gross observations which Ussing and his collaborators set about to dissect by the application of isotopic tracers. Disregarding for the time being the net movement of water which in any case is negligible when the skin is bathed on both sides with solutions of equal osmotic pressure, Ussing showed that the forces acting on an ion to cause movement across the intervening membrane would produce a flux from one surface to the other described by the equation

$$M = -\frac{d\bar{a}}{dx}\frac{ART}{GNof}\exp - \frac{zF}{RT}\psi,$$

where $M$ is the unidirectional flux across the membrane, $\bar{a}$ the chemical activity of the species under consideration, $A$ the area of the membrane available for penetration by the species under consideration, $G$ is a frictional coefficient describing the interaction between the membrane and a diffusing ion, $No$ is Avogadro's number, $z$ the valence of the ion, $\psi$ the electrical potential, and $f$, $F$, $R$, and $T$ have their usual meaning.[3] The equation contains several unknowns which are not determinable, and, therefore, it is not too useful in this form. However, if attention is directed to the ratio of fluxes from one side of the membrane to the other, the indeterminate quantities cancel out leaving upon integration the relatively simple relationship,

$$\frac{M_{1.2}}{M_{2.1}} = \frac{f_1C_1}{f_2C_2}\exp\left[\frac{zF}{RT}(\psi_1 - \psi_2)\right],$$

which contains only experimentally determinable quantities, except perhaps the activity coefficients, which, however, also cancel out when the solutions on the two sides of the membranes are essentially the same.

The skin was set up between two chambers, each containing solutions vigorously stirred and oxygenated by a stream of air or oxygen. When the inside solution

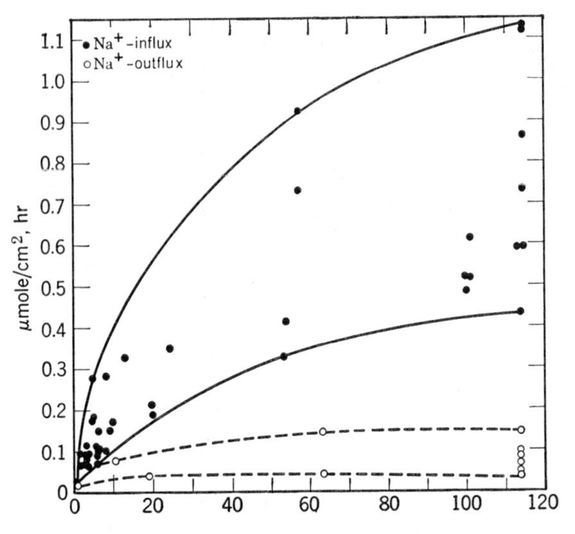


FIG. 1. Sodium fluxes across isolated frog skin as functions of outside sodium concentration [from H. H. Ussing, Cold Spring Harbor Symposia Quant. Biol. **13**, 193 (1948)].

was Ringer's solution and the outside various dilutions of Ringer's, it was found[11] that the influx of sodium always was greater than the outflux so long as the outside concentration was not less than about 1 mM/l. The influx of sodium increased in nonlinear fashion as outside sodium was increased; there also was some small increase in outflux but this always remained small (Fig. 1.).

These, of course, represent merely some quantification of what was already well known—that sodium could be transported inward against combined chemical and electrical gradients and was, therefore, an active process. However, it was found that the chloride fluxes corresponded well to the predictions of the flux-ratio equation and that chloride movements could be considered entirely passive.[12] The sodium influx was generally greater than that of chloride so that there could be no question of bulk movements of elements of outside solution to the inside.[13] In general, the potential was higher when the sodium influx was greater and higher when chloride influx was decreased as could be achieved by treating the skin with $10^{-5}$ $MCu^{++}$ solutions. It was proposed that the potential was produced by the transport of sodium and that the passive movements of other ions tended to reduce the potential.[14]

TABLE I. Unidirectional fluxes, net flux, and electric current in short-circuited frog skin [from H. H. Ussing, *The Relation between Active Ion Transport and Bioelectric Phenomena* (Instituto de Biofisica, Universidade do Brasil, Rio de Janeiro, 1955)].

| | μa/cm² | | | |
|---|---|---|---|---|
| | $Na_{in}$ | $Na_{out}$ | ΔNa | Current |
| I | 20.1 | 2.4 | 17.7 | 17.8 |
| II | 11.1 | 1.5 | 9.6 | 9.9 |
| III | 40.1 | 0.89 | 39.2 | 38.6 |
| IV | 62.5 | 2.2 | 60.3 | 56.8 |
| V | 47.9 | 2.5 | 45.4 | 44.3 |

Studies of the short-circuited frog skin served to show that only sodium was actively transported and that, when the potential across the skin was kept at zero, there was a remarkable correspondence between the current flowing through the skin and the net movement of sodium ions[14] (Fig. 2 and Table I).

These observations, then, were compatible with the hypothesis that the only active process involved was the transport of sodium ions from outside to inside. No other ion, except—to a small extent—lithium could substitute for sodium,[4] and, in particular, potassium could not, although the complete absence of potassium from the bathing solutions very markedly reduced the transport of sodium—a circumstance to be mentioned again later.

These studies, then, have defined the behavior of the frog skin, but the question remains as to what kind of process might be involved in the transport of sodium

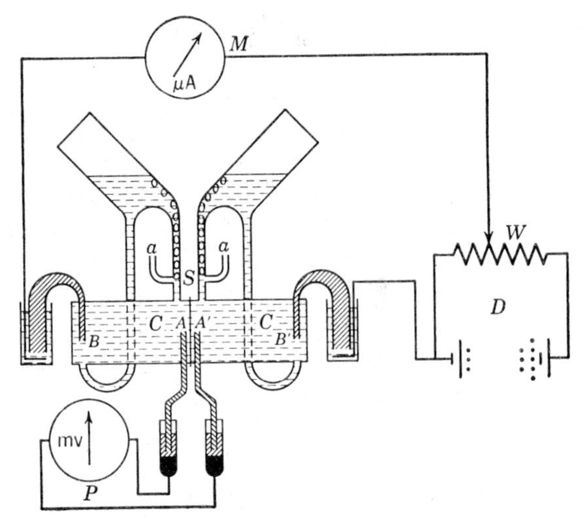


FIG. 2. Apparatus used for determining sodium fluxes and current in short-circuited frog skin [from H. H. Ussing and K. Zerahn, Acta Physiol. Scand. **23**, 110 (1951)].

which, of course, is the major question in the field of active transport at present. Two additional types of study involving the amphibian skin have yielded information pertinent to this question.

One of the favorite theories of ion transport has involved the so-called redox or electron-transport pump, variations of which have been suggested by a number of investigators. These hypotheses are based on the assumption that oxidation and reduction of cytochrome iron are spatially separated and that hydrogen ions formed in the reductive step are available for exchange for some cation from outside the cell. The electrons then are passed along the chain, the cations taken up possibly moving with them, and the electrons finally are donated to oxygen to yield hydroxyl ions. Such a scheme has the stoichiometric limitation that only four electrons and four univalent ions can be transported per oxygen molecule consumed. By careful measurement of oxygen consumption and current flow (net sodium transport) in the frog-skin system, both Zerahn[15] and Leaf and Renshaw[16] have been able to show that more than four ions are transported per oxygen consumed. The average figure obtained by Leaf and Renshaw (Fig. 3) was close to seven and some of the ratios were in excess of ten. Thus, even if all of the oxygen consumed were utilized in ion transport—a doubtful assumption—the ratio of transport to oxygen consumption is too great to be attributed to a redox pump mechanism. (Even higher ratios, going up to 20, can be obtained, if one divides the *change* in transport by the *increase* in oxygen consumption when sodium transport is increased in one of several ways.)

An alternative type of mechanism is suggested by other recent observations of Koefoed-Johnsen and Ussing.[5] The experiments were done with membranes in which maximum potentials were induced by reducing to a minimum the effective anion permeability. This

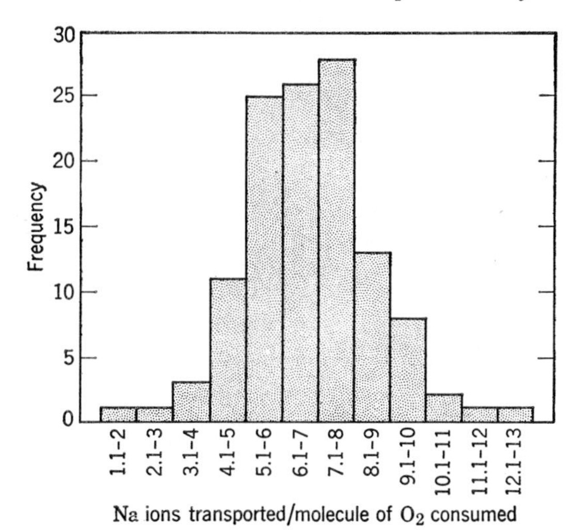


FIG. 3. Relationship of sodium transport to total oxygen consumption of isolated frog skin. Frequency distribution of 120 periods of simultaneous measurement [from A. Leaf and A. Renshaw, Biochem. J. **65**, 82 (1957)].

was done by treating the skin with $Cu^{++}$, or by replacing the chloride in the outside solution with sulfate to which the skin has a very low permeability. The concentrations of sodium and potassium on each side of the membrane were then varied systematically and the effect on the potential observed. Under these circumstances, the inside of the skin behaved very closely as if it were a potassium electrode, the potential across the skin decreasing approximately 60 mv with each tenfold increase in potassium concentration. Changing the potassium concentration on the outer surface of the skin had no effect, but changing the sodium concentration did, the potential increasing as sodium concentration was increased. This strongly suggested that the two surfaces of the cells making up the skin had quite different and specific properties—a fact that is apparent also from other properties. The two potentials thus behaved as diffusion or junction potentials at boundaries specifically permeable to single cations. The explanation offered is as follows: A sodium ion is extruded from the inner cell surface by forced exchange for potassium; that is, a charged carrier drops off a sodium ion and takes up a potassium ion. At the inner surface, by some exergonic reaction, the specificity of the carrier is changed and it gives up the potassium in exchange for another sodium ion. Thus, the sodium concentration inside the cell is reduced and the potassium concentration raised in a process which, in itself, produces no electrical potential. However, since the inner surface membrane is highly permeable to potassium, potassium diffuses out leading to a diffusion potential with the cell contents negative. The negativity of the cell contents tends to produce the uptake of a cation through the outer surface, but this surface being permeable only to sodium, it is a sodium ion which enters. The net effect is a cycling of potassium at the inner surface and a net movement of sodium ion from outer bathing solution to the inside of the skin. In the presence of chloride and a normal permeability to it, the potential is maintained at a lower level by the diffusion of chloride from outside to inside, and in the normal state the net movement of sodium and chloride, of course, must be very nearly equal.

This model is supported further by potential measurements made by Hoshiko and Engbaek[17] in Ussing's laboratory. By thrusting microelectrodes progressively through the skin, Hoshiko found the change to occur in two jumps which can be interpreted as representing changes occurring at the two surfaces of the active layer of cells.

This, then, provides a well-studied example of electrolyte transport by a membrane and it is generally hoped that, in principle, sodium transport is similar in other systems which are less accessible to isolation and study—such as the electrolyte transport in the renal tubules. Of course, here as elsewhere, very big questions remain—namely, what is the nature of the ion carrier? How does it derive its unusual specificity, and how is this specificity modified to provide the specific orientation of the transport process? Incidentally, mention should be made of the fact that the specificity for sodium on the outward trip (across the inner surface) in frog skin is not matched by a similar specificity for potassium on the inward trip, since varying the *p*H of the inside solution has an effect on the frog skin potential similar to that of changing potassium concentration,[11] suggesting the possibility that hydrogen and potassium may be interchangeable in this process. A similar situation appears to hold in the renal tubule.[18]

So much for the potential-generating and ion-transport properties of frog skin. There remains one aspect not mentioned yet—namely, the roles of diffusion and flow in the movement of water across the skin and the effect of such movements on the movement of solute. These phenomena again can be used to illustrate more-general phenomena characteristic of biological membranes. Actually, even in the conditions already discussed, there is some net movement of water inward in an amount such as to make the net transport of salt result in no net change in solute concentration. However, this is so small as to be justifiably neglected. Under other conditions, however, there may be appreciable net movements of water. How may such water movements be brought about and what might be their effects on the fluxes of water and various solutes? Aside from the possibility of active water transport, a phenomenon existence of which is doubtful and which certainly is not required to explain anything in connection with frog skin, movements of water might be the result of hydrostatic or activity gradients. There are no appreciable hydrostatic gradients involved in the skin system so one is left with gradients of water activity or osmotic pressure. If one has only osmotic gradients, does this exclude the possibility of hydrodynamic flow through pores? It has been claimed that this is the case.[19] However, Ussing has illustrated with a simple model a mechanism by which hydrodynamic flow can result from a gradient of solute concentration only.[13] There are undoubtedly more-rigorous approaches,[20] but this is an easily visualized model. Consider a pore permeating a membrane impermeable to some solute contained in a solution on one side and not on the other. Consider the boundary of the pore at the end facing the solution. Just inside the pore, which the solute cannot enter, is a pure water solution of molar fraction one while just across the boundary of the pore is a solution containing a finite concentration of solute and having a molar fraction of water something less than one. Now clearly there will be a net diffusion of water from just inside the pore into the solution. But since the loss of water from the end of the pore cannot change its molar fraction, there is no reason why water should diffuse from deeper in the pore. Nevertheless, there will be a net movement of water through the pore. This then must be the result of a decrease in pressure at the end of the pore from which water has been lost and, hence, there actually is a hydrostatic gradient causing the water movement and, of course, hydrostatic gradients can cause flow through pores. It has been reported by Mauro that gradients of osmotic pressure across artificial membranes produce flow equal to that produced by the equivalent hydrostatic pressure.[21]

The effect of flow through a pore is to accelerate the diffusion of molecules in the direction of flow and to retard it in the opposite direction. Since the solvent drag force is inversely proportional to the diffusion coefficient, the effect in inducing asymmetry should be greatest for molecules as large as possible which can still pass readily through the pores. Andersen and Ussing have used thiourea and acetamide to examine the phenomenon.[22] Thiourea or acetamide were added to the solution on both sides of the skin (in this case, toad skin because it showed more marked responses to the addition of posterior pituitary extracts) in equal concentration, but the molecules in the outside solution were labelled differently from those in the inside solution. It thus was possible to measure the flux in each direction. Experiments were also done with labeled water. The rate of net movement of water was varied by diluting the outside solution. When the solution on both sides was Ringer's, there was very little net movement of water and the fluxes of the test substance were very nearly symmetrical (Table II). When the outside solution was diluted, there was a net movement of water inward and a marked asymmetry of the fluxes of thiourea and acetamide. When the log of the ratio of inward permeability constant to the outward permeability constant was plotted against the net water transfer, a linear relationship was obtained as predicted theoretically and the ratio of the slopes of 2.3 fits closely with predicted ratios of 2.4 and 2.7 for acetamide and thiourea, respectively (Fig. 4).

Of particular interest was the effect of adding neurohypophyseal hormone (Table II) which greatly increased the net water movements and the asymmetry of the fluxes, but had only a very small effect on the total water flux. Thus, it seemed to increase the area available for flow with very little change in the area available for diffusion. Such an effect might be the result of changing the shape of pores without changing their total area. In any case, these experiments seem to provide a clear indication of the existence of pores and of the production of bulk flow by osmotic gradients.

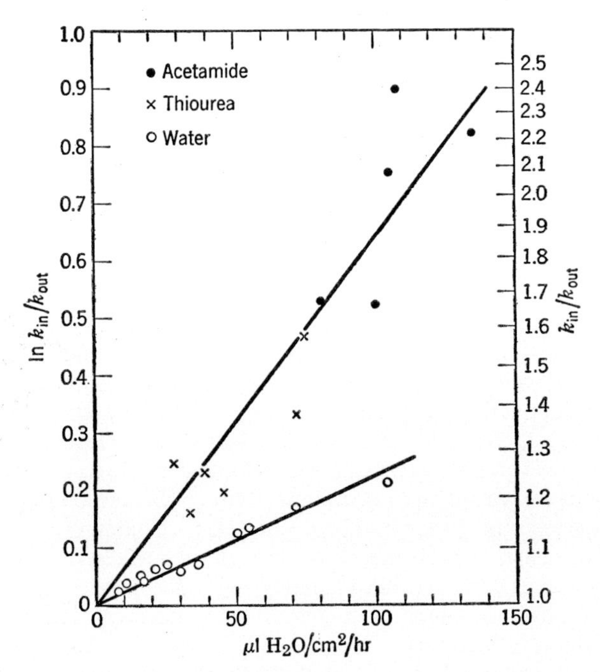


FIG. 4. Relationship between the rate of osmotically induced net water transfer (abscissa) and the logarithm of the ratio of inward to outward rate constants (ordinate) in isolated toad bladder [from B. Andersen and H. H. Ussing, Acta Physiol. Scand. **39**, 228 (1957)].

TABLE II. Effect of neurohypophyseal hormone on permeability of isolated toad skin. Inside bathing solution Ringer's solution, outside solution as specified. $K_{in}$=inward permeability coefficient (cm/hr), $K_{out}$=outward permeability coefficient (cm/hr), and $\Delta W$=net water transfer ($\mu$l/cm²/hr), [from B. Andersen and H. H. Ussing, Acta Physiol. Scand. **39**, 228 (1957)].

| | | | Before hormone | | | | After hormone | | | |
|---|---|---|---|---|---|---|---|---|---|---|
| Date | Outside medium | Test substance | $K_{in}$ ×10⁴ | $K_{out}$ ×10⁴ | $K_{in}/K_{out}$ | $\Delta W$ | $K_{in}$ ×10⁴ | $K_{out}$ ×10⁴ | $K_{in}/K_{out}$ | $\Delta W$ |
| March 26 | Ringer's | Thiourea | 6.63 | 4.72 | 1.40 | 0 | | | | |
| June 11 | | | 10.5 | 8.08 | 1.30 | 0 | 278 | 278 | 1.00 | 0 |
| June 14 | | | 11.5 | 9.75 | 1.18 | 0 | 106 | 109 | 0.97 | 2 |
| June 22 | | | 11.5 | 13.7 | 0.84 | 3 | 247 | 223 | 1.11 | 12 |
| June 23 | | | 12.4 | 14.1 | 0.88 | 2 | 64.7 | 62.3 | 1.04 | 1 |
| June 25 | | | 3.17 | 4.47 | 0.71 | 1 | 7.71 | 8.45 | 0.91 | 1 |
| | | | | | 1.05 | * | | | 0.98 | * |
| June 15 | | Acetamide | | | | | 537 | 516 | 1.04 | 14 |
| June 22 | | | | | | | 363 | 321 | 1.13 | 10 |
| March 25 | 1/10 Ringer's | Thiourea | 2.90 | 3.03 | 0.96 | 8 | 20.6 | 16.2 | 1.27 | 27 |
| April 6 | | | 6.48 | 6.51 | 1.00 | 14 | 17.4 | 14.4 | 1.21 | 45 |
| April 8 | | | 11.4 | 8.73 | 1.33 | 14 | 49.3 | 31.2 | 1.58 | 74 |
| April 26 | | | 4.00 | 4.02 | 1.00 | 8 | 7.52 | 6.42 | 1.17 | 33 |
| April 29 | | | 4.25 | 4.07 | 1.04 | 17 | 15.4 | 12.3 | 1.25 | 38 |
| May 3 | | | 9.86 | 6.52 | 1.52 | 14 | 156 | 113 | 1.38 | 71 |
| May 25 | | Acetamide | | | | | 251 | 121 | 2.08 | 105 |
| June 1 | | | | | | | 175 | 105 | 1.67 | 80 |
| June 6 | | | | | | | 299 | 180 | 1.66 | 100 |
| June 10 | | | | | | | 530 | 239 | 2.22 | 135 |
| June 29 | | | 129 | 100 | 1.29 | 24 | | | | |
| April 6 | | | 81.5 | 76.1 | 1.16 | 16 | 196 | 81.5 | 2.40 | 108 |
| April 26 | | Heavy water | 3730 | 3510 | 1.06 | 22 | 5550 | 4510 | 1.23 | 104 |
| May 3 | | | 4090 | 3890 | 1.05 | 20 | 4250 | 3750 | 1.13 | 50 |
| May 9 | | | 3630 | 3420 | 1.06 | 21 | 4270 | 3730 | 1.14 | 54 |
| May 16 | | | 4040 | 3760 | 1.07 | 28 | 4560 | 3850 | 1.18 | 71 |

They also make unnecessary any assumed active water transport by the amphibian skin. Similar phenomena also have been demonstrated to occur in some marine eggs,[23] and it is a reasonable conclusion that the earlier observation of unexplained high net water fluxes in intestinal loops has a similar basis.

The phenomenon of exchange diffusion is of interest with respect to the mechanism by which ions cross cell membranes. The concept was invented by Ussing[11] to rationalize the very high rate of isotopic exchange between the sodium in cells with that in their surroundings. Thus, with the potential inside negative relative to the outside and with the sodium concentration much higher outside, work is required to extrude sodium. If it is assumed that the inward movements are entirely by diffusion and that the outward movement proceeds at the expense of metabolic energy, then the energy requirements would exceed half of the total metabolic energy production of muscle in the resting state. Ussing proposed that sodium ions attached to a charged carrier might exchange for other sodium ions in the external solution or in the cell interior, the carrier making the round trip with sodium attached so that no net change in sodium concentration on either side of the membrane would occur. Yet, if the sodium on one side of the membrane were isotopically labeled, the process would result in movement of label.

The concept of exchange diffusion has been useful in the interpretation of a number of other phenomena some of which more directly are evidence for the idea of reversible combination of the same ion with a carrier so that exchange is greater than actual transport or movement of the free ion. For instance, Hogben[24] has studied the secretory membrane of the frog stomach using the Ussing technique. The stomach secretes hydrochloric acid, chloride against both electrical and concentration gradient, hydrogen ion with the electrical gradient which, however, is smaller than the very steep opposing concentration gradient. Sodium behaves passively. However, the striking phenomenon in connection with the phenomenon of exchange diffusion is the very large flux in the direction opposite to active transport (Table III). The flux from the secretory surface corresponding to the inside of the stomach, to the nutrient surface corresponding to the blood, is more than twice the membrane conductance as determined by the relationship between imposed current and developed potential. A considerable part of the chloride

TABLE III. Chloride fluxes in stomach mucosa of the bull frog [from A. Hogben, *Electrolytes in Biological Systems*, A. Shanes, editor (American Physiological Society, Washington, 1955), p. 176].

| Net chloride transfer $\mu$ Eq cm⁻² hr⁻¹ | | Net charge transfer $\mu$ Eq cm⁻² hr⁻¹ | |
|---|---|---|---|
| Flux N−S | 10.65 | Current | 3.05 |
| Flux S−N | 6.38 | $H^+$ secretion | 1.20 |
| | 4.27 | | 4.25 |

Mean conductivity: Start 3.21 m mhos; End 2.28 m mhos

TABLE IV. Na active transport and exchange diffusion in HK sheep red cells.

| External cation | Na outflux | Difference |
|---|---|---|
| Na+K | 3.5 | 2.7 E.D.[a] |
| Mg+K | 0.78 | |
| Mg+K+Stroph | 0.16 | 0.62 A.T.[b] |
| Mg | 0.16 | |

[a] E.D. = Exchange diffusion.
[b] A.T. = Active transport.

movement must occur in a form which does not contribute to the transfer of charge—presumably the same carrier complex by which the active transport is effected.

Another similar case involves the movement of chloride across the large intestine of the frog and has been studied by Cooperstein and Hogben.[25] In this case, there is no active transport, chloride fluxes being symmetrical when the electrical potential is zero. However, at various levels of potential, the flux ratio is lower than predicted from the electrochemical potential ratio. This could be explained if some part of the flux were owing to exchange diffusion and thus was not affected by the electrochemical potential gradient.

Finally, an example of quite a different sort involving sodium transport by the red blood cell. The red cell is rather unusual, in a number of respects, in that it is non-nucleated, lacks the machinery for oxidative metabolism, and, in general, transports sodium and potassium at rates several orders of magnitude lower than most body cells. Also, they contain a rather high chloride concentration, 60 to 70% of that in the plasma, whereas most other cells contain very little; since they have an extremely high permeability to chloride, with a half-turnover time, measured in milliseconds,[26] the presumption is that the distribution is passive and, therefore, that the electrical potential can be estimated from the chloride-distribution ratio. If this is true, the accumulation of potassium in this cell cannot be considered as a passive consequence of sodium extrusion, as is frequently considered the case in other cells. In any case, the slow rate of transfer and the ease of direct analysis make the red cell a suitable object for study and the relative simplicity of the red-cell metabolism somewhat facilitates the study of the metabolic correlates of ion transport.

Some data obtained with sheep red cells by Tosteson and Hoffman[27] in our laboratory illustrate the phenomenon of exchange diffusion (Table IV). When cells are incubated in normal plasma, an outflux of sodium of 3.5 mM/l/hr is observed. The removal of sodium and its replacement with magnesium very markedly depresses the outflux of sodium. Since the only type of process by which sodium leaves cells which we can visualize as requiring sodium on the *outside* is exchange diffusion, it appears that close to 80% of the sodium output from the sheep red cell is attributable to this process. When the potassium also is removed from the medium, sodium output is reduced even further to a level compatible with complete absence of active transport and which is the same low level reached when transport is inhibited with strophanthanthidin, a glycoside related to digitalis and a member of a group of powerful and apparently specific inhibitors of sodium transport. These results again suggest that the sodium extrusion is effected by a carrier-linked exchange for potassium. From the amount of glycoside required to inhibit potassium uptake by human red blood cells, Glynn[28] has estimated that there are less than 1000 transport sites on the surface of the human red blood cell, a disk 7 μ in diameter and perhaps 1 μ thick.

These similarities between membrane-transport phenomena in diverse systems and even in orders of vertebrates encourage one to believe that the processes involved fundamentally are limited in number and are not different in each cell type and species, and that the study of simpler systems may yield information applicable to more-complex and less-accessible ones.

## BIBLIOGRAPHY


[1] E. Overton, Vierteljahresschr. Naturforsch. Ges. Zürich **40**, 159 (1895).

[2] J. R. Pappenheimer, E. M. Renkin, and L. M. Borrero, Am. J. Physiol. **167**, 13 (1951).

[3] H. H. Ussing, Acta Physiol. Scand. **19**, 43 (1949).

[4] H. H. Ussing, Symp. Soc. Exptl. Biol. **8**, 407 (1954).

[5] V. Koefoed-Johnsen and H. H. Ussing, Acta Physiol. Scand. **42**, 298 (1958).

[6] A. Leaf, J. Anderson, and L. B. Page, J. Gen. Physiol. **41**, 657 (1958).

[7] A. Krogh, Skand. Arch. Physiol. **76**, 60 (1937).

[8] G. von Hevesy, E. Hofer, and A. Krogh, Skand. Arch. Physiol. **72**, 199 (1935).

[9] M. B. Visscher, E. S. Fetcher, Jr., C. W. Carr, H. P. Gregor, M. S. Bushey, and D. E. Barker, Am. J. Physiol. **142**, 550 (1944).

[10] F. Brunn, Z. ges. exptl. Med. **25**, 170 (1921).

[11] H. H. Ussing, Cold Spring Harbor Symposia Quant. Biol. **13**, 193 (1948).

[12] V. Koefoed-Johnsen, H. Levi, and H. H. Ussing, Acta Physiol. Scand. **25**, 150 (1952).

[13] H. H. Ussing, *The Relation between Active Ion Transport and Bioelectric Phenomena* (Instituto de Biofisica, Universidade do Brasil, Rio de Janeiro, 1955).

[14] H. H. Ussing and K. Zerahn, Acta Physiol. Scand. **23**, 110 (1951).

[15] K. Zerahn, Acta Physiol. Scand. **36**, 300 (1956).

[16] A. Leaf and A. Renshaw, Biochem. J. **65**, 82 (1957).

[17] T. Hoshiko and L. Engbaek, Abstr. Communications of the 20th International Physiological Congress, Brussels (1956), p. 443.

[18] R. W. Berliner, T. J. Kennedy, Jr., and J. Orloff, Arch. Int. Pharmacodynamie **97**, 299 (1954).

[19] F. P. Chinard, Am. J. Physiol. **171**, 578 (1952).

[20] O. Kedem and A. Katchalsky, Biochim. et Biophys. Acta **27**, 229 (1958).

[21] A. Mauro, Science **126**, 252 (1957).

[22] B. Andersen and H. H. Ussing, Acta Physiol. Scand. **39**, 228 (1957).

[23] D. M. Prescott and E. Zeuthen, Acta Physiol. Scand. **28**, 77 (1953).

[24] A. M. Hogben in *Electrolytes in Biological Systems*, A. Shanes, editor (American Physiological Society, Washington, D. C., 1955), p. 176.

[25] I. L. Cooperstein and C. A. M. Hogben, Federation Proc. **17**, 29 (1958).

[26] D. C. Tosteson, Abstr. Communications of the 20th International Physiological Congress, Brussels (1956), p. 892.

[27] D. C. Tosteson and J. Hoffman (personal communication).

[28] I. M. Glynn, J. Physiol. **136**, 148 (1957).

# 41

# Interaction Properties of Elongate Protein Macromolecules with Particular Reference to Collagen (Tropocollagen)*

FRANCIS O. SCHMITT

*Department of Biology, Massachusetts Institute of Technology, Cambridge 39, Massachusetts*

ORGANIZATION and the interaction of organized structures at all levels of complexity are of cardinal significance in life processes. This organization may be manifested within the protein or other biomolecule or may depend upon the interaction of many individual entities.

For purposes of biophysical and biochemical investigation of such interaction of organized systems, it is profitable to deal with relatively simple systems, preferably with pure substances. Such systems are available in materials which are primarily one-dimensional (fibrous), two-dimensional (membranous), or three-dimensional (crystalline). The present paper is concerned exclusively with the properties of highly elongate, fibrous macromolecular systems. In a study of the forces between such highly organized fibrous systems lies rich opportunity for fundamental biophysical research.

Many substances of crucial importance in cells and tissues occur as very thin (10 to 30 A), highly elongate (1000 to 5000 A) particles. Proteins (such as myosin, paramyosin, actin, fibrinogen-fibrin, collagen, and the nerve-axon protein), nucleic acids (DNA and RNA), and polysaccharides (such as cellulose, hyaluronic acid, chondroitin sulfate) are examples. Many of these substances are themselves polymers (as the protein macromolecules are polymers of many amino-acid residues) but, as monomers, these elongate macromolecules polymerize end-to-end and laterally to form fibrous structures.

These macromolecular systems lend themselves to detailed physicochemical, crystallographic, and electron-optical study. Polyelectrolyte theory may be applied fruitfully to them. In many cases, energy is coupled with the macromolecular system by interaction with "energy-rich" substances such as adenosine triphosphate (ATP), but the mechanism by which this available energy is caused to do work (mechanical, chemical, electrical, or osmotic) is still poorly understood.

## SOME TYPICAL BIOLOGICAL SYSTEMS IN WHICH FUNCTION DEPENDS UPON SPECIFIC INTERACTION OF ELONGATE MACROMOLECULES

It will help to identify the types of biophysical problems involved by mentioning a few typical examples.

### 1. Functions Involving Mechanical Properties

In this category, mention is made of but three types that may serve to illustrate fundamental problems: fibrous systems designed to afford high tensile strength; those which produce tension or undergo contraction; and those which, by forming gelled clots, occlude regions such as blood vessels and prevent bleeding.

An excellent example of a fibrous system that achieves high tensile strength (*ca* 100 000 lb/in.$^2$) by a lateral bonding of macromolecular polymers are the structures such as skin, tendon, and other forms of connective tissue involving the protein collagen. The macromolecules are synthesized within fibroblast cells, find their way into the intercellular tissue spaces, and eventually aggregate at the appropriate place and time to form fibers. The mechanism of fibrogenesis may be very complex, involving processes of activation and homeostatic control of such processes so as to facilitate fibrogenesis when needed (as in wound repair) and to prevent excessive fibrinogenesis (such as occurs in aging and in certain pathological processes as in atherosclerosis, and in certain rheumatoid and so-called "collagen diseases").

The second mechanical function, that of contraction or tension production, poses a problem as to whether the shortening occurs essentially as an intramolecular process of superfolding of polypeptide chains, or is rather an intermolecular process involving a rapid and reversible change in the affinity or interaction between two or more species of fibrous proteins leading to a shortening of the fibrous system without substantial change in the helical configuration of the intramolecular chains characteristic of the native macromolecules.

Another rather striking problem is well illustrated in the embryogenesis of fibrous structures such as striated muscle, in which the axial repeat (sarcomere length) is so precise as to give several orders of diffraction with visible light. This production of supermacromolecular patterns well may involve the specific aggregation of several species of macromolecules, each having lengths of the order of several thousands of Ångström units. Perhaps some day it will be possible to produce such super-repeating patterns by interacting several kinds of fibrous macromolecules (protein, nucleic acid, or polysaccharide) under appropriate conditions *in vitro*.

The third type of mechanical function mentioned in the foregoing is that of the clotting of fibrous protein,

* These studies were aided by a research grant from the National Institute of Allergy and Infectious Diseases, National Institutes of Health, Public Health Service, U. S. Department of Health, Education, and Welfare.

as in the transformation of the soluble fibrinogen in the blood into the insoluble fibrin of blood clot. This involves an enzymatic activation of the soluble protein monomers by means of an enzyme (itself activated by a series of interdependent processes) to produce "intermediate polymers" several thousand Ångström units long which then polymerize spontaneously to form the clot. The enzymatic activation itself is controlled by a highly complex system of kinases and antikinases by means of which a high degree of homeostatic control of this vital process is achieved.

### 2. Functions Involving Enzyme Action

As was emphasized by Engelhardt and Ljubimova,[1] large macromolecules may function enzymatically. This may occur either because the macromolecule as a whole acts like an enzyme or because a portion of the molecule is enzymatically active. Myosin, with which Engelhardt and Ljubimova were concerned,[1] was found to exert enzymatic action in splitting ATP. It is known now that only one component ("heavy meromyosin") is enzymatically active as ATPase. It remains to be seen how many other kinds of elongate macromolecules will be found to have enzymatic properties. It may be mentioned that, when an enzymatic group forms part of a large macromolecular complex, the configuration of the macromolecules or smaller molecules in the environment, for steric reasons, may strongly influence the availability of the enzymatic site. Such regulatory action well may be involved in muscle contraction.

### 3. Functions Involving the Maintenance of a Specific Linear Sequence of Chemical Groups as in Genetic Determiners

It is believed that in the linear sequence of nucleotide residues in the double helix of DNA is to be found the coding responsible for transmitting genetic information. The DNA occurs as highly elongate macromolecules extractable from the chromosomes by mild methods (such as by treatment with hydrogen bond breakers). It seems probable that the ability of such macromolecules to exert their specific controlling influence during development and differentiation must depend importantly upon their interaction with other constituents of the chromosomes such as histones, protamines, and protein macromolecules, and perhaps also with other nucleic-acid macromolecules.

### 4. Other Functions

Engelhardt[2] suggested that macromolecules may perform other types of work such as osmotic and electrical. These possibilities have not yet been explored thoroughly. The function of certain fibrous proteins, such as the axoplasmic protein of nerve, remains completely unknown.

Perhaps one may mention in this connection also the action postulated by Weiss[3] as important in determining the ordering of cells into typical tissues by means of the interaction of specific types of macromolecules at the surfaces of the cells that form the tissues. Presumably because of a complementarity or ordered type of interaction of the surface molecules, the cells are caused to aggregate in patterns characteristic of each tissue.

From this very brief description, it is obvious that crucial biological functions depend importantly upon the ways in which highly specifically structured macromolecules interact with one another and upon the manner in which environmental conditions change or regulate this interaction. It is the purpose of this paper to illustrate this specificity of interaction by a consideration of the properties of one particular type of macromolecule, collagen, chosen because of its specially favorable chemical and structural properties. Like many other biological macromolecules, collagen is very asymmetric (*ca* 14×2800 A). It has, however, the valuable property that, in its aggregation patterns, it forms fibrous structures characterized by highly specific band patterns as seen in the electron microscope. By analysis of these band patterns, it is possible to deduce the type of macromolecular interaction responsible for each characteristic pattern. From such studies, valuable lessons are learned that have direct application to other types of macromolecular systems in which no band patterns exist to serve as guides.

## BIOPHYSICAL AND BIOCHEMICAL PROPERTIES OF COLLAGEN

In contradistinction to most proteins, collagen (or perhaps more correctly the collagen class) possesses characteristic structural and chemical properties which permit its definitive identification. For present purposes, it is necessary only to sketch those properties necessary for an understanding of the internal structure, and the chemical properties of the collagen macromolecule. For recent excellent surveys of these properties, see Gustavson,[4] Highberger,[5] the CIOMS Symposium on Connective Tissue,[6] and that on Gelatin and Glue Research.[7]

Collagen occurs in dense fibrous tissue of high tensile strength, as in tendons, or less tightly woven tissue fabrics, as in skin, or in more sparse distribution as in loose connective tissue. The fibrous protein occurs in various hierarchies of fiber size, including the following: *fibers*, visible macroscopically or microscopically and having diameter of the order of micra; *fibrils* with widths of the order of a few hundred to several thousand Ångström units, observable in the dark-field microscope and resolvable in the electron microscope; the *protofibrils* which originally were defined as constituting "The unit columnar arrays which, when associated laterally, form the collagen fibril";[8] and the *collagen* (or "tropocollagen") *macromolecules* which constitute the monomeric units of the protofibrillar polymer.

For x-ray diffraction and for chemical-analytical

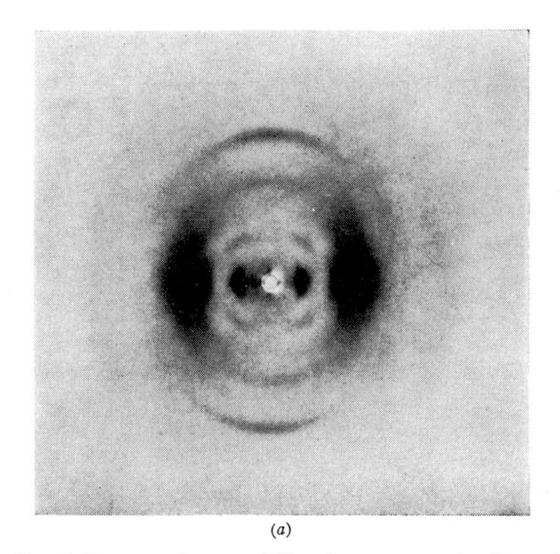

(a)

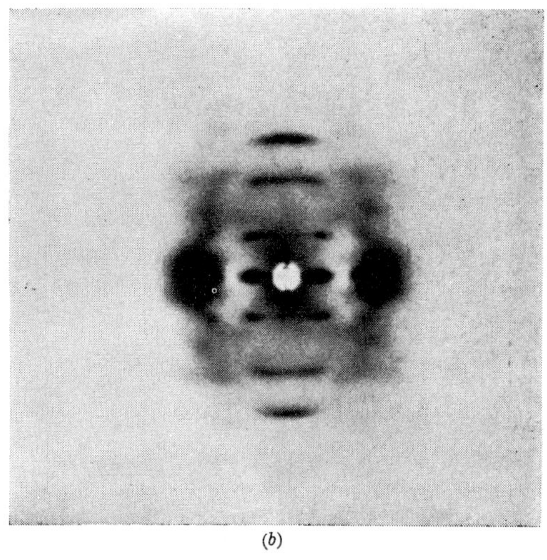

(b)

FIG. 1. Large-angle x-ray diffraction patterns of collagen from rat-tail tendon.[10] (a) Unstretched; (b) stretched 8% [from J. T. Randall, J. Soc. Leather Trades' Chemists 38, 362 (1954)].

studies, gross macroscopic fibers or whole tissues are used. For electron-microscopic investigation, the fibrous category of interest is the fibril which manifests a detailed and characteristic band pattern which, as is brought out below, results from a specific pattern of aggregation of the elongate native collagen macromolecules.

The collagen class of proteins, as Astbury[9] referred to them, is uniquely characterized by its amino-acid composition, its x-ray diffraction pattern, and its banded appearance in the electron microscope. These characteristics may be described briefly as follows.

The collagenous proteins differ from other proteins in that they contain the amino acids hydroxyproline and hydroxylysine. In mammalian collagen, about one-third of the amino-acid residues are glycine. Proline and hydroxyproline together make up almost another third, leaving approximately one-third for other amino-acid types. From a determination of the hydroxyproline and glycine content of a given preparation, one can make a good estimate of the collagen content.

Perhaps the most distinctive characteristic of collagen is its large-angle x-ray diffraction pattern (Fig. 1) which reflects the internal organization of the collagen macromolecule and is, therefore, characteristic of this class of proteins. Astbury[9] early called attention to the 2.86-A meridional reflection, which he considered to represent the length of the amino-acid residue, along the fiber axis, in a coiled polypeptide chain, and to the equatorial reflections at 10 to 15 A (depending upon the degree of hydration) which he attributed to the separation between main chains. With more refined technique, it has been possible to obtain far more reflections in the pattern and to achieve a higher degree of orientation by stretching fresh tendon. From such patterns, it has been possible for several groups of workers to agree that the diffractions are interpreted best in terms of a macromolecule containing three chains coiled in helical fashion about each other to form a coiled coil (see particularly the papers of Crick and Rich).[11–13] The proposed triple-stranded structure is shown schematically in Fig. 2.

It has been proposed also that there are only two types of three-stranded helical models of collagen structure, based on the so-called structure I and structure II, derived from a consideration of polyglycine, and compatible with the x-ray, infrared, analytical, and physicochemical data. In these models, the axial repeat occurs at 28.6 A. In collagen II, to agree best with the diffraction data, the OH groups of the hydroxyproline residues extend radially from the three chains, making it possible to form hydrogen bonds with CO groups of adjacent three-stranded macromolecules. In the collagen I structure, the hydrogen bonds from hydroxyproline are directed internally, bonding the three chains intramolecularly. The hydroxyproline content appears to be determinative of the denaturation temperature, which is a measure of the energy needed to disrupt the internal organization of the macromolecule. This fact tends to support the collagen-II type of structure. Rich (p. 50) has suggested that one type of structure might be convertible into the other and that this may result from application of stress to the fiber.

Although there is fairly general agreement that the collagen macromolecule is a three-stranded helix, it is not certain that the macromolecule is thus constructed over its entire extent. Gallop[14] suggests that as much as 30% may have a different configuration.

Treatment of soluble collagens with hydrogen-bond breakers like urea, or by heating, causes denaturation with the liberation of the constituent chains to form "parent gelatin." From the original macromolecule, having a weight of 360 000, there is formed, according to Doty and Nishihara,[15] one chain with a weight of

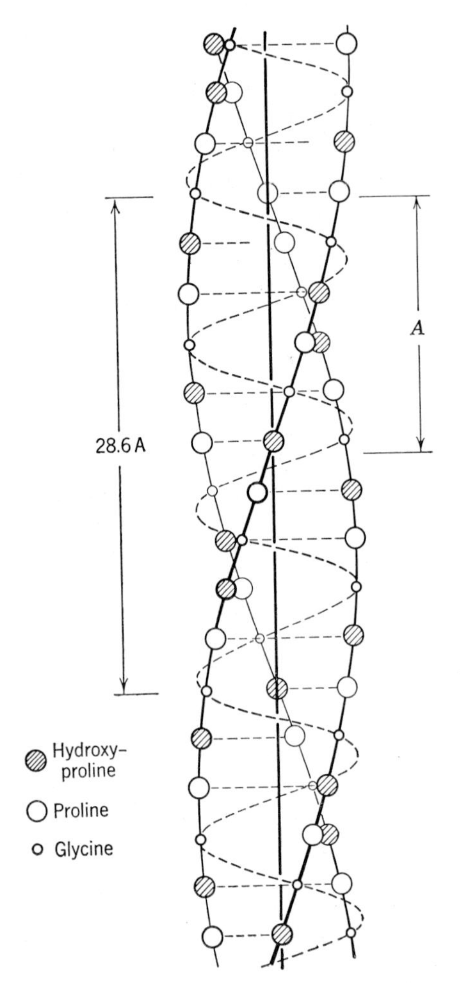


FIG. 2. Three-stranded helical structure of collagen macromolecules (courtesy A. Rich).

120 000 and another with a weight of 240 000. These authors believe that an alkali-labile ester bond links two chains of weight 120 000 to form the heavier chain obtained from denatured collagen. This is to be compared with the corresponding values of Orekhovitch and Shpikiter.[16]

In addition to the large-angle x-ray pattern, arising from the internal, presumably three-stranded, structure of the macromolecule, collagen also manifests a well-developed small-angle x-ray pattern (Fig. 3) consisting of many (*ca* 50) orders of a large axial repeat which Bear[18,19] showed to be 640 A in air-dried fibers and nearer to 700 A in moist fibers. Although all native collagen fibers from a wide variety of sources showed this axial repeat, its significance in terms of molecular structure was not obvious. The simplest early interpretation was that it represents the molecular length of the collagen molecules, an assumption that seemed to gain support from the fact that a similar axial repeat was observed in the band pattern observed in electron micrographs. Bear and Morgan[20] attempted to relate the positions of the intraperiod bonds observed electron optically with the characteristic intensities of the various orders of the small-angle x-ray pattern. As is shown in the following, the collagen macromolecule probably has a length of four times the 700-A period, i.e., about 2800 A.

The band pattern observed in high-resolution electron micrographs of teased collagen fibrils stained with phosphotungstic acid (PTA), or other heteropolyacid, is uniquely characteristic of collagen (see Fig. 4). This axial pattern repeats at about 700 A and contains a number of bands and interbands of characteristic density and position. It was suggested by Bear[21] that the bands represent regions of relative disorder due to the interaction of side chains of relatively large size while the interbands represent regions of relative order due to the interaction of the smaller side chains which are found in considerable abundance in collagen. Another interpretation of band structure depends upon the characteristic interaction of groups such as the guanidino groups of the arginine side chains with PTA, as suggested by Kühn *et al.*[22] It is noted from Figs. 4 and 5 that the band pattern—i.e., the intraperiod positions and the relative densities of the bands—is a polarized, asymmetric pattern. The significance of this pattern was discovered only after it became possible to take the native fibrils apart into their constituent macromolecules and to cause these to re-aggregate in characteristic and new band patterns. These results may now be described briefly.

## FORMATION OF ORDERED AGGREGATION STATES OF COLLAGEN BY PRECIPITATION FROM SOLUTION

A formidable difficulty in the characterization of the collagen molecules lay in the relative insolubility of collagen fibers. However, certain types of collagen, such as in rat-tail tendon and in the fish swim bladder, are

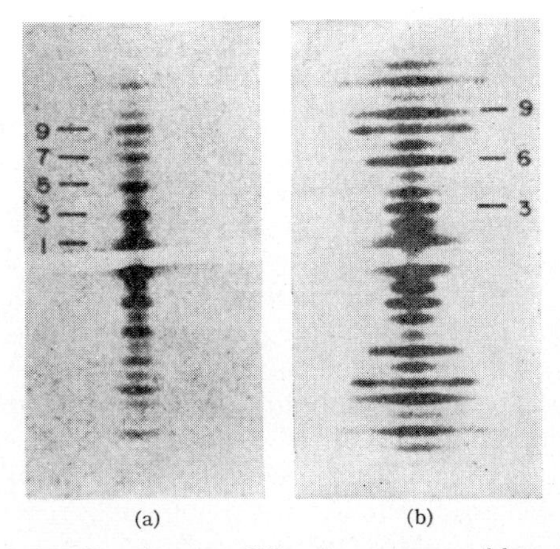


FIG. 3. Small-angle x-ray diffraction patterns of kangaroo-tail tendon collagen; (a) moist preparation; (b) after brief exposure to water and drying under tension.[17] Layer line indices are indicated [from R. S. Bear, O. E. A. Bolduan, and T. P. Salo, J. Am. Leather Chemists' Assoc. **46**, 107 (1951)].

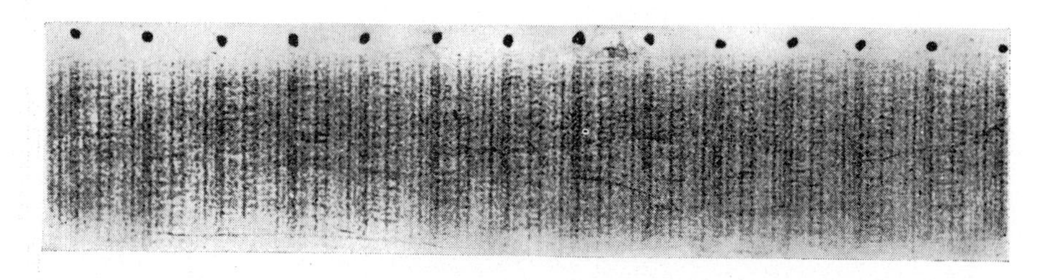


FIG. 4. Electron micrograph of calfskin collagen reconstituted from solution, stained with phosphotungstic acid. Band pattern is of the native type. Axial repeats marked (courtesy A. J. Hodge).

soluble in dilute acid. From the classical early work of Zachariades, Nageotte, Fauré-Fremiet, Wyckoff, and Corey, and others, it is known that, by appropriate adjustment of the *p*H and ionic strength of such acid solutions, the collagen can be precipitated reversibly in fibrous form. Examined in the electron microscope after staining with PTA, the reprecipitated fibrils were found to possess structure, the type of which depends upon the conditions of precipitation. With increasing ionic strength, the band pattern may be that characteristic of native fibrils (period $\cong$700 A), it may be about one-third this value, or the precipitate may have tactoidal appearance, showing no bands at all. These different forms can be produced reversibly from acid solutions of highly purified collagen; presumably the different ordered states depend only upon the collagen and require no additional organic material.

When certain types of extracts are made from connective tissue or when certain organic substances, particularly highly negatively charged substances, are added to the collagen solutions and the conditions are adjusted appropriately, a new modification is found which manifests an axial repeat or identity period about four times that of normal collagen (i.e., about 2600 to 3000 A) and which, therefore, are called "long-spacing" types. Two such forms, called "fibrous long-spacing" (FLS) and "segment long-spacing" (SLS) are shown diagrammatically in Fig. 5. The FLS modification is produced routinely by addition of $\alpha$-1 acid glycoprotein to an acetic-acid solution of collagen, followed by dialysis against water. The SLS modification is produced routinely by addition of ATP to the acid solution of collagen; the precipitate forms directly without further adjustment of conditions.

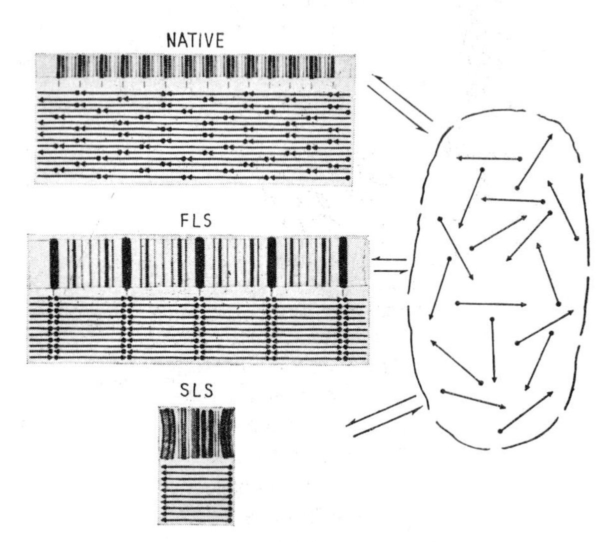


FIG. 5. Diagrammatic illustration of patterns of aggregation of tropocollagen macromolecules in native, FLS, and SLS types. Polarization of macromolecules indicated by arrow.

It is noted that the FLS type has a symmetrical, nonpolarized band structure while the SLS has an asymmetrical, polarized pattern of banding.

Each of the five band patterns described may be produced reversibly from an acid solution of collagen. The particular patterns produced depend for their specificity upon the collagen rather than upon the other substances added or conditions imposed; rather these substances and conditions serve to evoke the structure inherently characteristic of the collagen itself.

## MACROMOLECULAR MONOMER OF COLLAGEN—THE "TROPOCOLLAGEN" HYPOTHESIS

The structures described in the foregoing, discovered in collaboration with Gross and Highberger, were interpreted as follows (see summaries of this work by Schmitt, Gross, and Highberger;[23] Schmitt;[24] Gross;[25] and Highberger.[5] It is assumed that the long-spacing (about 2800 A) represents the length of the native collagen macromolecule which has a three-stranded helical internal structure, as deduced from the large-angle x-ray pattern (see the foregoing). The long, thin macromolecules were given the term "tropocollagen" (TC), because they are capable of "turning into" or forming the native collagen structure, and also to distinguish them from various other collagen fractions (such as procollagen) previously described.

The TC macromolecules are assumed to be essentially identical in structure and composition and to be themselves polarized in the sense of the linear sequence of amino-acid residues in the constituent intramolecular strands.† This is indicated by the arrows on the TC

† The unit of native collagen structure is referred to as a macromolecule rather than as a molecule because it appears to be composed of several covalent polypeptide chains bonded together by hydrogen bonds.

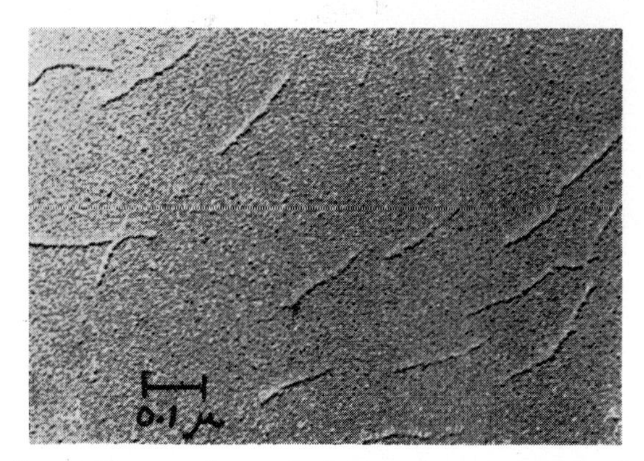


FIG. 6. Electron micrograph of tropocollagen macromolecules prepared by the method of Hall[28,29] (courtesy C. E. Hall).

macromolecules in Fig. 5. The hypothesis assumes that the various types of ordered patterns of TC aggregation occur by virtue of relatively stable bonding between terminal groups on the side chains of laterally adjacent macromolecules. Each type of ordered aggregation type represents a particular pattern of interacting side chains. In the SLS form, it is assumed that the TC macromolecules are essentially in register with respect to their ends and are "pointing" all in the same direction; i.e., they are in parallel array. The SLS pattern, therefore, provides a molecular "fingerprint" of the sequence of amino-acid residue types along the TC macromolecule—information not deducible by examination of the band pattern of native fibrils. The FLS is assumed to be formed by an antiparallel packing of TC in which the macromolecular ends are approximately in register (see Fig. 5).

Since the axial repeating pattern of native fibrils, both as seen in electron micrographs and as measured in the small-angle x-ray pattern, is about a quarter of that of the length of the TC macromolecules, it was assumed that the latter are arranged in parallel array but are displaced in the axial direction by one-quarter of a length in adjacent macromolecules (see Fig. 5). A specific suggestion somewhat along the same line has been proposed by Tomlin and Worthington.[26]

This concept of the structure and properties of the native macromolecule of collagen was deduced from the electron-optical observations of the various ordered aggregation types observed. The hypothesis received confirmation from the physicochemical studies by Boedtker and Doty[27] performed on solutions which were highly monodisperse with respect to the monomer macromolecules (achieved by centrifuging out the larger aggregates). These data indicated that the macromolecules behave like rigid rods with dimensions about 14×2800 A and molecular weight about 360 000. Previous estimates of other workers about particle sizes were considerably greater, probably because their preparations were heterodisperse, containing polymers of collagen as well as the monomers.

Finally, the tropocollagen macromolecules were visualized directly in the electron microscope by a method developed by Hall.[28] This consists in depositing the molecules upon the atomically smooth surface of freshly cleaned mica by spraying a very dilute solution of the protein. After drying, this surface is shadowed by evaporation of platinum at a small angle. The metalized layer then is backed with a thin collodion supporting film, stripped from the mica, and examined in the electron microscope at high resolution. From such electron micrographs (see Fig. 6), Hall[29] found the fibrous particles to be about 15 A in width but their lengths to be somewhat smaller than had been predicted by the physicochemical data of Boedtker and Doty. Subsequently, with improved technique, Hall and Doty[30] obtained a weight-average length of 2820 A, in good agreement with the physicochemical data and with the lengths determined in this laboratory on the same solutions used by Hall and Doty by conversion to the FLS modifications and measurement of the axial period (average value was 2700 A).

The problem of the nature of the precursor of fibrous collagen in the fibrils of connective tissue has been the subject of much investigation. Orekhovitch *et al.*[31] suggested that the fraction soluble in citrate buffer ($\mu \cong 0.2$, $p\mathrm{H} \cong 3.5$) is such a precursor and, therefore, gave the material the name "procollagen." However, from turnover studies or the incorporation of $C^{14}$-labeled glycine, Harkness *et al.*[32] suggested that the precursor is to be found in the material soluble in slightly alkaline buffer, with a much shorter half-life than citrate-soluble collagen. This conclusion was confirmed by Jackson[33] using other methods. The possibility that tropocollagen macromolecules soluble in neutral salt solutions (Gross, Highberger, and Schmitt,[34]) may be the precursor of fibrous collagen has been discussed in some detail by Gross.[35] Orekhovitch and Shpikiter[16] have concluded that procollagen and tropocollagen are, in fact, identical. The possibility that collagen, as synthesized in the fibroblasts, requires activation before it is capable of being incorporated into fibrous tissue has been much investigated, but the details of the process remain to be disclosed.

## FRAGMENTATION OF TROPOCOLLAGEN MACROMOLECULES BY SONIC IRRADIATION

The discovery by Nishihara and Doty[36] that sonic irradiation of tropocollagen rapidly reduces the viscosity without substantial reduction in optical rotation suggested that the irradiation fragments the macromolecules into shorter pieces, actually into halves and quarters, which retain the triple-chain helical structure characteristic of the native macromolecules. This possibility was confirmed by Hodge and Schmitt[37] by electron-microscopic examination of irradiated collagen. The loci along the macromolecules which undergo scission could be determined with considerable precision

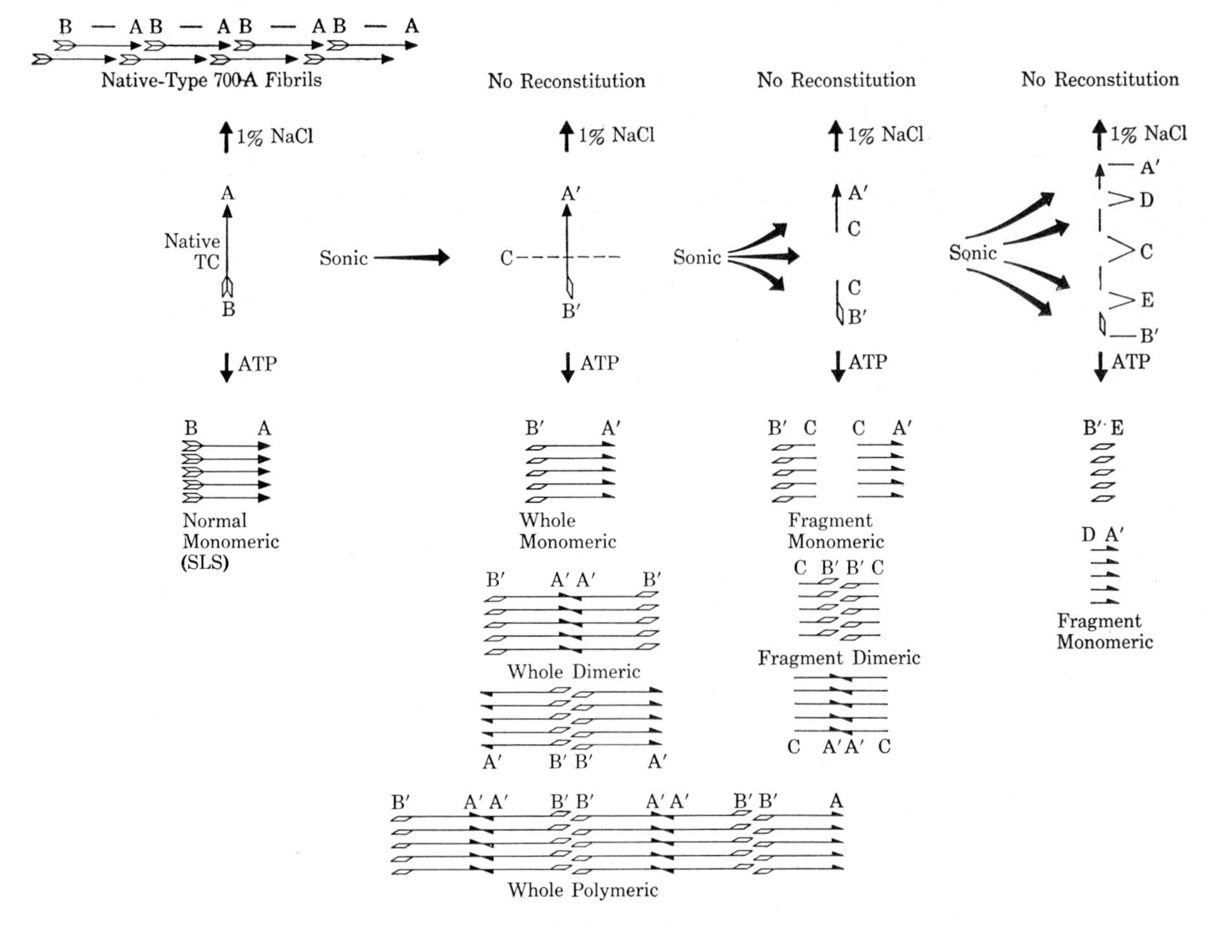


Fig. 7. Diagrammatic illustration of the chief effects of sonic irradiation on the tropocollagen macromolecules [from A. J. Hodge and F. O. Schmitt, Proc. Natl. Acad. Sci. U. S. 44, 418 (1958)].

by reference to the band patterns of the SLS-type aggregates produced by the addition of ATP to the acid solutions after irradiation.

It was discovered that sonic irradiation produces profound effects in addition to that of scission of the macromolecules into smaller fragments. The most striking of these is an alteration of "end regions," produced by relatively short periods of irradiation, without change in the length of the macromolecules. The results are shown schematically in Fig. 7, wherein the native TC macromolecules are represented by an arrow with A and B ends, indicative of the asymmetric distribution of amino acid residues reflected in the SLS type of aggregation pattern. It is thus possible to tell at a glance which is the A and B end of any particular SLS; in addition to the specific band pattern at each end, the polarization of the TC is shown at once by the position of the broad, slightly off-center interband (labeled *F-G* by Schmitt, Gross, and Highberger[23]). As was indicated earlier, the formation of the native type (700-A axial repeat) involves the formation of protofibrils that are actually linear polymers of TC by end-to-end interaction of the A-B type; lateral aggregation of such protofibrils occurs in a manner such that adjacent protofibrils are displaced axially with respect to one another by a quarter of a macromolecular length (*ca* 700 A). It is this A-B type of interaction of macromolecular ends that is first affected by sonic irradiation. In the diagrammatic representation, the altered ends are designated A′ and B′.

Following irradiation, sufficient to prevent the formation of native-type fibrils (tested for by dialysis *vs*

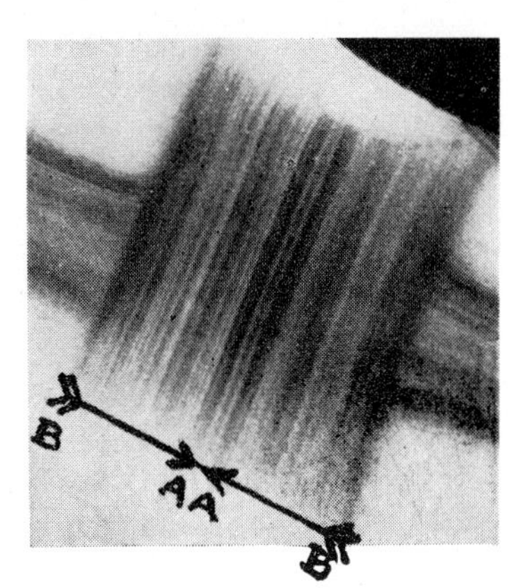


Fig. 8. Dimeric aggregation form of tropocollagen macromolecules of the type

B′–A′–A′–B′.

Produced by sonic irradiation of calfskin collagen for 20 min [from A. J. Hodge and F. O. Schmitt, Proc. Natl. Acad. Sci. U. S. 44, 418 (1958)].

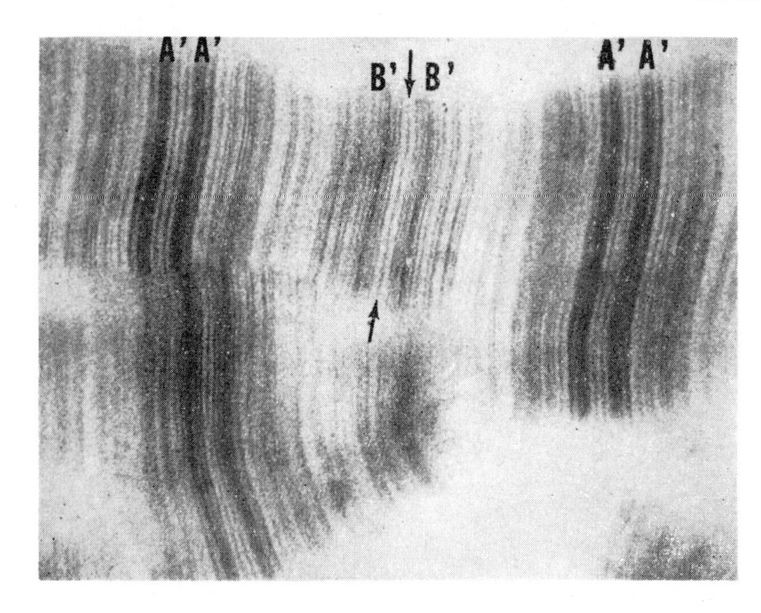


FIG. 9. Whole polymeric aggregation types of SLS aggregates from a solution of calfskin collagen treated with sonic irradiation for 240 min. Locus of A′ and B′ ends of macromolecules labeled. Arrow points to dense band at the junction between macromolecules at the B′ end [from A. J. Hodge and F. O. Schmitt, Proc. Natl. Acad. Sci. U. S. 44, 418 (1958)].

1% NaCl), i.e., by alteration of macromolecular ends, two pronounced changes are found in the ATP precipitates: (1) an increased side-to-side interaction, producing a highly exaggerated lateral aggregation into long ribbons of SLS forms; and (2) a progressive increase in the amount of A′−A′ and B′−B′ types of interaction. As a result, the formation of dimeric and polymeric forms is favored (see Figs. 8 and 9). With longer irradiation, the macromolecules are fragmented, the locus of the scission being indicated by the band pattern of the fragments.

It is noteworthy that end-to-end polymerization of scission products never involves ends produced by the fragmentation (such as those designated C, D, or E in Fig. 8). Apparently, the original ends of macromolecules are different from those produced by sonic scission. From the high density of bands (i.e., regions which combine preferentially with phosphotungstic acid) in end regions of SLS, it seems clear that certain amino-acid side chains (possibly the guanidino groups of arginine, as suggested by Kühn, Grassmann, and Hofmann,[22] or the ϵ-amino groups of lysine) may be concentrated in the end regions of the macromolecule.

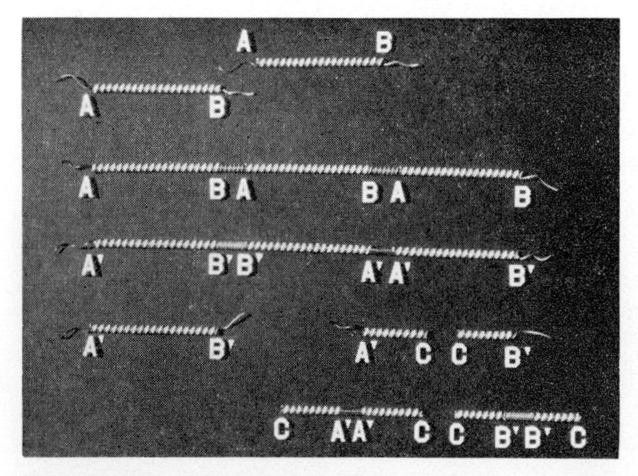


FIG. 10. Diagrammatic illustration of suggested end-to-end interaction of normal collagen macromolecules by formation of mutual coiling of terminal chains (at A-B junctions). Abnormal A′−A′ and B′−B′ junctions in sonic irradiated preparations. Scission of chains at C illustrates effect of longer periods of sonic irradiation. Note that no C—C linkage occurs [from A. J. Hodge and F. O. Schmitt, Proc. Natl. Acad. Sci. U. S. 44, 418 (1958)].

Because of the special properties of the end regions in end-to-end polymerization, it is important to obtain evidence concerning the structure in these regions. Clues are afforded by a study of the band fine structure in A′−A′ and B′−B′ linkages in the dimeric and polymeric forms. As shown in Fig. 9, the first bands at the A′ ends are separated by a region, about 100 A long, which is a typical interband (i.e., shows no dense band, hence presumably contains relatively few side chains reacting with the phosphotungstic-acid "stain"). In the case of the B′−B′ junctions, however, the separation between the first bands is about 180 A, and a darkly staining band occurs in the middle of the junctional region.

This behavior is highly suggestive concerning the nature of macromolecular ends and of end-to-end polymerization of macromolecules as follows: (1) chain appendages may occur at both ends of the native TC macromolecule; (2) these appendages may have lengths of about 100 A and 200 A at the A and B ends, respectively; (3) the amino-acid composition of the terminal chain appendages resembles typical interband regions—i.e., lacking concentrations of basic amino-acid side chains thought to characterize the band regions (except for a portion of the B end as mentioned in the foregoing).

Such considerations led Hodge and Schmitt[37] to suggest that normal end-to-end polymerization of TC monomers may involve a specific type of coiling of the terminal chains at A and B ends about each other to form a highly ordered, possibly helical, structure (see

Fig. 10). The "coiling energy" of such interaction in fact may be represented by the difference between the thermal shrinkage and denaturation temperatures (found by Doty and Nishihara[15] to be constant and equal to 29°C in three different types of collagen). It seems probable that such end-chain interaction involves primarily rather weak, hydrogen bonding. This would be consistent with the findings of Gross[25] that thermal gelation of salt solutions of collagen are inhibited by hydrogen-bond breakers, such as urea. In addition to such weak bonds, more stable bonds may be formed between terminal chains in the collagens of certain kinds of connective tissue, such as those which resist acid solution.

## DISCUSSION

It is apparent from the behavior of TC macromolecules already described that the interaction between these macromolecules and the various types of ordered structures that result from such interaction is determined by two important factors:

(1) The specific properties and reaction potentialities that are built into the macromolecules by virtue of the linear sequence of amino-acid residues in the covalent chains, the number of chains in the macromolecule and the specific type of coiling of these chains.

(2) The chemical environment of the macromolecules which may evoke one or another of the various types of interaction patterns made possible by virtue of the internal structure of the macromolecules.

A few fibrous proteins other than collagen have been investigated along similar lines and results consistent with the foregoing conclusions obtained. Thus, Hodge (p. 409) found paramyosin to be a long (*ca* 1400 A) macromolecule capable of forming an FLS type of structure when packed in antiparallel array, though in the native fiber giving rise to a 145-A repeat or a larger period five times this long. Tropomyosin also appears to behave somewhat similarly (Hodge[38]). Also, the repeat pattern in fibrin fibrils is considerably less than the length of the fibrinogen molecules (Hawn and Porter[39] and Hall[40]), and it has been suggested that the latter are staggered with respect to molecular ends (Ferry *et al.*[41]).

Glimcher, Hodge, and Schmitt[42] presented evidence in support of the concept that the initiation of mineralization involves the nucleation of inorganic crystals by a precise juxtaposition of groups in the organic matrix to form specific stereochemical arrays. In the nucleation of hydroxyapatite in calcification, only when the purified TC macromolecules were precipitated as native-type fibrils (640-A repeat) did the system induce the formation of hydroxyapatite crystals when exposed to otherwise stable solutions of calcium and phosphate ions. If the TC macromolecules were allowed to assume other types of aggregation patterns, no nucleation occurred. In the most general case, an example is seen here of how specific macromolecular interaction serves to govern very fundamental processes, such as that of the deposition of inorganic material of specific crystalline form.

If macromolecules also possess enzymatically active sites, it is obvious that even more-complex interaction behavior may occur, particularly if products of the enzyme reaction themselves strongly influence the behavior of one or more macromolecular species in the system (e.g., ATPase in myosin).

Crucial clues to many vital biological processes, such as those briefly mentioned at the beginning of this paper, eventually may be found if such biophysical and biochemical properties of elongate macromolecules are kept in mind.

## BIBLIOGRAPHY


[1] W. A. Engelhardt and M. N. Ljubimova, Nature **145**, 668 (1939).

[2] W. A. Engelhardt, Bull. Acad. Sci. USSR Ser. biol. **5**, 182 (1945).

[3] P. Weiss, Int. Rev. Cytol. **7**, 391 (1958).

[4] K. H. Gustavson, *The Chemistry and Reactivity of Collagen* (Academic Press, Inc., New York, 1956).

[5] J. H. Highberger in *The Chemistry and Technology of Leather*, F. O'Flaherty, W. T. Roddy, and R. M. Lollar, editors (Rheinhold Publishing Corporation, New York, 1956), p. 65.

[6] R. E. Tunbridge, editor, *Connective Tissue: A CIOMS Symposium* (Blackwell Scientific Publications, Oxford, England, 1957).

[7] G. Stainsby, editor, *Recent Advances in Gelatin and Glue Research* (Pergamon Press, London, 1958).

[8] F. O. Schmitt, C. E. Hall, and M. A. Jakus, J. Cellular Comp. Physiol. **20**, 1 (1942).

[9] W. T. Astbury, J. Int. Soc. Leather Trades' Chemists **24**, 69 (1940).

[10] J. T. Randall, J. Soc. Leather Trades' Chemists **38**, 362 (1954).

[11] F. H. C. Crick and A. Rich, Nature **176**, 780 (1955).

[12] A. Rich and F. H. C. Crick, Nature **174**, 915 (1955).

[13] F. H. C. Crick and A. Rich, see reference 7, p. 20.

[14] P. Gallop (personal communication, 1958).

[15] P. Doty and T. Nishihara, see reference 7, p. 92.

[16] V. N. Orekhovitch and V. O. Shpikiter, Science **127**, 1371 (1958).

[17] R. S. Bear, O. E. A. Bolduan, and T. P. Salo, J. Am. Leather Chemists' Assoc. **46**, 107 (1951).

[18] R. S. Bear, J. Am. Chem. Soc. **64**, 727 (1942).

[19] R. S. Bear, J. Am. Chem. Soc. **66**, 1297 (1944).

[20] R. S. Bear and R. S. Morgan, see reference 6, p. 321.

[21] R. S. Bear in *Advances in Protein Chemistry*, M. L. Anson, K. Bailey, and I. T. Edsall, editors (Academic Press, Inc., New York, 1952), Vol. VII, p. 69.

[22] K. Kühn, W. Grassmann, and U. Hofmann, Naturwissenschaften **44**, 538 (1957).

[23] F. O. Schmitt, J. Gross, and J. H. Highberger in *Fibrous Proteins and Their Biological Significance*, Symposia Soc. Exptl. Biol. **9**, 148 (1955).

[24] F. O. Schmitt, Proc. Am. Phil. Soc. **100**, 476 (1956).

[25] J. Gross, J. Biophys. Biochem. Cytol. **2**, 261 (1956).

[26] S. G. Tomlin and C. R. Worthington, Proc. Roy. Soc. (London) **A235**, 189 (1956).

[27] H. Boedtker and P. Doty, J. Am. Chem. Soc. **78**, 4267 (1956).

[28] C. E. Hall, J. Biophys. Biochem. Cytol. **2**, 625 (1956).

[29] C. E. Hall, Proc. Natl. Acad. Sci. U. S. **42**, 801 (1956).

[30] C. E. Hall and P. Doty, J. Am. Chem. Soc. **80**, 1269 (1958).

[31] V. N. Orekhovitch, A. A. Tustanowsky, K. D. Orekhovitch, and N. E. Plotnikova, Biokhimija **13**, 55 (1948).
[32] R. D. Harkness, A. M. Marko, H. M. Muir, and A. Neuberger, Biochem. J. **56**, 558 (1954).
[33] D. S. Jackson, see reference 6, p. 62.
[34] J. Gross, J. H. Highberger, and F. O. Schmitt, Proc. Natl. Acad. Sci. U. S. **41**, 1 (1955).
[35] J. Gross, see reference 6, p. 45.
[36] T. Nishihara and P. Doty, Proc. Natl. Acad. Sci. U. S. **44**, 411 (1958).
[37] A. J. Hodge and F. O. Schmitt, Proc, Natl. Acad. Sci. U. S. **44**, 418 (1958).
[38] A. J. Hodge. Proc. Natl. Acad. Sci. U. S. **38**, 850 (1949).
[39] C. V. Z. Hawn and K. R. Porter, J. Exptl. Med. **86**, 285 (1947).
[40] C. E. Hall, J. Biol. Chem. **179**, 857 (1949).
[41] J. D. Ferry, S. Katz, and J. Tinoco, J. Polymer Sci. **12**, 509 (1954).
[42] M. J. Glimcher, A. J. Hodge, and F. O. Schmitt, Proc. Natl. Acad. Sci. U. S. **43**, 860 (1957).

# 42

# Molecular Biology of Mineralized Tissues with Particular Reference to Bone*

MELVIN J. GLIMCHER†

*Department of Biology, Massachusetts Institute of Technology, Cambridge 39, Massachusetts*

## INTRODUCTION

THERE are several reasons for presenting the topic of biological mineralization. In the first place, the organization of these tissues, from the macromolecular level to the level of the macroscopic tissue elements, offers superb examples both of structural order in biological systems and of the relation between tissue architecture and tissue function. Secondly, the *process* of mineralization illustrates the way in which the physiological function of a tissue or organ may be interpreted on the basis of the molecular structure and macromolecular organization of its components, and in terms of fairly well-characterized physicochemical phenomena. Thirdly, many of the unsolved phenomena in mineralization provide challenging problems, particularly for those with training in the physical sciences.

Certain tissues of the organism must perform a variety of mechanical functions, such as the maintenance of the form and shape of the organism against the forces of gravity, and the protection of certain delicate organs (or the organisms themselves) by enclosing them in rigid vaults. As sites of attachments for muscles and by virtue of articulations, they also provide the organism with a system of movable but structurally rigid levers. Thus, certain organisms have not only definite form and shape, but also flexibility and a means for locomotion and prehension.

Nature has devised a number of different ways of differentiating these specialized tissues. In the case of certain insects, the polymeric chitinous-protein procuticle becomes highly crosslinked providing the structural properties required for the exoskeleton. In the case of certain *Elasmobranchii*, such as the shark, a fibrous, gel-like structure, cartilage, provides both resiliency and structural rigidity. A third method, the subject of this paper, is the deposition of a substantial amount of inorganic crystals within an organic matrix: *tissue mineralization*. This process is widespread in biology (Table I), both in the plant and in the animal kingdoms, ranging from the most primitive to the most highly ordered species. Classical examples are the exoskeletons of certain marine mollusks (the shells of clams, oysters, etc.) and the endoskeletons of the vertebrates (calcified cartilage, and bone). Also, a combination of methods is used, such as the crosslinking of the chitinous-protein shell of the lobster and the collagen matrix of bone, both of which are also mineralized.

The inorganic crystals not only serve to confer new structural properties on the tissues, but also provide a storehouse of inorganic ions which may be used to help maintain the constancy of the ionic environment of the organism.

Space does not permit the discussion of the molecular biology of all of the mineralized tissues. Illustrated here are some of the major points and problems in one representative tissue, bone.

## COMPOSITION AND MOLECULAR STRUCTURE OF THE MAJOR COMPONENTS OF BONE

Analytically, on a dry-weight basis, bone consists roughly of 65 to 70% of the inorganic crystals of the calcium-phosphate salt, apatite, and 30 to 35% of organic matrix of which collagen makes up the major fraction (95 to 99%).[1,2] The collagen in bone appears to be *structurally* similar to that in other tissues, although it is probably highly crosslinked.[3] The other components in the organic matrix include a number of ill-defined proteins, as well as the acid mucopolysaccharides, and constitute part of the ground substance. Some of the mucopolysaccharides probably are present as mucoproteins (noncollagenous protein complexes). The exact anatomical location of these components at a macromolecular level is not certain, and their state of aggregation and polymerization is not well known.

Although it had been known for nearly two hundred years that bone contains calcium and phosphate, it was not until 1926 that DeJong[4] demonstrated by x-ray diffraction that the crystal structure was similar to that of the apatites, more specifically hydroxyapatite [$Ca_{10}(PO_4)_6(OH)_2$], the structure of which is illustrated in Fig. 1. The exact nature of the apatite in bone, however, is still being debated.[5–11] This is partly because of the very broad x-ray diffraction reflections resulting from the extremely small crystal size, preventing crystallographic differentiation between a number of very similar proposed structures.

Further difficulties arise because the stoichiometry

* These studies were aided by research grants E-1469 (C1) from the National Institute of Allergy and Infectious Diseases, and A-2317 from the National Institute of Arthritis and Metabolic Diseases of the National Institutes of Health, Public Health Service, U. S. Department of Health, Education, and Welfare, and by research grant 12A from the Orthopedic Research and Education Foundation. Many of the major concepts involved have already been dealt with in an earlier paper [M. J. Glimcher, A. J. Hodge, and F. O. Schmitt, Proc. Natl. Acad. Sci. U. S. 43, 860 (1957)]. The detailed evidence and documentation for these concepts are currently in the press.

† Special Postdoctoral Research Fellow of the National Heart Institute, National Institutes of Health, Public Health Service, U. S. Department of Health, Education, and Welfare.

TABLE I. Examples of biologically mineralized tissues.

| Species | Tissue mineralized | Crystal chemistry | Mineral form | Major organic matrix components |
|---|---|---|---|---|
| Plants | Cell wall | $CaCO_3$ | Calcite | Cellulose, Pectins, lignin |
| Radiolariens | Exoskeleton | $SrSO_4$ | Celestite | (?) |
| Diatoms | Exoskeleton | Silica | (?) | Pectins |
| Mollusks | Exoskeleton | $CaCO_3$ | Calcite, aragonite | Chitin, protein |
| Arthropods | Exoskeleton | $CaCO_3$ | Calcite, aragonite | Chitin, protein |
| Vertebrates | Endoskeleton | | | |
| | Bone | $Ca_{10}(PO_4)_6(OH)_2$ | Hydroxyapatite | Collagen |
| | Cartilage | $Ca_{10}(PO_4)_6(OH)_2$ | Hydroxyapatite | Collagen, acid mucopolysaccharides |
| | Tooth | | | |
| | Dentin | $Ca_{10}(PO_4)_6(OH)_2$ | Hydroxyapatite | Collagen |
| | Cementum | $Ca_{10}(PO_4)_6(OH)_2$ | Hydroxyapatite | Collagen |
| | Enamel | $Ca_{10}(PO_4)_6(OH)_2$ | Hydroxyapatite | Eukeratin |

of the bone mineral (and many synthetically prepared apatite crystals) departs from the theoretical value of hydroxyapatite (lower Ca/P ratio). Because of their extremely small size and, therefore, large surface area, attempts have been made to explain this nonstoichiometry on the basis of the surface adsorption of excess phosphate on stoichiometric hydroxyapatite.[12,13] Aside from a number of theoretical objections to this explanation, direct experimental evidence has not verified this hypothesis.[11,14]

Other investigators have felt that lattice defects account for the aberrant stoichiometry.[6,11] In most instances, these investigators have proposed that calcium atoms were missing from the atomic structure (either internally or on the surface) and were replaced either by other cations, water, or by protons (as hydrogen bonds between the oxygens of orthophosphate groups and protons). The latter have been referred to as defect apatites.[11] However, defect crystals in which there are vacancies in certain lattice positions are ones in which the order of magnitude of these "holes" or "defects" is possibly one per thousand or ten-thousand

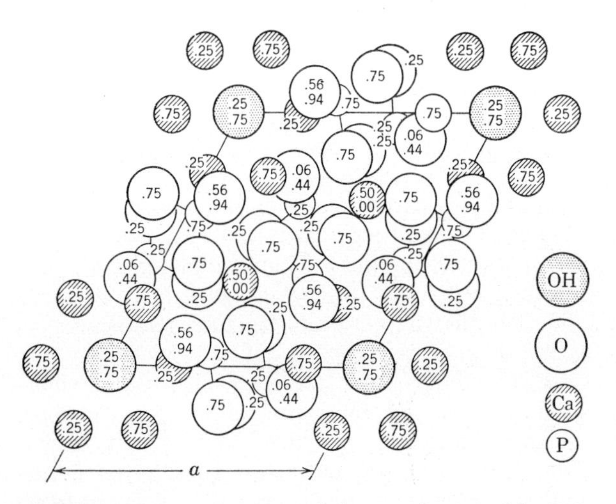


FIG. 1. Atomic arrangement of the constituents of hydroxyapatite projected on the 001 plane. The numbers refer to the fractional height in the unit cell of the atoms in the plane perpendicular to the paper ($c$-axis), as reported by Carlström[5] [from A. S. Posner, Clin. orthoped. **9**, 5 (1957)].

atoms. To satisfy the requirements in this particular case, however, one or even two calcium atoms out of every ten would have to be absent from the lattice structure. It is very doubtful that so many calcium atoms could be absent or substituted for without some structural change in the lattice.

Such structural changes have been postulated in the case where water or hydronium ions replace Ca atoms in the lattice structure. Experiments have shown (1) that a considerable amount of water is lost both from bone apatite crystals and *in vitro* prepared apatite crystals when subjected to progressively increasing temperatures after initial dehydration,[8–10,15] and (2) that apatite crystals precipitated from aqueous solutions have associated with them an amount of water many times greater than that adsorbed by initially dry crystals from a vapor phase of water.[16] This "excess" water cannot be separated from the crystals by mechanical centrifugation at 80 000 g.[16]

Since the structure of hydroxyapatite allows for no water of crystallization, several explanations have been offered to account for this "excess" water.

Thus, the substitution of water (or hydronium ions) for Ca atoms, would not only explain the low Ca/P ratios, but would also be consistent with the data on water discussed above.

The objection that such a relatively large replacement of Ca atoms would lead to structural changes in the lattice (*vide supra*) is acknowledged by one group which believes that there *are* structural differences between hydroxyapatite and the bone apatite crystals, and refers to the latter as α-tricalcium phosphate or hydrated tricalcium phosphate without specifying either the exact differences in the structural parameters or the position of the water in the new structure.[8–10,17]

Considering the water as specifically replacing collumnar calcium atoms in the apatite lattice, it has been proposed also that certain low-ratio apatites are layered structures consisting of "sheets" of hydroxyapatite held together by hydrogen bonds between the phosphate groups and water.[18]

This layered apatite structure referred to as octocalcium phosphate (OCP) gave an almost identical x-ray diffraction pattern as that of "pure" hydroxyapatite except for several additional reflections,[5,17] presumably thought to be owing to the superlattice effect of the layered water.

On the other hand, the "excess" water (particularly in the case of *in vitro* precipitated apatite crystals in aqueous solutions) has also been postulated to be the result of a large hydration shell composed of about 60 molecular layers of water "bound" to the crystals because of an electrical-field asymmetry of the crystal surfaces.[16] This amount of water is many times greater than that adsorbed by initially dry crystals from a vapor phase of water and consists of many more molecular layers of water than usually considered possible for surfaces to bind as a result of electrical-field effects.

It appears possible from the data on the replacement of calcium atoms by water (or hydronium ions) that, when apatite crystals are precipitated in aqueous solution, part of the "excess" water may be accounted for by its incorporation *into the crystal lattice*, resulting in a phase change of the solid.

Another factor in determining the amount of water associated with apatite crystals, particularly those precipitated in aqueous solutions, is the size, shape, and interaction properties of the crystals which are discussed later in this paper (page 386). Recently, however, single crystals of OCP have been prepared, and it has been shown to be a distinct compound whose structure, while closely related, is not identical to hydroxyapatite.[19] These investigators also felt, however, that OCP was also a layered structure with the layers separated by water molecules.

Dehydration of OCP caused a characteristic x-ray reflection (18.4 A) to shift to progressively lower spacings and to disappear completely upon elimination of approximately two-thirds of the water. The crystals then showed a typical apatite pattern, indicating that, with loss of a certain amount of water, the OCP structure is unstable and a true phase change occurs from OCP to apatite.

Octocalcium-phosphate (OCP) crystals prepared in the laboratories of A. S. Posner, American Dental Association,[20] Research Division, National Bureau of Standards, have been well characterized by x-ray diffraction and have shown minor differences both in the lattice spacings and intensities of the reflections as compared to those of hydroxyapatite. These OCP crystals have been examined in this laboratory by electron-microscopy and electron diffraction. The electron-diffraction patterns were indistinguishable from those of hydroxyapatite indicating that, under the conditions of vacuum and temperature in the electron microscope, a phase change had occurred from an OCP lattice to an apatite lattice by the dehydration process, confirming the report[19] referred to earlier.

Another problem is the determination of the position and state of aggregation of carbonate, which has been postulated to exist in the crystal lattice *per se* (as a carbonate apatite) as a separate phase (calcium and magnesium carbonate) or adsorbed on the crystal surfaces as carbonate ions.[7–10,21,22]

These examples illustrate that, although the evidence is quite substantial that the lattice structure of the inorganic crystals in bone is either that of hydroxyapatite or of something similar, the intimate details such as the relationship between OCP, other hydrated calcium phosphate compounds, defect apatites, hydroxyapatite, and the structure and nonstoichiometry of the bone crystals and synthetically prepared "apatite" crystals is obviously not yet clear.

## ANATOMICAL STRUCTURE

Figure 2(a) demonstrates the gross appearance of the upper end of a femur. The honeycomb-like appearance of the head and metaphysis is referred to as the *spongiosa* and is composed of delicate spicules of bone called *trabeculae.* The bone in the cortex of the shaft is much more densely packed and is referred to as *compact bone.* In both cases, the structure of the adult bone consists of a series of layers or lamellae. In the cortex of many long bones, the lamellae are further arranged concentrically around a central canal forming hollow "cylinders" containing small blood vessels. These small cylinders of bone are the basic units of such compact bone and are called *osteones* or *Haversian Systems* (Fig. 3). They are longitudinally oriented in the general direction of the long axes of the long bones (Fig. 4).

The collagen fibers in a lamella are arranged in small bundles which encircle the canal in continuous spirals crossing one another and resulting in a trellis-like arrangement.[23,24] Although the general direction of the bundles in the same lamella is similar, it varies from one lamella to the next, giving a characteristic appearance when viewed in polarized light.[25] This arrangement of the fibers within any one lamella and in consecutive lamellae imparts maximum structural properties to the tissue.

The inorganic crystals of apatite are deposited in this highly organized and ordered matrix of collagen and ground substance. Their distribution can be visualized by the use of microradiography (Fig. 5). Note the inhomogeneity, not only with respect to the entire section, but even within any one Haversian system.

The trabeculae, or "bone girders" of the spongiosa are oriented functionally in that they closely parallel the trajectories of maximum stress [compare Figs. 2(a) and 2(b)]. This allows the structure to resist mechanical stress and strain in the most efficient manner and with the greatest economy of material in accordance with sound engineering principles.

The *size, number, disposition,* and *orientation* of trabeculae of the spongiosa also can *change* in response to altered mechanical demands. This functional adaptation of bone to mechanical stress was emphasized first by Wolff in 1892 (Wolff's law),[26] yet, to date, no adequate explanation of the mechanism has been offered.

The structural order evident in the arrangement of the macroscopic trabeculae and in the microscopic Haversian systems of bone is also evident at a lower order of magnitude. In Fig. 6, an electron micrograph of a longitudinal section of cortical bone, the collagen fibrils are well oriented and the 640-A axial repeat of the fibril is accentuated in many areas by the specific location of the dense inorganic crystals. The crystals are well oriented with their long axes parallel to the collagen fibrils and intimately related to them. Electron diffraction (Fig. 7) confirms that it is the crystallographic *c*-axis of the inorganic crystals which is parallel to the fibril axis of the collagen.

Most workers have felt that the inorganic crystals of adult bone were in the ground substance between the collagen fibrils,[27,28] although more recent work indicates that some of the crystals might be within the fibrils.[29] Interpretation of electron micrographs has been difficult because of the dense packing of the collagen fibrils (Fig. 5), making it impossible to say with certainty whether the crystals were within or on the surface of the collagen fibrils, or just generally associated and oriented with them by virtue of the over-all organization, orientation, and dense packing of the fibrils in the tissue. Our laboratory has recently resolved this issue by studying bone in which the collagen fibrils are widely separated, and it is possible to see the relationship between individual fibrils and individual crystals by high-resolution electron microscopy.

Figure 8, an electron micrograph at low magnification, indicates the regular arrangement of the collagen fibrils in alternate layers and the relative looseness in the packing of the individual fibrils. It is quite obvious that the crystals are *within the fibrils* and not in the intervening ground substance. Figures 9–12 are longitudinal and cross sections at higher magnifications, and more clearly emphasize the position of the inorganic crystals *within the collagen fibrils*. That this is not an initially random crystal precipitation between the fibrils, later reorganized within the fibrils by recrystallization, is evident from an examination of electron micrographs of the earliest stages of calcification in

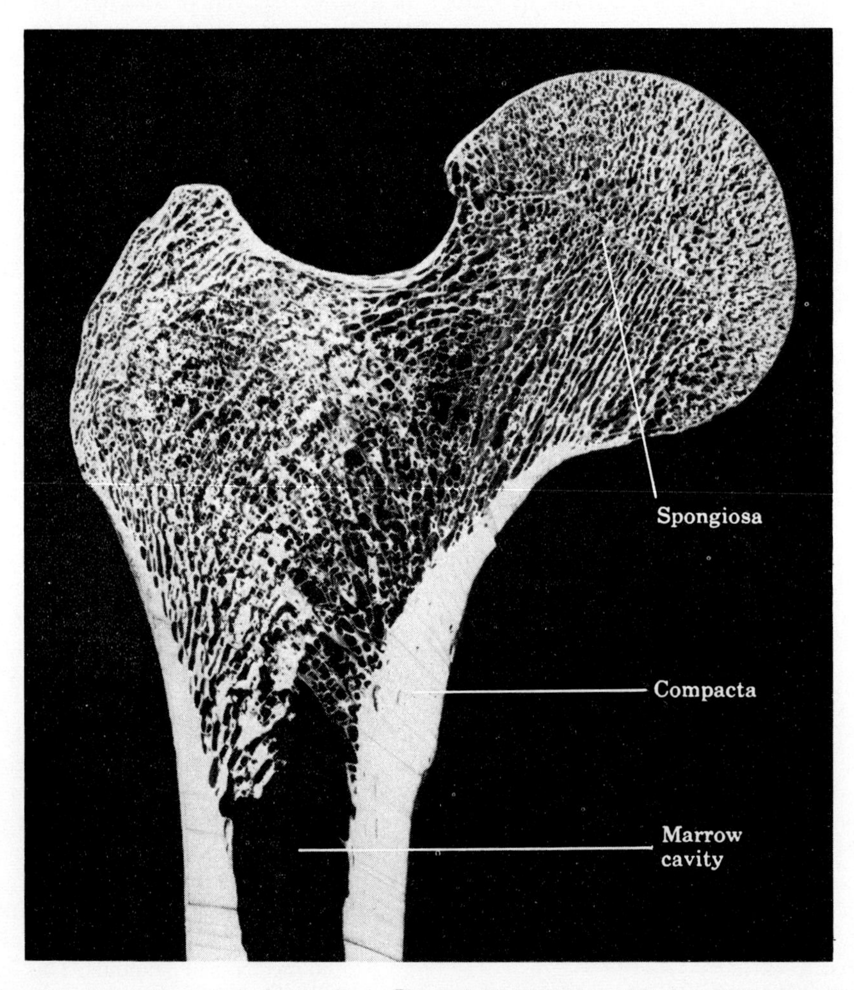


FIG. 2(a). Coronal section of the upper end of a human femur [from D. W. Fawcett, in *Histology*, R. Greep, editor (The Blackiston Company, New York, 1954), p. 133; reproduced from original kindly supplied by the author].

embryonic bone.[30,31] Figures 13 and 14 are electron micrographs of the initial stage of calcification in embryonic bone. These clearly show that the earliest crystals are regularly spaced and deposited *within* the collagen fibrils.

The location, distribution, and orientation of the crystals, by virtue of their position within the fibrils, provide the most efficient arrangement for effectively resisting mechanical stresses.

The extremely small size of the crystals (200 to 400 A ×15 to 30 A) and the fact that one and probably two of their dimensions consist of only a few unit cells, result not only in a tremendous surface area, but also in a large percentage of the atoms being in surface or near-surface positions. This easy accessibility to the crystal interior allows for a rapid exchange of ions between the crystals and the interstitial fluids.

In summary, starting with the major components themselves (collagen and apatite) which are ordered materials, the organization of the tissue, from a macromolecular level to the arrangement of the macroscopic trabeculae, constitutes a highly ordered, well-organized structure, superbly adapted both mechanically and chemically to perform its biological functions.

## MECHANISM OF CALCIFICATION

### Introduction

The many theories of calcification may be classified, in general, into two groups. The first, referred to as the "booster" theory,[16] proposed that a specific enzyme (or enzymes) exists in the areas undergoing calcification and that it splits off inorganic phosphate from an organic substrate.[32] This local "boosting" of the concentration of phosphate then exceeds the level of spontaneous precipitation and results in crystallization.

A second concept, first proposed in 1921[33] and recently revived, suggests that the *organic matrix* of calcifiable tissues initiates the formation of crystals. Recent biochemical evidence[34] and the inadequacy of the "booster" theory in explaining the precise localization of the

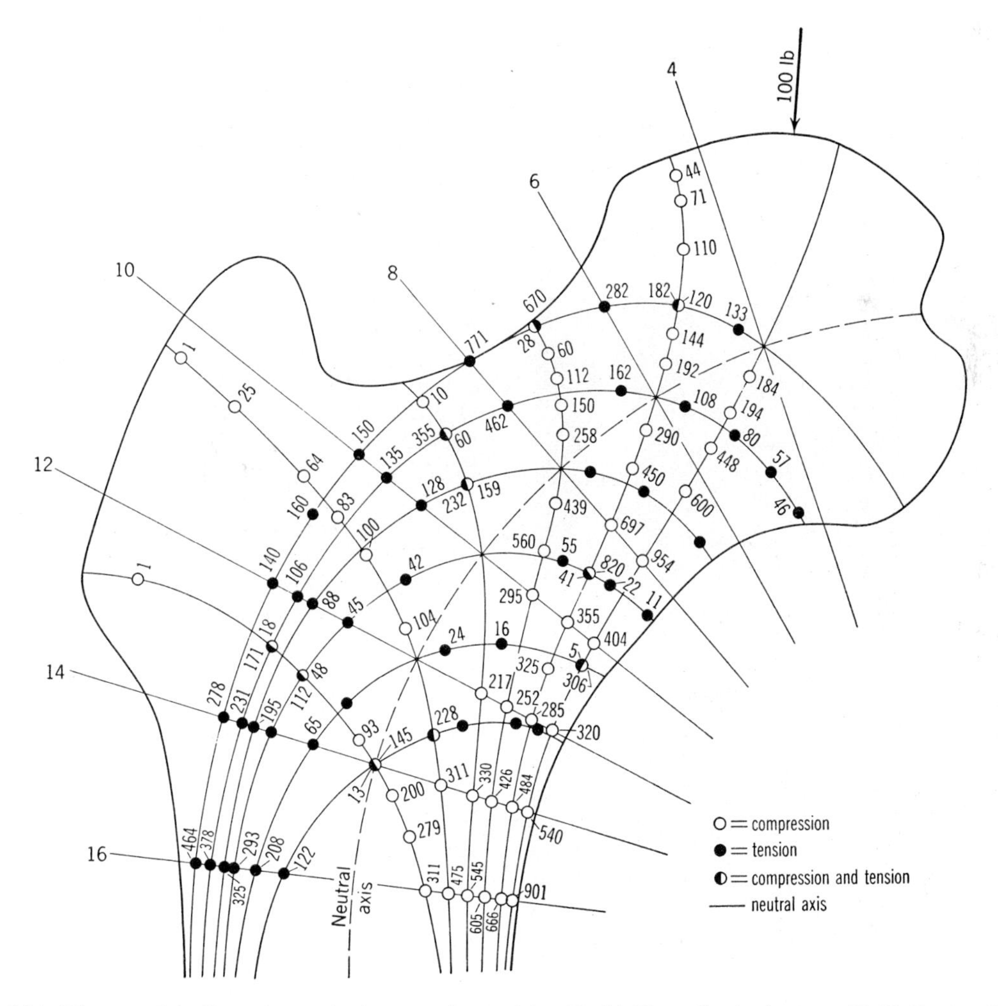


FIG. 2(b). Diagram of the lines of stress in the upper femur (after Koch) [from *Gray's Anatomy*, W. H. Lewis, editor (Lea and Febiger, Philadelphia, 1942), twenty-fourth edition, Fig. 245].

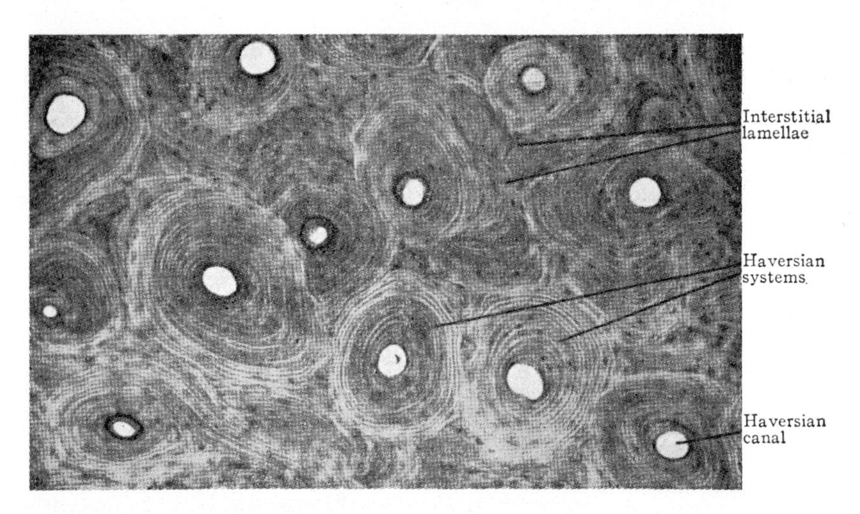


FIG. 3. Cross section of decalcified compact bone ×80. Note the alternating dark and light layers in the Haversian systems and interstitial lamellae attributable to consecutive changes in collagen-fiber orientation [from D. W. Fawcett in *Histology*, R. Greep, editor (The Blackiston Company, New York, 1954), p. 134; reproduced from original kindly supplied by the author].

crystals at an ultrastructural level strongly support the role of the organic matrix in calcification. Opinion as to what substance (or substances) in the organic matrix is responsible for the induction of crystallization and the nature of the *mechanism* of induction has varied.[35–42]

Because of the complexities in biological mineralization, it is important that the physicochemical nature of the mineralization phenomenon be clearly defined before proceeding to the experimental evidence and hypothesis

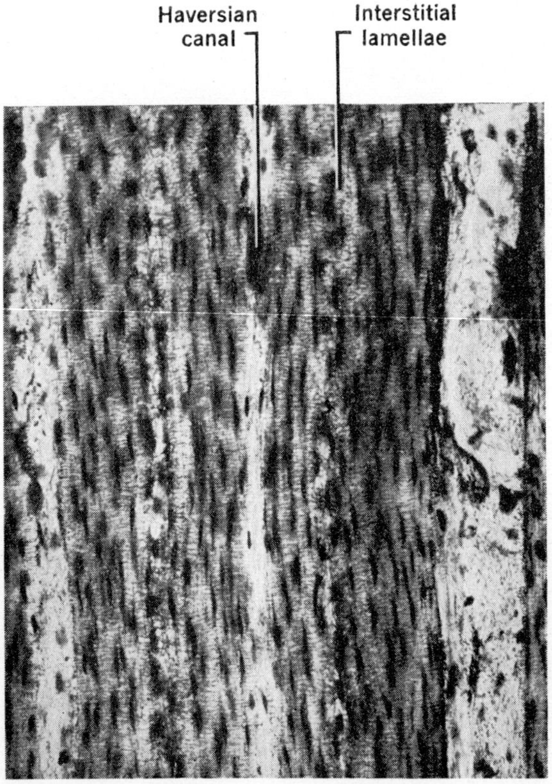

Haversian canal

Interstitial lamellae

FIG. 4. Longitudinal section of compact bone illustrating Haversian systems (original magnification ×120) [from J. P. Weinmann and H. Sicher, *Bone and Bones, Fundamentals of Bone Biology* (C. V. Mosby Company, St. Louis, 1947); reproduced from original kindly supplied by the authors].

which follow. Crystallization, or the formation of inorganic crystals from solutions where none previously existed, represents a *phase change*. This physical change in state can be divided arbitrarily into *crystal nucleation*, the process of forming the initial fragments of the new phase, and *crystal growth*, the subsequent growth of these fragments into clearly defined crystals. In addition, the phenomena of recrystallization (the growth of large crystals at the expense of smaller ones), may also play a role even after the solid state has been achieved. Failure to distinguish these interrelated but separate phenomena and failure to differentiate between the many *regulatory* processes controlling them and their underlying *mechanisms*, can be confusing. A brief review of some of the thermodynamic and kinetic principles related to *phase transition* is given in the Appendix at the end of this paper.

Certain misconceptions concerning the formulation of apatite crystals from solution are clarified by the phenomenological description in the Appendix regarding the manner in which phase changes occur.

It has been assumed[6,16,43,44] that, in the precipitation of apatite crystals, the formation of brushite ($CaHPO_4\cdot 2H_2O$) must occur first, later to be hydrolyzed or otherwise converted to apatite [$Ca_{10}(PO_4\cdot 6(OH_2)$]. This assumption has been based on the incorrect premise that an 18-body collision is required for hydroxyapatite formation whereas only a two-body collision is required for brushite formation.

Several conceptual errors need clarification. In the first place, when calcium and phosphate ions in solution aggregate to form a crystal, the change in state which occurs is a *physical change in state*—i.e., a change in the state of aggregation of the ions from a solution phase to a *solid* phase. The empirical formulas of the solids in such cases [$CaHPO_4\cdot 2H_2O$ *or* $Ca_{10}(PO_4)\cdot 6(OH_2)$] do not refer to a molecule or molecules of brushite (containing 2 ions and 2 water molecules) or of apatite (containing 18 ions), but merely represent the *ratios* of all of the constituent ions of the *solid phase* in terms of smallest whole numbers. The ions are part of a crystalline solid containing *thousands* or usually *millions* of ions arranged in a definite spatial configuration characteristic for the particular crystal.

Secondly, even in chemical reactions where a physical change of state does not occur, neither the reaction *order* nor the reaction *mechanism* can be deduced simply on the basis of the chemical formulas of the reactants or the reaction products. In the nucleation of new phases, nuclei probably arise by the stepwise addition of single molecules, atoms, or ions[45,46], and there is no obvious association between the empirical formula of the solid and the size and composition of the nucleus or the order of the reaction.

Inspection of the empirical formulas of various possible solids in no way permits one to predict the *probability* that the formation of *one* particular solid is *more likely* than another. This, of course, will depend on the relative amount of work which is necessary for the formation of a cluster of a particular composition, size, shape, and configuration under the specific conditions of the experiment, i.e., *p*H, temperature, etc.

Some simple examples illustrate the foregoing points. The formation of ice from water may be written $H_2O$(liquid)→$H_2O$(solid). It is obvious that one is not dealing with single "molecules" of ice, but with a *solid* containing many water molecules in a specific steric

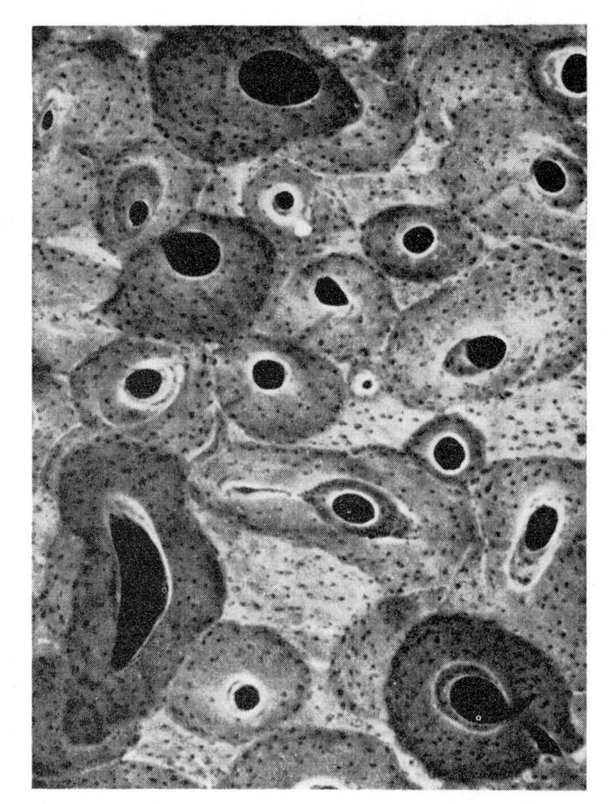

FIG. 5. A microradiogram of a thin cross section of normal cortical bone. Note the variation in density (indicating inorganic-crystal concentration) within any one Haversian system, as well as within the section as a whole. Reproduced from original kindly supplied by A. Engström, Karolinska Institutet, Stockholm, Sweden.

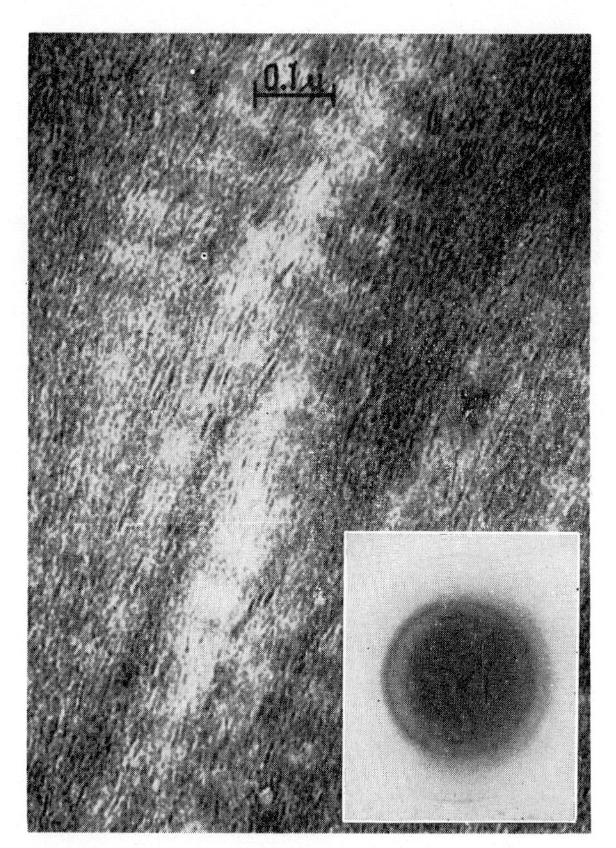


FIG. 6. Electron micrograph of a longitudinal section of compact bone. Note the accentuation of the characteristic axial repeat of the collagen fibrils in some areas by the small inorganic crystals, whose long axes approximately parallel the collagen-fiber axis. ×100 000.

FIG. 7 (lower right-hand corner). Selected-area electron diffraction of the specimen in Fig. 6, showing characteristic pattern of apatite. Arcing of the 002 and 004 reflections indicates that the crystallographic *c*-axes of the crystals are oriented parallel to the collagen-fiber axis, and, therefore, correspond to the long axes of the crystals.

configuration. It is also apparent that neither the *size* of the critical cluster nor the *reaction kinetics* can be deduced, either from the chemical formula or from the equation which describes the change in state. Physicochemical evidence[46] suggests that the nucleus which initiates this phase change is composed of about 80 to 100 water molecules. These nuclei do not form by the simultaneous collision of 80 to 100 water molecules, but are believed to arise as the result of the stepwise addition of single water molecules. The reaction is, therefore, bimolecular in *mechanism*, and of the 80th to 100th *order*.

As a second example, the formation of sodium chloride crystals from solution may be written as $Na^{+}$ (ion aqueous) $+Cl^{-}$ (ion aqueous)→NaCl (solid). This is a physical change in state—i.e., a change in the state of aggregation of the sodium and chloride ions from ions in solution to ion constituents of a crystal. In this solid, there are no NaCl "molecules," but an array of $Na^{+}$ and $Cl^{-}$ ions, with each sodium ion equally shared by six chloride ions and each chloride ion by six sodium

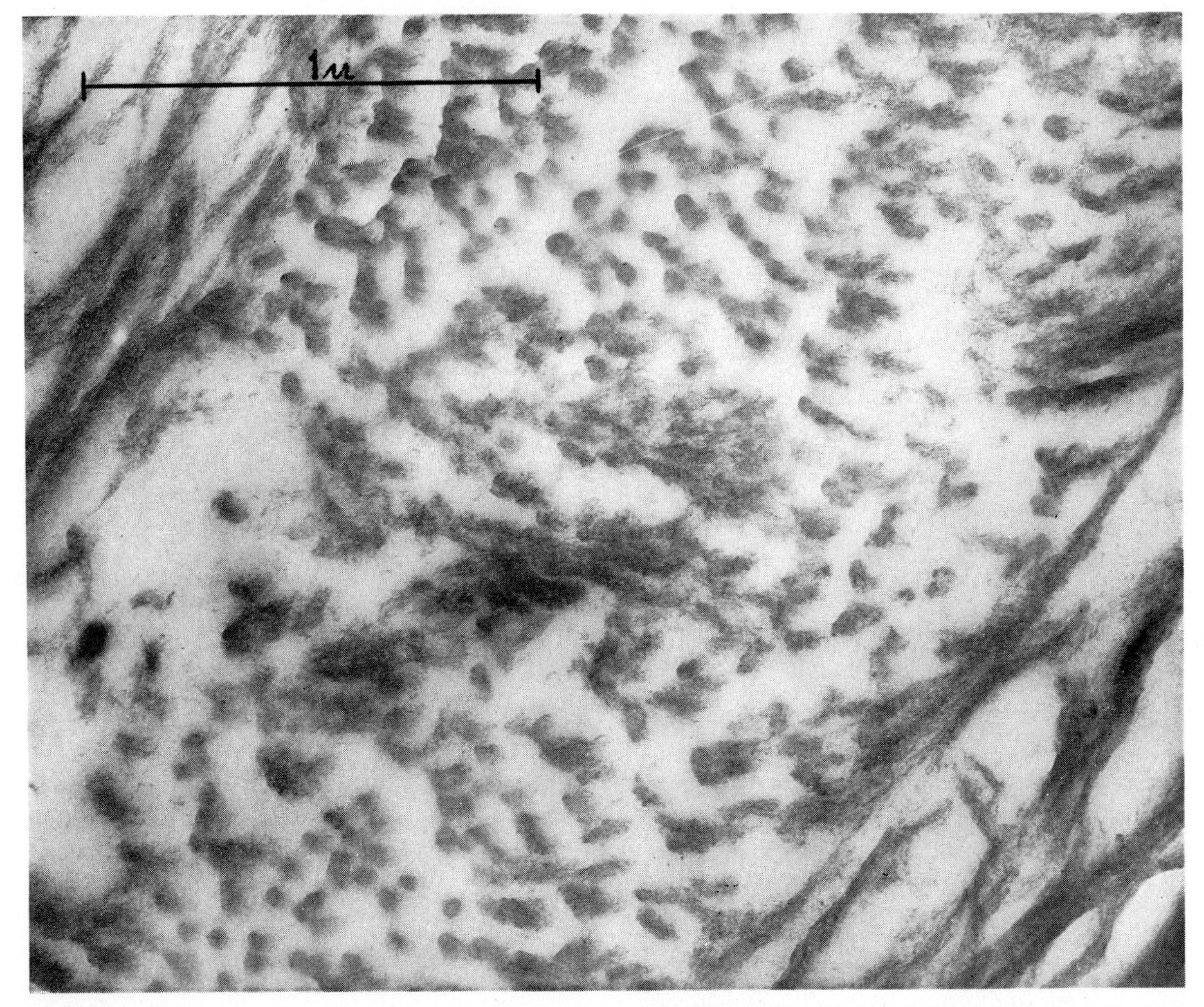


Fig. 8. Electron micrograph of a cross section of fish bone. Note the alternating direction of the collagen fibers in consecutive layers and the packing of the fibrils. Even at this low magnification it is obvious that the dense inorganic crystals are *within* the fibrils and not in the intervening ground substance. ×58 000.

ions. The chemical formula of the solid (NaCl) obviously does not indicate that this change occurs as a result of a 2-body collision.

There appears to be no doubt that, under *certain* physicochemical conditions (low *p*H, 6.0–6.2), one *can* precipitate the calcium phosphate solid, brushite, and that this can in turn be hydrolyzed or otherwise converted to apatite under the proper physicochemical conditions (raising the *p*H, for example). However, one cannot infer that, when a calcium-phosphate salt is precipitated, the formation of brushite is more probable or more likely than apatite, simply because its *empirical* formula contains fewer atoms and ions.

## MECHANISM OF CRYSTAL INDUCTION

The paper by F. O. Schmitt (p. 349) notes that not only is the collagen macromolecule a "crystalline" protein, but also the collagen *fibril*—by virtue of its *ordered aggregation* of such macromolecules, and high degree of structural regularity, as seen both by electron microscopy and by low-angle x-ray diffraction—may be considered "crystalline."

These facts and the obvious association between the apatite crystals and the collagen fibrils in normally calcified tissues recommended investigation of the possibility that this intimate anatomical relationship was evidence that either the macromolecular *structure* of collagen or the macromolecular *aggregation state* of the collagen fibrils was responsible for the *induction* of calcification,[40] by acting as a catalytic heterogeneity for the nucleation of apatite crystals. Other studies had also suggested a role for collagen[38,39] or collagen chondroitin sulphate complexes in calcification.[35,36]

For such a mechanism to be operative, however, the solutions from which the crystals are formed must be in metastable equilibrium. It is, therefore, important to establish the state of equilibrium of the fluids immediately surrounding the fibrils in relation to the formation of this new phase (apatite crystals) *in vivo*. Although it has been demonstrated that the bulk of the extra-

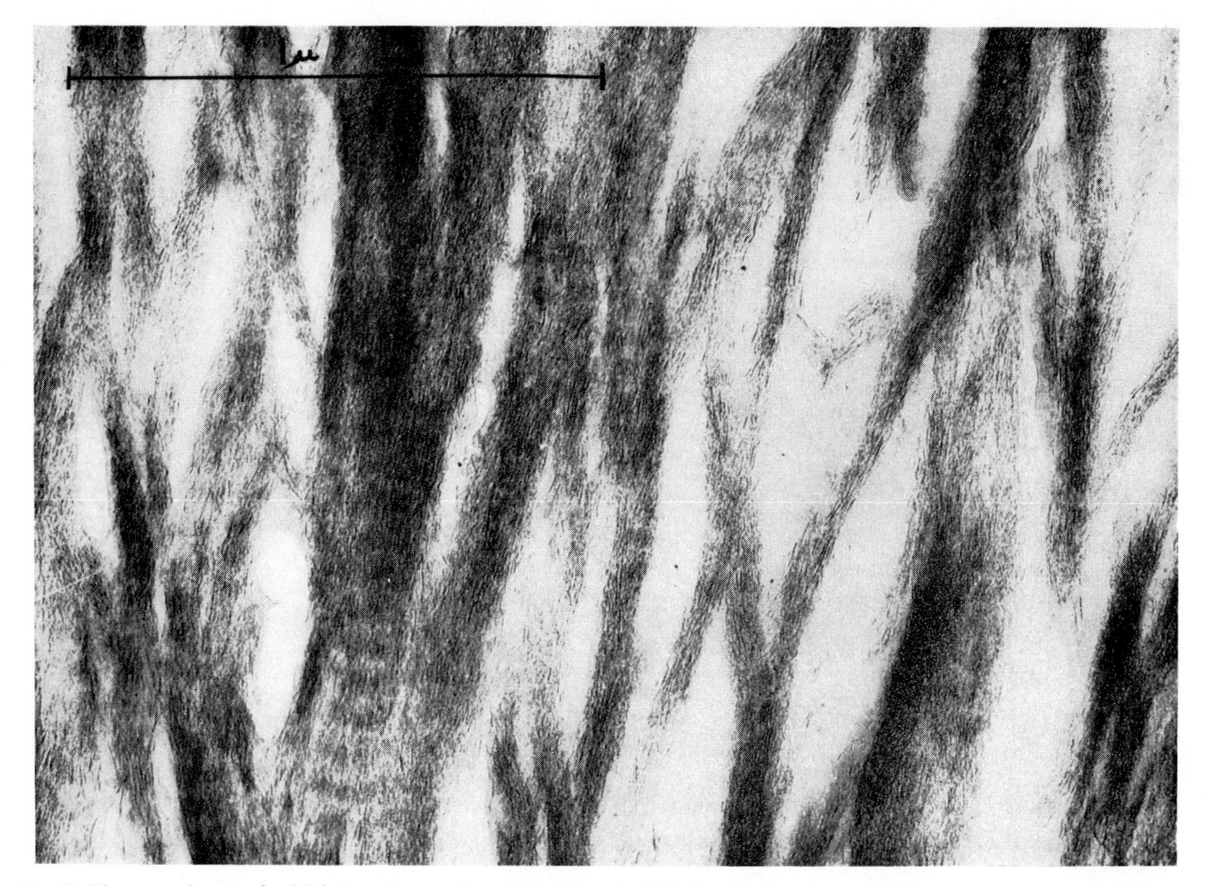

FIG. 9. Electron micrograph of fish bone in a region of loosely packed fibrils such as in Fig. 8. Despite variations in individual fibril direction, long axes of inorganic crystals are parallel to the individual collagen fibrils with which they are associated by virtue of their position within the fibrils. ×75 000.

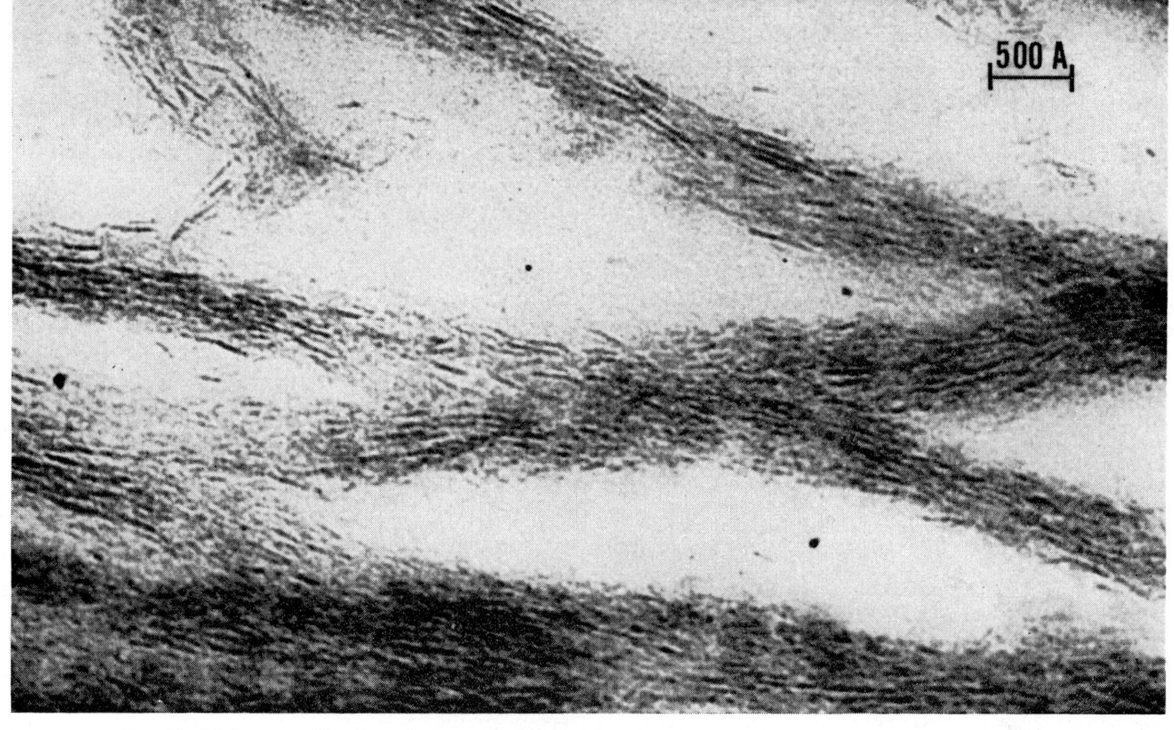


FIG. 10. Higher magnification of an area in Fig. 9. The crystals appear to be rod-shaped, approximately 200 to 400 A long and 15 to 40 A wide. ×225 000.

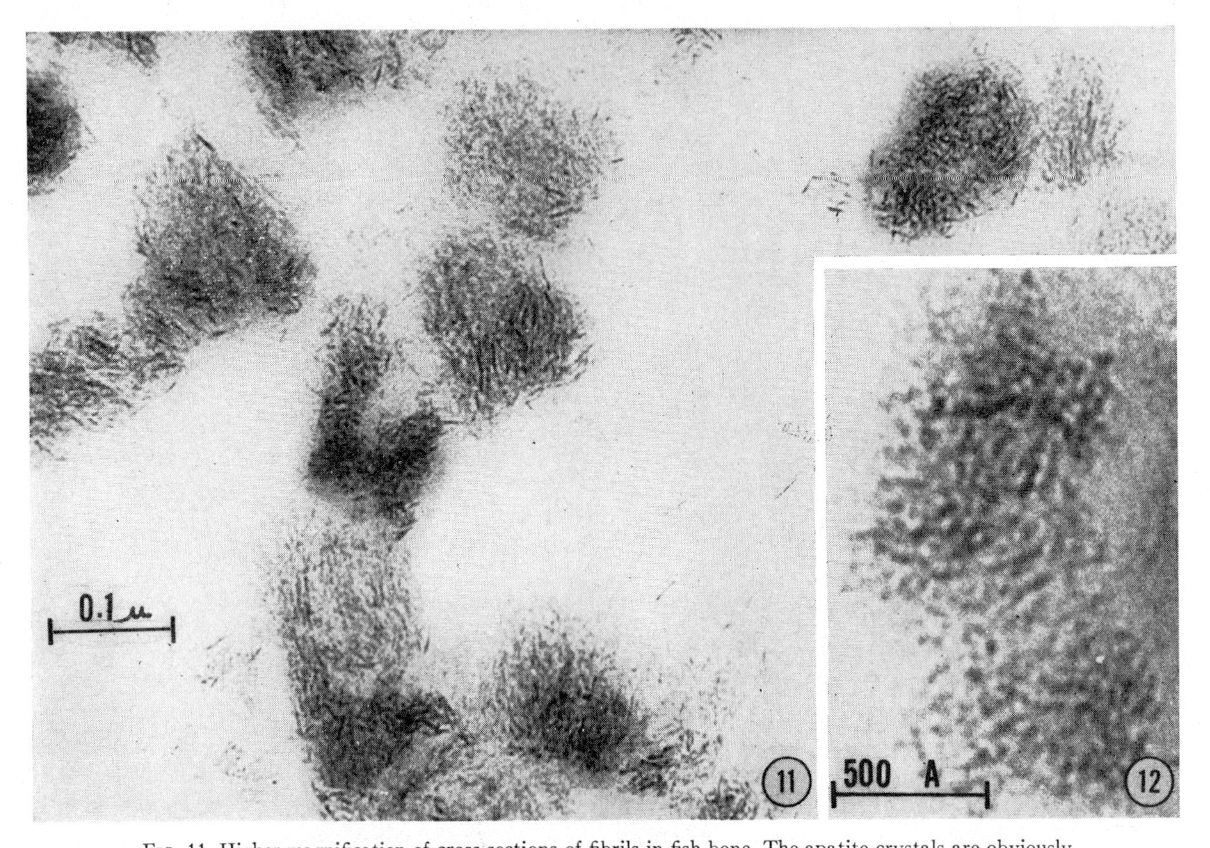


FIG. 11. Higher magnification of cross sections of fibrils in fish bone. The apatite crystals are obviously within the collagen fibrils. ×165 000.

FIG. 12. Cross section of two fibrils in an area similar to Fig. 11. The crystals have appearance of rods viewed on end and appear to be hexagonally packed. ×430 000.

cellular interstitial fluid is metastable with respect to the formation of apatite crystals,‡ the *local* values immediately surrounding the collagen fibrils in any particular tissue might be markedly different because of active cellular-controlled compositional changes, or active or passive transport and diffusion phenomena.

Although it is informative to examine the data with regard to the state of equilibrium of the unaltered extracelluar fluids, not only with reference to the mechanism of induction but also particularly with regard to the *regulation* and *control* of this process in other non-mineralized tissues, the *immediate objective*, however, was to establish whether or not collagen *could* induce the formation of apatite crystals from solutions which *were* metastable, and to determine the nature of this specificity (if any) and its mechanisms. The experiments discussed in the following were conducted at the Massachusetts Institute of Technology in collaboration with A. J. Hodge and F. O. Schmitt, and were designed to answer these questions.§

The collagen of many connective tissues can be dissolved in a number of weak acids and neutral buffers yielding viscous solutions of the macromolecules (see F. O. Schmitt, p. 349). These can subsequently be reaggregated and reconstituted into fibrils with the typical and characteristic axial repeat and intraperiod fine structure of native fibrils. By appropriate treatment of the connective tissue prior to dissolving the collagen and by several recrystallizations, the collagen fibrils can be prepared relatively pure, with only minute traces of ground-substance constituents.

‡ The ion product ($a_{Ca^{++}} \times a_{HPO_4^{-}}$) in the serum of many vertebrates is considerably higher than the same product in solutions of equivalent ionic strength, *p*H, temperature, etc., where equilibrium has been approached either through precipitation or through dissolution of apatite crystals.[38,39] Since this product appeared to be critical in determining whether or not precipitation of a solid phase occurred, it was concluded that serum and interstitial fluid were supersaturated with respect to the bone mineral. While these thermodynamic data are not conclusive, direct experimental evidence supports this supposition. Inorganic solutions with ion products ($a_{Ca^{++}} \times a_{HPO_4^{-}}$) similar to serum were stable for indefinitely long periods of time, but showed rapid separation of *more* solid when exposed to bone mineral[47] or to synthetic apatite crystals.[48] It would appear, therefore, that the unaltered extracellular interstitial fluid is metastable with respect to the formation of apatite crystals.

§ Although we are fully aware of the limitations of a model system which is admittedly a great deal simpler than the process as it occurs *in vivo*, it has enabled us to characterize and distinguish what appears to be the *mechanism* of crystal induction as well as several closely allied phenomena in the process of calcification and to gain an insight into their mechanisms. Many of the minute details and a quantitative description of these phenomena and the manner in which they are biologically *regulated* will, of course, have to be gathered from the intact organism. However, one must remember that it is difficult even to know *which* details to search for if the basic underlying mechanisms are not at least clearly defined conceptually.

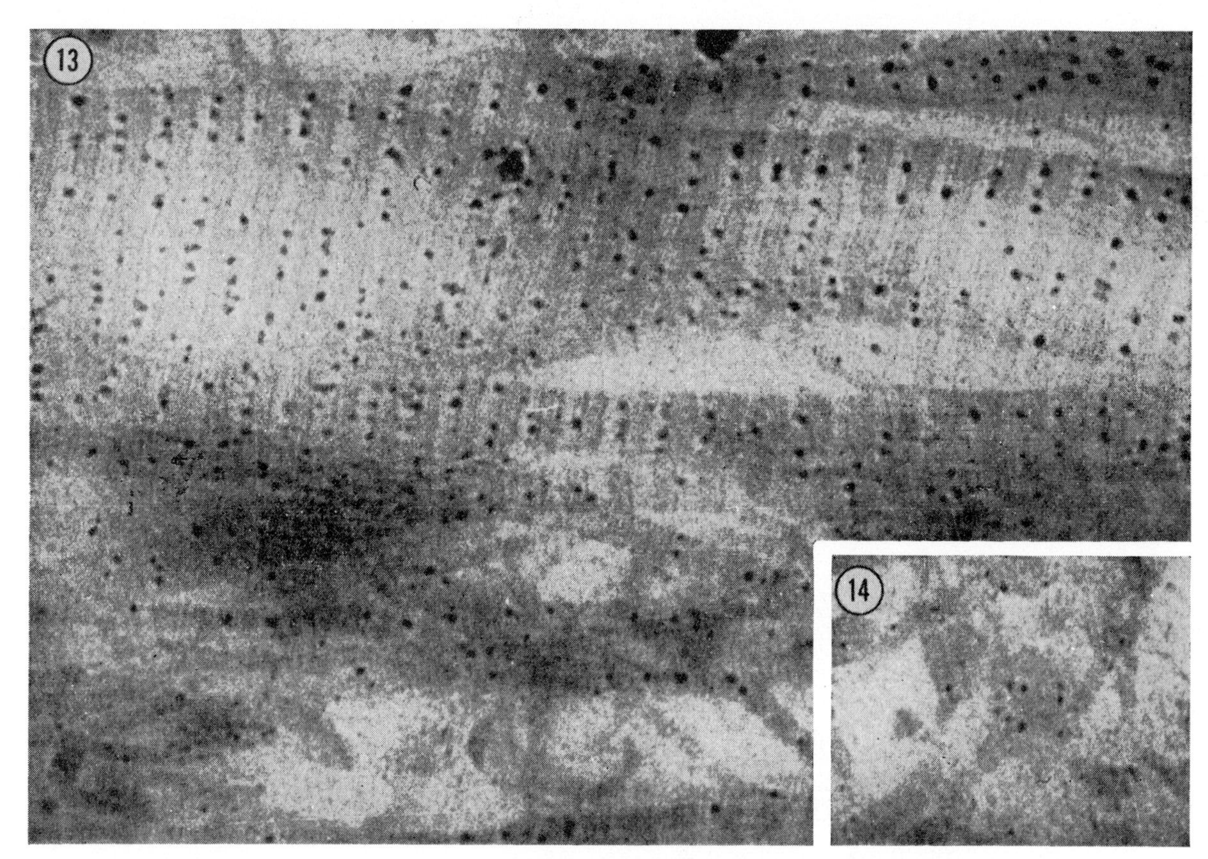


FIGS. 13 and 14. Electron micrographs of avian embryonic bone in the earliest stages of calcification. Note the regular and periodic arrangement of the dense apatite crystals within the collagen fibrils [from S. Fitton-Jackson, Proc. Roy. Soc. (London) **B146**, 270 (1957); reproduced from originals kindly supplied by the author]. ×100 000.

The first experiments were carried out by exposing such reconstituted 640-A axial-repeat collagen fibrils to calcium-phosphate (Ca-P) solutions shown previously to be metastable with respect to the formation of apatite crystals. Because of the technical difficulties involved, and because the thermodynamic properties of calcium-phosphate solutions and their relation to the formation of solid phases are not well-enough characterized, it has not yet been possible to procure quantitative data on nucleation rates, crystal growth rates, free energy of nucleation, etc. The more qualitative methods used were designed to detect the ability of the test materials to *induce* the formation of apatite crystals, and depended mainly on the identification of the formed crystals. They did not distinguish between nucleation rate, crystal growth, recrystallization, etc.

The results of these experiments demonstrated that native-type, 640-A axial-repeat, reconstituted collagen fibrils, prepared from normally uncalcified tissues such as rat-tail tendon, calf skin, guinea-pig skin, fish swim bladder, etc., were able to nucleate apatite crystals from metastable calcium-phosphate solutions. Figure 15 is an x-ray diffraction pattern of such a calcified collagen preparation showing the typical diffraction rings of apatite.

Ironically, we were not able to prepare reconstituted collagens from bone, presumably because of its highly crosslinked nature. However, bone could be decalcified under a variety of conditions, which maintained the 640-A structure of many of the collagen fibrils. This

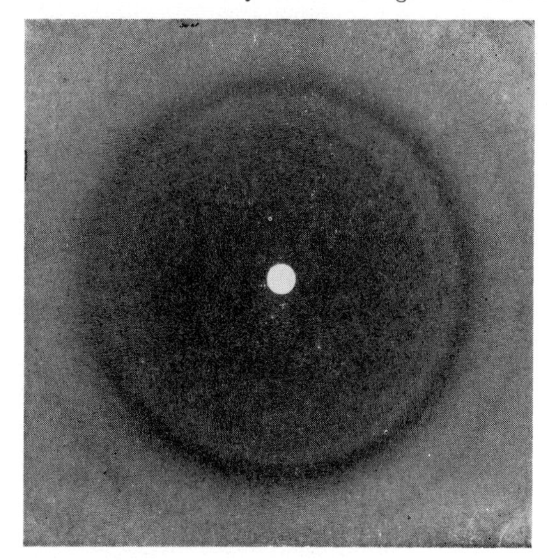

FIG. 15. X-ray diffraction pattern of *in vitro* calcified collagen. Note the lack of crystal orientation (evidenced by complete rings) and the broadness of the diffraction lines attributable primarily to small crystal size.

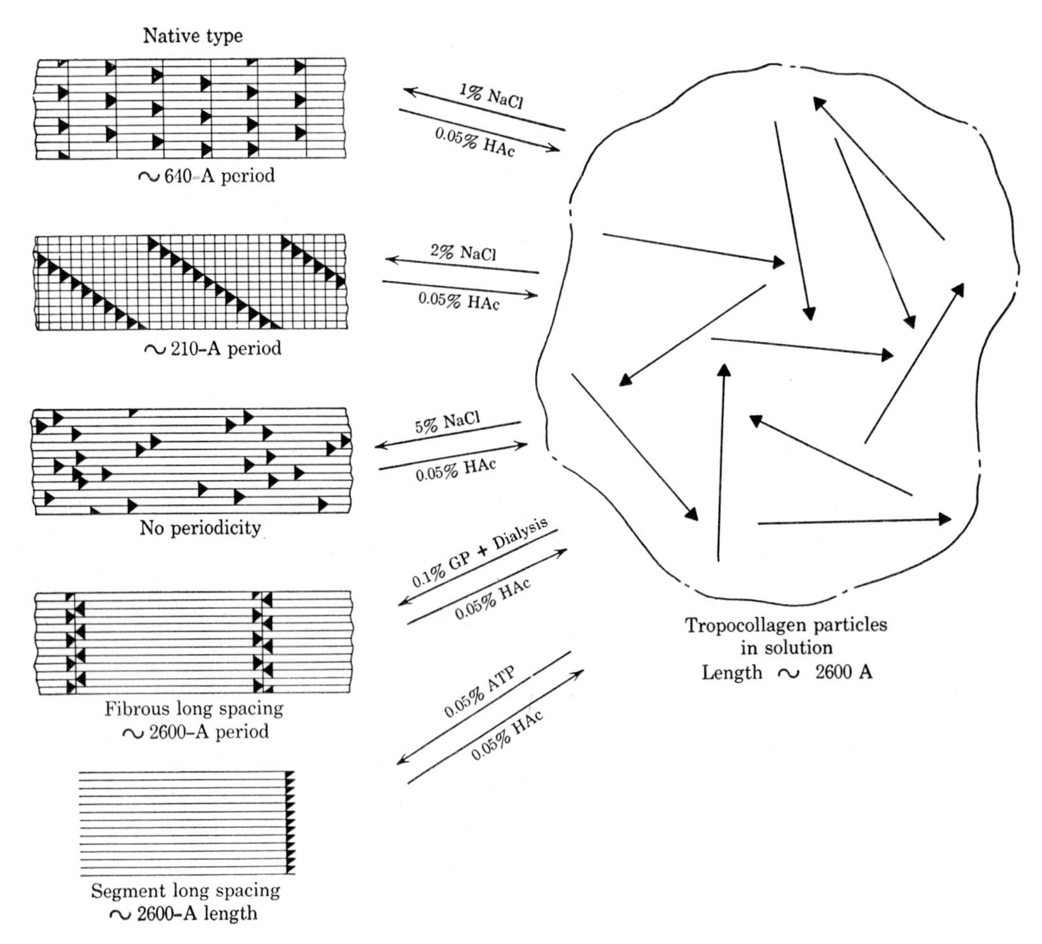


FIG. 16. Modes of aggregation of collagen macromolecules *in vitro* [from F. O. Schmitt, Proc. Am. Phil. Soc. **100**, 476 (1956)].

preparation was also able to induce crystallization of apatite. Although it is very difficult to make quantitative comparisons, it appeared that many of the decalcified bones were not grossly as potent in this respect as the reconstituted collagens, presumably because of changes in the structure of many fibrils and to alteration of certain important functional groups during decalcification.[49–51]

Since, in nucleation phenomena, a number of discontinuities are able to induce phase changes, the next problem was to determine whether or not this property was specific for collagen and, if so, where the specificity resided. Paramyosin fibrils, another well-organized and ordered fibrous protein structure, showing a regular periodic pattern by electron microscopy and low-angle x-ray diffraction, were obtained from the adductor muscles of clams, and used in similar experiments under identical physiochemical conditions. They were not able to induce this phase change.

A number of other fibrillar forms can be reconstituted *in vitro* from the same solution of collagen macromolecules (see F. O. Schmitt, p. 349). These include fibrils with 220-A axial periods, structureless fibrils with no discernible band pattern, fibrous long-spacing (FLS), and segment long-spacing (SLS) (Fig. 16). These are *different* ordered-aggregation states of the *same* macromolecules and reflect differences in the packing and intermolecular geometry of the fibrils. These differences are, of course, also reflected in differences in the *kinds* of amino-acid side chains which interact, and in differences in their *stereochemical relations* to one another. In addition, the isolation of the collagen macromolecules and linear polymers of collagen macromolecules (protofibrils), both in solution and in the solid state, provided a unique system with which to determine the steric nature of the specificity.

Figure 17 is a schematic representation of the basis for these experiments. Using SLS as a typical example with which to compare the native-type fibril, it is apparent that within any region of the fibril, although the *same* macromolecules and, therefore, the *same* amino acids are present, the *steric* relations between amino-acid side-chain groups are different.

The macromolecules, protofibrils, and the various aggregation states were exposed to metastable Ca-P solutions. Neither the macromolecules, linear aggregation of macromolecules, nor any of the fibril forms, *other than those with the native-type 640-A axial repeat,*

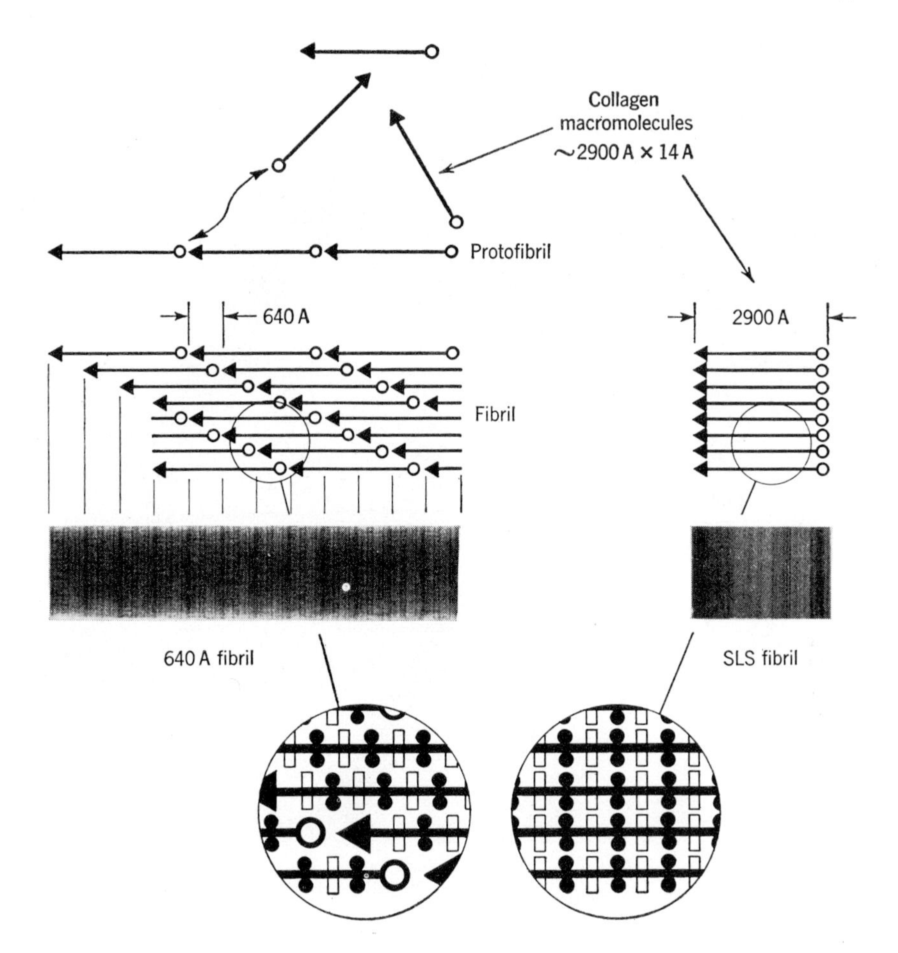


FIG. 17. The steric differences between the reactive side-chain groups of adjacent macromolecules in the native and SLS forms of collagen diagrammatically portrayed.

were able to initiate this phase change under identical physicochemical conditions. Since it was possible to pass reversibly from one fibril form to another and to demonstrate nucleating ability *only* while in the native 640-A form, it was evident that failure to initiate crystal formation was the result of the unfavorable configurations of the other types of reconstituted collagen fibrils and not the result of denaturation or other changes in the macromolecules produced in aggregating them from solution, or of the nature of the metastable solutions.

These experimental findings indicated that the process of *induction of crystallization* was a heterogeneous nucleation of apatite crystals from metastable Ca-P solutions by the collagen fibrils, and was *not* dependent on the macromolecules *per se* or on single linear adlineations of macromolecules, *but on groups of macromolecules and protofibrils polymerized laterally and longitudinally in a highly specific fashion characteristic of native collagen.* Thus, it would appear that the particular type of aggregation and packing of macromolecules characteristic of native-type fibrils creates highly specific steric relationships between reactive amino-acid side-chain groups from adjacent macromolecules within the fibril which serve as centers for nucleation.

Additional confirmation of the necessity of a *specific juxtaposition* of *certain* reactive *groups* was obtained by subjecting aliquots of reconstituted native-type collagen fibrils and demineralized bone to a number of physical and chemical agents, without altering the reactivity of the amino-acid side chains, and demonstrating that the ability to induce crystallization had been lost.

These included the effect of heat, acid, and alkali. In the case of heat, no change was noted *until* the thermal-shrinkage temperature was reached. Thermal shrinkage of collagen results in a phase change as regards its state of aggregation similar to gelatinization.[52,53] This disrupts both the molecular structure and the macromolecular-aggregation state of collagen.

In the case of acid-treated demineralized bone, or alkaline-treated reconstituted fibrils, although the collagen fibrils undergo a good deal of swelling and distortion, leading to loss of the typical 640-A repeat by low-angle diffraction and electron microscopy, the *macromolecular structure* remains intact. In both cases, the amino-acid side chains are not *chemically* altered, but only their *steric* interrelations are changed.

With reference to the phenomenon of nucleation, three distinctly different aspects of the relation between collagen and apatite must be considered. The first is the stereochemical relationship between the reactive side-chain groups which constitute a nucleation center. The second is the demonstration that there are *preferred*

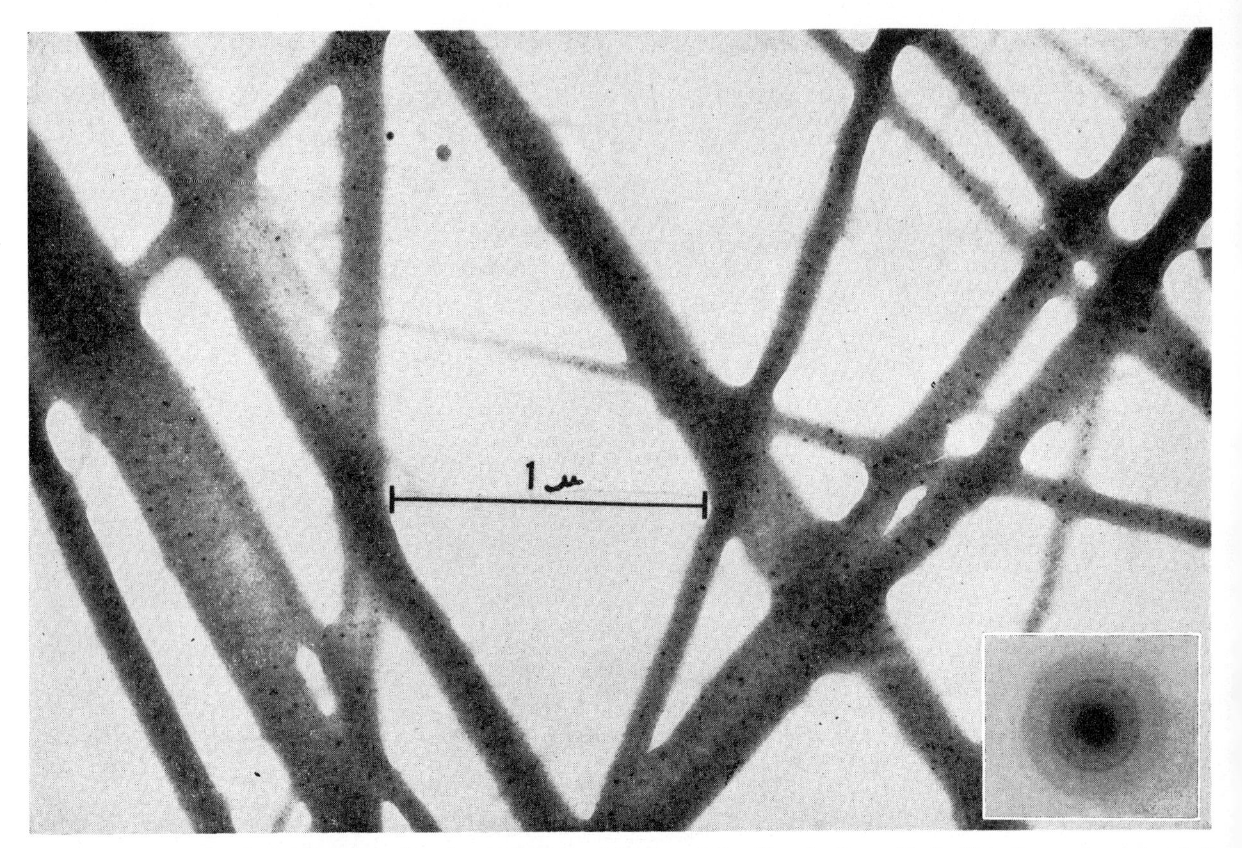


FIG. 18. Unstained, unshadowed preparation of an early stage of *in vitro* calcification of collagen fibrils. Note the regular and periodic distribution of the crystals along the collagen-fibril axis corresponding to the intraperiod fine structure of the fibrils and occurring primarily once per axial period.

FIG. 19 (lower right-hand corner). Selected-area electron-diffraction pattern of preparation similar to Fig. 18, showing the characteristic apatite reflections. There is no evidence of preferred crystal orientation, even though the diffraction pattern was obtained from an area where the fibrils were well oriented. ×42 000.

nucleation centers and the determination of their location with respect to the known electron-microscopic intraperiod fine structure of the collagen fibrils. The third is the nature of the chemical groups in the nucleation centers, and the nature of the intermolecular forces between these groups and the mineral ions.

An undue emphasis seems to have been placed on the *absolute* value of *640 A*, characteristic of native-type fibrils, and its relation to nucleation and other calcification phenomena. It should be apparent, however, that there is nothing specific about this *absolute* value, since it is merely a visible manifestation of the particular aggregation state of the collagen macromolecules and acts as a "fingerprint" for the recognition of the native-type fibrils. Its importance lies in the fact that when the macromolecules are arranged in a specific three-dimensional array so that this characteristic electron-density distribution occurs along the fibril, *certain regions* are created within the fibril whose reactivity and steric relations permit them to act as sites of heterogeneous nucleation. It is this stereochemistry *between reactive side-chain groups within the nucleating regions* which is directly related to the mechanism of nucleation, and this has no direct relation to the *absolute* value of 640 A.

In attempting to correlate the molecular structure of collagen and the property of nucleation, therefore, one must distinguish clearly between the structural characteristics of the *macromolecules* and the higher-ordered configurations which are the result of the specific *arrangement* of the *macromolecules* in the *fibril*.

Unfortunately, unlike the case of nucleation of inorganic crystals by other inorganic crystals, a direct correlation cannot be made between the wide-angle diffraction spacings of collagen and the known lattice spacings of apatite, as has been attempted by one group of workers.[16] The wide-angle diffraction pattern of collagen cannot be treated as arising from simple Bragg planes as in inorganic crystals, but must be interpreted on the basis of the theory of helical diffractors.[54,55] Except for the 10- to 12-A equatorial spacings, indicating the distance between the centers of adjacent macromolecules, the wide-angle reflections are characteristic of the triple-chain helical-coiled structure of the *macromolecule* and give *no* information as to the interatomic distances and configurations of the side-chain groups in the *fibril*, *which* information *is* directly related to the nucleation phenomena. It should be clearly understood that not only do the 640-A-type

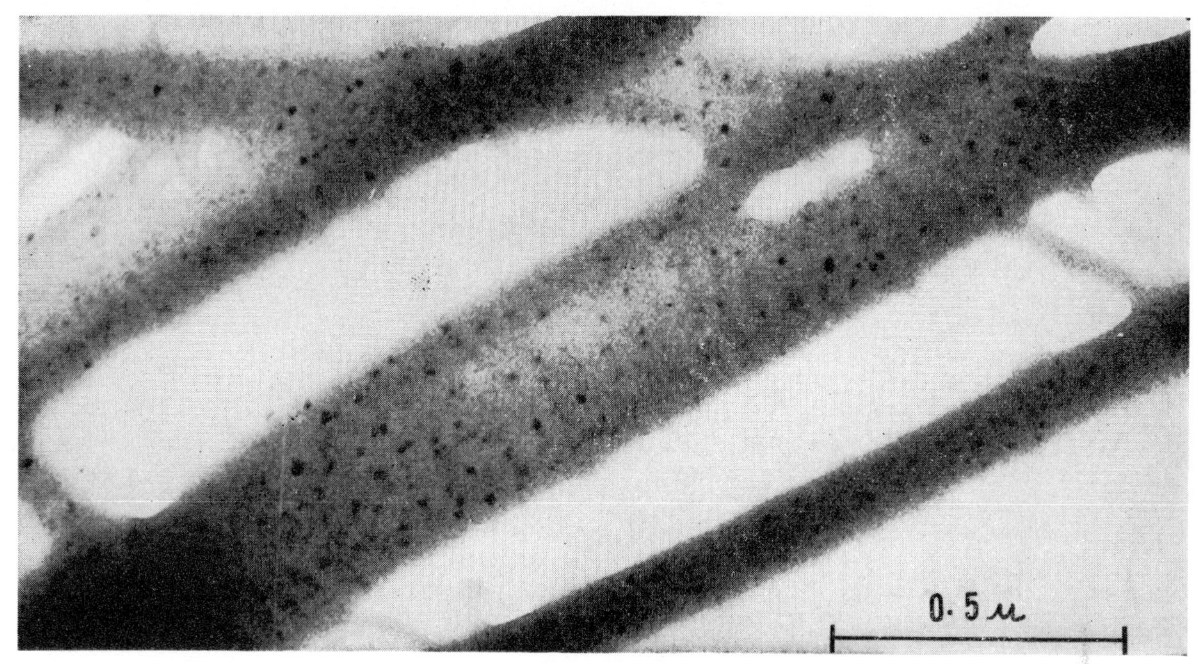


FIG. 20. Higher magnification of an area in Fig. 18. ×80 000.

fibrils give the characteristic wide-angle x-ray diffraction pattern, but also the macromolecules themselves and all other fibrillar forms, as well as cold gelatin films.

To date, neither the exact sequence nor stereochemistry of the amino acids in the collagen macromolecule are well enough known to permit determination of this specific configuration, although some general statements may be surmised from other data presented below.

As regards the second point—the demonstration of specific nucleation centers and their location—we have attempted to elucidate this aspect by several different methods. Direct visualization by electron microscopy and correlation of the changes in electron density, either in electron micrographs or from low-angle x-ray diffraction patterns, have been tried.[56] Interpretations based on electron-micrograph density changes and low-angle x-ray diffraction density changes are fraught with technical and theoretical difficulties, and no definite statement is as yet possible from these data.

Electron micrographs taken during the course of time-sequence studies of *in vitro* calcification (Figs. 18–20) and during the earliest stages of *in vivo* calcification of embryonic bone (Figs. 13, 14) provide direct evidence that there are specific regions in the fibrils which act as nucleation centers. These show that, during the earliest stages of calcification *in vitro* and *in vivo*, small dense particles, varying in size from 20 to 150 A, are deposited *regularly spaced* along the fibrils. Selected-area electron-diffraction of these particles in both cases revealed that they are crystals of apatite (Fig. 20).

Since the smallest crystals visualized in electron micrographs represent the *summation* of nucleation and crystal growth, interpretations based on such visual observations must be viewed with some caution. It is entirely possible that regions most favorable for crystal growth may be quite different from those where nucleation occurs. Also, there may be different areas capable of nucleation but differing in their degree of catalytic potency.

In addition, it should be pointed out that the nucleating center is not represented by an entire "band" or "interband" in the collagen fibrils, but only by *regions* within such transverse sections of the fibril. This can be deduced easily if one considers that it takes a certain number of repeating groups from adjacent macromolecules in just the right configuration to make a nucleation center, and that statistically there will only be a finite number of such groups with the proper steric configurations within any one transverse section.

Although it has not been possible to identify the exact location of the nucleation centers with respect to the intraperiod fine structure of the collagen fibrils, observations thus far indicate that these centers do correspond to certain of the so-called band (electron dense) regions. This would correlate well with related chemical and low-angle diffraction studies[57,58] which have shown that the band regions are the most reactive sites for many electron stains, tanning agents, etc., and that these areas probably contain many of the amino acids with long, reactive, polar side-chain groups.

In this respect, it is interesting to note the recent demonstration of the direct correlation between the degree of mineralization and the reactivity of the ε-amino groups of lysine in bone and tooth during demineralization,[50] since lysine is considered to be a major constituent of the band regions.[57–60]

Whereas these experiments have indicated the *steric*

*specificity* in the heterogeneous nucleation of apatite crystals by native-type collagen fibrils, they do not provide information concerning the manner in which these steric factors operate or concerning the nature and importance of the chemical interaction between collagen and the mineral ions, and the specific amino acids and mineral ions involved.

In this regard, certain general principles may be stated. Thus, while it is obvious that some sort of chemical interaction must occur between certain reactive groups in the collagen fibrils and the appropriate ions for nucleation to occur, it is imperative that the nature of the forces involved be such that the ions are still capable of interacting with the other constituent ions of the crystal lattice. For, if either calcium or phosphate ions were very *strongly bound* by the collagen fibrils, collagen would act as a *demineralizer* similar to chelating agents, unless a second mechanism were invoked which *released* the ions after they were bound.

There are several distinct ways in which the steric factors could operate to induce crystallization. The simplest case would be by facilitating the local concentrations of calcium and/or phosphate ions without requiring that the ions themselves be arranged in any particular steric fashion. This local increase in ion concentration would lead to crystallization by exceeding the metastable limit.

Another possibility is that the precise array of groups necessary for nucleation imparts a specificity for the *selective adsorption* or binding of the calcium or phosphate ions, either randomly or in specific steriochemical fashion, and that they then interact with the other constituent ions of the lattice to produce the first fragments of the new phase.

A third method would be the most specific of all, and would statistically require the fewest number of atoms. In this case, the precise array of reactive groups would sterically closely approximate certain low-index planes of apatite in a fashion similar to that proposed by Turnbull *et al.*,[61–63] and discussed in the Appendix. Thus, calcium, phosphate, and hydroxyl ions, either singly or in combination, depending upon which atomic plane of apatite is involved, would be "lightly" bound in such a *geometric configuration* that they would constitute a reactive nucleus.

Of course, one must consider the possibility that certain *clusters* of ions are bound by the reactive groups in the collagen structure. This is really a semantic issue, however, since the binding of such clusters, which presumably have the configuration of the bulk-solid phase, would require similar steric considerations.

Experiments in which nucleation failed to occur, after native-type reconstituted fibrils and demineralized bone were alternately exposed many times to solutions containing *either* calcium or phosphate ions in concentrations equal to or several times greater than those of metastable solutions from which nucleation readily occurred, confirmed the supposition that the intermolecular forces between collagen and the inorganic ions must be just strong enough to affect their interaction energies without *firmly* binding them to the collagen structure.

As to the specific organic groups and mineral ions involved in calcification, most investigators have felt that the initial step in calcification involves the presence in the organic matrix of an anionic group which combines with calcium ions. The anionic site has been variously assigned to chondroitin sulfate,[37] to chondroitin sulfate-collagen complexes,[35,36] or to a phosphorylated polysaccharide.[64] In addition to the fact that the experimental evidence indicates that the nucleating centers consist of reactive amino-acid side-chain groups in the collagen fibril, other considerations cast some doubt on the hypothesis of the *primary* role of calcium-binding in crystal induction.

In the first place, the apatites are *phosphate* salts, and the structural characteristics of the apatite lattice are primarily attributable to the phosphate groups, not to the calcium atoms which can be replaced by a number of other cations (Sr, Pb, etc.) without changing the major features of the crystal structure and symmetry. Since the phosphate groups are the "backbone" of the lattice, their role in the formation of the initial crystal structure would appear to be equally as important, if not more important, than that of the calcium ions.

The findings of Solomons *et al.*,[50] referred to earlier, which showed a direct correlation of the available $\epsilon$-amino groups of lysine (and hydroxylysine) and the degree of mineralization in bone and tooth also is very suggestive, and may indicate that the primary collagen-mineral-ion interaction is between the $\epsilon$-amino groups of lysine and the phosphate ions. Phosphorylation of $NH_2$ groups has been proposed also by Polonovski and Cartier[65] as the initial step in calcification. Experiments, now under way in this laboratory, in which specific amino-acid groups are being blocked singly and in combination, should provide the necessary data for the interpretation and elucidation of the actual molecular mechanism of the nucleation process.

## LOCALIZATION, REGULATION, AND INHIBITION OF CRYSTALLIZATION

The phenomena of *localization*, *regulation*, and *inhibition* of the physicochemical mechanism which initiates crystallization both in normally mineralized and unmineralized tissues are different but closely related. On the basis of the hypotheses and experimental data presented, it would appear that all collagenous matrices are *inherently capable* of nucleating apatite crystals from metastable solutions. Since collagen is the major fibrous protein of all of the connective tissues (skin, tendon, ligaments, etc.), the questions arise: Normally, why do all of these tissues *not* calcify? Why *do* they calcify in certain pathological conditions? Also, even under normal circumstances, apatite crystals are deposited in tissues which do *not* contain collagen fibrils (enamel), and in abnormal states in other noncollagenous tissues as well.

It is impossible to answer all these questions definitively at the present time, but the controlling factors and the circumstances under which they may be operative can be developed within the framework of the physicochemical concepts of solution metastability and heterogeneous nucleation sites.

### Localization of the Crystals to Specialized Tissues

It is not suggested that collagen is the *only* organic compound capable of nucleating apatite crystals. This is quite unlikely, as, in all nucleation processes, many materials can act as nucleation catalysts with varying degrees of potency. In the case of tooth enamel, the apatite crystals are closely associated with another structural and "crystalline" protein, a eukeratin,[66–68] and it is likely that a similar mechanism for crystal induction exists here.

### Regulation and Inhibition

Since the native collagen fibrils of most connective tissues do *not* calcify under normal conditions, one of several general situations or a combination of them must exist. Either the collagen in those tissues *not* normally mineralized is different from that of bone or calcified cartilage; or the degree of metastability of the extracellular fluids immediately surrounding and within the collagen fibrils in the various tissues is different; or other local phenomena increase the catalytic potency of the collagen fibrils in the normally mineralized tissues.

Since *reconstituted* native-type collagen fibrils from a wide variety of tissues normally *not* calcified *were* able to initiate crystal formation *in vitro* in our experiments, *failure* to mineralize, if attributed to the collagenous component, would involve subtle *structural* differences between native and reconstituted fibrils, as yet unresolved by the physical and chemical methods employed to date. However, it is possible that, during the extraction and reconstitution of the fibrils, parts are lost of the collagen macromolecules which normally inhibit calcification *in vivo*. There are no data available with which to evaluate the feasibility of this suggestion.

With respect to the second possibility, that differences in the degree of solution metastability account for the specific localization and regulation of calcification, two different points of view are possible. If one assumes that the degree of metastability of the unaltered interstitial fluid is sufficiently great so that collagen *can* induce perceptible rates of nucleation, some mechanism must be operative in the normally unmineralized tissues which *prevents* crystallization. If the degree of metastability of the unaltered interstitial fluids is *not* sufficiently high, either an *increase* in the degree of metastability or an increase in the catalytic potency of the collagen fibril is necessary for crystallization even in the normally mineralized areas, and a *minimal* protective mechanism is required in the normally uncalcified regions.

These variations in the degree of metastability could result either (a) from cellular-controlled compositional changes in the interstitial fluids, or (b) from the presence of other substances in the tissue which actively or passively controlled diffusion and specific ion transport and transfer, or competed with the mineral ions for active sites in the collagen fibril.

Cellular-controlled compositional changes of the interstitial fluids include variations in the calcium and/or phosphate concentrations, calcium to phosphate ratios, *p*H, ionic strength, ion complexes, etc. These could be mediated directly by the cells or by substances secreted by the cells, such as enzymes. For example, the demonstration that phosphorylative glycogenolysis can produce a local increase in phosphate concentration in epiphyseal cartilage[34] illustrates the role of such a device in the *regulation* of calcification in *normally mineralized tissues*, either by merely making more phosphate available, or by participating in another enzymatic cycle[69] which actively transfers phosphate ions to specific groups in the collagen fibril.

As to another method of *regulating* the degree of metastability (particularly in *normally unmineralized tissues*) by controlling diffusion, ion transport etc., the possible role of the ground substance and certain of its components have also been investigated.

Although much research has been done on the ground substance of connective tissues, its exact composition, state of aggregation, and anatomical distribution are not clear. Prominent among its components are the various acid mucopolysaccharides. These are thought to exist in the tissues as complexes with noncollagenous proteins.[70,71] In the past, most investigators have postulated that, of the ground-substance components, the chondroitin sulfates specifically, either alone or in combination with collagen, have played a role in the initiation of calcification.[35–37]

That it is not solely the *amount* of the acid mucopolysaccharides which determines whether calcification is initiated is obvious from the fact that hyaline cartilage—one of the richest sources of this material—is normally uncalcified, whereas bone, which does calcify, contains extremely small amounts of these substances.

Differences in the *kinds* of mucopolysaccharides present is not a plausible explanation, since the various acid mucopolysaccharides found in adult bone or in the growing ends of bone (bone, epiphyseal cartilage, etc.) are present in other tissues.[72]

It is instructive to review the findings as to the nature of the ground substance in cartilage, since a transition occurs from the normally uncalcified hyaline portion to the normally calcified epiphyseal region over a relatively short anatomical distance.

When tissue sections of hyaline cartilage, epiphyseal cartilage, and bone are stained with metachromatic dyes or by the Hotchkiss procedure, definite differences are noted.[43] The staining characteristics of hyaline cartilage gradually change as the epiphyseal cartilage is approached. This change is also evident in bone which is being actively deposited or resorbed. Since metachromatic dyes react with negatively charged, high molecular-weight compounds, these staining characteristics have been linked primarily to the sulfated mucopolysaccharides in such connective tissues. The change in metachromasia and in the staining properties of these tissues by the Hotchkiss procedure has been interpreted as evidence for the depolymerization of the anionic mucopolysaccharides, as the zone of calcification is approached and reached.[41–43]

Although there is obviously a number of other possible reasons for this change in staining properties, the important point is that *there is some alteration*, either in the amount, state of aggregation, or reactivity of charged groups, etc., in the ground substance, and that this change accompanies calcification. Analytical data[73] which demonstrate a very marked loss of organic sulfate during cartilage calcification and during the formation of bone matrix indicate that it is the *depolymerization* and subsequent *removal* of these compounds which are related to calcification.

In attempting to assign a specific role to the ground substance in calcification, two properties stand out as important. One is that its state of aggregation is probably that of a gel in which the mucopolysaccharides exist complexed with a protein moiety as a mucoprotein.[70,71] Such a physical state of aggregation might well limit the diffusion of interstitial fluids and ions to the collagen fibrils.

The anionic groups of this mucoprotein are free and reactive,[74] which accounts for the second property of importance—the large cation-binding capacity of the ground substance. Under normal circumstances, therefore, the mucoproteins may help to inhibit calcification by limiting the available mineral ions, both by diffusion and by selective cation (calcium) binding. It is also possible, but less likely (since collagen and the chondroitin sulfates, for example,[75] do not react with collagen above *p*H 4.0), that the reactive groups of the mucopolysaccharide portion of the mucoprotein compete with the mineral ions for positions in the collagen fibrils, but it is possible that the noncollagenous protein portion of the mucoprotein may interfere with the process by just such a mechanism.

Depolymerization and removal of these compounds by eliminating the diffusion barrier, decreasing the cation-binding capacity of the ground substance, and possibly "freeing" some of the reactive groups of the collagen from their interaction with the protein moiety of the mucoprotein, would allow the mineral ions to react with the collagen. The cation-binding capacity of the remaining depolymerized mucoprotein components would also most likely be decreased, since it has been shown that metallic cations are more strongly bound to high molecular-weight acids and bases than to their monomeric compounds.[76]

*In vivo*, the rapid depolymerization of the mucopolysaccharide-protein complexes and the subsequent decrease in their cation-binding properties might well lead to a local release of free cations including calcium, so that the resultant *increase* in calcium-ion concentration might then actually *aid* the initiation of crystallization by increasing the degree of metastability and making additional calcium ions available.

This hypothesis as to the role of the ground substance is supported by a series of *in vitro* experiments, conducted in our laboratory, in which native collagen-rich tissues, such as rat-tail tendon, calf skin, guinea-pig skin, etc., *failed* to mineralize under physicochemical conditions identical with those used in testing *reconstituted* native-type fibrils *from these same tissues*. In addition, when these tissues were treated so as to *extract* many of the components of the ground substance, including the chondroitin sulfates, either directly or by enzymatic depolymerization, these same native tissues readily mineralized.

Further support for this hypothesis comes from the experimental observation that hyaline cartilage, rich in the acid mucopolysaccharide-protein complexes *fails* to mineralize *in vitro* but selectively *removes* $Ca^{++}$ from the calcium-phosphate mineralizing solutions. This property has been shown to be related to the sulfate groups of the chondroitin sulfates.[77]

This "shielding" of reactive sites in the collagen fibrils by ground substance components in intact tissues has also been noted in the tanning and dying of skins, a process which depends on the interaction of certain dyes, complex metallic ions, etc., with specific groups in the collagen. It was found that such interactions were markedly facilitated when the tissues were first tested by procedures designed to extract the ground substance components, and even more so by the prior depolymerization of the chondroitin sulfates by testicular hyaluronidase.[78] The latter method, in addition to being more specific, was also carried out under milder conditions so that the collagen fibrils were presumably less distorted and less swollen.

Figure 21 is a schematic representation of many of the experiments described. It illustrates both the specificity of the collagen macromolecular-aggregation state in *initiating* calcification, and the *possible* role of the ground substance as *a factor* in inhibiting it.

Since nucleation rate is so markedly dependent on the degree of metastability (p. 390), the organism has the dual problem of keeping the metastability of the extracellular fluids at a sufficiently high level for collagen to induce a perceptible rate of nucleation in normally mineralized tissues, and at the same time not so high that aberrant calcification cannot be safely controlled. It seems likely that a compromise is obtained by maintaining the degree of metastability in unaltered extracellular fluid *just* below the point where collagen is very effective. The rate of nucleation in bone and cartilage could then be increased and controlled by very small increases in the local concentration and/or active

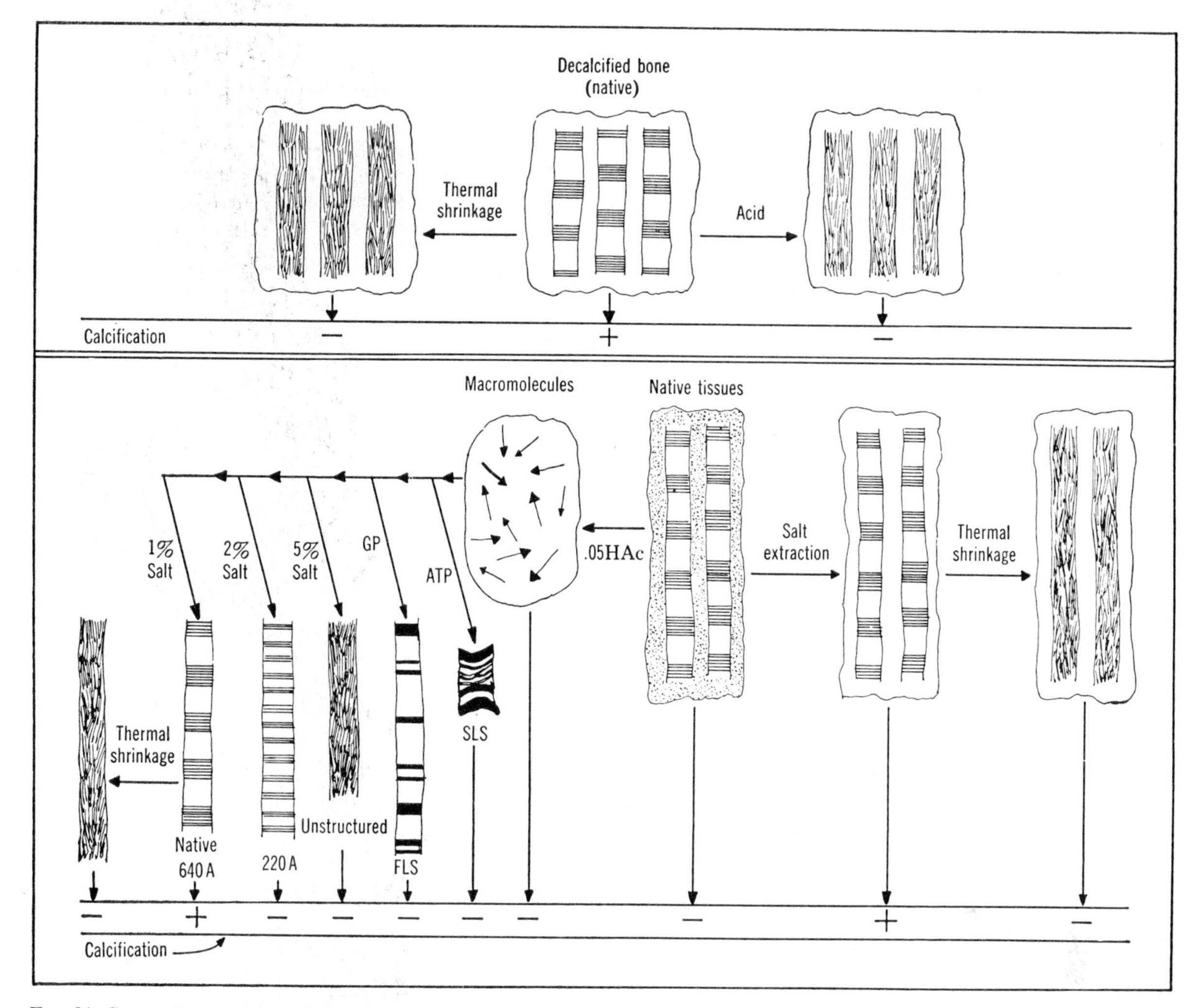


FIG. 21. Composite and diagrammatic illustration of the experiment demonstrating the specificity of the macromolecular aggregation state of native-type collagen fibrils in calcification, and the possible role of the ground substance (heavily stippled regions surrounding fibrils in native tissues) in inhibiting and controlling it. Enzymatic treatment of the native tissue is not shown, since it is not yet certain whether depolymerization is itself effective, or whether depolymerization and subsequent removal from the tissue is necessary.

transfer of mineral ions by enzyme mechanisms, such as the one mentioned earlier for phosphate ion, and the prevention of pathological calcification assured by decreasing the degree of metastability by utilizing the properties of the ground substance, as enumerated in the foregoing.

There is every reason to expect that, like other functions of such vital importance to the organism, mineralization is under the control of many factors, delicately balanced to provide *biological* and *cellular regulation* of the physiochemical *mechanism* which initiates crystallization.

## COLLAGEN-INORGANIC CRYSTAL RELATIONSHIPS

### Location of the Inorganic Crystals with Respect to the Collagen Fibrils

The location of the crystals within the collagen fibrils supports the theory of heterogeneous nucleation as presented, since the probability of arranging a number of side-chain groups in the proper steric configuration is statistically higher *within* the fibrils, and where the density, intermolecular forces, and interaction energies are highest.

The longitudinal position of the crystals tends in many instances to be localized to certain regions accentuating the 640- to 700-A axial repeat of the collagen fibrils. It is difficult to say exactly where these regions are in relation to the intraperiod fine structure of the fibrils, although density considerations and shadowed preparations indicate that they are most likely in those regions where the density of the bands is greatest. This is especially true in the less heavily calcified fibrils. In many other fibrils, however, especially where calcification is quite high, no such localization is apparent. It would appear, therefore, that—although *initially* there are preferred regions in the fibril for nucleation and for crystal growth—as the fibril continues to calcify, crystals are deposited throughout the fibril.

X-ray diffraction evidence that the average size of the crystals is approximately 220 A led some workers to

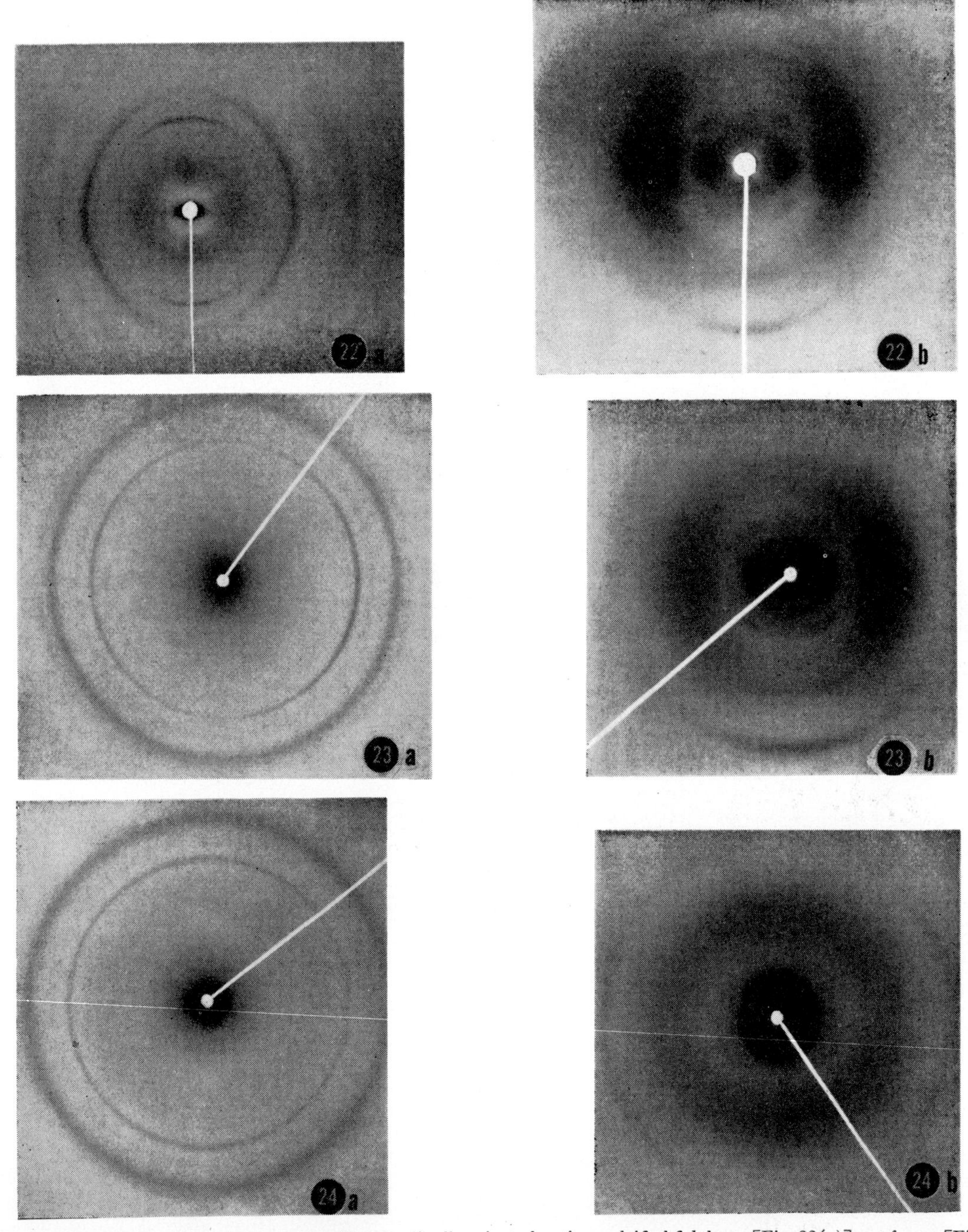


FIGS. 22, 23 and 24. X-ray diffraction patterns of longitudinally oriented native, calcified fish bone [Fig. 22(a)], rat femur [Fig. 23(a)], and embryonic metatarsal rudiment [Fig. 24(a)], showing apatite orientation; same specimens [Figs. 22(b)–24(b)] decalcified, showing collagen orientation.

speculate that three such crystals were aligned longitudinally per axial period of the collagen fibril.[79] Such an arrangement would result in the structure having an axial repeat of 220 A. This has not been borne out by direct visualization of the fibrils and the inorganic apatite crystals by electron microscopy.

## Inorganic Crystal-Collagen Fibril Coorientation

Earlier workers have demonstrated by electron microscopy that the long dimension of the inorganic crystals, while closely parallel to the fibril axis in bone, is for the most part randomly oriented in calcifying cartilage.[80] Since the process of orientation was felt to be a true-oriented overgrowth directed by certain crystallographic planes in the collagen fibrils (epitaxy)[5,16] and directly related to the process of mineral phase induction, it was suggested that *possibly* different mechanisms were involved in the initiation of calcification in these two closely related tissues.[16]

This was based also on electron-microscopic evidence

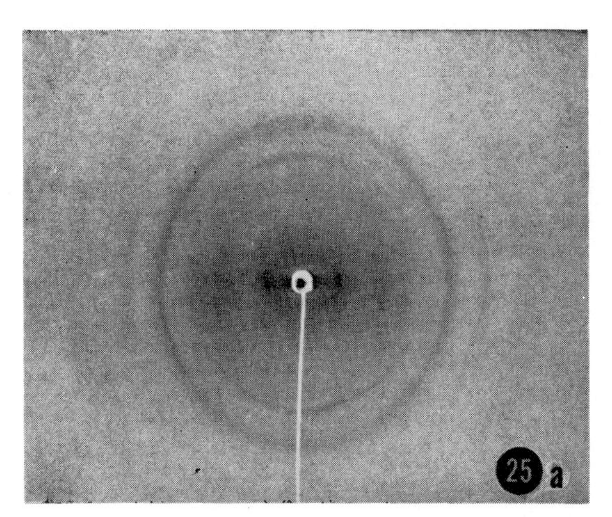


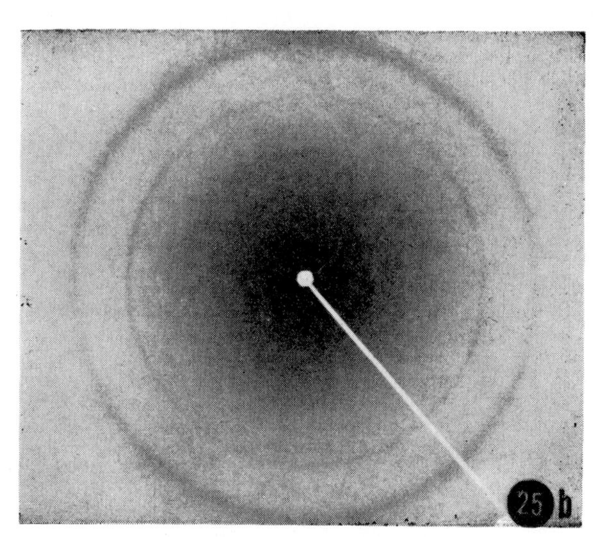


FIG. 25 X-ray diffraction patterns of *in vitro* (a) recalcified fish bone and (b) rat-femur bone. Note that, although there is *some* orientation of the apatite crystals as evidenced by the arcing of the 00*L* reflections, it is *not* as prominent as in the original native, calcified bone [compare Fig. 22(a) with Fig. 25(a), and Fig. 23(a) with Fig. 25(b)].

that, in the zone of provisional calcification in cartilage, the size, appearance, and organization of the collagen fibrils were markedly different from those in bone.[80] More specifically, the collagen fibrils in this calcifying zone of cartilage are quite thin (50 to 250 A) and do not show any visible interperiod fine structure. The fibrils are, in addition, widely separated and randomly oriented. Correlations of this type, however, depend both on an intimate knowledge of the *mechanism* of crystal orientation and its relation to the *induction* of crystallization, and on information concerning the macromolecular organization of the *apparently* "structureless" collagen fibrils. Furthermore, several recent studies also have demonstrated that the embryonic *bone* of some species (and even early postfetal bone in others) does not show any preferred orientation of the inorganic crystals[81–83] (as deduced by x-ray diffraction, based primarily on the arcing of the 00*L* reflections of apatite), and that crystal orientation varies not only from bone to bone, but also in different areas of the same bone.[82,83]

As in other problems in calcification, the term "orientation" must be defined carefully. Orientation of the inorganic crystals has been related to the long axis of the entire bone and to the various hierarchies of collagen structure. These definitions obviously are not identical, and the orientation of the crystals with respect to the *collagen fibrils with which they are associated* is the relationship of interest.

The evaluation and interpretation of such co-orientation by x-ray diffraction evidence alone is limited and may be quite misleading. The difficulties stem both from the basic nature of the method and from the failure to interpret the results in the light of the gross and microscopic arrangements of the collagen fibers in the tissue.

An x-ray beam, regardless of how small a collimator is used, integrates the orientation of the collagen and crystals over an enormous volume as compared with the order of magnitude discussed here. Since the x-ray diffraction data are summations from many crystals and fibers, it is not possible to discern *individual inorganic crystal-collagen fibril* relationships, particularly if the collagen fibrils are themselves randomly dispersed or if account is not taken of the higher-ordered organization of the structures in the tissue.

Electron microscopy, on the other hand, gives a direct result if the crystals can be visualized with respect to individual fibrils. In cases where small local areas are in question, selected-area electron diffraction—since it integrates over a much smaller volume than x-ray diffraction—also can be of great help.

In order to evaluate whether the marked variation in crystal orientation as deduced by x-ray diffraction is indicative of varying degrees of collagen fibril-inorganic crystal orientation, or whether it is the result of the geometric factors in tissue organization and of x-ray technique, a number of specimens showing the complete spectra of orientation was examined by electron microscopy, electron diffraction, and x-ray diffraction. Figures 22–24 are x-ray diffraction patterns of fish bone, rat bone, and 16-day-old embryonic chick bone, calcified and decalcified. In the case of fish bone, there is marked orientation of both the apatite crystals and the collagen fibrils. The rat bone shows less orientation of both the apatite crystals and the collagen fibrils, and in the embryonic bone, no orientation of the apatite is present and there is very little evidence of collagen orientation.

Electron micrographs and selected-area electron diffraction of the two extreme cases (fish bone and embryonic chick bone) clarified the issue. Figure 6 demonstrates that the fish bone consists of extremely well-oriented collagen fibrils with the inorganic crystals lying with their long axes (crystallographic *c*-axis) parallel to the collagen-fibril axis. Figure 26 is an electron micrograph of embryonic bone which reveals the general randomness of the collagen fibrils. Within very

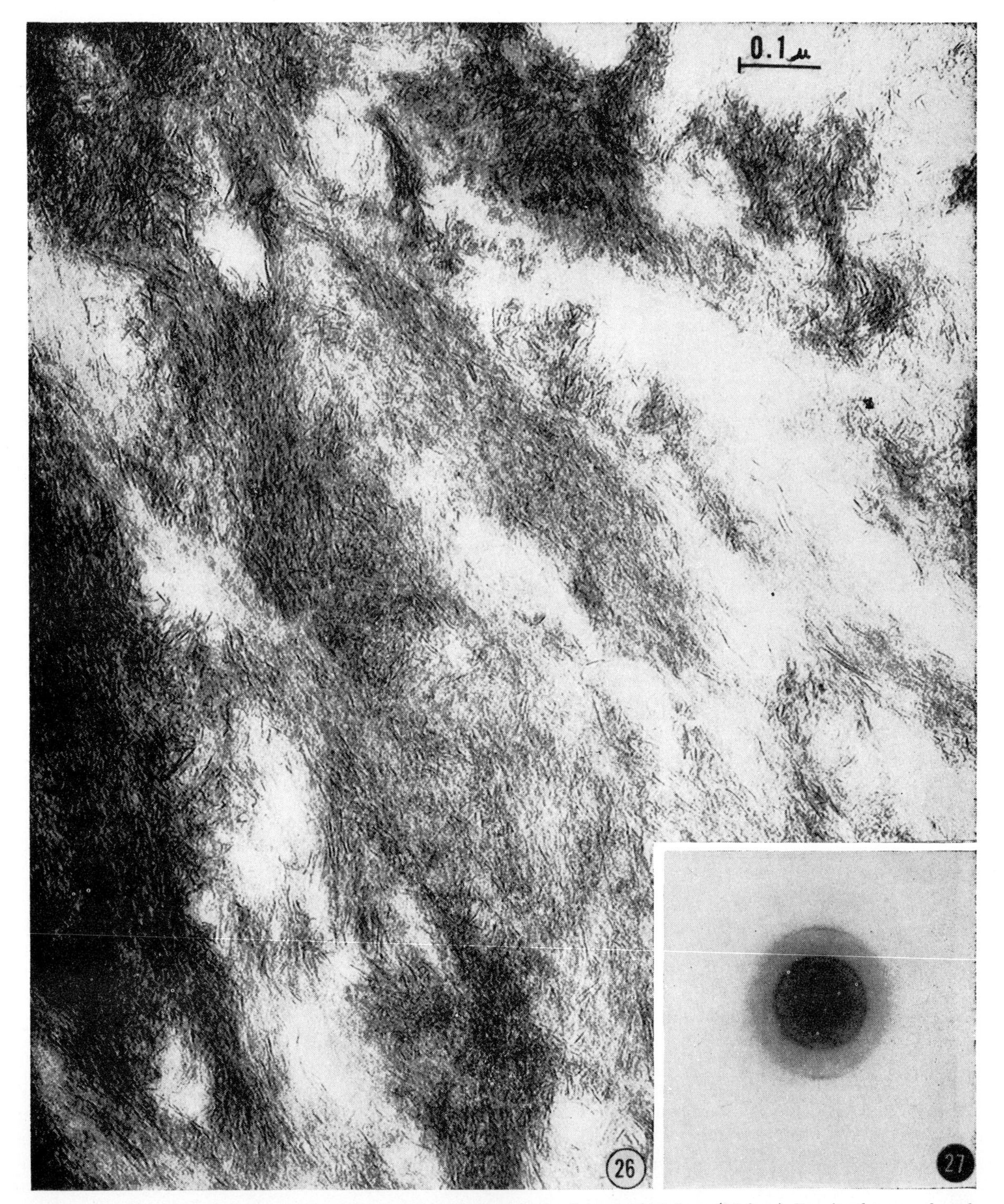


FIG. 26. Electron micrograph of a section of the metatarsal rudiment of embryonic chick bone (16 days). Despite the general randomness of the collagen fibrils, the crystals are oriented in local regions corresponding to the individual collagen-fibril directions. The size and shape of the crystals are similar to those in adult bone. ×40 000.

FIG. 27. Selected-area electron diffraction of specimen shown in Fig. 26. The preferred orientation of the crystals is evident by the arcing of the 002 and 004 reflections.

local areas, however, *the same general parallel alignment of crystals and the collagen fibrils is present*, confirmed by selected-area electron diffraction (Fig. 27). The orientation of individual crystals with *the fibrils with which they are associated* is also apparent in certain adult bones where the collagen fibrils are clearly sepa-

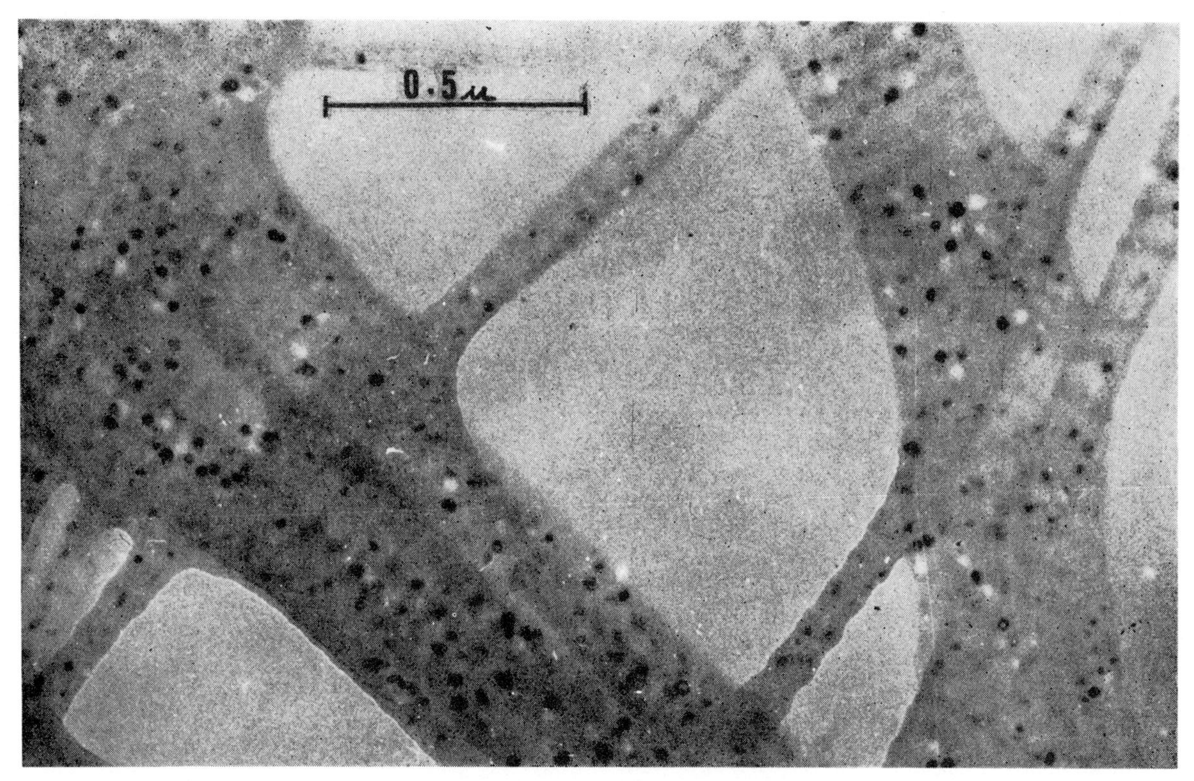


FIG. 28. Unstained, unshadowed preparations of *in vitro* calcified collagen at an early stage. Although the crystals are situated regularly spaced in certain regions of the fibrils, the haphazard arrangement of the crystal reflections (white spots) indicates lack of preferred *orientation* of individual crystals with the fibrils with which they are associated. ×70 000.

rated, as shown in Figs. 8 and 9. It is clearly evident, therefore, that x-ray diffraction evidence *alone* cannot be used as a criterion in assessing or interpreting crystal orientation, it may, as in this case, be quite misleading. Thus, the complete lack of preferred crystal orientation as evidenced by x-ray diffraction data in the embryonic bone is simply the result of the randomness of the collagen fibrils themselves, and, as far as the crystals within any one fibril with which they are associated are concerned, they are well oriented similar to adult bone.

### Mechanism of Crystal Orientation

As further experimental results unfolded, it became apparent that the reasons for crystal orientation were twofold: *the size and habit of the inorganic crystals, and the location of the crystals within the fibrils.*

Electron micrographs taken during time-sequence studies of *in vitro* calcified reconstituted collagen have shown a striking similarity in the appearance of the inorganic crystals to those of embryonic bone during the initial stages of calcification (Figs. 13, 14, 18, and 20). In both cases, the crystals appeared dot-like with none of the axes elongated in any direction, and in both cases selected-area electron diffraction (Fig. 19) showed *no* preferred crystal orientation from areas where the collagen fibrils were relatively well oriented. On the other hand, in later stages of embryonic bone, once crystal growth (and/or recrystallization) has occurred, leading to *small* asymmetric crystals *within* the fibrils, the crystals do become oriented with their long axes approximately parallel to the collagen fibrils *with which they are associated* (Figs. 26 and 27).

Although it was impossible to obtain single-crystal patterns from the *in vitro* calcified specimens, individual diffraction spots, rather than complete rings, were obtained in many instances from small local areas with relatively few crystals and in regions where there were several well-oriented fibrils.

In order to obtain more information about the relationship of individual crystals to the collagen fibrils, recourse was had to another experimental device. The objective aperture of the electron microscope was removed and the specimen photographed slightly out of focus. The resultant crystal reflections, seen as white dots in Fig. 28, are from a definite set of crystal planes, confirmed by measuring the distance from the crystals to the reflections. One easily can see the haphazard patterns of these reflections, which clearly indicate that the individual crystals *are not oriented with their crystal axes parallel to the fibril axes, but are randomly oriented.* These data indicate that the *initial* crystals *in vivo* and *in vitro* have not grown in an epitatic fashion and that their *subsequent* orientation must be attributable to other factors.

A detailed electron-microscopic examination of calcification in cartilage has been made by Robinson *et al.*[80]

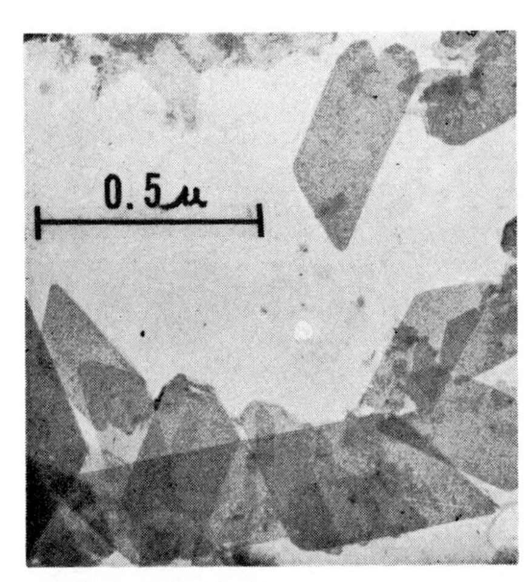


FIG. 29. Electron micrograph of *in vitro* apatite crystals precipitated under conditions similar to those of the *in vitro* collagen-nucleation studies. Note large hexagonal-type crystals. ×50 000.

They observed that, in the initial stages of cartilage calcification, the crystals *are* directly related to the collagen fibrils and are propagated in the intervening interfibrillar areas. The crystals are not preferentially oriented, except for a few which lie directly on, or possibly in, the fibrils.

The other point concerning calcifying cartilage—that is, the lack of the usual interperiod fine structure of the collagen fibrils—appears to be attributable to the very small size of the fibrils and possibly to the large amount of ground substance surrounding them. The question of collagen-fibril size is also important from the standpoint of the number of collagen macromolecules that must be polymerized laterally in order to constitute a nucleus.

With these factors in mind, reconstituted collagen fibrils were prepared, approximately 50 to 100 A in diameter and not showing cross-striations when stained with phosphotungstic acid (PTA). Well-oriented fibers of this material, however, demonstrated the characteristic 640-A low-angle x-ray diffraction pattern, indicating the *macromolecular organization* to be similar to the larger fibrils. Preparations of these fibrils were able to initiate crystallization of apatite *in vitro* from metastable solutions in experiments similar to those described.

The data may be summarized as follows. In calcifying cartilage, embryonic bone, and *in vitro* calcified collagens, crystallization is initiated in *direct relation to the collagen fibrils*. During the *first stages* of mineralization in embryonic bone, reconstituted collagens, and calcifying cartilage, the inorganic crystals are *not* oriented with respect to the collagen-*fibril axis*. In both embryonic bone and reconstituted collagens, the crystals are strikingly similar: dot-like with *none of the crystallographic axes elongated.*

Further crystal growth results in assymetric crystals with one of their axes elongated. In bone, this occurs primarily *within the fibrils* and is *associated with orientation of the crystals*. That the orientation of the crystals is not a function of the over-all compactness or organization of the collagen fibrils in the tissue, but is dependent upon the *location* of the crystals within the fibrils, is shown by the observations on embryonic bone and on certain adult bone. In the former (Fig. 26), despite the general randomness of the fibrils, and in the latter (Figs. 8 and 9), although the collagen fibrils are clearly separated and somewhat randomly directed, the inorganic crystals are well oriented with their long axes parallel *to the collagen fibril axis with which they are associated.* In calcifying cartilage, however, because of the very small diameter of the fibrils, crystal growth is propagated primarily *between* the widely spaced fibrils, and these crystals, therefore, *incapable* of being oriented by the fibrils, are randomly oriented.

The conclusions are reached that: (1) the basis for the orientations of the inorganic crystals is the asymmetric growth of small crystals *within the collagen fibrils, between tightly packed, longitudinally oriented chains of macromolecules* (*protofibrils*), which necessarily results in the crystals being similarly oriented; (2) the process of crystal orientation, therefore, is unrelated to the *mechanism of induction* of crystallization; (3) since crystal orientation is a function of the *size* and *shape* of the *crystals* and their position *within the fibrils* and not part of the crystal induction mechanism, the *lack* of preferred orientation in cartilage in no way suggests that the *mechanism* of crystal formation (heterogeneous nucleation by the collagen fibrils) is different in cartilage from that in bone.

In fact, the theory of heterogeneous nucleation of apatite crystals by collagen fibrils may be more convincing in cartilage than in bone, since the collagen fibrils are so closely packed (particularly in compact bone) that there is very little space for crystal formation *except* within the fibrils, whereas in cartilage, the fibrils are relatively far apart and randomly oriented with large amounts of intervening interfibrillar ground substance available for crystallization. Despite this, calcification does *not* begin randomly in the tissue, but is initiated in *direct relation* to the very thin fibrils and is then propagated throughout the intervening ground substance.

## CRYSTAL HABIT AND SIZE

A good deal of disagreement exists as to the exact habit and size of the apatite crystals in bone. Electron micrographs obtained in our laboratory, in agreement with those of Speckman and Norris,[84] and Fernández-Morán and Engström,[85] show what *appear* to be rod-shaped particles, the majority varying in thickness from 15 to 30 A and in length from 200 to 400 A. It is possible, however, that these represent extremely thin, lathe-like crystals, several of which are stacked together. Perfect cross sections of well-oriented fibrils are obtained with great difficulty, but a number of fibrils in many fields appear to show a dot-like appearance in cross section consistent with a rod-like habit of the crystals (Fig. 12).

Because the size of the crystals approaches the limit of resolution of the electron microscope, particularly in a tissue technically difficult for sectioning such as bone, and because of other factors such as overlap, etc., it is not possible to be absolutely certain of the exact size and shape of the inorganic crystals. In any event, the issue is somewhat semantic: when does a rod become a plate? There may be variations in the habit of the crystals from true rods to a *somewhat* lathe-like habit where one of the faces is *slightly* larger than the other. But from many cross sections, longitudinal sections and oblique sections, one can say with certainty that the crystals are *not* the large hexagonal plates (500×250×100 A) reported by some workers on the basis of the examination of blended or autoclaved bone.[27,28]

On the other hand, the apatite crystals formed by simple precipitation in the test tube,[86] or crystals nucleated and *grown in the presence of collagen*, are usually hexagonal plates quite unlike those occurring in bone.[87] Occasionally, lathe-like crystals can be formed in the test tube, but again these are several orders of magnitude larger than the bone crystals.[86] Thus, although the early crystals nucleated by collagen *in vitro* are similar to those nucleated *in vivo*, further crystal growth appears to be quite different.

This difference in crystal habit and size between bone crystals and precipitated calcium phosphate crystals might be explained on the basis of the mechanical factors resulting from crystal growth within the fibrils, or by an active process controlled by the structure of collagen such as has been suggested by others.[79] It is also possible that it is entirely unrelated to the structure of collagen.

The position of the bone crystals within the collagen fibrils suggests that the collagen fibrils play a primary role in determining crystal size and shape. The finding, however, that collagen fibrils were *not* able to alter the size or the habit of the crystals under the physicochemical conditions of the *in vitro* experiments casts some doubt on the validity of this hypothesis. This is also confirmed by an examination of electron micrographs of calcifying cartilage which reveals that the crystals lying in the interfibrillar space and *anatomically unrelated to the collagen fibrils* are the same size and shape as those within the collagen fibrils of bone.

Since it is a well-established fact that the size and shape of crystals can be altered by changes in the physicochemical environment—including the addition of a number of substances bound on certain crystal faces and limiting their further growth—studies were carried out on the precipitation and growth of apatite crystals from solution under a wide variety of experimental conditions.

Figure 29 is an electron micrograph of a calcium phosphate precipitate in which the physicochemical conditions were similar to the *in vitro* collagen-nucleation experiments. These large, hexagonal plates are somewhat larger than those usually seen under these circumstances, but serve as a control to illustrate the progressive change in size and habit that can be produced *in vitro* in the absence of collagen. X-ray diffraction of this preparation revealed a typical apatite pattern without any evidence of the presence of octocalcium phosphate or brushite which also crystallize in large plate-like form.

Figure 30 is another preparation in which an attempt was made to alter crystal size and shape. One can easily see the marked difference in the size of the crystals, although they are still in the form of hexagonal plates. Note that where the crystals overlap there is a suggestion of a dense "rod," particularly if viewed without reference to the neighboring crystals.

Figures 31–33 are electron micrographs of apatite preparations where the crystal size was further reduced, and it appears that the habit has also been altered so that a number of rod-like crystals have formed. It is impossible, however, to tell whether these are truly rods or represent extremely small, thin plates stacked together or supporting each other and viewed along their edges. The "rods" are now approximately the size of the bone crystals.

Figure 34 is an electron micrograph of another preparation in which the crystals are even thinner (many ~10 A or less) and very similar in size and appearance to the bone crystals, and in which there is no apparent evidence of hexagonal plates. If one examines the elec-

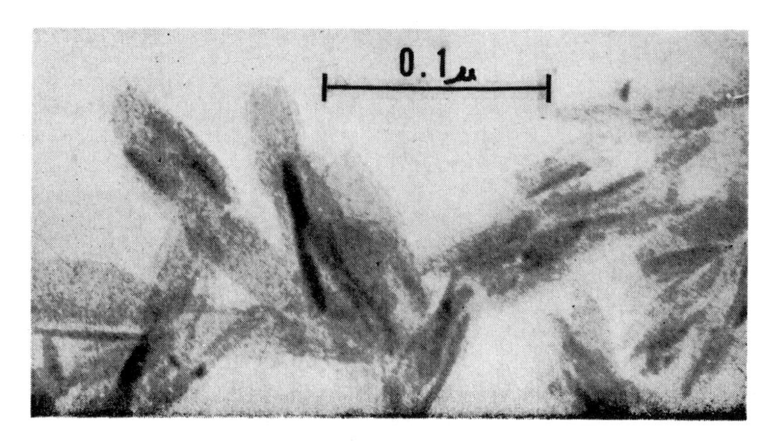


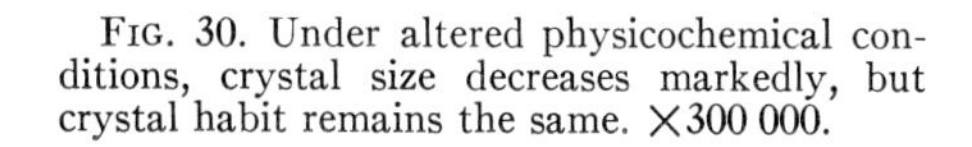

FIG. 30. Under altered physicochemical conditions, crystal size decreases markedly, but crystal habit remains the same. ×300 000.

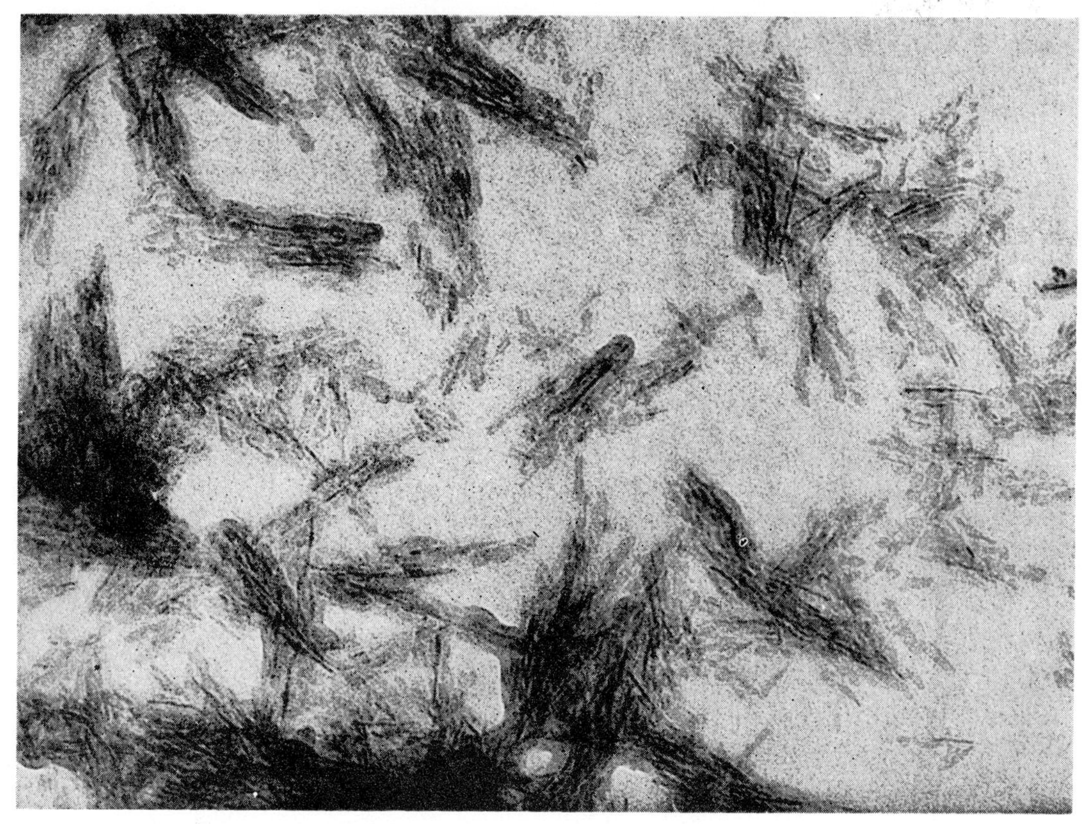

FIG. 31. Further alteration in the physicochemical environment produces not only a further decrease in crystal size but also, for some, a change in the crystal habit to a rod-like appearance. ×180 000.

tron micrographs carefully, however, it is obvious that there is a minimal but definite background electron density which may indicate that these crystals are extremely thin plates, some of which are stacked together or are upright and supporting each other, and, when viewed along their edges, give the appearance of rods. The fact that many of the rods appear to be bent in several directions may indicate that this is the correct interpretation. This same appearance has been seen in very lightly calcified collagen fibrils of bone.

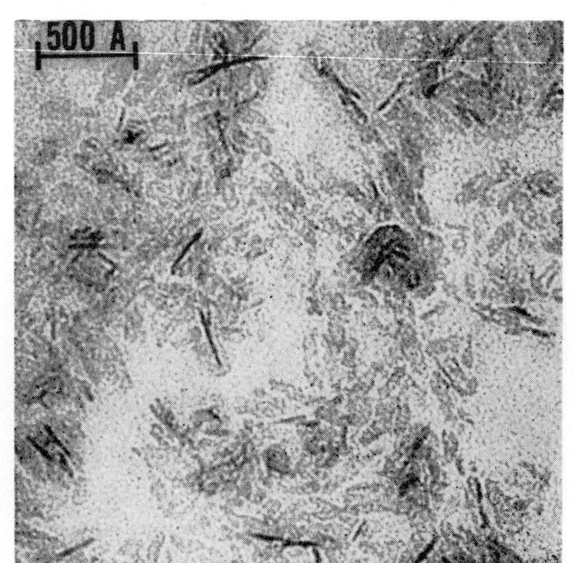


FIG. 32. Higher magnification of an area in a sample similar to Fig. 31, where the "plates" predominate. ×260 000.

Although it is still impossible to state dogmatically the exact size and shape not only of the bone crystals but also of many *in vitro* precipitated crystals, these experiments show that *it is possible to grow from solution under controlled physicochemical conditions apatite crystals whose size, shape, and appearance in electron micrographs and whose x-ray diffraction characteristics are similar to those of bone crystals.*

The facts (1) that *in vitro* collagen *per se* is not able to modify the crystal habit or size, (2) that *in vivo* (calcified cartilage) crystals similar in size and shape to the bone crystals lie in the interfibrillar spaces *between* the collagen fibrils and anatomically unrelated to them, and (3) that crystals *can* be grown *in vitro* similar to the size and habit of bone crystals in the absence of collagen, indicate that the *size* and *shape* of the *apatite crystals are primarily controlled by other physicochemical factors in vivo* and *not by the collagen fibrils.*

Obviously, the conditions under which crystal growth occurs within longitudinally oriented and closely packed fibrils may exert some secondary influence on the crystal size and habit, both mechanically and by selectively facilitating and inhibiting the diffusion of ions; but when the other physicochemical conditions are not met, crystal growth is *not altered* by the collagen fibrils.

Under such conditions, the crystals ordinarily tend to grow to a relatively large size as compared with normal bone crystals, and since such crystal growth cannot take place within the fibrils, further seeding, recrystallization, etc., occurs *outside* the fibrils. The results of experiments in which crystals initially nucleated within the collagen fibrils *in vitro* proceeded to recrystallize and grow outside of the fibrils, substantiates this conclusion, and clarifies the observations related to crystal orientation in *in vitro* calcified collagens including bone.

Thus, *in vitro* calcified, reconstituted collagens showed *no* preferred inorganic crystal orientation by x-ray diffraction, even though the collagen was moderately oriented. On the other hand, specimens of bone recalcified *in vitro* did show some preferred inorganic crystal orientation, but not as much as similar samples of native bone, as shown in Figs. 22(a), 25(a) and 23(a),25(b). In the case of the recalcified bone, the degree of orientation of the crystals was related to the degree of orientation of the collagen matrix [Figs. 22(b) and 23(b)].

Electron micrographs of heavily calcified preparations of *in vitro* calcified collagens demonstrated that the majority of the crystals were in the interfibrillar space

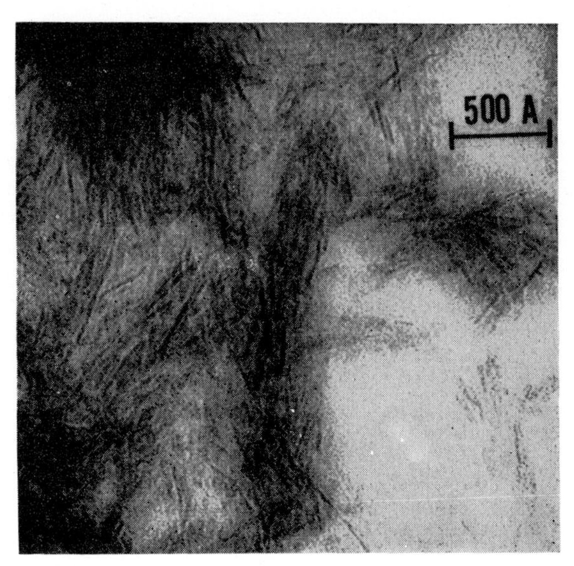


FIG. 33. Higher magnification of an area in a sample similar to Fig. 31, where the rod-like crystals predominate. ×260 000.

and *not* within the fibrils. In addition, the crystal habit and size were unlike that of native bone and resembled the large *in vitro* precipitated hexagonal plate-like crystals. Because of size and shape, crystal growth was more favorable in the interfibrillar regions rather than within

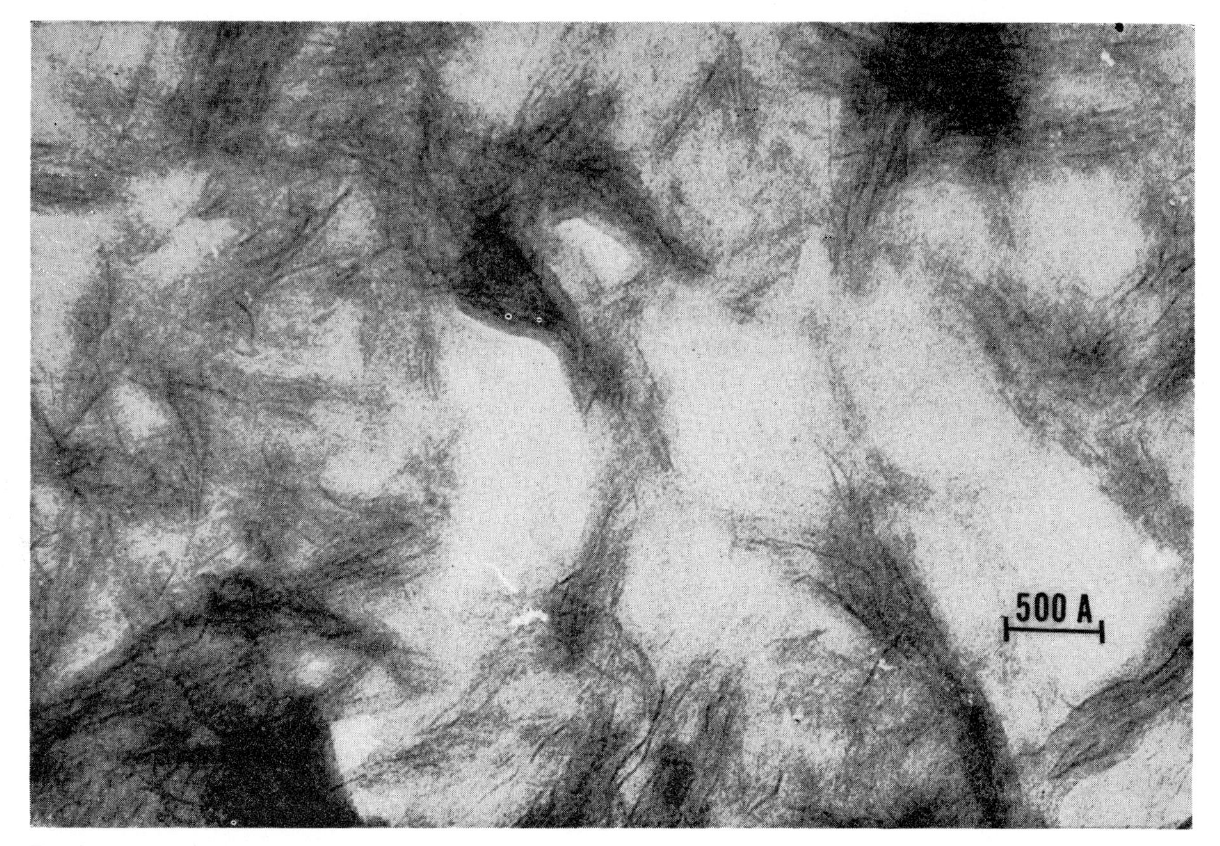


FIG. 34. *In vitro* precipitated apatite crystals. Note the rod-like appearance without evidence of plate formation, except for the minimal background electron density. Crystals vary from 8 to 20 A in width, and many appear to be "bent" similar to areas of lightly calcified bone. ×260 000.

the fibril. These crystals were, therefore, in no way capable of being oriented by *the individual fibrils.*

In samples of reconstituted collagens, the interfibrillar spaces were large, and the over-all organization of the fibrils was also not able to influence the orientation of the crystals (similar to cartilage). In the case of demineralized bone, however, the collagen fibrils are more closely packed and well oriented. Under these circumstances, the elongated plate-like crystals mechanically orient with their flat, thin faces between adjacent fibrils and their long dimensions approximately parallel to the collagen fiber axis. This arrangement gives a co-oriented x-ray diffraction pattern of apatite and collagen but not as good a one as that in native bone, where the majority of the crystals are within the fibrils.

Experiments are now underway in which the physicochemical conditions during crystal *growth* as well as crystal *induction* are being carefully controlled in an attempt to grow the crystals primarily within the collagen fibrils similar to that of native bone.

### Crystal Nonstoichiometry

As mentioned in the beginning of this paper, one of the perplexing problems in the study of the nature of the bone crystals or of *in vitro* precipitated apatites has been their nonstoichiometry. Recall that, in bone and in some *in vitro* precipitated crystals, the average width of the rods or of the stacked plates (or both) varied from 10 to 30 A. This dimension corresponds to the *a*- or *b*-axis of the unit cell which is approximately 9.43 A. Thus, the crystals are composed of 1 to 3 unit cells in this dimension. Although the unit cell of hydroxyapatite may be represented as in Fig. 1, it must be remembered that it depicts only *conceptually* the *spatial relations* of the constituent atoms and molecules, and that many of the individual atoms and molecules are shared by adjoining unit cells in an actual crystal.

Therefore, the stoichiometry of the solid phase may be represented by the structure of the unit cell *only* when there is a very large number of unit cells comprising the crystal, so that the unit cells or the portions of them making up the surface are not statistically significant. When the entire crystal is composed (in any dimension or dimensions) of only a few unit cells, however, and if the planes on which crystal growth ceases are relatively uniform along the length of the crystal, *it is not possible to obtain the theoretical stoichiometry.* The *actual* stoichiometry in these cases will be determined by the atomic planes in which crystal growth ceases. If the crystal surface planes on which crystal growth ceases vary a good deal, it is possible that a fortuitous combination might occur which would result in a theoretical stoichiometry, but this is highly unlikely. In *in vitro* precipitated crystals, however, there is ordinarily a very wide range of crystal size and shape (and probably hydration), and the statistical summation of the calcium and phosphorus content of such an heterogeneous sample, whether stoichiometric or not, would not be very meaningful. Although there is obviously a number of other previously mentioned factors such as internal lattice substitution, surface adsorption, vacant lattice sites, etc., which may influence the stoichiometry, the point is that, *if none of these existed, it still would not be possible to obtain stoichiometric hydroxyapatite* with crystals the size of those obtained in bone or prepared *in vitro.*

The larger "defect" apatite crystals reported by Posner and Perloff[11] of course poses another problem, but it is possible that such large crystals represent a somewhat different structure than hydroxyapatite and are similar to OCP.

The small size of the crystals also may be an important factor in determining the amount of water associated with apatite crystals precipitated from aqueous solutions as discussed on p. 361. This "excess" water is many times greater than that adsorbed by initially dry crystals from a vapor phase of water and consists of many more molecular layers of water than that usually considered possible for a surface to bind because of electrical-field effects.

In addition to the fact that some of the water may enter the lattice structure itself, since the minute apatite crystals in aqueous suspension tend to stick together and form stacks and bundles, the very large surface tension and capillarity effects between such crystal surfaces could well "trap" a large amount of water (in addition to a bound monolayer) and still resist separation from the crystals by mechanical centrifugation.

### Closing Comments

As regards other mineralized biological tissues, one can see from Table I that the various crystals are anatomically related to specific organic matrices. Note the wide variety both of inorganic crystals and of organic matrices in which mineralization occurs. In fact, in some marine mollusks, calcium carbonate exists in two of its three possible crystallographic forms (calcite and aragonite) *in the same shell,* in well-demarcated regions which *border* on each other.

Another example of such an intimate chemical and ultrastructural relationship between the inorganic and organic phases is that which occurs in oyster shells. Electron micrographs have shown a close and ordered relationship between the calcium carbonate crystals and the organic matrix of conchiolin.[88–90] Here too, calcification is initiated by small seed-like crystals which are initially deposited in a regular fashion in the organic matrix and appear similar to synthetically grown crystals from *supersaturated* solutions.[90]

Although the mechanisms by which mineralization is initiated and regulated in the organism are undoubtedly far more complex than those in the relatively simple model system described, the process of the heterogeneous nucleation of inorganic crystals by highly specific regions in the organic matrix as the result of a characteristic stereochemical array of certain reactive groups is probably a fundamental one, not only for the calcification of bone and cartilage but also for biological mineralization in general.

## APPENDIX A. SOME THERMODYNAMIC AND KINETIC CONSIDERATIONS IN PHASE TRANSFORMATIONS‖

Crystallization is a specific case of a more general phenomenon in which a change in state involves a phase transformation. A complete consideration of phase transformations would encompass (1) the conditions under which a system can or cannot persist in a distinct, physically homogeneous state (a single phase); (2) a *mechanism* for the formation of new phases from such systems; and (3) a quantitative description of the time-dependence of the process.

Since the primary purpose of this exposition, however, is to provide a background of some basic concepts in phase transformations in order to understand better the general phenomenon of crystallization and of biological mineralization, attention is directed primarily to some general considerations of the first two aspects.

### Equilibrium and the Stability of Phases¶

In most general terms, Gibbs[91] has defined equilibrium as a state independent of time, that is, a state in which *all* of the sensible properties which describe the system do not vary with time. Thus, in formulating both the necessary and sufficient conditions of equilibrium in terms of thermodynamic functions (e.g., energy, entropy, free energy, etc.), Gibbs considered equilibrium with respect to *all possible variations* of the state of the system, and *not* with respect to *certain* variations only. In practice, however, it is necessary to idealize most systems by imposing certain conditions of restraint and then by studying them with respect to *certain* possible variations only. In the case at hand, the possible variation of interest is the formation of entirely new phases from an initially homogeneous phase.

The usual criteria of equilibrium [for example, that for all possible variations in the state of an isolated system $(\delta E)_S \geqq 0, (\delta S)_E \leqq 0, (\delta F)_{p,T} \geqq 0$, etc.] do not give the information necessary for the interpretation of phase transformations, since many systems which meet these general criteria of equilibrium may vary widely in their *relative* tendency or ability to form new phases. On the basis of this *relative stability*, Gibbs[91] distinguished four kinds of equilibrium: (1) stable, (2) neutral, (3) unstable, and (4) metastable equilibria.

‖ The reader is referred to the following article for a more comprehensive review of the subject: D. Turnbull, Advances in Solid-State Phys. 3, 225 (1956).

¶ The material presented here has been compiled from a series of lecture notes from a graduate course in Chemical Thermodynamics, based on the writings of J. Willard Gibbs, given at the Massachusetts Institute of Technology, Department of Physical Chemistry, by Professor James A. Beattie. I would like to thank Professor Beattie for permission to use some of this material, including Figs. 35 and 36, and for many helpful suggestions and criticisms in the preparation of the Appendix.

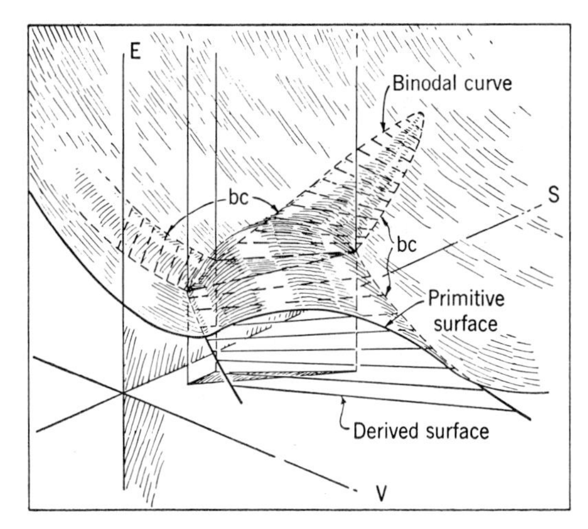


FIG. 35. Artist's (P. Lund) conception of the Gibbs energy-entropy-volume surface for a one-component system. The dotted lines on the primitive surface connect points of common tangency (neutral equilibrium), and the construction of a derived surface from such points is illustrated also. The orthogonal projection of the plane triangle on the S-V plane, representing the triple point, is shown. bc represents binodal curves.

The fundamental principles can, with respect to the stability of phases, conveniently and easily be visualized by reviewing certain pertinent features of the Gibbs energy-entropy-volume thermodynamic surface. For purposes of exposition, consider the special case of a pure substance when effects produced by gravity, electricity, capillary tensions or distortion of solid phases are absent, or may be neglected.

For all possible infinitesimal changes of state involving only expansion work in a closed system, the first and second laws of thermodynamics state that

$$dE = TdS - pdV. \qquad \text{(A1)}$$

Integration of (1) gives the equation

$$E = E(S,V). \qquad \text{(A2)}$$

Thus, the relations between the energy, entropy, and volume may be represented by a surface in space whose rectangular coordinates are represented by the volume, energy, and entropy of the system (Fig. 35). This surface, called the *primitive surface*, represents equilibrium states of a homogeneous phase. At any point on the primitive surface, the equilibrium temperature and pressure are defined by the slopes of the surface at that particular point:

$$\left(\frac{\partial E}{\partial S}\right)_V = T, \left(\frac{\partial E}{\partial V}\right)_S = -p \qquad \text{(A3)}$$

and the tangent plane at the point given by

$$E = E_0 + T_0(S - S_0) - p_0(V - V_0). \qquad \text{(A4)}$$

The intersection of the tangent plane with various other planes and the coordinate axes give various thermodynamic functions. For example, its intersection with the $E$ axes gives $F_0$, the Gibbs free energy of the substance at the point $P_0$.

States of the system consisting of two or more phases in equilibrium (heterogeneous equilibrium) can also be represented in $E$-$S$-$V$ space, since the system as a whole has a definite energy, entropy, volume, free energy, etc., as well as temperature and pressure. In heterogeneous equilibrium, the substance in each of its aggregation states has the same temperature, pressure, and molal free energy. Thus, planes tangent to points on the primitive surface representing, for example, one mole of the substance in each of these aggregation states will have the same slope $(-p,T)$ and the same intercept in the $E$-axes (free energy). Hence, these points have a common tangent plane.

In the case of two phases (say liquid and vapor) in equilibrium at a definite temperature and pressure, there are two points on the primitive surface, one representing one mole of liquid and the other representing one mole of vapor at this pressure and temperature which have a common tangent plane. Lying along the straight line connecting the points are the states of mixtures of liquid and vapor in equilibrium at this particular temperature and pressure. By taking all of the lines determined by such tangent planes for varying temperatures and pressures, another surface may be generated called the *derived surface* which represents states of heterogeneous equilibrium. One may conceive of this derived surface as being produced by rolling a double tangent plane (tangent to the primitive surface at two points) on the primitive surface and by connecting the successive pairs of *conjugate points* of tangency by a series of straight lines (Fig. 35).

In the case of three phases in equilibrium (i.e., liquid, vapor, solid), there are three points on the primitive surface with a common tangent plane, and states of the substance at the triple point (mixtures of all three phases in equilibrium) are presented in $E$-$S$-$V$ space by points in a plane triangle (Fig. 35). The derived surfaces, therefore, include all of the plane triangles representing three phases in equilibrium, and all of the developable surfaces representing two phases in equilibrium. It is a continuous ruled surface (generated by the motion of a straight line), but it is *not* cylindrical.

## Properties of the Thermodynamic Surfaces Which Indicate the Stability of Thermodynamics Equilibrium

Consider a specified amount of a pure substance $A$, immersed in a large medium $M$, at constant pressure $p_0$ and temperature $T_0$ with an initial energy, entropy, and volume $(E',S',V')$ and which undergoes the following change of state

$$A(E',S',V')=A(E'',S'',V'')$$

under conditions where the action of the system on the media is substantially reversible. It can easily be shown from the first and second laws of thermodynamics that the following relation must hold:

$$(E''-T_0S''+p_0V'')\leqq(E'-T_0S'+p_0V'). \quad \text{(A5)}$$

With reference to the thermodynamic surface, the terms enclosed in parentheses are the vertical distances of the points $(E'',S'',V'')$ and $(E',S',V')$ representing the final and initial states of the system above a plane passing through the origin having the slopes $-p_0,T_0$. Geometrically, it defines the conditions under which changes in state can occur spontaneously: where this vertical distance representing the final state is *less* or at most *equal* to the distance representing the initial state. Changes in state which result in an increase of this distance are, therefore, *not* possible. This principle is crucial to an understanding of phase changes, for it clearly delineates the regions in $E$-$S$-$V$ space (and, therefore, the equilibrium of such states) where phase transitions are possible.

In examining the possible equilibrium states of a pure substance in a medium at temperature $T_0$ and pressure $p_0$ (represented by points on the primitive surface), two possible categories of change must be considered. The behavior of systems may be related to (1) *continuous changes*—that is, changes considered with respect to nearby or adjacent states, and (2) *discontinuous changes*—that is, in relation to states at a finite distance from the point in question. The *curvatures* of the primitive surface determine the former, while the *over-all relation* of the tangent plane to the surface determines the latter.

### Stable Equilibrium

If the primitive surface falls *above* the fixed tangent plane, *except* at the single point of contact (representing the initial state), the state of the system represented by that point is "absolutely stable" toward a phase change when in a medium of constant pressure and temperature. That is, there are no points in $E$-$S$-$V$ space for which Eq. (5) is satisfied, and there are, therefore, no possible changes in state which can occur. Any "unnatural" perturbations of the system (such as, a local fluctuation in density) would necessarily lead to changes in state represented by points in $E$-$S$-$V$ space where the distance of any such point above the tangent plane would be greater than that of the original point. Since Eq. (5) has shown that such changes in state are *not* possible, natural processes will occur which will return

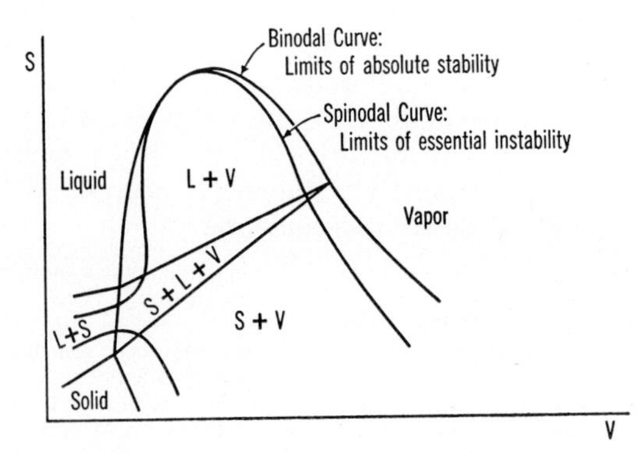


FIG. 36. Orthogonal projection of the limits of absolute stability (binodal curve) and limits of essential instability (spinodal curve) on the S-V plane for a substance having one solid phase.

the system to its original state. It can be shown that that part of the primitive surface representing such stable states of equilibrium is concave upward in both of its principal curvatures.

The points of tangency of the rolling double-tangent plane representing two phases in equilibrium trace the *binodal curves* (Fig. 36) or the limits of *absolute stability*. The tangent plane to points on the primitive surface which fall *outside the binodal curves* has the primitive surface entirely above it except at the single point of contact. That portion of the primitive surface outside the binodal curves represents states of *absolute stability*. This part of the primitive surface (referred to as the *surface of absolute stability*) together with the derived surfaces constitute the *surface of dissipated energy*. The tangent plane to a point on the surface of *absolute stability* is always below the surface of dissipated energy except at the point of tangency.

Solutions in such states of equilibria would be absolutely stable with regard to the *formation* of new phases. The mechanical analog of such a system may be represented by a marble in a hemispherical bowl fitted with a cover, with the conditions of restraint that the marble cannot sink through the bowl or be otherwise removed from it [Fig. 37(a)].

## Neutral Equilibrium

If the primitive surface does not fall anywhere below the fixed tangent plane but meets it at more than one point, the equilibrium of such states of the system is considered *neutral*. In such cases, if the system in such an equilibrium state is changed from its original state to a new state represented by another point of tangency to the primitive surface, the distances of both points above the fixed tangent plane (at constant $T$ and $p$) obviously will be equal (zero). Therefore, such systems will have no tendency to pass into one of the other states

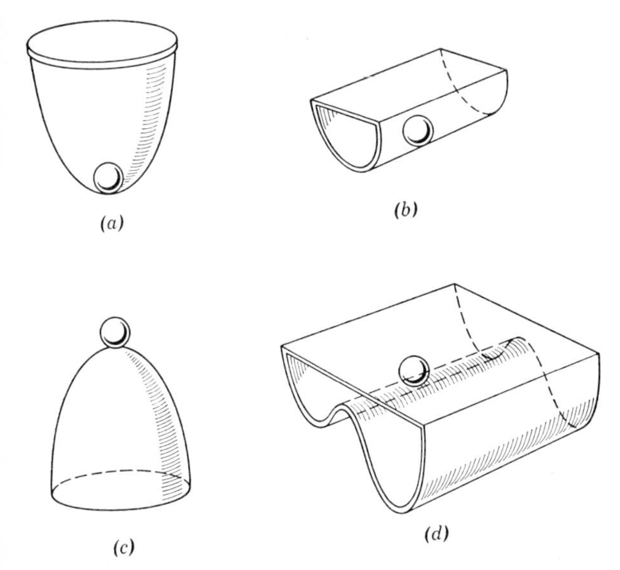


FIG. 37. Diagrammatic illustration of the four types of equilibria based on their relative stability.

(as represented by other points of common tangency) or to return to their original state if so displaced. However, such systems are stable with respect to continuous changes in state similar to systems in stable equilibrium.

A marble in a horizontal trough is an example of such a state of neutral equilibrium [Fig. 37(b)]. Although the displacement of the marble up the sides of the container eventually results in the return of the marble to its original position at the bottom of the trough, there are a number of positions of the marble along the bottom of the trough where no spontaneous tendency to change exists and where no spontaneous tendency for a return of the marble occurs if such horizontal displacements are made. Solutions in such states of neutral equilibrium are, therefore, also stable with regard to the *formation* of new phases.

## Unstable Equilibrium

If the primitive surface be continuous, there must necessarily exist regions between the binodal curves where the curvature of the primitive surface is concave *downward* in at least one of its principal curvatures. Points *on* the binodal curves (representing states of neutral equilibrium) separate such states of stability and instability with respect to *discontinuous* changes. The lines on the primitive surface dividing the portion which is concave *upward* in both of its principal curvatures from the portion which is concave downward in one or both of its principal curvatures represent the limit of *essential instability* or the *spinodal curves* (Fig. 36).

In such regions, it is obvious that, where part of the surface falls below the fixed tangent plane, it is possible to change the initial state of the system such that the point representing the final state is now below the fixed tangent plane. In this case, according to Eq. (5) natural processes occur which cause the system to continue to change (represented by moving the point further from the tangent plane) until a state is reached which is entirely different from the initial state [Fig. 37(d)]. Such states of equilibrium are *unstable* with respect *both* to continuous and discontinuous changes. That part of the primitive surface which lies inside the spinodal curve represents such states of *unstable* equilibrium. This type of equilibrium is rarely—if ever—attained in practice, and solutions in such a precarious state of equilibrium would hardly remain so for very long periods.

## Metastable Equilibrium

The primitive surface which lies *between* the spinodal and binodal curves is concave upward in both of its principal curvatures and is *stable* with respect to *continuous changes* (adjacent states). However, it is so situated with respect to the over-all thermodynamic surface that other states do exist a finite distance away, where the primitive surface falls *below* the tangent planes drawn through points on the primitive surface which lie between the binodal and spinodal curves. Such systems which are *stable* with respect to *continuous changes* (adjacent states) and *unstable* with respect to *discontinuous* changes (distant states) are said to be in *metastable equilibrium*. If the conditions necessary for such a discontinuous change *were not* present, the equilibrium would remain stable *indefinitely*. But, for example, if small portions of the same substance in one of the more stable states of aggregation, represented by points below the tangent plane, are introduced or otherwise caused to form by very small disturbances (perhaps ones that cannot be detected experimentally), the equilibrium would be destroyed and a change of state (a phase change) would occur [Fig. 37(d)]. Solutions in metastable equilibrium, therefore, although *capable* of remaining stable indefinitely, *can* form new, more-stable phases under certain conditions without changing the entire system.

With reference to the thermodynamic surface, it is thus apparent, since points *on* the surface of dissipated energy represent *stable* or *neutral* equilibrium and since points *below* the surface of dissipated energy have no physical significance, that points on the primitive surface which lie *above* the surface of dissipated energy representing states of *unstable* or metastable equilibrium delineate the regions in $E$-$S$-$V$ space from which phase changes can occur at constant temperature and pressure.

On the basis of these conditions, it can be shown that a single analytical expression representing both the necessary and sufficient conditions of equilibrium for a system in a medium of temperature $T_0$ and pressure $p_0$ is

$$\delta(E-T_0S+p_0V)\equiv\delta F=0 \qquad \text{(A6)}$$

for all variations of the state of the system. The system is absolutely *stable* with respect to first-order changes if $\delta F$ is a minimum, and it is *unstable* if $\delta F$ is a maximum. Finally, if Eq. (6) holds, but for *some finite changes*, the inequality $\Delta F<0$ exists, the state is *metastable*.

Although the foregoing discussion is limited to systems of one component, Gibbs (reference 91, pp. 100–115) has extended his treatment to a study of the internal stability of homogeneous fluids of many components.

## Mechanisms and Time Dependence of Phase Transformations

The discontinuous change of metastable or unstable systems resulting in the formation of the initial fragments of a new, more stable phase is called *nucleation*. When such a phase-change occurs in the interior of a metastable or unstable system in the absence of structural impurities, it is referred to as *homogeneous nucleation*, whereas a phase-change initiated *by* and *on* foreign inclusions extraneous to the system is called *heterogeneous nucleation*.

### Homogeneous Nucleation

Classical theories of homogeneous nucleation have assumed that phase-changes occur by the formation of intermediaries as a result of local fluctuations of certain properties (such as density) in initially homogeneous phases in metastable or unstable equilibrium.[46,92–94]

These intermediaries consisting of clusters of molecules (or ions) of the initial phase vary in size (and possibly composition, shape, and structure). They are referred to as "embryos" and are considered to exist as a true heterogeneous system with the mother phase. The cluster of critical size, composition, structure, etc., capable of *further* spontaneous growth (with a net decrease in free energy) and, therefore, capable of initiating the formation of the new phase, is called a *nucleus* (Fig. 38).

The embryos and nuclei are considered to arise by the stepwise addition of single molecules, i.e., a *bimolecular mechanism*.[46] The *order* of the reaction, therefore, is equal to the number of molecules in the nucleus.

As first proposed by Volmer and Weber,[93] the theory considered that nuclei can exist in a state of unstable equilibrium with the mother phase, making it possible, at least in principle, to define the nucleus in terms of reversible thermodynamics. Considering surface effects, the free energies of formation of spherical clusters can be calculated as a function of the radius (Fig. 38). It is obvious from Fig. 38 that the over-all free energy goes through a maximum which corresponds to the critical size cluster (nucleus) of radius, $r^*$, which can grow spontaneously with a net decrease in free energy. From fluctuation theory, the probability of nuclei formation and, therefore, also of their concentration can be approximated since they are proportional to $\exp(-\Delta F^*/kT)$, where $\Delta F^*$ is equal to the free energy or the work of formation of a nucleus corresponding to the maximum as shown in Fig. 38. The condensation velocity or nucleation rate is then evaluated by computing the collision frequency of single molecules with the nuclei. This free-energy barrier is, therefore, similar to the activation energy of ordinary chemical reactions in permitting the derivation of a *nucleation rate*.††

Becker and Döring[45] treated the problem differently. They assumed a size distribution of "embryos" and evaluated the rate of condensation by the solution of a set of equations relating how the number of embryos of any particular size changed with time. By assuming a steady state, the mechanism of change was, therefore, simply the kinetic process of unit condensation and unit evaporation.

In both of the cases, expressions for nucleation rate were quite similar [$J\sim\exp(-1/\ln S^2)$], where $J$ is the nucleation rate and $S$ the supersaturation ratio $p/p_e$ where $p$ is the actual vapor pressure and $p_e$ the equilibrium vapor pressure), and indicated a very marked

†† In more-complicated systems, such as occurs in the case of inorganic crystallization, the various-sized embryos may have a number of compositional and structural differences which also involve free-energy changes, and, in particular instances, there may be important effects owing to strain energy as well. Thus, over-all free-energy barrier to nucleation will be provided by a combination of the chemical-, interfacial-, and strain-free energies.

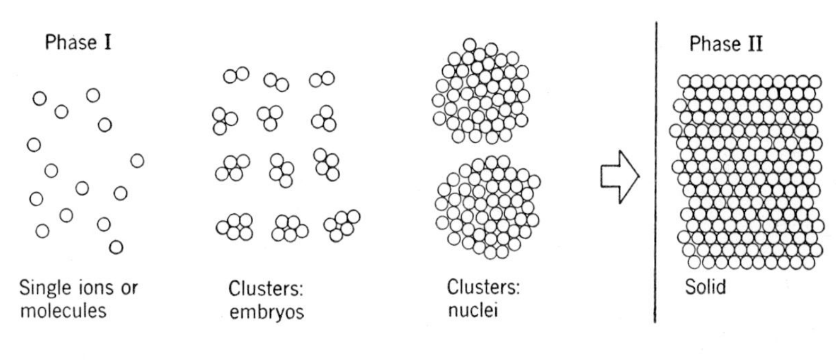


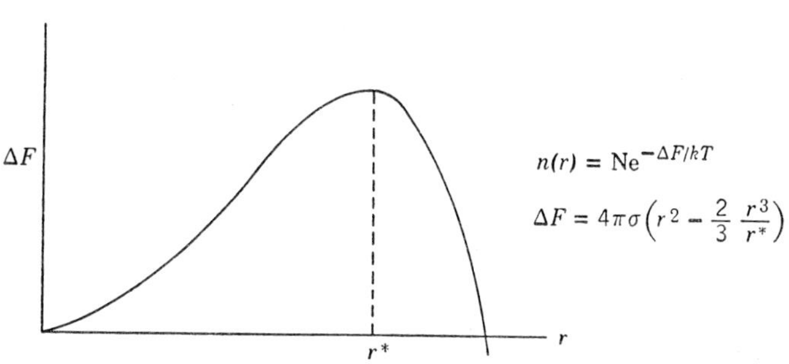


FIG. 38. Schematic illustration indicating mechanism of nucleation as proposed principally by Volmer *et al.*[46,93,94] and by Frenkel.[92]**

dependence of nucleation rate upon the degree of supersaturation, particularly near the critical supersaturation ratio (i.e., where the nucleation rate rather suddenly becomes sufficiently large for ready measurement). For example, in the case of water vapor condensing to liquid drops, the time that must elapse for the appearance of the first drop at a supersaturation ratio of 4, is 0.1 sec, whereas increasing the supersaturation ratio to 5 decreases the time to $10^{-13}$ sec, and decreasing the ratio to 3 increases the time to $10^3$ years![46]

Although based on the theory of Volmer and Weber (fluctuation theory), it is possible that homogeneous nucleation can occur at any level of metastability. Certain theoretical considerations (as well as experimental findings) indicate that this is not possible, except at the limit of essential instability or from systems in *unstable* equilibrium rather than in *metastable* equilibrium. The difficulties arise conceptually from the definitions of equilibrium (a state independent of time), of metastability (stable with respect to continuous changes in state, but unstable with respect to certain discontinuous changes in state), and of what constitutes a discontinuous change in state, and from the applicability of $e^{-\Delta S/k}$ to predict the probability of fluctuations which are large enough to be considered a new phase. Thus, Frenkel[92] considers that the density fluctuations which lead to the formation of embryos (heterophase) transcend the limits usually considered in the ordinary statistical theory of homogeneous systems wherein ordinary fluctuations (homophase) lie "within the limits compatible with the preservation of a given aggregation state" (Frenkel,[92] p. 375).

** While this is usually considered the mechanism of nucleation for most systems, there is recent evidence that, in the nucleation of solid-state transformations[95] the nucleii arise by a cooperative phenomenon and not by discrete atomic jumps.

From these considerations, it would appear that nucleation from true metastable systems occurs only as a result of the introduction of heterogeneities and not spontaneously. This would also seem to be Gibbs' interpretation of the stability of metastable systems, since he states:

". . . the mass in question must be regarded as in strictness stable with respect to the growth of a globule of the kind considered, since *W*, the work required for the formation of such a globule of a certain size (viz, that which would be in equilibrium with the surrounding mass), will always be positive. Nor can smaller globules be formed, for they can neither be in equilibrium with the surrounding mass, being too small, nor grow to the size of that to which *W* relates. If, however, by *any external agency* [Ital.: Ed.] such a globular mass (of the size necessary for equilibrium) were formed, the the equilibrium has already (page 243) been shown to be unstable, and with the least excess in size, the interior mass would tend to increase without limit except that depending on the magnitude of the exterior mass" (Gibbs,[91] pp. 255–256).

## Heterogeneous Nucleation

The discontinuous change in state which results in nuclei formation by the introduction of a foreign inclusion may be the result of a number of different effects. Thus, either by virtue of nonspecific surface forces or by very specific interactions between certain groups on the surface and the molecules or ions of the initial phase, such heterogeneities may adsorb or bind the molecules or ions of the initial phase on their surface; by acting either as a core whose surface is now composed of the molecules of the initial phase or by binding the molecules or ions in a particular steric array, they may form a cluster of critical size, shape, and configuration necessary for nucleation. Since the critical-size cluster (assuming no structural or compositional changes) varies with the supersaturation ratio, nucleation rates for heterogeneous nucleation will, as in the case of homogeneous nucleation, be strongly dependent upon and markedly dominated by the degree of metastability of the system.

With regard to crystallization, it is of interest to note the findings of Turnbull and Vonnegut[63] who have proposed a crystallographic theory of crystal nucleation. The theory emphasizes the importance of *geometric* and *structural* factors of nucleation catalysts with regard to their catalytic potency and ability to initiate phase-changes. Based on a number of their own observations as well as on experiments of others, they have postulated that those substances acting as effective nucleation catalysts have atomic arrangements and lattice spacings on certain low-index planes which are very similar to those of the crystal being nucleated.

The potency of the catalyst is formulated to be directly proportional to the reciprocal of the disregistry between the lattice parameters of the catalytic surface and the forming crystal on these planes. Thus, in the formation of ice crystals from supersaturated droplets of water vapor for example, the most potent nucleation catalyst was found to be AgI[96,97] whose lattice structure and atomic arrangement are remarkably similar to ice. Similarly, in the case of the nucleation of hydrated sodium-sulfate ($Na_2SO_4 \cdot 10H_2O$) by sodium-tetraborate ($Na_2B_4O_7 \cdot 10H_2O$) crystals, it was noted that both crystals belong to the same space group and that the disregistry was very small between the two types of crystals on certain basal planes.[98] That some heterogeneities will not act as catalysts at all is probably related to such specific interactions and geometrical factors.

Although many considerations such as specific interactions, types of chemical bonds, etc., as well as relatively nonspecific factors such as surface dislocations and defects, influence the potency or ability of the catalyst in any particular case, the importance of *geometric* and *structural* factors should be kept in mind because of their possible implications for the theory of biological mineralization which is presented.

## Other Related Considerations

Whereas *nucleation*—referring to the process of *forming* the infinitesimally small fragments of the new phase—can *in concept* be separated from the subsequent *growth* of these fragments into macroscopic particles of the new phase, *nucleation rate*—which is usually determined by measuring the appearance of the macroscopic particles of the new phase—*operationally* is not completely independent of the kinetics of such growth.

Therefore, the *operational* definition of *nucleation rate* is not completely consistent with the *theoretical* concept of nucleation rate, since the latter strictly speaking refers only to the rate at which the nuclei are forming and not to the subsequent growth of the nuclei to crystals, nor to the phenomenon of recrystallization.

Furthermore, measurements of nucleation rate, particularly in the case of heterogeneous nucleation have given little or no evidence of the molecular or atomic sequence of the nucleation phenomena itself—i.e., the mechanism of nuclei formation—which is of particular importance in the case of biological mineralization, where certain of these steps may be enzymatically controlled and regulated, for example.

Moreover, in the case of crystallization, the number and size of crystals is determined by the relative rates of crystal growth and nucleation rate once nucleation has been initiated. Considered independently, crystal growth may vary in just the opposite fashion as does nucleation rate with respect to certain variables such as temperature. In addition, the growth of crystal nuclei is dependent not only on diffusion rate (i.e., supersaturation ratio) but also on the presence of surface defects (e.g., growth by screw dislocations) even at low-supersaturation ratios.[99]

An interpretation of the entire phenomenon of the crystallization of a specific mineral would, therefore, require not only a study of nucleation phenomena but also considerations of the factors influencing crystal growth, crystal size, crystal habit, etc., for a particular crystal structure. Furthermore, the conditions influencing *recrystallization* (the growth of large crystals at the expense of smaller ones), might also be quite important in the over-all process even after the solid state had been achieved.

## BIBLIOGRAPHY


[1] J. E. Eastoe and B. Eastoe, Biochem. J. **57**, 453 (1954).

[2] J. E. Eastoe in *The Biochemistry and Physiology of Bone*, G. H. Bourne, editor (Academic Press, Inc., New York, 1956), p. 81.

[3] Author's unpublished observations.

[4] W. F. DeJong, Rec. trav. Chim. **45**, 445 (1926).

[5] D. Carlström, Acta Radiol. Suppl. 121 (1955).

[6] W. F. Neuman and M. W. Neuman, Chem. Revs. **53**, 1 (1953).

[7] P. Cartier, Bull. soc. chim. biol. **30**, 65 (1948).

[8] M. J. Dallemagne and H. Brasseur, Acta Biol. Belgica **2**, 440, 445 (1952).

[9] H. Brasseur, M. J. Dallemagne, and J. Mélon, Bull. soc chim. France **16**, 109 (1949).

[10] M. J. Dallemagne, C. Fabry, and A. S. Posner, J. physiol. (Paris) **46**, 325 (1954).

[11] A. S. Posner and A. Perloff, J. Research Natl. Bur. Standards **58**, 279 (1957).

[12] D. Carlström, Biochim. et Biophys. Acta **17**, 603 (1955).

[13] S. B. Hendricks and W. L. Hill, Proc. Natl. Acad. Sci. U. S. **36**, 731 (1950).

[14] J. H. Weikel, Jr., and W. F. Neuman, U. S. Atomic Energy Report, University of Rochester, UR-228 (1952).

[15] R. Kunin, K. L. Elmore, and D. L. Johnson, TVA Report, unpublished. Data reported by W. F. Neuman and M. W. Neuman in *The Chemical Dynamics of Bone Mineral* (The University of Chicago Press, Chicago, 1958), p. 39.

[16] W. F. Neuman and M. W. Neuman, *The Chemical Dynamics of the Bone Mineral* (The University of Chicago Press, Chicago, 1958).

[17] M. J. Dallemagne and J. Mélon, Arch. de Biol. **57**, 79 (1946).

[18] P. W. Arnold, Trans. Faraday Soc. **46**, 1061 (1950).
[19] W. E. Brown, J. R. Lehr, J. P. Smith, and A. W. Frazier, J. Am. Chem. Soc. **79**, 5318 (1957).
[20] A. S. Posner (personal communication, 1958).
[21] J. W. Gruner and D. McConnell, Z. Kristallog. **97**, 208 (1937).
[22] D. McConnell in *Metabolic Interrelations*, Transactions 4th Conference, E. C. Reifenstein, Jr., editor (Josiah Macy, Jr., Foundation, New York, 1952).
[23] L. Huber and C. Rouiller, Experientia **7**, 338 (1951).
[24] C. Rouiller, L. Huber, E. Kellenberger, and E. Rutishauser, Acta Anat. **14**, 9 (1952).
[25] W. J. Schmidt, Handbuch biol. Arbeitsmethoden, **A5**, T10, 435 (1934).
[26] J. Wolff, *Das Gesetz der Transformation der Knochen* (Quarto, Berlin, 1892).
[27] R. A. Robinson, J. Bone and Joint Surg. **34A**, 389 (1952).
[28] R. A. Robinson and M. L. Watson, Anat. Record **114**, 383 (1952).
[29] R. A. Robinson and M. L. Watson, Ann. New York Acad. Sci. **60**, 596 (1955).
[30] S. Fitton-Jackson, Proc. Roy. Soc. (London) **B146**, 270 (1957).
[31] H. Sheldon and R. A. Robinson, J. Biophys. Biochem. Cytol. **3**, 1011 (1957).
[32] R. Robinson, *The Significance of Phosphoric Esters in Metabolism* (New York University Press, New York, 1932).
[33] E. Freudenberg and P. György, Biochem. Z. **118**, 50 (1921).
[34] A. B. Gutman and T. F. Yü, in *Metabolic Interrelations*, Transactions 1st Conference (Josiah Macy, Jr., Foundation, New York, 1950), p. 11.
[35] A. E. Sobel and M. Burger, Proc. Soc. Exptl. Biol. Med. **87**, 7 (1954).
[36] A. E. Sobel, Ann. N. Y. Acad. Sci. **60**, 713 (1955).
[37] P. S. Rubin and J. E. Howard in *Metabolic Interrelations*, Transactions 2nd Conference (Josiah Macy, Jr., Foundation, New York, 1951), p. 155.
[38] B. S. Strates, W. F. Neuman, and G. J. Levinskas, J. Phys. Chem. **61**, 279 (1957).
[39] B. S. Strates and W. F. Neuman, Proc. Soc. Exptl. Biol. Med. **97**, 688 (1958).
[40] M. J. Glimcher, A. J. Hodge, and F. O. Schmitt, Proc. Natl. Acad. Sci. U. S. **43**, 860 (1957).
[41] M. B. Engel, N. R. Joseph, and H. R. Catchpole in *Metabolic Interrelations*, Transactions 5th Conference, E. C. Reifenstein, Jr., editor (Josiah Macy, Jr., Foundation, New York, 1953), p. 105.
[42] I. Gersh in *Connective Tissues*, Transactions 2nd Conference (Josiah Macy, Jr., Foundation, New York, 1952), p. 11.
[43] F. McLean and M. R. Urist, *Bone, An Introduction to the Physiology of Skeletal Tissue* (The University of Chicago Press, Chicago, 1955).
[44] F. McLean in *Metabolic Interrelations*, Transactions 2nd Conference (Josiah Macy, Jr., Foundation, New York, 1950), discussion p. 101.
[45] R. Becker and W. Döring, Ann. Physik **24**, 719 (1935).
[46] M. Volmer, *Kinetik der Phasenbildung* (Edwards Brothers, Ann Arbor, 1945).
[47] W. F. Neuman in *Metabolic Interrelations*, Transactions 2nd Conference (Josiah Macy, Jr., Foundation, New York, 1950), p. 32.
[48] Author's unpublished results.
[49] D. S. Jackson (personal correspondence, 1958).
[50] C. C. Solomons and J. T. Irving, Biochem. J. **68**, 499 (1958).
[51] B. F. Martin and F. Jacoby, J. Anat. **83**, 351 (1949).
[52] P. Doty and T. Nishihara in *Recent Advances in Gelatin and Glue Research*, G. Stainsby, editor (Pergamon Press, London, 1958), p. 92.
[53] P. J. Flory and R. R. Garrett, J. Am. Chem. Soc. **80**, 4836 (1958).
[54] W. Cochran, F. H. C. Crick, and V. Vand, Acta Cryst. **5**, 581 (1952).
[55] F. H. C. Crick, Acta Cryst. **6**, 685 (1953).
[56] Many of the low-angle $x$-ray diffraction studies were carried out in cooperation with Dr. A. Posner, Am. Dental Assn., Research Division, National Bureau of Standards, and with Dr. L. Bonar, Department of Biophysics, Massachusetts Institute of Technology.
[57] R. S. Bear, Advances in Protein Chem. **7**, 69 (1952).
[58] O. E. A. Bolduan, T. P. Salo, and R. S. Bear, J. Am. Leather Chem. Assn. **46**, 124 (1951).
[59] K. Kühn, W. Grassmann, and U. Hofmann, Naturwissenschaften **44**, 538 (1957).
[60] W. Grassmann, U. Hofmann, K. Kühn, H. Hörmann, H. Endres, and K. Wolf in *Connective Tissue*, Symposium organized by Council for International Organizations of Medical Sciences, R. E. Tunbridge, editor (Blackwell Scientific Publications, Oxford, 1957), p. 157.
[61] D. Turnbull, J. Appl. Phys. **21**, 1022 (1950).
[62] D. Turnbull, J. Chem. Phys. **20**, 411 (1952).
[63] D. Turnbull and B. Vonnegut, Ind. Eng. Chem. **44**, 1292 (1952).
[64] V. DiStefano, W. F. Neuman, and G. Rouser, Arch. Biochem. Biophys. **47**, 218 (1953).
[65] M. Polonovski and P. Cartier, Compt. rend. **232**, 119 (1951).
[66] R. J. Block, M. K. Horwitt, and D. Bolling, J. Dental Research **28**, 518 (1949).
[67] R. J. Block, Ann. N. Y. Acad. Sci. **53**, 608 (1951).
[68] D. B. Scott, Ann. N. Y. Acad. Sci. **60**, 575 (1955).
[69] H. R. Perkins and P. G. Walker, J. Bone and Joint Surg. **40B**, 333 (1958).
[70] J. Shatton and M. Schubert, J. Biol. Chem. **211**, 565 (1954).
[71] S. M. Partridge and H. F. Davis, Biochem. J. **68**, 298 (1958).
[72] K. Meyer, E. Davidson, A. Linker, and P. Hoffman, Biochim. et Biophys. Acta **21**, 506 (1956).
[73] M. A. Logan, J. Biol. Chem. **110**, 375 (1935).
[74] M. Schubert and D. Hamerman, J. Histochem. Cytochem. **4**, 159 (1956).
[75] J. Einbinder and M. Schubert, J. Biol. Chem. **185**, 725 (1950).
[76] H. P. Gregor, Trans. New York Acad. Sci. Ser. II, **18**, 667 (1956).
[77] E. S. Boyd and W. F. Neuman, J. Biol. Chem. **193**, 243 (1951).
[78] D. Burton and R. Reed, Discussions Faraday Soc. **16**, 195 (1954).
[79] D. Carlström, A. Engström, and J. B. Finean in *Fibrous Proteins and Their Biological Significance*, Symp. Soc. Exptl. Biol. **9**, 85 (1955).
[80] R. A. Robinson and D. A. Cameron, J. Biophys. Biochem. Cytol. **2**, Suppl. 253 (1956).
[81] S. M. Clark and J. Iball, Nature **179**, 94 (1957).
[82] S. M. Clark and J. Iball, Progr. in Biophys. and Biophys. Chem. **7**, 225 (1957).
[83] G. Wallgren, Acta Paediate. **113**, Suppl. (1957).
[84] T. W. Speckman and W. P. Norris, Science **126**, 753 (1957).
[85] H. Fernández-Morán and A. Engström, Biochim. et Biophys. Acta **23**, 260 (1957).
[86] M. L. Watson and R. A. Robinson, Am. J. Anat. **93**, 25 (1953).
[87] Author's unpublished results.
[88] C. Gregoire, J. Biophys. Biochem. Cytol. **3**, 797 (1957).
[89] T. Tsujii, D. G. Sharp, and K. M. Wilbur, J. Biophys. Biochem. Cytol. **4**, 275 (1958).
[90] N. Watabe, D. G. Sharp, and K. M. Wilbur, J. Biophys. Biochem. Cytol. **4**, 281 (1958).
[91] J. W. Gibbs, *The Collected Works of J. Willard Gibbs* (Longmans, Green and Company, New York, 1928), Vol. I.
[92] J. Frenkel, *The Kinetic Theory of Liquids* (Clarendon Press, Oxford, 1946, and Dover Publications, New York, 1955).
[93] M. Volmer and A. Weber, Z. physik. Chem. **119**, 277 (1926).
[94] M. Volmer, Z. Elektrochem. **35**, 555 (1929).
[95] M. Cohen, Transactions Metallurgical Soc. AIME **212**, 171 (1958).
[96] V. J. Schaefer, Chem. Revs. **44**, 291 (1949).
[97] B. Vonnegut, J. Appl. Phys. **18**, 593 (1947).
[98] M. Telkes, Ind. Eng. Chem. **44**, 1308 (1952).
[99] F. C. Frank, Advances in Phys. **1**, 91 (1952).

# 43
# Structure of Muscle Cells*

H. Stanley Bennett

*Department of Anatomy, University of Washington School of Medicine, Seattle 5, Washington*

IT IS my task to talk about the microscopic aspects of muscle structure. Muscle cells are specialized structurally for the function of contractility. In a muscle fiber, the ratio between deforming force and elongation can change rapidly and reversibly. This change characterizes the phenomenon of contractility. Contractility involves changes in the viscoelastic behavior of some fibrous protein structures in the muscle. In the relaxed state, a muscle fiber subjected to tensile stress along its axis will elongate, and a characteristic stress-strain diagram can be prepared. If a corresponding curve is taken while the muscle is contracted, the slope of the curve is much steeper. The shift from one elastic state to another can occur in a few milliseconds.

Contractile structures of this sort are found in a great many different forms. Inoué (p. 402) discusses certain features of contractile mechanisms other than those called muscle. Now, consider primarily a specialized class of biological contractile mechanisms called muscle. The boundary between muscle and nonmuscle is often not easy to draw. There are protozoa, for example, which have contractile filaments within certain portions of their complicated cells. Many of these contractile structures have properties of muscle.

Metazoans may likewise contain such dual-purpose cells. For example, some of the cells which surround the blood vessels of the earthworm serve the functions both of lining the channel and of controlling the size of the lumen. These endothelial cells contain contractile filaments which resemble those of muscle cells elsewhere in the organism. As another example, in hydra, a small metazoan organism, certain cells lining the gut cavity have an important role in the digestion and in the absorption of the food consumed. A part of these cells contain contractile filaments, and this part could be regarded as being similar to muscle. One might cite many other examples of dual or triple function cells, one of whose properties is that of contractility.

But even the well-defined muscle cells found in the limbs of a vertebrate organism may display more than one physiological function. In addition to the main function of contractility, the muscle fibers of the mammals have a very important role in the production of heat. A good fraction of the body temperature is generated by combustion in the muscle cells. If the heat losses from the body are sufficient, a very special type of contractile response known as shivering occurs. This is characterized by very rapidly repeated contractions which do not accomplish any useful work but which generate a good deal of heat, thus helping to restore a satisfactory thermal balance. In addition, there are very important electrical characteristics of muscle. The phenomenon of excitation of muscle is accompanied by electrical changes. These electrical changes involve voltage transients of 20 to 100 mv, depending on the muscle. These are generally only of local importance as far as that muscle is concerned. But, in an aqueous medium, these electrical transients can escape into the surrounding fluid and be detected by sensitive electrodes. In electric fishes, the muscle becomes modified so that the electrical transients become very highly directional and strengthened. In most cases, the contractility of such specialized muscle fibers is lost. In such a fish, each modified muscle cell is polarized, forming an electroplax. Considerable numbers of these electric generating units may be stacked up in series and arranged in parallel. In the most highly developed forms, the electric organ can produce formidable power. The torpedo ray, for example, can deliver a current of 60 amp at 80 v as a result of series and parallel summation of many of these specialized muscle cells. In the case of the electric eel (*Electrophorous electricus*), more of these cells are summed in series and potentials can achieve values of 300 or more volts. Currents in this case, with fewer generating units in parallel, are of the order of 0.3 to 0.5 amp. Thus, by suitable electrical coupling (by giving a suitable vector to the electrical discharge of the muscle and connecting the modified muscle units in series and in parallel), rather astonishing electrical effects can be produced. In addition, many fish have very sensitive electric sensing organs and can broadcast electric pulses (some of them small) to explore the surrounding water. By sensing these pulses with their special electric receivers, they test the conductivity of the water and use this information for navigation, for escaping enemies, and for finding prey.

Muscle fibers proper contain not only a contractile mechanism itself, but also other structures which control and fuel the contractile devices and render the whole apparatus a workable and effective part of the organism. The unit which accomplishes all of these coordinated physiological functions is a muscle fiber. These are really muscle cells, but since many of them are long and thread-like, they are called fibers. There are many kinds of muscle fibers. They vary from 10 to 100 $\mu$ in diameter and from 20 $\mu$ to several centimeters in length. The smaller muscle fibers usually contain only one nucleus.

* The preparation of the material presented in this report was assisted by Grant B-401 from the National Institutes of Health, Public Health Service, U. S. Department of Health, Education, and Welfare, and by a grant from the Muscular Dystrophy Associations of America, Inc.

Many of the larger fibers, however, contain very many nuclei, up to hundreds.

Muscles also vary greatly with respect to the fraction of their resting length by which they can shorten. At one extreme are the flight muscles of insects, such as those of the fly or wasp. The change in length between the relaxed state and contracted state in this type of muscle is relatively small, of the order of 2 to 5%. At the other extreme, one might cite the proboscis retractor muscles in certain marine worms, where the changes upon shortening are of the order of 90% of the original length. In vertebrate skeletal muscle such as is found in humans, the changes in length between the contracted and relaxed state do not exceed 20% under ordinary circumstances.

For over a century, it has been realized that an explanation of contraction of muscle depended on an understanding of its molecular components and in a characterization of the actions and interactions of these components in terms of the contractile process. This concept of the participation of molecular units in contraction was explicitly stated in 1859 by Kühne,[1] when he made the first attempt to obtain from muscle proteins which might be important in the contractile process. He pressed out muscle juice and obtained from it a crude protein preparation which he called "myosin."

In recent years, with improvements in chemical and physical techniques, there have been a number of attempts to relate the molecular components of muscle with the ultrastructural anatomy and the actual process of contraction. Two important hypotheses have been proposed (see Morales, p. 426):

The first may be called the one-filament hypothesis, and the second the two-filament hypothesis. In the first case, contraction is considered to be the result of a change in the configuration of a unit consisting of a single filament. Shortening of many of these unit filaments in parallel and in series would occur when a muscle cell contracts. So far as I am aware, the clearest and earliest explicit statement of this hypothesis can be laid to Meyer.[2] He proposed a number of models in which attraction and repulsion of charged groups on a helix were responsible for contraction. Astbury,[3] from his x-ray diffraction work, elaborated on this concept. The view that the unit of contractility resides in a single filament dominated the literature until very recent years.

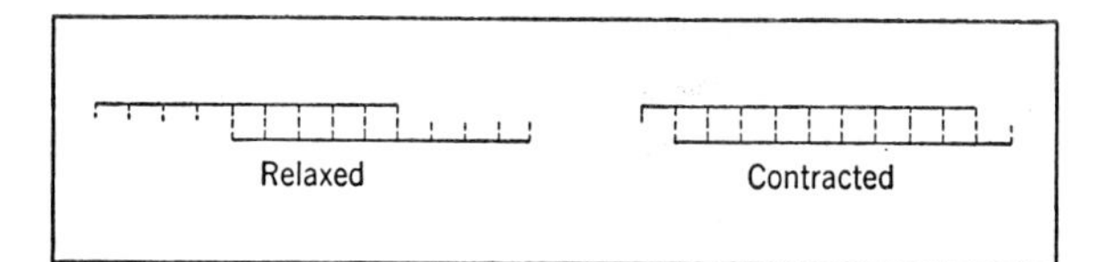


FIG. 1. Diagram of a conceptual model of the simplest possible muscle, based on the two-filament hypothesis of Huxley and Hanson. Each filament, represented by a horizontal solid line, is fitted with a series of binding sites, shown as short vertical dashed lines. The filaments constituting a pair overlap to a certain extent, and are bound to each other by a number of binding groups. In the relaxed state, the length of the overlapping portions is relatively short, and only a few of the binding sites are in contact. In contraction, the interaction between the binding sites changes, more binding sites become engaged, and one filament is drawn up alongside the other, increasing the proportion of overlap, shortening the total length of the system, and exerting tension in a direction parallel to that of the filaments. The filaments are of molecular dimensions.

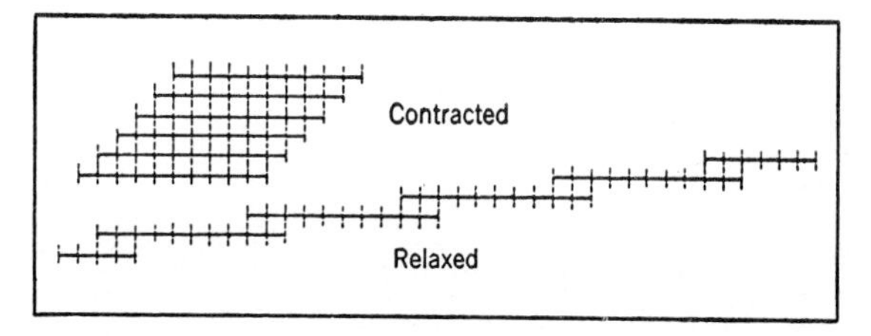


FIG. 2. Diagram of a conceptual model of a modification of the two-filament hypothesis designed to account for a muscle which can contract to a small fraction of its resting length. The concept holds that in the relaxed state a series of myofilaments, each fitted with two sets of bonding sites capable of interacting with those of fellow filaments, overlaps a neighboring filament to a slight extent at each end. Upon contraction, the length of overlap between adjacent filaments is caused to increase as each filament is pulled lengthwise along its neighbor by a change in the interaction between binding sites. The assembly thus behaves like a multiple set of extension ladders. The terminal extremities of the two terminating members of the series of extended relaxed filaments, however far separated, can be pulled by the contractile process to a distance apart limited only by the length of the longest filament of the series.

The second hypothesis involves the view that the unit structure in contractility consists of two parallel, overlapping and closely spaced filaments. One of the filaments moves parallel to the other during the contractile process. Thus, the over-all length of the systems is decreased, although each of the individual filaments maintains its original length (Fig. 1). This two-filament hypothesis, which has been most explicitly stated by Hanson and H. E. Huxley[4] and H. E. Huxley,[5] has been considered from the energetic and physiological point of view rather elaborately by A. F. Huxley.[6] It is now an important topic of discussion amongst muscle physiologists. This two-filament hypothesis can be regarded as a reversible special version of the concept of aggregation of macromolecules and protofibrils which F. O. Schmitt discusses in connection with collagen (p. 349). He points out that there is evidence that collagen macromolecules can aggregate in various spatial relationships with respect to one another. As applied to muscle, the theory would state that the protein filaments in muscle could shift reversibly from one position of parallel alignment to another.

According to this hypothesis, the simplest muscle would consist of two filaments (Fig. 1). An elaboration of this would permit one to build an analogous model to fit a fiber which contracted to only a small fraction of its relaxed length. As mentioned previously, such muscles are found in some invertebrates. Such a muscle might be composed of a series of filaments arranged in staggered array, as in Fig. 2. Upon contraction, these filaments would simply slide up on each other like the units of a multiple or extension ladder. This provides a way in which the two-filament hypothesis can be elabo-

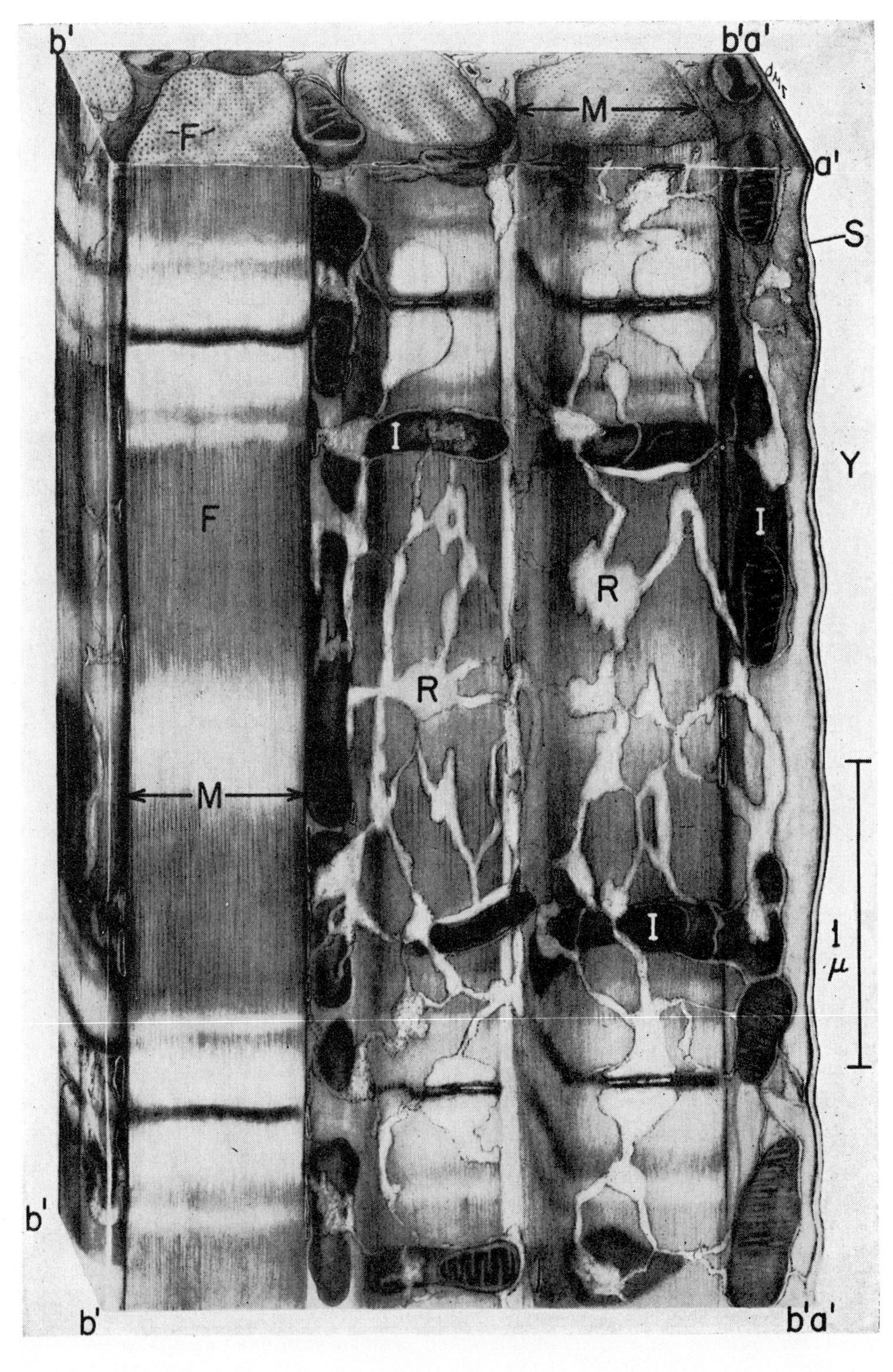


FIG. 3. A diagram to show some of the structural features of a typical striated muscle fiber, such as might be found in the limb muscle of a vertebrate. The figure represents a semitransparent block cut out of the muscle. *Y* designates extracellular space. The sarcolemma *S* is the plasma membrane of the muscle cell. Many mitochondria *I* are interspersed between the myofibrils *M*. Within the myofibrils are the myofilaments *F*, which are of molecular dimensions. Forming laceworks surrounding the myofibrils are elements of the sarcoplasmatic reticulum *R*, also called sarcotubules.

rated to account conceptually for some of the more extreme examples of muscular contraction. There is no experimental evidence, however, to support this model for muscles which contract to only a small fraction of their resting length.

On the basis of electron micrographs, the two-filament hypothesis is supported by the visualization in striated muscle of two sets of filaments.[5] In the muscles which Huxley studied, one type of filament is thicker than the other (Figs. 4 and 5). This is not an essential part of the concept, however, and it would not detract from the theory if, in some muscles, all of the filaments were found by electron microscopy to have geometrically similar dimensions.

In addition to the contractile elements, muscles contain a number of other important components. Some of these are represented for one type of muscle in Fig. 3. The muscle is surrounded by a membrane which corresponds to the plasma membrane of the muscle and is called the sarcolemma (*S*). The concentrations of constituents of the fluid encompassed by the plasma membrane differ from those outside. If one places an electrode inside the muscle cell and connects it through a suitable recording device to an electrode in the fluid outside the muscle cell, one finds a potential difference between the inside and outside of the muscle. This potential difference has been related to the difference in ionic concentrations across the sarcolemma. Fairly satisfying relationships, based on the differential concentrations of potassium and sodium, have been derived to account for this potential difference. In electron micrographs, the sarcolemma (the structure responsible for separating these two ionic compartments) is of the order of 100 A thick. A measured 100-mv potential difference between the inside and the outside of the muscle fiber means that there is a voltage gradient of the order of 100 000 v/cm across the sarcolemma.

Katz (p. 466) presents some of the features of membranes of this sort in his discussion of nerves and neural conduction. If one applies a quick transient electric pulse to certain portions of the sarcolemma of many muscles, one gets an electrical disturbance which sweeps in all directions over the muscle fiber without loss of amplitude. Such electrical excitability resides in large portions of the sarcolemma, but is not found in the immediate region where nerves reach the muscle, nor is it found in the region where the tendons attach to the muscle fiber. From this difference in electrical behavior, one can conclude that the molecular organization of the sarcolemma varies in different portions. There is, however, no satisfactory correlation of these presumed differences in molecular structure with images of the sarcolemma in electron micrographs, nor is it known in any detail what these differences might be physiochemically.

The sarcolemma can be traced continuously around both ends of the muscle. In general, in most vertebrate muscle and in some insect muscle, the fibers of the tendons are bonded in some way to the outer surface of the sarcolemma. Certain noncontractile intracellular filaments continuous with the myofilaments also connect with the sarcolemma, but with its inner surface, opposite the tendon attachments. This arrangement is capable of transmitting the tension generated within the fiber to some extrafibrillar unit, such as bone.

Most muscle fibers, but not all, are equipped with a nerve supply which, in more elaborate fibers, serves a number of different functions. In some types of vertebrate muscle, such as skeletal muscle, there is a main motor nerve which supplies each fiber. When this nerve is excited, the muscle responds by contracting. In addition, in many cases there are sensory organs located in the tendons. These sense organs are similar to strain gauges in that they can respond to changes in tension and send impulses to the central nervous system. These signals can then be processed and correcting directions relayed back to the muscle fiber. This provides a type of feedback loop which regulates the strength and temporal nature of the muscular contraction. In certain muscles, much more elaborate types of regulatory devices are present. One is known as a muscle spindle. This is a complex structure embedded amongst ordinary muscle fibers and consisting of a very delicate muscle fiber inside a connective tissue capsule. This special fiber is fitted with a nerve supply containing motor and sensory components. The motor component can actuate the contractile mechanism of the very fine fiber within the spindle capsule and thus vary the bias on this sensory device. The sensory fibers reach nerve endings associated with the fiber and pick up signals which result from changes in tension. These are sent to the central nervous system. These examples do not exhaust the types of regulatory mechanisms which operate in muscle fibers. In certain cases, rather different mechanisms are found.

One can divide the contents of the muscle cell into two components: the contractile portion and the noncontractile portion. The latter is called the sarcoplasm. This sarcoplasm contains mitochondria, nuclei, fat droplets, glycogen, and other structures. The fat and the glycogen serve primarily as fuel. Among the other structures, muscle contains an elaborate internal system of membranes (Fig. 3, *R*). These have been studied by Sjöstrand,[7] by Andersson,[8] by Porter and Palade[9], and others. These membranes often define tubular structures. This has led Sjöstrand to speak of them as "sarcotubules." The structures were seen with the light microscope by Retzius (1881, 1890) many years ago and termed the "sarcoplasmatic reticulum."[10,11] Thus, one can choose between two terms, each proposed by an eminent Swedish anatomist.

This internal system of membranes has, in general, not enjoyed much attention. It has been ignored completely by biochemists, and only recently has received the gaze of physiologists. Indications at the moment are that it may have a role in transmitting an excitatory signal from the sarcolemma to the contractile elements within the fiber.

Many of the features introduced above are diagramed in Fig. 3, which represents a semitransparent block cut out of a piece of striated muscle. It shows, in a rather over-simplified way, many of the features of muscle structure discussed in the foregoing. *V* represents the intercellular space outside the fiber. The double lines *S* represent the sarcolemma or plasma membrane of the muscle. This has been severed by a knife in the directions $a'$–$a'$. The flesh itself has been cut along $b'$–$b'$. The myofilaments *F* are represented by the very small dots in the cross sections at the ends of the block and by the very fine striations seen on longitudinal section. In this

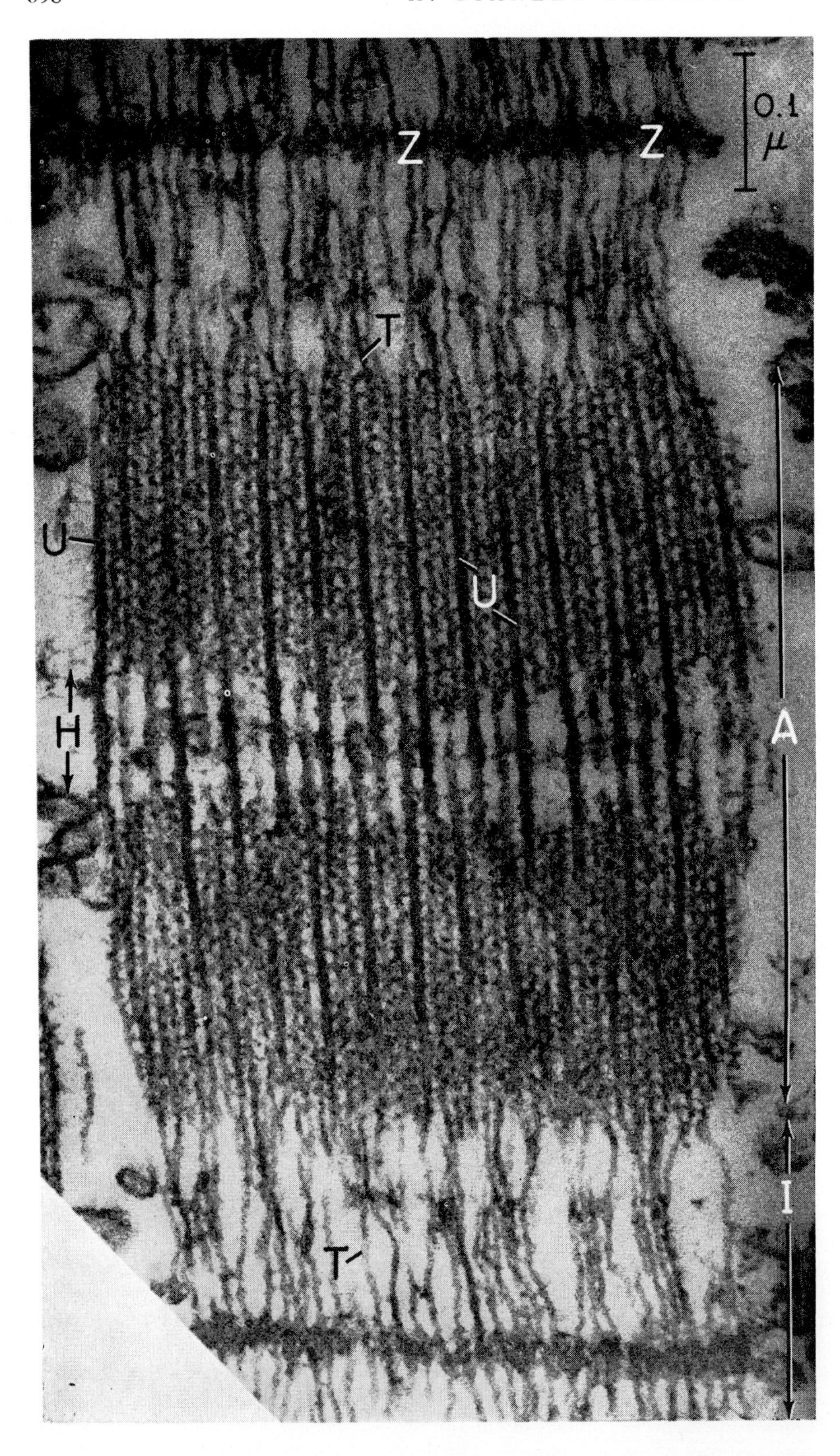


Fig. 4. An electron micrograph showing a single sarcomere from the striated muscle of a rabbit. This very thin section is cut nearly parallel to the axis of the myofilaments. Thin filaments *T* and thick filaments *U* can be distinguished and their overlap is displayed. *A*, *I*, *H*, and *Z* designate the corresponding cross bands of the striped muscle repeating unit, the sarcomere. Electron micrograph courtesy of H. E. Huxley.

muscle, the myofilaments are organized in groups, the myofibrils *M*. Myofibrils are found in many muscles but not in all. They are approximately 0.5 to 2 μ in diameter. In striated muscles, one finds a longitudinal repeating pattern, which F. O. Schmitt (p. 349) discusses earlier. This repeating pattern usually varies in relaxed muscle from about 1½ to 2 μ. However, in invertebrate striated muscle, the length of the unit may extend to

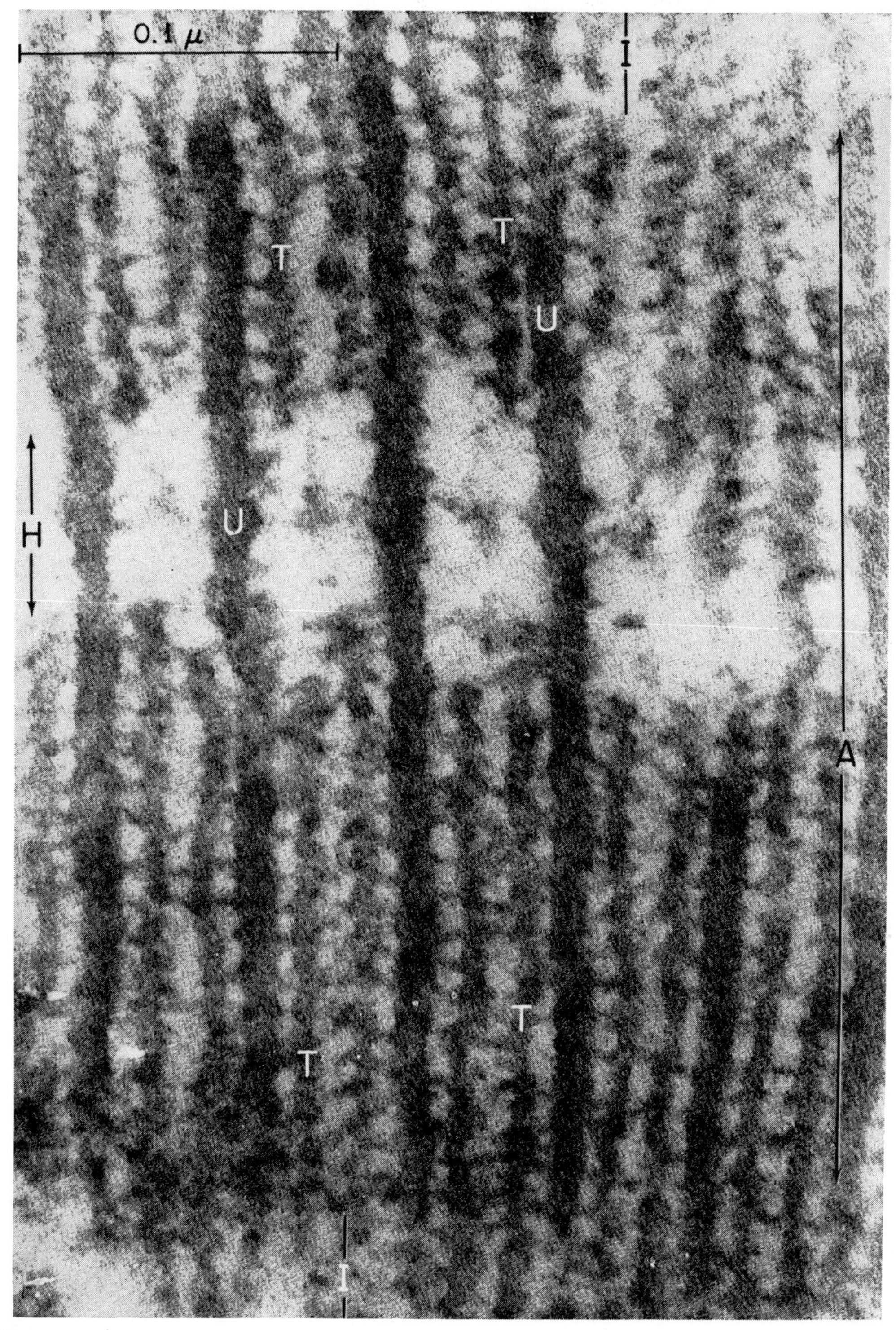


FIG. 5. A high-power detail of a portion of Fig. 4. The thin filaments *T* can be seen interdigitating with the thick filaments *U*, overlapping throughout the *A* band, except for the region of the light *H* band in the center. Slender bridges can be seen connecting the thick and thin filaments. These are believed to represent interaction sites, such as those represented by vertical short dashed lines in Figs. 1 and 2. Electron micrograph courtesy of H. E. Huxley.

10 or 15 μ, and in certain extreme cases (in certain unusual worms) repeating patterns of 50 or more μ can be encountered.

In most muscle, one finds many glycogen granules scattered about in the sarcoplasm. These are not shown here. In addition, particularly in certain phases of

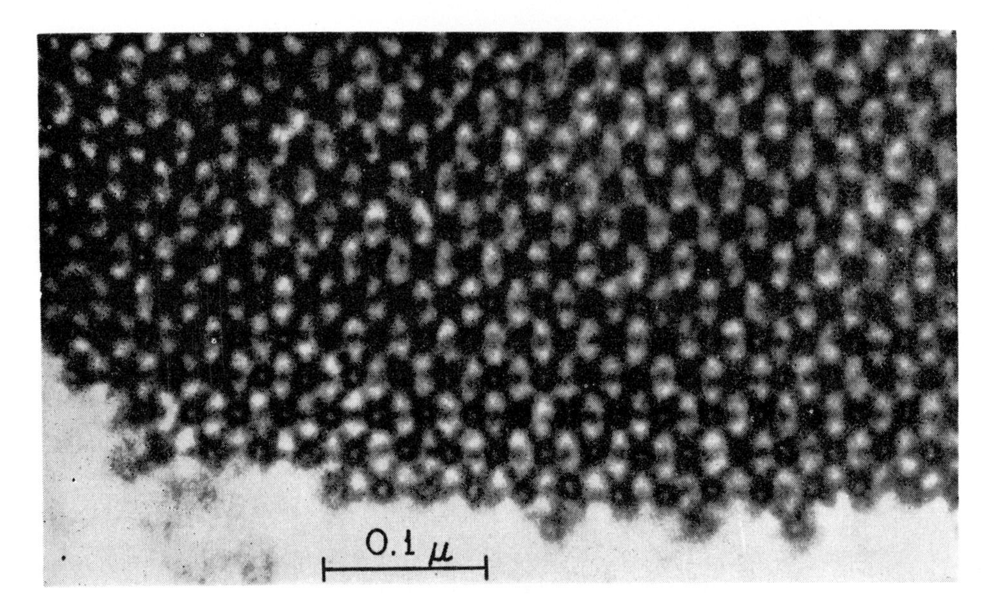


FIG. 6. An electron micrograph showing a cross section of a portion of a myofibril of the muscle of a fly. The thick myofilaments appear in cross section as circles arranged in hexagonal array, connected to each other by bridges forming equilateral triangles. In the middle of each bridge is a dense spot which may represent cross sections through thin filaments. Electron micrograph courtesy of A. J. Hodge.

physiological activity, one can find some intramuscular fat. If the animal is starved somewhat, a good deal of fat will appear in heart and other muscles. Under those conditions, mitochondria are closely associated with the fat droplets as though they were burning the lipid as fuel. In migrating salmon and in migrating geese, which feed extensively before their long journeys, the muscle fibers and adjacent cells accumulate very large amounts of fat. It is thought that the distance which a flock of geese can cover in a single flight is determined by the amount of fuel they can carry. It is much more economical to carry it in the form of fat than in the form of glycogen, since they can burn the hydrogen in the fat.

Turning back to Fig. 3, the mitochondria *I* can be clearly seen. Close to them and intertwining with the myofibrils are elements of the sarcoplasmatic reticulum of Retzius, *R*. In certain places, it can be seen that this reticulum comes in close contact with the mitochondria. It forms an extensive plexus here and there within the sarcoplasm of the muscle fiber. Thus, in gross outline it forms a network appearance. This gave rise to the concept that it was a reticulum. But if one cuts it in cross section, one sees that it is composed of membranes which form tubules which are very often flattened. Sjöstrand's term of "sarcotubule" is also appropriate, as it describes a second important feature of the structure. Retzius suggested in 1881 that this reticulum might pick up a signal from the sarcolemma and convey that signal deep into the interior of the fiber, thus exciting myofibrils a long way from the surface. Experimental justification of this concept, however, was a long time in coming. Only within the last two or three years A. F. Huxley and Taylor (see A. F. Huxley[6]) have brought forth evidence that there is indeed a conducting pathway which conforms to these specifications. Huxley and Hodgkin are now exploring the possibilities that the conduction might in fact reside within this system of tubules or membranes.

Robertson[12,13] has studied the sarcolemma in the electron microscope at high resolution. He has observed that it conforms in structure to the "unit membranes" found in other cells, presenting two peaks in density about 55 A apart, separated by a region of lesser density. On the outer surface of the sarcolemmatic unit membrane is a cloud of material which gives the chemical reactions associated with polysaccharides. It is believed to contain some glycoproteins. This type of polysaccharide-rich coating is found very frequently in association with plasma membranes, so its presence in association with the sarcolemma of muscle is not exceptional.

There is reason to think that there are probably two layers of lipid and some protein in the unit membrane comprising the sarcolemma. Various attempts have been made to fit these layers into the patterns of density seen in Robertson's micrographs. As Sjöstrand and Hodge point out, precise information about this point is not available. My own view is that all of the models concerned with the actual disposition of lipid within these membranes are so uncertain that one of our major tasks is to make a detailed map of the true arrangements of the proteins and lipid in biological membranes. I am doubtful of the usual presentation of the lipoprotein membrane structure as presented by Davson and Danielli,[14] where the lipid molecules are presented with their polar groups out and their nonpolar groups in.

Figure 4 is one of H. E. Huxley's elegant electron micrographs representing the arrangements of myofilaments in a myofibril. It provides some of the evidence for the hypothesis there are two interacting types of filaments differing from each other in structure. One, the thin filament *T*, is envisioned as starting at the *Z* band and extending a certain distance. The second type, the thick filament *U*, occurs in the middle of the repeating unit, which is called the sarcomere. Contraction is envisioned as involving an interaction between these two filaments, so that traction is exerted and the ends of the large filaments approach the *Z* band. Figure 5 shows the interdigitation of these filaments at higher power. In cross sections, the large filaments are arranged in a hexagonal pattern, and the small filaments are arranged in special relationship to them. In muscle from the rabbit, the small filaments are arranged so that each one is located in the center of a triangle formed by the large ones. One can see that, in such a case, if a section is cut through the muscle sufficiently thin to encompass only one layer, a plane can be found which will accommodate the thick filaments and two thin ones. Figures 4 and 5 are examples of such sections visualized in the electron microscope. Of importance are certain bridges which can be seen transversely extending between the two types of filaments. These may represent sites of interaction between the two filaments.

Hanson and Huxley[4] have proposed that the thick filaments contain the protein myosin and the thin filaments contain the protein actin. Huxley interprets contraction of the muscle as a change in the bonding occurring through reaction sites at the bridges, taking place in such a way that the actin and myosin filaments slide along each other.

Figure 6 is an electron micrograph by Hodge.[15] It represents a cross section of insect muscle. The large filaments and the cross bridges connecting them in a hexagonal pattern can be seen clearly. There are no small filaments lying in the center of the triangles formed by the large filaments in this particular type of muscle. However, in many cases, one can see a dense spot in the middle of the line connecting adjacent large filaments. It can be shown that these represent cross sections of small filaments which, in this type of muscle, are arranged in the centers of lines connecting individual large filaments rather than in the middle of triangles bounded at the corners by the large filaments. Huxley, however, believes that this is a trivial difference and that in both cases the interactions between the small and large filaments are the basis of muscular contraction.

Whether or not one thinks of contractility as residing in one filament or in the interaction between two filaments, these micrographs of Huxley and of Hodge have gone far in showing the geometrical framework within which the contractile mechanism resides.

## BIBLIOGRAPHY


[1] W. Kühne, Arch. Anat. Physiol. u. wiss. Med., Leipzig 564–642, 748–835 (1859).

[2] K. H. Meyer, Biochem. Z. **214**, 253 (1929).

[3] W. T. Astbury, Proc. Roy. Soc. (London) **B134**, 303 (1947).

[4] J. Hanson and H. E. Huxley, Symposia Soc. Exptl. Biol. **9**, 228 (1955).

[5] H. E. Huxley, J. Biophys Biochem. Cytol. **3**, 631 (1957).

[6] A. F. Huxley, Progr. in Biophys. and Biophys. Chem. **7**, 255 (1957).

[7] F. S. Sjöstrand, Intern. Rev. Cytol. **5**, 455 (1956).

[8] E. Andersson in *Electron Microscopy*, Proceedings of the Stockholm Conference, September, 1956 (Academic Press, Inc., New York, 1956), p. 208.

[9] K. R. Porter and G. F. Palade, J. Biophys. Biochem. Cytol. **3**, 269 (1957).

[10] G. Retzius, Biol. Untersuchungen **1**, 1 (1881).

[11] G. Retzius, Biol. Untersuchungen, Neue Folge, **1**, 51 (1890).

[12] J. D. Robertson, J. Physiol. **140**, 58 (1957).

[13] J. D. Robertson, Anat. Record **130**, 440 (1958).

[14] H. Davson and J. F. Danielli, *The Permeability of Natural Membranes* (Cambridge University Press, New York, 1943), p. 64.

[15] A. J. Hodge, J. Biophys. Biochem. Cytol. **1**, 361 (1955).

# 44

# Motility of Cilia and the Mechanism of Mitosis*

SHINYA INOUÉ

*Department of Biology, The University of Rochester, Rochester 20, New York*

THE movement of cilia and chromosomes are interpreted in terms of the microscopic and fine structures in cells. Section 1 discusses structure and motility of cilia; Sec. 2, microscopic structure of the mitotic spindle; Sec. 3, centers of organization in spindle and cilia; and Sec. 4, on the physicochemical nature of the spindle; and Sec. 5, relation of cilia and chromosome movement to muscle contraction.

## 1. STRUCTURE AND MOTILITY OF CILIA

Many small organisms—bacteria and protozoa, sperm and embryos of larger organisms—are propelled by beating their thin whip-like cilia or flagella (used interchangeably here). Also, ctenophores even a foot long and several inches wide can swim about or adjust their gravitational orientation by coordinated beating of their ciliary bundles. When the cell or organism is fixed, as in the gills of mussels and clams and in the human trachea, cilia create a current capable of pumping a considerable quantity of liquid or mucous material.[1–3]

The length of cilia may range from a few microns to several millimeters, and there may be one to several thousand cilia per cell but their diameter is quite constant, usually between a tenth and a half micron. Where a number of cilia occur in a row, adjacent cilia beat slightly out of phase with each other and a regular propagating wave is observed.

At the base of each cilium is found a characteristic bulbous enlargement, the basal granule. Further, proximal to the basal granule, rootlets sometimes are found which may function as anchorage or serve to conduct impulses. The membrane of the cilium appears continuous with that of the cell body.

Examined with an electron microscope (Fawcett and Porter,[4] and also dicussed in the following), cilia reveal a characteristic inner fibrillar pattern of amazing uniformity. Near the periphery of the cilium, there are usually nine (or occasionally a multiple thereof) fibrils, each composed of two filaments (or tubules?) some 100 to 200 A in diameter. Surrounded by the outer nine are two additional central fibrils of somewhat smaller diameters (Figs. 1 and 2). The nine outer fibrils reach the base of the cilium and appear to merge laterally into a hollow tube to make up the basal granule. The two central filaments also extend the whole free length of the cilium but apparently do not reach the basal granule. At the tip of the cilium, the outer nine and inner two fibrils are said to merge. In cilia with distinct directional beating, the line intersecting the two central filaments lies at right angles to the direction of beat.

The pattern of nine plus two has been found in cilia from a wide variety of cells, in fact, in practically all cilia observed with adequate resolution. The eleven fibrils are unlikely to be artifacts formed during preparation for electron microscopy, as sperm-tail flagella macerated in distilled water also show the frayed eleven fibrils in dark field illumination.

Given this structure, how does one explain the mechanism of ciliary beat? The pattern of beat may be relatively simple, as shown in Fig. 3(a). There is a recovery stroke in which the limp cilium stiffens from the base up, and an effective stroke where bending is mostly at the base and the rest of the cilium acts as if it were stiff. The same flagellate organism which swims forward by this

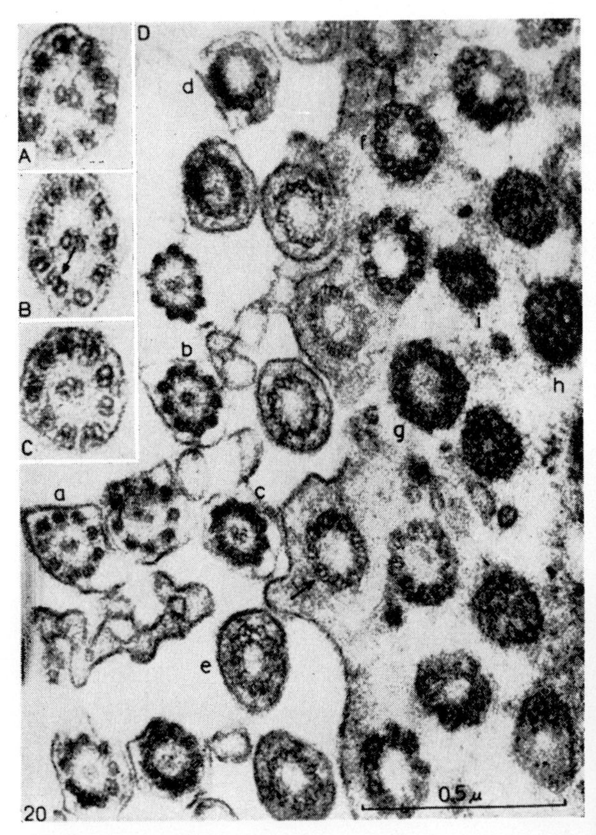


FIG. 1. Electron micrograph of cross sections of rat-trachea cilia. Compare with interpretive diagram, Fig. 2 [from J. Rhodin and T. Dalhamn, Z. Zellforsch. u. mikroskop. Anat. 44, 345 (1956)].

* Original work appearing in this paper was supported in part by the Public Health Service, U. S. Department of Health, Education, and Welfare (G-3002-C), and by the American Cancer Society.

beat may swim backward, also sidewise or circularly, as shown in Fig. 3(b).[2,5] Bradfield[6] postulates that waves of contraction proceed along the length of the outer nine fibrils with a message perhaps traveling in advance along the two inner ones. The waves of contraction may be started at the base by a commutator-like device which would result not in a synchronous contraction but in waves with various phase lags. It is more likely, however, that the outer fibrils may be the conductive elements, the inner two at least partaking a more active function in beating. For, in certain sensory cells (see following) and at the embedded base of each cilium[4,7] where one would expect conduction but not contraction, one in fact finds the outer nine fibrils and not the inner two.

It was tacitly assumed in the foregoing that contraction of the fibrils was the basis for cilia beat. Actually, the only evidence for contraction of cilia components (at the molecular level) appears to lie in the x-ray diffraction studies of Astbury *et al.*[8] There they find in flagella collected from bacteria, in addition to an $\alpha$-protein pattern (which is characteristic of many fibrous proteins, such as keratin, myosin, elastin, etc., and is believed to reflect the fundamental spacing of the polypeptide backbone), a folded $\beta$ pattern. This supercontracted pattern they believe reflects the folding of a fraction of the polypeptide chains responsible for contraction. The protein isolated from this bacterial flagella

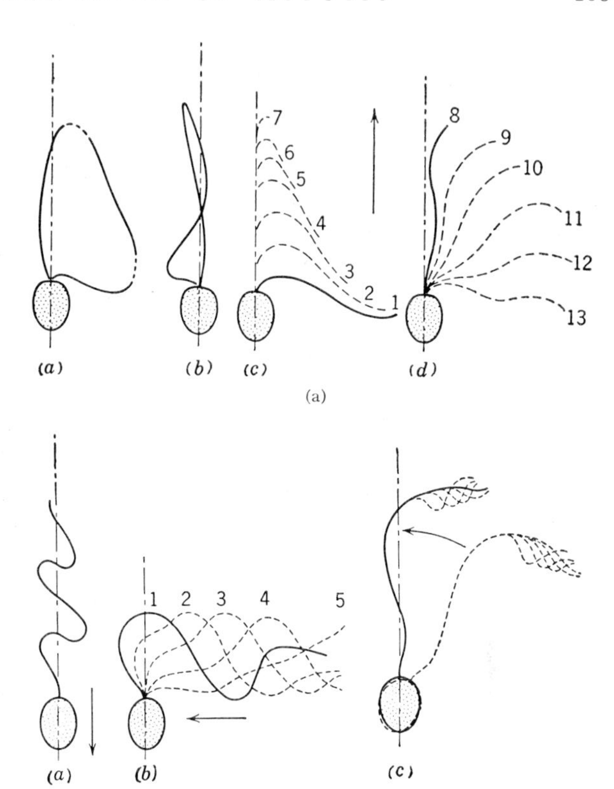


FIG. 3. Various patterns of beat of a *Monas* flagellum. Arrows indicate directions of swimming [from M. Hartmann, *Allgemeine Biologie* (Gustav Fischer Verlag, Stuttgart, 1953), fourth edition; after Krijgsman[5]].

preparation lacks sulfur-containing amino acids and appears dissimilar to any of the muscle components so far known.

Aside from localized contraction, the mechanism of ciliary beat has been interpreted by other schemes also, such as local swelling, reciprocal pumping of a liquid in and out of the cilium, or discontinuous flow of material through the cilium.[1,3,9] The exact site of the motor function in the cilium is also disputed, but nevertheless there exist certain observations (Secs. 3 and 5) which link the structure and function of these minute structures to other specialized motile structures of the cell.

DeRobertis, Sjöstrand, and others[10,11] made an interesting discovery related to the structure of cilia in the filaments of the retinal-rod cells and of the sensory-hair cells of the inner ear. These fibrous elements, believed for some time to be derived from embryonic cilia, also show a fine structure similar to that of cilia described in the foregoing. In these apparently non-motile fibers, the same outer nine fibrils are found, but the inner two are missing.

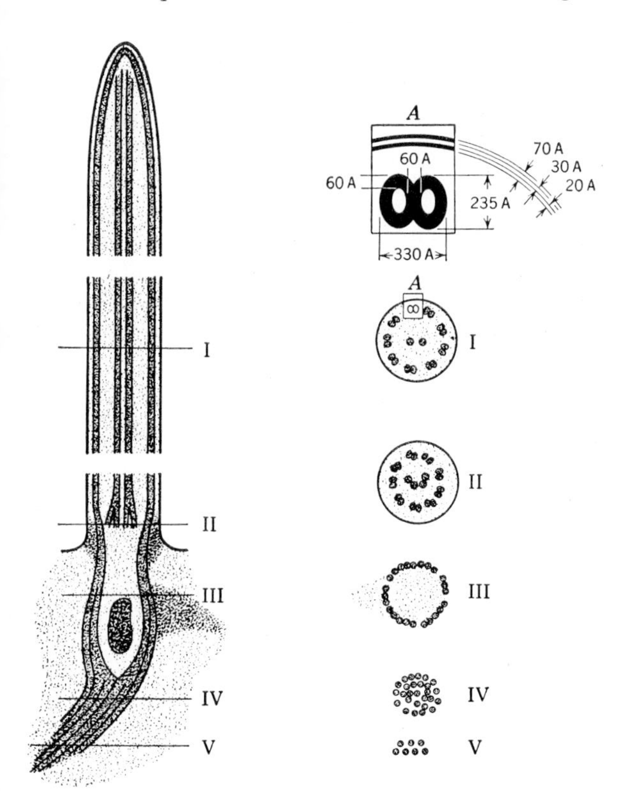


FIG. 2. Interpretation of electron micrograph (Fig. 1) showing fine structure of cilia [from J. Rhodin and T. Dalhamn, Z. Zellforsch. u. mikroskop. Anat. 44, 345 (1956)].

## 2. MICROSCOPIC STRUCTURE OF THE MITOTIC SPINDLE

Cilia may beat as frequently as a hundred cycles per second. Chromosomes, on the other hand, move extremely slowly, the maximum velocity being a few

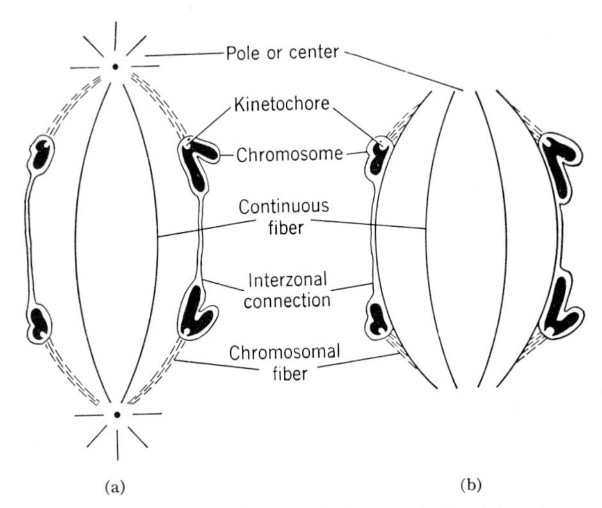


FIG. 4. Schematic diagram of mitotic spindle, (a) with centrioles and (b) without (modified from Schrader[19]).

microns per minute at anaphase (for reviews of mitosis see references 12 to 21). However, certain elements of the mitotic apparatus responsible for chromosome movement may be identical to, or at least have properties common to, portions of cilia. Before discussing this problem in Sec. 3, the structure of representative mitotic apparatuses are described.

Using Schrader's description,[19] the following terms are used. A dot or rod-like structure, the "centriole," is found at the poles of the mitotic spindle of many animal cells and occasionally in plant cells. The centrioles are morphologically the focal points for the spindle fibers and the astral rays [Fig. 4(a)]. In an average plant cell [Fig. 4(b)], both the centrioles and the astral rays are missing and the spindle fibers show less tendency to converge at the poles. Between the two spindle poles lie "continuous fibers." From the "kinetochores," a specific region of the chromosomes "chromosomal fibers" extend to or toward the spindle poles.

For half a century, the reality of these fibers in *living* cells has been disputed, for, with very few exceptions, they could be seen only in cells after fixation and staining. Recently, however, with improvements of the polarizing microscope, the author has been able to show these fibers clearly in living cells of many animals and plants by virtue of their positive birefringence (strength of birefringence $10^{-3}-10^{-4}$).[22,23] In animal cells, the fibers converge and their birefringence is stronger adjacent to the chromosome kinetochores and near the centrioles [Fig. 5(a)]. In plant cells, the situation is similar at the kinetochore region, but toward the "poles" the birefringence is weaker and the fibers are more diffuse [Fig. 5(b)]. During anaphase movement, the birefringence of the chromosome fibers persists and always is strongest adjacent to the kinetochores and the centrioles. The birefringence of the continuous fibers falls once and then rises again after complete separation of the chromosomes. The midregion containing fibers with strong secondary birefringence becomes the phragmoplast of plant cells [Fig. 5(c)]. Within the phragmoplast, small granules align, fuse, and become the cell plate, which divides the original cell into two. In animal cells, the cytoplasm generally cleaves inward at right angles to the spindle remnant forming two new cells with one nucleus each. Time-lapse motion pictures of these processes have been made by the author using a special polarizing microscope and were shown at the meeting.

The material of the mitotic apparatus has been isolated from sea urchin and other eggs by Mazia, Dan, and their collaborators in quantities sufficient for chemical analyses.[24–26] Amino-acid composition of their protein fraction (molecular weight *ca* 45 000) shows, unlike the bacterial flagella protein, a fair content of sulfhydryl-containing amino acids.

Although the spindle isolated after alcohol treatment by Mazia and Dan is stable, the fibers of the mitotic spindle in living cells are apparently extremely labile. The spindle fibers may disappear by slight mechanical agitation or by low-temperature treatment of the cell, only to reform in the course of a few minutes (see Sec. 5, also Carlson[13,27], Chambers,[28] Inoué,[29] and Östergren[30]).

### 3. CENTERS OF ORGANIZATION IN SPINDLE AND CILIA

Although the apparent velocity and very probably the stability of the spindle fibers differ by orders of magnitude from the cilia, the fibrous elements of the two structures may be formed or organized in a very similar fashion. The argument follows.

(A) The birefringence of the spindle fibers and astral rays is strongest adjacent to the kinetochores and centrioles throughout metaphase and anaphase (Inoué[23] and Schmidt[31]). The fibers are arranged radially (within restricted cones, in the case of the kinetochores) from these centers, and the growth of the spindle (at least in animal cells) takes place by lengthening of the fibers joining the centers. This strongly suggests that both the centrioles and the kinetochores are centers of fiber orientation.

(B) Growth of the axial filament of the sperm-tail flagellum starts from the basal granule, which same structure during the last mitosis acted as a centriole of the spindle.[21,32] In certain protozoa, flagella and the mitotic spindle both grow simultaneously from common giant centrioles.[14,32]

(C) In abnormal divisions of snail spermatocytes, some chromosomes lose their kinetochores and cannot partake in mitosis. The Pollisters[33] have shown that a clear correlation exists between the number of such chromosomes and the number of supernumerary basal granules, which migrate to the cell periphery and form the same number of extra sperm tails. Also, those kinetochores earlier dissociated from chromosomes form small extra astral rays while the spindle for the next division is formed.

(D) With the electron microscope, centrioles of the mitotic apparatus have been shown to exhibit the same general structure and dimensions as that earlier described for the cilia basal granules, namely, a cylindrical structure containing nine groups of rods or tube-like elements.[34–37] This same structure is observed for the basal granule of sperm-tail flagellum.[38,39]

Thus, it appears that basal granules of cilia, kinetochores, chromosomes, and centrioles of the mitotic apparatus all act as centers of fibrous organization in cells. The basal granules, centrioles, and kinetochores, may be in fact identical structures, taking on different functions at different loci within the cell (see also Meves[40]).

### 4. ON THE PHYSICOCHEMICAL NATURE OF THE MITOTIC SPINDLE

Electron-microscope studies have revealed a characteristic fine structure in cilia (Sec. 1). The possible identity of their basal granules to centrioles of the spindle also was strengthened (Sec. 3). However, this technique as yet has revealed rather little of the structure and behavior of spindle fibers and kinetochores.[25,41,42] The lack of success, I believe, is attributed to the difficulty or impossibility of preserving the spindle material in a reasonably native form after fixation and electron bombardment. In contrast, the polarizing microscope enables one to observe birefringence of spindle fibers in actively dividing cells (Secs. 2 and 3). Although the fine structure cannot be resolved directly, measurement of birefringence allows one to interpret the changes taking place in the microscopically unresolvable domain. This section describes further polarization-optical observations which may shed some light on the physical chemistry of the mitotic spindle.

The spindle of the egg cell (of a marine worm *Chaetopterus*) can be stretched if the cell is flattened very gently. The length of the spindle is then found to be strictly proportional to the diameter of the compressed egg.[22] This relation is explained by the attachment of the spindle poles through astral rays[43,44] to the cortical-gel layer. Immediately upon stretching, the spindle is thinner and more pointed at the poles, while in a minute or two it grows fatter and the birefringence increases. When the egg is compressed suddenly, the link between spindle poles and the cortical gel is apparently broken and the spindle shortens as it loses its birefringence (Fig. 6).

At a given length, the birefringence of the spindle fibers is a function of temperature.[29] With abnormally low temperature (4° to 6°C), the spindle birefringence is abolished completely. When the temperature is raised, the birefringence returns. The loss of birefringence with low temperature is rapid (less than half a minute), but when the temperature is raised the birefringence and structure of the spindle fluctuate until, after several minutes, they reach an equilibrium specific for the new

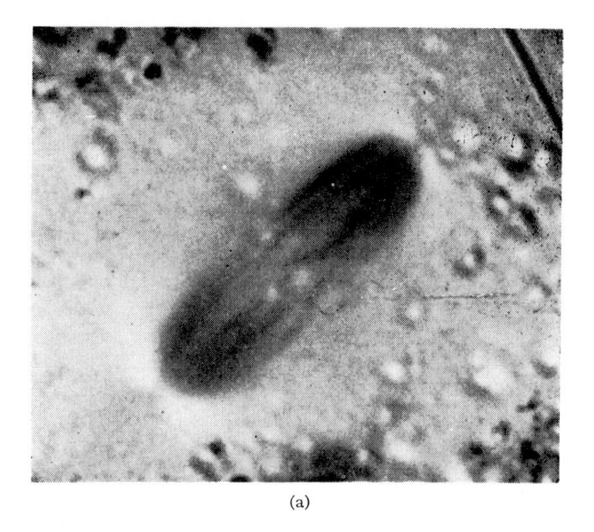

(a)

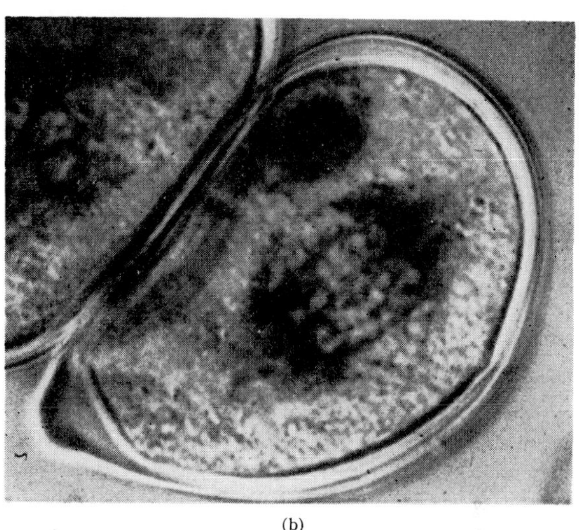

(b)

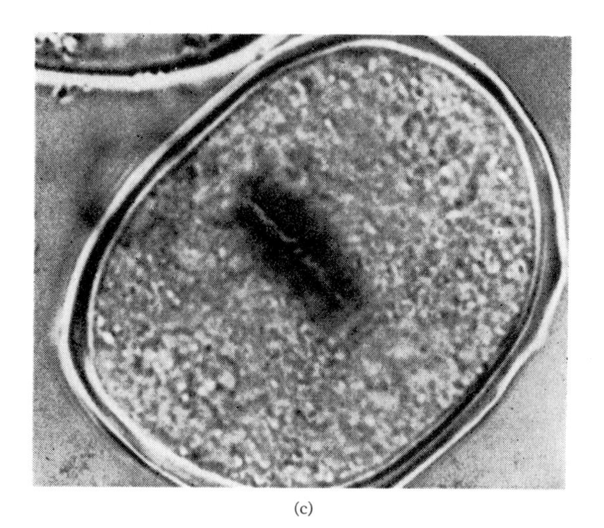

(c)

FIG. 5. Birefringent spindle fibers in living cells. Photographs are printed as negatives and show spindle fibers parallel to spindle axis black. (a) *Chaetopterus pergamentaceous* metaphase; (b) *Lilium longiflorum* early anaphase; (c) the same, phragmoplast with early cell-plate formation (modified from Inoué[23]).

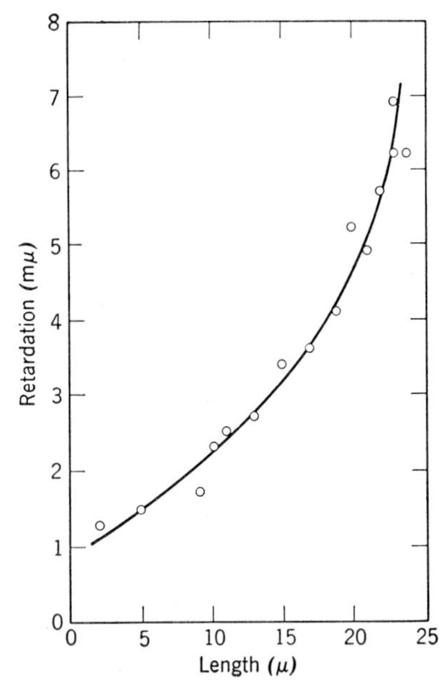


FIG. 6. Retardation of *Chaetopterus* spindle plotted against length [from S. Inoué, J. Exptl. Cell Research Suppl. 2, 305 (1952)].

temperature. The native spindle is, therefore, in a temperature-sensitive equilibrium.

The equilibrium birefringence at various temperatures is plotted in Fig. 7. If it is assumed that the birefringence of spindle fibers is directly proportional to the amount ($B$) of material oriented in that region, that only the equilibrium constant [$k(T)$] between oriented and nonoriented material is influenced by temperature, and that the total amount ($A_0$) of the orientable material in the same region remains constant, then the equilibrium is expressed by

$$A_0 - B \overset{k(T)}{\rightleftarrows} B.$$

$A_0$ was assumed constant since the spindle in the *Chaetopterus* egg is in metaphase equilibrium and also

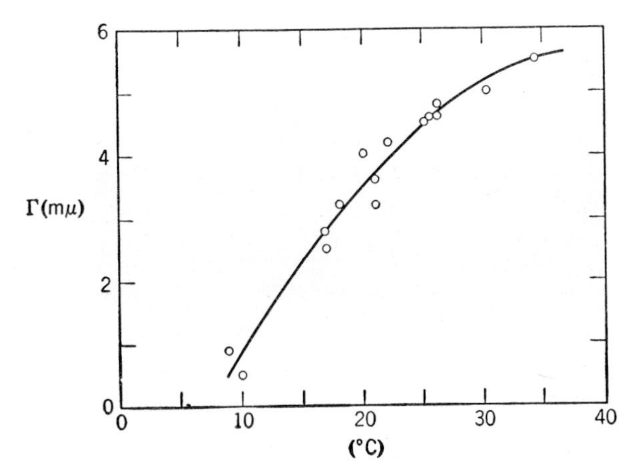


FIG. 7. Equilibrium retardation (Γ) of *Chaetopterus* spindle at various temperatures.

because cells which already have entered metaphase can go through division even in the presence of metabolic inhibitors such as cyanide and carbon monoxide.[45]

$A_0$ then is determined as the asymptote of the curve in Fig. 7. Were these assumptions warranted, one should expect a linear relationship between log $B/(A_0-B)$ and $1/T$°K. Figure 8 shows this plot. From the slope and intercept we (Morales[46] and Inoué) calculate the evolution of 28 kcal of heat per mole reacted, while at 25°C a free-energy change of $-1.8$ kcal/mole and an entropy increase of 100 eu/mole is observed.

The very low free-energy change agrees with the proposed lability of the spindle structure (weak gel with small number of active hydrogen bonds?), while

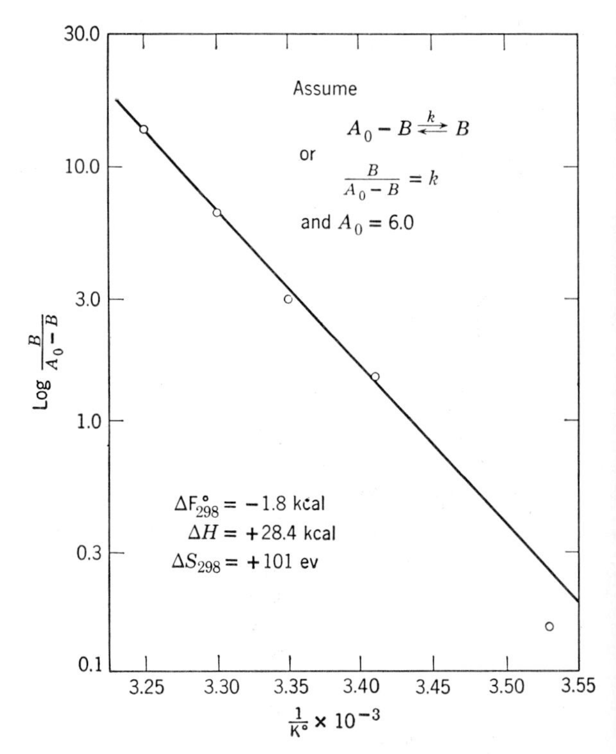


FIG. 8. Log plot of spindle reaction equilibrium *vs* inverse absolute temperature.

the high heat of reaction and the high positive entropy which nearly cancel each other explain the apparent decrease of entropy (increase of spindle birefringence) at higher temperatures. At high temperature, the still-unoriented protein molecules presumably absorb a considerable amount of heat, thus, for example, releasing bound water which could have prevented their orientation. The melting and randomizing of the bound water then could account for the large increase in entropy (also see Anderson[12]).

It appears that the spindle fibers are regions with high degrees of orientation, although very labile, expressing the orienting influence of the kinetochores and centrioles. As Östergren's earlier observation on the chromosome behavior of a plant *Luzula* also suggests,[30] spindle fibers are undoubtedly almost fluid in nature and are not stably crosslinked gels as the term "fibers" may imply. With dehydrating agents (e.g., alcohol) and in an acidic environment the crosslinking is probably enhanced until the spindle is finally "fixed."

On this basis and from observations described in Sec. 2, anaphase movement of chromosomes may be explained by local reduction in the quantity of oriented material and the consequent shortening of chromosomal fibers. The orienting forces of kinetochore and centriole must be just as actively at work throughout this process. The continuous fibers, either actively elongating or at constant length, could function as supports to counteract the pulling action of the chromosomal fibers. This mechanism of contraction of the chromosomal fibers is similar to that postulated by the author for the action of low concentrations ($<10^{-3}M$) of colchicine.[22] It is, however, in distinct contrast to the mechanism suggested by Swann[47,48] whose microscope lacked the power of resolution for detecting individual spindle fibers.[49]

### 5. RELATION OF CILIA BEAT AND MITOSIS TO MUSCLE CONTRACTION

The sole evidence for molecular contraction in cilia appears to lie in the x-ray diffraction data on isolated bacterial flagella (Sec. 1). Hypotheses involving mechanisms other than contraction (e.g., differential swelling) also have been postulated, but discriminating experiments are lacking.

Anaphase movement of chromosomes was explained by an orientation equilibrium of spindle fibers (Sec. 4). The action of centrioles and kinetochores—the center of foci of orientation in the spindle—appears similar to that of basal granules of cilia during fibrogenesis (Sec. 3).

Occasionally, cilia are resorbed or re-formed, but the process is much slower than the formation and disappearance of the mitotic spindle at each cell division. It appears that the cilia fibrils are quite stable while the molecules in spindle fibers probably are barely crosslinked (Sec. 4). In comparison, the contractile material in muscle may have a stability lying in between that of cilia and spindle fibers. The primary function of muscle and cilia is repeated rapid contraction, while with the spindle it is a single successful partition of the chromosomes into two new cells.

Mechanisms of muscle contraction are discussed in other papers of this symposium. It is interesting that one of the most widely discussed (and rather widely accepted) current hypotheses is that involving the creeping of two sets of filaments past each other.[50,51]

Regardless of the exact mechanism of choice, one may not overlook the muscle model systems which can contract and produce the same force per cross section as live muscles. This is true in muscle fibers extracted with 50% chilled glycerol, and in an oriented gel fiber formed by mixing two purified muscle proteins, actin and myosin. In either case, contraction is induced specifically by the addition of ATP (adenosine triphosphate) in the presence of magnesium and potassium ions.[52–54]

Hoffmann-Berling has shown further that motility can be induced in glycerol-extracted cells other than muscle, again by the addition of ATP. Thus, he was able to induce glycerinated sperm-tail flagella to undergo prolonged beating, chromosomes to separate, and extracted dividing cells to complete formation of their cleavage furrow (motion picture commercially available[54,55]). These models respond to approximately the same concentration of ATP as muscle models.

To what extent the movements induced by ATP in various cells reflect the same molecular mechanisms still is not clear. For example, the elongation of the central spindle, apparently responsible for the separation of chromosomes in the cell model, is not prevented by the same poisons to which the muscle and cilia model are very sensitive. Furthermore, although the organic triphosphate ATP (and ITP) specifically induces movements in extracted cells, the responding proteins show significant difference in their amino-acid compositions (see Secs. 1 and 2). It is nevertheless encouraging that movements closely resembling those found in living cells can be induced by the same reagent in cells from which much of the complex structures and materials have been removed.

In conclusion, evidence for the long-sought molecular folding is still weak, and no unifying molecular mechanism has been found for cilia beat, anaphase chromosome movement, and muscle contraction. Developments in recent structural and physicochemical analyses are, however, encouraging, and with a concerted intelligent approach, one may acquire before too long a much clearer understanding of the mechanisms underlying these cellular movements.

### BIBLIOGRAPHY


[1] J. Gray, *Ciliary Movement* (The Macmillan Company, New York, 1928).

[2] M. Hartmann, *Allgemeine Biologie* (Gustav Fischer Verlag, Stuttgart, 1953), fourth edition.

[3] L. V. Heilbrunn, *An Outline of General Physiology* (W. B. Saunders Company, Philadelphia, 1952), third edition.

[4] D. W. Fawcett and K. R. Porter, J. Morphol. **94**, 221 (1954).

[5] B. J. Krijgsman, Arch. Protistenk. **52**, 478 (1925).

[6] J. R. G. Bradfield, Symposia Soc. Exptl. Biol. **9**, 306 (1955).

[7] D. R. Pitelka and C. N. Schooley, J. Morphol. **102**, 199 (1958).

[8] W. T. Astbury, E. Beighton, and C. Weibull, Symposia Soc. Exptl. Biol. **9**, 282 (1955).

[9] W. J. Schmidt, *Protoplasma-Monographien* (Gebrüder Borntraeger, Berlin, 1937), Vol. II.

[10] E. De Robertis, J. Biophys. Biochem. Cytol. **2**, 319 (1956).

[11] F. S. Sjöstrand, J. Cellular Comp. Physiol. **42**, 45 (1953).

[12] N. G. Anderson, Quart. Rev. Biol. **31**, 243 (1956).

[13] J. G. Carlson, Science **124**, 203 (1956).

[14] L. R. Cleveland, J. Protozool. **4**, 230 (1957).

[15] A. Hughes, *The Mitotic Cycle, the Cytoplasm and Nucleus during Interphase and Mitosis* (Academic Press, Inc., New York, 1952).

[16] H. Ris in *Analysis of Development*, B. H. Willier, P. A. Weiss, and V. Hambürger, editors (W. B. Saunders Company, Philadelphia, 1955), p. 91.

[17] F. Schrader, Biol. Bull. **67**, 519 (1934).

[18] F. Schrader, "A critique of recent hypotheses of mitosis," *Symposium on Cytology* (Michigan State College Press, East Lansing, Michigan, 1951).

[19] F. Schrader, *Mitosis* (Columbia University Press, New York, 1953), second edition.

[20] C. P. Swanson, *Cytology and Cytogenetics* (Prentice-Hall, Inc., Englewood Cliffs, New Jersey, 1957).

[21] E. B. Wilson, *The Cell in Development and Heredity* (The Macmillan Company, New York, 1928), third edition.

[22] S. Inoué, Exptl. Cell Research. Suppl. 2, 305 (1952).

[23] S. Inoué, Chromosoma **5**, 487 (1953).

[24] D. Mazia, Symposia Soc. Exptl. Biol. **9**, 335 (1955).

[25] D. Mazia, Advances in Biol. and Med. Phys. **4**, 70 (1956).

[26] D. Mazia and K. Dan, Proc. Natl. Acad. Sci. U. S. **38**, 826 (1952).

[27] J. G. Carlson, Chromosoma **5**, 199 (1952).

[28] R. Chambers in *General Cytology*, E. V. Cowdry, editor (The University of Chicago Press, Chicago, 1924), p. 235.

[29] S. Inoué, Biol. Bull. **103**, 316 (1952).

[30] G. Östergren, Hereditas **35**, 445 (1949).

[31] W. J. Schmidt, Chromosoma **1**, 253 (1939).

[32] K. Bělař, Ergeb. Fortschrift. Zool. **6**, 235 (1926).

[33] A. W. Pollister and P. F. Pollister, Ann. N. Y. Acad. Sci. **45**, 1 (1943).

[34] M. Bessis and J. Breton-Gorius, Bull. microscopic appl. **7**, 54 (1957).

[35] E. de Harven and W. Bernhard, Z. Zellforsch. u. mikroskop. Anat. **45**, 378 (1956).

[36] J. Rhodin and T. Dalhamn, Z. Zellforsch. u. mikroskop. Anat. **44**, 345 (1956).

[37] Ch. Rouiller and E. Fauré-Fremiet, J. Ultrastructure Research **1**, 289 (1958).

[38] M. H. Burgos and D. W. Fawcett, J. Biophys. Biochem. Cytol. **2**, 223 (1956).

[39] D. W. Fawcett, Intern. Rev. Cytol. **7**, 195 (1958).

[40] F. Meves, Verhandal. Anat. Ges. (Jena) Halle, 152 (1902).

[41] P. R. Gross, Trans. N. Y. Acad. Sci. **20**, 154 (1957).

[42] K. R. Porter in *Harvey Lectures* (Academic Press, Inc., New York, 1957), Vol. LI, p. 175.

[43] J. C. Dan, Physiol. Zool. **21**, 191 (1948).

[44] S. Inoué and K. Dan, J. Morphol. **89**, 423 (1951).

[45] M. M. Swann, Quart. J. Microscop. Sci. **94**, 369 (1953).

[46] Thermodynamical analysis of the data was carried out in 1958 by M. Morales of Dartmouth College Medical School. The author is deeply indebted for his contribution.

[47] M. M. Swann, J. Exptl. Biol. **28**, 417 (1951).

[48] M. M. Swann, J. Exptl. Biol. **28**, 434 (1951).

[49] A. F. Hughes and M. M. Swann, J. Exptl. Biol. **25**, 45 (1948).

[50] J. Hanson and H. E. Huxley, Symposia Soc. Exptl Biol. **9**, 228 (1955).

[51] A. F. Huxley, Progr. in Biophys. and Biophys. Chem. **7**, 255 (1957).

[52] T. Hayashi and R. Rosenbluth, J. Cellular Comp. Physiol. **40**, 495 (1952).

[53] A. Szent-Györgyi, *Chemistry of Muscular Contraction* (Academic Press, Inc., New York, 1951), second edition.

[54] H. H. Weber, *The Motility of Muscle and Cells* (Harvard University Press, Cambridge, 1958).

[55] H. Hoffmann-Berling, Fortschr. Zool. **11**, 142 (1958).

# 45
# Fibrous Proteins of Muscle*

Alan J. Hodge
*Department of Biology, Massachusetts Institute of Technology, Cambridge 39, Massachusetts*

A FULLY coherent and integrated picture of the structure of muscle in terms of its various components and the changing relationships accompanying or responsible for contraction is still a long way off, as becomes evident later. An adequate description of muscle contraction requires knowledge sufficient to answer questions such as the following: First, which are the components of the myofibril actually participating in the contraction? Second, how does the structure of the individual components and their interrelationships one with another change during shortening? Third, how is the mechanical work produced at the expense of free energy derived from the hydrolysis of adenosine triphosphate (ATP)? The first of these questions is, at present, the only one which can be answered with any certainty. Although many models have been proposed for the contraction process, the physical changes accompanying contraction remain a subject for speculation, largely because of the difficulty in applying appropriate physical methods, such as x-ray diffraction and optical rotation, to the problem. The electron microscope has been employed with striking success in elucidating the details of muscle fine structure but, because of certain inherent limitations, it has been much less useful in following the structural changes accompanying contraction. A final solution, it would seem, can be accomplished only by a careful and rigorous synthesis of data obtained from light and electron microscopy, x-ray diffraction, optical rotation, and other physical and chemical techniques.

Before passing on to a consideration of the individual muscle proteins, it seems appropriate to review briefly some aspects of the progress which has been made since Engelhardt and Ljubimova[1] demonstrated the ATPase activity of "myosin." Later, Szent-Györgyi and his collaborators showed that myosin is a complex of two proteins, actin and L-myosin, and that the association of these components is influenced by the presence of ATP.[2,3] This discovery led to an extended period in which the mechanical and other properties of actomyosin threads were intensively studied. The discovery and development of the glycerine-extracted model system provided further stimulation for the study of artificial model systems in general. Further progress in narrowing the gap between intact muscle and the various extracted actomyosin systems was made by the discovery of substances able to influence contraction in one way or another. Of these, perhaps the best-known is the "relaxing factor" of Marsh.[4] The more-refined physicochemical methods developed in recent years have been applied to advantage in characterizing the various muscle proteins, but they have been relatively unsuccessful in elucidating the molecular changes accompanying contraction phenomena. Furthermore, it is difficult to extrapolate from the behavior of macromolecules in solution to their behavior in a complex paracrystalline multicomponent system like the myofibril. There exists an urgent need for the development of techniques which will allow the determination of configurational changes both at the molecular and macromolecular levels during contraction in intact muscle. X-ray diffraction is the only method currently applicable to this problem, and its application is fraught with very great technical difficulties, stemming largely from the short duration of a twitch and from the very high source intensities required for an adequate diffraction record at low angles. The application of optical-rotation methods appears to be an attractive possibility provided that the physical difficulties arising from the small dimensions and high degree of orientation of the contractile components can be overcome.

Following the pioneering work of Astbury in classifying the fibrous proteins into the $\alpha$ class (kmef group, now known to possess the $\alpha$-helical configuration[5–7]), and other classes which need not be considered here, Bear[8] showed that the x-ray diffraction patterns of various types of muscle fall into two distinct groups as judged by their low-angle diffraction patterns: (1) the Type I or paramyosin pattern corresponding to an axial repeat of 725 A found only in certain invertebrate muscles possessing the so-called "catch mechanism"; and (2) the Type II diffractions corresponding to axial spacing of about 400 A which are found in all muscles. As becomes evident later, this classification based on x-ray spacings is probably far more reliable as a criterion than some of the biochemical properties such as solubility and amino-acid composition, which have been employed in recent years as bases for nomenclature of the muscle proteins. At first, it was thought that this 400-A periodicity arose from a single component. However, as is seen, x-ray diffraction studies on purified components of striated muscle indicate that most, if not all of them, exhibit periodicities of about 400 A. Furthermore, the macromolecules of many of these Type II components have lengths of about this magnitude, and electron micrographs of myofibrils frequently show an axial spacing of similar dimension. It seems very likely, therefore, that, in the myofibril, the various

* This paper contains descriptions of original work by the author (on tropomyosin and paramyosin) which was aided by a research grant, E-1469 (C1), from the National Institute of Allergy and Infectious Diseases, National Institutes of Health, Public Health Service, U. S. Department of Health, Education, and Welfare.

FIG. 1. Crystal of LMM showing the sharp axial banding observed in the electron microscope in the absence of an "electron stain." The very dense transverse striations spaced about 420 A apart are probably owing to binding of salts by the end regions of the LMM molecules [from D. E. Philpott and A. G. Szent-Györgyi, Biochim. et Biophys. Acta **15**, 165 (1954)]). ×200 000.

components interact with one another in orderly fashion by means of reactive sites repeating at distances of about 400 A along the fiber axis.

The results of Huxley and Hanson[9] indicate that intermolecular rearrangement is an important part of the mechanism of contraction in striated muscle. How-

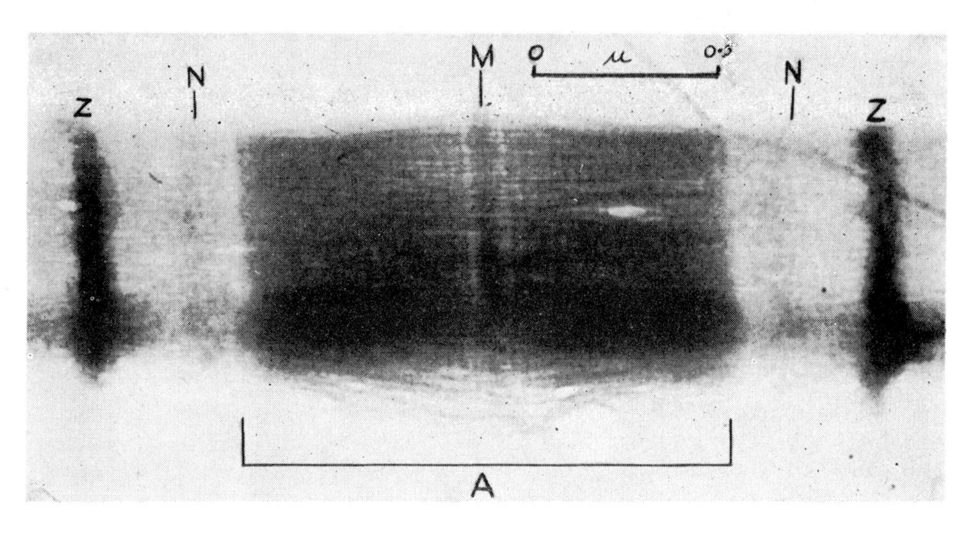


FIG. 2. Electron micrograph of a myofibril isolated from formalin-fixed toad muscle and stained with phosphomolybdic acid, showing the regular 400-A cross-striation in the form of fine, transversely oriented, dense lines [from M. H. Draper and A. J. Hodge, Australian J. Exptl. Biol. Med. Soc. 27, 465 (1949)]. ×64 000.

ever, the results obtained so far do not rule out the possibility that configurational changes are involved as well—the results of detailed correlative x-ray diffraction and other measurements must be awaited before any definite conclusions can be drawn, and in any case, the results obtained with striated muscle are not necessarily applicable to all types of contractile systems.

## ACTIN

This component of the actomyosin complex was isolated by Straub.[10] It appears to be a necessary component of the contractile system in muscle and comprises 15 to 20% of the structural proteins in rabbit striated muscle. An outstanding characteristic of actin is its ability to undergo polymerization (the $G-F$ transformation) to form well-defined fibrous elements, a process accompanied by the dephosphorylation of strongly bound ATP. The monomeric form of actin is one of the most difficult muscle proteins to characterize from a physicochemical point of view since the presence of salt favors the formation of the fibrous form. However, it seems likely that the monomeric unit has a molecular weight of about 60 000 (Table I, from work by Cohen and A. G. Szent-Györgyi[11]) with physical dimensions rather more uncertainly defined. The work of Astbury[22,23] and of Selby and Bear[24] indicates that actin exhibits an axial-repeat period of about 400 A in conformity with the other proteins of vertebrate striated muscle and of smooth muscle, both vertebrate and invertebrate. Actin has a pronounced tendency for complex formation with myosin, but is devoid of enzymatic properties.

TABLE I. Dimensions of fibrous-muscle proteins [from C. Cohen and A. G. Szent-Györgyi, paper read at International Biochemistry Congress, Vienna (1958)].

| | M.W. | Est. length (A) | Est. axial ratio |
|---|---|---|---|
| Actin | 57 000[12]–74 000[13] | 290[13] ? | 12[13] ? |
| Myosin | 450 000[14–17] | 1600 | 50 |
| HMM | 230 000[18]–330 000[19] | 400[18] | 15–20[18] |
| LMM | 96 000[18]–140 000[19] | 550[18] | 30–40[18] |
| LMM fr. 1 | 110 000 –120 000 | | 30–40 |
| Tropomyosin | 53 000[20] | 385[20] | 25[20,16] |
| Paramyosin | 134 000[21] | 1400[21] | 80[20] |

## MYOSIN

Myosin is the major constituent of most types of muscle, e.g., it accounts for 55 to 60% of the structural proteins of rabbit skeletal muscle. It has ATPase activity which is calcium activated and magnesium inhibited. It is a globulin, precipitating at low ionic strength, and can be extracted from muscle at neutral $p$H with solutions of ionic strength higher than about 0.6. Characterization of the myosin molecule has been a subject of controversy for many years, but there seems now to be general agreement that the molecule has a length of about 1600 A, an axial ratio of about 50, and a molecular weight of about 420 000 (Table I).

Brief exposure of myosin solutions to proteolytic enzymes such as trypsin,[25–27] chymotrypsin,[28] and subtilizin[29] degrades the myosin molecule into two fairly well-defined components, heavy meromyosin (HMM) and light meromyosin (LMM). The well-known properties of the myosin molecule are divided between these fragments. LMM behaves physically very much as does myosin. It precipitates at low ionic strength and has a molecular weight of about 100 000. However, it has no ATPase activity and does not combine with actin. On the other hand, HMM has all of the ATPase activity of the myosin from which it was derived, combines with actin in the same proportions as does myosin, but differs from myosin in being soluble in solutions of low ionic strength. Reported molecular weights range from 230 000 to 320 000 (Table I). HMM, like myosin, does not form ordered precipitates. LMM, on the other hand, under appropriate conditions, forms highly ordered needle-shaped crystals exhibiting a very striking axial period of about 420 A in the electron microscope, even in the unstained condition, as shown in Fig. 1 from work by Philpott and Szent-Györgyi.[30] It seems very likely that

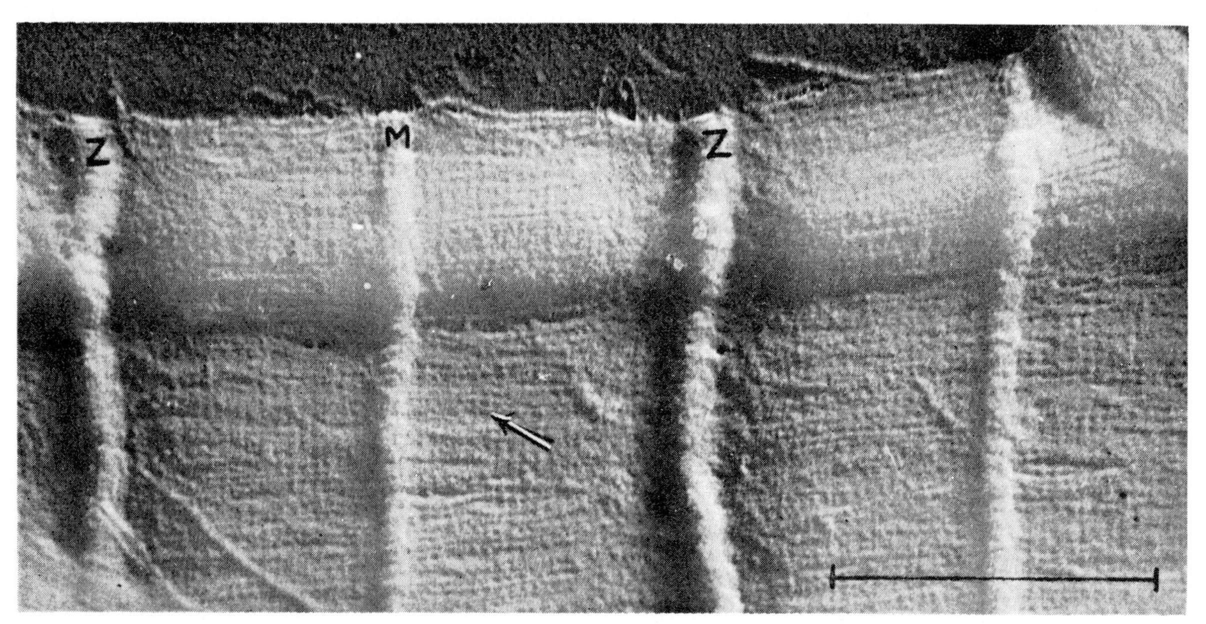


FIG. 3. Semicontracted myofibril of toad muscle, shadowed with platinum, showing the regular axial period apparently associated with the presence of transverse bridges (arrow) [from M. H. Draper and A. J. Hodge, Australian J. Exptl. Biol. Med. Soc. 27, 465 (1949)]. ×45 000.

the fine cross-striations evident in these crystals arise from a selective binding of inorganic ions at specific sites located about 400 A apart (possibly at the ends of the LMM "molecules"). This result is in good agreement with the fact that isolated myofibrils, when shadow-cast or stained with phosphotungstic acid (PTA) or phosphomolybdic acid, exhibit a sharp and regular cross-striation[31,32] (Figs. 2 and 3) with an axial period of about 400 A in the relaxed condition, and it agrees with the evidence obtained by Draper and Hodge[33,34] concerning the distribution of bound mineral in myofibrils after electron-induced microincineration in the electron microscope.† These authors found that the mineral residue from well-washed myofibrils obtained from formalin-fixed muscles is present as fine cross-striations with an axial spacing of about 400 A (Fig. 4), a result suggesting that the various inorganic and organic components of muscle interact one with another by means of reactive sites or groups located about 400 A apart. Further evidence[36] for this view is provided by the regular cross striations observed in thin sections of muscle (Fig. 5). There is in addition, some electron-microscopic evidence to suggest that this fundamental period may shorten during contraction,[37] a result which, if borne out by further investigation, strongly suggests that contraction must involve a configurational change in at least some of the macromolecular components of the myofibril. On the other hand, the sliding filament model, in which contraction is achieved by progressive interdigitation of actin and myosin filaments, does not require such configurational changes, and it should be noted that no changes have so far been observed in the wide-angle x-ray pattern of muscle during contraction. However, such configurational changes

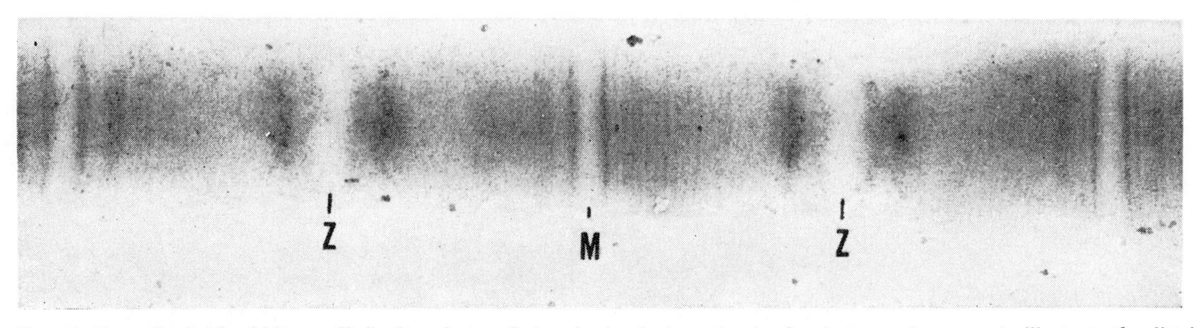


FIG. 4. Formalin-fixed rabbit myofibril after electron-induced microincineration in the electron microscope to illustrate the distribution of the bound mineral residue within the sarcomere in the form of regularly spaced, fine transverse striations spaced about 400 A apart [from M. H. Draper and A. J. Hodge, Nature 163, 576 (1949)]. ×30 000.

† The work of Leisegang[35] provides a rationale for the microincineration of organic material induced by high beam intensities in the electron microscope.

FIG. 5. Thin longitudinal section of rabbit muscle fixed in buffered $OsO_4$ and stained with PTA, showing the regular cross-striation present in all bands of the sarcomere [from A. J. Hodge, H. E. Huxley, and D. Spiro, J. Exptl. Med. **99**, 201 (1954)]. ×40 000.

would be difficult to detect if, as seems possible, a sequential shortening of small contractile units (perhaps 400 A long) is involved in the mechanism of contraction.

A number of recent experimental investigations point toward the rather disquieting possibility that the myosin "molecule" may be an artifact of the extraction methods used to isolate this protein. Thus, although it has so far proved necessary to employ proteolytic enzymes such as trypsin, chymotrypsin, and subtilizin in order to obtain the meromyosins from preparations of myosin, there is as yet no convincing evidence to indicate that the splitting of peptide bonds is necessary for the liberation of the meromyosins from the parent macromolecule. Middlebrook[29] was unable to demonstrate the presence of C-terminal groups in the ratios expected from the known specificities of trypsin and chymotrypsin. Furthermore, the degradation of the meromyosins by these enzymes is a much slower process than their production from the parent macromolecule, a result indicating the presence of a highly sensitive region within the myosin molecule, and the results of turnover-rate studies indicate a metabolic independence of the two meromyosins. Thus, Velick[38] has shown that the turnover rate of phenylalanine is about

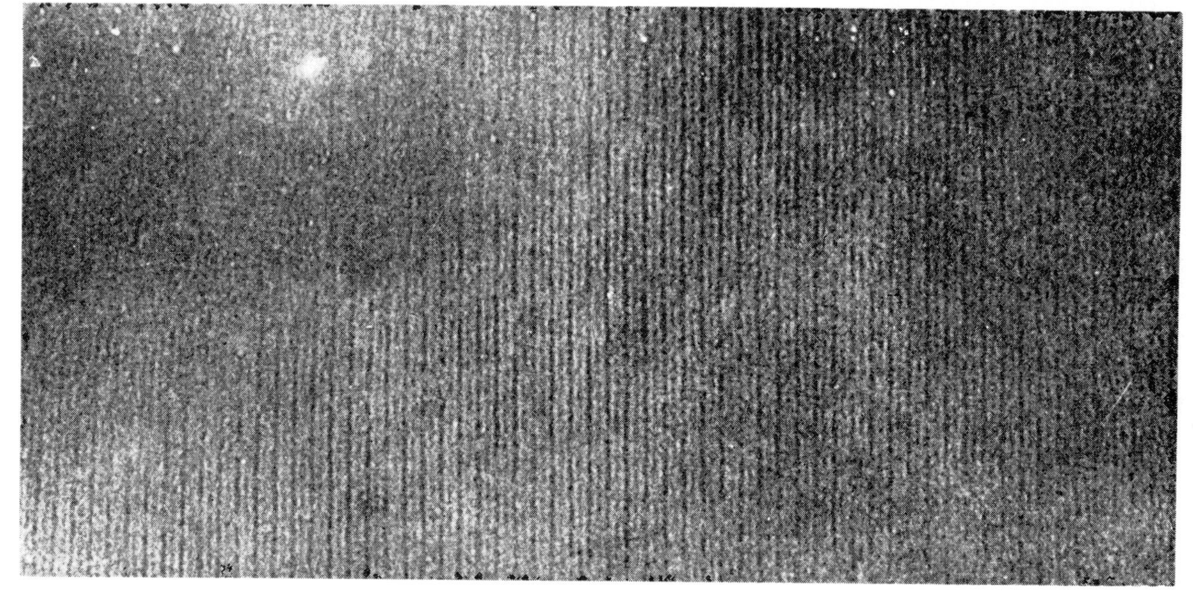

FIG. 6. Electron micrograph of a crystal of rabbit tropomyosin deposited on a supporting film and stained with PTA in *p*H 4.2 phosphate buffer. The main spacing is about 200 A, and a definite intermediate line is present. ×125 000.

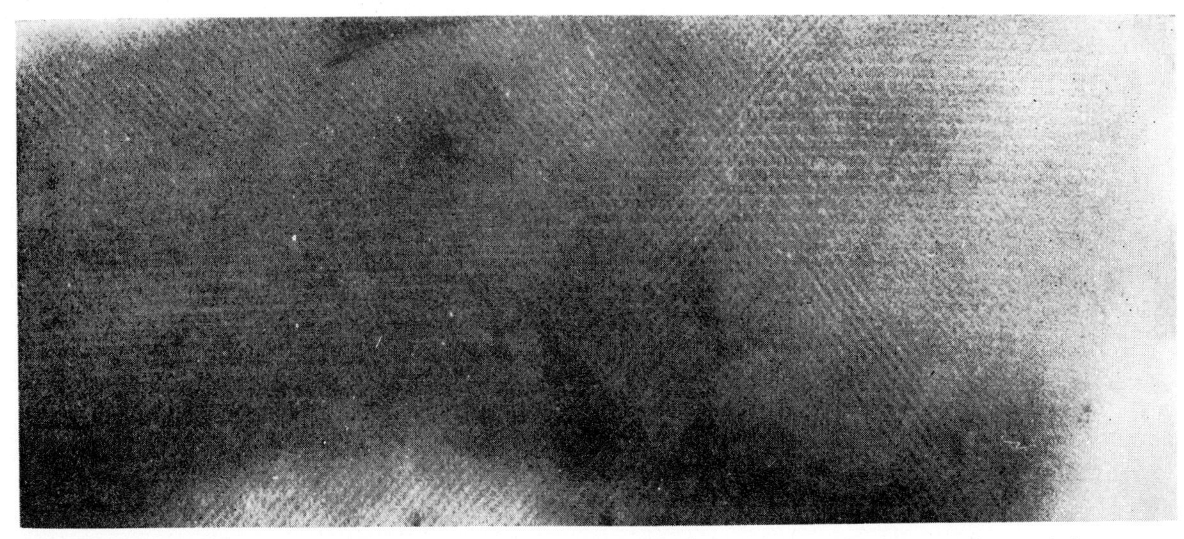

FIG. 7. Crystal of rabbit tropomyosin, stained with PTA, showing the crossed-grid appearance characteristic of many of these crystals in electron microscopic preparations. ×70 000.

five times more rapid in LMM than in HMM. Similarly, Schapira *et al.*[39] have observed that the incorporation of glycine is more rapid for LMM than for HMM. These results, on their face value, may be interpreted as indicating that the meromyosins represent precursors of the myosin macromolecule. However, there is also evidence derived from careful experiments with fluorescent antibodies to suggest that at least a part of the meromyosins in the myofibril are distributed independently of each other and in a highly characteristic pattern. This is discussed later. If nothing else, the sum total of the foregoing results must sow a seed of doubt concerning the almost universally accepted assumption that myosin *per se* is a well-defined macromolecular species which by some relatively simple interaction with other species such as actin is able to bring about the phenomenon of contraction.

### TROPOMYOSIN

This protein component, which is universally distributed in small concentrations in a wide variety of muscles, was discovered by Bailey[40] and is remarkable in that it is the first and probably the only fibrous protein thus far obtained in a truly three-dimensional crystalline form. The crystals of tropomyosin are plate-like and have an unusually high degree of hydration (80 to 90%). This latter property probably results from the fact that tropomyosin, unlike many of the fibrous proteins of muscle (which tend to form one-dimensional paracrystalline arrays), exhibits a strong tendency for the formation of three-dimensional crossed-grid net-

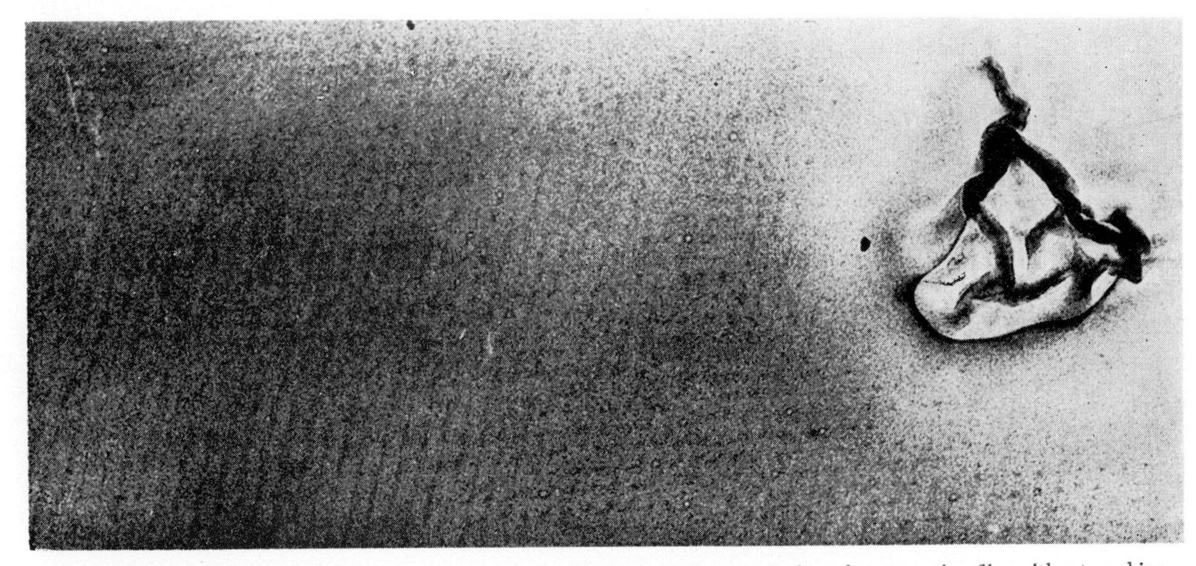

FIG. 8. Rabbit tropomyosin crystal from an ammonium-sulfate suspension mounted on the supporting film without washing or staining. The points and lines of high density probably correspond to regions where salt is accumulated. ×135 000.

works. A crystal of tropomyosin can, in fact, be regarded as a very highly ordered gel. Tropomyosin has no enzymatic activity and, as its name implies, was at one time thought to be a precursor of myosin itself. However, there is no substantial evidence for this view, and indeed, the role played by tropomyosin in muscle function remains a subject for speculation.

Tropomyosin, the water-soluble component originally defined by Bailey,[40] appears to be present in small amounts in all muscles (e.g., it accounts for about 4 to 5% of the fibrous proteins of rabbit striated muscle) and is present even in those invertebrate muscles that exhibit Type I low-angle diffraction patterns (i.e., those that have paramyosin as well as the actomyosin system). It is of interest to note that the tropomyosin molecule with a molecular weight of about 53 000 (Table I), has a length of about 400 A, as estimated from physicochemical evidence, a result in keeping with the lengths of all of the other components and subunits in the Type II system so far described.

Examination of tropomyosin crystals in the electron microscope in this laboratory has proved instructive and the spacings observed appear compatible with physicochemical estimates of the molecular length. A correlation with low-angle x-ray diffraction is currently being carried out.[41] The results obtained when a preparation of small tropomyosin crystals is placed on a grid, stained with buffered phosphotungstic acid (*p*H 4.2) or allowed to dry without staining are illustrated in Figs. 6–9. A frequently observed pattern is one comprising striations spaced about 200 A apart (often with intraperiod lines) or a crossed-grid network with periods of the same magnitude in two directions (Figs. 6 and 7). In unstained crystals (Fig. 8), the periodic structure appears rather as a pattern of dots, strongly suggesting that inorganic ions are occluded or bound at, or near, the sites of interaction of the long thread-like macromolecules arranged in an ordered gel structure. The square two-dimensional pattern with spacings of about 400 A (Fig. 9) affords yet another example of the remarkable capacity of tropomyosin to form open, ordered network structures, a property which may well have some significance in relation to the transverse bridges known to connect the longitudinal filamentous elements in striated muscle at regular intervals along the fiber axis. The regular outlines of the tropomyosin crystals become evident if they are fixed, embedded, and sectioned by conventional procedures (Figs. 10 and 11). However, dimensional analysis is rather more uncertain than in the cases already mentioned, since it is technically difficult to define the plane of the section in relation to a particular crystallographic plane. Nevertheless, the three-dimensional ordered net structure of the crystals is often clearly discernible (Fig. 12). This type of net structure is reminiscent of that observed in collagenous tissues such as Descemet's membrane[42] and in the pharyngeal structures of certain protozoa.[43] It may represent an important structural principle in ordered biological systems and may not be irrelevant to the process of contractility itself.

Fig. 9. Open, two-dimensional, net structure observed in a rabbit tropomyosin crystal preparation. The spacing of about 400 A corresponds to the length of the tropomyosin molecule as determined by physicochemical techniques (Table I). ×75 000.

As has been seen, there exists a rationale for an orderly interaction of the muscle components so far described in terms either of their lengths being about 400 A or of there being subunits of this length. It seems clear, therefore, that the total x-ray pattern (both meridional and equatorial) of whole muscle must contain information concerning the precise state of interaction of the macromolecular components for any given state of the contractile system, and offer the

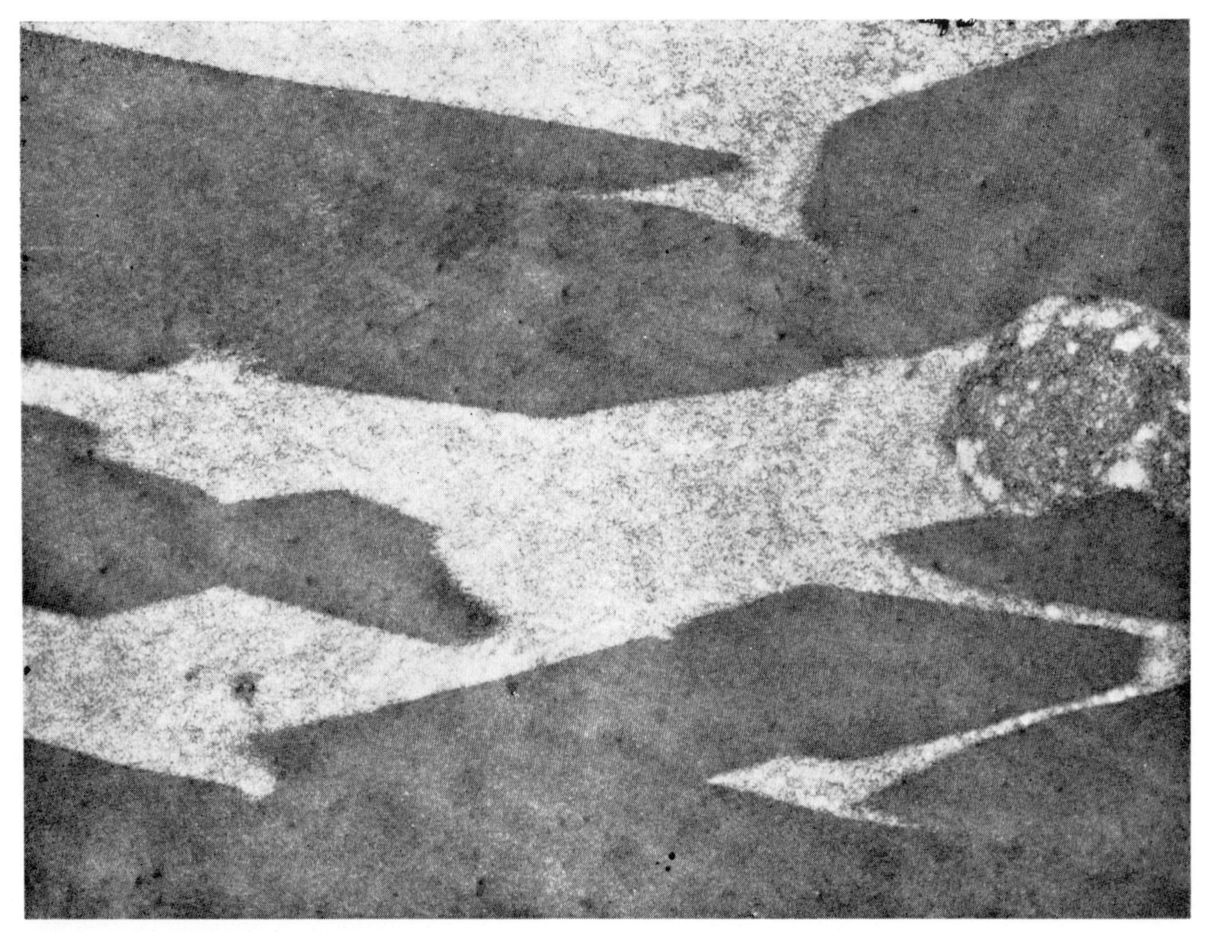

FIG. 10. Electron micrograph of a thin section of rabbit tropomyosin crystals, fixed in buffered, osmium-tetroxide solution, stained with PTA, and embedded in *n*-butyl methacrylate. Note the regular outlines of the crystals and the precise cross-striation corresponding to a spacing of about 200 A. ×20 000.

possibility of following the physical changes accompanying contraction. However, as already noted, there are severe technical problems to be overcome in any such investigation.

### PARAMYOSIN

The term paramyosin was introduced by Hall *et al.*[44] to describe a major component of certain specialized molluscan and annelid muscles (Table II),[45] which possess the ability to maintain a rigor-like contracture for long periods of time and which were described in the older literature as having a "catch mechanism." It is this protein that is predominantly responsible for the Type I low-angle diffraction pattern obtained by Bear[8] from these muscles, a highly characteristic pattern comprising reflections corresponding to a fundamental repeat period of 725 A, with every fifth reflection highly accentuated. The pattern can be interpreted formally (see Bear and Selby[46]) in terms of a net structure of the type shown in Fig. 13. "Paramyosin fibrils," which are easily isolated in native form from the adductor muscles of certain marine mollusks such as the clam, *Venus mercenaria*, exhibit a characteristic band structure with a period of 145 A in the electron microscope; together with a spot pattern, the symmetry of which is such that the true repeat period of this composite pattern is 5×145 A=725 A (Fig. 13). Since this spot

TABLE II. Relative intensities of the myosin and paramyosin x-ray diffraction systems exhibited by various muscles [from F. O. Schmitt, R. S. Bear, G. E. Hall, and M. A. Jakus, Ann. New York Acad. Sci. **47**, 799 (1947)].[a]

| Muscle | Myosin diffractions | Paramyosin diffractions |
|---|---|---|
| *Mytilus* adductor | + | +++++ |
| *Venus* adductor, w | ++ | ++++ |
| *Anodonta* adductor, w | ++ | +++ |
| *Anodonta* adductor, t | +++ | +++ |
| *Mya* adductor | +++ | +++ |
| *Venus* adductor, t | ++++ | ++ |
| *Pecten* adductor, w | ++ | ++ |
| *Phascolosoma* retractor | + | + |
| Dog retractor penis | + | ... |
| *Pecten* adductor, t | + | ... |
| *Thyone* retractor | + | ... |
| Frog sartorius | + | ... |

[a] Rough visual estimates of the relative intensities of the myosin and paramyosin diffractions on the patterns of the various muscles are indicated by +; w and t refer, respectively, to the "white" and "tinted" components of the muscles which possess them.

FIG. 11. Higher magnification view of two tropomyosin crystals in the same preparation as Fig. 10, showing the regular open-network lattice characteristic of these crystals. ×95 000.

Pattern is frequently absent in these native isolated fibrils, and because the major component of these fibrils can be extracted and reconstituted into fibrils in which no spots are visible but which exhibit band patterns with fundamental repeat periods of 725 A (Fig. 16), it seems likely that the spots reflect the presence of one or more additional components within the fibrils. In any case, it is proposed here to restrict the term paramyosin to the major component of these native fibrils. A composite structure for the native paramyosin fibrils is further indicated by some recent work[47] in which transverse sections of adductor muscle show that the fibrils are built up of a number of thin layers.

In 1952, the author[48,49] found that paramyosin could be quantitatively precipitated from acid solutions in the form of ordered tactoidal fibrils with an axial period of about 1400 A and with symmetrical intraperiod band structure, (Fig. 14) rather than the "polarized" band structure characteristic of the native 725-A period. At that time, Schmitt *et al.*[50] were in the process of analyzing the band structures of the newly discovered segment long-spacing (SLS) and fibrous long-spacing (FLS) forms derived from soluble collagen (asymmetric and symmetric band patterns, respectively), and had deduced that the collagen macromolecule was probably four times the length of the axial period (*ca* 700 A)

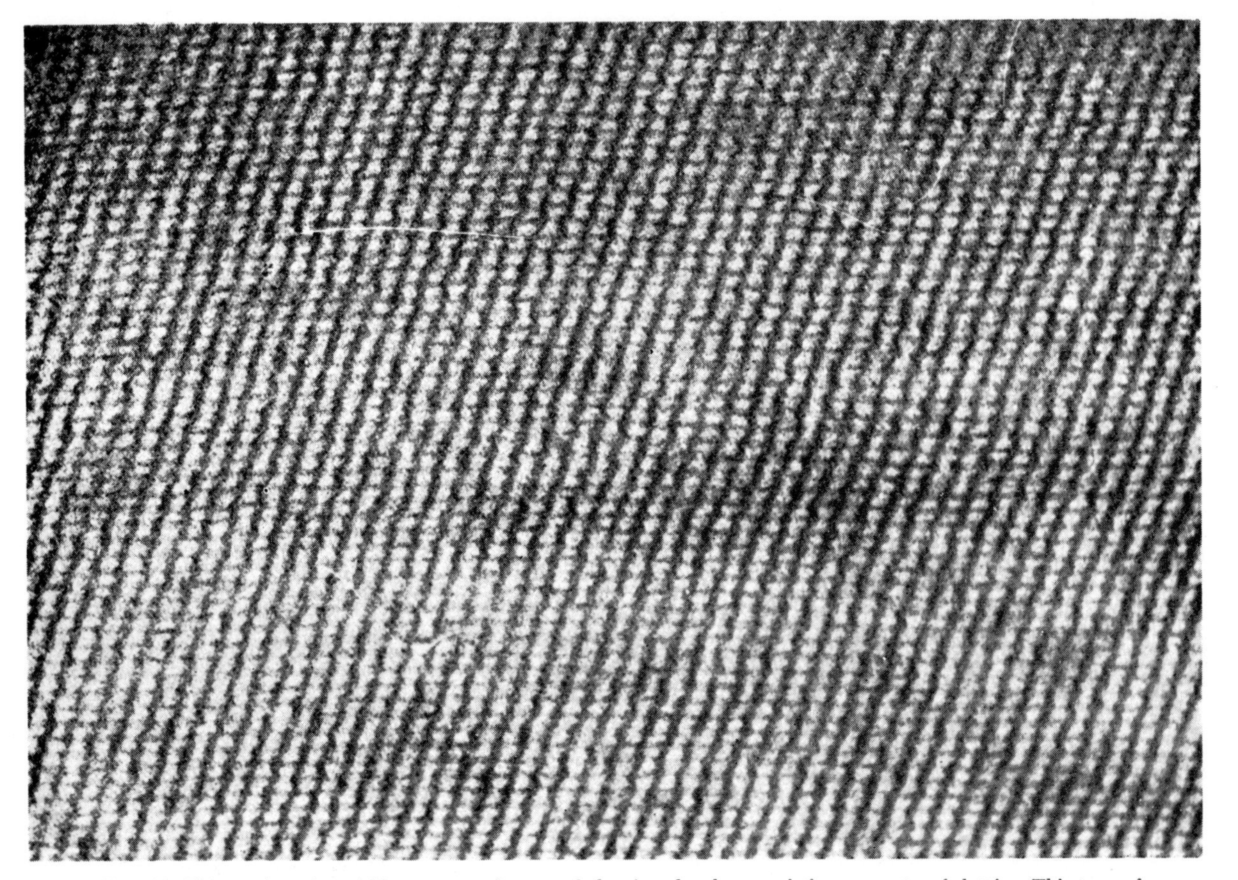

FIG. 12. Thin section of a rabbit tropomyosin crystal showing the characteristic open-network lattice. This type of structure probably accounts for the very high degree of hydration of the tropomyosin crystals. ×190 000.

displayed by native-collagen fibrils. According to their concept, the SLS forms arose by a parallel packing of the macromolecules with like ends in register, the FLS by an antiparallel packing, thus accounting for the symmetrical band structure of FLS. Consequently, by analogy with the collagen picture, it seemed likely that the macromolecules of paramyosin were about 1400 A long, and that they were packing in antiparallel array to form an FLS-type banded structure. This prediction was borne out[48,49] by the results of sedimentation, diffusion, viscosity, and light-scattering investigations, which were necessarily rather crude because of the low ionic strengths required to keep the protein in solution under these acid conditions. However, the data were sufficiently good to indicate the presence in the solutions of particles with lengths between 1200 and 1500 A and with axial ratios of about 70. Thus, the combined electron-microscopical and physicochemical evidence established the length of the paramyosin macromolecule (or at least of the kinetic unit in solution) with some certainty as being about 1400 A. The more recent and more accurate measurements of Kay,[21] carried out on paramyosin in neutral solution at high ionic strength, appear to confirm this value for the length of the paramyosin macromolecule.

Bailey[51] obtained paramyosin in purified form, and, on the basis of similarities in the amino-acid compositions of tropomyosin and paramyosin, was led to call the latter "insoluble tropomyosin."[52] Similarly, Kominz *et al.*[53] refer to paramyosin as "tropomyosin A" to distinguish it from "tropomyosin B" (i.e., the original tropomyosin of Bailey). This profusion of nomenclature seems unwarranted, and it would seem advisable at the present time to retain the term paramyosin, since there exist definite differences in solubility properties, crystalline form, and x-ray and electron-microscopic periodicities which clearly differentiate this protein from the original water-soluble tropomyosins of Bailey. Paramyosin has no enzymatic activity, according to Szent-Györgyi,[54] does not combine with actin, and there is as yet no evidence to indicate a possible function for it in relation to the specialized properties of the "catch" muscles.

A remarkable variety of fibrous structures can be formed from paramyosin solutions under various conditions of *p*H and ionic strength. Those so far obtained include fibrils with axial periods of *ca* 70, 145, 725, and 1800 A. (This last, together with one of 360 A, was first observed by Locker and Schmitt.[55]) In the case of collagen, it can be shown that the native type, 700-A

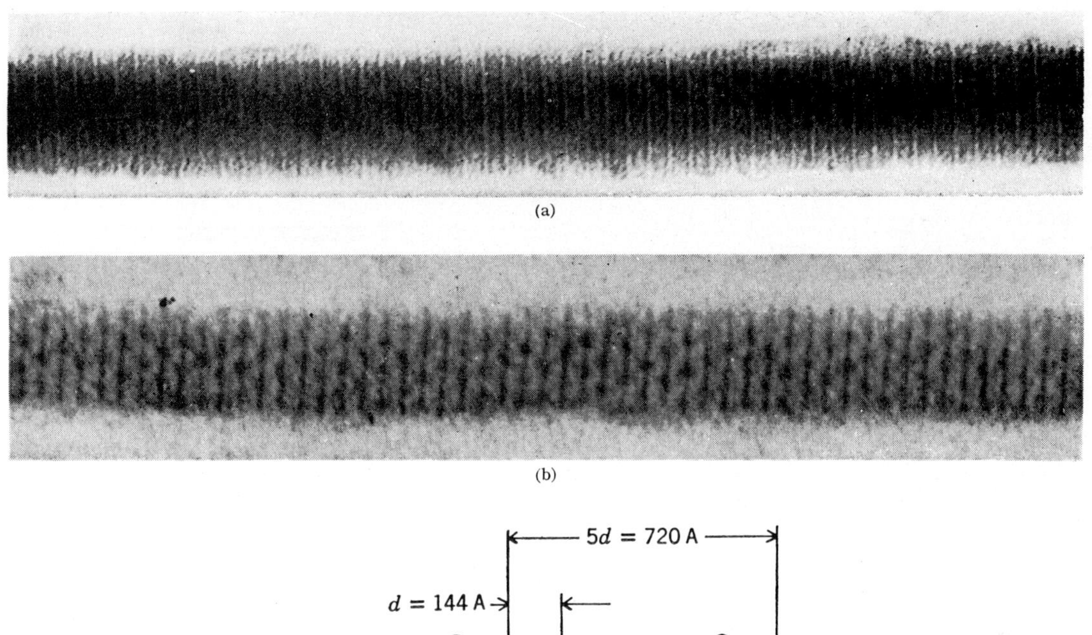

(a)

(b)

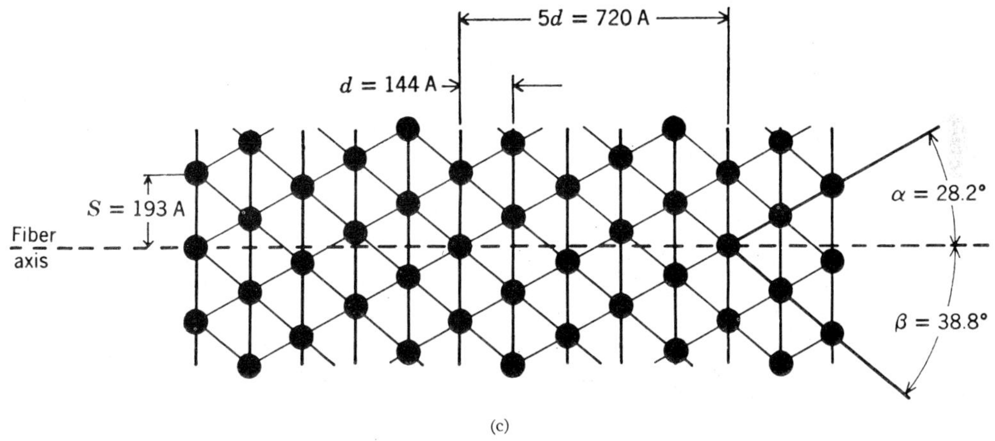


(c)

FIG. 13. Native "paramyosin fibrils" isolated from the adductor muscles of *Venus mercenaria* in dilute salt solutions. These usually show (a) a transverse cross-striation with an apparent period of 145 A,[48,49] (b) a pattern consisting of the 145-A cross-striation and a superposed dot pattern of symmetry such that the true repeat period is 5×145 A=725 A [from F. O. Schmitt, R. S. Bear, C. E. Hall, and M. A. Jakus, Ann. N. Y. Acad. Sci. **47**, 799 (1947)]. (c) shows the node pattern derived from electron micrographs. This formal net structure is in good agreement with the results of low-angle x-ray diffraction studies [from C. E. Hall, M. A. Jakus, and F. O. Schmitt, J. Appl. Phys. **16**, 459 (1945)].[8] (a) and (b) ×190 000.

periodicity can be obtained by a parallel packing in which the tropocollagen macromolecules (2800 A long) are staggered with respect to one another by one-quarter of their length. The key to such a synthesis of band structure is the availability of the SLS band structure, which is in effect a "fingerprint" of the tropocollagen macromolecule. Attempts to produce segment-type structures from paramyosin solutions have so far proved unsuccessful, but it seems likely that parallel, antiparallel, and staggered arrangements involving axial displacements of one-tenth the molecular length (i.e., 1400 A/10) or integral multiples thereof could explain the various band patterns encountered thus far (Figs. 13–18). A feature of interest in relation to these band patterns is the frequent occurrence of transitions from one type of packing to another. Figure 18 clearly shows such a transition from the 1800-A type to the 145-A spacing, Fig. 17 a rhythmic transition from the 145-A spacing to a *ca* 70-A spacing and vice versa.

## X-RAY DIFFRACTION AND ROTATORY DISPERSION

All of the proteins considered up to this point, with the exception of actin, give wide-angle x-ray diffraction patterns in which the prominent feature is the 5.1-A meridional reflection, indicating that at least a significant proportion of the polypeptide chains possess the α-helical configuration observed by Pauling and Corey.[5,6] This is true also of the meromyosins,[11] LMM giving a good pattern, with HMM being more variable since it is rather easily denatured. However, the x-ray diffraction method does not lend itself to estimation of the noncrystalline but helical parts of the molecule. The helix content of proteins in aqueous solution can best be estimated at the present time by rotatory-dispersion measurements, and the equation by Moffitt[56] describing the rotatory dispersion of the α-helix can be used for this purpose with the muscle proteins, subject to certain theoretical limitations, and on the assumption that the

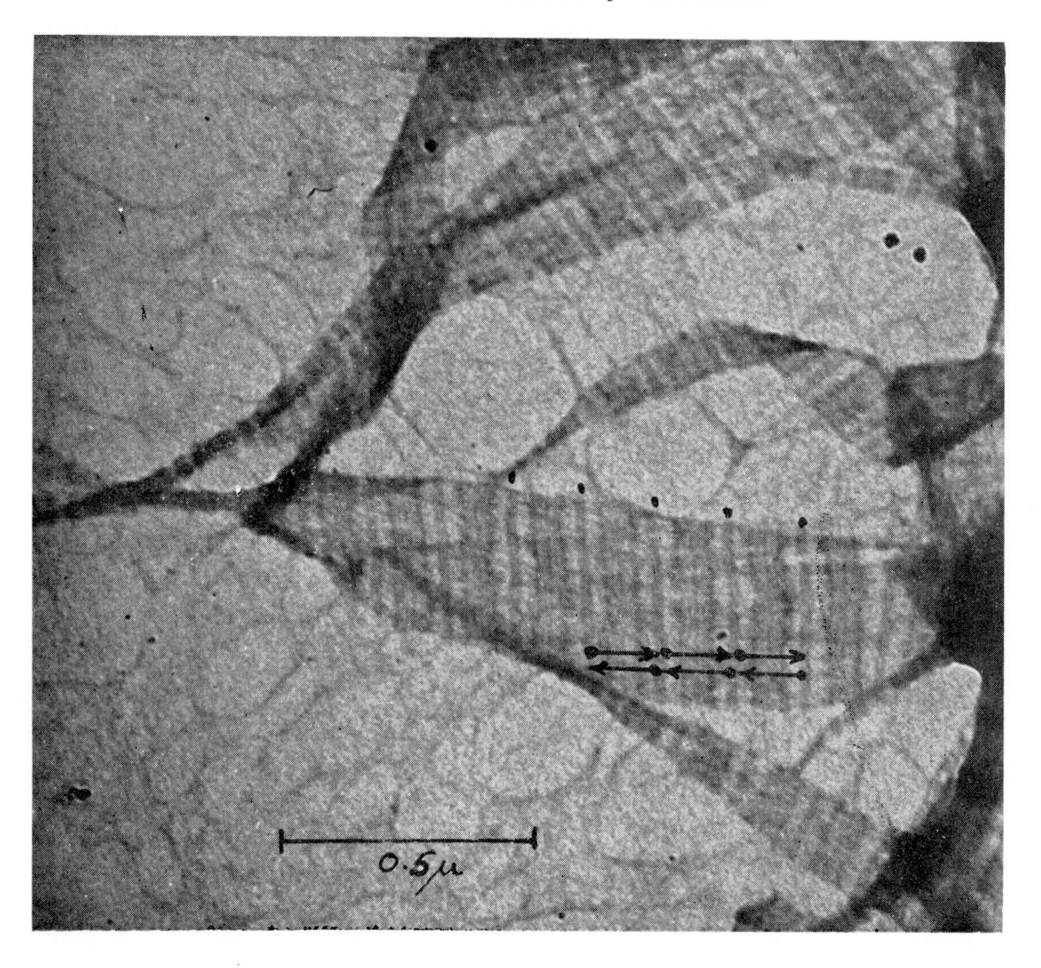


Fig. 14. Fibrils reconstituted from an acid solution of paramyosin showing an axial repeat period of about 1400 A. Note that the band structure within each repeat period is arranged symmetrically. This is interpreted as arising from an antiparallel packing of rodlike molecules about 1400 A long [from A. J. Hodge, Proc. Natl. Acad. Sci. U. S. **38**, 850 (1952)]. ×70 000.

entire helical content is in the right-handed α-helical configuration. Table III, from the work of Cohen and Szent-Györgyi,[11] shows the estimated α-helix content in aqueous solutions of the muscle proteins and fragments under discussion. It can be seen from this table also that the extent of the helical regions seems to be dependent upon the proline content. The proline contents of the highly helical muscle components (LMM fr. I, tropomyosin, and paramyosin) are very low, while those of myosin and HMM (which exhibit the lowest helix content) are correspondingly high. On a statistical basis, the data indicate that each proline residue prevents about 20 other residues from participating in α-helix formation.

Table III. Helix content and proline and cystine concentration of muscle proteins [from C. Cohen and A. G. Szent-Györgyi, paper read at International Biochemistry Congress, Vienna (1958)].

| Protein | Wt. % helix | Equ. cystine in $10^5$ g | Wt. % proline | No. of nonhelical residues per proline residue |
|---|---|---|---|---|
| LMM fr. 1 | 100 | 0 | 0.22 | |
| Tropomyosin | 94 | 1.65 | 0.35 | 15 |
| Paramyosin | 91 | 0 | 0.21 | 36 |
| LMM | 74 | 0.65 | 0.97 | 23 |
| Myosin | 56 | 0.6 | 2.08 | 18 |
| HMM | 45 | 1.2 | 2.87 | 16 |

## LOCALIZATION OF THE MUSCLE PROTEINS IN THE MYOFIBRIL

The commonly accepted distribution of actin, myosin, and tropomyosin in striated muscle as deduced from extraction experiments of a more or less selective nature is shown in Fig. 19, reproduced from a review by Perry.[57] However, there are indications that the picture is not quite so simple, especially in relation to the distribution of myosin and to the changes in the density distribution within the sarcomere accompanying shortening. According to the concept of Huxley and Hanson,[9] the actin and myosin are in the form of two separate sets of interdigitating filaments. Shortening is accomplished not by contraction of the filaments themselves, but rather by a mutual sliding action. As the *I* bands disappear, the ends of the myosin filaments pile up against the *Z* band to produce the well-known contraction bands ($C_z$ bands). However, there are many indications in the literature (e.g., Hodge[58]) which suggest that at least some of the material contributing to *A*-band density in the relaxed myofibril is capable of actual migration to the *Z* bands (and possibly to the *M* bands)

FIG. 15. Electron micrograph of a large flat paramyosin "crystal" obtained by reducing the ionic strength of an approximately neutral paramyosin solution, stained with PTA. The axial period is about 145 A. ×190 000.

independently of any sliding of formed myosin elements. Possibly, such substances act as moderators, activators, or inhibitors in controlling the orderly interaction of the actin, myosin, and other macromolecular components.

There is evidence to indicate that the myosin content of the *A* band is not homogeneous with respect to its extractibility. Thus, on extraction with Guba-Straub-ATP solution, a band of appreciable density remains on both sides of the *M* band[59] (Fig. 20), which can be removed only by further extraction with a pyrophosphate solution.[60] Furthermore, it has been shown that appreciable amounts of components other than myosin are extracted from the *A* band by these procedures.[61] The distribution of birefringence in the sarcomere as found by Inoué and Szent-Györgyi[62] using high-resolution polarization-optical methods coupled with extraction and replating experiments is also at variance with the simple distribution of actin and myosin as postulated by Huxley and Hanson.[63]

The polarization-optical results are more in harmony with the very interesting results of Holtzer and Marshall[64] using the technique of "fluorescent antibody analysis" to determine the localization of the various muscle components. The distributions of actin and myosin found with this technique are essentially in accord with that shown in Fig. 19. However, when using antibodies specific for LMM and HMM, the very curious result emerges that these two components appear to be localized independently of each other within the $A+H$ region of the sarcomere. The LMM appears to be located preferentially in the *A* bands proper, the HMM in a complementary pattern consisting of two bands, one on each side of and adjacent to the *M* band. More recently, Holtzer and Szent-Györgyi[65] have confirmed these observations and demonstrated that there is no physical hindrance to penetration of the antibody into the myofibril, by carrying out experiments involving extraction with KI solutions. In the control, this treatment removes most of the protein of the myofibril. However, pretreatment of a myofibril with a specific antibody prevents the extraction of the particular protein under consideration. The independent but somewhat overlapping distributions found for LMM and HMM by the above methods have also been confirmed by Inoué and Szent-Györgyi,[62] who first selectively extracted for myosin and then exposed the

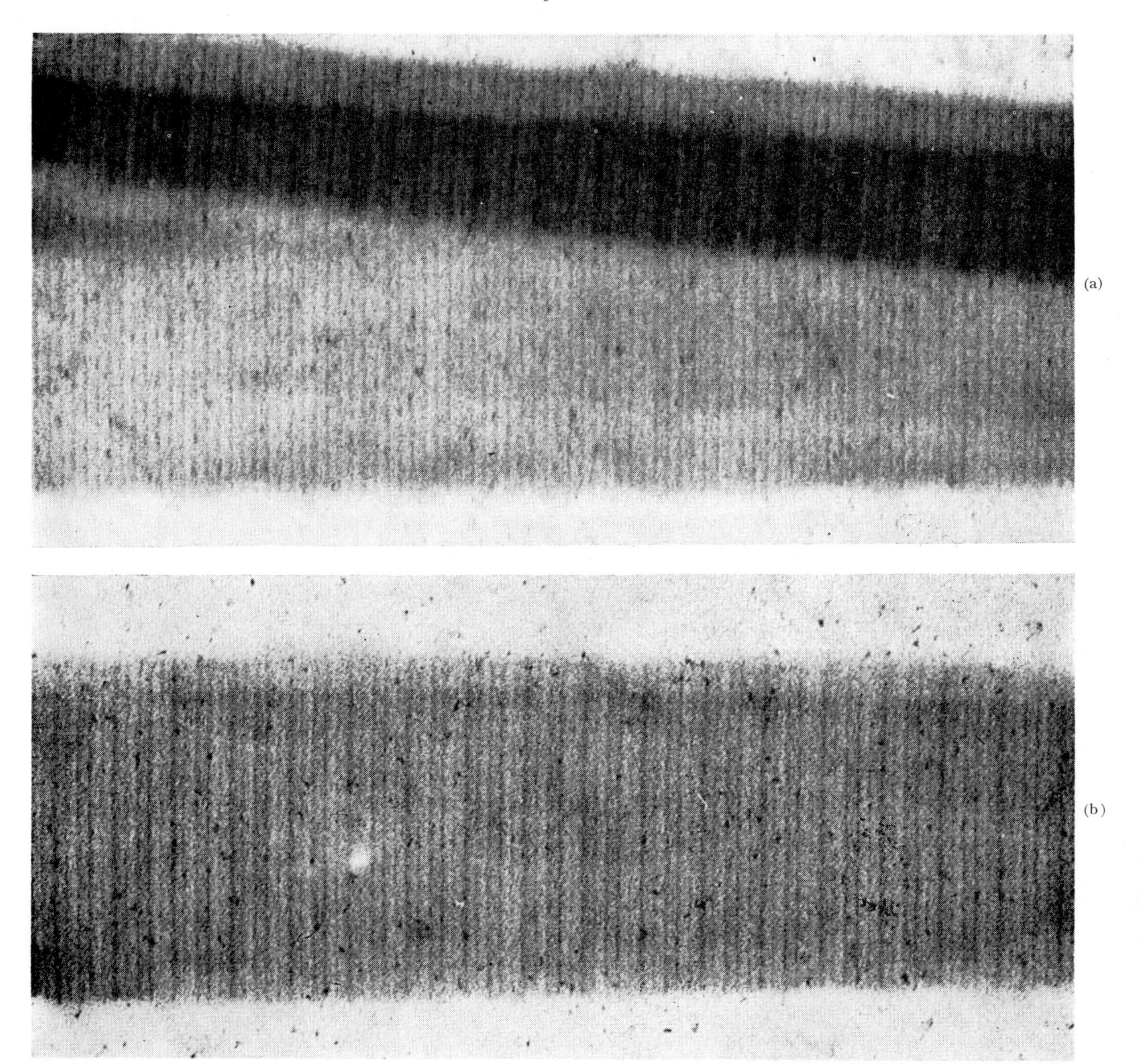


Fig. 16. Two fibrils from the same preparation as Fig. 15. The upper one (a) shows a simple period of about 145 A with a superposed apparently isomorphous structure having a repeat of 725 A, the lower one (b) shows a more complex pattern in which the fundamental repeat period is 5×145 A=725 A. Stained with PTA. Similar patterns have been observed by Hanson *et al.*[66] ×125 000.

myofibrils to LMM and HMM solutions before polarization-optical examination. They found birefringent bands appearing in positions consistent with the fluorescent-antibody results. This apparent independence of at least some of the LMM and HMM within the myofibril recalls the observations of metabolic independence mentioned earlier, and raises the question of whether or not some kind of interaction or reaction of LMM with HMM is involved in the process of contraction.

## PERSPECTIVES

The sliding filament model of muscle is based, to a considerable extent, on the observation in the electron microscope of two sets of filaments in the *A* bands, identified by Huxley and Hanson as actin and myosin. However, these filaments have been observed, of necessity, only in thin sections of muscle subjected to the procedures involved in fixation, dehydration, and embedding, so that the original state in the living, resting muscle fiber is a matter of inference. The danger of extrapolation from such observations is illustrated by comparing the low-angle x-ray diffraction data with the results of electron microscopy for muscle fixed (a) in the fresh state and (b) after glycerination. Living muscle gives a low-angle equatorial x-ray diffraction pattern consisting of two reflections corresponding to rod-like elements about 450 A apart in a hexagonal array,[67]

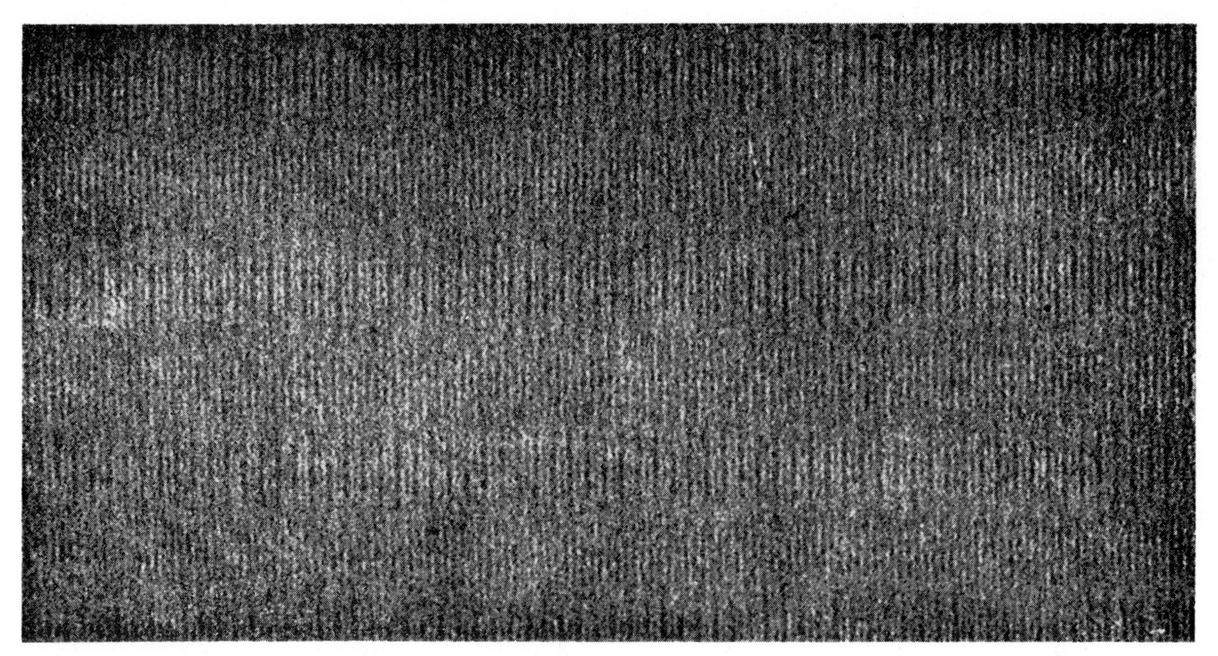

FIG. 17. Paramyosin crystals from the same preparation as Fig. 15, showing a rhythmic transition from a 145-A repeat structure to a *ca* 70-A repeat structure and vice versa. The author is grateful to J. L. Farrant for his suggestion that this pattern could result from the superposition (with small angular displacement of fiber axes) of two paramyosin crystals with 145-A spacings similar to that shown in Fig. 15. ×160 000.

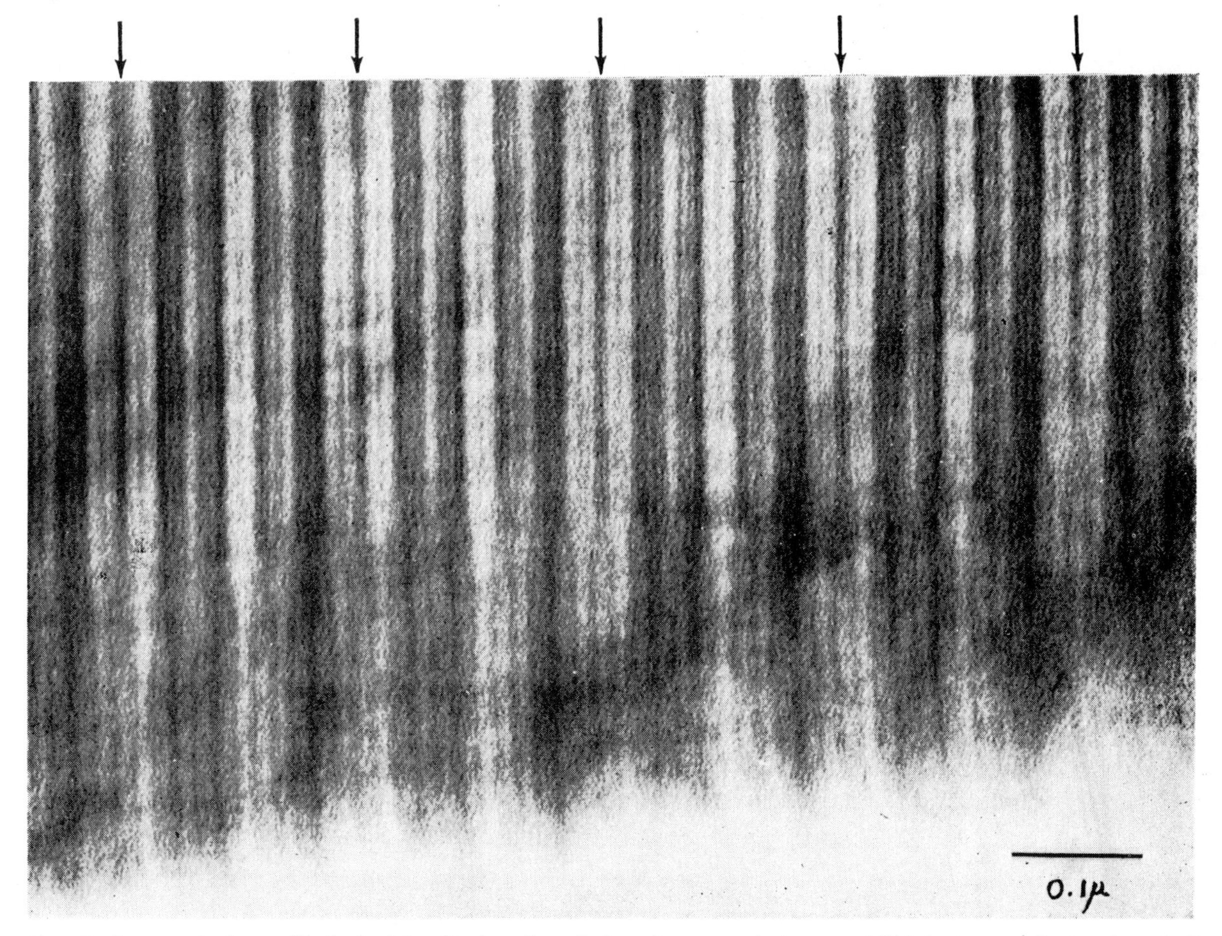


FIG. 18. Paramyosin "crystal" obtained by dilution of a solution of paramyosin in neutral high ionic strength phosphate buffer and stained with PTA, showing a complex band pattern of the type described by Locker and Schmitt[55] with an axial repeat period of about 1800 A (indicated by arrows at top), and a region of transition from the 1800-A repeat structure to one in which the 145-A spacing is accentuated. ×180 000.

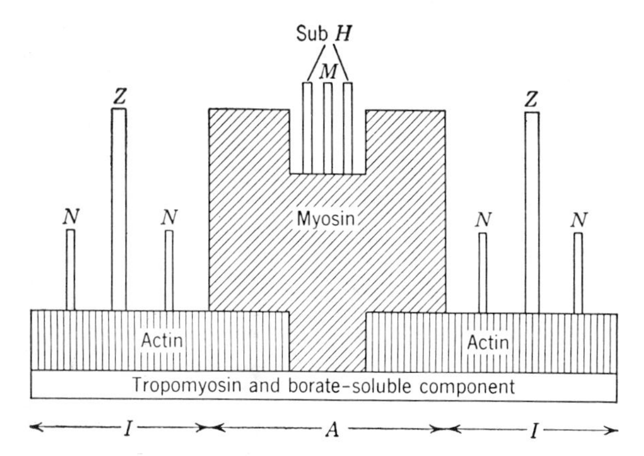


FIG. 19. Diagram illustrating the distribution of the protein components of the myofibril within the sarcomere. The height of the shaded areas represents the protein density at any point along the myofibril axis [from S. V. Perry, Physiol. Rev. **36**, 1 (1956)].

FIG. 20. Electron micrographs of thin sections through (a) intact glycerinated fibril and (b) glycerinated fibril after extraction with Guba-Straub-ATP solution [from J. Hanson and H. E. Huxley, Nature **172**, 530 (1953)].

with a strong first order and with a relatively weak second order. If the ATP is removed from the muscle (i.e., in glycerinated muscle or muscle in rigor), a striking reversal of the intensities of these two reflections takes place, the second order becoming very strong.[67] This indicates that, in the absence of ATP, a region of high electron density is present at a position intermediate between the primary rod-like elements, while, when ATP is present, the electron density between these elements is relatively uniform. The electron microscope results are at variance with this, for the two sets of filaments can be observed equally well in muscle fixed in the fresh state or after glycerination.[37] It thus is necessary to conclude that one (or more) of the processes involved in fixation, dehydration, and embedding causes the muscle to go into a rigor-like state, and consequently, that the density distributions observed in electron micrographs do not correspond to those present in the living, resting muscle fiber. The results clearly demonstrate the need for a thorough correlation by x-ray diffraction of the procedures used in the preparation of specimens for the electron-microscopic examination of muscle.

In the sliding-filament model, the two sets of filaments are postulated to remain at constant length on the basis of evidence derived largely from observations of $A$-band lengths during shortening. Thus, this mechanism does not require configurational changes of the protein components, neither at the $\alpha$-helix nor at the macromolecular level. No direct evidence on this point has been reported. However, Huxley[67] found no change in the axial spacing measured from the low-angle x-ray diffraction pattern on passive stretching of the muscle, a result which has been interpreted as possibly indicating that no change would accompany shortening. On the other hand, some as yet inconclusive evidence from electron-microscopical observations[32,37] suggests that there is a correlation of the axial period both with the sarcomere length and with the type of band pattern observed during shortening. Such a variation in the axial period would strongly imply the occurrence of configurational changes during shortening. It is clear that further evidence of a more direct nature is urgently needed in order to settle this matter.

As has been seen, the application of analytical and degradative methods has proved to be of great importance in solving many of the problems associated with muscle structure and function. However, it would seem that the time is ripe for an approach to muscle which considers the whole machine as a complex mixed macromolecular "crystal," capable of changes in physical state in order to convert available chemical energy into mechanical work. Any completely adequate theory of contraction must be able to take into account the interactions between the many and varied components of the myofibril and other contractile units.

## BIBLIOGRAPHY

1 W. A. Engelhardt and M. N. Ljubimova, Nature **144**, 668 (1939).

2 Albert Szent-Györgyi, Acta Physiol. Scand. **9**, Suppl., 25 (1945).

3 Albert Szent-Györgyi, *Chemistry of Muscular Contraction* (Academic Press, Inc., New York, 1951), second edition.

4 B. B. Marsh, Nature **167**, 1065 (1951).

5 L. Pauling and R. B. Corey, Nature **171**, 59 (1953).

[6] L. Pauling and R. B. Corey, Proc. Roy. Soc. (London) **B141**, 21 (1953).
[7] F. H. C. Crick, Acta Cryst. **6**, 689 (1953).
[8] R. S. Bear, J. Am. Chem. Soc. **67**, 1625 (1945).
[9] H. E. Huxley and J. Hanson, Symp. Soc. Exptl. Biol. **9**, 228 (1955).
[10] F. B. Straub, Studies Inst. Med. Chem. Univ. Szeged **2**, 3 (1942).
[11] C. Cohen and A. G. Szent-Györgyi. Paper read at International Biochemistry Congress, Vienna (1958).
[12] W. F. H. M. Mommaerts, J. Biol. Chem. **198**, 445 (1952).
[13] T. C. Tsao, Biochim. et Biophys. Acta **11**, 227 (1953).
[14] K. Laki and W. R. Carroll, Nature **175**, 389 (1955).
[15] A. Holtzer and S. Lowey, J. Am. Chem. Soc. **78**, 5954 (1956).
[16] W. F. H. M. Mommaerts and B. B. Aldrich, Science **126**, 1294 (1957).
[17] J. Gergely and H. Kohler, Conference on the Chemistry of Muscular Contraction, Tokyo (1957).
[18] A. G. Szent-Györgyi, Arch. Biochem. Biophys. **42**, 305 (1953).
[19] J. Gergely, H. Kohler, W. Ritschard, and L. Varga, Abstract of Meeting of the Biophysical Society, Boston (1958), p. 46.
[20] T. C. Tsao, K. Bailey, and G. S. Adair, Biochem. J. **49**, 27 (1951).
[21] C. M. Kay, Biochim. et Biophys. Acta **27**, 469 (1958).
[22] W. T. Astbury and L. C. Spark, Biochim. et Biophys. Acta **1**, 388 (1947).
[23] W. T. Astbury, Nature **160**, 388 (1947).
[24] C. C. Selby and R. S. Bear, J. Biophys. Biochem. Cytol. **2**, 71 (1956).
[25] J. Gergely, J. Biol. Chem. **200**, 543 (1953).
[26] E. Mihályi, J. Biol. Chem. **201**, 197 (1953).
[27] E. Mihályi and A. G. Szent-Györgyi, J. Biol. Chem. **201**, 211 (1953).
[28] J. Gergely, M. A. Gouvea, and D. Karibian, J. Biol. Chem. **212**, 165 (1955).
[29] W. R. Middlebrook, Abstract of Meeting of Biophysical Society, Boston (1958), p. 46.
[30] D. E. Philpott and A. G. Szent-Györgyi, Biochim. et Biophys. Acta **15**, 165 (1954).
[31] C. E. Hall, M. A. Jakus, and F. O. Schmitt, Biol. Bull. **90**, 32 (1946).
[32] M. H. Draper and A. J. Hodge, Australian J. Exptl. Biol. Med. Soc. **27**, 465 (1949).
[33] M. H. Draper and A. J. Hodge, Nature **163**, 576 (1949).
[34] M. H. Draper and A. J. Hodge, Australian J. Exptl. Biol. Med. Soc. **28**, 549 (1950).
[35] S. Leisegang, *Proceedings of the Third International Conference on Electron Microscopy*, London, July 1954 (Royal Microscopical Society, London, 1956), p. 176.
[36] A. J. Hodge, H. E. Huxley, and D. Spiro, J. Exptl. Med. **99**, 201 (1954).
[37] A. J. Hodge, J. Biophys. Biochem. Cytol. **2**, Suppl., 131 (1956).
[38] S. F. Velick, Biochim. et Biophys. Acta. **20**, 228 (1956).
[39] G. Schapira, G. Broun, J. C. Dreyfus, and J. Kruh, Compt. rend. soc. biol. **150**, 944 (1956).
[40] K. Bailey, Biochem. J. **43**, 271 (1948).
[41] A. J. Hodge, A. G. Szent-Györgyi, and C. Cohen (paper in preparation).
[42] M. A. Jakus, J. Biophys. Biochem. Cytol. **2**, Suppl., 243 (1956).
[43] C. Rouiller and E. Fauré-Fremiet, J. Ultrastructure Research **1**, 1 (1957).
[44] C. E. Hall, M. A. Jakus, and F. O. Schmitt, J. Appl. Phys. **16**, 459 (1945).
[45] F. O. Schmitt, R. S. Bear, C. E. Hall, and M. A. Jakus, Ann. New York Acad. Sci. **47**, 799 (1947).
[46] R. S. Bear and C. C. Selby, J. Biophys. Biochem. Cytol. **2**, 55 (1956).
[47] G. F. Elliott, J. Hanson, and J. Lowy, Nature **180**, 1291 (1957).
[48] A. J. Hodge, "Studies on paramyosin; the *in vitro* reconstitution and transformation of periodic structure," Ph.D. thesis, Massachusetts Institute of Technology (1952).
[49] A. J. Hodge, Proc. Natl. Acad. Sci. U. S. **38**, 850 (1952).
[50] F. O. Schmitt, J. Gross, and J. H. Highberger, Proc. Natl. Acad. Sci. U. S. **39**, 459 (1953).
[51] K. Bailey, Publ. Staz. Zool. Napoli **29**, 96 (1956).
[52] K. Bailey, Biochim. et Biophys. Acta **24**, 612 (1957).
[53] D. R. Kominz, F. Saad, and K. Laki, Nature **179**, 206 (1957).
[54] A. G. Szent-Györgyi (personal communication, 1958).
[55] R. H. Locker and F. O. Schmitt, J. Biophys. Biochem. Cytol. **3**, 889 (1957).
[56] W. Moffitt, J. Chem. Phys. **25**, 467 (1956).
[57] S. V. Perry, Physiol. Rev. **36**, 1 (1956).
[58] A. J. Hodge, J. Biophys. Biochem. Cytol. **1**, 361 (1955).
[59] J. Hanson and H. E. Huxley, Nature **172**, 530 (1953).
[60] H. E. Huxley and J. Hanson, Nature **173**, 973 (1954).
[61] A. G. Szent-Györgyi, D. Mazia, and Albert Szent-Györgyi, Biochim. et Biophys. Acta **16**, 339 (1955).
[62] S. Inoué and A. G. Szent-Györgyi (unpublished data).
[63] H. E. Huxley and J. Hanson, Biochim. et Biophys. Acta **23**, 229 (1957).
[64] H. Holtzer and J. M. Marshall, results presented at the Meeting of the Biophysical Society, Boston (1958).
[65] H. Holtzer and A. G. Szent-Györgyi, (unpublished data).
[66] J. Hanson, J. Lowy, H. E. Huxley, K. Bailey, C. M. Kay, and J. C. Ruegg, Nature **180**, 1134 (1957).
[67] H. E. Huxley, Proc. Roy. Soc. (London) **B141**, 59 (1953).

# 46
# Mechanisms of Muscle Contraction

MANUEL F. MORALES
*Department of Biochemistry, Dartmouth Medical School, Hanover, New Hampshire*

THIS article discusses current thought on the fundamental problem of muscular contraction, i.e., on the nature of the molecular transducer which converts chemical free energy into mechanical work. Structural observations (Bennett, p. 394; Inoué, p. 402; Hodge, p. 409) have greatly influenced this thought, but most current hypotheses are based to an even greater extent upon the discovery by Engelhardt and Ljubimova that the essential features of muscle action can be reproduced *in vitro*, with fairly pure materials extracted from muscle. This work provided such a far-reaching simplification that many researchers have been impelled to discard the system as a whole and to concentrate on figuring out how the *in vitro* system works. Fortunately, not everyone feels this way.

The simplest system which shows muscle-like contractility is one wherein a thread spun from a protein extracted from muscle is suspended in a neutral medium of low ionic strength. If one places, in this medium, mM $Mg^{++}$ and mM ATP, the thread contracts and does external work. Reciprocally, the thread catalyzes the hydrolysis of ATP to ADP and P—an exergonic reaction[1–4] which is considered to "drive" many biological processes. The question is: How—in a molecular sense—are the hydrolysis and the shortening "coupled"? To give an impression of current thought on this question, some theories or "models" of the process are summarized and passing mention is made of the observations which inspired them. These particular models were chosen, in part because they are among the more reasonable ones, and in part because they were developed from ideas and techniques discussed by Doty (p. 107), Zimm (p. 123), and Rice (p. 69).

In 1948, Kirkwood[5] suggested that the contractile element in muscle might be a flexible polyelectrolyte whose charge is modulated by phosphorylation of its serine residues by ATP. Actually, his was a scheme for relaxation rather than one for contraction, but his basic idea was all-important. At the time, Botts was making a thermoelastic analysis of threads spun from contractile protein, and she came to the conclusion that both the energy and the entropy decreased with increasing length.[6,7] This is the behavior expected of a polyelectrolyte bearing a net charge. Later, impressed by the high charge which ATP has in neutral solution, by the essentiality of $Mg^{++}$ for contraction, and by the fact that $Mg^{++}$ could bind so tightly to the protein as to shift its isoelectric point (I.P.) beyond *p*H 9, we suggested[6,7] that the contractile element of a thread might be a positively charged polyelectrolyte whose charge could be reduced merely by the adsorption of $ATP^{4-}$ in the *first* step of the ATPase process (Fig. 1). This neutralization would lead then to contraction, because, in shortening, the element would—for various reasons—gain entropy. Such a proposition relegated the *hydrolytic* step of the ATPase process to a minor role, a feature that greatly provoked biochemists; however, a conjoint thermodynamic analysis with Hill[8], and later a statistical-mechanical investigation of our model by Hill[9] convinced us that the model is quite acceptable from the point of view of energetics.

The model in question purported to be a model only of the contractile element in muscle. To make the model fit the facts about the *in situ* organization, other assumptions had to be made. In 1946, Schmitt and his associates[10] provided evidence that, in the sarcomeres of uncontracted muscle, the filamentous elements apparently were already extended. Since it was known[11] that these elements exert no tension in passively stretched muscle, I suggested[12] that, at rest, the contractile elements from either end of the sarcomere must interdigitate, but that cross-bonds between such elements must remain absent until the moment of excitation. The main concern, however, is—and was—to see if the model of the *contractile element* is correct, for it is in that device that transduction occurs. In its favor, we and others have mobilized additional evidence of which the following is representative.

Consider the Michaelis-Menten equation which relates the steady-state rate of substrate degradation to the concentration of substrate. It has been shown[13] that the enzymatic activity of the muscle protein system obeys this equation very well. This makes it possible to extract from experimental data the two Michaelis-Menten parameters. The one of interest here is the reciprocal Michaelis constant;

$$\bar{K} = k_1/(k_{-1}+k_2),$$

in the scheme,

$$E+S \underset{k_{-1}}{\overset{K_1}{\rightleftarrows}} X \overset{k_2}{\rightarrow} E+P.$$

There exist simple ways[14] of ascertaining whether $k_{-1} \gg k_2$; i.e., whether $\bar{K}$ can be considered identical with the equilibrium constant of substrate-to-enzyme binding. When this identity is legitimate,

$$\Delta F^0_{\text{binding}} = -RT \ln \bar{K}.$$

Many factors are known which influence the effectiveness with which the ATP structure can bring about contraction; for example, variations in ionic strength, in $[Mg^{++}]$, $[Ca^{++}]$, susbtitutions in the ring (thus generating other members of the nucleotide family), or elimination of the ring (TPP). As a rule, these variations also influence the ATPase activity, and, therefore, one can obtain also a set of $\bar{K}$ values from chemical measurements alone. When one compares *mechanical* measurements of contractant effectiveness with values of $\bar{K}$, there is a fairly good correlation, but in many instances there is no correlation whatever with maximum rate of ATPase activity (as there should be if the tension generator were "geared" to ATP hydrolysis).

Although there are other indications of polyelectrolyte behavior in contractile protein—for instance, the fact that *anions* totally unrelated to ATP, e.g., $I^-$, $SCN^-$, $Fe(CN)^{4-}{}_6$, also bring about contraction[15,16]—one additional experimental foundation of the model can be discussed briefly. To do this, and to prepare for the description of other models, the term "contractile protein" has to be defined. The best preparation, from the point of view of work performance, is obtained by extracting directly from muscle mince, according to a well-known recipe, a system referred to as "myosin-B." A contractile system, however, can be obtained also by complexing two separately extracted proteins. One protein is fibrous, exhibits ATPase activity, and is called simply "myosin." The other is globular, contains bound nucleotide, and is called "actin." The contractile complex of the two is called "actomyosin." For the moment, the question is left open as to whether or not myosin-B is identical with actomyosin. Up to this point, myosin-B has been considered in its precipitated (e.g., thread) form, at low ionic strength. If the ionic strength is raised to a few tenths molar, myosin-B dissolves, only then becoming accessible to the traditional tools of physical biochemistry. Electrophoretic examination shows that in such a state the myosin-B particles are anionic. The dissolved system still responds to ATP, by reducing its viscosity, and by lowering its optical turbidity, etc. It has been thought generally that, under these conditions, the underlying structural change is a dissociation of particles into actin and myosin. On this account, the interpretation which we placed on our early light-scattering experiments[17]—that the myosin-B particles were inflating on ATP adsorption—met with something less than global enthusiasm. New investigations[18,19] were carried out in our laboratory, and also while guests in that of Schachman; this new work was sparked principally by Gellert and von Hippel.

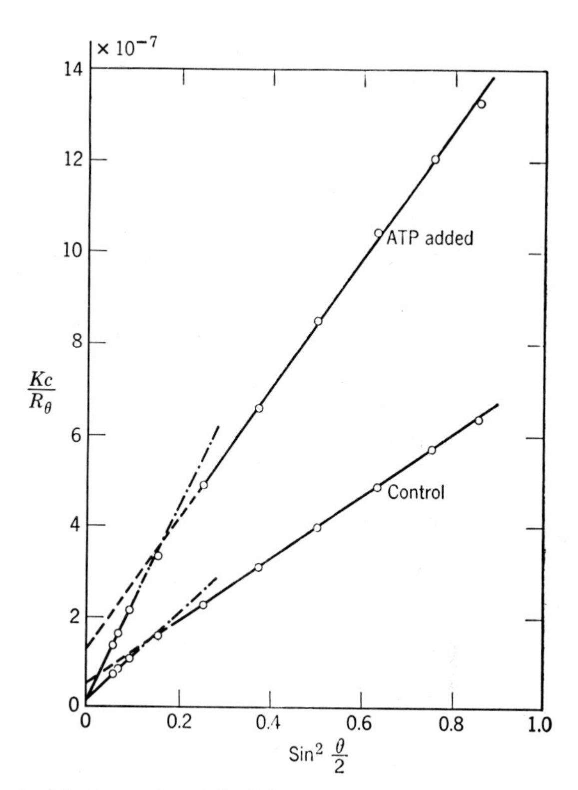


FIG. 2. Zimm plot of the light scattered from a myosin-B solution without and with ATP. Extrapolation from the points obtained at small angles gives an intercept which does not change on ATP addition, thus indicating an essentially constant weight-average molecular weight. ATP addition does increase the slope, however, thus indicating (see text) that at least some particles are "inflating."

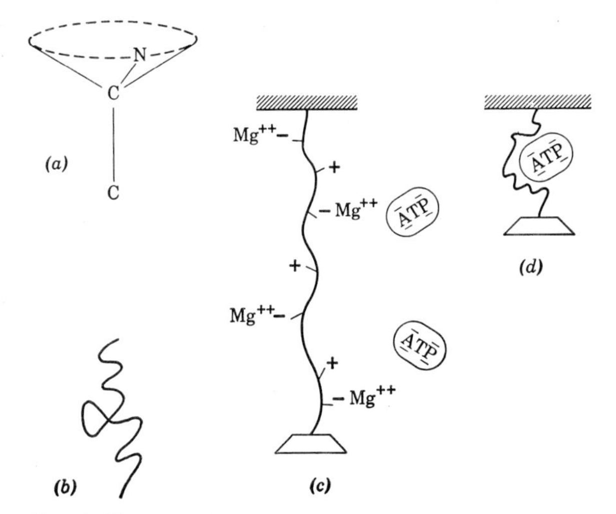


FIG. 1. Elements of a polyelectrolyte model of contraction. It is assumed that the molecular chain has some internal flexibility owing to a free rotation at at least some points (a). In the absence of other effects, the chain then assumes a coiled form which maximizes its configurational entropy (b). However, if the chain bears a net electrostatic charge, it will tend to extend out owing to repulsion between charges. Mg-myosinate might be so extended by a net positive charge in some of its regions (c). If so, adsorption of quadrivalent anions of ATP in the *first* step (complex formation) of the enzymatic hydrolysis would discharge the chain. An effectively neutral chain (d) would shorten toward the configuration of the unperturbed coil (a).

Figure 2, from this work, shows a "Zimm plot" (cf. Doty, p. 61) of data obtained on pristine protein, using the best techniques possible. Only the zero-concentration points are shown because there is ample evidence that the second virial coefficient is negligible. On a plot of this type, the reciprocal intercept gives the weight-average molecular weight ($\bar{M}_w$) and that the initial slope gives a higher order average of the square of a particle dimension. Accepting this data at face

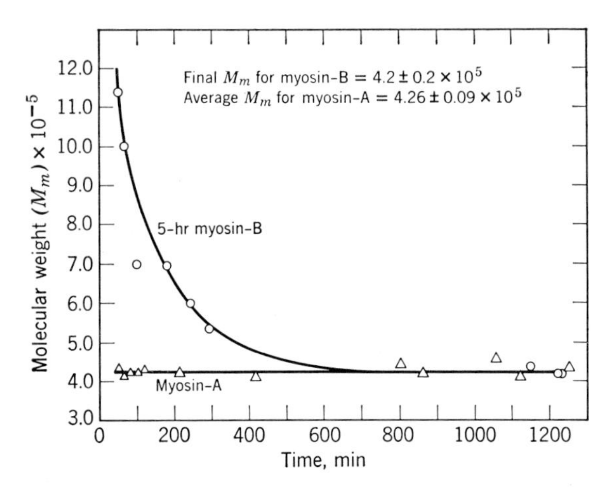


FIG. 3. Molecular weight of myosin (lower curve) and 5-*h* extracted myosin-B (upper curve), measured at the meniscus by the "Archibald" approach to sedimentation equilibrium technique, as a function of time of sedimentation at 4197 rpm and *ca* 5°C. Initial protein concentrations, 0.5 g/100 ml.

value, it is concluded that the addition of a saturating amount of ATP to this system *inflates particles at constant molecular weight.* (The Yang plot, which is more sensitive to changes in $\bar{M}_w$, leads to the same conclusion.) A similar result is obtained using $OH^-$ instead of ATP. Thus, when the protein is in the precipitated form and is cationic, the addition of ATP anions causes contraction; but, when it is anionic, the addition of ATP anions causes inflation. I feel that this constitutes some evidence for the idea that the interaction of ATP with this protein is that of an ion and of a polyelectrolyte. However, conclusions drawn from the light-scattering behavior of a polydisperse system can be very tenuous. It could be, for example, that the solution contains large ATP-inert particles which dominate the scattering at small angles, and smaller particles which simply dissociate (say, into myosin and actin), but whose influence is felt only at larger angles. This actually is suggested in our extrapolations from high-angle measurements. If the amounts and dimensions of these two hypothetical classes of particles were just right, the surviving heavy fraction might be so enriched with *inert* particles that its *average* dimension might appear to increase, even if the particles themselves are not inflating. To block this possibility some excursions in the ultracentrifuge were necessary.

With the ultracentrifuge, we soon found that the myosin-B system was readily separable into light and heavy components. This made it possible to apply the molecular-weight technique developed by Kegeles and Schachman on the basis of the Archibald equations. In Fig. 3 the $\bar{M}_w$, from meniscus measurements, is plotted as a function of sedimentation time. It decreases from very high values (such as are given by light scattering) to an asymptotic value of 4.2×10^5 g. On the other hand, if the same technique is applied to purified *myosin*, from the start one obtains a constant value of 4.2×10^5 g. This result immediately identifies the "light component" in myosin-B as *myosin* and reveals that the molecular weight of myosin is 4.2×10^5 g (until now, this quantity has been very controversial). To explore the heavier and more polydisperse components which generate no schlieren boundary, we turned to the ultraviolet absorbance method. This method gives $C(x)$ rather than $C'(x)$. Figure 4 shows two sample graphs constructed from densitometer measurements on a cell containing myosin-B saturated with ATP; both are plots of absorbance against distance in the direction of centrifugation—the top one corresponding to some time during the acceleration phase and the bottom one to many minutes at speed. By the time the lower measurements had been taken, the heavier components had plummeted to the bottom of the cell, and myosin (the "light component") was beginning to be swept out. From curves such as these, the concentration of myosin was measured. Extrapolating backward in time, allowing for sectorial dilution, the myosin concentration was subtracted from the total concentration to get the concentration of the "heavy components" at the earlier time; for instance, the time at which the upper plot was made. By doing this at various times, the "clearance curve" of the heavy components was inferred, and from it the weight distribution of the various components also, both in the presence and in the absence of saturating concentrations of ATP. The conclusion from these manoeuvers was that the (5-hr extracted) myosin-B system contains about 65% myosin and 35% heavier components, the latter distributed roughly into two classes. On addition of ATP (or PP), about one-third of the total heavy components depolymerizes (yielding

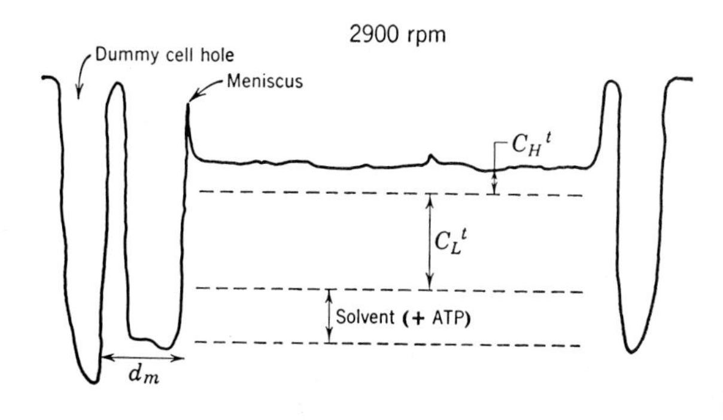


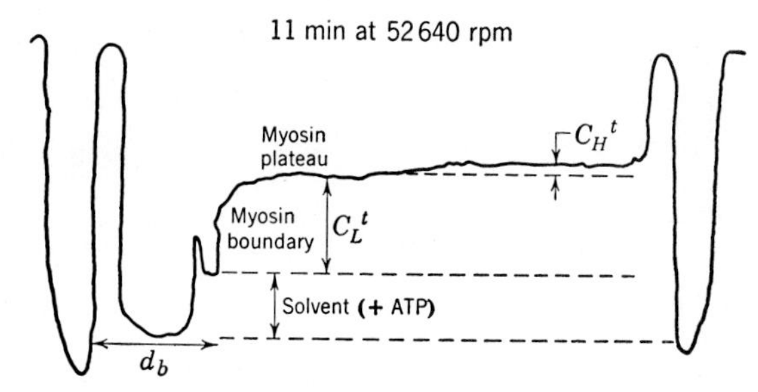


FIG. 4. Absorption of ultraviolet light by an ultracentrifuge cell filled with a solution of myosin-B and ATP, observed at different stages of sedimentation.

certainly myosin, and possibly actin), while two-thirds seem to remain intact. Based upon these determinations, the light-scattering data were re-examined to determine if the inflation suggested by those measurements might have been apparent only. It turns out very clearly that the inflation must be real, so I feel that one is entitled to say that the nature of the interaction between ATP and the myosin-B system in solution is understandable as an ion-polyelectrolyte effect.

The polyelectrolyte model is a special case of a *class* of models in which it is assumed that the adsorption of ATP—and not its enzymatic cleavage—causes a change in the elasticity of a mechanically continuous protein element. In the same class, for example, are the proposals of Pryor,[20] Flory,[21,22] and Laki[23] in which it is assumed that on adsorption of ATP a "crystallized," perhaps helical, configuration becomes unstable and passes into a random coil. Also in the class is the proposal of Astbury,[24] which envisions passage from the $\alpha$-configuration into the "super-contracted" configuration.

Another model which also could be constructed from materials found in muscle, but which would operate on a radically different principle has been described by H. Huxley and Hanson and by A. Huxley and Niedergerke[25] and was cast very recently into chemical language by Weber.[26] This model frankly is designed to agree, on the one hand, with certain observations on the organization of proteins within the fibril, and, on the other hand, with the widespread belief that the "coupling" of ATP hydrolysis for the purpose of "driving" another reaction always involves phosphorylated intermediates. The central feature of the Huxleys-Hanson theory[25] of the intrafibrillar organization of proteins is that it adopts, first of all, the "actomyosin" interpretation of the myosin-B system, and considers that over-all shortening is achieved by the motion of actin filaments *relative* to myosin filaments, without

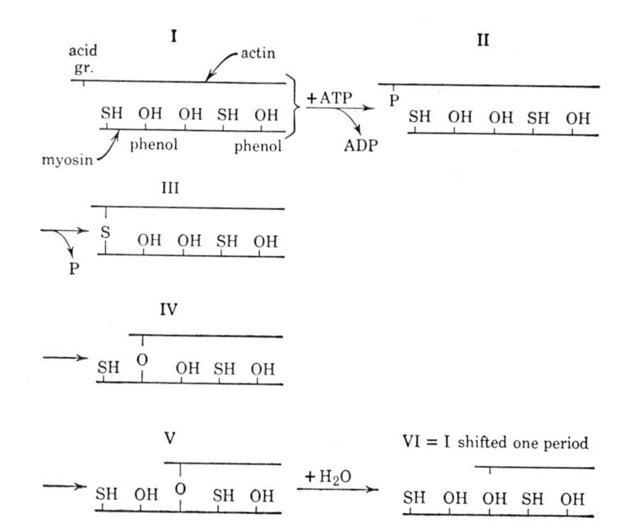


FIG. 5. Weber's[26] proposed chemical scheme to account for the sliding motion assumed in the Huxley-Hanson theory of contraction.

invoking the intrinsic contraction of either type of filament. How this sliding motion could result from the interaction of the filaments with ATP is shown in Fig. 5.

In schemes of this type it is necessary to explain how it is that the filaments slide in one direction only and not in the other, since obviously the reaction scheme is quite symmetrical. Weber does not allude to this point, but it has been well-recognized by Huxley,[27] who *formally* invoked a spatially asymmetric probability of reaction without pretending to give it a chemical basis.

Key support for the Weber scheme is considered to reside in certain tracer experiments of his colleagues, the Ulbrechts.[28] They reported that myosin-B—but not reconstituted actomyosin, which is also contractile—under conditions in which contraction occurs, catalyzed the back incorporation of labeled phosphorus from ADP into ATP. This suggested existence of the reaction,

$$\text{ATP}+\text{actomyosin} \rightleftarrows \text{actomyosin}\cdot\text{P}+\text{ADP}.$$

Since the Ulbrechts found, however, that fibrils from which myosin had been extracted catalyzed the exchange just as well, Weber reasoned that it must be the *actin* partner which is phosphorylated first; that is,

$$\text{ATP}+\text{actin} \rightleftarrows \text{actin}\cdot\text{P}+\text{ADP}.$$

To get the desired translation, he then assumed that P is extruded with the formation of an actin-myosin bond, which bond then migrates across the enzymatic site. At this point, water attacks the bond and the process starts anew with another ATP. This interesting scheme deserves close attention, for, if correct, it should harmonize much chemical and structural data. At the same time, some of its weaknesses must be considered.

Confirmation of the basic Ulbrecht experiment is not yet forthcoming; indeed, in some quarters, there is concern lest the observed exchange might be the result of contamination by granular ATPases. The plausibility of *actin* phosphorylation by ATP is also in serious doubt. Strohman[29] recently demonstrated phosphorylation of actin by the creatine kinase system, but attempts to phosphorylate actin with ATP itself have shown only exchange of whole nucleotide molecules, and no real phosphorylation. Likewise, Levy and Koshland[30] failed to find $O^{18}$ incorporation into ATP when ATP was incubated with actin in the presence of labeled water. Evidence for a transphosphorylation in just the *reverse* direction (i.e., myosin to actin) actually is suggested in the Levy-Koshland work. These authors found that the orthophosphate produced by myosin ATPase was enriched twofold over what it should have been, had the incorporation of $O^{18}$ occurred only at a single hydrolytic step. Still another purely biochemical difficulty arises from considering the speed required by the proposed hydrolytic reaction in order to account for the observed rate of shortening. From the model by Weber, one would expect an individual enzymatic site to be, say, 5 A or $5\times10^{-4}\,\mu$ long. For esterases which are actually

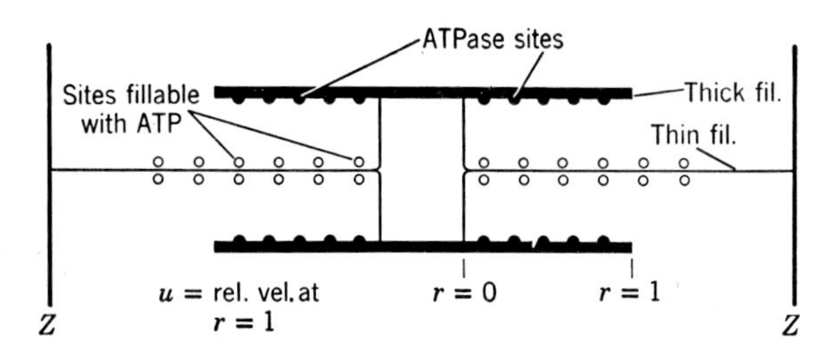


FIG. 6. Scheme for the Podolsky model of contraction. The mechanism for phosphorylating the thin filament (e.g., the creatine kinase-creatine phosphate system) is not included in the diagram.

*much faster* than myosin-B ATPase, $(V_m/E_0)$ at 25°C is about $10^2$ sec$^{-1}$. Therefore, the relative velocity of filaments moved by the esterase activity should be about $5\times10^{-2}$ $\mu$sec$^{-1}$. Since the myosin filaments are supposed to be closing on the actin filaments from either side, the predicted shortening velocity of the sarcomere is $10^{-1}$ $\mu$sec$^{-1}$. For an unloaded 3-cm frog muscle contracting at 0°C, Hill[31] reported a shortening speed of 5 cm sec$^{-1}$. In such a muscle, the sarcomeres, which are 2 $\mu$ long, would shorten at a speed of $5/3\times2$ $\mu$sec$^{-1}$, or 3.3 $\mu$sec$^{-1}$, at 25°C this speed easily could reach 10 $\mu$sec$^{-1}$, or 100 times the speed expected from a fast totally unencumbered esterase. To these difficulties should be added two others which stem from the observed length—tension behavior of muscle and myosin-B threads. Both living muscle (single fibers[8]) and myosin-B threads are able to contract until they are 20 to 30% of their original length. But, assuming the forces which bring about the interfilament motion to be short-ranged, it is hard to see how the system ever would contract to less than 50% of its original length. At the other extreme, a living single muscle fiber which is passively stretched and then excited to contraction continues to generate tension until it has been stretched to about twice its rest length. Already at 1.4 times rest length the overlap between the presumably interacting filaments has ceased!

Although I have shown a measure of dissatisfaction with the sliding filament model, I wish at the same time to avoid an impression of blissful contentment with the coiling class of models typified by our polyelectrolyte scheme. To mention but two serious difficulties with coiling models, consider that if myosin is invoked as the polyelectrolyte, one must face up to the difficulty that present measurements indicate that one mole of ATP may influence maximally as much as $10^5$ g of myosin. This seems too large a sphere of influence, even for Coulombic forces, and it is the difficulty which has led Szent-Györgyi[32] to invoke very unorthodox means of energy excitation and transmission. Another important shortcoming of the coil model ( a shortcoming recently removed for the sliding model[27]) is that, until now, it has never been shown to be consistent with certain heat-work generalizations established by Fenn and Hill for living muscle. In this situation, the proposal recently put forth by Podolsky[33] is most intriguing. His model is basically a variant of the polyelectrolyte model, but its modifications allow it to agree, on the one hand, with some structural observations, and, on the other, with the Fenn-Hill relationships.

Podolsky commences with two empirical equations evolved by Hill,[31] from measurements by Fenn and Hill. These equations describe the heat and work exchanges of a muscle contracting in the neighborhood of its rest length, bearing a load $t$ and shortening at the constant velocity $v$. The first equation is the "force-velocity" relation (Fig. 6):

$$(\tau+a)(v+b)=5ab;\ a,b,\ \text{constants.} \tag{1}$$

The second equation asserts that $-\dot{Q}$, the rate at which the muscle produces heat, is a linear function of $v$; i.e.,

$$-\dot{Q}=ab+av. \tag{2}$$

From the first law of thermodynamics, it is known that for a system at constant pressure which performs work $w$ in addition to work of expansion,

$$\Delta H=Q-w. \tag{3}$$

If this be a chemical system proceeding under conditions in which $\Delta H$ can be expressed as the extent of reaction, $\nu$, multiplied by a (constant) "heat per equivalent of reaction," $h$, then,

$$h\dot{\nu}=\dot{Q}-\dot{w}. \tag{4}$$

At this point, one has two choices, since Eq. (1) can be solved for either $\tau(v)$ or $v(\tau)$. If one takes $\tau(v)$, combines it with Eq. (2), and substitutes into Eq. (4), one obtains,

$$\dot{\nu}=\dot{\nu}(v,\ \text{constants}); \tag{5}$$

if one takes $v(\tau)$, combines it with Eq. (2), and substitutes into Eq. (4) one obtains,

$$\dot{\nu}=\dot{\nu}(\tau,\ \text{constants}). \tag{6}$$

Thus, Hill's equations seem to force one to assume that the reaction rate of the net driving reaction depends either upon the shortening velocity or upon the load (analytically, of course, these are equivalent conclusions.) To Podolsky, it has seemed easier to interpret Eq. (5) physically, since a dependence of reaction rate on shortening rate can arise if the two participants in the chemical reaction are attached to structures that are in motion relative one to another. Specifically, he has elaborated the model shown in Fig. 6. At rest, the "thin filament" is supposed to be a positively charged polyelectrolyte at equilibrium length. On excitation, there goes into operation a system capable of depositing on this filament the negative ions of ATP. The thin filament thus develops tension and begins shortening, drawing inwards the two $Z$ membranes. At the same time, the ATPase sites on the thick filaments are free to pick the bound ATP molecules off the thin filament, so that the net number of bound ATP molecules (and, therefore, the tension in the thin filament) will be a balance between the charging rate and the ATPase rate. The ATPase rate now must be considered more closely. The thick filaments do not move. There is, therefore, a *relative* velocity between every point on the thin filament and the point on the thick filament vertically opposite to it. This velocity is zero at the central ends of the thin filaments (where they are supposed to be attached rigidly to the thick filaments), and it rises to a maximum, say $u$, at the outer ends. If $r$ is a fraction which rises linearly from 0 at the central (fixed) ends of the thin filaments to unity at the outer edges of the thick filaments, it is simplest to assume that, at any position along the thick filament, the relative velocity between opposite points on the two filaments is $ru$. Since both $Z$ membranes are closing in on the center, each with velocity $u$,

$$2u/\delta=v \text{ i.e., } u=u(v, \text{constants}), \tag{7}$$

where $\delta$ is the sarcomere rest length and $v$ is the shortening velocity of the whole fiber, expressed in muscle lengths per second. The *chemical* basis for the dependence of reaction rate on shortening rate is introduced in this way (Fig. 7): It is assumed that, as an ATP molecule bound to a thin filament, there can be defined a probability of reaction, $P$, which depends essentially upon a velocity constant (conventionally defined) and upon the time spent in transit, which in turn depends upon $ru$. If the distance between bound ATP molecules on the thin filament be $l$, then an ATPase site, at position corresponding to $r$, will be presented with passing ATP molecules $ru/l$ times per second; therefore, per second it will carry out $(ru/l)P(ru)$ fruitful reactions, and the entire filament will carry out,

$$\int_0^1 (ru/l)P(ru)\lambda dr \tag{8}$$

fruitful reactions, where $\lambda dr$ is the number of ATPase sites in the element $dr$. Multiplying expression (8) by the number of filament units per ml of tissue, $\dot{\nu}$ ($u$, constants) is obtained, and, finally, by means of relation (7),

$$\dot{\nu}=\dot{\nu}(v, \text{constants}). \tag{9}$$

Thus, on the basis of the Poldolsky model, one can reach the functional dependence (9) which is that called for by the Fenn-Hill relations [Eq. (5)]; moreover, the explicit version of Eq. (9) can be fitted very closely to

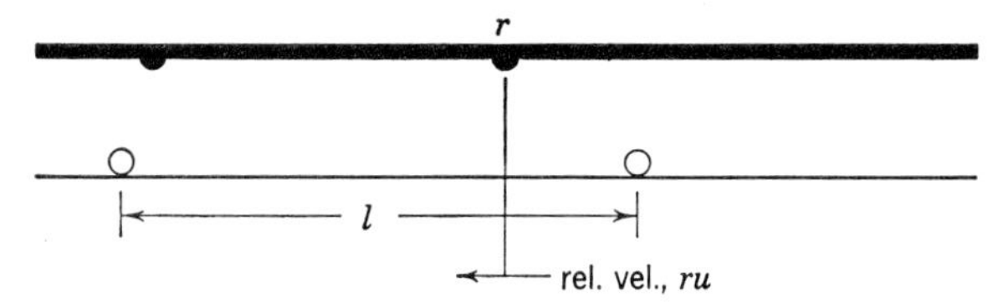


FIG. 7. The motion of an ATP affixed to a thin filament relative to an ATPase site on a thick filament.

the experimentally determined version of Eq. (5). The obvious merits of the Podolsky suggestion are three: there is retained what seems to this author the best molecular transducer by far—i.e., the coiling polyelectrolyte (or any of its thermodynamic analogs); there is invoked a structural organization consistent with currently popular ideas; and it is shown that the heat-work behavior of the model would be very like that observed in living muscle.

To what Podolsky actually has written, I should like to add some opinions of my own. If, indeed, his model is to enjoy consistency with current views on protein deployment, then it must be assumed that his thick filaments are myosin (or its subunits), and that his contractile thin filaments are actin. It is, therefore, perfectly reasonable to assume the existence of ATPase sites on the thick filaments. However, it is not at all obvious that actin could function in the polyelectrolyte sense required by the model in its present form. As already mentioned, actin is a globular protein of monomer weight, about $6\times10^4$ g. It has an interesting content of bound adenine nucleotide. It is well known that, under certain conditions, the globular (G) form of actin polymerizes into necklace-like aggregates of fibrous (F) actin. As normally prepared, the nucleotide of G-actin is ATP, and of F-actin it is ADP, but the precise mechanistic connection between dephosphorylation and the G-F transformation is unknown. Recently, Oosawa and his collaborators[34,35] carried out an elegant study of the G $\rightleftarrows$ F transformations. They showed them to be cooperative processes, and concluded that the F $\rightarrow$ G transformation is assisted by (a) hydrolysis of bound nucleotide—i.e., $H_2O+G(ATP) \rightarrow F(ADP)+P$— (b) addition of divalent cations such as $Mg^{++}$, and (c) increase of ionic strength. Naturally, these observations suggest that in G-actin the monomers are kept from aggregating by the electrostatic repulsion between their negative charges; in other words, the dispersed form of actin is charged more negatively than the aggregated form. To the extent that these transformations depend upon the state of the bound nucleotide, they could be driven by interaction with enzyme systems. On the basis of the recent work by Strohman,[29] transphosphorylation from the creatine-phosphate, creatine-kinase system would *disperse* the actin, and transphosphorylation from the dispersed actin to myosin ATPase would *condense* the actin. But this is just opposite to what the Podolsky model requires, since in his model the form of actin with the bound ATP would be assumed to be the *condensing* form, while the form of actin from which the myosin ATPase has removed a P would be assumed to be the *dispersing* form. One way out of this difficulty might be to suppose that *in situ* the actin has sufficient adsorbed cations so that the G$\rightleftarrows$F transformations are inverted from what they are in free solution, e.g., that the G$\rightarrow$F condensation is prevented by electrostatic repulsions between *positive* charges on the monomers. Again, not enough is known about intracellular-ion distributions to reject this possibility. To complete this description on a favorable note, I will add that the Podolsky model enjoys a substantial advantage over any coiling model in which it is assumed that myosin is the responding substance, since, if actin is assumed to be the responding substance, the volume of material which would have to be influenced by one molecule of ATP is much smaller and correspondingly more plausible.

Nothing I have read or thought about has weakened my intuitive conviction that the tension-generating device in excited muscle will prove to be a mechanically continuous structure, because any alternative device is apt to run afoul of vector analysis; nor am I ready to relinquish my faith in Coulombic interactions as the most rapidly generated and the most long ranged of the forces that the transducer could employ. On the other hand, I feel utterly dismayed over the experiments which have *not* been done in this field. For example, x-ray diffraction, intrinsic birefringence, and optical-rotation techniques have been available for some time, but one has been content to speculate about small-scale configurational change without trying to establish its occurrence or nonoccurrence in the *in vitro* system. Everything points to the transcending importance of both $Ca^{++}$ and $Mg^{++}$ in the fundamental process—indeed, to the likelihood that it is these ions, not ATP, which are moved in excitation; yet, there is very little certain information on where these ions reside in the fiber. One debates nowadays whether or not ATP or creatine phosphate are dephosphorylated in the twitch, but, so far as I know, little effort is being put into devising optical methods to follow dephosphorylation, and, at present, there are not even the means for blocking one of these reactions and not the other. However, ignorance in these matters will be dispelled only by hard work and certainly not by prolonging this gloomy epilogue.

*Added note*: Since this manuscript was prepared, W. T. Astbury has informed us that his associates have demonstrated the appearance of the cross-$\beta$ x-ray diffraction pattern when ATP shortens a film of myosin-B, thus strongly suggesting a real configurational change. Also, F. Buchthal has come forth with electron micrographs which show a wide gap (of nonoverlap) between thin and thick filaments, at muscle lengths at which tension can still be readily demonstrated.

## BIBLIOGRAPHY


[1] L. Levintow and A. Meister, J. Biol. Chem. **209**, 265 (1954).

[2] M. Morales, N.M.R.I. Memorandum Report 54-2 (1954).

[3] M. Morales in *International Symposium on Enzymes*, O. H. Gaebler, editor (Academic Press, Inc., New York, 1956), p. 325.

[4] R. J. Podolsky and M. Morales, J. Biol. Chem. **218**, 945 (1956).

[5] J. Riseman and J. G. Kirkwood, J. Am. Chem. Soc. **70**, 2820 (1948).

[6] D. J. Botts and M. Morales, J. Cellular Comp. Physiol. **37**, 27 (1951).

[7] M. Morales and D. J. Botts, Arch. Biochem. Biophys. **37**, 283 (1952).

[8] T. L. Hill and M. Morales, Arch. Biochem. Biophys. **37**, 425 (1952).

[9] T. L. Hill, Discussions Faraday Soc. **13**, 132 (1953).

[10] C. E. Hall, M. A. Jakus, and F. O. Schmitt, Biol. Bull. **90**, 32 (1946).

[11] R. W. Ramsey and S. F. Street, J. Cellular Comp. Physiol. **15**, 11 (1940).

[12] M. Morales, Biochim. et Biophys. Acta **2**, 618 (1948).

[13] L. Ouellet, K. J. Laidler, and M. Morales, Arch. Biochem. Biophys. **39**, 37 (1952).

[14] M. Morales, J. Am. Chem. Soc. **77**, 4169 (1955).

[15] L. Churney, Federation Proc. **13**, 26 (1954).

[16] K. Laki and W. J. Bowen, Biochim. et Biophys. Acta **16**, 301 (1955).

[17] J. J. Blum and M. Morales, Arch. Biochem. Biophys. **43**, 176 (1953).

[18] M. F. Gellert, P. H. von Hippel, and M. Morales, J. Am. Chem. Soc. (to be published).

[19] P. H. von Hippel, H. K. Schachman, P. Appel, and M. Morales, Biochim. et Biophys. Acta **28**, 504 (1958).

[20] M. G. M. Pryor, Progr. in Biophys. and Biophys. Chem. **1**, 216 (1950).

[21] P. J. Flory, Science **124**, 53 (1956).

[22] P. J. Flory, J. Am. Chem. Soc. **78**, 5222 (1956).

[23] K. Laki, Discussion at Gordon Conference on Muscle (1958).

[24] W. T. Astbury, Proc. Roy. Soc. (London) **B134**, 303 (1947).

[25] H. Huxley and J. Hanson, and A. F. Huxley and R. Niedergerke, Nature **173**, 971, 979 (1954).

[26] H. H. Weber, *The Motility of Muscle and Cells* (Harvard University Press, Cambridge, 1958).

[27] A. F. Huxley, Progr. in Biophys. and Biophys. Chem. **7**, 257 (1957).

[28] G. Ulbrecht, M. Ulbrecht, and H. J. Wustrow, Biochim. et Biophys. Acta **25**, 110 (1957).

[29] R. C. Strohman, Federation Proc. **17**, 157 (1958).

[30] H. M. Levy and D. E. Koshland, J. Am. Chem. Soc. **80**, 3164 (1958).

[31] A. V. Hill, Proc. Roy. Soc. (London) **B126**, 136 (1938).

[32] Albert Szent-Györgyi, *Bioenergetics* (Academic Press, Inc., New York, 1957).

[33] R. J. Podolsky in Symposium on Cardiac Contractility Ann. N.Y. Acad. Sci. (to be published).

[34] S. Asakura, K. Hotta, N. Imai, T. Ooi, and F. Oosawa in *Proceedings of the Tokyo Conference on Muscle Chemistry*, S. Ebashi, editor (Igaku Shoin, Tokyo, 1957).

[35] F. Oosawa, S. Asakura, K. Hotta, N. Imai, and T. Ooi (to be published).

# 47
# Quantitative Studies on Mammalian Cells *in Vitro**

T. T. Puck

*Department of Biophysics, University of Colorado Medical Center, Denver 20, Colorado*

The major advances of the science of genetics have come about through the use of two fundamental types of operations. The first and oldest consists in mating of selected, multicellular plants or animals to produce large numbers of offspring, among which the distribution of various inherited characters is examined. This procedure has produced that body of knowledge known as classical genetics and which, beginning systematically with the studies of Mendel, has delineated the processes which sustain and modulate biological heredity in all living forms. This experimental approach, which essentially examines genetic mechanisms in the germ cells, has been extremely successful in elucidating hereditary phenomena in organisms as diverse as *Drosophila* and *Zea mays*, but is difficult to apply to the slowly and less extensively reproducing organisms like the mammals. Thus, while some notable advances have been accomplished, it has not been generally possible to obtain sufficiently large numbers of progeny from selected mammalian matings to provide populations adequate for measurement of many important genetic events.

The second major class of genetic operations involves a more recent technique, which consists in study of the independent micro- and ultramicro-organisms—the free-living cells and the viruses. Here the standard unit is a single cell or particle, and the fundamental operation involves examination of the distribution of genetic traits among the progeny of asexual reproduction. The principal power of this method lies in the vast multiplication factor which it affords. Instead of hundreds or thousands, millions or billions of progeny, all arising from a single individual, can be rapidly produced and quickly scanned for a large variety of genetic characters, so that even rare events can be quantitatively examined. In the last twenty years, such genetic studies on microorganisms have produced a whole new field of knowledge. Sexual methods of genetic analysis are not excluded from these operations but can sometimes be arranged in a step preceding the clonal multiplication of the single cells. In addition, however, new methods for nonsexual genetic analysis have been found which promise to be exceedingly productive. Among the fundamental results which have issued from use of single-cell techniques applied to microorganisms during the last two decades are included: (a) the first systematic demonstration of the direct relationship between genes and enzymes; (b) delineation of many specific, amebolic pathways involved in gene-controlled biosyntheses; (c) demonstration of the direct exchange of genetic materials between cells previously considered to multiply only mitotically; (d) introduction of genes into cells by means of temperate viruses or pure deoxyribonucleic acid; (e) the most accurate measurement of gene-mutation rates; (f) demonstration and quantitative analysis of the processes involved in enzyme induction among genetically competent cells in response to specific environmental stimuli; (g) delineation of the characteristics of linear inheritance in bacteria; (h) systematic use of mitotic crossing-over in certain organisms to localize genes on their chromosomes; and (i) the most detailed mapping at the level of molecular dimensions of the linear sequence of genetic determinants, as accomplished in a bacteriophage chromosome.

An index of the fruitfulness of this latter approach to the analysis of genetics and related metabolic processes is afforded by the fact that the great majority of the discussion devoted to genetic processes in this biophysical conference has been concerned with developments arising directly from microbial genetics.

In our laboratory, a program was initiated, attempting to develop a methodology that would make possible studies of mammalian cells by means of the techniques of microbial genetics. If successful, such a methodology would provide a new avenue for exploration of mammalian genetic processes among the somatic cells of multicellular organisms, which would complement studies on the genetics of the germ cells achieved by conventional mating studies. In addition, it would afford all of the many kinds of analyses which an unlimited number of progeny provides. Such techniques might facilitate elucidation of the nature of any changes in the genome and phenome, respectively, which are responsible for the characteristic behavior of the various differentiated tissue cells; localization of different genes on the various chromosomes; provision of mutated cells containing specific biochemical blocks for the delineation of various enzymatic steps through quantitative study of specific biosyntheses, like that of hormones and antibodies in specialized cells; search for processes like sexual genetic exchange and transduction in somatic cells, and investigation of the genetic biochemistry of normal and malignant cells.

Participating in this program have been a series of devoted co-workers: Steven Cieciura, Harold Fisher, and Philip Marcus carried out studies which formed the basis of their doctoral theses; and D. Morkovin, G. Sato, and N. C. Webb joined this program in a postdoctoral capacity. Chromosome studies have been

* Contribution No. 79 from the Florence R. Sabin Laboratories.

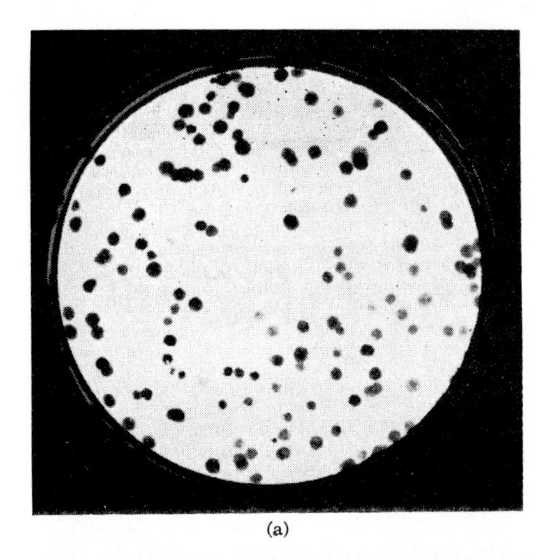

(a)

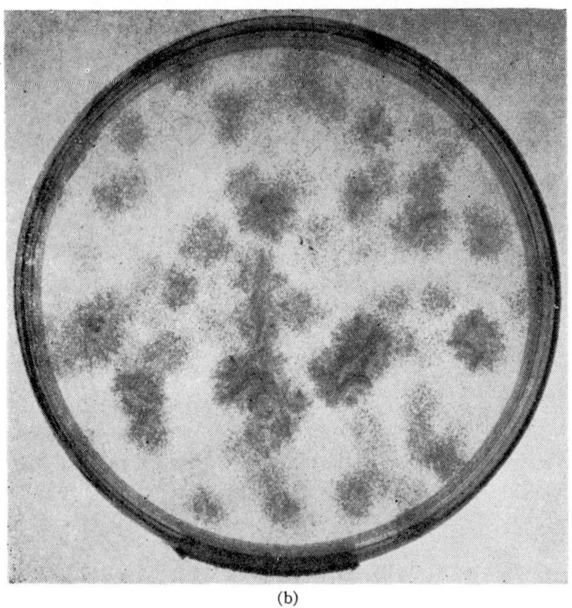

(b)

FIG. 1. (a) Petri dish which was seeded with 100 single cells of the S3 strain of the HeLa culture, an aneuploid which originated in a human cancer. (b) Petri dish seeded with single cells of a euploid human strain. These cells tend more to spread and migrate, and so the colonies tend to run together unless a smaller number of cells is plated or a larger plate is employed.

carried out initially in a collaborative arrangement with Dr. Chu and Dr. Giles at Yale University, and more recently with J. H. Tjio, who joined our laboratory last year. Dr. Arthur Robinson has conducted those aspects of these studies which involved work on human patients.

The first step in this program, which, of course, has drawn a great deal on previously existing tissue-culture methodology, was to develop means for plating single mammalian cells, under conditions such that every single cell develops into a discrete, macroscopic colony.[1] This is essential in order to quantitate growth of single cells of a large population, and to make possible isolation of mutant clones. A simple, quantitative means for growth of single cells into colonies was achieved by a procedure, identical in principle to, and only slightly more complex in practice than, the plating method which constitutes the basis of quantitative microbiology. Figure 1(a) illustrates a representative plating in which 100 cells of the S3 HeLa clone, plated in a Petri dish in nutrient medium, have attached to the glass and produced, within the limits of sampling uncertainty, a discrete macroscopic colony from each cell inoculated. The counting of such colonies is completely objective and highly reliable, so that this methodology permits all of the types of experiments characteristic of quantitative bacteriology to be applied to mammalian cells. Cells taken from organs as diverse as skin, liver, spleen, bone marrow, testis, ovary, kidney, lung, and others, from man and other mammals, and from individuals of a large variety of ages, all respond similarly. It can be concluded that at least large numbers of cells exist in virtually every organ of the mammalian body which retain the potentiality of growth as independent microorganisms, if provided with an adequate physical and chemical environment.

By and large, two general types of cellular and colonial morphologies result from such plating. That illustrated in Fig. 1(a) has been called "epithelial-like" because the cells form compact, tight colonies with relatively smooth, well-defined edges. In our experience so far, the cells that tend to grow stably in this fashion have been polyploid, and usually aneuploid. The other type of colony, which has been called "fibroblast-like" but should perhaps be referred to simply as "stretched," is illustrated in Fig. 1(b), which shows the tendency of the colonies to grow with rough edges, and of the cells to line up as elongated parallel structures. This stretched, needle-like conformation is more characteristic of the euploid cells grown by the methods developed in our laboratory. However, the molecular environment strongly influences the morphology of cells grown in this manner.[1]

The course of developments in microbial genetics has demonstrated the enormous utility of quantitative single-cell plating for the isolation of stocks with rare genetic markers that permit analytical experimentation. Unless each cell can grow in isolation to form a colony, it becomes impossible to apply the highly discriminating screening methods by which millions of cells are subjected to a stressful situation permitting growth of only occasional rare mutants, for otherwise, the exceedingly rare mutant which though potentially is capable of reproduction is not part of a large reproducing population and may not be enabled to express its growth potentiality. An effective single-cell plating technique permits ready recognition of even very rare genetic events, and quantitation of their frequency.

Isolation of mutant clones from such plates is a straightforward process for which a variety of different mechanical methods have been developed. We have established mutant clones which have arisen spontaneously or have been induced as a result of x-irradiation. Among these are included forms with divergent nutri tional requirements for colony formation, changed colonial morphologies, resistance to destruction by specific viruses like Newcastle disease virus, and differences in chromosomal constitutions (Fig. 2). These mutants have been grown for months or years in continuous culture, during which they have produced astronomical numbers of progeny, and have exhibited stability with respect to their identifying characteristics which is in every way comparable to that of the familiar mutants in bacteria and molds.[2,3]

Attention was next devoted to the development of a defined chemical medium which would promote growth in high efficiency of single mammalian cells, and so make available means for isolating mutants with specific nutritional markers. Many laboratories have contributed studies elucidating small molecular requirements of mammalian cells in tissue cultures utilizing massive cell inocula.[4–6] In studying the growth requirements of individual, isolated cells, we found that single cells of the S3 clonal strain of the HeLa cell form colonies with 100% plating efficiency, when a chemically defined small-molecular medium is supplemented with two macromolecular serum fractions each migrating as a single moving boundary in an electric field. The two proteins of mammalian serum which, under the conditions of these experiments, complete the growth requirements of single cells are serum albumin and an $\alpha_1$-globulin with a molecular weight of approximately 45 000, which appears to be a glycoprotein in composition. Both substances are necessary for growth of single cells under the specific conditions of our procedures. The functions of these proteins are still under study, but that of the $\alpha_1$-globulin has been found to be fulfilled completely by the protein fraction named fetuin, an $\alpha_1$-globulin which constitutes 45% of calf fetal serum protein.[7]

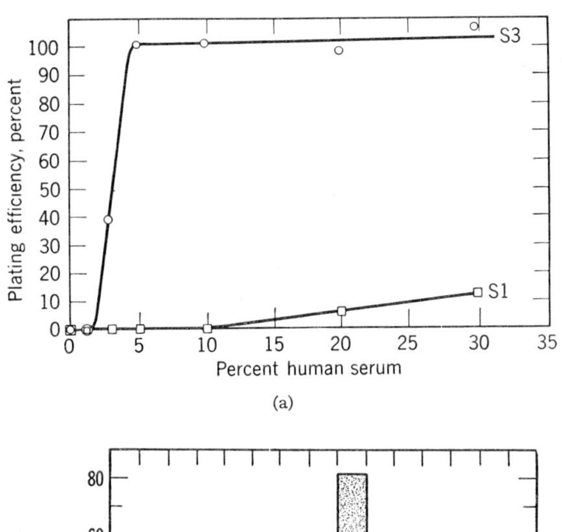


(a)

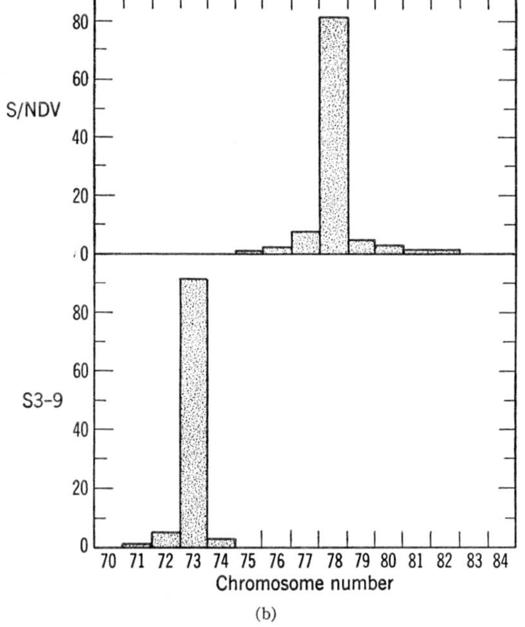


(b)

FIG. 2. Demonstration of some representative mutant strains isolated from the HeLa population. (a) Plating efficiency curves in the presence of different amounts of human serum of 2 mutants found to occur spontaneously in the original HeLa population. The cellular and colonial morphologies of these two strains are identical but their growth requirements are different. (b) Distribution of chromosome number in 2 HeLa clones. The clone at the top (*S/NDV*) has the same chromosome number as S3, but is characterized by its resistance to destruction by Newcastle disease virus.

Albumin and fetuin have been purified by repeated precipitation to the point where preparations are approximately 98% homogeneous electrophoretically and ultracentrifugally, although the ultimate purity of such preparations is, of course, difficult to specify with certainty, particularly since fetuin has been demonstrated to be heterogeneous immunologically. A typical electrophoretic pattern is presented in Figs. 3(a), and 3(b) demonstrates the colonial development achieved from inoculation of single cells into a medium containing the purified albumin and fetuin, amino acids, growth factors, salts and glucose. The plating efficiency equals that in serum-containing medium. Experiments on mutants of S3 which exhibit different molecular nutritional requirements are now in progress.

It seems likely that fetuin and albumin may not be required as true metabolites, but rather as conditioning factors which permit establishment of the necessary physicochemical relationships between the cell and surrounding medium. Thus, the fetuin fraction exercises a specific action on the cell membrane, since trypsinized cells added to a medium, complete except for the presence of fetuin, remain as rounded spheres, unattached to the glass surface. With the addition of fetuin to such a medium, the cells rapidly attach to the glass and stretch out in a highly characteristic configuration [Figs. 4(a) and 4(b)]. As little as 1 mg/cc of fetuin can be detected by means of this action. Fetuin has been known to be a powerful antitryptic agent, and at least part of the function of fetuin seems to be antiproteolytic in nature, because this substance prevents the tryptic loosening and rounding of glass-attached cells. The participation in the attachment reaction of a protein not included in the fetuin fraction has been claimed by some workers.[8]

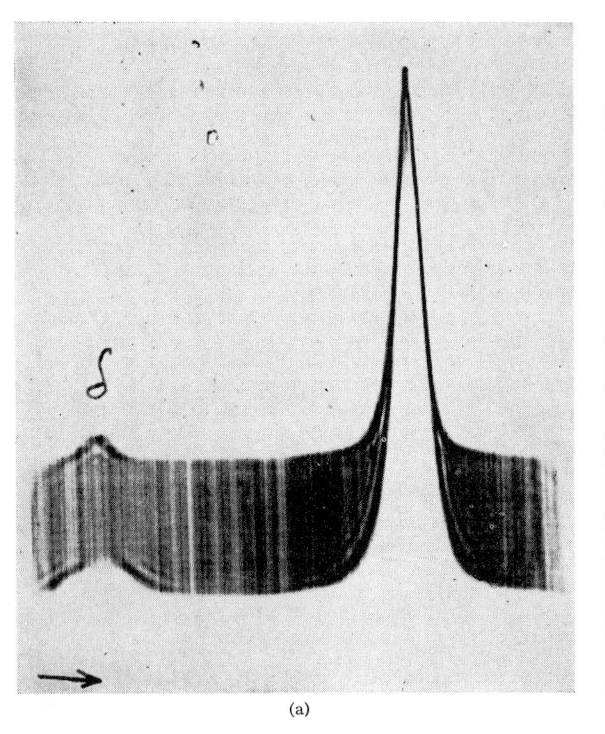
(a)

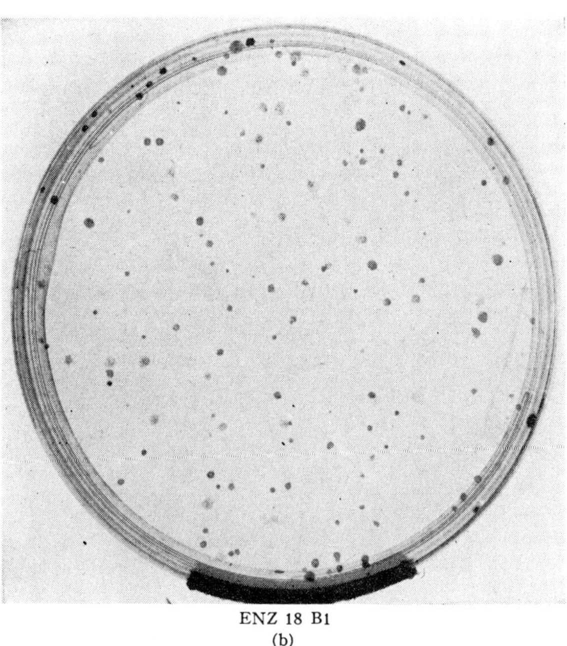

(b)

Fig. 3. (a) Electrophoretic pattern of fetuin purified by repeated, fractional $(NH_4)_2SO_4$ precipitation from fetal calf serum. (b) Colonies developing on plate seeded with 100 S3 cells, in a medium containing known small molecular components, and purified fetuin and human serum albumin.

In further development of tools for quantitative study of the growth and genetics of the mammalian cell *in vitro*, it became necessary to attempt to control the lability of karyotype which had been demonstrated to affect cells cultivated by standard tissue-culture techniques. Recent studies have shown that virtually every mammalian cell which has been established as a stable tissue-culture strain has a chromosomal constitution radically different from that of the parental species from which it originated.[9,10] Euploid cells were generally found either to stop growth after a very short period in culture, or else to proliferate, but with drastic changes in chromosome number and structure. In order to achieve conditions eliminating this proclivity for aneuploid chromosomal constitution in cell cultures (which renders genetic studies of questionable significance), methodologies were required for simple and reliable monitoring of chromosomal constitution of cells grown *in vitro*. Such methods were developed by Axelrad and McCulloch,[11] by Rothfels and Siminovitch,[12] and by Tjio and the author in our laboratories.[13] When the karyotypes of such cultures could be checked with ease and accuracy, it was possible to find growth conditions which permit karyotype integrity to be maintained. These conditions were secured by development of a screening test for media and regulation of other conditions so as to remove conditions leading to mitotic inhibition, which might induce polyploidy. By careful control of the processes involved in cell dispersion, by regulation of the *p*H and temperature during cell incubation, and by use of pretested media which incorporate all the needed nutritional requirements for cell growth and exclude toxic substances, it became possible to grow cells from normal tissues of man and other mammals, for periods now approaching a year, and for numbers of progeny which have exceeded $10^{25}$ without change in chromosomal constitution.[14] Figure 5 shows the chromosome complement of typical human male and female cells. These methodologies have permitted characterization of each of the human chromosomes, and particularly of the $X$ and $Y$ sex-determining chromosomes.[15]

In order to employ human genetic markers with biochemical, immunologic, and pathologic characteristics, a method was devised by which cells from any individual could be simply and reliably introduced into stable growth *in vitro*. A small piece of skin, approximately 10 to 50 mg in mass, is excised from the ventral part of the forearm, dispersed by trypsinization, and grown in the medium which contains fetal calf serum which was shown to maintain stable karyotype. This method has proved itself reliable in securing actively growing cultures from any individual.[14] It has made possible confirmation of the fact that the human karyotype does, indeed, consist of 46 chromosomes, as first reported by Tjio and Levan.[16] Studies are now under way on cells of individuals with phenyl ketonuria and other genetic diseases, in an attempt to establish

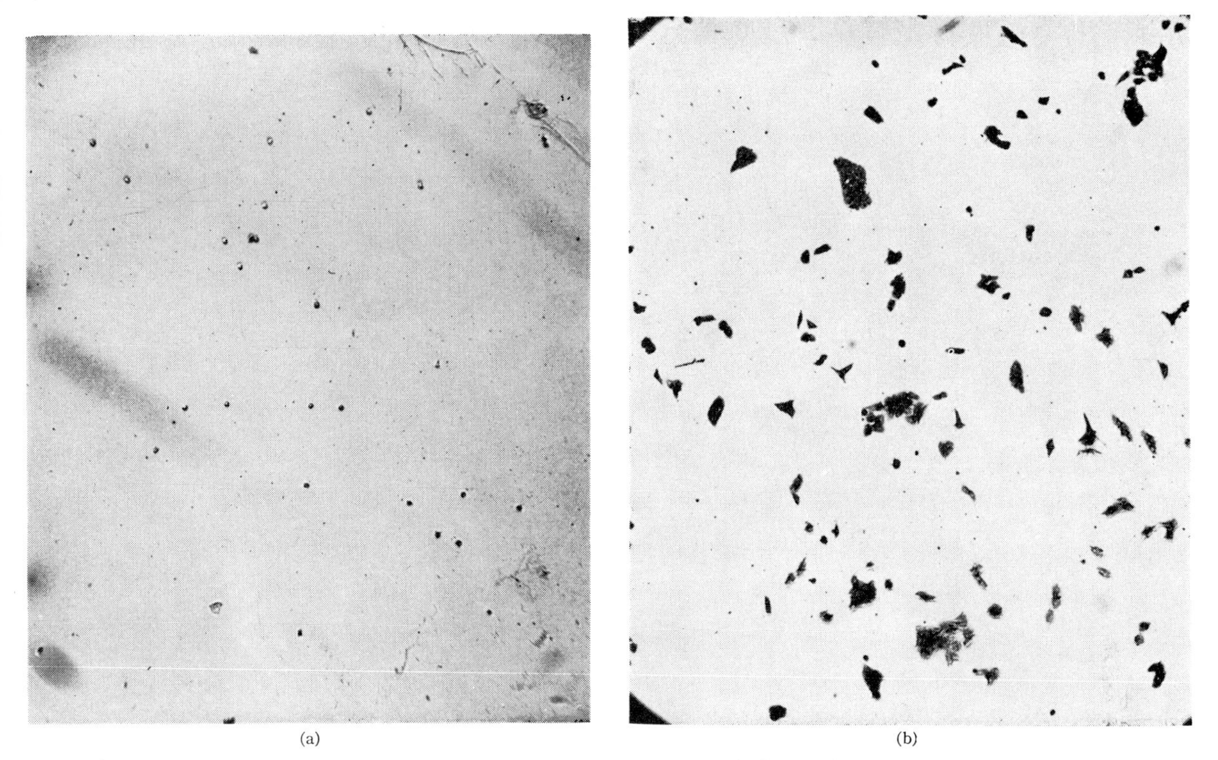


FIG. 4. (a) Photomicrograph demonstrating the rounded condition of cells, which will shortly be released from their bond to the glass in the absence of any "stretching factor." (b) Photomicrograph demonstrating the stretched condition of the cells in an adequate concentration of purified fetuin.

cytogenetic and biochemical differences between these and normal cells in the *in vitro* cultures. These methodologies are also being employed clinically to determine the chromosomal constitution of persons with various degrees of clinical hermaphroditism and to compare the results so obtained with those utilizing the sex chromatin method devised by Barr and his associates.[14] Thus, an anatomically female patient suffering from ovarian dysgenesis (Turner's syndrome) was shown to possess only 45 chromosomes.

In contrast with the normal human chromosome constitution shown in Fig. 5, Fig. 6 presents that of the aneuploid S3 HeLa cell, a clonal stock which we developed from the culture which Gey and his associates established from a human carcinoma of the cervix.[17] The chromosome number of this clonal strain is 78, which represents a hypotetraploid condition. Studies are in progress comparing the different morphology and behavior of cells which differ in chromosomal number and constitution. Thus, mutant clones of animals like the Chinese hamster have been isolated, which have stemline chromosome numbers of 22 and 23, respectively.

As one example of the kinds of quantitative studies made possible by these techniques, the remainder of this discussion is devoted to analysis of their application to study of the action of high-energy radiation on mammalian cells. Ionizing radiations are capable of disrupting any chemical bond in any molecule of any cell which has absorbed such energy. On the basis of such qualitative considerations alone, one might expect an enormous variety of cellular pathologic actions. Radiobiologic studies on mammalian systems carried out by many laboratories over many years have more than borne out this expectation, demonstrating a bewildering variety of pathological consequences attending various kinds of exposure to radiation of mammals, including phenomena as diverse as gastrointestinal disturbance, drop in the level of blood white cells, epilation, and carcinogenesis. One of the most critical problems in this field involves determination of the degree to which chemical disorganization attending exposure to a given dose of ionizing radiation affects the genome or the phenome, respectively, of the irradiated cell. A quantitative answer to this question would greatly simplify attempts to understand the enormously complex series of events attending irradiation of mammalian cells and of whole animals. In addition, the answer to this question is essential in order to determine the degree to which exposure of mammalian populations to various doses of ionizing radiations will result in random mutations transmissible to succeeding generations.

Physiologic damage to cells in the form of reversible lag periods in cell multiplication, changes in cell permeability and inhibition of many specific enzymatic activities, have been demonstrated in many organisms.

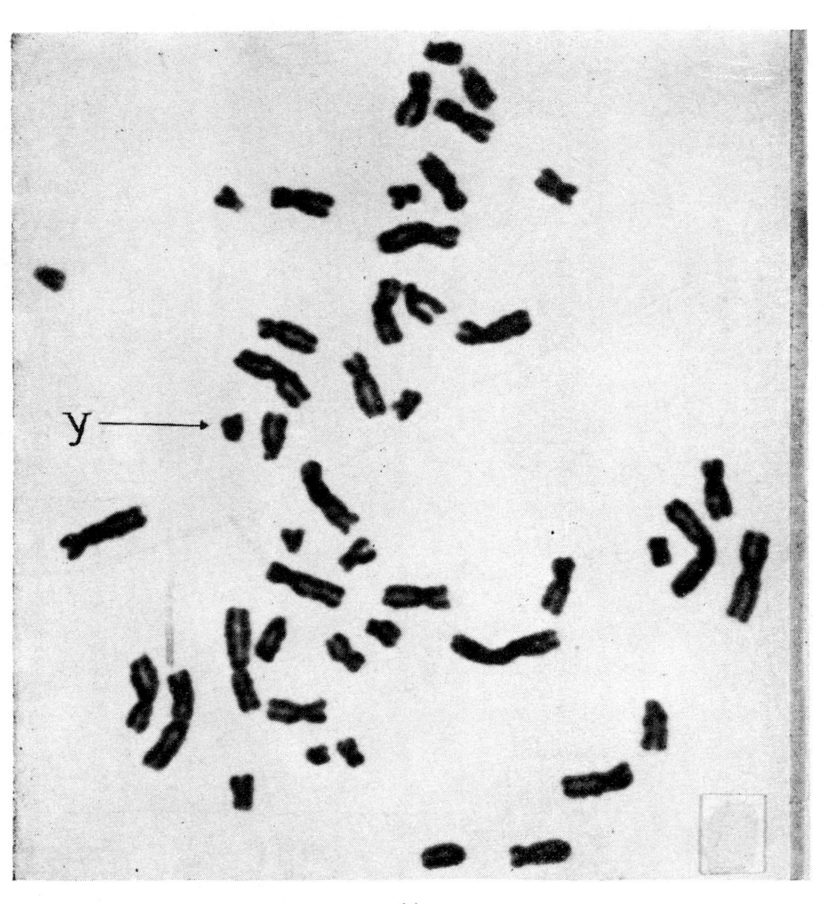


(a)

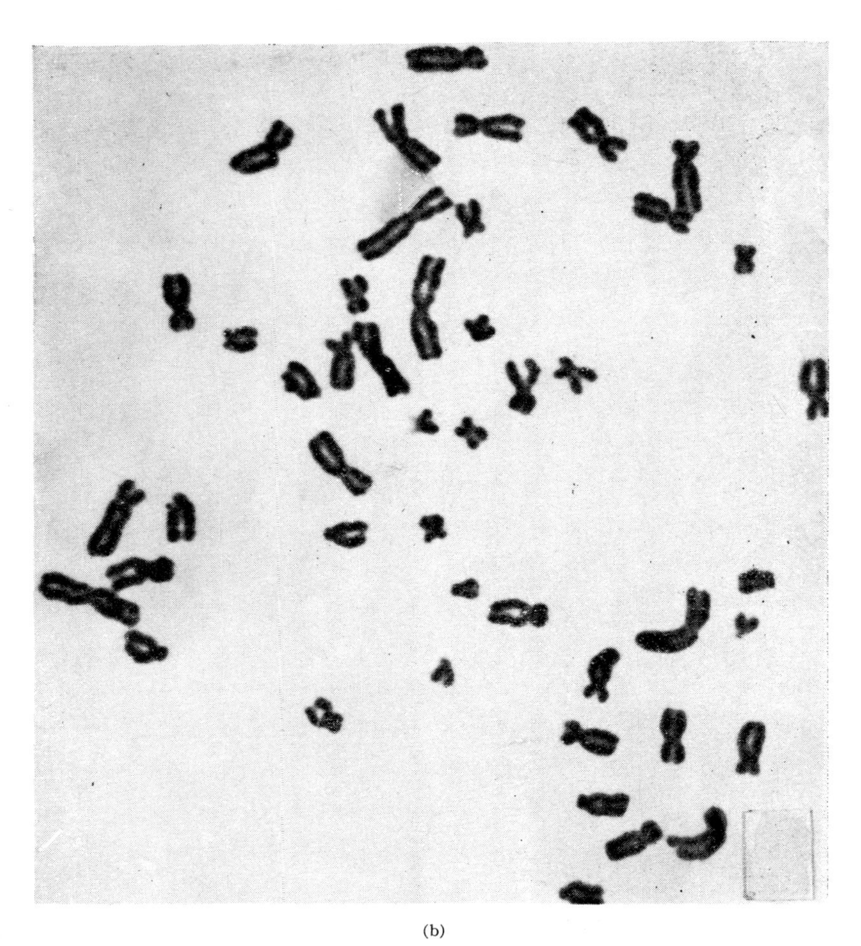

(b)

FIG. 5. (a) Chromosome constitution of normal human male cell.[13] (b) Chromosome constitution of normal human female cell.[13]

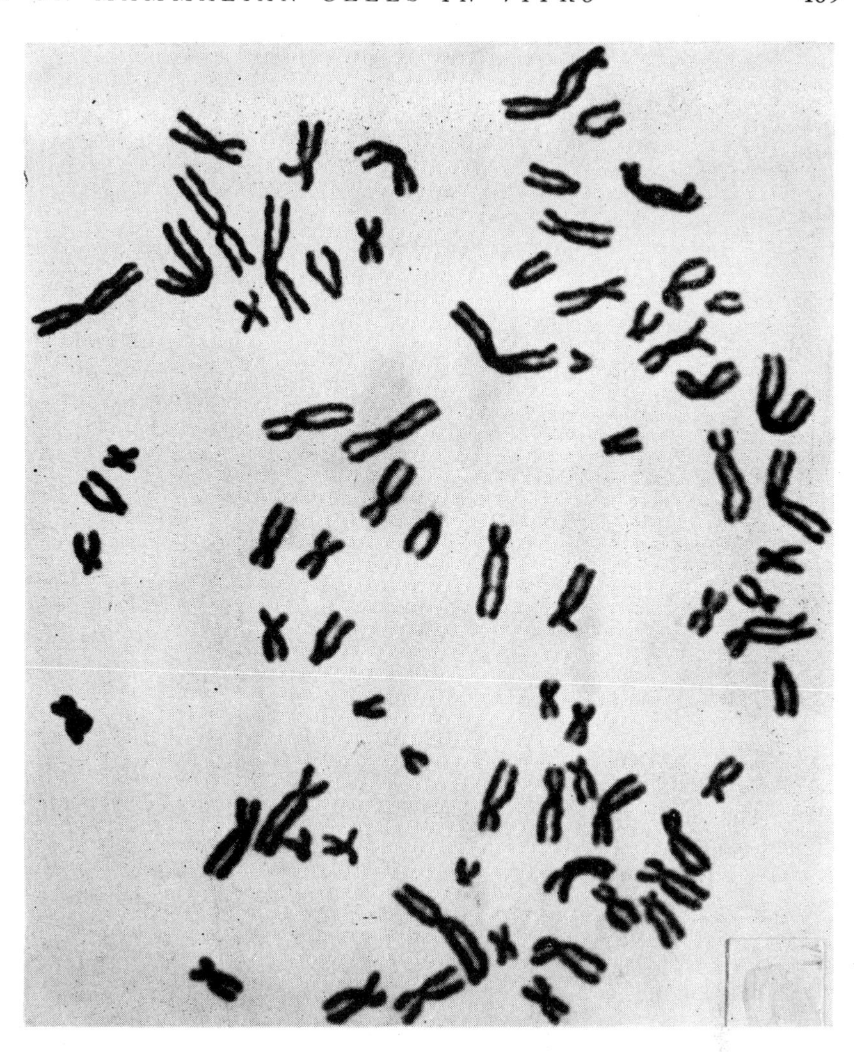

FIG. 6. Chromosome complement of a typical S39 HeLa cell, a subclone of the S3 strain, of human malignant origin, with a stemline number of 78.

In a variety of nonmammalian forms it had also been shown that the cell nucleus is much more sensitive to damage by irradiation than the cytoplasm. Genetic and cytologic examination of radiation effects, particularly in the simpler organisms, has revealed that changes of the genetic structure arising from irradiation include single gene mutations, chromosome breaks, production of areas deficient in chromatin, and occasionally elongated and amorphous chromosomes, these latter presumably arising from cells which had been irradiated during mitosis. At the site of fragmentation, a chromosome exhibits sticky surfaces which can re-attach so that the chromosomes may reconstitute themselves into a configuration more or less closely resembling the normal. In all probability, however, a defect, even though invisible, will persist at the site of breakage unless this happened to lie in a genetically inactive region. All or part of any chromosome which has suffered such a "hit" may, through imperfect restitution, fail to be incorporated in the mitotic apparatus and be lost in subsequent cell divisions. Such a condition can cause genetic imbalance that may destroy the ability of a cell exhibiting it to reproduce indefinitely. In addition, if an unrestituted chromosome divides, the two sister chromatids may unite at their broken ends to form an anaphase bridge that can prevent completion of mitosis. If multiple chromosome hits occur in the same cell, all of the foregoing damaging actions characteristic of single breaks are intensified. Moreover, abnormal restitution of the sticky, broken ends of different chromosomes may occur, producing a variety of complex translocations, such as dicentric chromosomes which can destroy the cell's reproductive capacity.[18]

Earlier studies of the relative sensitivities of the mammalian genome and phenome, respectively, to damage by high-energy radiations at first seemed to favor the thesis that for mammalian cells, in contrast to cells of other organisms, physiologic rather than genetic effects may be most important in the dose range up to and including the mean lethal dose for the whole organism, which is about 400 to 500 r. This supposition appeared to be supported by several kinds of evidence: studies in which massive cell inocula were irradiated in tissue culture with progressive doses of irradiation revealed that permanent loss of cell proliferation did

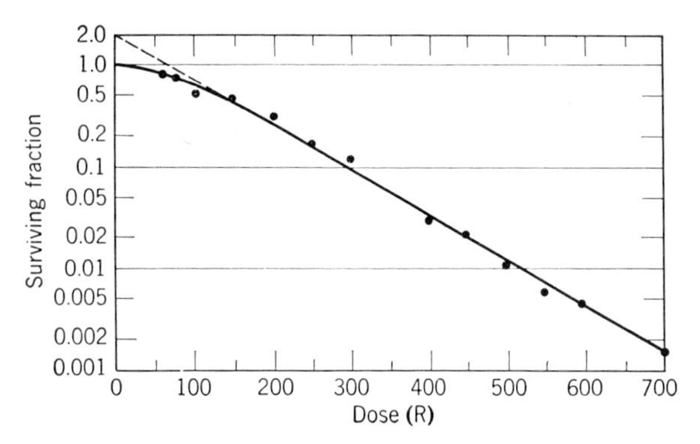


FIG. 7. X-ray survival curve of the ability of single S3 HeLa cells to form colonies as a function of the dose, $D$. The points can be fitted either by the equation $N/N_0=1-[1-e^{-D/96}]^2$ or by $N/N_0=e^{-D/77}(1+D/77)$, the latter relationship being preferable, since it is based on the model that reproductive death for this polyploid cell requires two or more effective hits anywhere in the chromosome complement. A more refined and comprehensive equation which expresses the contributions to genetic death of haploid, diploid, or polyploid cells of any species by single-hit and multiple-hit radiation damages is presented elsewhere.

not occur until many thousands of roentgens had been administered. Presumably because of the way in which tissue-culture thinking in the past was dominated by the concept that cell proliferation is a property of a cellular community, experiments like this were interpreted as an index of the dose needed to inactivate the cellular reproductive mechanism. Moreover, irradiation of single cells of bacteria, yeasts and protozoa, which might be considered as models for the mammalian cell, indicated that in all these forms, inactivation of reproduction required thousands of roentgens.[19] Hence, since death of a whole mammal can be effected by an exposure of only 400 to 500 roentgens, it appeared possible that mammalian-cell interaction with irradiation proceeds in a manner fundamentally different from that of other forms. While this thesis has been vigorously opposed, especially by geneticists, among whom H. J. Muller has been particularly active, it is still occasionally affirmed in the current scientific literature that very high doses in the neighborhood of many thousands of roentgens are needed to achieve irreversible damage to mammalian cells *in vitro*[20] and the ability of ionizing radiation to produce mutagenesis in mammals is still being challenged.[21]

With the development of the techniques which have been here described, it became evident that the use of cell proliferation from massive cultures cannot be used as an end point to quantitate irreversible effects of radiation on cell reproductive capacity unless the results are calculated as a survival function based on the number of cells irradiated. Since even a single cell remaining intact can proliferate eventually to produce any degree of outgrowth whatever, the dose at which a large but unspecified population fails to produce recognizable outgrowth is meaningless in terms of the dose needed on the average to inactivate the cellular reproductive apparatus. Moreover, since it is characteristic of irradiated cells to exhibit reversible mitotic lags, and to continue to multiply for one or even several generations, but then to produce a microcolony of sterile progeny, use of the mitotic index of the population as a measure of irreversible effects on cellular reproduction is not reliable. However, the use of the single-cell plating procedure makes possible precise determination of survival curves for mammalian cells by procedures exactly like those which had been used for cells of *E. coli*. In this way it is possible to measure accurately the dose required for inactivation of the reproductive mechanism of the individual cells constituting a population. The first such curve obtained is shown in Fig. 7, and approximates a two-hit relationship, with an initial shoulder, followed by a linear exponential drop, which is maintained for doses up to at least 1800 r. Of major importance is the fact that the mean lethal dose, $D_0$, which is obtained from the slope of the linear part of the curve, was only 96 r. The determination of this curve is completely objective and has been verified by this time in a number of laboratories.

Several features of the behavior of such irradiated cells pointed to damage to the genome as the primary process responsible for reproductive death.[19] The small value of the mean lethal dose for the S3 HeLa cell made it possible to rule out of consideration single-gene mutation as the dominant radiobiological process. The hypothesis was advanced that cell death is a consequence of chromosomal damage resulting from irradiation, and this proposal appeared to give good quantitative fit with the behavior of the system. This interpretation was supported by the following facts. After x-irradiation of single cells with approximately 6 mean lethal doses, the survivors were allowed to form colonies which were then picked and grown into new cell stocks. Examination showed at least 4 out of 5 such strains to be mutated from the original form, exhibiting morphological or nutritional differences, which persisted after long-term growth. Moreover, each of these four strains was revealed to possess a chromosomal constitution grossly altered from that of the original, unirradiated strain (Fig. 8). The 2-hit nature of the curve for the S3 cell was explained on the basis that reproductive death in such cells requires, in the main, at least two independent hits, and therefore appears to be a result of aberration produced by interaction between two separately damaged chromosomes. Other aneuploid cells of human origin, like S3, and with chromosome numbers far in excess of the normal 46, gave x-ray survival curves similar to that of S3.[22]

The fate of the cells whose ability to multiply indefinitely has been destroyed by irradiation is of interest. While some may multiply for a few generations all appear to retain the ability to metabolize effectively and to carry out complex biosyntheses of macromolecular structures like that of a specific virus, added to the system after irradiation. Under proper environmental conditions, cells of either aneuploid or euploid constitution may, after irradiation, produce giant forms. Thus, these cells which have been destroyed reproductively by irradiation, have maintained intact a large proportion of their other metabolic activities. At least small distortions of these functions probably also exist, but may require subtle means for their demonstration.

A pattern of events similar in some ways, but different in others, was obtained on irradiation of human cells with normal chromosome complements. All of these cells originated in normal human tissues and possessed the typical, elongated configuration commonly associated with euploid human cells.[14] Determination of chromosome constitution on several of these revealed them to have the normal diploid number. All of them behaved identically, and exhibited a mean lethal dose of about 50 r, corresponding to an even greater radiosensitivity than the HeLa cell. The hit number of these survival curves is less than that of the aneuploid HeLa cell, and lies somewhere between 1 and 2, a greater accuracy not yet being available for these cells because of their great radiation sensitivity. Hence, the conclusion may be drawn that euploid human cells are even more radiosensitive than the malignant, aneuploid HeLa.

As a further test of the hypothesis that cell reproductive death, as a result of x-irradiation, arises directly from chromosomal damage, a series of experiments was carried out in which euploid human cells were irradiated, after which the chromosomes were delineated and examined cytologically for direct evidence of aberrations. A preliminary report by Bender,[24] counting

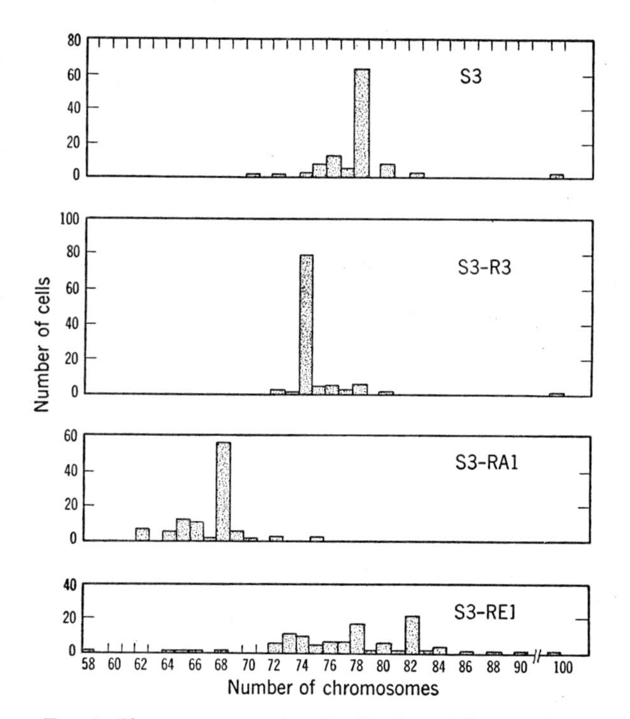


FIG. 8. Chromosome number distribution of the clonal strain S3 (top), and of 3 typical subclones isolated from among the survivors of radiation with 500–700 r (S3R3, S3RA1, and S3RE1). These chromosome analyses were carried out by Chu and Giles[23] in a collaborative study with our laboratory.

TABLE I. Chromosome aberrations at various radiation doses. Single-hit aberrations are chromosomal defects caused by a single ionizing event, and include a complete break in one chromatid only; a break in both chromatids at the same point presumably reflecting a break in the chromosome before it has doubled or a traverse of both chromatids by the same ionizing particle; an achromatic region in which the continuity of the chromosome is uninterrupted, but chromatin has disappeared from a particular area; or the presence of one or more greatly elongated chromosomes trailing "sticky" streamers. Multihit complexes comprise chromosomal aberrations involving interaction between two or more independent hits to 1 or more chromosomes, such as translocated dicentrics and ring chromosomes.

| Dose (roentgens) | No. of mitoses scored | Single-hit aberrations: Single chromatid break | Double chromatid break | Achromatic regions | Presence of one or more "sticky" chromosomes | Multihit aberrations |
|---|---|---|---|---|---|---|
| 0 | 116 | 22 | 1 | 1 | 1 | 1 |
| 10 | 3 | 0 | 0 | 0 | 0 | 0 |
| 20–25 | 33 | 6 | 1 | 0 | 0 | 0 |
| 40–50 | 20 | 37 | 4 | 0 | 2 | 1 |
| 75 | 101 | 113 | 23 | 7 | 2 | 14 |
| 150 | 26 | 26 | 5 | 8 | 4 | 10 |
| Totals | 299 | 204 | 34 | 16 | 9 | 26 |

chromosome breaks induced in euploid human cells grown *in vitro*, suggested that approximately 300 r were required to produce an average of one visible chromosome break per cell. This figure is almost six times higher than the mean lethal dose for these cells. In an attempt to obtain an estimate of the roentgen efficiency of chromosome breaks for euploid human cells, valid under our conditions of study, experiments were carried out in our laboratory in which the conditions of growth of the cells were brought as nearly as possible to optimal values in order to minimize some of the uncertainties owing to mitotic lags. In addition, however, an independent method of arriving at the chromosome-breaking efficiency was employed, based on the fact that the appearance of aberrations due to abnormal chromosome restitutions, can occur only in appreciable numbers at doses beyond the mean lethal dose. Such abnormal restitutions will not disappear, as can single chromosome breaks which can reseal without leaving any visible trace.[25]

Figure 9 presents typical sets of mitotic figures obtained when cells with normal chromosome constitution taken from various tissues of normal human subjects are irradiated *in vitro* at different dose levels. It is obvious that, with cells irradiated with 50 r, single chromosome breaks are readily evident [Fig. 9(b)]. When the dose is increased to 75 r, two effects appear: the number of single breaks is increased; in addition, however, new types of aberrations appear which indicate that, at this dose, multiple chromosome hits in individual cells have become a frequent occurrence. Examples of such complex anomalies formed through interaction of several radiation-damaged chromosomal sites are presented in Figs. 9(c) and 9(d). A summary of this data is presented in Table I. From these figures,

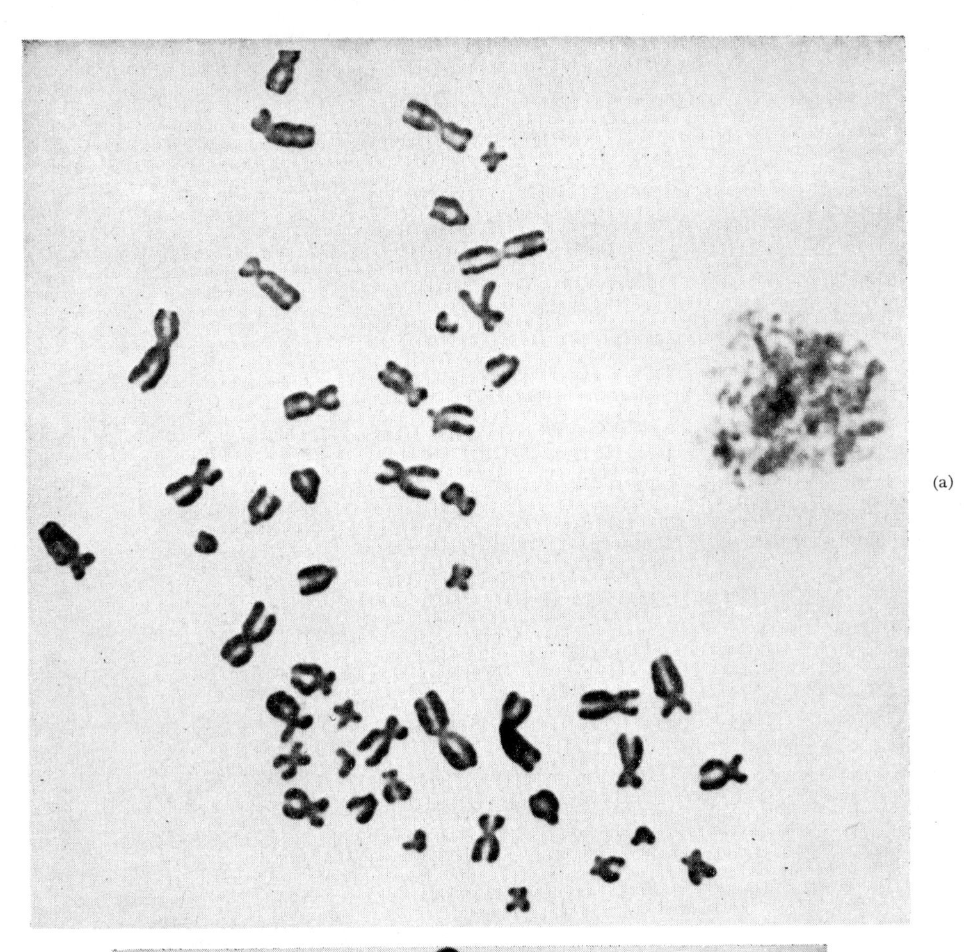

(a)

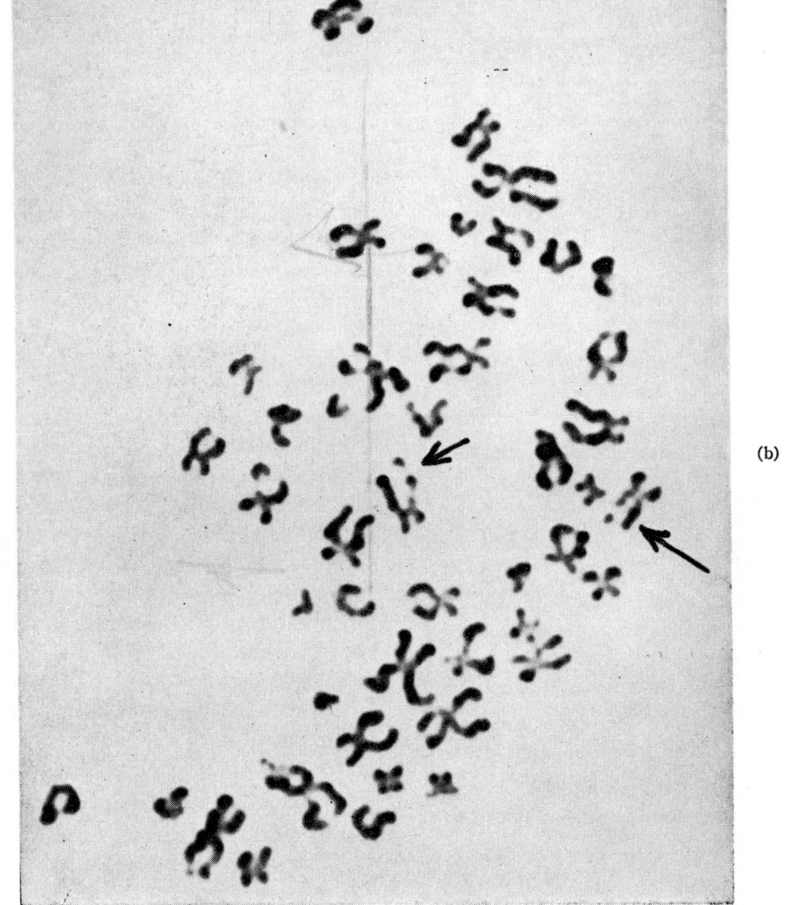

(b)

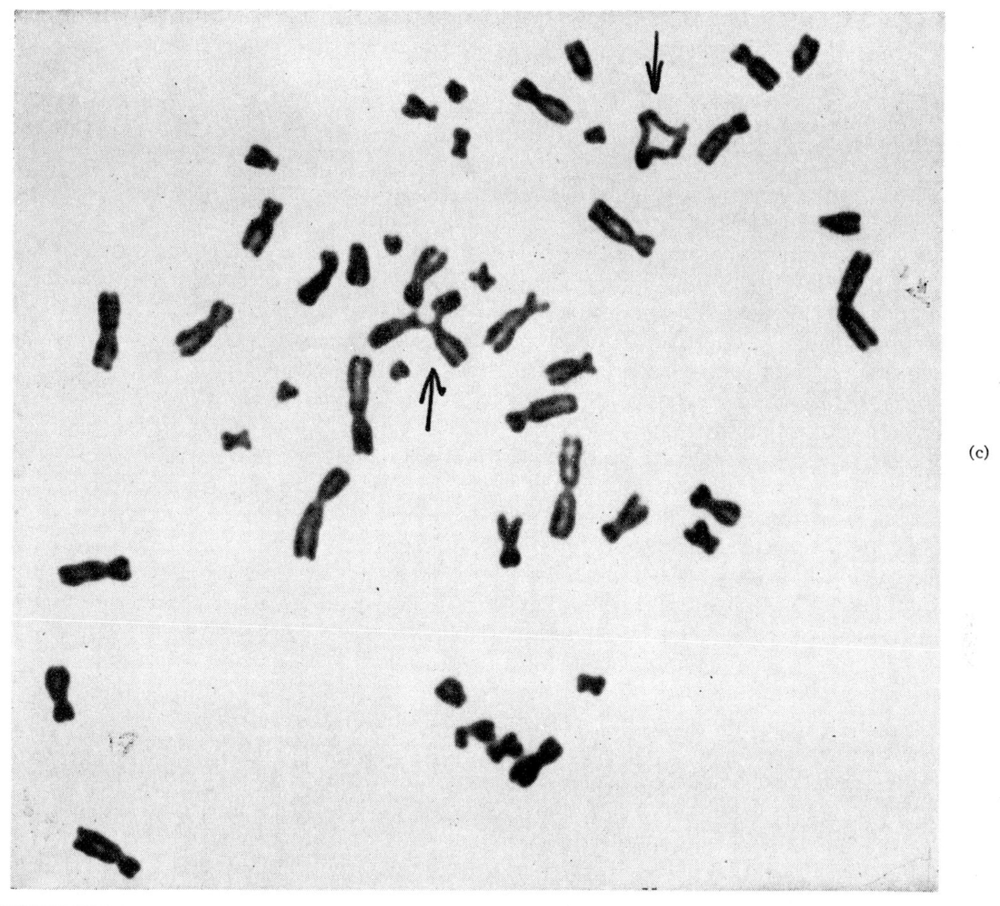

(c)

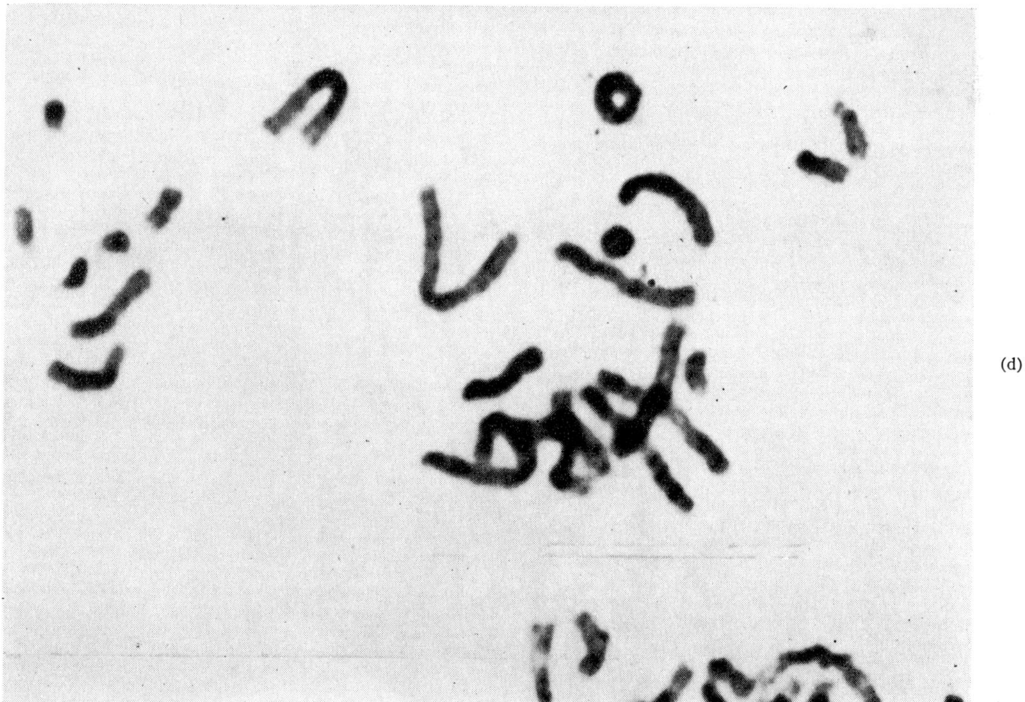

(d)

FIG. 9. Typical kinds of chromosome abnormalities observed when euploid human cells are irradiated *in vitro* with 230-kv x-rays. (a) Unirradiated cell. (b) Cell irradiated with 50 r. Breaks and deletions appear as indicated by arrows. (c) Chromosomes of cell irradiated with 75r, showing a translocation in the center, and a dicentric. (d) Portion of a mitotic figure of a cell after 150-r irradiation, demonstrating formation of ring chromosomes.

one can safely conclude that the dose needed to introduce an average of one visible chromosome break per cell is no more than 40 to 60 r, a value which agrees within the limits of experimental uncertainty with that obtained as the mean lethal dose for colony formation of these cells.

The conclusion may be drawn that the genome of the euploid human cell is extraordinarily susceptible to chromosome damage by ionizing radiation, and that the average dose needed to introduce one chromosome break per cell is similar to the mean lethal dose to the cell's reproductive function, i.e., approximately 50 r. Since this value is far less than that required for the inactivation of various enzymatic activities in mammalian cells,[26] it appears to establish the fact that the cellular genome of man is far more sensitive to radiation damage which is, at least in part, presumably irreversible, than are other of its structures. While studies are currently in progress to determine whether cells *in vivo* will behave in the same way, experiments with radioprotective agents added to the medium suggest that only relatively small changes in the cellular $D_0$ value can be accomplished by changes in the cellular environment.

These considerations afford an understanding of the action of ionizing radiation on mammalian cells in terms of a primary effect—the initiation of a chromosome break, which requires on the average an exposure of a normal human cell to 50 r or less, and a secondary effect—the production of complex chromosomal aberrations through abnormal restitution of the fragments resulting from multiple breaks occurring within the same cell. The first of these involves a variety of alternative possibilities: The cell may continue reproduction with no recognizable cytogenetic change, though with perhaps a gene mutation at the site of the original chromosome break; it may suffer loss of all or part of a chromosome which could impair its ability to reproduce; or it may undergo formation of a chromosome bridge at anaphase which will certainly end its reproductive ability. When multiple chromosome fragmentations leading to complex translocations occur, the cell is much more likely to be incapacitated with respect to reproduction. Different cells should display 1-hit, multiple-hit, or intermediate types of survival curves, depending on the degree to which these different processes operate in causing destruction of the ability to multiply.

This analysis appears to offer an explanation for the difference in radiobiologic response exhibited by various mammalian cells. For example, normal diploid cells of different species, differing in their total mass of chromosome material, should be expected to exhibit differences in radiation sensitivity roughly though not exactly paralleling their chromosomal volumes, since the cell with greater chromosomal mass will offer greater opportunity for a given radiation dose to produce ionization directly or indirectly effective within this region. In accordance with this expectation, we have found that cells of the chick, whose DNA content is less than one-quarter that of human cells, possess a mean lethal dose for x-rays approximately 4 to 8 times greater than that of euploid cells of man. In Table II is presented a comparison of the $D_0$ values and DNA content of diploid cells from a variety of different species. Despite relatively large uncertainties in the individual determinations of the mean lethal dose and of the DNA content, and the large range of values encompassed, it is evident that these very different diploid cells exhibit a remarkably constant value for the product of these 2 variables. Other aspects of these relationships are considered elsewhere.

These considerations afford insight into the effects of various chromosome conditions like polyploidy on the radiosensitivity of mammalian cells from the same species. With increasing numbers of chromosome sets within a cell, its sensitivity to killing by a 1-hit process should decrease. Thus, a haploid cell will lose its ability for indefinite multiplication as a result of damage to any single gene essential to replication. Diploid cells which have a double gene set presumably will require loss or aberration of larger portions of a chromosome before the genetic imbalance necessary for inhibition of multiplication results. Since such chromosome losses readily occur as a result of 1-hit processes following exposure to ionizing radiations, such cells will often exhibit curves with a hit number close to unity, but may in some cases be multiple hit, if the gene distribution among the chromosomes is such that death rarely follows a single-hit process. The $D_0$ value of such curves will be influenced by the volume of each chromosome, the magnitude of the breaks which are produced, and the degree to which 1-hit lethal and 2-hit lethal processes occur when cells are exposed to various types of irradiation under various conditions. Cells with higher degrees of ploidy will be much less susceptible to reproductive killing by loss of part or all of any chromosome because of the smaller imbalance produced when larger multiples of the chromosome set are present. In such cells, the major lethal process will involve interaction between

TABLE II. Comparison of radiation and the DNA content for diploid cells of three different living forms. The data indicate that, despite individual variations in $D_0$ values and DNA contents of over a hundred-fold, the product of these two variables remains reasonably constant, varying only by a factor of about two.

| Diploid cell type | $D_0$ value[a] | DNA content (in picograms per cell) | Product: $D_0 \times$DNA content (roentgens $\times$picograms) |
|---|---|---|---|
| Yeast | about 8000 r[b] | 0.05[b] | 400 |
| Chick | about 300–400 r[c] | 2.5[d] | 900 |
| Man | 50–60 r[e] | 8.3[d] | 500 |

[a] The $D_0$ value is taken as the dose needed to reduce the reproducing fraction of the cell population to 37%, in the region where the survival curve is linearly exponential.
[b] See reference 27.
[c] See reference 25.
[d] See reference 28.
[e] See reference 22.

two or more chromosomes to produce malformations like anaphase bridges. Hence, such cells will usually display a hit number in the neighborhood of 2. While the larger volume of the chromosome set in such cells might at first be considered to make them more vulnerable than diploid cells to the accumulation of hits that can lead to cell death, this factor may be more than counterbalanced by the relatively low probability with which two simultaneously broken chromosomes will restitute abnormally, so as to form a nondividing configuration (like a dicentric) as opposed to a union still permitting normal mitosis. Thus, the HeLa S3 cell, an aneuploid human malignant cell with 78 chromosomes, exhibits a survival curve which is approximately 2-hit, as opposed to the more nearly 1-hit curve of the euploid cell, and exhibits a $D_0$ value of 96 r instead of 50 r characteristic of the normal diploid cell. These parameters—the degree of cell ploidy; the distribution among the different chromosomes of regions whose loss may lead to cell death through imbalance; and the separate probability of formation of chromosomal aberrations which mechanically prevent unlimited division by 1-hit and 2-hit processes distributed among the various chromosomes—would appear to afford explanation of the difference in radiosensitivities of different cells. Quantitative evaluation of these parameters in selected normal and malignant cell systems is now under study.

These considerations also can offer explanation for our recent findings that the S3 cell which is enormously more sensitive than bacterial cells to killing by x-rays, is much more resistant to ultraviolet irradiation. Ultraviolet, while a highly effective lethal and mutagenic agent for bacteria, is known to be much less efficient than x-rays in producing chromosome breaks. While haploid bacterial cells can be killed effectively by point mutations and so are readily inactivated by ultraviolet light, the need for chromosome breakage to occur before the multiploid animal cell is inactivated would render it resistant to the action of ultraviolet irradiation.

Application of these aspects of the action of ionizing radiation at the cellular level, to analysis of the underlying mechanism involved in the pathologies arising from acute, whole body irradiation of animals leaves little doubt that many of the most important features of the mammalian radiation syndrome are due to the cumulative actions on the individual body cells of the same kind as those described in the preceding paragraphs:

(a) Knowledge of the $D_0$ value for individual cells for the first time provides clear explanation why the mean lethal dose for total body irradiation of mammals should lie in the range of 400 to 600 r. Exposure of the whole body to a dose of this magnitude would inactivate the reproductive mechanism of more than 99% of the cells with radiosensitivities like those of the cells studied *in vitro*. The body might recover from smaller doses which would still leave sufficient intact cells which could reproduce at a rate rapid enough to compensate for the losses. The mean lethal dose to the whole animal represents a point of loss of reproducing cells so extensive as to provide widespread damage to the integrities of physiological function which depends on cell reproduction and which constitutes a stress which cannot ordinarily be withstood.

(b) This formulation is in accord with the failure to find any specific biochemical lesion resulting from animal irradiation in the mean lethal range. Significant depression of individual enzyme functions have been found to require much higher doses than that needed to kill the whole animal. While no specific enzyme function could be implicated, many studies showed that DNA synthesis was greatly reduced as a result of irradiation in the dose range here considered.[29] The theory here considered would predict such behavior. DNA synthesis can proceed normally only when cell reproduction progresses. Destruction of this function through production of chromosome aberrations must eventually depress DNA synthesis, even though no enzyme system of the cells or body fluids has been significantly altered by the irradiation.

(c) Similarly, the failure to find any accumulation of toxic material in the body fluids after exposure to radiation in the mean lethal range is to be expected. While at higher doses significant amounts of toxic products might be produced as a result of chemical structural alterations resulting from ionizations produced in the tissue, doses of several hundred roentgens only will alter significantly structures with a size of the order of magnitude of the chromosomes, and will produce biological changes only if the uninterrupted integrity of such structures is vital to normal body operations as is indeed the case with the genetic apparatus of the individual cells.

(d) It follows that one would expect those tissues to be most radiosensitive which must maintain the highest rate of cell mitosis, since these will suffer the most immediate loss of their specific functions, on which the body depends. This correlation between radiosensitivity and mitotic rate has been recognized as one of the earliest generalizations to emerge from radiobiologic experience. However, while the rapidly dividing tissues are indeed most radiosensitive from the point of view of their immediate contribution to embarrassment of the whole organism through failure to maintain their specific functions, it is necessary to regard virtually all of the body cells as equally sensitive to chromosomal damage. Each normal cell nucleus has an equal probability of accumulating similar chromosomal injuries. Such aberrations will remain largely latent in each cell until it comes to reproduce. Hence, the cells of the more rapidly dividing tissues are simply the first to make evident the injuries which must be regarded as equally numerous in the more slowly multiplying regions of the body. This fact is of vital importance in understanding genetic transmission of radiation damage, and the ability for tumors to develop many years after a radiation experience.[30]

(e) Similarly, our picture accounts in straightforward fashion for the characteristic lag period between radiation exposure and the development of symptoms, which has long been recognized as typifying mammalian injury by ionizing radiation. The cell which has suffered damage which is primarily only chromosomal, will continue to function effectively for a period, since its full complement of enzymes and other structures presumably remains largely unaltered. Not until the cell reaches the point where it should normally divide will it begin to exhibit grossly deviant behavior from the normal. Moreover, we have shown that such cells may even multiply for several generations, before they and their progeny cease reproduction.[19]

(f) The sequence of events which ultimately leads to death of any acutely irradiated animal may thus be explained as follows. Chromosomal aberrations introduced among all of the cells exposed will first reveal themselves in those cells of the most actively mitotic tissues. Division will be prevented in most of such cells irradiated with several hundred roentgens, with the result that the functions supplied by these tissues will fail and the body will be threatened. As other more-slowly dividing tissues reach the point where they too should produce cell proliferation, they will add the results of their failure to those which have already introduced distortion of normal body physiology. In contrast to the actively dividing structures like the bone marrow, and the epithelial linings of the gastrointestinal tract, tissues like nerve and muscle in which cell division rarely occurs, can continue to exhibit normal function even after exposure to many thousands of roentgens. It is of interest that cell reproductive death has also been identified as the major factor in the mammalian radiation syndrome by Quastler and his co-workers, using experimental approaches and techniques completely different from those employed here.[31]

In this connection, an alternative hypothesis has been proposed that the more-rapidly dividing tissues are more radiosensitive only because cells in mitosis are in a more sensitive condition and hence are damaged more readily by irradiation. This proposal does not fit the facts, since even in the most rapidly dividing tissues, no more than 3% of the cells are in mitosis at any one time. Hence, if this were the true explanation, it could account at most for a negligible reduction in cell viability of such tissues. However, in early embryonic development, where high mitotic frequencies often are achieved and where phased cell reproduction may occur to bring many cells into mitosis simultaneously, this factor might easily play an important role.

(g) These considerations also explain quite naturally how it is possible to save animals irradiated with doses in the mean lethal range, by injection of viable bone-marrow cells, though not by cellular fragments or extracts. Such cells recolonize the irradiated tissue which is being depleted through impaired cell-reproductive function. While other tissues have presumably suffered equal diminution in the percent of cells capable of reproduction, these require replenishment less rapidly and hence provide more time for the surviving cells to recolonize the tissue, provided the bone-marrow functions can be maintained. It may be expected that animals might be saved from lethal effects of even higher doses, by additional inoculation of viable cells from other tissues which, as the dose increases progressively, would become critical in their failure to maintain functional integrity of the body.

(h) Similarly, the interesting experiments first carried on by Patt and his co-workers,[32] in which animals were subjected to low temperatures immediately after irradiation, become explicable. Such animals failed to develop any of the symptoms of radiation injury until after their temperature was restored, after which the entire sequence of pathogenesis was initiated as though the irradiation had just occurred at the time their body temperatures had been raised back to normal. At the low temperature, mitosis is inhibited so that the body does not produce the conditions which can bring into expression the latent damage to the cellular genetic apparatus. On rewarming, normal cell reproduction is again initiated, so that each time a genetically damaged cell comes to mitosis its injury becomes functional and contributes to embarrassment of a normal function.

(i) One might expect, then, to find some rough correlation between the $D_0$ obtained for euploid cells of different animals as measured by the survival curves here described, and the mean lethal dose for the whole animal. While the latter figure must, of necessity, be influenced by many complex interactions of an unpredictable kind, it is of distinct interest to find that such a correlation is at least suggested by the small amount of currently available data, as shown in Table III.

Studies of the action of radioprotective agents on x-ray survival curves of S3 cells are completely consistent with the interpretations here developed. The compound, 2-mercaptoethyl guanidine, which has been demonstrated to raise the MLD for mice from 900 to a maximum of 1450 r,[35,36] also exercises significant radio-

TABLE III. Data demonstrating that possibly some parallelism may exist between the mean lethal dose for whole warm-blooded organisms ($LD_{50}$) and the $D_0$ values of their single, euploid cells as determined *in vitro*.

| | X-ray $LD_{50}$ for whole animal | X-ray $D_0$ of euploid cell *in vitro* |
|---|---|---|
| Man | 400– 500 r | 50– 60 r |
| Chinese hamster | 825–1190 r[a] | 160 r |
| Fowl | 1000 r[b] | 300–400 r |

[a] See reference 33.
[b] See reference 34.

protection on S3 HeLa cells plated *in vitro*. The shape of the survival curve remains approximately 2-hit as it was in the absence of the compound, but $D_0$, the mean lethal dose for cell reproduction, is raised from 96 r to a maximum of 160 r, a value which agrees far better than perhaps could be expected with the degree of protection achieved in the whole animal. All of the data on the kinetics of the radioprotective action of this compound on single plated cells are consistent with the interpretation that the presence of this material lowers the effective dose of x-rays which reaches the sites within the cell whose inactivation results in loss of the ability to reproduce indefinitely.

(j) Finally, these data support the cellular-genetic interpretation of the great effectiveness of relatively low doses of radiation in suppressing antibody formation in the mammalian body. The dose range needed to inhibit antibody production significantly lies in the region of 50 to 150 r,[37] a value which, through its close correspondence with the mean lethal dose for euploid cell reproduction, suggests this action to reflect the effect of radiation in preventing multiplication of antibody-producing cells. By contrast, specific macromolecular biosynthetic mechanisms appear essentially undamaged even after irradiation of mammalian cells with thousands of roentgens.[15] Thus, the great radiosensitivity of antibody production suggests that the mechanism of new antibody production involves the need for cell multiplication, rather than simply antibody synthesis by pre-existing cells.

The fact that some mammalian cells exhibit survival curves for the reproductive function which are multiple-hit in character must not be taken as an indication that a threshold exists for the induction of mutations by high-energy radiations. On the contrary, all of the evidence here presented leads to the conclusion that, while a cell displaying a 2-hit survival curve does display a threshold for killing by high-energy radiation, the production of single chromosome breaks, and the attendant alteration of genetic material at the site of such a break, goes on even when doses insufficient to kill are applied. From the point of view of the entire organism and its progeny, it may be far better for any cell to be permanently inactivated by multiple chromosomal aberrations than to suffer only a single chromosomal change, which then may be handed on to the offspring, usually as a recessive genetic defect.

The data here presented make possible an estimation as to whether background radiation may contribute to the processes of aging in man. Since 50 r is the average dose needed to produce 1 visible chromosome break per cell by the techniques here employed, one may well expect that damage on a smaller scale which cannot be seen by ordinary microscopy results from even smaller doses. Thus, one can calculate that, with a background exposure of 0.1 r per year, in 70 years the accumulated dose is sufficient to have produced gross chromosome damage in one-seventh of all of the body cells, and probably more subtle effects in a much larger number. This would appear not to be a negligible process, and thus makes it probable that the gradual attrition of body functions which constitute aging, may be, to a very significant degree, the reflection of accumulation of cellular genetic injuries in cells and their descendants originating from background irradiation. By extension of the methodologies here discussed, it is proposed to attempt analysis of physiologic and genetic differences in the behavior of cells taken from aging animals, and to examine as well the effects of the molecular constituents of body fluids from such subjects on physiologic and genetic behavior of standard cell strains *in vitro*.

Discussion of the use of quantitative methods in measuring mammalian cell growth and genetics would be incomplete without at least mention of the outstanding development by Dulbecco and his associates[38] in achieving a plaque technique for precise enumeration of single particles of mammalian viruses exactly as has been current in bacteriophage studies. Thus, it becomes possible also to quantitate virus and cell interaction in animal systems, so that one may expect equally profound insights to arise from such studies as has been the case in the interaction of bacteriophages with their bacterial hosts.

The thesis which has been the purpose of this discussion is that newer developments promise to provide well-defined systems for quantitative exploration of physical and physicochemical mechanisms involved in growth, genetics, and differentiation in mammalian-cell systems. There is every indication that in the space of a few years it will be possible to come to grips with specific molecular aspects of these basic mechanisms in as intimate a fashion as is now current in the biophysical approach to microbial systems.

## BIBLIOGRAPHY


1 T. T. Puck, P. I. Marcus, and S. J. Cieciura, J. Exptl. Med. **103**, 273 (1956).

2 T. T. Puck and H. W. Fisher, J. Exptl. Med. **104**, 427 (1956).

3 T. T. Puck, J. Cellular Comp. Physiol. (to be published).

4 H. Eagle, J. Biol. Chem. **214**, 839 (1955).

5 V. J. Evans, J. C. Bryant, W. T. McQuilkin, M. C. Fioramonti, K. K. Sanford, B. B. Westfall, and W. R. Earle, Cancer Research **16**, 87 (1956).

6 J. F. Morgan, H. J. Morton, and R. C. Parker, Proc. Soc. Exptl. Biol. Med. **73**, 1 (1950).

7 H. W. Fisher, T. T. Puck, and G. Sato, Proc. Natl. Acad. Sci. U. S. **44**, 4 (1958).

8 I. Lieberman and P. Ove, J. Biol. Chem. **233**, 637 (1958).

9 R. C. Parker, L. N. Castor, and E. A. McCulloch, "Cellular biology, nucleic acids and viruses," special publication of New York Acad. Sci. **5**, 305 (1957).

10 A. E. Moore, C. M. Southam, and S. Sternberg, Science **124**, 127 (1956).

11 A. A. Axelrad and E. A. McCulloch, Stain Technol. **33**, 67 (1958).

12 K. H. Rothfels and L. Siminovitch, Stain Technol. **33**, 73 (1958).

13 J. H. Tjio and T. T. Puck, J. Exptl. Med. **108**, 259 (1958).

[14] T. T. Puck, S. J. Cieciura, and A. Robinson, J. Exptl. Med. **108**, 945 (1958).
[15] J. H. Tjio and T. T. Puck, Proc. Natl. Acad. Sci. U. S. (to be published).
[16] J. H. Tjio and A. Levan, Hereditas **42**, 1 (1956).
[17] G. O. Gey, W. D. Coffman, and M. T. Kubicek, Cancer Research **12**, 264 (1952).
[18] J. H. Muller in *Radiation Biology*, A. Hollaender, editor (McGraw-Hill Book Company, Inc., New York, 1954), p. 351.
[19] T. T. Puck and P. I. Marcus, J. Exptl. Med. **103**, 653 (1956).
[20] C. M. Pomerat, Ann. New York Acad. Sci. **71**, 1143 (1956).
[21] T. H. Ingalls, H. R. Morrison, and L. I. Robbins, New Engl. J. Med. **2581**, 252 (1958).
[22] T. T. Puck, D. Morkovin, P. I. Marcus, and S. J. Cieciura, J. Exptl. Med. **106**, 485 (1957).
[23] E. K. Y. Chu and N. M. Giles, J. Natl. Cancer Inst. **20**, 383 (1958).
[24] M. A. Bender, Science **126**, 974 (1957).
[25] T. T. Puck, Proc. Natl. Acad. Sci. U. S. **44**, 772 (1958).
[26] Z. M. Bacq and P. Alexander, *Fundamentals of Radiobiology* (Butterworths Scientific Publications, London, 1955), pp. 228–262.
[27] C. A. Tobias, Natl. Acad. Sci.-Natl. Research Council, Publ. No. **18**, 46 (1956).
[28] I. Leslie in *The Nucleic Acids*, E. Chargaff and J. N. Davidson, editors (Academic Press, Inc., New York, 1955), Vol. II, p. 1.
[29] Z. M. Bacq and P. Alexander, see reference 26, pp. 257–258.
[30] C. L. Simpson and L. H. Hempelmann, Cancer **10**, 42 (1957).
[31] H. Quastler, Radiation Research **4**, 303 (1956).
[32] H. M. Patt and M. N. Swift, Am. J. Physiol. **155**, 388 (1948).
[33] G. Yerganian, Federation Proc. **14**, 1371 (1955).
[34] Z. M. Bacq and P. Alexander, see reference 26, p. 220.
[35] D. G. Doherty (personal communication).
[36] R. Shapira, D. G. Doherty, and W. T. Burnet, Radiation Research **7**, 22 (1957).
[37] H. T. Kohn, J. Immunol. **66**, 525 (1951).
[38] R. Dulbecco and M. Vogt, J. Exptl. Med. **99**, 167 (1954).

# 48
# Interactions between Cells

Paul Weiss
*The Rockefeller Institute, New York 21, New York*

INTERACTIONS among cells are the means by which the cell community of the organism establishes and maintains its organizational harmony. They are so numerous and varied that it would take more space than is allotted here just to draw up a reasonably comprehensive list. Yet, knowledge about them is still very scanty. Fascinating progress has been made in the study of some of the biochemical and biophysical components of the living system. But knowledge seems to have grown the faster, the smaller the sample of the living system that was taken under investigation. The major advances were made down at the molecular level. The task of dealing with the larger cellular level, therefore, involves a re-entry into areas of uncertainty, and the course from the introductory article on cell dynamics (Weiss, p. 11) to the present article marks a full circle from relative ignorance through knowledge back to relative ignorance. The study of cells in interaction deflates complacent notions that all the major facets of cell life are truly understood even in principle.

"Interactions between cells" covers practically everything that is going on in organisms, and obviously this account must confine itself to a few crucial examples. There are essentially two ways for cells to interact: either a cell elaborates a diffusible substance which affects another cell at some distance, or a cell transmits an effect to another cell by direct contact. Since mediation by diffusible agents, as in the hormone system, is more widely studied and better understood than are the contact interactions, the emphasis here is on the latter. A more detailed account can be found in the author's recent review article on "Cell Contact."[1]

First, a few words on what is meant by "contact." It has been very easy to define contact between cells in terms of observations with the ordinary light microscope. If microscopically two cell borders came so close as to merge into a single line, there was contact, whereas a microscopic space in between signified lack of contact. Conversely, any interruption of the microscopic outline between two cells was interpreted as "protoplasmic continuity." But evidently, such questions of "contiguity" *vs* "continuity" between cells are merely questions of the resolving power of the instruments at hand. It is not surprising, therefore, that the introduction of electron microscopy has brought a marked change of outlook.

In the first place, electron microscopy has revealed a higher degree of organization at cell surfaces than previously assumed. As an example, one can cite the larval amphibian skin referred to in the introductory article (Weiss, p. 11). The underside of the epidermal cells, which rest on the basement membrane, is dotted at regular intervals with submicroscopic bodies, about 1200 A high, consisting each of two electron-dense round plates connected by a lighter neck.[2] The outer one of the plates forms part of the cell surface. This surface itself is separated from the underlying basement membrane by a gasket-like granulated film of a few hundred Ångströms thickness. The cell surface thus is actually a mosaic of patches of supramolecular order and different chemical and physical properties, which have been partly identified. This was done by applying various enzymes to fragments of live skin before they were fixed in osmium tetroxide and prepared for electron microscopy. In pancreatic lipase, the dark plates lose their osmiophilia, indicating that normally they have a high lipid content. In distilled water, the neck of these bodies swells and breaks, leaving unconnected double plates. Thus, the neck region is hydrophilic. Salivary amylase dissolves the film through which the epidermal cell is attached to the basement lamella, suggesting abundance of carbohydrate. It also erodes the parts of the cell surface between the dense bodies, but leaves the latter intact and protruding, thus identifying them as solid bodies.

Such observations demonstrate that it would be quite mistaken to consider a cell as being equal and uniform over its whole surface. Characteristically, this surface pattern is confined sharply to the portion of the cell that is in immediate apposition to the basement membrane, thus indicating a direct contact interaction between cell and substratum. This supposition is confirmed by experiments in which the contact between the two structures was first broken and then restored, as follows. When part of the skin is injured, the epidermal cells near the wound roll off. The dense bodies, which seem to serve as suckers to attach the cells to their substratum, become detached and are resorbed within a few days. They are re-formed during wound healing as the epidermal cells migrate over the defect, but are precisely confined to that fraction of cell surface now in fresh contact with the substratum. In other words, the cell has a mechanism for producing this specialized apparatus which responds to the interaction between cell surface and the underlying carbohydrate-rich film. As a result, an "induction" on the submicroscopic level occurs. The cell is induced along a geometrically and physically defined surface to display a specific fraction of its synthetic repertory. Other instances of contact induction are given later in this paper.

When two cells are in contact, the problem becomes more complicated. At the contact points between two epidermal cells, single dark plates are present, the well-known nodes of Bizzozero. Electronmicroscopically, they show as a single dark plate in each of the contacting cells with a lighter space in between.[1] Odland[3] has recently scanned this region densitometrically in superior electron micrographs and has found an additional dark line in the center of the light interspace. This would indicate that the molecular array separating the two cell surfaces is not random, but has some degree of spatial order. It may be that the cell surfaces are linked by macromolecules normal to the surfaces of some 100 A in length, which is the order of width of the light space. Gaps of similar dimensions have been found between practically all cells that make intimate connections with each other, including synapses between nerve cells. It would seem best, therefore, to define two cells as being in contact, operationally speaking, when they are separated by a space whose contents are not subject to random perturbation. Depending on the dimensions, several factors might operate to limit perturbation of the interspace. For example, macromolecular compounds moving out of the cells may form bridges between organized parts and establish special submicroscopic attachments or interaction points between the adjoining cell surfaces. Evidence from tissue culture, obtained over a decade ago,[4] indicates that certain cells, and perhaps all cells, when left to their own devices, surround themselves with characteristic colloidal exudates, which must be taken into account when cell-to-cell relations are analyzed. Such surface coronas, extending for unknown distances into the cellular environment, might be major factors in immune reactions, cellular aggregations into colonies, phagocytosis of specific substances, selective associations among nerve fibers, and the like. It is equally possible that cell surfaces react with each other specifically without such mediators. A few examples are cited in the following.

In experiments on the development of nerve fibers, it has been found that fibers of the same type have a selective affinity to one another—motor fibers applying themselves to motor fibers, and sensory fibers to sensory fibers, and within each class, each subgroup to its corresponding type.[5] Only by virtue of such specific grouping is it possible for a surgeon, for instance, to expose the spinal cord and sever discriminately a bundle of pain fibers or of fibers for deep perception. Such fibers would not run in common bundles unless they had grown out that way, and since they do not develop all at the same time, the older fibers must have served as guides to latecomers of the same character.

The same principle applies, even more subtly, to the relations among the cells within the central nervous system on which the coordination of neural functions depends. Our current concern with the nervous system as a system centers mostly around such matters as geometric properties, electric parameters and time constants—length of interconnections within the network, number and distribution of branches, temporal characteristics of the individual units, synaptic resistances, etc. We generally ignore the fact that there is great biochemical diversity within the system far beyond the gross distinctions of cholinergic and adrenergic portions, which makes it work as something more than a monotonic network of conducting fibers. Developing nerve cells have highly selective discriminatory affinities by which they "recognize" each other and their surroundings. This introduction of the anthropomorphic term of "recognition" is not anything one need apologize for so long as physicists speak of electrons "seeing" each other. Recognition of one cell by another may be based on conforming charge distributions, on conforming molecular groupings, or on as yet wholly unsuspected mechanisms. The problem of explaining such recognition is of course encountered in many other biological phenomena, e.g., in enzyme-substrate relations and in antibody-antigen reactions. With a few notable exceptions,[6] affinity reactions among somatic cells, however, have received little attention and even less methodical study, even though there is ample evidence for the widespread occurrence of such selective behavior.

To cite a specific example, cells can "locate" their proper destinations in the body even if they are deprived of their customary routes for getting there. This has been shown[7] by letting embryonic cells of known destination be distributed at random through the blood stream of a host embryo. In order to be able to identify the injected cells, one chooses cells which carry a marker. We chose the precursor cells of pigment cells which, when introduced into nonpigmented breeds of chicks, would reveal their origin by synthesizing black melanin. It was found that after an injection of such cells into the blood stream, about 8% of the host chicks later carried scattered patches of black pigment cells in their feathers and skin. These pigment cells were always situated in the precise positions where such cells would have resided in the donor animals, and in no case was a pigment cell ever observed to have become lodged where pigment cells do not normally belong. One must conclude, therefore, that donor propigment cells had colonized only specified areas in the host embryos where such cells normally end up, which, in view of the random dissemination of the dissociated cells in the host body, implies the existence of a mechanism for highly selective localization. As few as one or two propigment cells, when arriving by chance at an appropriate site, "recognize" it and settle down to proliferate and differentiate into pigment cells. Cells that miss encountering within the embryo an environment favorable to their type fail to become lodged and to differentiate, and presumably are resorbed.

On the other hand, outside the embryo, on the yolk sac, where specific sites for embryonic cells are lacking, clumps of injected cells get stuck, and their fate led to a further significant extension of these observations. It was discovered[8] that, where random mixtures of embryonic cells from a variety of sources (nerve, muscle, cartilage, glands, etc.) had become lodged on the yolk sac, they gave rise not to indiscriminately mixed structures, as might have been expected, but to quite harmoniously organized organ complexes. Some degree of self-sorting according to types must have taken place among these scrambled cells after they had become trapped.

Further conclusive evidence of "self-sorting" was found in tissue cultures in which random mixtures of suspensions of cells of different embryonic origins —for instance, cartilage and kidney cells—had been combined.[9] These cultures developed compound structures in which islands of pure cartilage were sharply demarcated from blocks of pure kidney tissue, indicating that cells had become re-associated like-to-like. Thus, even *in vitro* cells, which have acquired in their prior embryonic history some biochemical differential related to their subsequent development as either kidney cells or cartilage cells, can assort themselves according to type. Since the cells make contact only with their surfaces, one must assume that some property associated with differentiation into kidney cells or into cartilage cells has imparted type-specific markers to their surfaces for mutual recognition. But how such "recognition" leads to active sorting remains obscure.

How do mixed cell populations achieve this orderly reassortment? Do like cells "attract" each other? Or do they "recognize" each other only after chance encounters? A clue came first from some observations on selective wound healing after the grafting of various kinds of epithelia into skin gaps in the flanks of amphibian embryos.[10] The first coverage of a wound is effected by the migration of epithelial cells of the wound edge over the raw surface. When these advancing cells reach a graft, fusion of the two fronts occurs only if the graft contains an epithelium of a type which normally borders on skin epidermis. In that event, the advancing cells stop in their tracks, so to speak, merge into a uniform sheet with the graft, and no further proliferation occurs. Implants of skin, or of cornea or oral lining—which are normally continuous with skin—are accepted in this manner. But, when a fragment of lung or gall bladder or esophagus is implanted in a skin wound, the migrating skin epidermis is not halted on contact with those foreign-type cells. The epidermal edges glide over or under the graft and continue until they meet with the skin advancing from the opposite side.

Results of this kind have long been known in surgery and lead to the conclusion that, whether we have a plausible explanation for the phenomenon or not, cells of identical or conforming constitution (as epidermis, cornea, and oral lining) will recognize each other as akin and stay put; whereas cells of unlike character develop a reaction which will cause them to separate.

With these facts as background, we proceeded to study the contact reactions between different kinds of cells directly *in vitro*,[1] by taking phase-contrast time-lapse motion pictures of encounters among cultured cells.* If the cells that make contact are all of the same kind—all liver cells, for example—they aggregate and stay together. The free borders of roaming cells are in constant motion; but wherever two like cells touch each other, that portion of their surfaces becomes quiescent. The bond between them, however, is not static; there is no cement which sticks them together as in an immunological precipitin reaction. On the contrary, the cells continually glide over one another, constantly changing their positions relative to each other and, by the time others have joined them, their positions in the group.

On the other hand, if cells of two different types are cultured together (e.g., lung and liver or lung and kidney), the result is quite different. As the loose cells stray about, they collide at random. There is not the slightest indication that they would react to each other's presence unless or until their free borders touch. Neither do like cells "attract" each other, nor do unlike cells avoid each other. Mutual recognition and consequent discriminatory behavior are decidedly contact responses. All cells, whether alike or not, make primary contact indiscriminately. But, whereas like pairs thereafter draw closer together and remain joined, unlike pairs separate secondarily by reciprocal withdrawal of those parts of their borders that had been in fleeting touch. While the mechanism of identification undoubtedly resides in the surface, the interior of the cell seems to be engaged in the withdrawal reaction, giving the phenomenon the general appearance of a "reflex" response with a "sensory" and a "motor" component. The "reflex time" is of the order of $10^3$ sec, varying both with the strangeness of the confronted cells and with their respective rates of motility. In conclusion, self-sorting of scrambled cell types results from chance collisions with matching combinations holding on to each other, whereas nonmatching ones do not last.

"Matching" combinations involve either cells of the same type or of complementary types, the latter being types that normally cooperate in the building and functional operation of complex organs. For instance, an active mutual adhesion between nerve processes and enveloping sheath cells has been demonstrated,[11] and our motion pictures reveal a similar marked tendency of macrophages to confine their excursions to within the borders of the large flattened lung cells with which they have been jointly explanted.

The discriminatory response is strictly cell-type specific, but it is not species specific. That is to say, cartilage cells of the chick will reject association with liver or kidney cells of the same chick, but they will combine readily with cartilage cells of a mouse. This fact, derived earlier from reaggregation experiments,[12] has now

* These moving pictures, made in collaboration with A. C. Taylor and A. Bock, were shown at the Study Program in Biophysical Science in conjunction with the lecture on which this article is based.

been proven visually by the motion pictures. On the other hand, the same cell type may undergo progressive alterations with age, so as to make more mature and less mature cells of the same strain less acceptable to each other than cells of the same age would be. Furthermore, there are signs that the differentials among cell types, which underly their discriminatory responses to each other, may be subject to gradations in accordance with their ontogenetic relationships.

The details of this principle still remain to be worked out. But what has emerged thus far is enough to force a substantial revision of former excessively static pictures of the organism as a fixed framework of parts shifted into predestined places by a rigidly prescribed system of tracks and schedules and then immured by fibrous cements. Although, grossly, this picture still holds, it gains much greater flexibility in detail from the realization that the eventual pattern of combination, association, and segregation among cell types in typically ordered arrays is not just passively arrived at, but is actively insured and guarded—and restored after disturbance—by a system of subtle mutual conformances and nonconformances with which the various cell types have been endowed as a corollary of their ontogenetic differentiation.

The principle of the self-linking of cells into matched groups by contact affinities and disaffinities has an important bearing on the explanation of the establishment and maintenance of order in the networks of the nervous system. The building up of central and peripheral nerve cables by the selective attachment of nerve fibers of a given type to others of like specificity has already been referred to in the foregoing. The same principle holds for interneuronal relations, as well as for the relations between neurons and their non-neural receptor and effector organs, on an even subtler scale. Since this is one of the most compelling, yet at the same time least recognized, instances of specificity and selectivity in cell-to-cell interactions, it is well to recapitulate briefly the relevant experiments, which I started almost forty years ago (see a recent review[13]) and which have more recently been amplified by my student, Sperry.[14]

During development, countless connections are established among individual neurons and between neurons and peripheral sense organs and muscles. The latitude inherent in the primary developmental mechanisms of neurogenesis[5] is so great that it would rule out any such microprecision in the details of the neuronal circuitry as is usually postulated as basis for coordinated functions. On the other hand, there is equally conclusive evidence to show that the nervous system does emerge from its embryonic phase with a large list of ready-made coordinated performances, which, since learning by trial and error could have played no part in their formation, must be explicable in terms of developmental interactions. The dilemma was solved by the demonstration of secondary processes which adjust the details within the gross primary system of connections to the unique arrangements of each individual specimen. Although nerves from the same central sources may in different individuals end up in different muscles, the muscles themselves then send a specifying influence back into the centers with the information on just what muscle of what name lies at the end of what line.

The test experiment consisted of transplanting a supernumerary muscle (or a whole set of limb muscles) near a normal limb and hitching it to a random branch of the local nerve supply. This amounted to "tapping" the communication network by inserting an extra receiver. What was observed then was that the transplanted muscle responded in the course of normal activities of the animal always at the precise time and with the same strength as the muscle bearing the same name in the normal limb. Thus, each muscle is called up, as it were, by the central system by its proper name, and if there are several muscles of the same name present in a common innervation district, they all respond simultaneously when the particular code is called. Since the test muscle was inserted arbitrarily, it is clear that it must have established its name-specific correspondences in the nerve centers secondarily. The same correspondence between periphery and centers was also confirmed for the sensory part of the system, first for proprioceptive[15] and tactile[16] sensations and later for the visual field.[14]

It all boils down to the following. Even though it is impossible to distinguish different skeletal muscles by present biochemical methods, the nervous system can tell them apart very well. Whatever the telling difference between muscles, it must be something that can make itself known in a retrograde manner over the motor nerve fiber in such a way that the central apparatus thereby gains exact information about its peripheral terminations. Motor fibers are initially blank and unspecified. They acquire specific identity only after they penetrate identified muscles at the periphery, and the same holds for sensory neurons in regard to their endings. Somehow, the acquired detailed specificity is then transmitted still further back from the primary to secondary neurons, and so on up the lines.

By this device, the neuronal population becomes progressively specialized into subunits whose member cells can henceforth regulate their own mutual relations by virtue of their acquired distinctive properties. Neurons of corresponding or complementary specificities thus become joined into cooperative systems. The proposition that this implies only linkage of a static morphological order all the way through the nerve centers[14] does not seem to go far enough if one is to explain either the aforementioned experimental results or the general problems of neural coordination.[13] But, if one adds the assumptions: that synaptic contact between neurons is merely an enabling, but not a decisive, condition for impulse transmission; further, that actual transmission requires specifically matched states of two apposed surfaces; and, finally, that the specificities underlying this conformance are subject to modification by both peripheral and central influences; one comes closer to a satisfactory concept. The nervous system would then emerge as a vast system of resonance circuits, the elements of which would be linked by conforming molecular surface configurations, partly permanently, partly variably in response to changing central and peripheral influences.

From these remarks, it can readily be seen that compatibilities and incompatibilities along the boundaries between neurons are as significant as discriminatory devices as are the signs of surface "recognition" described earlier in this article for other cell types. Indeed, the direct demonstration of cell discrimination in the motion pictures should help to make the corresponding property of neurons acceptable, if not more palatable, to those who had hoped to be able to maintain their oversimplified faith in the essential identity of all neurons so long as they were faced only with the more indirect earlier evidence. In reopening the whole question of specificity in the nervous system, which of late has been lying dormant, our observations on cell encounters in tissue culture thus assume added significance, as they also open the way to a more direct practical study of the nature of the specificities concerned.

The reference to the progressive diversification within nerve-cell populations points logically to the more elementary question of how any primitive cell types, including neurons, acquire their distinguishing characteristics in the first place. To discuss the problem of divergent differentiation would go far beyond the scope of this article. There is one aspect, however, that fits into the context. Differentiation occurs essentially by dichotomies in the courses of transformation of cells with basically identical physical and chemical endowments.[17] Any cell strain is, by virtue of such endowments, initially capable of effecting a limited variety of qualitatively different reactions. Every step of differentiation implies that from that multiple repertory one definite course has become activated in some members of the group, and an alternative course in the other members, the two courses being mutually exclusive. Sooner or later, the different reactions lead to manifest diversity. Development is composed of long series of such steps. Present evidence is that each step has its own mode of triggering mechanism, and that, contrary to earlier illusions, there is no single master agent that could be held generally accountable for the various and successive dichotomous changes. Formally, however, one can distinguish two classes—one in which the differentiating activity plays entirely within the bounds of a group of similar cells, referable perhaps to differential interactions among the members of the group depending on their relative positions; and another in which the reaction of a given cell group is decisively influenced by an extraneous cell group.

This second class is usually termed "induction"; again, different steps of "inductive" influences need have little in common but the name. Some of them operate over distances, others only between adjacent tissues. Evidently, these latter raise the issue of whether true contact interactions are involved. The answer seems to be that in some instances and for certain steps of differentiation, intimate contact between the interacting cells is essential, whereas, in other cases, such direct contact is dispensable, the interaction being mediated by diffusible substances.

Some cases in which direct contact is necessary may be cited as examples. When devitalized (frozen-dried and rehydrated) cartilage is transplanted into the vicinity of a limb bone in amphibian larvae, the implant can induce the formation of new cartilage in contact with its surface.[1] Thus, competent living host cells build on cartilage to the dead cartilage model as bees would build on new honeycomb to an artificial wax honeycomb presented to them. This is a particularly interesting case, because the bodies of the cells in the implant cannot have been the inducing agents directly, as they are enclosed and insulated in the cartilaginous matrix; therefore, the inductive stimulus to surrounding host cells must have originated in the exposed surface of that matrix. Another set of experiments showed the same thing for bone.[1] A frozen-dried and rehydrated metatarsal bone of a rat, inserted into the leg of a host rat, induced there new bone, including new bone marrow, in contact with the old bone. That contact is essential has been confirmed from other sources.[18] Such observations support the conclusion that extracellular materials may play important roles in certain inductive interactions.[19]

The next case is of special interest to students of collagen. The stroma of the cornea consists of layers of collagen in a characteristic regular arrangement. When frozen-dried cornea was transplanted into a corneal defect in rabbits, the grafted stroma as such persisted for many months. In addition, however, it induced along its inner side the formation of several new layers of typical stroma.[20] Thus, the architecture of the dead stroma matrix has in some fashion been transmitted to the induced layers, so that they assumed the characteristic pattern of collagenous corneal lamellae. This result again proves that not only cells but cellular products likewise can influence other cells inductively.

The intimacy of submicroscopic cell contact, in the sense of the introductory remarks, which may be required for "contact inductions" has not yet been clarified. One of the few clues is the fact that, during the period of inductive interaction between eye cup and prospective lens epithelium, the attachment between the interacting cell layers is so firm as to resist mechanical separation.[21] But, it is still undecided whether the specific transmission across the cell borders is simply an orienting or sorting influence on pre-existing molecular populations[22] or whether it involves the actual passage of substance,[23] perhaps prepared by molecular reorientation.[1] There is a vast field here for exact studies on the relations between physical configuration and chemical activities. These relations are bound to remain in obscurity so long as one focuses attention on those cell-to-cell influences that are mediated not by contact, but by remote control through diffusible agents. The latter, although equally real, can tell only part of the story of cellular interaction.

In summing up, to state that all of the specific interactions among cells exemplified in this article are based on specific properties of the interacting cells is a truism. But, it explains why so little is known about the molecular phenomena governing cell-to-cell relations. There just has not been enough preoccupation with the molecular basis of cellular specificity. Unquestionably, cell "recognition" and selective response must eventually be reduced to terms of properties of molecules and molecular populations. But, so far, there has been very little tangible progress in that direction. There have been a few hypothetical suggestions to explain cell-to-cell conformances, none of them quite tenable in the light of recent observations. I had proposed[24,25] that some sort of steric fitting (e.g., by corresponding charge distributions) between complementary molecules might produce bonds or linkages between like cells, whereas unlike cells without conforming molecules in their surfaces to interlock could establish no connections. Yet, this supposition is now clearly contradicted by the fact that like cells, while remaining together, are not fastened but move continuously around each other. This leaves us for the time being without substantial explanation of the phenomena of selective aggregation as shown in the films.

The only thing that seems clear is that these phenomena point to the same general area of specificity that covers immune reactions,[26] enzyme-substrate interactions, pairing of chromosomes, fertilization, parasitic infection, and phagocytosis. The latter clearly involves selective uptake of particles of suitable molecular surface organization. The problem is no different from that met in cell-to-cell encounters, except for the great disproportion of size between the members of the pair in phagocytosis. Presumably, the selective ingestion of macromolecules belongs in here, too, with the discrepancy of size even more pronounced. The opposite extreme is found in a cell which spreads out selectively over a large body of proper substratum, where the cell is very small in comparison to the partner which it tries to engulf. So, from phagocytosis and the uptake of macromolecules, through cell-to-cell contacts with specific "recognitions," up to the problem of selective adhesion of a cell to its substratum, one deals with a continuous spectrum of problems of the same nature.

This very fact holds promise of progress, as advances in any one sector of that broad area will shed light on other sectors. On the other hand, cleverly getting around the acknowledgment of specificity in any one sector, whether by theoretical constructs or by unrealistic models, will not strip the other sectors of their aspect of specificity, and, hence, will not relieve the intellectual discomfort engendered by our inability to squeeze the broad subject of specificity into the limited conceptual framework which we have erected from the study of fragmentary vital phenomena lacking that aspect. Specificity, as a real and basic control mechanism of cell behavior, can no longer be relegated to a corner, but must be placed in a central position in cellular and molecular biology. The less that is known about it, the greater is the challenge. If there has been some dodging in the past, it was because of some concealed hope that the whole thing might yet in the end turn out to have been an illusion born of inadequate penetration. Exactly the opposite has happened. The more penetrating the analysis, the more cogent has the evidence of specificity become; witness the motion pictures of cell discrimination. There seems little doubt that once the reality and universality of the problem are generally acknowledged, and the promising techniques at our disposal for its disciplined study are recognized, progress can be rapid. But there is even less doubt that, if the existence of the problem continues to be widely ignored, or even denied, some of the major clues to the understanding of living systems will remain missing.

**BIBLIOGRAPHY**


1. P. Weiss, Intern. Rev. Cytol. **7**, 391 (1958).
2. P. Weiss and W. Ferris, Exptl. Cell Research **6**, 546 (1954).
3. G. F. Odland, J. Biophys. Biochem. Cytol. **4**, 529 (1958).
4. P. Weiss, J. Exptl. Zool. **100**, 353 (1945).
5. P. Weiss in *Analysis of Development*, B. H. Willier, P. Weiss, and V. Hamburger, editors (Saunders Company, Philadelphia, 1955), p. 346.
6. J. Holtfreter, Arch. exptl. Zellforsch. Gewebezücht. **23**, 169 (1939).
7. P. Weiss and G. Andres, J. Exptl. Zool. **121**, 449 (1952).
8. G. Andres, J. Exptl. Zool. **122**, 507 (1953).
9. A. Moscona, Proc. Soc. Exptl. Biol. Med. **92**, 410 (1956).
10. J. J. Chiakulas, J. Exptl. Zool. **121**, 383 (1952).
11. M. Abercrombie, M. L. Johnson, and C. A. Thomas, Proc. Roy. Soc. (London) **B136**, 448 (1949).
12. A. Moscona, Proc. Natl. Acad. Sci. U. S. **43**, 184 (1957).
13. P. Weiss, Symp. Soc. Exptl. Biol. **4**, 92 (1950).
14. R. W. Sperry, Growth Symposia **10**, 63 (1951).
15. F. Verzar and P. Weiss, Pflüger's Arch. ges. Physiol. **223**, 671 (1930).
16. P. Weiss, J. Comp. Neurol. **77**, 131 (1942).
17. P. Weiss, J. Embryol. Exptl. Morphol. **1**, 181 (1953).
18. F. C. McLean and M. R. Urist, *Bone* (The University of Chicago Press, Chicago, 1955).
19. C. Grobstein in *Aspects of Synthesis and Order in Growth*, D. Rudnick, editor (Princeton University Press, Princeton, 1954), p. 233.
20. I. H. Leopold and F. H. Adler, A.M.A. Arch. Ophthalmol. **37**, 268 (1947).
21. M. S. McKeehan, J. Exptl. Zool. **117**, 31 (1951).
22. P. Weiss, Quart. Rev. Biol. **25**, 177 (1950).
23. J. Brachet, *Chemical Embryology* (Interscience Publishers, Inc., New York, 1950).
24. P. Weiss, Growth **5**, 163 (1941).
25. P. Weiss, Yale J. Biol. Med. **19**, 235 (1947).
26. A. Tyler, Growth **10**, Suppl., 7 (1947).

# 49
# Molecular Organization of the Nerve Fiber*

Francis O. Schmitt

*Department of Biology, Massachusetts Institute of Technology, Cambridge 39, Massachusetts*

IN the context of these studies in the biophysical sciences, it is the purpose of this contribution to identify some of the basic problems of nerve that may be effectively studied at the molecular level by the methods of biophysics and biophysical chemistry, and to describe certain nerve structures, such as the myelin sheath, which provide great scope for the application of such methods and which have important application also to such general biological problems as the nature of cell membranes.

The primary function of neurons is the transfer of information from one region to another, the message being propagated along the fiber from cell body to axon terminals by means of a local disturbance or propagated bioelectric impulse. To many, particularly to those studying the activity of the larger nerve units in the processing of information by the central nervous system or in the behavior of the organism as a whole, the individual neuron is but the structural and functional unit in a highly complex network whose basic function involves the all-or-nothing, on-or-off responses of the units. Thus, in neurophysiology, as well as in bioelectrical studies generally, the most elementary problem which could be posed to the molecular biologist or biophysicist would doubtless be the elucidation of the mechanism by which the impulse is propagated along nerve fibers—i.e., the nature of the changes in the properties of the surface film of the axon that underlie excitation and propagation of the action wave.

If this were in fact the only problem to be dealt with, the present contribution would be very short indeed. Little is known about the molecular composition of the axon surface membrane and even less is known about changes that occur in it during propagation of the impulse. However, for several reasons, I do not wish to take such a limited point of view. In the first place, the axon surface membrane does not exist as a separate entity. It is but part of the neuron which, as a complex reacting system, contains many constituents. Even if one were interested only in bioelectric processes, it would be impossible to understand the phenomena going on in the surface membrane without reference to the structural and biochemical properties of the fiber as a whole. Moreover, nerve performs many functions other than that of propagating impulses. Some of these functions are known, some can be guessed, while others are completely unknown.

For example, somewhere in the neuron there are synthesized certain highly active substances, such as acetyl choline and adrenergic substances, which have transmitter functions; i.e., they transmit the impulse across the junction from the nerve fiber to the muscle fiber or other end organ. The site of synthesis of this material in the neuron, its mode of transfer from this site to the nerve endings (possibly enclosed there in vesicles) and its transmission to the innervated tissue are at present very poorly understood. Certain neurons also produce neurosecretions that can be identified cytologically and histochemically and can be demonstrated to pass from the nerve cell body down the axon to the axon terminals. Possibly other hormonal substances are similarly produced and transmitted, but have thus far escaped detection.

Neurons also manifest certain properties, especially during embryogenesis or regeneration, that are highly specific. Thus, each motor fiber grows out and eventually innervates its appropriate muscle. This innervation is not only highly specific with respect to the muscle fiber which is "sought out" and innervated, but also it is necessary for the stability and maintenance of the normal properties of the muscle fiber. Cutting the nerve fiber to a muscle may cause the latter to change its physiological and pharmacological properties markedly, or even to degenerate.

Finally, it is probable that certain aspects of nerve function may involve chiefly the nerve cell or axoplasm without obvious relationship to the axon surface film. Such processes would probably not lend themselves to study or detection by bioelectric methods. For example, as is discussed in more detail below, certain constituents, such as the fibrous protein, seem to be present in all neurons and to represent one of the characteristic features of neurons generally. However, there is still not the foggiest notion as to the biochemical and physiological function of this protein or protein complex.

It is desirable, therefore, that the biophysical and biochemical investigation of the neuron be not limited to bioelectric processes, but be systematic and exhaustive so as to reveal "what is there," particularly at the molecular and macromolecular level. One has only to consider the history of other fundamental biological problems, such as that of muscle contraction, to realize the importance of such a systematic biochemical and biophysical approach. Only when significant advances in knowledge of the composition and molecular organi-

* These studies were aided by a research grant (B-24) from the National Institute of Neurological Diseases and Blindness, of the National Institutes of Health, Public Health Service, U. S. Department of Health, Education, and Welfare; by a contract between the Office of Naval Research, U. S. Department of the Navy, and the Massachusetts Institute of Technology (NR 101-100); and by grants from the Trustees under the wills of Charles A. King and Marjorie King, and from Lou and Genie Marron.

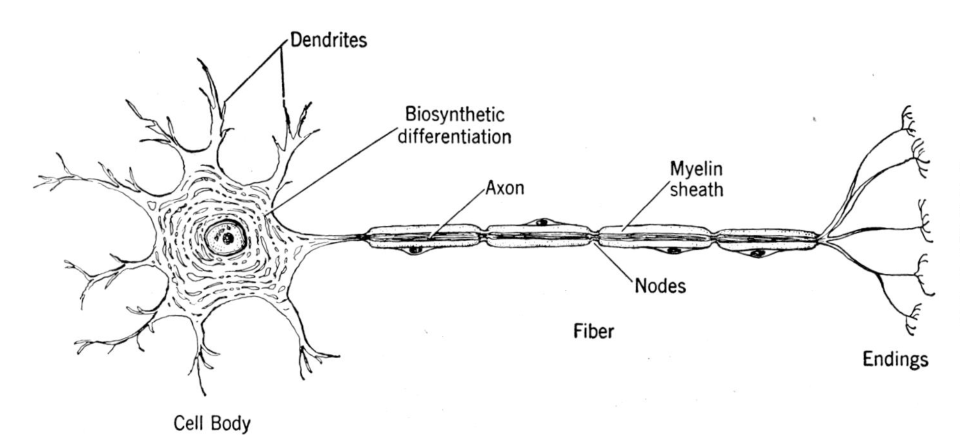


FIG. 1. Diagrammatic representation of a myelinated nerve fiber. Note cell body with dendrites and cytoplasmic organelles, axon, myelin sheath with Schwann cells and nodes of Ranvier, and terminal twigs making synaptic connection with peripheral tissue.

zation of muscle and its fibrous proteins were obtained, could the fundamental problem of the molecular mechanism of contractility be fruitfully attacked. By comparison with muscle, despite much research extending over a century, knowledge of the composition and molecular organization of nerve is still primitive. Moreover, impulse propagation in the nerve fiber is a subtle phenomenon which, unlike contractility in muscle, is not revealed in structural alterations thus far demonstrable even in the electron microscope.

The investigation of the molecular biology of nerve is, therefore, very challenging, particularly to young biophysical scientists not yet steeped in traditional concepts of neurophysiology. This article, which is necessarily limited in space, is meant to be provocative rather than exhaustive in its treatment of basic matter. For more extensive information, the reader is referred to reviews by Schmitt.[1–3]

## I. NEURON

The unit of the nervous system, the neuron, consists of a nerve cell body with its dendritic arborization and synaptic connections with the terminal twigs of other neurons; of an axon which is the thin, cylindrical outgrowth of the nerve cell; and of the endings or terminal twigs which make apposition with the innervated cell or tissue through the synapse (see Fig. 1). The neuron has physiological as well as morphological polarization. The impulse passes from presynaptic fiber to nerve cell, down the axon and across the terminal synaptic membrane to the innervated tissue, but not in the reverse direction. For present purposes, it is convenient to consider the cell body, axon, and endings separately.

### (A) The Cell Body

The nerve cell body has the task of spinning out the axon in development and of regenerating it in the event of injury. Since the volume of axoplasm to be biosynthesized in a long fiber may be hundreds and, in some cases, thousands of times that of the cell body, it is not unexpected that the metabolism of the nerve cell is very high. To accomplish its impressive task of biosynthesis, the cell is equipped with the cytological differentiation found by electron microscopists characteristically in highly active, secreting gland cells. Particularly prominent are the membrane-limited structures (endoplasmic reticulum, ergastoplasm) richly studded with RNA-rich ribosomes (see Palay[4]). These basophilic structures are identical with the Nissl substance, which has long been studied by cytological methods and by microabsorption spectroscopy and which is known to reflect the synthetic activity of the nerve cell. With improved spectrophotometric methods, such as those developed by Hydén, it has been possible to follow quantitatively the changes in RNA, protein, lipid, and dry weight of regenerating neurons and to correlate these changes with the peripheral growth of the axon (see Brattgård *et al.*[5]). Despite the increase in protein, lipids, and water during the axon outgrowth, the amount of RNA appears to remain constant. However, changes occur in the state of aggregation of RNA; small, finely dispersed particles are formed in the process of "chromatolysis," long known to occur during regeneration. When contact is re-established between the outgrowing axon and the peripheral innervated cell, there is characteristically an increase of RNA and of cell volume until the original values are restored. According to these authors, the only type of cell that can compete with the nerve cell in biosynthetic activity and with the degree of cytological differentiation for such processes, particularly that involving the RNA system, is the exocrine cell of the pancreas of small animals. The neuron is visualized as "an enormous gland cell structure whose lively protein metabolism serves the specific nerve function."[5]

The high metabolic activity of nerve cells (as distinguished from nerve fibers), long known to be characteristic also of normal, unregenerating cells (as in the brain), may be owing in part to the constant synthesis of axoplasm which—according to Weiss and Hiscoe[6] and supported recently by Waelsch[7] and by Ochs and Burger[8]—passes continuously, during the life of the neuron, as a plastic, gelled mass down the axon at a slow rate (*ca* 1 mm/day). The manner in which axoplasm disappears peripherally, e.g., by oxidation or other metabolic process, is not known. If nerve cells are in fact constantly synthesizing axoplasm, it becomes important to establish the purpose served by such a metabolically expensive process. It is possible that one such purpose may be the maintenance of the turgor of the axon over its long extent to the endings. One constituent that would also be constantly synthesized in such a process is the fibrous protein of the axon. When the function of this protein is discovered, light may be shed also on the apparent necessity for its constant synthesis.

The metabolism and the cytological differentiation of the nerve cell are responsive not only to the need for active biosynthesis, as in the regeneration of a severed axon, but also to physiological activity of the neuron. This apparent energy dissipation is in contrast to the highly conservative nature of the process of impulse propagation in the peripheral fiber.

The evidence for the highly developed membrane-limited structures in the cytoplasm of living nerve cells and for the presence of fibrous protein oriented with long axes parallel to the distinguishing directions of the nerve cell is not merely cytological and electron microscopical, but derives also from polarization-optical studies of fresh nerve cells.[9]

### (B) Axoplasm and the Axon

As a differentiated type of nerve-cell cytoplasm, axoplasm is a highly complex system. Noteworthy is the low absorption at about 2600 A which characterizes the region peripheral to the axon hillock, roughly at the junction of the nerve-cell cytoplasm with the axon. This is in agreement with the relative scarcity in the axoplasm of ribosomes, as seen in the electron microscope, and of basophilic material, as seen cytologically. Membranous material, possibly consisting chiefly of fragments of the membrane-rich cytoplasmic system, have been observed in squid axoplasm in this laboratory by Maxfield[10] during the process of protein fractionation and physicochemical analysis of axon proteins. Peripheral axoplasm is low in materials absorbing at 2600 A (see von Muralt[11]), consistent with the apparent lack of substantial biosynthesis in this region of the neuron.

Among the particulates commonly present in cytoplasm, only the mitochondria of axoplasm have been subjected to special study. These have properties similar to those isolated from other tissues (Foster[12]). In thin sections of lobster fibers, they have been observed to occur preferentially near the Schwann-cell surface, suggesting the possibility that metabolic energy utilization may occur in this region (Geren and Schmitt[13]).

Although axoplasm is a highly complex, specially differentiated form of nerve-cell cytoplasm, it is highly desirable that the main features of its composition and macromolecular organization be established, not merely for the light it might throw upon the energy coupling involved in active ion transport and in other processes related to impulse propagation, but also as clues to the discovery of other processes carried out by nerves, possibly unrelated to, and not predictable from, those which may be studied by electrical methods.

For such studies of axoplasm, the squid giant fiber is uniquely suited. Axoplasm, relatively uncontaminated by sheath constitutents and other nonaxonic material, can be readily obtained from these fibers. From axoplasm obtained from *Loligo pealii*, the New England squid, analyses have demonstrated (Koéchlin[14]) the presence of previously unknown substances, particularly isethionic acid, which turns out to be the major anion in the acid-base balance of the squid nerve fiber. The fibrous protein of axoplasm has also been studied in this material (Maxfield,[10] Maxfield and Hartley,[15] Schmitt[16]).

More recently, the large squid, *Dosidicus gigas*, that abound in the waters of the Humboldt current off the western coast of South America, have been utilized for chemical investigations. Deffner and Hafter[17] have demonstrated the presence in this axoplasm of some fifteen free amino acids, as well as cysteic amide (a sulfonate closely related to isethionic acid and taurine, which are also present in relative abundance) and other low molecular weight constituents (see Table I), including some five peptides whose composition is now being investigated and whose function is not known. Using high-capacity methods of electrophoretic fractionation and large amounts (10 to 20 g) of freeze-dried axoplasm, these substances are being isolated in quantities sufficient to permit determination of their composition and biological properties.

Among the macromolecular constituents of axoplasm, the fibrous protein is of special interest. Polarization-optical analysis (Bear *et al.*[18]) demonstrated (1) that the fibrous protein exists in the normal fresh axon where it is oriented with long axes parallel to the fiber axis, (2) that the intrinsic birefringence of the protein macromolecules is similar in magnitude to that of other fibrous proteins, and (3) that the partial volume occupied by the oriented fibrous protein is probably relatively small. Electron-microscopic observations[19,20] revealed that the fibrous protein occurs characteristically as thin (*ca* 100 to 200 A) filaments of indefinite length and with no demonstrable axial discontinuities or periodic structure. P. F. Davison and E. W. Taylor in this laboratory have made recent physicochemical studies (unpublished) on fresh axoplasm of *Dosidicus* shipped iced from Chile to Cambridge. From these, it appears that the fibrous protein may have a molecular weight in the order of millions and may be very long and thin. However, the possibility cannot be excluded that the protein as it occurs in the axoplasm so far available (i.e., suspended in cold saline and studied some days after removal from the squid) may in fact be a high polymer, of which the monomer may have a length of the order of 2500 A. In addition, the material which has so far been isolated may represent a complex of two or more proteins. Only when this protein or protein complex has been isolated and characterized chemically and structurally will it be

TABLE I. Dialyzable substances in squid axoplasm. Expressed as micromoles per gram dry weight.

| Substance | Acidic Loligo | Acidic Dosidicus | Neutral Loligo | Neutral Dosidicus | Basic Loligo | Basic Dosidicus |
|---|---|---|---|---|---|---|
| Aspartic acid | 73.0 | 74.8 | | | | |
| Glutamic acid | 19.6 | 26.2 | | | | |
| Glycine | | | 10.7 | 10.4 | | |
| Alanine | | | 7.8 | 9.3 | | |
| Serine | | | 3.7 | 1.0 | | |
| Leucine and isoleucine | | | 2.7 | 0.2 | | |
| Valine | | | 2.2 | 0.5 | | |
| Threonine | | | 1.9 | 0.3 | | |
| Proline | | | 1.0 | 0.1 | | |
| Tyrosine | | | 0.7 | 0.3 | | |
| Phenylalanine | | | 0.6 | 0.15 | | |
| Methionine | | | 0.4 | 0.15 | | |
| Arginine | | | | | 3.2 | 4.1 |
| Lysine | | | | | 2.4 | 0.2 |
| Ornithine | | | | | 1.8 | 0.3 |
| Isethionic acid | 220.0[a] | | | | | |
| Taurine | | | 98.0 | 31.0 | | |
| Cysteic acid amide | | | 4.4 | 0.4 | | |
| Glycocol, betaine | | | + | + | | |
| Glycerol | | | + | + | | |
| Homarine | | | 17.1 | 18.6 | | |
| Peptides | | | + | + | | |
| Fumaric acid | | | | | | |
| Succinic acid | 15.0[a] | | | | | |
| K | | | | | 344.0[a] | + |
| Na | | | | | 65.0[a] | + |
| Ca | | | | | 7.0[a] | + |
| Mg | | | | | 20.0[a] | + |
| $PO_4$ | 16.6[a] | 16.1 | | | | |
| $SO_4$ | + | + | | | | |
| Cl | 140.0[a] | + | | | | |

Reaction at physiological $p$H ($\sim$6.7)

[a] Values taken from Koéchlin.[14]

possible to determine its biological role. Meanwhile, as it is purified and fractionated, it will be subjected to screening tests intended to provide clues regarding its biological activity.

### (C) The Axon Surface Membrane

Electron microscopy permits one to visualize directly the axon surface membrane, though admittedly this is possible only in fibers which have been fixed, embedded, and cut into ultrathin sections. This membrane, or some component of it, presumably corresponds to the irritable membrane whose physical properties before, during, and after the passage of the nerve impulse have been studied by electrophysiologists. Some of the functions that must be subserved by membrane constituents are shown schematically in Fig. 2.

At high resolution, the membrane appears as a double-edged structure with two dark edges separated by a less dense area (see Robertson[21,22]). The total thickness is about 60 to 70 A, the dark and light lines having about equal thickness. The apparent thickness of the dense regions in the membrane may be reduced as methods are improved. A satisfactory interpretation of this structure, in terms of the lipid and protein components of the membrane, must await a detailed study of model systems of lipids and lipid-protein systems. The consensus at present seems to be that the dark regions represent the aqueous, protein-containing interfaces. The less dense regions represent the lipid phase with its preponderant aliphatic hydrocarbon chains. This point of view is not shared by all. Stoeckenius[23,24] believes that the osmium-tetroxide deposits electron-dense material in the region of the unsaturated double bonds of the lipids rather than at the protein, aqueous interface.

Whatever may be the eventual interpretation of the structure of the axon surface membrane, it is perhaps safe to say that the membrane includes a bimolecular

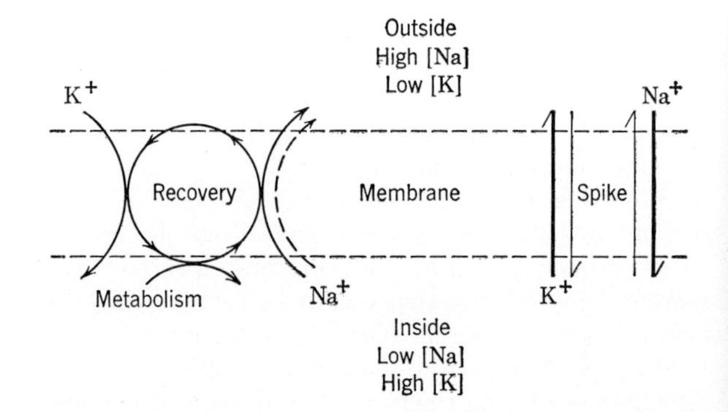


FIG. 2. Diagrammatic illustration of ion movements through the nerve membrane [after A. L. Hodgkin and R. D. Keynes, J. Physiol. **128**, 28 (1955)]. Ion movements occurring during the action wave are shown at the right; active transport, involving coupling of metabolic energy, is shown on the left. The excitable membrane pictured is presumably identical with the limiting surface membrane of the axon, or some portion thereof.

leaflet of mixed lipids and at least a monolayer of protein or other hydrophilic, osmiophilic material on either side of the lipid layer. This represents the "unit" membrane structure, thought by many competent electron microscopists to be characteristic of biological membranes generally. In addition to this structure, which might be thought of as the "floor space" of the cellular factory, one supposes that there are mounted upon or within this membrane the enzymes responsible for the special molecular and biochemical processes which underlie the rapid change in cation permeability at the time of impulse propagation and the "pumping" of sodium ions from axoplasm against an activity gradient into extracellular space. Obviously, much research is required before the details of the molecular organization of the membrane can be established. Meanwhile, in describing the membrane structure, it is perhaps desirable to avoid terms such as "triple" membrane (so-called because of the two dense and one less dense regions); this might cause investigators of the nature of the true "physiological" membrane (i.e., that deduced from indirect, chiefly electrical, methods) to identify one or another component of the fixation artifact seen in the electron microscope as a self-sufficient entity, ignoring the fact that the membrane must almost certainly be considered as a complex system of interacting, interdependent components. Precisely in such situations are the methods of molecular biology different from those of descriptive morphology.

### (D) Nerve Endings and the Synapse

At axon endings, electron microscopy reveals an accumulation of particulates, particularly mitochondria. Also characteristic is the presence of small (*ca* 200 to 500 A) vesicles, appearing sometimes in large quantity. It has been suggested that these vesicles contain acetyl choline, adrenalin-like materials, and other physiologically and pharmacologically active substances such as histamine and serotonin. If transmitter substances occur in packets and if such packets were presented at the end organ, perhaps as a secretion across the membrane, certain electrophysiological studies of synaptic potentials might thus find explanation (Katz, p. 524). At present, the origin of the vesicles is still unknown, as are the details of the mechanism by which they or their contents find their way into the endings.

Of considerable theoretical significance is the problem of the nature of the relationship between the limiting membranes of the nerve terminals and of the muscle or other innervated cell. If, as current consensus indicates, there is a space of at least a few hundred Ångström units between these membranes, the problem of the mechanism whereby transmitter substances pass from neuron to innervated tissue is a different one than if the two membranes were strictly in molecular contact. This is still a matter for discussion. The factors involved in determining membrane separation by electron microscopy are considered below in more detail.

## II. SATELLITE-CELL DIFFERENTIATIONS AND RELATIONSHIP TO THE AXON

Extending to the farthest reaches of the body, portions of the nerve axon may be very remote from the metabolic center near the nucleus of the cell body. This would constitute an inherently very unstable system. However, the axon is probably seldom actually remote from metabolic centers because it is everywhere surrounded by thin satellite cells, the Schwann cells, which are themselves metabolically active. These cells may well have intimate biochemical relationship with the axon. Such cell-to-cell biochemical relationship and interdependence have in fact been demonstrated *in vitro*. Puck,[25] for example, has shown that a monolayer of cells of a certain type, which are themselves unable to obtain all of their nutritional requirements from the medium, can be stabilized by depositing upon them a layer of cells which can provide or utilize the required factor. These act as "feeder" cells to the underlying cells and constitute a stable system. It seems possible that the evolutionary advance to the higher organisms, with their extensive nervous system, depended upon the development of the process whereby satellite cells come to envelop outgrowing axons, thus permitting the fibers to extend to the extremities of the body as stable neurons.

The satellite cells may function in relation to the axon in other ways also. In any homogeneous group of nerve fibers, the velocity of impulse propagation is a direct function of the diameter of the fiber. In invertebrate fibers, to achieve relatively high velocity of propagation (30 to 40 m/sec), the axon diameter becomes relatively enormous (0.5 to 1.0 mm in the squid). However, in myelinated fibers in which the axon is covered to some 99% by myelin, impulse-propagation velocity may be high (50 to 100 m/sec) in fibers of modest diameter (10 to 20 $\mu$). In these cases, impulse propagation is believed to be saltatory in nature rather than continuous; i.e., the membrane changes responsible for excitation occur in very restricted regions (at the nodes of Ranvier), and internodal segments respond as entire units. The investigation of the molecular organization of the myelin, which acts as internodal insulation, and of its mode of formation by the Schwann cell, present a most rewarding opportunity for the methods in the molecular biology of nerve.

### (A) Myelin

From polarization-optical analyses, it was shown by Schmidt[26] that the lipid molecules in the myelin sheath are oriented with their paraffin chains extending radially, whereas the protein components are oriented with their distinguishing directions perpendicular to this direction, i.e., tangential. The lipid-protein layers must be thin with respect to the wavelength of light, constituting a Wiener mixed body.[27]

X-ray diffraction studies supported this view and

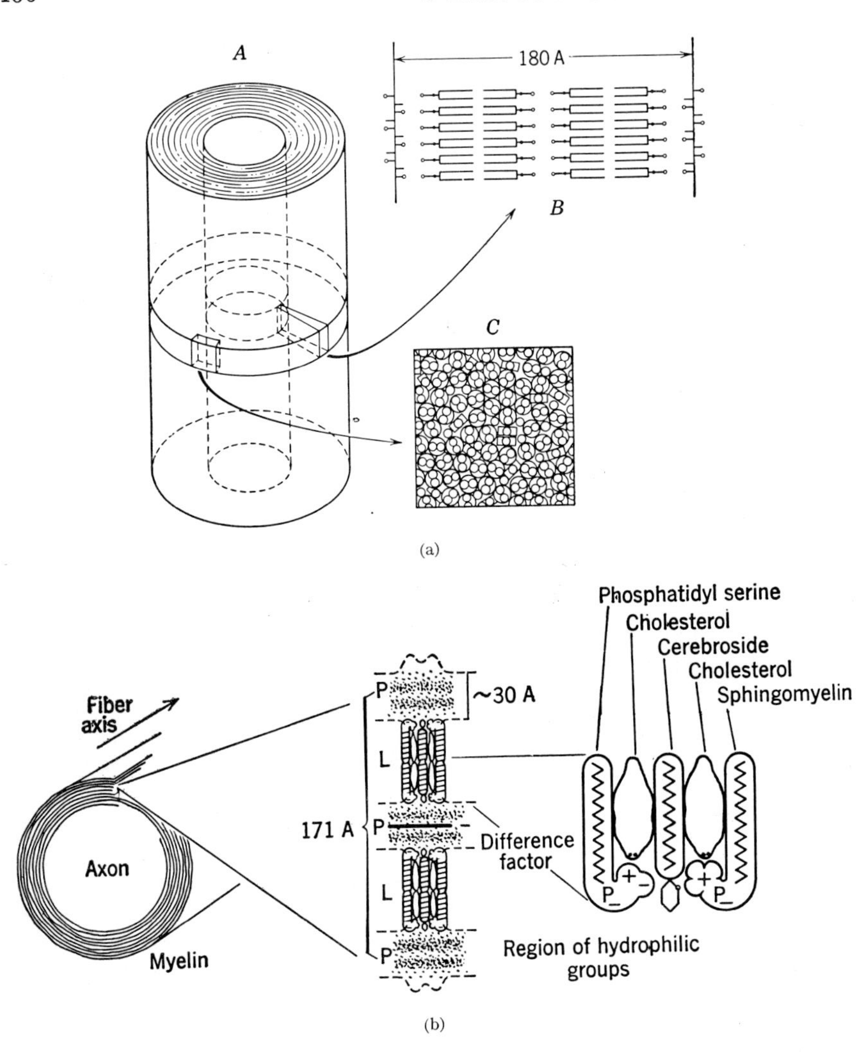


FIG. 3. (a) Schematic representation of molecular organization of the nerve myelin sheath (after Schmitt, Bear, and Palmer[28]). *A*, concentric lipid-protein layered structure deduced from polarization optics and from small-angle x-ray diffraction; *B*, unit lipid-protein structure in the radial direction (perpendicular to the planes of the layers); *C*, section through the molecular chains in the planes of the layers. (b) Diagram summarizing molecular organization of nerve myelin from x-ray diffraction data, as suggested by Finean.[34]

small-angle diffraction showed that the lipid-protein layers have a characteristic thickness of about 171 A in cold-blooded animals and about 185 to 190 A in warm-blooded animals (Schmitt *et al.*[28]). On the basis of extensive studies of the individual lipid types both dry and in aqueous systems, these authors showed that the identity period in the radial direction must include two bimolecular leaflets of mixed lipids, and that the protein and water components must occupy approximately 25 A each. Several alternative models involving bimolecular leaflets flanked by monolayers of protein were proposed (see Fig. 3). It was pointed out[29] that nerve myelin, being susceptible to x-ray diffraction analysis and having a lipid-protein architecture similar to that of cell membranes, might provide a model for the study of such membranes which, because of their thin paucimolecular nature, could not be similarly studied.

The concentric-layered structure of myelin was confirmed by direct visualization in the electron microscope (see Fernández-Morán[30] and Sjöstrand[31]) after development of the technique of thin sectioning for electron microscopy. The thickness of the layers was somewhat less than that demonstrated by x-ray diffraction for the fresh fibers. This difference was shown to be owing to the fixation, dehydration, and embedding required in the electron-optical technique; nerves were subjected to diffraction analysis after each step in the preparative technique, and the thickness of the layers was determined by electron-microscopic examination of thin sections of the same specimens.[32–34]

After the main features of myelin architecture had been determined, the question arose as to how such a structure is produced by the Schwann cell, in collaboration with the axon. A clear-cut answer to this question was provided by Geren,[35,36] at least in the case of peripheral fibers. From observations of embryonic sciatic nerves in chick embryos,[37] Geren proposed her membrane theory which is schematically illustrated in Fig. 4. According to this ingenious theory, now verified by the observations of many investigators (e.g., Robertson[38]), the Schwann cells, after enveloping the outgrowing axons and causing their respective surface membranes to adhere firmly, wrap themselves many times around the axon by a continuous infolding of the outer Schwann-cell surface membrane. The mechanism of myelination in the central nervous system is less clear. Although various authors have seen the myelin layers as a spiral (in transverse section), as in peripheral fibers, others (Luse[39] and DeRobertis *et al.*[40]) have obtained evidence which they interpret as indicating that lipid-protein layers are formed in the cytoplasm of the oligodendrocyte, perhaps by the fusion of vesicles 200 to 800 A in diameter, which condense into the compact myelin layers. The myelin layers are hence not concentric but are spirally wrapped about the axon. After many wrappings (or few, depending upon the fiber type) the Schwann-cell cytoplasm is pressed from among the layers which become condensed into the smectic, paracrystalline material of the finished myelin. This process requires a prolific biosynthesis of lipid-protein material. Where this is synthesized in the cell and how it comes to be incorporated in the cell membrane is not yet known. The myelination process would seem to provide an ideal system in which to study the molecular mechanism of lipid-protein synthesis and of membrane formation in general.

From Geren's membrane theory, it immediately becomes obvious why there are *two*, rather than one, bimolecular layers of lipids in the x-ray identity period in the radial direction (see Schmitt, Bear, and Palmer[28]). As the outer Schwann-cell membrane infolds, the outer surfaces of the membrane remain in contact. As the layers condense, with the expulsion of Schwann-cell cytoplasm, the membrane surfaces which faced the Schwann-cell cytoplasm also adhere, fuse, and form the darkly staining line seen in the electron micrographs (see Fig. 5). The alternate less-dense line represents the fusion of membrane surfaces which originally faced outward to extracellular space. Finean,[34] who has continued the x-ray analysis in a systematic fashion over the last decade, has proposed a molecular model of myelin in which the cholesterol molecules fit in between the phospholipid and cerebroside molecules in characteristic fashion (see Fig. 3). Further refinements in the x-ray and electron-microscopic techniques, together with the study of various lipid-protein models, may provide important new information about membrane structure which will be of vital importance in the rapidly expanding field of molecular biology.

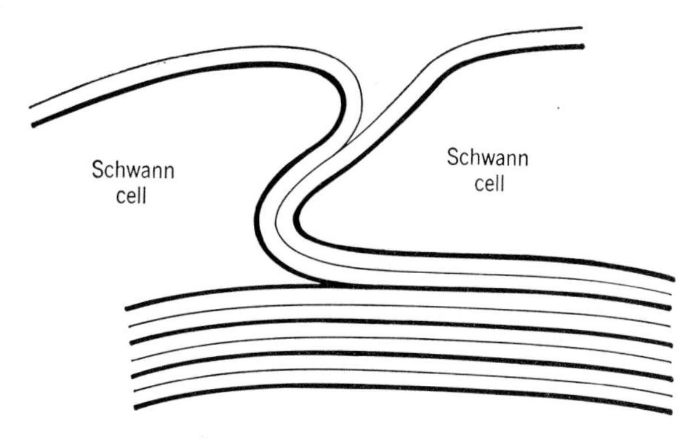


FIG. 5. Illustrating enfolding of Schwann-cell surface membranes to form compact myelin, after the theory of Geren.[35] Note differentiation between inner and outer surfaces of enfolding membrane; in the double membrane, this gives rise to Finean's "difference factor" and explains why the full repeat distance in the radial direction involves two bimolecular leaflets of lipo-protein (two membranes).

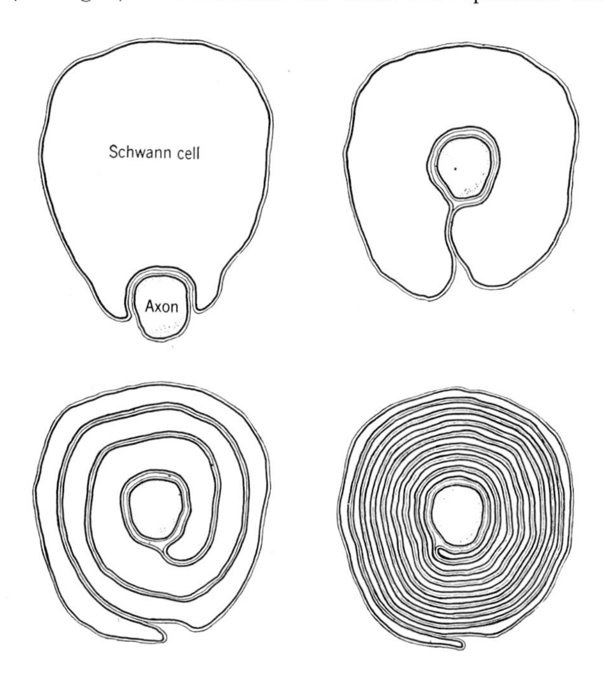


FIG. 4. Schematic representation of the membrane theory of the origin of nerve myelin, (after Geren[35]). Four stages in the engulfment of the axon by the Schwann cell and the wrapping of many double layers which, after condensation, form the compact myelin.

As Schwann or satellite cells envelop the outgrowing axon and start to wrap their surfaces in multiple folds about it, adjacent Schwann cells come into contact and form the nodes of Ranvier. This nodal region of the finished, myelinated fiber has not yet been exhaustively studied in the electron microscope. In view of the critical nature of this region, suggested by the theory of saltatory conduction, its ultrastructure merits careful study. According to a preliminary report by Robertson,[41] the Schwann cells on either side of the node form finger-like processes which interdigitate closely over the node. The region between the processes, presumably containing extracellular fluid, is of the order of several hundred Ångström units thick. To this extent, the axon surface membrane is exposed to extracellular fluid at the node, according to Robertson.

The incisures of Schmidt-Lantermann presented a problem because, if they are continuous channels between the axon and the exterior through the myelin, they would offer conducting paths for current. However, electron microscopy has demonstrated that this is not the case.[42,43] The incisures are produced by a systematic shift in the relationship of all of the helically wrapped membranes, but each membrane is continuous across the incisural gap. There are thus as many layers traversing the incisures as occur in the compact region of the myelin, and the incisures are therefore not regions through which ions and currents can freely flow into the axon.

## III. SATELLITE-CELL-AXON RELATIONSHIPS IN "UNMYELINATED" FIBERS

It has long been known to neurohistologists, particularly to those of the Spanish school, that nerve fibers are everywhere covered by satellite cells. The fact that these cells and their possible functional role—in the mature as well as in developing and regenerating fibers—have not greatly interested physiologists, or even histologists, is to some extent owing to their small size (*ca* 0.1 to 1 in most invertebrate fibers) or to the fact that myelin (which has been much studied by histologists and physiologists) was not recognized as satellite-cell substance (i.e., a compact helical layering of double surface membranes). The fact that, early in the development of histology, a sharp distinction was made between "myelinated" and "unmyelinated" fibers (primarily on the basis of staining reactions), led some physiologists to the mistaken notion that a substantial fraction of the animals' nerve fibers are in effect naked axons. From this, it is obvious why the cellular covering of the axon attracted little interest except to those studying nerve development or regeneration, where the satellite-cell covering is unquestionably of functional significance.

Polarization-optical studies (see Schmitt and Bear[29]), based largely on a reinterpretation of the mechanism of the so-called metatropic reaction discovered by Göthlin,[44] strongly suggested that at the surface of all nerve fibers, vertebrate or invertebrate, myelinated or unmyelinated, oriented lipid-protein material is present which has qualitative similarity of molecular organization though quite variable quantitatively from one fiber type to the next. This led to the notion that "all nerve fibers are to some extent myelinated."[29]

This matter could be clarified only when the region at the surface of nerve fibers—i.e., the monolayer of satellite cells—could be observed at high resolution by the electron microscope. Details of the discoveries thus brought to light are beyond the scope of this paper, but a few of the more significant aspects at the molecular level, particularly insofar as they bear on basic physiological concepts, are considered.

Investigation with the electron microscope of the detailed structure of the satellite cells and their relation to the axon is not yet a decade old, and it is, therefore, not unexpected that knowledge of the subject is still very primitive. No doubt, special methods of fixation and preparation will have to be developed to permit investigation of the most fundamental aspects of the problem, such as the possible variation of intercellular relationship, as manifested morphologically by the apparent spaces or channels between axon surface and satellite cells and between satellite cells themselves, with function.

The typical invertebrate, unmyelinated (but not unsatellited!) nerve fibers, such as those of the lobster leg or claw nerve, are surrounded by satellite cells which are very thin (*ca* 0.1 μ). Except at the region of the nuclei, there is relatively little space between the two parallel surfaces of the cells (i.e., facing the axon surface membrane and that facing the basement membrane, connective tissue investment), although mitochondria and some of the typical equipment of the cytoplasm of active cells may be seen. The axon surface is characteristically highly contorted, and axoplasmic mitochondria are seen with the long axes of their ellipsoids oriented into the folds of these contortions, suggestive of energy utilization at this interface (see Fig. 6). The outpocketings of the satellite cells into the axon may manifest all of the stages to be expected if particulates (mitochondria?) were actually being formed and extruded into the axon by filling such outpocketing with satellite-cell cytoplasm (including its particulates) and pinching them off into the axoplasm (see Geren and Schmitt[13,45]).

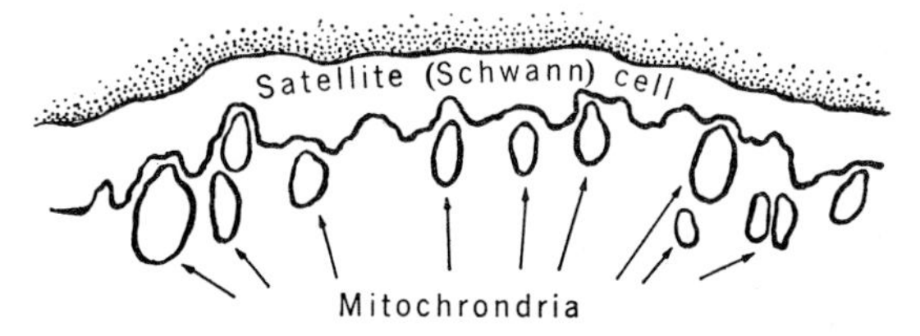


Fig. 6. Diagrammatic illustration of thinness of satellite cell in lobster fiber and of intimate relationship between mitochondria and axon surface membrane (after Geren and Schmitt[45]).

Such very thin cells would be expected to show only a very weak birefringence and metatropic effect, since only two lipid-protein surface membranes are involved. Nevertheless, this optical effect was observed and correctly interpreted, except that the layers were simply assigned to the region at the surface of the axon, rather than to the membranes of the satellite cells.[46]

The satellite cells of squid giant fibers are somewhat thicker (*ca* 1 μ) than those of lobster, though the ratio of thickness to axon diameter may actually be less. In any histological section of the giant fibers, as many as two or three nuclei may be present. If one were to chart the boundaries and domains of the satellite cells upon the axon surface membrane (e.g., by slitting the fiber and flattening out the membrane), one would see that many cells are involved in the formation of the cellular monolayer that surrounds the axon.

The most distinguishing feature of the cytoplasm of these cells is the presence of many (a half-dozen or more) osmiophilic, dense membranous layers lying in planes predominantly parallel with the cell surface. Whether these layers are related to those present in typical, metabolically active cells ("endoplasmic reticulum"), or whether they have a special structure and function in satellite cells remains to be established by more-detailed examination. However, if one assumes that the layers have the composition and ultrastructure typical

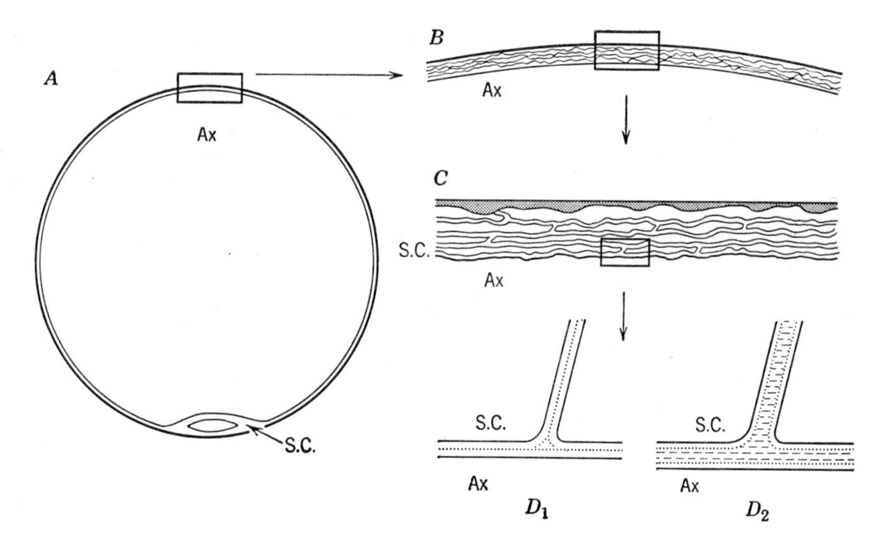


FIG. 7. Relationship of axon (Ax) and Schwann cell (S.C.) in the squid giant fiber (after Schmitt and Geschwind[2]). *A*, cross section of giant fiber showing very thin Schwann cell surrounding axon. ×125. *B*, enlarged segment portion of Schwann cell, ×3000; *C*, same as *B*, showing intracytoplasmic layers, ×9000; and $D_1$ and $D_2$, possible relationships at the molecular level of axon surface membrane and Schwann-cell membrane. $D_1$ presumes layers are in molecular contact. (Water layer does not exceed about 15 A in thickness.) $D_2$, showing thicker water layers (100 to 200 A) between adjacent membranes. ×150 000.

of lipid-protein membranes generally (hence, also of myelin), it is possible to account for the polarization-optical properties described by Bear *et al.*[47] and ascribed by them to a metatropic layer at the outer surface of the satellite cells—a kind of myelin-like layer. Geren and Schmitt[45] have shown that the optical effects can satisfactorily be accounted for on the basis of the intracytoplasmic dense layers present in the satellite cells. There is no differentiated myelin-like layer at the outside of these cells, only the amorphous basement membrane.

One of the structural details that must await further investigation for clarification, but that may be of great ultrastructural and perhaps functional significance, is the nature of the boundaries between satellite cells and the axon surface membrane, between adjacent satellite cells and between intracytoplasmic membranes. Of particular importance is the thickness of the aqueous regions between these membranes, because ions passing from extracellular space to axoplasm and in the reverse direction must traverse such channels.

The situation is illustrated schematically in Figs. 7 and 8. If the adhesion between the surface membranes were as close as in the myelin sheath, there would be no substantial channels for ion diffusion. The aqueous phase would have a thickness of the order of 12 A according to the x-ray diffraction analysis of myelin struc-

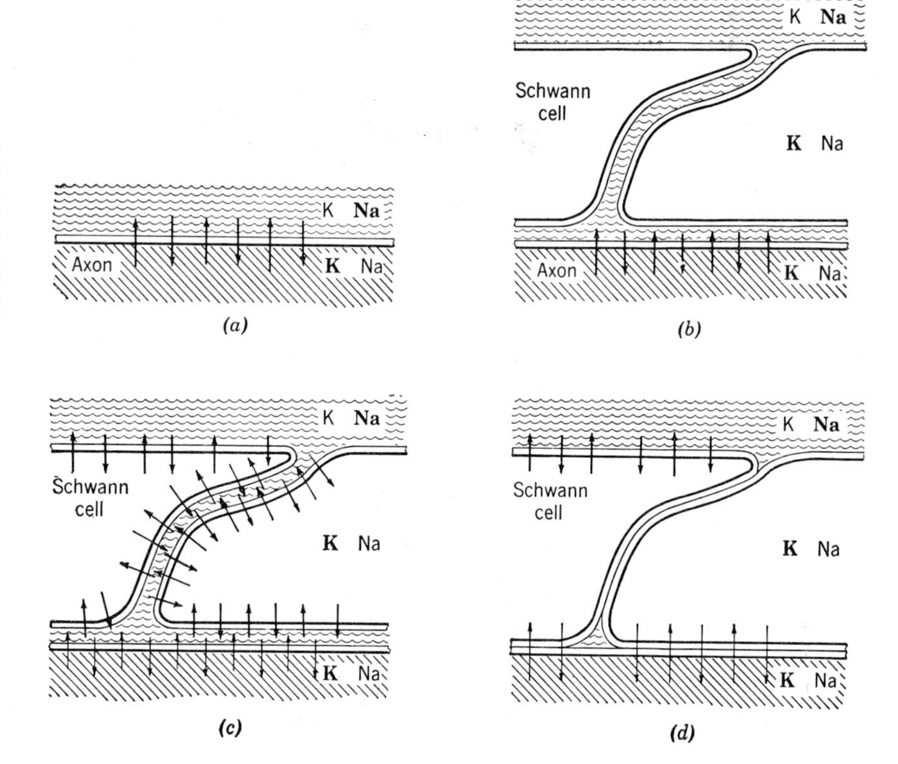


FIG. 8. Possible diffusion pathway for solutes passing between axoplasm and extracellular fluid (after Schmitt[3]). (*a*) bare-axon membrane is exposed to extracellular fluid (*b*) diffusion channel is 150 to 200 A thick, filled with extracellular fluid. Active interchange occurs only between axon surface membrane and fluid in channels; passive diffusion occurs in other channels. (*c*) same as (*b*) except that active interchange of metabolites and solutes occurs across all membrane surfaces; satellite cells participate actively in processes involving interchange between axon and extracellular fluid. (*d*) satellite cells and axon membranes are in molecular contact; water channel not more than 10 to 15 A thick with structure comparable to that of compact myelin.

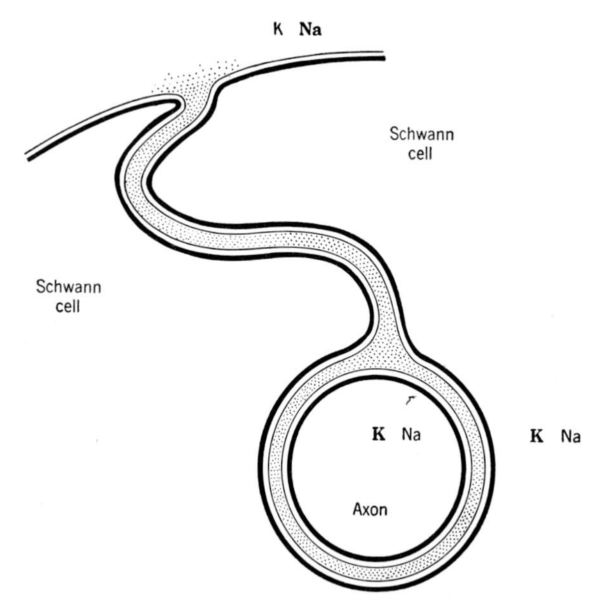


Fig. 9. Diagrammatic representation of relationship of very thin *C* fibers to Schwann cell (after Robertson[48]).

ture.[28] On the other hand, if these intercellular relationships are similar to those found by Robertson[48] and by electron microscopists to obtain between cells generally, the aqueous channels, communicating with the frank extracellular fluid of the ground substance of the surrounding connective tissue, would have a thickness of about 150 A.

Robertson[21,22,48] has found such a membrane relationship in the case of the vertebrate "*C*" fibers. In these nerves, the very thin (*ca* 0.1 $\mu$) axons pass through the satellite (Schwann) cell tube enclosed in a fold of the surface membrane of the satellite cell in a manner analogous to that assumed by Geren's membrane theory to explain the origin of myelin (see also the earlier works of Gasser[49–52]).

In analyzing the functional significance of the intermembrane relationship and the thickness of the aqueous regions between limiting membranes *in life*, several factors must be considered. Indeed, this situation illustrates certain indeterminacies that characterize the electron microscope approach to molecular biology generally.

From the properties of the constituent membranes, including the charge on the membrane molecules, a separation of the order of 100 to 200 A between the membranes might be expected.[53–56] This equilibrium distance will vary with the ionic strength (a statistical physicochemical concept that is difficult to apply to such thin capillary spaces). If the equilibrium distance between surface membranes is responsive to the ionic strength of the environment, it is obvious that a factor of importance is the ionic strength of the fluid used to fix the nerves. Another factor is the osmotic pressure normally characterizing the intercellular fluid. If this is not identical with that of the intracellular fluid, then when the system is placed in the fixative (which presumably quickly destroys membrane semipermeability and such vital equipment as "ion pumps"), an osmotic transfer of water may be expected, and this might importantly influence the distance between surface membranes as observed in the electron microscope after fixation and sectioning.

It has become evident from electron-microscopical studies of metabolically active cells not only that the metabolic activity involves the presence of membranes separated by small distances, but also that the products of the metabolism, presumably including ions, must find their way into the aqueous space between the membranes. The presence of these metabolites in turn must affect the equilibrium separation between the membranes and the properties of the intercellular fluid within these thin capillary spaces. Such factors must be taken into consideration in dealing realistically with these spaces which are the channels of communication between the axon surface membrane (presumably the seat of the rapid alterations responsible for the propagation of the impulse in the nerve fibers) and frank extracellular space. According to Frankenhaeuser and Hodgkin,[57] if such intercellular channels are the route of diffusion of potassium ions from axoplasm to the exterior, the electrical data would require a channel thickness of about 300 A in a structure similar to that shown by Geren and Schmitt[13] assuming a process of passive diffusion calculable on the basis of properties characteristic of macroscopic bulk phases. Whether or not additional properties, deriving from the microscopic dimensions of the intermembrane system and from the presence of solutes, particularly metabolic products of the satellite cells, require modification of this view remains for further analysis to determine. That the satellite cells of the squid giant fiber are indeed metabolically active has been shown by the measurements of oxygen consumption of the sheath cells after removal of the axoplasm,[58] and from an assay of enzyme activities of sheath cells and axoplasm.[59]

## BIBLIOGRAPHY

[1] F. O. Schmitt in *Metabolism of the Nervous System*, D. Richter, editor (Pergamon Press, London, 1957), p. 35.

[2] F. O. Schmitt and N. Geschwind, Progr. in Biophys. and Biophys. Chem. **8**, 166 (1957).

[3] F. O. Schmitt, Exptl. Cell Research **5**, Suppl., 33 (1958).

[4] S. L. Palay in *Ultrastructure and Cellular Chemistry of Neural Tissue*, H. Waelsch, editor (Paul B. Hoeber, Inc., and Harper and Brothers, New York, 1957), p. 31.

[5] S. O. Brattgård, J. E. Edström, and H. Hydén, Exptl. Cell Research **5**, Suppl., 185 (1958).

[6] P. A. Weiss and H. P. Hiscoe, J. Exptl. Zool. **107**, 315 (1948).

[7] H. Waelsch, J. Nervous and Mental Diseases **126**, 33 (1958).

[8] S. Ochs and E. Burger, Am. J. Physiol. **194**, 499 (1958).

[9] P. Chinn, J. Cellular Comp. Physiol. **12**, 1 (1938).

[10] M. Maxfield, J. Gen. Physiol. **37**, 201 (1953).

[11] A. von Muralt, *Die Signalübermittlung im Nerven* (Birkhäuser, Basel, 1946).

[12] J. M. Foster, J. Neurochem. **1**, 84 (1956).

[13] B. B. Geren and F. O. Schmitt, Proc. Natl. Acad. Sci. U. S. **40**, 863 (1954).

[14] B. A. Koéchlin, J. Biophys. Biochem. Cytol. **1**, 511 (1955).
[15] M. Maxfield and R. W. Hartley, Biochim. et Biophys. Acta **24**, 83 (1957).
[16] F. O. Schmitt, J. Cellular Comp. Physiol. **49**, Suppl. 1, 165 (1957).
[17] G. Deffner and R. Hafter, Biochim. et Biophys. Acta (to be published).
[18] R. S. Bear, F. O. Schmitt, and J. Z. Young, Proc. Roy. Soc. (London) **B123**, 505 (1937).
[19] F. O. Schmitt, J. Exptl. Zool. **113**, 499 (1950).
[20] F. O. Schmitt and B. B. Geren, J. Exptl. Med. **91**, 499 (1950).
[21] J. D. Robertson in *Ultrastructure and Cellular Chemistry of Neural Tissue*, H. Waelsch, editor (Paul B. Hoeber, Inc., and Harper and Brothers, New York, 1957), p. 1.
[22] J. D. Robertson, J. Biophys. Biochem. Cytol. **4**, 349 (1958).
[23] W. Stoeckenius, Exptl. Cell Research. **13**, 410 (1957).
[24] W. Stoeckenius in *Proceedings of the Fourth International Congress on Electron Microscopy* (Springer-Verlag, Berlin, to be published).
[25] H. W. Fisher and T. T. Puck, Proc. Natl. Acad. Sci. U. S. **42**, 900 (1956).
[26] W. J. Schmidt, *Die Doppelbrechung von Karyoplasma, Zytoplasma und Metaplasma* (Gebrüder Borntraeger, Berlin, 1937).
[27] O. Wiener, Abhandl. sächs. Ges. Wiss. math. phys. Kl. **32**, 509 (1912).
[28] F. O. Schmitt, R. S. Bear, and K. J. Palmer, J. Cellular Comp. Physiol. **18**, 31 (1941).
[29] F. O. Schmitt and R. S. Bear, Biol. Revs. **14**, 27 (1939).
[30] H. Fernández-Morán, Exptl. Cell Research **1**, 309 (1950).
[31] F. S. Sjöstrand, Experientia **9**, 68 (1953).
[32] J. B. Finean, F. S. Sjöstrand, and E. Steinmann, Exptl. Cell Research **5**, 557 (1953).
[33] H. Fernández-Morán and J. B. Finean, J. Biophys. Biochem. Cytol. **3**, 119 (1957).
[34] J. B. Finean, Exptl. Cell Research **5**, Suppl., 18 (1958).
[35] B. B. Geren, Exptl. Cell Research **7**, 558 (1954).
[36] B. B. Geren in *Cellular Mechanisms in Differentiation and Growth*, D. Rudnick, editor (Princeton University Press, Princeton, 1956), p. 213.
[37] B. B. Geren and J. Raskind, Proc. Natl. Acad. Sci. U. S. **39**, 880 (1953).
[38] J. D. Robertson, J. Biophys. Biochem. Cytol. **1**, 271 (1955).
[39] S. A. Luse, J. Biophys. Biochem. Cytol. **2**, 777 (1956).
[40] E. DeRobertis, H. M. Gerschenfeld, and F. Wald, J. Biophys. Biochem. Cytol. **4**, 651 (1958).
[41] J. D. Robertson, J. Physiol. **135**, 56*P* (1957).
[42] M. Luxoro, Proc. Natl. Acad. Sci. U. S. **44**, 152 (1958).
[43] J. D. Robertson, J. Biophys. Biochem. Cytol. **4**, 39 (1958).
[44] G. F. Göthlin, Kgl. Svenska Vetenskapsakad. Handl. **51**, 1 (1913).
[45] B. B. Geren and F. O. Schmitt in *Symposium on Fine Structure of Cells* (Interscience Publishers, Inc., New York, 1955), p. 251.
[46] R. S. Bear and F. O. Schmitt, J. Cellular Comp. Physiol. **9**, 275 (1937).
[47] R. S. Bear, F. O. Schmitt, and J. Z. Young, Proc. Roy. Soc. (London) **B123**, 496 (1937).
[48] J. D. Robertson, J. Biophys. Biochem. Cytol. **3**, 1043 (1957).
[49] H. S. Gasser, Cold Spring Harbor Symposia Quant. Biol. **17**, 32 (1952).
[50] H. S. Gasser, J. Gen. Physiol. **38**, 709 (1955).
[51] H. S. Gasser, J. Gen. Physiol. **39**, 473 (1956).
[52] H. S. Gasser, Exptl. Cell Research **5**, Suppl., 3 (1958).
[53] K. J. Palmer and F. O. Schmitt, J. Cellular Comp. Physiol. **17**, 385 (1941).
[54] K. Hess and H. Kiessig, Chem. Ber. **81**, 327 (1948).
[55] H. L. Booij and D. Bungenberg-de Jong in *Protoplasmatologia Handbuch der Protoplasmaforschung* (Springer-Verlag, Vienna, 1956).
[56] J. Th. G. Overbeek, unpublished lecture, "Physical chemical properties of layered systems," given before the Electron Microscope Society of America (10 September 1957).
[57] B. Frankenhaeuser and A. L. Hodgkin, J. Physiol. **131**, 341 (1956).
[58] R. R. Coelho, J. W. Goodman, and M. B. Bowers (manuscript in preparation).
[59] N. R. Roberts, O. Lowry, E. J. Crawford, and R. Coelho, J. Neurochem. **3**, 109 (1958).

# 50
# Nature of the Nerve Impulse

BERNARD KATZ
*Department of Biophysics, University College, London, W. C. 1, England*

## INTRODUCTION

THE mechanism by which signals are rapidly transmitted over long distances in the body has formed one of the principal preoccupations of biophysicists for well over a century. Since the days of Du Bois-Reymond (about 1850),[1] it was known that nerves not only are excited by electric currents but also produce, in the course of their activity, electric currents of their own. His pupil, Hermann,[2] suggested that the propagation of a signal along a nerve fiber (or axon) involves a process of recurrent electric stimulation from point to point. Hermann also called attention to the apparent cable properties of nerve fibers, although he recognized that, by passive cable transmission alone, nerve signals could not travel over any appreciable length. Later, under the influence of Nernst, Bernstein[3] developed the basis of the modern membrane theory. According to Bernstein, the key to our understanding of the nerve impulse is to be found in the properties of the surface membrane of the axon, and especially its selective and changeable permeability (or conductance) to ambient ions. His theory has been modified in important details, but in essence it still stands, and the advances which have been made in this field during the last two decades have confirmed the usefulness of Bernstein's basic concept.

The view which now has emerged is (a) that the electrical events in a nerve fiber are governed by the differential permeability of its surface membrane to sodium and potassium ions, and (b) that these permeabilities themselves depend upon the electric field across the surface. The interaction of these two factors leads at a certain critical threshold level to excitation, that is, to a regenerative release of electrical energy from the axon membrane, and the propagation of this change along the fiber in the form of a brief, all-or-none, electrical impulse (the so-called spike or action potential).

As F. O. Schmitt has pointed out (p. 455), when a physiologist speaks of the axon membrane, he usually has in mind an abstraction rather than a microscopically identified structure. This point, *viz.*, the relation between fine structure and functional properties, is discussed later in this paper; here the main concern is to indicate what evidence there is for such physiological "abstractions."

## RESTING NERVE FIBER

For the purpose of discussing the principal features of the nerve signal, the structural picture can be reduced to that of a long cylindrical tube with a surface membrane which separates two aqueous solutions of equal osmolarity but of different chemical composition. In the external medium, more than 90% of the osmotic balance is made up of sodium and chloride ions; while inside the cell, these ions account for less than 10% of the solutes, potassium taking the place of sodium, and various organic anions (which presumably are synthetized within the cell itself) taking the place of chloride. Our main concern is with the concentration differences of sodium (about 10 times higher outside) and potassium (about 30 times higher inside) across the cell surface. With the use of very fine KCl-filled micropipettes, it has been possible to penetrate the fiber surface without serious damage and to measure the electric PD between inside and outside. There is a potential difference across the membrane of some 60 to 90 mv (inside negative), while there are no detectable potential differences within the interior of the normal resting cell.

Observations of this kind have been made on a variety of nerve and muscle fibers. Most of the evidence described in the following has been obtained from the giant axon of the squid whose large size (nearly 1 mm thick) allows one to introduce an assembly of fine electrodes along the inside to analyze the ionic content of a single fiber and to observe the movement of radio-tracers within selected regions of such fibers (see Hodgkin,[4,5] also Huxley[6] and Katz[7] for review and references).

It should be mentioned perhaps that the nerve fibers of squid and other cold-blooded animals can be isolated and kept in a suitable salt solution. They remain in a functioning condition for many hours during which they can be made to propagate many thousands of impulses of approximately the same voltage and velocity as initially, *in situ* and with the blood circulation intact. Such fibers remain capable, to some extent, of replenishing the losses from their chemical stores which are drawn upon during periods of activity. Nevertheless, it is true that isolated tissues and, in particular, nerve fibers (which, after all, are only peripheral stumps of a cell) are no longer in a steady state, and their electrochemical accumulator gradually runs down.

It is of interest that muscle fibers can be used to study propagation of impulses as well as nerve. Muscle fibers, unlike severed nerve axons, are self-contained cell units which possess a built-in apparatus of distributed nuclei (and which, incidentally, do not possess investing layers of satellite cells). The surface membrane of a muscle fiber has certain important properties in common with that of a nerve: it provides a selective barrier between cytoplasm and surroundings, and it serves to conduct electric excitation rapidly over the whole length from the nerve-muscle junction to the tendon. This is an essential mechanism for the skeletal muscle, for the action potential provides the local stimulus to the contractile process at each point of the fiber, and a high speed of propagation is needed to elicit an efficient synchronous twitch. As far as has been ascertained, the properties of the vertebrate-muscle impulse are exactly analogous to those of the nerve action potential.

## MAINTENANCE OF THE STEADY STATE

The first question to consider is: What are the mechanisms responsible for the maintenance of a steady state in a resting (i.e., nonstimulated) nerve or muscle cell? In particular, how are the electrical and ionic concentration differences kept up between interior and surroundings?

*Is potassium chemically bound?* It has been suggested from time to time that one does not require selective membrane properties in order to explain the preferential accumulation of potassium in the cytoplasm. It would be sufficient, for example, if the cellular proteins had a special chemical affinity to potassium rather than to sodium. This idea of chemical binding of potassium seemed unlikely on several grounds: it would be difficult to account for the osmotic pressure and electric conductivity of the cell contents unless at least a large proportion of the intracellular potassium was present in the form of free ions. The most direct evidence was obtained by Hodgkin and Keynes[8] in the following experiment. A region of an isolated *Sepia* axon (about 150 μ thick) was bathed in a droplet containing radiopotassium. After a quantity of the labeled ions had entered the axoplasm, the external droplet was washed away. It may be noted that the mixing between applied tracer and intracellular potassium takes many hours, but the exposure time was sufficient to build up an adequate amount of radioactivity within the axon. Subsequently, the spread of this patch by diffusion along the interior of the fiber was determined; at the same time, a potential difference was applied to the ends of the fiber, and the speed with which the labeled patch moved toward the cathode was measured. The experiment showed that the diffusion coefficient and the electric mobility of the tracer ions were only slightly less inside the cell than in free aqueous solution. There was clearly a highly resistive surface barrier, for the entry of the externally applied potassium into the axoplasm was very slow; but once the tracer ions had passed this barrier, they continued to behave as free ions.

## BERNSTEIN MEMBRANE THEORY

The experiment of Hodgkin and Keynes seems to rule out the suggestion of chemical binding of potassium and provides direct evidence for the existence of a resistive surface membrane as envisaged by Bernstein more than 50 years ago. He postulated that the nerve membrane was selectively permeable to potassium alone, and impermeable to sodium, chloride, and the internal anions. This was an attractive hypothesis, for it would explain not only the maintenance of steady ionic concentration gradients, but also the existence of the resting potential as resulting from a potassium-concentration cell. Moreover, it was possible to predict an inverse logarithmic relation between the external potassium concentration and the membrane PD, and over a fairly wide range there was satisfactory agreement between theory and experiment. Bernstein went one step further. He postulated that the membrane permeability to the other ions increased during electric excitation so that the impulse should be accompanied by a sharp increase of the membrane conductance and a drop toward zero of the membrane potential (or rather to the low level of a liquid junction PD between axoplasm and external fluid). For many years, these ideas were accepted as a reasonable explanation, and the evidence, rather inaccurate and indirect as most of it was, tended to support it. More recently, it was strengthened by the observation of Cole and Curtis[9] that the rising phase of the action potential indeed is accompanied by a rapid 40-fold increase of the ionic membrane conductance.

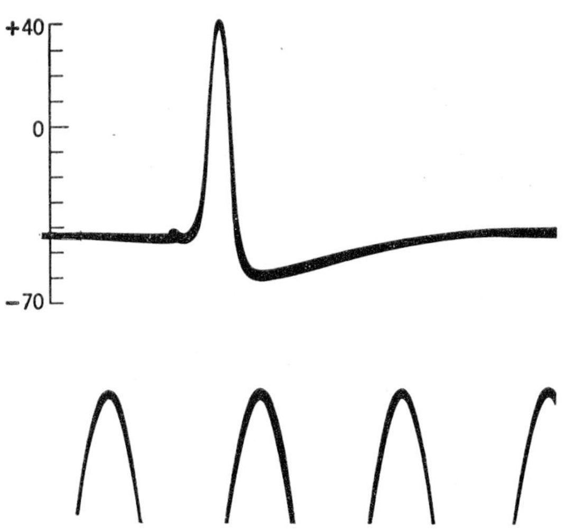


FIG. 1. Action potential recorded with an internal electrode from a squid giant axon. The scale shows the internal potential in millivolts, relative to the outside bath. The time marks are at 2-msec interval [from A. L. Hodgkin and A. F. Huxley, Nature **144**, 710 (1939)].

The flaws began to appear soon after, when increasing use was made of the critical methods introduced by Cole and by Hodgkin, *viz.*, the single axon technique and the use of intracellular recording, and also when improved chemical and tracer methods were employed in observing ion movements. In 1939, Hodgkin and Huxley[10] (see also Curtis and Cole[11]) discovered that the membrane potential during the nerve impulse does not move simply toward the zero level (Bernstein's predicted "depolarization"), but reverses substantially to a level of 40 to 50 mv—inside positive (see Fig. 1). This was clearly incompatible with the second part of the Bernstein theory just now outlined. Two years later,

Boyle and Conway[12] demonstrated that muscle fibers are permeable to chloride as well as to potassium, and shortly afterward it became clear that even sodium can enter the fibers, though apparently it encounters a much higher membrane resistance than does potassium.

### SODIUM PUMP

This raised an altogether new problem, for the demonstration of even a small permeability to *sodium* made it impossible to regard the physical properties of the surface barrier as sufficient to maintain the concentration and potential gradients in a steady state. Potassium and chloride appear to be very nearly in electrochemical equilibrium, their large concentration differences being balanced approximately by the membrane potential. But, in the case of sodium, both driving forces act in the same direction and tend to shift the ion into the interior. If the membrane resistance toward sodium is not infinite, then some mechanism entirely different from the ones discussed so far is required to keep its internal concentration at a steady low level.

It was at this stage that the postulate of a metabolic sodium pump was introduced—that is, some secretory process in which an energy-yielding reaction of the cell is utilized to perform the work of a Maxwell demon, moving sodium ions uphill and outward through the membrane as fast as they leak into the cell in the direction of the electrochemical gradient. There was nothing intrinsically implausible about the suggestion, for the permeability of the resting cell surface to sodium is normally so low that the leakage rate remains very small, and the work required of the pumping process amounts to only a small fraction of the energy which is continuously being made available by the metabolism of the cell.[13]

That there is, in fact, a direct relation between cellular metabolism and sodium extrusion was first shown by Hodgkin and Keynes[14] who observed a cessation of the efflux of tracer sodium when the axon was treated with certain metabolic inhibitors (dinitrophenol, azide, cyanide). This was a reversible change and one which specifically affected the extrusion rate of sodium, not its rate of leakage into the cell. More recently, Caldwell and Keynes[15] have shown that "pumping" is temporarily resumed by the cyanide-inhibited axon if a dose of adenosinetriphosphate or arginine phosphate is injected into the axoplasm.

At one time, it was thought that the sodium pump operated by expelling sodium alone, and in doing so was capable of separating electric charges at the fiber surface, moving them against the existing potential gradient, and so contributing directly to the build-up of the resting emf. This idea has been made unlikely by the findings (a) that the extrusion rate of sodium is unaffected by the potential gradient against which such an electrochemical pump would have to work, and (b) that the stoppage of sodium extrusion by metabolic poisons has no immediate effect on the resting (or action) potential, which runs down very slowly as a result of the gradual diminution of the ionic concentration gradients. It appeared, therefore, that the ion pump works on a principle of electroneutrality, either taking sodium out in company with some anion (e.g., bicarbonate or phosphate), or through a cation-exchange process. There is evidence that the latter is the prevailing mechanism, and that extrusion of sodium and internal uptake of potassium are linked closely in the pumping process. The most suggestive piece of evidence is the finding[14] that the extrusion rate of sodium becomes reduced greatly when potassium is withdrawn from the bath solution, and that, in general, the rates of potassium entry and of sodium efflux change simultaneously and in a parallel manner.

To summarize present views of the ionic maintenance mechanisms of the resting cell, one imagines that the cellular metabolism is responsible for the upkeep of ionic concentration differences between cell and surroundings, (1) by synthetizing large organic anions which cannot diffuse through the membrane, and (2) by providing the energy for an active ion-exchange mechanism, expelling sodium and accumulating potassium ions, possibly one for one. The cell membrane has a generally extremely low ionic permeability so that even in the complete absence of pumping action it will take many hours before the Na/K concentration gradients run down. The ionic permeability of the resting membrane, apart from being small, is also differentially very selective, that to potassium being much higher than to sodium, so that the PD across the resting membrane approximates to that of a potassium-concentration cell, though it does not quite reach this level (it would do so only if the sodium permeability were negligible).

The evidence for the existence of a cation pump is impressive; there is little doubt that the required Maxwell demons are being paid for; the account seems satisfactory not only in terms of available calories, but even in the form of the available currency (made up of ATP and other acceptable phosphates). But, as in most instances of specific cellular utilization of metabolic energy yields, there is no theory at present to explain how the so-called driving reaction is geared to the particular purpose of expelling sodium and piling up potassium. This is one of the familiar challenging gaps waiting to be bridged, even if only by a temporary working hypothesis.

### CABLE PROPERTIES OF NERVE

It has been seen from the tracer experiments that the surface membrane presents a formidable barrier to the diffusion of ions, which agrees with the conventional picture of the membrane consisting of, or containing, a thin layer of insulating lipid material. It may correspond to the "double contours" of 50- to 100-A thickness which have been seen in electron micrographs of various cellular or intracellular surface structures.

The conclusions which have been reached from chemical and tracer measurements are in general agreement with what is known about the electrical properties of nerve and muscle fibers. By studying alternating-current impedance, or by analyzing the attenuation and temporal distortion along the fiber surface of a small voltage-step signal applied at one point between inside and outside, it had been shown that the interior of axons and muscle fibers behaves as a conducting cylinder of slightly higher electrolytic resistivity than the outside fluid but is separated from the outside by a surface layer of very low conductance (of the order of $10^{-4}$ to $10^{-3}$ $ohm^{-1}$ $cm^{-2}$) and of high capacitance. The fibers possess, therefore, electrical properties which are analogous to those of a cable, and one may ask to what extent these properties can be used for the purpose of signal transmission. Now, taken as a passive transmission line, the axon would be of little use, for its cable losses are very great: its surface leakage and the resistivity of its core are some $10^8$ times greater, and its sheath capacity about $10^6$ times higher than those of an ordinary commercial cable. In fact, a weak applied square-pulse signal (i.e., an electric pulse of subthreshold intensity which fails to excite the inherent relay mechanism of the axon membrane) fades out and becomes badly blunted within a few millimeters of its origin. It is clear, therefore, that the cable properties of the nerve are quite insufficient to serve the propagation of a message over the required distances, unless the deficiencies of the cable are made up by a special boosting process. This is indeed the case, and the sole purpose of the excitatory mechanism is to regenerate and to reamplify the signal at each point of the line, and so to insure its forward passage without attenuation or distortion of its brief wave form.

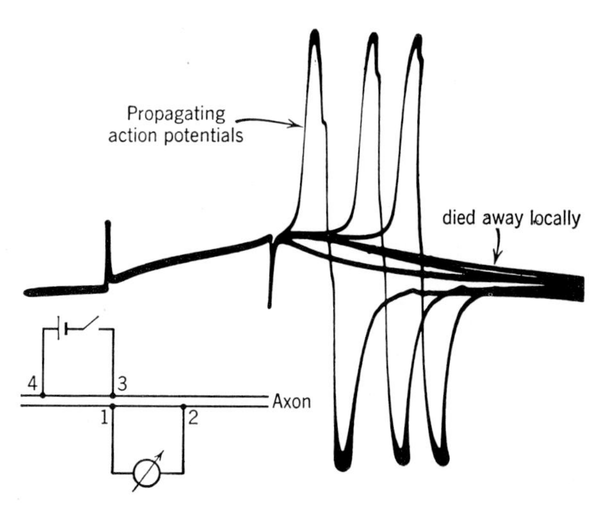


FIG. 3. Local and propagated potential changes, recorded at the site of stimulation from a single crab nerve fiber. Inset diagram shows arrangement of recording (1 and 2) and stimulating (3 and 4) electrodes, all four being placed on the axon surface, 1 and 3 close together. The records show, superimposed, the local potential changes produced by 6 successive pulses of threshold strength (duration of pulse *ca* 2 msec, beginning and end marked by brief artifacts). The potential change either died out locally or flared up after a variable delay into a large propagating action potential (about 100 mv peak-to-peak). Its diphasic nature is due to the wave of surface negativity passing first point 1, then point 2.

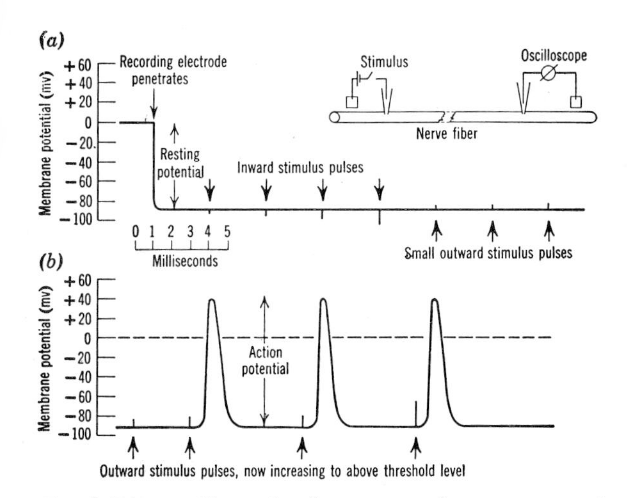


FIG. 2. Diagram illustrating the response of a nerve or muscle fiber to an electric stimulus applied at a distant point. The inset diagram shows the arrangement of stimulating and recording electrodes. In (a), a recording micropipette is inserted into the fiber interior, showing the resting potential across the cell membrane. Vertical scale indicates level of potential at the tip of the micropipette relative to the outside bath. Current pulses are applied through the distant stimulating electrodes, first inward, then outward through the membrane. No response is produced until, in (b), the outward pulses exceed a critical threshold when an all-or-none action-potential wave is recorded.

It is worth emphasizing that the presence of a cable-like mechanism, insufficient though it is for long-distance signaling, forms an essential link in the process of impulse propagation: it allows the action potential which has arrived at one point to impart a stimulus to the next region and to excite its latent relay mechanism. Moreover, the passive cable spread is of great importance for the integration of converging messages which takes place in the cells of the central nervous system. These cells become excited, or inhibited, by the interaction and summation of many subthreshold electrical changes which are imposed on different points of their surface membrane, all within a short range of the central cell body. The effective summation of such converging electrical influences is possible, because over short distances (a fraction of a millimeter) the cable properties of the cell are quite adequate, and the subthreshold potential changes do not suffer much attenuation. They can, therefore, sum up to the firing point at which an impulse emerges, capable of traveling the whole length of the axon process.

## PROCESS OF EXCITATION

What is the nature of the excitatory relay mechanism by which the impulse is boosted to its full strength? The process can be demonstrated by recording the electrical events at the point of the fiber at which an impulse is initiated by an applied electric stimulus (Figs. 2–4). When a square pulse of current is passed inward through the axon membrane, the membrane potential is displaced

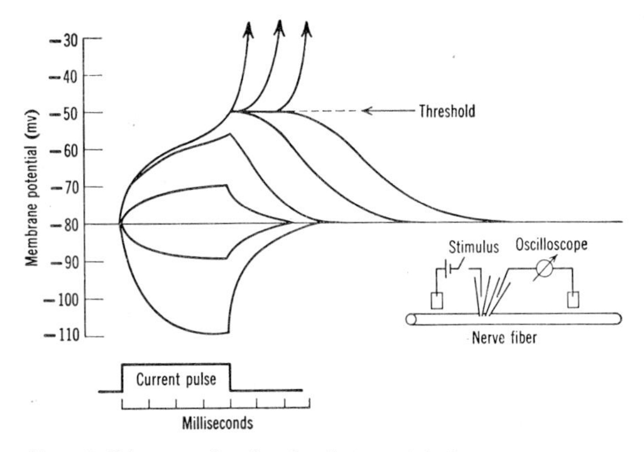


FIG. 4. Diagram showing local potential changes across axon membrane, due to rectangular current pulses (4-msec duration, varying direction and amplitude). At threshold, the membrane potential is displaced to a level at which it is in a state of unstable equilibrium, either flaring up into a propagating signal or subsiding locally.

from its resting level of, say, −60 mv inside to a higher level of internal negativity. The potential change takes a certain time to develop and to decay; this time course depends upon the resistance and capacity of the fiber, as in an ordinary leaky and capacitative cable structure. Currents of this direction, which increase the membrane PD, do not excite, but simply produce local potential changes approximately proportional to current strength, and fading-out along the fiber with a decrement of about 50% per 1 or 2 mm, characteristic of its cable coefficients.

When the polarity is reversed and weak current intensities are employed, very similar local potential changes of the opposite direction (reducing the membrane PD, or partially depolarizing) are observed. However, as the current strength is increased the picture alters. The membrane potential changes progressively, more than in proportion to the applied current, and a point is reached at which the depolarization becomes regenerative and, after withdrawal of the current, the decay of the effect greatly retarded. A slight further increase in current strength causes a flare-up of the potential change which now passes through an automatic cycle; it rises to the full height of the propagating action potential and then returns to the initial level within 1 or 2 msec. This event (known as the spike) travels at constant speed (about 25 m/sec in the squid fiber at 20°C) down the axon, without loss of amplitude anywhere; it leaves behind a short refractory period of a few msec, after which the axon is again excitable and capable of repeating the event. It is interesting that even below the firing point (or threshold), some local "smoldering" seems to go on. Indeed, there is a good analogy between the local electrical events in a nerve fiber and any other form of regenerative chain reaction, in which the initially applied stimulus is reinforced by the reaction which it produces. It would be possible, for instance, to obtain a very similar family of kinetic curves, if one used an explosive gas mixture (say hydrogen and oxygen), applying or withdrawing heat instead of current, and plotting on a suitable time scale the temperature of the gas instead of membrane potential. Well below the ignition point, the temperature changes along an exponential time course and equilibrates at a level determined by a balance between heat leakage to the surroundings and rate of heat supply. Close to the ignition point, some gas molecules combine and produce extra heat, but not at a sufficient rate to keep the reaction going when the applied heat source is switched off. At the ignition point, the rate of reactive heat production just balances the leakage to the cooler surroundings and, for a while, maintains the system in unstable balance from which either an explosive flare-up or a return to the ambient stable temperature occurs. Apart from this formal analogy, there is, of course, little resemblance between an explosive exothermic process and the much more subtle and repeatable release of electric energy by the nerve membrane.

## PROPERTIES OF THE NERVE IMPULSE

Several questions arise at this point: (1) What is the nature of the regenerative factor in a nerve fiber which causes an initially applied potential change to amplify itself? (2) What brings the action potential down to the original baseline? (3) What causes the system to become temporarily inexcitable? (4) How does the axon recover from its impulse activity?

### Regenerative Factor

It has been shown[16,17] that the sodium conductance of the membrane depends upon the electric field across it; the sodium conductance is very low at the normal level of the resting potential, but increases, temporarily, when this potential difference is reduced. *How* this comes about is not known and remains one of the outstanding problems, but the statement that sodium permeability is raised by a depolarization of the axon membrane is based on very strong evidence. The consequences of this effect are far-reaching: suppose a partial depolarization is produced by passing current outward through the membrane. The consequent rise in sodium permeability causes an increased leakage of these ions down their concentration gradient into the fiber. As a result, positive change is transferred to the interior and the membrane potential is diminished further. This is the regenerative step which causes the permeability to sodium to rise again and more sodium ions to enter. At a certain level of the membrane potential, the threshold, this process becomes progressive and flares up into the action-potential wave. Like the ignition point of the explosive-gas analogy, threshold is a point of unstable equilibrium at which restoring and regenerative tendencies just balance each other. The regenerative factor responsible for the automatic ascent of the action potential is the progressive relation between sodium-conductance change and fall of membrane potential; the restoring factor is to be found in the potassium (and chloride) conductances which tend to reset the membrane potential to the initial stable baseline.

Once the threshold level is exceeded, the sodium conductance becomes rapidly the dominant factor and the regenerative entry of sodium would continue until the membrane potential swings well to the other side, and the interior attains sufficient positive charge to balance the inward concentration gradient and prevent any further net influx of sodium ions. In other words, if the initial excitatory process continued, a new equilibrium level would be reached corresponding to that of a sodium-concentration cell. This is, indeed, the accepted explanation for the reversal of the membrane potential during the peak of the impulse. The level which is attained falls somewhat short of the sodium potential, because the initial permeability change is only transient: within about a millisecond, the high, specific sodium conductance is switched off and apparently is converted into a state of high potassium conductance.

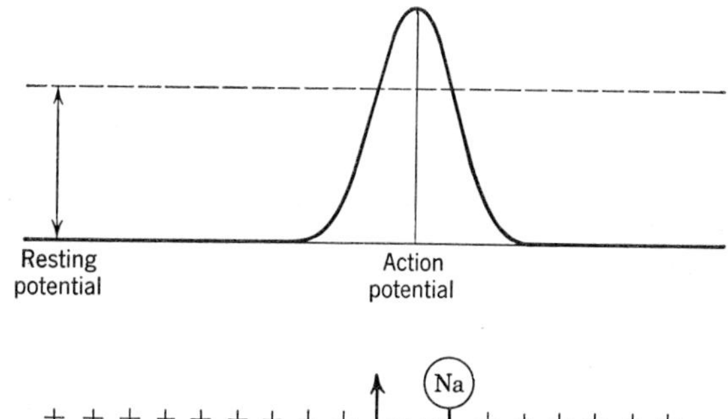


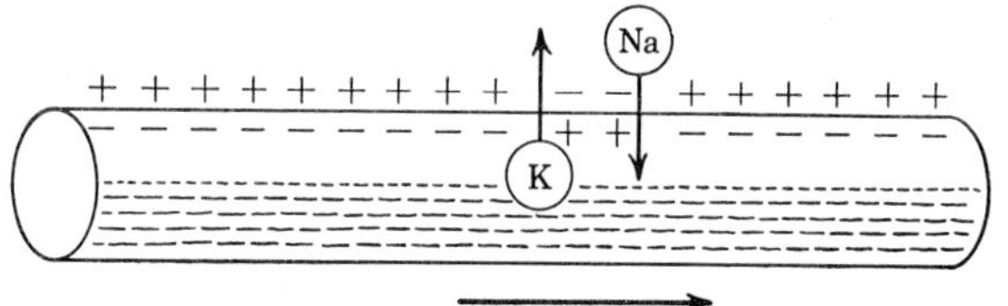


FIG. 5. Diagram of action potential traveling (in direction of arrow) along the axon. During the rise of the wave, sodium ions enter the fiber and charge the interior positively. During the decline, potassium ions leave the fiber and restore the initial membrane potential.

## Restoring Factor

This secondary changeover, from sodium to potassium permeability, ensures a rapid return of the potential to its original baseline, for potassium ions now will leave the fiber rapidly in the direction of their electrochemical gradient. Such electric restoration would be effected even without a secondary increase of potassium conductance, provided the initial sodium-permeability rise is cut off (a process known as inactivation). In this event, however, the decline of the action potential would take a relatively long time and proceed with the time constant of the resting-fiber membrane. In the squid axon, it would take several milliseconds at the ordinary leakage rates of potassium (and chloride) to restore the membrane potential to the resting level. By opening up the potassium channels, this process becomes greatly accelerated, and a few milliseconds later the axon is ready to fire again.

The work of Hodgkin and Huxley has shown that a depolarization imposed on the axon membrane (by the so-called voltage-clamp method) produces these two successive permeability changes (Na, then K), the intensity of both changes increasing similarly with the amplitude of the enforced voltage step. During the normal action potential, the sodium-conductance rise is regenerative and self-reinforcing, while the potassium-permeability change is restorative and gradually cuts itself down as the membrane potential is caused to return to the original stable level.

It should be noted that the electrical return of the system during the impulse is *not* brought about by a reversal of the ionic movement (which would mean a forced expulsion of the quantity of sodium which had entered during the ascending phase of the action potential), but by the leakage of an equivalent quantity of potassium ions (see diagram in Fig. 5). Both sodium and potassium are moving downhill during the impulse; the quantities, however, are so small that no detectable change of the internal concentration would occur during one impulse, and there is plenty of time for the pumping process to restore the chemical situation in resting periods later on.

## Nature of the Refractory Period

The secondary permeability effects which bring about the decline of the action potential are also responsible for the short period of inexcitability (refractoriness) which follows the propagation of each impulse. Clearly, during the state of inactivation—i.e., while the regenerative dependance of sodium permeability on membrane potential is in abeyance—an applied electric current fails to excite; secondly, a state of very high potassium permeability implies that the restoring factor is greatly strengthened and tends to oppose any displacement of the membrane potential from the stable potassium-equilibrium level. Hodgkin and Huxley have determined the rate constants of these changes and their time course of decline following a restoration of the normal membrane potential. It was found that the period of recovery of excitability after the impulse is simply a measure of the time by which the return of the normal permeability relations lags behind the return of the membrane potential.

## Chemical Recovery

There are several phases of restoration: the recovery of the membrane potential is followed by a return of excitability, but the system has not yet reverted completely to its initial state; in fact, a minute fraction of the intracellular potassium store has been lost and replaced by sodium. The downhill movement of sodium and potassium ions across the membrane has provided the energy for the nerve signal. The amount of ions which exchange during the impulse is extremely small; it has been determined for a series of impulses in tracer experiments, and by chemical analysis of single fibers, and was found to exceed the minimum coulomb requirement by a factor of 3. The transfer of charge needed to alter the membrane potential by 100 mv, with a capacity of 1 $\mu f$ $cm^{-2}$ is $10^{-7}$ coul which requires a net transfer of about $10^{-12}$ M of univalent ions: the net exchange of sodium and potassium per impulse is about 3 to 4 times as much.[16] These quantities produce only a minute internal concentration change per single impulse (about 0.0001% in the large squid axon, but considerably more in small fibers whose volume/surface ratio is correspondingly less). Even if the ion pump were not active, a squid axon could conduct some hundred thousands of signals before its accumulated ion store would be exhausted. As in many other forms of brief, high-intensity biological action, a period of impulsive activity is completed at the expense of small but cumulative chemical losses; these are made good during periods of rest by continuous chemical recovery mechanisms which Connelly discusses (p. 475).

A rough qualitative picture has been given here of the conclusions derived from the extremely accurate quantitative work of Hodgkin and Huxley and others, and reference is made to their papers for a description of the experimental evidence on which these conclusions are based.[17–20] Summarizing briefly, it has been found that the resting potential is related to the external potassium but independent of sodium concentration, while the peak of the action potential depends on the sodium-concentration ratio in the way predicted by the Nernst theory. In experiments in which the axon membrane was subjected to a voltage clamp, it was found that a sudden stepwise depolarization (e.g., to zero membrane potential) was followed by a surge of positive inward current followed by a maintained outward current. The first, transient, component (in which current flows "in opposition to Ohm's law" indicating the presence of the required negative resistance, or regenerative factor) could be identified as owing to the inrush of sodium, for it could be made to reverse in polarity by reversing the concentration gradient of sodium ions. With a variety of sodium concentrations, the reversal was shown to occur invariably at the electrochemical equilibrium potential for sodium ions. The second, maintained, component was identified by tracer-efflux measurements as a maintained high-intensity outward current of potassium ions. The rate constants of the several successive conductance changes were determined, and from this information the properties of the impulse, its excitatory threshold, its amplitude, duration, and conduction velocity, could be synthetized with a high degree of precision.

While this series of investigations has been brought to a decisive conclusion, this should, of course, not conceal the fact that the nature of the permeability relations and the reason why membrane potential and ionic conductances are linked in such a specific manner remains an altogether unsolved problem at the present time.

## SOME SPECIAL CASES

### Medullated Nerve Fibers

The picture presented so far describes the situation in the giant axon of the squid, on which most of the crucial experiments have been carried out. But substantially the same conclusions apply to all other types of nerve fibers and to most types of skeletal-muscle fibers which have been studied. There is, however, an important structural difference between the so-called nonmedullated nerves, and the medullated (myelinated) axons of vertebrates which possess relatively thick, segmented sleeves of myelin broken at regularly spaced intervals (the nodes of Ranvier about one or a few millimeters apart) at which a small area of axon membrane is exposed to the outside fluid. In the ordinary nonmedullated fibers, the propagation of the signal depends upon the existence of a continuous relay mechanism which is built in along the whole line at many scattered random points of the axon membrane. This makes up for the deficiencies of the poor cable transmission: without it, the action potential would be attenuated rapidly in the region ahead. Experimentally, this can be shown to be the case by poisoning or anaesthetizing a narrow region of the axon cylinder and so abolishing locally the regenerative mechanism. Provided, however, the attenuated signal which spreads to the other side of the blocked region exceeds the threshold depolarization (which amounts to only $\frac{1}{10}$ to $\frac{1}{5}$ of the full amplitude of the spike), the relay process will start again and the impulse will jump the block.

In medullated nerves, transmission of signals normally occurs in jumps.[21,22] The segmented myelin sheath provides a relatively good, low-capacity insulation around the axon, and the relay mechanism is concentrated and restricted to the discrete nodes of Ranvier where the axon membrane is exposed. There is, in other words, a rather close miniature analogy to a submarine cable with relay stations interposed at the nodes. In the internodal region, the fiber behaves as a relatively good cable with only slight capacitative losses which are made good at the next node. The final result is a great improvement in speed and economy of signaling: the ionic losses are restricted to relatively few microscopic areas of the fiber surface, and the propagation velocity of the impulse is some 10 times greater than for a nonmedullated axon of the same size. To achieve the speed which medullated axons attain *without* the provision of a myelin sheath, the only alternative way for nature would have been to reduce the internal core resistance; that is, to increase the fiber size. Considering the need for a large number of separate signaling channels within a confined space (there are one million medullated axons in the human optic nerve alone), the great advantages of myelination are obvious.

## Other Tissues

Most of the tissues, nerve or skeletal muscle, whose function depends upon the rapid propagation of an all-or-none impulse, appear to make use of a regenerative entry of sodium ions in the course of excitation; there is, however, some doubt as to whether the secondary increase of potassium permeability is a similarly wide-spread phenomenon, or whether it may be a more or less dispensable effect. In some tissues, e.g., certain medullated axons, the potassium resistance and the electric time constant of the resting membrane appear to be sufficiently low to insure a rapid decline of the action potential once the sodium conductance has dropped back towards the initial low level.

In heart muscle, the action potential has a different shape, of "flat top" and long duration. The initial fast rise is probably governed by a sodium mechanism just as in nerve, but the subsequent changes differ at least in quantity. It may be that the maintained plateau of depolarization arises from the increased sodium conductance not being switched off completely, and from a failure of the potassium efflux to rise above its steady rate. It is interesting to consider the usefulness of this long flat-topped heart potential: its duration apparently governs the systolic contraction of the heart muscle, and because of the relatively low frequency of its regular beat, there is no need for rapid electric recovery of the membrane potential.

In skeletal muscle, a quantitative difference from nerve was found in the large capacity of the surface membrane (5 to 10 $\mu f/cm^2$ as against 1 $\mu f/cm^2$ in nerve). There is a correspondingly larger transfer of charge during the action potential, and recent tracer measurements by Hodgkin and Horowicz[23] have shown that the amounts, per unit surface area, of sodium and potassium exchange are about 3 times greater than in the nerve fibers. There appears to be less temporal overlap of the two ionic permeability changes; indeed, resistance measurements indicate that there are two discrete phases of high membrane conductance during the action potential. This would enable the sodium ions which enter to produce a relatively larger net effect. It also seems that the delayed rise of potassium conductance is shortlived and stops before the action potential has returned to the baseline, giving rise to a characteristic long tail (the so-called negative afterpotential) of the muscle spike.

There are finally some tissues (e.g., crustacean muscle and certain conducting plant cells) in which action potentials of the usual propagating all-or-none type can be elicited by electric stimulation, but sodium entry does not seem to be responsible for their regenerative ascent. In many of their features, these action potentials resemble those of the nerve axon very closely. It appears that the underlying events are similar and involve ionic permeability changes, but that the specific channels which are being opened during the depolarization of the membrane are different. For example, in certain algae, a regenerative exit of chloride from the cytoplasm takes the place of sodium entry,[24] while in crustacean muscle, entry of divalent cations possibly may be responsible for the rise of the action potential.[25]

## Structure and Function

To return to some of the problems raised by F. O. Schmitt (p. 455): Can one try now to identify our physiological mechanisms with any one of the structures revealed by the electron microscope? The question has been raised, for instance, whether ion permeabilities are entirely to be attributed to the axon membrane, or could they be properties of Schwann cells which envelope the axon? Insofar as the observations have been repeated on skeletal-muscle fibers, which are not invested with satellite cells, and insofar as they corroborate the main conclusions derived from the nerve experiments, one may feel that the presence of a Schwann-cell envelope does little to mask or mimic the electrical properties of the axon membrane. It is important also to recall that the cellular investment of the axon is not complete, but there are gaps through which the axon surface communicates with the extracellular fluid. Although these channels may be narrow (only a few hundred Ångström units wide), they are also short (a few microns long) and probably add only a small resistance in series with the axon membrane. There is evidence for such an external nonreactive resistance of about 5 ohms in series with a resting membrane of approximately 1000 ohms $\times cm^2$. Recent work of Frankenhaeuser and Hodgkin[26] suggests that the presence of this external bottleneck gives rise to a slow accumulation of potassium ions on the surface of the squid axon during a period of prolonged impulse activity. This is associated with cumulative changes in the membrane potential, but these are rather minute by comparison with the main features of the spike. Nevertheless, one must bear in mind the possibility that the distribution of current flowing through the axon membrane may be modified locally by Schwann cells which cling to its surface and by the presence of discrete channels between them.

It may seem a little disappointing that, with all of the high-resolving power which has been attained in recent years, both in the study of fine structure and of the functional mechanism of nerve, there is still such a conspicuous lack of correlation between these two fields of work. One of the reasons is undoubtedly that the details which are revealed in their most striking form by such different techniques are not necessarily related to the same features. In the electron micrograph, the almost complete investment of an axon by a Schwann cell, or the almost intimate contact between the two surfaces, may be the most impressive phenomenon, while from the point of view of electric current distribution the appearance of a fine gap or crack in this investing layer (which is difficult to find and might easily be discarded as an artifact) may be a much more important feature.

A second reason for the apparent discrepancy is that there must be structural counterparts of functional aspects other than those with which the student of the nerve impulse is directly concerned. The purpose of this paper is to describe some efforts that have been made to elucidate the nature of the nerve signal. But, in addition, there is built into the nerve a machinery for chemical recovery processes, and for the maintenance of the axon and its enzymic apparatus, which one associates with the so-called trophic influence of the nerve-cell body and its nucleus. Little is known about this process, but there are many indications that satellite cells are actively involved, both in trophic maintenance and in the disintegrative processes which follow the severance of the axon from the cell body.

The elementary membrane structure which has been seen in electron micrographs, e.g., of permanganate-fixed axons, is a double contoured-layer 50 to 100 A thick. It is not known what fraction of this layer is composed of the insulating lipid material which we hold responsible for the ion-impermeable and capacitive properties of the axon membrane. Nor can one be certain that improvements in microscopic resolution will not show further subdivisions and even thinner membrane elements. And ultimately, the physiologist is interested not so much in the lipid bulk phase of the membrane which occupies most of it, and which seems never to change nor to take part in the transport of ions, but in those very sparse molecular patches—possibly at continuously varying sites, at which ionic gaps or channels are thought to be formed, and where "carrier molecules" are believed to be active in transferring sodium and potassium ions from one side to the other.

## BIBLIOGRAPHY


[1] E. Du Bois-Reymond, *Untersuchungen über thierische Elektricität* (G. Reimer, Berlin, 1848–1860), Vols. I, II.

[2] L. Hermann, *Untersuchungen zur Physiologie der Muskeln und Nerven* (Hirschwald, Berlin, 1868).

[3] J. Bernstein, Pflüger's Arch. ges. Physiol. **92**, 521 (1902).

[4] A. L. Hodgkin, Proc. Roy. Soc. (London) **B148**, 1 (1958).

[5] A. L. Hodgkin, Biol. Revs. Cambridge Phil. Soc. **26**, 339 (1951).

[6] A. F. Huxley in *Ion Transport across Membranes*, H. T. Clarke, editor (Academic Press, Inc., New York, 1954), p. 23.

[7] B. Katz, Symposia Soc. Exptl. Biol. **6**, 16 (1952).

[8] A. L. Hodgkin and R. D. Keynes, J. Physiol. **119**, 513 (1953).

[9] K. S. Cole and H. J. Curtis, J. Gen. Physiol. **22**, 649 (1939).

[10] A. L. Hodgkin and A. F. Huxley, Nature **144**, 710 (1939).

[11] H. J. Curtis and K. S. Cole, J. Cellular Comp. Physiol. **19**, 135 (1942).

[12] P. J. Boyle and E. J. Conway, J. Physiol. **100**, 1 (1941).

[13] R. D. Keynes, J. Physiol. **114**, 119 (1951).

[14] A. L. Hodgkin and R. D. Keynes, J. Physiol. **128**, 28 (1955).

[15] P. C. Caldwell and R. D. Keynes, J. Physiol. **137**, 12P (1957).

[16] R. D. Keynes and P. R. Lewis, J. Physiol. **114**, 151 (1951).

[17] A. L. Hodgkin and A. F. Huxley, J. Physiol. **116**, 449, 473, 497 (1952).

[18] A. L. Hodgkin and A. F. Huxley, J. Physiol. **117**, 500 (1952).

[19] A. L. Hodgkin and A. F. Huxley, J. Physiol. **121**, 403 (1953).

[20] A. L. Hodgkin, A. F. Huxley, and B. Katz, J. Physiol. **116**, 424 (1952).

[21] A. F. Huxley and R. Stämpfli, J. Physiol. **108**, 315 (1949).

[22] I. Tasaki, *Nervous Transmission* (Charles C. Thomas, Springfield, 1953).

[23] A. L. Hodgkin and P. Horowicz, "Movements of Na and K in single muscle fibres," J. Physiol. (to be published).

[24] C. T. Gaffey and L. J. Mullins, "Ion fluxes during the action potential in *Chara*," J. Physiol. (to be published).

[25] P. Fatt and B. L. Ginsborg, J. Physiol. **142**, 516 (1958).

[26] B. Frankenhaeuser and A. L. Hodgkin, J. Physiol. **131**, 341 (1956).

# 51
# Recovery Processes and Metabolism of Nerve

C. M. Connelly
*The Rockefeller Institute, New York 21, New York*

## INTRODUCTION

THE prime function of peripheral nerve is the conduction of impulses. In biological terms, the nerve impulse is the means of transmission of an item of information from one point in an organism to another. In terms of physics and chemistry, the impulse is a spatially propagating, transient disturbance in the state of a complex reaction system; it might be detected by, or it might be described in terms of, any of those conceivably measurable changes which are a part of it or to which it gives rise. In thermodynamic terms, the impulse is a dissipative process in which the free energy of the nerve and its environment is decreased, principally by a flow of ions down gradients of their electrochemical potential. In this discussion, attention is focused upon processes by which resting nerve maintains itself in a steady state, ready to function, and upon processes which serve to restore nerve to its resting state after it has conducted impulses. A central question is how energy-yielding processes may be coupled to those requiring energy. In an approach to a more specific formulation of this question, the discussion emphasizes correlations between changes in rate of oxidative metabolism and electrochemical manifestations of ionic movement.

In resting nerve there are at least two processes which may be presumed to require energy. These are the maintenance of structural integrity and the maintenance of the ionic-distribution characteristic of the resting state. About the former very little can be said. Perhaps the cell bodies from which the excised nerve fibers have been severed are the primary site of synthetic reactions underlying such maintenance. During activity and recovery therefrom energy demand may be increased in two ways: there must be an acceleration of ion transport processes in order to reverse the exchange of ions which occurs during the passage of impulses and, secondly, there may be a dissipation of chemical energy associated with the permeability cycle that allows the ionic exchange to take place.

Insofar as nerve at rest is in a steady state and following activity returns to the same steady state, the over-all process of impulse conduction involves no net external work and results only in the conversion of chemical energy to heat.

The following discussion indicates that oxidative metabolism in nerve can serve as an adequate energy source. Brink[1] has recently reviewed the evidence that the biochemistry of peripheral nerve of the frog resembles that of other animal cells; the nerve appears to contain and utilize the Meyerhof-Embden pathway of carbohydrate breakdown, the Krebs tricarboxylic acid cycle, and the cytochrome chain of enzymes. The energy turnover appears to be mediated by the usual system of phosphate compounds (adenosine phosphates and creatine phosphate) which are replenished principally by oxidative phosphorylation carried out by mitochondria. In what follows, it will be assumed (1) that any processes which rely upon metabolic energy do so by chemical reactions coupled to the breakdown of adenosine triphosphate, or its equivalent, and thereby increase the intracellular level of phosphate acceptor; (2) that the kinetics of oxygen uptake by nerve are some reflection or measure of the changes in concentration of phosphate acceptors at the mitochondria. A particularly dramatic experiment directly supporting the first of these assumptions is that in which Caldwell and Keynes[2] injected ATP into a metabolically poisoned squid axon and observed a partial restoration of the rate of sodium extrusion.

Firstly, the kinetics of the increases in oxygen utilization associated with the conduction of impulses is described and compared with the heat measurements of Hill, and then it is shown how the kinetics are modified under circumstances in which rates of ion transport have presumably been affected. Secondly, data on the ionic fluxes across the nerve surface are examined and the estimated energy requirements of transport processes are compared with energy available from oxidative metabolism. Thirdly, some observations are described of prolonged positive afterpotential (post-tetanic hyperpolarization) which appears to be closely related to ionic transport processes.

Most of the observations which follow come from experiments on the excised sciatic nerve of the frog. This preparation has some advantages over the giant axon of the squid in that its metabolic and electrical properties remain more nearly constant over an experimental period of ten to fifteen hours. It has, however, the disadvantage of being a bundle of fibers of several different types and there are complications arising from the existence of an appreciable extracellular space. The minimum environment required by frog nerve to maintain function and a satisfactory steady state at 20°C is a balanced bathing solution containing sodium, potassium and calcium salts, and dissolved oxygen. It should be emphasized that many or most of the electrical and metabolic properties of nerves appear to be basically similar from one species of animal to another, and similar also, in fact, to the properties of other excitable tissues such as skeletal and cardiac muscle.

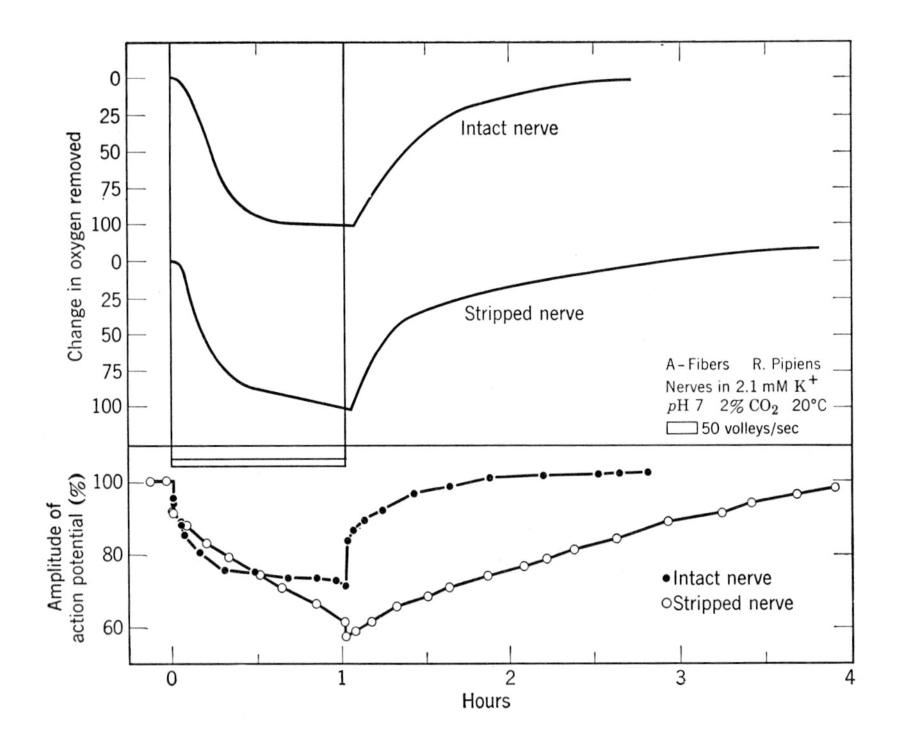


FIG. 1. (Upper) Relative time courses of the increases in oxygen uptake resulting from activity in intact frog sciatic nerve (perineurium not removed) and in stripped nerve (perineurium removed). Measurements made in a flow respirometer[3] with a polarized oxygen cathode. Polarographic current is a linear measure of dissolved oxygen remaining in solution after it has flowed past the nerve, a downward deflection in the trace, as shown, indicating an increase in oxygen uptake. (In all tetani in this and the following figures, only type A fibers stimulated.) (Lower) Time courses of changes in amplitudes of compound action potentials of the two nerves, measured during tetanus and occasionally by test shocks during recovery.

## OXIDATIVE METABOLISM AND HEAT PRODUCTION

The oxygen uptake by resting frog nerve is usually 30 to 40 mm$^3$/g(wet)/hr or about 1.5 $\mu$moles/g(wet)/hr. When a nerve is tetanized, its rate of oxygen uptake increases and approaches a new steady level in about 30 min, closely following an exponential time course with a time constant of 5 to 8 min, Fig. 1, upper curve.[3] The amplitude of the increase is greater the higher the frequency of tetanus, at low frequencies, and approaches a maximum limiting value (of about 1 $\mu$mole/g (wet)/hr) at about 100 volleys/sec,[4] as shown in Fig. 2. At the end of tetanus, the rate of uptake declines slowly toward the resting value along an exponential whose time constant is usually 15 to 25 min. [After tetani at frequencies lower than about 10/sec, recoveries may be more rapid than this (see Fig. 11 in an article by Connelly *et al.*[5]).]

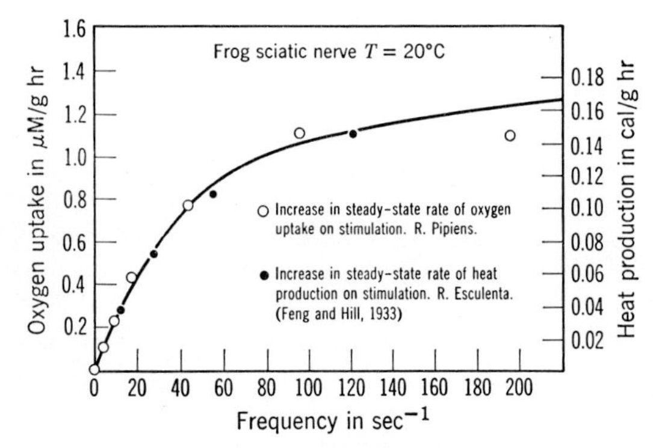


FIG. 2. Oxygen uptake and heat output for active nerve in relation to frequency of conducted impulses. Perineuria not removed. Ordinate scale symbol $\mu$M means micromole [from F. Brink, D. W. Bronk, F. D. Carlson, and C. M. Connelly, Cold Spring Harbor Symposia **17**, 53 (1952)].

The over-all increases in heat production by tetanized frog nerve show kinetics similar to those of the increases in oxygen consumption.[6] The quantitative comparison of steady-state heat rate and oxygen consumption as a function of frequency is shown in Fig. 2. The ratio of the right-hand to the left-hand scale is within 15% of the accepted value of the calorific equivalent of oxygen, 5 cal/cc. The agreement is quite satisfactory, in view of the difference in species.

A detailed analysis of the heat production of nerve during short tetani reveals the presence of a component whose onset and termination are abrupt and correspond to the beginning and end of the tetanus.[7] This "initial heat," only a few percent of the total heat associated with the tetanus, appears to have no counterpart in the time course of the increase in oxygen uptake, the curve of which rises linearly from the beginning of the tetanus.[4] In recent elegant work, Hill and his coworkers[8] have resolved the initial heat associated with a single volley in crab nerve into a positive phase (heat production) followed by a smaller negative phase (heat absorption). These phases may result from the heats of dilution or mixing of the sodium and potassium ions exchanged during the impulse, but other events of the permeability cycle have not been excluded as contributing causes.

The second curve of Fig. 1 illustrates the time course of increase in rate of oxygen uptake by a nerve trunk from which the perineurium, or connective-tissue sheath, has been removed. The striking difference between this and the upper curve is that the recovery is characterized by two components, the first lasting no more than about 30 min and the second extending over several hours.

The lower part of Fig. 1 shows the accompanying changes in the amplitudes of the compound action potentials of the two nerves, the stripped nerve showing much less rapid recovery than the unstripped nerve. The possibility that these differences in the metabolic and electrical behavior of stripped and intact nerve might result principally from differences in the extracellular level of potassium ions was tested by the series of experiments illustrated in Figs. 3 and 4. These compare the kinetics of the changes in oxygen uptake and action potentials, respectively, of stripped nerves bathed in solutions containing different concentrations of potassium ion. The three lower curves of Fig. 3 possess fast components of about the same relative magnitude and time course whereas the slow components of these curves show gradation from effectively no recovery in K-free solution, to a slow almost linear recovery in 2.1 mM $K^+$, to a somewhat more rapid curvature characteristic of an exponential of about a 40-min time constant, in 5 mM $K^+$. In the top curve (8.5 mM $K^+$), the over-all recovery is even more rapid than any observed in intact nerve (in 2.1 mM $K^+$), with an exponential time constant of only 11 min instead of 15 to 25. It is not possible to distinguish two phases of recovery in this case and it is not clear whether the slower phase observed in the lower curves has been speeded up to merge with the rapid phase or whether it has decreased in magnitude to zero. In Fig. 4, one sees that the lower the concentration of potassium in the bathing solution, the more rapid is the decline in height of the action potential during tetanus and the less rapid is its recovery. Recovery of action potential is incomplete in K-free solution, as the recovery of oxygen uptake was seen to be. In 8.5 mM $K^+$, however, not only does the action potential not decline in amplitude during tetanus but it shows about 13% increase in amplitude during the post-tetanic period.

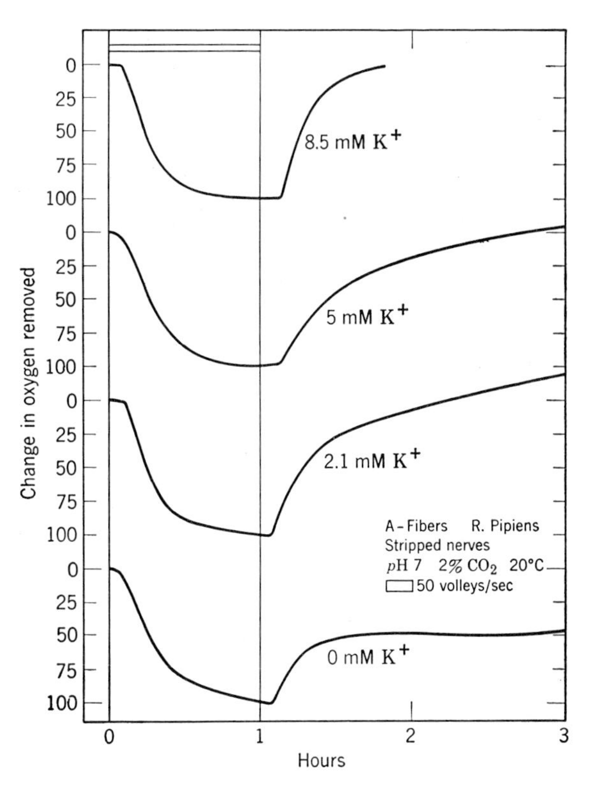


FIG. 3. Relative time courses of the increase in oxygen uptake by stripped nerve bathed in Ringer's solutions containing different concentrations of $K^+$. Ordinate as described for Fig. 1.

A possible key to the interpretation of these observations is the finding of Hodgkin and Keynes[9] (see also Hodgkin's Croonian Lecture[10] for an extended discussion of the movements of ions in giant nerve fibers) that the efflux of sodium from *Sepia* fibers is increased in a solution containing more potassium than does sea water and is depressed by K-free solution to one-third or one-

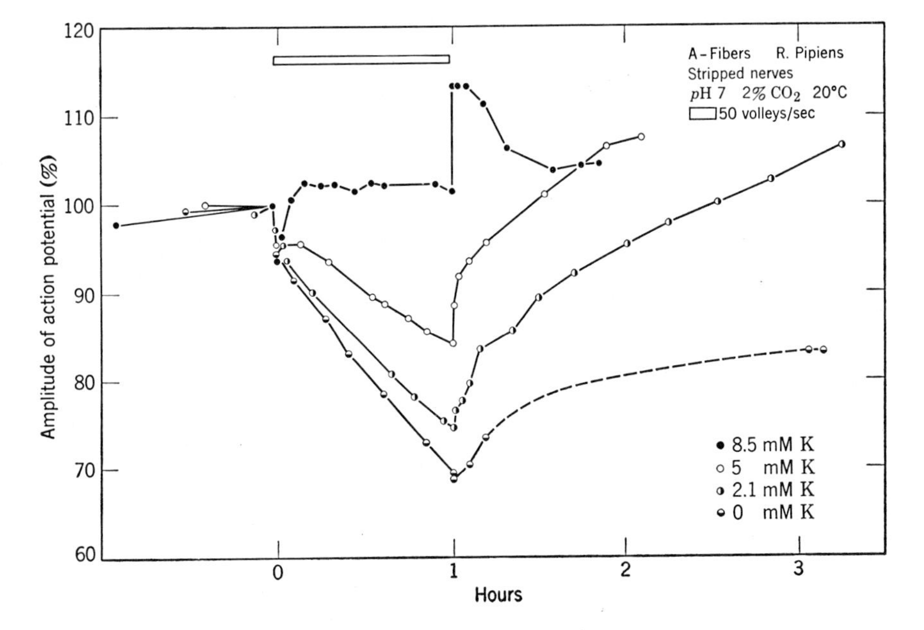


FIG. 4. Time courses of changes in amplitudes of compound action potentials of the four nerves of Fig. 3.

quarter of the efflux into sea water. If, in frog nerve, the rate of sodium extrusion is similarly dependent upon the level of potassium in the external solution, the slow component of recovery respiration may be interpreted as reflecting the energy demand of the transport machinery. The fast component, on the other hand, might result either from cessation at the end of tetanus of energy demand by the excitability cycle, or from a rapid reduction in rate of ion transport, at the end of tetanus, because of relaxation by diffusion of the internal ionic gradients maintained near the nerve surface by the entrance of sodium and by the exit of potassium during each impulse. The following considerations suggest that additional factors may affect the kinetics of the fast component.

The principal locus of ionic flow during an impulse in a myelinated fiber appears to be across the membrane at the node.[11,12] If ionic recovery takes place principally across the nodal membrane, the question arises as to whether or not phosphate compounds participating in this recovery are rephosphorylated by mitochondria close to the node or by those farther down the internodal space. Thus, longitudinal diffusion of phosphate acceptors in the internode may determine in part the kinetics of changes in rate of oxygen uptake. These considerations apply, as well, to the kinetics of the rising phase of uptake that occurs during the tetanus. The fact that the time constant of this rising phase in intact nerve is as long as it is (5 to 8 min) and does not vary markedly with tetanus frequency[4] is consistent with the idea that the kinetics of this phase are restricted in part by configurational parameters.

The differences in the kinetics of the changes in oxygen uptake and in the action potentials of intact and stripped nerves in 2.1 mM $K^+$ Ringer's appear, then, to be due to restrictions imposed by the perineurium on the diffusion of ions between the extracellular space and the bathing solution. The differences arise from the secondary effects of a changing ionic composition around the fibers of the intact nerve, especially as regards the concentration of potassium.

A most important and revealing experiment is that of measuring the respiration of nerve bathed by a solution in which lithium has replaced sodium ion. It is known that lithium can substitute for sodium in the conduction process. As soon as lithium Ringer's is substituted for sodium Ringer's, the resting respiration begins a slow decline. Tetanization produces almost no increase in rate; there may result a slowing or interruption in the decline of the resting respiration, but over all the effect of activity during exposure to lithium Ringer's is certainly not more than 5% of the effect measured in sodium Ringer's. The observation suggests strongly that lithium ions are extruded from frog nerve very slowly or not at all. Swan and Keynes[13] have reported that lithium ions are extruded from frog muscle much less rapidly than are sodium ions. Another conclusion from this experiment is that energy demand originating in the excitability cycle is less than 5% of the total. Therefore, it appears unlikely that the fast recovery components of the three lower curves of Fig. 3 result from the cessation of such a demand.

## ION TRANSPORT AND ENERGY REQUIREMENTS

In Table I are shown data from the work of Hurlbut and Asano on water distribution, ion distribution and fluxes, and the net ion changes which result from activity or exposure to an altered environment. The water distribution refers to intact nerve. The assignment of water and ions is based upon the assumption that the nerve consists of two compartments, intracellular and extracellular, the latter occupied by bathing solution. The internal concentrations of ions are based upon measurements in which the extracellular space had been washed free of sodium and potassium ions and include a small correction for sodium lost during this washing.[14] The fluxes of sodium and of potassium were measured as outfluxes from resting nerve which had previously been equilibrated with radioactive ions.

At the bottom of Table I the sodium gained and the potassium lost during activity and during exposure to oxygen-free and K-free environments are compared. It is apparent that in each case the sodium gain is nearly equal to the potassium loss. Asano and Hurlbut[15] have shown that ionic recovery does take place in the three or four hours following activity (50 volleys/sec, 1 hr, in 2 mM $K^+$ Ringer's). Figure 5 indicates that recovery from rather large ionic shifts takes place in such a way that the potassium movement in one direction balances

TABLE I. Frog nerve.

| | |
|---|---|
| Extracellular water content | 45% wet weight[a] |
| Intracellular water content | 29% wet weight[a] |
| Dry weight | 26% wet weight[a] |

| | Concentration, Intracellular (mmole / kg intracell water) | Concentration, Extracellular (mM) | Flux (mmole / kg intracell water, hr) |
|---|---|---|---|
| Sodium | 47[a] | 116 | 23[b] |
| Potassium | 159[a] | 2 | 10–28[b] |

| Net changes resulting from | Activity 50 volleys/sec 1 hr | Asphyxia 5 hr | K-free Ringer's 5 hr |
|---|---|---|---|
| Sodium gain (mmole / kg intracell water) | 18[c] | 35[a] | 31[b] |
| Potassium loss (mmole / kg intracell water) | 21[c] | 41[a] | 34[b] |

[a] Reference 14.
[b] W. P. Hurlbut and T. Asano (unpublished observations).
[c] Reference 15.

the sodium movement in the other. The rates of ionic recovery during the first hour or two (Fig. 5) produce concentration changes of about 12 mM (intracellular) per hour. This net ionic exchange that is measured during recovery is presumably superimposed upon the normal fluxes, measured in resting nerve, of about 23 mM (intracellular) per hour for sodium and probably about the same for potassium.

For a two-compartment system, the thermodynamic expression for the energy required to move one mole of sodium from inside ($i$) to outside ($o$) and one mole of potassium in the opposite direction is

$$RT\left(\ln\frac{Na_o}{Na_i}+\ln\frac{K_i}{K_o}\right),$$

where $Na_o$, . . . etc., are ion concentrations (strictly activities). This expression has a value of about 3000 cal/mole for the concentrations of sodium and potassium given in Table I. If the energy available from the hydrolysis of one mole of ATP to ADP, under intracellular conditions, is 7000 to 12 000 cal, then energetically it should be possible for some 2 to 4 sodium ions to be extruded for each molecule of ADP produced. Assuming, conservatively, that one ADP is produced for each sodium ion transported, it may then be asked whether the observed oxygen consumption appears to be capable of providing sufficient energy, via phosphorylated intermediates, to transport ions at the rates observed. For frog nerve at rest, the oxygen uptake is about 1.5 $\mu$mole/g (wet)/hr and the sodium flux about 6.6 $\mu$mole/g (wet)/hr. Thus the ratio Na:O is $6.6/(2\times1.5)=2.2$. If, in the functioning cell, mitochondria maintain an average P:O ratio (phosphate acceptors phosphorylated to oxygen atoms reduced) of about 3, as isolated mitochondria do,[16] then it would seem that oxidation provides adequate energy for transporting one sodium ion per ATP broken down. If the Na:P ratio is actually unity, the comparison implies that only about 30% of the resting respiration is available for processes other than ion transport.

An estimate of the increase in rate of sodium extrusion during a period of activity can be made from other observations of Asano and Hurlbut.[15] They found that after one hour of activity at 50 volleys/sec the net gain of sodium averaged 32 $\mu$mole/g (dry) when the nerve was bathed with K-free solution and only 16 when the Ringer's contained 5 mM $K^+$. This is an equivalent difference in rate of 4.2 $\mu$mole/g (wet)/hr. If, in the two solutions, about the same amount of sodium entered the nerve during each impulse, the observed difference may be attributed to effective, restorative transport taking place during activity. The ratio of this difference in rate to the increase in respiration during activity in 5 mM $K^+$ is approximately $Na:O=4.2/(2\times.8)=2.6$. This figure suggests that, in the case of active nerve also, there may be little energy utilized for purposes other than ion transport.

It should be emphasized that only one plausible line of thought has been followed in making these estimates and comparisons and that assumptions without direct experimental support have been invoked. Alternative lines of reasoning may ultimately prove to be more acceptable than the one outlined here.

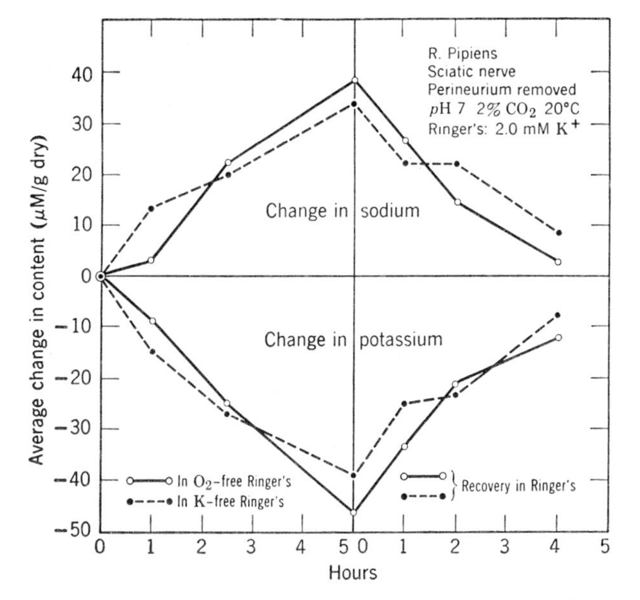


Fig. 5. Time courses of the changes in ionic contents of stripped nerves during and after exposure to oxygen free or potassium-free Ringer's solution. (Partially unpublished observations of W. P. Hurlbut and T. Asano.) Ordinate scale symbol $\mu$M means micromole.

### POST-TETANIC HYPERPOLARIZATION

Experiments which indicate that there are measurable electrical changes associated with the ionic transport events of recovery[17] are now described.

A stripped frog nerve is mounted in a plastic chamber so as to pass through five compartments (separated by grease-seals) containing oxygenated solutions. The first and fifth compartments are connected via calomel half-cells to a stable dc amplifier (chopper amplifier, time constant about 1 sec). The first compartment usually, and the second, fourth, and fifth compartments always contain a 5 mM $K^+$ Ringer's, in the experiments to be described. The nerve is stimulated before it enters the first compartment and the impulses are blocked in the third which contains a choline chloride (sodium-free) Ringer's. By this arrangement, it is possible to follow the changes in potential developed by an active or recovering region of nerve, the steady potential of an inactive region serving as a reference.

Figure 6 shows the potential changes observed during and following a 25-min period of activity at 50 volleys/sec. The downward deflection at the beginning of the tetanus is the time average of the negative action potentials. Once each minute the tetanus was interrupted for 5 sec and the recorded potential showed a positive deflection which reached its peak within this

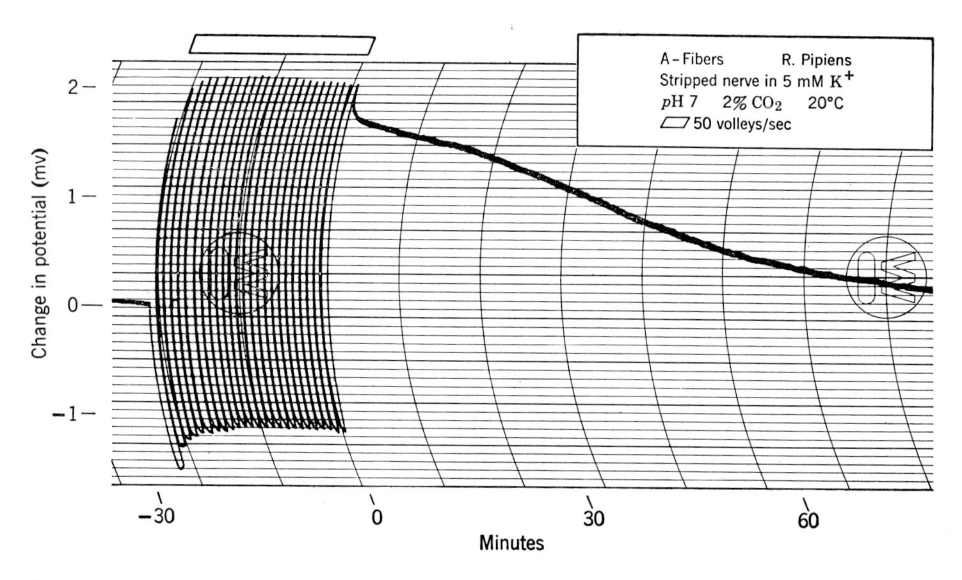


FIG. 6. Changes in average membrane potential recorded during and after 25-min tetanus. Tetanus interrupted for 5 sec each minute [from C. M. Connelly, "Post-tetanic hyperpolarization in frog nerve," in *Proceedings of the National Biophysics Conference, 1957* (Yale University Press, New Haven, to be published)].

interval. The amplitude of this hyperpolarization (net increase in membrane potential) reached a maximum after about 4 min and thereafter changed little. After the end of the tetanus, the hyperpolarization lasted for more than an hour; its recovery time course cannot be described as exponential.

Figure 7 shows how the prolonged positive after-potential is affected by the level of $K^+$ in the bathing solution (in the first compartment). A stimulation of 25 min in a solution containing 5 mM $K^+$ was first carried out as a control, and after recovery the solution was changed either to 8.5 mM $K^+$ (upper) or to K free (lower). The potential changes recorded during tetanus under the modified conditions have been superimposed on the control observations. The amplitude of the hyperpolarization varies appreciably with the level of potassium; recovery is rapid in high potassium and appears to be very slow and prolonged in K-free solution. During recovery from activity, the kinetics of oxidative recovery and the kinetics of the positive after-potential are both affected in much the same way.

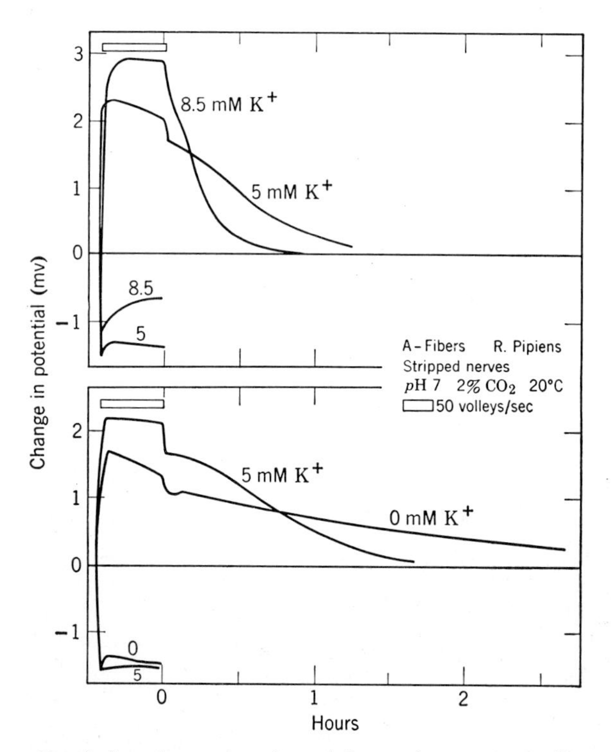


FIG. 7. Superimposed tracings of changes in average membrane potential recorded during and after 25-min tetani in Ringer's solutions containing different levels of $K^+$. Traces shown during period of tetanus are envelopes of changes similar to those shown in Fig. 6. In each of the two experiments the first tetanus was in Ringer's with 5 mM $K^+$, and the second tetanus carried out after 30 to 40 min exposure to the second solution. Zeroes of potential adjusted to superimpose at beginning of tetanus.

Other characteristics of this after-potential have been examined. It has been found that the observed amplitude of hyperpolarization approaches a maximum, or "saturation level" as the frequency of tetanus is increased above about 25/sec.[17] This maximum is about 2 mv. Depolarization produced by introducing isotonic potassium chloride into the first compartment averages 10 mv. Thus, maximum hyperpolarization corresponds to an increase in membrane potential of about 20%.

Post-tetanic recovery may be rapid or long-lasting depending on duration and frequency of tetanus. At frequencies above about 25/sec the effect of duration is striking, as illustrated in Fig. 8 which shows, superimposed, the recoveries from four tetani of different durations, at 50 volleys/sec. The longer the tetanus, the longer recovery takes, as if continued activity resulted in an accumulation of something, the final dissipation of which has associated with it an emf. Areas under the total hyperpolarization-time curves (i.e., area above the zero-potential line, including the periods of tetanus and recovery) have been measured as a function of duration of tetanus. Within experimental limits, area varies linearly with duration suggesting that the magnitude of hyperpolarization is approximately a linear measure of the rate of a recovery process.

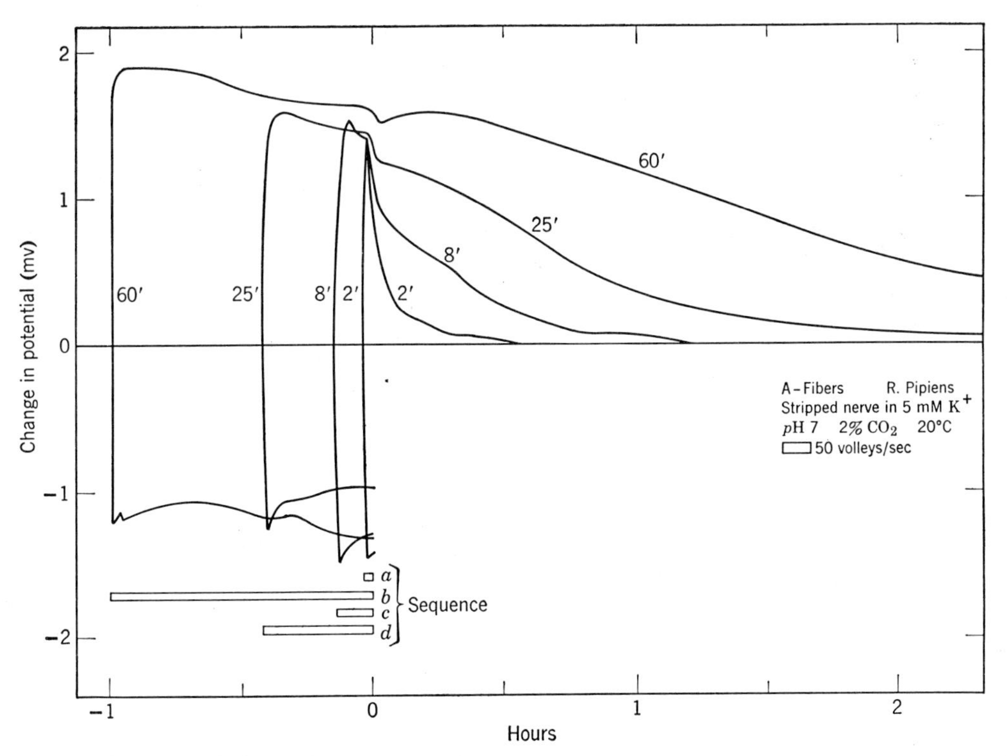


FIG. 8. Superimposed tracings of changes in average membrane potential recorded during and after a series of tetani of different durations. Traces shown during period of tetanus are envelopes of changes similar to those shown in Fig. 6 [from C. M. Connelly, "Post-tetanic hyperpolarization in frog nerve," in *Proceedings of the National Biophysics Conference, 1957* (Yale University Press, New Haven, to be published)].

A variety of observations appears to be consistent with the idea that the recovery process involved is the outward transport or extrusion of sodium ions. One experiment which has a direct bearing on the question is illustrated in Fig. 9. The positive afterpotential associated with a tetanus is effectively eliminated upon complete substitution of lithium for sodium in the bathing Ringer's. The parallelism between this result and that described earlier on the effect of lithium Ringer's on activity respiration furnishes strong support for the idea that the activity respiration and the positive after-potential both have their origin in the process involving the outward transport of sodium coupled to the inward movement of potassium.

One further point should be discussed. Ritchie and Straub,[18] in studying the positive afterpotential of

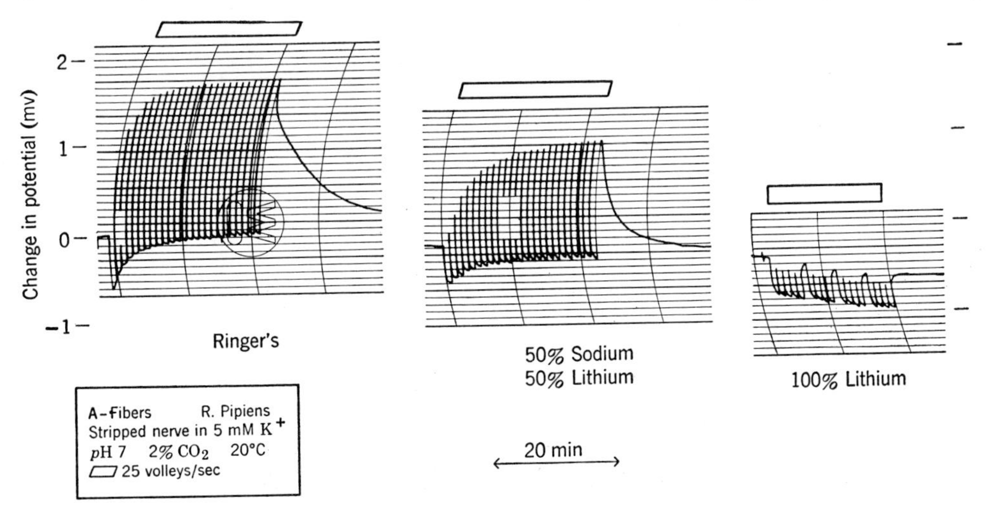


FIG. 9. Changes in average membrane potential recorded during and following tetani in solutions having different proportions of sodium and lithium ions. Potential zero common to the three records. During intervals (about 30 min) between records, nerve was bathed in the next solution. First and second records, 25-min tetani interrupted for 5 sec each minute. Third record, 20-min tetanus with 13 interruptions of 5 sec and 3 interruptions of 1 min each [from C. M. Connelly, "Post-tetanic hyperpolarization in frog nerve," in *Proceedings of the National Biophysics Conference, 1957* (Yale University Press, New Haven, to be published)].

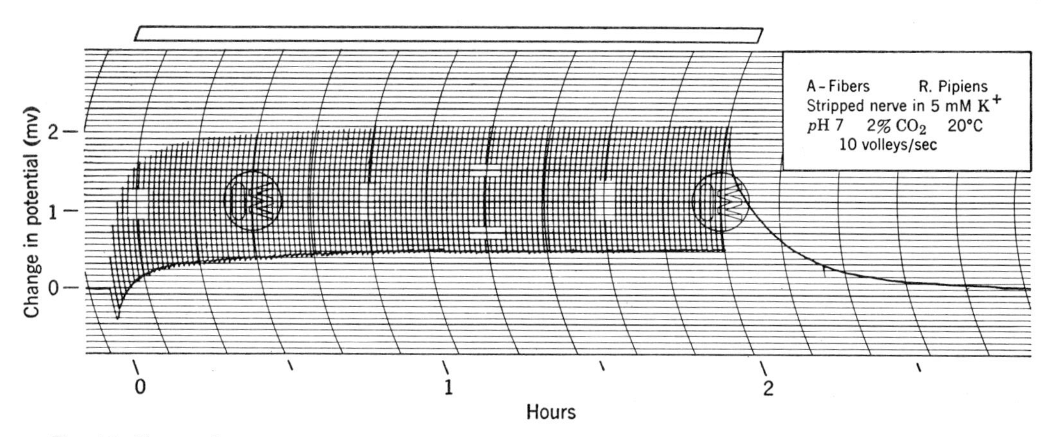


FIG. 10. Changes in average membrane potential recorded during and after 2-hr tetanus at 10 volleys/sec. Interrupted 5 sec each minute.

mammalian C-fibers, came to the conclusion that their observations could be explained solely by an electroneutral pump operating to restore the normal sodium-potassium balance following activity. The after-positivity was described as a variation of membrane potential in response to the variation in the concentration of potassium in the fluid immediately outside the nerve membrane. This happens as follows: during recovery, the pump produces a net flow of potassium inward across the membrane. By this action, the concentration of potassium at the outside surface of the fiber is lowered below its value in the body of the bathing solution. Since it is known that a decrease in external concentration of potassium does produce an increase in membrane potential, the conclusion appears to be unassailable, qualitatively. One must agree that any ion pumping system, whether electroneutral or inherently electrogenic, must have this property. It is to be taken

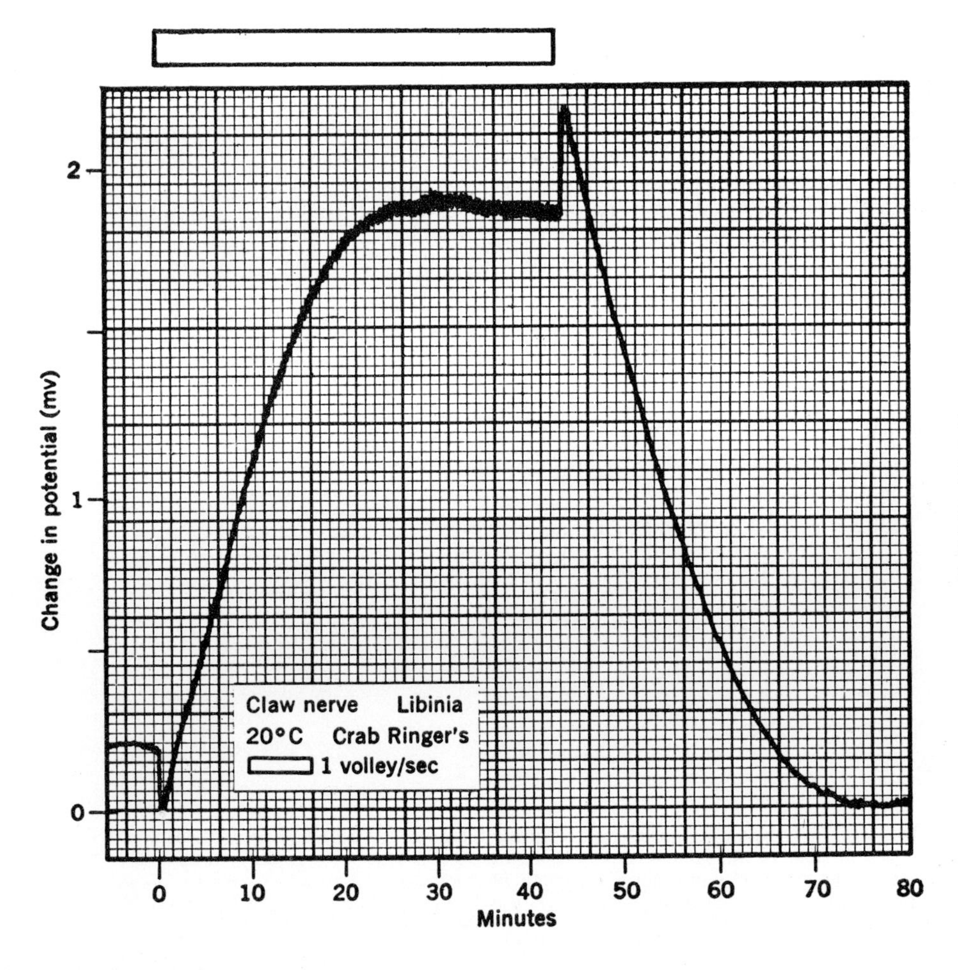


FIG. 11. Changes in average membrane potential of spider-crab nerve recorded during and following 43-min tetanus at 1 volley/sec. Not interrupted during tetanus. Measured depolarization of spider-crab nerve in isotonic potassium chloride about 30 mv.

for granted that the transport system in frog nerve produces some potential by this mechanism. On the other hand, Fig. 10 illustrates an experiment in which the Ritchie and Straub mechanism is not sufficient to explain the observed variations of potential. During a tetanus, the average membrane potential (exclusive of the action potentials) should be either negative or zero, according to the Ritchie and Straub scheme. The average should be negative during periods in which there is a net loss of potassium from the fibers and would approach zero if the system approached a steady state (i.e., $K^+$ pumped back inside during pulse interval $\equiv K^+$ lost/impulse). The lower envelope of the record in Fig. 10 shows that the membrane became hyperpolarized four minutes after the beginning of the tetanus and remained so for the remainder of it. With correction for the apparent depolarization introduced by averaging the action potentials, the statement could be made that hyperpolarization began at the beginning of the tetanus and reached a final average level (in the 100-msec interval between pulses) of almost one millivolt. Thus, the ionic transport system of frog nerve appears to be inherently electrogenic (positive outward).

A similar statement, based on similar evidence applies to unmyelinated limb nerve of the spider crab, shown in Fig. 11. Here the membrane also develops a net hyperpolarization during low-frequency tetani.

If one takes as a working hypothesis the proposition that the magnitude of hyperpolarization is proportional to the rate of sodium extrusion, how can the shapes of the recovery curves of Fig. 8 be explained? The kinetic behavior of a simple reaction mechanism showing saturation kinetics is portrayed in Fig. 12. It is the classical enzyme-substrate reaction of Michaelis, the rate of which is given by the second expression in the figure, where $K_M$ is the concentration of substrate that produces half-maximal rate. The curve describes the rate of reaction as a function of time as an initially large concentration of substrate decreases. If nerve were to extrude the sodium remaining within it with the kinetics of this mechanism, the hyperpolarization should follow this curve exactly, starting at a point appropriate to the

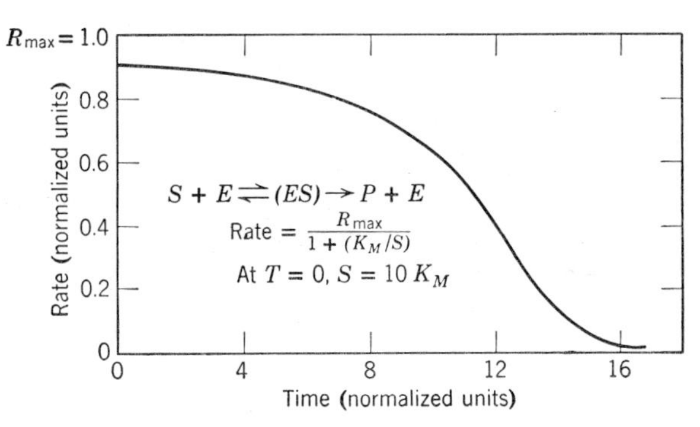


FIG. 12. Calculated time course of rate of disappearance of substrate with Michaelis type of enzyme-substrate reaction mechanism. $K_M$ is concentration of substrate producing half-maximal rate [from C. M. Connelly, "Post-tetanic hyperpolarization in frog nerve," in *Proceedings of the National Biophysics Conference, 1957* (Yale University Press, New Haven, to be published)].

amount of sodium accumulated during the tetanus. But, the recovery curves of Fig. 8 cannot be superimposed by displacement along the time axis, and, therefore, do not conform to this model.

If sodium ions enter the active fiber at the nodes and are also extruded principally from the nodes, then ions not extruded immediately after entry must tend to diffuse into the myelin-insulated internodal space. To determine the possible effect of diffusion in the internode on the kinetics of ionic recovery, the analog model shown in Fig. 13 was constructed. The model corresponds to one-half a node and to the adjacent one-half internodal space. A pentode, whose plate current-plate voltage curve is approximately a hyperbolic saturation curve of the Michaelis type, simulates the ion-extruding mechanism at the node; a filter network of 10 *RC* units is the analog of the diffusion field in the internode. Positive charges correspond to sodium ions; potentials at various points in the plate and filter circuits to sodium concentrations; plate current to rate of sodium extrusion; and injection of a constant current to the plate circuit simulates a tetanus. Seconds in the model correspond to minutes in the nerve. Figure 14 shows superimposed records of the output of the analog circuit, with parameters chosen to duplicate as closely as possible the

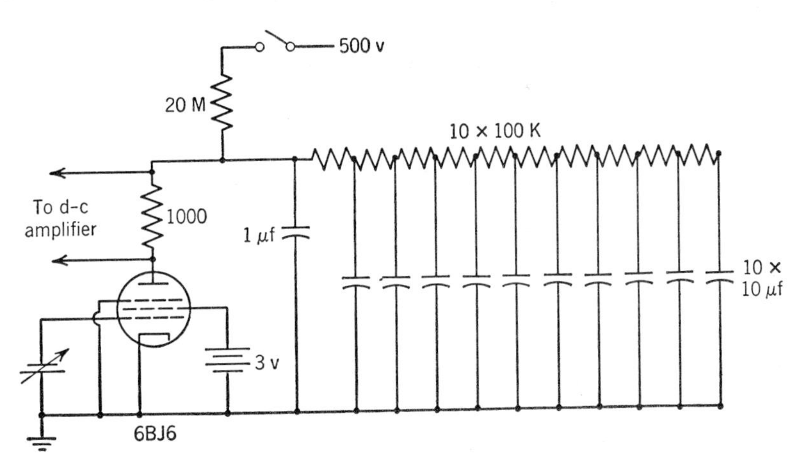


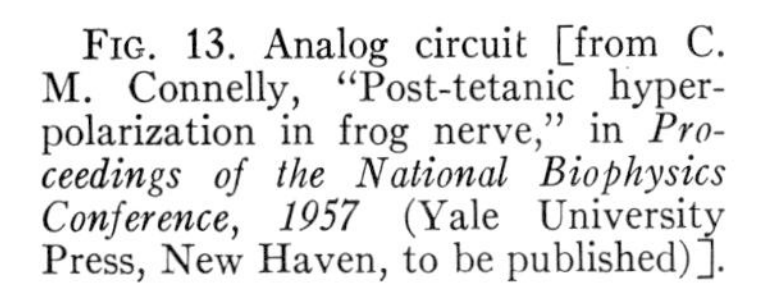
FIG. 13. Analog circuit [from C. M. Connelly, "Post-tetanic hyperpolarization in frog nerve," in *Proceedings of the National Biophysics Conference, 1957* (Yale University Press, New Haven, to be published)].

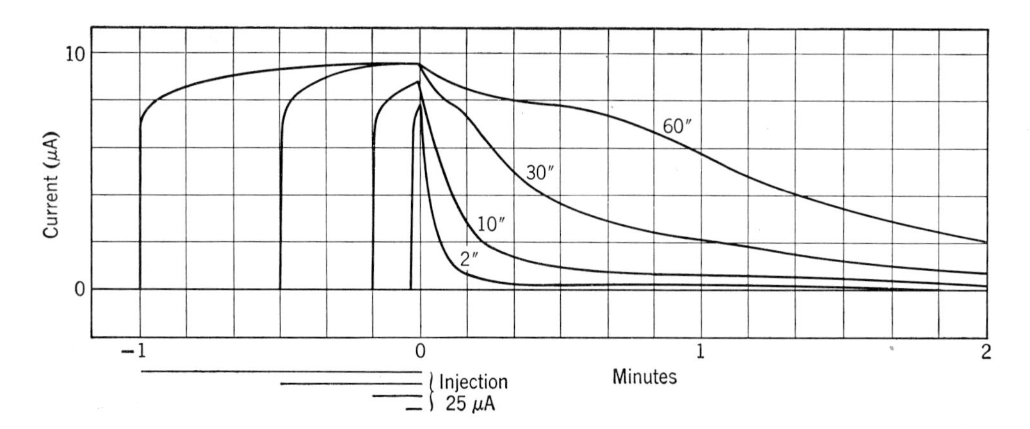


FIG. 14. Superimposed records of changes in plate current of analog circuit of Fig. 13, during and after 25-μA injections of different durations. Grid bias adjusted to give maximum plate current of 10 μA [from C. M. Connelly, "Post-tetanic hyperpolarization in frog nerve," in *Proceedings of the National Biophysics Conference, 1957* (Yale University Press, New Haven, to be published)].

observations of Fig. 8. The time constant of the network in this case is about twice as large as the corresponding time constant for the diffusion of sodium chloride in the internodal space (estimated assuming free aqueous diffusion and node-to-node spacing of 2 mm). It is encouraging that the shape of the recovery curve changes more or less in the proper manner as the duration of tetanus is increased. This oversimplified analysis tends to support the hypothesis that the positive after-potential has its origin in a saturable, ion extruding mechanism operating at the nodes.

## CONCLUDING REMARKS

The events of recovery may be described tentatively as follows: Sodium ions enter a fiber during activity. The increase in internal concentration accelerates a saturable ion-transport process in which sodium ions and high-energy phosphate compounds are obligatory participants. Sodium ions are liberated to the exterior, and potassium in the exterior medium participates in the over-all cycle, being transferred to the interior. The operation of the mechanism generates an emf, directed positively outward. Oxidative metabolism maintains the supply of high-energy phosphate compounds.

The observations outlined in this discussion have emphasized the coupling between oxidative metabolism as an energy source and ionic transport processes as an energy sink. The molecular mechanisms of active ionic transport across membranes are almost a complete mystery. Proposed mechanisms range from shuttling carriers to micro-pinocytosis to enzymatic modification of pore configuration or charge distribution. This is an outstanding problem in cellular and molecular biology today.

## BIBLIOGRAPHY

[1] F. Brink, in *Metabolism of the Nervous System* edited by D. Richter (Pergamon Press, New York, 1957), p. 187.
[2] P. C. Caldwell and R. D. Keynes, J. Physiol. **137**, 12*P* (1957).
[3] F. D. Carlson, F. Brink, and D. W. Bronk, Rev. Sci. Instr. **21**, 923 (1950).
[4] F. Brink, D. W. Bronk, F. D. Carlson, and C. M. Connelly, Cold Spring Harbor Symposia **17**, 53 (1952).
[5] C. M. Connelly, D. W. Bronk, and F. Brink, Rev. Sci. Instr. **24**, 683 (1953).
[6] T. P. Feng and A. V. Hill, Proc. Roy. Soc. (London) **B113**, 356 (1933).
[7] A. V. Hill, Proc. Roy. Soc. (London) **B113**, 345 (1933).
[8] B. C. Abbott, A. V. Hill, and J. V. Howarth, Proc. Roy. Soc. (London) **B148**, 149 (1958).
[9] A. L. Hodgkin and R. D. Keynes, J. Physiol. **128**, 28 (1955).
[10] A. L. Hodgkin, Proc. Roy. Soc. (London) **B148**, 1 (1957).
[11] I. Tasaki, *Nervous Transmission* (Charles C. Thomas, Springfield, Illinois, 1953).
[12] R. Stämpfli, Physiol. Rev. **34**, 101 (1954).
[13] R. C. Swan and R. D. Keynes, Abstr. Comm. 20th Int. Physiol. Congr. 869 (1956).
[14] W. P. Hurlbut, J. Gen. Physiol. **41**, 959 (1958).
[15] T. Asano and W. P. Hurlbut, J. Gen. Physiol. **41**, 1187 (1958).
[16] B. Chance and G. R. Williams, Advances in Enzymol. **17**, 65 (1956).
[17] C. M. Connelly, "Post-tetanic hyperpolarization in frog nerve," Proc. Natl. Biophys. Conf. (1957) (to be published).
[18] J. M. Ritchie and R. W. Straub, J. Physiol. **136**, 80 (1957).

# 52
# Sensory Performance of Organisms*

Walter A. Rosenblith

*Department of Electrical Engineering and Research Laboratory of Electronics, Massachusetts Institute of Technology, Cambridge 39, Massachusetts*

THIS paper is concerned with those aspects of organized neural complexity that permit organisms, such as man, to deal successfully with their sensory environment. Examples presented below illustrate how man, as an organism, detects, orders, and identifies events in his environment to which his sense organs are sensitive. The roles that stimulus intensity and time play in these operations are emphasized, as opposed to those aspects of sensory quality (pitch and hue, for example) for which one might expect the underlying mechanisms to differ rather considerably in going from one sensory modality to another.

Studies of communication processes in the nervous system should be based on a realistic view of the way in which the total organism reacts when it reaches selectively into its surroundings—be it under self-instruction or under instructions from others—to process stimuli that have informational value. This emphasis upon the organism's behavior in communication tasks leads to a preoccupation with certain dependent and independent variables. It leads to an inquiry into the "operating characteristics" of the organism in order to be able to specify, albeit in a statistical manner, its maximum sensitivity, its resolving power, its dynamic range, its characteristics in the frequency and time domains, its information-handling capacity, and so forth.

The data presented in this paper come mainly from *contemporary* psychophysics. Subjects are not asked to introspect, but, rather, to indicate by standardized motor responses (most often verbal responses) whether they judge a stimulus to be present or not, whether they judge two stimuli to differ, whether they can order a set of stimuli or identify members of the set.

Assume that one is dealing with well-instructed cooperative subjects who are skilled in the execution of the expected response and who are familiar with the set of stimuli that will be presented to them.†

Assume also that there is agreement on a program of stimulus presentation and on a method of response analysis. Thus, one must decide whether the admissable response categories shall be simply: "yes" and "no," "same" and "different," "more" or "less," or whether they should include the set of natural numbers. There must also be rules for dealing with false responses (false-alarm rate), and, finally, one must decide whether or not to quantify the temporal aspects of responding in addition to recording the mere emission of responses.

Such considerations may seem unnecessary details that should be left to methodologists, but unless there is a realistic understanding of the measurement and quantification problems in fields like psychophysics and neurophysiology, one can hardly evaluate the store of knowledge they have produced, or understand the relation of such knowledge to the knowledge that exists in the several areas of biophysics.

## INFLUENCE OF INSTRUCTIONS UPON THRESHOLD OF DETECTION

The task of detecting the presence of a stimulus includes the classical absolute threshold as well as the generalized "masked" threshold—i.e., detection of the stimulus against a background of sensory stimulation. Such "background noise" refers necessarily to the sense modality under study. Even the best isolation booths do not reduce to zero all unwanted sensory influx other than that under the experimenter's control.

It has long been known that instructions significantly influence both the shape and the anchor points of "frequency-of-seeing" or "frequency-of-hearing" curves. A recent study by Smith and Wilson[2] has forcefully illustrated this dependence upon the judgmental criteria that are suggested to observers. Ten observers were asked to report the presence of a tone in noise. The three curves of Fig. 1 show how the identical group of observers performed this task under three different sets of instructions. Examples of the instructions that were designed to induce a "conservative" or a "liberal" attitude in the reporting of signals follow:

Conservative groups:

"Keep in mind that it is important for you to be sure you hear the tone. In many cases, there will be no tone, and you will be making a mistake if you indicate that you hear one. None of the tones is really easy to hear, but don't push the switch unless you are sure that you heard the tone. If you are in real doubt as to whether or not you heard a tone, assume that there was none."

* This work was supported in part by the U. S. Army (Signal Corps), the U. S. Air Force (Office of Scientific Research, Air Research and Development Command), and the U. S. Navy (Office of Naval Research).

† Experiments parallel to those considered here could be conducted with monkeys or even with rats or pigeons, if appropriate behavioral techniques were used. However, few experimenters[1] have been willing to train their animals thoroughly enough, or to reformulate the required response in a way that was compatible with the animal's behavioral repertory. The short-cut of employing highly motivated and intelligent subjects permits experimenters to bypass a lengthy learning process.

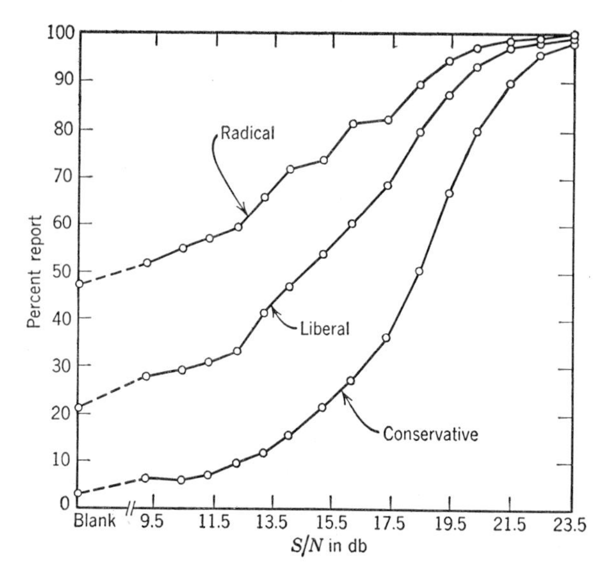


FIG. 1. Cumulative percentage of instances in which the presence of an 800-cps tone was reported as a function of signal/noise ratio. Attitude towards listening tasks is varied by means of instructions (see text). Each of 10 observers made approximately 150 judgments per point.[2]

Liberal groups:

"Keep in mind that all of these tones are hard to hear, and that you will rarely be absolutely sure that you heard something. But if you think you heard something, probably you did—report it. If you are very sure that you didn't hear anything—then don't touch the switch."

The results illustrate clearly that one cannot hope to predict accurately from a knowledge of signal-to-noise ratio alone the performance of a group of observers, even if the task calls for nothing more than the reporting of the presence of a single class of signals against the background of an unvarying noise.

Statistical models of the detection process have been formulated[3] in which performance is predicted not only on the basis of stimulus values and organismic variables; there are additional parameters that are sensitive to instructions and to amounts of reward or punishment. Such models may account for behavior such as is depicted in Fig. 1. A physiological interpretation of this sensitivity of the organism to nonstimulus variables remains to be given, though some physiological phenomena that provide grist for speculation have been reported in recent years.

When tasks that involve some ordering of stimuli are assigned to organisms, three problems are encountered. These stimulus-ordering operations may be listed under the headings of (a) differential sensitivity, (b) identification, and (c) psychophysical scaling.

## DIFFERENTIAL SENSITIVITY

Such labels as just-noticeable difference (jnd), differential thresholds, and difference limens have been used to describe the changes in stimulus characteristics that make it possible for an observer to distinguish, with various degrees of certainty, between two stimuli. All of these measures of differential sensitivity are statistical in character and the jnd increases in size if a listener is required to identify correctly the more intense of two sounds in 90% of the trials than if he is required to be correct in 75% of the trials only.

More than a century ago, Weber formulated, on the basis of his own experiments, a generalization according to which the jnd between two stimuli gets larger as the stimuli get larger. This assertion has been called Weber's law, and the expression‡

$$\Delta I/I=k \tag{1}$$

has been written to express this relative constancy of human differential sensitivity. Here, $I$ is the intensity of the reference stimulus, $\Delta I$ represents the difference along the intensity dimension that is discriminable (in 75% of the trials, for example), and $k$ is a constant. Weber's law states, therefore, that the jnd will remain proportional to the intensity of the reference stimulus. Today, there exist Weber functions (see Fig. 2) that cover a wide enough dynamic range to provide a fair test of the validity of Weber's law. The data indicate that, at least for auditory and visual stimuli, this relation breaks down as the absolute threshold is approached. Both frequency and intensity discrimination worsen as the stimulus becomes weaker. The existence of the so-called achromatic and atonal intervals[7] is further evidence along these lines. A better fit to the data of Fig. 2 is then provided by an equation of the form.

$$\Delta I=k(I+N_i); \tag{2}$$

$N_i$ has such a value that for low values of $I$, $\Delta I$ depends primarily upon $N_i$, while for medium and high values of $I$, $N_i$ becomes negligible.

The introduction of $N_i$ can be interpreted in different ways. Years ago, both Fechner and Helmholtz suggested that Weber's law be modified to acknowledge the presence of intrinsic interfering stimulation that could not be eliminated. In contemporary jargon, one might say that $N_i$ corresponds to the "internal noise level" of the organism. This expression should, however, not be taken too literally. Several theoretical models exist in which a generalized noise term is a crucial feature of the discriminating organism. Thus far, modifications of Weber's law have aimed at agreement with the data near the absolute threshold. There is some evidence that the presence of extremely intense sensory stimulation leads again to some deterioration of discrimination. It is easy to see that the model could be further modified

‡ The symbolic notation employed in writing this Weber fraction or Weber ratio seems to emphasize the intensitive aspects of a stimulus. Actually, what is involved is not necessarily some aspect of energy flux, but rather loose usage of the term "intensity." Thus, fractions have been determined for lifted weights, length of line, pressure on the skin, and so forth, and comparisons between sense modalities have been made on the basis of these numbers.

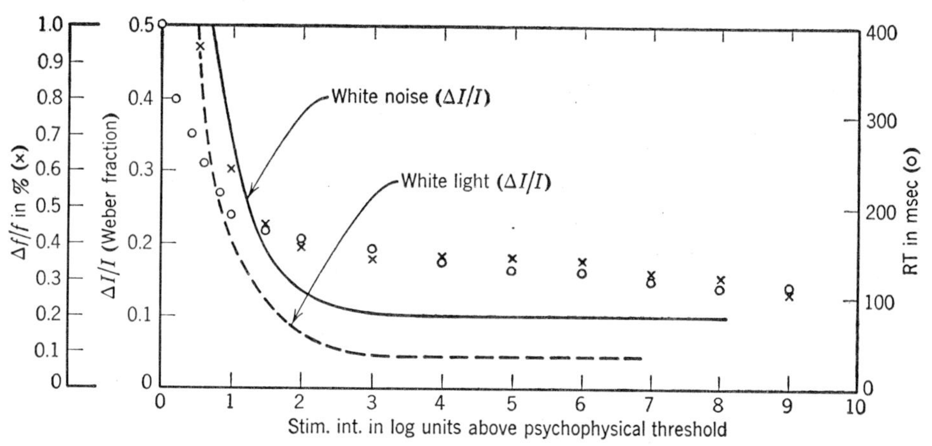


FIG. 2. Just-noticeable differences and reaction times as a function of stimulus intensity. Data on frequency discrimination of a 1000-cps tone after Shower and Biddulph[4]; on intensity discrimination for white light and white noise after Boring, Langfeld, and Weld[5]; data on reaction time after Chocholle.[6]

by the introduction of another term corresponding to an equivalent noise level that would only come into play for large values of $I$.

One further comment regarding the amount of invariance that it is realistic to expect in psychophysical data. Figure 3 summarizes the findings from more than half a dozen investigators, all of whom were trying to measure man's ability to discriminate pitch. In different experiments, different subjects, several psychophysical methods, several methods of stimulus presentation, and even several response criteria were employed. While it does not seem possible to read from this graph a value for the jnd for frequency, it is possible to point out that practically all of the data lie within the two dashed lines; i.e., their total spread is approximately a decade. The compilers of handbooks or of "Critical Tables on Sensory Performance" would thus be well advised to emphasize those aspects of psychophysical judgments that exhibit what might be called "relative invariance." Instead, they often insist upon quoting numerical values (sometimes with several significant figures), perhaps with the forlorn hope that the users of such tables will take these values with a grain of *ceteris paribus*.

## SPAN OF ABSOLUTE JUDGMENT

From data on just-noticeable differences, estimates have been made of the total number of distinguishable sounds, lights, or smells. Usually, the authors of these estimates have simply divided the range over which an organism is sensitive by the corresponding average jnd. Thus, the "auditory area"[9] has been estimated to represent several hundred thousand distinguishable sounds, and the number of distinguishable colors has been estimated to be of the same order.

Several unjustified assumptions underlie these extrapolations. Subjects may find it quite easy to detect a difference between two stimuli without being able to identify either of them. In everyday situations, an individual rarely has difficulties in distinguishing between two other individuals even though he cannot tell which one is Jones and which one is Smith. Absolute identification requires much more information than differential discrimination.

Under the influence of the mathematical theory of communication,[10,11] with its concept of channel capacity, many experiments have been carried out to determine man's capacities to process selective information. By considering man as a communication channel with stimuli for inputs and responses for outputs, it has been possible to estimate maximal rates of transmission through him. The amount of input information can be varied either by varying the amount of information per stimulus or by varying the rate at which stimuli are presented. As Quastler[12] has pointed out, the maximal rate of transmission is limited by two factors: (a) people cannot emit more than 5 to 9 responses per second, and (b) people get confused when they have to discriminate

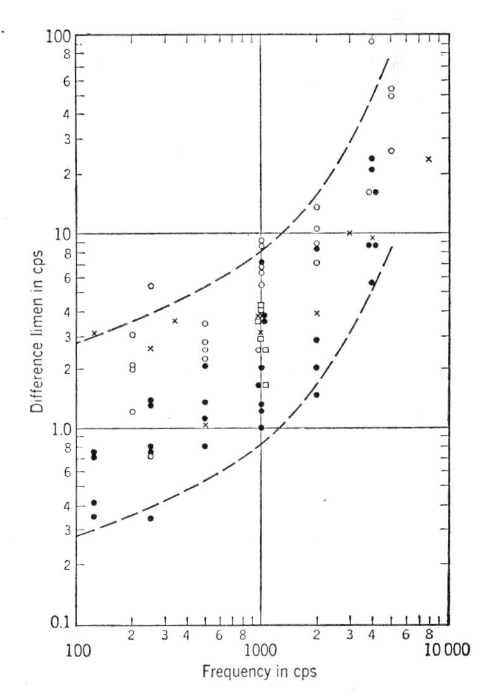


FIG. 3. Data on pitch discrimination reported by several investigators. Different symbols identify different experimental procedures (see Rosenblith and Stevens[8]).

rapidly among too many alternatives. Thus, even under optimal conditions, transmission rates of more than 25 bits per second are hardly ever observed. The type of motor response required and the degree to which stimuli and responses are "compatible"[13] determine the particular figure that will be obtained for the transmission rate.

A different aspect of man's information handling capacity comes into focus if subjects are no longer required to respond as quickly as possible. They now need simply to tell which of several alternative stimuli occurred on a particular stimulus presentation. Given knowledge of the probability of occurrence for the different stimuli and of the accuracy with which the subjects perform, one can calculate the amount of information in these absolute judgments. This experimental paradigm has been much used in connection with "simple" stimuli, such as pure tones or monochromatic lights. The results from these experiments in several sense modalities have led to the emergence of what Miller[14] has called "the magical number 7 plus or minus 2." This number represents the number of alternative stimuli that subjects can accurately identify as long as only one aspect ("dimension") of the stimulus is varied. This magical number thus corresponds more or less to the upper and lower bounds for the "span of absolute judgment." More-precise estimates of performance figures will need to take account—in addition to individual differences—of such experimental conditions as the range of stimulus variation, the spacing of stimuli over a given range, and the number of available response categories. Subjects seem also to perform better if knowledge of results is fed back to them after each response.

Sensory communication in everyday situations is, however, not restricted to making unidimensional judgments. The sounds, shapes, tastes, and smells upon whose successful recognition one's sensory commerce with the environment depends are basically "multidimensional displays." Laboratory experiments with such displays indicate that the span of absolute judgments increases in size as an increasing number of stimulus aspects are varied. Pollack and Ficks[15] obtained 7.2 bits per judgment when they presented to their subjects sounds that varied along six different acoustic "dimensions." This performance corresponds to correct identification of approximately 150 different categories. Comparison of the Pollack-Ficks data (1.2 bits/judgment/dimension) with the characteristic figure for unidimensional judgments (2.3 bits/judgment) shows that, while each new dimension helps to transmit more information, these increments tend to diminish as the dimensionality of the display increases. But it is not known where the asymptote lies for this relation between bits per judgment and stimulus dimensionality. Thus, organisms seem much better in absorbing information from the environment by making several crude distinctions simultaneously, instead of by making extremely precise discriminations of a single aspect of a sensory stimulus. This finding will not surprise those who are familiar with the facts of speech communication where the precise spectral composition of speech sounds is rather unimportant in conveying information. Instead, man's ability to recognize speech sounds seems to depend upon the detection of the presence or absence of "distinctive features"[16] that are thus rather analogous to the dimensions of the acoustic display.

## PSYCHOPHYSICAL SCALING

Among all the attempts to establish orderly relations between verbal responses and sensory stimuli, Fechner's law is probably the best known. By the use of indirect methods in which he used the just-noticeable difference as his scale unit, Fechner concluded that psychological magnitude varied as the logarithm of stimulus magnitude. There has been much argument about the measurability of sensation, about the validity of Weber's law, about the legitimacy of the Fechnerian integration of Weber's law which Fechner used to derive his own law; and yet, by and large, Fechner has been correct in predicting that his psychophysical edifice would stand because his detractors would not be able to agree on how to tear it down. In recent years, S. S. Stevens,[17] whose concern with measurement and scaling had led him to investigate systematically many of the fundamental assumptions in psychophysics, has produced a series of painstaking and ingenious studies of how subjects order stimuli by *direct* methods. Stevens did not accept the assumption according to which the unit of resolving power§ is the natural unit for experiments in which observers are asked to order a stimulus continuum by assigning numbers to it and in which these numbers are to reflect the judged or subjective magnitude of the stimuli. Stevens[18] divides perceptual continua into two general classes: continua that have to do with *how much* are called prothetic; continua that deal with *what kind* and *where* (position) are called metathetic. Well-known examples of prothetic continua are loudness and brightness; pitch is an example of a metathetic continuum. On the basis of experiments involving more than a dozen prothetic continua, Stevens concludes: ". . . there is a general psychophysical law relating subjective magnitude to stimulus magnitude . . . this law is simply that equal stimulus ratios produce equal subjective ratios. On numerous perceptual continua, direct assessments of subjective magnitude seem to bear an orderly relation to the magnitude of the stimulus. To a fair first-order approximation the ratio scales constructed by "direct" methods (as opposed

§ This indirect derivation of the scale of psychological magnitude parallels in some sense the extrapolated number of distinguishable sounds or colors. In both instances, the just-noticeable difference is the key concept of quasi theoretical deductions without regard for operational compatibility.

to the indirect procedures of Fechner) are related to the stimulus by a power function of one degree or another."[18]

Figures 4 and 5[19,20] present two samples of the numerous results obtained. In Fig. 4 there are plotted the magnitude estimates of loudness made by listeners who had not been given standard sounds of fixed intensity for comparison purposes. The eight different intensities were presented irregularly, and the subject estimated the loudness by numbers of his own choosing. These numbers were multiplied by appropriate factors to make each subject's estimate at 80 db the same, namely, 10. The points represent the medians of these "normalized" judgments, and the vertical lines represent the interquartile range of the judgments. By plotting the estimated magnitudes on a logarithmic scale *versus* the logarithm of stimulus intensity, the slope of the straight line is the exponent of Stevens' power function

$$\Psi = kS^n. \tag{3}$$

Here, $\Psi$ is the psychological magnitude, $S$ is the stimulus magnitude, and $k$ and $n$ are constants. For the stimuli that have been studied, the approximate value of $n$ varies from 0.3 for loudness and brightness‖ to 4 to 5 for electric shocks. The exponent for vibratory stimulation applied to the finger has a value of approximately 1. It is interesting to speculate that these exponents stand in inverse order to the dynamic ranges for the relevant sense modalities. The dynamic range for hearing and vision spans approximately 12 logarithmic units, the range for tactile vibration spans perhaps 3 to 4 logarithmic units, and the range for electric shocks spans only about 1 logarithmic unit. Does the fact that

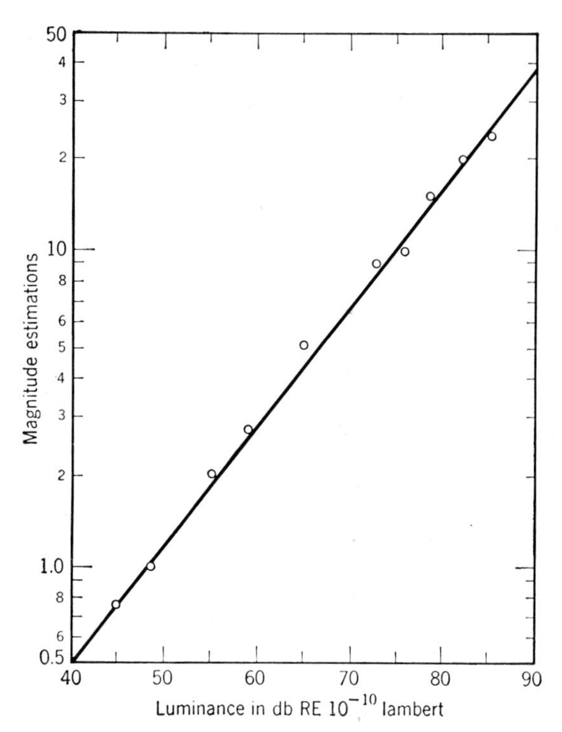


FIG. 5. Magnitude estimates of brightness of a luminous spot; the median estimates of ten observers are plotted.[20]

the product of the exponent multiplied by the dynamic range yields numbers of the same order of magnitude have any deeper significance? Does it reveal anything about the functioning of the sense organ, about the functioning of the nervous system, or does it rather reveal a certain constancy in verbal behavior—i.e., about the way in which people use numbers between the threshold of detection and the threshold of pain?

Another recent study by Stevens[21] deserves to be reported in this context. Once scales of subjective intensity have been established for several sensory continua, can these scales be used to predict what people will do when asked to match directly the loudness of a noise against the subjective intensity of a vibratory stimulus or an electric shock to the finger? Cross-modality matches were made between each modality and each of the other two. To a good approximation, the predicted functions were confirmed by direct experimentation, and the internal consistency of the direct approach to scaling has been further validated by this cross-modality study.

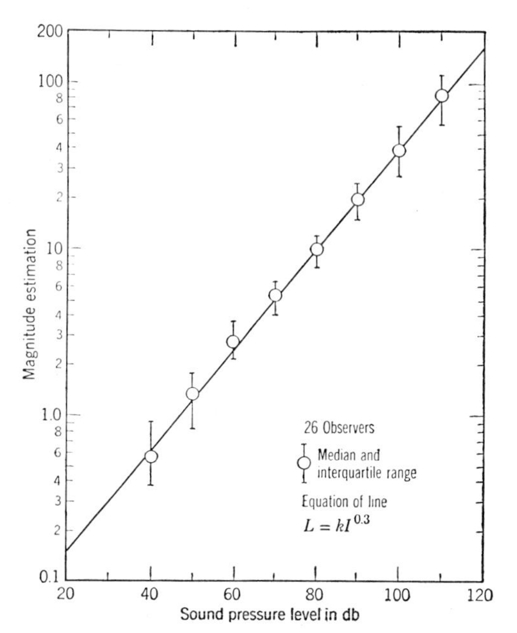


FIG. 4. Magnitude estimates of loudness made with no fixed standard.[19]

‖ The exponent is 0.6 if stimulus magnitude is expressed in terms of sound pressure instead of sound energy.

## TEMPORAL CHARACTERISTICS OF SENSORY PERFORMANCE

Since Helmholtz's time, experiments on reaction time have been used to make inferences on the processing of sensory signals in the nervous system and on the duration of so-called mental processes such as discrimination, recognition, and choice. It was established that reaction times to sounds, electric shock, and touch were faster than reaction times to visual stimuli; also, that reaction

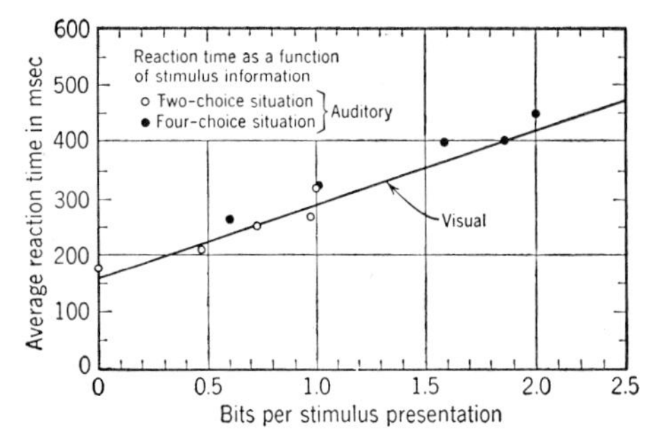


FIG. 6. The straight line represents the best fit to data obtained by Hyman[23] from one of his subjects in response to visual stimuli; the open and closed circles represent reaction-time data obtained by Albert[25] from one of his subjects in response to auditory stimuli; here, the stimulus set was composed of either two or four elements.

times decrease—at first rapidly and then more slowly—as the intensity of a sensory stimulus is raised above threshold (Fig. 2).

In addition to these data on simple reaction times, other experiments have attempted to assess the effect of task complexity on the speed of more discriminative reactions. It has been shown that reaction time increases as the stimulus ensemble increases in size and as its members become harder to distinguish from each other. This whole topic had lain more or less dormant for several decades until it became obvious that the application of certain concepts from information theory might make it possible to explore another aspect of man's capacity to handle information. A series of experiments[22–24] indicated that, at least over a certain range, choice time (or disjunctive reaction time) was proportional to the average amount of information transmitted per stimulus-response event. This finding applies equally well to visual as to auditory stimuli (see Fig. 6), and there is sufficient agreement between the data obtained by different experimenters that it is possible to talk about a characteristic figure for the "rate of gain of information" whose value lies near 6 bits per second. It is not possible to account for these time intervals in terms of delays on either the afferent or the motor sides of the nervous system. Enough is known about the latencies of evoked electric activity to permit one to say that these events consume only a fraction of the time involved in even the fastest reaction. The discrepancy becomes much greater, of course, upon considering sensory stimuli that are barely above the absolute or masked threshold or reaction times that involve complex discrimination. Thus, one must infer that a large fraction of the total reaction must be spent in "central information processing."

Recent experiments by Davis[26,27] and others throw some new light on these inferences. Davis presented to his observers two visual stimuli in succession and asked them to respond to each stimulus in turn. He then varied the interval between his two stimuli randomly between 50 and 500 msec. As long as the separation was more than 250 msec, the reaction time to either stimulus was approximately 200 msec; but when the time interval between the two stimuli fell below 250 msec, the second reaction time increased sharply. The second reaction time reaches its maximum value, approximately 400 msec, for an interval of 50 msec. The same phenomenon occurs when one stimulus is visual and the other is auditory, no matter what their order. Next, the subjects were instructed not to respond to the first of two lights. The results show little improvement over the delays that occurred when motor responses were made to both signals; the only difference was a slight shortening of the second reaction time, by about 20 to 30 msec. In a final experiment, the subjects were asked to make a "response" without a signal having occurred. Following the pressing of the key, a light came on (after an interval that was again randomly varied between 50 and 500 msec) to which the subjects were required to respond. This time there was no additional delay for any time interval. The subject was just as efficient in responding to a signal that came 50 msec after he pressed the first key as he was in dealing with a signal that came 500 msec afterwards. It would thus appear that the actual performance of the response does not contribute much to the delays that are observed, but that these delays are imputable to the processing of a signal that is still going on when the second signal is presented. There is further evidence that emphasizes the amount of time that central processes play in the handling of sensory information. In hearing and in vision, there is almost perfect integration of energy at the absolute threshold, as long as the stimulus is being presented for time intervals that are shorter than 200 msec. There is a whole class of phenomena in both hearing and vision in which a strong succeeding stimulus interferes with the "proper" analysis by the nervous system of a preceding weaker stimulus. The shorter the time interval between the two stimuli and the greater the discrepancy in intensity, the more severe is the interference. Since the delay times on the afferent side of the nervous system are longer for weak stimuli than for strong ones, the more intense stimulus has a chance to "catch up." However, the time intervals involved are sufficiently long that these differences in delay times along the ascending pathway are hardly the whole story. There is the further fact that interference with the more complex discrimination seems to be effective over longer time intervals than with more primitive tasks (such as simple detection of a preceding stimulus).

The preceding illustrations all point to the conclusion that the analysis of sensory information is a time-consuming task that occupies an organism's central nervous system in relation to the amount of information to be processed. This is hardly surprising; yet, it is often overlooked in practical situations.

The foregoing briefly reviews man's sensory performance as it appears when examined by the quantitative methods of contemporary psychophysics. Organisms are able to deal with the sensory bombardment that impinges upon them by being satisfied to make crude discriminations, unless specifically instructed to make precise ones (at the cost of spending more time making them and of neglecting other sensory events than those under analysis). The flexibility of organisms in switching from one sense modality to another, from one aspect of sensory display to another, is bought at the expense of relatively low rates for the transmission of information. The mere existence of organisms seems to entail a high, internal noise level that prevents responses to environmental stimuli from being anything but statistical in character. As long as thresholds of tolerance are not approached, the more energy a sensory stimulus delivers to an organism, the higher is the probability that the signal will be processed in the central nervous system with both discrimination and dispatch. One may paraphrase this situation by saying that the higher the energy level of a sensory signal, the greater the redundancy with which the representation of this signal will be handled in the central nervous system. These are some of the considerations that led us to select certain aspects of the electrical activity of the nervous system for quantitative study.

## BIBLIOGRAPHY


[1] D. Blough, J. Comp. and Physiol. Psychol. **49**, 425 (1956).

[2] M. H. Smith and E. Wilson, Psychol. Monogr. Gen. and Appl. **67**(9), 1 (1953).

[3] W. P. Tanner and T. G. Birdsall, J. Acoust. Soc. Am. **30**, 922 (1958).

[4] E. G. Shower and R. Biddulph, J. Acoust. Soc. Am. **3**, 275 (1931).

[5] E. G. Boring, H. S. Langfeld, and H. P. Weld, *Foundations of Psychology* (John Wiley and Sons, Inc., New York, 1948).

[6] R. Chocholle, Ann. oto-laryngol. **71**, 379 (1954).

[7] I. Pollack, J. Acoust. Soc. Am. **20**, 146 (1948).

[8] W. A. Rosenblith and K. N. Stevens, J. Acoust. Soc. Am. **25**, 980 (1953).

[9] S. S. Stevens and H. Davis, *Hearing: Its Psychology and Physiology* (John Wiley and Sons, Inc., New York, 1938).

[10] C. E. Shannon and W. Weaver, *The Mathematical Theory of Communication* (University of Illinois Press, Urbana, 1949).

[11] G. A. Miller, Am. Psychol. **8**, 3 (1953).

[12] H. Quastler, editor, *Information Theory in Psychology* (Glencoe Free Press, Urbana, Illinois, 1955).

[13] P. M. Fitts and R. L. Deininger, J. Exptl. Psychol. **48**, 483 (1954).

[14] G. A. Miller, Psychol. Rev. **63**, 81 (1956).

[15] I. Pollack and L. Ficks, J. Acoust. Soc. Am. **26**, 155 (1954).

[16] R. Jakobson and M. Halle, *Fundamentals of Language* (Moulton Company, The Hague, 1956).

[17] S. S. Stevens, Science **127**, 383 (1958).

[18] S. S. Stevens, Psychol. Rev. **64**, 153 (1957).

[19] S. S. Stevens, J. Acoust. Soc. Am. **27**, 815 (1955).

[20] S. S. Stevens and E. H. Galanter, J. Exptl. Psychol. **54**, 377 (1957).

[21] S. S. Stevens, "Cross-modality validation of subjective scales for loudness, vibration, and electric shock" (J. Exptl. Psychol., to be published).

[22] W. E. Hick, Quart. J. Exptl. Psychol. **4**, 11 (1952).

[23] R. Hyman, J. Exptl. Psychol. **45**, 188 (1953).

[24] E. R. F. W. Crossman and J. Szafran, Experientia Suppl. IV, 128 (1956).

[25] A. E. Albert, *Analysis of variance in reaction time experiments.* B. S. Thesis, Department of Mathematics, Massachusetts Institute of Technology, June, 1956 (unpublished).

[26] R. Davis, Quart. J. Exptl. Psychol. **8**, 24 (1956).

[27] R. Davis, Quart. J. Exptl. Psychol. **9**, 119 (1957).

# 53
# Biological Transducers and Coding*

Otto H. Schmitt

*Departments of Zoology and Physics (Biophysics), University of Minnesota, Minneapolis 14, Minnesota*

In this discussion of biological transducers and coding, no attempt is made to describe in detail the specific codes in which diverse animal (and plant) transducers operate, for this attractive topic in comparative physiology and biophysics is discussed at length in other sections of this book by Bullock, Hartline, Katz, and others. Nor is an attempt made here to correlate over-all organismic behavior with transducer functions for that is, to a large extent, the burden of the communication by Rosenblith. Instead, the niche between organismic behavior and detailed transducer codings is examined to determine whether or not it is possible to develop new and useful analytical tools by which one can better formalize the behavior of individual transducer elements and their interrelationships, so as to approach a deeper and more intuitive understanding of the resulting whole organism's stability, decision making, intelligent behavior, and homeostatic performance.

While it is impossible to develop the point of view adequately in a few pages of text, it is the firm conviction of the author that biophysical theory, where it has developed at all, has developed too closely in the footsteps of physical theory and has not yet evolved mathematical models and analytical techniques specifically tailored to fit biological phenomenology. Instead of adapting physical-science terminology and theoretical formulations to its needs, it has adopted them as though they were "true" rather than merely convenient approximate descriptions of experimental experience. This has been especially true in the studies of biological transductive processes where the biologist, armed only with experimental facts, has for the first time come up against the new and highly formalized jargon of information theory and automatic-control engineering. Through modesty or lack of mathematical dexterity, he usually tries to make-do with these mental tools as furnished him ready-made by physics and engineering. It is not the author's purpose to deny that one must start with available techniques, but rather to propose that one must quickly amend and supplement these with uniquely biophysical hypotheses and an appropriate calculus, if the problems of this field are to be reduced to a complexity where intuitive insight into them can readily be gained.

It is necessary, however, to set up some ground rules before launching into a discussion of biological transducers, their transfer functions and other possible means for studying their organization and coding. As there are no uniformly adopted definitions for transducers in biology and biophysics, an explanation is probably necessary for the term biological transducers. The usage of the author may be somewhat different from that of others in the biophysical field, and perhaps different even from that of other authors contributing to this volume.

What, then, is a biological transducer? The term transducer is borrowed from the engineering sciences, where it is usually defined so as to include those devices which "lead across" or transform energy from one form into another, and thus translate signals from one energy modality—such as sound—into another—such as a varying electrical current. While the physical-science definitions do not usually specifically exclude gross-energy converters such as steam boilers, storage batteries, and electric light bulbs from being considered as transducers, it is tacitly assumed that only devices which serve a communication or signal-converting function will be called transducers. Thus, a motor turning the rudder or shifting the ailerons of an airplane in response to instructions from the autopilot is clearly regarded as a transducer; one driving a washing machine is not. Similarly, a resistance-wire strain gauge is considered a transducer in its function of converting mechanical position changes into electrical-current variations, but its incidental function of converting electrical power into heat is not usually thought of as transduction.

Biological transducers, then, should logically be the biological counterparts of the physical sciences' transducers and should comprise all of those biological units and systems which perform transformations of energy modality within the context of a communication or informational transformation function. Thus, biological transducers should include not only the familiar sensory transducers such as the rods and cones of the retina, the thermal, olfactory, and gustatory receptors, the vestibular and cochlear receptors, and the many proprioceptive organs, but they should also include numerous synaptic and related intermediary units involved in biological-data processing and storage, and the output or motor transducers as well. Nor is there any reason to exclude humorally mediated transducers. The concept of a transducer fits the function of a cortical neuron or of the motor end plate at least as well as it does that of the muscle-spindle strain gauge.

It is convenient to use the terms afferent and efferent

* Much of the original work reported in this section was supported by grants H513 from the National Institutes of Health, Public Health Service, U. S. Department of Health, Education, and Welfare, by contract N6onr-24609 with the Office of Naval Research, U. S. Department of the Navy, and by grants from the Minnesota Heart Association and the University of Minnesota Graduate School.

in reference to biological transducers to distinguish incoming from outgoing transduction functions, and it seems desirable to adopt directly from engineering the terms "passive" and "active" transducers, to distinguish between those where the energy of an incoming signal provides directly the output energy and those transducers, very common in biological systems, where the incoming energy triggers or controls the release of locally supplied energy into an output modality.

Transducers that operate in one direction only are called unilateral transducers, while those that operate in either direction are called bilateral. The quite special transducers that not only operate in either direction, but also obey the reciprocity theorem, are called reciprocal transducers.

The linear transducer is one where the output can be related through linear-differential or algebraic equations to the input function. It is a type of transducer not often found in biological nature, but a form to which most experimental transducer results are approximated, as its behavior is easily formulated and understood mathematically. The linear transducer does deserve special mention, for it is out of the notion of a linear transducer that the transducer transfer-function idea arose.

The transfer function of a linear transducer is usually expressed simply as the ratio of its output to its input signal with frequency as a parameter. This ratio is usually a complex number which may be dimensionless in the case of a transduction between similar modalities, or may carry within it the ratio of the dimensions in which output and input are expressed. It is valuable to remember that, around any properly described, closed, biological control loop, transfer functions must cancel dimensionally. Thus, a mathematical description of a biological loop system can be tested much as elementary physics equations are tested dimensionally by cancellation of dimensions and by reduction to mass, length, and time units.

In studying biological transducers, it is often desirable to regard the transfer function as a much more general operator than merely a linear complex ratio, for many of the merits of transfer-function analysis remain even where the transfer operator is a very elaborate time- and pattern-dependent code.

It is evident from these definitions and the associated discussion that the foregoing concept of biological transduction is very inclusive. The term "biological transducer," is used here to include all of those organs and processes wherein the purpose of the reaction involved is to convey, store, or process information about the environment and organism for its preservation and effective operation. Included, therefore, are all of those processes whereby information is transduced or led across some kind of a transformative exchange, with no restriction, for example, to sensory receptor organs. In this definition, genetic determination, many processes of differentiation and growth, as well as reflex and higher level abstraction mechanisms, become specifiable in transductive terms. It may be felt that this is too inclusive a definition, for it obviously includes a great deal of biochemistry, hormonal control, enzyme dynamics, growth processes, and central nervous whole-organism behavior control. It can be limited, however, and made meaningful by adding the loose criterion that has entered into some European engineering literature to distinguish between what are called large energy processes and small energy processes, and by recognizing that this distinction is one of convenience rather than one of fundamental importance.

Behind this categorization is the idea that, while every energy turnover process, whether power-plant generation of electricity or ATP utilization in muscle, involves a certain amount of direction, and while every informational biological activity, even so explicitly directive as the axonal propagation of an impulse, requires the expenditure of energy, one can usually choose one of these as being of dominant importance in the particular process in question.

One often finds in complex control systems that up to about 10 or 15% of the total energy consumed will go into processes that may legitimately be considered as primarily informational. This fits reasonably well with biological systems where, if one considers the human as an example, something like 10 to 15 w of power goes into the total nervous-system maintenance and operation, whereas the total average power turnover for a normal-size person of average activity is somewhat over 100 w. It is sound economical biological design that one should put perhaps 10 to 20% of the available metabolic energy to work for learning about the environment, deciding what to do about it, and issuing proper instructions to do something appropriate.

Now, to turn to a different aspect of biological transducers and coding, it is rather hard to break from tradition to frame variables conforming to codes in which biological processes are described by the organism, as against those terms in which one is used to describing similar phenomena in the physical sciences. In engineering analysis, one very firmly separates time as a variable apart from space coordinates, whereas one is forced to conclude that the organism uses in almost every transducer a conveniently mixed code and pays very little attention to whether it separates out temporal and spatial aspects of pattern.

Figure 1 illustrates one of the classical sensory-receptor code patterns (that of the carotid-sinus pressure receptor) to show that these receptors do not slavishly report on the value of a function being transduced or on the time derivative of this function exclusively. Instead, they answer the biologically important questions: "What is the present state of affairs?" "What is changing?" and "What has been the recent past experience?" These reports correspond loosely to engineering transducer language in which function, time derivative, and integral transducers are involved, but in

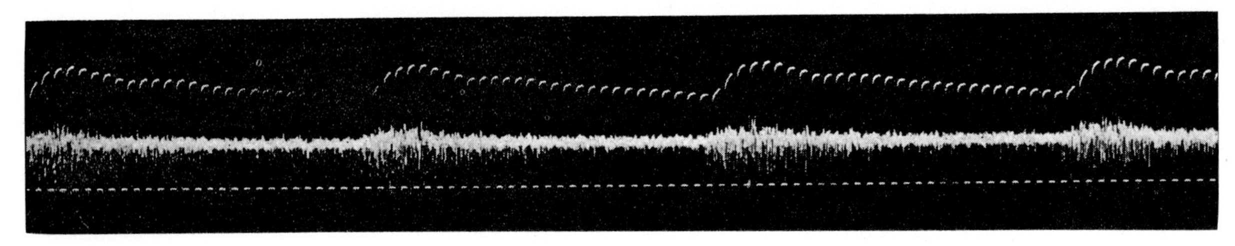

FIG. 1. Carotid sinus pressure transducer response (dog). Central record represents multifiber nerve response. Upper curve envelope is directly measured blood pressure referred to lower base line as zero. Mean blood pressure 150 mm Hg, interruption rate on pressure measuring trace 60/sec. Each pattern represents one heart beat (records by courtesy John W. Trank).

engineering transducers these factors usually are rather carefully separated out or very cautiously combined. Figure 1 is a record of the action potentials recorded from a few nerve fibers leading from the carotid-sinus end organs in a dog. This organ is concerned with reporting on the mean blood pressure in the animal and also on the details of pressure during the heartbeat. In this case, the average pressure is an important item and its absolute value remains important. Therefore, this must be a relatively nonaccommodating organ in the sense that it must continue to report the present state of affairs.

On a short time basis, it has a large "derivative" term, and thus is modulated from a high frequency almost to cutoff during each beat. Its steady-state response as reflected in the mean frequency, and especially in static records like those of Fig. 2 for a similar organ in a rabbit, is shown to convert static pressure almost linearly into pulse frequency. In this static sense, one might call this carotid-sinus organ a linear pulse frequency-modulation transducer. One must however point out the danger of specifying a transducer as "linear" or "nonlinear" on this basis by referring to the two graphs accompanying Fig. 2 where data from the same impulse records are plotted in (a) as a frequency *vs* pressure-transfer relationship and in (b) as a pulse interval *vs* pressure graph. Note how the choice of variable determines "linearity" or "nonlinearity" in this casual sense. One must, therefore, define quite carefully on a theoretically sound basis what kind of nonlinearity is meant when that term is used.

This transducer an engineer might still consider reasonably well-behaved, but now consider the photoreceptors of Fig. 3 which use an amplitude, nonlinear heavily time-influenced code. These transducers operate on a pulse-code modulation characteristic much like that which would be anticipated for a leaky, resistance-condenser coupling decay characteristic. Frequency is very high initially and decays for constant illumination so that frequency increases, but not linearly, with intensity, and decays with time for any given intensity. In addition to this general pattern of response, some animal photoreceptors also give oppositely polarized responses so as to remark pointedly that the light is decreasing or going out; an event worthy of mention, even though in the negative sense. Some sensory transducers, such as those in the tongue concerned with measuring temperature, use their nerve frequency spectra doubly, once in an ascending sense, once in a descending, the range in use being determined by a separate signal.

It was only about fifteen years ago that it was discovered that "contrast" or accentuation of detailed changes in a one- or multi-dimensional field of transduced data could be greatly enhanced by systematic suppression in a readily specifiable manner by each region of every adjacent region. In fact, it was only in about 1945 that spatial, maximum, and minimum finding networks were first produced, temporal peak predictors having come only a little earlier.

Now it is found that a very similar system exists in optical, acoustic, and probably other sense-modality, biological transducer systems, so that much of the pattern finding is done before data reach the central nervous system.

Where there is ample signal energy to operate many transducers at substantial output, as in foveal bright-light vision, this detailed differential discrimination is common, while in threshold signal performance, adjacent units are often pooled to improve statistics. There is an interesting auditory case reported where noise heard in one ear can be used to subtract out coherent noise from noise plus signal in the other ear, to permit recognizing signal otherwise submerged below recognition. In this case, two ears cooperate in a field-subtractive type of selective filtering.

It is often stated that the sensory codes are essentially logarithmic in the sense that equal fractional increases in intensity of stimulation rather than absolute increases are required to evoke correspondingly detectable increases in response. While this generalization has been elevated to the status of a "law" and is often roughly true over a decade or two, it is usually noticeable that there is a threshold region where logarithmicity is not obeyed and that there is a bending over of even the logarithmically plotted line at extremely high intensities. In long-range senses, there will usually be one or more automatic-gain control systems operating to extend the working range. It will interest engineers and physicists to know that biological transducers often operate in pseudo-instability, much like that of the superregenerative radio receiver in the absence of signal, and share

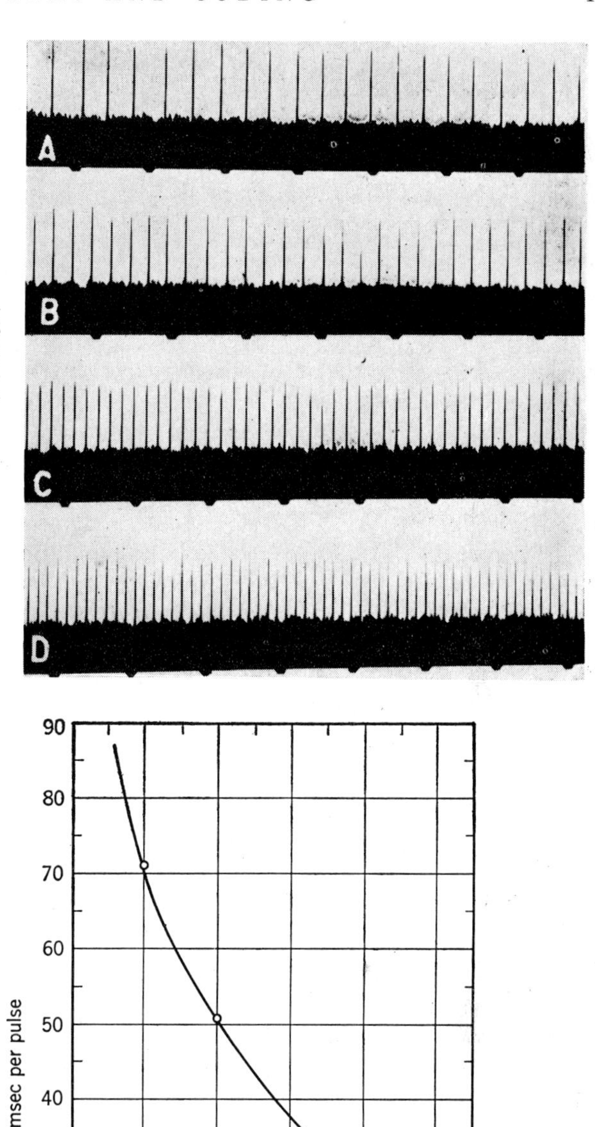


FIG. 2. Steady-state response of carotid sinus pressure transducer [from D. W. Bronk and G. Stella, Am. J. Physiol. **110**, 708 (1934)]. Data from original records of this figure are replotted in Fig. 2(a) as frequency *vs* pressure and in 2(b) as pulse interval *vs* pressure to emphasize importance of descriptive variable chosen in determining whether transducer is "linear" or "nonlinear."

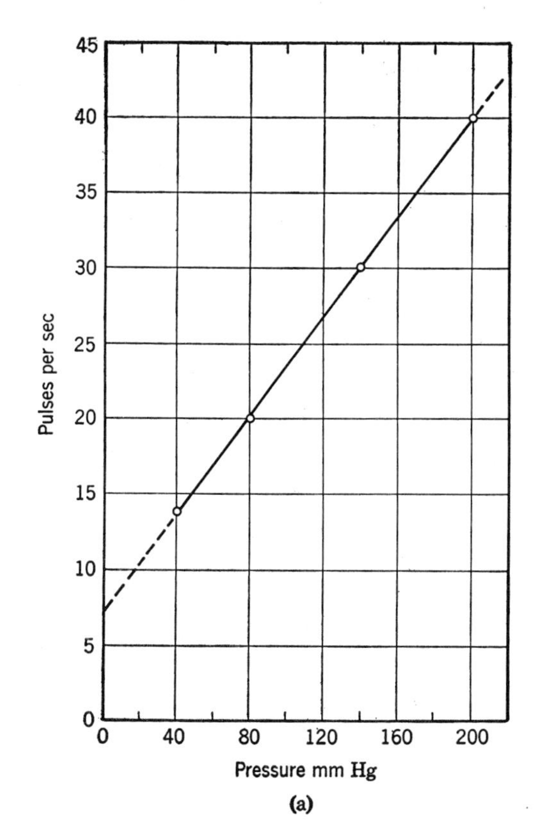


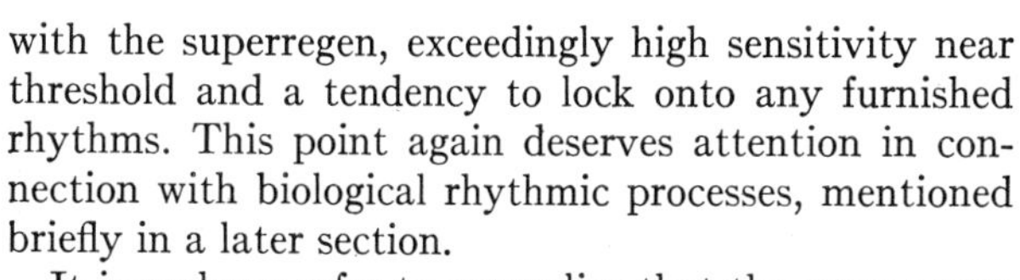

with the superregen, exceedingly high sensitivity near threshold and a tendency to lock onto any furnished rhythms. This point again deserves attention in connection with biological rhythmic processes, mentioned briefly in a later section.

It is perhaps safer to generalize that the sense-organ receptors which report in neural pulse-code modulations usually spread the available frequency band width roughly equally, in terms of importance, over the continuum of intensity time space in which they have to report, than it is to assume that they adhere to the Weber-Fechner ideal of true logarithmic response. The accuracy with which a sense organ can report over a given neural channel depends on the length of time in

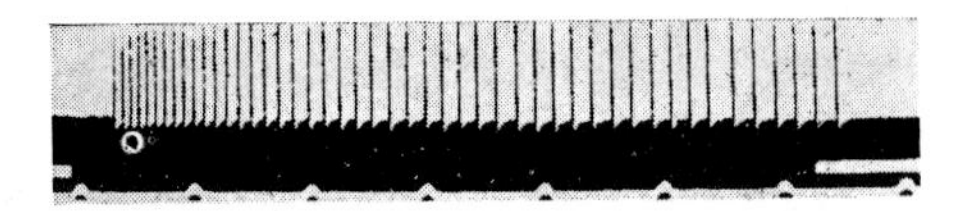

FIG. 3. Record showing adaptation of a sense organ (photoreceptor). Stimulus was maintained at constant strength throughout, but interval between discharges steadily diminished. Signal indicates duration of stimulus application and time is in 0.2-sec intervals [record of H. K. Hartline cited from D. W. Bronk, Research Publs., Assoc. Research Nervous Mental Disease **15**, 60–82 (1934)].

which it is allowed to report for a given discriminating mechanism. In general, the organism must have its report quickly, to avoid vacillatory behavior and sluggishness, yet reasonably accurately over a wide range. Thus, the neural codes turn out, for the narrow band width in frequency of pulses available, to be apt compromises between adequate and quick reporting.

One is familiar with the interestingly interchangeable way in which temporal and dimensional variables enter into the ordinary wave equation describing the progression of a pattern or wave of form $f$ in space $x$ and time $t$. Equations (1) show this relationship for one dimension:

$$\begin{aligned} F &= f(x-ut) \\ \partial F/\partial x &= f'(x-ut) \\ \partial F/\partial t &= -uf'(x-ut) \\ \partial V/\partial t &= -u\partial V/\partial x, \end{aligned} \tag{1}$$

but it is easy to see how it can be generalized for more dimensions. There is a linear relationship between the position of a particular feature of the wave in space and the passage of time.

As a practical illustration of this equivalence, it is recognized that one frequently interchanges mentally the picture of a nerve action potential in the space-time world. Usually, one measures the potential variations at one place along the nerve as a function of time, but, without feeling any strain, one substitutes the picture of a wave of identical form progressing smoothly down the length of the nerve as a fixed pattern.

It will certainly be important to discover the mechanisms within neural net systems whereby such interchanges become easy and in which this kind of spatial-temporal pattern analysis can be accomplished with the available neuronal networks and fields. It must be expected in this seemingly messy and complicated situation that the usual biological optimization of representation will apply. One must expect near-logarithmic compression of past and future in this representation, and similarly one must expect that spatial features will be compressed in their representation as they become farther removed pattern-wise.

It is attractive to think of the neural net as a multidimensional analytical space in which distance along partially excited, short neuron extensions corresponds to temporal or spatial separation. Figure 4 shows how the phase velocity of an unpropagated nerve impulse goes on at near-normal speed in a passive axon, ready at any time to be reinforced into partial or full excitation by surrounding fields. A corresponding lower-velocity process easily can be anticipated for the short neuronal extensions in the central nervous system. In this connection, it is worth mentioning that application of readily available, distributed negative-resistance networks to *in vitro* nerve preparations allows exploration of this type of reinforcement.

Consider now some basic differences in approach to informational coding that must apply in biological studies as contrasted to the conventional engineering analysis of comparable systems. In an engineering system, the coding is generally superimposed from without by the machine design, so that there is real justification for dissociating the message from the code. In biological systems, the organism must create the complex of a code plus the messages in this code with which it will function.

It consequently becomes possible in the biological case to compare meaningfully the cost of a very elaborate code, in which a one- or two-bit message can tell one to record the complete contents of the *International Critical Tables* or of the *Encyclopedia Britannica*, with a very elaborate message in a simple code whereby one dictates one or the other of these works at length. In a genetic plus sensory-motor biological code complex, the organism seeks to minimize the combination of these two aspects.

Where there is to be very frequent repetition of similar messages, it becomes economical to devise fairly elaborate codes in which these messages can be shortened and sharpened in discrimination. For the very rare or

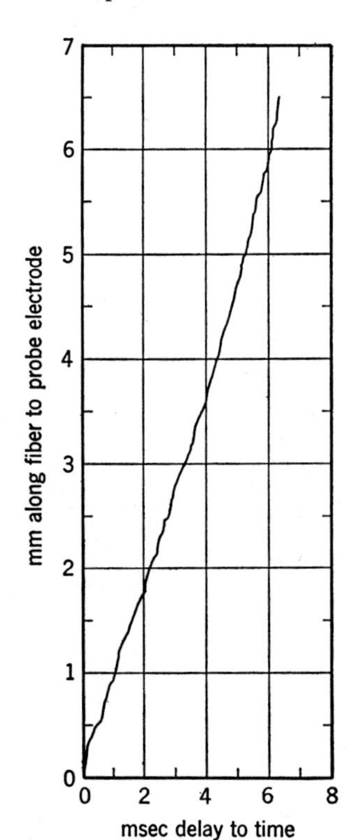


FIG. 4. Phase velocity of an unpropagated subthreshold sinusoidal wave in living axon. Data for squid axon, mean diameter 0.41 mm, 300 cps, 21°C. Phase velocity 10 meters/sec varying slightly with taper of fiber. Record is directly traced from machine drawn curve without smoothing.

one-time situations, it is economical to spell out the message in detailed, simple versatile code so long as it is not essential that the communication be completed in a very short time. Undoubtedly, some of the complex innate behavioral responses of lower and higher organisms have to be spelled out rather expensively in informational, genetic, and developmental symbols, as they are essential to survival and can be left only slightly sketchy.

It is an almost universal observation that, wherever a specific biological organic structure or organization functionally need not be built to a firm pattern, it will be left to grow as dictated by local circumstances. This relieves the informational system of that much superfluous burden. If it is necessary only that there be a substantial number of muscle fibers or connective-tissue units in a particular structure, it is very improbable that an exact number will be present, while it is equally certain that a nine-plus-two or a correspondingly important topological or chemically significant structure will be firmly designed-in. The informational design, once having been paid for genetically, will be used widely, thus minimizing the necessity for superfluous duplicate instruction. Biological organisms are probabilistic mechanisms, and their transducer codings reflect their willingness to take a statistical chance for survival in a risky world, especially if the odds are good. Where an item of information is especially important, it will usually be protected by replicate and nonreplicate redundancy. Such redundancy may become too expensive in some cases and may have to be sacrificed. This may be the case in some genetic transformations where cells and indeed the continuity of life for the organism become susceptible, briefly, to minor external mutational influences.

To avoid further digression from specific, experimentally accessible material, some of this free-wheeling thinking should be tied down to mechanisms and bits of analysis where there is a chance for specific solutions to specific problems. This set of analytical tools is still rather dull, but undoubtedly can be sharpened by use and extended to the effective analysis of the behavior of complex informational systems prevalent in animals.

As a starting point, consider some of the ways in which one can elaborate on the standard notion of a transfer function, which forms the usual stock-in-trade of the servo-engineer and electronic designer when he is trying to understand the behavior of control-communication and computer systems. The transfer function is usually represented as a black box into which a signal is put and out of which a related signal emerges. Normally, it is very tightly hemmed in with formal restrictions of allowable behavior not unrealistic, in most cases, for amplifiers and physical transducers, but usually very unbiological. As a result, many a biologist has been led down the garden path when he has attempted to use this concept, for, unlike the engineer, he is unable to make his system conform to the limitations of the model on which he bases his reasoning.

In the conventional cases, a transfer-function black box will be classed as either passive or active. If passive, it will require that there be put in as a signal all of the power required to actuate the box. It merely gives out a proportion of this power modified in some internally determined manner. Alternatively, it may have a local power source, as in an amplifier, and thus actively give an enlarged or otherwise modified more-powerful or effective output. In either case, the transfer box is usually assumed to be linear, except for special cases. The output is, thus, in terms of magnitude at any one frequency, a kind of phase-shifted, differently scaled replica of the input, although the relationship may be different at different frequencies. For the simple cases then, the performance of the box is adequately represented by describing the ratio of the output to the input as a function of frequency when driven by a sine-wave input variation. Output is presumed proportional to input, but is acknowledged to vary in relative magnitude and also to advance or retard in relative phase as a function of frequency. Consequently, for many purposes, a complex plane phase-amplitude plot of this transfer function is considered to convey adequately the behavior of the box.

Usually, the transfer-function box is used as a building block to construct functional block diagrams, often including loops like those in Fig. 5(a) to confer a measure of stability absent in a system that is not back-compared. Occasionally, a long, compound loop system like that of Fig. 5(b) will be used, because it has still greater stability than the simpler loop systems for a given available gain, but it is also more likely to become violently unstable if certain limits are exceeded.

Note that the feedahead loop illustrated in Fig. 5(c), while not intrinsically as stable as the feedback loop, has an advantage in being capable of introducing exactly compensatory control corrections without having infinite gain, whereas the feedback loop can only approach this ideal asymptotically. Feedahead can be combined with feedback to give a very desirable compound operation and is now gaining widespread acceptance in engineering applications.

Figure 6 illustrates a more typical, biological-transfer loop system which characteristically includes feedahead as well as feedback loops and almost always displays control parameter dependency upon signal. It should be emphasized that the traditional transfer-function box clearly carries with it the notion of signal pathways firmly fixed spatially; that is to say, there is some kind of a discrete-circuit-pathway diagram characteristic of the system. There may be switches, but these merely direct activities between alternative discrete pathways. Each transfer box is a point-to-point function; whatever enters, comes in from a source at a specifiable place in the diagrammatic model, and what comes out, comes out at a specified place or places. Such function dia-

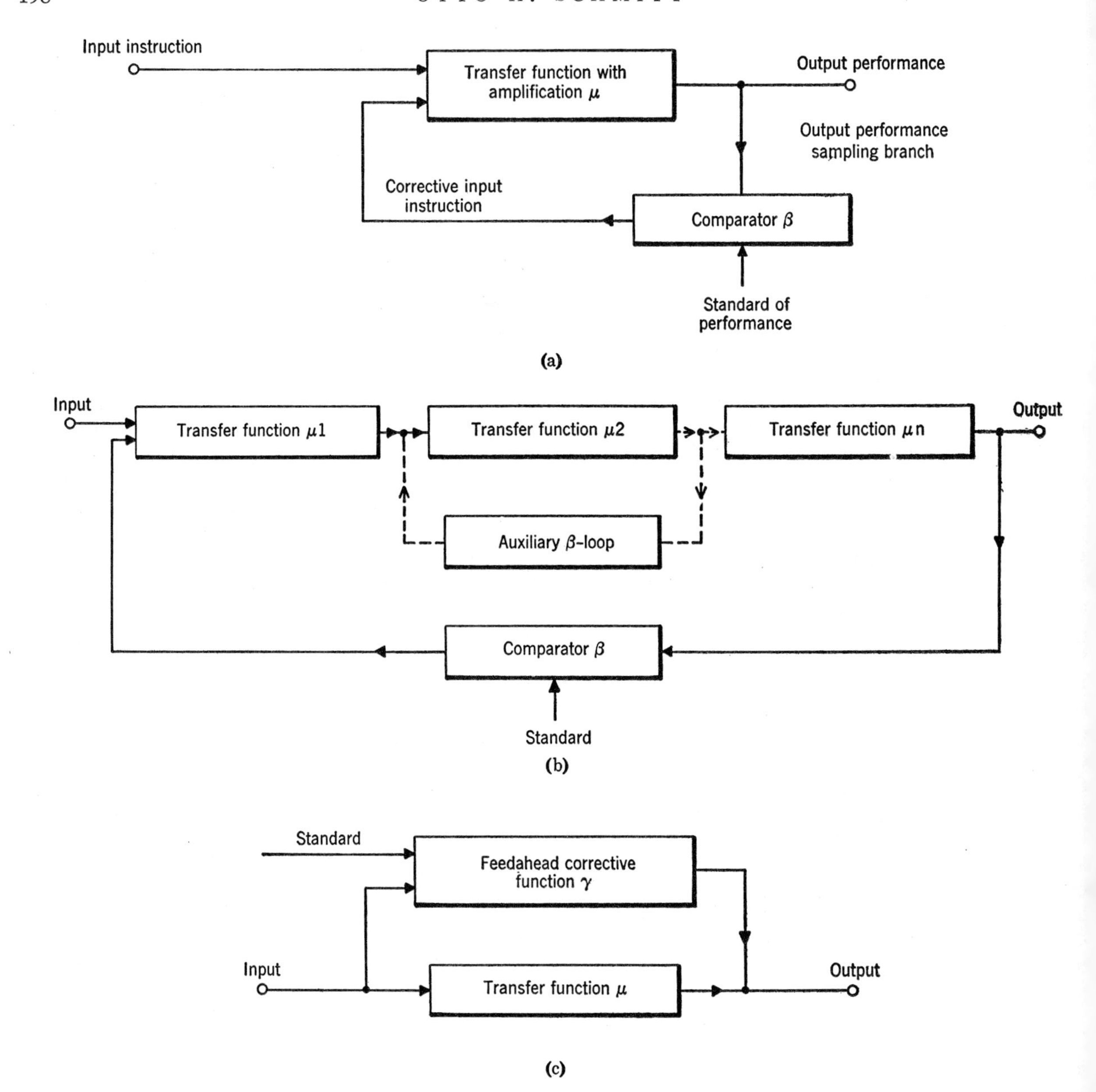


Fig. 5.

grams are hardly ever generalized to anything more complicated than a substantial interconnected block of such units, possibly with systems of switching arrayed in a logical, or occasionally in a statistically determined pattern.



Fig. 6.

In all fairness, it should be pointed out that there i another and quite different set of packages, sometime also called transfer functions, where the black box, in stead of performing linear operations, now perform logical operations. These units, which figure prominentl in digital-computer design, do something specific at a output when received signals in some predetermine logical set appear or fail to appear at its input. Typica signal sets might represent both, neither, or, *a* before *b* time coincidence, equality, inequality, etc. These tw quite different kinds of transfer-function black boxe can, in fact, be considered as epitomes, respectively, o the functional organs of analogical-computer systems and of digital-computer systems.

Biology does not have *a priori* knowledge of a real difference between analogical and digital computers, and consequently is not hampered by this distinction. In fact, this difference is much more of the experimenter's making than it is an intrinsic property of the system being analyzed.

Digital systems are really those where enough probabilistic configurations of a system are lumped redundantly into a few significant categories to make the whole system approximate the all-or-nothing pattern characteristic of Aristotelian logical arguments. Analog systems are those where the available system configurations are dispersed over a significance range and, if numerous enough, resemble a value continuum. In the systems where energy redundancy is reduced to a minimum and nearly all of the available time band width of an information-handling system is utilized, the conventional thinking about digital and analog distinction becomes hazy and a new and more general type of analysis has to be substituted.

It is perhaps unfortunate that the term "analog system," which originally implied that one physical variable was being substituted for another for computational purposes, has come also to be used to describe any system utilizing a continuous range of variables in its operation. One is thus left with the anomaly that an analogical computer may be digital in makeup and that circuits performing quite direct control operations, and not particularly analogical physically to anything else, will be called analog-control systems.

It would be unfair to suggest that the modern engineer is not aware of the importance of nonlinear systems or to accuse him of thinking that digital and analogical systems are intrinsically different. He does, however, tend to build apparatus so as to conform essentially with one or the other of these because these analyses are familiar and easy, and he most likely has not been forced to cope with distributed transfer-function systems, particularly those that are highly re-entrant.

A little futher on, the behavior of two importantly different types of re-entrant systems is discussed, but first some places in biology where one can use fragments of available analyses are examined and embroidered on so as to make them more biological in their characteristics. In this way, they may be made applicable to the increasingly important cases where there are biological-transducer elements, particularly neural nets, which are simultaneously a continuum and a system of tight topological connectivities. This is a complicated way of saying that they are to an extent switchboard connected, but are simultaneously a system of diffuse field interactions. In this case, these field interactions mean simultaneously the chemical field, e.g., the hormonal or other chemical concentration-gradient field, and equally important, the electric field contributed by the surrounding active elements.

There will almost certainly be found in this complex an understanding of the efficient large-scale behavioral control of higher organisms, and mathematical techniques will have to be expanded to give facility and simplicity in handling systems which are hopelessly complex if treated as telephone exhanges gone berserk or soaked in electrically leaky saline.

One cannot offer more than an introduction to this kind of thinking at this time—first because it is complicated, and second because most of the answers are still unknown—but several extensions of ordinary field theory can be stated that are not presently being used in neural-net analysis.

There is one experimental and theoretical approach to this area that should be explored. This is the approach based on cardiac transfer-impedance theory, a technique used successfully in our laboratory in examining the superficially accessible action potentials of heart muscle, i.e., the spatial electrocardiogram. This theory, developed to handle biophysical heart data, will in all probability prove even more valuable in examining neural field interaction than it has in the purpose for which it was intended. Strangely enough, the key notion of reciprocity inherent in this theory was beautifully stated by Helmholtz, one of the early biophysicists, just about a century ago. It was not applied to biophysical problems until quite recently, and then it was introduced in another disguise.

The following demonstrates how the transfer-function notion can be spatialized. Consider Eq. (2), the definition of transfer impedance in its usual engineering form for a linear passive network:

$$\text{output voltage} \equiv \text{input current} \times \text{transfer impedance}. \tag{2}$$

Such a network may include any combination of capacitive, resistive, or inductive components that do not contribute external energy. For any combination of such linear components, one can determine the current applied between any two terminals in the system and the resulting potential difference developed between any other two terminals. The ratio of this voltage to the current will be dimensionally an impedance, and will have a constant value at any specific frequency if the system is linear.

Substitute for the passive circuit a biological tissue, nervous or otherwise, which is electrically at least approximately isotropic and linear although not necessarily homogeneous. These qualifications are included to avoid tensor notation with its complications when simple vector expressions are probably adequate for the present state of the art. In the low-frequency range which is of interest neurophysiologically, the brain approximates such a system of linear electrical components, except for microregions. Consequently, if any two points are picked as terminals through which to apply current, and any other two points are picked to

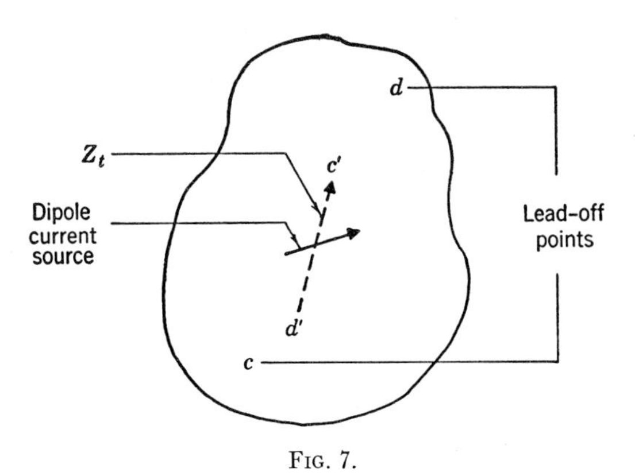


FIG. 7.

read off the potential difference, one finds that the voltage and current will be in a constant linear relationship which is formally described by their transfer-impedance ratio. If one now brings the two current points close together, they can be treated as a dipolar source-sink combination, and such a dipole is the kind of source that is encountered in biological tissue.

Because of the relatively short, length constants involved in muscle, nerve, and other biological current-generating systems, one can generally assume quite safely that source and sink are dipolar on a millimeter basis, and that macroscopic linearity prevails. In an isotropic medium, another generalization can be added. One can observe that, so long as the source dipole remains essentially in one region, it with its orientation becomes vectorial in the sense that the voltage transferred to a specified pair of terminals will obey a cosine relationship with respect to some specifiable preferred axis.

Thus, the dipole can be thought of as projecting itself, to a constant scale, on an effective axis not necessarily going through the recording points. In the diagram (Fig. 7), lead-off points $c$ and $d$ may be thought of as effectively represented by a vector on an axis $c'd'$ at the source. The magnitude of potential realized at the lead-off electrodes will be proportional to the magnitude of the current source times its distance of separation (i.e., its current dipole moment) multiplied by the cosine of the spatial angle between it and a reference line $c'd'$.

It is seen at once that this has become a vector dot product system and that the $c'd'$ reference has become a reference vector characteristic of a particular lead system. One, therefore, vectorializes $Z$ in the alternating-current version of the Ohm's law relationship $E=IZ$. $I$ becomes a dipole current moment (amperes times separation in centimeters). $Z$ becomes a transfer impedance which is dimensionally ohms per centimeter in magnitude, but assumes the spatial attributes of a vector. Voltage, of course, remains a scalar quantity which is consistent with its representation as the scalar product of the transfer impedance with the current dipole moment.

$$V=\mathbf{I}_m \cdot \mathbf{Z}_t.$$

With $Z$ vectorialized, one now has a function capable of being distributed spatially throughout a tissue system; that is to say, the transfer function has been successfully distributed in a neural field space. This function must necessarily be a vector-point function—that is, a quantity having specifiable magnitude and direction at every point throughout the space involved. By evaluating this function, one can characterize the interaction of a source anywhere in the field space with a receptive region anywhere else in the field, and can quantitatively predict their mutual electrical coupling.

It is at this point that Helmholtz's reciprocity theorem comes to bear. One must be rather careful in stating the theorem for a distributed medium, but it can be put in the following form. For any given system of lead-off electrodes, there is a spatial vector-point function which will relate the potential difference measured in this lead to the current dipole moment introduced anywhere in the medium in any orientation. The potential will be $V_t=\mathbf{I}_m \cdot \mathbf{Z}_t$ where $V_t$ is the led-off voltage, $\mathbf{I}_m$ the dipole-source moment, and $\mathbf{Z}_t$ the transfer-impedance function. Upon applying an adaptation of the reciprocity relationship, it is found that the electric field or potential gradient $\mathbf{E}_t$ anywhere in the medium owing to introduction of current $I$ into the former lead-off electrodes will be exactly $\mathbf{Z}_t I$, so that the spatial transfer-impedance function $\mathbf{Z}_t$ is identical in the two cases for any electrode or medium configuration.

What is the value of this $\mathbf{Z}_t$ function and its reciprocity property? In the first place, it permits one to predict, by applying currents to finite leads and measuring the resulting potential distribution with microelectrodes, what potential will be set up as a field by microsources such as neurons. Model measurements and *in vivo* measurements of transfer impedances by the reciprocal method now become feasible and valuable. While these measurements have not yet been done to any great extent on brain tissues, it is possible to illustrate the kind of results to be expected by comparison with results obtained in the electrocardiographic field.

Figure 8 is an example of such an application where a set of orthogonal leads was developed experimentally by linear combination of partial derivatives of the transfer-impedance function. Knowing that transfer impedance functions must obey the superposition theorem as well as the reciprocity rule, it was possible to develop leads sensitive, for practical purposes, *only* to sagittal, vertical, or horizontal sources of current irrespective of position within the heart tissue.

Inversely, it was possible to show theoretically and experimentally that fields from neurons in certain configurations can add up so as to be quite strongly concentrated at substantial distances from the sources while in other cases they die out even more rapidly than the inverse cube law. Measurements of transfer impedance fields via models give surprising insight into such complicated field patterns. The analysis indeed

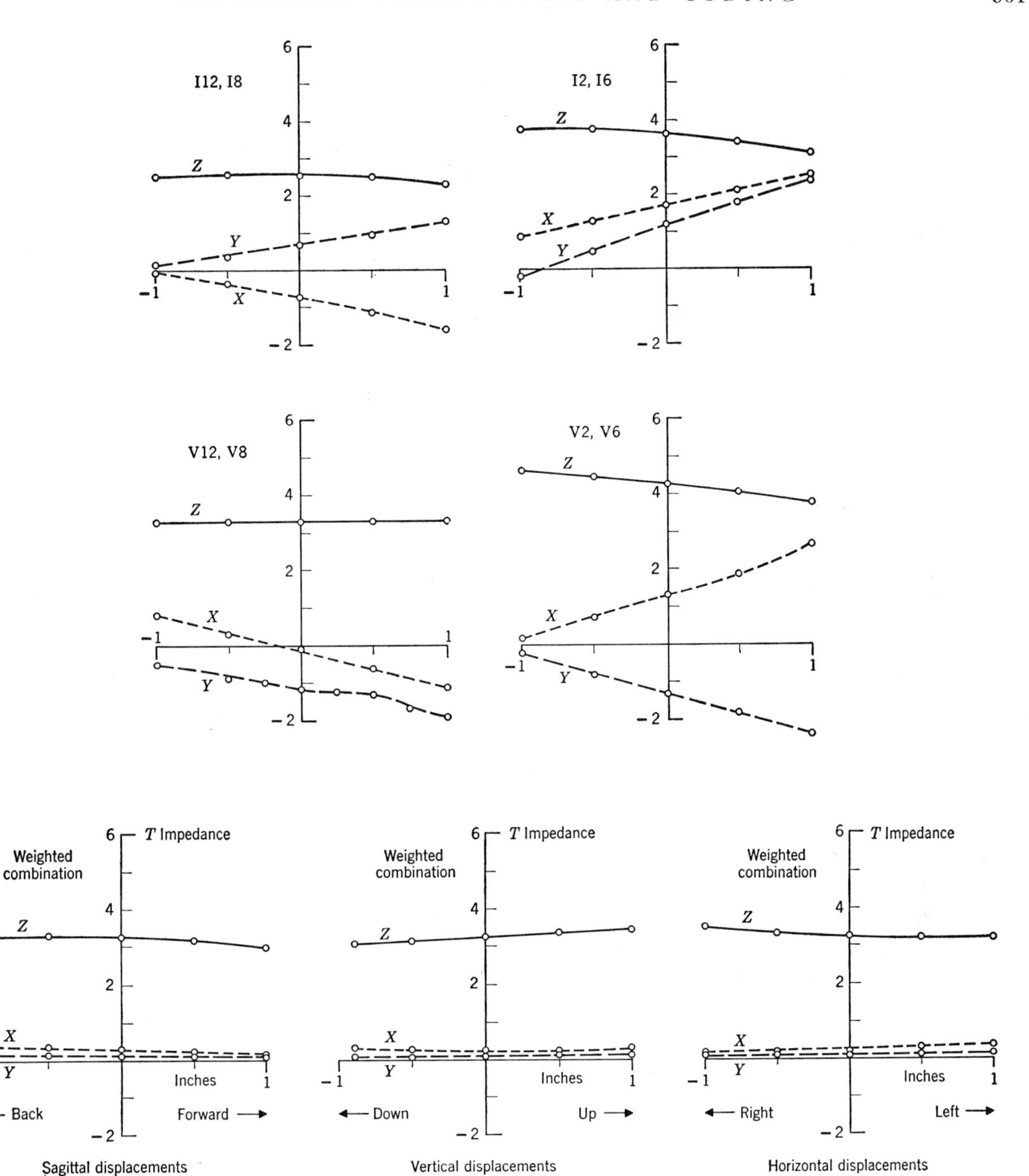


FIG. 8. Orthogonalization of transfer impedance fields. For any tissue system to which a specified lead system has been applied, there a characteristic associated transfer-impedance field which takes the form of a continuous spatial vector-point function. By examining e principal partial derivatives of this vector function for several different lead systems applied to the same tissue, it is usually possible devise weighted combinations of leads which yield nearly orthogonal functions in a finite source region so that potentials picked up om the combined leads report uniformly and orthogonally on sources in the corrected region, and stimuli applied through the leads imulate the source region uniformly and in a constant known direction. Alternatively, highly distorting leads can be devised which ill record and stimulate selectively. The patterns shown here as examples were developed to yield a uniform sagittal, or front to back, ad for the human chest to be used in electrocardiography. Ideally, the transfer impedance should be large and constant in $Z$ and should ave negligible $X$ and $Y$ contamination.

esembles a kind of degenerate directional antenna de- ign procedure.

Spatial transfer-function analysis is very attractive n studying field stimulation of neurons, for it gives an ntuitive insight into the remarkable time sensitivity of stimulation patterns. If one remembers that the conduction time at constant velocity is equivalent to distance, and that the inverse equivalence is also valid, then one has in a neural net a most interesting mechanism for the spatial-temporal localization of a pattern,

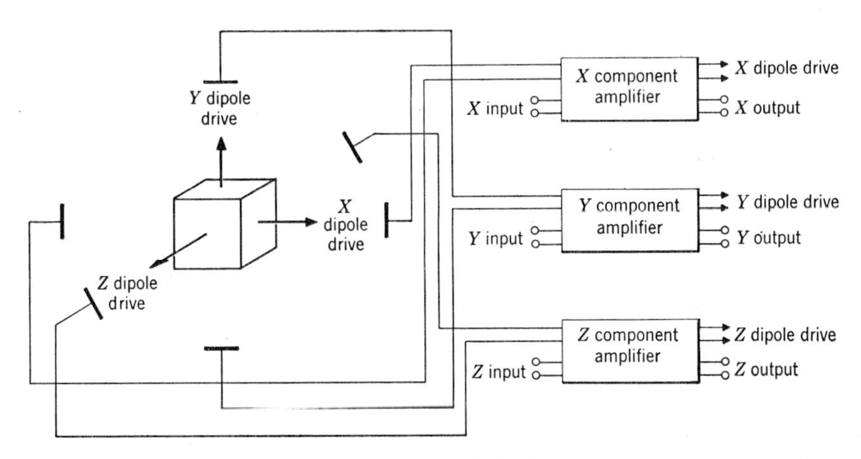


FIG. 9. System for automatic synthesis of equivalent concentrated dipole-current moment sources. A conductive model of the tissue system is fitted with a lead system equivalent to that on the experimental animal or person, and a vector dipole source is installed in the model at the chosen locus for which the equivalent vector source is to be calculated. A system of high gain amplifiers is looped around the model lead system to the dipole, and the amplified signal from the biological subject is inserted as a forcing function. In order to maintain all the amplifier inputs at near-zero potential simultaneously, the amplifier system must generate a dipole source which just cancels the complex of lead potentials and thus one which duplicates, in a negative sense, the desired source. The solution is, of course, unique only for a specified dipole source locus and electrode configuration as can be demonstrated by rearranging the electrodes on model and subject or by moving the dipole source about in the model and noting the continually altering resultant pattern of source dipole moment.

and hence an ideal medium for distributed probabilistic field effects.

Another closely related extension of transfer-function analysis was introduced by Burger in Holland a few years ago. In this approach, an image space is developed as a complementary space to the real field space. In the image space, vectorial summation of superimposed field contributions becomes beautifully simple, even though working out the shape of the image space is a more complicated process.

There is a very interesting feedback method for synthesizing spatial excitation sources (Fig .9). If one builds a conductive three-dimensional scaled model of a tissue system and immerses in it an orthogonal, triple dipolar source at a place corresponding to a supposed bioelectric source, this artificial source will excite a transfer-impedance pattern similar to that of the original living system. If each component of the dipole is fed from a high-gain amplifier which is fed, in turn, from one of a set of three reasonably orthogonal lead sets corresponding to the dipole orientation, the system then becomes an automatic synthesizer of excitation sources in terms of the actual picked up potential. Explained in behavioristic terms, the feedback system tries constantly to null itself in all directions simultaneously and can do so only by creating a source equivalent to the real but inaccessible one. To what extent this general scheme can be extended beyond three loops to synthesize more complex sources, and to what extent noise can be used as a tracer to establish source localization, are not yet known, but this approach seems to be very promising.

Space does not permit amplification of many additional aspects of the biological-transducer function which are of great importance, but the transductive aspects of animal "orientation" behavior must not be totally omitted. Animal orientation includes especially the navigational processes whereby an organism "steers" itself toward a desirable goal literally as well as figuratively. It is well known that pigeons home, that fish almost unerringly find their native stream after months at sea, that birds migrate to very specific targets, and that bees communicate to their fellows the range and azimuth of food depots. Birds are even believed to utilize astronavigation, as suggested by recent experiments. All of these processes are biological-transducer problems at a higher level and ought to be investigated on this basis, as well as by the traditional types of experiment. If a moth knows which way to fly to find another moth a kilometer away, this is a transduction and coding problem, and one should seek not only the sensory transducers involved but also the transfer functions describing the reduction of massive sensory data to net decisive instructions for a flight plan.

Another set of biophysical phenomena classed as animal orientation is the family of rhythmic or oscillatory processes found in almost all biological organisms for which one has yet to discover a biophysical mechanism. It is well known that animals and plants exhibit strongly rhythmic properties with periods ranging from 1 or 2 msec in the case of some nerve processes to periods of a month, a year, or even several years in other cases. These rhythmic processes often have very nearly, but not exactly, a one-day period. Again, the period may coincide with the succession of high and low tides in a particular place, with the lunar month, or with the terrestrial year. While these periodicities are related to those of the normal environment, many of them persist for many cycles in experiments where the organism is carefully shielded from all known environmental clues to these periods.

Some of the simple biorhythms like the cricket chirping rate have a reasonable chemical-process temperature coefficient but others have an almost exactly zero temperature coefficient. Some of these temperature-independent periodicities can be shifted in phase by a systemic shock such as dunking in cold water. Others can be "pulled," as a crystal oscillator can be pulled slightly off frequency by coupling to another resonator of similar frequency. None of the basic oscillatory mechanisms has been identified nor has an adequate hypothetical master oscillator been devised which will conform with experimentally determined properties of these natural oscillations.

Crabs change color on a tidal-periodic basis; some bacteria luminesce periodically on the basis of when it should be day or night; some flies emerge from pupae on an integral diurnal but not fixed illumination-period basis; some poisons change toxicity on a diurnal basis; and cilia beat rhythmically. No one seems to have found the transducer mechanisms by which these rhythms are loosely but adequately coupled to the environment. Surely, there should be a search for mechanisms of periodicity hinted at by the three major classes of physical oscillating systems. In any case, one should be embarrassed to have failed completely in turning up even approximate explanations for these phenomena that are of undoubted prime biological importance at every level from the intracellular to that of the entire organism.

Finally, there are two items referred to earlier—bivalent codes and hierachical re-entrant control systems. It is evident that biological control systems are, in general, organized in a series of semi-autonomous hierarchically arranged levels where each subordinate system left to itself will achieve a more primitive, less flexible, but still effectual, type of homeostatic control. Single isolated cells get on somehow in tissue culture, and a heart will beat in parts if it fails to receive coordinating excitations from its pacemaker. What are the mechanisms and the general control principles whereby several, often contradictory, courses of motor behavior are coordinated so as to remain unified and influenced by the several operator controls, yet not uselessly vacillatory or indecisive in action? Evidently, whole patterns of feedback and feedahead control are coordinatedly included or excluded from taking dominance in effective control. What does an animal do when exposed to unfamiliar, conflicting sensory information such as might originate from a minor malfunction of one of its transducer mechanisms? Generally, it chooses one set of the feasible interpretations of the available data and ignores incompatible parts of the others. Sometimes, as in the case of Von Holst's fishes (where the fish, used to having 1*g* gravity downward and light from above, are experimentally exposed to light from the side and to a centrifugally increased apparent *g*), they seek a weighted compromise. Very seldom do they fail completely, as do our rockets, from some single small malfunction.

In summary, one may conclude that there is at present no unified, formally manipulable system in the biophysical sciences for examining and describing systematically the functional characteristics of biological transducers and for studying their aggregate behavior when combined into functional ensembles. There is ample mathematical experience in the engineering fields to permit a start in this direction, but existing techniques fall far short of adequacy, and it is certain that new, uniquely biological theory and techniques for transductive-function manipulation and code specification will have to be developed.

# 54

# Initiation of Nerve Impulses in Receptor and Central Neurons

THEODORE HOLMES BULLOCK

*Department of Zoology, University of California, Los Angeles 24, California*

THE objective of this paper is twofold: to point out some of the problems which represent the present state of physiological analysis of some sense organs as detectors and transducers, and to point out a current view of the complex chain of events between transducing a stimulus and initiating a nerve impulse. It will be argued that this chain is similar in receptor and central neurons and that information about either one is relevant to the other.

Biophysicists long have been interested in sensory receptors, since they present in such a conspicuous form the conversion of physical events of the inanimate world into physiological events in the organism.[1] Some of the best-known achievements of biophysical science lie in this realm. Indeed, a vast literature exists on sensory capacities, analysis of the parameters of reception, and quantitative description of the subvarieties of unit receptors within modalities[2]. There continue to appear in the literature new kinds of receptors physiologically identified, and there remains a host of histologically known sense organs and diffuse receptors, differentiated visibly but not yet defined physiologically (e.g., pectines of scorpions, esthetes of chitons, osphradia of snails and clams, etc.).

This situation, virtually unique among organ systems, is not surprising if one thinks of behavior as the principal attribute in which the great wealth of animal types is differentiated, if one ascribes behavior to differentiation in the central nervous system and remembers the intimate and perhaps causal correlation between differentiation in the central nervous system and that in the sense organs. One has learned to expect startling discoveries in each new batch of journals—polarized-light detection, wind-speed indicators, gyroscopic-deflection sensing, infrared directional devices, olfactory separation of optical isomers, hydrostatic-pressure detection in supposedly gas-free organisms, ultrasound reception—not to speak of systems achieving a complex degree of analysis of signals such as the ultrasonic-pulse-reflection analysis in bats, the amplitude-modulation frequency analysis of the ears of crickets, the integration of proprioceptive information in the strike of a praying mantis, and of we do not know what information in many cases such as orientation of fish to turbulence and to other unseen fish. This list furthermore omits mention of eyes of various kinds and their fantastic achievements, of which Hartline writes (p. 515).

Since a systematic survey is patently impossible, I offer here only a small selection of some recent instances of progress in the *identification and characterization of receptors* with respect to their capacities and properties. A closer look is taken later at the *parameters of impulse initiation.*

## ELECTRORECEPTORS

Recent work[3–6] has opened up a new sensory modality in the sensitivity to normally occurring electric fields, found in many species of fish of at least three unrelated families, the Gymnotidae, the Gymnarchidae, and the Mormyridae (Fig. 1). These fish have electric organs which emit pulses of low voltage—of the order of 1 v—in some species in brief, low-frequency bursts, and in other species continuously, hour after hour, with characteristic frequencies and pulse form and duration, diagnostic of the species (Fig. 2). The frequencies commonly lie between 60 and 400/sec, but in some cases exceed 1000/sec. The pulse duration—from the whole animal, representing activity of many electroplates—ranges from 10 msec to less than 0.2 msec. What is significant for us is that the behavioral evidence clearly shows in some species the use of these signals in orientation with respect to objects, apertures, and other emitting fish in the immediate environment. Clearly, the fish detects alterations in the pattern of the electric field in the water surrounding it (Fig. 3). Marked signs of agitation are elicited by a wire brought within range or by the discharges of other fish. Conditioned-reflex experiments show the ability to detect the presence of a stationary magnet outside the aquarium and to discriminate between conductors and nonconductors in the aquarium. The fish respond to the movement of a small electrostatic charge such as that produced by combing one's hair with a vulcanite comb. Sensitivity to imposed electric currents has been known, but what is of interest for us is the demonstration that a normally developed sensibility exists which is apparently used in nature to detect and to analyze in a complex way naturally occurring electric fields.

The sense organ is presumably the lateral line, heretofore regarded as a special form of mechanoreceptor. We will have to make room in our lists of modalities for a new category—electroreceptors. It is perhaps more curious that this modality is not more widespread. Muscle-action potentials can be recorded in the water at some distance from fish as well as from other animals but so far the tests outlined in the foregoing, when applied to fish of other families than these three, hav

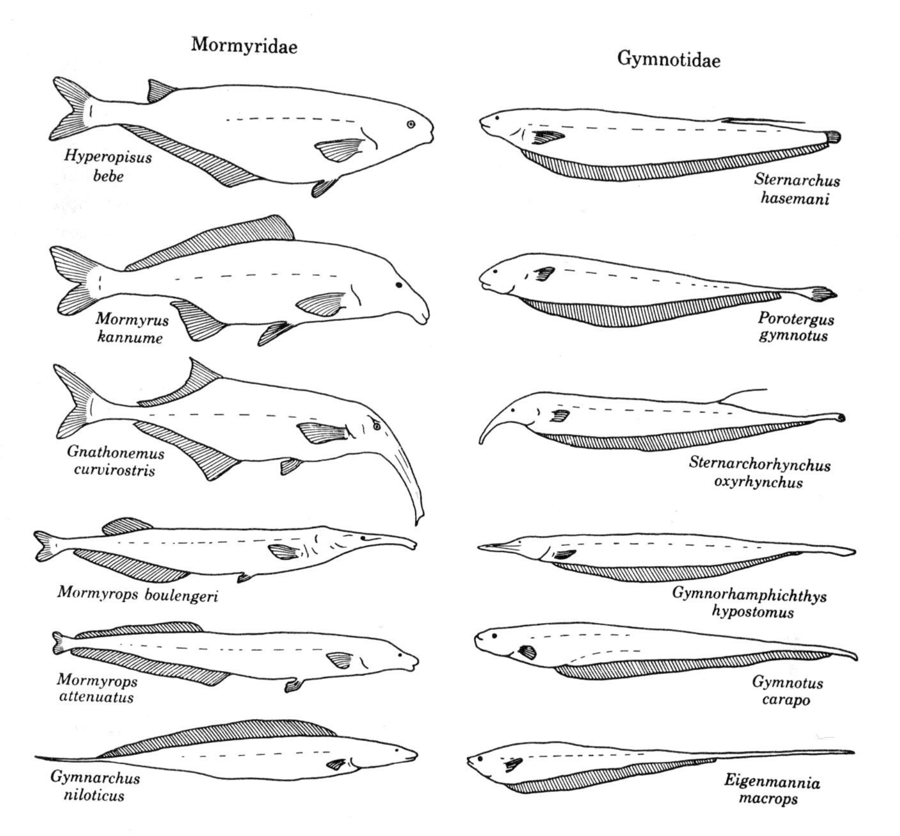


FIG. 1. Examples of continuously discharging, low-voltage electric fish. Representatives of the Mormyridae, Gymnotidae, and Gymnarchidae [from H. W. Lissman, J. Exptl. Biol. **35**, 156 (1958)].

indicated no such sensitivity (with the possible exception of the *Siluridae*). One may conjecture that it is perhaps not so much the deficiency in peripheral sensibility as the lack of central apparatus for making use of the signals, which distinguishes the ordinary from the electric fish. All electric fish share, among other things, the enormous development of the valvulae cerebelli, the so-called mormyro-cerebellum.

The sensitivity of the receptor has not been measured directly but Lissman and Machin[3] have calculated the intensities of the electric fields which are effective in altering behavior, for example in conditioned response tests to electrostatic charges or magnets outside the aquarium. The field in the water around the fish must be changed by about 0.003 $\mu$v/mm at threshold, in *Gymnarchus niloticus*. This corresponds to a current through the fish of $2\times10^{-5}$ $\mu$A/cm$^2$, some hundred thousand times smaller than a still subthreshold stimulus current density across the membrane of a squid axon (2 mv across $10^3$ ohms/cm$^2$). The problem of the possible meaning of such high sensitivity to electrical events already has been raised by Terzuolo and Bullock.[7] Our evidence suggests that even in ordinary nerve cells the sensitivity, while not nearly as high as in these electroreceptors, is several orders of magnitude higher than the usually recognized threshold changes of 10 to 20 mv required to excite a silent nerve fiber. In the first place, this high sensitivity is manifested only as an alteration of frequency of an already discharging cell. In the next place, to be sensitive to minute changes in membrane potential, both the critical potential for spike initiation and the rate of rise of the prepotential must be extraordinarily stable—and must be localized in a limited part of the cell. The basic problem of high electrical field sensitivity—of special interest because it may be the one case where one does not need to search for a transducing mechanism as we do in photoreceptors, mechanoreceptors, etc.,—is the matter of *stability*. What are the requirements in terms of stability of the threshold and of the prepotential such that a given channel can provide a useful signal within a reasonable time

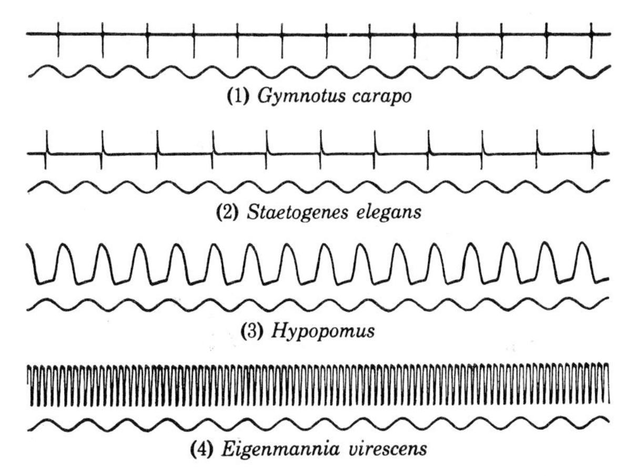


FIG. 2. Examples of types of electric pulses produced by four species of the Gymnotidae. Both pulse duration and pulse frequency are different and characteristic for each species. The fish are continuously discharging day and night. Time marker 50 cps [from H. W. Lissman, J. Exptl. Biol. **35**, 156 (1958)].

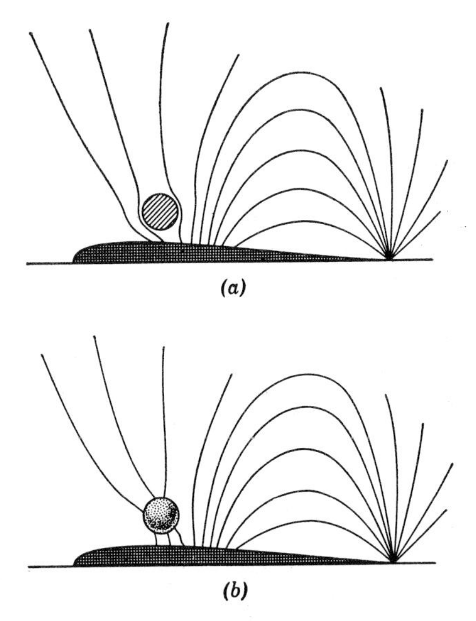


FIG. 3. The influence of objects upon the electric field around a fish. (a) An object of low conductivity, and (b) one of high conductivity. The fish detects the distortion of the field and reacts to both classes of objects [from H. W. Lissman and K. E. Machin, J. Exptl. Biol. **35**, 451 (1958)].

(e.g., 1 sec)? At another level is the important problem of whether or not the high electric field sensitivity in the modulation of ongoing discharge plays a role in normal situations in the nervous system. If so, it permits effects not only of changes in electric fields normally present but also of changes in the chemical milieu which may act through the same mechanism, by changing the frequency of the spontaneous discharge for a given membrane potential at the critical locus.

## MECHANORECEPTORS

Turning to the realm of mechanoreceptors, an attractive problem exists in the wide variety of arthropod hair-like exoskeletal projections (Fig. 4), each of which receives a single distal process of a primary sensory neuron and is specialized for signaling deflection of the hair in a certain direction.[8–10] In the lobster statocyst (an equilibrium sense organ where this is especially clear and for which recent analyses have been made), there are several specific types of hairs, differentiated for sensitivity to movement in one or another direction, or for sensitivity to a maintained position in one or another plane, or for sensitivity to vibration. Since the stimulus seems likely to be limited to a very restricted region—namely, the point of articulation of the hair near its base—one is faced with the general problem of transducing a mechanical event into a physiological one in perhaps a more simple and discrete form than in the more familiar mechanoreceptors such as stretch, touch, and auditory organs. Although the sensitivity must be high in terms of displacement, it cannot be estimated accurately, owing to the fact that, at the base of the long hair, where the relative movement that must act as a stimulus occurs, the mechanical disadvantage is maximal and the small threshold movements of the tip of the hair must be reduced enormously in absolute magnitude. Fundamental questions remain, regarding the *ultrastructure and the local events* in the region of the mechanical deformation of the sensory nerve ending.

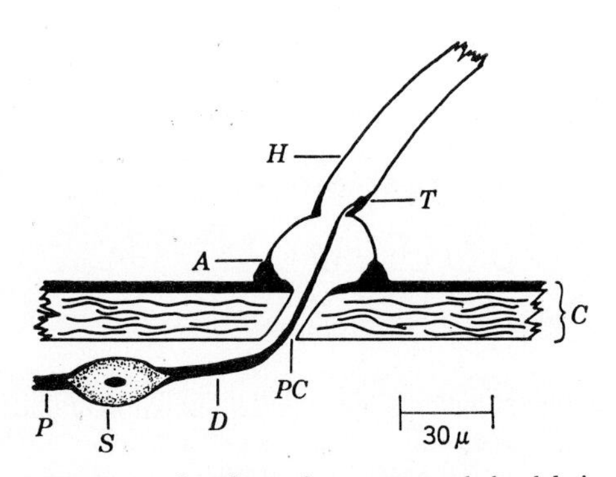


FIG. 4. Diagram of arthropod sensory exoskeletal hair of the type innervated by a single primary sensory neuron. The distal process of the sense cell ends near the region of articulation of the hair [from M. J. Cohen and S. Dijkgraaf in *The Physiology of Crustacea*, T. H. Waterman, editor (Academic Press, Inc., New York) (to be published)].

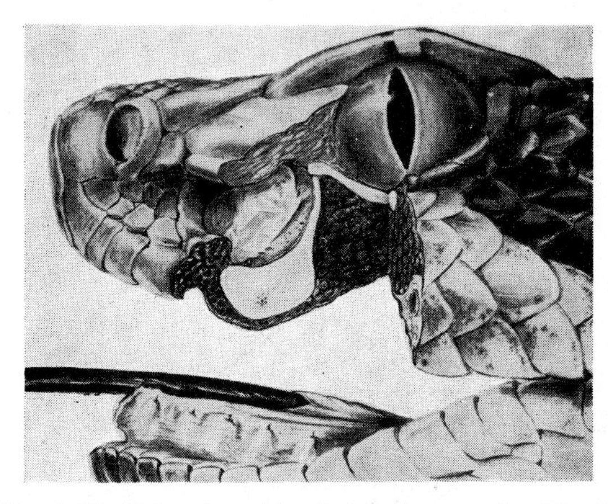

FIG. 5. The infrared sensitive facial-pit organ of rattlesnakes, *Crotalus*. Here, a wedge has been removed to show the deep pit and the sensitive surface—the thin membrane at its back (shown crinkled but normally smooth), with an air chamber both in front and behind it. This membrane, 10 to 15 μ thick, is heavily innervated and vascular [from T. H. Bullock and F. P. J. Diecke, J. Physiol. **134**, 47 (1956)].

As an example of the specialization of which arthropod mechanoreceptors are capable, Burkhardt and Schneider[11] have found that units in the Johnston organ of the antenna of flies (*Calliphora*) are hardly less sensitive than the human ear to sound frequencies between 150 and 250 cps, the range of the wing-beat frequency. Even beyond this range the units follow the sound frequency faithfully, giving the animal sensory information corresponding to each wing beat with a delay of only about 1 msec. The speed of flight is controlled by this sense organ. For some other recent studies in this area, see references 12–17.

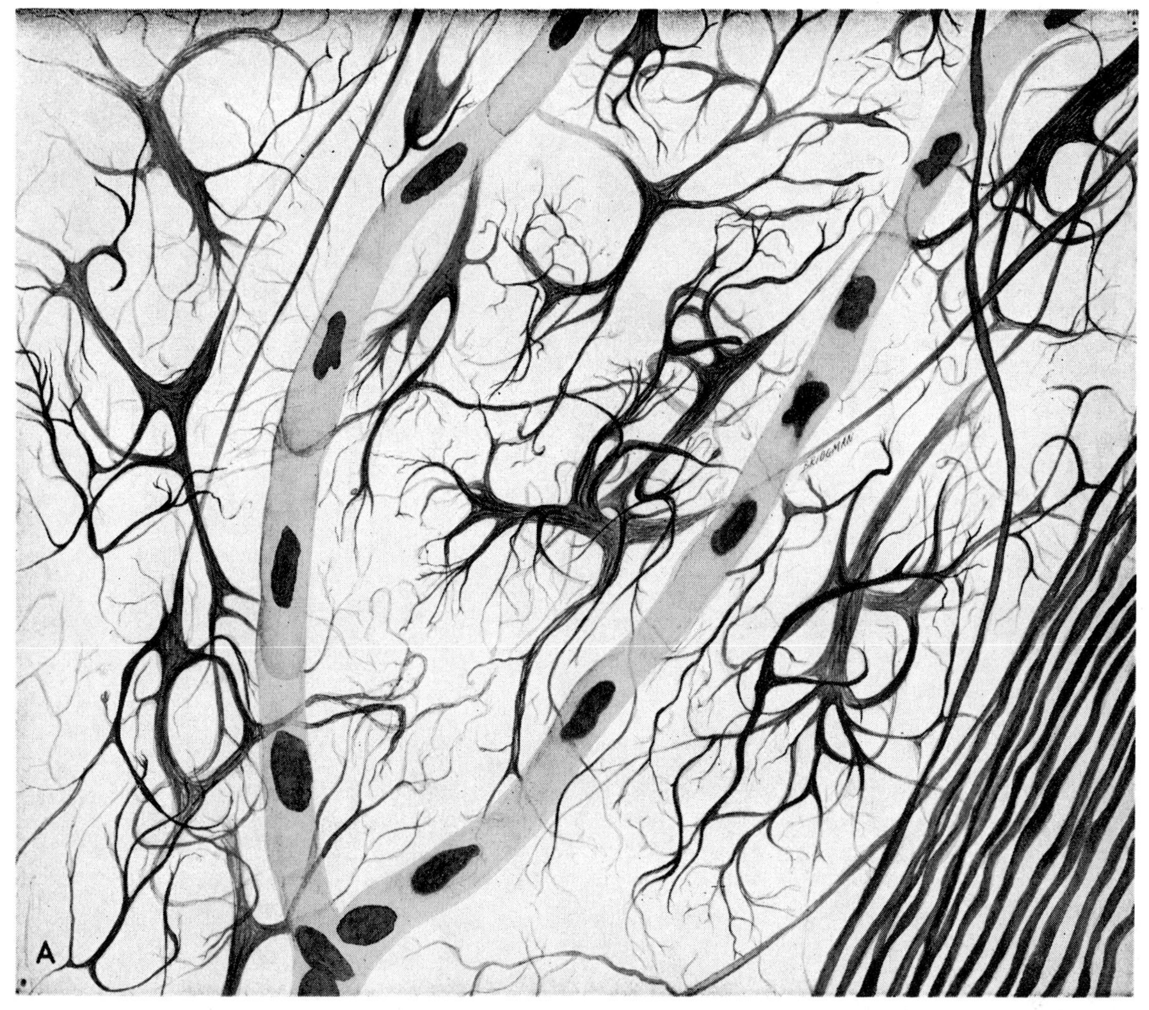


FIG. 6. The sensory endings in the pit-membrane of a rattlesnake. This is a faithful drawing from a silver-stained whole mount of the 10 to 15 $\mu$ membrane. The nerve fibers end freely in palmate expansions with branching processes. 500 to 1500 such expanded endings occur per mm$^2$. Apparently all are of one functional type—so-called warm receptors. Width of picture, 150 $\mu$ [from T. H. Bullock and S. W. Fox, Quart. J. Microscop. Sci. **98**, 219 (1957)].

## TEMPERATURE RECEPTORS

Referring briefly to temperature receptors, there is some information regarding the extraordinarily developed long-infrared-radiation detectors found in rattlesnakes and other pit vipers in the thin membrane at the base of the facial pit (Figs. 5 and 6)[18,19]. This structure, richly provided with a special form of free nerve ending and specialized in other ways—such as, for the directional estimation of radiant sources—responds to very small doses, of the order of $10^{-11}$ small calories in $\frac{1}{10}$ sec on the area of the terminal ramification of one nerve fiber (2000 $\mu^2$). The evidence indicates that this response is due to the change in temperature of the tissue, which both by indirect and direct methods is estimated to be of the order of 0.001°C, close to the value already found for the human.[20] Expressed as $Q_{10}$, the frequency of nerve impulses in a single fiber increases with temperature with a $Q_{10}$ of about $10^{30}$, a figure which offers considerable room for speculation about *high amplification with preservation of reasonable stability* (Fig. 7). Much more could be said about each of these cases with respect to physiological properties and characteristics of response. It is the purpose here only to call attention to the variety of opportunities and problems presented by a few recently studied receptors.

## CHEMORECEPTORS

In the area of chemoreceptors, I choose only two recent reports. Hodgson and Roeder[21] have discovered labellar hairs in various flies in which two primary sensory neurons send distal processes to the chemoreceptive area at the tip of the hair. These two neurons have different modes of sensitivity: one responds only to sugars; the other to salts, acids, and alcohols. Not only is this preparation of interest because of the opportunity for the physiological *study of unit receptors of different chemical specificity*, but also, because of its additional sensitivity to mechanical stimuli and to tem-

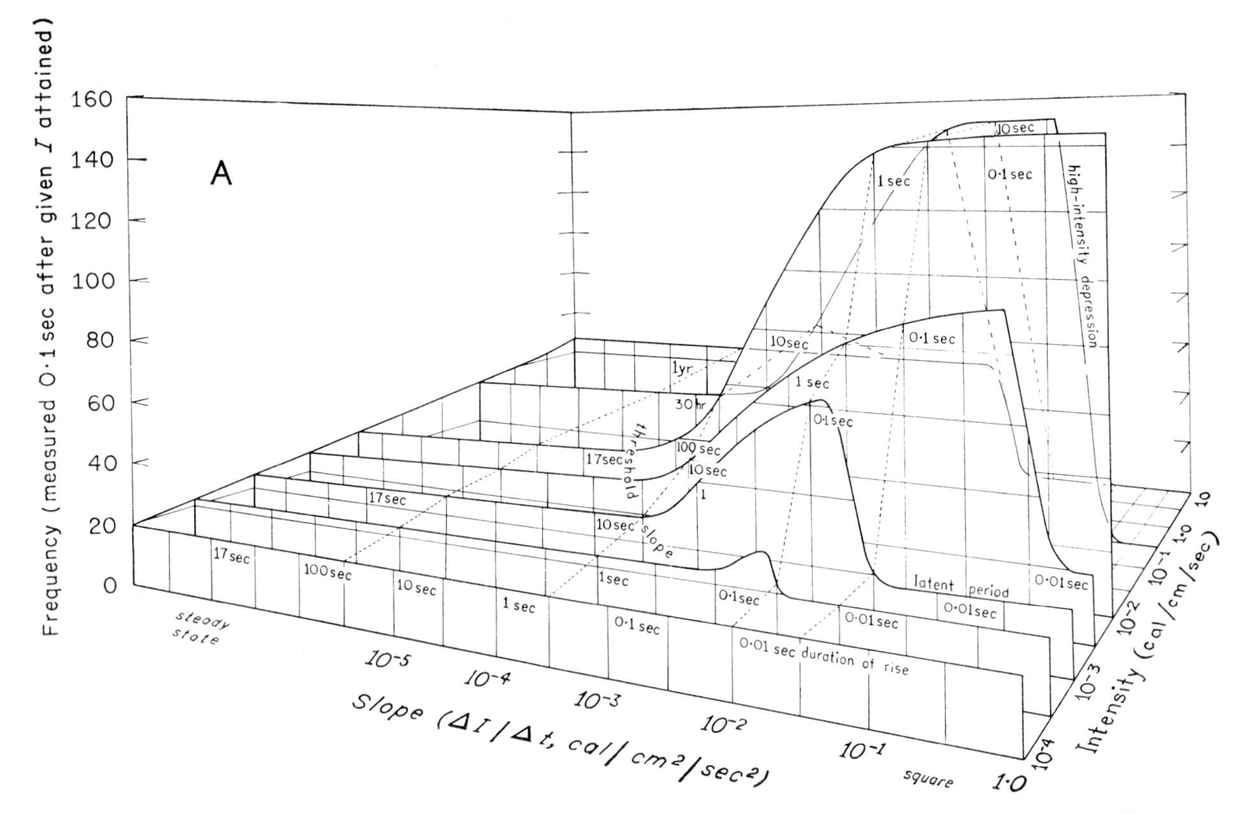

FIG. 7. Response of typical receptor unit of rattlesnake infrared sense organ, measured as nerve impulse frequency at different rates of increase of radiation and at different flux levels [from T. H. Bullock and F. P. J. Diecke, J. Physiol. **134**, 47 (1956)].

perature changes as small as $\frac{1}{10}$°C, it presents in a clear form the general *problem of analyzing the world through receptors which are not unambiguously* detecting one aspect of the environment only. Furthermore, these labellar hairs are remarkable, as shown by Dethier,[22] in that complete behavioral rejection or acceptance responses of the intact animal occur to stimulation by a microdrop which can reach only one single sensory neuron. Schneider[23] has discovered an electrophysiological response in the antennae of silk moths (*Bombyx mori*) in the males only, which is highly specific apparently to the naturally occurring odorous material produced by the female. This odorous material attracts the males from great distances. In microelectrode records from the isolated antenna, picked up extracellularly from many units, Schneider finds a slow wave of several millivolts which is negative for some substances and positive for others. Spikes are superimposed preferentially on the negative phase of the electroantennogram. As in the preceding case, there is here an apparently peripheral filtering of the normally specific stimulus, which filtering is achieved by a *specific chemical sensitivity* of a receptor and again one is faced with high sensitivity. de Vries[24] has calculated that at threshold there are far fewer stimulating molecules impinging upon the sensory epithelium in man than there are olfactory receptor cells in the same area.

## ABSOLUTE RECEPTION

Extending the remarks by Rosenblith (p. 485), it is highly desirable that further attention be given the problems raised by the sensory reception of absolute stimulus values as opposed to reception where comparison can be made by the receptors with a status in the recent past. Besides body temperature, blood pressure, and the like in man, there are many other indications of reception of values which may be called absolute, e.g., preferred temperatures in cold blooded animals, levels of light which day after day induce a given response, $CO_2$ concentration reactions, and pitch recognition. Even in many humans lacking phenomenal ability, the cross-modality subjective-intensity matching and many psychophysical phenomena illustrate scales not dependent upon relative stimulation as measured against a just-preceding level. One faces here the problem of stability and drift correction—not only the control loops, intra- or inter-cellular, but the standard of reference for detecting error.

## SEQUENCE OF LABILE COUPLINGS LEADING TO IMPULSE INITIATION

Consider now what makes a neuron fire—namely, the specification of the distinguishable processes within the neuron leading to an explosive firing of a propagated all-or-none nerve impulse. The interesting conclusion

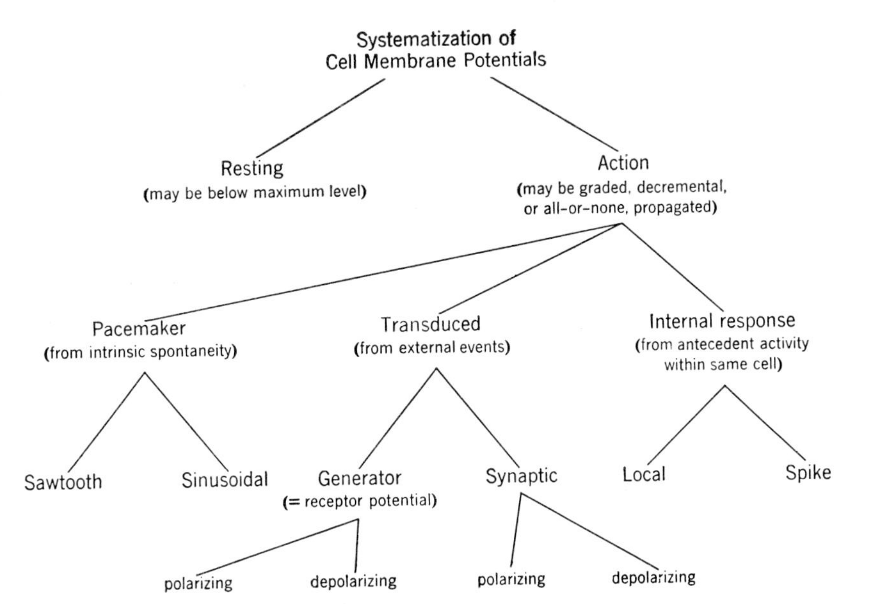


FIG. 8. The types of nerve cell membrane potentials.

permitted by current evidence is that *the determination of firing is not merely the build-up of an adequate stimulus* or its transduced and amplified resultant to a critical level, the threshold, but rather it is a sequence of labile couplings between graded (analog) events, each occurring in a limited fraction of the neuron. Normal firing is preceded by a series of steps, with alternative pathways. The several processes are separate, partly in space and partly in time, but causally are interconnected. Some of these processes are reflected in membrane-potential changes and some are not.

The *processes reflected in the membrane potential* include slow shifts of the average membrane potential, synaptic, generator, local, and pacemaker potentials. Figure 8 makes it clear that generator and synaptic potentials are parallel types of transduced potentials representing amplified response to external events whether from a presynaptic cell, from a non-nervous sense cell, or from the external environment. It also points out that both of these classes of response may occur either in polarizing or depolarizing directions. Depolarizing responses are commonly called excitatory because they are likely to increase the probability of firing. Polarizing, or as they are sometimes called, hyperpolarizing, responses generally decrease the probability of firing and are called inhibitory.

First among the processes which determine the initiation of impulses in receptor and central neurons, we should list what is actually no doubt a whole class of different and complex processes by which the level of the so-called *resting potential* is under the influence of the milieu and nonspecific agents (nonspecific to a given cell, though possibly specific to a cell type). Hormones, inorganic and organic constituents of the medium, deformation of the cell (even in neurons other than specific mechanoreceptors), and other factors are known to have some effects, although in most cases their importance in normal physiology cannot be assessed yet. In addition to these, we now can recognize as an important class the general shift in membrane-potential level accomplished through specific nervous pathways. For example, Otani and Bullock[25] found in certain cells of the 9-celled cardiac ganglion of lobsters that certain presynaptic fibers exert an influence, but without any discrete synaptic potentials. With repetitive stimulation they cause a slow, smooth shift of membrane potential (Fig. 9) as seen by an intracellular electrode in the soma. Terzuolo[26] has shown that certain neurons in the spinal cord of the cat similarly respond to stimulation of certain parts of the cerebellum by a shift of membrane potential level. This absence of discrete synaptic potentials even with a small number of incoming pathways possibly means that the main factor in determining the slow, smooth shift of potential is the distance of the synapses from the soma. If the distance is considerable, individual deflections would be smoothed out by the spread through leaky cables. But, if one believes that the membrane potential level at the basis of the axon or some such limited locus is crucial to the determination of cell firing, any such properties of the dendrites and soma—such as the electrotonic conduction of slow potential changes, the smoothing out of potential changes, or the discrimination against rapid deflections—would be of decisive importance in spreading the influence of the decrementally propagating activity of the much-branched dendritic processes to the region in which the spike originates. This may be one reason for having long dendrites. Among sensory neurons, there are many whose distal process divides to supply a considerable number of receptor regions of the periphery (Fig. 10, from the work of Meyer[27]) and in which the histology is suggestive that these loci are not alternative sources of full-fledged impulses but may, in some cases, make graded contributions to the probability of an impulse arising at some

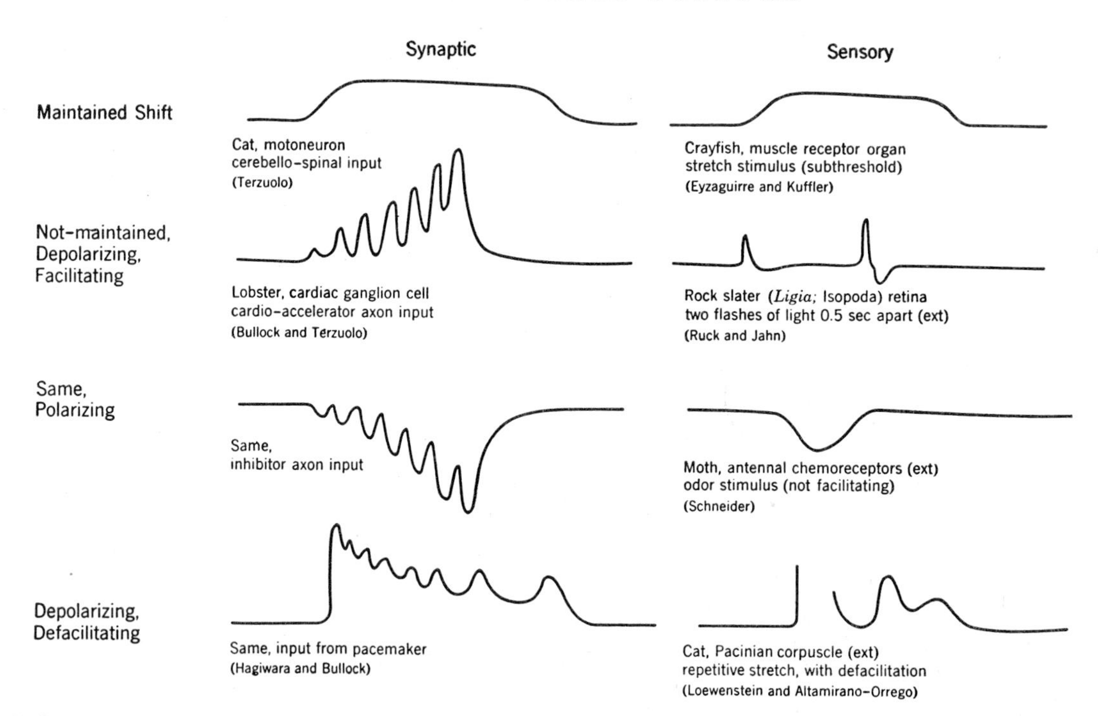


FIG. 9. Some of the types of subthreshold potentials, to show the correspondence between synaptic and sensory types. Hand-drawn approximations of original records, intracellular (except when marked ext=external); time scale is not uniform; upward deflection is depolarization or, in external recording, negativity of closer electrode.

point, such as a confluence of main branches of the distal process. So much for the processes that act upon the spike probability through the so-called resting potential.

More typically, or at least more familiarly, *the response to an external event is the transient deflection*, called a synaptic potential or a generator potential according to whether the external event is presynaptic or whether it is a sensory stimulus (Fig. 9). These responses arise in restricted portions of the neurons and have their significance in the effects they produce in other regions after nonregenerative spread—with decrement and phase delay. They may be polarizing or depolarizing, and this alternative may be decided by the nature of the external event or, in certain circumstances by the

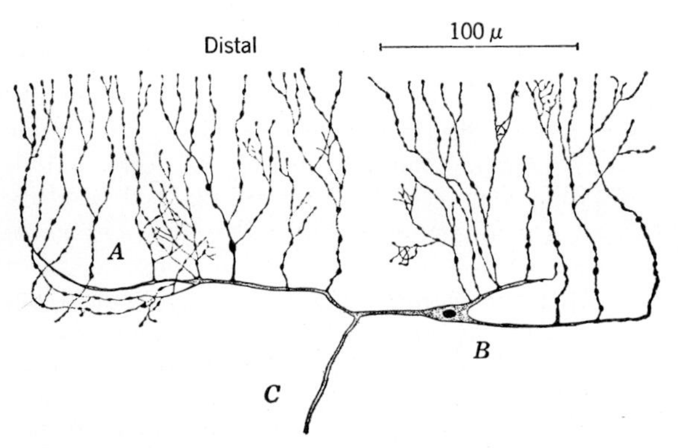


FIG. 10. Sensory nerve cell from the leg of a centipede, *Lithobius*; methylene blue. *A*—lattice of fine terminal fibers, *B*—cell body, *C*—axon [from G. F. Meyer, Zool. Jahrb. Anat. 74, 381 (1955)].

level of the membrane potential obtaining at the moment. Some normally polarizing, hence inhibiting, events are reversed readily to a depolarizing response which can be shown to increase the probability of firing and are thus at least in a measure excitatory. Such change in the response to a given input owing to different levels of membrane potential is not purely a laboratory phenomenon resulting from experimentally imposed polarization, but occurs normally, for example, in the aforementioned situation where Terzuolo found the membrane level of spinal motoneurons shifted by cerebello-spinal pathways, thereby altering the response to dorsal-root inputs. (In some accounts, emphasis has been laid on the fact that an inhibitory input drives the membrane potential from either side toward a certain equilibrium value and does not cause any synaptic potential if the membrane is already at that level. Even under the latter condition inhibition can result, because the lowered impedance tends to clamp the membrane. But this can inhibit only in the near vicinity. Upon spike-initiating regions or synaptic regions at some distance, the sign and magnitude of the potential change will determine the sign and magnitude of the effect.)

In at least some cases, there intervenes between the synaptic or generator potential and the critical level at the spike-initiating locus (which is a very restricted part of the neuron at some distance from the principal synaptic and generator loci) *an intermediate graded potential* which is usually called the *local potential* (Fig. 11, from the work of Bullock and Terzuolo[28]). This is presumed to arise in an electrically excitable membrane adjacent

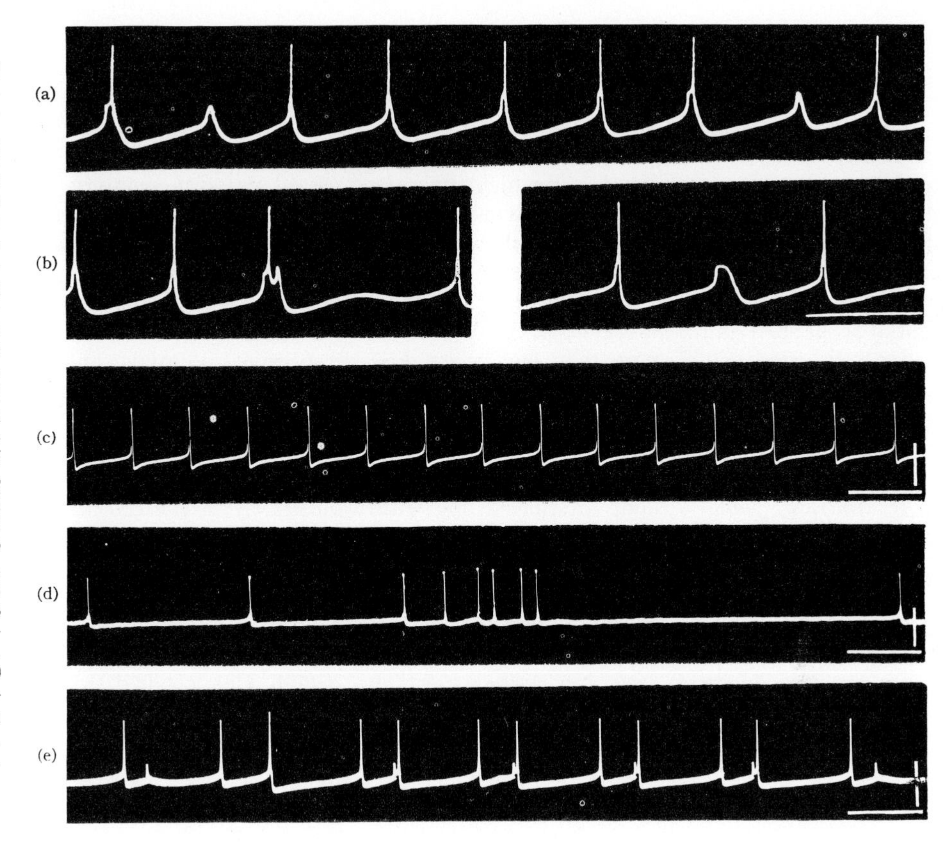


FIG. 11. Examples of true spontaneous activity as seen by an electrode inside a nerve cell. Cardiac ganglion cell of the lobster, *Panulirus* [(a) and (b)] or the crab, *Cancer* [(c) (d) (e)]. (a) and (b) show a large pacemaker potential, presumably arising nearby, and another prepotential, regarded as a local potential, before the spike. The local potential may fail to elicit a spike and then can alone cause repolarization. Note the failure of the local potential to arise following the third spike in (b) with instead an undulation leading to a new cycle. (c), (d), and (e) show different forms and permutations of pacemaker potential and repolarization. Scales: (a) and (b)—500 msec; (c), (d), and (e)—50 mv, 200 msec [from T. H. Bullock and C. A. Terzuolo, J. Physiol. **138**, 341 (1957)].

to the synaptic or generator membrane, but distinct from it, and as a consequence of the local circuits from antecedent activity of the same cell. In this respect, the local potential is like the spike potential, which also arises by electrical excitation from the local circuits in neighboring regions of the same neuron. Indeed, local potentials can be made to have graded amplitudes all the way to the full amplitude of the spike potentials, and they differ only in being nonregenerative and, therefore, in spreading decrementally for distances of the order of a millimeter only. In some sensory neurons, we may conjecture that such local potentials occur in different branches of the distal process and sum in the stem process to determine the initiation of a spike. The experiments of Katz[29] on the summating subthreshold deflections in sensory terminals of muscle spindles may be so interpreted. Situations where distal sensory processes branch over a considerable area are quite common.

The nonregenerative property of some neuronal membranes is important in those cases where it results in the *inability of spike potentials, once initiated, to invade* those regions of the neuron. This preserves these regions from explosive depolarization and from subsequent strong repolarization, and thus enhances their integrative capacity. For example, the soma of the large cardiac ganglion cells in the lobster and perhaps the dendrites on some other types of neurons never experience an explosive all-or-none event. This characteristic is very likely true of most neuronal membranes, that is to say, of the vast forests of fine branching processes making up neuropiles. If this is so, one must think of all-or-none spike potentials as a peculiar development of a limited portion of the neuron whose function is the faithful propagation of signals over long distances rather than integration.

The familiar local and spike potentials are depolarizing in direction, but there are a few instances indicating that membranes are capable of equivalent ***regenerative potentials in the opposite direction.*** For example, during the plateau of an essentially completely-depolarized potential in the Purkinje tissue of the mammalian heart, during a heart beat, a threshold stimulus in the polarizing direction sets up a regenerative repolarization which grows explosively and restores a resting potential [Weidmann[30]]. Hisada[31] has recorded action potentials of the same polarity—that is, increasing polarization across the cell membrane of the protozoan *Noctiluca*.

Finally, we have *spontaneous activity* manifested in *pacemaker potentials* in some neurons including many sensory neurons. A background of continuous spike discharge in the absence of known stimulation or under steady-state stimulation may be called spontaneity. Of course, such activity depends upon permissive conditions of the metabolism and of the milieu of the cell. If these conditions are normal ones, however, steady-state firing is spontaneous in the sense that the origin of the intermittent activity lies within the neuron, rather than in the environmental mechanisms. Such activity has been explored in several cases (Fig. 11), and the significant finding from the present point of view is the *localization of the pacemaker process to a limited portion of the neuron* different from that in which spikes arise, and perhaps also different from that in which synaptic or generator activity occurs primarily. Of course, these regions must be within shouting distance of each other, for their normal significance lies in the summing of their effects to produce a threshold change in the membrane potential at the spike-initiating region. In lobster cardiac ganglion cells, evidence of more than one pacemaker locus is seen in the same neuron. This is possible because these loci apparently occur in different major processes of the cell, each of which has its own spike-initiating region; two different rhythms of spikes arise and can be seen in the same soma without interfering one with another since they cannot invade the soma. What this says is that the absolute dimensions of neurons are not accidental but are fixed by engineering requirements specified by the electrotonic parameters of the membrane.

There are very likely at least *two distinct types of spontaneous potential* change. The rarer is approximately sinusoidal. Spikes tend to occur during the phase of maximum depolarization. Several spikes may occur during this phase, but they are not necessary for and do not accelerate the repolarizing phase. In contrast, the more common type is similar to a relaxation oscillation, in that the pacemaker potential itself is a steadily depolarizing potential which must be interrupted by some new process with a threshold whose net result is to repolarize, thereby starting a new cycle. This threshold process is typically the spike potential, but it may be preceded by and indeed may be substituted by a local potential which is capable of repolarizing the membrane during its recovery phase, thus setting the stage for another pacemaker potential to begin.

There are other *permutations and complexities in the processes of spontaneous impulse initiation* as is suggested by the high rhythmicity of some receptors and the low rhythmicity of others. In fact, it may be characteristic that, in one and the same unit, rhythmicity is relatively high at high frequencies of discharge and low at low frequencies. Tokizane and Eldred[32] have found two distinct populations of stretch receptor fibers in the dorsal roots of cats. One population is consistently more rhythmic at any given average frequency than the other; that is, the standard deviation of intervals is smaller for a given average interval. They believe that the more-regular units come from flower-spray endings in the muscle spindles and that the less-regular units belong to annulo-spiral endings.

The various possibilities for understanding the *origin of nonrhythmicity* or randomness-within-limits of successive intervals cannot be discussed here. Nor can one discuss the interesting question of the central problem created by nonrhythmicity, the *distinction of signals from noise*—that is, of weak true stimuli from random changes in the frequency of firing. These problems have been discussed elsewhere (see references 33–35).

These separate processes reflected in membrane potential shifts, in synaptic, generator, local, pacemaker, and spike potentials are *sequentially coupled* in complex ways: partly because there are several alternative sequences; partly because much depends on the particular microanatomy of a given neuron, the spatial relations of the separate loci, and the possibilities of spread of the respective potentials; and partly because of the labile character of the coupling constants. The constants or transfer functions well may be nonlinear.

However, these do not exhaust the processes of which there is evidence that determine the initiation of impulses within the cell. Turning to *processes not manifested by membrane potential* changes—at least by changes detectable with the same methods used for measuring the just-discussed events—we are faced mainly with indications of excitability changes. These are brought to attention only by responses to subsequent stimuli. One is called *facilitation* (Fig. 9). The cell can be initially at a certain membrane potential and respond to a given arriving impulse (in this case, a single presynaptic fiber) with a small synaptic potential. After repetition, the response to the same presynaptic impulse is much greater even when the membrane potential is allowed to return to the initial value between stimulations. Another type, which may be seen in the same neuron may be called *defacilitating*; the response is less to successive presynaptic impulses the closer together they come. Here also, at the same membrane potential level, the cell or presynaptic terminals are altered in response or excitability according to the recent history. This is not a laboratory artifact but is an essential part of the normal mechanism of burst formation in the 9-celled cardiac ganglion of the lobster. Here, the large follower neurons, driven by a burst of arriving presynaptic impulses from the pacemakers, characteristically give the form of synaptic potential response shown in Fig. 9. A large, initial synaptic potential is followed by a series of small ones during the high-frequency portion of the burst and then by growing amplitude responses as the frequency of the arriving presynaptic impulses declines.

Here we have to deal with the *time course of excitability of graded, subthreshold events*. There must be a separate curve for each of the subthreshold events previously discussed and for each locus and type of synapse. The last two instances, of facilitation and defacilitation, recorded in the same neuron were both excitatory synaptic potentials, from different presynaptic pathways (one from the pacemaker in the cardiac ganglion, and the other from the cardioacceleratory nerve from the central nervous system). The same neuron also has facilitating inhibitory synapses. Each of these synaptic responses must have its own dependence of response upon input and of this relation (response per input) upon time after preceding activity. The local potentials also may have a separate time dependence of excitability and so may the spike-initiating locus where the excitability is measured by a sharp threshold. We note that the classical spike threshold, so easily measured any place in the axon, is normally a significant parameter at a very limited region only in the whole neuron. Beyond this region, the safety factor is normally sufficiently high that the spike threshold is without great significance.

Another degree of freedom not predicted by the membrane potential is the presence or absence of after-effects—either persistent response after the input ceases or of overshooting rebound (post-excitatory inhibition or post-inhibitory excitation). Cases of each kind are known for receptors and for central neurons.

Still another line of evidence for processes determining spike initiation, which cannot be seen in the usual membrane potential measurements, is that of significant *alteration of the frequency of already-active neurons by weak electric fields* in the surrounding medium, already mentioned. In a preliminary analysis, Terzuolo and Bullock[7] estimated the intensity of the voltage gradient in the saline surrounding a neuron (stretch receptor of the crayfish) when an imposed polarization was of a magnitude just sufficient to cause a noticeable change in the maintained frequency of firing. The results, as already mentioned, were that very weak fields (of the order of 1 $\mu v/\mu$ in the external medium around the cell) were sufficient. There will be a potential change in a given direction (for example, an excitatory direction) in certain parts only of the neuron, and since this will be graded in magnitude geographically, there must be a very limited region only wherein the imposed field actually exerts its effect. It can be concluded, therefore, that, even in the absence of average change in membrane potential, very localized regions may be critical in determining the firing frequency and may have an extremely high sensitivity to small voltage gradients. Thus, old ideas are confirmed in a certain sense that, among other things, the neuron is to a significant degree under the influence of differences of potential between one part and another of the surface of the neuron.

The general picture then is quite different from that of the axon. Grundfest[36] has given arguments for believing that adjacent patches of neuronal membrane are, respectively, electrically excitable and electrically inexcitable in a certain meaning of these terms. It may be emphasized that the evidence cited here is also strongly indicative of a *neuronal membrane consisting of a patchwork* of different kinds of subthreshold-response capacities. The responsiveness or the excitability—these two are different, but ordinarily are not easy to distinguish experimentally—vary separately one from another; some are manifest by electrical events, others are not.

One more degree of complexity must be added: the separate processes do not simply sum algebraically to determine the level of probability of impulse initiation, but at least some of them apparently interact: the presence of an increased magnitude of one subthreshold process may alter the magnitude of one of the others. For example, in the presence of a large generator potential, excitatory synaptic potentials may not simply add but will themselves be altered in size. The same may be true for inhibitory and pacemaker potentials. These interactions have been little studied, but there is no doubt that they will be found, in some cases, to contribute importantly to the integration of input to determine output (see, for example, the evidence of multiplicative neural events adduced by Hassenstein and Reichardt[37]).

## SUMMARY

A selection of sense organs upon which recent progress has been made is reviewed in order to point out problems and opportunities of biophysical interest. The electroreceptors in electric fish, mechanoreceptor hairs in arthropods, infrared receptors in pit vipers, and specific chemoreceptors in flies and moths are the examples presented.

The general problem of absolute reception as opposed to reception of change is emphasized in terms of stability, drift correction, and reference standard.

The long series of subthreshold events which may intervene between transducing a stimulus and initiating a spike is reviewed. These include events which are reflected in the membrane potential and others which are not. Specific shifts of membrane potential, pacemaker, generator, synaptic, local and spike potentials, and some subvarieties are sequentially coupled in labile and perhaps nonlinear ways. They occur, in general, in restricted regions of the neuron which interact in complex ways not only because of the complex coupling constants between successive steps but also because of the profound influence of the anatomical distribution of the differently responding types of membranes.

Events not reflected in the membrane potential are the excitability cycles and the presence or absence of aftereffects. In addition to the spike threshold, each graded process preceding the spike has a curve of response against input and of this relation against time after preceding activity.

It is probably important that these several processes not only are sequentially related but that they interact—the amplitude of one may alter the responsiveness of another. Spike initiation is, therefore, potentially a highly derived and integrative result.

## ACKNOWLEDGMENTS

Permission to reproduce figures from the *Journal of Physiology*, *The Journal of Experimental Biology*, and *The Quarterly Journal of Microscopical Science* is gratefully acknowledged.

## BIBLIOGRAPHY

[1] R. Granit, *Receptors and Sensory Perception* (Yale University Press, New Haven, 1955).

[2] T. H. Bullock, Federation Proc. **12**, 666 (1953).

[3] H. W. Lissman and K. E. Machin, J. Exptl. Biol. **35**, 451 (1958).

[4] H. W. Lissman, J. Exptl. Biol. **35**, 156 (1958).

[5] H. Grundfest, Progr. in Biophys. and Biophys. Chem. **7**, 1 (1957).

[6] C. W. Coates, M. Altamirano, and H. Grundfest, Science **120**, 845 (1954).

[7] C. A. Terzuolo and T. H. Bullock, Proc. Natl. Acad. Sci. U. S. **42**, 687 (1956).

[8] M. J. Cohen, J. Physiol. **130**, 9 (1955).

[9] M. J. Cohen and S. Dijkgraaf in *The Physiology of Crustacea*, T. H. Waterman, editor (Academic Press, Inc., New York) (to be published).

[10] C. A. G. Wiersma and E. Boettiger (to be published).

[11] D. Burkhardt and G. Schneider, Z. Naturforsch. **12B**, 139 (1957).

[12] C. Eyzaguirre and S. W. Kuffler, J. Gen. Physiol. **39**, 87 (1955).

[13] L. H. Finlayson and O. Lowenstein, Proc. Roy. Soc. (London) **B148**, 433 (1958).

[14] E. Florey, Z. Naturforsch. **10B**, 591 (1955).

[15] E. Florey, Z. Naturforsch. **11B**, 504 (1956).

[16] S. W. Kuffler in Submicroscopic Organization and Function of Nerve Cells, H. Fernández-Morán and R. Brown, editors, Exptl. Cell Research Suppl. 5, 493 (1958).

[17] W. Loewenstein and R. Altamirano-Orrego, J. Gen. Physiol. **41**, 805 (1958).

[18] T. H. Bullock and F. P. J. Diecke, J. Physiol. **134**, 47 (1956).

[19] T. H. Bullock and S. W. Fox, Quart. J. Microscop. Sci. **98**, 219 (1957).

[20] J. D. Hardy and J. W. Oppel, J. Clin. Invest. **16**, 533 (1937).

[21] E. S. Hodgson and K. D. Roeder, J. Cellular Comp. Physiol. **48**, 51 (1956).

[22] V. Dethier, Quart. Rev. Biol. **30**, 348 (1955).

[23] G. Schneider, Z. vergleich. Physiol. **40**, 8 (1957).

[24] Hl. de Vries, Progr. in Biophys. and Biophys. Chem. **6**, 207 (1956).

[25] T. Otani and T. H. Bullock, Physiol. Zool. (to be published).

[26] C. A. Terzuolo, (personal communication, 1958).

[27] G. F. Meyer, Zool. Jahrb. Anat. **74**, 381 (1955).

[28] T. H. Bullock and C. A. Terzuolo, J. Physiol. **138**, 341 (1957).

[29] B. Katz, J. Physiol. **111**, 261 (1950).

[30] S. Weidmann, Ann. N. Y. Acad. Sci. **65**, 663 (1957).

[31] M. Hisada, J. Cellular Comp. Physiol. **50**, 57 (1957).

[32] A. Tokizane and E. Eldred, (personal communication, 1958).

[33] T. H. Bullock in *Recent Advances in Invertebrate Physiology*, B. T. Scheer, editor (University of Oregon Press, Eugene, 1957), p. **1**.

[34] R. Fitz-Hugh, J. Gen. Physiol. **40**, 925 (1957).

[35] R. Fitz-Hugh, J. Gen. Physiol. **41**, 675 (1958).

[36] H. Grundfest, Physiol. Rev. **37**, 337 (1957).

[37] B. Hassenstein and W. Reichardt, Z. Naturforsch. **11B**, 513 (1956).

55

# Receptor Mechanisms and the Integration of Sensory Information in the Eye

H. K. Hartline

*The Rockefeller Institute, New York 21, New York*

THE processing of sensory information begins in the sense organs themselves. It is in them that the first steps take place in the transformation of external influences into the patterns of nervous action that regulate the activity of an animal in its complex environment. The fundamental nature of the receptors and the design of the accessory structures in which the receptors are deployed determine how external information flows into the organism. Thus, sense organ and receptor mechanisms determine the character of the neural activity that is passed on to higher neural centers. In addition, the first steps in neural integration take place within the sense organs, for in many of them the receptors interact with one another. As a result of both of these actions, patterns of sensory nerve fiber activity transmitted to the higher centers are more than mere replicas of the temporal and spatial patterns of external stimuli. Certain significant features of the stimulus patterns are accentuated at the expense of less important fidelity of representation. This can be clearly illustrated in the analysis of the first steps of the visual process, with which this paper deals.

One of the great contributions of biophysics in the last century was the precise description of the human eye as an optical instrument. The high degree of perfection of our eyes enables us to exploit many of the peculiar advantages of luminous energy as a source of information; their shortcomings set limits to our visual performance. The vertebrate scheme of optical imagery by a lens system is not the only one that is used by animals; compound eyes also have been evolved—made up of small optical units, each having a narrow entrance angle and each pointed in a different direction so that all cover the entire field of view. They too have both advantages and disadvantages, one of the advantages being that short wavelengths can penetrate to their receptors. In either case, retinal receptors arranged in a mosaic receive light in varying amounts from the various parts of the animal's surroundings. The mechanism of the visual receptor units that compose the retinal mosaic determines many of the properties of vision.

The photoreceptor offers certain advantages over many other receptors in the study of sensory mechanisms, for in it the very first step in the transducer mechanism for translating the stimulus into nervous action is beginning to be well understood. This is a consequence of the general principle that electromagnetic radiation, to produce a permanent effect on a material system, must yield some of its energy to the system. Consequently, the action spectrum of the visual apparatus is simply the manifestation of the absorption spectrum, or a portion of it, of the primary photosensitive material in the visual receptors. It is the absorption spectrum of the primary visual pigment that sets the rather indistinct limits to the extent of the visible region within the electromagnetic spectrum and that determines quantitatively the relative effectiveness of different wavelengths of visible light. There is now very good agreement between the measurements of the absorption spectrum of photolabile pigments extracted from the retina and the "action spectrum" of vision for several animal forms, especially for man.[1,2]

The fact that one can identify the photosensitive material of the visual receptor makes it possible to say whereabouts in the receptor cell the first act of the visual process takes place. In the vertebrate eye, the visual pigment "rhodopsin" is known to be concentrated entirely in the outer segments of the retinal rods. This identifies the outer segments as the locus of the initial step in the visual receptor process. Rhodopsin can be extracted from suspensions of the outer segments of retinal rods, and its absorption spectrum, after appropriate correction, agrees well with the distribution of spectral sensitivity of rod vision. It is clearly the primary photosensitive substance of the rods. A number of visual pigments related to rhodopsin are now known. One of them, iodopsin, is the corresponding photosensitive substance of the retinal cones.[3] The biochemistry of the visual pigments constitutes an extensive and elegant chapter of modern biochemistry that cannot be discussed in detail in this paper (cf. Wald[4]).

Rhodopsins are known to be conjugated proteins, the prosthetic group being a carotenoid called retinine. Retinine is an aldehyde, the corresponding alcohol being vitamin A, and is known in a number of isomeric forms. The first act of light apparently is to produce an isomerization of the carotenoid group while it is still attached to the protein.[5] Retinine is then split off the protein molecule by subsequent reactions that are independent of light, and may be converted reversibly into vitamin A.

After photolysis, visual pigments can regenerate. Otherwise, one would have one look at the world and then be forever blind. The kinetics of photolysis and regeneration of visual pigments has been studied extensively, both *in vitro* and recently in the living retinas of experimental animals and human subjects.[6] Many

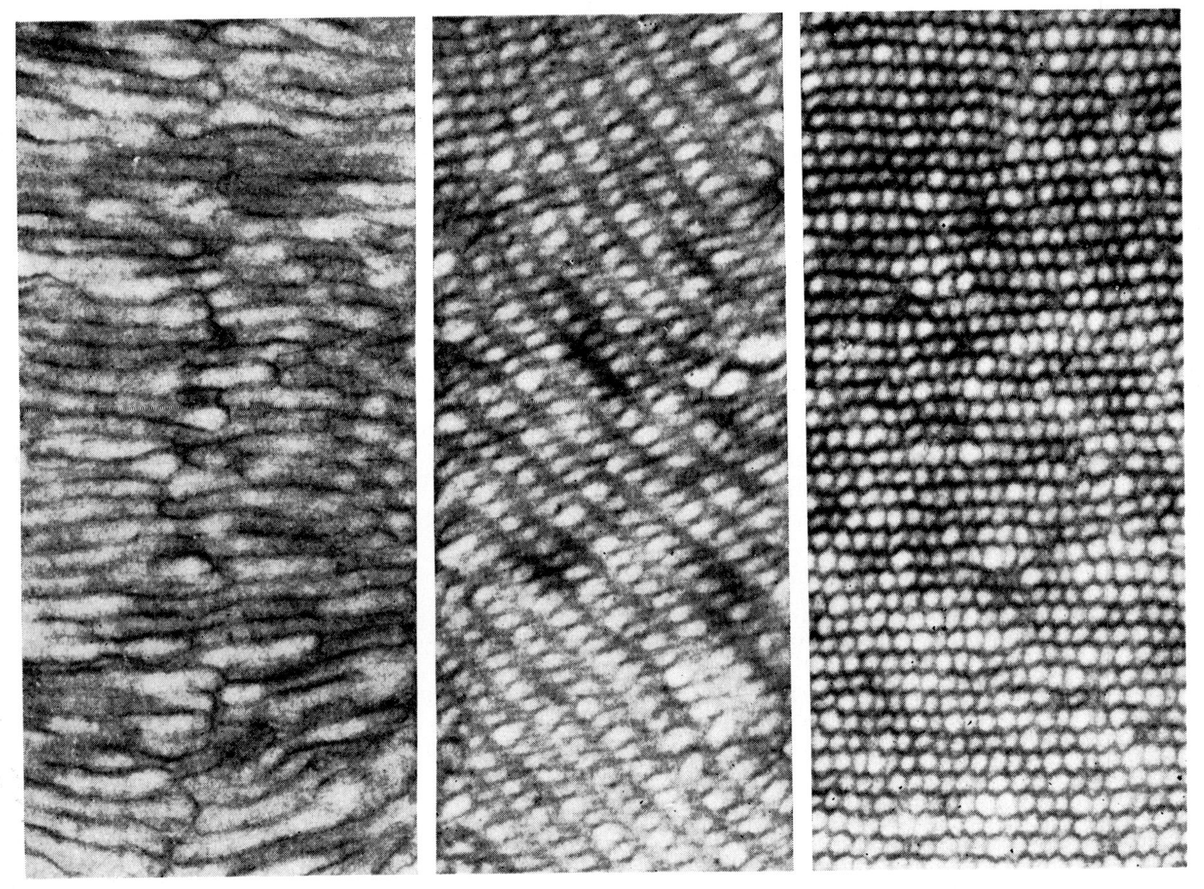

FIG. 1. Electron micrographs of rhabdom of an ommatidium of an arthropod compound eye (*Limulus*), showing honeycomb-like arrangement of osmium-staining membranes. Left: plane of section perpendicular to optical axis of ommatidium. Center: oblique section. Right: section in an axial plane. Height of figure approx 2 μ. Courtesy W. H. Miller.

receptor properties, such as the loss of sensitivity in the light and its recovery during dark adaptation, can be explained qualitatively by these elementary biochemical processes in the visual receptors. Moreover, a simple model of the photochemical system of the receptor was used by Hecht to explain quantitatively many psychophysical measurements of vision.[7] His formulations remain the most comprehensive and successful theoretical treatment of visual-receptor physiology, although his model is oversimplified and the theory needs reworking in light of recent developments in biochemistry and physiology.

The primary photosensitive pigment of the visual receptor is present in a structured system. Electron-microscope studies show a profusion of osmium-staining membranes in visual receptors. Sjöstrand[8] has shown that the outer segments of the receptors of the vertebrate retina have a lamellar structure. The rod outer segment is thus a stack of thin plates crowded with rhodopsin. In the arthropods, instead of a lamellar system, the part of the receptor cell (the rhabdom) that presumably contains the visual pigment is composed of myriads of microvilli densely packed to form a honeycomb-like structure. Figure 1 is an electron micrograph of the rhabdom of an ommatidium of an arthropod compound eye.[9] In the vertebrates, the outer segments of the retinal rods and cones have been shown to be derivatives of cilia.[10,11] In the arthropods, where cilia are extremely rare, there is no evidence of a ciliary derivation. In some mollusks, however, there is a different system of membranes and again the structures are derived from cilia.[12] Exactly how the visual pigment is arranged within any of these membranous structures is not known, though there have been speculations on this point.

Visual receptors have evolved into light detectors that are so sensitive that they work at the limit set by the quantum nature of light. A human observer is able to see a flash of light that contains only about 100 quanta, measured at the cornea of the eye. After correction for losses in transmission through the ocular media and failure of the visual purple to be present in sufficient amount in the retina to absorb all of the quanta that fall on it, this figure comes down to something of the order of 10 quanta.[13] This aspect of visual physiology has been extensively studied and is well reviewed in a recent article by Pirenne.[14] Obviously, it is of great significance to visual performance, especially at low

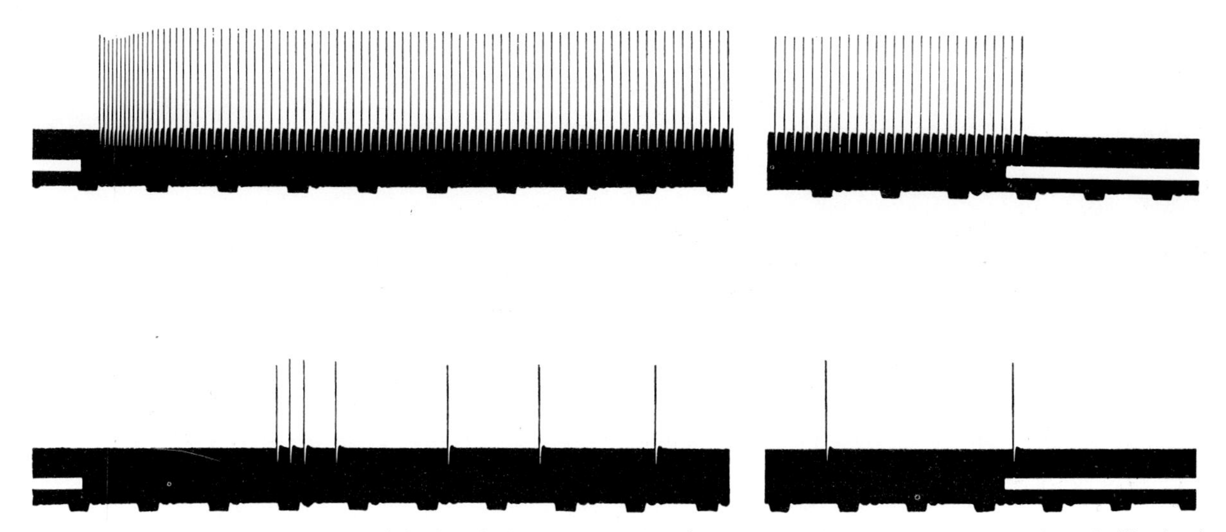

FIG. 2. Oscillograms of the electrical activity in a single optic-nerve fiber (eye of *Limulus*) in response to prolonged steady illumination of the facet of the eye innervated by the fiber.[15] Each spike is the "action potential" associated with the passage of a nerve impulse in the fiber. For the top record, the intensity of stimulating light was $10^4$ times that used for the bottom record. Signal of exposure to light blackens out the white line above the time marks. Time marked in $\frac{1}{5}$ sec.

illuminations. In the short "action time" of the retina, so few quanta are needed that the retinal image, though visible, is too "grainy" to be seen with high resolution. Also, at threshold, seeing is uncertain. Indeed, the statistical uncertainty at threshold can be explained almost entirely by the fact that a very few quanta suffice to excite a response. Nevertheless, the visual threshold is sharp enough so that it is quite certain that a human observer cannot see just one quantum, although exactly how many are needed is still a matter of controversy. The small amount of light that is just visible can be seen if it is spread over a retinal area containing about five hundred rods. This must mean that near threshold there is almost no chance of any one rod receiving more than one quantum, and that the cooperative activity of several rods is necessary to reach the threshold of vision. Thus, a single quantum of light absorbed within the stack of plates comprising the outer segment of a rod is sufficient to excite that rod, causing it to transmit some kind of nervous influence that can sum with similar influences from several other rods to reach the threshold for a behavioral response. In this retinal summation, one has an example of the simplest kind of neural integrative action, exerted at the very threshold of vision.

The end result of receptor excitation is the generation of nervous influences in its attached nerve fiber. It has not yet been possible to record the neural activity of the receptors (rods and cones) of the vertebrate retina, but some invertebrate eyes afford an opportunity to record optic activity that appears to be very close to the action of the primary receptors. The eye of the common horseshoe crab, *Limulus*, is particularly favorable for the study of the action of single receptor units. This eye is a coarsely facetted compound eye. Individual receptor units corresponding to each facet (ommatidia) can be separately illuminated, and the electrical activity of the optic nerve fiber from such a unit can be recorded.[15]

The neural activity recorded from one of the receptor units of the eye of *Limulus* consists of trains of uniform nerve impulses similar in all respects to the sensory discharges observed in nerve fibers in all of the higher animal forms (Fig. 2). As in all receptors, the higher the intensity of the stimulus, the higher the frequency of the impulses with which the receptor responds. In the visual receptor, it is noteworthy that frequency changes over a relatively small range for a large range of light intensity: the dynamic range of a single receptor is five or six orders of magnitude. Roughly, the relation between frequency of discharge and intensity of light is a logarithmic one (Fechner's Law). Thus, the transducer mechanism of the visual receptor covers a large range and is adapted to signal the ratios of stimulus values. In our own visual experience, values of light and shade stay more or less fixed, no matter what the ambient level of illumination, over a large range. In such situations, stimulus ratios stay constant and the visual receptors yield approximately a fixed difference in the frequency for a given ratio of stimulus values, even though the absolute differences may vary widely.

Another important receptor property is illustrated in Fig. 2. The discharge of nerve impulses begins at a high frequency when the light is turned on, but the frequency of the discharge subsides in a fraction of a second to a considerably lower level, which is then maintained with only slight diminution as long as light continues to shine on the receptor. This sensory adaptation is manifested by all other receptors, some to a far greater extent than others. As a result of sensory adaptation, receptors provide a somewhat distorted report of the stimulus events, such as to accentuate any sudden change. Sensory trans-

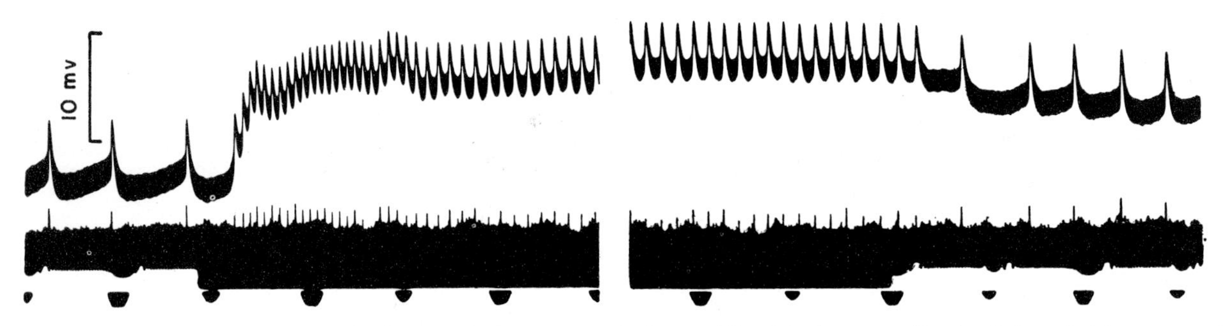


FIG. 3. Electrical responses to illumination of a single receptor unit (ommatidium) in the eye of *Limulus*.[15] Recorded (upper trace) by a micropipette electrode (tip diam 1 μ) in the neuron of the ommatidium, and simultaneously (black edge just below upper trace) by a pair of wire electrodes over which was slung the nerve bundle from the ommatidium containing the neuron's axon. At the beginning of the record, the microelectrode base line records the resting level of the electrical polarization of the cell membrane, at a potential about 50 mv negative to the solution bathing the outside of the cell. When the ommatidium was illuminated (black band above the time marks), there was a partial depolarization of the neuron (potential becoming less negative: rise in the base line) accompanied by an increase in the frequency of the spike-like deflections, each one of which was synchronous with the discharge of an impulse in the nerve bundle (small spikes on the black edge). Time marked in $\frac{1}{5}$ sec.

ducers are not concerned so much with a faithful representation of the world as with a useful one, and it is especially useful to the organism to accentuate the *changes* that occur in external conditions. If the illumination on a visual receptor unit is given a small increment, the receptor response consists of a modulation of the discharge of impulses in which there is an exaggeration of frequency changes at the onset and again when the increment is turned off. This permits the receptor to signal small changes, and still possess a large dynamic range. But what may be even more important, the suddenness of the changes enhances their stimulating effectiveness. Thus, the inherent properties of the sensory receptors determine how the patterns of neural activity they generate will represent the stimulus events. This is a first step in the processing of information for use by the organism.

As yet not much is known about the nature of the excitatory processes following the initial photochemical reaction in visual receptors until one comes near the end of the receptor process. In the eye of *Limulus*, it has been possible to learn a little about the actual production of nerve impulses in the axon of the excited neuron in the receptor unit. By the use of a micropipette electrode penetrating the sensory structure of the ommatidium, changes in the electrical polarization of the cell membrane of the neuron in the ommatidium have been recorded.[15–17] These changes are associated with the trains of impulses initiated by this cell when the receptor unit is illuminated (Fig. 3). When light is turned on, the cell membrane becomes somewhat depolarized, and simultaneously there is a speeding up of the discharge of impulses in its axon. Such depolarization is referred to as a "generator potential,"[18] in the belief that the nerve impulses are generated by local electric currents flowing as a result of the difference in potential between the axon and the depolarized cell body (or more probably, in the present case, the depolarized dendritic process of the cell which penetrates the rhabdom of the ommatidium). The degree of depolarization depends on the intensity of the stimulating light and in turn determines the frequency of the relaxation oscillations of the membrane of the initial segment of axon in the region where it leaves the cell body and from which the propagated impulses take off. The discharge of trains of impulses by depolarized neurons is a familar process in neurophysiology. For the photoreceptor, the question is how the initial photochemical reaction produces the depolarization and the ensuing "generator potential." About this, almost nothing is known.

As discussed in the foregoing, the receptor itself by its inherent properties does a certain amount of processing of the information from the outside world. It is concerned with the report only of certain aspects of the physical stimulus that acts on it, and it is not necessarily a high-fidelity recording device. Built as it is, it selects certain features of the stimulus pattern for accentuation. The next step in the processing of sensory information in the visual system concerns the distribution of light over the entire population of visual receptors. A retina, whether in a vertebrate or an arthropod, is more than a mosaic of independent detecting elements. In the vertebrates, it is well known that the retina is a highly organized nervous center. It is really a part of the brain closely applied to a mosaic of sensory receptors. The first step in the neural analysis of the pattern of the retinal image requires the intercomparison of what happens in the various differently stimulated receptors, and a modification of the pattern of neural activity to accentuate important features of the spatial distribution of light over the receptor mosaic. Evidently, it is profitable to do this close to the point where the information is being picked up. In the vertebrate retina, the early neurons in the visual pathway are spread out in correspondence with their associated receptors, and many of the processes in the first step of neural integration apparaently can be done most effectively in the retina itself. This is not a simple process; patterns of activity observed in the optic-nerve fibers in the verte-

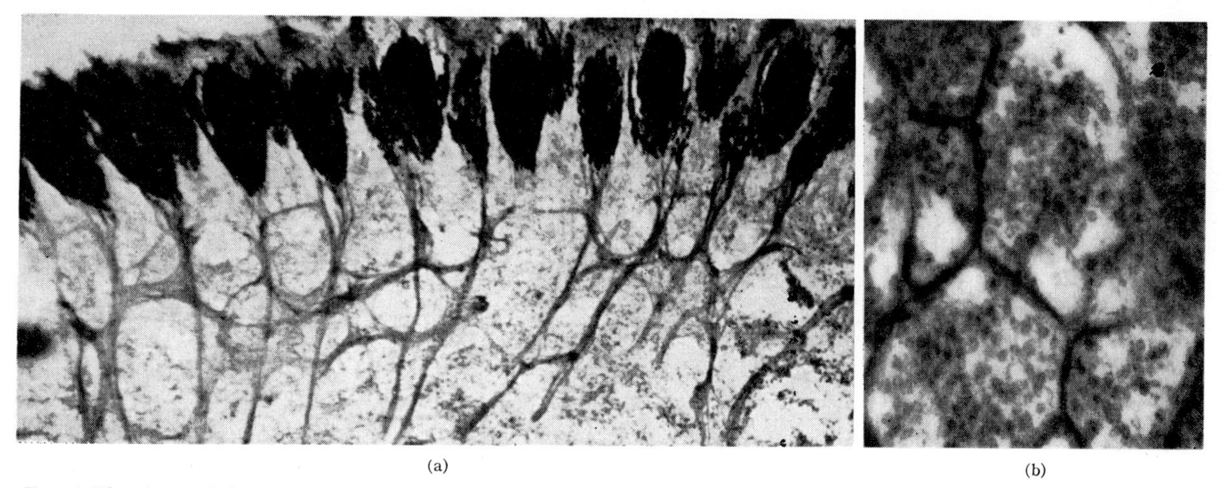


FIG. 4. The plexus of the compound eye of *Limulus*. (a) Light micrograph of a section cut through the eye in a plane perpendicular to its external surface (cornea removed), showing on its upper border a row of the heavily pigmented ommatidia, from each of which emerges a small bundle of nerve fibers (stained with silver by Samuel's method) that contains, together with small fibers, the axon of ommatidium neuron. Connecting these bundles are festoons of fibers, with clumps of neuropile that appear at this magnification as condensations in the meshes of the plexus. Width of figure = 2.2 mm. Photograph by W. H. Miller [from H. K. Hartline, H. G. Wagner, and F. Ratliff, J. Gen. Physiol. **39**, 651 (1956)]. (b) Electron micrograph of a portion of a clump of neuropile in the plexus, showing a few outlines of the fibers composing the clump, within which are numerous small circular outlines interpreted as synaptic vesicles. Width of figure = 1.2 $\mu$. Photograph by W. H. Miller.[19]

brate retina are very complex, and their analysis is difficult. In simpler visual systems, integrative processes can be more readily analyzed. The eye of *Limulus* again affords a good oportunity for such studies.

The neural structure of the compound eye of *Limulus* is much simpler than that of the vertebrate retina or the eyes of more highly developed arthropods, but it is nevertheless a retina: the units of the receptor mosaic are interconnected by a network of nerve fibers [Fig. 4(a)]. The nerve fibers from the ommatidia branch profusely on their way out of the eye to form the optic nerve. Festoons of these branches connect each receptor unit with its neighbors. There are no nerve-cell bodies in this plexus of interconnections, as in more complex retinas, but there are numerous knots composed of a felt-work of very fine branchlets closely intertwined. The fibers in these clumps of "neuropile" are packed with "vesicles" typically present in synapses [Fig. 4(b)]. Evidently, the clumps of neuropile are synaptic regions, where influences are transmitted from one set of

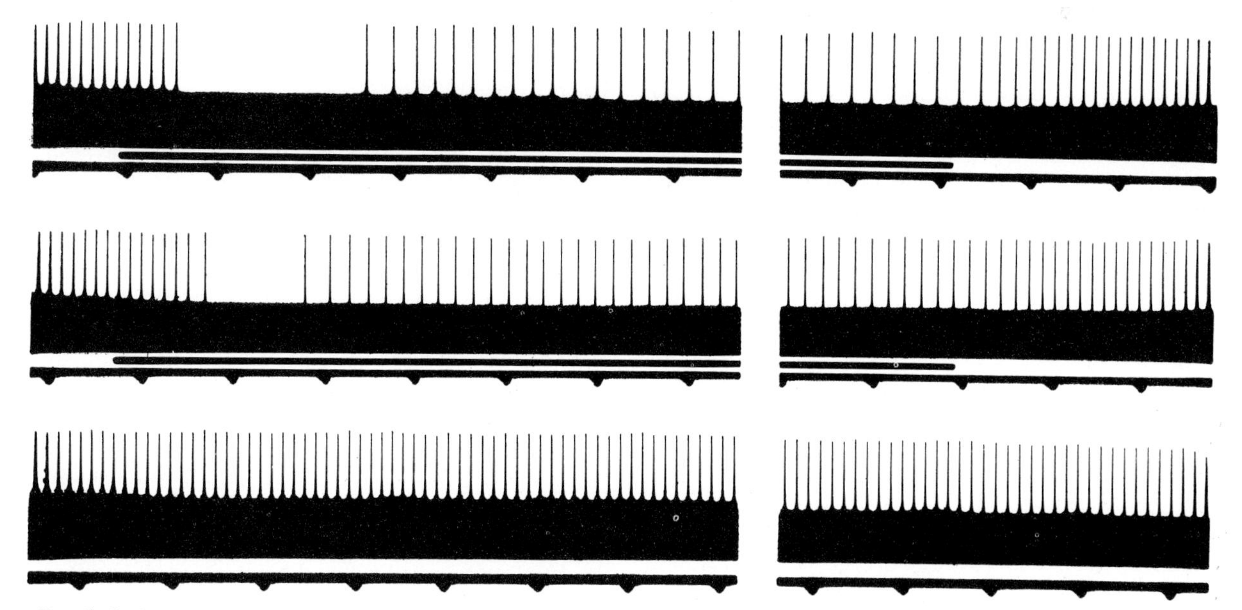

FIG. 5. Oscillograms of nerve action potentials, showing inhibition of the impulses in a single optic-nerve fiber of *Limulus*. The ommatidium of the eye from which the fiber arose was illuminated steadily at a fixed intensity, beginning 3 sec before the start of each of the records; adjacent ommatidium were illuminated during the interval signalled by the blackening out of the white line above the time marks, in the upper two records. In the top record, the intensity of the illumination of adjacent receptors was ten times that used in the middle record. Bottom record is a control (no adjacent illumination). Time in $\frac{1}{5}$ sec. Experimental arrangement as in Fig. 6(a).

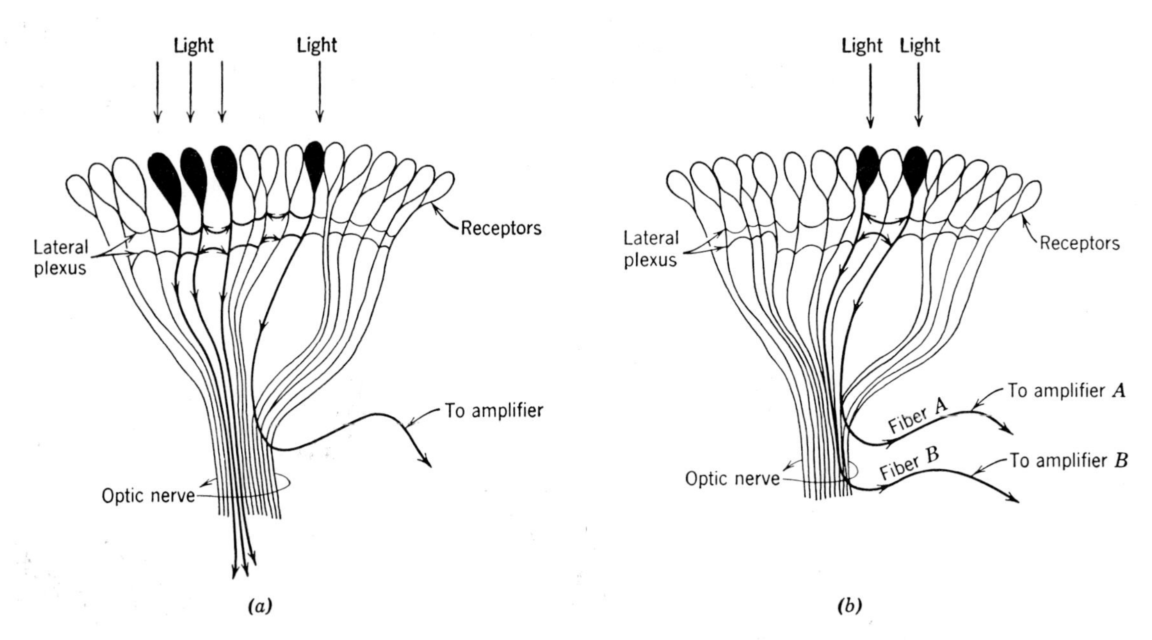


FIG. 6. Schematic diagrams of experimental arrangements used. (a) Experiment of Fig. 5 (inhibition of test ommatidium by illumination of nearby ommatidia). (b) Experiments of Figs. 7 and 8 (mutual inhibition of two ommatidia that were close to one another).

branches to another.[19] Based on this structural organization is a simple functional organization: each ommatidium tends to inhibit the activity of its neighbors. This influence is indeed exerted over the plexus of nerve fibers, for, by cutting the interconnecting branches to an ommatidium, the influence of its neighbors on it can be abolished.

The inhibition that is exerted on an ommatidium by its neighbors is illustrated in Fig. 5. In the experiment from which these records were taken, the discharge of impulses was recorded in a single optic nerve fiber in response to illumination, by a small spot of light, of the facet of the ommatidium from which that fiber arose [Fig. 6(a)]. During steady illumination of that one ommatidium alone, a steady discharge of impulses resulted (bottom record). When, during steady and continuous illumination of this "test" ommatidium, light was caused to shine also on other ommatidia in neighboring regions of the eye (top and middle records of Fig. 5), the frequency at which the test ommatidium discharged impulses was reduced.

Figure 5 also shows that strong illumination of the adjacent region produced a greater depression of frequency than weak illumination. It has also been shown that the magnitude of the inhibition exerted on an ommatidium is greater the larger the number of neighboring ommatidium that are stimulated. Thus, the inhibitory influences from many neighbors can combine to increase the net effect they produce. Also, the inhibition exerted on an ommatidium by its neighbors is greater the closer they are to it. Ommatidia that are separated by a distance exceeding 4 or 5 mm have no effect on one another.

Inhibition in the eye of *Limulus* is exerted mutually by the receptor units.[20] Each ommatidium, being a neighbor of its neighbors, inhibits them as well as being inhibited by them. This is shown in Fig. 7, obtained by recording activity simultaneously in the optic nerve fibers from two independently illuminated ommatidia, close to each other in the eye [Fig. 6(b)]. The frequency of each receptor unit was lower when both were illuminated together than when each was illuminated by itself. When this experiment is performed using various intensities on the two receptors in various combinations, it has been shown that the amount by which the steady frequency of discharge of each receptor unit is lowered depends on the degree of concurrent activity in the other, and is indeed a linear function of the frequency of its discharge (Fig. 8). Thus, the response of one receptor unit is determined by the excitation furnished by the stimulating light shining on it, diminished by the inhibitory influence from the second receptor, which in turn depends on the resultant of the excitation furnished by its own stimulus and the inhibition exerted on it by the first. This mutual interdependence of any two neighboring receptor units may be described by a pair of simultaneous equations, linear in the frequencies of the discharges.

When more than two interacting receptors are illuminated simultaneously, each is subject to the combined inhibitory influences from all of the others. The law that determines how the inhibitory influences from several active receptor units combine in affecting the activity of a neighboring unit has been found by experiment: if the influences on a given unit are measured by the reduction they produce in its frequency of nerve impulse discharge, the combined effect of all of the other units is simply given by the sum of the influences

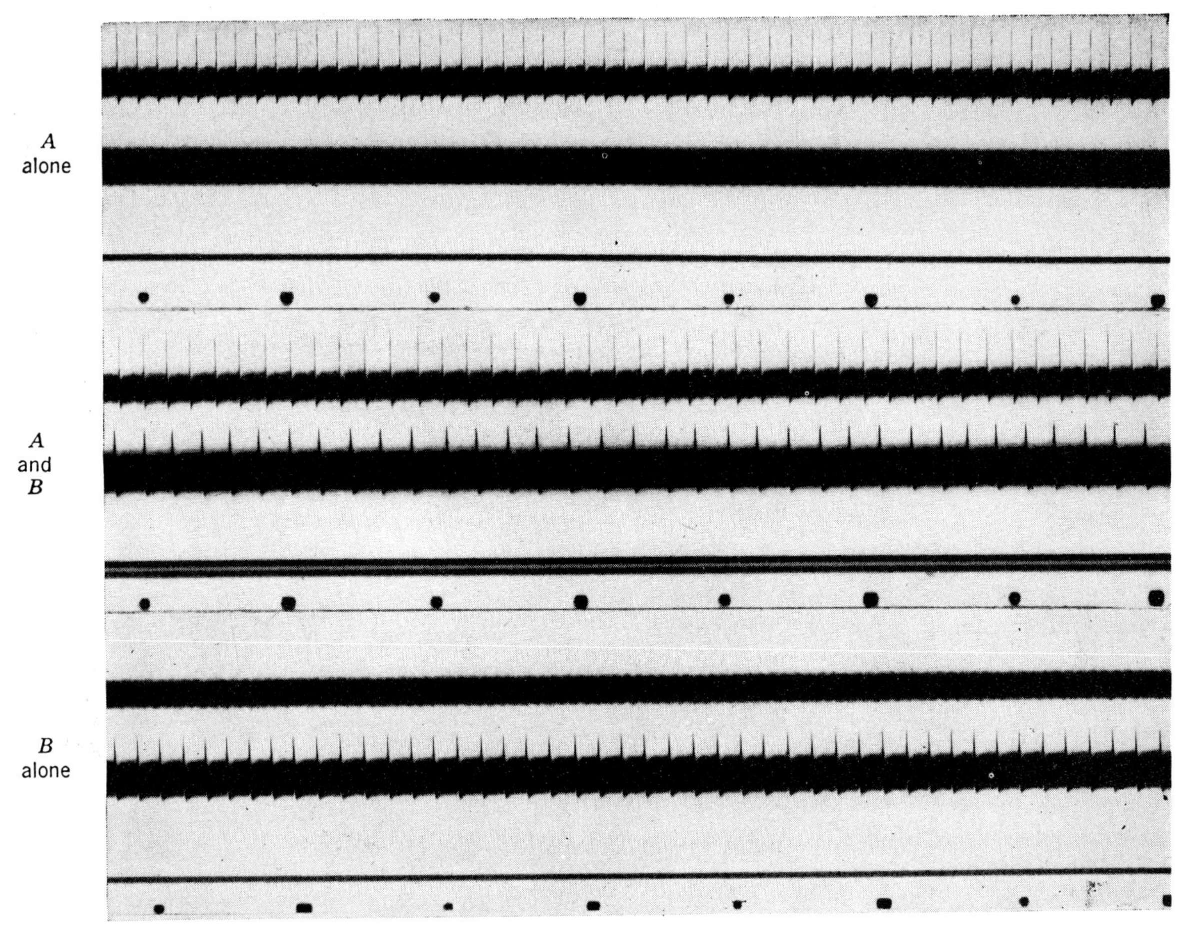


FIG. 7. Mutual inhibition of two ommatidia close to one another in the eye of *Limulus*, steadily illuminated at fixed intensity on each. Experimental arrangement as in Fig. 6(b). Time (black dots) in $\frac{1}{5}$ sec.

exerted by each.[21] The responses of a set of $n$ interacting receptor units, measured by the frequencies of their optic nerve discharges, are therefore expressed by a set of $n$ simultaneous linear equations,

$$r_p = e_p - \sum_{j=1}^{n} K_{pj}(r_j - r_{pj}^0) \quad p = 1, 2 \cdots n.$$

In these equations, $r_p$ stands for the response of the $p$th unit (measured by its steady frequency of impulse discharge) when it is illuminated steadily together with the other units. Its excitation, $e_p$, is measured by the response it has when it is illuminated alone. Each constant $K_{pj}$ is the coefficient of the inhibitory action of the $j$th receptor on the $p$th (usually less than 0.2) and each constant $r_{pj}^0$ is the threshold of that action. Terms for which $j=p$ are usually omitted. The equations as written apply only to those units and that range of activity for which $r_j$ is not less than $r_{pj}^0$. As a rule, the closer the interacting elements are to one another, the larger the $K$'s and the smaller the $r^0$'s. Exceptions are often found, however, and it is not yet possible to state the statistical law governing the effects of distance on the inhibitory interaction.

If $N$ small groups of receptors are considered, each group uniformly illuminated and assumed to consist of receptors with similar properties exerting equal actions, the foregoing set of equations may be reduced to $N$ simultaneous equations with lumped coefficients representing the group interactions. Applied to three interacting receptors or receptor groups, the theory outlined in the foregoing can account quantitatively for a number of effects that have been observed with various experimental configurations of retinal illumination. Thus, if a test receptor is located midway between two groups of receptors that are themselves too far apart to interact, the combined inhibitory action of these two on the test receptor is equal to the sum of the separate actions of each, unless the test receptor itself has an appreciable effect upon them. If the two groups are close together and both are near the test receptor, their combined inhibitory effect will be less than the sum of their separate effects since they inhibit one another mutually when both are active. If a group of receptors is too far from a test receptor to influence it directly, it may nevertheless affect its response by inhibiting a second group located close to the test receptor, thereby releas-

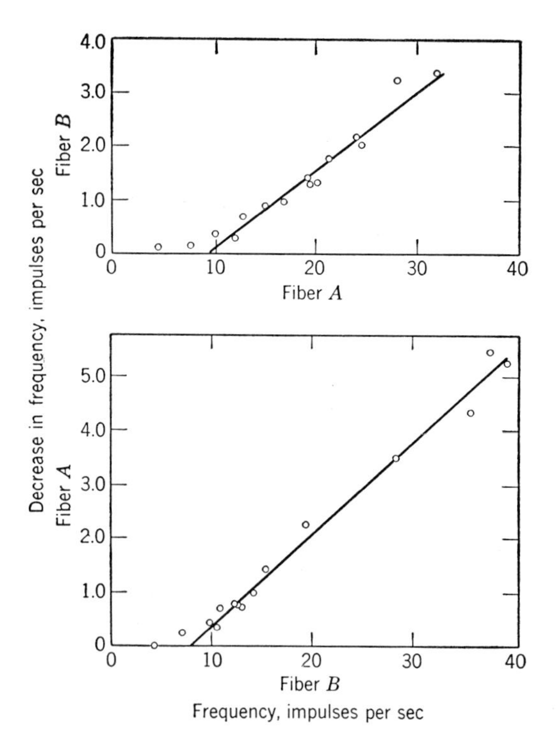


FIG. 8. Mutual inhibition of two ommatidia close to one another in the eye of *Limulus*, illuminated independently at various levels of intensity in various combinations. Amount of inhibition (decrease in frequency) of each ommatidium plotted as a function of concurrent level of response (frequency) of the other. From an experiment similar to that of Fig. 7, experimental arrangement as in Fig. 6(b).

ing the test receptor from the inhibition exerted by the second group. Such "disinhibition" illustrates how indirect effects may be exerted beyond the limits of direct influence and, in principle at least, extend over the entire mosaic of interdependent receptor units.

The inhibitory interaction just described may be considered a simple integrative mechanism that takes place at or close to the level of the receptors themselves. Because of it, patterns of optic nerve fiber activity yield a distorted representation of the patterns of incident illumination. This distortion, however, serves a useful function, for it is clear that inhibitory interaction must enhance contrast: brightly lighted elements in the receptor mosaic inhibit the dimly lighted ones more than the latter inhibit the former. If, as in the eye of *Limulus*, mutual inhibition is greater between close neighbors than distant ones, contrast will be greatest near regions of steep intensity gradients and borders and edges in the retinal image will be "crispened." Phenomena of border contrast are illustrated in our own vision by the light and dark bands bordering a penumbra (Mach bands), and by the fluted appearance of an optical step-wedge or of shadows cast by multiple light sources. Inhibitory interaction is probably one of the mechanisms in our own visual systems that gives rise to these phenomena.[22]

Direct experimental demonstration of the "crispening" of the contours by inhibitory interaction can be made, using the eye of *Limulus*.[23] The discharge of impulses is recorded from a "test" receptor near the center of the eye as the eye is caused to scan slowly a pattern of illumination containing a gradient of intensity. When all of the receptors are masked except for the one from which activity is being recorded, a faithful representation is obtained of the distribution of intensity in the image viewed. But when the mask is removed, so that all of the receptors view the pattern, maxima and minima in the frequency of the test receptor discharge occur, corresponding to the regions bordering the gradient. These resemble in form and location the "Mach bands" seen by a human observer viewing the same pattern.

Interaction is known to take place in other sense organs. In the ear, von Békésy[24] has suggested that inhibitory interaction may be important in increasing pitch discrimination. Indeed, Galambos and Davis[25] have demonstrated inhibition of the activity of single auditory nerve fibers by tones differing in frequency from those used to excite the fibers. Also, von Békésy[26] has demonstrated "contrast" effects in the tactile stimulation of the skin, suggesting inhibitory interaction over considerable distances over the surface of the body.

In the higher nervous centers, integrative processes of great complexity take place. In the retina of the vertebrate eye, which—even though located in the peripheral sense organ itself—is nevertheless a nervous center of high order, there is an intricate interplay of excitatory and inhibitory interactions.[27] As a result, diverse and labile patterns of optical nerve fiber activity are generated.[28,29] In the vertebrate retina, to a much greater degree than in the primitive retina of *Limulus*, the patterns of afferent nervous activity are greatly modified to accentuate significant features of information about the environment. The process of neural integration is well begun by the time the afferent messages are transmitted to still higher centers in the brain.

## BIBLIOGRAPHY

[1] H. J. A. Dartnall, *The Visual Pigments* (Methuen and Company, Ltd., London, 1957).

[2] G. Wald and P. K. Brown, Science **127**, 222 (1958).

[3] G. Wald, P. K. Brown, and P. H. Smith, J. Gen. Physiol. **38**, 623 (1955).

[4] G. Wald, Am. Scientist **42**, 73 (1954).

[5] R. Hubbard and A. Kropf, Proc. Natl. Acad. Sci. U. S. **44**, 130 (1958).

[6] F. W. Campbell and W. A. H. Rushton, J. Physiol. **130**, 131 (1955).

[7] S. Hecht, Physiol. Rev. **17**, 239 (1937).

[8] F. S. Sjöstrand, J. Cellular Comp. Physiol **42**, 15 (1953).

[9] W. H. Miller, J. Biophys. Biochem. Cytol **3**, 421 (1957).

[10] E. DeRobertis, J. Biophys. Biochem. Cytol. **2**, 209 (1956).

[11] K. R. Porter in *Harvey Lectures* (Academic Press, Inc., New York, 1957), Vol. LI, p. 175.

[12] W. H. Miller, J. Biophys. Biochem. Cytol. **4**, 227 (1958).
[13] S. Hecht, S. Schlaer, and M. H. Pirenne, J. Gen. Physiol. **25**, 819 (1942).
[14] M. H. Pirenne, Biol. Revs. Cambridge Phil. Soc. **51**, 194 (1956).
[15] H. K. Hartline, H. G. Wagner, and E. F. MacNichol, Cold Spring Harbor Symposia Quant. Biol. **17**, 125 (1952).
[16] E. F. MacNichol, Am. Inst. Biol. Sci. **1**, 34 (1956).
[17] T. Tomita, Japan. J. Physiol. **6**, 327 (1956).
[18] R. Granit, *Sensory Mechanisms of the Retina* (Oxford University Press, London, 1947).
[19] F. Ratliff, W. H. Miller, and H. K. Hartline, Ann. N. Y. Acad. Sci. **74** (2), 210 (1958).
[20] H. K. Hartline and F. Ratliff, J. Gen. Physiol. **40**, 357 (1957).
[21] H. K. Hartline and F. Ratliff, J. Gen. Physiol. **41**, 1049 (1958).
[22] G. A. Fry, Am. J. Optom. and Arch. Am. Acad. Optom. **25**, 162 (1948).
[23] F. Ratliff and H. K. Hartline, J. Gen. Physiol. (to be published).
[24] G. von Békésy, Physik. Z. **19**, 793 (1928).
[25] R. Galambos and H. Davis, J. Neurophysiol. **1**, 287 (1944).
[26] G. von Békésy, J. Acoust. Soc. Am. **30**, 399 (1958).
[27] R. Granit, *Receptors and Sensory Perception* (Yale University Press, New Haven, 1955).
[28] H. K. Hartline in *Harvey Lectures* (Academic Press, Inc., New York, 1942), Vol. XXXVII, p. 39.
[29] S. W. Kuffler, J. Neurophysiol. **6**, 37 (1953).

# 56
# Mechanisms of Synaptic Transmission

BERNARD KATZ
*Department of Biophysics, University College, London, W. C. 1, England*

## INTRODUCTION

THE propagation of an impulse along a nerve or muscle fiber is brought about by two coupled processes: (i) cable transmission, which allows an electric potential change to spread along a short distance, but with rapid attenuation, and (ii) a boosting mechanism by which the full signal strength is regenerated at each point. If either of these processes is interfered with, the signal will be blocked and will fade out locally. The cable mechanism depends upon the continuity of the fiber structure, with a relatively low-resistance core and high-impedance surface layer. During the impulse, sufficient current must be able to flow forward along the inside of the axon and outward through the resting membrane to stimulate it. If one were to close the core with a high-resistant transverse membrane, or to place a low-resistance shunt across the fiber surface, transmission would be impaired and probably would fail at that point. The very reason why thousands of axons packed together within one nerve bundle can conduct their messages independently, without mutual interference, rests on the absence of structural continuity, and so of an effective cable connection between them.

The purpose of this paper is to consider what happens at "synapses," the points of contact between one nerve cell and the next, or between nerve and muscle fiber. There is no sign of cytoplasmic continuity between the different cell units. Electron-microscope evidence shows that the membranes of the synapsing cells are arranged in close proximity, though in general they do not seem to fuse or to come into intimate contact. The electron micrographs, however, do not reveal anything about the electrical properties of the contacting surfaces; and one has no means of guessing intuitively whether or not an effective cable linkage exists across the synapse.

## ELECTRICAL AND CHEMICAL TRANSMISSION

There are, in principle, two basically different modes of synaptic transmission, electrical and chemical. *Electric* transmission implies that in spite of apparent morphological complexities, an effective local-circuit connection exists which allows sufficient current to pass from one cell to the next to restimulate it. In other words, the transmission is one continuous process without any essential change at the synapse. There must, of course, be *some* difference; for all synapses have the property of functioning only in one direction unlike nerve fibers which can conduct impulses with equal facility in either direction despite the fact that in normal life, because of their terminal synaptic connections, they are used for one-way traffic only.

*Chemical* transmission implies the intervention of an entirely different process specific for the synaptic area. It presupposes that the ordinary cable-connection is interrupted at the contact point between the cells, and replaced by the agency of a chemical mediator. A specific chemical stimulant which is synthetized and stored inside the nerve terminal is liberated by the nerve impulse. When the substance is released, it attaches itself to special receptor molecules in the surface of the contacting postsynaptic (or effector) cell. This chemical combination leads to a membrane change which gives rise to a local depolarization of the effector cell. When the depolarization exceeds the threshold level, a new action potential is set up which then travels along the cell in the manner previously described. Thus one has, interposed between two separate waves of propagated electric activity, a secretion of a specific substance from one cell, and a chemoreceptor reaction in the surface of the next cell.

Now, it is not possible to predict without thorough experimental examination which of these two modes of transmission occurs at a particular type of synapse. Various attempts have been made to generalize by drawing analogies from the few cases which have been explored; and in recent years the view has become prevalent that transmission is llkely to be chemical at all synapses, but that a variety of substances are being employed in different cases, and only a few of them, like acetycholine and noradrenaline, have so far been identified. This view, however, seems a little too sweeping, and the discussion is begun, therefore, by quoting an example of electric transmission which has just been brought to light by the work of Furshpan and Potter.[1] This may well be an exception to the rule, but it provides a definite warning against too much generalization in this field.

Furshpan and Potter used a "giant" synapse in an abdominal ganglion of the crayfish cord. This is a contact point between a very large nerve fiber which runs through the central nervous system of the crayfish and a somewhat smaller motor axon which emerges from the cord to supply the flexor muscles of the "tail." This synapse was chosen for two reasons: (a) because the large size of the two contacting cells made it possible for a pair of microelectrodes (one to pass current and one to measure membrane potential) to be introduced into each of them; (b) it seemed *a priori* that electric transmission might be feasible at this synapse, more so than at other types where minute nerve endings usually terminate in contact with a huge cell (which implies, from the electrical point of view, a poor mismatch, very little current being available from the high-impedance terminals to discharge the large surface of the postsynaptic cell). Furshpan and Potter were able to show that the membrane contact of this special synapse acts as a good electric rectifier, allowing current to pass relatively easily from the presynaptic to the postsynaptic cell, but not in the reverse direction. In other words, at this particular synapse, an adequate cable connection exists between the interior of the two cells, in the *normal* direction of impulse travel alone. Provided the internal potential of the prefiber was *higher* than that of the postfiber, electric current could flow across and influence the membrane potential of the adjoining cell. The result is that a depolarization such as occurs during the impulse can be transmitted in the normal *orthodromic* direction, but not the other way. And, conversely, a local hyperpolarization, produced experimentally by passing a current inward through the fiber membrane was found to be transmitted only in the *antidromic* direction (from postfiber to prefiber) but not the other way. Thus, at the giant synapse of the crayfish, one has a case—so far, the only known example—of electric transmission, in which the action current generated by the arrival of an impulse in the presynaptic cell is passed on without finite delay and can directly depolarize and thereby excite the postfiber. One-way transmission is owing to the valve-like one-way resistance of the synaptic contact membranes.

There are only a few giant synapses available in nature allowing such a direct experimental approach to both sides of the junctional region. In most cases, the presynaptic nerve endings are too small to be tackled with intracellular electrodes and their electrical behavior has to be inferred from a more indirect approach. It is of great interest, however, that at another giant synapse, in the stellate ganglion of the squid, Bullock and Tasaki and Hagiwara[2,3] obtained evidence of a different kind; they observed a definite local delay in the propagation of the electrical change, indicating a stoppage of the local-circuit transmission at the junction; there was no detectable transfer of subthreshold cable signals in either direction. These observations provide another fair warning against attempts to generalize about synaptic mechanisms.

If one now takes an entirely different case—namely, the skeletal neuromuscular junction—one finds here one of the few examples of a synapse where chemical transmission has been firmly established. It was shown by Dale and his colleagues that a specific cholinester, almost certainly identical with acetylcholine, is released from the active motor-nerve endings. This substance is a very potent local stimulant and, provided it is applied rapidly to the junctional end-plate region of a muscle fiber, causes a local depolarization of the cell membrane and sets up propagated impulses and contraction in the fiber. The chemical effect is localized to the synaptic area of the muscle surface; it is at these points exclusively that a number of chemical blocking agents, like curare, act (apparently by a competitive attachment to the acetylcholine receptors). By histochemical methods, a high local concentration of a specific enzyme, acetylcholinesterase, has been found at the same point; the apparent purpose is to hydrolyze the transmitter substance within a very short time after it has exerted its action. Electrical studies have shown that there is an irreducible delay of 0.5 to 1 msec between the arriving nerve impulse and the start of the so-called end-plate potential (which is the local depolarization in the muscle fiber produced by the transmitter substance). There is no cable transfer of electric current, of either polarity, directly from the nerve axon to the muscle fiber. When potential changes are imposed on the terminal portion of the motor nerve, these changes do not spread beyond the nerve terminal, but can be shown to increase the rate at which acetylcholine is being released from the nerve endings and so, indirectly, to influence the membrane potential of the muscle fiber.

The most direct way of establishing chemical transmission by nerve impulses would be to show that a substance is released, on nerve stimulation, into the circulating fluid, and when applied to a remote effector cell produces the same, excitatory or inhibitory, action. This has been achieved in a few cases, notably the classical experiment of Otto Loewi in which he demonstrated the role of acetylcholine as the transmitter of nervous inhibition of the heart beat. Usually, such a direct demonstration is not feasible because of the enormous dilution of the transmitter agent, on the way from its primary point of release and action to the assaying tissue. Indeed, this discrepancy between (i) the amount of acetylcholine (ACh) which has to be applied artificially to stimulate a muscle, and (ii) the much smaller quantities which are released into the perfusion fluid from stimulated nerve endings, has been used as an argument against the validity of the chemical-transmitter theory. By a method of microscopic ionophoresis, much more effective applications have been made recently, and it has been possible to show that as little as $10^{-16}$ g equiv of ACh can give rise to an effective, superthreshold, depolarization of the muscle fiber. This is still a few hundred times more than the amount per impulse recovered in the earlier perfusion experiments, but it must be remembered that even with the best micropipettes one cannot reduce the average diffusion distance to that of the natural synaptic contact. The remaining quantitative discrepancy is entirely within the range to be expected and hardly can be used as a theoretical counterargument. The ionophoretic microtechnique has shown a number of other interesting results; it has confirmed the extremely critical localization of the chemoreceptors at and within

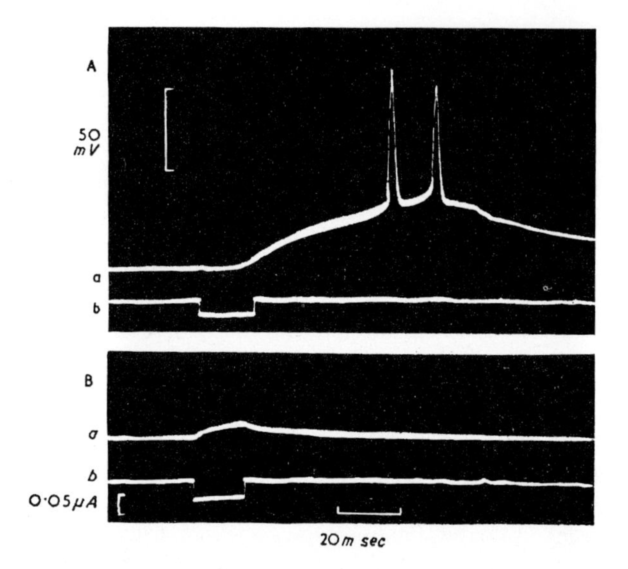


FIG. 1. External and intracellular application of acetylcholine to a motor end plate [from J. del Castillo and B. Katz, J. Physiol. **128**, 157 (1955)]. Intracellular recording of membrane-potential changes from the junctional region of a frog muscle fiber. In *A*, an ACh-filled micropipette was placed on the outside of the end plate, and a quantity of ACh was released by passing a brief outward-directed current pulse through the pipette (registered in *b*). It produced the effect shown in trace *a*: a depolarization developing after a diffusion delay and culminating in two spikes. Between records *A* and *B*, the ACh-pipette entered the muscle fiber. An outward pulse produces now a small, direct potential change, owing to the passage of current through the fiber membrane. (If, for comparison, a KCl-filled micropipette is used, no potential change is produced by the pulse until the pipette has entered the fiber, when the effect is identical with that recorded in *B*, a.)

the neuromuscular junction. Moving the tip of the pipette by several microns can substantially reduce the effect of a given dose. Furthermore, it has been possible (Fig. 1) to insert the tip into the interior of the muscle cell, and so apply the acetylcholine alternatively to the external and internal side of the postsynaptic end-plate membrane.[4] The result showed that a depolarization was produced only by external, not by intracellular application; and this was observed with acetylcholine as well as carbamylcholine (a substance of similar action, but not destroyed by the local cholinesterase), and the same result was found for the blocking action of curarine. It seems that the first chemical attachment on to the receptor molecules must take place at the external surface of the end-plate membrane, which is, of course, the side facing the nerve endings from which the acetylcholine emerges under normal conditions.

Before discussing other peculiarities of the neuromuscular junction, the principal features and problems inherent in chemical transmission in general may be considered briefly.

There are two main steps interposed between the arrival of a presynaptic and the departure of a postsynaptic impulse: (i) the process by which the arriving impulse releases the transmitter substance, from its storage place inside the terminal into the narrow cleft between the contacting cells—this is a special case of what has been called "neuro-secretion"; (ii) there is the process by which the transmitter substance becomes attached to specific molecules in the postsynaptic cell surface and causes its electric membrane properties to change—this is a special example of *chemoreceptor* action, that is a process analogous to that occurring in our various chemical sense organs where a minute concentration change of some specific substance is registered in the form of sensory nerve impulses. As an intermediate step, one should consider also the mechanism by which transmitter molecules are transported across the small synaptic gap; however, the path length is only a fraction of a micron, and the time taken up by simple diffusion over such a short distance is well within the range of the observed synaptic latency.

### EXCITATORY AND INHIBITORY CHEMORECEPTOR ACTION

To begin with, consider the second step, that is, the chemoreceptive mechanism. How do transmitter substances alter the membrane potential? Only a very incomplete answer to this question can be given. In general, the primary action leads to the opening of some ionic permeability channel in the membrane. Depending upon the size or specificity of this ionic channel, the membrane potential either tends to *fall* toward a low level, well beyond the firing threshold of an impulse, or it may become stabilized in the vicinity of the resting level or even tend to rise somewhat (hyperpolarize). In the first case, excitation ensues; in the last cases, one obtains an opposite, inhibitory, action (see Eccles[5]). But it may be noted that, underlying all of these changes, there is a common primary effect—namely, an increase of some ionic conductance.

For example, at the motor end plate there is evidence that acetylcholine causes the membrane permeability to increase, simultaneously, to several monovalent cations (e.g., sodium, potassium, ammonium) and possibly opens up an indiscriminate aqueous channel to all small ions on either side of the membrane. The result is a depolarization which has a "null point" at about 10 to 20 mv, negative inside, which corresponds to the level of a free-diffusion or liquid-junction potential between cytoplasm and external fluid. The effect is to short-circuit and depolarize the surrounding muscle membrane beyond the level at which a new impulse arises which then travels rapidly along the whole length of the muscle fiber. The methods by which these conclusions were reached have been described elsewhere;[6,7] briefly, they consisted in measuring the current/voltage relation across the end-plate membrane, and observing the particular level of the membrane potential at which the electromotive effects of ACh reversed, with normal as well as with altered composition of the ionic environment.

While ACh has a depolarizing and excitatory effect at the end plate, it produces the opposite action, that is hyperpolarization or stabilization of the resting potential, in the regions of the heart muscle onto which it is released by impulses in the vagus nerve. Here also, the basic effect is an increase of ionic conductance, but the channel which is being opened is restricted to potassium, and does not include sodium. As a result the membrane tends to move towards, or to be held at, the potassium equilibrium potential which is usually somewhat greater than the existing resting potential (hence, a hyperpolarization).

Somewhat similar changes appear to be associated with the excitatory and inhibitory synapses in the motor neurons of the spinal cord.[5] At these junctions, the transmitter substances are unknown, but the "null points" of the potential changes which they produce have been determined and correlated with the existing ionic concentration gradients. Here also, an inhibitory hyperpolarization occurs which appears to be associated with an increase of membrane conductance to various small ions but excluding sodium; while excitation apparently arises from an indiscriminate "short circuit" in which sodium as well as the other small ions are allowed to pass.

Although the transmitter effects are fairly well understood in terms of ionic conductance changes, and the subsequent steps leading to excitation or inhibition present no special problem, the molecular mechanism by which the chemical attachment, e.g., of acetylcholine, to the receptive sites of the membrane alters its permeability is still far from understood. At one time it was thought possible that $ACh^+$ ions might produce a local depolarization without permeability change, simply by being able to move very rapidly through specific channels into the interior of the muscle fiber. This idea had to be abandoned when it became clear that the transfer of Coulombs during the end-plate depolarization exceeds the charge on the available ACh ions by several orders of magnitude. The observation shown in Fig. 1—namely, that a positive quantity of charge applied directly into the interior of the cell is much less effective in depolarizing the fiber membrane than an equivalent charge of ACh ions applied on the outside—illustrates this point rather clearly: most of the externally released ACh ions will diffuse away and have no chance of penetrating, or even colliding with, the end-plate surface. Yet under its influence much more positive charge enters the fiber than with the direct intracellular discharge. The conclusion is that the relatively few ACh ions released from the nerve terminal cause a vastly greater quantity of other, ambient, ions to flow through the end-plate membrane, and so achieve the great amplification of local current which is needed to transmit an impulse from the minute nerve endings to the much larger muscle cell.

Regarding the molecular combination between ACh and receptors and the subsequent chain of events, all that can be said at present is that a study of various chemical inhibitors (e.g., del Castillo and Katz[8]) suggests the presence of a 2- or 3-stage process whose kinetics resemble those of many enzyme-substrate reactions. Substances like tubo-curarine appear to act as competitive inhibitors, by interfering with the initial site of attachment, without themselves leading to the next phase which involves a change in the physical membrane properties.

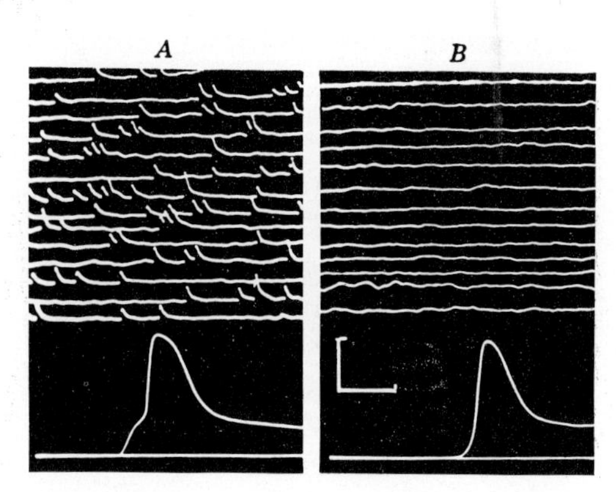


FIG. 2. Spontaneous miniature end-plate potentials. *A:* A recording microelectrode was placed inside a frog muscle fiber at the nerve-muscle junction. *B:* The electrode was placed 2 mm away into the same muscle fiber. The upper portions were recorded with slow speed and high amplification and show the occurrence of spontaneous small potential changes, restricted to the junctional region (calibrations: 3.6 mv and 47 msec). The lower portions show the response to a nerve impulse with fast-speed and low-gain recording (calibrations: 50 mv and 2 msec). The stimulus was applied to the nerve at the beginning of the trace; response *A* (at the end plate) shows the step-like initial end-plate potential which leads up to the propagating wave; response *B* shows only the propagated action potential, delayed by conduction over a distance of 2 mm [from P. Fatt and B. Katz, J. Physiol. **117**, 109 (1952)].

### Quantal Nature of Acetylcholine Release

An interesting feature which has emerged during a detailed study of the vertebrate nerve-muscle junction is that the release of ACh from the motor nerve terminals occurs in discrete packets or quanta each containing a large number of molecules. Even in the absence of a nerve impulse, such packets are released "spontaneously" at infrequent random intervals (Fig. 2). The arrival of an impulse at a cell junction apparently causes a few hundred events to be synchronized within a fraction of a millisecond instead of going on at a leisurely average rate of about 1/sec.[9]

The evidence for this state of affairs was obtained soon after it became possible to apply intracellular recording electrodes to the motor end plate. If a recording electrode is inserted into a resting muscle fiber well away from its junctional region, one observes a steady resting potential of about 90 mv, negative inside. But as one approaches the end plate with the recording probe, a characteristic form of spontaneous activity shows up which consists of an intermittent random

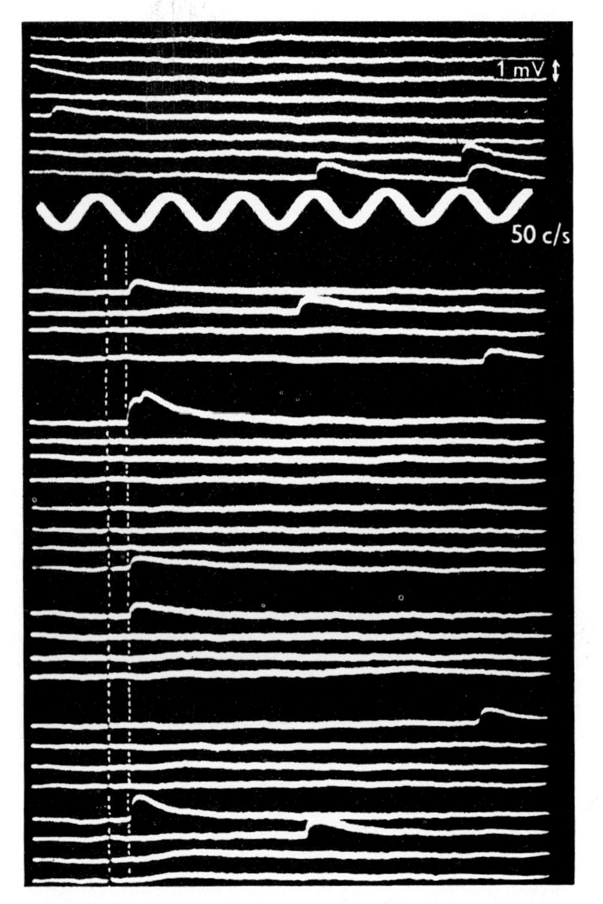


FIG. 3. Quantal units of end-plate response, recorded intracellularly from a muscle fiber in a calcium-deficient and magnesium-rich medium. The top portion shows a few spontaneous potentials. The lower part (below the 50-cps time signal) shows, in addition, the responses to single nerve impulses. Stimulus and response latency are indicated by a pair of dotted lines. There was a high proportion of failures, and only 5 single unit-responses to twenty-four impulses [from J. del Castillo and B. Katz, J. Physiol. **124**, 560 (1954)].

discharge of minute potential changes of standard size and time course. Each deflection is a transient depolarization of the order of 0.5 mv, with a rapid (1 msec) rise and a slower decay, lasting altogether about 20 msec. It resembles in many respects the end-plate potential (e.p.p.)—that is, the immediate depolarization of the end plate produced after the arrival of a nerve impulse, but differs from it in its much smaller size (about 1%) and its spontaneous random occurrence. Fatt and I called it the miniature end-plate potential. It resembles the e.p.p. in its time course, its restricted localization to the innervated region of the muscle fiber, and its pharmacological reactions. It is reduced in size by curare, and its amplitude and duration increases when the local hydrolysis of ACh is prevented by a potent anti-esterase. In both respects, the miniature potential behaves exactly like the depolarizations produced by an applied dose of ACh, and we believe, therefore, that the spontaneous discharges arise from local random impacts of ACh on to the end-plate receptors. The source of these impacts is evidently a spontaneous release or leakage of ACh from the motor-nerve terminal where the substance is stored, for the miniature potentials vanish in the course of experimental nerve degeneration, at a time when neuromuscular transmission fails.

We considered the possibility that random molecular diffusion of ACh from the motor-nerve ending might be responsible for the minature e.p.p.'s. If this were true, then the same type of discharge ought to be elicited, at vastly increased frequency, by applying ACh to the fluid surrounding the end plate. This, however, is not the case: the depolarization which one then observes is continuously graded in size and time course, depending upon the dose and length of the diffusion path. It is clear that the effects of single molecular impacts of ACh must be far below the resolving power of our re-

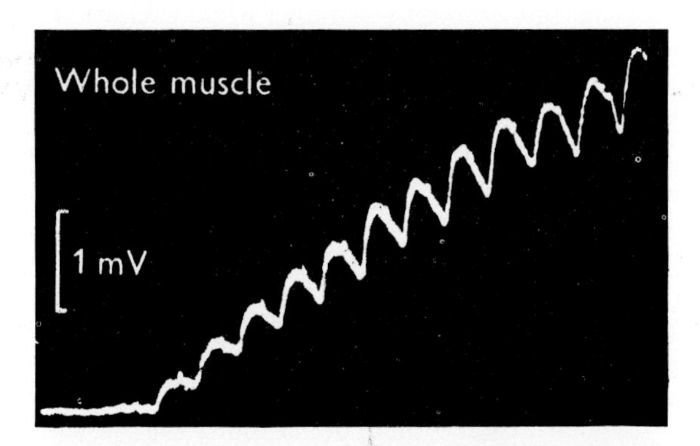


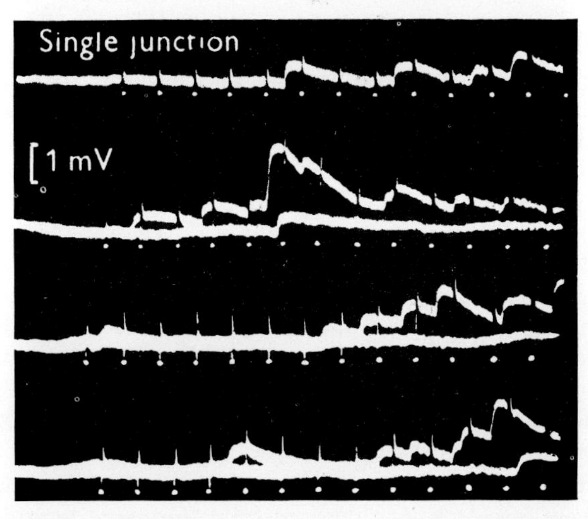


FIG. 4. Statistical properties of the end-plate response. The nerve-muscle preparations were blocked by a high magnesium- and calcium-deficient medium. The nerve was stimulated at 100 shocks per sec which produces a progressively increasing end-plate response. The upper record was obtained from the surface of a sartorius muscle showing the "smooth" average population response of a few hundred end plates. In the lower part, the response of a single end plate is recorded intracellularly, showing the quantal fluctuations of the response. Stimuli indicated by dots. Note spontaneous potentials on the superimposed "base lines" [from J. del Castillo and B. Katz, J. Physiol. **124**, 574 (1954)].

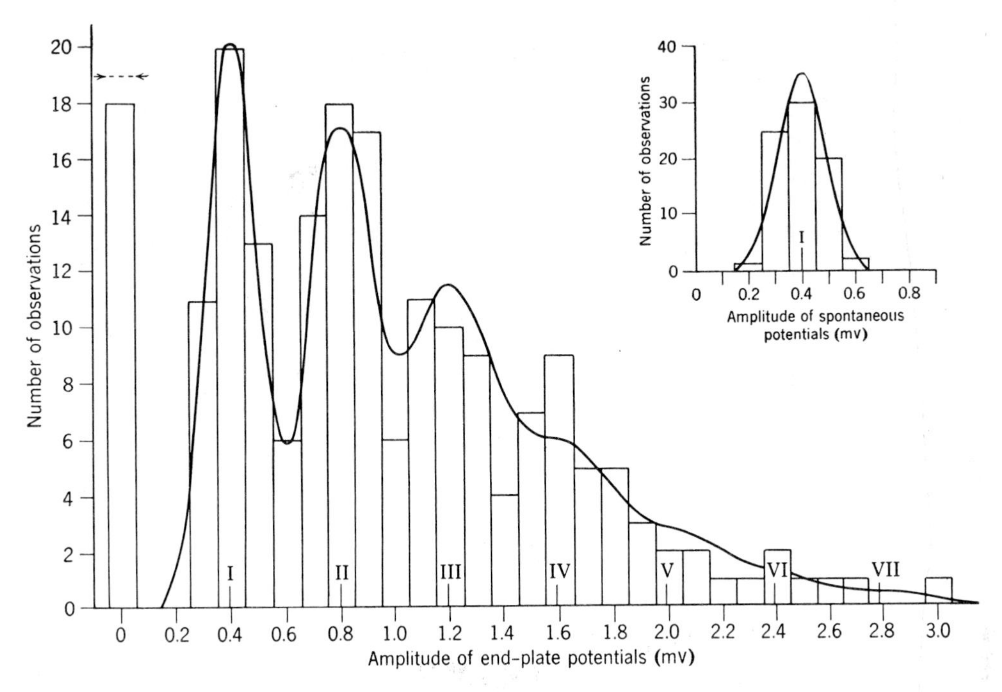


FIG. 5. Histograms of e.p.p. and spontaneous potential amplitudes (inset), from a mammalian end plate blocked by magnesium. Peaks of e.p.p. amplitude distribution occur at 1, 2, 3, and 4 times the mean amplitude of the spontaneous miniature potentials. A Gaussian curve has been fitted to the latter and used to calculate the theoretical distribution of e.p.p. amplitudes (continuous curve). Arrows indicate expected number of failures (zero amplitude) [from I. A. Boyd and A. R. Martin, J. Physiol. **132**, 74 (1956)].

cording equipment, and conversely that a discrete miniature e.p.p., with its standard size and brief time course, must be due to a synchronous package of ACh molecules, may be hundreds or thousands, discharged in the immediate vicinity of our end-plate receptors.

That this packet is the basic coin in which the trans-

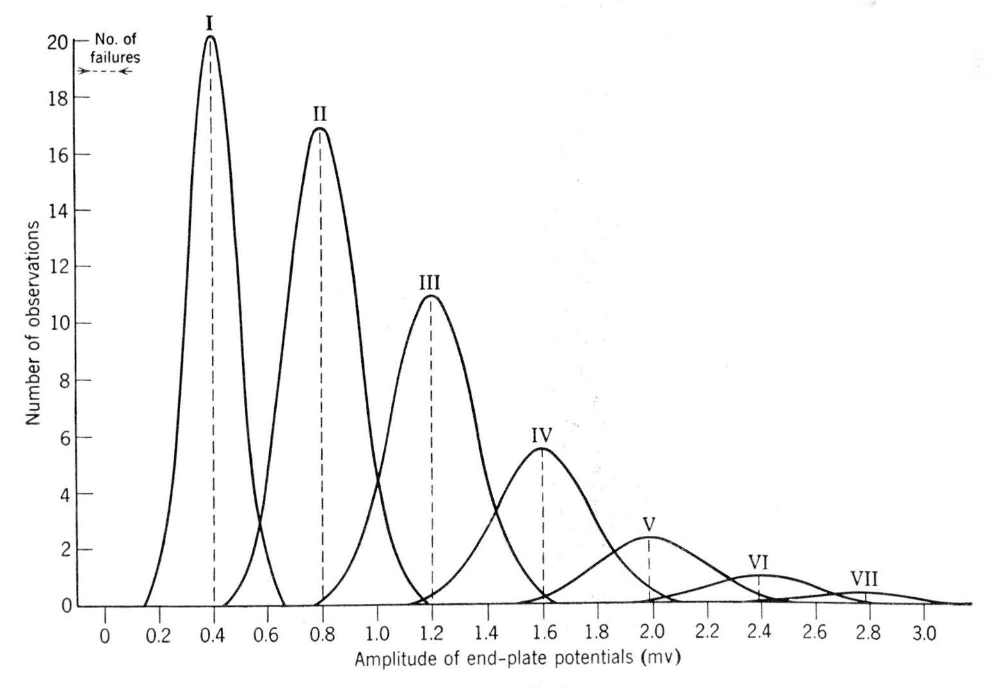


FIG. 6. Method of obtaining the continuous theoretical curve in Fig. 5. A Poisson distribution was calculated, for a mean value $m$ = mean amplitude of e.p.p. responses/mean amplitude of spontaneous potentials. The calculated numbers of each Poisson class have been distributed along Gaussian curves, corresponding to multiples of the spontaneous potentials (see Fig. 5). Algebraic summation of ordinates gives the continuous curve of Fig. 5 [from I. A. Boyd and A. R. Martin, J. Physiol. **132**, 74 (1956)].

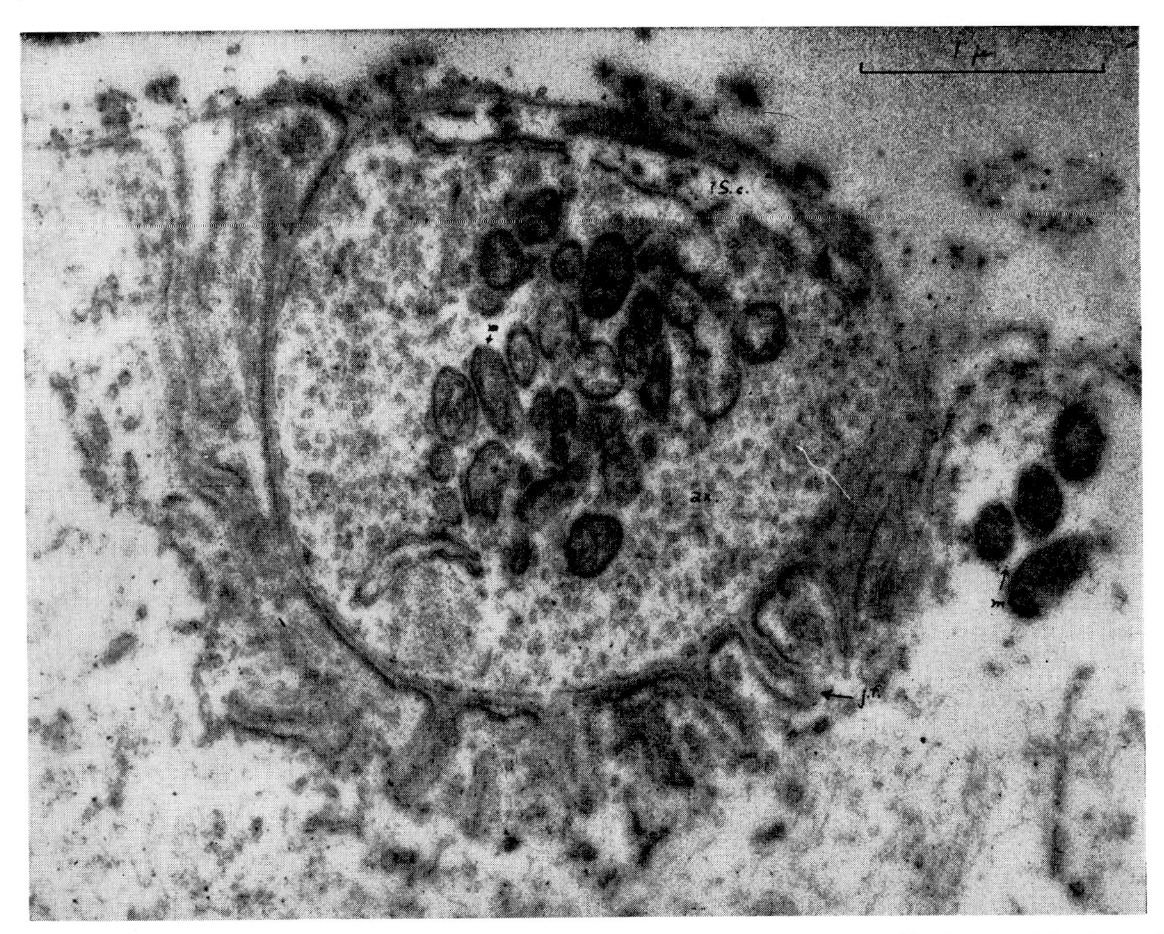

FIG. 7. Electron micrograph of a reptilian neuromuscular junction. Diameter of the motor-nerve ending is approx 2.5 $\mu$; it contains, in addition to mitochondria (large dark particles), many small vesicles of a few hundred A units diameter [from J. D. Robertson, J. Biophys. Biochem. Cytol. 2, 381 (1956)].

mitter is normally delivered from the nerve endings, has been shown in several kinds of experiments. The release of ACh by an impulse depends upon a number of "co-factors" which can be varied experimentally: among these perhaps the most important are the concentrations of calcium and magnesium in the surrounding fluid. Calcium is an essential adjuvant, magnesium an inhibitor of the release mechanism. By lowering the calcium and raising the magnesium concentrations, the quantity of ACh liberated during an impulse, and the size of the resulting e.p.p., can be progressively reduced towards zero. The point of interest is that, during such an experiment, the e.p.p. at each individual end plate is found to be diminished in discrete steps, which correspond to the dropping out of individual miniature potentials, one by one. With a suitable ratio of Ca/Mg concentrations the liberation can be reduced to a small number of ACh packets;[10] in this condition, the size of the end-plate potential during successive nerve impulses has been found to fluctuate in a characteristic stepwise manner, corresponding to a Poisson-wise distribution of the number of units released in each instance.[11,12] The unit-step is identical with the spontaneously occurring miniature potential. Examples are shown in Figs. 3–5, Fig. 6 showing the method whereby Fig. 5 is obtained; detailed studies of this effect on a variety of vertebrate nerve-muscle junctions have made it certain that transmission is brought about by a summation of many of these quantal units of activity.

A further point of interest is that the size of the unit parcel of ACh which is delivered from the nerve endings, spontaneously or in response to an impulse, is relatively constant at all cell junctions and apparently quite unaffected by the many changes which may be imposed on the system in the course of an experiment. On the other hand, the probability of release of any one parcel in a given time interval—that is, the frequency of the miniature potentials—can be altered by several orders of magnitude, for example, by electrically depolarizing the nerve ending with a steady current, or by changing the chemical composition of the environment. The action of the nerve impulse itself can be described as causing a momentary enormous increase in the frequency of the miniature potentials (by a factor of nearly $10^6$, provided a high Ca/Mg ratio exists in the surrounding medium).

The basis of this multimolecular quantum of ACh release is not yet known; an attractive suggestion is that the transmitter substance is stored, within the nerve endings, in minute intracellular corpuscles from which it is discharged at the surface in an all-or-none manner.[6,7] Electron micrographs[13,14] have revealed a mass of fairly densely packed so-called vesicles inside the nerve terminals, and it is conceivable that these are the intracellular bags in which ACh is being stored prior to its release (Fig. 7). It is possible to imagine mechanisms by which a collision between such particles and certain critical spots in the nerve membrane could bring about sudden liberation of the vesicular contents straight into the synaptic cleft. But while it is easy to present speculations which are compatible with the existing evidence, to put them on a firm experimental basis will be a much more difficult task.

## BIBLIOGRAPHY

[1] E. J. Furshpan and D. D. Potter, Nature **180**, 342 (1957).
[2] T. H. Bullock and S. Hagiwara, J. Gen. Physiol. **40**, 565 (1957).
[3] S. Hagiwara and I. Tasaki, J. Physiol. **143**, 114 (1958).
[4] J. del Castillo and B. Katz, J. Physiol. **128**, 157 (1955).
[5] J. C. Eccles, *The Physiology of Nerve Cells* (The Johns Hopkins Press, Baltimore, Maryland, 1957).
[6] J. del Castillo and B. Katz, Progr. in Biophys. and Biophys. Chem. **6**, 121 (1956).
[7] B. Katz, Bull. John Hopkins Hosp. **102**, 275 (1958).
[8] J. del Castillo and B. Katz, Proc. Roy. Soc. (London) **B146**, 369 (1957).
[9] P. Fatt and B. Katz, J. Physiol. **117**, 109 (1952).
[10] J. del Castillo and B. Katz, J. Physiol. **124**, 560 (1954).
[11] J. del Castillo and B. Katz, J. Physiol. **124**, 574 (1954).
[12] I. A. Boyd and A. R. Martin, J. Physiol. **132**, 74 (1956).
[13] J. D. Robertson, J. Biophys. Biochem. Cytol. **2**, 381 (1956).
[14] G. E. Palade, Anat. Record **118**, 335 (1954).

# 57

# Some Quantifiable Aspects of the Electrical Activity of the Nervous System (with Emphasis upon Responses to Sensory Stimuli)*

WALTER A. ROSENBLITH

*Department of Electrical Engineering and Research Laboratory of Electronics, Massachusetts Institute of Technology, Cambridge 39, Massachusetts*

IN assessing the sensory performance of higher organisms (Rosenblith, p. 485), a certain number of characteristics emerge that suggest problems for a quantitative study of the electrical activity of the nervous system. In considering organisms engaged in communication tasks, it has been seen (a) that their performance is statistical in character; (b) that they need more time to handle more information; (c) that their capacity to discriminate and their speed of reaction depend on stimulus intensity; and (d) that their repertory of absolute identifications ("span of absolute judgment") is relatively small and seemingly based upon the ability to make several rather crude discriminations simultaneously, i.e., to classify environmental sensory inflow into some rather gross categories. One is thus led to examine the electrical phenomena in the nervous system in their statistical aspects and to look for over-all, i.e., relatively gross patterns of electrical behavior such as those exhibited by populations of neurons at various levels of a sensory system, rather than to attach disproportionate significance to the behavior of isolated members of such populations. By choosing this quasi-thermodynamic approach, it is not intended to deprecate the complementary view that derives its inspiration from statistical mechanics. But, it must be emphasized that we do not know how to sample a population of neural units adequately (it is far from easy to define a "typical" unit when experimenting with a single microelectrode), and that there is much to be learned of the laws of interaction between units that presumably belong to a population.

It is from this viewpoint that data are presented in this article on (a) electrical responses evoked by sensory stimuli and on (b) so-called on-going activity; the latter may be thought of as reflecting an organism's "state" as it reacts selectively to sensory stimulation.

## SCHEMATIC FLOW CHART OF A MODEL SENSORY SYSTEM

It is obviously impossible to deal here realistically with the neuroanatomy of those structures in the nervous system that become involved when a sensory stimulus is presented. It may, therefore, be expedient to start the discussion by presenting, in Fig. 1, a highly oversimplified view of a model sensory system. *Pour fixer les idées*, consider the auditory system, which is perhaps the one for which the greatest amount of quantitative data is available.

A stimulus impinges upon the organism: The appropriate transducer (here located in the inner ear) is activated, and electrical activity is generated in the region of the hair-cell-neural junction.[1] If the stimulus is a "transient," certain subsequent electrical phenomena at the various higher levels of the nervous system are comparatively easy to follow, and both amplitude and temporal characteristics of responses at various locations can be measured with accuracy.

Once the electric signal has entered the organism's nervous system, it is possible for it to ascend the classical afferent pathway to the cortex. The route can be represented schematically by straight segments in the nature of nerve tracts; these alternate with circles that symbolize the main relay stations of the pathway. Along the axons of the longitudinal tracts, the transmission of signals is predominantly in the form of conducted nerve impulses (all-or-none activity or "spikes"). Transmission is one-to-one in the sense that a given axon exhibits identical patterns of "spikes" along its entire length. The relay stations exhibit activity that has both "discrete" and "continuously graded" aspects. It is in these regions that both excitatory and inhibitory synaptic junctions play an important integrative role by permitting interactions of the many-to-one and perhaps also one-to-many type.

One of the more significant features of the flow chart of Fig. 1 is that it depicts, in parallel with the classical afferent pathway, another route that leads—via the brain-stem reticular formation—to the cortex. In his recent book,[2] Magoun characterizes the reticular system as follows:

"From recent study of the brain stem . . . the concept has developed of a major non-specific or transactional mechanism in the brain, paralleling the specific sensory and motor systems of classical neurology and richly interconnected with them. This non-specific neu-

* This work was supported in part by the U. S. Army (Signal Corps), the U. S. Air Force (Office of Scientific Research, Air Research and Development Command), and the U. S. Navy (Office of Naval Research).

ral system is distributed through much of the central core of the brain stem and, as spokes radiate from the hub of a wheel to its peripheral working rim, so functional influences of this central system can be exerted in a number of directions: caudally upon spinal levels to influence both reflex and other spinal activity; rostrally and ventrally upon hypothalamic and pituitary mechanisms through which influences can be exerted upon visceral and endocrine functions; cephalically upon the diencephalic and rhinencephalic brain, where affect and emotion now reign instead of in the heart; and, more cephalically and dorsally still, upon the neocortex of the cerebral hemispheres which, with its interconnected thalamic and basal ganglionic masses, subserves higher sensory motor and intellectual performance. The influences of this non-specific system in the brain stem are thus brought to bear upon most other portions and functions of the central nervous system, either to diminish or to raise the level of their activity, or to interrelate or integrate their several performances." . . . "It does this as a reflexion of its own internal excitability, in turn a consequence of both afferent and corticifugal neural influences, as well as of the titer of circulating humors and hormones which affect and modify reticular activity."

Relatively few of these properties are depicted in Fig. 1. Yet there can be seen collaterals that branch off from the afferent pathway to furnish the reticular system with information regarding the signals that are being processed along the more direct and specific route. There is also an indication that this system receives impulses from all sensory modalities, i.e., both from the *milieu extérieur* and the *milieu intérieur*. And there are indications that the flow of neural signals for even a single sensory modality is not just one-way in the reticular system, i.e., directed toward the cortex; there are important descending pathways whose functional significance is just beginning to be understood.

It is perhaps somewhat surprising to learn that neuroanatomists have given descriptions of the reticular system before the beginning of the twentieth century, but that it was only after Bremer's experiments of the midthirties, and in particular after the now classical Moruzzi-Magoun experiments[3] on the "Brain stem reticular formation and activation of the EEG" (1949), that this system began suddenly to loom large in the understanding of all integrated behavior. Previous to that period (and even in a large number of experiments performed since then, in the fifties), the use of centrally acting anesthetics has resulted in differential interference with reticular activity. Under these circumstances, most explanations of the role which the nervous system plays in perception or in the handling of sensory information are based on evidence from, neurophysiologically and behaviorally speaking, badly distorted preparations.

Among the more dramatic findings that have forced one to give up oversimplified views of the organization

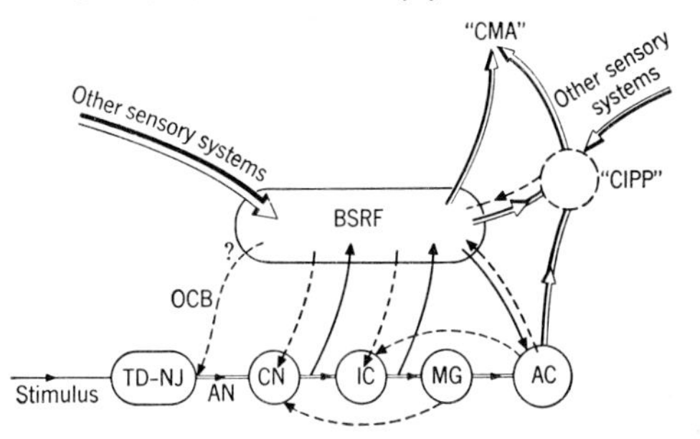


FIG. 1. For the sake of simplicity, this diagram presents only a gross outline of the neuroanatomy of the auditory system. Such important way stations as the cerebellum, the superior olive, the lateral lemniscus, and other regions of the brain from which responses to acoustic stimuli have been recorded have been left out; the still incompletely known and complex organization of the brain-stem reticular formation (BSRF) has not even been suggested. Furthermore, no attempt has been made to indicate connections that correspond to acoustic reflexes of short latency: middle ear muscles contract about 10 msec after stimulus delivery, while eye blinks occur about 35 to 40 msec after delivery of a strong click; such reflexes represent simple forms of integrated behavior that do not necessarily involve the higher levels of the auditory system.

In order to appreciate the number of neural elements present along the afferent pathway, Chow's[4] numerical estimates of the auditory system of rhesus monkey are reproduced here. There are about $10^5$ cells in the cochlear nucleus (CN), about $4\times10^5$ cells in the inferior colliculus (IC) and also in the medial geniculate (MG); the auditory cortex (AC) is estimated to contain about $10^7$ cells; these numbers, which represent orders of magnitudes only, may be compared with the $3\times10^4$ fibers in the auditory nerve (AN) and the $2.5\times10^4$ hair cells in the inner ear, i.e., at the transducer neural junction (TD-NJ). The foregoing data are the results of histological investigations; hence, they do not permit any direct inference concerning the number of neural units that will be either directly excited or inhibited in the presence of a stimulus of given intensity.

The remaining symbols are to be identified as follows: OCB stands for the olivo-cochlear bundle (depicted in dashed lines like the other descending pathways) whose influence upon the activity of the auditory nerve has been investigated by Galambos,[5] CIPP stands for the central information-processing pool of neurones, whose existence, though neither neuroanatomically nor neurophysiologically established, is conceptually useful. The relations of this hypothetical construct to the sensory-projection areas and association areas remain to be worked out to give a more realistic account of the role that the brain and its various components play in the organism's handling of sensory information. CMA stands for the inferred control unit of the motor activity that takes place in connection with the stimulus-related response.

and functioning of sensory systems have been the following: (a) sectioning of the classical auditory ascending pathway was shown to leave intact an animal's ability to be aroused by sound; (b) animals whose auditory cortex had been ablated were shown to be capable of simple frequency and intensity discriminations, though they were not capable of distinguishing more-complex patterns such as simple melodies. Experiments of this sort should, however, not be interpreted to mean that the anatomical structures that can be removed or interfered with play no role in the normal performance of sensory tasks. There are certainly different ways of assessing impairment of performance, and the mere statement that an animal is still capable of carrying out a given task, after ablation, does not force the conclu-

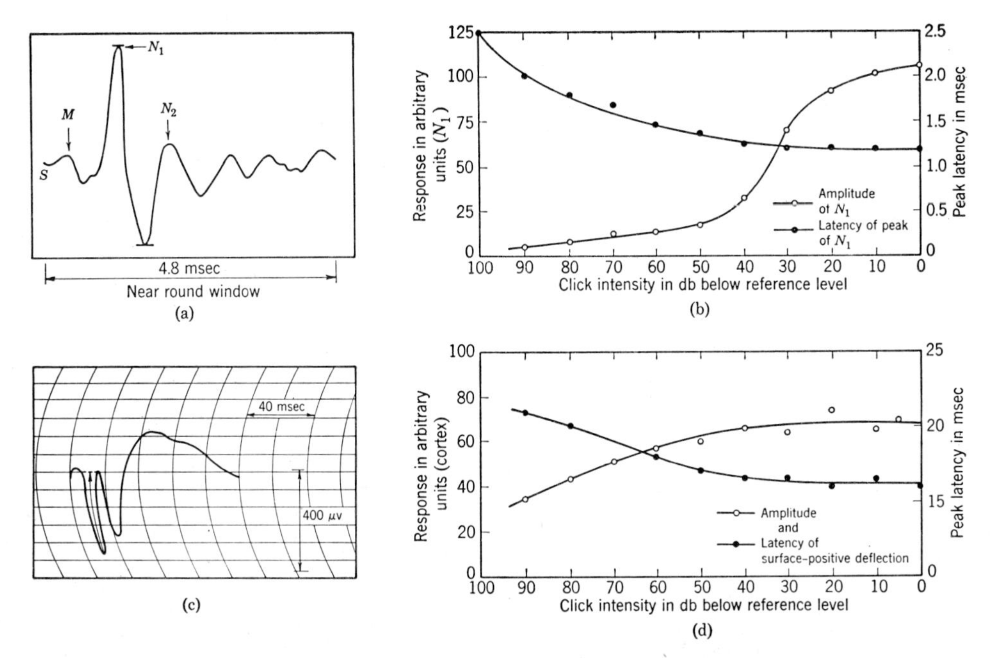


FIG. 2. (a) A typical response to a click of medium intensity recorded near the cat's round window of the inner ear. $M$ is a so-called microphonic potential which is non-neural in origin. $N_1$ is the earliest and most prominent neural component that can be recorded at this location. It represents the summated action potentials from the auditory nerve. $S$ indicates the occurrence of the electrical pulse that generates the acoustic click. (b) An example of how the amplitude and latency of $N_1$ vary as the intensity of the click is varied. The latency values include a delay of approximately 0.5 msec that corresponds to the time that it takes for the sound to travel from the earphone diaphragm to the eardrum. (c) An average response to a click recorded from a location on the auditory cortex of an unanesthetized cat. The averaging operation was carried out by an electronic computing device.[14] Surface positive deflections are plotted downward. (d) Intensity functions for the baseline-to-peak amplitude and the peak latency of the surface positive component of the cortical response to clicks in an anesthetized cat.

sion that there is no sensory deficit. More generally, one is impressed with the fact that the nervous system of higher animals seems to be characterized by enough flexibility (which some people have metaphorically referred to as a "safety factor" or "redundancy") so as to permit the performance of certain rather basic discriminations in a variety of ways.

For a long time students of the nervous system have wondered how its flexibility manages to deal with sensory stimuli, when they are in the focus of the organism's attention in contrast to situations in which they seem to be of no importance. Adrian formulated this question most incisively at the 1954 Symposium on *Brain Mechanisms and Consciousness*[6]:

"The operations of the brain seem to be related to particular fields of sensory information which vary from moment to moment with the shifts of our attention. The signals from the sense organs must be treated differently when we attend to them and when we do not, and if we could decide how and where the divergence arises, we should be nearer to understanding how the level of consicousness is reached. The question is whether the afferent messages that evoke sensations are allowed at all times to reach the cerebral cortex or are sometimes blocked at a lower level. Clearly we can reduce the inflow from the sense organs as we do by closing the eyes and relaxing the muscles when we wish to sleep and it is quite probable that the sensitivity of some of the sense organs can be directly influenced by the central nervous system. But even in deep sleep or coma there is no reason to believe that sensory messages no longer reach the central nervous system. At some state therefore on their passage to consciousness the messages meet with barriers that are sometimes open and sometimes closed. Where are these barriers, in the cortex, the brain-stem, or elsewhere?"

The schematic flow diagram of Fig. 1 suggests, in agreement with the beautiful experiments that have been carried out since 1954, that the barriers that Adrian refers to are at no single place in the nervous system, but that they are multiple. This is clearly not the place to review in detail the experiments carried out by Hernández-Peón[7] and many others.[8–13] Suffice it to report that all of these experiments indicate that the descending pathways exercise powerful gating, screening, and regulating influences. This central or centrifugal control of the afferent sensory inflow has been demonstrated to be effective across sense modalities (for example, the presence of a mouse or of fish odors reduces the response to a click at the level of the cat's cochlear nucleus) or during "habituation" (which presents itself as a reduction in the amplitude of evoked responses for stimuli that have been repeated over and over again). This effect of habituation can, however, be suddenly abolished by pairing the "boring" stimulus judiciously with a few electric shocks. From the available evidence (these dramatic effects of attention and habituation are, for instance, not found in anesthetized animals), one is led to conclude that the selective processing of the sensory inflow is under the control of the brain stem or of descending pathways that are structurally or functionally closely related to the reticular system. Much remains to be learned about the precise ways in which the informational aspects of a sensory signal affect its progress in the nervous system, but it is already clear that sensory systems cannot be characterized by invariant sensitivity and invariant "tuning."

### EVOKED RESPONSES ALONG THE AFFERENT PATHWAY

Having discussed the schematic flow chart and some of the phenomena that led to its formulation, it is now necessary to admit that most of the quantitative data that exist deal with electrical events along the classical afferent pathway. This admission should give one a more realistic appreciation of what the phenomena are for which there exist functional relations of the stimulus-response type. What follows should, therefore, be merely considered as an introduction to some problems of recording, data-analysis, and model-making in relation to the electrical activity found at various levels of the nervous system.

The left side of Fig. 2 depicts samples of responses to clicks from populations of neural units at the level of the auditory nerve and at the level of the auditory cortex. As indicated, it is possible to make quantitative measurements along both the voltage and the time axes of these displays. Thus, one can measure amplitudes of deflections (either peak-to-peak or baseline-to-peak) and "latencies" (i.e., time intervals that have elapsed since the delivery of the stimulus) of extreme values or zero crossings. Naturally, the characteristics of the recording equipment must be chosen appropriately. It would be futile if one were to attempt to measure accurately either the latency or the amplitude of rather sharp deflections on the ink tracings taken by the customary EEG recording device.

When such amplitudes or latencies are measured, one must ask whether one shall be satisfied with average values or whether one shall need to know entire distributions in order to be able to give a rational account for the neuronal events. As has been shown by McGill,[15] Macy,[16] Frishkopf[17] and others, evoked responses are afflicted with variability, especially in regard to amplitude and particularly at the higher neural centers unless the animal is rather heavily anesthetized. It is important to realize that this variability of evoked responses to identical stimuli is present in spite of the fact that the gross electrodes that record the activity of reasonably large neural populations are already performing an averaging process. Some studies of the variability of evoked responses in anesthetized preparations have been undertaken[16–18] and have yielded results that

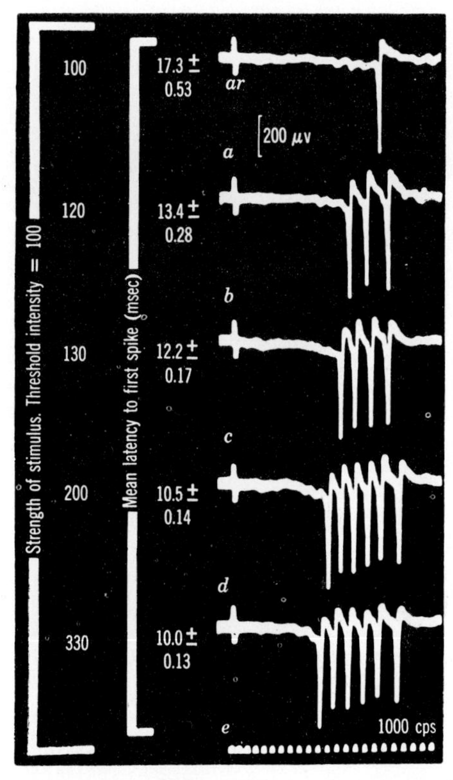


FIG. 3. Responses of single units in tactile thalamic region to transient electrical stimulation of forepaw. Position of stimulation is unchanged for all records. Numbers in column at extreme left give a measure of stimulus strength. Numbers in second column indicate (in msec) mean latency to first spike and standard error of the mean for each stimulus intensity. The records in the right-hand column depict the modal value of the number of spikes that are recorded for each intensity. The short vertical line at the left of each response trace indicates the delivery of the electrical stimulus [from J. E. Rose and V. B. Mountcastle, Bull. Johns Hopkins Hosp. 94, 238 (1954)].

are suggestive from the viewpoint of probabilistic models.

On the basis of amplitude and latency measurements, one can plot intensity functions, i.e., functions that relate the average amplitude of characteristic deflections and their average latency to the intensity of the stimulus. Figures 2(b) and 2(d) illustrate such functions. These and similar graphs permit one to state the following generalization: As the intensity of a transient sensory stimulus is increased, the average amplitude of evoked responses increases (or at least does not decrease), and the average latency of either the onset or the peak of the deflection decreases. [Parallel generalizations can, of course, be formulated for single neural units (see Figs. 3 and 4, from Rose and Mountcastle[19]).] The detailed shape of these intensity functions varies for different locations in the nervous system and for different stimuli, and it is, therefore, hardly appropriate to search for a unique representation of stimulus intensity in the nervous system.

It must be noted that the quantification process just described runs into difficulties if stimuli of appreciable

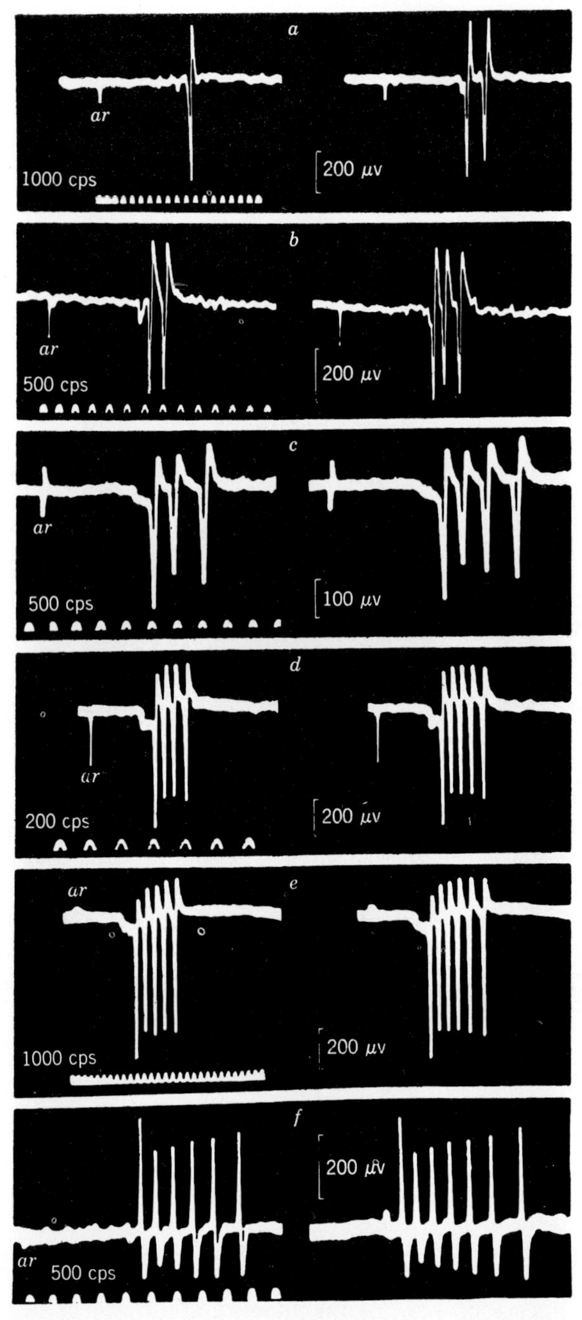


FIG. 4. Each pair of records comes from the same unit. All six units were activated by electrical stimulation of different locations on the skin. The stimulus artefact (*ar*) indicates the instant of stimulus delivery [from J. E. Rose and V. B. Mountcastle, Bull. Johns Hopkins Hosp. **94**, 238 (1954)].

duration are used. Thus far, it has not been possible to suggest, especially for the higher neural centers, meaningful measures of the evoked activity on the basis of visual inspection only. It is by no means certain that simple intensity functions will be obtained for all instants during which the stimuli are "on"; the interaction of excitatory and inhibitory phenomena, as well as various adaptation phenomena, can be counted upon to complicate the picture.

Before dealing with microelectrode data on the behavior of single units at various way stations of the nervous system, mention should be made of the series of mathematical models developed by McGill,[15] Frishkopf,[17] Goldstein,[20] and Macy.[16] These models, which are all probabilistic in character, relate certain aspects of gross-electrode responses (i.e., responses to which populations of neural units have contributed) to stimulus parameters. All of these models make certain, physiologically not unreasonable, assumptions regarding the behavior of single units in the auditory nerve or at the auditory cortex. From these postulated properties and from certain assumptions regarding the interaction between various members of a population of neural units, the behavior of population responses (i.e., evoked responses recorded by gross electrodes) can be predicted and compared with experimentally available data. The response parameters that have been investigated include average amplitudes, variability in amplitude, wave form of a population response, and temporal recovery characteristics. The stimuli were clicks or bursts of noise: these were presented either singly or in pairs (or even triplets), either in the absence or in the presence of background noise. While it is obviously premature to forecast the ultimate success of such model-making, it is safe to assert that the models have demonstrated their usefulness in summarizing sizable quantities of experimental data and in suggesting systematic experimentation in an area wherein workers seem only too often satisfied by demonstrating the existence of phenomena. These models force one, furthermore, to face an issue that may well be critical for the understanding of the organization of the nervous system: What is the relationship between the behavior of the components that can be studied (single cells), the behavior of subsystems (made up of more or less homogeneous populations of single units), and the behavior of an entire sensory system? For it is hardly reasonable to expect that the knowledge of the functioning of the brain will be revealed *in toto* in the absence of a working knowledge of principles of organization of functional units of the brain.

## PATTERNS OF ELECTRICAL ACTIVITY IN SINGLE UNITS OF SENSORY SYSTEMS

Over twenty years ago, first Blair and Erlanger[21] and then Pecher[22] showed that, under carefully controlled conditions, a single nerve fiber will sometimes respond and sometimes not respond when the same stimulus is presented repeatedly. They furthermore showed that, at least to a first approximation, spike responses in fibers that belonged to the same nerve trunk were statistically independent. In 1950,[23,24] it was shown that the probability of response to a click of a single unit in the cochlear nucleus of the cat increases from "almost never" to "almost always" as the stimulus energy was multiplied by a factor of approximately 100 (or 20 db);

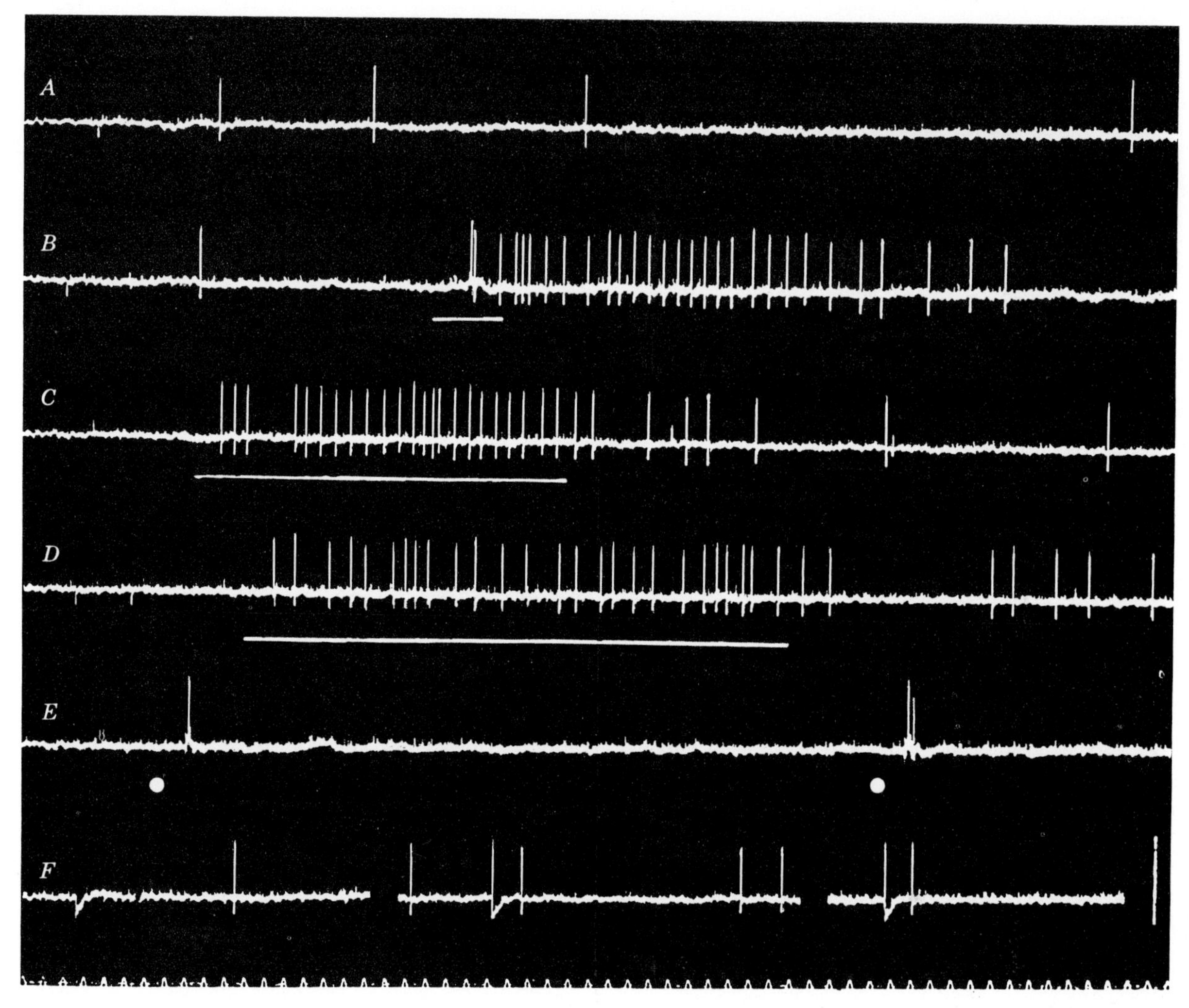


FIG. 5. Microelectrode recording from a unit in the medial part of the pontine reticular formation of a cat: *A*, spontaneous activity; *B*, tapping the ipsilateral forelimb; *C*, rubbing the animal's back; *D*, touching the cat's whiskers; *E*, hand claps; *F*, single shocks delivered to the animal's sensorimotor cortex. Delivery of stimuli is indicated by white horizontal lines (*B*,*C*,*D*), dots (*E*), and artifacts that displace the record's baseline (*F*). Time calibration at the bottom of figure indicates 10-msec intervals [from M. Palestini, G. F. Rossi, and A. Zanchetti, Arch. Ital. Biol. **95**, 97 (1957)].

at the same time, the latency of the spike response that is elicited decreases substantially. Since then, a considerable amount of such data has become available for different structures in the nervous system.[25,26]

Rose and Mountcastle[19] have given some of the most beautiful records that are available today. These records illustrate the finesse and the subtlety with which the nervous systems can "code" stimulus intensity into response patterns of single cells. These workers stimulated the skin of cats electrically and recorded the activity of single neurons in the thalamic relay nucleus of the cat's tactile system. Figure 3 shows how the number of spike responses increases and the latency of the first spike decreases as the strength of the transient peripheral stimulus is increased. Measurements of intervals between spikes yield such interesting results as: (a) Units start to fire at a rate which is higher, the larger the number of spikes in a train of repetitive discharges. (b) The interspike interval increases throughout a train and firing ceases when the interval exceeds a value such as 2 msec.

Figure 4 shows, in each horizontal subdivision, pairs of responses from the same unit to identical stimuli. These discharges are again all-or-none in character and, in spite of the fact that their height (amplitude) varies within a given train, there is overwhelming evidence that all spikes in a single train come from the same unit. This figure furnishes striking evidence for the view according to which the variability in response patterns of single units to identical stimuli renders their statistical description imperative.

Figures 3 and 4 represent locations in the modality-specific classical afferent pathway. The next record, taken in Moruzzi's laboratory (Fig. 5, from Palestini *et al.*[27]), illustrates, in contrast, the response pattern of a single unit in the brain stem.[28] This unit was responsive

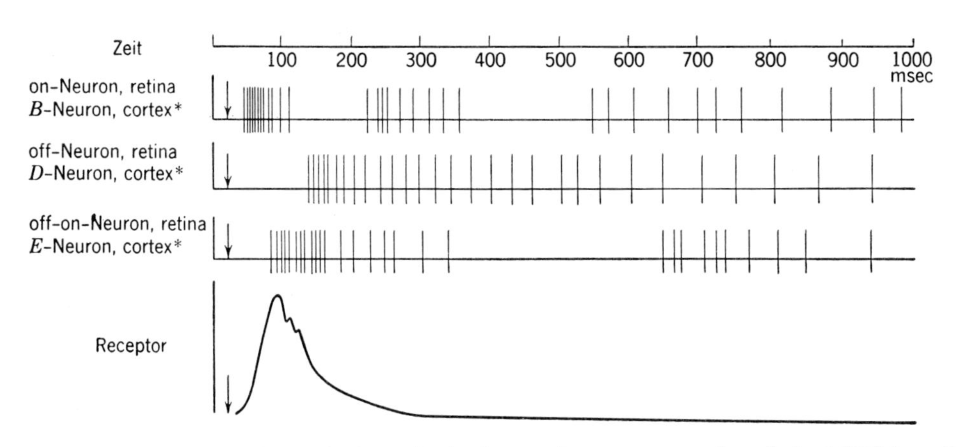


FIG. 6. Typical patterns of unitary discharges in the cat's visual cortex in response to a short flash of 3000 Lux; the bottom trace shows a typical electroretinogram that is observed at the same time. Arrow indicates delivery of flash; the *B*-, *D*-, and *E*-neurons reflect, to a certain degree, patterns of retinal excitation (on, off, and on-off). The pattern of activity of the *B*-neuron—i.e., an initial burst followed by a pause of about 0.1-sec duration—is found in other sensory systems (after O. J. Grüsser and A. Grützner[29]).

to all test stimuli that were tried, although the pattern of spike activity was rather different for the different modalities.

The next microelectrode record (Fig. 6) demonstrates that some of the response patterns that can be observed in sensory systems are discernible over time intervals whose length is comparable to behavioral reaction times. Jung and his group[30] distinguish four different types of neurons in the visual cortex on the basis of their activity patterns. Figure 6 describes schematically three of these typical patterns. The fourth pattern, which corresponds to the so-called *A*-neuron, is not represented. Actually, more than one-third of the units in the visual cortex from which records have been obtained fail to yield visually detectable changes in their spontaneous activity after delivery of an optical stimulus.† There is, however, some evidence that these units may exhibit some kind of responsiveness, provided a form of optical stimulation is used that attracts the animal's attention.

From the highly selected evidence that has been presented in Figs. 3 to 6, a certain number of conclusions impose themselves regarding the strategy of future model-oriented research in this area. New recording devices and new techniques of analysis are needed. One must be able to record simultaneously the firing patterns of not one or two cells but of significant assemblies and subassemblies of cells. Gross electrodes permit the recording of potentials that are very coarse spatio-temporal averages of both spike and continuously graded

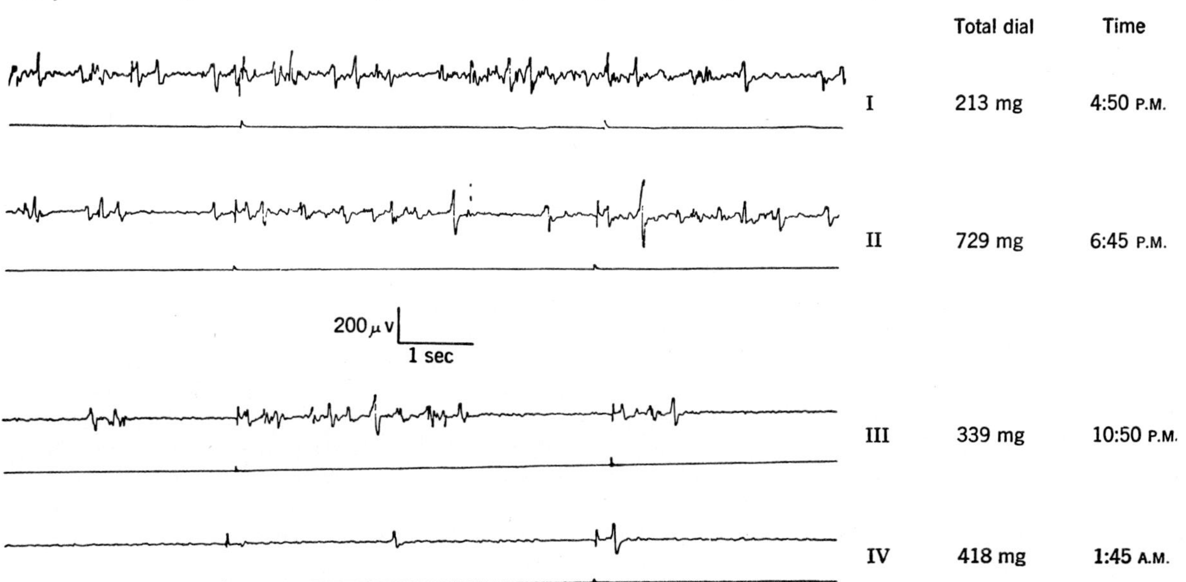


FIG. 7. Four tracings (I, II, III, IV) of electrical activity by a gross electrode from the auditory cortex of a cat. These records were taken during the presentation of clicks. The delivery of the stimuli is indicated on the accompanying lower traces. Depth of anesthesia increases from I to IV, as the amount of dial administered is increased.[32]

† Erulkar, Rose, and Davies[31] found, likewise, that about one-third of the units in the auditory cortex was unresponsive to acoustic stimulation.

activity. The time has now come for solid-state recording devices that will permit the exploration of what might be called the "domain-structure" of sensory systems. It would, however, be foolish to expect that the data from 10 or 50 inputs could be handled adequately without the help of digital computers of appreciable capacity. One cannot be satisfied with visual inspection of records if one wants to find out whether unit $X$ responds most of the time only if units $Y$ and $Z$ are inhibited or after units $A, B,$ and $C$ have fired in a certain order and separated by certain time intervals.

## DETECTION OF EVOKED RESPONSES IMBEDDED IN BACKGROUND ACTIVITY

In recent years, as neurophysiologists attempted to study cortical responses to sensory stimuli in animals that were, physiologically speaking, in a "more normal" state, two problems arose: First, it became more difficult to detect the evoked responses that were now imbedded in a much livelier background activity of the cortex than had been observed in the previously used, deeply anesthetized preparations. Second, it became important

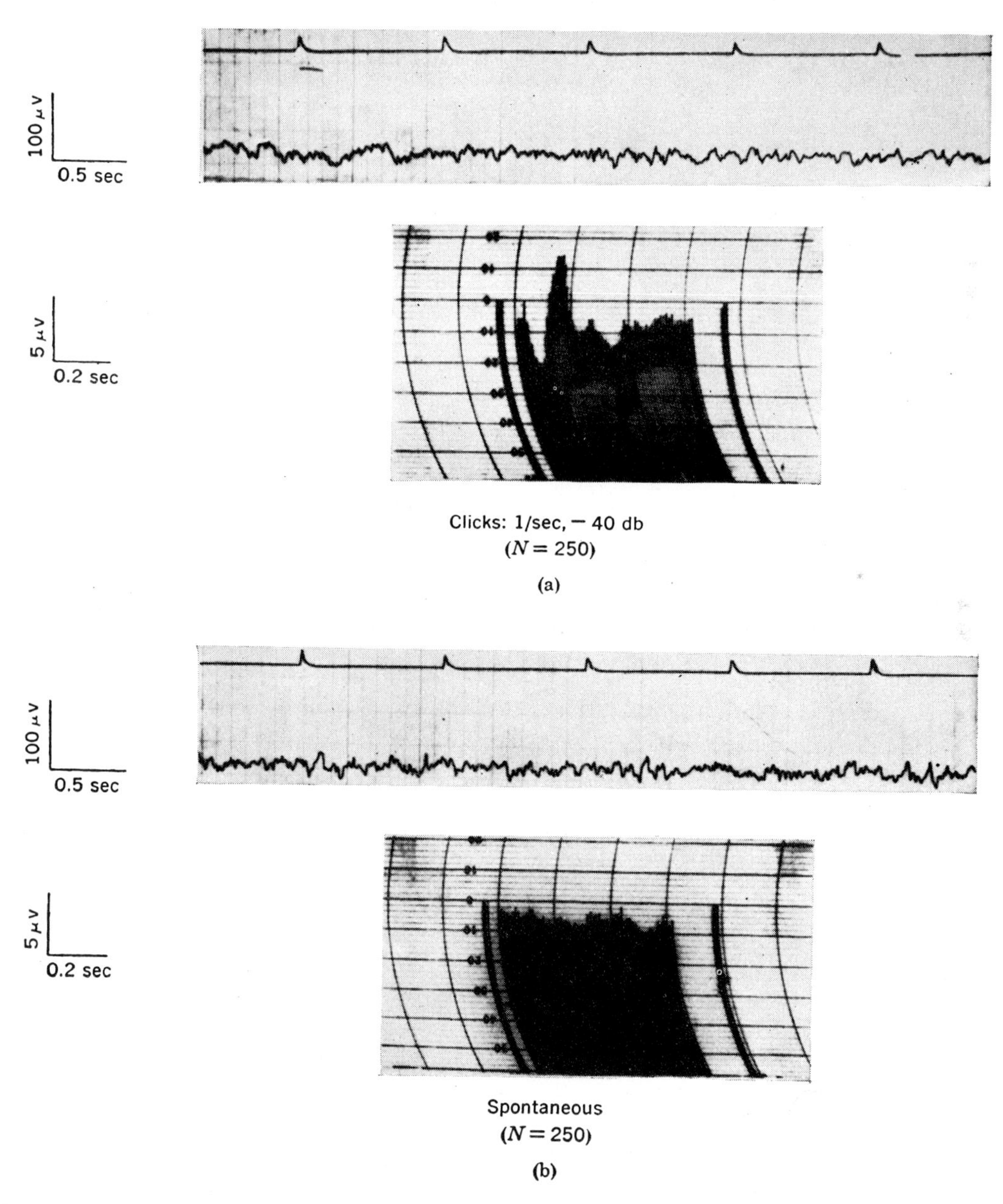


FIG. 8. The upper records of (a) and (b) show ink traces obtained from the skull (vertex) of human subjects (a) in the presence of clicks and (b) in the absence of clicks. The pulse channel, recorded simultaneously, indicates in (a) the instants at which clicks were delivered and in (b) serves as a comparable time reference. The lower records are the average wave forms obtained, by a process of cross correlation of the signal with a series of brief pulses,[14] from a sample of 250 intervals of which 5+ are shown in the ink records directly above. The average wave forms are the envelope of a series of closely spaced lines (10 msec).[33]

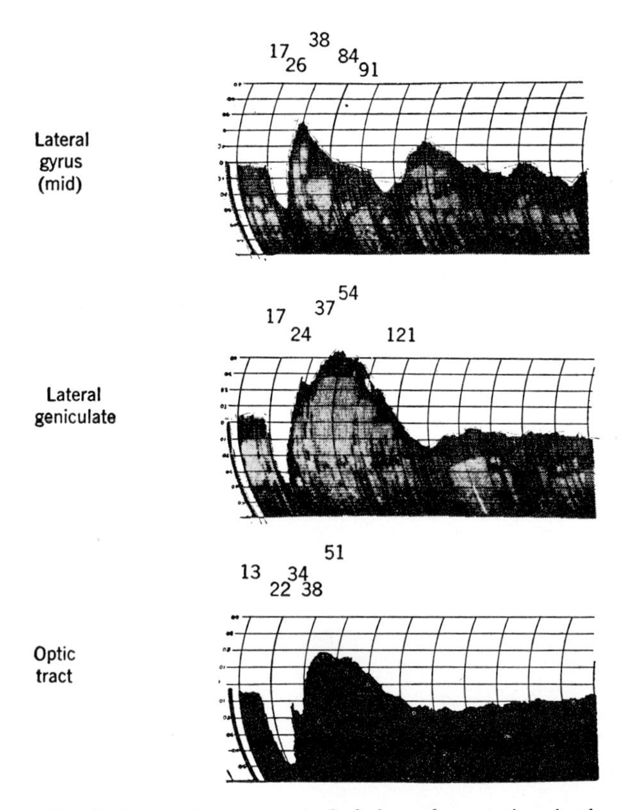


FIG. 9. Averaged responses to flash from three stations in the visual pathway of a cat; simultaneous recordings under moderate pentobarbital anesthesia with reference lead on the back of the neck. The average response is the envelope of a series of pen deflections separated by intervals of 1 msec. The delivery of the flash coincides with the first of the continuous series of pen deflections. The numbers above each average response are the latencies of characteristic points in the particular wave form [from M. A. B. Brazier, Acta Physiol. Pharmacol. Neerlandica 6, 692 (1957)].

to find ways of characterizing this background activity in a more than impressionistic manner.

Figure 7 illustrates the first problem by demonstrating how additional doses of anesthetic make it much easier to detect the "signal" (the evoked response) in the much diminished background activity that can, however, hardly be called "noise." Raab and Kiang[32] were actually looking for correlations between measures of the activity that preceded the delivery of a stimulus and the size of the evoked response. Their precise results are not of particular import here. It is, however, important to note that the interaction between evoked and background activity of the nervous system is not a problem for which a simple or even a single solution can be assumed.

Having a probabilistic bias in the interpretation of the electrical activity of the nervous system, members of our laboratory have attempted for several years to treat the obtained data in a manner that will yield average responses to sensory stimuli. The reasoning was, qualitatively, somewhat as follows: In the handling of sensory information, organisms behave as if they were acting on the basis of estimates of activity averaged over substantial regions of the nervous system; such a weighted averaging process could be carried out by comparing the outputs of a large number of neural elements. Since there are at present no devices capable of performing this task (even if one knew exactly where and how to perform it), one chooses to average over an ensemble of responses to repeated identical stimuli in order to bring out certain typical aspects of the behavior of the nervous system. Once the "correlator system" (referred to below) was available, it became possible to use it with comparatively slight modifications for the detection of evoked responses.[14] Figure 8 illustrates how this averaged display yields average results that would be hard to foretell on the basis of visual inspection either of ink tracings or of tracings of single responses photographed from the screen of an oscilloscope. Electronic computation of the sort used in our laboratory necessitates the recording of data onto FM magnetic tapes. Such tapes not only contain, in general, the specific responses that are set off by delivery of the stimuli but also offer, especially when multichannel recording is used, a much more representative sample of the organism's electrical activity. Appropriate programming permits the carrying out of different analyses of the recorded data as desired by replaying the tapes as often as is necessary.

Figures 9 (from Brazier[34]) and 10 offer just two examples of how the Evoked Response Detector (ERD) modification of the correlator system was used in the study of the visual[34] and auditory[35] systems. Other problems investigated include: cortical off-responses in the auditory system,[36] the effect of anesthesia upon the wave form of evoked cortical responses,[37] as well as responses to photic stimulation in man.[38]

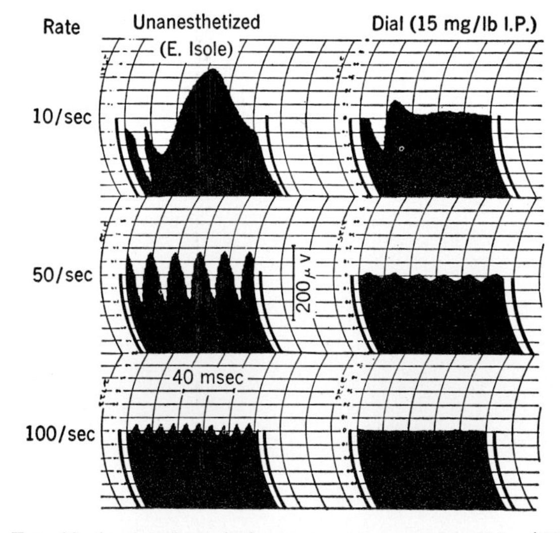


FIG. 10. Averaged cortical responses to repeated clicks (25-db threshold) indicate that the auditory cortex in nonanesthetized preparations is capable of "following" at higher rates. Note, also, the difference in wave form of response in anesthetized and nonanesthetized preparations for 10 clicks/sec. The evoked-response detector averaged the following number of responses at the different rates: 600 responses at 10/sec, 3000 at 50/sec, and 6000 at 100/sec.[35]

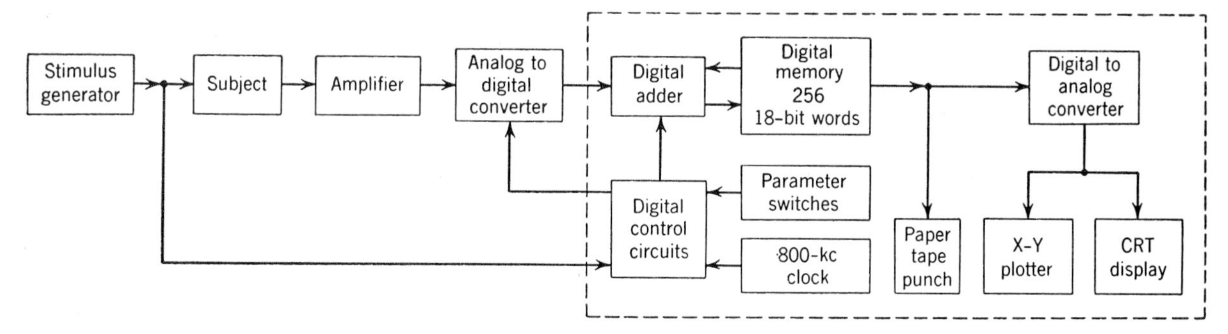


FIG. 11(a) Block diagram of average response computer (ARC-1).

More recently, we have been able to use (thanks to the cooperation of some colleagues at the Lincoln Laboratory) a considerably more powerful special-purpose digital computer, the ARC-1 (Average Response Computer[39]) in these investigations. Perhaps the most significant feature of this transistorized computer is that it computes "in real time" (Fig. 11). Thus, a display can be obtained instantaneously of the average of all response traces including the one just recorded from the animal's or subject's head. This immediate access to knowledge of results offers advantages, especially when as complex a system as the nervous system is being investigated. Without entering into the realm of science fiction, one can foresee experiments in the near future in which an organism and a computer operate in a "closed loop," i.e., in which the organism's response behavior alters the programming of sensory stimuli that the organism is instructed to deal with. With the help of ARC-1, our laboratory has continued (see Fig. 8) a rather exciting study of evoked potentials in man. It has thus been possible not only to record extracranially what appear to be cortical responses to acoustic clicks in many experimental subjects,[40] but also to demonstrate that these electrical responses appear at stimulus intensities that are quite comparable to the subject's absolute psychophysical threshold for clicks (see Fig. 12). Furthermore, it has been noted that the responses disappear near the subject's masked threshold. These findings suggest numerous sensory experiments in which convergent neurophysiological and psychophysical data can be obtained from human beings and are thus in tune with an era of experimentation in which the electrical activity of behaving animals will be recorded by implanted electrodes and broadcast by transistorized transmitters.[41]

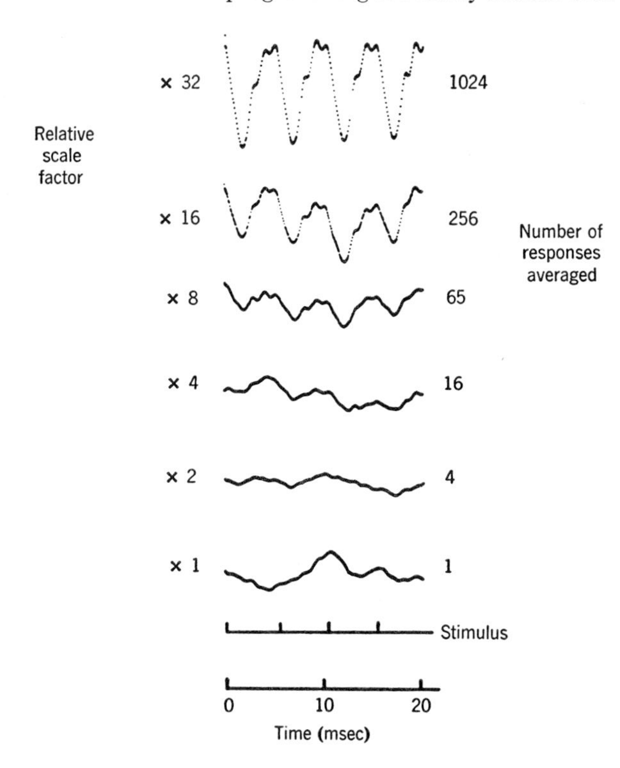


FIG. 11(b). Emergence of an average response from background activity as the number of averaged responses is increased.[39]

## ATTEMPTS AT QUANTIFICATION OF ONGOING ACTIVITY

Up to this point, this paper has emphasized specific electrical responses to specific sensory stimuli, although it has been pointed out that these responses are, at all levels of the nervous system, in a nontrivial sense functionally related to the existing background activity.§

Trained electroencephalographers are clearly capable of arriving at diagnostic judgments on the basis of ink tracings of the EEG. Attempts have been made to "quantify" (here the term is used in its broadest meaning) various aspects of brain-wave activity in experimental situations. These attempts at quantification have ranged from a classification of the recorded complex visual patterns into broad classes (corresponding either to characteristics of the EEG wave form or to such states as *arousal* or *drowsiness*, or to some pathological condition) to a running Fourier analysis of short-time (several seconds) samples of the EEG activity. Given our lack of theoretical understanding of the physiological mechanisms that underlie the EEG[42] and the multiplicity of purposes for which EEG data have been used, it is not surprising that no ideal method

§ This background activity has been designated as *spontaneous* or *ongoing* activity. The precise conditions under which it is observed in pure form are extremely hard to define.

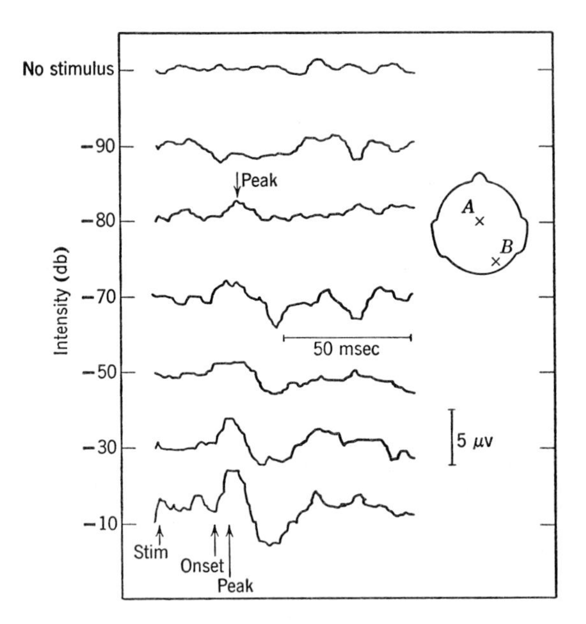


FIG. 12. Computed average responses to monaural periodic clicks obtained from scalp electrodes for different stimulus intensities. Each trace represents the wave form of the average response to 400 individual presentations of identical click stimuli delivered at a rate of 1.5/sec. The subject's (W. P., 10 September 1957; awake, eyes open) psychophysical threshold for these clicks, as determined during the same experiment, lay at approximately −85 db. Upward deflection indicates that electrode *A* is positive with respect to electrode *B* [from C. D. Geisler, L. S. Frishkopf, and W. A. Rosenblith, Science **128**, 1210 (1958)].

of analysis has been found. Several years ago Wiener—intent on testing certain of his ideas regarding the presence of a "clock" in the brain[43]—interested Brazier's group at the Massachusetts General Hospital and the Communications Biophysics group in the Research Laboratory of Electronics, MIT, in applying correlation analysis to the EEG. A "correlator system"[44] was built, and a great quantity of experimental data has by now been processed.[45] There developed a considerable amount of discussion regarding the mathematical nature of the EEG time series, regarding the usefulness of correlation analysis of the EEG (since it discards so much of the information contained in the raw data), and even regarding the ease with which a correlation function of the EEG can be interpreted in a really quantitative manner. Whatever the merits of these issues, any data-reducing scheme should prove of some usefulness if the final display of the dependent variable satisfies several requirements: (a) it should be stable for a given organism when data are taken under comparable conditions; (b) it should be sensitive to important changes in the organism's environment; (c) it should perhaps be possible to classify a whole population on the basis of a limited number of displays.

Figures 13 to 15 illustrate some of the properties of the autocorrelation function as such a display. Figure 13 demonstrates that the display is capable of reflecting a threefold classification of the EEG ink tracings into

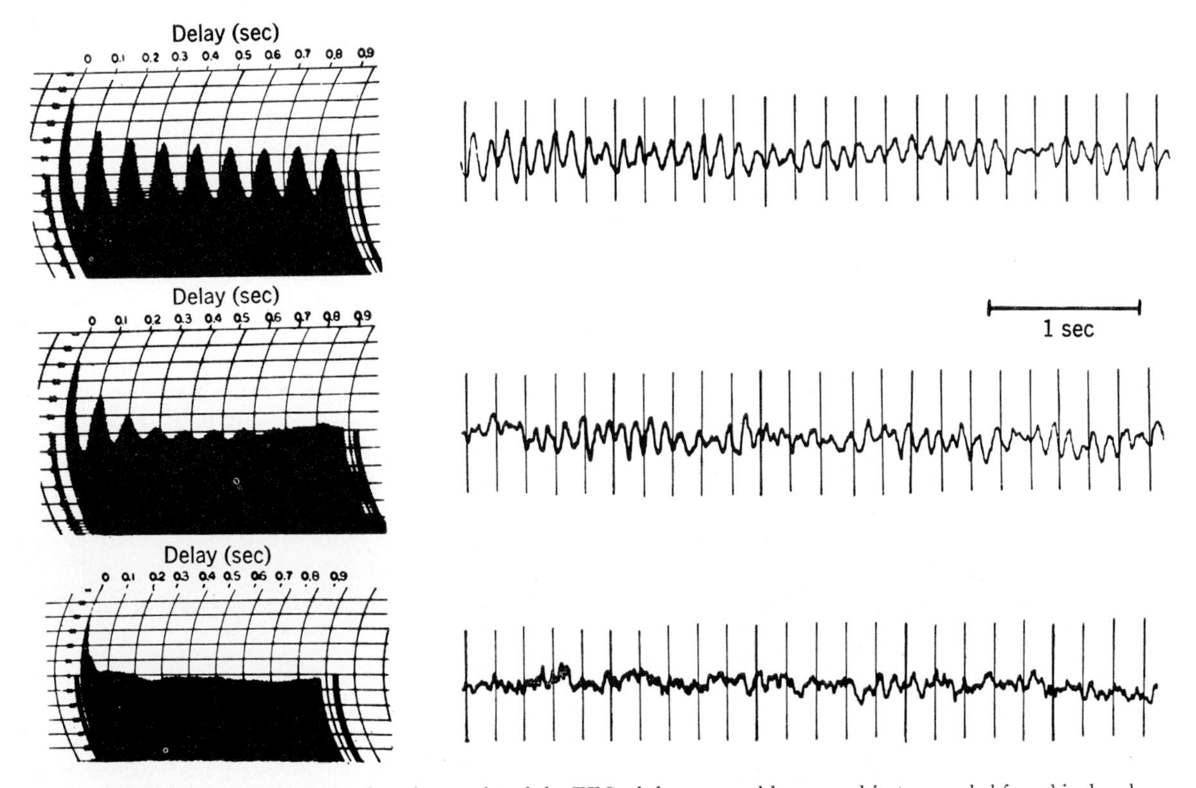


FIG. 13. Left: Autocorrelograms of 1-min samples of the EEG of three normal human subjects recorded from bipolar electrodes (right parietal to right occipital) on the scalp. Right: Sample of ink records from these same subjects [from J. S. Barlow, M. A. B. Brazier, and W. A. Rosenblith, "The application of autocorrelation analysis to electroencephalography," in *Proceedings of the National Biophysics Conference, 1957* (Yale University Press, New Haven, to be published)].

"high alpha," "medium alpha," and "no alpha." It remains to be seen to what extent a much larger population could be accurately classified on the basis of a certain number of such displays for various locations on the skull.

Figure 14 asks the question: How stable is such a display for a given person over a prolonged period of time? Although it is not possible at present to evaluate the similarity of these two displays in precise numerical terms, they appear to be quite comparable if the conditions of recording the EEG are reasonably well controlled and the time sample from which the correlation

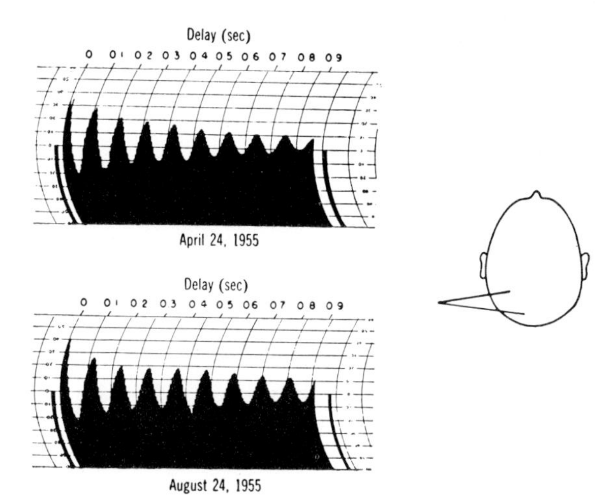


FIG. 14. Autocorrelograms of the EEG (1-min sample) of a normal human subject recorded four months apart [from J. S. Barlow, M. A. B. Brazier, and W. A. Rosenblith, "The application of autocorrelation analysis to electroencephalography," in *Proceedings in the National Biophysics Conference, 1957* (Yale University Press, New Haven, to be published)].

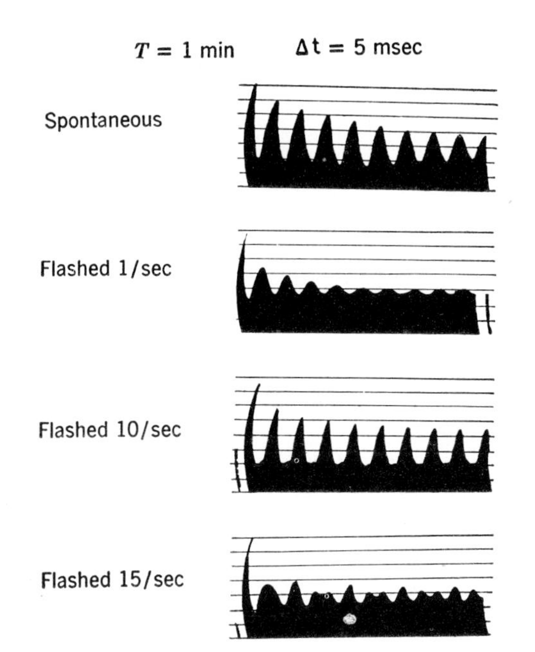


FIG. 15. Autocorrelograms computed for same subject in absence of controlled sensory stimulation ("spontaneous") and in presence of different rates of periodic photic stimulation.

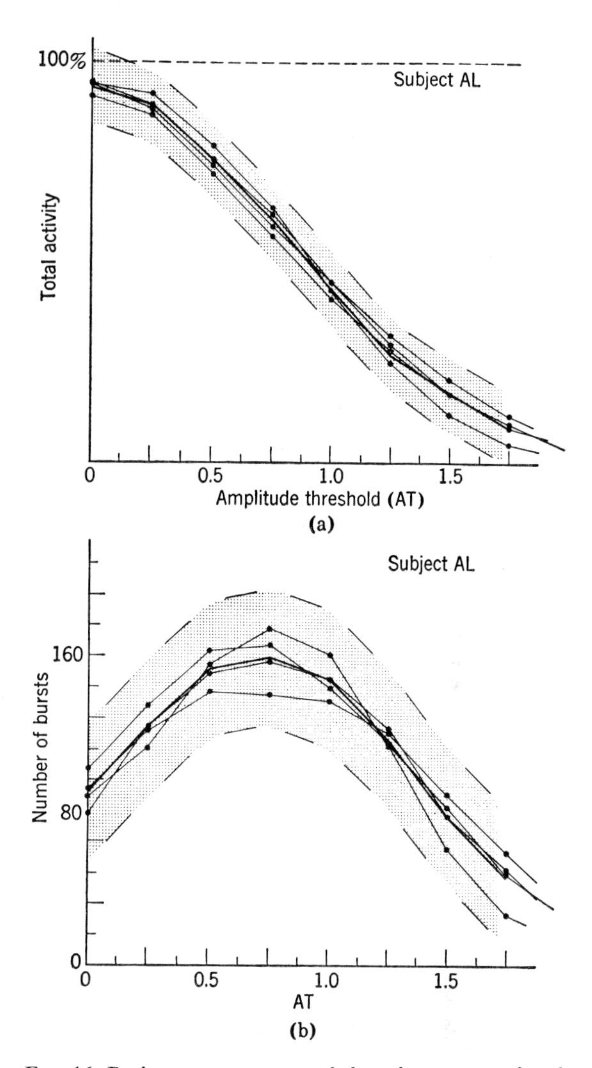


FIG. 16. Brain waves were recorded on four consecutive days (at same time of the day) from an awake subject sitting with closed eyes in an anechoic chamber. The displays show the four sets of results in fine lines, with the mean of the four runs as a heavy solid line. Each analyzed run lasted 3 min. For details of computer program and definitions of such terms as *amplitude threshold*, *total activity*, and *number of bursts*, see Farley *et al.*[46] The shaded region is bounded by the estimated confidence limits for the variation of the displays.[46]

function is computed is sufficiently long. Figure 15 demonstrates that the autocorrelational display is sensitive to photic stimulation.

It is not possible to examine here the extent to which the findings obtained with the aid of correlation (both autocorrelation and crosscorrelation) analysis support a particular model of brain function. One must, of course, admit that any method of data analysis implies a model of the process that is being investigated. The research worker has a burden of proof in the sense that he must show that the implied model fits the process well enough for his purpose. The fact remains that, in the absence of systematic computational analysis, most questions regarding the mathematical nature of the signal cannot even be raised in a meaningful manner. In addition, one should not underestimate the heuristic

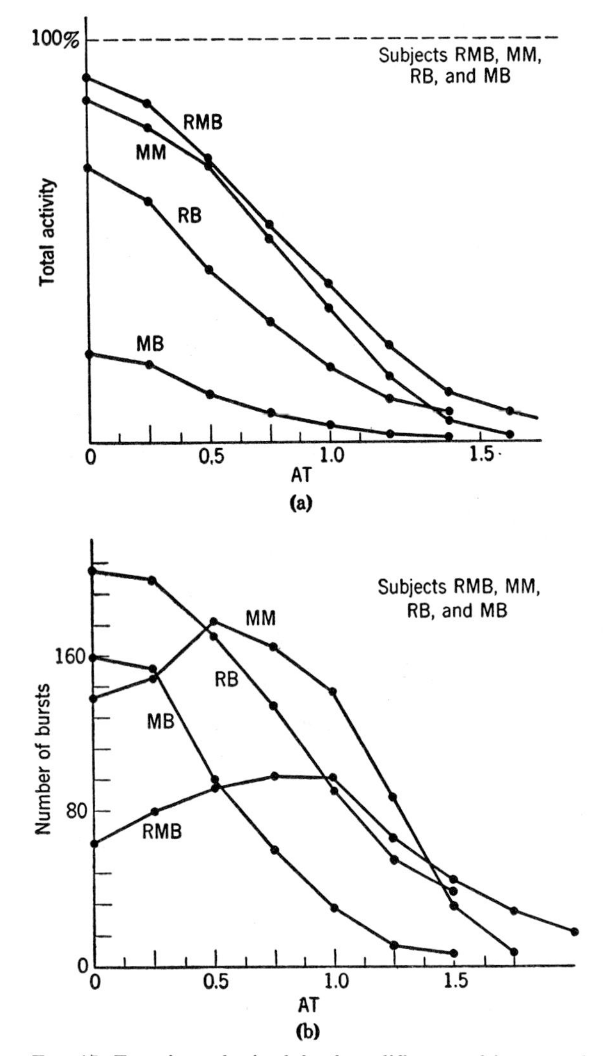


FIG. 17. Functions obtained for four different subjects on the basis of single 3-min runs. Note that subjects RMB and MM do not differ much in their total activity profile, but they are clearly distinguished in the display depicting the number of bursts.[46]

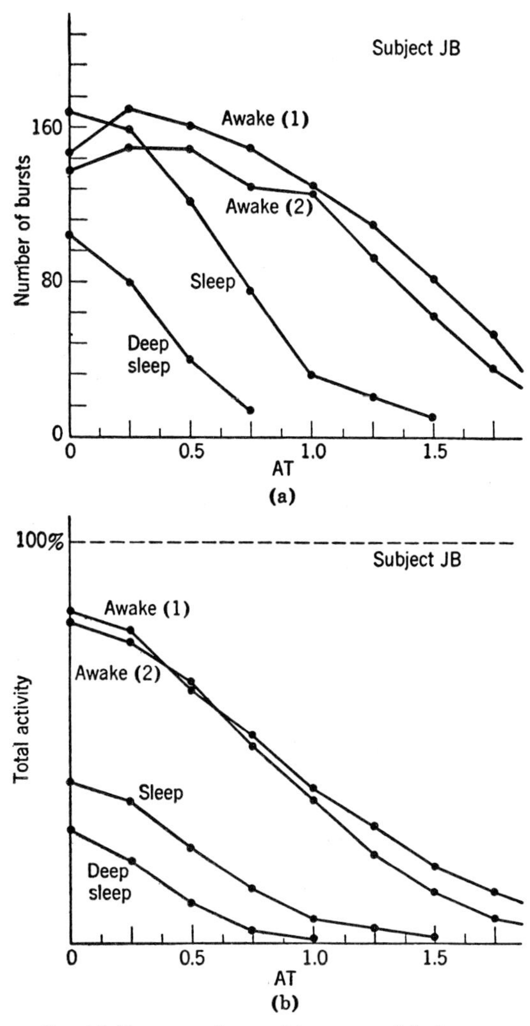


FIG. 18. Four runs for a subject, two while he was awake and two in two different states of sleep.[46]

benefits that derive from the use of a display that is in itself "anschaulich."

Displays of correlation functions of ongoing electrical activity are just one way of trying to describe an organism's "state" in terms of pseudo-state-variables. A report entitled "Computer Techniques for the Study of Patterns in the Electroencephalogram"[46] represents perhaps the most far-reaching effort in this direction. With the help of a rather flexible general-purpose digital computer equipped with a large memory, the attempt was made to "recognize the pattern of rhythmic bursts" in a given individual's EEG. Once a program for this purpose had been written, the attempt was made to distinguish different subjects from each other, as well as different "states" in the same subject.

Figures 16 to 18 show some of the results that have been obtained. Figure 16 illustrates that it is possible to obtain quite reproducible displays for the same subject if precautions are taken to obtain the data under fairly constant conditions. It was even possible to estimate confidence limits for the variability of the display for this subject and to compare displays from other subjects with these limits in order to assess significant differences. The initial results by Farley *et al.*[46] indicate that even a limited number of displays is capable of separating subjects into statistically distinguishable classes. Increasing the number of displays by increasing the number of pseudo-state-variables should make it possible to remove existing "degeneracies."

Figure 18 shows how these displays in a given subject are affected by his going to sleep. The influence of drugs or of sensory stimulation could be studied in this manner. Some evidence has already been obtained that these displays are sensitive to even short periods of sensory deprivation.

It is, however, appropriate to utter here a caveat regarding this and other methods that aim at quantifying data on the electrical activity of the nervous system. The multivariate nature of the nervous system prevents one from taking too much stock in any given number or even in a computed curve. Somehow one must learn to derive quantities that characterize the essentials of neural activity in a fashion that is perhaps somewhat analogous to the dimensionless parameters of engineering systems. To make this point more concrete: Farley *et al.*[46] have shown that there are individuals whose total activity profile when they are awake bears some resemblance to the profile of another individual while the latter is asleep. However, whenever a person falls asleep there are changes in his activity profile. In other words, one must learn to assess the significance of change in relation to certain base lines and must refrain from attaching too much value to isolated pieces of numerical information. Unfortunately, economical mathematical descriptions of patterns and sequences are not yet available. These considerations can perhaps be carried somewhat further by noting that Fig. 13 illustrated how widely the autocorrelogram from apparently normal individuals may vary. It would thus seem unwise to draw too far-reaching conclusions regarding, for instance, a person's psychological make-up on the basis of a single display or even a set of summary displays computed from the electroencephalogram.

## ACKNOWLEDGMENTS

The author owes a debt of gratitude to his colleagues and associates in MIT's Research Laboratory of Electronics. In connection with the data presented in this paper, he is particularly indebted to J. S. Barlow, M. A. B. Brazier, R. M. Brown, B. G. Farley, M. Z. Freeman, L. S. Frishkopf, C. D. Geisler, M. H. Goldstein, Jr., N. Y.-S. Kiang, J. Macy, Jr., C. E. Molnar, W. T. Peake, D. H. Raab, T. T. Sandel, and T. F. Weiss. To some extent, their views are reflected in the discussion, although the author must bear sole responsibility for the conclusions that he has drawn.

## BIBLIOGRAPHY

[1] H. Davis, Physiol. Rev. **37**, 1 (1957).
[2] H. W. Magoun, *The Waking Brain* (Charles C. Thomas, Springfield, Illinois, 1958).
[3] G. Moruzzi and H. W. Magoun, Electroencephalog. and Clin. Neurophysiol. **1**, 455 (1949).
[4] K. L. Chow, J. Comp. Neurol. **95**, 159 (1951).
[5] R. Galambos, J. Neurophysiol. **19**, 424 (1956).
[6] E. D. Adrian, F. Bremer, and H. H. Jasper, editors, *A C.I.O.M.S. Symposium: Brain Mechanisms and Consciousness* (Blackwell Scientific Publications, Oxford, 1954).
[7] R. Hernández-Peón, H. Scherrer, and M. Jouvet, Science **123**, 331 (1956).
[8] S. W. Kuffler and C. C. Hunt, Research Publ. Assoc. Nerv. Mental Dis. **30**, 24 (1952).
[9] K.-E. Hagbarth and D. I. B. Kerr, J. Neurophysiol. **17**, 295 (1954).
[10] D. I. B. Kerr and K.-E. Hagbarth, J. Neurophysiol. **18**, 362 (1955).
[11] R. Granit, *Receptors and Sensory Perception* (Yale University Press, New Haven, 1955).
[12] R. Galambos, G. Sheatz, and V. G. Vernier, Science **123**, 376 (1956).
[13] M. Jouvet and J.-E. Desmedt, Compt. rend. **243**, 1916 (1956).
[14] J. S. Barlow, Electroencephalog. and Clin. Neurophysiol. **9**, 340 (1957).
[15] W. J. McGill, "A statistical description of neural responses to clicks recorded at the round window of the cat," Ph.D. thesis, Harvard University (1952).
[16] J. Macy, Jr., "A probability model for cortical responses to successive auditory clicks," Ph.D. thesis, Massachusetts Institute of Technology (1954).
[17] L. S. Frishkopf, Tech. Rept. 307, Research Laboratory of Electronics, Massachusetts Institute of Technology (1956).
[18] K. S. Harris, "Serial characteristics of neural responses to acoustic clicks recorded at the auditory cortex," Ph.D. thesis, Harvard University (1954).
[19] J. E. Rose and V. B. Mountcastle, Bull. Johns Hopkins Hosp. **94**, 238 (1954).
[20] M. H. Goldstein, Jr. and N. Y.-S. Kiang, J. Acoust. Soc. Am. **30**, 107 (1958).
[21] E. A. Blair and J. Erlanger, Am. J. Physiol. **106**, 524 (1933).
[22] C. Pecher, Arch. Int. Physiol. **49**, 129 (1939).
[23] W. A. Rosenblith, J. Acoust. Soc. Am. **22**, 792 (1950).
[24] W. A. Rosenblith in *Proceedings of the Symposium on Information Networks* (Polytechnical Institute of Brooklyn, 1954), p. 223.
[25] L. S. Frishkopf and W. A. Rosenblith in *Symposium on Information Theory in Biology* (Pergamon Press, New York, 1958), p. 153.
[26] V. E. Amassian and H. J. Waller in *Reticular Formation of the Brain* (Little, Brown and Company, Boston, 1958), p. 69.
[27] M. Palestini, G. F. Rossi, and A. Zanchetti, Arch. Ital. Biol. **95**, 97 (1957).
[28] G. F. Rossi and A. Zanchetti, Arch. Ital. Biol. **95**, 199 (1957).
[29] O. J. Grüsser and A. Grützner, v. Graefe's Arch. Ophthalmol. **160**, 65 (1958).
[30] R. Jung, O. Creutzfeldt, and O. J. Grüsser, Deut. med. Wochschr. **82**, 1050 (1957).
[31] S. D. Erulkar, J. E. Rose, and P. W. Davies, Bull. Johns Hopkins Hosp. **99**, 55 (1956).
[32] D. H. Raab and N. Y.-S. Kiang, Quart. Progr. Rept., Research Laboratory of Electronics, Massachusetts Institute of Technology (October 15, 1955), p. 76.
[33] L. S. Frishkopf and C. D. Geisler, Quart. Progr. Rept., Research Laboratory of Electronics, Massachusetts Institute of Technology (January 15, 1957), p. 123.
[34] M. A. B. Brazier, Acta Physiol. Pharmacol. Neerlandica **6**, 692 (1957).
[35] M. H. Goldstein, Jr., N. Y.-S. Kiang, and R. M. Brown, J. Acoust. Soc. Am. **31**, 356 (1959).
[36] N. Y.-S. Kiang and T. T. Sandel, Quart. Progr. Rept., Research Laboratory of Electronics, Massachusetts Institute of Technology (July 15, 1957), p. 139.
[37] N. Y.-S. Kiang and M. H. Goldstein, Jr., Quart. Progr. Rept., Research Laboratory of Electronics, Massachusetts Institute of Technology (July 15, 1957), p. 142.
[38] M. A. B. Brazier in *Reticular Formation of the Brain*, H. H. Jasper *et al.*, editors (Little, Brown and Company, Boston, 1958), p. 151.
[39] W. A. Clark, Jr., Quart. Progr. Rept., Research Laboratory of Electronics, Massachusetts Institute of Technology (April 15, 1958), p. 114.
[40] C. D. Geisler, L. S. Frishkopf, and W. A. Rosenblith, Science **128**, 1210 (1958).
[41] G. E. Forsen, "An FM transmitter for recording the electrical activity of the nervous system," S.M. thesis, Massachusetts Institute of Technology (June, 1957).
[42] M. A. B. Brazier, J. Nerv. Mental Dis. **126**, 303 (1958).
[43] N. Wiener, Scientia **52**, 1 (September, 1958).
[44] J. S. Barlow and R. M. Brown, Tech. Rept. 300, Research Laboratory of Electronics, Massachusetts Institute of Technology (1955).
[45] J. S. Barlow, "Autocorrelation and crosscorrelation analysis in electroencephalography," Eleventh Annual Conference on Electrical Techniques in Medicine and Biology," 19–21 November 1958, Minneapolis (manuscript in preparation).
[46] B. G. Farley, L. S. Frishkopf, W. A. Clark, Jr., and J. T. Gilmore, Jr., Tech. Rept. 337, Research Laboratory of Electronics, Massachusetts Institute of Technology, and Tech. Rept. 165, Lincoln Laboratory, Massachusetts Institute of Technology (1957).

# 58

# Imitation of Pattern Recognition and Trial-and-Error Learning in a Conditional Probability Computer*

ALBERT M. UTTLEY
*National Physical Laboratory, Teddington, Middlesex, England*

A CONDITIONAL probability computer, appropriately connected with input and output devices, can "recognize" patterns and "learn" by trial and error. The principle of such a system developed at the National Physical Laboratory is described briefly in this paper, together with some comments on the role of engineering in the study of biological problems.

An engineer generally starts with a specification, a set of rules of behavior for a machine, and then tries to design a machine that possesses that behavior. He designs an airplane, for example, to fly at certain speeds, to go certain distances, and to carry certain loads; from these requirements he deduces the required structure.

Nature is the engineer that constructs the difficult bridge between structure and behavior of living systems. In psychology, one discusses the behavior of an organism and wonders what kind of structure could cause it. In physiology and anatomy, on the other hand, one usually studies the structure and wonders what it is doing, what function it is supposed to serve.

We believe, in all humility, that the engineer can perhaps help to bridge the gap between structure and behavioral function in biological systems. Using his known methods, the engineer can most fruitfully start with observed behavior and work toward a possible elucidation of structure. Some engineers have tried to work in the opposite direction: to think up some properties of units, give them random connections, and find out what they will do. We do not think that that approach has been fruitful. We have considered, specifically, the problems of pattern recognition and learning, and have tried to design computers to do these things in the hope that the structure of the computers might be suggestive of the organization of the nervous system.[1]

This approach differs in one respect from that of Shannon and many others, who have also studied the design of machines to do certain things which animals and men can do. Their approach has been to program a universal computer. Such an approach should lead to a deep mathematical understanding of behavior, but it is not likely to lead to a specific structure which will give clues to the physiologist, because a universal machine can do anything.

The first step toward synthesizing pattern recognition is to devise a method for distinguishing one *set* of signals from another, corresponding to sets of signals in nerve fibers. The signals might be continuously variable, or they might be discrete and binary; the latter type is much easier to handle, and is used in this study. The machine must have a component unit for each signal-input channel to indicate if that channel is active; and a unit for each possible pair of channels to indicate, by means of a coincidence circuit, whether or not that pair of channels is active. In general, one needs a unit for each possible set of channels. Such an arrangement is called a classification system; an example is shown in Fig. 1.[2]

Next, the system must be made to recognize patterns in *time*; this requires the introduction of delay circuits. Suppose that there are only two input channels, and that each is put through a fixed delay. If these two inputs and their delayed versions are used as four inputs of a classification system, the system will recognize all 12 "tunes" which can be played on a two-note piano if one allows only two temporal ideas: "before" and "after." The connections and recognition possibilities of such a system are shown in Fig. 2.

A classification system can be built with quite *random* connections. The reason is that all possible arrangements of connections between units and channels are required. If there are sufficient units to accommodate the desired inputs, and if they are connected entirely at random to the inputs, a classification system will arise automatically.

Next, the system must be given *plastic* behavior, in which past events can influence future responses. This property is introduced through the computation of *conditional probability*. Suppose that there are two

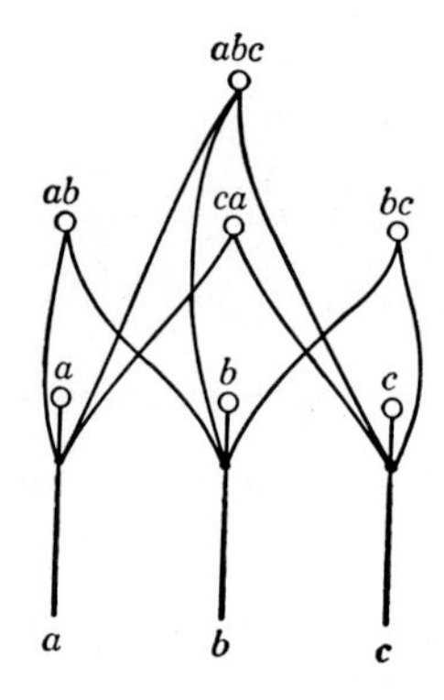

FIG. 1. A classification system for three inputs has seven units.

* This paper is published by permission of the Director of the National Physical Laboratory.

events, $A$ and $B$, which occurred as follows on nine occasions:

$A$: $I\ I\ I\ I\ I\ O\ O\ O\ O$ ($I$=occurrence,
$B$: $O\ O\ I\ I\ I\ I\ I\ I\ I$ $O$=nonoccurrence)

One now asks: "If the future is like the past—if one accepts the inductive principle—and if $A$ occurs, what is the chance of $B$?" In the past, $A$ occurred on five occasions, and on three of these $B$ also occurred. The corresponding probability of $B$, given $A$, is three out of five. Conversely, the probability of $A$, given $B$, is three out of seven. A machine to make these calculations must be able to count the five, in an $A$ unit; the seven, in a $B$ unit; and the three, in a coincidence-counting $AB$ unit. Thus, to possess plastic behavior, the machine needs an additional feature: each unit must be able to *count*, even if in a very nonlinear way. The system will then be capable of inductive inference.[3]

Another important matter which comes easily out of this approach is the "context" in which events occur. Suppose that, in the foregoing example, an event $C$ occurred on the last five occasions:

$A$: $I\ I\ I\ I\ I\ O\ O\ O\ O$
$B$: $O\ O\ I\ I\ I\ I\ I\ I\ I$
$C$: $O\ O\ O\ O\ I\ I\ I\ I\ I$

This history now has two different contexts ($C$ occurs and $C$ does not occur), and one can ask: "In the context $C$, what would be the probability of $A$, given $B$?" In this context, $B$ occurred five times and $A$ once, so the three-out-of-seven chance has changed to one-out-of-five. Similarly, the probability of $B$, given $A$, in the context $C$, is unity. The system can be extended indefinitely to infer one thing from another in all sorts of different contexts.[4]

There is a practical difficulty in the design of counters: after a time they get full. The engineer has to put a leak into the counter to avoid this dilemma—and a very interesting property is thereby produced. The leaky counter gives a time weighting in which past events matter less than recent ones. The computer can then *adapt* fairly quickly to new conditions.

Finally, the system must be able to *calculate ratios* of

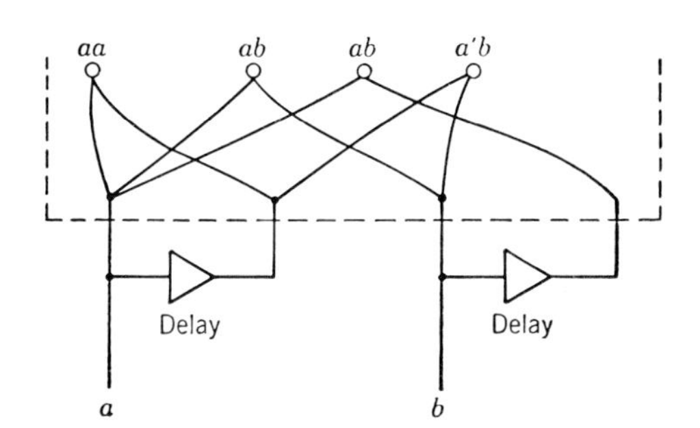


FIG. (2a). A classification system for 2 inputs which distinguishes "before" and "after." Four of the 15 units are shown.

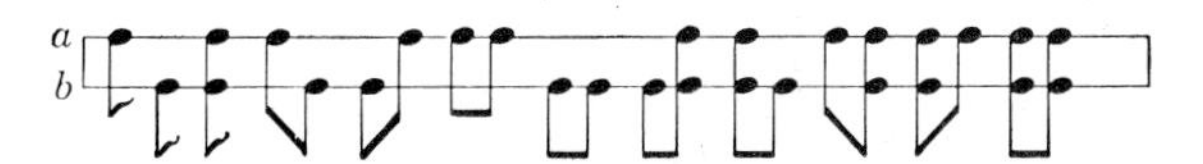


FIG. 2(b). The 12 tunes which can be distinguished by the system of Fig. 2(a). For the first 3 tunes—$a$, $b$, and $ab$—there are duplicate units.

numbers. Division is done most easily by using logarithms. It turns out that a leaky counter produces something very close to a logarithmic count, so the required ratios can be obtained simply by taking the differences between the states of two units. (The "state" is just the stored voltage in the counter.) Clearly, the effect of an active counting unit on one to which it is connected must depend on their relative state.

This relationship suggests a very important feature of a system that is capable of learning. The old idea that "facilitation" of a path depends on the number of times that it is used is not adequate; it is as illogical as the story of the Irishman and his boots—he couldn't get them on until he had worn them a few times. It is well known that a novel stimulus may, initially, produce no reaction from an animal, but eventually produce a response owing to trial-and-error learning. If initially a path is not made at all, how could it possibly be made on the principle that the more often it is used, the better it is?

The difficulty is resolved by a comparison of the states of two units. In neural terms, if a probability is to be computed, the numerator must be stored on one side of a synapse and the denominator on the other, so that a comparison can be made.

In summary, a machine to simulate pattern recognition and learning must be able to:

(a) distinguish sets of signals, through the use of coincidence counting;

(b) recognize time patterns, by use of delay circuits;

(c) count occurrences of signals and sets of signals;

(d) store information regarding past occurrences, at least for a certain length of time; and

(e) calculate ratios to determine relative states of units with stored counts of occurrences.[5]

These properties are incorporated in the machine shown in Fig. 3.

## CONCLUDING REMARKS

Mathematical models can be very useful for studying complex systems and functions, including those of living organisms. The utility of a model, and especially the confidence with which one can draw inferences from its behavior, can be greatly enhanced by building a physical analog of the model—a machine. In the process of building the machine and making it work, one may find shortcomings or omissions in the theoretical model. More important, one may find that the machine is capable of doing more than was anticipated. The system reported here, for example, was designed to do pattern

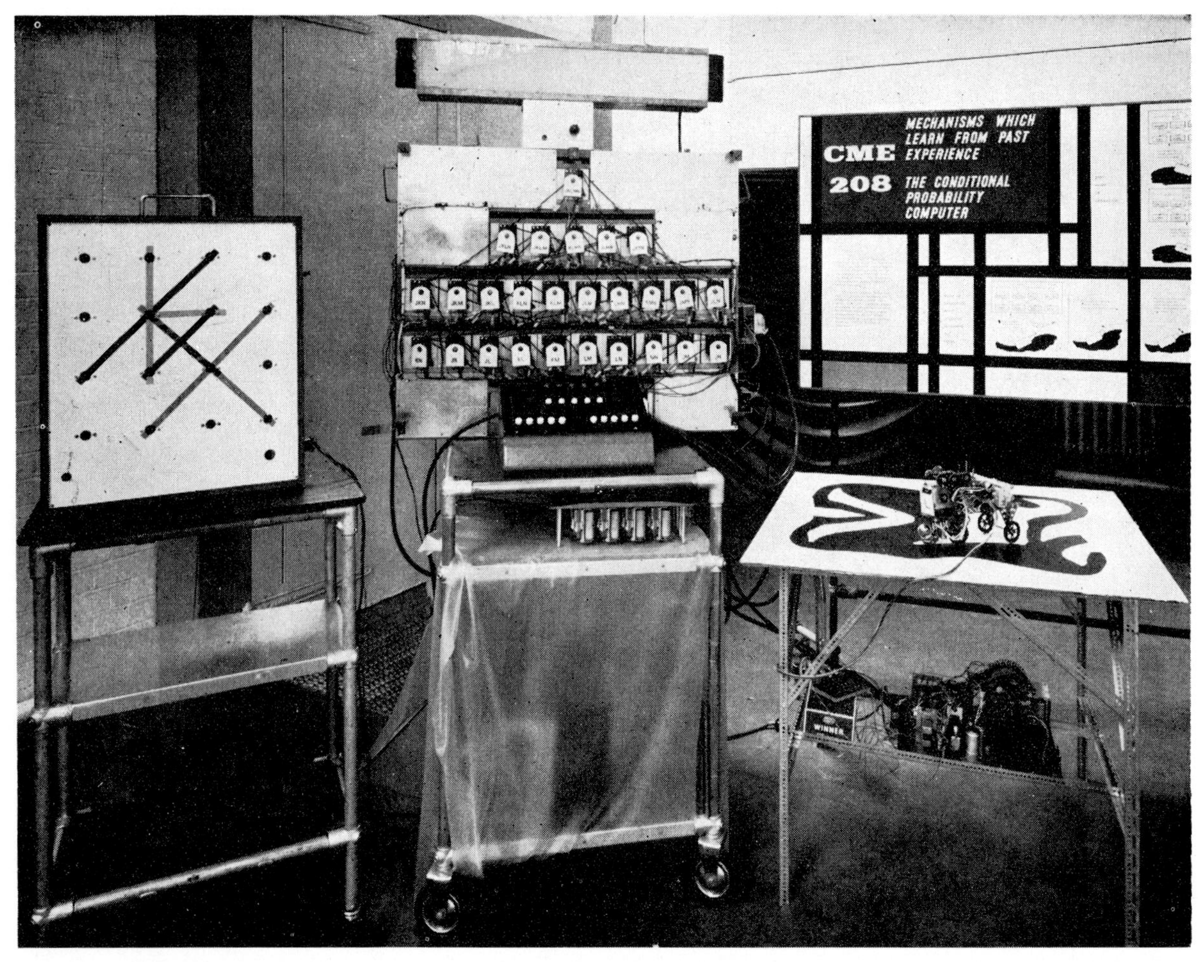


FIG. 3. A conditional probability computer, built at the National Physical Laboratory, capable of pattern recognition and trial-and-error learning. On the left is a model "retina" of photo cells to receive T-shaped shadow patterns. On the right is an electrically driven "vehicle" that learns, with the help of its computer "brain" which is shown in the center of the picture, to move efficiently along the black-white border of its "highway."

discrimination, and it turned out to be capable of trial-and-error learning as well.

Another important reason for making models is that people in different specialties use different languages—as is amply evident in a conference such as this Study Program. People try to talk about the same thing and hardly understand one another. A model is a sort of universal language. No matter what your field is, if you look at a physical model and see it working—see its parts working—you can get something out of it. In particular, the physiologist and anatomist can look at an engineer's model of a psychologist's observation and consider whether the structure of the model bears any resemblance to the structure of the living organism.[6]

One final comment: engineers use a very limited number of types of vacuum tubes to do a very large number of different things. The nervous system appears to be similar. There must be an enormous number of things neurons can do with just a few basic properties. The computation of conditional probabilities is, perhaps, just one of them.

## BIBLIOGRAPHY

[1] A. M. Uttley, *Conditional Probability as a Principle in Learning*, Namur Cybernetics Conference (1956) (Gauthier-Villars, Paris) (to be published).

[2] A. M. Uttley, Electroencephalog. and Clin. Neurophysiol. **6**, 479 (1954).

[3] A. M. Uttley in *Automata Studies*, C. E. Shannon and J. McCarthy, editors, Ann. of Math. Studies **34**, 253 (1956).

[4] A. M. Uttley, Electroencephalog. and Clin. Neurophysiol. **6**, 277 (1954).

[5] A. M. Uttley, "The design of conditional probability computers," Information and Control (to be published).

[6] A. M. Uttley, *Conditional Probability in a Nervous System*, N. P. L. Symposium (1958) (Her British Majesty's Stationery Office, London) (to be published).

59

# Chemical Specificity in Biological Systems

WALTER KAUZMANN

*Department of Chemistry, Princeton University, Princeton, New Jersey*

THE importance of specificity in biology has been mentioned or implied in a great many contributions presented in this volume, and many examples have been discussed. It is worthwhile, however, to examine the concept of specificity by itself, and it is advisable to review some of the more important physical and chemical factors that are currently believed to be involved in the phenomenon.

In this paper are considered some typical examples of biological specificity involving more or less well-defined chemical systems as one of the specific interacting agents. The molecular basis of specificity is discussed with special emphasis on the concept of complementarity. Finally, some words are devoted to a review of the important and old-fashioned idea of specificity as a form of organization.

Serological specificity was defined by Landsteiner[1] as "the disproportional action of a number of similar agents on a variety of related substrata," and this definition is readily extended to other fields in which biological specificity manifests itself. Boyd[2] points out that this definition "would also include many chemical reactions, as might be expected if serological specificity were fundamentally chemical in nature, which we now believe," and again there is no need to hesitate in applying the remark to biological specificity in general. Indeed, specificity is an important aspect of chemistry; qualitative and quantitative analysis in organic and inorganic chemistry depends in a large degree upon "disproportionate actions" of similar agents on "related substrates" (e.g., the precipitation of silver halides in water and the separation of silver chloride from the bromide and iodide by treatment with ammonia, $H_2O$ and $NH_3$ being regarded as similar agents and the silver halides being related substrates).

It is important to emphasize at the outset, too, that biological specificity is very variable in degree. Certain systems show a wide range of specificity whereas other systems show extremely narrow selectivity. What is likely to impress the chemist about biological specificity is the sharpness with which closely similar substances can be distinguished in some instances, but this sharpness is not at all universal.

It is worth mentioning that the concept of specificity is involved in the actions of inhibitors as well as in those of normal substrates, since many closely related biological systems are poisoned to decidedly different degrees by substances having similar chemical compositions. The drug industry thrives on this fact—as does mankind in general.

## EXAMPLES OF CHEMICAL SPECIFICITY IN BIOLOGICAL SYSTEMS

### 1. Antibody-Antigen Interactions

Because of the pioneering work of Landsteiner,[1] the field of immunochemistry provides one with probably the most easily understood examples of biological specificity, and it has made possible a fruitful method of attack on the relationship between specificity and chemical structure. When certain foreign substances (antigens) are introduced into an organism, the organism reacts by producing proteins (antibodies) which are capable of reacting specifically with the antigens that gave rise to them. Antigens are invariably substances of rather high molecular weight, so that it is difficult to alter their chemical composition in a systematic fashion. Landsteiner was able to show, however, that new antigens may be prepared by combining small chemical groups (such as substituted aromatic rings) with proteins. These new antigens give rise to antibodies whose specificity depends on the nature of the conjugated groups. The small chemical group is called a hapten. Let $w$, $x$, $y$, $z$ represent a series of chemically related haptens, let $Aw$, $Ax$, $Ay$, $Az$ represent antigens produced from protein $A$ by conjugation with these haptens, and let $Bw$, $Bx$, $By$, $Bz$ represent the antibodies arising from injection of $Aw$, $Ax$, $Ay$, $Az$ into a rabbit. It is clear that a great deal might be learned about the chemical factors involved in immunological specificity by comparing the interactions of the different antibodies with the different antigens. Furthermore, it is found that a low molecular-weight compound, $aw$, containing the hapten group $w$, is able to inhibit the interaction between $Aw$ and $Bw$. Evidently, $Bw$ contains sites which are capable of reacting with the hapten groups on $Aw$, and the haptens in $aw$ can compete for these sites. The study of this competition using a series of related haptens makes possible a more quantitative study of antibody-antigen specificity.

Some typical results of this general approach are reported in a paper by Landsteiner and Lampl.[3] Metaaminobenzoic acid and metasulfanilic acid were coupled by diazotization of the amino group to the proteins of horse serum, and antibodies to these modified proteins were obtained by injection into rabbits. The resulting rabbit antisera then were mixed with samples of chicken sera which had been diazotized to aniline or to various aniline derivatives. Evidence for interaction consisted in the formation of a precipitate. (The use of horse serum in the production of antibodies and of chicken serum in the antibody-antigen test

TABLE I. Effects of structural changes in the hapten on antibody-antigen interaction.[a]

| Hapten present on horse-serum antigen used in preparing antisera. | Hapten present on chicken-serum antigens used to test interaction with antisera. | | | | | |
|---|---|---|---|---|---|---|
| | –N=N–C₆H₅ | –N=N–C₆H₄–$CO_2^-$ (meta) | –N=N–C₆H₃(Cl)–$CO_2^-$ (meta) | –N=N–C₆H₃($CH_3$)–$CO_2^-$ (meta) | –N=N–C₆H₄–$CO_2^-$ (para) | –N=N–C₆H₄–$SO_3^-$ (meta) |
| –N=N–C₆H₄–$CO_2^-$ (meta) | 0 | +++ | ++++ | +++ | 0 | + |
| –N=N–C₆H₄–$SO_3^-$ (meta) | 0 | 0 | | 0 | 0 | ++++ |

[a] 0 = no precipitate; +, +++, ++++ = increasing amount precipitate.

eliminates any contribution of the nonhapten portion of the antigen to the antibody-antigen interaction.) The results are shown in Table I. It is seen that:

1. No interaction will occur if an acidic group is not present on the hapten of the chicken-serum antigen.
2. No interaction will occur unless the acidic group is meta to the amino group (i.e., $CO_2^-$ and $SO_3^-$ must be meta to the point of coupling of the hapten to the protein).
3. Replacement of a hydrogen atom on the benzene ring by a chlorine atom or a methyl group has little effect on the interaction.
4. The antibodies are readily able to distinguish between a carboxyl group and a sulfonic-acid group on the antigen.

On the basis of many experiments of this type, it could be concluded that the nature of the charged groups of the hapten is of decisive importance in determining the specificity. Haptens containing sulfonic-acid groups do not tend to compete effectively for sites on antibodies which are generated by carboxyl-containing haptens, and vice versa. The introduction of an arsenic-acid group into an antigen makes for a particularly strong interaction between the modified antigen and the antibodies which it induces. Introduction of an uncharged methyl, halogen, methoxyl, or nitro group into a hapten has a much smaller effect on the specificity; strong cross reactions are observed between antigens containing a given hapten and antibodies induced from antigens containing haptens that differ from the given hapten merely through the presence of one of these uncharged groups. On the other hand, antibodies are found to be able to distinguish very well between two haptens which are optical isomers or which are (*cis*-)(*trans*-) isomers.

The important inference to be drawn from these studies is that the specificity of the antibody-antigen reaction must depend upon a complementarity between the surfaces of the antibody and antigen molecules. This complementarity presumably involves an apposition of positive and negative electric charges, if a charge is present in the antigen. It also involves a rather close matching of the contours of portions of the surfaces of the antibody and antigen molecules—a lock and key relationship that had been visualized already by Ehrlich early in the development of immunology. Small deviations in the perfection of this complementarity are tolerated (e.g., replacement of a methyl group by a chlorine atom, or even of a hydrogen atom by a methyl group). Forces of all types are presumably in action across the surface of contact of the antibody and antigen, but electrostatic forces seem to be particularly important. One can visualize these relationships by means of semischematic drawings such as Fig. 1, where a portion of the antibody molecule is supposed to envelop the hapten, with electric charges suitably located on the antibody surface so as to interact favorably with the opposite charge on the hapten.

A paper by Pressman, Siegel, and Hall[4] illustrates the possibility of obtaining quantitative information about complementarity by making use of the hapten-inhibition phenomenon. Samples of ovalbumin were diazotized to ortho-, meta-, and para-aminobenzoic acid. Three antigens were produced in this way, consisting of ovalbumin molecules bristling with azobenzoate groups having the three isomeric configurations shown in Fig. 2. The three antigens are referred to as "$X_o$-ovalbumin," "$X_m$-ovalbumin," and "$X_p$-ovalbumin." The three corresponding antibodies were prepared by injecting these antigens into rabbits, bleeding the rabbits, and separating the globulin fraction (which contains the antibodies) by ammonium-sulfate precipitation. The addition of suitable proportions of $X_o$-ovalbumin to anti-$X_o$-globulin gave a visible precipitate, the amount of which could be measured. When low molecular-weight haptens [such as benzoate, *o*-chlorobenzoate, *m*-chlorobenzoate, *p*-chlorobenzoate, *o*- (*p*′-hydroxyphenylazo) benzoate, and various chlorinated derivatives of *o*- (*p*′-hydroxyphenylazo) benzoate] were added, the amount of the precipitate was decreased because of the competition of the small haptens for the complementary site on the antibodies. By varying the concentration of low molecular-weight hapten and observing the change in the amount of precipitate produced, it was possible to compare the affinities of the different hapten groups for the sites on the anti-$X_o$-globulin.* These relative affinities could be expressed in terms of the differences between the mean free energies of combination of the haptens for the sites on the anti-$X_o$-globulin.

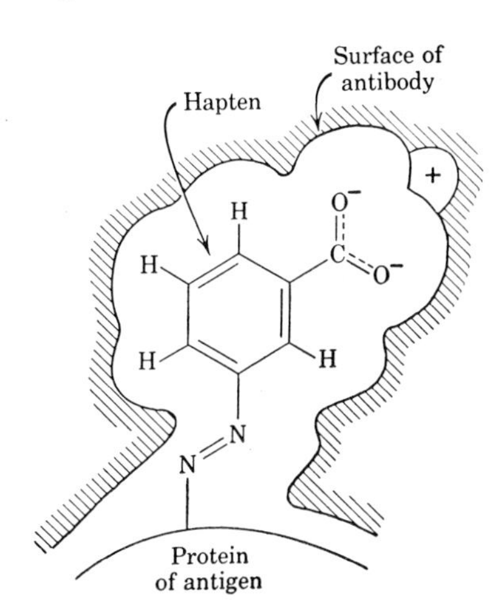


FIG. 1. Diagrammatic representation of complimentarity of the hapten-antibody relationship.

TABLE II. Relative affinities of various haptens for anti-$X_{o,m,p}$-globulins.

| Hapten | $\Delta F_{rel.}$ cal/mole of hapten |
|---|---|
| For anti-$X_o$-globulin: | |
| Benzoate | (0) |
| *o*-chlorobenzoate | −580 |
| *m*-chlorobenzoate | 0 |
| *p*-chlorobenzoate | 430 |
| *o*-(*p*′-hydroxyphenylazo) benzoate | −1700 |
| 3-Cl, 2-(*p*′-hydroxyphenylazo) benzoate | −1100 |
| 4-Cl, 2-(*p*′-hydroxyphenylazo) benzoate | −1100 |
| 5-Cl, 2-(*p*′-hydroxyphenylazo) benzoate | −1300 |
| 6-Cl, 2-(*p*′-hydroxyphenylazo) benzoate | −1500 |
| For anti-$X_m$-globulin: | |
| Benzoate | (0) |
| *o*-chlorobenzoate | 680 |
| *m*-chlorobenzoate | −300 |
| *p*-chlorobenzoate | 480 |
| *m*-(*p*′-hydroxyphenylazo) benzoate | −1600 |
| 2-Cl, 3-(*p*′-hydroxyphenylazo) benzoate | −560 |
| 4-Cl, 3-(*p*′-hydroxyphenylazo) benzoate | −350 |
| 5-Cl, 3-(*p*′-hydroxyphenylazo) benzoate | −980 |
| 6-Cl, 3-(*p*′-hydroxyphenylazo) benzoate | −500 |
| For anti-$X_p$-globulin: | |
| Benzoate | (0) |
| *o*-chlorobenzoate | 960 |
| *m*-chlorobenzoate | −300 |
| *p*-chlorobenzoate | −560 |
| *p*-(*p*′-hydroxyphenylazo) benzoate | −1700 |
| 2-Cl, 4-(*p*′-hydroxyphenylazo) benzoate | −940 |

Similar experiments were performed on the $X_m$-ovalbumin-anti-$X_m$-globulin system and on the $X_p$-ovalbumin-anti-$X_p$-globulin system. As a result, it was possible to observe the effects that different groups in different positions in the hapten have on the ability of the hapten to combine with a given antibody. Some results for a series of chlorobenzoates and chloro-*p*-hydroxyphenylazobenzoate haptens are shown in Table II. The numerical values of $\Delta F_{rel}$ listed in this table are the differences between the average free energy for the combination of the hapten in question with the antibody sites and the average free energy of combination of the unsubstituted benzoate ion with the same sites. It is seen that, with each of the three globulins, the *p*-hydroxyphenylazo group greatly increases the affinity as compared with the benzoate ion (makes $\Delta F_{rel}$ more negative), presumably because it is able to take advantage of some of the binding forces associated with the portion of the antibody site that the organism intended to be occupied by the azo group and by the side chain of the protein to which the azo group is attached (see Fig. 3). This simple picture also shows

* The detailed mathematical analysis by Pauling, Pressman, and Grossberg[5] of the dependence of the amount of precipitate on the concentration of the low molecular-weight hapten showed that there must be a considerable variability in the affinities of the hapten for different sites on the antibody molecules. That is, the antibody sites at which the haptens are bound are not all alike. This heterogeneity can be described adequately by means of a Gaussian distribution of the free energies of binding about a mean value which is characteristic of each hapten, the spread in the free energies on either side of the mean value amounting to 1000 to 2000 cal/mole of hapten. The complication introduced by heterogeneity can be avoided by talking in terms of the mean free energy of binding for a given hapten at the sites on the anti-$X_o$-globulin.

orthobenzoate derivative
= "$X_o$-ovalbumin"

metabenzoate derivative
= "$X_m$-ovalbumin"

parabenzoate derivative
= "$X_p$-ovalbumin"

FIG. 2. Configuration of haptens used in studying antibody-antigen complementarity.

why the introduction of a chlorine atom in place of a hydrogen atom at various positions in the benzene ring in the simple benzoate ion tends to be unfavorable (makes $\Delta F_{rel}$ more positive by several hundred calories), if the slightly bulky chlorine atom is not able to go into the position intended for the azo group of the original antigen. When the much larger *p*-hydroxyphenylazo group is present in the hapten, it occupies the position intended for the protein of the original antigen; the introduction of a chlorine atom into such a hapten then invariably decreases its affinity for the antibody site. The amount of the decrease in affinity is different, depending upon the location of the chlorine atom on the benzene ring. This reveals something about the nature of different portions of the surface of the hapten-binding site on the antibody. It is clear that experiments of this kind allow one to probe the various regions of the antibody-combining sites and to obtain quite detailed information on the factors that determine antibody-antigen specificity.

## 2. Enzyme-Substrate Interactions

Enzymes are well known to be rather particular about the substrates whose reactions they choose to catalyze. The degree of this specificity varies widely, however. The enzyme urease evidently catalyzes only the hydrolysis of urea and appears to have no noticeable effect on any urea derivative or other substance that has yet been tested. Fumarase adds water to the double bond of the fumarate ion to produce L-malate ion and no other double-bonded organic molecule has been found which is attacked by this enzyme. On the other hand, certain lipases will hydrolyze ester bonds containing widely different organic groups attached to either side of the bond.

It is generally agreed that enzymes act by adsorbing the substrate molecule or molecules at some "active site" on the enzyme molecule in such a way that the appropriate bonds in the substrate show an increased reactivity. It is difficult to avoid thinking in terms of the hypothesis that the enzyme-substrate interaction, like the antibody-antigen interaction, depends upon a complementarity in structure between the two substances, the substrate molecule fitting into a characteristically shaped impression in the surface of the enzyme molecule. The increased reactivity may arise because in the adsorbed state the bonds involved in the

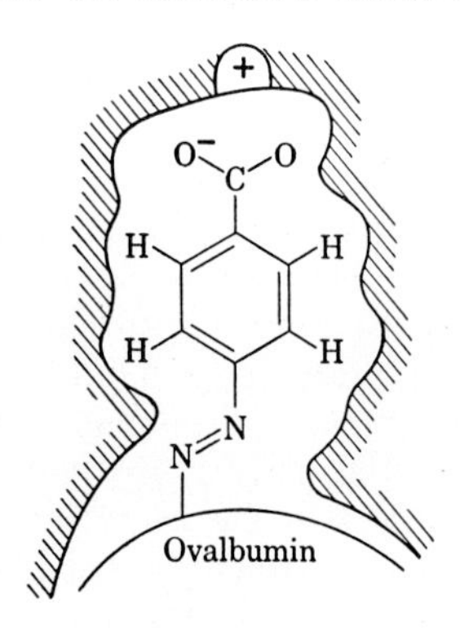


Interaction of $X_p$-ovalbumin with anti-$X_p$-globulin

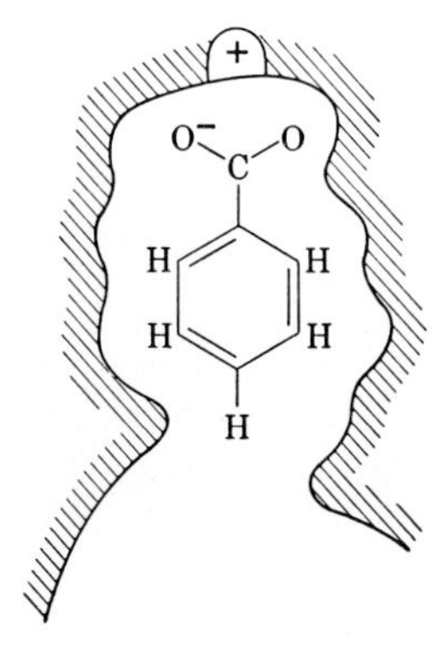

Interaction of benzoate ion with anti-$X_p$-globulin

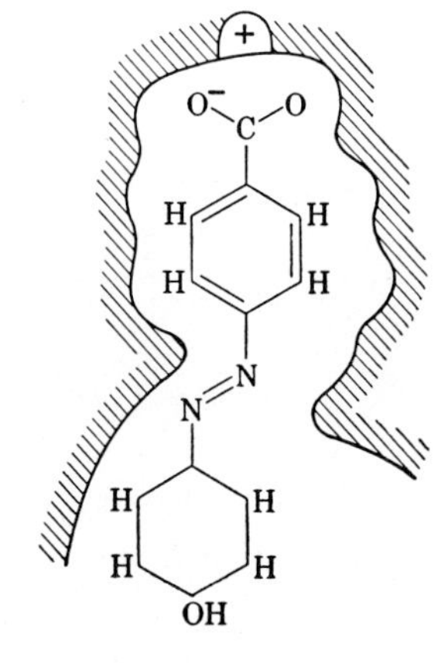

Interaction of *p*-hydroxyphenylazobenzoate with anti-$X_p$-globulin

FIG. 3. Hapten inhibition of anti-$X_p$-globulin.

reaction are placed under strain or because they are brought into a favorable position for reaction. This simple picture seems to be consistent in a general way with the wide range of specificity found among enzymes. Highly specific enzymes presumably contain a highly characteristic impression of the substrate molecule into which it is not possible to fit even closely related molecules capable of undergoing the same reaction as the substrate. Enzymes showing a wide range of specificity, on the other hand, must interact with only the region of the substrates in the immediate vicinity of the bonds that are involved in the reaction. This picture also explains why substances closely related to a given substrate, but incapable of undergoing the reaction that is catalyzed by the enzyme, often act as powerful inhibitors (e.g., malonic acid and malic acid strongly inhibit the conversion of succinic acid to fumaric acid by succinic dehydrogenase). These inhibitors must fit into the site intended for the substrate but, like the dog in the manger, they neither take advantage of the catalytic capabilities of the site nor permit access to the true substrate.

The hypothesis of structural complementarity also explains the high degree of stereospecificity shown by practically all enzyme reactions in which the potential substrates are capable of existing as stereoisomers. (E.g., the dehydration of malic acid in the presence of fumarase to form fumaric acid is absolutely specific for L-malic acid, D-malic acid being unaffected by fumarase to any detectable degree; similarly, maleic acid, the *cis* isomer of fumaric acid, is not hydrated by fumarase to any detectable degree. Most peptidases will not hydrolyze peptide bonds involving D-amino acids.) Suppose that the active center of an enzyme is surrounded by a mold into which all of the chemical groups in the vicinity of a double bond of the *trans* isomer of a compound will fit. It is easy to understand why this mold will not be able to accommodate the *cis* isomer of the compound, which has an entirely different external shape. The specificity to optical isomers is no less easy to understand in terms reflecting the common experience that a left hand will not fit into a glove made to cover its mirror image, the right hand.

It is interesting to examine some typical enzyme systems in order to observe the types of behavior that can be accommodated to the complementarity concept. One finds that many systems can be interpreted very easily in these terms, but one finds also that there are certain instances in which the concept does not present a very convincing picture of what must be going on.

### (*a*) *Peptidases*

Bergmann studied the action of various peptidases on simple peptides and was able to show that there is a high degree of specificity among these enzymes which depends on the nature of the groups located on either side of the peptide bond that is split. He was able to

TABLE III. Bergmann's specificity rules for the hydrolysis of the peptide bond in $R_2$—NH—$CHR_1$—CO—NH—$R_3$.

| Enzyme | $R_1$ | $R_2$ | $R_3$ |
|---|---|---|---|
| Pepsin | glutamic or aspartic | not H | aromatic (tyrosine or phenylalanine) |
| Trypsin | cationic (lysine or arginine) | not H | nonspecific; may be H |
| Chymotrypsin | aromatic (tyrosine or phenylalanine) | nonspecific | nonspecific |

formulate certain general specificity requirements for the action of pepsin, trypsin, and chymotrypsin, which are summarized in Table III. Further work has shown, however, that many of these requirements are not met by all substrates of these enzymes, and the specificity picture has become somewhat more complex (see, for example, the discussion in Dixon and Webb[6] and Neurath and Schwert[7]). To illustrate these complexities, the chief points of attack on ribonuclease by various peptidases are shown in Fig. 4. Inspection of this figure and Table III will show that the specificity discovered by Bergmann for trypsin is invariably maintained: the points of attack by trypsin are always at the peptide bond next to either arginine or lysine, with the carbonyl end of the bond that is hydrolyzed belonging to the cationic amino acid. It does appear, however, that the bond adjacent to one of the cationic amino-acid residues of ribonuclease (the lysine at position 44 from the lysine end of the chain) may not be attacked by trypsin. The points of attack on ribonuclease by chymotrypsin are frequently adjacent to aromatic side chains (usually at the carbonyl side), as demanded by the Bergmann rules. Unexpected exceptions are found, however, in the ready opening of the leucine-threonine linkage between positions 35 and 36, and of the alanine-lysine linkage between positions 103 and 104. A weak ability of chymotrypsin to split peptide bonds involving methionine had been noted in simple peptides (chymotrypsin also hydrolyzes a cystine-serine bond in insulin). These exceptions are somewhat puzzling from the point of view of the complementarity concept, because the groups that are involved can hardly be said to resemble in general shape the aromatic groups that the Bergmann rule demands for chymotrypsin. More complex factors seem to operate than the mere binding by the enzyme of the amino-acid residue next to the peptide bond that is being hydrolyzed. The slight ability of methionine and cystine to replace an aromatic ring in some substrates for chymotrypsin is especially puzzling and might suggest that an electronic mode of interaction may be involved rather than one involving steric complementarity.

The specificity rule for pepsin proposed by Bergmann on the basis of its action on simple peptides turns out to be violated frequently when pepsin acts on long

```
                                   :                                        :
Lys.Glu.Thr.Ala.Ala:Ala.Lys|Phe|Glu.Arg|Ser.Thr.Ser.Ser.Asp.His.Met:Glu.Ala.Ala.
                P? P       T  C       T                             C
                              P

       NH2            NH2 NH2                      NH2                          :
Ser.Ser.Asp. Ser.Tyr|Cys.Asp. Glu. Met.Met.Lys|Ser.Arg|Asp. Leu|Thr.Lys|Asp.Arg:Cys.
                    C                         T       T        C       T       T
                                                                       S

                         NH2                      NH2                          NH2
Asp.Val.Pro.Lys.Thr.|Phe|(Glu,Val,Leu,Ser,His)|(Glu, Asp,Ala,Val)|(Ala,Val,Cys,Ser,Glu)
                    P   C                      P                 P
                    S   P
                        S

  NH2:                 NH2          NH2 NH2            NH2
Lys|Asp.:Val.Ala.Cys.Lys|Asp.Gly.Thr.Asp. Glu.Cys.Tyr|Glu.Ser.Tyr.|Ser.Thr.Met|Ser.
  T    C                T                            C            C           C
                                                                              S

                                                               NH2
Ileu.Thr.Asp.Cys|Arg|Glu.Ser.Thr|Ser.Gly.Lys|Tyr.Pro.Asp. Ala.Cys.Tyr|Lys|Thr.Thr.
                S   T           S           T                        C   T

NH2 NH2              :              (NH2)
Asp. Glu. Ala|Lys|His: (Ileu,Ileu,Val)|(Asp,Glu,Pro)Gly.Cys.Ala|Tyr|Val(Val,Pro,His)Phe|
             C   T   C                P                         S   C                  C
                                      S                             S                  P

Asp.Ala.Ser.Val
```

FIG. 4. Points of hydrolysis of ribonuclease by chymotrypsin (C), pepsin (P), trypsin (T), and subtilisin (S). Dotted lines represent points of slower hydrolysis.

polypeptide chains. This is evident from Fig. 4, where the attack on an alanine-alanine bond may be noted. Pepsin also attacks an alanine-serine bond in ACTH and a leucine-valine linkage in insulin. In general, however, pepsin does show a tendency to attack either bonds involving glutamic or aspartic acids by themselves, or bonds involving aromatic amino acids by themselves—in partial agreement with Bergmann's rules. It appears that the true range of specificity of pepsin does not reveal itself until one allows it to act on relatively long polypeptides.

It is interesting that trypsin and chymotrypsin are able to hydrolyze esters and suitably activated carbon-carbon bonds as well as peptides. The specificity requirements on the amino acids adjacent to the ester or carbon-carbon bond are similar to those observed for peptides with the same two enzymes. Here one has a particularly striking example of the importance of the environment of a reacting bond, rather than the bond itself, in manifesting the specificity of the enzyme.

### (b) *Cholinesterase and Acetylcholinesterase*

The reaction

$$\underset{\text{(acetylcholine)}}{(CH_3)_3N^+-CH_2-CH_2-O-CO-CH_3} + H_2O \rightarrow \underset{\text{(choline)}}{(CH_3)_3N^+-CH_2-CH_2-OH} + CH_3-CO-OH$$

is catalyzed by two distinct groups of enzymes which have interestingly different specificity requirements. One of these groups of enzymes, the acetylcholinesterases is inhibited by the substrate acetylcholine at higher substrate concentrations, so that its activity *vs* substrate-concentration curves pass through a maximum. The other group of enzymes, known as cholinesterases, is not so inhibited, and the activity *vs* substrate-concentration curves follow the normal Michaelis-Menton behavior. Both kinds of enzymes can act on many substrates related (in some instances rather distantly) to acetylcholine, but mention is made only of the interesting and opposite effects of replacing the acetyl group with other acyl groups containing longer hydrocarbon chains. In mammalian brain acetylcholinesterase, the replacement of acetyl with butyryl greatly lowers the activity and does not remove the inhibition at high substrate concentrations. In cholinesterase, on the other hand, the same substitution greatly increases the activity of the enzyme. Here one has an example of two enzymes which catalyze the same reaction but whose specificity patterns are markedly different, showing that considerably different types of complementarities must be possible for a given substrate.

### (c) *Glycosidases*

The hydrolysis of $\alpha$- or $\beta$-alkyl- or aryl-glycosides to form free sugars plus alcohols or phenols is catalyzed by a group of enzymes that show interesting specificities. In general, a given glycosidase is highly specific to the sugar moeity, and to whether the glycoside has the $\alpha$- or $\beta$-configuration, but it is relatively unconcerned about the nature of the aglucone residue. For instance, $\alpha$-glucosidase will hydrolyze $\alpha$-methylglucoside, $\alpha$-phenylglucoside, and many other $\alpha$-glucosides, but it will not cause the hydrolysis of any $\beta$-glucoside or any $\alpha$-glycoside of a sugar other than glucose. Here one has examples of very high specificity toward a group that lies on one side of the chemical bond that is attacked, and a very low specificity to the other side.

*(d) Stereospecificity in the Transfer of Hydrogen to DPN in Dehydrogenase Reactions.*

Vennesland[8] and Westheimer have shown by means of ingenious experiments with deuterium that, in the reaction

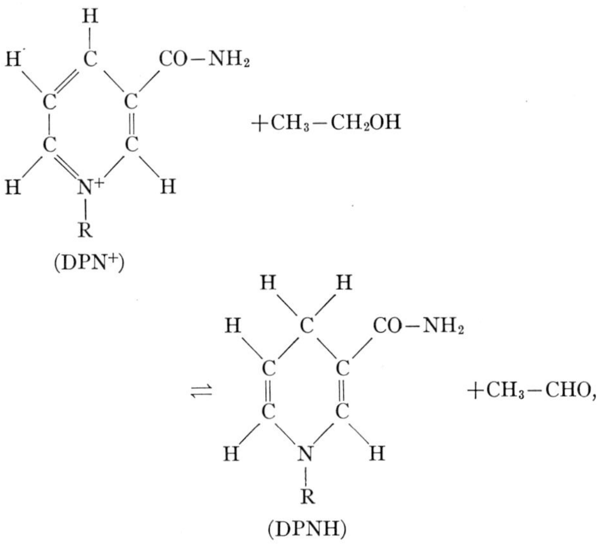

```
       H
       |
H      C      CO-NH2
 \   //  \   /
   C       C
   |       ||          +CH3-CH2OH
   C       C
 /   \\  /   \
H      N+      H
       |
       R
    (DPN+)

                 H     H
                  \   /
           H        C        CO-NH2
             \    /   \    /
               C        C
       ⇌       ||       ||          +CH3-CHO,
               C        C
             /   \    /   \
           H       N        H
                   |
                   R
                (DPNH)
```

both the addition of the hydrogen atom to the pyridine ring of $DPN^+$ in the forward direction and the addition of hydrogen to acetaldehyde in the reverse reaction are stereospecific. Thus, if the equilibrium

$$CH_3-CH_2-OD+DPN^+ \rightleftharpoons CH_3-CHO+DPNH+D^+$$

is carried out in $D_2O$, it is found that no additional deuterium atoms are introduced into either $DPN^+$, DPNH, alcohol, or aldehyde. Similarly, no normal hydrogen is introduced when the reaction

$$CH_3-CD_2-OH+DPN^+ \rightleftharpoons CH_3-CDO+DPND+H^+$$

is equilibrated in $H_2O$. This proves that the hydrogen atom is transferred directly from the carbon atom of the alcohol to the $DPN^+$ (and vice versa) and does not go through the solvent. Furthermore, the hydrogen atom introduced into the DPNH must lie on one definite side of the pyridine ring. The ring must lie, therefore, in a definite orientation on the enzyme surface, and relative to the alcohol molecule, in the complex of the enzyme with its two substrates. That the orientations of the ethanol and aldehyde molecules also must be fixed rigidly relative to the pyridine ring is shown by the fact that the reaction

$$H^+ + DPND + CH_3-CHO \rightarrow DPN^+ + CH_3-CHD-OH$$

produces a single optical isomer of ethanol-1*d*. Here is an especially clear example of the detailed stereochemical specification of the structure of an enzyme-substrate complex. It must take advantage of a large fraction of the possible complementarity that can exist between the substances concerned in the reaction.

*(e) The Koshland-Stein Rule*

Koshland and Stein[9] have proposed the general principle that "if the enzyme requirements for a substrate having an oxygen bridge, i.e., R—O—Q, show high specificity for R and low specificity for Q, than R—O cleavage occurs in the enzyme reaction." This rule has been found to be valid for invertase, alkaline phosphatase, acid phosphatase, trypsin, and chymotrypsin. The simplest interpretation of the rule in terms of complementarity would be that the group with the highest specificity would be presumed to have the closest contact with the enzyme; the enzyme then would be expected to have the greatest activating effect on the bond adjacent to this group.

*(f) Existence of Identical Amino-Acid Sequences in the Active Centers of Enzymes Having Diverse Specificities*

It has been found recently that the active centers of thrombin, chymotrypsin, trypsin, and phosphoglucomutase all contain the sequence aspartic acid-serine-glycine. These enzymes have markedly different specificities, the first three catalyzing the hydrolysis of peptide bonds, whereas phosphoglucomutase catalyzes the transfer of a phosphate group from one glucose ring to another. The specificity of these enzymes must clearly reside in regions quite distinct from the center that activates the chemical bonds which are disrupted.

### 3. Binding of Small Molecules to Proteins (Other Than Hapten-Antibody Interactions)

Serum albumin is remarkable for its ability to adsorb a wide variety of small molecules, particularly (though not at all exclusively) those bearing negative charges (acetate ion, halide ions, methyl orange, and many other anionic dyes, detergents, etc.). The most striking feature of this adsorption, however, is the lack of any reasonably marked specificity. Changes in the structures of the adsorbed molecules do have some effects on the strength of the binding. These changes in affinity are not, however, as large as one would expect to find if the surface of the serum albumin molecule were covered with a mosaic of rigid binding sites having a reasonably high degree of complementarity to the necessarily limited number of types of substances which such a mosaic might potentially bind. (For example, enantiomorphic isomers of a dye are adsorbed to almost the same extent.) In order to account for this, Karush[10] has proposed the concept of "configurational adaptability," according to which the serum-albumin molecule is considered to possess a degree of internal flexibility.

As a result, the surface conformation can be changed so as to establish a measure of complementarity with the surface of almost any molecule that approaches it, and there is no need to deal with a limited number of fixed types of potential binding sites. A strong point in support of this concept is the fact that serum albumin does possess a high degree of internal flexibility as compared with other proteins, which also show a far smaller affinity for small molecules.[11] In a sense, the serum-albumin molecule behaves like a lump of putty onto which one can stick an unlimited number of different shapes.

The concept of configurational adaptability is important because it shows that complementarity need not imply specificity. Furthermore, it raises the possibility that, in some instances, the specific complimentarity structure may be developed only when the substrate is present.

## SPECIFICITY AND ORGANIZATION

Green has set forth[12] reasons for believing that, in certain biochemical processes involving a sequence of steps, each catalyzed by a different enzyme, the enzyme molecules must occupy fixed positions in space which are located so as to facilitate the successive reactions. He speaks of "an organized mosaic of enzymes in which each of the large number of component enzymes was uniquely located to permit efficient implementation of consecutive reaction sequences." A primary example of this mosaic is the cyclophorase system, which includes all of the enzymes for the citric-acid cycle, fatty-acid oxidation, oxidative phosphorylation, and terminal electron transport. This system of enzymes has been shown to be located exclusively in the mitochondria. In view of the striking mitochondrial structures which have been revealed by electron microscopy [see the papers by Sjöstrand (p. 301) and Fernández-Morán (p. 319) in this volume], there is a certain temptation to accept this theory.

Dixon and Webb[13] have disputed the necessity for assuming organization in this spatial sense. As an alternative, they go back to the point of view presented twenty-five years ago by Hopkins[14] in these words: "The organizing potentialities inherent in highly specific catalysts have not, I believe, been adequately appraised in chemical thought. Highly specific catalysts determine just what particular materials, rather than any others, shall undergo reaction. They select from their environment. The specific catalyst determines which among possible paths the course of change shall follow. It has directive powers . . . ." Dixon[15] speaks of "organization by specificity" as an alternative to Green's spatial organization of enzyme systems. The concentration of the cyclophorase enzymes into the mitochondria is ascribed to the need to reduce the distance between the enzymes involved in consecutive reactions. The cell is able in this way to reduce the "transit times" required for the products of one enzymatic step to diffuse to the enzyme molecules involved in the next step. According to this view, there is no need to assume any regular arrangement of the different enzymes. It is necessary only to have an arrangement (a random one would do) with a sufficiently small mean distance of separation between the enzymes involved in the different enzymatic steps.

It is interesting in this connection to estimate the order of magnitude of the transit time for a typical metabolite to move a given distance from one enzyme to another by a diffusion mechanism. It is well known that if $x^2$ is the mean square distance moved in time $t$ by a particle whose macroscopic diffusion constant is $D$, then

$$x^2 = 2Dt.$$

Typical metabolites have diffusion constants of the order of 1 cm²/day, or $10^{-5}$ cm²/sec. Thus, it is found that, if the enzymes required for two successive steps of a reaction sequence are 1000 A apart, the transit time is 10 μsec. If the separation is 100 A, the transit time is 0.1 μsec. It would appear, therefore, that the arrangement of enzymes of a reaction sequence in layers whose repeat distance is a few hundred Ångström units, without any regard to the serial positions in space of the enzymes involved in successive reaction steps, would give a more than adequately short transit time for most biochemical requirements. Specificity alone should be adequate to cope with problems of organization at this level. This is not to say, of course, that the living cell actually is organized in this manner. For reasons unknown, it may have chosen to solve its transit time problems in a less straightforward manner.

## BIBLIOGRAPHY


1 K. Landsteiner, *The Specificity of Serological Reactions* (Harvard University Press, Cambridge, 1945), revised edition.

2 W. C. Boyd, *Fundamentals of Immunology* (Interscience Publishers, Inc., New York, 1956), third edition.

3 K. Landsteiner, and H. Lampl, Biochem. Z. **86**, 343 (1918).

4 D. Pressman, M. Siegel, and L. A. R. Hall, J. Am. Chem. Soc. **76**, 6336 (1954).

5 L. Pauling, D. Pressman, and A. L. Grossberg, J. Am. Chem. Soc. **66**, 784 (1944).

6 M. Dixon and E. C. Webb, *Enzymes* (Academic Press, Inc., New York, 1958), p. 251.

7 H. Neurath and G. W. Schwert, Chem. Revs. **46**, 69 (1950).

8 B. Vennesland, J. Cellular Comp. Physiol. **47** (suppl. I), 201 (1956).

9 D. E. Koshland and S. S. Stein, J. Biol. Chem. **208**, 139 (1954).

10 F. Karush, J. Am. Chem. Soc. **72**, 2705 (1950).

11 W. Kauzmann in *Symposium on the Mechanism of Enzyme Action*, W. D. McElroy and B. Glass, editors (The Johns Hopkins Press, Baltimore, Maryland, 1954), p. 70.

12 D. E. Green, Symposia Soc. Exptl. Biol. **10**, 30 (1957).

13 See reference 6, p. 242.

14 F. G. Hopkins, Proc. Roy. Soc. (London) **B112**, 159 (1932).

15 M. Dixon, *Multienzyme Systems* (Cambridge University Press, London, 1949).

# 60
# Blood Coagulation—A Study in Homeostasis

David F. Waugh

*Department of Biology, Massachusetts Institute of Technology, Cambridge 39, Massachusetts*

In this discussion, a brief account is to be made of one of the mechanisms involved in homeostasis, a word which means that the chemical and physical processes occurring within the fluid matrix or internal environment of an organism are integrated so that a stable physiological level of each is maintained. Thus, if one process deviates from its stable level, other mechanisms are shifted slightly or considerably, as the occasion demands, in an attempt to bring this process and the entire system back to its normal state. Of course, everyone appreciates the fact that the closely related set of activities occurring in one physiological system are influenced by the sets of activities occurring in other systems. For example, the system discussed herein, the blood coagulation system, is influenced markedly by stress (as, of course, are the other physiological systems.) The mechanisms in one physiological system, then, are not at liberty to establish levels of activity which are optimal for that system as an isolated system: the levels are a compromise derived from the activities of the organism as a whole.

It should be made clear at the start that the information presented here is also the result of a compromise: it represents the activities in a chosen individual only in a general way and is a simplification most parts of which can be challenged. This situation stems from the fact that, while there is general agreement about many of the coagulation mechanisms, others have been recognized as possibilities only recently and an understanding of details is incomplete. Extensive use has been made of certain references in preparing this discussion: the books by Wintrobe[1] and Biggs and MacFarlane,[2] the review articles by Brinkhous, Langdell, and Wagner,[3] and Seegers,[4] and the symposium edited by Brinkhous.[5]

Applied to the blood vascular system the word homeostasis would include first, the hemostatic mechanisms—that is, the mechanisms which are involved in sealing a wound which has opened the usually closed circulatory system—and second, the more mysterious set of mechanisms which are involved in maintaining the thin fragile walls of the capillaries in such a state that the proteins and formed elements (red and white cells) of the blood do not escape from the circulation and pass into the tissue spaces. The first set of mechanisms, the hemostatic, involve the closing of small vessels through vascular constriction, plugs of platelets, and coagulation in which the fluid plasma is transformed into a gel. We can be concerned only with coagulation, where it is found that an extensive set of interlocking processes exist in what is quite evidently a poised equilibrium such that a displacement of the equilibrium in the proper direction leads to a rapid autocatalytic development of the clotting enzyme thrombin from its inactive precursor prothrombin. A general scheme of the features of the coagulation system to be discussed here is shown in Fig. 1. The explosive development of thrombin swamps out, locally, all of the anticlotting mechanisms and a clot develops, the latter being the result of a change in the structure and in the state of aggregation of the protein fibrinogen. The clot is confined locally by a set of slower reactions, some of which are autocatalytic and controlled by thrombin development and most of which are backed up by large capacities. These reactions nullify the clot-developing system at several points.

The trigger for the mechanisms which displace the poised equilibrium is apparently disruption of a boundary which isolates key substances, either within small blood elements, the platelets, or within the tissue cells. On wounding, the intermixing of blood and tissue juices and the presence of rough surfaces at the site of injury, which cause platelet decomposition, are sufficient to give the rapid development of thrombin described in the following.

One might venture that if hemostasis were the only objective of the coagulation system it could be much simpler than it is. Many of the extraordinary complications are probably associated with the fact that coagulation must occur continuously at low levels in order to preserve the normal structural and permeability characteristics of the capillaries. The physiological necessity for slow continuous clotting is referred to later in an attempt to justify this statement. It is just mentioned now, however, to focus attention at the start on the possibility that one is dealing with a system which, on the one hand, must operate at a low but precisely controlled level of activity but which, on the

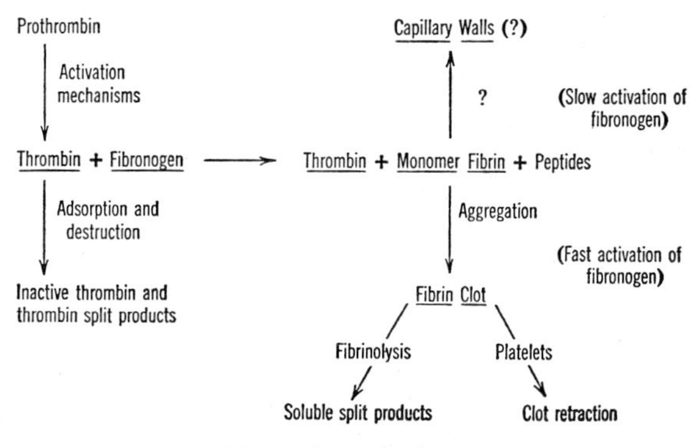


Fig. 1. General scheme.

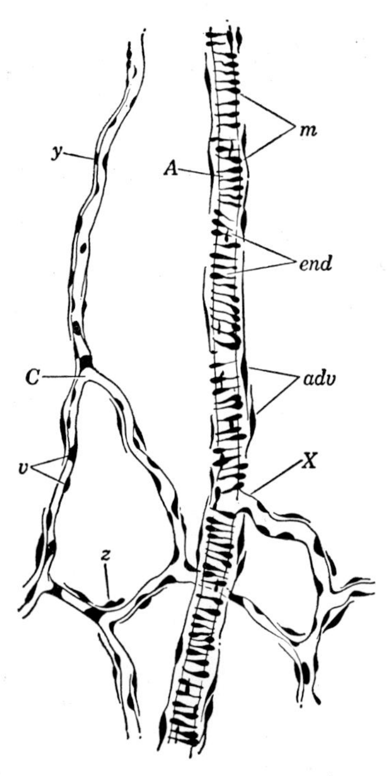


FIG. 2. Small artery, *A*, and capillaries, *C*, from the mesentary of a rabbit; *M*, muscle cells in the media; *adv*, adventitia; *X*, origin of a capillary from the artery; *y*, pericyte; *z*, perivascular histiocyte; *end*, *v*, endothelial nuclei [from A. Maximow and W. Bloom, *A Textbook of Histology* (Saunders Company, Philadelphia, 1942), fourth edition].

other hand, can be triggered into explosive activity which is not controlled but must be confined.

Blood flows within a set of tubes of varying size, the larger tubes (arteries, arterioles, veins, venules) being covered by substantial layers of muscle and connective tissue. The smallest vessels, the capillaries, are our chief concern. These fragile vessels (Fig. 2) have a wall which is constructed of a single layer of flat endothelial cells cemented at their boundaries and invested by a thin network of connective-tissue fibers. The caliber of the capillaries is close to the diameter of a single red cell, and is thus about 8 $\mu$ in man. It is believed that water and solutes generally pass through the cytoplasm of the endothelial cells. White cells may migrate through the substance of the endothelial cells or through the boundaries between cells.

Blood[1] itself is a fluid 40% of whose volume is made up of formed elements (red cells, white cells, etc.) and 60% of a straw-colored fluid plasma containing seemingly an endless variety of compounds, many in the process of being transported from one organ to another. The most important of the formed elements in this discussion is the platelet, a disk-shaped body about 2 $\mu$ in diameter. The platelets can become sticky and adhere to each other forming a mass which itself is of importance in stopping the flow of blood from capillaries. In addition the platelets, as mentioned, normally act as unavailable depots for important coagulation components. The platelet is fragile. If it comes in contact with a surface foreign to the circulation (i.e., glass) it adheres, spreads out, and bursts liberating its contents.

Within the plasma are found a series of proteins and lipoproteins all involved directly or indirectly in clotting. These proteins are described as they come up for discussion, the current series includes prothrombin, thrombin, and fibrinogen. Prothrombin, in the purified preparation of Seegers *et al.*,[6] appears as a molecule of $M\sim 66\,000$ and as a prolate ellipsoid may be described as having a length of 119 A and a diameter of 34 A.[7] This molecule can dissociate on dilution into halves. It is quite stable in the pure state and stable in plasma when prothrombin activators are absent. Prothrombin, as its name implies, has no action whatsoever on fibrinogen.

Thrombin is one of the derivatives of prothrombin. It can be obtained from prothrombin by a variety of techniques, the important one in this instance being the normal biological-activation process involving a variety of substances. In this activation process, the molecular weight of prothrombin does not appear to change radically. In the concentrated solutions required for physical studies, however, thrombin may be associated. Our current studies suggest that the minimum molecular weight of thrombin is about 22 000. One unit of thrombin is defined as that amount which clots plasma or a solution of purified fibrinogen corresponding to plasma in 16 sec. Since preparations of thrombin containing 2000 units/mg have been made, it is clear that a single unit is represented by about $5\times 10^{-4}$ mg. There is sufficient prothrombin in one milliliter of plasma to give 300 units of thrombin, corresponding to the small amount of about 0.15 mg of prothrombin per milliliter of plasma. The total possible number of thrombin units is, however, staggering. Immediate activation of a few percent of the prothrombin would coagulate blood in seconds. In the coagulation process, thrombin acts as an enzyme which splits an acidic peptide from fibrinogen (Fig. 1).

Fibrinogen has a molecular weight of 340 000 and is an asymmetric protein of axial ratio about 10.[8] It has an interesting substructure[9] which cannot be discussed here for lack of space. In pure form, it is a fairly stable protein which survives rough handling. After thrombin has split off an acidic peptide, the reactivity of the original protein is altered, the fibrin monomer produced by this enzymatic alteration having now a marked capacity to interact in a relatively specific fashion with other fibrin monomers. At the same time, a small amount of stabilizing plasma globulin is incorporated.[10] The interaction leads predominantly to end-wise association which produces fibrin fibrils or fibers having asymmetries orders of magnitude greater than that of the original fibrinogen molecule. During growth of the set of fibrin fibrils there occurs a brief interval of time during which the total volume changes from being greater to being less than the sum of the volumes dominated by each fibril. The fibrils start to overlap and to crosslink where they come into contact. In this way a space filling network of strands develops.[11] At first, the strand network is tenuous and easily ruptured, but as old strands become thicker and as new strands are incorporated into the network, the clot progressively develops rigidity. As is seen later, this is not the end of the story, for, in the presence of platelets, a spontaneous squeezing out of fluid may occur (clot retraction). This is an important activity of the clot in forming a compact closure over a wound. If retraction and drying of the clot do not occur, an enzyme plasmin may be activated from its precursor.[12] Apparently, the enzyme plasmin has been devised specifically to dissolve fibrin.

The first portion of this general scheme to consider is the transformation of prothrombin into thrombin. This phase of the over-all problem has received a great deal of attention, the natural result of the fact that discoveries in clinics have revealed an ever-increasing series of patients whose plasmas have lacked one or another of the activators necessary for the prothrombin-thrombin conversion, and of the fact that prothrombin has been prepared in relatively pure form, first by Seegers and more recently by others, and its activation has been studied under a variety of conditions.

The scheme which is shown in Fig. 3 is a simplified version of experience in the field.[1–5] The central theme is the conversion of prothrombin into thrombin. The top half of the figure depicts clotting events which can occur when blood is drawn and every effort is made to keep the blood free of contamination with tissue components. Before proceeding, attention is directed to the central role assigned here to thrombin not only as the clotting enzyme but also as an agent which feeds back and promotes the formation of more of itself. As can be seen, this feedback enters at several levels into the set of interactions and, in order to promote the explosive

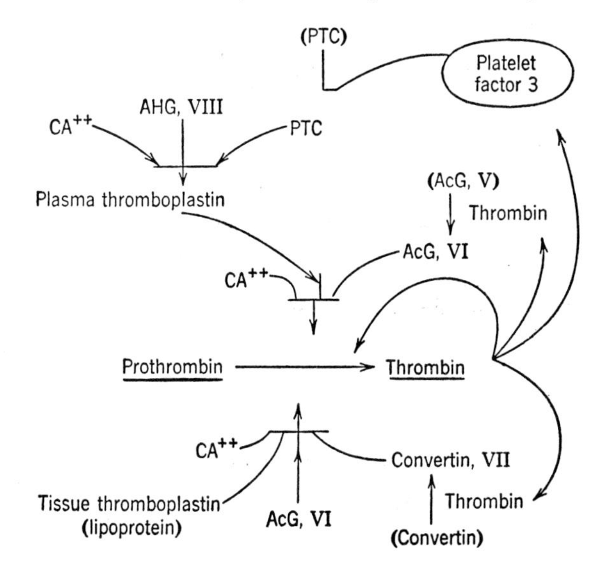


FIG. 3. Illustrative scheme for mechanisms involved in prothrombin activation.

liberation of thrombin, what is necessary, so far as this portion is concerned, is platelet decomposition and a nudge in the direction of thrombin production.

To start with, on being drawn, the blood comes into contact with a foreign surface and certain of the platelets respond by rupturing and liberating a series of compounds, the one singled out being platelet factor 3, a lipoprotein. This factor combines with or activates an inactive form of a substance which is probably a protein called plasma thromboplastin component (PTC). The inactive forms of components have been placed in parentheses, the Roman numerals indicate the accepted number in the "factor" series. Plasma thromboplastin component now interacts with antihemophilic globulin (AHG) and calcium ion to give plasma thromboplastin. This is not sufficient in itself to produce rapid prothrombin conversion, at least one additional step being necessary—namely, an interaction with the active form of accelerator globulin (AcG, VI). A rapid conversion of prothrombin now takes place. Note how thrombin is instrumental in converting inactive accelerator globulin to an active form, that it is involved in platelet alterations and may be involved directly in the conversion of prothrombin. All of the elements in this scheme are present in plasma.

The lower part of the Fig. 3 shows a sequence of events involving tissue components (tissue thromboplastin). Tissue thromboplastin short circuits platelet contributions at the start but requires calcium ion, the active form of accelerator globulin, and a new protein factor, convertin, which is produced from a precursor by the action of thrombin (cf. Goldstein and Alexander in reference 5, p. 93). It is generally held that the tissue thromboplastin-convertin cycle gives an immediate thrombin production which then operates on itself and on the upper mechanisms of Fig. 3 to yield a fast production of thrombin.

During the rapid development of thrombin (upper part of Fig. 3), all of the prothrombin may not be converted to thrombin. In the scheme presented, this is most readily understood by the fact that, during prothrombin activation, plasma thromboplastin, and thus at least some of the components which give rise to plasma thromboplastin, are either inactivated or consumed, as is true also of tissue factors. This is related to the confinement mechanisms which are outlined in the following.

Another aspect of the thrombin-production feedback mechanism which should be mentioned is that the platelets are thought to liberate a substance which is an antiheparin, and since, as is seen, heparin is intimately involved in the destruction of thrombin and thromboplastin, removal of heparin constitutes an additional positive factor in thrombin production.

The scheme of Fig. 3 has presented one possible sequence of events. The situation is probably more complicated than this, as is evident from a consideration of the following. First, at least two possible factors have been omitted from the scheme. These are the Hegeman factor and PTA or plasma thromboplastin antecedent. Second, the number and sequences of reactions leading to the development of plasma thromboplastin are not clear. Not only is it possible that the details of factor interactions as given are incorrect, but also there is known to be opportunity for substitution under the proper conditions. For example, the platelets liberate a material which can substitute for the active form of accelerator globulin.

It is in this series of reactions that one finds the proven genetically controlled clotting defects. An abnormality in antihemophilic globulin is the defect of classical hemophilia, which was described over 150 years ago. It is a sex-linked recessive and manifests itself in varying deficiency. Seegers *et al.*[13] have evidence to show that this abnormality may be owing to inhibitors which block AHG. Lack of accelerator globulin (V), although it may be acquired through liver damage, may also be of hereditary origin; however, the inheritance is uncertain. Lack of PTC (hemophilia B, Christmas disease, etc.) is an hereditary disease which is a sex-linked recessive, similar to AHG deficiency.

There are two important series of reactions which are involved in confining coagulation to a locality and in removing unwanted clots. In approaching the mechanisms which confine clot development, it is necessary to realize that in wounds the hemostatic mechanisms are operating in the relatively stagnant or slowly flowing blood which occurs near the constricted ends of cut vessels. Whatever blood pressure is operating on these cut vessels will tend to work the coagulating system to the exterior. Thus, the problem is one of handling back-diffusion and any back-flow of active clotting agents. Of course, if such agents are carried into the general circulation, there occurs a relatively extensive dilution.

Confinement of clot development and the shutting off of coagulation reactions involve a series of antieffects which operate on the basis of adsorption, stoichiometric combination, and enzyme action. Some mechanisms are not well established, although a high probability can be assigned to their existence.

The first rather interesting reaction which is known to have importance in limiting clot development is the adsorption of thrombin to fibrinogen and fibrin.[3,4,12,14] In a static blood volume, such an effect allows complete local clotting to be accomplished before thrombin can diffuse out of the locality.

There then follows a series of reactions which have been grouped in Table I according to whether or not they are thrombin dependent. The second group is thrombin dependent; the equations mean that thrombin, alone or with other materials, will destroy three key procoagulants: accelerator globulin, prothrombin, and antihemophilic globulin.

What is implied by the three equations listed in the third group is that for every coagulant in the active form there is a factor more or less specifically designed to effect its inactivation. When the mechanisms of the first three groups of Table I are coupled with the many observations that materials such as AHG, PTC, AcG, platelet factor 3, etc., appear to be *consumed* in activating prothrombin, it is apparent that the explosive development of thrombin can be sharply terminated in the region of the clot.

The presence of sizable amounts of free thrombin always, however, constitutes a dangerous situation and two mechanisms, one rapid and the other slow, have apparently been devised to cope with the situation. The first of these is indicated in the first equation under group 4 and is the rapid reaction. The experiments designed to prove the existence of this inactivating factor are assailable, but essentially the suggestion is that, when normal prothrombin is activated to thrombin in plasma, there arises at the same time an ephemeral factor which can inactivate the resulting thrombin.

The greatest capacity for inactivating thrombin is distributed throughout the plasma itself. This activity is indicated in the second reaction under group 4, the responsible substance or substances being referred to as normal antithrombin. There is present in plasma sufficient normal antithrombin to inactivate 1.5 times the thrombin which could appear if all of the prothrombin were activated to thrombin. This places the capacity of antithrombin at about 450 units/ml. Under ordinary conditions, normal antithrombin acts relatively slowly, as is evident from the time disappearance of thrombin added to plasma which has been diluted fifty-fold with saline (Fig. 4, curve 1, see Waugh and Fitzgerald[15]). It is certain that antithrombin is itself inactivated in the process of destroying thrombin; the reaction is thus stoichiometric. A reversal of the thrombin-antithrombin complex has so far *not* been accomplished. It should be emphasized once more that the capacity of plasma antithrombin is about 450 units/ml. This large plasma-antithrombin capacity to a certain extent compensates for the small rate constant of the second-order thrombin-antithrombin reaction, the result being that when one unit of thrombin is present per milliliter plasma, inactivation will take place at the rate of 0.05 units

TABLE I. Confinement and shut-off mechanisms.

| | |
|---|---|
| Group 1. | Thrombin+fibrin $\rightleftarrows$ thrombin fibrin |
| Group 2. | AcG, VI $\xrightarrow{T}$ inactive AcG, VI |
| | Prothrombin $\xrightarrow{\text{T, factors}}$ inactive prothrombin |
| | AHG $\xrightarrow{T}$ inactive AHG |
| Group 3. | Antithromboplastin+thromboplastin → inactive product |
| | Antiplatelet factors+platelet factors → inactive products |
| | Anti AcG VI+AcG VI → inactive product |
| Group 4. | Thrombin+inactivating factor → inactive thrombin |
| | Thrombin+antithrombin → inactive (thrombin antithrombin) |

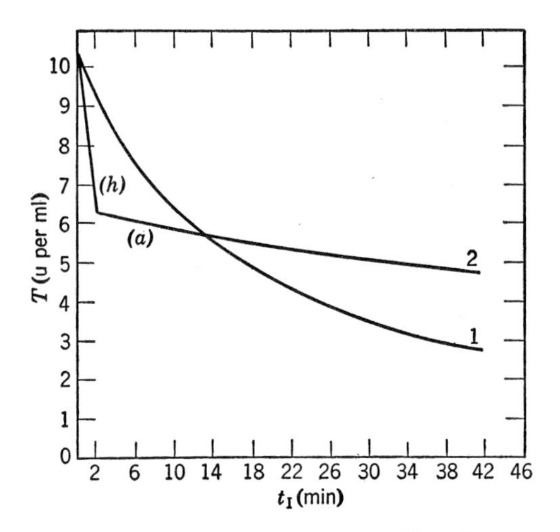


FIG. 4. Inactivation of thrombin by antithrombin. Curve 1 refers to normal antithrombin and curve 2 to the effects which are produced by traces of heparin.

thrombin per second. This rate of inactivation will increase the clotting time under conditions of fast activation but will not act with sufficient rapidity to prevent clotting.

The use of heparin, the highly acidic (sulfated) mucopolysaccharide, as an anticoagulant is well known. At this time, attention is directed only to the fact that heparin, at the small concentrations at which it exerts a strong anticoagulant action, acts through the normal antithrombin mechanism.[15] Figure 4 indicates the important points of this action—namely, that heparin speeds up the rate of normal antithrombin action by a factor of over 50 and at the same time decreases the efficiency of inactivation so that the antithrombin capacity drops by a factor of 1.5 to 2.0. Heparin, a product of the liver, is present in most tissues of the body and can be liberated and appear in the blood stream during anaphylactic shock. The normal concentration of free heparin in the blood, if it is present at all, is so low as to be difficult to measure. From the standpoint of hemostatic feedback mechanisms, the possibility that traces of heparin determine the rate at which normal antithrombin acts is most interesting. Heparin has been mentioned earlier as an important factor in the destruction of plasma thromboplastin.

It is now clear that the shut-off mechanisms attack any active components which leave the clot area. Under more favorable circumstances of time it would be necessary at this juncture to examine some of the concepts concerning the necessity of having threshold levels of procoagulants before the activating mechanisms are effective. There is evidence that such is the case and, if so, threshold levels must be intimately associated with the shut-off mechanism. That is, they would prevent an extensive spread of the explosive development of thrombin under conditions where activators would escape and be carried into the blood stream.

Consider now the clot structure itself. The developing clot has platelets adhering to its fibers, particularly at the fiber junctions. These platelets may be the loci at which such junctions were formed in the first place, it being understood that purified fibrinogens will form firm clots without platelets being present, thus that platelets are not necessary for fibrin development. In any event, the presence of the platelets gives rise in the normal course of events to an expulsion of the fluid portions of the clot, thus to clot retraction. As pointed out at the start, retraction gives rise to a relatively fine-meshed protective wound cover. It is now quite certain that platelets or platelet derivatives (e.g., serotonin) are necessary in the retraction process. The opinion has been expressed that the fibers of the clot are "contracting" in this process of clot retraction. Another alternative, which is the more likely, is that a process of zippering together is initiated at the fiber junctions and that this zippering process extends gradually throughout the clot.

Another important action takes place during clotting. This is the adsorption of an inactive enzyme called profibrinolysin[12] on to the strands of the clot. On activation, profibrinolysin gives rise to the enzyme fibrinolysin whose specificity is directed against fibrin, and to an extent fibrinogen. Fibrin is attacked via the splitting of peptide bonds, the fragments produced being unable to reassociate to form fibrin or a clot structure under the conditions existing in plasma. Not much is known about the normal mechanism for the activation of profibrinolysin, current methods for activation involving such techniques as the shaking of plasma with chloroform or the addition of chemical compounds, for example, streptokinase which is obtained from the corresponding cocci. The focus of attention in this discussion is the fact that fibrinolysin activity is known to be generated within the liquor sanguinis itself: after accidental death, by fear, stress, and severe exercise, after epinephrine injections and at times in "normal" individuals. With an appreciation of the importance of the coagulation mechanisms in everyday physiology, it is not surprising to find an interest in the possibility that a chemical or physical treatment can be found which activates only the fibrinolysin which is adsorbed to the fibrin clot. Another statement, which should come as no surprise, is that there occurs in the blood a potent antifibrinolysin activity which so far has prevented the therapeutic use of fibrinolysin preparations.

At the onset, reference was made to the possibility that a slow, continuous coagulation was necessary in the maintenance of capillary structure. Immediately thereafter, the factors involved in fast clotting and the control of the fast-clotting process were outlined. Return now to the problem of slow clotting, realizing that many of the complications of clotting may occur as a necessity for maintaining and controlling a system acting slowly and continuously, but poised for rapid action.

No absolute proof can be offered in which coagulation factors or the products of coagulation, for example monomer fibrin, can be directly implicated in capillary

TABLE II. Hemorrhage owing to clotting defects.

| | |
|---|---|
| I. Platelet deficiency: from unknown causes or symptomatic | |
| II. Defects in activating components | |
| (a) Antihemophilic globulin | (AHG) |
| (b) Plasma-thromboplastin component | (PTC) |
| (c) Plasma-thromboplastin antecedent | (PTA) |
| (d) Accelerator globulin | (AOG) |
| III. Prothrombin deficiency: lack of vitamin K, caused by drugs, by disease or is congenital | |
| IV. Fibrinogen deficiency: congenital or through liver damage | |
| V. Anticoagulants: heparin, enzymes (shock) | |

properties such as fragility and permeability. The evidence is indirect and can be summarized briefly by pointing out that interferences in the normal clotting mechanisms often lead to internal hemorrhage, a situation in which blood cells and plasma leak out through the capillaries into the tissue spaces. A few of the coagulation defects which lead to hemorrhage are shown in Table II (see also Wintrobe,[1] p. 285).

First, it is seen that a deficiency in platelets may lead to hemorrhage. This deficiency can arise either through an inherent platelet deficiency or through an external agent, disease for example, which reduces the numbers of platelets.

Second, deficiencies in procoagulation components have the same effect. Here one finds included antihemophilic globulin, plasma-thromboplastin component (mentioned earlier), possibly plasma-thromboplastin antecedent, and accelerator globulin.

Third, a deficiency of prothrombin, whether it be induced by vitamin-K deficiency, by drugs, by disease, or is congenital.

Fourth, a deficiency in fibrinogen: congenital or as a result of liver damage.

Fifth, drugs which increase antithrombin action such as heparin and circulating anticoagulant enzymes which may appear in shock.

The common denominator for this impressive list is active coagulation. In considering the physiological necessity for slow coagulation, the author has been impressed by the difficulty with which the normal anticoagulation mechanisms remove or inhibit the last traces of thrombin. This is a situation so frequently encountered that one is tempted to think in terms of traces of highly resistant thrombin. Such a conclusion would be premature since the alternative that the coagulation system also has a mechanism which preserves slight traces of thrombin is attractive: such a system would not only provide for slow continuous clotting but would supply the traces of thrombin which might spark the initial stage of autocatalytic activation. I feel that the slow-clotting mechanism will emerge as a mechanism warranting close attention in the future.

## BIBLIOGRAPHY


[1] M. M. Wintrobe, *Clinical Hematology* (Lea and Febiger, Philadelphia, 1956), fourth edition.

[2] R. Biggs and R. G. MacFarlane, *Human Blood Coagulation and its Disorders* (Charles C. Thomas, Springfield, Illinois, 1957), second edition.

[3] K. M. Brinkhous, R. D. Langdell, and R. H. Wagner, Ann. Rev. Med. **9**, 159 (1958).

[4] W. H. Seegers, Advances in Enzymol. **16**, 23 (1955).

[5] K. M. Brinkhous, editor, *International Symposium on Hemophilia and Hemophiloid Diseases* (University of North Carolina Press, Chapel Hill, North Carolina, 1957).

[6] W. H. Seegers, R. I. McClaughry, and J. L. Fahey, Blood **5**, 421 (1950).

[7] F. Lamy and D. F. Waugh, Physiol. Rev. **34**, 722 (1954).

[8] S. J. Shulman, J. Am. Chem. Soc. **74**, 5706 (1952).

[9] J. E. Fitzgerald, N. S. Schneider, and D. F. Waugh, J. Am. Chem. Soc. **79**, 601 (1957).

[10] L. Lorand and A. Jacobsen, J. Biol. Chem. **230**, 421 (1958).

[11] W. A. Scheraga and M. Laskowski, Jr., Advances in Protein Chem. **12**, 1 (1958).

[12] S. Shulman, N. Alkjaersig, and S. Sherry, J. Biol. Chem. **233**, 91 (1958).

[13] W. H. Seegers, B. H. Landaburu, R. R. Holburn, and L. M. Tocantins, Proc. Soc. Exptl. Biol. Med. **95**, 583 (1957).

[14] D. F. Waugh and B. J. Livingstone, J. Phys. & Colloid Chem. **55**, 1206 (1951).

[15] D. F. Waugh and M. A. Fitzgerald, Am. J. Physiol. **14**, 627 (1956).

# 61
# Hormone Regulation

DeWitt Stetten, Jr.

*National Institute of Arthritis and Metabolic Diseases, National Institutes of Health, Public Health Service, U. S. Department of Health, Education, and Welfare, Bethesda 14, Maryland*

In higher organisms, whether animal or vegetable, there exist channels through which fluids are transmitted from one tissue to another. Into these fluids or humors are delivered many compounds elaborated in one or another tissue. It is to be expected that, in many cases, these compounds will exert recognizable physiological effects in other tissues to which they may be hydraulically transmitted. Among the host of compounds which meet these criteria, a limited number have been set apart under the special category of hormone or "chemical messenger." It is the purpose here to review briefly what these hormones are, whence they arise, whither they go, and how they accomplish their missions. Although hormonal mechanisms are well described in invertebrates and indeed in plants, these remarks are restricted to mammalian endocrinology.

To speak of the mission of a compound is to lapse into teleological jargon. Teleology, however, has been defined as a woman with whom no scientist cares to be seen in public, but with whom all scientists consort in private. It will simplify my task if you will forgive my teleological indiscretions in this paper which permits me to dodge the otherwise necessary cumbersome circumlocutions.

What then is a hormone? It is a compound devoid of intelligence but containing nonetheless in its structure certain information. While there are no Koch's postulates by which one may judge whether a compound is or is not a hormone, there are certain accepted standards to which most respectable hormones conform.

1. In many but not all cases, hormones arise in tissues which appear to be specialized for their production. In the mammal, these organs include the neurohypophysis, the adenohypophysis, the thyroid, the parathyroids, the islets of Langerhans of the pancreas, the adrenal medullae, the adrenal cortices, the testes, the ovarian follicles, and the ovarian corpora lutea. Other organs, such as the pineal body, the hypothalamus, the thymus, the carotid bodies, etc., are not included in the present list of so-called endocrine organs in view of their relatively uncertain status and lack of specific and convincing information.

It should not be supposed, however, that all known hormones arise in tissues anatomically specialized for their production. The word "hormone" was coined actually to describe an as yet unidentified principle termed secretin, discovered by Bayliss and Starling, which appears to arise in the intestinal mucosa, to enter the blood stream, and to stimulate, upon its arrival there, the secretion of juices by the pancreas. A second but distinct material, arising from the same source, cholecystokinin, stimulates the contracture of the gall bladder, and the injection of bile into the duodenal lumen. Similarly, there is partial evidence for the existence of renal hormones affecting vascular tone, hepatic hormones affecting uptake of blood glucose by muscle, and indeed a hormone liberated by working skeletal muscle which affects the metabolism of resting muscle in the same animal.

Whereas a number of the hormones which arise in the specialized endocrine glands, e.g., the pituitary, the adrenal, etc., have been purified and characterized structurally, this has not been accomplished for hormones arising from such tissues as liver or intestinal mucosa. Consequently, the structural information presented is confined to products of the so-called glands of internal secretion.

2. A second characteristic of hormones is that they normally circulate and exert their regulatory roles at low molar concentrations. On the basis of published estimates of the normal concentration of insulin in blood, it would appear that this agent is effective at concentrations of about $10^{-11}$ to $10^{-12}$ molar, and epinephrine, glucagon, and thyroxin all appear to exert recognizable effects at similar levels of concentration.

3. Some, but not all, hormones exert their effects upon highly specific target organs. Thus, adrenocorticotropic hormone of the pituitary operates essentially uniquely upon the adrenal cortex, while the activity of thyrotropic hormone is restricted largely to the thyroid. The sole convincingly demonstrated action of glucagon is upon the liver. On the other hand, epinephrine exerts actions upon liver and muscle, insulin has been shown to operate in many tissues including skeletal muscle, adipose tissue, and mammary gland, whereas the effects of adrenocortical steroids and thyroxin may be demonstrated in a wide variety of tissues.

4. If a hormone is to exert a regulatory effect, clearly the rate of its release into the blood stream must be variable and should be determined by some function of the mechanism which it regulates. Such variation in rate of discharge has been demonstrated for certain but not for all hormones. For example, one may cite insulin, which regulates the level of glucose in the blood and the release of which by the islet cells of the pancreas is determined directly by the concentration of glucose in the pancreatic blood supply. Analogously, ACTH, released by the pituitary, specifically stimulates the adrenal cortex to discharge steroid hormones, which, if injected, cause a prompt suppression of ACTH produc-

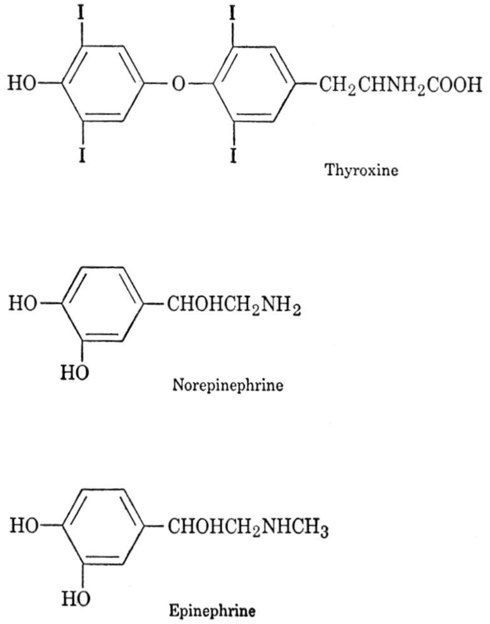

Thyroxine

Norepinephrine

Epinephrine

FIG. 1. Structures of certain phenolic hormones.

tion. Many other such examples of negative feedback, more or less well-documented, might be cited.

It may be profitable to consider briefly some of the structural types which are represented among the hormones. These include a variety of substances, among the simplest of which are certain phenols (Fig. 1). Under the regulation of thyrotropic hormone derived from the anterior pituitary gland, a protein, thyroglobulin, is elaborated in the thyroid gland, which contains the unique amino acid thyroxin, 3,5,3′,5′-tetra-iodothyronine. This amino acid, accompanied by tri-iodothyronine, is discharged into the bloodstream where it circulates, largely protein-bound, in the plasma. Epinephrine and norepinephrine derived from adrenal medulla are related closely both metabolically and structurally. There is a considerable amount of information

Estrone

| Compound | Substituent at C Numbers 17 | 16 |
|---|---|---|
| α or β Estradiol | —OH | —H |
| Estriol | —OH | —OH |

FIG. 2. Structures of certain $C_{18}$ steroid hormones (N. B. Equilin and equilenin represent successive states of oxidation of B ring of estrone).

at hand defining the biosynthetic origins of these compounds. Tyrosine contributes the phenolic portions, in each case. With these, as well as with certain other hormones, secondarysites of origin exist. The synthesis of thyroxin-like materials in the totally thyroidectomized animal[1] is perhaps not surprising in view of the fact that tyrosine-containing proteins, after incubation with iodine in alkali, in the absence of enzymes, exhibit thyroxin activity.

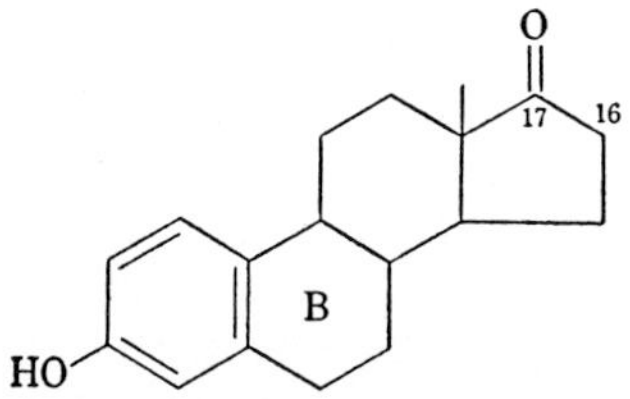

Steroidal hormones comprise a large and much-studied group. They often are classified structurally according to the number of carbon atoms contained. Thus, the $C_{18}$ steroids include the estrogens (Fig. 2), estrone, α- and β-estradiol, estriol, as well as the less familiar equilin and equilenin. All of these compounds have phenolic—i.e., acidic—properties. The methyl group born on carbon atom 10 of most other steroids is lacking in these compounds, as a consequence of the unsaturation. Arising predominantly in the Graafian

Testosterone

| Compound | Substituent at C Numbers 3 | 11 | 17 |
|---|---|---|---|
| Adrenosterone | =O | =O | =O |
| Androstenedione | =O | —H | =O |
| Androstanediol* | —OH | —OH | =O |

*Double bond at $\Delta^4$ is reduced

FIG. 3. Structures of certain $C_{19}$ steroid hormones.

follicle of the ovary, under regulation by pituitary gonadotropin, estrogens are generated also by testis and possibly also by adrenal cortex. They are detoxified normally in the liver by conjugation with glucuronic or sulfuric acids, and the loss of this function in hepatic failure results in feminizing symptoms and signs.

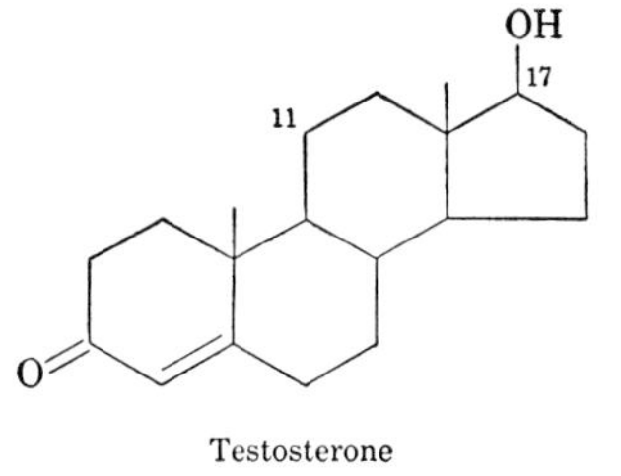

The $C_{19}$ steroids include (Fig. 3) the androgens, testosterone and its congeners. There are a number of steroids with androgenic activity and these may arise in the adrenal cortex as well as in the testis. The $C_{21}$ steroids (Fig. 4) include not only the several adrenocortical steroids but also progesterone, the hormone of the ovarian corpus luteum. The generation of these hormones is under direct control of specific anterior pituitary hormones. Testicular production of androgens is regulated by pituitary gonadotropin, adrenocortical activity is determined by pituitary adrenocorticotropin, and progesterone production depends upon a luteinizing hormone of the pituitary gland.

The structural similarities of the several steroid hormones are explained by the fact that they are all biogenetically closely related. Cholesterol appears to be a parent compound to all of them, and many of the individual enzyme-catalyzed steps whereby they arise have been elucidated. It is, of course, of great interest that specific and important biological activities of a wide variety are associated with steroids. In addition to the steroid hormones, other biologically important steroids are vitamin D, the cardiac aglycones of the digitalis group of drugs, the toad poisons, and certain carcinogens. What it is in the steroid structure which imparts to the members of this class these various activities is not known.

The remaining groups of hormones about which there is structural information are polypeptides. Of these, insulin (Fig. 5) is probably the best-studied. Beef insulin, whose chemical structure was given by Sanger,[2] is made up of two polypeptides, an *A*- or acidic chain comprising 21 amino acids and a *B*- or basic chain containing 29 amino acids. The intramolecular disulfide bridges have been described previously by Oncley (p. 30), and their integrity is essential to the physiological activity. The C-terminal alanine in the *B*-chain is apparently not essential, however, and indeed the eight C-terminal amino acids of this chain may be eliminated without complete inactivation. The question of what constitutes the essential center of physiological activity of insulin, if indeed there is such, remains unanswered. When this structure first was published, attention was attracted to the pentapeptide closed ring in the *A*-chain, particularly since rings of similar size were found by du Vigneaud[3,4] to be present in both of the hormones of the posterior pituitary gland, vasopressin and oxytocin. The curious run of three adjacent aromatic amino acids in the *B*-chain also was pointed out.

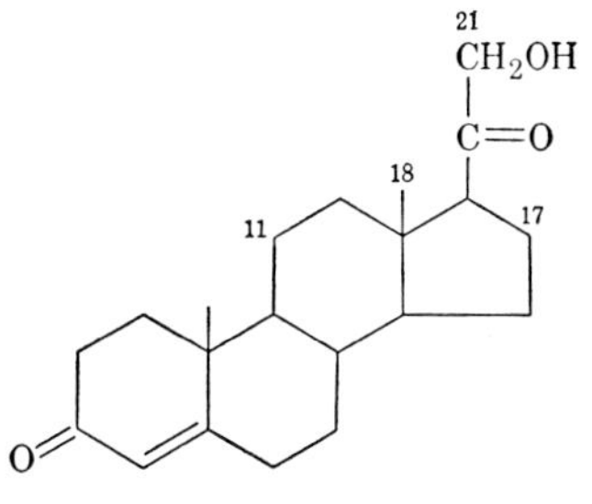

21 $CH_2OH$
C=O
18
11
17
O

Deoxycorticosterone

Substituent at C Numbers

| Compound | 11 | 17 | 18 |
|---|---|---|---|
| Corticosterone | —OH | —H | $-CH_3$ |
| 11-Dehydrocorticosterone | =O | —H | $-CH_3$ |
| 17-Hydroxycorticosterone | —OH | —OH | $-CH_3$ |
| Cortisone | =O | —OH | $-CH_3$ |
| Aldosterone | —OH | —H | —CHO |

FIG. 4. Structures of certain $C_{21}$ steroid hormones (N. B. Progesterone differs from deoxycorticosterone only in that $-CH_2OH(21)$ is reduced to $-CH_3$.

No evidence has been presented implicating either of these sites specifically in insulin function.

The posterior pituitary hormones present striking similarities and, as might be expected, some degree of cross-reactivity physiologically. There is some reason to believe that these compounds may be parts of a larger polypeptide arising in the hypothalamus and transmitted down the pituitary stalk, to be stored in and released from the posterior lobe of the pituitary gland.

Attention is directed now to amino acids 8, 9, and 10 in the *A*-chain of insulin. Insulins of several species have been prepared; all are very similar in physical properties and are essentially identical in physiological assay, about 27 units per mg. Chemical differences do exist,

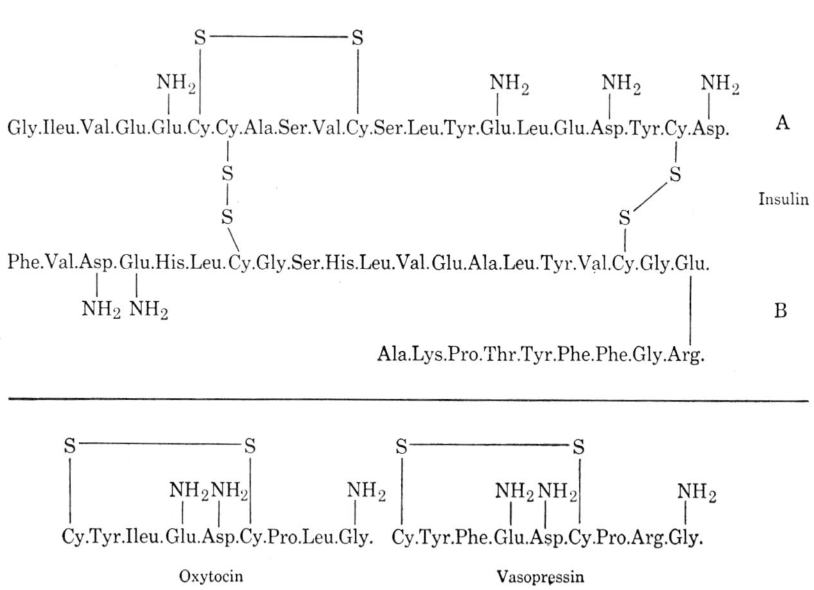

S—S
$NH_2$ $NH_2$ $NH_2$ $NH_2$
Gly.Ileu.Val.Glu.Glu.Cy.Cy.Ala.Ser.Val.Cy.Ser.Leu.Tyr.Glu.Leu.Glu.Asp.Tyr.Cy.Asp. A
S S
S S Insulin
Phe.Val.Asp.Glu.His.Leu.Cy.Gly.Ser.His.Leu.Val.Glu.Ala.Leu.Tyr.Val.Cy.Gly.Glu.
$NH_2$ $NH_2$ B
Ala.Lys.Pro.Thr.Tyr.Phe.Phe.Gly.Arg.

S—S S—S
$NH_2$ $NH_2$ $NH_2$ $NH_2$ $NH_2$ $NH_2$
Cy.Tyr.Ileu.Glu.Asp.Cy.Pro.Leu.Gly. Cy.Tyr.Phe.Glu.Asp.Cy.Pro.Arg.Gly.
Oxytocin Vasopressin

FIG. 5. Structures of bovine insulin and the hormones of the posterior pituitary.

Insulin–A–Chain*

| 1 — — 7 8 9 10 — — — — 20 | |
|---|---|
| ...Cys.Ala.Ser.Val... | Beef |
| ...Cys.Thr.Ser.Ileu.. | Pig † |
| ...Cys.Ala.Gly.Val... | Sheep |
| ...Cys.Thr.Gly.Ileu.. | Horse |
| ...Cys.Thr.Ser.Ileu.. | Whale † |

*Remainder of A chain and entire B chain (1 through 29) are invariant

FIG. 6. Species variations in insulin structure.

however, but thus far the sole differences that have been detected reside within these 3 amino acids (Fig. 6). Although the sample studied is confessedly too small to permit firm conclusions to be drawn, a few suggestive findings have turned up. In each position, 8, 9, or 10, only one of 2 amino acids has been found. Thus, in position 8, there is either alanine or threonine; in position 9, serine or glycine; in position 10, valine or isoleucine. If this binary replacement proves general, and no other alterations are found, only 8 insulins are possible. There may be, however, an additional restriction. On comparison of positions 8 and 10, it appears that, whenever the former is alanine, the latter is valine; whenever the former is threonine, the latter is isoleucine. Should this represent a firm link in the coding to these 2 positions, only 4 insulins would occur, and indeed all types now are known. It is not surprising, then, that pig and whale insulins prove to be identical (Fig. 6†).

Insulin is generated in the $\beta$ cells of the islets of Langerhans. Its function, as is discussed later, is to favor the utilization of blood glucose by muscle, to restrict glucose formation by liver. It is noteworthy, therefore, that the release of insulin by $\beta$ cells is favored by a high concentration of blood glucose.

Intimately associated anatomically with the $\beta$ cells are the $\alpha$ cells of the islets which are the source of glucagon. This polypeptide (Fig. 7) was for many years an unsuspected contaminant of the best available insulin from American commercial sources although, because of differences in manufacturing procedure, it was lacking in Danish insulin. Chemically and physiologically it is

$NH_2$ (on Glu 3)

His.Ser.Glu.Gly.Thr.Phe.Thr.Ser.Asp.Tyr.Ser.Lys.
1 2 3 4 5 6 7 8 9 10 11 12

$NH_2$ (on Glu 20)

Tyr.Leu.Aop.Ser.Arg.Arg. Ala.Glu.Asp.Phe.Val.
13 14 15 16 17 18 19 20 21 22 23

$NH_2$ (on Glu 24) $NH_2$ (on Asp 28)

Glu.Try.Leu.Met.Asp.Thr.
24 25 26 27 28 29

FIG. 7. Structure of glucagon.

unrelated to insulin.[5] The sole sequential similarity between insulin and glucagon that has been pointed out is the series 22–24 in glucagon (Phe. Val. Glu $NH_2$) which resembles Phe. Val. Asp $NH_2$ (1–3 chain B) of insulin. Whereas both glucagon and insulin affect the level of blood glucose, they do so by influencing quite different processes, and appear to be antagonists only on very superficial inspection.

The last group of hormones about which there is useful structural information are the polypeptides of the anterior pituitary gland. There are at least six of these, thyrotropin, adrenocorticotropin, somatotropin (growth hormone), follicle stimulating, luteinizing, and melanophore stimulating hormones. Other activities have been suspected in the past and may be revealed in the future. Complete structural information is available, for certain species, in regard to corticotropin and melanotropins. The formulas for certain of these have been given by Oncley (p. 30). Here, attention is directed only to certain salient points of similarity and dissimilarity (Fig. 8). Amino acids 4 through 10 of all corticotropins studied are identical with amino acids 7 through 13 of all known melanotropins. This identity of a series of seven amino acids is sufficiently startling

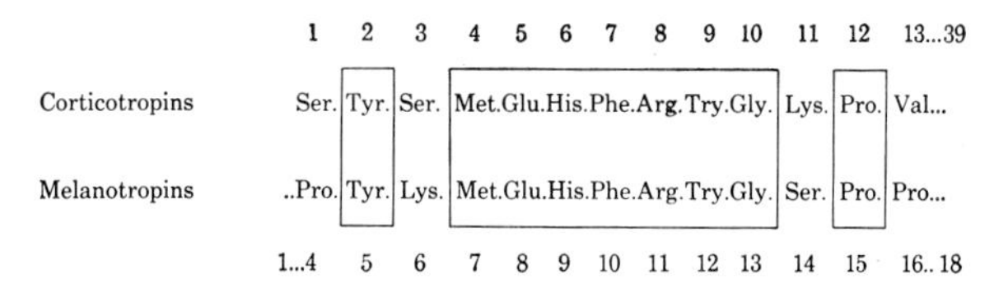


FIG. 8. Structural relationships of ACTH and melanophore stimulating hormone.

to attract attention. Li[6] has pointed out that the identical sequences would have been eleven rather than seven, but for a curious confusion about positions 3 and 11 of corticotropins, or 6 and 14 in melanotropins. An interchange between the serine and lysine in the appropriate positions of either of these peptides would prolong the identical sequences from seven to eleven.

This observation, as well as the previously mentioned species variations in insulin, suggests the occurrence of certain blunders in the transmission of the coded information pertaining to amino-acid sequences. Such blunders, if indeed so they be, may themselves be informative. Thus, the present figure suggests that the code is such as to permit interchange of information about 2 amino acids 8 positions remote in a polypeptide.

To date, accumulating knowledge of hormone structure has, for the most part, revealed little about how these compounds function. Only within the past few years have reasonable hormone effects been observed in biological preparations disorganized below the cellular level. Several of these now may be mentioned.

Thyroxin has been suspected for some time of exerting an effect similar to that of 2,5-dinitrophenol, a suspicion dating back to the early clinical observations of the

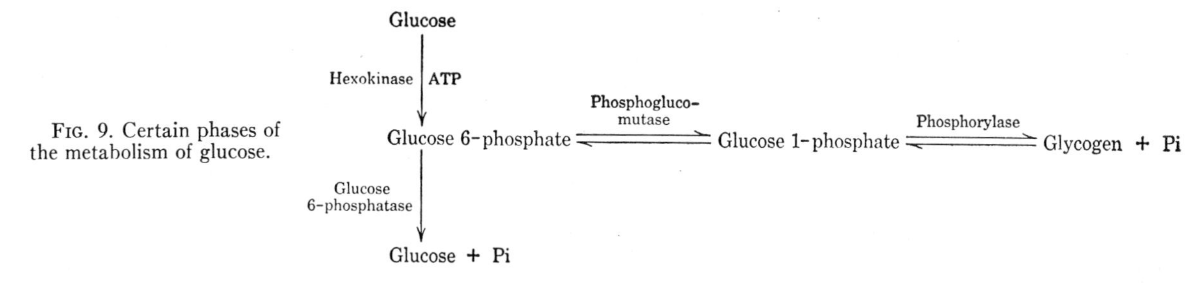


FIG. 9. Certain phases of the metabolism of glucose.

toxicology of the latter drug during World War I. When it was ascertained that dinitrophenol effected an uncoupling of oxidative phosphorylation, permitting respiration to proceed without obligatory and coincident phosphorylation, a similar effect was attributed to, and soon found with, thyroxin. This effect has been demonstrated with isolated mitochondria by Lehninger.[7] Such uncoupling, and consequent escape of glycolysis from the restraining influence of the Pasteur effect, would explain many of the clinical manifestations of thyrotoxicosis.

Estrogens have been shown by Villee[8] to activate the isocitric-acid dehydrogenase of placenta and this effect has been demonstrated in homogenized preparations (cell-free). More recently, Talalay[9] has indicated that the enzyme of placenta directly affected is a transhydrogenase which regenerates the TPNH required by the dehydrogenase in question. It is possible that, in this transhydrogenase reaction, the estrogen serves as a self-regenerating co-substrate.

Two hormones about which there is partial mechanistic information are epinephrine and glucagon. It is perhaps remarkable that these two substances, wholly unrelated structurally and biogenetically, should both exert, at equimolar concentrations, virtually identical effects. These effects, elucidated by Sutherland,[10] relate to the activation of hepatic phosphorylase. This enzyme, required for glycogen breakdown and probably also for its synthesis, is itself enzymatically inactivated with the loss of inorganic phosphate. Its reactivation also may be accomplished enzymatically, but this requires a novel cofactor, an anhydride of AMP. The synthesis of this cofactor, by yet another enzyme, proceeds only in the presence of epinephrine and/or glucagon. The two hormones, though acting identically in liver, are quite different in their effects upon muscle glycogen. Here, whereas epinephrine is highly active, glucagon is without effect.

Insulin, which has received the attention of many excellent investigators since its isolation, is in a more confused state than some of the other hormones which have been considered. The dramatic capacity to lower blood sugar apparently is the result of effects upon two processes. To evaluate these two effects, one should consider briefly some of the sources and fates of blood glucose (Fig. 9). Here are represented in abbreviated form some of the reactions of concern. In the absence of insulin, as in the severely diabetic human or as in the pancreatectomized animal, the conversion of extracellular glucose into intracellular glucose 6-phosphate, an essential first step in glucose utilization, is impaired. This impairment is especially demonstrable in muscle tissue and characteristically is seen in the isolated diaphragm. The basis for the impairment is still somewhat obscure but the bulk of current evidence favors the view, first propounded by Levine,[11] that insulin favors the entry of glucose into the intracellular compartment of muscle. Insulin demonstrably performs this function in relation to a number of other sugars. Possibly significant is the fact that the sugars most responsive to insulin in this regard are configurationally identical with glucose about carbon atoms 1, 2, and 3.

A second pertinent insulin effect is defined best by the fact that the activity of glucose 6-phosphatase is greatly enhanced in the liver of the diabetic animal, an effect reversible by administration of insulin. The importance of this enzyme is revealed by the fact that it catalyzes the reaction responsible for the majority of glucose contributed to the blood by animal tissues.

Consider the dilemma of the diabetic animal in the light of these two defects. With the conversion of glucose to glucose 6-phosphate restricted, and with the hydrolysis of this ester exaggerated, there results a curtailment of glucose phosphate needed for a variety of important processes, including glycolysis, oxidation, generation of reduced coenzymes, preparation of essential intermediates. Many, but certainly not all, of the late consequences of untreated diabetes are referable directly to this combination of biochemical defects. The question of which of the two defects is most significant is in a state of flux at the present time. Data from Hastings[12] and others suggest that the increase in utilization of glucose by muscle occurs much more promptly in response to insulin than does a decrease in glucose 6-phosphatase activity. On the other hand, Weinhouse and his collaborators[13] have data indicating that the reverse is true.

In relation to this figure, one can consider also the purported "antagonism" between insulin and glucagon. Glucagon, as mentioned previously, effects favorably the activity of hepatic phosphorylase, promotes breakdown of liver glycogen, and produces a resulting rise in blood glucose. Insulin, by its actions, produces a fall in blood glucose. If one's vision is restricted to the glucose of the blood, these two hormones act antagonistically. The blood glucose, however, is really relatively un-

$$HT \rightleftharpoons H + T$$

$$K = \frac{[H]\ [T]}{[HT]}$$

$$Q = [T] + [HT]$$

$$Q = K\frac{[HT]}{[H]} + [HT]$$

$$\frac{[H]}{[HT]} = \frac{K}{Q} + \frac{[H]}{Q}$$

or

$$\frac{1}{[HT]} = \frac{1}{[H]} \cdot \frac{K}{Q} + \frac{1}{Q}$$

FIG. 10. Expected arithmetical relationship between concentrations of free hormone, H, of bound hormone, HT, and of acceptor sites on target organ, Q.

important, except insofar as it nourishes the several tissues of the body, and the major consuming tissue is skeletal muscle. With respect to the nutrition of skeletal muscle, glucagon and insulin, far from being antagonists, are synergists. Glucagon supplies to the blood the glucose which muscle requires, and insulin facilitates the entry of this glucose into the cells so that it may be utilized.

It was mentioned at the outset that many endocrines, insulin among these, are physiologically active at remarkably low concentrations. The finding of Stadie[14] that several tissues responding to insulin *in vitro* are endowed with the capacity to bind insulin to their surfaces, and that the response to insulin is, under appropriate circumstances, proportional to the amount of insulin bound, is perhaps relevant. If the target organ has the capacity to bind the hormone to itself, local concentrations of hormone may be built up far in excess of the concentrations occurring in the blood. The dose-response relationship for several hormones is compatible with this idea in that, whereas the response is approximately proportional to dosage at low dosage levels, for many hormones a maximal response is attained asymptotically at high dosage levels. This suggests that something, possibly the acceptor site on the target organ, is being saturated with hormone. This situation may be treated arithmetically[15] by assuming a reversible association between hormone (H) and target-organ acceptor site (T). If K is the dissociation constant and Q is the total concentration of acceptor sites (Fig. 10), the problem may be treated as Lineweaver and Burk treated the Michaelis-Menten problem, to yield the final expression on the figure. If now it be assumed tentatively that the concentration of hormone H is proportional to dosage and that the response is proportional to the concentration of bound hormone HT, the reciprocals of these two quantities should be related linearly. In a limited number of cases where this relationship has been tested, the expected linear relationship indeed has been found. Values for K and Q can be estimated from the slopes and intercepts of such plots. The usefulness of the relationship diminishes insofar as some quantity other than the concentration of bound hormone limits the observed response.

It is perhaps unfortunate to terminate this distinguished series upon such an unsatisfactory note. In earlier papers on macromolecular structure and interaction, biological fine structure, information storage and retrieval, etc., one was left with the idea that there is a narrowing gap between the biological description and the physical model, or even explanation. In the matters presently under consideration, the gap is so wide that I, for one, am unable to see across it. Of such a gap, one may hope, but one can not know, that it is being narrowed. Certainly many experiments will have to be performed before the hormones, their origins, their modes of action, and their interrelationships can be defined in physical terms.

## BIBLIOGRAPHY


1. M. E. Morton, I. L. Chaikoff, W. O. Reinhardt, and E. Anderson, J. Biol. Chem. **147**, 757 (1943).
2. F. Sanger, L. F. Smith, and R. Kitai, Biochem. J. **58**, vi (1954).
3. V. du Vigneaud, C. Ressler, and S. Trippet, J. Biol. Chem. **205**, 949 (1953).
4. V. du Vigneaud, H. C. Lawler, and E. A. Popenoe, J. Am. Chem. Soc. **75**, 4880 (1953).
5. W. W. Brommer, L. G. Sinn, A. Staub, and O. K. Behrens, J. Am. Chem. Soc. **78**, 3858 (1956).
6. C. H. Li, Advances in Protein Chem. **12**, 269 (1957).
7. A. L. Lehninger in *Enzymes: Units of Biological Structure and Function*, O. H. Gaebler, editor (Academic Press, Inc., New York, 1956), p. 217.
8. C. A. Villee and D. D. Hagerman, Endocrinology **60**, 552 (1957).
9. P. Talalay and H. G. Williams-Ashman, Proc. Natl. Acad. Sci. U.S. **44**, 15 (1958).
10. T. W. Rall, E. W. Sutherland, and J. Berthet, J. Biol. Chem. **224**, 463 (1957).
11. R. Levine, M. S. Goldstein, B. Huddlestun, and S. F. Klein, Am. J. Physiol. **163**, 70 (1950).
12. A. E. Renold, A. B. Hastings, F. B. Nesbett, and J. Ashmore, J. Biol. Chem. **213**, 135 (1955).
13. D. F. Dunn, B. Friedmann, A. R. Maass, G. A. Reichard, and S. Weinhouse, J. Biol. Chem. **225**, 225 (1957).
14. W. C. Stadie, N. Haugaard, and M. Vaughan, J. Biol. Chem. **200**, 745 (1953).
15. DeW. Stetten, Jr., Science **124**, 365 (1956).

# Author Index

A number in parentheses and the number immediately preceding it designate, respectively, a reference number in one of the papers and the number of the page on which that reference appears. A few papers contain an unnumbered bibliography instead of the numbered references; for these cases, the numbers in parentheses in this index designate the order number of the entry in the bibliography. Italicized numbers designate the pages on which are given source information (e.g., journal, volume, date, etc.).

# Subject Index

The letters F and T after numerals indicate figures and tables, respectively.

E

J



K



L

O

T